院士文库　厦门大学专辑

# 蔡启瑞 院士论文选集

## 【下册】

厦门大学出版社
XIAMEN UNIVERSITY PRESS
国家一级出版社
全国百佳图书出版单位

谨以此文集

热烈庆贺蔡启瑞先生百岁华诞
暨厦门大学催化学科创建五十五周年

《蔡启瑞院士论文选集》编辑小组

厦门大学化学系催化科学与工程研究所

物理化学研究所

固体表面物理化学国家重点实验室

醇醚酯化工清洁生产国家工程实验室

厦门大学化学化工学院

2013年11月

蔡启瑞院士

# 《院士文库·厦门大学专辑》出版说明

院士,是国家的宝贵财富,是推动科技进步和经济社会发展的重要力量。

在全国1 000多位"两院"院士中,厦门大学有12位。他们是:化学化工学院的蔡启瑞、田昭武、张乾二、黄本立、万惠霖、赵玉芬、郑兰荪、田中群;生命科学学院的 唐仲璋 、唐崇惕、 林鹏 ;海洋与地球学院的焦念志。而在计算机、材料科学、海洋生物等研究领域,厦门大学还引进了新的人才机制,先后聘请了10余位双聘院士。

几十年来,这些院士辛勤耕耘于科学园地,孜孜努力于科研创新,不仅为国家培养了大批专业人才,而且为我国科学技术的繁荣与发展做出了突出的贡献。他们的科学精神,他们的聪明智慧,他们的创新成果,不仅是厦门大学的宝贵财富,也是全体教育、科研工作者学习的榜样。

1931年,著名教育家梅贻琦在出任清华大学校长时曾经这样说过:"所谓大学者,非谓有大楼之谓也,乃有大师之谓也。"回顾厦门大学创办与发展的历程,人们不能不对此感同身受,钦服之至。

86年前,在我们国家和民族处于贫穷落后、灾难深重的年代,陈嘉庚先生基于"教育救国"的理念,毅然倾资创办了厦门大学。他在发起人会议上慷慨陈述:"今日国势危如累卵,所赖以维持者,惟此方兴之教育与未死之民心耳。""民心不死,国脉尚存,以四万万之民族,决无甘居人下之理。"

为了不甘居人下,为了实现"南方之强"的目标,厦门大学在创办之初,就十分注重招揽名师执教,并把"研究高深学术,养成专门人才,阐扬世界文化"作为自己的三大任务。一时之间,群贤毕至,名流云集,如文学家鲁迅,动物学家何博礼,植物学家钟心煊,数学家姜立夫,化学家刘树杞,物理学家朱志涤等等。这些国内第一流的名师为厦大的初创和人才培养奠定了良好的基础。

此后,一代代的名师前赴后继,悉心传道、授业、解惑,培养出包括物理学家谢希德、经济学家许涤新、化学家卢嘉锡、数学家陈景润、遗传学家方宗熙、水生物学家伍献文等在内的一大批很有影响力的专业人才,他们为国家的进步和科学的发展做出了不可磨灭的贡献。

"名师出高徒"。名师的传承、交流、融汇正是一所国际高水平大学生生不息的源泉。今天,厦门大学化学化工学院能够拥有8位院士,在全国高校化学化工学院中名列前茅,能够在物理化学的三个分支学科——催化化学、电化学、结构与量子化学领域,形成自己的创新优势和研究特色,并蜚声海内外;而在生物学研究领域,3位院士能够在寄生虫及红树林研究方面独树一帜,这不能不说与名师间的传承效应、群体效应有很大的关系。另一方面,"双聘院士"的引进,不仅弥补了厦大在一些研究领域的薄弱环节,而且为不同高校和科研机构之间的学术交流提供了一个很好的"平台"。毫无疑问,这些院士的创新精神和学术影响力,已远远超出了自己的专业领域,而成为不同领域科学工作者不可或缺的科学素养。

厦门大学出版社历来把弘扬科学精神和出版优秀的学术著作,作为自己矢志追求的目标。

为了展示"两院"院士国际领先的学术水平和求实探索的科学精神,同时也为了向学界提供更为系统、完整的专业论著,厦门大学出版社决定倾力编辑、出版一套《院士文库》丛书;而首编便是即将呈现在读者面前的《院士文库·厦门大学专辑》。

该专辑所选论著有的发表时间较为久远,有的作者已经去世。在编辑出版时,我们既注重整套专辑及丛书风格的统一,又注重时代痕迹的保留。为此,在重新录入时,对书眉、标题字体以及参考文献的格式加以统一;但对发表在上个世纪不同杂志上的论文则依然保留了当时的简化字、字符、量纲以及体例,尽量使其原汁原味。

希望本文库的出版能对相关学科的科研起到一定的推动作用,尤其能使后辈学人从中汲取科学的营养,领略院士们的治学精粹,为学术的传承与创新"牵线搭桥",为新一代大师的不断涌现推波助澜。

如是,则读者幸甚,作者幸甚,编者幸甚!

《院士文库》编委会

2007 年 10 月

# 《蔡启瑞院士论文选集》代序

恩师蔡启瑞先生 1913 年农历十一月初六出生于福建省同安县（今厦门市翔安区）马巷镇番薯市五甲尾一个华侨店员家庭。

在恩师百岁高寿的 2013 年，厦门大学在 4 月 6 日 92 周年校庆庆典上，将首次设立的厦门大学最高奖"南强杰出贡献奖"颁给了恩师，以表彰恩师为国家和人民以及学校和科学所做出的卓越贡献，颁奖辞赞曰："蔡启瑞先生，中国科学院院士，德高望重的物理化学家、分子催化专家。在他心里，国家民族为重，个人利益为轻。为了祖国的召唤，他执意回国；为了国家的需要，他毅然转行。催化学科，他是奠基人；物化研究，他是引领者；工科发展，他是开拓者。他呕心沥血，携手攀登，他在厦大领衔创建了中国高校第一个催化教研室、厦大第一个国家重点实验室、福建省首个国家工程实验室，圆了几代人梦寐以求的'化学梦'，奠定了厦大化学学科的一流地位。他为人平和，谦逊礼让，如清泉般透彻。他以身作则，提携后辈，像泰山般厚道。古人赞曰：'仁者寿！'先生以百岁的实践证明古人之云然也！"

在恩师百岁高寿的 2013 年，我们怀着感恩和崇敬之心，迎来了《蔡启瑞院士论文选集》的正式出版。论文集收录了：从恩师署名的 380 篇有关论文中选出的 225 篇全文，论文（著）总目，专利目录（发明专利 19 项，实用新型 2 项，已授权 18 项），主要活动年表（学习、教学、科研和学术活动，以及主要社会职务和主要奖项），指导研究生名单，个人照、工作照、活动照和生活照。特别令人欣喜的是，在老科学家（蔡启瑞）学术成长资料采集工程小组的努力下和厦门大学美洲校友会的支持下，论文集首次收录到恩师作为厦门大学第 12 届毕业生、在张怀朴教授指导下于 1937 年 6 月 11 日完成的厦门大学理学学士学位论文 *ELECTROMETRIC DETERMINATION OF THE HYDROLYSIS OF ZINC AND CADMIUM NITRATES*（《硝酸锌和硝酸镉水解的量电法测定》），以及在马克（E. Mack，Jr）、哈里斯（P. M. Harris）和纽曼（M. S. Newman）教授的指导下于 1950 年 3 月完成的美国俄亥俄州立大学（Ohio State University）化学领域的哲学博士学位（Ph. D.）论文 *A STUDY OF MAC-RO-RING CLOSURE IN HETEROGENEOUS REACTIONS：SURFACE FILMS OF HIGH POLYMETHYLENE DICARBOXYLIC ACIDS AND GLYCOLS*（《多相反应中大环闭合的研究：高聚亚甲基二羧酸和二元醇的表面膜》）。

因篇幅所限，论文集仅是恩师部分学术成就的反映。论文集的宗旨在于给后人以启示，为后人之所用。为此，论文集特别将厦门大学化学系催化教研室和物理化学研究所催化研究室撰写的《祝贺蔡启瑞教授从事化学工作五十年》（《卢嘉锡/蔡启瑞教授从事化学工作五十年纪念册》，1986 年），厦门大学化学系催化教研室、物理化学研究所催化研究室和化工系工业催化教研室撰写的《我国分子催化的奠基人之一蔡启瑞教授》（《庆贺蔡启瑞教授八秩华诞》，1994 年），《祝贺我国著名物理化学家,中国科学院院士蔡启瑞教授九十华诞暨执教五十八年》(《化学学报》,2004年,第62卷,第18期)以及经恩师蔡启瑞先生亲自审订的《20世纪中国知名科学家学术成就概览·化学卷·第一分册》中的"蔡启瑞"篇(科学出版社,2011年,第1版)转载于论文集部首,以便更好和简要地反映恩师的主要成就。更详细的资料可参阅今后可能出版的老科学家(蔡启瑞)学术成长资料采集工程的研究报告《蔡启瑞传》。

恩师蔡启瑞先生一生平和朴实、谦逊礼让、学风正派、为人正直、淡泊名利,是学术界公认的德高望重的学术大师。学如流水行云,德比松劲柏青,探赜索隐老而弥笃,立志创新志且益坚,这些科教界名流的题词嘉勉是对恩师学识和师德的赞许,是对恩师的大胆假设、小心求证、不迷信权威、勇于创新的科学研究素质的评价,是对恩师学术道德和为人风范的写照。

在恩师百岁高寿的2013年,论文集得以顺利出版,要感谢厦门大学化学系催化科学与工程研究所、论文集编辑小组以及厦门大学出版社同仁们的辛勤工作,还要感谢厦门大学特批的出版基金资助。因恩师蔡启瑞先生正在住院康复中,不便亲自写序,嘱生代笔,学生师从恩师几十载催化研究,受益良多,代序之言中不妥之处,还请不吝指教。

廖代伟  万惠霖

2013年4月于厦门大学化学楼

# 目　录

## （下册）

■ 本文原载：Thermocimica Acta 274(1996)，pp. 289～301.

# Computer Simulation of Derivative TPD

Yun-Hang Hu①，Hui-Lin Wan，Khi-Rui Tsai，Chak-Tong Au

*(Department of Chemistry and State Key Laboratory for Physical Chemistry of Solid Surface，Xiamen University，Xiamen 361005，People's Republic of China)*

Received 8 August 1994；accepted 12 August 1995

**Abstract**　In this paper，the advantages of employing a Derivative Temperature-Programmed Desorption (DTPD) curve in TPD analysis are demonstrated. Based on a series of theoretical DTPD curves obtained by computer simulation with double assumption of zero signal noise and no temperature gradients across the sample，a comparison is made between the TPD and DTPD curves，and it is found that the approach can (a) estimate desorption order，(b) raise resolving power，and (c) eliminate baseline drift. The equations for calculating kinetic parameters from DTPD curves are also presented. The results show that these equations are valid.

**Key words**　Computer simulation　Derivative TPD curve　TPD

## 1. Introduction

Elucidation of the interaction of reactants with catalyst surfaces is of primary importance in heterogeneous reaction systems. In the past twenty years, although techniques such as AES, XPS, HREELS, etc., have become popular, some traditional methods such as TPD and TPR have been found to be indispensable[1]. Generally speaking, an ultra-high vacuum is required in studies of the bonding between adsorbates and surfaces when modern techniques are used and in situ studies under reaction conditions are very difficult. Temperature-programmed desorption (TPD), however, can be conducted under reaction conditions and has wide application in catalytic research[1-5].

Although the TPD technique is simple, it is rather difficult to extract all the information present in the TPD curve. In order to exploit TPD fully, many researchers have undertaken a theoretical approach[6-14]. Generally speaking, TPD diagrams can be used qualitatively to obtain information about the number of adsorption forms on the surface and their relative stability. When catalyst surfaces are being analyzed, one finds that in most cases there are more than one kind of binding site and the desorption of adsorbants might produce overlapping peaks. It is generally accepted that better resolution between overlapping peaks can be achieved by using lower heating rates; however, the method is found to be ineffective in a number of cases[15].

Quantitative analysis of TPD curves allow calculation of the Arrhenius kinetic parameters for desorption[1,2,8,10,15]. Although there are better methods for calculating these parameters from TPD

---

① Corresponding author.

curves with a single-peak profile[1,2,8,10,15], it is difficult to analyze overlapping peaks quantitatively. There is a "simplex method" for resolving overlapping peaks[13]. However, before using such a method, it is necessary to estimate the number of peaks present in the profile from the peak shape. Therefore, whether in a qualitative or quantitative way, it is very important to improve the resolving power of the TPD curve.

Derivative spectra have been widely applied in UV, IR, AES and ESR to enhance resolving power[16-18]. In the present work, in order to improve resolving power of the TPD curves, we propose the adoption of a derivative temperature-programmed desorption (DTPD) curve. In the following sections, we describe simulation of the DTPD curves by computer, discuss the characteristics of the DTPD curves, and present the equations for calculation of kinetic parameters from the DTPD curves.

# 2. computer simulation of theoretical DTPD curves

## 2.1 The definition of DTPD

The curve of $-d\theta/dT \sim T$ is generally called a TPD curve. Hence the curve of $-d^2\theta/dT^2 \sim T$ is the first DTPD curve, and the curve of $-d\theta^3/dT^3 \sim T$ the second. Higher-order DTPD curves can similarly be defined.

## 2.2 Mathematical equations

On the surface of a catalyst, although there might be many kinds of active sites, one can assume that each kind is homogeneous, i. e. the desorption of each adsorbent conforms to the model

$$-\frac{d\theta}{dt} = A\exp\left(-\frac{E}{RT}\right)\theta^n \tag{1}$$

and with a constant heating rate, $dT/dt = \beta$, Eq. (1) becomes

$$-\frac{d\theta}{dT} = \left(\frac{A}{\beta}\right)\exp\left(-\frac{E}{RT}\right)\theta^n \tag{2}$$

with $A$ the pre-exponential factor, $T$ temperature, $E$ activation energy, $\beta$ heating rate, $n$ order, $\theta$ surface coverage, $t$ time, and $R$ the gas constant. Eq. (2) can be rearranged and integrated

$$-\int_{\theta_0}^{\theta}\frac{1}{\theta^n}d\theta = \frac{A}{\beta}\int_0^T \exp\left(-\frac{E}{RT}\right)dT \tag{3}$$

where $\theta_0$ is the initial surface coverage. Writing

$$F(T) = \frac{A}{\beta}\int_0^T \exp\left(-\frac{E}{RT}\right)dT \tag{4}$$

and if $X = E/RT$, one can get

$$F(T) = \frac{EA}{R\beta}\int_0^{(E/RX)} \exp(-X)d\left(\frac{1}{X}\right) \tag{5}$$

Because in most chemical desorption, $X \gg 1$, Eq. (5) can be expressed as[19]

$$F(T) = \frac{EA}{R\beta}\left[\frac{\exp(-X)}{X^2}\left(1 - \frac{2!}{X} + \frac{3!}{X^2} - \frac{4!}{X^3} + \cdots + (-1)^{m-1}\frac{m!}{X^{m-1}}\right)\right] \tag{6}$$

where m is a positive integer.

From Eqs. (3) and (4)

$$\theta = \theta_0 \exp[-F(T)] \qquad (n=1) \tag{7a}$$

$$\theta = [(n-1)F(T) + \theta_0^{1-n}]^{(1/1-n)} \qquad (n \neq 1) \tag{7b}$$

From Eqs. (7) and (2)

$$-\frac{d\theta}{dT}=\frac{A}{\beta}\exp\left(-\frac{E}{RT}\right)\exp[-F(T)]\theta_0 \qquad (n=1) \qquad (8a)$$

$$-\frac{d\theta}{dT}=\frac{A}{\beta}\exp\left(-\frac{E}{RT}\right)[(n-1)F(T)+\theta_0^{(1-n)}]^{(n/1-n)} \qquad (n\neq 1) \qquad (8b)$$

Eq. (2) can be differentiated to give

$$-\frac{d^2\theta}{dT^2}=-\left(\frac{d\theta}{dT}\right)\left[\left(\frac{n}{\theta}\right)\left(\frac{d\theta}{dT}\right)+\frac{E}{RT^2}\right] \qquad (9)$$

Eq. (9) can be differentiated to give

$$-\frac{d^3\theta}{dT^3}=\left(-\frac{d^2\theta}{dT^2}\right)\left[\frac{E}{RT^2}-\frac{2n}{\theta}\right]\left(\frac{-d\theta}{dT}\right)-\left(\frac{d\theta}{dT}\right)\left[\frac{n}{\theta^2}\left(-\frac{d\theta}{dT}\right)^2+\frac{2E}{RT^3}\right] \qquad (10)$$

If the parameters $A, E, n, \beta$ and $\theta_0$ are known, theoretical TPD, first DTPD and second DTPD curves can be simulated by using Eqs. (6)-(10). According to this principle, computer programs were written. A series of theoretic curves have been simulated.

# 3. Characteristics of DTPD

## 3.1 Characterization of DTPD

Figs. 1 and 2 are respectively the TPD, first DTPD and second DTPD curves of first and second-order desorption. One can see that after differentiation, the single peak of the TPD curve, is converted into one positive peak and one negative peak in the first DTPD curve; while there are two positive peaks and one negative peak in the second DTPD curve. Moreover, the peaks of the DTPD curves are sharper than those of the original TPD curves. For first-order desorption (Fig. 1), the maximum height of the positive peak is smaller than the maximum height of the negative peak (absolute value) in the first DTPD curve; and the maximum height of the positive peak at the lower temperature is smaller than the one of the positive peak at the higher temperature in the second DTPD curve. For second-order desorption (Fig. 2), the shapes are markedly different: the absolute values of the maximum heights of the positive and negative peaks are approximately the same in the first DTPD curve; and in the second DTPD curve, one sees a group of symmetrical peaks. Therefore, from the characteristics of the DTPD curves, one can determine the order of desorption fairly easily.

## 3.2 Separation of overlapping peaks

As shown in Fig. 3, overlapping peaks corresponding to two kinds of state are difficult to distinguish in the TPD curve. However, it is obvious in the first DTPD curve that there are two positive peaks (at 267 and 288 K) and two negative peaks (at 280 and 304 K). Because each state would have one positive and one negative peak in the first DTPD curve, the first DTPD curve with two positive peaks and two negative peaks indicates the existence of two states. The second DTPD curve shows three positive peaks (at 257, 284, and 309 K) and two negative peaks (at 275 and 299 K). Because each state would have two positive peaks and one negative peak in the second DTPD curve, from the second DTPD curve with three positive peaks and two negative peaks, one can conclude that there are two states present. Therefore, both the first and second DTPD curves reveal that there are two states of desorption present in the peak profile in Fig. 3. Hence the DTPD curves can reveal the number of states present.

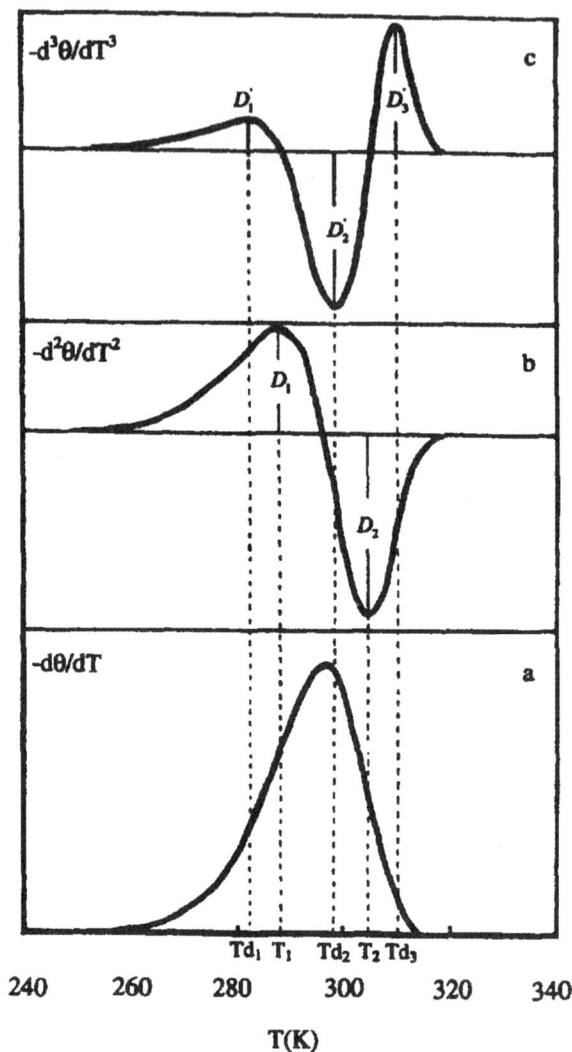

**Fig. 1** Theoretical DTPD curves: $n=1$, $A=10^{13}\ \mathrm{s}^{-1}$, $E=83.684\ \mathrm{kJ\ mol}^{-1}$, $\theta_0=1$, $\beta=10$
$\mathrm{Kmin}^{-1}$. a, TPD curve; b, first DTPD curve; c, second DTPD curve.

### 3.3 Elimination of baseline drift

As shown in Fig. 4, when there is a linear baseline drift, the TPD curve is obviously affected. However, in the first DTPD curve, the baseline is raised slightly and in the second DTPD curve there is no change at all. When there is a second-order-function baseline drift, there are changes in the TPD and the first DTPD curves. However, in the second DTPD curve, the baseline is only raised, indicating that as far as the elimination of baseline drift is concerned, DTPD curves have advantage over TPD curves, while the second DTPD curves have advantage over the first DTPD curves. This is because any function can be expressed as

$$-\frac{\mathrm{d}\theta}{\mathrm{d}T}=C_0+C_1T+C_2T^2+C_3T^3+\cdots+C_mT^m \tag{11}$$

On differentiation

$$-\frac{\mathrm{d}^2\theta}{\mathrm{d}T^2}=C_1+2C_2T+3C_3T^2+\cdots+mC_mT^{m-1} \tag{12}$$

one can see that the constant term is eliminated and the linear term has become a constant term,

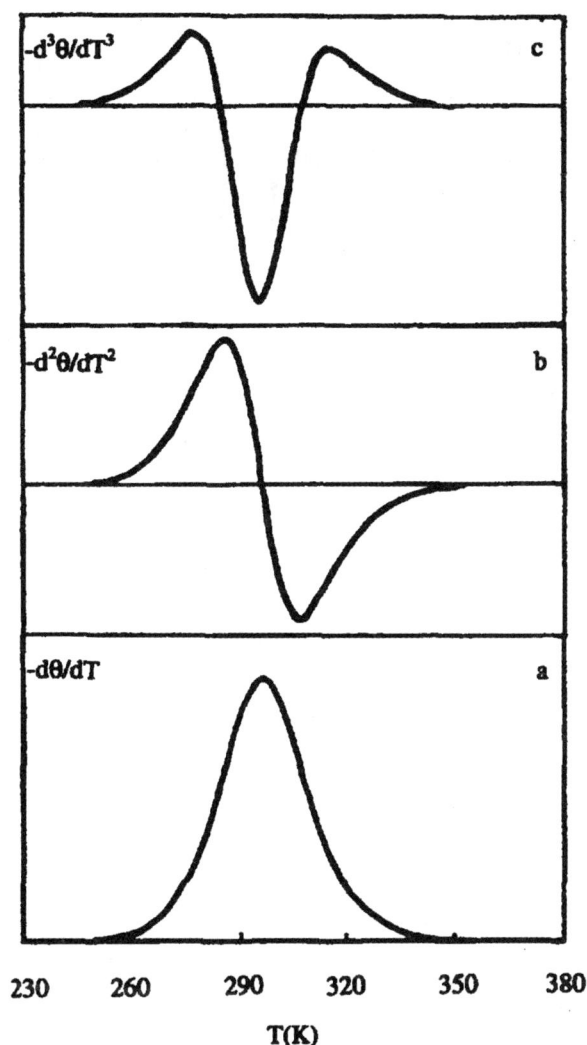

**Fig. 2** Theoretical DTPD curves: $n=2$, $A=10^{13}\,s^{-1}$, $E=83.\,684$ kJ mol$^{-1}$, $\theta_0=1$, $\beta=10$ Kmin$^{-1}$. a, TPD curve; b, first DTPD curve; c, second DTPD curve.

resulting in the elimination of the linear baseline drift. Eq. (12) can be differentiated to give

$$-\frac{d^3\theta}{dT^3}=2C_2+6C_3 T+\cdots+(m-1)mC_m T^{m-2} \tag{13}$$

One can see that the $(C_0+C_1 T)$ term of Eq. (11) is eliminated, i. e. the second-order term has become the constant term, indicating that the second-order-function baseline drift is eliminated. It can be seen that higher-order DTPD curves can eliminate lower-order-function baseline drift.

## 4. Equations for calculating desorption kinetic parameters from DTPD curves

According to Murray and White's expression[20], Eq. (2) can be written as

$$\ln\left(\frac{\theta_0}{\theta}\right)=\frac{A}{\beta}\left(\frac{RT^2}{E}\right)\left(1-\frac{2RT}{E}\right)\exp\left(-\frac{E}{RT}\right)(n=1) \tag{14a}$$

$$\frac{1}{\theta}\left(1-\frac{\theta}{\theta_0}\right)=\frac{A}{\beta}\left(\frac{RT^2}{E}\right)\left(1-\frac{2RT}{E}\right)\exp\left(-\frac{E}{RT}\right)(n=2) \tag{14b}$$

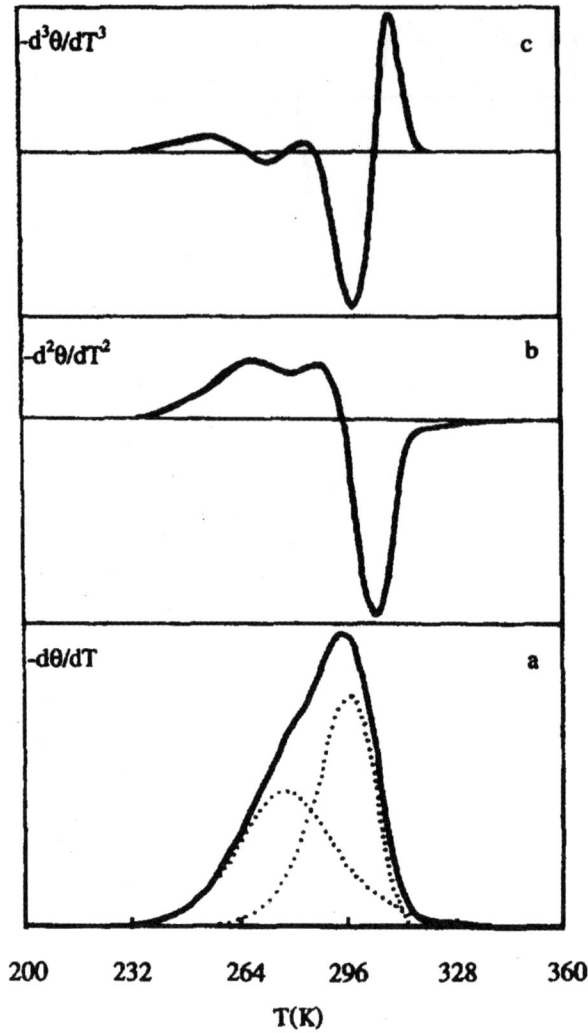

**Fig. 3** **Theoretical DTPD curves of overlapped peaks. State 1: $n=1$, $A=10^{13}$ s$^{-1}$, $E=83.684$ kJ mol$^{-1}$, $\theta_0=1$, $\beta$ $=10$ Kmin$^{-1}$. State 2: $n=2$, $A=10^{10}$s$^{-1}$, $E=63.353$ kJmol$^{-1}$, $\theta_0=1$, $\beta=10$K min$^{-1}$. a, TPD curve; b, first DTPD curve; c, second DTPD curve.**

In general, $2RT/E \ll 1$, and Eq. (14) can be simplified as

$$\ln\left(\frac{\theta}{\theta_0}\right)=\frac{A}{\beta}\left(\frac{RT^2}{E}\right)\exp\left(-\frac{E}{RT}\right)(n=1) \tag{15a}$$

$$\frac{1}{\theta}\left(\frac{\theta}{\theta_0}\right)=\frac{A}{\beta}\left(\frac{RT^2}{E}\right)\exp\left(-\frac{E}{RT}\right)(n=2) \tag{15b}$$

From Eqs. (2) and (15), one can get

$$-\frac{d\theta}{dT}=-\frac{E\theta}{RT^2}\ln\left(\frac{\theta}{\theta_0}\right)(n=1) \tag{16a}$$

$$-\frac{d\theta}{dT}=-\frac{E\theta}{RT^2}\left(\frac{\theta}{\theta_0}-1\right)(n=2) \tag{16b}$$

From Eqs. (16) and (9), one can get

$$-\frac{d^2\theta}{dT^2}=-\theta\ln\left(\frac{\theta}{\theta_0}\right)\left[\ln\left(\frac{\theta}{\theta_0}\right)+1\right]\left(\frac{E}{RT^2}\right)^2(n=1) \tag{17a}$$

$$-\frac{d^2\theta}{dT^2}=-\theta\left(\frac{\theta}{\theta_0}-1\right)\left[2\left(\frac{\theta}{\theta_0}\right)-1\right]\left(\frac{E}{RT^2}\right)^2(n=2) \tag{17b}$$

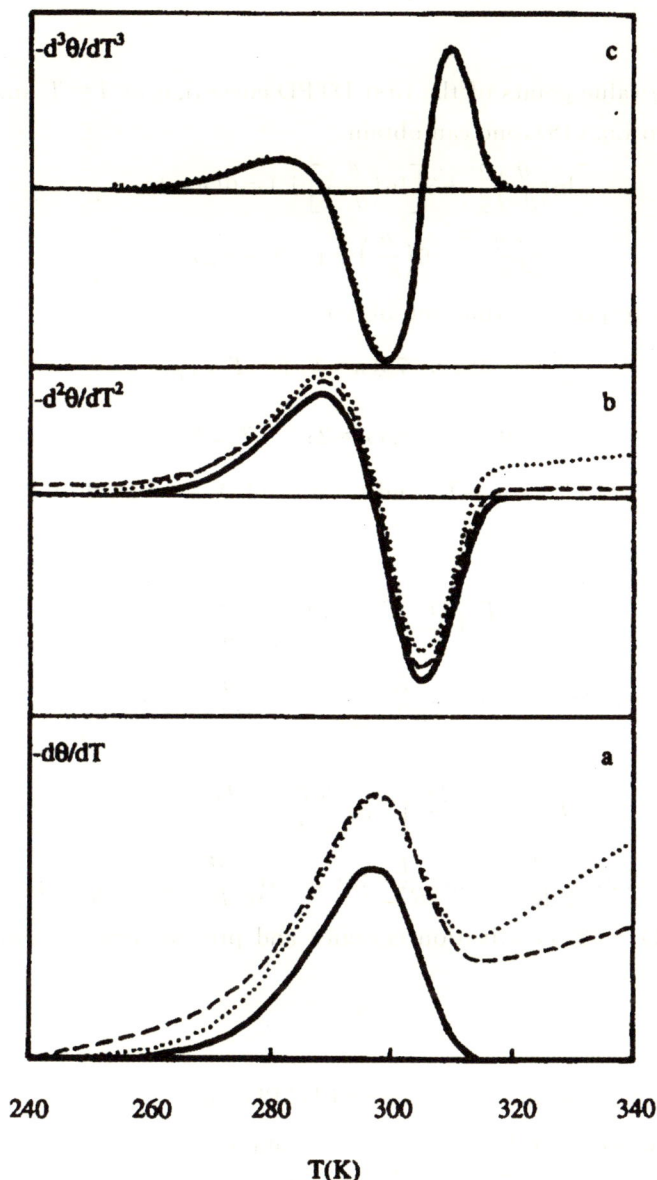

**Fig. 4** Theoretical DTPD containing baseline drift: $n=1$, $A=10^{13}\,\text{s}^{-1}$, $E=83.\,684\ \text{kJ mol}^{-1}$, $\theta=1$, $\beta=$ 10 K min$^{-1}$. a, TPD curve; c, second DTPD curve; —, no baseline; ---, containing linear baseline drift; ⋯, containing second-order-function baseline drift.

Eq. (17) can be differentiated to give

$$-\frac{\mathrm{d}^3\theta}{\mathrm{d}T^3}=-\theta\ln\left(\frac{\theta}{\theta_0}\right)\left[\left(1\mathrm{n}\left(\frac{\theta}{\theta_0}\right)\right)^2+31\mathrm{n}\left(\frac{\theta}{\theta_0}\right)+1\right]\left(\frac{E}{RT^2}\right)^3\ (n=1) \qquad (18\text{a})$$

$$-\frac{\mathrm{d}^3\theta}{\mathrm{d}T^3}=-\theta\left(\frac{\theta}{\theta_0}-1\right)\left[6\left(\frac{\theta}{\theta_0}\right)^2-6\left(\frac{\theta}{\theta_0}\right)+1\right]\left(\frac{E}{RT^2}\right)^3\ (n=2) \qquad (18\text{b})$$

From similar procedures, one can get

$$-\frac{\mathrm{d}^4\theta}{\mathrm{d}T^4}=-\theta\ln\left(\frac{\theta}{\theta_0}\right)\left[\left(1\mathrm{n}\left(\frac{\theta}{\theta_0}\right)\right)^3+6\left(\ln\left(\frac{\theta}{\theta_0}\right)\right)^2+7\ln\left(\frac{\theta}{\theta_0}\right)+1\right]\left(\frac{E}{RT^2}\right)^4\ (n=1) \qquad (19\text{a})$$

$$-\frac{\mathrm{d}^4\theta}{\mathrm{d}T^4}=-\theta\left(\frac{\theta}{\theta_0}-1\right)\left[24\left(\frac{\theta}{\theta_0}\right)^3-36\left(\frac{\theta}{\theta_0}\right)^2+14\left(\frac{\theta}{\theta_0}\right)-1\right]\left(\frac{E}{RT^2}\right)^4\ (n=2) \qquad (19\text{b})$$

## 4.1 First DTPD

Because at the extreme value points of the first DTPD curve, i. e. at $T=T_1$ and $T_2$, $(\mathrm{d}/\mathrm{d}t)(-\mathrm{d}^2\theta/\mathrm{d}T^2)=0$ (Fig. 1), according to Eq. (18), one can obtain

$$\left[\ln\left(\frac{\theta}{\theta_0}\right)\right]^2+3\left[\ln\left(\frac{\theta}{\theta_0}\right)\right]+1=0(n=1) \tag{20a}$$

$$6\left(\frac{\theta}{\theta_0}\right)^2-6\left(\frac{\theta}{\theta_0}\right)+1=0(n=2) \tag{20b}$$

From Eqs. (20a) and (20b) respectively, one can obtain

$$\frac{\theta}{\theta_0}=0.68, 0.073(n=1; T=T_1, T_2) \tag{21a}$$

$$\frac{\theta}{\theta_0}=0.79, 0.21(n=2; T=T_1, T_2) \tag{21b}$$

At the extreme values of $(-\mathrm{d}^2\theta/\mathrm{d}T^2)_{T=T_1}=D_1$ and $(-\mathrm{d}^2\theta/\mathrm{d}T^2)_{T=T_2}=D_2$ (Fig. 1), from Eqs. (15), (17) and (21), one can obtain

When $n=1$

$$E=\frac{RT_1^2}{0.4}\left(\frac{D_1}{\theta_0}\right)^{1/2} \text{ or } \frac{RT_2^2}{0.56}\left(-\frac{D_2}{\theta_0}\right)^{1/2} \tag{22}$$

$$A=0.38\left(\frac{\beta E}{RT_1^2}\right)\exp\left(\frac{E}{RT_1}\right) \text{or } 2.62\left(\frac{\beta E}{RT_2^2}\right)\exp\left(\frac{E}{RT_2}\right) \tag{23}$$

When $n=2$

$$E=\frac{RT_1^2}{0.31}\left(\frac{D_1}{\theta_0}\right)^{1/2} \text{ or } \frac{RT_2^2}{0.31}\left(-\frac{D_2}{\theta_0}\right)^{1/2} \tag{24}$$

$$A=0.27\left(\frac{\beta E}{\theta_0 RT_1^2}\right)\exp\left(\frac{E}{RT_1}\right) \text{or } 3.76\left(\frac{\beta E}{\theta_0 RT_2^2}\right)\exp\left(\frac{E}{RT_2}\right) \tag{25}$$

According to Eqs. (22)−(25), activation energies and pre-exponential factors can be calculated from first DTPD curves.

## 4.2 Second DTPD

Because at the extreme value points of the second DTPD curve, i. e. at $T=T_{d_1}$, $T_{d_2}$ and $T_{d_3}$, $(\mathrm{d}/\mathrm{d}T)(-\mathrm{d}^3\theta/\mathrm{d}T^3)=0$ (Fig. 1), according to Eq. (19), one can obtain

$$\left[\ln\left(\frac{\theta}{\theta_0}\right)\right]^3+6\left[\ln\left(\frac{\theta}{\theta_0}\right)\right]^2+7\ln\left(\frac{\theta}{\theta_0}\right)+1=0 \qquad (n=1) \tag{26a}$$

$$24\left(\frac{\theta}{\theta_0}\right)^3-36\left(\frac{\theta}{\theta_0}\right)^2+14\left(\frac{\theta}{\theta_0}\right)-1=0 \qquad (n=2) \tag{26b}$$

The two equations can be resolved to give

$$\frac{\theta}{\theta_0}=0.85, 0.26, 0.011 \qquad (n=1; T=T_{d_1}, T_{d_2}, T_{d_3}) \tag{27a}$$

$$\frac{\theta}{\theta_0}=0.91, 0.50, 0.092 \qquad (n=2; T=T_{d_1}, T_{d_2}, T_{d_3}) \tag{27b}$$

At the extreme values of $(-\mathrm{d}^3\theta/\mathrm{d}T^3)_{T=T_{d_2}}=D_1'$, $(-\mathrm{d}^3\theta/\mathrm{d}T^3)_{T=T_{d_2}}=D_2'$, $(-\mathrm{d}^3\theta/\mathrm{d}T^3)_{T=T_{d_3}}=D_3'$ (Fig. 1) Eqs. (15), (18), and (27) can be simplified

When $n=1$

$$E=\left(\frac{RT_{d_1}^2}{0.42}\right)\left(\frac{D_1'}{\theta_0}\right)^{1/3} \text{ or} \left(\frac{RT_{d_2}^2}{0.75}\right)\left(-\frac{D_2'}{\theta_0}\right)^{1/3} \text{or} \left(\frac{RT_{d_3}^2}{0.73}\right)\left(\frac{D_3'}{\theta_0}\right)^{1/3} \tag{28}$$

$$A=\frac{0.16\beta E}{RT_{d_1}^2}\exp\left(\frac{E}{RT_{d_1}}\right)\frac{1.34\beta E}{RT_{d_2}^2}\exp\left(\frac{E}{RT_{d_2}}\right) \text{or} \frac{4.51\beta E}{RT_{d_3}^2}\exp\left(\frac{E}{RT_{d_3}}\right) \tag{29}$$

When $n=2$

$$E=\left(\frac{RT_{d_1}^2}{0.35}\right)\left(\frac{D_1'}{\theta_0}\right)^{1/3} \text{ or }\left(\frac{RT_{d_2}^2}{0.50}\right)\left(\frac{D_2'}{\theta_0}\right)^{1/3} \text{ or }\left(\frac{RT_{d_1}^3}{0.35}\right)\left(\frac{D_3'}{\theta_0}\right)^{1/3} \tag{30}$$

$$A=\frac{0.10\beta E}{\theta_0 RT_{d_1}^2}\exp\left(\frac{E}{RT_{d_1}}\right) \text{ or }\frac{\beta E}{\theta_0 RT_{d_2}^2}\exp\left(\frac{E}{RT_{d_2}}\right) \text{ or }\frac{9.87\beta E}{\theta_0 RT_{d_3}^2}\exp\left(\frac{E}{RT_{d_3}}\right) \tag{31}$$

By using Eqs. (28)—(31), desorption activation energies and pre-exponential factors can be calculated from second DTPD curves.

## 4.3 The test of the equations by simulated DTPD curves

**Table 1  Quantitative results of the first derivative TPD curves simulated by computer**

| Theoretical | | | Calcd. [a] | | | |
|---|---|---|---|---|---|---|
| $n$ | $\lg A$ | $E/\text{Jmol}^{-1}$ | $\lg A_1$ | $E_1/\text{J mol}^{-1}$ | $\lg A_2$ | $E_2/\text{Jmol}^{-1}$ |
| 1 | 13 | 41842 | 13.19 | 42269 | 13.69 | 44583 |
| 1 | 13 | 62763 | 13.19 | 63353 | 14.00 | 67156 |
| 1 | 13 | 83684 | 13.20 | 84538 | 13.98 | 89232 |
| 1 | 13 | 104605 | 13.20 | 105697 | 14.00 | 111647 |
| 1 | 13 | 12026 | 13.21 | 127070 | 14.00 | 133961 |
| 1 | 13 | 146447 | 13.20 | 148041 | 13.99 | 156146 |
| 1 | 13 | 167368 | 13.20 | 169155 | 14.00 | 178544 |
| 1 | 5 | 83684 | 5.17 | 84776 | 6.04 | 95734 |
| 1 | 8 | 83684 | 8.18 | 84688 | 9.03 | 92337 |
| 1 | 10 | 83684 | 10.19 | 84700 | 11.02 | 90981 |
| 1 | 15 | 83684 | 15.21 | 84772 | 15.98 | 88617 |
| 2 | 13 | 41842 | 13.13 | 42043 | 14.64 | 46541 |
| 2 | 13 | 62763 | 13.13 | 63135 | 14.63 | 69633 |
| 2 | 13 | 83684 | 13.12 | 84023 | 14.63 | 92797 |
| 2 | 13 | 104605 | 13.13 | 105170 | 14.62 | 115680 |
| 2 | 13 | 125526 | 13.13 | 126233 | 14.62 | 138798 |
| 2 | 13 | 146447 | 13.13 | 147196 | 14.64 | 162146 |
| 2 | 13 | 167368 | 13.13 | 168284 | 14.62 | 184992 |
| 2 | 5 | 83684 | 5.12 | 83969 | 6.62 | 101287 |
| 2 | 8 | 83684 | 8.11 | 83951 | 9.62 | 96671 |
| 2 | 10 | 83684 | 10.13 | 84128 | 11.62 | 94676 |
| 2 | 15 | 83684 | 15.11 | 83960 | 16.64 | 91814 |

[a] $A_1$ and $A_2$ are respectively the calculated pre-exponential factors from the positive and negative peaks of the first derivative TPD curves. $E_1$ and $E_2$ are respectively the calculated desorption activation energies from the positive and negative peaks of the first derivative TPD curves.

According to Eqs. (22)—(25), activation energies and pre-exponential factors were calculated by using the positive peak and the negative peak of the simulated first DTPD curve. The results are shown in Table 1. From Table 1, it can be seen that the calculated activation energies and pre-exponential factors from the positive peaks are consistent with the actual values. However, the calculated results from the negative peaks have some deviation. This is because certain approximate treatments introduced in the calculation are more appropriate at low temperature than at high temperature. Therefore, in the analysis of first DTPD curves, positive peaks should be used.

Any of the three peaks in the second DTPD curve is suitable for the calculation of the desorption kinetic parameters. According to Eqs. (28)－(31), the activation energies and pre-exponential factors were calculated using the positive and negative peaks of the simulated second DTPD curve. The results are shown in Table 2. From Table 2, it can be seen that the calculated activation energies and the pre-exponential factors from positive peaks at low temperature are consistent with the actual values. However, calculated results from negative and positive peaks at higher temperature have some deviation. This is also because certain approximating treatments introduced in the calculation are more appropriate at low temperature than at high temperature. Therefore, in the analysis of second DTPD curves, positive peaks at low temperature should be used.

**Table 2  Quantitative results of the second derivative TPD curves simulated by computer**

| Theoretical | | | Calcd. [a] | | | | | |
|---|---|---|---|---|---|---|---|---|
| $n$ | $\lg A$ | $E/\mathrm{Jmol}^{-1}$ | $\lg A_1$ | $E_1/\mathrm{J\ mol}^{-1}$ | $\lg A_2$ | $E_2/\mathrm{Jmol}^{-1}$ | $\lg A_3$ | $E_3/\mathrm{J\ mol}^{-1}$ |
| 1 | 13 | 41842 | 12.51 | 40424 | 13.70 | 43779 | 14.17 | 45248 |
| 1 | 13 | 62763 | 12.52 | 60692 | 13.68 | 65525 | 14.23 | 68152 |
| 1 | 13 | 83684 | 12.50 | 80776 | 13.69 | 87433 | 14.23 | 90780 |
| 1 | 13 | 104605 | 12.51 | 101099 | 13.70 | 109325 | 14.23 | 113413 |
| 1 | 13 | 125526 | 12.51 | 121342 | 13.70 | 131130 | 14.22 | 135974 |
| 1 | 13 | 146447 | 12.51 | 141468 | 13.71 | 153163 | 14.23 | 158765 |
| 1 | 13 | 167368 | 12.51 | 161740 | 13.71 | 175033 | 14.23 | 181410 |
| 1 | 5 | 83684 | 4.49 | 77529 | 5.65 | 90496 | 6.23 | 98216 |
| 1 | 8 | 83684 | 7.49 | 79232 | 8.68 | 89065 | 9.23 | 94266 |
| 1 | 10 | 83684 | 9.49 | 79922 | 10.69 | 88316 | 11.22 | 92442 |
| 1 | 15 | 83684 | 14.51 | 81186 | 15.72 | 87261 | 16.23 | 90023 |
| 2 | 13 | 41842 | 12.32 | 39859 | 13.87 | 44181 | 15.13 | 47976 |
| 2 | 13 | 62763 | 12.31 | 59822 | 13.87 | 66219 | 15.12 | 71793 |
| 2 | 13 | 83684 | 12.29 | 79567 | 13.86 | 88161 | 15.13 | 95743 |
| 2 | 13 | 104605 | 12.30 | 99567 | 13.87 | 110325 | 15.14 | 119693 |
| 2 | 13 | 125526 | 12.29 | 119451 | 13.85 | 132070 | 15.12 | 143234 |
| 2 | 13 | 146447 | 12.30 | 139535 | 13.86 | 154117 | 15.12 | 167079 |
| 2 | 13 | 167368 | 12.30 | 159468 | 13.87 | 176289 | 15.13 | 191063 |
| 2 | 5 | 83684 | 4.38 | 76261 | 5.87 | 92040 | 7.2 | 108572 |
| 2 | 8 | 83684 | 7.36 | 78186 | 8.87 | 90149 | 10.18 | 101776 |
| 2 | 10 | 83684 | 9.33 | 78889 | 10.86 | 89132 | 12.16 | 98760 |
| 2 | 15 | 83684 | 14.30 | 80132 | 15.87 | 87759 | 17.12 | 94408 |

[a] $A_1$, $A_2$ and $A_3$ are respectively the calculated pre-exponential factors from the positive peak at low temperature, and the negative and positive peaks at high temperature of the second derivative TPD curves. $E_1$, $E_2$ and $E_3$ are respectively the calculated desorption activation energies from the positive peak at low temperature, and the negative and positive peaks at high temperature of the second derivative TPD curves.

# Acknowledgement

We thank the Natural Science Foundation of China (NSFC) for supporting this research.

# References

[1]S. Bahtia, J. Beltramini and D. D. Do, Catalysis Today, **7** (1990) 1.

[2]R. J. Cvetanovic and V. Amenomiya, Catal. Rev. -Sci. Eng., **6** (1972) 21.

[3]Y. P. Arnaud, Appl. Sci., **62** (1992) 21.

[4]B. Fastrup, M. Muhler, H. Nygard Nielsen and L. Pleth Nielsen, J. Catal., **142** (1993) 135.

[5]A. Rochefort, F. L. Peltier and J. P. Boitiaux, J. Catal., **145** (1994) 409.

[6]J. A. Konvalinka, J. J. F. Scholten and J. C. Rasser, J. Catal., **48** (1977) 365.

[7]P. I. Lee and J. A. Schwarz, J. Catal., **73** (1982) 272.

[8]J. L. Falconer and J. A. Schwarz, Catal. Rev. -Sci. Eng., **25** (1983) 14.

[9]J. S. Riech and AT. Bell, J. Catal., **85** (1984) 143.

[10]Y. H. Hu, Petrochemical Technology of China, **15** (1986) 167.

[11]J. M. Criado, P. Malet and G. Munuera, Langmuir, **3** (1987) 973.

[12]T. Ioannides and X. E. Verykios, J. Catal., **120**(1989) 157.

[13]Y. H. Hu, A. M. Huang, J. X. Cai and H. L. Wan, Chem. J. Chin. Univ., **13** (1992) 952.

[14]Y. H. Hu, H. L. Wan and K. R. Tsai, Chem. J. Chin. Univ., **14** (1993) 238.

[15]P. Malet, in J. L. G. Fierro (Ed. ), Studies in Surface Science and Catalysis, Vol. 57B, Elsevier Science, Amsterdam, 1990, p. B333, and references cited therein.

[16]R. N. Hager, Jr., Anal. Chem., **45** (1973) 1131 A.

[17]J. Alvarer and M. C. Asensio, in J. L. G. Fierro (Ed. ), Studies in Surface Science and Catalysis, Vol. 57A, Elsevier Science, Amsterdam, 1990, p. A79, and references cited therein.

[18]M. Che, B. Canosa and A. R. Gonzalez-Elipe, J. Phys. Chem., **90** (1986) 618.

[19]A. W. Smith and S. Aranoff, J. Phys. Chem., **62** (1958) 684.

[20]M. P. White, J. Trans. Br. Ceram. Soc., **54** (1955) 15.

■ 本文原载:J. Chem. Inf. Comput. Sci. 36(1996),pp. 1178~1182.

# Computer-Aided Molecular Design of Catalysts Based on Mechanism and Structure

Dai-Wei Liao[①], Zun-Nan Huang, Yin-Zhong Lin, Hui-Lin Wan,
Hong-Bin Zhang, Khi-Rui Tsai

(*Department of Chemistry, Institute of Physical Chemistry and State Key Laboratory for Physical Chemistry of the Solid Surface, Xiamen University, Xiamen, Fujian 361005, China*)
Received September 14, 1995[①]

**Abstract**    A bridge theory including a mathematical expression for the process of catalysis is proposed. A catalytic reaction can be expressed as ‖reactantmatrix‖ × ‖bridgematrix‖=‖productmatrix‖. The physicochemical bases of computer-aided molecular design of catalysts are discussed. Five variations on both structure and mechanism are identified as important in the process of catalyst design. An expert system of molecular design of catalysts (ESMDC) including a knowledge base and a reasoning scheme has been developed and is described.

## 1. INTRODUCTION

The catalytic process has been playing an increasingly important role in many fields of modern science and industry. Development of a commercial industrial catalyst however is a task requiring much time (7－10 years) and money because thousands of experimental tests are involved. As a consequence, the task of developing more effective catalysts has come into focus in recent years. Trimm[1], a pioneer in catalyst design, has analyzed the causes of success and failure in practical design of some industrial catalysts published before 1980 and has suggested a scientific basis for catalyst design. Cusumano[2], Zamaraev[3], and Likolobov[4] have each provided a perspective on the molecular design of industrial catalysts, and, since 1970, significant work on catalyst design has also been published by Liao[5], Hegedus[6], and Graziani[7].

More recently, advances in computer power have led to the development of reliable bases and useful tools for the design of catalysts at the molecular level, but, to date, no practical expert system for molecular design of catalysts has been reported. The main difficulties facing computer-aided molecular design of catalysts are the absence of a uniform mechanism and mathematical representation for all catalytic reactions. It is also uncertain which variations in structure and mechanism are important to catalyst design. Thus it is a problem to decide on the knowledge rules and reasoning process to be used

---

① Abstract published in *Advance ACS Abstracts*, September 1,1996.

for a specific catalytic reaction.

In the next section of this paper, we propose a bridge theory with a mathematical expression for the catalytic process. In section 3, the physicochemical bases of molecular design of catalysts and catalytic processes from the periodic properties of elements and compounds are discussed, and, in section 4, using the oxidative coupling of methane (OCM) as an example, a practical expert system, ESMDC1995, including a knowledge base and a reasoning algorithm, for the design of metal oxide catalysts is described.

# 2. BRIDGE THEORY AND MATHEMATICAL EXPRESSION FOR CATALYSIS

The development of catalysis theory can be regarded as having gone through several periods: (1) an initial hypothesis of "magic" catalytic force, (2) the early pure chemical, physical, and physicochemical theories of catalysis, (3) recent theory of catalysis, and (4) the modern theory of catalysis. All these various theories and hypotheses attempted to explain the phenomenon of catalysis, but a uniform theory of catalysis is still elusive. Modern techniques however have provided a deep insight into the nature of catalysis and catalysts, but the search for a general and common principle of catalysis remains a paramount problem.

An individual catalytic reaction can be described at the molecular or the atomic level. A hypothesis of "frequency catalysis" involving a relationship between motion, frequency, and energy was proposed by Liao who also recently proposed a new hypothesis for microcatalysis and information catalysis in the life science[8]. Catalysis, in nature, is both a physical and chemical process. A complete explanation of catalysis however also requires a mathematical view, and efforts to provide this have been reported by Balaban[9], Ugi[10], Koca, and Kvasnicka et al[11]. Who have examined the applicability of graph theory to the chemical problem. A program, COMSYCAT, developed by Likolobov and co workers[12], is a mathematical model of the structural chemistry implied by the sequence $\{X + M*\} \rightarrow \{ZM*\} \rightarrow \{Y + M*\}$ in which the molecular system $\{X+M*\}$ is described as a structural bond-electron matrix. Other attempts to clarify the catalysis problem have been made. Hoffmann has attempted to build bridges between inorganic and organic chemistry, Fukui has studied reaction pathway, and Taube has examined the rate at which electrons are transmitted through the bridge of a metal ligand. In this paper, we develop a bridge theory for which we provide a mathematical representation which deals with three distinct species.

In the bridge theory, the catalyst is treated as a bridge between reactants and products. Any reactant or catalyst can be represented as a mathematical set, and any component of either is defined as an element of the set to which it belongs. All reactants and catalysts can be defined as a product of these sets. Thus the bridge representation of a catalytic process corresponds to a map. The mathematical meaning of a map is a rule, $\phi$. Only one element $d$ of a set $D$ can be obtained by the operation of a rule upon any element of a set $(a_1, a_2, \cdots, a_n")(a_i \in A_i)$, and it can be expressed as follows:

$$\phi; (a_1, a_2, \cdots, a_n) \rightarrow d = \phi(a_1, a_2, \cdots, a_n)$$

The reverse catalytic process can be expressed as $\phi^{-1}$ in bridge representation. An active center and an active adsorption can be also represented as a subset and a mapping, respectively.

With this preamble, we can use matrix algebra to explain the mechanism of catalytic reactions and in the design of catalysts. This done by introducing the concept of both set and group into the mathematical representation of catalysis. A bridge matrix can be defined as consistent with the following expression：

||reactant matrix||×||bridge matrix||＝||product matrix||

The key step in determination of the matrix representing the target species is the determination of the bridge matrix given that the following expression is thermodynamically reasonable：

||species A|| ||species Bll···||species M||＝||species 1|| ||species 2||···||species N||

For example，if suppose species 1 is a target product，we can use the following expression to represent the catalytic process：

||species A|| ||species B||···||bridge matrix||＝||target species 1||

where the bridge matrix includes matrix of active center, promoter matrix, and matrix of reaction conditions and so on.

For an easy understanding of the above matrix expression，here we show a mathematical example. For example，if we use a matrix

$$\left\| \begin{matrix} 1 & 1 \\ 2 & 1 \end{matrix} \right\|$$

for representing a $H_2$ molecule and a matrix

$$\left\| \begin{matrix} 7 & 3 \\ 14 & 3 \end{matrix} \right\|$$

for a $N_2$ molecule and

$$\left\| \begin{matrix} 42 & 42 \\ 42 & 42 \end{matrix} \right\|$$

for a $NH_3$ molecule，it can be shown that the best bridge matrix is

$$\left\| \begin{matrix} 6 & 7 \\ 3 & 14 \end{matrix} \right\|$$

because

$$\left\| \begin{matrix} 1 & 1 \\ 2 & 1 \end{matrix} \right\| \left\| \begin{matrix} 7 & 3 \\ 14 & 3 \end{matrix} \right\| = \left\| \begin{matrix} 7 & 14 \\ 6 & 3 \end{matrix} \right\| \text{ and } \left\| \begin{matrix} 7 & 14 \\ 6 & 3 \end{matrix} \right\| \left\| \begin{matrix} 6 & 7 \\ 3 & 14 \end{matrix} \right\| = \left\| \begin{matrix} 42 & 42 \\ 42 & 42 \end{matrix} \right\|$$

It is not so simple，of course，for a catalytic process，in which various codes are used to represent matrix elements and both algebraic and logic operations are used to calculate them. Likholobov and his co-workers suggested an alternative coding in which the matrix representations for CO, $H_2$, and metal surface are

$$\left\| \begin{matrix} 2 & 2 \\ 2 & 4 \end{matrix} \right\|, \left\| \begin{matrix} 0 & 1 \\ 1 & 0 \end{matrix} \right\| \text{ and} ||1||$$

respectively.

How is a catalytic process expressed in a mathematical method of bridge theory? consider the synthesis of ammonia on an iron catalyst，$N_2 + 3H_2 = 2NH_3$，as an example. All molecules of nitrogen and all molecules of hydrogen can be expressed as a set，$\boldsymbol{N_2} = \{N_2^a, N_2^b, \cdots, N_2^z\}$ and $\boldsymbol{H_2} = \{H_2^a, H_2^b, \cdots, H_2^\beta\}$，respectively. Let us suppose that the catalyst is simply composed of only element Fe which is expressed as a set $\boldsymbol{Fe_n} = (Fe_n^a, Fe_n^b, \cdots, Fe_n^\gamma)$. The promoter is not considered for the moment. The set $\boldsymbol{Fe_n}$ is a subset of a set $\boldsymbol{Fe}$. We can note $\phi: N_2^a \rightarrow N_2^a$ and $\phi: H_2^a \rightarrow 2H^a$ or $H_2^a$ for chemisorption of $N_2$ and $H_2$ molecule，respectively. For a complete set，however，we can note $\phi_1: N_2 \rightarrow N_2$ and $\phi_1: H_2 \rightarrow 2H$ or $H_2$，respectively. Both set $\boldsymbol{N_2}$ and set $\boldsymbol{H_2}$ do not map fully to $\phi_1$ which corresponds to the set $\boldsymbol{Fe}$. Only for the map $\phi_2$ corresponding to the subset $\boldsymbol{Fe_n}$, the set $\boldsymbol{N_2}$ and set $\boldsymbol{H_2}$ are fully mapped one by one. In chemical language，nitrogen and hydrogen can be actively chemisorbed and combined with each other only if they are adsorbed on an active center of $Fe_n$ cluster. So we have $\phi_2: N_2 \rightarrow N_2$ and $\phi_2^{-1}: N_2 \rightarrow N_2$ as well as $\phi_2 : H_2 \rightarrow 2H$ or $H_2$ and $\phi_2^{-1} : 2H$ or $H_2 \rightarrow H_2$. The adsorption and desorption are in equilibrium.

Define a merge set $N_2 \bigcup H_2 = \{N_2^a + H_2^a, N_2^b + H_2^b, \cdots, N_2^a + H_2^\beta\} =$ set $NH_3 = \{NH_3^a, NH_3^b, \cdots,$ $NH_3^\delta\}$. Suppose the set $Fe_n$ is an empty subset corresponding to the set $N_2$ and $H_2$. We can note that $N_2 = \{N_2^a, N_2^b, \cdots, N_2^a, Fe_n^a, Fe_n^b, \cdots, Fe_n^\gamma\}$ and $H_2 = \{H_2^a, H_2^b, \cdots, H_2^\beta, Fe_n^a, Fe_n^b, \cdots, Fe_n^\gamma\}$. Therefore, we have a merge set $NH_3 = \{NH_3^a, NH_3^b, \cdots, NH_3^\delta, Fe_n^a, Fe_n^b, \cdots, Fe_n^\beta\}$. Mathematically, this means that the catalyst is unchanged after the reaction. If the symbol "$\circ$" is used to express a set operation, we can note $(N_2, H_2) \rightarrow 1/2N_2 \, 3/2H_2 = 1/2N_2 \circ 3/2H_2$. Also, $(N_2, H_2, Fe_n) \rightarrow 1/2N_2 \, 3/2H_2 Fe_n = (1/2N_2 \circ 3/2H_2) \circ Fe_n$ can be expressed as the following matrix expression:

$$
\frac{1}{2}
\begin{vmatrix}
N_{2,a1} & N_{2,a2} & \cdots & \cdots & N_{2,ai} \\
N_{2,b1} & N_{2,b2} & \cdots & \cdots & N_{2,bi} \\
\cdots & \cdots & \cdots & \cdots & \cdots \\
\cdots & \cdots & \cdots & \cdots & \cdots \\
N_{2,n1} & N_{2,n2} & \cdots & \cdots & N_{2,ni}
\end{vmatrix}
\frac{3}{2}
\begin{vmatrix}
H_{2,a1} & H_{2,a2} & \cdots & \cdots & H_{2,ai} \\
H_{2,b1} & H_{2,b2} & \cdots & \cdots & H_{2,bi} \\
\cdots & \cdots & \cdots & \cdots & \cdots \\
\cdots & \cdots & \cdots & \cdots & \cdots \\
H_{2,n1} & H_{2,n2} & \cdots & \cdots & H_{2,ni}
\end{vmatrix}
$$

$$
\begin{vmatrix}
Fe_{2,a1} & Fe_{2,a2} & \cdots & \cdots & Fe_{2,ai} \\
Fe_{2,b1} & Fe_{2,b2} & \cdots & \cdots & Fe_{2,bi} \\
\cdots & \cdots & \cdots & \cdots & \cdots \\
\cdots & \cdots & \cdots & \cdots & \cdots \\
Fe_{2,n1} & Fe_{2,n2} & \cdots & \cdots & Fe_{2,ni}
\end{vmatrix}
=
\begin{vmatrix}
NH_{2,a1} & NH_{2,a2} & \cdots & NH_{2,ai} \\
NH_{2,b1} & NH_{2,b2} & \cdots & NH_{2,bi} \\
\cdots & \cdots & \cdots & \cdots \\
\cdots & \cdots & \cdots & \cdots \\
NH_{2,n1} & NH_{2,n2} & \cdots & NH_{2,ni}
\end{vmatrix}
$$

The mathematical expression including map, group, and matrix is useful and helpful for the computer-aided design of catalysts.

# 3. PHYSICOCHEMICAL BASES FOR MOLECULAR DESIGN OF CATALYSTS

Ioffe et al[13]. have carried out excellent work in application of pattern recognition to catalytic research. More recently, Kuntz et al.[14] summarized their experience with the DOCK program in the drug discovery process and the challenges that lie ahead for structure-based design. We have investigated molecular design of metal oxide catalysts for oxidation coupling of methane (OCM). Based on the periodic properties of elements and compounds, five physicochemical variations, namely, (1) geometric topology, (2) electronic structure, (3) magnetic property, (4) other physicochemical properties, and (5) catalytic active model, are identified as important for the computer-aided molecular design of catalysts.

The variations of (1) to (4) can be represented and evaluated with a series of relative periodic properties of elements and compounds which are components of a catalyst. The following properties, as suggested by Trimm[1], are usually important and useful in the process of design of oxide catalysts: radius of ion, acidity, and basicity, affinity for gases, the parameter $d\%$, electronic structure of $d^0$ and $d^{10}$, coordination model, electronegativity, ionization potential, work function, magnetic susceptibility, lattice parameter, surface energy, adsorption heat, valence state, partial charge of oxygen, stability, melting point, entropy of fusation, melting heat, and formation heat, etc.

The exchange of ions with similar radius, cation with cation, or anion with anion may have an effect on a significant change for both catalytic activity and selectivity. An active catalyst for OCM is mostly composed of the alkali and alkaline earth metal compounds and, also, rare earth metal compounds or other metal compounds, especially with variable valence. A pair of matching cations usually has a close or similar radius, as in the case of $Li^+$ (0.068 nm) vs $Mg^{2+}$ (0.074 nm) and $Na^+$ (0.098 nm) vs $Ca^{2+}$ (0.104 nm). The replacements between cations or between anions with different valence states, in this case, may take place and cause such change on the surface lattice as the generation of holes and so on.

In general, acidity is favorable for dehydration, but basicity is favorable for dehydrogenation. It is a continuous property and generally related to catalytic activity. Also, acidity favors carbonization, but basicity favors protection against carbonization. The oxide with less partial charge on oxygen has stronger acidity. The electronegativity of metal ions also corresponds to their acidity.

If a metal element has an affinity for a gaseous molecule, generally, the metal and its compounds can catalyze the catalytic reaction involving the gaseous molecule. The magnitude of affinity usually can be expressed as heat of adsorption. Obviously, a gaseous molecule with larger affinity for a metal is more active.

The parameter $d\%$ usually is related to the adsorption and catalytic selectivity. It represents the contribution of $d$ orbitals to $dsp$ hybridized orbitals for a transition metal. An metal oxide with electronic structure of $d^0$ or $d^{10}$ usually has good selectivity for catalytic oxidative reaction.

A coordination model of adsorbed species on surface of solid catalyst generally relates to the activity of a catalyst. The adsorption model usually can be divided into two types, namely, associate and dissociate or molecular and atomic. The chemisorbed species may be electrically neutral, positive, or negative.

The direction and pathway of transfer of both electrons and charges in catalysis usually may be correlated with the electronegativity of the catalyst components, and the electronegativity can be divided into electronegativity of elements, orbitals, ions, or groups. The ionization potential also has an effect on the transfer of electrons in catalysis and is related to the electronegativity and work function. The orbitals of an element with a larger ionization potential are more occupied. The exit of electrons from surfaces of metal with less work function into vacuum is facilitated.

Magnetic susceptibility, a very important factor in magnetic catalysis, is often related to $f$ electrons. Paramagnetism or diamagnetism are also important parameters in magnetic catalysis. The lattice parameter is an important parameter in molecular design of catalysts based on structure. Catalytic activity and selectivity generally relate to the stereo structure of catalysts. The geometric topology usually has a considerable effect on catalysis.

The catalytic active model is the most complex and difficult in the process of catalyst design. It mostly depends on the type of a catalytic reaction and its catalysts. Fortunately, the development in the field of catalysis in the past decades has provided a lot of useful information. We can find almost all types of catalytic reactions and catalysts as well as their mechanisms from various handbooks and literatures as a reference for the catalyst design. However, the individual knowledge and experience of experts in catalysis are often the most important and decisive factor.

The first four variations provide physicochemical bases for catalyst design based on structure, but the last variation provides the bases of design based on mechanism. A good design should be based on both structure and mechanism.

# 4. CONSTRUCTION OF AN EXPERT SYSTEM FOR MOLECULAR DESIGN OF CATALYSTS

Based on above physicochemical bases in both structure and mechanism and the mathematical expression method, an expert system for the molecular design of heterogeneous oxide catalysts has recently been developed in our group. Using this system, we can predict and theoretically prepare a catalyst for a desired catalytic reaction catalyzed by oxidic catalysts. We also can find information useful to an explanation of the mechanism of a known catalytic reaction. Figure 1 shows the system flow diagram for computer-aided molecular design of catalyst components. The system structure of our Export System of the Molecular Design of Catalyst (ESMDC) is shown in Figure 2.

Now, we take the design of composition of metal oxide catalysts for OCM as an example. If the alkali metal cation $Li^+$, $Na^+$, and $K^+$, respectively, is selected as a reference ion, the recommended matching ions in catalyst for OCM are as follows. (1) For alkali metal ion $Li^+$, we can find that alkaline earth metal ion $Mg^{2+}$, rare earth metal ions $Lu^{3+}$, $Yb^{3+}$, $Sc^{3+}$, transition metal ions $Ti^{3+}$, $Ti^{4+}$, $Ti^{2+}$, $V^{3+}$, $V^{2+}$, $V^{4+}$, $Nb^{4+}$, $Nb^{5+}$, $Ta^{5+}$, $Cr^{2+}$, $Cr^{3+}$, $Mo^{4+}$, $W^{4+}$, $W^{6+}$, $Mn^{3+}$, $Mn^{2+}$, $Fe^{2+}$, $Fe^{3+}$, $Ru^{4+}$, $Os^{3+}$, $CO^{3+}$, $CO^{2+}$, $Rh^{3+}$, $Rh^{4+}$, $Ni^{2+}$, $Ir^{4+}$, $Pt^{4+}$, $Cu^{2+}$, $Zn^{2+}$, etc., other metal ions $Al^{3+}$, $Ga^{3+}$, $Ga^{2+}$, $Sn^{4+}$, $As^{3+}$, $Pb^{4+}$, $Bi^{5+}$, $Sb^{5+}$, and anions $O^{2-}$, $F^-$ may be a good matching ion. (2) For alkali metal ion $Na^+$, alkaline earth metal ions $Ca^{2+}$, $Sr^{2+}$, rare earth metal ions $La^{3+}$, $La^{4+}$, $Ce^{3+}$, $Ce^{4+}$, $Y^{3+}$, $Pr^{3+}$, $Nd^{3+}$, $Pm^{3+}$, $Sm^{3+}$, $Eu^{3+}$, $Gd^{3+}$, $Dy^{3+}$, $Th^{4+}$, $Pa^{3+}$, $Pa^{4+}$, $Nb^{3+}$, $Pu^{3+}$, $Am^{3+}$, transition metal ions $Mn^{2+}$, $Ti^{2+}$, $Zr^{4+}$, $Cr^{2+}$, $Hf^{4+}$, $Ag^+$, $Cu^+$, $Hg^{2+}$, other metal ions $In^{3+}$, $Cd^{2+}$, $Ti^{3+}$, $Sn^{2+}$, $Sb^{3+}$, $Te^{4+}$, and anions $Cl^-$, $Br^-$ may be a good matching ion. (3) For alkali metal ion $K^+$, alkaline earth metal ions $Ba^{2+}$, $Sr^{2+}$, $Ra^{2+}$, rare earth metal ions $Ac^{3+}$, transition metal ions $Ag^+$, other metal ions $Bi^{3+}$, $Tl^{3+}$, $Pb^{2+}$, and anions $F^-$, $O^{2-}$ may be a good matching ion. Again, if the $La^{3+}$ is selected as a reference ion, we can find only the $Ca^{2+}$ ion as a recommended matching ion among the alkaline earth metal ions. The selection should follow some physicochemical rules in both structure and mechanism. Figure 3 shows the procedure of system design for combination of different ions with specific properties to form a complex oxide with a defect lattice.

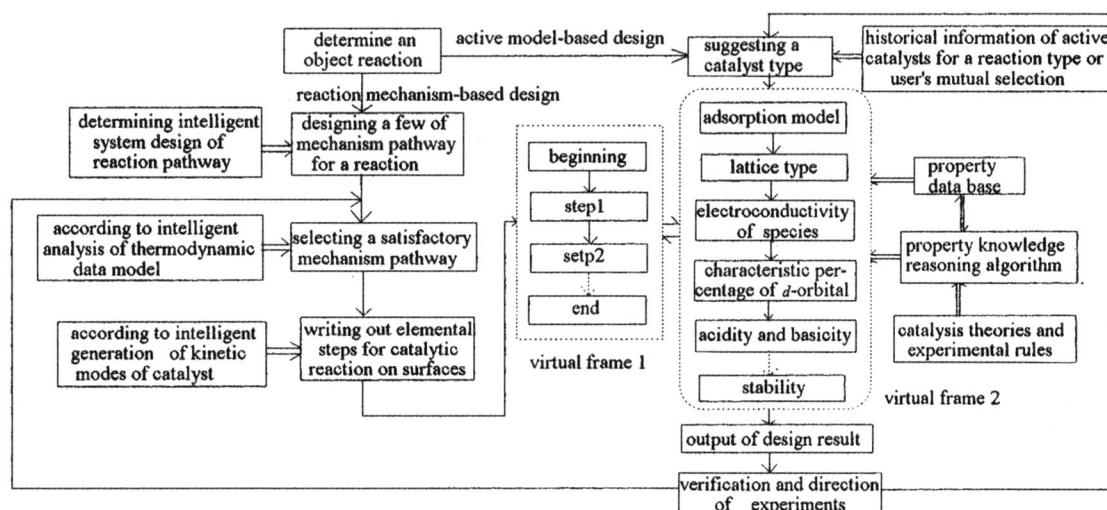

Figure 1　System flow diagram for computer-aided molecular design of catalyst components.

The system includes a user-computer interface, a knowledge collection scheme, a property data base, a knowledge rule base, an explanation scheme, a memory scheme, and a reasoning scheme. The general knowledge base is divided into two bases. The property data base is composed of periodic properties of elements and compounds. The knowledge rule base is composed of the knowledge rules from the field of catalysis.

The collection and expression of data and knowledge should be consistent with the reasoning performance. The knowledge in catalysis can be classified as follows: (1) data type knowledge which involves periodic properties of elements and compounds, etc. and can be collected mostly from various

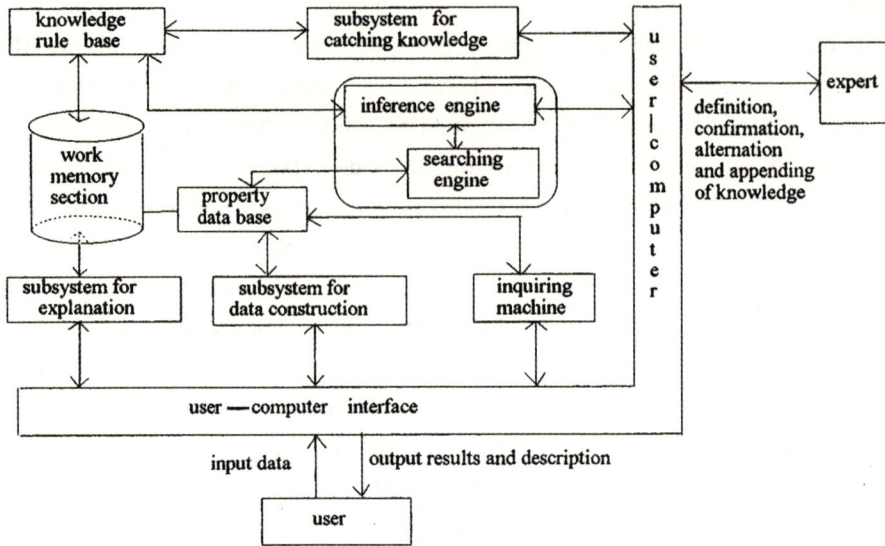

**Figure 2 System structure of ESMDC.**

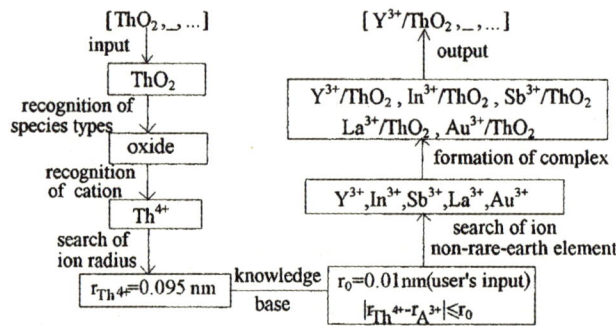

**Figure 3 Procedure of system design for combination of different ions with specific properties to form complex oxide with defect lattice.**

handbooks such as CRC and literatures and can be expressed in a frame structure, (2) descriptive type knowledge which involves the principle, mechanism, and structure, etc. in catalysis and can be collected from the facts and experimental results in literature and experience of experts, (3) solution type knowledge which is an ambiguous and blurring judgment and can be collected from literature and experience, and (4) process type knowledge, which is a knowledge of a control scheme, is a comprehensive knowledge of principles, facts, and experiences and forms a reasoning frame. The last three types of knowledge can be expressed as a generative expression and predicate logic expression. For example, we can write down the following rules of knowledge:

[Rule 1] (Catalyst has active phase) $\wedge$ (Catalyst has promoter) $\rightarrow$ (The catalyst is an active catalyst)

[Rule 2] (component has defect structure) $\wedge$ (component has $CaF_2$ type lattice) $\rightarrow$ (The component is an active phase)

[Rule 3] (complex, active phase) $\wedge$ {(complex, promoter) $\vee$ (complex, stabilizator) $\vee$ (complex, inert diluter) $\vee$ (complex, quencher)} $\rightarrow$ (It is an active complex catalyst)

[Rule 4] (complex, oxide) $\wedge$ cation (oxide, stable valence) $\rightarrow$ (It may be an active complex catalyst)

[Rule 5] If $(0.79/1.32) \leqslant (r_M^{n+}/r_o^{2-}) \leqslant (1.34/1.33)$ then the oxide $M_2O_n$ can form a lattice $CaF_2$

**Table 1  Some Output of Catalyst component Design for OCM**

| | |
|---|---|
| with a $CaF_2$ lattice, n-type metal cation with stable valence melting point $> 500$ ℃, p-type | $CaO, ZnO, SrO, CdO, Sc_2O_3, La_2O_3, Bi_2O_3$ |
| to be able to form a defect CaF2 lattice, matching ions with radius of $\pm 0.01$ nm, metal cation with variable valences, $T_{min}$ of phase transformation $> 500$ ℃ | $Na^+/MnO, Cu^+/MnO, Sc^{3+}/ZrO_2, Rh^{3+}/ZrO_2, Sb^{3+}/ZrO_2, Ir^{3+}/ZrO_2, Au^{3+}/ZrO_2, Sc^{3+}/CeO_2, Y^{3+}/CeO_2, In^{3+}/CeO_2, Sb^{3+}/CeO_2, Au^{3+}/CeO_2, Y^{3+}/ThO_2, In^{3+}/ThO_2, Sb^{3+}/ThO_2, La^{3+}/ThO_2, Au^{3+}/ThO_2, Sc^{3+}/UO_2, Y^{3+}/UO_2, In^{3+}/UO_2, Sb^{3+}/UO_2, Au^{3+}/UO_2$ |
| to be able to form a defect CaF2 lattice, matching ions with radius of $\pm 0.01$ nm, both n-and p-type electroconductivity, carbonate modified with alkaline-earth metal | $Ca^{2+}/Sm_2O_3/SrCO_3, Mn^{2+}/Sm_2O_3/BaCO_3, K^+/PbO/BaCO_3, Cd^{2+}/Sm_2O_3/SrCO_3, Sn^{2+}/Sm_2O_3/CaCO_3, In^+/PbO/MgCO_3, Mn^{2+}/Gd_2O_3/BaCO_3, Rh^{2+}/Gd_2O_3/CaCO_3, Hg^+/PbO/BaCO_3, Cd^{2+}/Gd_2O_3/BaCO_3, Sn^{2+}/Gd_2O_3/CaCO_3, Tl^+/PbO/MgCO_3$ |
| to be able to form a defect $Ce_2O_3$ lattice, matching ions with radius of $\pm 0.005$ nm, mp $> 500$ ℃, ionic type electroconductivity, modified with a metal ion being able to form superoxide | $Sc^{3+}/ZrO_2/Na^+, Na^+/SnO/Cs^+, Cu^+/SnO/Rb^+, Sc^{3+}/CeO_2/Ra^{2+}, In^{3+}/CeO_2/Sr^{2+}, Sb^3+/CeO_2/Ba^{2+}, Au^{3+}/CeO_2/K^+, Rh^{3+}/PbO_2/Sr^{2+}, Ir^{3+}/PbO_2/K^+, Y^{3+}/ThO_2/Ba^{2+}, In^{3+}/ThO_2/Ra^{2+}, Sb^{3+}/ThO_2/Na^+, Au^{3+}/ThO_2/K^+, Sc^{3+}/UO_2/Ba^{2+}, In^{3+}/UO_2/K^+, Sb^{3+}/UO_2/Ba^{2+}, Au^{3+}/UO_2/Cs^+$ |

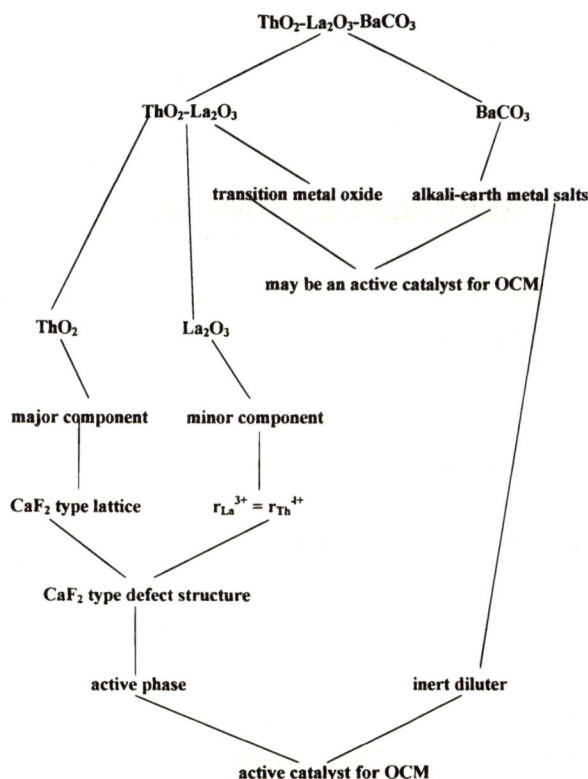

**Figure 4.  An example of generative reasoning.**

[Rule 6] If a complex catalyst is composed of a reducible metal oxide of IV, VA, or IIB group and an alkali metal oxide, then the catalyst may be an active catalyst for OCM

where the symbol $\wedge$ and $\vee$ represented logic "and" and logic "or", respectively. We can, in this way, write down many knowledge rules for the knowledge base.

The reasoning scheme is the most difficult key problem in the molecular design of catalyst since the catalysis is extremely complex. Different reasoning schemes are designed for different knowledge expression scheme in catalysis, respectively. They include generation reasoning, reasonable inference, graph searching, deductive reasoning, and algorithm reasoning, etc. Figure 4 shows an example of generative reasoning. Table 1 shows some output of catalyst component design for OCM.

## 5. CONCLUSION

A new and common method for explaining and mathematically expressing a catalytic reaction is proposed in the bridge theory which is favorable for computer-aided molecular design of catalysts. According to the physicochemical and mathematical bases for catalyst design, a practical expert system based on both structure and mechanism, is described with catalyst for OCM as an example. An advanced program for molecular design of catalysts has been developing. We believe that the molecular design of catalysts should play a more and more important role in developing new catalysts and catalytic processes and describing the mechanism of a known catalytic reaction as well as preparing high effective catalysts.

## ACKNOWLEDGMENT

This research is funded by the National Science Foundation of China and the Grant from the Department of Education of China. We greatly appreciate Dr. G. W. A. Milne for his editing of the manuscript and assisting in the publication of this paper.

## REFERENCES AND NOTES

[1]Trimm,D. L. Design ofIndustrial Catalysts; Elsevier Sci. ;New York,1980.

[2]Cusumano, J. A. In IUPAC Perspectives in catalysis—A Chemistry for the 21st Century Monograph;Thomas,J. M.,Zamaraev,K. I.,Eds. ;Blasckwell Sci. ;Oxford,1992;pp 1—34.

[3]Zamaraev, K. I. In IUPAC Perspectives in catalysis—A Chemistry for the 21st Century Monograph;Thomas,J. M.,Zamaraev,K. I.,Eds. ;Blasckwell Sci. ;Oxford,1992;pp 35—66.

[4]Likholobov, V. A. In IUPAC Perspectives in catalysis—A Chemistry for the 21st Century Monograph;Thomas,J. M.,Zamaraev,K. I. Blasckwell Sci. ;Oxford,1992;pp 67—90.

[5]Liao,D. W. Chemistry1982,(3),38,44—51(in Chinese).

[6]Hegedus, L. L. et al. Catalyst Design-Progress and Perspectives; John Wiley & Sons;New York,1987.

[7]Proceedings of Workshop on Catalyst Design. In Advances in Catalyst Design; Graziani,M., Rao,C. N. R.,Eds. ;World Sci. ;New Jersey,1991.

[8]Liao,D. W. et al. Study on Micro-catalysis in Life Science (1) Separation of Ginseng Saponins and Quantum Chemical Calculation of Panaxdiol; (2) Isolation of Polysaccharides and Information Catalysis,to be published in J. Xiamen University (N. S. ).

[9]Balaban,A. T. Chemical Application of Graph Theory; Academic;London,1976.

[10]Ugi,I. ;Wochner,M. J. Mol. Struct (THEOCHEM) 1988,165,229.

[11]Kvasnicka, V. ;Pospichal,J. Int. J. Quant. Chem. 1990,38,253.

[12]Likholobov,V. A. et al. Homogeneous and Heterogeneous Catalysis;VNU Sci. ;Utrecht,1986.

[13]Ioffe,I. I. Application of Pattern Recognition to Catalytic Research;John Wiley & Sons;New York,1988.

[14]Kuntz,I. D. ;Meng,E. C. ;Shoichet,B. K. Acc. Chem. Res. 1994,**27**,117.

■ 本文原载:Catalysis Today 30(1996),pp. 67～76.

# Constituent Selection and Performance Characterization of Catalysts for Oxidative Coupling of Methane and Oxidative Dehydrogenation of Ethane

Hui-Lin Wan[①], Zi-Sheng Chao, Wei-Zheng Weng, Xiao-Ping Zhou,
Jun-Xiu Cai, Khi-Rui Tsai

(Department of Chemistry and State Key Laboratory for Physical
Chemistry of the Solid Surface, Xiamen University, Xiamen 361005, China)

**Abstract**   The catalytic property of $BaF_2$-doped tetragonal LaOF catalysts for oxidative coupling of methane and oxidative dehydrogenation of ethane was evaluated in detail by independently varying the concentration of $BaF_2$ in the catalysts. The results indicate that addition of $BaF_2$ into the tetragonal LaOF significantly improves the catalytic performance of LaOF for both reactions,and the best results are observed over the $BaF_2$/LaOF catalysts with $BaF_2$ content of ca. 10 mol%. The XRD analysis of the fresh catalysts suggests that,for the catalysts with $BaF_2$ content less than 15 mol%, most of the $Ba^{2+}$ ions together with the accompanying $F^-$ and $O^{2-}$ may be dispersed in the LaOF,forming a $Ba^{2+}$-doped LaOF phase in which some of the $La^{3+}$ lattice points might be replaced by the less positively charged $Ba^{2+}$,leading to the formation of an F-center,anion vacancies or $O^-$ species in the LaOF lattice which will be favorable to the adsorption and activation of $O_2$. The results of in situ Raman characterization of the surface dioxygen species over the LaOF and 10% $BaF_2$/LaOF catalysts suggest that the improvement of alkane conversions on a $BaF_2$-doped LaOF may be explained by the relative abundance of the dioxygen adspecies over the catalyst. While one of the possible reasons for the significant improvement of the selectivities may have resulted from the favorable effect of $F^-$ ions for the isolation of the surface oxygen adspecies.

**Key words**   Oxidative coupling of methane   Oxidative dehydrogenation of ethane   Alkaline-earth fluorides doped lanthanum oxyfluoride catalysts   Defective fluoride structure   Anionic vacancy   Raman spectroscopy

## 1. Introduction

Since the early work of Keller and Bhasin[1] on the oxidative coupling of methane (OCM),there has been an extensive effort throughout the world to develop an economically feasible catalytic process for converting methane to $C^{2+}$ hydrocarbons. Many catalyst systems have been developed for this reaction and most of them contain the oxides with basic properties[2] such as the oxides of alkali metals,alkaline

---

① corresponding author.

earth metals and lanthanides. However, it is now evident that oxidative coupling catalysts need not be strongly basic oxides[3]. Some experimental evidences have shown that there is no simple correlation between the catalyst basicity and OCM activity[4]. A complex relationship exists between surface acidity/basicity and the catalytic activity/selectivity, and surface basicity alone cannot control OCM catalytic properties[5]. Moreover, catalysts with strong basicity were easily poisoned by the $CO_2$ generated in the reaction, so that higher reaction temperature was required to maintain an adequate level of activity. Many investigators have found that the catalytic property, especially the selectivity to ethylene, can be significantly improved by chlorine present either in the form of a chloride component built into the catalyst or as a volatile chlorinated compound (organic or inorganic) in the reactant feed[2,6]. Recently, Lunsford et al. have reported a $Cl^-$ promoted $Li^+$-MgO system which shows good catalytic performance for both OCM and ODE (oxidative dehydrogenation of ethane)[7,8]. Based on their studies, Lunsford suggested that the presence of $Cl^-$ ions at an appropriate level modified the catalyst so that it no longer functioned as a strongly basic oxide catalyst and would not be easily poisoned by $CO_2$. Therefore, under steady-state conditions a less basic catalyst may be more active, particularly with respect to ODE, since ethane has a weaker C-H bond than methane.

In addition to chloride, it will also be interesting and of fundamental significance to have a detailed investigation on the influences of other halides, especially fluoride, on the catalytic properties of the OCM and ODE catalysts. The general reasons of considering fluoride first instead of bromide or iodide are as follows: (i) alkaline-earth and rare-earth fluorides are usually more stable than the corresponding chlorides, bromides or iodides, so the loss of fluorine during the OCM and ODE reactions may be less compared to the other halides; (ii) $F^-$ ($r=1.33$ Å) and $O^{2-}$ ($r=1.32$ Å) have similar ionic radii, and anionic exchange between metal fluoride and metal oxide may easily take place, leading to the formation of lattice defects, such as anionic vacancies that are requisites for the adsorption and activation of $O_2$ over compound catalysts with stable cationic valencies; (iii) the fluoride-doped OCM and ODE catalysts would be less basic than the undoped metal oxides, thus they will be favorable to the prevention of $CO_2$ inhibition as well as to the improvement of the catalytic activity; (iv) the distribution of $F^-$ ions on the catalyst surface will have a positive effect on the dispersion of the surface active oxygen species that will be helpful to decrease the deep oxidation and improve the $C^{2+}$ selectivity. Following these clues, a series of metal-fluoride promoted metal-oxide catalysts were developed, many of the catalysts demonstrated a satisfactory catalytic property for both OCM and ODE reactions[9-13]. XRD characterizations of the catalysts clearly indicated that anionic exchange did happen between the metal-oxide and the metal-fluoride phases, as have been proven by the observation of the lattice expansion/contraction of oxide/fluoride phases and the formation of the new oxyfluoride compounds in the catalysts containing rare earth elements as well as transition metals[9-13]. Among these oxyfluoride compounds, the LaOF phase has been identified in several catalysts prepared with $LaF_3$ as a starting material (followed by calcination which removes part of the $F^-$), including the $LaF_3$/SrO system that shows good catalytic property for OCM[12] and the $LaF_3$/$BaF_2$ system that is good for ODE[11,12]. As has been indicated in the literatures, LaOF has the fluorite-type structure or its superstructure[14,15], depending on the arrangement of the anions, it can form the stoichiometric rhom-bohedral phase (LaOF) as well as the nonstoi-chiometric tetragonal phase (expressed by the formula $LaO_xF_{3-2x}$, $0.7<x<1$) and cubic phases (observed in the LaOF-$La_4O_3F_6$ region)[15]. In the latter two cases, the excess of $F^-$ ions occupy some of the interstitial positions as Frenkel defects. This result is consistent with the fact that many compounds with fluorite-like structure, such as alkaline-earth halides, $ZrO_2$ etc. contain anion Frenkel defects and anionic vacancies[16], and indicates that tetragonal or cubic LaOF is actually a compound with defective fluorite structure. It has been generally accepted that one of the essential conditions for a compound with stable cationic valency to be a good OCM catalyst is the presence and the mobility of the anionic vacancies in the lattice, so as to adsorb and activate $O_2$ during the reaction. Among the normally used OCM catalyst systems of oxide or composite-oxide, there is a series of catalysts with good catalytic performance in which the oxides of fluorite structure or defective fluorite structure, such as the cubic C-type structure, have been adopted as the hosts of catalysts[2,17,18]. correspondingly, for the above mentioned fluoride-promoted OCM catalyst system, it is also possible to develop a new series of catalysts with good catalytic properties for oxidative dehydrogenation of light alkanes, in which LaOF of fluorite structure or its superstructure is employed as a fluoride-containing host. By proper doping of LaOF with metal fluorides of lower cationic valency such as alkaline-earth fluorides, it may be possible to create a LaOF of defective fluorite structure with higher concentration of anionic defects including F-centers and anionic vacancies, as a result, the catalytic properties of LaOF may also be improved.

In the present study, the catalytic properties of $BaF_2$-doped tetragonal LaOF catalyst for OCM and ODE reactions were evaluated in detail by independently varying the concentration of $BaF_2$ in the catalyst, with attention being given to the correlations between catalytic properties and the bulk composition and structure of the catalysts. The results of photocurrent measurement with LaOF or $BaF_2$/LaOF catalyst as a working electrode as well as exsitu and insitu Raman spectroscopic characterization of the oxygen species performed at 650℃, 550℃ and 25℃ over the LaOF and $BaF_2$/LaOF catalysts are presented. These results will provide us with more information for the understanding of the nature of the promotion effects of the $F^-$ ions in metal fluoride/metal oxide system, especially for the catalyst system in which alkaline-earth fluoride or oxide is employed as the dopant of lanthanum oxyfluoride.

# 2. Experimental

LaOF was prepared by grinding equal molar ratios of $LaF_3$ and $La_2O_3$ in a mortar. The mixture was then pressed to pellets under a pressure of 300 $kg/cm^2$, followed by calcination at 900℃ for 4 h. XRD measurement proved that the resulting material was tetragonal LaOF. $BaF_2$/LaOF catalysts were prepared by mixing appropriate amounts of $BaF_2$ and LaOF according to the same procedures used for LaOF preparation. After calcination at 800℃ in air for 4 h, the pellets of LaOF and $BaF_2$/LaOF were crushed and sieved to grain size of 20 to 40 mesh.

The catalytic reactions were performed in a plug-flow fixed-bed quartz reactor (i. d. 0. 80 cm) under atmospheric pressure. For OCM reaction, 0. 50 ml of catalyst was loaded in the middle part of the reactor. For ODE reaction, 2. 0 ml of catalyst was loaded in the middle part of the reactor and the rest of the reactor was filled with quartz sand of grain size of 20 to 40 mesh. No diluent gas was used in all reactions. A Shang Fen 102-GD gas-chromatograph (operated at room temperature) equipped with thermal conductivity detector was employed to analyze the gaseous effluent. A 5 Å molecular sieve column was used to analyze $O_2$, CO and $CH_4$ and a Porapak Q column to analyze $CO_2$, $C_2H_4$ and $C_2H_6$. The conversion of alkane ($X_{alkane}$) and selectivity ($S_i$) were calculated from the equations: $X_{alkane} = (\sum A_i \times F_i)/[\sum (A_i \times F_i) + A_{alkane} \times F_{alkanc}]$ and $S_i = (A_i \times F_i)/\sum (A_i \times F_i)$, respectively, where $A =$ peak area of carbon-containing species and $F =$ correction factor of response and carbon number.

XRD analysis of the catalyst samples was carried out on a Rigaku Rotaflex D/max-C XRD system using Cu $K\alpha(\lambda = 1.5406$ Å) radiation. The specific surface area of the catalysts was measured by the BET method at liquid nitrogen temperature on a Sorpmatic—1900 system with $N_2$ as adsorbate.

Photocurrent measurement was carried out in 0. 1 M NaF aqueous solution on a conventional three-electrode system, with pt disk, SCE and LaOF or $BaF_2$/LaOF as the counter, reference and working electrodes, respectively. The electrodes were arranged in an electrolytic cell with quartz window. A deuterium lamp(DLPS-3,30W) was used as the light source. The monochromatic light modulated with a parc Light Chopper Model 197 was illuminated to the working electrode through the quartz window and the potential of the cell was controlled by a Hokuto Denko Dual Potentiostat/Galvanostat HR101B. The photocurrent were amplified and recorded by a pqrc 5206 two-phase lock-in analyzer.

The Raman spectra were recorded on a Jobin Yvon U-1000 Raman spectrometer at a resolution of 4 cm$^{-1}$ using a self-designed insitu Raman cell which could be heated up to 750℃. The temperature of the cell was measured by a thermal couple mounted close to the sample. In the experiments, the sample was first thermally treated at 750℃ under $H_2$ for 2 h to remove the surface carbonate residue, followed successively by purging with He at 750℃, and exposure to $O_2$ at same temperature. The $O_2$-treated sample was then cooled down gradually under $O_2$ atmosphere from 750℃ to 25℃ within a period of 3 h, and the Raman spectra were recorded at specified temperature, i. e. 650, 550 and 25℃, respectively during the cooling period. At each specified temperature point, temperature of the Raman cell was maintained constant for about 15 min to insure that the Raman measurement was conducted under thermostatic condition.

# 3. Results and discussion

## 3. 1  Catalytic performance

The catalytic performance of the LaOF-based catalyst for OCM and ODE reactions are listed in Tables 1 and 2. The data for OCM and ODE were collected after 10 and 6 h on stream, respectively. It can be seen from Table 1 that LaOF is an active catalyst for OCM at 780℃, however, its selectivity for $C_2$ hydrocarbons was only about 46. 6%. When 5 mol% of $BaF_2$ was added to the LaOF, the catalytic performance improved significantly. With the increase of $BaF_2$ content from 5 to 10 mol%, both $CH_4$ conversion and $C_2$ selectivities increased, and the highest $CH_4$ conversion and $C_2$ selectivity were obtained over the catalysts with $BaF_2$ contents of 10 to 15 mol%. The catalyst could run for about 100 h maintaining the $C_2$ yield at about 18 to 19%. Since OCM reaction is a strongly exothermic reaction, the management of the large amount of heat released has also attracted much attention. Generally, it may be acceptable to carry out the reaction at lower conversion of $CH_4$ and higher $C_2$ selectivity[3]. As shown in Table 1, the $CH_4$ conversion and $C_2$ selectivity over the $BaF_2$/LaOF catalyst could be controlled by varying the $CH_4$:$O_2$ ratio. When the $CH_4$:$O_2$ ratio increased to 6:1 and 9:1, $C_2$ selectivities of 81. 2 and 84. 6% with $CH_4$ conversions of 19. 5 and 16. 5% were obtained over 10%$BaF_2$/LaOF catalyst at 770℃.

As can be seen from Table 2, LaOF is also an active catalyst for ODE at 660℃, but its selectivity for ethylene formation is limited to 58. 5%. When $BaF_2$ was added to LaOF, both ethane conversion and ethylene selectivity increased significantly, and the catalysts with $BaF_2$ content close to 10 mol% gave the highest ethane conversions and ethylene selectivities under the conditions of 660℃ and GHSV= 2700 h$^{-1}$. Even higher ethane conversions can be obtained at higher GHSV. However, there is no simple correlation between the ethylene selectivity and the GHSV which may indicate that a hot spot exists in the reactor, especially at higher GHSV. The life-span test with a 14% $BaF_2$/LaOF catalyst showed no decrease in catalytic activity and ethylene selectivity during 26 h on stream. However, taking in consideration that the life time of the same catalyst for OCM was higher than 100 h, and that the reaction temperature of OED was much lower than that of OCM, it is expected that the catalyst will demonstrate even better stability for ODE reaction.

**Table 1　The OCM performance of the BaF$_2$/LaOF catalysts with different BaF$_2$ content [a]**

| content of BaF$_2$ (mol%) | conversion of CH$_4$(%) | Selectivity (%) | | | | | Yield of C$_2$(%) |
|---|---|---|---|---|---|---|---|
| | | CO | CO$_2$ | C$_2$H$_4$ | C$_2$H$_6$ | C$_2$ | |
| LaOF | 25.0 | 12.6 | 40.8 | 27.2 | 19.4 | 46.6 | 11.7 |
| 5% BaF$_2$/LaOF | 26.4 | 4.7 | 38.6 | 31.9 | 24.8 | 56.7 | 15.0 |
| 7% BaF$_2$/LaOF | 26.8 | 9.0 | 32.7 | 35.5 | 22.8 | 58.3 | 15.6 |
| 10% BaF$_2$/LaOF | 28.7 | 3.1 | 29.6 | 44.7 | 22.6 | 67.3 | 19.3 |
| 10% BaF$_2$/LaOF [b] | 19.5 | 0 | 18.8 | 41.2 | 40.0 | 81.2 | 15.8 |
| 10% BaF$_2$/LaOF [c] | 16.5 | 0 | 15.5 | 23.5 | 61.0 | 84.5 | 13.9 |
| 15% BaF$_2$/LaOF | 27.9 | 3.0 | 29.1 | 43.7 | 24.2 | 67.9 | 18.9 |
| 18% BaF$_2$/LaOF | 27.4 | 8.2 | 27.9 | 40.3 | 23.6 | 63.9 | 17.5 |
| 20% BaF$_2$/LaOF | 26.9 | 9.9 | 27.0 | 41.6 | 21.5 | 63.1 | 17.0 |
| 30% BaF$_2$/LaOF | 23.8 | 12.8 | 28.0 | 39.0 | 20.2 | 59.2 | 14.1 |

[a] Reaction conditions: feed=CH$_4$ : O$_2$=4 : 1, temperature=780℃, GHSV=15000 h$^{-1}$.

[b] Reaction conditions: feed=CH$_4$ : O$_2$=6 : 1, temperature=770℃.

[c] Reaction conditions: feed=CH$_4$ : O$_2$=9 : 1, temperature=770℃.

**Table 2　The ODE performance of BaF$_2$/LaOF catalysts with different BaF$_2$ content [a]**

| content of BaF$_2$ (mol%) | Temp. (℃) | $X_{O_2}$ [b] (%) | conv. of C$_2$H$_6$(%) | Selectivity(%) | | | | Yield of C$_2$H$_4$(%) |
|---|---|---|---|---|---|---|---|---|
| | | | | CO | CH$_4$ | CO$_2$ | C$_2$H$_4$ | |
| Quartz sand | 700 | 31.7 | 3.0 | 0 | 0 | 25.8 | 74.2 | 2.2 |
| | 720 | 30.2 | 5.2 | 0 | 0 | 20.7 | 79.3 | 4.1 |
| LaOF | 660 | 0 | 44.6 | 13.3 | 4.0 | 24.2 | 58.5 | 26.1 |
| 6% BaF$_2$/LaOF | 660 | 0 | 54.7 | 10.3 | 3.6 | 17.1 | 69.0 | 37.7 |
| 8% BaF$_2$/LaOF | 660 | 0 | 55.2 | 2.9 | 3.8 | 19.3 | 74.0 | 40.8 |
| 10% BaF$_2$/LaOF | 660 | 0 | 57.8 | 8.4 | 4.3 | 17.6 | 70.7 | 40.9 |
| 10% BaF$_2$/LaOF [c] | 660 | 0 | 75.5 | 11.8 | 8.6 | 12.7 | 66.9 | 50.5 |
| 12% BaF$_2$/LaOF | 660 | 0 | 50.2 | 8.1 | 3.1 | 20.3 | 68.5 | 34.4 |
| 14% BaF$_2$/LaOF | 680 | 0 | 53.4 | 9.8 | 3.6 | 19.7 | 66.9 | 35.7 |
| 26% BaF$_2$/LaOF | 660 | 0 | 52.4 | 9.3 | 3.5 | 20.8 | 66.4 | 34.8 |
| 30% BaF$_2$/LaOF | 660 | 0 | 46.2 | 8.0 | 3.5 | 24.5 | 64.0 | 29.6 |
| 30% BaF$_2$/LaOF [d] | 640 | 0 | 80.8 | 8.7 | 8.8 | 11.7 | 70.8 | 57.2 |

[a] Reaction conditions: feed=C$_2$H$_6$ : O$_2$=67.7 : 32.3; GHSV=2700 h$^{-1}$.

[b] $X_{O_2}$=the molar percentage of O$_2$ in the effluent.

[c] GHSV=6000 h$^{-1}$.

[d] GHSV=11600 h$^{-1}$.

## 3.2　XRD and BET specific surface area measurement

The phase composition of LaOF and BaF$_2$/LaOF with different BaF$_2$ content were determined by XRD, and the results were listed in Table 3. It can be seen that the tetragonal, LaOF, which has a defective fluoride structure and can be expressed by the formula LaO$_x$F$_{3-2x}$, (0.7 < $x$ < 1)[14,15], were detected in the LaOF and BaF$_2$/LaOF prepared with different BaF$_2$ contents. However, for the BaF$_2$/LaOF with BaF$_2$ content less than 15%, XRD could hardly detect the BaF$_2$ phase, only the strongest diffraction line of $BaF_2$ with $d \approx 3.53$ could be found. These results suggest that when the $BaF_2$ content was 15 mol% or less, most of the $BaF_2$ was dispersed in the LaOF, forming a $BaF_2$-or BaO-containing LaOF phase. Under the circumstances, some of the $La^{3+}$ lattice points might be replaced by the less positively charged $Ba^{2+}$, leading to the formation of F-centers, anionic vacancies or $O^-$ species in the LaOF lattice to maintain electroneutrality. Comparing with the lattice of standard LaOF, the lattice of LaOF in the prepared samples (LaOF or $BaF_2$/LaOF) slightly contracted. A possible explanation for the phenomenon is that the latter one may contain higher concentrations of anionic vacancies. These vacancies may result from partial replacement of the $F^-$ by $O^{2-}$ ions due to slow hydrolysis of LaOF in the calcination process (for the cases of LaOF and $BaF_2$/LaOF) and partial substitution of the $La^{3+}$ in the lattice by the $Ba^{2+}$ ion (for the case of $BaF_2$/LaOF). When the content of $BaF_2$ was above 15 mol%, the LaOF lattice could not accommodate all the $BaF_2$, and thus a separate $BaF_2$ phase was formed. In this case, some of $Ba^{2+}$ in the $BaF_2$ lattice might also be substituted by $La^{3+}$, leading to the formation of a lattice contracted $BaF_2$ phase taking in consideration the more positive charge and a smaller ionic size of $La^{3+}$ than $Ba^{2+}$.

**Table 3　XRD analysis results of $BaF_2$, LaOF and $BaF_2$/LaOF catalysts**

| Samples (fresh catalyst) | $d(\text{Å})(I/I_0)$ | | | | | | | | | |
| --- | --- | --- | --- | --- | --- | --- | --- | --- | --- | --- |
| $BaF_2$(standard) | 3.579(100) | | 3.100(27) | | 2.193(79) | | | 1.870(51) | | |
| LaOF(standard) | | 3.35(100) | | 2.90(25) | | 2.06(60) | 2.05(33) | | 1.76(22) | 1.75(44) |
| LaOF(prepared) | | 3.348(400) | | 2.892(18) | | 2.051(33) | 2.044(24) | | 1.756(10) | 1.743(34) |
| 5%$BaF_2$/LaOF | 3.534(5) | 3.341(100) | | 2.840(16) | | 2.048(35) | | | 1.752(11) | 1.742(29) |
| 10%$BaF_2$/LaOF | 3.537(8) | 3.346(99) | | 2.890(21) | | 2.051(39) | 2.044(25) | | 1.755(10) | 1.743(34) |
| 15%$BaF_2$/LaOF | 3.537(12) | 3.346(100) | | 2.890(19) | | 2.051(38) | 2.043(22) | | 1.753(9) | 1.743(27) |
| 18%$BaF_2$/LaOF | 3.562(21) | 3.346(99) | | 2.888(19) | 2.178(12) | 2.051(36) | 2.044(23) | 1.856(10) | 1.755(9) | 1.743(27) |
| 20%$BaF_2$/LaOF | 3.559(29) | 3.348(100) | 3.081(10) | 2.890(20) | 2.178(18) | 2.052(35) | 2.044(23) | 1.857(11) | 1.753(9) | 1.743(26) |
| 30%$BaF_2$/LaOF | 3.565(45) | 3.346(100) | 3.087(11) | 2.888(21) | 2.178(27) | 2.052(38) | 2.044(23) | 1.860(19) | 1.753(10) | 1.743(27) |

The data of the specific surface area of the catalysts were listed in Table 4. As shown in the Tables 1 and 2, the selectivities to $C_2$ (for OCM) or ethylene (for ODE) over $BaF_2$/LaOF catalysts were higher than those over LaOF, while the specific surface area of the $BaF_2$/LaOF catalyst was larger than that of LaOF. Thus the increase in the selectivities over $BaF_2$/LaOF catalysts cannot be attributed to a reduction in surface area. It may result mainly from the isolation effect of the $F^-$ ions to the surface oxygen adspecies as a result of dispersion and/or formation of the alkaline-earth fluorides on the surface of the $BaF_2$/LaOF system.

## 3.3　Photocurrent measurement and Raman spectroscopic characterization of the oxygen adspecies over the LaOF and $BaF_2$/LaOF

The experiment of photocurrent measurement with LaOF or $BaF_2$/LaOF as working electrode shows that, in the range of 0.30 to 1.0 V of applied potentials, the cathodic and anodic photocurrents were detected on the LaOF and 10% $BaF_2$/LaOF electrodes, respectively, and the intensities of photocurrent in the latter system were found to be about 100 times as high as those in the former one

(Fig. 1). These results suggest that LaOF is a solid electrolyte with 'p-type' conductivity, after modifying with 10% $BaF_2$, the conductivity of the sample changes to 'n-type', implying the possible formation of F-center or more anionic vacancies, or decrease of the concentration of interstitial $F^-$ ions in the 10% $BaF_2$/LaOF system as the result of partial substitution of the $La^{3+}$ in the LaOF lattice by $Ba^{2+}$. Incidentally, this result indicates that p-type conductivity may not be a requisite attribute for a good OCM catalysts.

. **Table 4    The BET specific surface area of LaOF, $BaF_2$ and $BaF_2$/LaOF catalysts**

| | BaF₂ content (mol%) | | | | | | | | | | |
|---|---|---|---|---|---|---|---|---|---|---|---|
| | 0 | 5.0 | 8.0 | 10.0 | 12.0 | 15.0 | 18.0 | 20.0 | 26.0 | 30.0 | 100 |
| Specific surface area (m²/g) | 2.9 | 11.0 | 5.2 | 6.5 | 4.7 | 7.5 | 6.3 | 5.1 | 4.4 | 4.3 | 2.7 |

The Raman spectroscopic characterizations of the oxygen adspecies under various temperatures were carried out over the LaOF and 10% $BaF_2$/LaOF catalysts. Fig. 2a shows the room temperature Raman spectrum of an $O_2$-treated 10% $BaF_2$/LaOF. Before the sample was exposed to $O_2$ at room temperature, it had been treated with $H_2$ at 750 ℃ for 2 h followed by cooling down under He atmosphere. No Raman peak was observed in the region of 700—1500 $cm^{-1}$ over the $H_2$ treated 10% $BaF_2$/LaOF before exposure to $O_2$. As can be seen in Fig. 2a, after exposure to $O_2$, several Raman peaks attributable to the surface dioxygen adspecies were observed in the wavenumber range of 738—1462 $cm^{-1}$. The peak at 738 $cm^{-1}$ is close to the low end of the wavenumber region for the peroxide species, and may be assigned to the surface $O_2^{2-}$ species[19]. The Raman peaks at 1016, 1042, 1080, 1154 and 1198 $cm^{-1}$ are close to the known IR bands of adsorbed $O_2^-$ (1015—1160 $cm^{-1}$) observed and attributed by Zecchina et al.[20] over the diluted MgO-CoO solid solution, based on the isotope substitution experiments and the comparison with homogeneous analogues, and that (1180 $cm^{-1}$) reported by Davydov et al.[21] on $O_2$-adsorbed $TiO_2$, and may tentatively be assigned to the surface $O_2^-$ species with different adsorption modes in different chemical environments. In order to explain the IR peaks of surface $O_2^-$ species with lower stretching frequencies, such as the peak at 1015 $cm^{-1}$, Zecchina et al.[20] have suggested a bridged $\mu$-superoxo adsorption mode. This suggestion could actually be correlated to that proposed by Madix et al.[22] and Biloen et al.[23] in which low-frequency peroxide species on Ag (110) may have resulted from the possible increased donation from the $O_2$ $\pi$ orbitals to the empty orbitals of the central metal besides the electron back-donation into the $\pi^*$ orbitals of the oxygen molecule. The preliminary results of isotope substitution experiment performed on 20% $BaF_2$/LaOF by Zhou et al.[24] indicated that, after exchanging with $^{18}O_2$, the peaks at 738 and 1198 $cm^{-1}$ shifted to ca. 700 and 1126 $cm^{-1}$, respectively. The observed isotope shifts for these two peaks are in good agreement with theoretically expected values of the difference between the stretching frequencies of $^{16}O-^{16}O$ and $^{18}O-^{18}O$ species. The peaks at 1388 and 1462 $cm^{-1}$ are lower than the frequency range of adsorbed neutral $O_2$ species (ca. 1550 $cm^{-1}$) but higher than that of the superoxide species, and can be assigned to the weakly adsorbed species ($O_2^{\delta-}$) between adsorbed $O_2^-$ and $O_2$. In order to get rid of the possible interference of the surface carbonate species to the assignment of the oxygen adspecies, the $CO_2$ adsorption experiment was also performed. After the above $O_2$-treated 10% $BaF_2$/LaOF was heated at 750℃ under one atmosphere of $CO_2$ followed by cooling to room temperature under $CO_2$, the corresponding Raman spectrum is shown in Fig. 2b. As can be seen, in addition to the peaks of adsorbed dioxygen species, two bands at 826 and 1058 $cm^{-1}$ with medium intensities were also observed. These bands obviously result from the surface carbonated species. comparing with the Raman spectra of a working OCM catalysts of the composite metal-oxide type such as Th-La-$O_x$[25], the band intensity of the surface carbonate species over the 10% $BaF_2$/LaOF is much lower, implying that basicity of the $BaF_2$/LaOF system is much weaker compare to that of the oxide system. This property is important for an OCM or ODE catalyst to give a lower 'light off' temperature. The similar $CO_2$ adsorption experiment was also performed over the LaOF catalyst; however, the peaks of surface carbonate species were hardly detectable, a possible explanation is that the surface of the LaOF is less basic than the $BaF_2$ doped LaOF, leading to decreased adsorption of $CO_2$ on LaOF.

Figs. 3-5 show the in situ and ex situ Raman spectra of oxygen adspecies on LaOF (a) and 10% $BaF_2$/LaOF (b) recorded at 650, 550 and 25℃, respectively, during the period when the samples were cooling from 750℃ to room temperature under $O_2$ atmosphere. Before the samples were exposed to $O_2$ at 750℃, similar treatment procedures as above were also followed inorder to clean the surface. As shown in Figs. 3 and 4, the Raman spectra recorded at 650℃ and 550℃ were not well resolved. Two broad bands with maxima at ca. 780−800 $cm^{-1}$ and 1040−1060 $cm^{-1}$, respectively, can be identified, and the intensity of the bands at 550℃ was a little higher than that at 650℃. The bands with maxima at ca. 800 $cm^{-1}$ (LaOF) and 780 $cm^{-1}$ ($BaF_2$/LaOF) were closed to the frequencies of peroxide signal, and might be

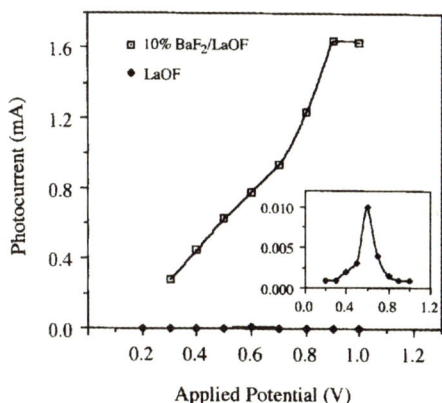

Fig. 1 Plots of photocurrent intensities vs. applied potentials. Cathodic and anodic currents were detected on the LaOF and $BaF_2$/LaOF electrodes, respectively.

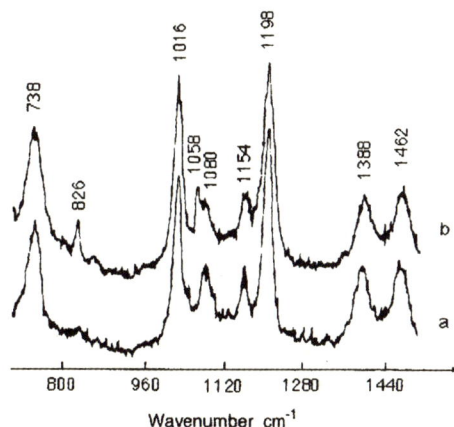

Fig. 2 Room temperature Raman spectra of the catalysts, (a) $O_2$ adsorbing 10% $BaF_2$/LaOF. (b) After treatment of (a) with 1 atmosphere of $CO_2$ at 750℃ followed by cooling to room temperature under $CO_2$.

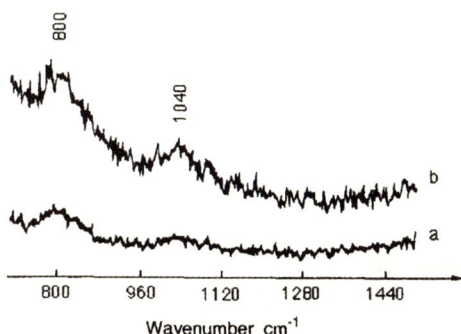

Fig. 3 Raman spectra of (a) LaOF and (b) 10% $BaF_2$/LaOF after exposure to $O_2$ at 750℃ followed by cooling to 650℃ under $O_2$.

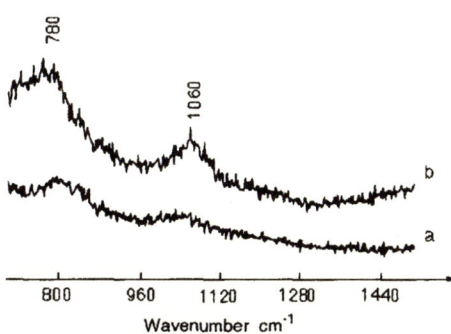

Fig. 4 Raman spectra of (a) LaOF and (b) 10% $BaF_2$/LaOF after exposure to $O_2$ at 750℃ followed by cooling to 550℃ under $O_2$.

assigned to the adsorbed $O_2^{2-}$ species[19]. The Raman signals of surface peroxide species over $La_2O_3$, $Na^+$ or $Sr^{2+}$ modified $La_2O_3$ and $Ba^{2+}/MgO$ with frequencies between 823 and 878 $cm^{-1}$ have been observed by Knozinger and Lunsford et al. at 700℃ [3,26]. Based on the above discussion, the bands centered at ca. 1050 $cm^{-1}$ may be attributed to the $O_2^-$ adspecies[20]. comparatively, the spectrum recorded at room temperature (Fig. 5) are much better resolved, and it shows exactly the same peaks as we have seen in Fig. 2a. As can be seen, the surface carbonate peaks at 828 and 1058 $cm^{-1}$ are not observed in Fig. 5, indicating that the possibility of coexistence of the surface carbonate species with the adsorbed dioxygen species in Figs. 3 and 4 can be excluded

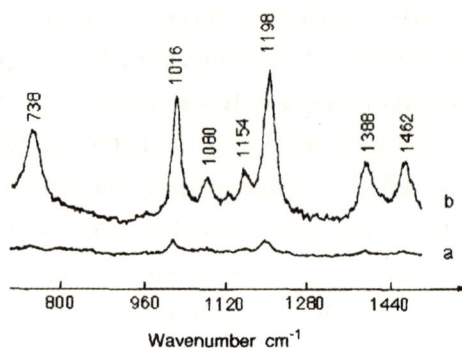

Fig. 5 Raman spectra of (a) LaOF and (b) 10% $BaF_2/LaOF$ after exposure to $O_2$ at 750℃ followed by cooling to room temperature under $O_2$.

or beyond the detection limit of Raman spectroscopy. It is also clear to see from Figs. 3-5 that intensities of bands of the oxygen adspecies ($O_2^{2-}$ and $O_2^-$) over the 10% $BaF_2/LaOF$ are significantly higher than those over the LaOF. This accounts for the improvement of the alkane conversion over the $Ba^{2+}$ doped LaOF catalysts, and the increase in the surface concentration of oxygen adspecies ($O_2^{2-}$ and $O_2^-$) may also be explained by the possible formation of F-centers or higher concentration of anionic vacancies in the $BaF_2/LaOF$ system compared with the LaOF system.

# 4. Conclusions

Based on the above results, we may conclude that addition of $BaF_2$ into the tetragonal LaOF significantly improves the catalytic performance of LaOF for both OCM and ODE. The best results were observed over the $BaF_2/LaOF$ catalysts with $BaF_2$ content of ca. 10 mol%. Under this condition, most of the $Ba^{2+}$ ions together with the accompanying $F^-$ and $O^{2-}$ may be dispersed in the LaOF, forming a $Ba^{2+}$-doped LaOF phase in which some of the $La^{3+}$ lattice points might be replaced by the less positively charged $Ba^{2+}$, leading to the formation of F-centers, anion vacancies or $O^-$ species in the LaOF lattice which will be favorable to the adsorption and activation of $O_2$. The results of in situ Raman characterization of the surface dioxygen species over the LaOF and 10% $BaF_2/LaOF$ catalysts suggest that the improvement of alkane conversions on a $BaF_2$ doped LaOF may be explained by the relative abundance of the dioxygen adspecies over the catalyst. While one of the possible reasons for the significant improvement of the selectivities may have resulted from the favorable effect of $F^-$ ions for the isolation of the surface oxygen adspecies. Though hydrolytic loss of $F^-$ from such catalyst system was detected by analysis, this type of catalyst may still be promising especially as catalyst operating for ODE at lower temperature (600℃). The studies of this type of catalysts illustrate many important points on the activity/selectivity/stability controlling factors of OCM and ODE catalysts.

# Acknowledgements

This Work is supported by the National Natural Science Foundation of China.

# References

[1]G. E. Keller and M. M. Bhasin,J. Catal.,**73** (1982) 9.

[2]A. M. Maitra,Appl. Catal. A:Gen.,**104** (1993) 11.

[3]J. H. Lunsford, in H. E. Curry-Hyde and R. F. Howe, Nature Gas conversion II, Elsevier Science,1994,p. 1.

[4]J. A. Lapszewicz and X. Z. Jiang,Catal. Lett.,**13** (1992) 103.

[5]V. R. Choudhary and V. H. Rane,J. Catal.,**130** (1991) 411.

[6]O. V. Krylov,Catal. Today,**18** (1993) 209.

[7]J. H. Lunsford,P. G. Hinson,M. P. Rosynek,C. Shi,M. Xu and X. Yang,J. Catal.,**147** (1994) 301.

[8]D. Wang,M. P. Rosynek and J. H. Lunsford,J. Catal.,**151** (1995) 155.

[9]S. Zhou,X. Zhou,H. Wan and K. R. Tsai,Catal. Lett.,**20** (1993) 179.

[10]X. Zhou,W. Zhang,H. Wan and K. Tsai,Catal. Lett.,**21** (1993) 113.

[11]X. P. Zhou,S. Q. Zhou,F. C. Xu,S. J. Wang,W. Z. Weng,H. L. Wan and K. R. Tsai,Chem. Res. Chin. Univ.,**9**(3) (1993) 269.

[12]X. P. Zhou,S. Q. Zhou,W. D. Zhang,Z. S. Chao,W. Z. Weng,R. Q. Long,D. L. Tang,H. Y. Wang,S. J. Wang,J. X. Cai,H. L. Wan and K. R. Tsai,Preprints,Div. Petro. Chem. Inc.,**39**(2) (1994) 222.

[13]X. P. Zhou,Z. S. Chao,W. Z. Weng,W. D. Zhang,S. J. Wang,H. L. Wan and K. R. Tsai,Catal. Lett.,**29** (1994) 177.

[14]A. F. Wells,Structural Inorganic Chemistry,5th edn.,Clarendon Press,Oxford,1984,pp. 252, 482.

[15]K. Niihara and S. Yajima,Bull. Chem. Soc. Jpn.,**44** (1971) 643.

[16]Z. Zhang,X. Verykios and M. Baerns,Catal. Rev.,Sci. Eng.,**36**(3)(1994) 507.

[17]Y. Amenomiya,V. I. Birss,M. Goledzinowski,J. Galuszka and A. R. Sanger,Catal. Rev. Sci. Eng.,**32** (1990) 163.

[18]Y. D. Liu,G. D. Lin,H. B. Zhang,J. X. Cai,H. L. Wan and K. R. Tsai,Preprints,Fuel Chem. Div.,**37**(1) (1992) 356.

[19]M. Che and A. J. Tench,Adv. Catal.,**33** (1983) 1.

[20]A. Zecchina,G. Spoto and S. coluccia,J. Mol. Catal.,**14** (1982) 351.

[21]A. A. Davydov,M. P. Komarova,V. F. Anufrienko, N. G. Maksimov,Kinet. Katal.,**14** (1973) 1519.

[22]B. A. Sexton and R. J. Madix,Chem. Phys. Lett.,**76**(2) (1980) 294.

[23]C. Backx,C. P. M. De Groot and P. Biloen,Surf. Sci.,**104** (1981) 300.

[24]X. P. Zhou and C. T. Au,unpublished results.

[25]Y. D. Liu,H. B. Zhang,G. D. Lin,Y. Y. Liao and K. R. Tsai,J. Chem. Soc.,Chem. commun., (1994) 1871.

[26]G. Mestl,H. Knozinger and J. H. Lunsford,Ber. Bunsenges. Phys. Chem.,**97** (3) (1993) 319.

■ 本文原载：Catalysis Letters 42(1996)，pp. 15～20.

# Preparation of Supported Gold Catalysts from Gold Complexes and Their Catalytic Activities for CO Oxidation

You-Zhu Yuan [a,b], Kiyotaka Asakura [c], Hui-Lin Wan [b], Khi-Rui Tsai [b],
Yasuhiro Iwasawa [a,①]

(ᵃ Department of Chemistry, Graduate School of Science,
The University of Tokyo, Hongo, Bunkyo-ku, Tokyo 113, Japan
ᵇ Department of Chemistry and State Key Laboratory for Physical Chemistry
of the Solid Surface, Xiamen University, Xiamen 361005, PR China
ᶜ Research Center for Spectrochemistry, Faculty of Science,
The University of Tokyo, Hongo, Bunkyo-ku, Tokyo 113, Japan)

Received 12 March 1996; accepted 27 August 1996

**Abstract**     A phosphine-stabilized mononuclear gold complex Au（PPh₃）（NO₃）（**1**）and a phosphine-stabilized gold cluster $[Au_9(PPh_3)_8](NO_3)_3$（**2**）were used as precursors for preparation of supported gold catalysts. Both complexes **1** and **2** supported on inorganic oxides such as $\alpha\text{-}Fe_2O_3$, $TiO_2$, and $SiO_2$ were inactive for CO oxidation, whereas the **1** or **2**/oxides treated under air or CO or 5% $H_2$/Ar atmosphere were found to be active for CO oxidation. The catalytic activity depended on not only the treatment conditions but also the kinds of the precursor and the supports used. The catalysts derived from **1** showed higher activity than those derived from **2**. $\alpha\text{-}Fe_2O_3$ and $TiO_2$ were much more efficient supports than $SiO_2$ for the gold particles which were characterized by XRD and EXAFS.

**Key words**     supported gold catalysts     CO oxidation     gold monomer and cluster     EXAFS

## 1     Introduction

Catalysis of small Au particles supported on inorganic oxides has been intensively studied in the literature[1−11]. It has been known for a long time that gold surfaces are capable of catalyzing the oxidation of carbon monoxide to carbon dioxide[3] and also that the nature of the support has a decisive influence on the catalytic activity of gold[4]; magnesia-and alumina-supported gold were more active by more than one order of magnitude for oxygen transfer between CO and $CO_2$ than silica-supported gold. Cant and Fredrickson studies the reactions of CO with $O_2$ and NO over gold and silver sponges[5]. On gold, the oxidation of CO by oxygen was about 40 times faster than that by nitric oxide. Small Au particles could adsorb CO and also increased the amount of oxygen adsorbed on the supporting oxide.

---

①     To whom correspondence should be addressed.

Vannice and coworkers studied[6] CO oxidation over differently pretreated $Au/TiO_2$ and $Au/SiO_2$ catalysts. After a high-temperature reduction at 773 K, $Au/TiO_2$ became very active for the oxidation at 313 K as compared to $Au/SiO_2$. The high activity of the $Au/TiO_2$ catalyst was attributed to synergistic interaction between gold and titania.

Nevertheless, it is very difficult to obtain the gold particles in a state of dispersion as high as those of platinum and palladium because the melting point of Au particles with a diameter of 2 nm is lowered to 573 K due to the quantum size effect[7]. This is mainly because the melting point of gold at 1336 K is much lower than those of Pt (2042 K) and Pd (1823 K). Haruta et al. prepared highly dispersed gold particles on various metal oxides by coprecipitation or deposition[8-10]. They found that gold supported on $TiO_2$, $\alpha$-$Fe_2O_3$, $Co_3O_4$, NiO, $Be(OH)_2$, and $Mg(OH)_2$ was very active for low-temperature CO oxidation. For $Au/TiO_2$, $\alpha$-$Fe_2O_3$, and $Co_3O_4$, the turnover frequencies for CO oxidation were almost independent of the kind of supporting oxides and increased sharply with a decrease in diameter of the gold particles below 4 nm. Recently, Hug et al. [11] studied $Au/ZrO_2$ prepared by coprecipitation and observed significant deactivation in long-term tests for CO oxidation although the catalyst was highly active at the beginning of the reaction.

However, there is no report hitherto on catalysis of supported gold complexes or cluster compounds, although the preparation and characterization of supported homonuclear naked gold clusters ($Au_x$) by a molecular beam technique have been reviewed[12]. We attempted to obtain supported gold catalysts by using well-characterized gold complexes as precursors by extending the knowledge of the nature of gold cluster compounds[13] and surface organometallic chemistry[14,15], and the way of preparation of highly dispersed metal catalysts by using metal complexes and clusters[16]. Here we report the preparation and characterization of gold catalysts derived from a mononuclear gold complex $Au(PPh_3)(NO_3)$ (**1**) and an enneanuclear gold cluster compound $[Au_9(PPh_3)_8](NO_3)_3$(**2**) supported on $TiO_2$, $\alpha$-$Fe_2O_3$ and $SiO_2$, and the catalytic performance of these samples in CO oxidation.

# 2   Experimental

## 2. 1   Synthesis of Au (PPh$_3$) (NO$_3$) (*1*) and [Au$_9$(PPh$_3$)$_8$](NO$_3$)$_3$(*2*)

Gold complexes 1 and 2 were synthesized according to the literature[17]. The total yields (based on Au) for **1** and **2** were above 80% and about 40%, respectively. The white crystal of **1** was light-and temperature-sensitive, and showed IR bands (KBr) for $v(NO_3)$ at 1495 and 1268 $cm^{-1}$. UV-visible spectroscopy of **2** exhibited $\lambda_{max}$(EtOH) at 442, 374 (sh), 351 (sh), and 313 nm, which coincides with that in the literature[13].

## 2. 2   Preparation of TiO$_2$-, SiO$_2$-, and $\alpha$-Fe$_2$O$_3$-supported *1* and *2*

$TiO_2$(Aerosil P25, surface area 50 $m^2/g$) and $SiO_2$ (Aerosil 300, surface area 300 $m^2/g$) were pretreated at 673 K under vacuum for 1 h before use as supports, $\alpha$-$Fe_2O_3$ was obtained by precipitation of $Fe(NO)_3$ with $Na_2CO_3$ in aqueous solution, followed by washing, drying under vacuum at room temperature overnight, and finally calcination at 673 K for 2 h. These inorganic oxides were evacuated at room temperature in situ before supporting the Au complexes. The supports were impregnated with a $CH_2Cl_2$ solution of the An complex while vigorously stirring, followed by evacuation of the solvent at room temperature for 5 h. The $CH_2Cl_2$ solvent was carefully dried over 5A molecular sieve for at least 12 h. The supported Au samples were then stored in sealed glass ampoules (under vacuum) at room temperature. The loading of Au was always 3. 0 wt%.

## 2. 3  Characterization of the Au catalysts

The supported Au catalysts were characterized by X-ray diffraction (XRD) and extended X-ray absorption fine structure (EXAFS). XRD patterns were measured on a Rigaku powder X-ray diffractometer with Cu $K_\alpha$ radiation over the $2\theta$-range of 20 to 80°. The mean particle size of the gold crystallites was estimated by using the Scherrer equation and the half-widths of selected diffraction lines of Au(111) at $2\theta=38. 2°$ for Au/SiO$_2$ and Au/$\alpha$-Fe$_2$O$_3$, and of Au(200) at $2\theta=44. 4°$ for Au/TiO$_2$.

Au $L_3$-edge EXAFS spectra of the samples were measured at room temperature in a transmission mode at beamline 10B of the Photon Factory in the National Laboratory for High Energy Physics (KEK) (Proposal No. :95G200). Cluster **2** was used as a reference compound in analyzing the EXAFS data of the samples **2**/ TiO$_2$ and **2**/SiO$_2$. The parameters extracted from the EXAFS data for Au foil and the cluster **2** are summarized in table 1.

## 2. 4  Catalytic CO oxidation

Catalysts were pretreated under several conditions before catalytic runs. **1** and **2**/Fe$_2$O$_3$ and **1** and **2**/ SiO$_2$ were calcined at 673 K under an O$_2$ flow. For the TiO$_2$ support, several pretreatments including heating at 673 K under CO or O$_2$ and reduction at 773 K under H$_2$ were examined. Kinetic measurements were carried out in a fixed-bed flow reactor with a computer-controlled auto-sampling system by using 200 mg of catalyst powder. The reaction gas consisting of 1. 0% CO balanced with air through a molecular sieve column was passed through the catalyst bed at a flow rate of 34 ml/min (SV=10 000 h$^{-1}$). The reaction products were analyzed by a gas chromatograph using a column of Unibeads C for CO$_2$ and a column of 5A molecular sieve for CO and O$_2$. The material balance in the catalytic reactions was checked from the concentrations of CO$_2$ and CO, both being in good coincidence with each other under all the conditions tested. Since temperature increase and pressure change at the surface are estimated to be 0. 15 K and 0. 01 mbar for 100% conversions, respectively, there seem to be no heat and mass transfer limitations.

# 3  Results and discussion

## 3. 1  EXAFS and XRD measurements

Fig. 1 shows the EXAFS spectra for **2**/TiO$_2$ before treatment. Figs. 1a, 1b and 1c present the EXAFS oscillation, its associated Fourier transform, and curve fitting, respectively. By comparison with the EXAFS Fourier transform for **2** in fig. 1d, the peaks in fig. 1b are straightforwardly assigned to Au-P and Au-Au bonds. The best-fit results are listed in table 2. The data demonstrate that the cluster framework [Au$_9$(PPh$_3$)$_8$]$^{3+}$ in **2**/TiO$_2$ is almost the same as that for the unsupported **2**. By heating the incipient **2**/TiO$_2$ to 573 K under vacuum, the EXAFS Fourier transform (fig. 2b) changed to exhibit only one peak due to the Au-Au bond. The Au-Au bond distance and the coordination number were determined to be 2. 88 and 11. 9 Å, respectively, by curve-fitting analysis (fig. 2c and table 2). The lack of an Au-P bond and the coordination number for the Au-Au bond suggest decomposition of the cluster framework to form metallic gold particles on TiO$_2$ surface at 573 K under vacuum.

The EXAFS analysis for the incipient **2**/SiO$_2$ at 293 K is shown in table 3. The data imply that a phosphine ligand still coordinates to an Au atom but the cluster framework is partially fragmented on

SiO$_2$, judging from a decrease in the coordination number of Au-Au bonds from 4.4 to 3.0. After heating the sample to 573 K under vacuum, the incipient supported clusters were decomposed to metallic Au particles, as shown in table 3.

**Table 1  Crystallographic data and curve-fitting results for Au L$_3$-edge EXAFS data of 2 and Au foil[a]**

| Sample | Crystallographic | | | EXAFS | | | | |
|---|---|---|---|---|---|---|---|---|
| | shell | N | r(Å) | N | r(Å) | $\Delta\sigma^2$(Å$^2$) | $\Delta E$(eV) | Rt(%) |
| Aufoil | Au-Au | 12 | 2.88 | 12.0 | 2.88 | 0.0000 | 0.7 | 0.04 |
| [Au$_9$(PPh$_3$)$_8$]$^{3+}$ | Au-Au | 4.4 | 2.78 | 4.4 | 2.79 | 0.0004 | 2.9 | 0.35 |
| | Au-P | 0.9 | 2.30 | 0.9 | 2.31 | 0.0000 | 1.9 | 0.35 |

[a] N: coordination number, r: interatomic distance, $\sigma$: Debye-Waller factor, $\Delta E$: difference in the origin of photoelectron energy between reference and sample, R$_f$: residual factor in the curve fitting. $\Delta k$: 3.00—14.56 Å$^{-1}$ (Au foil), 3.43—13.50 Å$^{-1}$ (Au-Au and Au-P for 2); $\Delta r$: 1.95—3.18 Å (Au foil), 1.55—3.00 Å (Au-Au and Au-P for 2).

**Table 2  Curve-fitting results for the Au L$_3$-edge EXAFS data of 2/TiO$_2$**

| T(K) | Au-P | | | | Au-Au | | | | R$_f$ (%) |
|---|---|---|---|---|---|---|---|---|---|
| | N | r(Å) | $\Delta\sigma^2$(Å$^2$) | $\Delta E$(eV) | N | r(Å) | $\Delta\sigma^2$(Å$^2$) | $\Delta E$(eV) | |
| 293 | 1.0 | 2.33 | 0.0030 | 6.1 | 4.0 | 2.77 | 0.0016 | 3.5 | 3.3 |
| 573 | | | | | 11.9 | 2.88 | 0.0000 | 0.2 | 1.1 |

**Table 3  Curve-fitting results for the Au L$_3$-edge EXAFS data of 2/SiO$_2$**

| T(K) | Au-P | | | | Au-Au | | | | R$_f$ (%) |
|---|---|---|---|---|---|---|---|---|---|
| | N | r(Å) | $\Delta\sigma^2$(Å$^2$) | $\Delta E$(eV) | N | r(Å) | $\Delta\sigma^2$(Å$^2$) | $\Delta E$(eV) | |
| 293 | 1.1 | 2.28 | 0.0023 | 3.6 | 3.0 | 2.81 | 0.0000 | 10.5 | 1.7 |
| 573 | | | | | 11.7 | 2.88 | 0.0004 | 0.4 | 0.1 |

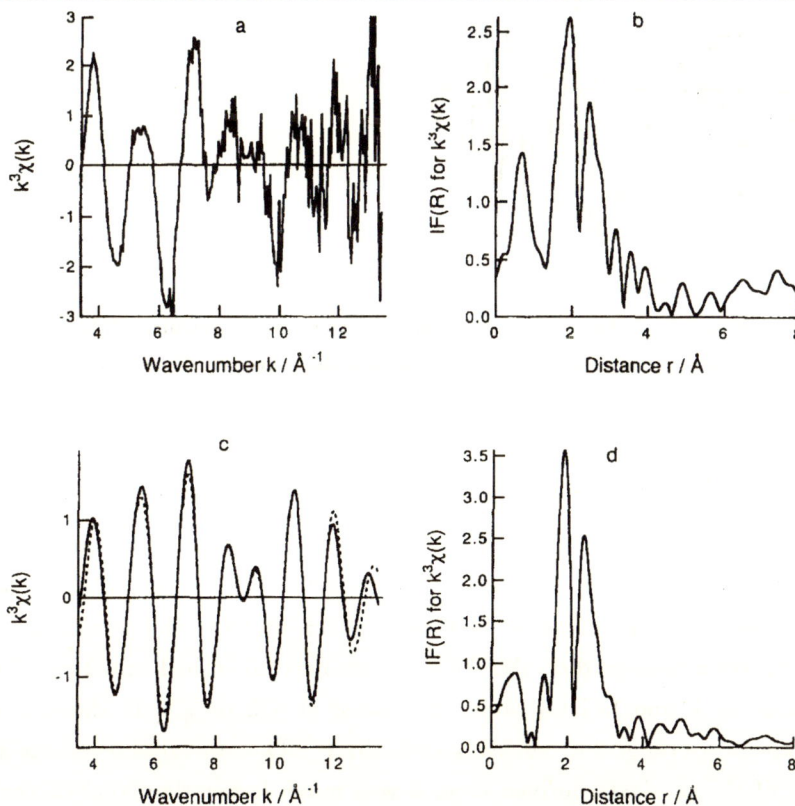

**Fig. 1   The EXAFS data for 2/TiO$_2$ at Au L$_3$-edge measured at 293 K in vacuum, (a) $k^3$-weighted EXAFS oscillation; (b) its associated Fourier transform; (c) curve-fitting analysis (inversely Fourier transformed) for Au-P and Au-Au; (——) observed; (···) calculated; (d) Fourier transform of the EXAFS oscillation for 2.**

X-ray diffraction patterns of the samples $1/TiO_2$ and $1/SiO_2$ before and after heat-treatments under air, 5% $H_2/Ar$, and CO atmosphere were measured to estimate the mean particle size of the Au crystallites. The patterns indicate that after the treatments the supported gold complexes were decomposed to form metallic particles on the oxides, while the incipient supported species showed no crystalline pattern, coinciding with the results from EXAFS. Similar decomposition to Au particles was observed in case of the samples $2/TiO_2$, $2/SiO_2$, $1/\alpha\text{-}Fe_2O_3$, and $2/TiO_2$ by XRD. For example, the Au particle sizes for $1/SiO_2$ (treated at 673 K in air), $1/TiO_2$, $2/TiO_2$ (both are treated at 773 K in 5% $H_2/Ar$), and $1/\alpha\text{-}Fe_2O_3$ (treated at 673 K in air) were estimated to be 160, 160, 180 and 120 Å as determined by XRD, respectively.

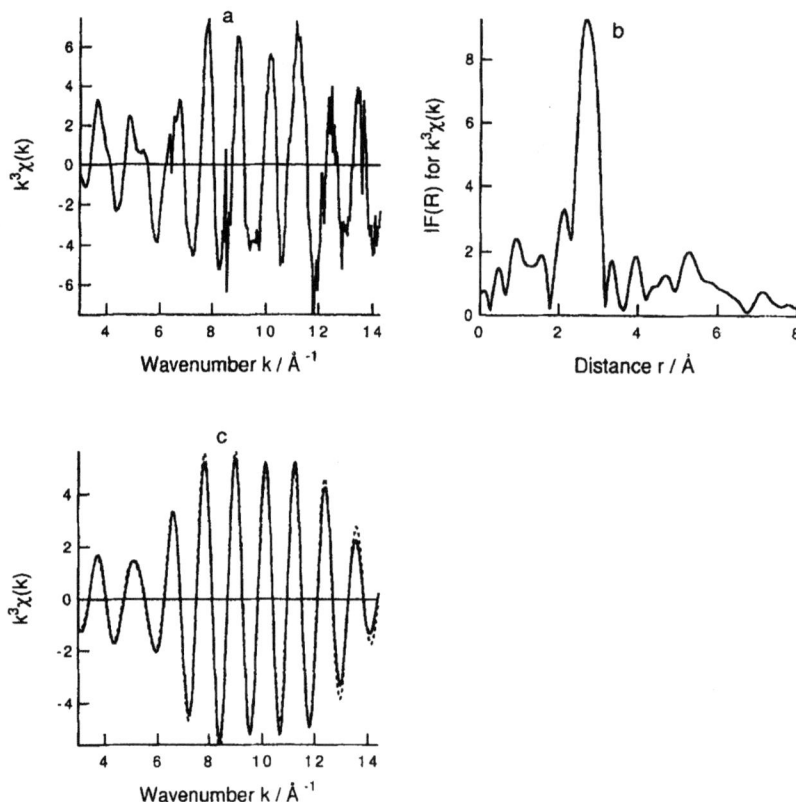

**Fig. 2** The EXAFS data for $2/TiO_2$ treated at 573 K in vacuum, (a) $k^3$-weighted EXAFS oscillation; (b) its associated Fourier transform; (c) curve-fitting analysis (inversely Fourier transformed) for Au-P and Au-Au: (——) observed; (···) calculated.

## 3.2 Catalytic activity for CO oxidation

No measurable activity of the incipient supported samples for CO oxidation at 313 K was detected. The catalytic activity was observed on the samples which were decomposed to form metallic gold particles on oxide supports except those on $SiO_2$ over which no $CO_2$ was produced at 313 K. Figs. 3, 4 and 5 depict CO conversion measured on the Au catalysts derived from **1** and **2** as a function of reaction temperature. The $1/\alpha\text{-}Fe_2O_3$ and $1/TiO_2$ catalysts treated at 673 or 773 K showed high CO oxidation activities in the temperature range 300~600 K, while the $1/SiO_2$ catalyst showed the activity only above 600 K. The activity of the catalysts derived from **2** was much lower than that of the catalysts derived from **1** as a common feature for $\alpha\text{-}Fe_2O_3$, $TiO_2$ and $SiO_2$ supports as shown in figs. 3, 4 and 5. Turnover frequencies (TOF) of some Au catalysts at 313 K are given in table 4. The TOF in this work is defined as the number of converted CO molecules per second divided by the total number of Au atoms included in the catalyst, because it is difficult to measure the accurate Au surface area by specific gas adsorption on XRD. The TOF of $1/TiO_2$, $2/TiO_2$ (pretreated at 773 K with $5\%$ $H_2/Ar$) and $1/Fe_2O_3$ (pretreated at 673 K in air), at 353 K are $0.0025$, $0.0004$, and $0.0082$ $s^{-1}$, respectively, as shown in table 4. The addition of **2** to $\alpha$-$Fe_2O_3$ had a little positive effect on the oxidation activity of $\alpha$-$Fe_2O_3$ (fig. 3). The activity of $1/\alpha$-$Fe_2O_3$ was higher than that of the catalyst prepared by an impregnation method using $HAuCl_4$ as precursor, but lower than that of the catalyst prepared by a coprecipitation method[8,9], possibly due to the different sizes of gold particles formed in these catalysts. The $TiO_2$-supported Au catalyst derived from $HAuCl_4$ was also less active than $1/TiO_2$, particularly at low temperatures as shown in fig. 4. The gold catalysts showed different activation energies: $1/TiO_2$ : 12kJ/mol, $2/TiO_2$ : 15 — 22 kJ/mol (the data are taken at $313 \sim 623$ K), $1/\alpha$-$Fe_2O_3$ : 9kJ/mol, and $2/\alpha$-$Fe_2O_3$ : 21 kJ/mol (calculated from the data at $303\sim473$ K), which were close to the activation energies ($16\sim35$ kJ/mol) reported for CO oxidation on Au catalysts prepared by coprecipitation[9]. The Au particle sizes for $SiO_2$ and $TiO_2$ supports were similar with each other, while $1/TiO_2$ catalyst exhibited much higher activity than the $1/SiO_2$ catalyst. The $1/\alpha$-$Fe_2O_3$ catalyst with an averaged particle size of 120 Å was more active than the $1/TiO_2$ catalyst with an averaged particle size of 160 Å. Thus the catalytic activity varied with the kind of support and the particle size. In this study, the catalysts derived from gold complexes **1** and **2** supported on $\alpha$-$Fe_2O_3$ among the three metal oxides were found to be most efficient for CO oxidation. There may exist synergistic interaction between Au particles and oxides which changes by the kind of precursors and the pretreatment conditions. The present study demonstrates importance of the kind of precursor for catalyst preparation in a synergistic mode with the oxide support[20], which would require precise characterization of the obtained catalyst at atomic or molecular scales.

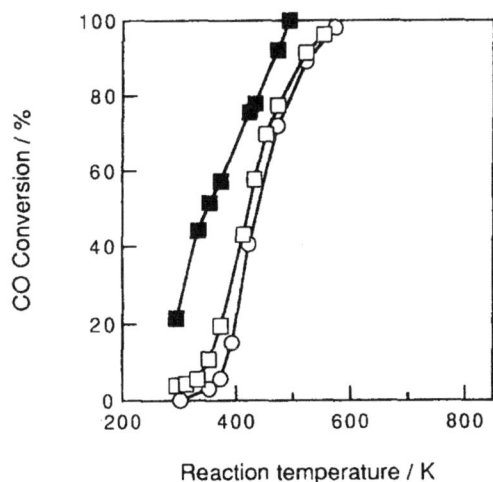

Fig. 3 Temperature dependence of CO oxidation activity of $1/\alpha$-$Fe_2O_3$ and $2/\alpha$-$Fe_2O_3$. The data at each temperature were measured after the catalytic reaction reached steady state; $1\%$ CO balanced with air at a flow rate of 34 mL/min (SV = 10000 $h^{-1}$). (■) $1/\alpha$-$Fe_2O_3$ treated under air at 673 K, (□) $2/\alpha$-$Fe_2O_3$ treated under air at 673 K, (○) $\alpha$-$Fe_2O_3$ calcined under air at 673 K.

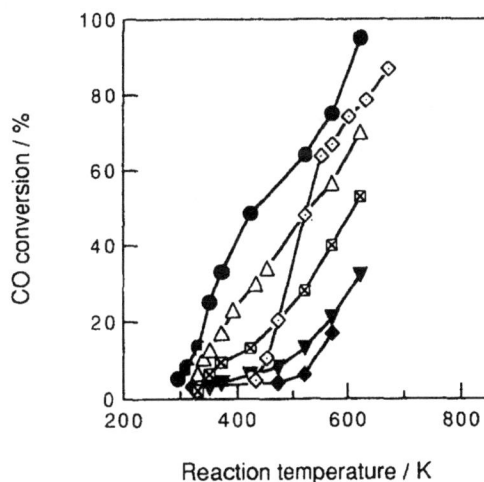

Fig. 4 Temperature dependence of CO oxidation activity of $1/TiO_2$ and $2/TiO_2$. The data at each temperature were measured after the catalytic reaction reached steady state; $1\%$ CO balanced with air at a flow rate of 34 mL/min (SV = 10000 $h^{-1}$). (●) $1/TiO_2$ treated under CO at 673 K, (△) $1/TiO_2$ treated under $5\%$ $H_2/Ar$ at 773 K, (⊠) $1/TiO_2$ treated under air at 673 K, (▼) $2/TiO_2$ treated under $5\%$ $H_2/Ar$ at 773 K, (◆) $2/TiO_2$ treated under CO at 673 K, (◇) $HAuCl_4/TiO_2$ calcined under air at 673 K.

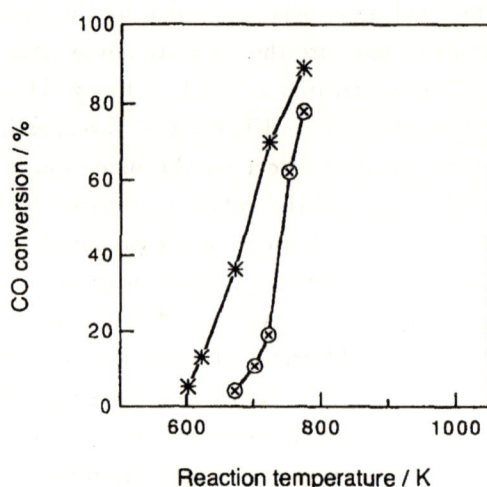

**Fig. 5** Temperature dependence of CO oxidation activity of 1/SiO$_2$ and 2/SiO$_2$. The data at each temperature were measured after the catalytic reaction reached steady state;1% CO balanced with air at a flow rate of 34 mL/min (SV=10000 h$^{-1}$). ( $*$ ) 1/SiO$_2$ treated under air at 673 K,($\otimes$) 2/SiO$_2$ treated under air at 673 K.

**Table 4  CO oxidation kinetics at temperature of 313 K [a]**

| Catalyst | Pretreatment | Activity TOF[c] (10$^{-3}$ s$^{-1}$) | Au crystalline size (Å) | Ref. |
|---|---|---|---|---|
| **1**/TiO$_2$ | 773K,5%H$_2$/Ar[b] | 2.5 | ~160 | this work |
| **1**/Fe$_2$O$_3$ | 673K,air | 8.2 | >120 | this work |
| **2**/TiO$_2$ | 773K,5%H$_2$/Ar[b] | 0.4 | ~180 | this work |
| 5%Au/Fe$_2$O$_3$ | 673K,air | 7.0 | 160 | [9] |
| 12%Au/Fe$_2$O$_3$[d] (Au/Fe=1/19) | 673K,air | >13 | 36 | [9] |
| 2.4%Au/TiO$_2$ (110m$^2$/g) | 673K,air | 83[e] | ~100 | [18] |
| 2.4%Au/TiO$_2$ (40m$^2$/g) | 673K,air | 83[f] | ~40 | [18] |
| 2.2%Pd/Al$_2$O$_3$ | | 0.08[g] | 37 | [19] |

[a] Measured with a gas flow rate 34 mL/min containing $P_{CO}$=7.6 Torr balanced by air.

[b] Heating at a rate with 5 K/min,keeping at final temperature for 4 h.

[c] TOF (turnover frequency) is defined as the number of reacted CO molecules divided by the total number of Au included in the catalyst.

[d] Prepared by coprecipitation,washed with hot water.

[e] Measured at 308 K.

[f] Measured at 273 K.

[g] $P_{CO}$=26 Torr,$P_{O_2}$=132 Torr.

# 4  Conclusion

We reported here the first example of supported gold catalysts prepared from a phosphine-stabilized mononuclear gold complex Au(PPh$_3$)(NO$_3$) (**1**) and an enneanuclear gold cluster compound [Au$_9$(PPh$_3$)$_8$](NO$_3$)$_3$ (**2**),which were active for CO oxidation after treatments under air or CO or 5% H$_2$/Ar atmosphere. These gold complexes can be used as precursors to develop a novel and relatively simple way to prepare supported gold catalysts. The catalytic activity of the gold particles on oxides depended on precursor,support,and treatment conditions. The catalysts derived from **1** showed a higher activity than those derived from **2**. Work is in progress to obtain much smaller gold particles on oxides and to find a new way to get stronger interaction between precursor and oxide for the genesis of high catalytic performance at low temperatures.

# References

[1]J. Schwank, Gold Bull. 1983, **16**: 103.

[2]A. G. Daglish and D. D. Eley. Proc. 2nd ICC, 1961, **2**: 1615.

[3]D. Y. Cha and G. Parravano. J. Catal, 1970, **18**: 200.

[4]S. Galvagno and G. Parravano, Ber. Bunsenges. Phys Chem, 1979, **83**: 894.

[5]N. W. Cant and P. W. Fredrickson. J. Catal, 1975, **37**: 531.

[6]S. D. Lin, M. Bollinger and M. A. Vannice. Catal. Lett, 1993, **17**: 245.

[7]P. Buffat and J. -P. Borel. Phys. Rev, 1976, **13**: 2287.

[8]M. Haruta, T. Kobayashi, H. Sano and N. Yamada. Chem. Lett, 1987: 405.

[9] M. Haruta, N. Yamada, T. Kobayashi and S. Iijima. J. Catal, 1989, **115**: 301.

[10]M. Haruta, S. Tsubota, T. Kobayashi, H. Kageyama, M. J. Genet and B. Delmon. J. Catal, 1993, **144**: 175.

[11]A. Baiker, M. Macijewski, S. Tagliaferri and P. Hug. J. Catal, 1995, **151**: 407.

[12]T. Castro, Y. Z. Li, R. Rifenberger, E. Choi, S. B. Park and R. P. Andres, ACS Symp Ser 1990, **437**: 329.

[13]K. P. Hall and D. M. P. Mingos, in: Progress in Inorganic Chemistry, 1984, **32**: 237.

[14]S. L. Scott and J. M. Basset. J. Mol. Catal, 1994, **86**: 5.

[15]A. Zecchina and C. Otero-Arean. Catal. Rev. Sci. Eng, 1993, **35**: 261.

[16]B. C. Gates. Chem. Rev, 1995, **95**: 511.

[17]F. Cariati and L. Naldini. J. C. S Dalton, 1972: 2286.

[18]S. Tsubota, M. Haruta, T. Kobayashi, A. Veda and Y. Nakahara, in: Preparation of Catalysts, 1991, **5**: 695.

[19]K. I. Choi and M. A. Vannice. J. Catal, 1991, **131**: 1.

[20]Y. Yuan, K. Asakura, H. Wan, K. Tsai and Y. Iwasawa. Chem. Lett, 1996, **9**, in press.

■ 本文原载:Chemistry Letters 1996,pp. 129~130.

# Structure and Catalysis of a SiO$_2$-Supported Gold-Platinum Cluster $[(PPh_3)Pt(PPh_3Au)_6](NO_3)_2$

You-Zhu Yuan[a,b], Kiyotaka Asakura[c], Hui-Lin Wan[b],

Khi-Rui Tsai[b], Yasuhiro Iwasawa[a]

(*aDepartment of Chemistry, Graduate School of Science,*

*The University of Tokyo, Hongo, Bunkyo-ku, Tokyo 113*

*bDepartment of Chemistry &.State Key Laboratory for*

*Physical Chemistry of the Solid Surface, Xiamen University, China*

*cResearch Center for Spectrochemistry, Faculty of*

*Science, The University of Tokyo, Hongo, Bunkyo-ku, Tokyo 113)*

Received October 19, 1995

**Abstract**   A novel catalyst of $[(PPh_3)Pt(AuPPh_3)_6](NO_3)_2$ **1** supported on SiO$_2$ was very active in H$_2$-D$_2$ equilibration with a TOF of 29.8s$^{-1}$, while it showed low activity for ethene hydrogenation and CO oxidation at 303 K. It was found that the catalysis of **1**/SiO$_2$ was not caused by platinum particle impurity but by the platinum atom which bound to the gold atoms in the cluster. The cluster framework of **1**/SiO$_2$ was stable during the reactions at 303 K. The change of Au-Pt bond in **1**/SiO$_2$ by heat treatment and the catalytic performance of produced cluster fragments were characterized by EXAFS, FT-IR and kinetic investigation.

Gold has the filled *d*-band located far bellow the Fermi level and is believed to be least useful for catalytic purpose, though small Au particles on α-Fe$_2$O$_3$, Co$_3$O$_4$ and TiO$_2$ exhibit high catalytic activity for low-temperature oxidation of CO[1]. There are some interesting examples, however, that evaporating gold onto platinum single-crystal surface displays markedly different activity and selectivity for conversion of *n*-hexene[2,3]. Conceivably, the "alkali-metal-like" $d^{10}s^1$ electronic configuration of Au atom leads to relatively tractable electronic structure compared to clusters of transition-metals with open *d* shells, which will attract much attention in catalysis of the transition-metal-gold clusters, especially ones containing catalytically important metals such as Pt and Pd. Those studies should provide a better understanding of metal-metal bonding and of the synergism often observed in bimetallic catalysis. Recently, Pignolet et al. reported a possible application of a phosphine-ligand stabilized Au cluster to heterogeneous catalysis by supporting on SiO$_2$ from surface organometallic viewpoints[4]. Independently, we have successfully obtained a new catalyst by supporting $[(PPh_3)Pt(AuPPh_3)_6](NO_3)_2$ **1** on SiO$_2$. The aim of this study is to examine the catalytic property of a Pt atom embedded in Au ensemble. The catalysis of **1**/SiO$_2$ for H$_2$-D$_2$ equilibration, ethene hydrogenation and CO oxidation has been investigated along with the surface structure characterization by means of FT-IR, EXAFS and TPR.

A dark-yellow microcrystalline of $[(PPh_3)Pt(AuPPh_3)_6](NO_3)_2$ **1** was synthesized according to the literature[5]. The framework of **1** is shown in Figure 1. Impregnation of **1** on SiO$_2$ (Aerosil 300) was conducted with a carefully dried ethanol solution of **1** in atmosphere of Ar (99. 9999%), followed by evaporation of the solvent under vacuum for 5 h at room temperature. The loading of **1** on SiO$_2$ was controlled to be 0. 5 Pt wt%. FT-IR studies on **1**/SiO$_2$ showed that all frequencies attributed to PPh$_3$ ligands were maintained in intensity when the sample was heated to 473 K at a rate of 4 K/min under vacuum, but the decrease in intensity was observed if temperature was over 473 K, especially above 520 K, indicating that a part of the ligands decompose to change the cluster framework when overheating.

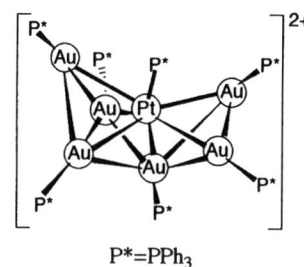

P*=PPh$_3$

**Figure 1** Framework of **1**.

EXAFS measurements have been carried out at room temperature to characterize **1**/SiO$_2$. The Fourier transforms of EXAFS oscillations at Au L$_3$-edge and Pt L$_3$-edge for **1**, **1**/SiO$_2$, **1**/SiO$_2$ treated at 473 K and **1**/SiO$_2$ treated at 773 K under vacuum, are shown in Figure 2.

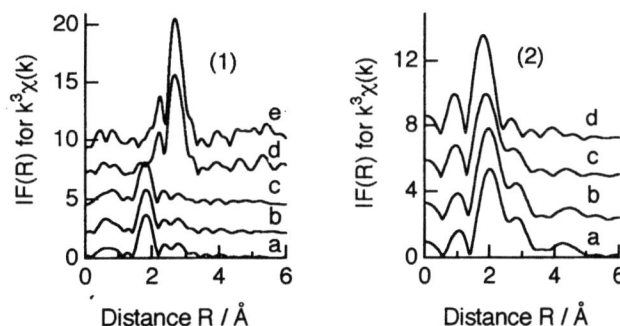

**Figure 2** Fourier transforms of the EXAFS oscillations at Au L$_3$-edge (1) and Pt L$_3$-edge (2); (a): **1**; (b): **1**/SiO$_2$; (c): **1**/SiO$_2$ treated at 473 K; (d): **1**/SiO$_2$ treated at 773 K; (e): Au foil.

The EXAFS fitting results are listed in Table 1. The EXAFS analysis revealed that there was no change in the cluster framework of **1** after deposition on the SiO$_2$ surface. However, cluster fragmentation irreversibly occurred when **1**/SiO$_2$ was treated at 473 K under vacuum. The bond numbers of Pt-Au and Au-Au(Pt) decreased to about 68% and 75% of the original cluster **1**, respectively. Moreover, the cluster framework of **1** was suggested to be completely destroyed to form gold particles and PtP$^\#$m (m≈3, P$^\#$ for phosphine species) when heated to 773 K. Change of the cluster framework in **1**/SiO$_2$ was proposed in Figure 3.

**Table 1** Curve-fitting results for the EXAFS data of **1**/SiO$_2$ [a]

| Sample | Au-P | | Au-Au(Pt) | | Pt-P | | Pt-Au | |
|---|---|---|---|---|---|---|---|---|
| | N | R/Å | N | R/Å | N | R/Å | N | R/Å |
| **1**/SiO$_2$ | 1. 0 | 2. 30 | 4. 0 | 2. 83 | 1. 0 | 2. 28 | 6. 0 | 2. 69 |
| | ±0. 1 | ±0. 01 | ±1. 0 | ±0. 02 | ±0. 2 | ±0. 02 | ±1. 0 | ±0. 03 |
| **1**/SiO$_2$[b] | 0. 6 | 2. 28 | 3. 0 | 2. 83 | 1. 3 | 2. 27 | 4. 1 | 2. 69 |
| | ±0. 1 | ±0. 02 | ±0. 9 | ±0. 02 | ±0. 3 | ±0. 02 | ±1. 0 | ±0. 03 |
| **1**/SiO$_2$[c] | — | | 10 | 2. 87 | 3. 3 | 2. 28 | — | |
| | | | ±1. 0 | ±0. 02 | ±0. 6 | ±0. 02 | | |
| **1**[d] | 1. 0 | 2. 30 | 4. 0 | 2. 83 | 1. 0 | 2. 28 | 6. 0 | 2. 69 |

[a] Fitting results were determined by comparing the EXAFS results with the crystallographic data of **1**; [4] [b] treated at 473 K; [c] treated at 773 K. [d] cluster **1**; Au-Au (Au foil): 2. 88 Å.

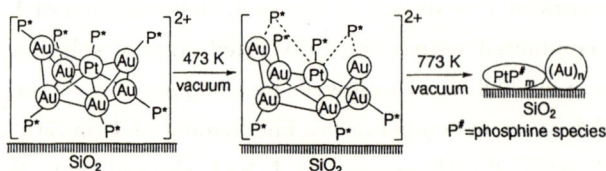

**Figure 3** Schematic framework transformation in 1/SiO₂ by heating. Dotted lines tentatively represent bonding of P* to Pt and Au.

$H_2$-$D_2$ equilibration on a series of catalysts related to $1/SiO_2$ was tested at 303 K in a fixed-bed flow reactor system equipped with a mass spectrometer. Ar was used as a diluent gas with a flow rate of 50 ml/min ($H_2 = D_2 = 2.0$ mL/min). The catalytic reaction rates are defined as turnover frequency (mol of HD)(mol of cluster)$^{-1}$(s)$^{-1}$, which are shown in Table 2.

From Table 2, it was observed that the catalytic activity of **1** under molecular solid-gas condition was promoted about 15 times by supporting on $SiO_2$[6]. The dramatic increase may be mainly attributed to a high surface area of the support. It was also found that the catalytic activity of $1/SiO_2$ was higher than that of the ones treated at 473 or 773 K. [$Au_9 (PPh_3)_8$]($NO_3$)$_2$/$SiO_2$ was inactive for $H_2$-$D_2$ equilibration, which implies that the Pt atom in **1** plays a key role in activation of $H_2$. This is contrasted to activation of $H_2$ on Pt catalysts which demands multimetal sites at Pt surface. Lower activity of $1/SiO_2$ treated at 473 K compared to $1/SiO_2$ is possibly referred to the increase in the number of Pt-P bonds and to the decrease in the number of Au-Pt bonds which may participate in the activation of $H_2$. The monometallic catalysts of $Pt(PPh_3)_4/SiO_2$ and that pretreated at 473 K were inactive (Table 2). In case of $1/SiO_2$ treated at 773 K, the $H_2$-$D_2$ equilibration may be caused by Au particles[7] or Pt sites with phosphine ligands. However, $Pt(PPh_3)_4/SiO_2$ treated at 773 K showed a remarkably high activity for $H_2$-$D_2$ equilibration, which may be ascribed to fragmentation of $Pt(PPh_3)_4$ to form metallic particles on $SiO_2$ because of irreversible $H_2$ adsorption (H/Pt = 0.32). No chemisorption of $H_2$ was observed with $Pt(PPh_3)_4/SiO_2$ and the one treated at 473 K. Increasing activity by treatment of $1/SiO_2$ at 773 K might be due to partial clusterization of Pt though no Pt-Pt bond was observed by EXAFS.

**Table 2** $H_2$-$D_2$ equilibration (à), ethene hydrogenation (b) and CO oxidation (c) over several catalysts at 303 K

| Catalyst | TOF/s⁻¹ | | |
|---|---|---|---|
| | (a) | (b) | (c) |
| **1** | 2.0 | — | — |
| 1/SiO₂ | 29.8 | $8 \times 10^{-4}$ | $7 \times 10^{-5}$ |
| 1/SiO₂ treated at 473 K | 11.1 | $5 \times 10^{-4}$ | $3 \times 10^{-5}$ |
| 1/SiO₂ treated at 773 K | 25.7 | $6 \times 10^{-4}$ | $1.5 \times 10^{-5}$ |
| Pt(PPh₃)₄/SiO₂ | 0.0 | 0 | — |
| Pt(PPh₃)₄/SiO₂ treated at 473 K | 0.0 | 0 | — |
| Pt(PPh₃)₄/SiO₂ treated at 773 K | 1299.1 | $7 \times 10^{-2}$ | — |
| [Au₉(PPh₃)₈](NO₃)₂/SiO₂ | 0.0 | — | — |

From pulse reactions, the following results were obtained: 1) About 80% decrease in the rate of $H_2$-$D_2$ equilibration was observed when ethene was mixed into a gas flow of $H_2$-$D_2$. When ethene was mixed with $D_2$, the HD formation rate was 1/20 of that in $H_2$-$D_2$ equilibration and a very weak signal in m/e 29 ($C_2H_3D$) was observed. It is, therefore, unlikely that $H_2$-$D_2$ equilibration on $1/SiO_2$ is caused by undetectable metallic platinum particle impurity. 2) No $H_2$-$D_2$ equilibration proceeded when CO pulse was admitted into a gas flow of $H_2$-$D_2$. By CO adsorption, the yellowish $1/SiO_2$ immediately turned to the orange-red one, which resembles the CO adduct of **1** in solution[4]. It is also deducible that the H$_2$-D$_2$ equilibration takes place on the coordination space enough for hydrogen dissociation on the Pt atom of **1**/SiO$_2$. Pt(PPh$_3$)$_4$/SiO$_2$ treated at 473 K was inactive, whereas, **1**/SiO$_2$ treated at 473 K catalyzed H$_2$-D$_2$ equilibration. The Au-Pt bonds may contribute to the formation of active Pt sites.

The rapid H$_2$-D$_2$ equilibration catalyzed by **1**/SiO$_2$ impelled us to carry out the ethene hydrogenation at C$_2$H$_4$/H$_2$ = 13 : 13 kPa in a closed circulating system. It was found that the hydrogenation of ethene to ethane proceeded over **1**/SiO$_2$ at an initial rate of $8 \times 10^{-4}$ s$^{-1}$ at 303 K (Table 2). TPR spectra of **1**/SiO$_2$ before and after ethene hydrogenation were almost the same each other, suggesting that no change occurred in the cluster framework of **1**/SiO$_2$ during the reaction.

We also performed CO oxidation reaction under CO/O$_2$ = 13 : 13 kPa at 303 K. The low reaction rate (TOF of $7 \times 10^{-5}$ s$^{-1}$) in Table 2 is due to strong CO adsorption as proved by the color change, which prevents O$_2$ adsorbing. The catalytic reaction mechanism on one Pt atom embedded in the Au cluster is not clear at present, but behavior of **1**/SiO$_2$ upon CO and O$_2$ adsorption related to structure characterized by FT-IR and EXAFS will be reported separately[8].

# References

[1]M. Haruta, S. Tsubota, T. Kobayashi, H. Kageyama, M. J. Genet, and B. Delmon. J. Catal, 1993, **144**:174.

[2]J. W. A. Sachtler and G. A. Somorjai, J. Catal. ,1983,**81**:77.

[3]J. W. A. Sachtler, J. P. Biberian, and G. A. Somorjai, Surf. Sci., 1983, **110**:43.

[4]I. V. Gubkina, L. I. Rubinstein, and L. H. Pignolet, Abst. of ACS Meeting, 1994, **208**:405.

[5] L. N. Ito, J. D. Sweet, A. M. Mueting, L. H. Pignolet, M. F J. Schoondergang, and J. J. Steggerda, Inorg. Chem., 1989, **28**:3696.

[6]M. A. Aubart, B. D. Chandler, R. A. T. Gould, D. A. Krogstad, M. J. Schoondergang, and L. H. Pignolet, Iorg. Chem., 1994, **33**:3724.

[7]S. Galvagno and G. Parravano, J. Catal, 1978, **55**:178.

[8]Y. Yuan, K. Asakura, H. Wan, K. Tsai, and Y. Iwasawa, to be published.

■ 本文原载:Chemical Research in Chinese Universities Vol. 12 No. 1,pp. 70~80,1996.

# Studies on Catalysis in Partial Oxidation of Methane to Syngas (I) *

## —Performance and Characterization of Ni-based Catalysts

Ping Chen, Hong-Bin Zhang①, Guo-Dong Lin, Khi-Rui Tsai

(Department of Chemistry & State Key Laboratory for Physical Chemistry of the Solid Surface, Xiamen University, Xiamen, 361005)

Received Feb. 27, 1995

**Abstract**  Highly active and selective Ni-based catalysts for partial oxidation of methane (POM) to syngas ($CO/H_2$) have been studied and developed. Spectroscopic characterization by XRD、XPS, EPR, *etc*. demonstrated that under the POM reaction conditions, the Ni-components of the catalysts investigated were reduced and enriched on the surface to form metallic $Ni^0$-phase. A comparative study of the first series of transition-metals showed that only Ni and Co have a high POM activity and selectivity, whereas the others (including Mn, Fe, Cu, *etc*.) give mainly complete combustion products, $CO_2$ and $H_2O$. The results favor the following viewpoints: the POM activity is related with the rapidly changeable valence transition-metal sites, $M^0/M^{2+}$ (e. g. $Ni^0/Ni^{2+}$), on the surface of the functioning catalysts; the transition-metal sites in zero-valence state seem to be responsible for the activation and dehydrogenation of methane by homolytic splitting of its C-H bonds on these sites, and the nature of rapidly changeable valence of the active sties is requisite for activation and rapid conversion of dioxygen.

**Key words**  Partial oxidation of methane  Syngas  Ni-based catalysts

## Introduction

With the development of a new generation of highly active catalysts, there has been renewed interest in the catalytic partial oxidation of methane (POM) to syngas as a potential commercial alternative to steam-reforming of methane in the past four or five years. In contrast to steam-reforming, the catalytic partial oxidation of methane to syngas,

$$CH_4 + \frac{1}{2}O_2 \rightarrow CO + 2H_2 (\Delta H = -36.0 \text{ kJ/mol}) \tag{1}$$

is mildly exothermic and more selective, and theoretically yields a $H_2/CO$ ratio of 2, which is lower than that obtained by steam-reforming, and thus is more applicable to the Fischer-Tropsch Synthesis (FTS) of higher hydrocarbons and the synthesis of methanol.

There have been a number of studies published recently on POM catalysts, Ashcroft *et al*. [1,2] first

---

＊ Supported by the National Natural Science Foundation of China.

① To whom correspondence should be addressed.

reported that the ruthenium compounds of pyrochlore type, $Ln_2Ru_2O_7$ (where Ln is a lanthanide) were extraordinarily active and selective catalysts for the POM reaction. After that, a few studies have been published on the POM over $Eu_2Ir_2O_7$[3], transition metal (i. e. , Ru, Rh, Pd, and Pt)-containing supported and mixed metal oxide catalysts[2,4], and supported Ni-catalysts[5−7]. More recently, two research groups have independently developed a new process for the POM[8−10], characterized by very short contact time between the reactants (methane and oxygen) and the catalyst. Moreover, Lunsford et al. [11] found that over $Ni/Yb_2O_3$, the oxidation of methane at high space velocities releases a sufficient amount of heat so that the thermal gradients within the catalyst bed may be over 300 ℃, and the observed concentrations of the products are somewhat less than those predicted by equilibrium at the hot spot temperature. However, quite limited amounts of the information about the nature of active sites and the mechanism of catalysis have been reported; many problems still remain to be solved.

In the present work, highly active and selective Ni-based catalysts, supported and mixed oxides, has been developed; spectroscopic characterization of catalysts and related systems have been performed using the methods of XRD, XPS, EPR etc. ; a comparative study on catalytic performance of the first series of transition metals for the POM reaction has been carried out; effect of alkali-earth-metal oxide-components on the activity and selectivity of Ni-catalysts was investigated; chemical behavior of methane on reduced and unreduced nickel catalysts was also examined. The results would provide significant implies for the understanding of nature of catalytically active site/phase and of mechanism of the catalysis.

# Experimental

## 1  Preparation of Catalysts

The supported catalysts were prepared by impregnating the corresponding nitrate dissolved in water on certain powdered supports of 80−100 meshes, followed by vacuum drying. The mixed metal oxides catalysts were prepared by thoroughly mixing and finely grinding the corresponding nitrates in the required proportion with deionized water to form a thick paste, followed by drying in air from room temperature to 473 K, Both types of catalysts were calcined in air at 1073 K for at least 5 hours. The mixed metal oxides catalysts, after the calcination, were crushed and sieved to particles of 80 − 100 meshes.

## 2  Evaluation of Catalysts

The evaluation experiment was performed in a fixed-bed continuous flow reactor-GC combination system operating at atmospheric pressure. The reactor is made of quartz tube of 4. 0 mm I. D. ; 100 mg of catalyst sample was used for each testing. Feed gases: methane, oxygen or air, all of them were in 99% purity. The reactants and products were analyzed by a on-line SQ-206 GC equipped with thermal conductivity (TC) detector and 601 carbon molecular sieve column, with hydrogen as the carrier gas. The catalytic POM-to-syngas reaction over the catalysts was carried out at the stationary state and under the following reaction conditions: feed, a gaseous mixture of $2CH_4/1O_2$; gas hourly space velocity (GHSV) at STP $10^5$ mL/h・g・catal. ; temperature 973~1173K, unless otherwise noted.

## 3  Spectroscopic Characterization of Catalysts

Samples used in spectroscopic measurements include those in three states: the oxidation state (precursors), the prereduced state, and the functioning state. Catalyst sample in functioning state was taken *in situ* after its operating under POM reaction conditions for 1 hour, followed by cooling it down to room temperature ana transferring it into a sample tube in nitrogen atmosphere for spectroscopic measurement. The prereduced catalyst was collected after it was prereduced by hydrogen for 1 hour, followed by the same procedures as that for the collection of functioning samples.

X-ray diffraction patterns were obtained by using a Rigaku D/Max-C X-ray Diffractometer with Cu $K\alpha$ radiation. The scanning rate was $8°/min$, scanning scope was from $15°$to $100°$. XPS measurements were done using a VG ESCA LAB MK-2 machine with Al$K\alpha$ radiation (1487 eV) and UHV ($1\times10^{-7}$ Pa), taking the Al($2p$) of $Al_2O_3$ at 74. 6 eV, or deposited carbon C($1s$) at 284. 6 eV, as the internal standard. EPR spectra of catalysts were recorded by a Brucker ER-200D ESR spectrometer operating in the X-band at room temperature or at liquid helium temperature (14K) when neccesary.

# Results and Discussion

## 1 POM Performance of Ni-based Catalysts

Table 1 summarizes the assay results of the POM activity of several Ni-based catalysts. The symbol "Ni/MO$_x$" represents the supported catalysts, and the symbol "Ni-M-O" is for those derived from precursors of composite oxide or mixed oxides. The results indicated that oxygen in the feedstream was completely consumed under the reaction conditions above 773K. It can be seen from Table 1 that the two catalysts developed by us, 4wt% Ni/ZrO$_2$ and Ni-Mg-O, show almost as high activity and selectivity as the noble metal catalysts reported, 1% Ru/Al$_2$O$_3$[1] and 1% Pd/Al$_2$O$_3$[1], do, especially under the reaction conditions of N$_2$-free feedstream of $2CH_4/1O_2$.

**Table 1    The performance of several Ni-based catalysts for POM reaction**

| Catalyst | $T/K$ | $CH_4/O_2/N_2$ | $C_{CH_4}$ (/%) | $S_{CO}$(C%) |
|---|---|---|---|---|
| Ni—La—O | 1050 | 2/1/4 | 92 | 92 |
| Ni—Mg—O | 1050 | 2/1/4 | 93 | 94 |
| | 1050 | 2/1/0 | 96 | 94 |
| | 973 | 2/1/4 | 86 | 87 |
| | 973 | 2/1/0 | 88 | 89 |
| Ni—Ca—O | 1050 | 2/1/4 | 93 | 93 |
| | 1050 | 2/1/0 | 90 | 90 |
| 4wt % Ni/Zr()$_2$ | 1050 | 2/1/4 | 94 | 95 |
| | 1050 | 2/1/0 | 93 | 97 |
| | 973 | 2/1/4 | 87 | 89 |
| | 973 | 2/1/0 | 88 | 89 |
| Ni—Sm—O | 1050 | 2/1/4 | 87 | 88 |

Fig. 1 illustrates the effect of temperature on $CH_4$-conversion and CO-selectivity, Both $C_{CH_4}$ and $S_{CO}$ increased with increasing temperature; as expected, high temperature favored the produccion of syngas.

It is clear that in order to achieve $CH_4$-conversion and CO-selectivity above 90% under the reaction conditions of $GHSV = 10^5$ mL/g $\cdot$ catal. $\cdot$ h and $2CH_4/1O_2$ ($N_2$-free) over the 4wt% $Ni/ZrO_2$ catalyst, temperature of approximate 1000K is required. On the other hand, the results of investigation about effect of $CH_4/O_2$ ratio on $CH_4$-conversion and CO-selectivity showed that as $CH_4/O_2$ ratio was increased, CO-selectivity was increased but $CH_4$-conversion dropped gradually. The syngas yield peaked near a $CH_4/O_2$ ratio of 2. When this ratio was over 2, there was an apparent deposition of carbon on the surface of catalyst.

From the experimental results of the effect of temperature on the conversion of methane, an apparent activation energy of 17.8 kJ/mol for the POM reaction over 4wt%$Ni/ZrO_2$ can be estimated. This value is quite close to the activation energy of methane decomposition on $Ni/SiO_2$ (25 kJ/mol)[12] and Rh-catalyst (20 kJ/mol)[13] reported.

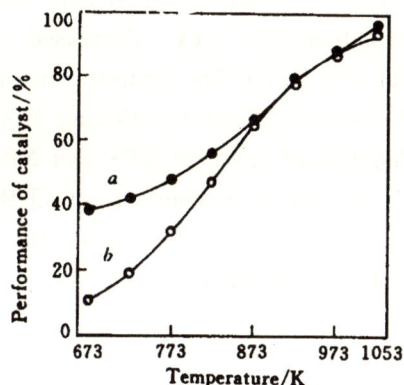

**Fig. 1** Effect of temperature on the performance of catalyst.

*a.* conversion of methane; *b.* selectivity to carbon monoxide; Catal.: 4wt% $Ni/ZrO_2$; $GHSV = 10^5$ mL/g $\cdot$ catal. $\cdot$ h.

## 2 Spectroscopic Characterization of Active Site/phase

### 2.1 XRD

As shown in Fig. 2 and Table 2, the precursor of Ni-Ca-O system (Fig. 2) did not form $ABO_3$ type composite oxide, but was present in the mixture of the two discrete oxides, NiO and CaO. In its functioning state, disappearance of the XRD feature of NiO is accompanied by the appearance of strong peaks of metallic nickel phase; whereas CaO remains unchanged and was more likely to serve as the support. The XRD results of the 2wt%$Ni/La_2O_3$ in the functioning state also revealed the reduction of NiO accompanied by formation of metallic Ni-phase.

**Table 2 Assignment of the XRD peaks observed on the Ni-Ca-O and 2wt% $Ni/La_2O_3$**

| Catalyst system | 2-THETA | $I/I_0$ | Phase |
|---|---|---|---|
| Ni—Ca—O | 43.24 | 100 | NiO |
| Precursor | 37.32 | 67 | NiO |
| | 62.72 | 46 | NiO |
| Ni—Ca—O | 44.30 | 100 | $Ni^0$ |
| Functioning state | 51.7 | 43 | $Ni^0$ |
| 2wt%$Ni/La_2O_3$ | 29.16 | 99 | $LaNiG_3$ and $La_2O_3$ |
| Precursor | 43.24 | 67 | NiO |
| 2wt%$Ni/La_2O_3$ | 29.96 | 99 | $La_2O_3$ |
| Functioning state | 47.3 | 44 | $La_2O_3$ |
| | 44.24 | 72 | $Ni^0$ |
| | 51.98 | 26 | $Ni^0$ |

### 2.2 XPS

The XPS measurements provided further information about valence-state of the Ni-species on the surface of the catalyst. The XPS spectra of Ni ($2p$) observed for 4wt%$Ni/ZrO_2$ catalyst are shown in

Fig. 3, which demonstrates that the Ni($2p$) binding energy (B. E. ) of the catalyst used (in functioning state) (about 853. 3 eV) decreased by about 1. 8 eV in comparison with that of unreduced sample (about 855. 1 eV). The computer-fitting results indicated the existence of Ni-species with mixed-valence on the surface of the functioning Ni-catalyst; relatively large amounts in $Ni^0$, and minor in $Ni^{2+}$, By combination of these of XPS and XRD (stated above) data, it can be seen that under the reaction conditions, the Ni-components of all these catalysts were reduced and enriched on the surface to form metallic $Ni^0$-phase. It is probably the metallic $Ni^0$ phase at the surface that is the catalytically active phase for the POM reaction.

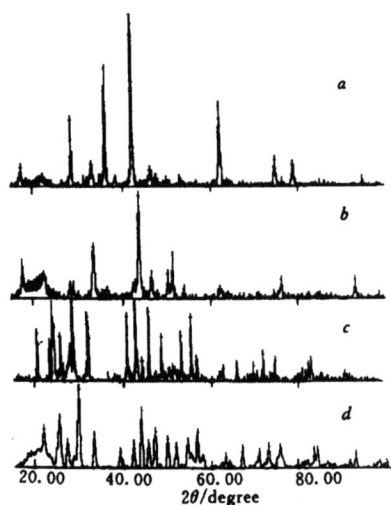

**Fig. 2** **XRD patterns of Ni—Ca—O catalyst in oxidation state (precursor)**
(a) ; Ni—Ca—O catalyst in functioning state (b); 2wt% Ni/La$_2$O$_3$ catalyst in oxidation state (c); 2wt% Ni/La$_2$O$_3$ catalyst in functioning state(d).

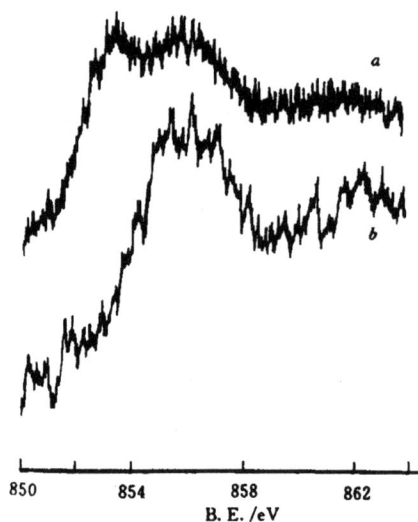

**Fig. 3** **XPS spectra of Ni (2p) for 4wt % Ni/ZrO$_2$ catalyst in functioning state (a); oxidation state (precursor) (b).**

## 2.3 EPR

The results of EPR observation at 14K of Ni-species of 2wt% Ni/Al$_2$O$_3$ demonstrated that EPR spectrum of the functioning catalyst almost coincided with that of the prereduced sample, but was obviously different from that of the unreduced (*i. e.* oxidation state) sample. Although the exact *g*-values of the separate peaks of these Ni species are difficult to determine (perhaps due to the characteristic of nickel), it is evident by the comparison of the overall appearance of these spectra that the Ni species in both functioning and prereduced samples are perhaps in the same valence state.

## 3 comparative Study of the POM Catalytic Performance of the First Series of Transition Metals

Table 3 shows the catalytic performance of several different transition metals for the POM reaction. Among Mn, Fe, Co, Ni, and Cu, only nickel and cobalt show a high activity and selectivity and very high productivity for the POM to syngas, whereas the others give complete combustion products, H$_2$O and CO$_2$. In comparison with nickel, cobalt requires a higher reaction temperature for reaching almost the same level of the activity as that of nickel catalyst.

**Table 3   The catalytic performance of several transition metals for the POM reaction** *

| Catalyst | $T$ | Conversion of | | Selectivity to | |
|---|---|---|---|---|---|
| | /K | $CH_4$(%) | $O_2$(%) | CO(C%) | $CO_2$(C%) |
| Mn—La—O | 1053—1173 | 24 | ~100 | — | ~100 |
| Fe—La—O | 1053 | 25 | ~100 | — | ~100 |
| Co—La—O | 1123—1173 | 95 | ~100 | 96 | 4 |
| Ni—La—O | 1053 | 92 | ~100 | 92 | 8 |
| Cu—La—O | 1053 | 30 | ~100 | — | ~100 |

\* Feedstream composition: $CH_4/O_2/N_2=2/1/4$; $GHSV=1\times10^5$ mL/g · catal. · h.

It is well known that all these transition metals can form oxides with the changeable valence cation. Indeed, the sites consisting of such cations can serve as the active sites for activation and conversion of dioxygen *via* alternate redox process. This can reasonably interpret the experimental fact that almost 100% dioxygen was converted over these transition metal-containing catalysts. However, it is difficult to rationalize why such significant difference in the activity and selectivity of methane conversion to CO exists among these transition metals.

By analysis of the standard electric potentials of these transition metals and comparison of their stable valence state in the functioning catalysts, some useful clues can be obtained. The standard electric potentials of these transition metals increased gradually from Ti to Cu as shown in Table 4[14]:

**Table 4   Standard electric potentials of some transition metals**

| Elements | Ti | V | Cr | Mn | Fe | Co | Ni | Cu |
|---|---|---|---|---|---|---|---|---|
| Potentials of $M^{2+}$—$M^0$(V) | −1.6 | −1.18 | −0.91 | −1.18 | −0.44 | −0.28 | −0.24 | 0.34 |

The front members in this transition series have stronger trend to lose their valence electrons, due to their lower standard electric potentials ( *i. e.* large negative potentials), and, thus, to be comparatively difficult to be reduced to $M^0$, so that they are easy to be stabilized at higher positive valence, in other words, they would form stable oxides under the condition with existence of certain amount of oxygen. For example, the XPS observation (Fig. 4) showed that the electron binding energy of iron remained almost unchanged in both oxidation state (precursor) and functioning state; whereas the most stable valence state of copper in the feedstream of rich-in-alkane/poor-in-oxygen was mostly in $Cu^0$, as evidenced by the experimental facts that the catalyst used is red-brown in color, and that on the internal surface of the reactor some red-brown deposit appears.

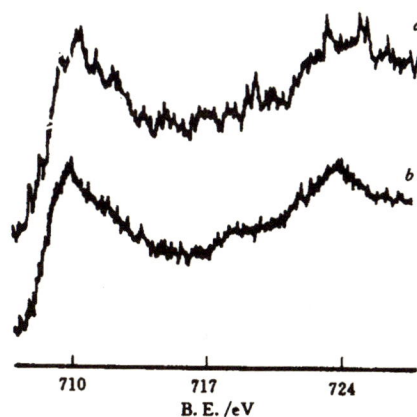

Fig. 4   XPS spectra of Fe ( $2p$ ) for Fe—Ti—O catalyst in oxidation state ( precursor ) ( *a* ); functioning state( *b* ).

Being different from the other members in the first transition series, the standard electric potentials of Ni and Co are relatively near to zero and close to each other; they most probably exist in a mixed valence state of $M^0/M^{2+}$ in the feedstream of $2CH_4/1O_2$ under the POM reaction conditions.

Thus, it seems that the POM activity and selectivity are more probably related with the changeable valence $M^0/M^{2+}$ sites. Under the POM reaction conditions, Ni-and Co-based catalysts possess a high concentration of the changeable valence $M^0/M^{2+}$ sites on their functioning surfaces, thus, showing a high POM activity and selectivity. It seems also that the activation of methane is more probably mainly *via* homolytic splitting of its C-H bond on $Ni^0$ or $Co^0$ sites and that oxygen-assisted dehydrogenation is unlikely the major pathway of methane activation in the POM-to-syngas reaction.

## 4  Effect of Alkali Earth Metal (AEM) Component on the Performance of Catalysts

Ni/MgO[15] and Co/MgO[16] systems are among the good POM catalysts reported. However, the effect of different AEM-components on the POM catalytic performance may be quite different. Table 5 shows the evaluation results of the POM performance of these Ni—AEM—O systems. As may be seen from Table 4, Ni—Mg—O and Ni—Ca—O systems have a very good catalytic activity and selectivity for the POM reaction, with the corresponding $C_{CH_4}$ and $S_{CO}$ both reaching 90% or more. However, the POM catalytic activity and selectivity decrease one by one from Ca to Ba, and, for Ni—Ba—O system, complete combustion products are mainly formed.

**Table 5  The performance of Ni—AEM—O catalysts for the POM reaction ***

| Catalysts | Conv. of CH$_4$ (%) | Select. to CO (C%) | Select. to CO$_2$ (C%) | Catalysts | Conv. of CH$_4$ (%) | Select. to CO (C%) | Select. to CO$_2$ (C%) |
|---|---|---|---|---|---|---|---|
| Ni—Mg—O | 93 | 94 | 6 | Ni—Ca—O | 93 | 93 | 7 |
| Ni—Sr—O | 75 | 80 | 20 | Ni—Ba—O | 30 | — | ～100 |

\* Feed gas composition: $2CH_4/1O_2/4N_2$; Temperature: 1050 K, GHSV: $10^5$ mL/g · catal. · h.

It is noted that all Ni-AEM-O systems do not form homogeneous composite oxide phases, as evidenced by our results of XRD measurement. In each of the catalyst precursors, there co-existed two separate oxide-phases, NiO and AEM-oxide, By the reduction of Ni-AEM-O binary oxides, the surface phase of $Ni^0$ on the AEM-oxide (AEO) supports, $Ni^0/$AEO, was developed. This indicates that the AEO components more likely play a role of support only. It has been found in the present work that the electron binding energy of the Ni-component remains almost unchanged in the precursor systems supported by the four different AEO (Fig. 5), implying that the difference in the influence of these AEO components on reducibility of Ni-component is not marked. However, the XPS results revealed that the abundance of nickel on the surface of these functioning catalysts decreased from Mg to Ba, i. e. , Mg ～Ca≫Sr＞Ba (Fig. 6). On the other hand, the XPS also demonstrated the formation of substantial amounts of carbonate species on the surface of Ni—Ba—O and Ni—Sr—O (Fig. 7).

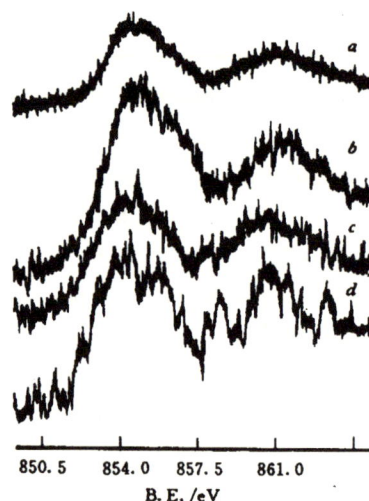

Fig. 5  XPS spectra of Ni (2p) for the precursors of the Ni—AEM—O catalysts.

a. Ni—Mg—O; b. Ni—Ca—O;
c. Ni—Sr—O; d. Ni—Ba—O.

The relative abundance of the surface carbonate species in these systems is in the order: Mg＜Ca≪Sr＜Ba, and in agreement with the sequence of their stabilities of the corresponding AEM-carbonates, The formation of carbonates would cause two possible effects on the performance of catalysts: first, it would lead to some chemical surroundings changes of nickel, and second, it may cover and deactivate catalytically active sites. Both factors may lead to the remarkable decrease in POM activity and selectivity of the catalysts.

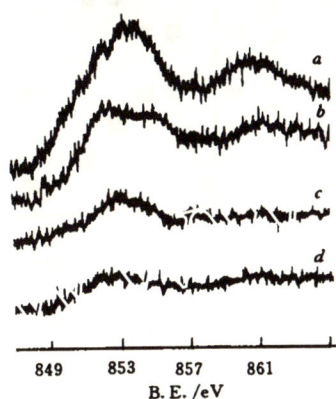

Fig. 6    XPS spectra of Ni(2p) for the functioning
Ni—AEM—O catalysts.
a. Ni—Mg—O; b. Ni—Ca—O;
c. Ni—Sr—O; d. Ni—Ba—O.

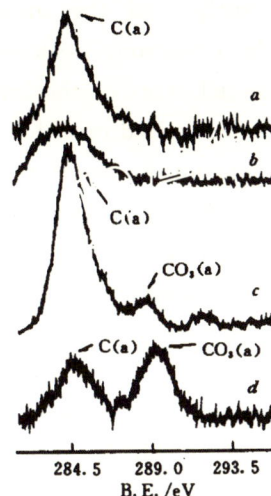

Fig. 7    XPS spectra of C(1s) for the functioning
Ni—AEM—O catalysts.
a. Ni—Mg—O; b. Ni—Ca—O;
c. Ni—Sr—O; d. Ni—Ba—O.

## 5  Chemical Behavior of Methane on the Surface of Reduced and Unreduced Nickel Catalysts

As early as 1957, it was reported that methane can be activated by homolytic splitting of its C-H bond on the surface of Pd in the reaction of $H_2$-$D_2$ isotope exchange[17]. On nickel catalysts, a similar process for activation of methane may be expected to occur. In order to confirm this, the following experiments have been performed: a sample of Ni-Ca-O catalyst was reduced by hydrogen at 1050K for 1 h, followed by switching the gas-supply to a stream of purified methane; the GHSV was monitored by soap-flow-meter of gas. The change of GHSV observed with the time indicated that there was an increasing pressure-drop in catalyst bed as the time passed due to deposition of carbon originated from decomposition of methane on the surface of reduced Ni-catalyst. Similar results could be also obtained at 773K, although the drop-rate of GHSV was slower than that at 1050 K.

In the other experiment, a sample of unreduced Ni—Ca—O catalyst was heated up to 1050K and continued to be heated at the same temperature for 10 minutes, followed by cleaning the surface using purified nitrogen for a few minutes, and then followed by switching the gas supply to purified methane. The change in composition of products with time was monitored by GC detector, which is shown in Fig. 8, This result demonstrated that in the first few minutes, the products contained mainly $CO_2$, $H_2O$, a large amount of unconverted methane and a little amount of CO. As time passed, composition of the products gradually varied; selectivity to CO gradually enhanced, whereas selectivity to $CO_2$ simultaneously dropped; conversion of methane increased in the first eight minutes and reached maximum in about 8 minutes and then decreased gradually in company with apparent dropping in the GHSV due to increment of pressure drop in the catalyst bed caused by the deposition of carbon, and, at last, conversion of methane was getting very low and only a very small amount of CO was detected.

The phenomena observed in the above experiment reflected the overall process of the interaction of methane with the Ni-catalyst being gradually reduced. It demonstrates that methane can react with lattice oxygen species on the surface of the unreduced Ni—Ca—O catalyst under the conditions associated with the POM reaction. In the initial stage of the interaction, only very little methane was converted probably *via* lattice-oxygen-assisted-dehydrogenation by heterolytic splitting of its C-H bond on the $Ni^{2+}$-$O^{2-}$ ion-pair site, with deep-oxidation products, $CO_2$ and $H_2O$, as the dominant products. As NiO-component was gradually reduced and the lattice oxygen on the surface was gradually consumed, the oxygen-assisted-dehydrogenation gradually became a minor pathway, simultaneously, with the development of $Ni^0$-sites, the homolytic splitting of C-H bond of methane molecule on $Ni_n^0$-site would become the major pathway for activation and conversion of methane. It may be inferred that under the POM reaction conditions, the latter

**Fig. 8** The variation of product composition with time during the interaction of methane with an unreduced Ni-Ca-O catalyst.

*a*. Conversion of methane; *b*. selectivity to CO; *c*. selectivity to $CO_2$.

reaction would proceed in a much higher rate than the former, with CO and $H_2$ as the major products, when the supply of oxygen was insufficient, it was also apt to lead to deposition of carbon on the surface of catalyst.

Therefore, it seems to us that there may exist two pathways for activation of methane on the Ni-catalyst in the POM reaction, *i. e.*,

(1)Dehydrogenation by homolytic splitting of C-H bond on reduced Ni-sites, namely, $Ni^0$-site,

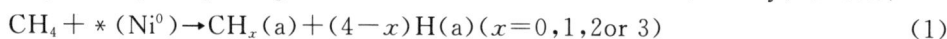

$$CH_4 + *(Ni^0) \rightarrow CH_x(a) + (4-x)H(a)(x=0,1,2 \text{ or } 3) \tag{1}$$

(2)Oxygen-assisted dehydrogenation by heterolytic splitting of C-H bond on oxidized Ni-sites, namely, $Ni^{2+}$-$O^{2-}$ ion-pair site,

$$CH_4 + *(Ni^{2+}\text{-}O^{2-}) \rightarrow CH_3(a) + OH(a) \tag{2}$$

Under the reaction conditions of rich-in-methane/poor-in-oxygen and high temperature, $Ni^0$ would be the predominant nickel-species on the surface of functioning catalyst, and the dehydrogenation by homolytic splitting of C-H bond would be the major pathway for the activation and conversion of methane, with syngas ($CO/H_2$) as the dominant product and also being easiiy ccmpanied by deposition of more or less carbon on the surface. However, under the reaction condition of rich-in-oxygen/poor-in-methane or of low temperature, $Ni^{2+}$-sites would co-exist with $Ni_n^0$-sites, and even become the dominant nickel-species on the surface of functioning catalyst, so that the oxygen-assisted dehydrogenation by the heterolytic splitting of C-H bond would be able to become the major pathway for activation of methane, thus, leading to a decrease in overall reaction rate and an increase in selectivity to deep oxidation products, $CO_2$ and $H_2O$.

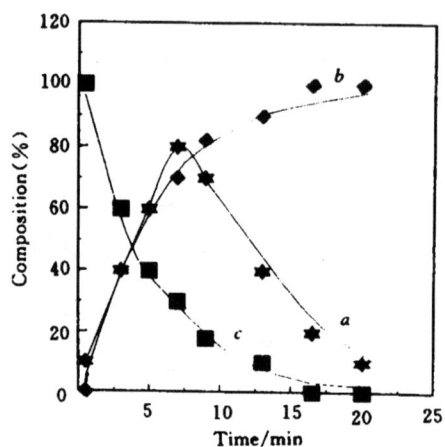

# References

[1]Ashcroft, A. T., Cheetham, A. K., Foord, J. S., et al. , Nature, 1990, **344**: 319.

[2]Vernon, P. D. F., Green, M. L. H., Cheetham, A. K., et al. , Catal. Lett. , 1990, **6**: 181.

[3]Jones, R. H., Ashcroft, A. T., Waller, D., et al. , Catal. Lett. 1991, **8**: 169.

[4]Vernon, P. D. F., Green, M. L. H., Cheetham, A. K., et al. , Catal. Today, 1992, **13**: 417.

[5]Dissanayake,D.,Rosynek,M. P.,Kharas,K. C. C.,et al. ,J. Catal.,1991,**132**:117.

[6]Choudhary,V. R., Rajput, A. M. and Rane, V. H.,J. Phys. Chem.,1992,**96**,8686;Catal. Lett. 1992,**16**:269.

[7]Vermeiren,W. J. M.,Blomsma,E. and Jacobs,P. A.,Catal. Today,1992,**13**:427.

[8]Hickman,D. A. and Schmidt,L. D.,J. Catal.,1993,**138**:267.

[9]Hickman,D. A.,Haupfear,E. A. and Schmidt,L. D.,Catal. Lett.,1993,**17**:223.

[10]Choudhary,V. R.,Rajput. A. M. and Prabhakar,B.,J. Catal.,1993,**139**:328.

[11]Dissanayake,D.,Rosynek. M. P. and Lunsford,J. H.,J. Phys,Chem.,1993,**97**:3644.

[12]Kuijpers,E. G. M.,Jansen,J. W.,Van Dillen,A. J.,Geus,J. W.,J. Catal.,1981,**72**:75.

[13]Brass,S. G. and Ehnich,G.,J. Chem,Phys.,1987,**87**:4285.

[14]cotton,F. A.,Advanced Inorganic Chemistry,3rd Ed.,New York,John Wiley & Sons,1972, 356.

[15]Takayasu,O.,Matsuura,I.,Nitta,K.,Yosliida,Y.,New Frontier in Catalysis,Proc. 10th ICC (Budapest,1992),Eds. Guczi,L. et al. ,Elsevier,1993,1951.

[16]Choudhary,V. R.,Sansare,S. D.,Mamman,A. S.,Appl. Catal.,A:General,1992,90,L1—L5.

[17]Burwell,R. L.,Shim,Jr. B. K. C. and Rowlinson,C.,J. Am. Chem. Soc.,1957,**79**:5142.

■ 本文原载：Catalysis Letters 1996，pp. 755～756.

# Supported Gold Catalysts Derived from Gold Complexes and As-Precipitated Metal Hydroxides, Highly Active for Low-Temperature CO Oxidation

You-Zhu Yuan[a,b], Kiyotaka Asakura[c], Hui-Lin Wan[b],

Khi-Rui Tsai[b], Yasuhiro Iwasawa[a]

( [a]Department of Chemistry, Graduate School of Science,
The University of Tokyo, Hongo, Bunkyo-ku, Tokyo 113
[b]Department of Chemistry and State Key Laboratory of Physical Chemistry for
Solid Surface, Xiamen University, Xiamen 361005, China
[c]Research Center for Spectrochemistry, Faculty of Science,
The University of Tokyo, Hongo, Bunkyo-ku, Tokyo 113)
Received May 24, 1996

Supported gold catalysts were prepared by attaching phosphine-stabilized gold complex and cluster on as-precipitated metal hydroxides $M(OH)_x$ ($M = Mn^{2+}$, $Co^{2+}$, $Fe^{3+}$, $Ni^{2+}$, $Zn^{2+}$, $Mg^{2+}$ and $Cu^{2+}$), followed by thermal decomposition and calcination. The obtained catalysts were remarkably active for CO oxidation at low temperatures below 273 K.

It is established that "ultra-dispersed metal particles" can be obtained by decomposition of molecular metal complexes including metal cluster compounds on inorganic oxide supports[1-5]. In these systems metal-support interface is more efficiently interacted to sometimes generate unique catalysis. However, it is scarcely known that supported small gold particles are obtained by decomposition of gold complexes or gold clusters, though catalysis of gold has accepted increasing interests[6-11]. Highly dispersed gold particles on metal oxides which show high catalytic activities for the oxidation of CO and $H_2$ at low temperatures have been reported by Haruta and his coworkers[6] who obtained dispersed gold catalysts by coprecipitation, deposition-precipitation[10,11], co-sputtering[12] and CVD method[13]. Our earlier works[14,15] on oxide-supported heteronuclear and homonuclear gold clusters stabilized by phosphine ligands showed that the gold compounds on oxide surfaces were readily decomposed to form metallic particles, but their catalytic activities for CO oxidation were lower than those of the catalysts prepared by coprecipitation of $HAuCl_4$ and $Fe(NO_3)_3$ due to the larger size of gold particles in the former systems. Here we report a new approach to prepare supported gold catalysts which show tremendously high activity for CO oxidation at low temperatures. Gold catalysts were obtained by supporting phosphine-stabilized gold complex $[Au(PPh_3)](NO_3)$ (**1**)[16] and cluster $[Au_9(PPh_3)_8](NO_3)_3$ (**2**)[17] on wet as-precipitated metal hydroxides $(MOH)_x$ originated from $Mn(NO_3)_2$, $Co(NO_3)_2$, $Fe(NO_3)_3$, $Mg(NO_3)_2$, $Zn(NO_3)_2$, $Ni(NO_3)_2$, $Cu(NO_3)_2$, $Ti(i\text{-}OC_3H_7)_4$, $VCl_3$, $Al(NO_3)_3$ and $Cr(NO_3)_3$.

Metal hydroxides were prepared by precipitation of metal nitrates (99.9% purity) with an aqueous

5.0 wt% solution of $Na_2CO_3$ (99.9% purity). Titanium hydroxide and vanadium hydroxide were obtained by hydrolysis of titanium-tetra-*iso*-propoxide (99.999%) and vanadium trichloride (99.9%) with an aqueous 5.0 wt% solution of $Na_2CO_3$, respectively. The precipitates were washed repeatedly with distilled water till the pH reached 7.0 and filtered. The obtained as-precipitated metal hydroxides without drying were impregnated with an acetone solution of **1** and a methanol solution of **2**, while vigorously stirring for 12 h, followed by evacuation for 5 h to remove the solvents. The gold content on each oxide was controlled to be 3.0 wt%. Decomposition and calcination of the Au complexes on the metal hydroxides were performed by heating to 673 K at a heating rate of 4 K/min at which temperature the samples were held for 4 h under a flow of air at 30 mL/min. The obtained samples are denoted as $1/M(OH)_x^*$ (x = 2 or 3), where asterisk stands for the as-precipitated metal hydroxides. For comparison, conventional samples were also prepared by impregnation of **1** on calcined metal oxides, followed by decomposition and calcination in the similar way to $1/M(OH)_x^*$, which are denoted as 1/oxide. Catalytic CO oxidation was carried out in a fixed-bed flow reactor equipped with a computer-controlled autosampling system by using 200 mg of catalyst powder. The reaction gas containing 1.0% CO balanced with air purified by a molecular sieve column was passed through the catalyst bed at a flow rate of 67 mL/min (SV=20000 $h^{-1}$). The reaction products were analyzed by a gas chromatgraph using a column of Unibeads C for $CO_2$ and a column of 5A molecular sieve for CO and $O_2$.

Table 1 shows CO conversion to $CO_2$ measured on the catalysts obtained from **1** supported on a variety of as-precipitated metal hydroxides. Among all catalysts the catalysts prepared by using the supports prepared from $Mn(NO_3)_2$, $Co(NO_3)_2$, $Fe(NO_3)_3$, $Mg(NO_3)_2$, $Zn(NO_3)_2$, $Ni(NO_3)_2$ and $Ti(i-OC_3H_7)_4$ were highly active for CO oxidation below 273 K. Especially $Au/Mn(OH)_2$ showed quite higher activity than $Au/MoOx$ catalysts ones previously reported. As typical examples, Figure 1 compares the catalytic activities for the CO oxidation of various iron-oxide-supported gold catalysts such as $1/Fe(OH)_3^*$, $2/Fe(OH)_3^*$, $1/Fe_2O_3$, $HAuCl_4/Fe(OH)_3^*$, $Au(PPh_3)Cl/Fe(OH)_3^*$ and $Fe(OH)_3^*$ alone. It was found that the catalytic activity of $1/Fe(OH)_3^*$ was remarkably high as compared with that of $1/Fe_2O_3$. The catalyst $1/Fe_2O_3$ was only active above 400 K, while the catalyst $1/Fe(OH)_3^*$ catalyzed CO oxidation even at 203 K. The $1/Fe(OH)_3^*$ showed better catalysis than $2/Fe(OH)_3^*$ for the CO oxidation in Figure 1.

**Table 1  Catalytic activities for CO oxdiation of various gold catalysts derived from $1/M(OH)_x^*$**

| Catalyst | $T_s$/K | $T_{50\%}$/K | $T_{100\%}$/K |
|---|---|---|---|
| $1/Mg(OH)_2^*$ | 203 | 250 | >373 |
| $1/Ti(OH)_4^*$ | <273 | 304 | 433 |
| $1/V(OH)_3^*$ | 383 | 649 | >773 |
| $1/Cr(OH)_3^*$ | 473 | 735 | >773 |
| $1/Mn(OH)_2^*$ | <203 | <203 | 273 |
| $1/Fe(OH)_3^*$ | <203 | 231 | 273 |
| $1/Co(OH)_2^*$ | <203 | <203 | 273 |
| $1/Ni(OH)_2^*$ | <203 | 230 | 273 |
| $1/Cu(OH)_2^*$ | <273 | 334 | 443 |
| $1/Zn(OH)_2^*$ | <203 | 248 | 273 |
| $1/Al(OH)_3^*$ | 373 | 606 | >633 |

$T_s$: temperature for CO oxdiation start; $T_{50\%}$: temperature for 50% conversion; $T_{100\%}$: temeprature for 100% converion.

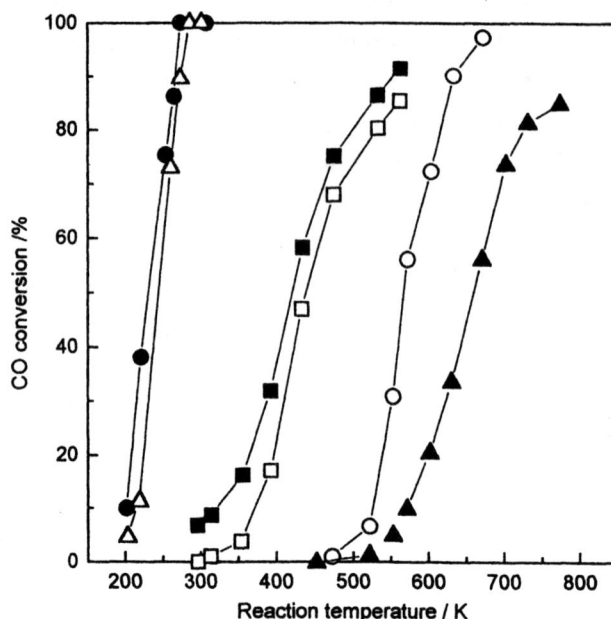

**Figure 1    Temperature dependence of CO oxidation activity of supported gold catalysts.**
—●—$1/Fe(OH)_3^*$ ；—△—$2/Fe(OH)_3^*$ ；—■—$1/Fe_2O_3$ ；—□—$Fe(OH)_3^*$ ；
—○—$HAuCl_4/Fe(OH)_3^*$ ；—▲—$Au(PPh_3)Cl/Fe(OH)_3^*$ .

TEM photographs showed that the gold particles in $1/Fe(OH)_3^*$ were nearly 4 times smaller on average than those in $1/Fe_2O_3$. The gold complexes **1** on oxides were decomposed to metallic gold particles at 573 K, while at this temperature the hydrated as-precipitated metal hydroxides ($M(OH)_x^*$) were dehydrated to form metal hydroxide anhydrides and partially converted to metal oxide. These transformations of the as-precipitated hydroxide supports may facilitate the gold-support interaction during the thermal decomposition, stabilizing the gold precursor on the surface of the metal hydroxides, and prevent the gold species aggregating on the support. On the other hand, there may exist weak interaction between the crystalline metal oxide and the gold particles in case of **1**/oxide, leading to the aggregation of gold particles during the calcination.

We have also tried to use $HAuCl_4$ and $Au(PPh_3)Cl$ as precursors which were impregnated on $Fe(OH)_3^*$ in the similar way. The catalysts obtained from $HAuCl_4/Fe(OH)_3^*$ and $Au(PPh_3)Cl/Fe(OH)_3^*$ showed much lower catalytic activities as compared to $1/Fe(OH)_3^*$ and $2/Fe(OH)_3^*$, and also $Fe(OH)_3^*$ alone as shown in Figure 1. The samples derived from $HAuCl_4/Fe(OH)_3^*$ and $Au(PPh_3)Cl/Fe(OH)_3^*$ gave a sharp XRD pattern at $2\theta = 38.2°$ for Au(111) growth, contrasting to a broad weak diffraction line observed with $1/Fe(OH)_3^*$.

The high activity for CO oxidation did not appear when **1** and **2** were supported on $Fe_2O_3$ treated with water vapor or water, and on $Fe(OH)_3$ commercially available. It is to be noted that the tremendous catalysis of the new supported gold catalysts for the low-temperature CO oxidation was achieved by choosing the suitable gold complexes as precursors and the as-precipitated metal hydroxides as supports.

# References

[1]G Maire, Stud Surf Sci, Catal, 1986, **29**: 509.

［2］B C Gates. Chem Rev,1995,**95**:511.

［3］Y I Yermakov, B N Kuznetsov and V A Zakharov. "Catalysis by Supported complexes."
Elsevier,Amsterdam,1981.

［4］J M Basset. J Mol Catal,1983, **21**:95.

［5］Y Iwasawa. "Tailored Metal Catalysts." Reidel,The Netherlands,1986.

［6］K P Hall and D M P Mingos. Prog Inorg Chem,1981, **32**:237.

［7］N W Cant and W K Hall. J Phys Chem,1971, **75**:2914.

［8］S Galvagno and G Parravano. J Catal,1978, **55**:178.

［9］A G Shastri,A K Datye,and J Schwank. J Catal,1984, **87**:265.

［10］M Haruta,S Tsubota,T Kobayashi,H Kageyama,M J Genet and B Delmon. J Catal,1993,
**144**:175.

［11］S Tsubota,U Atsushi,S Hiroaki. T Kobayashi and M Haruta. ACS Symp Ser,1994,**552**:420.

［12］T Kobayashi,M Haruta,S Tsubota and H sano,Sensors and Actuators,1990, **131**:222.

［13］M Okumura,T Tanaka,A Ueda and M Haruta. to be published.

［14］Y Yuan,K Asakura,H Wan,K Tsai and Y Iwasawa. Chem Lett, 1996:**129**.

［15］Y Yuan,K Asakura,H Wan,K Tsai and Y Iwasawa. to be published.

［16］A M Mueting,B D Alexander,P D Boyle,A L Casalnuovo,L N Ito,B J Johson and L H
Pignolet. Inorganic Synthesis,1992,**29**:280.

［17］F Cariati and L Naldini. J C S Dalton,1972:2286.

［18］S D Gardner,G B Hoflund,B T Upchurch,D R Schryer,E J Kielin,and J Schyer. J Catal,1991,
**129**:114.

［19］M Haruta, H Kageyama, N Kamijo, T Kobayashi and F Delannay. Successful Design of
Catalysts. Elsevier,Amsterdam,1988:33.

■ 本文原载：Catalysis Today 30(1996)，pp. 59～65.

# The Performance and Structure of Rare Earth Oxides Modified by Strontium Fluoride for Methane Oxidative Coupling

Rui-Qiang Long，Ya-Ping Huang，Wei-Zheng Weng，Hui-Lin Wan [①]，Khi-Rui Tsai

*(Department of Chemistry and State Key Laboratory for Physical Chemistry of the Solid Surface, Xiamen University, Xiamen, 361005, China)*

**Abstract** Strontium fluoride promoted rare earth（La, Nd, Sm, Eu, Gd, Dy and Y）oxides were more selective than the corresponding unpromoted rare earth oxides for the methane oxidative coupling to ethane and ethene. The XRD results indicated that the partial anionic or cationic exchanges and interaction between the oxide and fluoride phases took place in most of the catalysts studied in this paper, leading to the formation of new oxyfluoride phases. The possible formation of anionic vacancies in the lattice as the result of ionic exchange and interaction between the phases will be favorable to the activation of molecular oxygen and the improvement of catalytic performance.

**Key words**  Methane oxidative coupling   Rare-earth oxide   Strontium fluoride   Ionic exchange   Anionic vacancy

## 1  Introduction

Methane oxidative coupling（MOC）to $C_{2+}$ hydrocarbons has been a much investigated area in recent decade. A large number of catalysts has been found to be active and selective for the formation of ethane and ethene. Among these catalysts, the rare earth oxides（REO）, especially $Sm_2O_3$ and $La_2O_3$, have been extensively studied as MOC catalysts due to their high catalytic activities and selectivities as well as satisfactory thermal stabilities[1-4]. In order to further improve the catalytic performance of REO catalysts for MOC reaction, some promoters, usually alkali-metal or alkaline earth-metal oxides, have been added to the REO catalyst systems[5,6]. However, these alkali-metal or alkaline earth-metal oxides doped catalysts usually exhibit strongly basic property, and are easily poisoned by the $CO_2$ generated during the reaction, so that a higher reaction temperature is required to maintain an adequate level of activity. Besides the promotional effect of metal oxides mentioned above, the benefits of addition of chlorinated compounds to the MOC reaction have also been reported by many research groups[7-10]. Recently, Lunsford et al. [9,10] reported a $Cl^-$ promoted $Li^+$-MgO catalyst which showed good catalytic performance for both MOC and ODE（oxidative dehydrogenation of ethane）. In contrast to the unpromoted $Li^+$-MgO system, alkane conversions over the $Li^+$-MgO-$Cl^-$ catalysts with a Cl/Li ratio of

---

① Corresponding author.

0. 9 were almost unaffected by $CO_2$, since the presence of $Cl^-$ ions at an appropriate level modified the catalyst so that it no longer functioned as a strongly basic oxide catalyst and would therefore not be easily poisoned by $CO_2$. In addition to chlorinated compounds, We have recently found that metal fluorides also showed significant promotional effects on the oxide catalysts for the MOC and oxidative dehydrogenation of light alkanes ($C_2H_6$ and $C_3H_8$) reactions[11-13]. The beneficial effects resulting from addition of fluorides to oxides include (i) possible formation of lattice defects, such as anionic vacancies, which are requisite for the adsorption and activation of $O_2$ over the catalysts with stable cationic valencies, (ii) modification of the basicity of metal oxides, and (iii) isolation of the surface active oxygen species. In this work, the catalytic property of a series of $SrF_2$ promoted $Ln_2O_3$ catalysts with $Ln=La$, Nd, Sm, Gd, Eu, Dy and Y will be reported. Attention will be focused on the correlations between catalytic properties and the bulk composition and structure of the catalysts.

## 2　Experimental

The $SrF_2$ promoted catalysts were prepared by grinding different mole ratios of rare earth oxides ($La_2O_3$, $Nd_2O_3$, $Sm_2O_3$, $Gd_2O_3$, $Eu_2O_3$, $Dy_2O_3$ and $Y_2O_3$, purity $> 99.5\%$) with $SrF_2$ into fine powder. The mixture was then mixed with certain amount of deionized water to form a paste, followed successively by drying at 120℃ for 4 h and calcining at 900℃ (850℃ for $SrF_2/Sm_2O_3$) for 6 h. The resulting solid was crushed and sieved to 40~60 mesh particles. The pure rare earth oxides and $SrF_2$ used for the catalytic performance evaluation were also treated with the similar procedures as described above. Unless indicated elsewhere, the reagents used in the preparation were of analytical grade.

The catalytic reactions were carried out in a fixed bed quartz reactor (5.0 mm inside diameter) under the conditions of $GHSV=20000$ $h^{-1}$ and $CH_4/O_2=3$ (mole ratio). Methane (99.99%) and oxygen (99.5%) were used without further purification. In each experimental run, 0.20 mL of catalyst was used, and the gaseous effluent was analyzed at room temperature by an on-line Shang Fen 102GD gas-chromatograph equipped with thermal conductivity detector, with 5A molecular sieve column for $O_2$ and CO, and GDX 502 column for $CH_4$, $C_2H_4$, $C_2H_6$ and $CO_2$. The conversion of methane ($C_{methane}$) and selectivities of the products ($S_i$) were calculated from the equations:

$C_{methane}=(\sum A_i \times F_i)/[\sum(A_i \times F_i)+A_{methane} \times F_{methane}]$ and $S_i=(A_i \times F_i)/\sum(A_i \times F_i)$, respectively, where $A=$ peak area of carbon-containing species and $F=$ correction factor of response and carbon number.

The XRD measurements were carried out at room temperature on a Rigaku Rotaflex D/Max-C system with Cu K$\alpha$($\lambda=1.5406$ Å) radiation. The samples were loaded on a sample holder with depth of 1 mm. XRD patterns were recorded in the range of $2\theta=20\sim70°$.

## 3　Results and discussion

### 3.1　Catalytic performance evaluation

The catalytic performances of a series of $SrF_2$ promoted rare earth sesquioxide catalysts with different $SrF_2/Ln_2O_3$ mole ratios were evaluated and the results were summarized in Table 1. The blank reactor was found to have no activity for the reaction between $CH_4$ and $O_2$ at 750℃. For pure rare earth oxides, such as $La_2O_3$, $Nd_2O_3$, $Sm_2O_3$, $Eu_2O_3$ $Gd_2O_3$ and $Y_2O_3$ catalysts, 26%~31% of $CH_4$ conversions with $30\% \sim 43\%$ of $C_2$ selectivities were obtained. When a certain amount of $SrF_2$, which had almost no activity for MOC reaction, was added to the oxide catalysts, $CO_x$ ($CO + CO_2$) selectivities decreased significantly, while $C_2$ selectivities and yields were apparently improved under the same conditions, indicating that the addition of $SrF_2$ plays a significant promoting role for the MOC reaction with these REO as catalysts. Maximum $C_2$ yield of 19.6% was observed over $SrF_2/La_2O_3$ (1 : 4) and $SrF_2/Nd_2O_3$ (1 : 1) catalyst, respectively, which was about 9% higher than those over pure $La_2O_3$ and $Nd_2O_3$ under the same conditions. Comparatively, the promotional effect of $SrF_2$ on $Dy_2O_3$ was rather unpronounced, and similar results were also observed on the $SrF_2$ modified $Ho_2O_3$, $Er_2O_3$, $Tm_2O_3$ and $Yb_2O_3$ catalysts. This phenomenon may have resulted from the interaction between $SrF_2$ and $Ln_2O_3$ of which $Ln = Dy$, Ho, Er, Tm and Yb were relatively weak as compared with the other $SrF_2$ modified rare earth sesquioxide catalysts studied in this paper. It has been known that the conductivities of rare earth sesquioxides decrease with increasing atomic number (except for $Y_2O_3$), i.e. $La_2O_3$, $Nd_2O_3 > Sm_2O_3 > Eu_2O_3 > Gd_2O_3 > Dy_2O_3$ [14], suggesting that the mobilities of lattice oxygen may also decrease in this order. For those rare earth oxides with higher mobility of lattice oxygen, the anionic exchange between oxides and fluoride might be easier, which will be favorable to promote the interaction between the oxide and fluoride phases. From the data in Table 1, it is interesting to see that the decrease in $C_2$ yields of $SrF_2/Ln_2O_3$ catalysts follows the same order as the decrease in the conductivities of rare earth sesquioxides. In the next section, more discussion concerned with ionic exchange between rare earth sesquioxides and $SrF_2$ will be presented. As can be seen from Table 1, on both promoted and unpromoted REO catalysts, oxygen was almost completely consumed ($\geqslant 95\%$) in the reactions carried out at 750 ℃ or higher, which suggested that the results showed in Table 1 were obtained under an oxygen-limited condition.

**Table 1   Catalytic performance of the $SrF_2/Ln_2O_3$ catalysts for the oxidative coupling of methane**

| Catalyst (mole ratio) | Temperature (℃) | conversion (%) | | Selectivity (%) | | | | | $C_2$ Yield (%) |
|---|---|---|---|---|---|---|---|---|---|
| | | $CH_4$ | $O_2$ | CO | $CO_2$ | $C_2H_4$ | $C_2H_6$ | $C_2$ | |
| Blank experiment | 750 | no activity | | | | | | | |
| $SrF_2$ | 750 | 0.7 | 2.4 | 0.0 | 46.0 | 0.0 | 54.0 | 54.0 | 0.4 |
| $La_2O_3$ | 750 | 29.8 | 100 | 10.2 | 52.0 | 20.0 | 15.8 | 35.8 | 10.7 |
| | 700 | 29.4 | 100 | 10.5 | 52.3 | 21.5 | 15.7 | 37.2 | 10.9 |
| $SrF_2/La_2O_3$ (1 : 4) | 700 | 34.2 | 98.1 | 6.4 | 36.3 | 36.1 | 21.2 | 57.3 | 19.6 |
| $SrF_2/La_2O_3$ (1 : 1) | 750 | 35.6 | 99.2 | 10.6 | 34.3 | 31.4 | 23.7 | 55.1 | 19.6 |
| $Nd_2O_3$ | 750 | 27.2 | 99.1 | 7.2 | 53.5 | 20.1 | 19.2 | 39.3 | 10.7 |
| $SrF_2/Nd_2O_3$ (1 : 1) | 750 | 34.3 | 98.9 | 4.0 | 38.9 | 33.1 | 24.0 | 57.1 | 19.6 |
| $Sm_2O_3$ | 800 | 26.3 | 99.2 | 8.9 | 52.3 | 21.6 | 17.2 | 38.8 | 10.2 |
| $SrF_2/Sm_2O_3$ (1 : 1) | 800 | 34.0 | 99.5 | 3.9 | 40.3 | 33.1 | 22.7 | 55.8 | 19.0 |
| $Eu_2O_3$ | 750 | 26.3 | 98.4 | 13.8 | 53.1 | 21.8 | 11.3 | 33.1 | 8.7 |
| $SrF_2/Eu_2O_3$ (1 : 1) | 750 | 33.1 | 99.0 | 6.7 | 40.4 | 31.9 | 21.0 | 52.9 | 17.5 |
| $Gd_2O_3$ | 750 | 29.5 | 99.7 | 17.5 | 49.0 | 20.1 | 13.4 | 33.5 | 9.9 |
| $SrF_2/Gd_2O_3$ (1 : 1) | 750 | 34.4 | 99.5 | 5.9 | 39.5 | 32.0 | 22.6 | 54.6 | 18.8 |
| $Dy_2O_3$ | 750 | 31.3 | 99.5 | 12.4 | 45.0 | 25.0 | 17.6 | 42.6 | 13.3 |
| $SrF_2/Dy_2O_3$ (1 : 1) | 750 | 32.6 | 96.3 | 8.5 | 40.0 | 32.0 | 19.5 | 51.5 | 16.8 |
| $Y_2O_3$ | 750 | 27.2 | 98.3 | 16.9 | 50.2 | 20.1 | 12.8 | 32.9 | 9.0 |
| $SrF_2/Y_2O_3$ (2 : 1) | 750 | 33.6 | 94.4 | 8.3 | 35.1 | 34.1 | 22.5 | 56.6 | 19.0 |

Feed: $CH_4/O_2 = 3 : 1$, no inert gas for dilution, GHSV $= 20000$ $h^{-1}$. The data were obtained after 30 min on stream.

The stability of the catalytic performance of the $SrF_2$ promoted $Ln_2O_3$ catalysts varied with the rare earth elements. The results in Fig. 1 shows the catalytic performance of three $SrF_2$ promoted catalysts with respect to time on stream. The $SrF_2/La_2O_3$(1 : 4) catalyst was characterized by a little increase in $CH_4$ conversion and decrease in $C_2$ selectivity at the early period of reaction, as a result, the $C_2$ yield remained almost constant within a period of 36 h on stream. However, $CH_4$ conversion and $C_2$ selectivity over $SrF_2/Nd_2O_3$(1 : 1) catalyst gradually decreased over a period of 30 h on stream, leading to the decrease of $C_2$ yield from 19.6% to 15.2%. After that, both $CH_4$ conversion and $C_2$ selectivity remained almost unchanged in the following 15 h. The decrease of catalytic performance was also observed over the $SrF_2/Y_2O_3$(2 : 1) catalyst during a period of 31 h on stream. The loss of $F^-$ from the catalysts as HF during the reactions appears not to be the only reason for the decrease in catalytic performance of the catalysts, because there is no a corresponding relation between the relative loss amounts of $F^-$ from catalysts and the variations of catalytic performances.

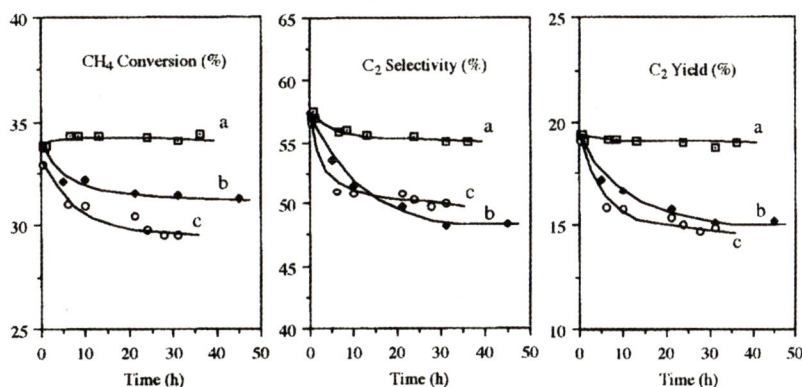

Fig. 1   Effect of time on stream over (a) $SrF_2/La_2O_3$(1 : 4) at 700 ℃, (b) $SrF_2/Nd_2O_3$(1 : 1) at 750 ℃, (c) $SrF_2/$
Y$_2O_3$(2 : 1) at 750 ℃ under the conditions of $CH_4 : O_2 = 3 : 1$ (no dilution gas) and GHSV=20000 h$^{-1}$.

## 3.2   Structure characterization

The XRD measurements indicated that new phases such as tetragonal and rhombohedral NdOF were formed in the fresh $SrF_2/Nd_2O_3$ sample (Fig. 2), suggesting that, during the process of the catalyst preparation, part of the $F^-$ ($r=1.33$ Å) in $SrF_2$ and $O^{2-}$ ($r=1.32$ Å) in $Nd_2O_3$ were substituted by $O^{2-}$ and $F^-$, respectively. Partial ionic exchange between $SrF_2$ and $La_2O_3$ phases and formation of the lanthanum oxyfluorides, such as tetragonal LaOF, may also happen in $SrF_2/La_2O_3$ system. Unfortunately, the diffraction lines of tetragonal LaOF [$d=3.35(100), 2.90(25), 2.06(60), 2.05(33)$, 1.76(22) and 1.75(44)] were overlapped with the characteristic diffraction lines of $SrF_2$ [$d=3.352$ (100), 2.900(25), 2.0508(80), 1.7486(52)] which made it difficult to identify the LaOF phase from a $SrF_2/La_2O_3$ catalyst. However, the formation of rhombohedral LaOF and $LaF_3$ have been detected by XRD in a similar 10% $BaF_2/La_2O_3$ catalyst[15]. For the other samples, XRD only detected cubic $SrF_2$ and rare earth sesquioxide phases, such as hexagonal $La_2O_3$, cubic and monoclinic $Sm_2O_3$, cubic $Eu_2O_3$, cubic $Gd_2O_3$, cubic $Dy_2O_3$, and cubic $Y_2O_3$ (Table 2). Another interesting result from Table 2 was that the content of cubic $Sm_2O_3$ in the $SrF_2/Sm_2O_3$(1 : 2) catalyst was a little more than that of monoclinic phase, while the pure $Sm_2O_3$ contained almost equal amounts of both phases, suggesting that cubic $SrF_2$ may play a certain role in stabilizing the cubic $Sm_2O_3$ at high temperature.

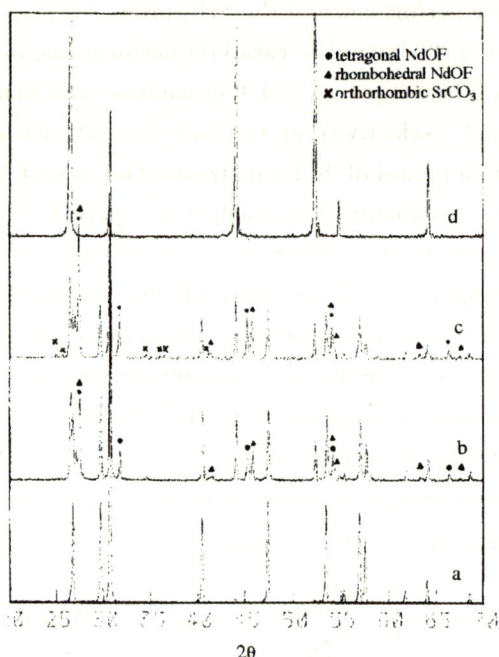

Fig. 2 ZX-ray powder patterns of (a) $Nd_2O_3$, (b) fresh $SrF_2/Nd_2O_3$ (1 : 1), (c) used $SrF_2/Nd_2O_3$ (1 : 1), (d) $SrF_2$ samples.

Table 2 The results of XRD analysis of the fresh $SrF_2/Ln_2O_3$ catalysts

| Catalyst | composition and structure [a] |
|---|---|
| $SrF_2/La_2O_3$ (1 : 4) | cubic $SrF_2$ (w); hexagonal $La_2O_3$ (s) |
| $SrF_2/La_2O_3$ (1 : 1) | cubic $SrF_2$ (s); hexagonal $La_2O_3$ (vs) |
| $SrF_2/Sm_2O_3$ (1 : 1) | cubic $Sm_2O_3$ (vs); cubic $SrF_2$ (s); monoclinic $Sm_2O_3$ (s) |
| $SrF_2/Eu_2O_3$ (1 : 1) | cubic $Eu_2O_3$ (s); cubic $SrF_2$ (m) |
| $SrF_2/Gd_2O_3$ (1 : 1) | cubic $Gd_2O_3$ (s); cubic $SrF_2$ (m) |
| $SrF_2/Dy_2O_3$ (1 : 1) | cubic $Dy_2O_3$ (s); cubic $SrF_2$ (m) |
| $SrF_2/Y_2O_3$ (2 : 1) | cubic $Y_2O_3$ (vs); cubic $SrF_2$ (s) |

[a] vs—very strong, s—strong, m—medium, w—weak.

When the $SrF_2/Ln_2O_3$ catalysts were exposed to $CH_4/O_2$ (3 : 1) at 750℃ for about 1.5 h, orthorhombic $SrCO_3$ resulting from the reaction between SrO and $CO_2$ (by-product of MOC reaction) was detected in the $SrF_2/Nd_2O_3$ (1 : 1) (Fig. 2) and $SrF_2/La_2O_3$ (1 : 4) catalysts (Table 3). This result provides further experimental evidence for the anionic or cationic exchanges between $SrF_2$ and the $Ln_2O_3$ phases. Besides, XRD also detected the formation of rhombohedral SmOF and YOF phases in the used $SrF_2/Sm_2O_3$ (1 : 1) and $SrF_2/Y_2O_3$ (2 : 1) catalysts, respectively (Table 3), which indicated that the exchange between $F^-$ and $O^{2-}$ was more favorable under the conditions of MOC reaction, and $H_2O$ generated in the reaction might play a certain role in promoting such exchange. These results were also in line with the observation that the NdOF content in a used $SrF_2/Nd_2O_3$ (1 : 1) catalyst was higher than that in the fresh (Fig. 2). Based on these results, it is reasonable to suggest that the anionic exchange between oxide and fluoride phases also happened more or less in a fresh $SrF_2/Sm_2O_3$ (1 : 1) or $SrF_2/Y_2O_3$ (2 : 1) catalyst, but the content of SmOF or YOF is probably too low to be detected by XRD. No formation of new phase was detected in the used $SrF_2/Eu_2O_3$ (1 : 1) catalyst.

**Table 3  The results of XRD analysis of the $SrF_2/Ln_2O_3$ catalysts after reacting about 1.5 h at 750℃**

| Catalyst | composition and structure [a] |
|---|---|
| $SrF_2/La_2O_3$ (1 : 4) | cubic $SrF_2$ (w); hexagonal $La_2O_3$ (s) |
| | orthorhombic $SrCO_3$ (w) |
| $SrF_2/Sm_2O_3$ (1 : 1) | cubic $Sm_2O_3$ (s); cubic $SrF_2$ (s); |
| | monoclinic $Sm_2O_3$ (vs); |
| | rhombohedral SmOF(w) |
| $SrF_2/Eu_2O_3$ (1 : 1) | cubic $Eu_2O_3$ (s); cubic $SrF_2$ (m) |
| $SrF_2/Y_2O_3$ (2 : 1) | cubic $Y_2O_3$ (vs); cubic $SrF_2$ (s); |
| | rhombohedral YOF (w) |

[a] vs—very strong, s—strong, m—medium, w—weak.

It has been generally accepted that one of the essential conditions for a compound with stable cationic valency to be a good MOC catalyst is the presence and the mobility of the anionic vacancies in the lattice, so as to adsorb and activate $O_2$ during the reaction[16]. According to the literature, the structure of cubic $Ln_2O_3$ compounds is closely similar to that of the fluorite but with 1/4 intrinsic oxygen vacancies[17]. The structure of hexagonal $Ln_2O_3$ can be described by slabs of $OLn_4$ tetrahedrons linked by three of their edges and forming a complex group cation $(LnO)_n^{n+}$ separated by ionic oxygens $O^{2-}$. The monoclinic structure is very similar to hexagonal, but the tetrahedrons are distorted[17]. "Genetic" vacancies also exist in the hexagonal structure of $Ln_2O_3$ and mono-clinic structure of $Sm_2O_3$. However, their concentrations are lower (ca. 17% of vacant positions in the oxygen sublattice) than those of cubic phase[18]. The tetragonal NdOF can be expressed by the formula $LnO_xF_{3-2x}$ ($0.7 < x \leqslant 1$)[19,20], whose structure is closely similar to that of the fluorite. The excess of $F^-$ ions can be accommodated in a fluorite-like structure and occupy some of the interstitial position as Frenkel defect[20]. The rhombohedral oxyfluorides are found to be of a single phase, which chemical composition is sharply defined and corresponds exactly to LnOF[19]. This structure is also a slightly distorted $CaF_2$-type structure. Taking into consideration the fact that many compounds with fluorite-type structure, such as alkaline earth halides, $ZrO_2$, etc. have anion Frenkel defects and anionic vacancies[21], it is reasonable to postulate that the similar anionic vacancies might also exist in the oxyfluoride compounds with fluorite-like structure. On the other hand, the ionic exchange between the oxide and fluoride phases of a $SrF_2/Ln_2O_3$ catalyst may also lead to the formation of anionic vacancies and other lattice defects such as F-center or $O^-$ species in order to maintain electroneutrality. The presence of anionic vacancies and F-center in the above catalyst will be favorable to the adsorption and activation of $O_2$ under the reaction condition.

# 4  conclusions

Based on the above results, it can be concluded that the extent of $F^-$ and $O^{2-}$ exchange in the $SrF_2/Ln_2O_3$ catalysts are in the order Nd, La > Sm, Y > Gd, Eu, Dy. For those rare earth oxides with higher mobility of lattice oxygen, the interaction between oxides and fluoride will take place easily, which will be favorable to the formation of new oxyfluoride compounds and lattice defects including anionic vacancies, F-centers and $O^-$ in the catalysts. The existence of anionic vacancies and the possible formation of F-centers of the oxyfluoride compounds will be favorable to the adsorption and activation of molecular oxygen under the MOC condition. The adsorption and activation of oxygen on the catalyst surface have been supported by the in situ FTIR spectroscopic observation of $O_2^-$ adspecies with characteristic vibration frequency near 1113 cm$^{-1}$ on the $O_2$-adsorbed $SrF_2/Nd_2O_3$ (1 : 1) and $SrF_2/La_2O_3$ (1 : 4) catalysts at 700 ℃ and 650 ℃, respectively[13,22]. On the other hand, the dispersion of $F^-$ on the surface of the catalysts will also be helpful to the isolation of the surface 'active oxygen' centers and the improvement of the $C_2$ selectivity. Therefore, the highest $C_2$ yield was obtained over $SrF_2$ promoted $Nd_2O_3$ and $La_2O_3$ catalysts among these catalysts, whereas, due to the weak interaction between $SrF_2$ and $Dy_2O_3$, the promoting effect of $SrF_2$ was not apparent for MOC reaction over $SrF_2/Dy_2O_3$ (1 : 1) catalyst.

# Acknowledgements

This work has been supported by the National Natural Science Foundation of China.

# References

[1]K Otsuka, K Jinno and A Morikawa. Chem Lett, 1985:499.

[2]K D Campbell, H Zhang and J H Lunsford. J Phys Chem, 1988, **92**:750.

[3]J H Lunsford. Catal Today, 1990, **6**:3.

[4]S Lacombe, C Geantet and C Mirodatos. J Catal, 1994, **151**:439.

[5]Z Kalenik and E E Wolf. in A. Holmen et al. (Eds.), Natural Gas conversion, Elsevier, Amsterdam, 1991:97.

[6]H Yamashita, Y Machida and A Tomita. Appl Catal A, Gen., 1991, **79**:203.

[7]A Z Khan and E Ruckenstein. J Catal, 1992, **138**:322.

[8]R Burch, G D Squire and S C Tsang. Appl Catal, 1988, **48**:105.

[9]J H Lunsford, P G Hinson, M P Rosynek, C Shi, M Xu and X Yang. J Catal, 1994, **147**:301.

[10]D Wang, M P Rosynek and J H Lunsford. J Catal, 1995, **151**:155.

[11]X P Zhou, W D Zhang, H L Wan and K R Tsai. Catal Lett, 1993, **21**:113.

[12]X P Zhou, S Q Zhou, S J Wang, J X Cai, W Z Weng, H L Wan and K R Tsai. Chem Res Chin Univ, 1993, **9**(3):264.

[13]X P Zhou, S Q Zhou, W D Zhang, Z S Chao, W Z Weng, R Q Long, D L Tang, H Y Wang, S J Wang, J X Cai, H L Wan and K R Tsai. Preprints, Division of Petroleum Chemistry, Vol. **39**(2), ACS, 1994. p. 222.

[14]K A Gschneidner, Jr and L Eyring. Handbook on the Physics and Chemistry of Rare Earth, Vol. 3, North-Holland, 1979, Chap. 27, p. 385.

[15]C T Au, H He, S Y Lai and C F Ng. Proc. 1st Global conf. Young Chinese Scientists on Catalysis Science and Technology, Tianjin, China. 12—15 Sept. 1995, p. 193.

[16]A G Anshits, E N Voskresenskaya and L I Kurteeva. Catal Lett, 1990, **6**:67.

[17]K A Gschneidner, Jr and L Eyring, Handbook on the Physics and Chemistry of Rare Earths, Vol. 5, North-Hol-land, 1982, **44**:322.

[18]E N Voskresenskaya, V G Roguleva and A G Anshits. Catal Rev Sci Eng, 1995, **37**(1):101.

[19]K Niihara and S Yajima. Bull Chem Soc Jpn, 1971, **44**:643.

[20]A F Wells. Structural Inorganic Chemistry, 5th edn., Oxford Univ. Press, NY, p. 483.

[21]Z Zhang, M Baems et al. Catal Rev Sci Eng, 1994, **36**(3):507.

[22]R Q Long, H L Wan, H L Lai and K R Tsai. Chem J Chin Univ, in press.

■ **本文原载**:《厦门大学学报》(自然科学版)第 35 卷第 5 期(1996 年 9 月),第 734～738 页。

# 氨合成铁催化剂上化学吸附物种
# 的 *in-situ* **FTIR** 谱*

廖代伟　　林种玉　　蔡启瑞

(物理化学研究所　化学系　固体表面物理化学国家重点实验室)

**摘　要**　应用原位傅里哀变换红外光谱方法,分别在 $400～450$ ℃、常压、高空速($12\,000～14\,400$ h. s. v. g.)的 $H_2$、$N_2/3H_2$ 或 $N_2$ 气氛的动态条件下,检测了双促进氨合成铁催化剂上的化学吸附物种,并分别在 $D_2$、$N_2/3D_2$ 或 $^{15}NH_3/H_2$ 气氛中,进行了常规同位素验证实验。结果表明,催化剂表面的主要含氮化学吸附物种是分子态的 $N_{2.ad}$,其 $v(N\text{-}N)=2\,036$ cm$^{-1}$(s),$2\,012$ cm$^{-1}$(w)和 $1\,935$ cm$^{-1}$(w),而不是原子态的 $N_{ad}$,其 $v(Fe\text{-}N)=1\,087$ cm$^{-1}$(vw),和 $NH_{ad}$. 其 $v(Fe\text{-}N)=886$ cm$^{-1}$(vw),表面有相当量的化学吸附氢存在,其 $v(Fe\text{-}H)=2\,056$ cm$^{-1}$(s),$1\,950$ cm$^{-1}$(ms),$1\,931$ cm$^{-1}$(m),$1\,902$ cm$^{-1}$(w),$915$ cm$^{-1}$(s,桥式)等。作为与报道过的激光拉曼光谱的互补研究,本结果支持了以缔合式途径为主、解离式途径为次的平行竞争缔合式合成氨催化反应机理。

**关键词**　原位傅里哀变换红外光谱　合成氨催化反应机理　化学吸附物种　同位素验证

**中国图书分类号**　O 643.12,O 647.32

铁催化剂上氮加氢合成氨是一个重要而典型的小分子多相催化反应。但是,关于这一催化过程的反应机理在分子水平上的认识至今还不一致。主要分歧在于究竟是解离式或是缔合式。显然,在氨合成反应条件下、铁催化剂表面上的化学吸附物种及其相对浓度的检测将为这一争论的圆满解决提供关键的证据。在这一方面,可以在一般催化反应条件下原位动态摄谱的互补红外与拉曼光谱学方法具有明显的优点。

Tamaru[1]、Brill[2] 和 Nakata 等[3] 分别就非促进或单促进铁催化剂上的化学吸附氮物种进行了红外光谱研究,报道了高温处理后冷至室温摄谱的静态结果。我们[4] 首先就 $400～450$ ℃、常压、高空速 $N_2/3H_2$ 合成氨反应条件下、双促进铁催化剂上的化学吸附物种进行了原位动态激光拉曼光谱研究,并报道了进一步的研究结果[5] 以及理论动力学[6] 和量子化学[7] 的研究结果。张鸿斌等[8] 也就双促进铁催化剂上的氨分解反应进行了原位动态激光拉曼光谱研究。

作为应用红外与拉曼光谱互补研究化学吸附物种方法的另一方面,本文将报道在上述合成氨反应条件下、就同一催化剂上的化学吸附物种所进行的原位动态傅里哀变换红外光谱(*in-situ* FTIR)的进一步研究结果。

# 1　实　验

催化剂为福州化工原料厂生产的双促进熔铁型合成氨工业催化剂 $A_{110\text{-}3}$。取研磨至粒度约 $2$ $\mu$m 的

---

*　本文 1995 年 8 月 11 日收到;国家自然科学基金和国家教育委员会留学回国人员资助项目。

催化剂约 150 mg,在 10 t/cm² 的压力和抽真空下重压两次成直径 13 mm 厚度 0.2 mm 的样品薄圆片,置于红外样品池内作为还原前催化剂样品。样品的还原和各种反应气体($H_2$、$D_2$、$N_2/3H_2$、$N_2/3D_2$、$N_2$、$^{15}NH_3/H_2$)的来源与净化如前文[5]所述。自制了适用于高温、常压、原位动态摄谱的红外样品池[9]。池内晶体光学片挡板(厚度 1 mm,相距 1 cm)的设置是为了使空速近似等于气体流经样品的线速度,并保护池两端的窗片。窗片与挡板根据检测范围的要求使用 $CaF_2$ 或 KBr 晶体光学片。

使用 Nicolet 5-DX 型傅里哀变换红外光谱仪进行检测。样品圆片用氢气流程序升温充分还原 48 h 以上,再用氩气流在 450～480 ℃排代 48 h 后动态或静态摄谱所得到的干涉图作为背景文件。扫描 960 次,透过率光谱图经 13 点平滑 1 次。所有吸附处理和动态摄谱均在 400～450 ℃、常压、高空速(12 000～14 400 $h^{-1}$)气体流通条件下进行,但部分静态摄谱则在切断气源并冷至室温或密闭加热到 400～450 ℃后检测。

## 2　结果与讨论

表 1 示出我们所观测到的化学吸附物种的主要红外(IR)谱峰振动频率及其归属,为便于比较,表中也列入前文[5]已报道的主要拉曼(R)振动频率及其归属。表中括号内值以及同位素观测值为动态处理静态摄谱频率。$v$ 示伸缩振动,$\delta$ 表示弯曲振动,数值后括号内的 b 表示桥式,s 表示强,ms 表示中强,m 表示中等,w 表示强度弱。同位素实验是在 400～450 ℃、分别用 $D_2$、$N_2/3D_2$ 或 $^{15}NH_3/H_2$ 气流动态充分还原处理样品后在室温或密闭加热到 400～450 ℃静态摄谱的。

**表 1　$A_{100-3}$ 催化剂上化学吸附物种的红外与拉曼振动频率(cm⁻¹)及其归属**

**Tab. 1　Infrared and Raman vibration frequency (cm⁻¹) and assignments of chemisorbed species on $A_{100-3}$ catalyst**

| 气体反应剂 | 归属 | IR | R | 归属 | IR | R |
|---|---|---|---|---|---|---|
| $N_2/3H_2$ | $v$(N-N)($N_{2,ad}$) | 2 036(s) | 2 040(s) | $v$(Fe-H)($H_{ad}$) | 2 056(s) | (2 056) |
| | | 2 012(w) | (2 013)(s) | | 1 960(ms) | 1 950(s) |
| | | 1 935(w) | 1 940(s) | | 1 931(m) | (1 928) |
| | | | | | 1 902(w) | 1 901(w) |
| | | | | | 910(s,b) | |
| | $v$(Fe-N)($N_{2,ad}$) | 450 | 443 | $v$(Fe-N)($N_{2,ad}$) | 469 | 465 |
| | | 434 | 423 | | | |
| $N_2/3D_2$ | $v$(N-N)($N_{2,ad}$) | 2 031 | | $v$(Fe-D)($D_{ad}$) | 1 432～1 444 | |
| | | 2 006(?) | 2 015 | | 1 382 | 1 381 |
| | | | 1 937～1 939 | | 1 354 | 1342 |
| $^{15}NH_3/H_2$ | $v$($^{15}N$-$^{15}N$)($^{30}N_{2,ad}$) | 1 980 | | $v$(Fe-H)($H_{ad}$) | 1 959 | 1 967 |
| | | 1 935 | 1 936 | | 1 908 | 1 917 |
| | | | 1 873 | | | |
| $H_2$ | | | | $v$(Fe-H)($H_{ad}$) | 2 052 | |
| | | | | | 1 952 | 1 950 |
| | | | | | 1 930 | 1 921～1 927 |
| | | | | | 1 906 | 1 901 |
| | | | | | 915(b) | |
| $D_2$ | | | | $v$(Fe-D)($D_{ad}$) | 1 430 | 1 414 |
| | | | | | 1 382 | 1 381 |
| | | | | | 1 364 | 1 365 |
| | | | | | 1 352 | 1 342 |
| $N_2$(或 $10N_2/H_2$) | $v$(Fe-N)($N_{ad}$) | 1 087 | 1 088 | $v$(Fe-N)($NH_{ad}$) | 886 | 890 |

如同前文[5]关于拉曼观测频率的归属理由,对于 $H_2/A_{110-3}$ 这一简单体系,显然可将 2 052 cm$^{-1}$、1 952 cm$^{-1}$、1 930 cm$^{-1}$ 和 1 906 cm$^{-1}$ 等红外峰归属于各种化学吸附氢物种的 $v$(Fe-H)。参照已知的桥式 $v$(M-H) 数据(900~1 100 cm$^{-1}$),可将 915 cm$^{-1}$ 归属于桥式 $v$(Fe-H)。比较 $N_2/3H_2/A_{110-3}$ 体系和 $H_2/A_{110-3}$ 体系的红外峰,同样可将 $N_2/3H_2/A_{100-3}$ 体系的 2 056 cm$^{-1}$、1 950 cm$^{-1}$、1 931 cm$^{-1}$、1 902 cm$^{-1}$ 和 910 cm$^{-1}$(桥式)归属于 $v$(Fe-H)。而较之 $H_2/A_{100-3}$ 体系多出的新峰 2 036 cm$^{-1}$ 则同样可归属于化学吸附 $N_{2.ad}$ 的 $v$(N-N),这一红外峰可能对应于 2 040 cm$^{-1}$ 的拉曼峰,因此是红外与拉曼活性的,它可能对应于单端基加侧基或多侧基直插式或斜插式多核配位吸附在 $Fe_n$ 簇的分子 $N_{2.ad}$ 物种。另一个位于 2 012 cm$^{-1}$ 的很弱的新峰看来也可归属于化学吸附 $N_{2.ad}$ 的 $v$(N-N),这一红外峰在静态摄谱时出现在 2 016 cm$^{-1}$,变得比 2 040 cm$^{-1}$ 峰还要强,在激光拉曼光谱实验中也观测到类似的现象。出现这一现象的可能原因稍后还将讨论,但这一吸附 $N_{2.ad}$ 物种也是红外与拉曼活性的,它可能对应于准双端基加侧基或多侧基桥型斜插式多核配位吸附在 $Fe_n$ 族的分子 $N_{2.ad}$ 物种。注意到对应于 1 940 cm$^{-1}$ 拉曼峰的 1 935 cm$^{-1}$ 红外峰强度很弱,因此 1 940 cm$^{-1}$ 拉曼峰所对应的化学吸附 $N_{2.ad}$ 物种可能主要是拉曼活性的,它可能对应于双端基加侧基或多侧基平躺式多核配位吸附在 $Fe_n$ 簇的分子 $N_{2.ad}$ 物种。在 $N_2/3H_2/A_{100-3}$ 体系的动态反应条件下,同样也没有观测到前文[5]在 $N_2$ 或 $10N_2/H_2/A_{100-3}$ 体系观测到并已合理地归属于原子态化学吸附氮的 $N_{ad}$(1 089 cm$^{-1}$)或 $NH_{ad}$(890 cm$^{-1}$)的红外峰,但对于 $N_2$ 或 $10N_2/H_2/A_{100-3}$ 体系则同样观测到 1 087 cm$^{-1}$ 和 886 cm$^{-1}$ 的红外峰。

用常规同位素实验初步验证了上述红外谱峰的归属。在 $D_2/A_{100-3}$ 或 $N_2/3D_2/A_{100-3}$ 体系,上述归属于 $v$(Fe-H) 的红外峰都几乎消失了或强度大大减弱(由于 $H_{ad}$ 排代不彻底),并位移至 1 430~1 444 cm$^{-1}$、1 382 cm$^{-1}$ 和 1 352~1 364 cm$^{-1}$ 等处。归属于 $v$(N-N) 的红外峰在 $D_2/A_{100-3}$ 体系同样观测不到,在 $N_2/3D_2/A_{100-3}$ 体系则仍被观测到,并有数 cm$^{-1}$ 的位移,这暗示氮的配位活化可能与邻近的化学吸附氢有关,但在 $^{15}NH_3/H_2/A_{100-3}$ 体系则消失了,而位移至 1 980 cm$^{-1}$ 和 1 935 cm$^{-1}$ 等处。$v$(Fe-D) 或 $v$($^{15}$N-$^{15}$N) 的位移值与按 $v$(Fe-D)=0.713 $v$(Fe-H) 或 $v$($^{15}$N-$^{15}$N)=0.966 $v$($^{14}$N-$^{14}$N) 的估计值是符合的。

业已提出的合成氨催化反应机理可以归纳为解离式机理和缔合式机理两大类。考虑到 1)解离式机理无法统一解释 $N_2$ 的化学计量数等于 1 和氘反同位素效应的实验事实,而若按缔合式机理则可以得到较圆满的统一解释[6];2)解离式机理关于氘反同位素效应的热力学解释的基本假设,即假设 $N_{ad}$ 或 $NH_{ad}$ 是表面最丰含氮化学吸附物种可能并不合理,而且缺乏充分的实验证据[5];3)已有间接或直接的实验证据说明在 350~450 ℃、$Fe_n$ 上有相当份量的分子态化学吸附氮的存在[5];4)由缔合式机理导出的动力学方程已被证明是符合大量工业生产中试数据的较好的数学模型[10];因此,缔合式机理可能是较合理的。

根据蔡启瑞提出的缔合式为主、解离式为次的铁催化剂上合成氨平行竞争反应机理,处于反应速度控制隘口步骤的化学吸附 $N_{2.ad}$ 的表面浓度必然要比快速加氢的 $N_{ad}$ 或 $NH_{ad}$ 来得大。本文和前文[5]的红外与拉曼光谱互补研究结果已经表明:在 400~450 ℃、常压、高空速 $N_2/3H_2$ 的动态反应条件下、铁催化剂表面存在着大量化学吸附氢,分子态化学吸附 $N_{2.ad}$ 的表面浓度远大于原子态化学吸附 $N_{ad}$ 或 $NH_{ad}$ 的表面浓度,这就为缔合式机理的合理性提供了相当可靠的实验证据。

在反应条件下,铁催化剂表面可能发生化学吸附 $N_{2.ad}$ 分子重新取向直至解离的过程[8]。但是由于大量化学吸附氢的存在,这一过程一般不可能进行到底。因为足够活化的化学吸附 $N_{2.ad}$ 分子无需解离,就已经有能力和许多机会与近邻的化学吸附氢物种发生对称或不对称加氢或氢解反应。随即再迅速进一步加氢成氨。而在局部活性氢缺乏的部位,极少数化学吸附 $N_{2.ad}$ 有可能进一步解离成 $N_{ad}$,而按解离式机理加氢成氨。

在动态反应条件下,对应于 $v$(N-N)=2 036 cm$^{-1}$(IR)或 2 040 cm$^{-1}$(R)以及 1 935 cm$^{-1}$(IR)或 1 940 cm$^{-1}$(R)的化学吸附 $N_{2.ad}$ 物种可能处于氮分子重新取向或加氢的溢口,因此,这两种物种的表面

浓度较大,因而谱峰强度相对较强。但在静态摄谱时,对应于 $v(\text{N-N}) = 2\ 012\ \text{cm}^{-1}$(IR)或 $2\ 013\text{cm}^{-1}$(R)的化学吸附 $N_{2\cdot ad}$ 物种可能变成处于氮分子重新取向或加氢的隘口,因而表面浓度较大,谱峰强度相对较强。这一现象还暗示了只有在特定位上配位的化学吸附氢物种才对氮加氢是有效的。

总之,我们的实验表明:催化剂表面的主要含氮化学吸附物种是分子态的 $N_{2\cdot ad}$,其 $[(\text{N-N}) = 2\ 036\ \text{cm}^{-1}(\text{s,IR}) \sim 2\ 040\ \text{cm}^{-1}(\text{s,R})$ 和 $1\ 935\ \text{cm}^{-1}(\text{w,IR})\ 1\ 940\ \text{cm}^{-1}(\text{s,R})]$,而不是原子态的 $N_{ad}$,其 $[(\text{Fe-N}) = 1\ 087\ \text{cm}^{-1}(\text{vw,IR}) \sim 1\ 088\ \text{cm}^{-1}(\text{vw,R})$ 和 $NH_{ad}$,其 $[(\text{Fe-N}) = 886\ \text{cm}^{-1}(\text{vw,IR}) \sim 890\ \text{cm}^{-1}(\text{vw,R})]$。由此,可以合理地推断,缔合式机理应是铁催化剂上合成氨催化反应的主要途径。这一机理是以分子氮的非解离化学吸附及其第一次缔合式加氢或氢解为反应速度控制步骤的。这也可由缔合式机理合成氨反应位能图[9]看出。

# 参考文献

[1] Okawa T, Onishi T, Tamaru K. Infrared and kinetic study of ammonia decomposition on supported iron catalysts: infrared observation of molecularly adsorbed nitrogen in ammonia decomposition. Z. Phys. Chem.,1977,**107**(2):239~243.

[2] Brill R,Jiru P,Schulz G. Infrared spectra of nitrogen-hydrogen adsorption complexes on an iron catalyst. Z. Phys. Chem.,1969,**64**(1~4):215~224.

[3] Nakata T, Matsushita S. Infrared studies of intermediates of ammonia synthesis. J. Phys. Chem. ,1968,**72**(2):458~464.

[4] 廖代伟,王仲权,张鸿斌等.氨合成铁催化剂上氮吸附态的研究(I)氨合成铁催化剂表面上吸附氮的激光 Raman 光谱和红外光谱.厦门大学学报(自然科学版),1982,**21**(1):100~103.

[5] Liao D W,Zhang H B,Wang Z Q,Tsai K R. Raman spectra of chemisorbed species on Ammonia synthesis iron catalysts. Scientia Sinica (Science in China),B,1987,**30**(2):246~255,也可见:中国科学,B 辑,1986,(7):673~680.

[6] 廖代伟.合成氨动力学方程和同位素效应.厦门大学学报(自然科学版),1995,**34**(2):204~208.

[7] Liao D W. EHMO study of the mode of coordination activation and hydrogenation of dinitrogen on $\alpha$-Fe (111) surface for ammonia synthesis J. Mol. Struc. (THEOCHEM),1985,**22**(121):101~107.

[8] Zhang H B, Schrader G L. Characterization of $NH_3$-Fe catalytic systems by laser Raman spectroscopy. J. Catal. 1986,**99**:461~471.

[9] 廖代伟.铁催化剂上的化学吸附物种和固氮成氨.[理学博士学位论文].厦门大学,1985.

[10] 刘德明,赵志良,黄目兴等.用序贯法判别"A"系催化剂上氨合成反应速率模型.化工学报,1979,(2):133~142.

# In-Situ FTIR Spectra of Chemisorbed Species on
# Iron Catalysts for Ammonia Synthesis

Dai-Wei Liao，Zhong-Yu Lin，Khi-Rui Tsai

（Inst. of Phys. Chem. Dept of Chem.，State Key Lab. for Phys. Chem. of the Solid Sur.）

**Abstract**　　Chemisorbed species on doubly promoted ammonia synthesis iron catalysts are investigated by *in-situ* Fourier transform infrared spectroscopy under dynamic conditions of $H_2$, $N_2/3H_2$ and $N_2$, respectively, with high space velocity （12 000～14 400 $h^{-1}$） at 1 atm and 400～450 ℃. Also, the general isotopic verifications are carred out with $D_2$, $N_2/3D_2$ and $NH_3/H_2$, respectively. The results indicated that, under above reaction condition, the most aboundant chemisorbed N-containing species are molecular $N_{2,ad}$ with $v(N\text{-}N) = 2\ 036\ cm^{-1}$（s）, $2\ 012\ cm^{-1}$（w） and $1\ 935\ cm^{-1}$（w）, rather than atomic $N_{ad}$ with $v(Fe\text{-}N) = 1\ 087\ cm^{-1}$（vw） and $NH_{ad}$ with $v(Fe\text{-}N) = 886\ cm^{-1}$（vw）. There are a lot of chemisorbed hydrogen with $v(Fe\text{-}H) = 2\ 056\ cm^{-1}$（s）, $1\ 950\ cm^{-1}$（ms）, $1\ 931\ cm^{-1}$（m）, $1\ 902\ cm^{-1}$（w） and $915\ cm^{-1}$（s, bridge）. As a complementary study for the laser Raman spectra previously reported, this paper supported the associative catalytic reaction mechanism of ammonia synthesis with both parallel competitive major associative pathway and minor dissociative pathway.

**Key words**　　In-situ FTIR　Catalytic reaction mechanism of ammonia synthesis　Chemisorbed species　Isotopic verification

■ **本文原载**:《科学通报》第 20 期第 41 卷,第 1919~1920 页。

# 丙烷氧化脱氢 VMgO 催化剂双相协同催化作用和活性位的研究*

方智敏　翁维正　万惠霖[①]　蔡启瑞

(厦门大学化学系　物理化学研究所　固体表面物理化学国家重点实验室,厦门　361005)

丙烷氧化脱氢制丙烯是有效利用低碳烷烃的一个重要催化过程。性能较好的 VMgO 催化剂引起了人们的兴趣和辩论。Kung 等人[1]认为活性相是正钒酸镁。Volta 等人[2]则认为焦钒酸镁是活性相,该相中稳定存在着与氧缺位形成有关的 $V^{4+}$ 离子,而正钒酸镁不存在 $V^{4+}$ 离子,有利于深度氧化。Delmon 等人[3]基于焦钒酸镁能够提高与之共存的正钒酸镁的选择性的实验事实,认为正、焦钒酸镁之间可能存在着协同作用。近来,我们采用柠檬酸盐法(C)、浸渍法(I)和硝酸盐法(N)制备了 3 个系列的 VMgO 催化剂,进行了性能评价,用 XRD,IR 和 LRS 方法测定了催化剂的相结构,并对其中钒的化学态进行了 EPR 和 XPS 表征。在此基础上对 VMgO 催化剂的双相协同催化作用和活性位进行了研究。

应用柠檬酸盐法制得了 3 种很纯的钒酸镁物相,即正钒酸镁 $Mg_3V_2O_8$、焦钒酸镁 $\alpha\text{-}Mg_2V_2O_7$ 和偏钒酸镁 $\beta\text{-}Mg_2V_2O_6$。在相同的丙烷转化率下,丙烯选择性顺序为: $\alpha\text{-}Mg_2V_2O_7 > Mg_3V_2O_8 > \beta\text{-}MgV_2O_6$,表明 $\alpha\text{-}Mg_2V_2O_7$ 为具有高丙烯选择性的活性相;在相同的反应条件下,$Mg_3V_2O_8$(转化率 C 25.1%,选择性 S 47.6%)比 $\alpha\text{-}Mg_2V_2O_7$(C 22.8%,S 55.4%)活性高但选择性低,$\beta\text{-}MgV_2O_6$(C20.3%,S49.6%)的催化活性最低,前两者可能分别更有利于丙烷的转化和丙烯的选择性生成。

浸渍法(以重质 MgO 为载体)制备的中低含钒量 2~60VMgO-I(即含 $V_2O_5$ 重量百分含量为 2%~60%,下同)催化剂具有较高的丙烷转化率和丙烯收率,500℃时 20VMgO-I 的丙烯收率(14.7%)比文献[1,2]报道的浸渍法制备的除 MgO 外只含有 $Mg_3V_2O_8$ 或 $\alpha\text{-}Mg_2V_2O_7$ 的 20VMgO 催化剂提高了 40%~50%,XRD,IR,LRS 和 EPR 测定结果表明,在该系列催化剂中比文献所用的较大钒含量范围内正钒酸镁和焦钒酸镁两相共存,因此,两相间可能存在互补性的协同作用。

柠檬酸盐法制备的 62,64,67VMgO-C 催化剂中 $Mg_3V_2O_8$ 和 $\alpha\text{-}Mg_2V_2O_7$ 双相共存,其转化率和选择性比 $Mg_3V_2O_8$ 和 $\alpha\text{-}Mg_2V_2O_7$ 都有不同程度的提高,其中 62VMgO-C 获得了已报道[1-3]的在可比条件下 VMgO 体系催化剂在 500℃的最高收率(15.5%)。

用机械混合 3 种不同摩尔比 3/1,1/1 和 1/3 的纯 $Mg_3V_2O_8$ 和 $\alpha\text{-}Mg_2V_2O_7$ 的方法制成双相共存的催化剂 62,64,67VMgO-M。在相同的丙烷转化率下,3 种催化剂的选择性均比 $Mg_3V_2O_8$ 和 $\alpha\text{-}Mg_2V_2O_7$ 高,这一结果进一步说明 $Mg_3V_2O_8$ 和 $\alpha\text{-}Mg_2V_2O_7$ 两相间可能存在着协同作用。

VMgO 催化剂的低温 EPR 研究结果表明,中低钒含量 VMgO-I 催化剂的 $V^{4+}$ 超精细结构由两套 8 条平行方向谱线和两套 8 条垂直方向谱线组成,表明 $V^{4+}$ 是以两种不同配位形式存在的,一种是处于正钒酸镁的分立 $VO_4$ 四面体中;另一种是处于焦钒酸镁的共角 $V_2O_7$ 四面体中。中低钒含量 VMgO-C 均只有一套 $V^{4+}$ 超精细分裂谱线,$V^{4+}$ 是处于正钒酸镁的分立 $VO_4$ 四面体中。性能评价结果表明,VMgO-I 和 VMgO-C 催化活性与 $V^{4+}$ 自旋浓度呈平行关系,暗示着包含 $V^{4+}$ 离子的某种微体系是丙烷氧化脱氢

---

\*　国家自然科学基金资助项目。

①　联系人。

VMgO 催化剂的一种活性位。VMgO-N 系列催化剂反应前后均基本没有观察到 V$^{4+}$ 信号,因此催化活性很低。

柠檬酸盐法制备的 2~60VMgO-C 的性能稍逊于相应的浸渍法催化剂,这是由于该制法具有分子内固相反应的均匀性和化学计量性,在此 V$_2$O$_5$ 含量范围内,除过量的 MgO 外,只得到一种正钒酸镁物相,不存在双相协同催化作用,但因能稳定存在 V$^{4+}$ 离子,催化性能仍较好。

硝酸盐法制备的 VMgO-N 系列催化剂虽然也在所用的 V$_2$O$_5$ 含量变化范围内检测到共存的正钒酸镁和焦钒酸镁相,但催化活件比另两个系列催化剂低得多,这是由于该系列催化剂具有不同的钒离子微环境而不利于稳定 V$^{4+}$ 离子的缘故。

以上结果表明,不同方法制备的 VMgO 催化剂具有不同的丙烷氧化脱氢催化性能;只有在合适的制备条件下,在钒酸镁相上生成了适量的 V$^{4+}$ 离子才具有较好的催化活性,因此含有适量 V$^{4+}$ 离子的钒酸镁可能是 VMgO 催化剂的活性相,其中焦钒酸镁的丙烯选择性最高;共存的正、焦钒酸镁可能存在着双相协同催化作用。

# 参考文献

[1]Chaar M. A.,Patel D.,Kung H. H.. Selective oxidative dehydrogenation of propane over V-Mg-O catalysts. J. Catal.,1988,**109**:463~467.

[2]Sam D. S. H,Soenen V.,Volta J. C.. Oxidative dehydrogenation of propane over V-Mg-O catalysts. J. Catal.,1990,**123**:417~435.

[3]Gao X. T.,Ruiz P.,Xin et al. Effect of coexistence of magnesium vanadate phases in the selective oxidation of propane to propene. J. Catal.,1994,**148**:56~67.

■ **本文原载**:《厦门大学学报》(自然科学版)第 35 卷第 1 期(1996 年 1 月),第 61～66 页。

# 低温催化裂解烷烃法制备碳纳米管*

陈　萍　　张鸿斌　　林国栋　　蔡启瑞　　翟和生

(化学系　固体表面物理化学国家重点实验室)(电镜室)

**摘　要**　实验发现,甲烷可在较低温度(723 K)下,在一种 Ni 催化剂上分解生成碳纳米管。透射电镜测试结果显示,通过此法制得的碳纤维几乎都具有管状结构。碳纳米管的外径,管长及产率明显地受催化剂的结构、性能,反应温度和原料气流速所支配。将所制得的产物浸泡在稀硝酸溶液中,溶去催化剂颗粒,经水洗,并于 473 K 温度下烘干,可达到将碳纳米管产物与催化剂分离、纯化的目的。

**关键词**　碳纳米管　Ni 催化剂　催化裂解　甲烷

**中国图书分类号**　O 643.35

碳纳米管的发现引起了人们极大的兴趣[1]。由于其管径大小为纳米级,量子限域及界面效应均较明显;其管壁结构类似石墨,导电性能好,比表面大,具有独特的物理、化学性质,因而是一种亟待进一步研究开发的新材料。

目前,国际上实验室内制备碳纳米管主要有两种方法:一是碳的电弧放电法[2],二是烃的催化裂解法[3]。前者所制得碳纳米管管直,结晶度高,但产率低,分离纯化困难;后者一般系将乙炔于 873 K 温度下在超细的 Ni-Cu 合金催化剂上分解,产率高,但结晶度较低。国内此类工作开展较少,主要的制备方法为高温(～1 273 K)热裂法,所得产物繁杂,分离纯化比较困难。

本研究组先前曾报道了在较低温度(～723 K)下采用催化裂解法由甲烷制得碳纳米管[4]。为进一步探讨碳纳米管的生成机理及可能用途,本文通过比较实验、程序升温、及透射电镜与扫描电镜等观测手段,考察了碳纳米管的生成条件及影响因素,发现碳纳米管的生成强烈地受催化剂的结构与性能,反应温度,原料气组成及其流速所支配。实验研究结果为增进对碳纳米管的生成机理的认识积累了有价值的信息。

# 1　实验部分

## 1.1　催化剂制备

硝酸镍与另一作为载体的硝酸盐经特定方法充分混合,烘干,于 973 K 下灼烧 5 h,得一膨松物,此即为前驱态催化剂,其比表面约 100 $m^2/g$。

## 1.2　反应系统

反应在常压固定床连续流动反应器上进行,反应器为内径 4 mm 的石英管。原料气和反应尾气组成由一在线 SQ-206 气相色谱仪热导检测器分析,色层分离柱为 GDX-502。程序升温及脉冲反应在色谱与

\* 国家"攀登"和自然科学基金资助项目。

反应系统联用装置上进行。程序升温反应条件为:以甲烷作载气,升温速率为 10 K/min;脉冲反应条件为:每次试验催化剂用量 12 mg,原料气体脉冲量 0.1 mL。

### 1.3　EM 观测

SEM 观测在日立 S-520 型扫描电子显微镜上进行,样品粘附在样品板上,镀金后进行测试;TEM 观测在日本产 JEM-100 CX 型透射电子显微镜上进行,样品粘附于小铜碗上,放大倍率示于相应图中。

# 2　结果与讨论

### 2.1　碳纳米管的制备

10 mg Ni 催化剂放置于石英反应管内,在氢气流中缓慢升温还原,至 873 K 稳定数分钟后,迅速调到设定的反应温度,导入甲烷,反应 0.5 h 后,逐渐冷却至室温,收集样品。碳纳米管的纯化,系将制得的样品经稀硝酸浸泡,溶去纳米管管端粘附的催化剂颗粒,经蒸馏水洗涤多次,于空气中 373 K 烘干 1 h,再于 473 K 下用 He 气吹扫 1 h,管内外的水分则可除去。由 TEM 观测可证实纯化后的碳纳米管试样,催化剂颗粒已除去,较为纯净。

为对比研究,CO 也用作原料,其相应碳纳米管的制备方法及操作程序与甲烷相同。

### 2.2　制备条件及影响因素

(1)反应气组成的影响

作为催化裂解法制备碳纳米管的原料气,国际上主要采用乙炔。本工作则选用甲烷和 CO 分别作为原料气,并发现在较低温度(~723 K)下则进行催化裂解或歧化;与国外反应条件相若,而甲烷和 CO 无疑丰富易得得多。碳纳米管的生成系由含碳反应物在催化剂上分解留下碳并按一定方式聚集成管状纤维。因而,包括像烃及 CO 等可在催化剂上裂解或歧化生成 C 物种的物料均有形成碳纳米管的可能。

为作比较研究,本文分别考查了 $CH_4$ 和 CO 在相同温度及催化剂上裂解或歧化生成碳纳米管的情况。873 K 下制备样品的 TEM 观测结果示于图 1. 结果显示,虽然二者都可生成碳纳米管,但其形貌有明显差别:由 CO 歧化生成的碳纳米管较 $CH_4$ 分解生成的细(平均外管径为 CO:15 nm,$CH_4$:40 nm),管壁较薄。除了二者分解反应有别(即甲烷:$CH_4 \rightarrow C + 2H_2$,CO:$2CO \rightarrow C + CO_2$)外,它们在分解积聚过程中对催化剂的重构作用似乎也有明显差异。SEM 测试结果(图 2)表明,经由 CO 还原活化的催化剂其颗粒明显地比 $CH_4$ 处理的小。这一点也可由示于图 1 的 TEM 图看出:CO 所生成的碳纳米管管端催化剂颗粒比 $CH_4$ 生成的碳纳米管管端的催化剂颗粒小,而催化剂颗粒的大小则直接控制碳纳米管的管径粗细。看来,CO 歧化并生成碳纳米管的过程对催化剂颗粒同时也产生较强的解裂作用,使其粒度变得更小。由此可以推测,CO 歧化并聚集成碳纳米管对催化剂颗粒的大小有选择性;只在其粒度小到一定程度的催化剂颗粒上,在催化剂上 CO 歧化留下的碳才能聚集成管。相对说来,$CH_4$ 对催化剂粒度的要求要宽得多(但其成管过程对催化剂颗粒也有细化作用);在较大的催化剂颗粒上,则生成粗管,而在小颗粒上则生成细管。但总体来说,$CH_4$ 还原处理的催化剂颗粒比 CO 处理的催化剂颗粒大 2 倍以上。当以 $CH_4$ 为原料气时,在大颗粒的催化剂上生成碳纳米管,其管壁较厚,暗示分解后的 C 可在较大的催化剂表面区域(即"碳管生长区")集结,并按一定方式排列后再一层接一层地堆积成管状结构。而当以 CO 为原料气时,这种可聚集 C 成管状结构的碳管生长区看来要小一些。也就是说,$CH_4$ 可以通过增加壁厚来维持碳纳米管的生长,而 CO 则往往导致催化剂先进一步碎裂细化,而后再在细颗粒的催化剂上结聚成管。这种对催化剂颗粒度的选择性反映出 $CH_4$ 和 CO 在分解、C 迁移和聚集成管过程中机制的差异。显而易见,两种物料的反应中间态碳物种就有差别:在表面上 CO 歧化只留下 C 物种,而在表面上 $CH_4$ 分解生成的中间物不仅有 C,也包含 CHx,后者在表面上迁移比 C 在表面或体相迁移扩散速率更快;这类物种可先迁移至碳管生长区附近,再脱氢为 C,而后聚集成管。

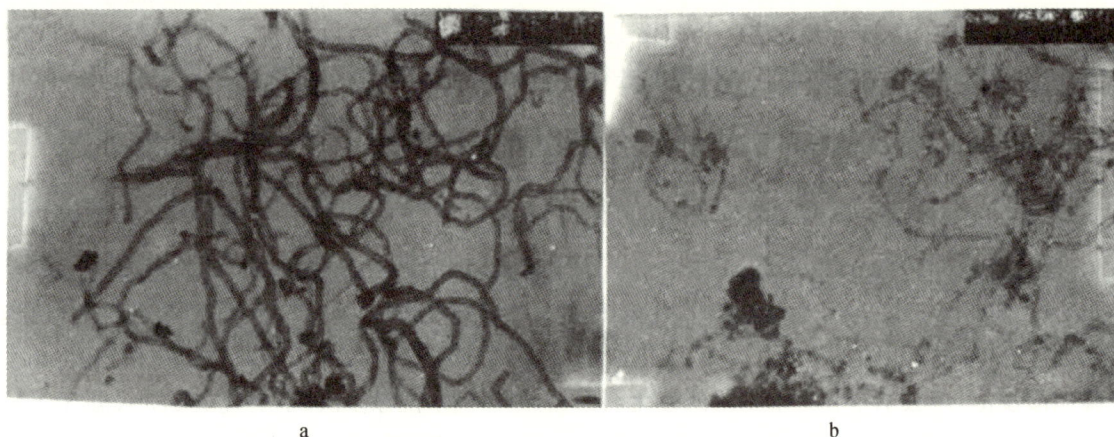

**图 1  873 K 下分别由 CH₄ 催化分解和 CO 歧化所制得碳纳米管的 TEM 图**

(a)CH₄(×27 000),(b) CO (×72 000)

**Fig. 1  TEM images of carbon nanotube produced at 873 K by catalytic decomposition of CH₄**

(a) (×27 000) and disproportionation of CO(b) (×72 000),respectively

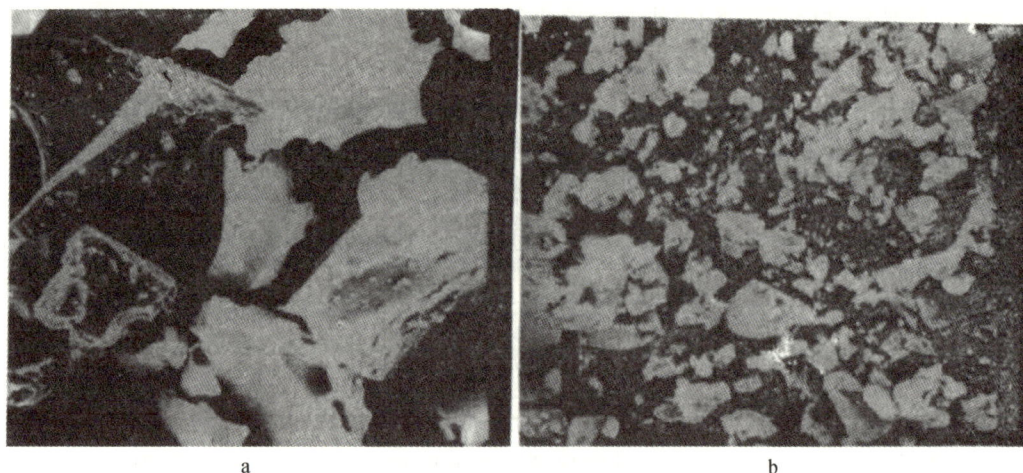

**图 2  Ni 催化剂的 SEM 图(×1 000);873 K 下分别经 CH₄(a),CO(b)处理**

**Fig. 2  SEM images (×1 000) of appearance of the Ni-catalysts after**

(a) methane-prereduction-treatment at 873 K,(b) CO-prereduction treatment at 873 K

(2)原料气流速的影响

在以甲烷为原料气制备碳纳米管的过程中,甲烷的空速大小对所生成的碳纳米管的形貌有着重大的影响。比较不同空速下收集之样品的 TEM 观测结果可以发现,当空速增大 4 倍后(图 3),所制得碳纳米管的管径变小;进一步观察可见,这些细的碳纤维的顶端均粘附有小的催化剂颗粒,其直径约 15 nm,不及较低空速下(参见图 1)相应之管端粘附催化剂颗粒的二分之一,且管壁较薄。由此看来,大空速原料气的导入有利于细管径的碳纳米管的生成,其原因很可能是在此状态下,CH₄ 对催化剂的细化作用得到加强。可以想象,大量甲烷的导入,就催化剂而言意味着在瞬间有大量的甲烷在其上活化分解生成大量的 C 并同时在较多的催化剂晶核上堆积成管状结构,因而可能导致催化剂颗粒更多地碎裂。而当甲烷空速较低,也即单位时间内与催化剂接触分解的甲烷量较少时,相应生成的碳量也较少,这有利于其在较少的晶核上作有序堆积,所制得的碳纳米管管径较粗,管壁也较厚。由此可见,催化剂颗粒的细化是在甲烷裂解、C 作有序堆积的过程中进行的。比较图 1 和图 3 还可发现,甲烷在催化剂上的活化裂解似乎与纳米级催化剂颗粒大小无太大关系,即在这两种情况下,都有大量的 C 生成。然而,所生成 C 的堆积结构则与其晶粒大小密切相关,并可能暗示某些特定晶面组成的多面体顶角才适合于管状碳纤维构造的形成。

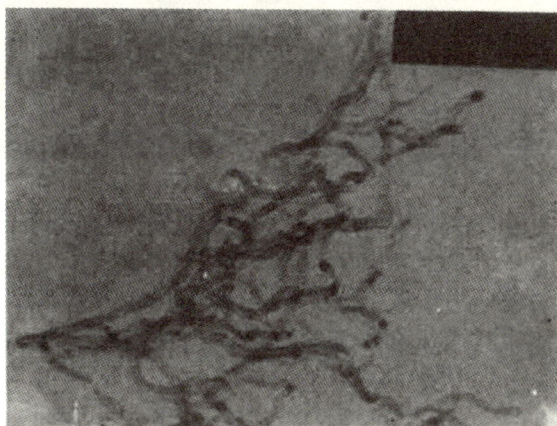

图 3 较高流速(4 000 mL/h)下 CH₄ 催化裂解
生成的碳纳米管 TEM 图(×100 000)

Fig. 3 TEM image (×100 000) of the carbon
nanotubes produced at higher GHSV
(4 000 mL/h) of CH₄ and 873 K

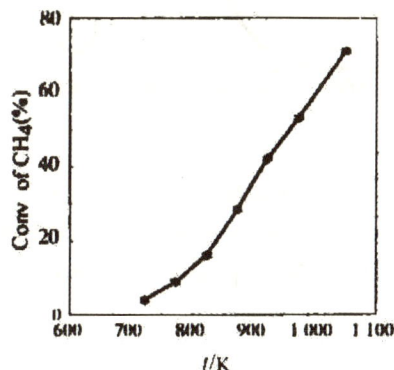

图 4 在 Ni 催化剂上不同温度下 CH₄ 脉冲反
应活性

Fig. 4 Activity of pulse reaction of CH₄ over the
Ni-catalyst at various temperatures

(3)温度效应

由脉冲实验结果(图 4)可以看出,温度越高越有利于甲烷的转化;由此似可预期,在其他条件相同的情况下,高温有利于碳纳米管产率的提高。然而实验结果并非如此。实验发现,碳纳米管的最大收率出现在 973 K 以下;而当温度高于 973 K 时,产率反而下降。表 1 示出 10 mg 催化剂与 CH₄ 反应 0.5 h 后碳纳米管的产率。对此催化剂进行 TPR 测量发现,当反应温度升至 973 K 时甲烷的分解达到一个峰值,而后急剧下降(见图 5),直至甲烷在催化剂上的分解几近停止,即催化剂明显失活。收集 1 023 K 产物样品进行 TEM 观测,发现大多数碳纳米管均从催化剂表面上剥离开来,而催化剂的表面覆盖着一层碳,这可从其与催化剂的电子透过率的差别来判定(见图 6)。表面碳的生成显然覆盖住催化剂活性表面,导致催化剂失活。

表 1 不同反应温度下碳纳米管的产率
Tab. 1 Yield of carbon nanotabe at several temperatures

| 温度(K) | 743 | 843 | 943 | 1 023 |
|---|---|---|---|---|
| 产率(mg/h·mg catal.) | 14 | 60 | 200 | 50 |

尽管文献上有关碳纳米管的生长机理尚未获定论,但至少如下观点是比较一致的,即与碳纳米管生长相关的催化剂颗粒其形状一般是具有一定对称性的多面体,碳纳米管往往是围绕多面体的顶角而生长,留下未被 C 覆盖的活性表面(多属一些晶面平面)区供作底物吸附、反应之用[1]。本文的实验结果支持上述观点,并进一步认为,底物分子 CH₄ 或 CO 在活性表面区吸附、分解留下表面碳物种,这些含碳物种将尽可能通过简捷的途径,包括在表面迁移或经催化剂颗粒体相扩散,到达碳纤维生长区(或生长点),

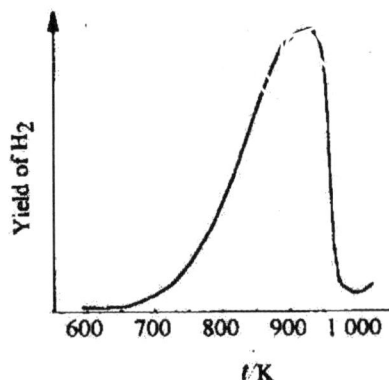

图 5　预还原 Ni 催化剂上 CH₄ 的 TPR 谱

Fig. 5　Temperature-Programmed-Reaction (TPR) of CH₄ on the prereduced Ni-catalyst

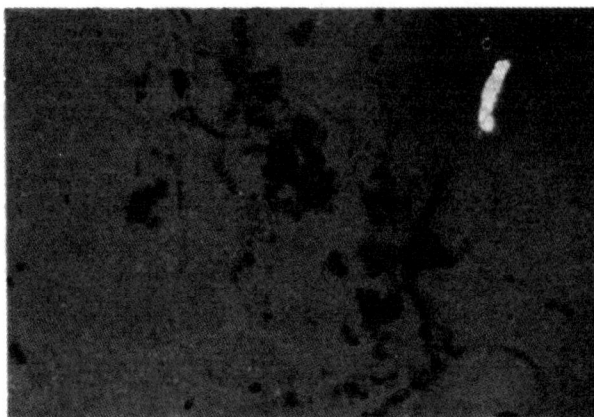

图 6　1 023 K 温度下 CH₄ 催化裂解生成的碳纳米管 TEM 图(×72 000)

Fig. 6　TEM image (×72 000) of the carbon nanotube produced from CH₄ catalytic pyrolysis at 1 023 K

进而以一定方式构建管状结构。这一总过程包含碳物种的生成、迁移和堆积成管状结构三个基本过程。由此可见,在较低温度(如 873 K)下,碳物种的生成速率较低,并可能低于其迁移及堆积成管状结构的速率,因而反应第一步生成的碳物种多数来得及迁移并以一定方式堆集成管状结构,所生成的碳纤维比较长而且较规整;而在较高温度(如 973 K 以上)下,碳物种的生成速率可能超过其迁移及堆积成管速率,第一步生成的碳物种有相当部分因来不及向碳管生长区迁移扩散而"就地"无规堆积并覆盖活性表面,从而导致催化剂失活和整个反应过程被阻化。

# 3　结　论

本文实验结果进一步支持以下观点,即碳纳米管的生长过程大体包含如下步骤:

(1)CH₄ 或 CO 在催化剂表面活化分解或发生歧化反应生成碳物种,并伴随着对催化剂颗粒产生不同程度的细化作用;

(2)表面碳在催化剂表面或体相迁移并向碳管生长区汇集;

(3)汇集的碳依照一定方式堆积成管状结构。

本文的实验结果表明,碳纳米管的生成受原料气组成、流速、及温度的影响。由甲烷生成的碳纳米管管径较粗,管壁较厚;而 CO 生成的碳纳米管则管径较细,管壁较薄。作为底物的 CH₄ 和 CO,在催化剂表面分解、反应并生成碳纳米管的过程中,伴随着对催化剂颗粒产生不同程度的细化作用;CO 只在较细的催化剂颗粒上生成碳纳米管。为获得较高的碳纳米管产率,适当高的反应温度和原料气空速固然必要,但过高的反应温度或过高的原料气空速将导致表面碳物种的生成速率超过其迁移、扩散并堆积成管状结构的速率,导致供底物分子吸附、反应的活性表面被来不及迁移而"就地"无规堆积的碳物种所覆盖,最终导致催化剂失活和整个反应过程被阻化。

# 4 参考文献

[1]Rodriguez N. M.. A review of catalytically grown carbon nanofibers. J. Mater. Res., 1993, **8** (12):3233~3250.

[2]Iijima S., Helical microtubules of graphitic carbon. Nature, 1991, (354):56.

[3]Baker R T K et al. Carbon, 1989, (27):315.

[4]陈萍,张鸿斌等.低温催化裂解烷烃法制备碳纳米管.高等学校化学学报,1995,16(11):1783~1784.

## Preparation of Carbon nanotube by Catalytic Pyrolysis of Methane at Lower Temperature

Ping Chen[1], Hong-Bin Zhang[1], Guo-Dong Lin[1], Khi-Rui Tsai[1], He-Shen Zai[2]

([1]Dept. of Chem. & State Key Lab. for Phys. Chem. of Solid Surf., [2]Lab. of Elec. Micr.)

**Abstract**   A new way for preparation of carbon nanotube by catalytic pyrolysis of methane on a Ni-catalyst at temperature of 723K is reported. TEM observation revealed that almost all carbon fibers formed by this method under the reaction conditions used were in the form of tubular structure. The results showed that the outer-diameter and the length of the carbon nanotube as well as the productivity were strongly governed by the structure-performance of the catalyst used, reaction temperature and feed gas flow-rate. The separation and purification of the carbon nanotube from the catalyst particles can be carried out by washing with aqueous solution of nitric acid followed by drying at 473 K.

**Key words**   Carbon nanotube   Methane   Catalytic pyrolysis   Ni-catalyst

■ **本文原载**：《厦门大学学报》（自然科学版）第 35 卷第 2 期（1996 年 3 月），第 220～225 页。

# 负载型水相膦铑配合物催化剂上
# 丙烯氢甲酰化制丁醛*

袁友珠　张　宇　杨意泉　林国栋　张鸿斌　蔡启瑞
（化学系　固体表面物理化学国家重点实验室）

**摘　要**　用连续流动加压微反-色谱装置评价了负载型水溶性膦铑催化剂对丙烯氢甲酰化制丁醛的反应活性，结果表明，催化剂活性与体系中水含量密切相关，适量的水可使催化剂活性明显提高，原位红外光谱考察结果显示，当催化剂吸附 CO 时在 2040、2012、1901 cm$^{-1}$ 等处出现一系列可归属于可逆吸附 CO 物种的伸缩振动带；通入原料气丙烯/CO/H$_2$ 时，很快可观测到～1714 cm$^{-1}$ 的醛基吸收峰。实验结果表明，水溶性膦铑配合物负载到固体表面后，在载体表面与体系中微量水形成液膜，该气/液膜（固）界面可能有利于丙烯氢甲酰化配位络合催化循环的进行。

**关键词**　丙烯　负载型水相催化剂　氢甲酰化　丁醛

**中国图书分类号**　O 621.256.1

在负载型油相催化剂（SLP 催化剂）作用下，使炼厂干气（主要含乙、丙烯）于气/固相进行连续氢甲酰化反应制取高附加值的丙醛和丁醛，这一催化过程已具备较大规模工业开发的价值[1]。该过程中所采用的 SLP 催化剂系将著名的 Wilkinson 催化剂 HRh(CO)(PPh$_3$)$_3$ 通过物理吸附负载到载体上（如高分子小球、SiO$_2$、γ-Al$_2$O$_3$ 等）来制备[2]。采用类似方法制备的负载型水相催化剂（SAP 催化剂），于气一液/固相体系中对长链烯烃氢甲酰化反应有较好的催化性能[3]，我们曾采用固定床微反系统对 C$_6$～C$_{11}$ 等长链烯烃在负载型水相膦铑催化剂（简称 SAP 铑催化剂）上的连续氢甲酰化进行过较深入研究[4]，并报道了碱金属促进的 SAP 铑催化剂的研究结果[5]。迄今文献中尚无在 SAP 催化剂上进行丙烯氢甲酰化制丁醛的研究报道。本文将考察丙烯在 SAP 铑催化剂上的氢甲酰化反应，并结合原位红外谱学方法研究 SAP 铑催化剂的氢甲酰化作用机理。

# 1　实　验

三磺化三苯基膦（TPPTS）按文献方法[6]制备，$^{31}$P 溶液 NMR 表明有约 5% 的三苯基膦氧化物。采用前文[5]的方法制备 SAP 铑催化剂。在固定床加压流动反应器—GC 联用装置进行催化剂活性评价，反应管为 Φ8 mm 不锈钢管，内装 80～100 目催化剂 0.5 g。原料气为 C$_3$H$_6$/CO/H$_2$＝1∶1∶1(v/v)。使用上分 102-GD 气相色谱仪（FID）分析氢甲酰化产物（色谱柱：分析醛为邻苯二甲酸二壬酯，柱长 2 m；分析烯和烷为 20% 三十烷/SiO$_2$，柱长 2 m）。

红外光谱仪为 Nicolet 740 FT-IR，仪器分辨率为 4 cm$^{-1}$，样品池窗片为 NaCl，扫描范围 4000～1300 cm$^{-1}$ 催化剂样品直接压片；所得 IR 图均扣除催化剂本底的贡献。

---

* 国家自然科学基金和中国石油化工总公司联合资助项目。

# 2　结果及讨论

## 2.1　载体的影响

载体 $SiO_2$,$\gamma$-$Al_2O_3$ 对 SAP 铑催化剂性能的影响见表1。由各种载体所制备的 SAP 铑催化剂,在丙烯氢甲酰化制丁醛的反应中,可检出～1％的加氢产物丁醇(包括正、异丁醇),而未检出丙烷,产物醛的选择性大于98％。由表1可见,同类载体中,比表面积大者所制备的催化剂活性高。但酸性载体($\gamma$-$Al_2O_3$)所制备的催化剂活性明显低于偏中性载体($SiO_2$)而得的催化剂。

**表 1　载体的影响**
**Tab. 1　Effects of support**

| 载体类型 | 比表面<br>($m^2/g$) | 转化率<br>（％） | 时空产率<br>($mmol/(g \cdot h)$) | n/b |
|---|---|---|---|---|
| $SiO_2$(l) | 208 | 10.22 | 16.8 | 1.5 |
| $SiO_2$(II) | 250 | 28.60 | 47.0 | 1.4 |
| $\gamma$-$Al_2O_3$ | 280 | 12.71 | 20.9 | 1.2 |

反应总压＝3.0 MPa,反应温度＝373 K,$C_3H_6$/CO/$H_2$＝1/1/1(v/v),

铑 0.02 mmol/g-$SiO_2$ or $\gamma$-$Al_2O_3$,L/Rh＝10(mol/mol).

## 2.2　反应压力和温度的影响

以 $SiO_2$(I)为载体制备的 SAP 铑催化剂在不同反应压力和温度下的催化性能见表2。由表2可见,随着压力升高,反应转化率提高,但将降低产物醛的 n/b;在333～393 K范围内,随着温度升高,催化剂性能也有类似的变化。由此可见,升高反应压力和温度对产率的提高有利,但大幅度提高反应压力或反应温度,将降低 n/b 比值;另外,过高的反应温度将不利于 SAP 催化剂的稳定使用。

**表 2　反应压力和温度的影响**
**Tab. 2　Effects of pressure and temperature**

| 压力<br>（MPa） | 温度<br>（K） | 转化率<br>（％） | 时空产率<br>($mmol/(g \cdot h)$) | n/b |
|---|---|---|---|---|
| 1.0 | 373 | 15.03 | 16.5 | 1.93 |
| 2.0 | 373 | 24.50 | 52.7 | 1.20 |
| 3.0 | 373 | 30.42 | 57.1 | 1.28 |
| 3.0 | 333 | 21.00 | 44.3 | 1.16 |
| 3.0 | 353 | 22.33 | 44.3 | 1.11 |
| 3.0 | 373 | 30.42 | 57.1 | 1.28 |
| 3.0 | 393 | 48.67 | 86.8 | 1.00 |

$C_3H_6$/CO/$H_2$＝1/1/1(v/v),GHSV＝1500 $h^{-1}$,其他条件同表1。

## 2.3　空速的影响

图1示出原料气空速对催化剂性能的影响。从图1可见,空速大于900 $h^{-1}$时,原料气在催化剂表面的接触时间过短,转化率随空速提高而下降;空速小于300 $h^{-1}$时,可能产物醛脱附太慢,部分催化剂活性点被占用,也不利催化剂活性提高。因此,水溶性配合物 HRh(CO)(TPPT)$_3$ 负载到 $SiO_2$ 上后,所得催化剂性能与多相催化剂类似,活性受接触时间和扩散速度影响。

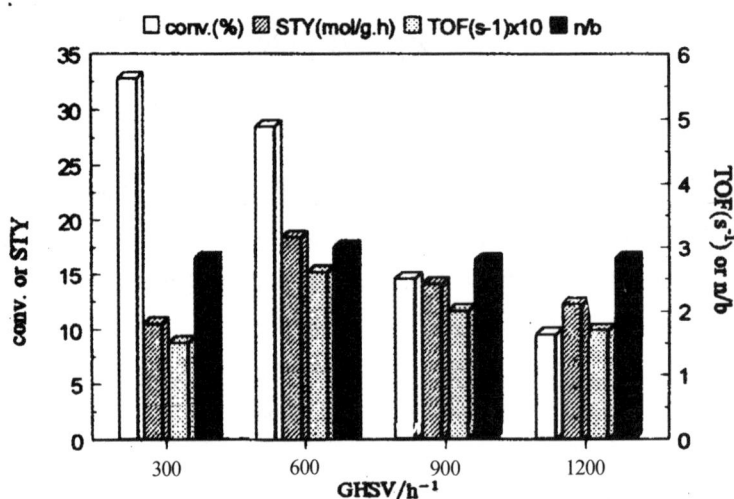

图1　空速对反应活性的影响

**Fig. 1　Effects of GHSV**

反应温度＝383 K,其他条件同表1

### 2.4　水对 SAP 膦铑催化剂丙烯氢甲酰化活性的影响

在 SAP 铑催化剂上进行 150 h 的烯烃氢甲酰化催化反应考察过程中发现,前 30 h 内的转化率和产物 n/b 可保持稳定;在往后的时间内催化剂活性有下降趋势。文献曾报道 SAP 铑催化剂活性连续 38 h 保持基本稳定的实验结果[7],但 Davis 等和刘海超等则分别报道 SAP 铑催化剂含水量为 4 wt％～12 wt％[8]和 25wt％～35wt％[9]时催化活性最高。由于稳态过程反应尾气带走催化剂上的水,我们设计了一个可控温的耐压鼓泡不锈钢管,通过原料气将水蒸汽带入反应体系来研究水分对催化剂性能的影响。实验结果示于表3。从表3可见,原料气带入体系的水量对催化剂活性的影响较大,当原料气通过 313 K 的水浴,催化剂时空产率为 24.9 mmol/(g·h);而当原料气未通过水浴情况下的时空产率仅为 7.3 mmol/(g·h),前者比后者高 3 倍多。进一步的实验结果表明,当反应原料气通过 313 K 水浴时,催化剂具有较好的活性稳定性。

表3　水对 SAP 膦铑催化剂丙烯氢甲酰化活性的影响[a]

**Tab. 3　Effects of water feed**

| 水温 | 转化率 | 时空产率 | n/b | 饱和水蒸汽压 |
| --- | --- | --- | --- | --- |
| (K) | (％) | (mmol/(g·h)) | | (Pa) |
| 干气 | 5.6 | 7.3 | 1.5 | |
| 303 | 14.3 | 18.4 | 2.2 | 4241 |
| 313 | 19.3 | 24.9 | 1.5 | 7374 |
| 333 | 9.3 | 12.0 | 1.5 | 19920 |

(a)干气通过盛水耐压容器鼓泡进入催化剂床层反应,反应总压＝1.0 MPa,$C_3H_6$/CO/$H_2$＝1/1/1(v/v),铑 0.02 mmol/g-$SiO_2$.GHSV＝900 $h^{-1}$,活性数据取自 4 h 后。

# 3　原位红外光谱结果

## 3.1　CO 在 SAP 铑催化剂上吸附的红外吸收光谱

我们先前已报道,SAP 铑催化剂的本底红外吸收光谱在 1980 $cm^{-1}$ 处出现可指认为配合物 HRh-

(CO)(TPPTS)$_3$ 的羰基配位基的 C-O 伸缩带[10]。当在 353~373 K,0.1~1.0 MPa 的实验条件下,向体系逐步引入 CO 时,如图 2 所示,在 1980、2012、2040、2108 cm$^{-1}$ 处出现若干 IR 吸收带。$v(CO) \approx 1980$ cm$^{-1}$ 带在 N$_2$ 气流中仍稳定存在,它应仍属于配合物 HRh(CO)(TPPTS)$_3$ 中的羰基配位基 C-O 伸缩带。另外三个谱带随着引入 N$_2$ 吹扫而逐渐消失,而当重新引入 CO 之后复又出现;前两个带(2012 和 2040 cm$^{-1}$)多半系产生自较弱地可逆吸附在 Rh 中心上的孪生配位羰基物种,而 2108 cm$^{-1}$ 处的吸收带很可能属于气相 CO 分子的振动—转动组合带。上述的实验结果可能暗示该吸附体系存在如下的平衡过程:

$$HRh(CO)(TPPTS)_3 \xrightleftharpoons[+TPPTS,-CO]{-TPPTS,+CO} HRh(CO)_2(TPPTS)_2$$

CO 分压的增加将有利于平衡向右移动,导致孪生配位的羰基物种浓度的升高(相应的 IR 吸收带强度增加)。

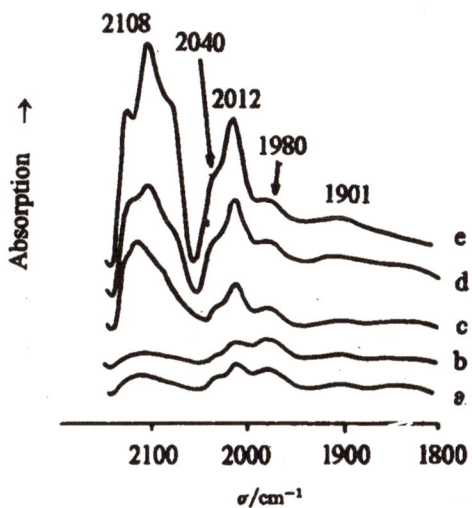

图 2  CO 在 SAP 铑催化剂上吸附的原位红外图

(a) 353 K,0.1 MPa;(b) 373 K,0.1 MPa;

(c) 373 K,0.3 MPa;(d) 373 K,0.6 MPa;

(e) 373 K,1.0 MPa

**Fig. 2**  *In-situ* IR spectra of CO adsorption on SAP-Rh catalyst

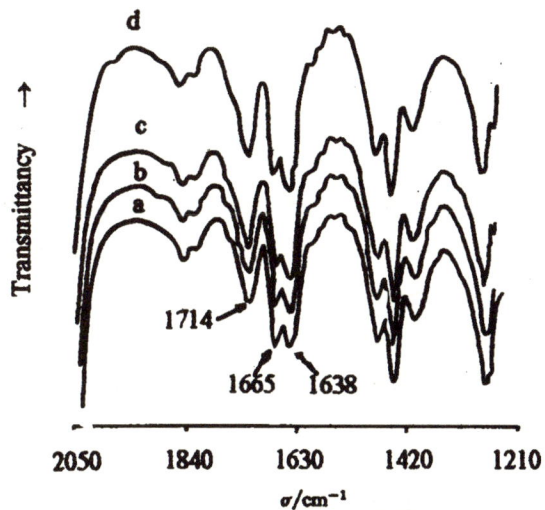

图 3  丙烯在 SAP 铑催化剂上氢甲酰化的原位红外图

(a) 1.0 MPa;(b) 1.0 MPa,2 min;

(c) 1.0 MPa,3 min;(d) 1.0 MPa,10 min

**Fig. 3**  *In-situ* IR spectra of propene hydroformylation over SAP-Rh catalyst at 373 K

### 3.2  丙烯权 SAP 催化剂上氢甲酰化的原位红外吸收光谱

SAP 催化剂上丙烯氢甲酰化的原位红外吸收光谱示于图 3。由图 3 可见,催化剂中一旦通入原料气丙烯/CO/H$_2$ 时,很快在 1714、1665、1638 cm$^{-1}$ 附近出现吸收峰,在纯粹 CO 吸附时这些峰并未观测到。因此,这些谱带可分别作如下归属:较宽的 1714 cm$^{-1}$ 带为在催化剂表面(液膜中)正/异醛分子的吸收,1665、1638 cm$^{-1}$ 两个带为配位在 Rh 中心上的反应中间物种正/异构酰化物。

## 4  催化机理及水的作用推测

从上述 SAP 铑催化剂的原位红外谱学结果可以推断,下列过程较容易进行:

$$HRh(CO)(TPPTS)_3 \xrightleftharpoons[+TPPTS]{-TPPTS} HRh(CO)(TPPTS)_2 \xrightleftharpoons[-CO]{+CO} HRh(CO)_2(TPPTS)_2 \xrightleftharpoons[+TPPTS]{-TPPTS} HRh-$$

$$(CO)_2(TPPTS) \underset{-CO}{\overset{+CO}{\rightleftharpoons}} HRh(CO)_3(TPPTS)$$

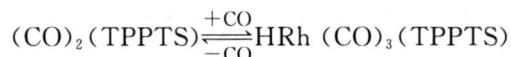

表明 SAP 铑催化剂的作用机理与 Wilkinson 提出的均相催化机理类似。另外,催化剂的丙烯氢甲酰化动力学考察结果显示产物醛的 n/b 较低(<3)。分析认为,水溶性膦配体受本身结构中的磺酸根与载体 SiO₂ 的-OH 基团间的相互作用(如氢键等),使铑膦配合物趋于平铺在 SiO₂ 上,铑中心的位阻减小,配体及配合物的结构因载体存在显得较为稳定。在所调查的反应温度下烯烃与 SAP 铑配合物催化剂的中心离子 Rh 的配位时容易产生快速的 1,2 取向配位,从而使产物醛的 n/b 较低。

在新鲜催化剂中,由于水溶性膦铑配合物负载在 SiO₂ 上并于常温真空抽干大部分水分后,水溶性配合物将均匀分布在 SiO₂ 表面,未能抽去的少量水份(为 10 wt%~15 wt%)与水溶性配合物在 SiO₂ 表面形成一层液膜,配体容易活动,载体的高比表面使其催化活性较高;当催化剂运转了一段时间后,催化剂中的水分已不足以使配合物成液膜,造成配位催化循环速率减慢,催化活性下降。因此,若在体系中引入适量的水,将有助于催化剂保持较高的活性。

# 5　参考文献

[1] 武戈,钱新荣. 炼厂干气综合利用方案评选. 化学进展,1994,**6**(2):161~170.

[2] 特木勒等. 分子催化,担载液相催化剂研究 Ⅱ. 炼厂干气中烯烃的醛化反应. 催化学报,1991,**12**(5):413~417.

[3] Davis M. E.. Supported aqueous-phase catalysis. Chemtech,1992:498~502.

[4] 袁友珠,陈鸿博,蔡启瑞. SiO₂ 负载的磺化三苯膦铑配合物催化高碳烯氢甲酰化. 应用化学,1993,**10**(4):13~17.

[5] Yuan Youzhu et al. The beneficial effect of alkali salt on supported aqueous-phase catalysts for olefin hydroformylation. Catal. Lett. ,1994,**29**:387~395.

[6] Kuntz E.. Catalytic hydroformylation of olefins. U. S. Pat. ,4248802(1981).

[7] Horvath I. T.. Hydroformylation of olefins with the water soluble HRh(CO)[P(m-C₆H₄SO₃Na)₃]₂ in supported aqueous-phase. is it really aqueous? Catal. Lett. ,1990,**6**:43~48.

[8] Davis M. E., Arhancet J. P., Hanson B. E.. Supported aqueous-phase catalysts. J. Catal. ,1990,**121**:327~339.

[9] 刘海超等. 负载水溶性铑—膦配合物催化 1-己烯氢甲酰化反应的研究. 分子催化,1994,**8**(1):22~27.

[10] 袁友珠等. SiO₂ 负载的磺化三苯膦铑配合物催化高碳烯氢甲酰化及反应中的氘逆同位素效应. 分子催化,1993,**7**(6):384~390.

## Hydroformylation of Propene to Butyl Aldehydes over Supported Aqueous-Phase Rh Catalysts

You-Zhu Yuan，Yu Zhang，Yi-Quan Yang，Guo-Dong Lin，Hong-Bin Zhang，Khi-Rui Tsai

(Dept. of Chem. & State Key Lab. for Phy. Chem. of the SolidSurf. )

**Abstract**　Hydroformylation of propene to butyl aldehydes over supported aqueous-phase Rh catalysts has been investigated in a fixed-bed reactor-GC system. The effects of the process parameters such as ratio of L(ligand)/Rh, type of support, reaction pressure and temperature as well as the water content in feedstream are studied. The feed rate of water vapor, which is introduced by bubbling feed-gases ($C_3H_6$/CO/$H_2$) through a bubbler containing water with adjusted temperature, is crucial for catalyst performance. The results indicate that the formation of the liquid film with a proper water content on the $SiO_2$ surface is the key factor for the dispersion and migration of the Rh complexes. In-situ IR spectroscopic investigation showed that when CO was admitted to the catalyst surface, several reversible IR bands at 1901, 2012, and 2040 $cm^{-1}$ have been observed, in addition to $\gamma_{(CO)} \sim$ 1980 $cm^{-1}$ for the absorption of complex HRh (CO) (TPPTS)$_3$. The IR band of butyl aldehydes at $\sim$1714 $cm^{-1}$ can be immediately observed when $C_3H_6$/CO/$H_2$ was induced into IR cell. These experimental results strongly suggest that the structure of complexes is stablilized during immobilization and the hydroformylation occurs on the interface of two immisible phase consisted of feed gas and liquid film.

**Key words** Propene  Supported-Aqueous-Phase Catalyst  Hydroformylation  Butyl aldehyde

■ **本文原载**:《生物化学与生物物理进展》第 23 卷第 1 期(1996 年),第 18～20 页。

# 固氮酶中的电子传递*

黄静伟　张鸿图　万惠霖　蔡启瑞

(厦门大学化学系,固体表面物理化学国家重点实验室,厦门 **361005**)

**摘　要**　提出改进的二步 ATP 驱动的电子传递机理,对还原剂和 MgATP 都充足或其中有一种不充足的情况下固氮酶体系的 EPR 信号变化作了合理的解释。

**关键词**　固氮酶　ATP 驱动　EPR 信号　电子传递

固氮酶是由 MoFe 蛋白($\alpha_2\beta_2$ 四聚体,分子量约为 230 ku)和 Fe 蛋白($\gamma_2$ 二聚体,分子量约为 65 ku)两个组分构成。随着 $N_2$ 和其他外源底物络合到 MoFe 蛋白的活性中心,电子从还原剂经由 Fe 蛋白传递到 MoFe 蛋白上,再由 MoFe 蛋白传递给底物,偶合着质子的传递,底物得到不同程度的还原。有关这一电子传递过程的研究一直是固氮酶研究的一个中心内容。本文提出改进的二步 ATP 驱动的电子传递机理,对还原剂($S_2O_4^{2-}$)和 MgATP 都充足或其中有一种不充足情况下固氮酶体系的 EPR 信号变化进行尝试性解释。

## 1　改进机理的提出

在 Hardy 等[1]证明生物固氮必须有 ATP 参加之后,Walker 和 Mortenson[2]在稀固氮菌——巴氏芽胞梭菌(*Clostridium pasteurianum*,*Cp*)酶溶液中观察到当还原剂($S_2O_4^{2-}$)和 MgATP 都充足时,少量的 Fe 蛋白能还原分子数量大得多的 MoFe 蛋白,使得 85% 的 MoFe 蛋白处在没有 EPR 信号的深度还原态,大部分 Fe 蛋白处于有 EPR 信号的还原态,大大降低了酶促活性。然而 Smith 和 Lowe[3]却在肺炎克氏杆菌(*Klebsiella pneumoniae*,*Kp*)*Kp*1:*Kp*2=1:1 的较浓固氮酶溶液中观察到稳态酶周转时,90% *Kp*1 处于无 EPR 信号的还原态,而 70% 的 *Kp*2 是没有 EPR 信号的氧化态。为了很好地说明这些实验现象,Cai 等[4]根据配位化学和络合催化作用原理,提出了二步 ATP 驱动电子传递的固氮酶催化反应机理。最近,Rees 等[5]成功地解析了 MoFe 蛋白和 Fe 蛋白单晶 X 射线衍射所得到的电子密度图,得到 MoFe 蛋白和 Fe 蛋白的三维空间结构,指出 $N_2$ 和其他外源底物是络合在固氮酶中的 M-簇合物(蛋白键合的 FeMo-cofactor)。M-簇合物是由 $MoS_3Fe_3$ 和 $FeS_3Fe_3$ 这二个缺口的类立方烷型簇合物通过三个非蛋白配体(二个硫和一个比硫轻的"Y")桥联而成。由于 M-簇合物是深埋在 MoFe 蛋白里,电子是由 Fe 蛋白先传递给处在 MoFe 蛋白表层的 P-簇合物(P-簇合物是由两个 $Fe_4S_4$ 簇合物通过二个半胱氨酸的巯基配体桥联而成),再传递给 M-簇合物。MgATP 可以络合到 Fe 蛋白或 MoFe 蛋白上[6],但单独的 Fe 蛋白或 MoFe 蛋白并不会使 MgATP 发生水解,只有当络合着 MgATP 的 Fe 蛋白与 MoFe 蛋白结合产生碱性基团时才能催化 MgATP 的水解[7]。动力学研究[8]表明:ATP 水解发生在 Fe 蛋白-MoFe 蛋白之

---
\* 国家基础性研究重大关键项目(攀登计划)资助课题。

间的电子传递之前。虽然 ATP 在 Fe 蛋白上的具体键合位目前还未证实,但已知 ATP 络合到 Fe 蛋白上会引起 Fe 蛋白的构型和氧化还原电位的变化,从而促进了 Fe 蛋白和 MoFe 蛋白的结合[9]。基因突变研究[10]发现体外 FeMo-co 的生物合成需要有 Fe 蛋白和 MgATP,但 Fe 蛋白并不必发生由于 MgATP 络合而诱发的构型变化,也不需要一般的 MgATP 水解及发生从 Fe 蛋白到 MoFe 蛋白的电子传递。ATP 水解的数量与电子传递的关系进一步证实,MgATP/2e>4[11]。根据上述这些研究结果,我们提出了改进的二步 ATP 驱动的电子传递机理,如图 1 所示。

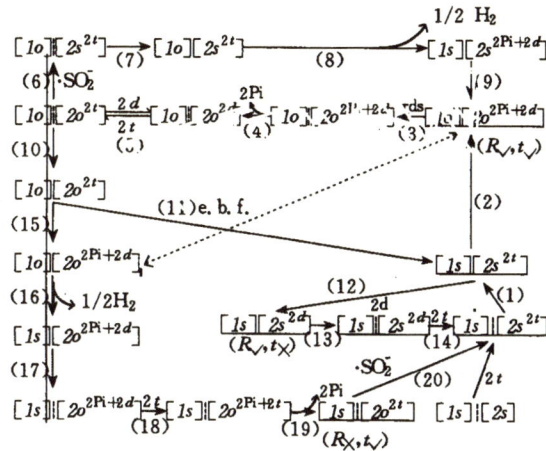

图 1  电子传递过程中固氮酶组分的氧化还原态和改进的二步 ATP 驱动的电子传递机理

$1s$:具有 EPR 信号的半还原态 MoFe 蛋白;$1o$:没有 EPR 信号的深度还原态 MoFe 蛋白;$2s$:具有 EPR 信号的还原态 Fe 蛋白;$2o$:没有 EPR 信号的氧化态 Fe 蛋白;$t$:ATP;$d$:ADP;Pi:磷酸盐;$R$:还原剂;√:表示充足供应;×:表示不充足供应;·$SO_2^-$:表示含有自由基的还原剂。

# 2  改进的 ATP 驱动的电子传递机理

## 2.1  第一步 ATP 驱动的电子传递

天然分离的 MoFe 蛋白和 Fe 蛋白分别具有 EPR 信号的半还原态($1s$)和还原态($2s$)。当 ATP 络合到 Fe 蛋白上时引起了 Fe 蛋白构型和氧化还原电位的变化,促进了 Fe 蛋白与 MoFe 蛋白结合为$[1s]$$[2s]^{2t}$。由于 Fe 蛋白和 MoFe 蛋白结合产生了碱性物种催化了 MgATP 的水解,产生 ADP 和 Pi(无机磷酸盐),同时放出大量的能量克服了电子从 Fe 蛋白传递给 MoFe 蛋白所需跃过的能垒,完成电子的传递,使 MoFe 蛋白变成没有 EPR 信号的深度还原态,而 Fe 蛋白自己也变成无 EPR 信号的氧化态,即得到$[1o][2o]^{2Pi+2d}$,这样就完成了 ATP 驱动的第一步电子传递,如图 1 中(1)→(2)。

## 2.2  第二步 ATP 驱动的电子传递

从 Fe 蛋白和 MoFe 蛋白结合物的结构上看,Fe 蛋白的 $Fe_4S_4$ 簇合物是被埋在 Fe 蛋白和 MoFe 蛋白结合的界面上,要使没有 EPR 信号的氧化态 Fe 蛋白被还原,Fe 蛋白和 MoFe 蛋白的结合物必须先解离,产生$[1o]$:$[2o]^{2Pi+2d}$。从动力学上来考虑,对于浓酶溶液,图 1(3)这一步是整个固氮酶电子传递过程的速率决定步骤,结果正如 Smith 等[3]所观察的。在 ATP 充足的情况下,ATP 会与 ADP 进行交换,产生络合着 ATP 的没有 EPR 信号的氧化态 Fe 蛋白。如果还原剂($S_2O_4^{2-}$)也很充分,则氧化态的 Fe 蛋白会被继续还原成有 EPR 信号的还原态,使得深度还原态的 MoFe 蛋白和还原态 Fe 蛋白缔合为$[1o]$$[2s]^{2t}$。对于稀酶溶液,图 1(7)是整个过程的速率决定步骤,该机理很好地说明了 Mortenson 等[2]在稀酶溶液中所观察到的实验结果。由于 ATP 和 Fe 蛋白络合在 MoFe 蛋白上,使得 MoFe 蛋白的构型发生变化,可能沟通了 MoFe 蛋白中 P-簇合物和 M-簇合物之间的电子传递路径,促使电子传递给络合在 M-簇合物上的底物。如果还原剂($S_2O_4^{2-}$)并不充分,则深度还原态的 MoFe 蛋白和氧化态 Fe 蛋白也会缔合为$[1o][2o]^{2+}$,同样会促进电子传递给底物。这两个子过程,如图 1 的(5)→(6)→(7)→(8)或(5)→(10)→(15)→(16)并不一定要求 ATP 发生水解,ATP 也可能是直接络合在 MoFe 蛋白上[10]。对于 ATP 不充足的情况,可能会使得深度还原态的 MoFe 蛋白和络合着 ATP 的氧化态 Fe 蛋白之间发生电子倒流并发生 ATP 与 ADP 的交换,见图 1(11)和(12),产生$[1s][2s]^{2+}$和$[1s][2s]^{2d}$。随着电子传递的循环往复进行和质子连绵不断的提供,使底物得到不同程度的还原。

## 参考文献

[1]Hardy R W F,D'Eustachio A J. Biochem Biophys Res Commun,1964;**15**:314.

[2]Walker M,Mortenson L E. Biochem Biophys Res Commun,1973;**54**:669.

[3]Smith B E,Lowe D J,Bray R C. Biochem J,1973;**135**:331.

[4]Cai Q R,Zhang H B,Lin G D,Advance in Science of China,1987;**2**:125.

[5]Rees D C,Kim J,Georgiadis M M et al. In:Stiefel E I eds. Molybdenum enzymes,cofactors,and model systems. Washington DC:American Chemical Society,1993:170.

[6]Burris R H. In:Gioson A H eds. Current perspectives in nitrogen fixation,Australia:Australian Academy of Science,1981:126.

[7]Wolle D,Dean D R,Howard J B. Science,1992;**258**:992.

[8]Thorneley R N F. In:Gresshoff P M eds. Nitrogen fixation:achievements and objectives. New York:Chapman and Hall,1990:103.

[9]Burgess B K. In:Stiefel E I eds. Molybdenum enzymes,cofactors,and model systems. Washington DC:American Chemical Society,1993:144.

[10]Gavini N,Burgess B K. J Biol Chem,1992;**267**:21179.

[11]Orme-Johnson W H. Science,1992;**257**:1639.

### The Electron Transport in Nitrogenase.

Jing-Wei Huang, Hong-Tu Zhang, Hui-Lin Wan, Qi-Rui Cai(Kui-Rui Tsai)

(Department of Chemistry, State Key Laboratory for Physical Chemistry of the Solid Surface, Xiamen University, Xiamen 361005, China).

**Abstract**    A modified mechanism of 2-step-ATP-driven electron transport is proposed, it can reasonably explain the change of EPR signal in the two components of nitrogenase when the supply of reductant($S_2O_4^{2-}$)and MgATP are enough or only one is enough.

**Key words**    nitrogenase    ATP-driven    EPR signal    electrontransport

■ **本文原载**:《厦门大学学报》(自然科学版)第 35 卷第 6 期(1996 年 11 月),第 890～899 页。

# 化学探针方法研究固氮酶 M-簇和 P-簇对的结构与功能关系[*]

万惠霖[1] [①]　黄静伟[1]　张凤章[2]　周朝晖[1]　张鸿图[1]　许良树[2]　蔡启瑞[1][①]

(¹ 化学系,固体表面物理化学国家重点实验室　² 生物学系)

**摘　要**　根据配位催化原理和化学探针思路,推断了野生菌固氮酶在酶促固氮反应中[Mo]位直接参与结合分子氮(N≡N);论证了固氮酶钼铁蛋白的 M-簇(Kim-Rees 模型)必须是活口的钼-铁-硫原子簇笼,对底物和抑制剂有分子识别能力,只有 N≡N 才能作为底物络合在[Mo-3Fe,3Fe]七核活性中心,而且必须有一条质子(和电子)接力传递链以达[Mo]位,N≡N 才能排去氢基配体而进到[Mo]位,(否则就只能象 HC≡CH 那样结合在[6Fe]位);此外还需要另一条质子接力传递链进到[Fe2]位才能使 N≡N 从外端 N 逐步还原加氢。其他十多种底物分子按形状和大小,只能在笼内[6Fe]位,或笼口[2Fe]位,或一部份在笼内、一部份在笼外还原加氢,这些底物只需要两条质子接力传递链中有一条不失效就行。阐明了 CO 不是底物而是所有外源底物酶促还原反应的抑制剂的原因。讨论了两条质子接力传递链及其支架基团的本质和进一步验证的方法。设计了一种基于其他底物或抑制剂对 HC≡CH 酶促还原加氘的竞争抑制来检验它们是在笼内或笼外结合的化学探针方法,并成功地用于验证高柠檬酸盐固氮酶 N≡N 确是在笼内强烈地抑制乙炔还原加氘(氢)的。

**关键词**　固氮酶活性位模型　活口 M-簇笼　分子识别底物　固氮与放氢机理　质子传递链分子探针

**中国图书分类号**　Q 502,O 643.31

固氮酶两组分——钼铁蛋白和铁蛋白 X-光晶体结构测定的成功[1~5],以及 M-簇(即钼铁蛋白内的铁钼辅因子,被认为是十多种底物的活性位所在之处)的 Kim-Rees 模型[1]和 P-簇对的 Chan-Kim-Rees 模型的提出,是继 60 年代初 Carnahan 和 Mortenson 等[6]用固氮酶萃取液加 MgATP 和 Na₂S₂O₄ 在细胞体外固氮成氨的成功和 1977 年 Shah 和 Brill[7]从钼铁蛋白提取铁钼辅因子的成功之后,固氮酶研究的又一次重大突破。

K-R 模型表示 M-簇笼是由两个开口类立方烷原子簇用三个螯型配体对口连接而成,其中二个螯型配体是 μ-S;另一个螯型配体"Y"可能也是 μ-S,或 μ-SH。M-簇分子式可表为 CysS$_\gamma$-FeS$_3$Fe$_3$($\mu$-S)$_2$($\mu$-"Y")Fe$_3$S$_3$Mo(His$^{\alpha442}$)($\mu$-homocitrate)。但不能排除一小部分 M-族的"Y"是 $\mu$-$\underline{O}$ 或 $\mu$-$\underline{N}$ 的可能性。鉴于固氮酶开始反应前有"氢爆发"现象,即每个 M-簇先放出一个 H₂(H₂ 与[Mo]之比约为 1:1)[8],然后才开始酶促还原反应,不能排除 M-簇笼中心有一个 $\mu_6$-$\underline{H}$(或每个[3Fe]口上有个 $\mu_3$-H,化学符号底下加划表示络合着的物种原子),即三桥连的双活口类立方烷的可能性。中国工作者[9,10]于 1973 年从分子

---

[*]　本文 1996 年 10 月 8 日收到;国家科委攀登计划共生固氮课题和国家自然科学基金资助项目。

[①]　通讯联系人。

轨道化学键理论的观点讨论了过渡金属分子氮配合物中 N≡N 的单核和多核配位络合模式,认为多核络合有利于固氮。接着又从稍为不同的角度,提出活口的和开口的两个类立方烷原子簇结构的活性中心模型(X-1 和 F-1 模型)[11,12],以及 N≡N、HC≡CH 等底物的多核络合活化模式;1978 年这两个模型又同时发展为大同小异的骈联双座双网兜型结构模型(F-2)[13]和骈联活口双立方烷结构模型(X-2)[14]。现在看来这些模型的思路和主要论据是有可取之处的;但由于 K-R 模型的簇骼结构类型在簇合物化学是前所未见的,F-1 和 X-2 模型没有考虑到两个开口或活口类立方烷可用三个硫桥对口连接,[Mo]在 M-簇笼底,而不是以[Mo]为核心骈联的。1978 以来到 K-R 模型发表前,国外工作者也提出了多种原子结构的模型,其中只有 Coucouvanis 等[15]提出的类硫镍铁矿石结构的模型是把[Mo]放在 MoFe$_7$S$_6$ 簇笼的底部的(但这是个实体的原子簇笼,难以看出它有多核络合底物分子的能力),其他模型都把[Mo]放在笼的中部。以上这些模型(包括 F-2、X-2 模型)都未能显示出固氮酶对不同底物有分子识别能力,因而未能完满地说明固氮酶十多种底物在活性位结合力($K_b \approx K_M^{-1}$,$(mmole/L)^{-1}$ 为单位)的大小次序和酶促还原产物选择性的某些特征[16]。

N≡N($K_b \approx 8-17$)≥(CH$_2$)(CH≡CH),(CH$_2$)(N≡N)($K_b \approx 10$)>HC≡CH($K_b \approx 2.5-7.2$;在 D$_2$O 中还原时,顺/反-d$_2$-烯比值约 99.5/0.5,无乙烷)≥CH$_3$N≡C($K_b \approx 1.0-5.0$,副产少量 C$_2$H$_4$ 等)≥C≡N$^-$,(HC≡N)($K_b \approx 1.0-2.5$,副产少量 C$_2$H$_4$ 等)≥N$_2$O,N$_3^-$,HN$_3$($K_b \approx 0.9-1.0$)≫CH$_3$C≡CH($K_b \approx 0.033$;在 D$_2$O 中还原时顺/反-d$_2$-烯比约 64/38)≫CH$_3$C≡N,C$_2$H$_5$C≡N($K_b \approx 0.002$)。

特别是未能说明为什么 N≡N 和环丙烯比 HC≡CH 和 CH$_3$N≡C 等底物络合得更牢?为什么 HC≡CH 结合力比 CH$_3$C≡CH 约大 200 倍,还原加氢顺式 d$_2$-烯选择性也高得多?

固氮酶两组分的晶体结构和 K-R 模型发表以来,固氮酶结构与功能关系成为国际上固氮酶研究的热点。已有多种 N≡N 配位模式被提出。如 Rees 等[1,2]提出 N≡N 络合在笼中部两个[3Fe]层之间的双端基六核配位 $\mu_6$($\eta^2$,$\varepsilon_3$,$\varepsilon_3'$)模式($\varepsilon_3$,$\varepsilon_3'$ 表示指示基团两端各与 3 核络合,侧基络合从总核数 $\mu_6$ 与端基络合总核数之差数看出,不另标明;以下配位模式标志仿此)有的提出 N≡N 在笼口[2Fe]位的双侧基[17]或双端基[18]二核配位,$\mu$-($\eta^2$)或 $\mu$-($\eta^2$,$\varepsilon_1$,$\varepsilon_1'$)模式[19],或络合在笼外[4Fe]面上的双端基四核配位 $\mu_4$($\eta^2$,$\varepsilon_2$,$\varepsilon_2'$)模式;或垂直[4Fe]面上在笼外以单端基四核配位络合的 $\mu_4$($\eta^1$,$\varepsilon_4$)模式[19],或以一端从[4Fe]面部分插入到笼内的络合模式[20];或 N≡N 在笼内以单端基络合在[Mo]、三侧基络合在[3Fe]层的 $\mu_1$($\eta^2$,$\varepsilon_1$)4 核配位模式[21];或 N≡N 在活口 M-簇笼内[3Fe]层下与[Mo-3Fe]络合的 $\mu_4$($\eta^2$,$\varepsilon_1$)配位模式[22],或在两个[3Fe]层之间、但偏近[Mo-3Fe]位的双端基七核配位 $\mu_7$($\eta^2$,$\varepsilon_4$,$\varepsilon_3'$)模式[23]。1992 年过后,晶体结构工作者比较一致地认为 M-簇笼中心无 S̲,但"Y"很可能也是 $\mu$-S̲[2]。看来这使许多工作者认为"Y"象其他两个 $\mu$-S̲ 一样在酶促反应中也是不能被移开的。按配位催化原理和化学探针思路,必须同时考虑 N≡N 及其他多种底物的络合活化模式;这就不难推断[Mo]位很可能是直接参与结合 N≡N 的,而且"Y"必须是 M-簇笼的活口[22]。本文简述化学探针方法研究固氮酶结构与功能关系的一些新进展。

# 1　化学探针方法进一步探讨固氮酶各种底物的络合活化模式

## 1.1　[Mo]直接参与结合 N≡N,M-簇笼是有分子识别能力的活口原子簇笼

固氮酶催化作用是一类含有电子与能量偶联传递的配位催化作用[14]。自然界许多不同来源的固氮微生物的固氮酶都含相同的铁钼辅因子,其固氮活性和产物选择性比后来发现的钒固氮酶和铁固氮酶都高得多。鉴于氮在金属钼单晶表面的原子吸附热(约 665 kJ[24])比在金属铁单晶表面的原子吸附热高 79~84 kJ,可以推测,Mo-N 单键键能也会明显地大于 Fe-N 单键键能,钼铁硫原子簇中的[Mo]与 N≡N 的配位键能也会高于[Fe]与 N≡N 的配位键能。在固氮酶的固氮反应中,[Mo]-位如直接参与多核络合

N≡N，必然有利于加强 N≡N 在活性中心的结合力，尤其有利于降低高位能中间态[Mo]-NNH，Mo-NNH$_2$ 的位能，从而加强酶促固氮反应对质子还原放氢反应竞争电子的能力。这很可能是钼固氮酶比钒固氮酶，尤其是比铁固氮酶，固氮效率高得多、放氢相对少得多[25]的主要原因。自然界选择钼为野生菌固氮酶中铁钼辅因子的金属组分之一，应该是让它直接参与结合 N$_2$ 的[22]。从 K-R 模型难以看出[Mo]在 M-簇笼外有结合 N≡N 的适当配位，更不用说含[Mo]的多核配位，但可以看出 N≡N 如能进入 M-簇笼内，它就有可能络合在[Mo-3Fe]、或[Mo-3Fe，3Fe]多核活性中心。这样的活性中心显然是 HC≡CH 和 CH$_3$N≡C 所不能达到的。这就容易理解为什么 N≡N 的结合力（K$_b$≈K$_M^{-1}$）大于 HC≡CH 和 CH$_3$N≡C。但如果 M-簇笼不是活口的，N≡N 就难以进入笼内；即使能进笼，还原产物 NH$_3$ 也难以出笼。因此"Y"必须是个活口。更直观的理由是，如果"Y"不是个活口，那么，从 K-R 模型不难看出，环丙烯，环偶氮卡宾、丙二烯和丙炔等底物分子在笼外就无任何配位立足点。如图 1a 所示，每个含[4Fe]面的 4Fe-4S 八员环的每个[Fe]位都是夹在两个高高隆起的 μ$_3$-S 和 μ-S 峰的，因此是这些底物分子都达不到的；这不符合环丙烯和环偶氮卡宾是结合力很强的底物和固氮反应的抑制剂[26]等已知的实验事实；仅这一点就足以推断"Y"必须是笼的活口。

关于活口"Y"的设想，不久前我们从 Seefeld 等[27]的工作获得了直接和完满的实验支持。Seefeld 等观察到 COS 是固氮酶的中等活性的底物，酶促还原时产生 CO 和 H$_2$S，而 CS$_2$ 是抑制剂。显然，COS 和 CS$_2$ 的 S 在酶活性位都有亲合力，COS 在活性位络合时析出 CO，S 还原为 H$_2$S；而 CS$_2$ 在活性位不能析出 CS，只起堵塞作用，因此成为抑制剂。从图 1a 不难看出，如果"Y"也是不能移开的，COS 就不能把庞大的 S 放置在任何一个[Fe]位上。唯一可以放置 S 的活性位是"Y"被移开后留下来的[2Fe]位。这正好说明在[2Fe]位上的 μ-S 在酶促反应中质子和电子流的作用下，是可以转化为 H$_2$S 而被移开的。"Y"所在的[2Fe]位是底物和反应产物的进出口，必然有足够宽敞的微环境使 μ-S 转化成 H$_2$S 而被置换；而其他二个 μ-S 则可能处在较紧束的微环境而不易还原加氢。

### 1.2　环丙烯、乙炔、丙炔等底物的配位模式和还原加氢反应

[2Fe]笼口有一定的张翕能力[22]，环丙烯、乙炔等底物分子可以平躺地部分进到笼内以双端基加双侧基 μ$_6$($\eta^2$，$\varepsilon_2$，$\varepsilon_2'$)模式配位络合在二个[3Fe]层的[6Fe]位[22]，而带"尾巴"($n$-R、CH$_2$＝CH—，或＝CH$_2$ 等)的一些底物分子，如 CH$_3$C≡CH，只能络合在笼口[2Fe]位(很可能按相似的取向和双端基 μ($\eta^2$，$\varepsilon_1$，$\varepsilon_1'$)配位模式[16,22])。所以乙炔在 M-簇的结合力比丙炔高两个数量级。络合在[6Fe]位的 HC≡CH 在氘水中酶促还原加氘时，由于夹在层间距仅约 0.3 nm 左右的二个[3Fe]层，中间体 HC＝CHD 的 CHD 不能按任何机理绕着碳碳键旋转，所以第二个 D 只能与第一个 D 加在同一边，即应该是 100%顺式加氘的；也不能进一步加氘成烷基，所以乙炔在[6Fe]位还原不会出乙烷。但乙炔总有一小部分(1%～2%，设为 1.5%)会象丙炔那样在活口[2Fe]位络合(Fig. 1b)和还原加氘，顺式加氘选择性可能略低于丙炔(后者顺反比约 64/36)[28]，但仍略高于 50/50；设其顺反比为 60/40。由此可估算出乙炔在氩气氛下和氘水中酶促加氘时，总的顺式加氘选择性应约为顺/反比≈99.4∶0.6 与 Hardy 等[29]观测到的顺、反 d$_2$—HDC＝CDH 的红外光谱峰(843 cm$^{-1}$、988 cm$^{-1}$)面积之比相近。

### 1.3　N≡N 的配位络合模式和酶促固氮及放氢反应机理

野生菌固氮酶(高柠檬酸盐固氮酶)酶促还原 N$_2$ 和乙炔的活性都很高，而且放氢反应不受 CO 所抑制；而 $nifV^-$(Kp 突变种，柠檬酸盐固氮酶)酶促还原 N$_2$ 的活性很低，而且放氢反应受 CO 抑制。但是放氢和还原乙炔的活性仍相当高；细胞体外生物合成的高柠檬酸盐固氮酶和柠檬酸盐固氮酶体系也是这样，酶促还原 N$_2$ 的相对活性，后者仅约为前者的 7%[31]。结合着高柠檬酸盐同系物或衍生物的多种固氮酶体系也有相似的规律[30,31]，前文[22]曾指出，在这些体系中，凡所结合的羟基多竣酸各有一个 γ-端羧酸根是指向[Mo]-ImH$^{α442}$的，它们的放氢活性都较高，而且对于 CO 抑制都不敏感。由此可以推断这些信

息小分子的 γ-端羧酸根都能参与支架一条从 P-簇对到[Mo]位的质子传递链；而柠檬酸盐螯型配体指向 ImH$^{\alpha 442}$ 的那一个 β-端羧酸根由于较短一个 $CH_2$ 链节，不能支架传递到 ImH$^{\alpha 442}$ 和[Mo]位的质子传递链。

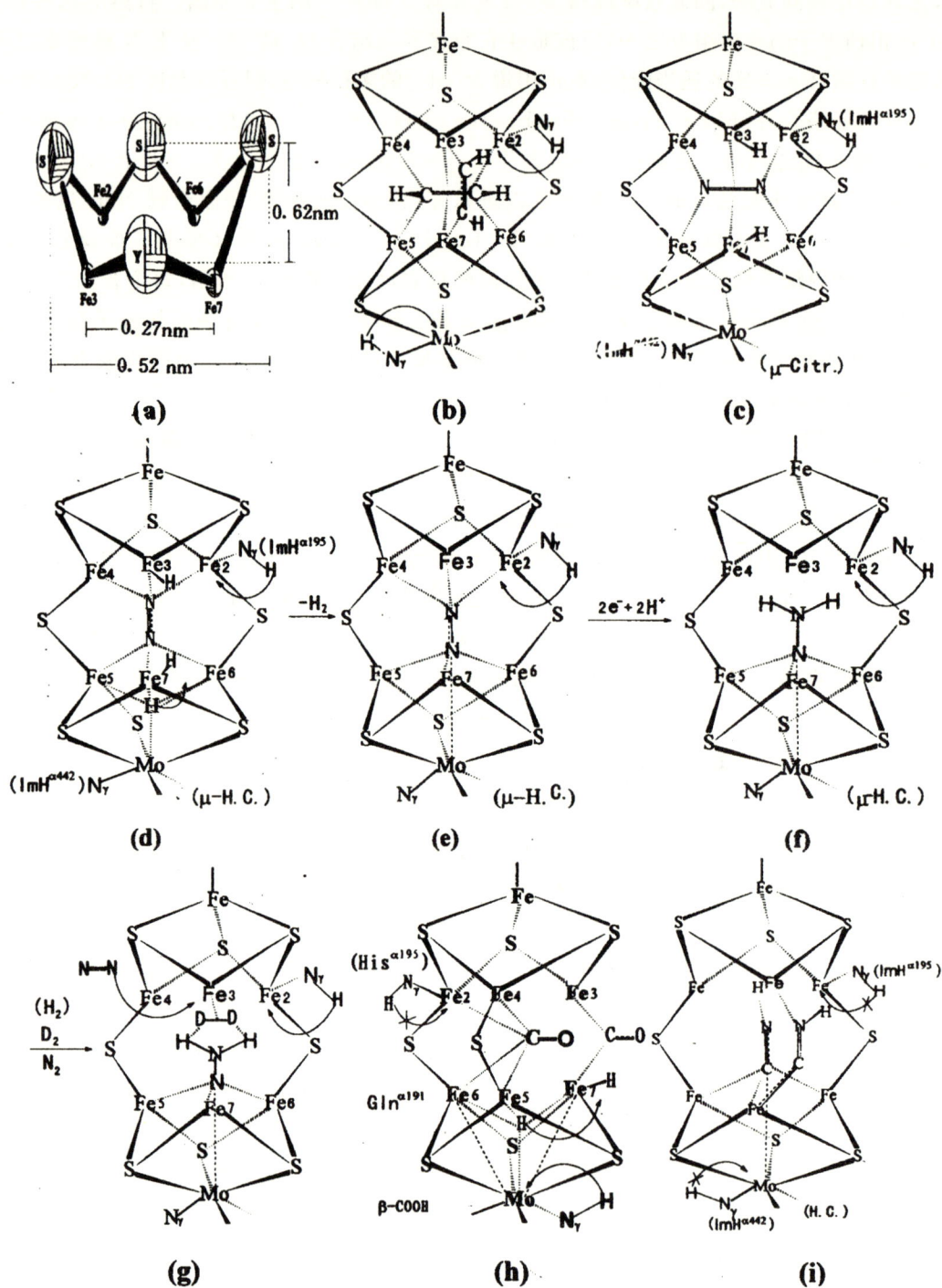

图 1　(a)固氮酶 M-簇的部分结构模式

(b)～(i)固氮酶对各种底物或抑制剂的络合及其作用机理

Fig. 1　(a)The partial structure model of nitrogenase M-cluster

(b)～(i)The coordination models of substrates or inhibitors in nitrogenase M-cluster and their mechanism.

因此高柠檬酸盐固氮酶能从[Mo-3Fe]位放氢,而柠檬酸盐固氮酶则不能。当电子和质子几乎同步传递到[Mo]位时,在[Mo]位生成 H 配体,使[Mo]位从本来的八面体六配位结构变成八面体单盖帽七配位结构;这很可能会引起[3S]层和[3Fe]层稍下沉约 0.03nm 左右,(但 Mo-S,Fe-S 核间距基本维持不变,而 Fe-Fe 和 S-S 等核间距则可稍变宽,Mo-Fe 核间距可稍变短)。这就使[3Fe]与[3Fe]的层间距扩大到 0.3 nm 左右,使 N≡N 能从靠近[2Fe]活口处竖着进入笼内[3Fe]层与[3Fe]层之间。[Mo]位上的 H 配体只要没有其他物种阻碍,应能溢流到笼内、外的一些[Fe]位;如笼口 [2Fe]的 2 个配位比较不饱和的[Fe7]和[Fe3]位。这样,竖着进到笼内的 N≡N 就能以一个 N 端排去[Mo-3Fe]位的 2 个 H 而析出 $H_2$,确切一点就是排去[Mo-Fe7]位上的 2 个 H,而放氢(图 1d),这个内 N 端是在[3Fe]层上络合着,而不是像以前[22]设想的那样内 N 在[3Fe]层下络合着的(因为后者 N-S 核间距太短),并同时与[Mo]位络合着(Mo-N 核间距约 0.22 nm,与 Mo-NγIm 相近),而外 N 端又与上面[3Fe]层络合着(Fe-N 核间距约 0.22 nm);这样,N≡N 就形成偏近[Mo-3Fe]一边的双端基七核配位 $\mu_7(\eta^2,\varepsilon_4,\varepsilon_3')$ 配位模式(图 1d,e),结合力必然大于结合在[6Fe]位的 $\mu_6-HC\equiv CH$。N≡N 再经两步还原加氢成为 $N-NH_2$(图 1f)。进一部还原加氢可能与 N-N 键的断裂同时发生,生成结合在[Mo-3Fe]位上的 NH(估计[Mo]位最多只能承担 NH 的一价,[3Fe]共承担 NH 的另一价),和结合在笼口 [2Fe]位的 $NH_2$。这样键能较有利。$NNH_2$ 虽有[Mo]垫底协助降低位能,但必然还很活泼,恰好能与疏松地络合在[Fe3]位的 D-D(或 H-H,Kubas 配合物[32])反应,产生 2HD(或 $2H_2$),但 $N_2$ 分压较高时 $N_2$ 会竞争 D-D 的配位(图 1g)。李季伦等[33]曾用实验证明,少量的 $N_2$ 对这副反应有促进作用,而在 $N_2$ 分压较高时又反而受到 $N_2$ 的抑制。我们[22]曾经指出,N≡N 在活口 M-簇笼内的络合活化和还原加氘恰好能说明这个 $N_2$ 酶促还原加氘所专有的副反应的机理。

柠檬酸盐固酶氮的 M-簇[Mo]位 ImH$^{α442}$附近无 γ-端羧酸根以协助支架质子传递链,所以[Mo-3Fe]位不能放氢。N≡N 显然不能从笼外使[Mo]改变配位结构,即不能从笼外使[3Fe]、[3S]层稍下沉,以使[3Fe]与[3Fe]层间距稍变宽,因此 N≡N 只能象 HC≡CH 那样 平躺地从笼口进到 M-簇笼内的[6Fe]位络合着,形成 $\mu_6(\eta^2,\varepsilon_2,\varepsilon_2')$ 配位模式(图 1b)。这样横躺的 N≡N 的 N 端距离[Mo]位仍然太远,在 0.36 nm 以上)。因此在柠檬酸盐固氮酶 N≡N 结合力比 HC≡CH 低得多,N≡N 酶促还原活性也低得多;体外生物合成的柠檬酸盐固氮酶的 $N_2$ 酶促还原相对活性仅约为高柠檬酸盐固酶的 7%[24]。但是柠檬酸盐固氮酶的放氢活性和乙炔还原活性却还不低[23,24],可见从 P-簇对到 M 簇还有其他质子传递链;这很可能是从 ImH$^{α195}$通过隧道效应进到笼内[Fe2]位的[22]。

高柠檬酸盐固氮酶 M-簇的[Fe2]位同样也有 ImH$^{α195}$。用基因点突变方法把 ImH$^{α195}$换成其他氨基酸残基,固氮活性就全失,而酶促放氢活性和还原乙炔活性还不低,但受到 $N_2$ 的强烈抑制[34];可见 $N_2$ 仍能有力地结合在笼内,但不能被还原;ImH$^{α195}$对于 $N_2$ 的酶促还原是绝对必需的(在此顺便指出,放氢反应受到 $N_2$ 的强烈抑制意味着这反应基本不是在 P-簇对上进行的)。这可根据 N≡N 的七核 $\mu_7(\eta^2,\varepsilon_4,\varepsilon_3')$ 络合活化模式来解释:N≡N 以其一端排去 2 H 后直接结合在[Mo]位及[3Fe]位,因此质子不能再从 ImH$^{α442}$传递到笼内[Mo]位;这样,N≡N 只能先从外端 N 还原加氢,而这时所需的质子源只有通过 ImH$^{α195}$传递到笼内[Fe2]位来获得。电子也可从这里与质子偶联传递进来,但不能排除相当一部分的电子流仍可通过 ImH$^{α442}$传递到笼内[Mo]位的可能性,电子可在笼内离域,也可从[Mo]位传梯给 N≡N 及 N=NH,同时从[Fe2]位接受质子。这或者更有利于酶促固氮反应对放氢反应竞争电子(如果 P-簇对上有放氢位,也会不利于 $N_2$ 竞争电子和质子)。

### 1.4　CO 及 CN⁻ 的络合模式

上面的一些事实表明 CO 不抑制固氮酶在[Mo-3Fe]位上的放氢反应,这意味着 CO 并不直接结合在[Mo]位。但 CO 能抑制柠檬酸盐固氮酶的酶促放氢反应,这意味着 CO 的结合位含[Fe2]。已知 CO 倾向于以 C 端配位络合,而不倾向于以侧基或 O-端络合。CO 与[Mo]的结合力大概和 CO 与[Fe]的结合力差不多。因此可以设想,CO 很可能是从笼口附近插进笼内、以 C-端络合在含[Fe2]的一个[4Fe]位的。如空间足够,CO 也能以 C 端络在笼口[2Fe]位。所以 CO 是所有外源底物的有效抑制剂。CO 不抑制[Mo-Fe7]位上的放氢,意味着由于微环境的约束 CO 不能在[Fe7]位成线状配体;$\mu_4$-CO 和 $\mu$-CO 本来就不会抑制[Mo-Fe7]位的放氢反应(图 1h)。$\mu_4$-CO 不能在笼内还原加氢成非线状的 HC=O,$\mu$-CO 也不易在笼口还原加氢成 HC=O,因为无适当助催剂(阳离子)作用于 HC=O 的 O-端以降低位能和活化能。所以 CO 不是固氮酶的底物。

C≡N$^-$、CH$_3$N≡C 和 CH$_3$CH$_2$N≡C 的酶促还原加氢都产生少量 C$_2$H$_4$ 等副反应产物,这种副反应显然不能在笼内进行;可见底物 C≡N$^-$ 主要是按 n-RN≡C(以及 n-RC≡CH,n-RC≡N)在笼口[2Fe]位的双端基配位模式还原加氢的[22]。HCN 结合力比 C≡N$^-$ 约低 3 个数量级,它主要是 C≡N$^-$ 的后备来源,而不是主要的底物物种[16,22],C≡N$^-$ 也能按双端基模式络合[15],C≡N$^-$ 有一定的机会进入笼内,象 N≡N 那样按双端基模式络合在[Mo-3Fe,3Fe]位,并质子化成为 C≡NH (图 1i)。但不能在笼内进一步还原加氢成非线状的 HC=NH;也难以进一步还原加氢成 C-NH$_2$,因为这使 C 端过于不饱和。C≡NH 在笼内会阻碍电子和质子偶联传递到[Mo]位及[Fe2]位,因而会使总电子流减少;所以会对 C≡N$^-$ 在笼口的还原加氢产生一定的自抑制作用[25]。少量 CO 的加入会排去笼内外的 C≡N$^-$(HNC,HCN),使放氢反应和总电子流恢复正常水平。N≡NO 和 N≡N—N$^-$ 都能以其一端伸人笼内,脱去 N$_2$,并有可能把笼内的 C≡NH 排出到到笼口[2Fe]位而起激活作用;O 或 NH 在笼内能被还原而脱去。可见 C≡N$^-$ 的自抑制作用是可逆的,不大可能发生在 P-簇对。

### 1.5 酶促反应中从 P-簇对到 M-簇的两条质子传递链的性质

这两条质子传递链每一条可能都是某些路易斯碱基团支架的,由若干个 H$_2$O 分子或 H$_3$O$^+$ 以氢键连接这些基团而串联构成的,而且可能是按氢键同步移位方式实现质子接力传递的[22]。与 ImH$^{\alpha195}$ 相邻的一个路易斯减支架基团可能是 $\mu$-(R)-高柠檬酸盐的 $\beta$-羧酸根,而不是 Gln$^{\alpha191}$,因为 Gln$^{\alpha191}$ 的谷氨酰胺残基变换成赖氨酸残基的 $\alpha$Q$^{191}$K 突变种固氮酶还有一定的酶促还原乙炔活性和放氢活性,而且后者对 CO 的抑制敏感[24]。从钼铁蛋白晶体结构的立体透视图[1,2]看,$\mu$-(R)-高柠檬酸盐的 $\beta$-羧酸根比较靠近 P-簇对的 S$\gamma$Cy$^{\alpha62}$;在活性酶分子中可能更靠近;ImH$^{\alpha195}$ 也被认为很可能移近[Fe2]而络合着;因此 ImH$^{\alpha195}$ 与 $\beta$-羧酸根之间、以及后者与 S$\gamma$Cy$^{\alpha62}$ 之间,有可能各以低能垒的氢健结合着一个 H$_2$O,以搭成一条氢键同步移位的质子接力传递链,最后通过隧道效应由 ImH$^{\alpha195}$ 进到笼内[Fe2]位。而 $\mu$-(R)-高柠檬酸盐的 $\gamma$-羧酸根并不像前文[22]所设想的那样靠近 S$\gamma$Cy$^{\alpha62}$;从 ImH$^{\alpha442}$ 到 $\gamma$-羧酸根,再到 P-族对的最靠近的一个保留的 S$\gamma$Cys,中间大概还需要一个支架基团,G(G 可能也是一个保留的氨基酸的竣酸根残基),中间可能以低能垒氢健支架着 3 个 H$_2$O(不能徘除 G 也是以氢连接在 S$\gamma$Cy$^{\alpha62}$ 与 $\beta$-羧酸根之间的 H$_2$O 的可能性)。这两条质子传递链和 G 的性质应该可以通过基因点突变和比较精确的晶体结构参数来定出。这是有深远意义的。

## 2 一个检验底物在笼内或笼外结合强度的化学探针实验方法
### ——底物对乙炔酶促还原活性和顺式加氘选择性的影响

根据上面推断,在野生菌活性固氮酶分子中 HC≡CH 在 M-簇笼内、笼外有着结合力大小悬殊、酶促还原顺式加氘选择性也悬殊的两个活性位,活口 M-簇笼有分子识别能力,只有 N≡N 能作为底物结合在笼内含[Mo]和多[Fe]的活性位,因此能对乙块在笼内活性位的结合产生强烈抑制,而基本不影响乙炔在笼外活性位的结合。

可以设想,如 HC≡CH 在氘水中还原加氘时有相当浓度的 $N_2$ 同时存在,则 HC≡CH 在笼内[6Fe]-位的还原加氘必定会受到强烈的竞争性抑制,而在笼口[2Fe]-位的还原加氘则基本不受 $N_2$ 影响,因此,还原产物 $C_2H_2D_2$ 中反式 DHC≡CDH 的百分比应该会明显增加,可用 FTIR 或 $^1$H-NMR 谱检测出来,从而可验证 N≡N 确是在笼内[Mo-3Fe,3Fe]-位对 HC≡CH 进行竞争抑制的,我们的实验结果(表1)证实了这一点。对于野生菌固氮酶,从上面所列 $K_b$ 值的大小看来,在常温常压下 N≡N 在水中的溶解度比乙炔约小 60~70 倍,但在笼内[Mo-3Fe,3Fe]-位的结合力约为乙炔在[6Fe]-位结合力的 3 倍左右(假如 N≡N 也是结合在[6Fe]-位的,则其结合力可能比乙炔小得多)。同理,可用其他底物对 HC≡CH 在氘水中还原加氘进行抑制试验,考察其对于乙炔顺式加氘选择性的影响程度,就可以协助推断该底物的结合位是在笼内,还是在笼口,还是两种位置都能结合。我们曾用此法设计实验考察了 $CN^-$、$N_3^-$、$NH_2NH_2$ 等底物对乙炔酶(Av)促还原加氢(氘)的抑制,得到了一些有趣的实验结果(将另文发表),并与用 $N_2O$ 的抑制作对比。如改用柠檬酸盐固氮酶($nifV^-$ 固氮酶)(根据周朝晖等[35]的钒、钼柠檬酸等复盐晶体结构的分析结果,我们有理由相信,柠檬酸盐固氮酶中的柠檬酸根配体也是以中间的羟羧基螯合在[Mo]-位的)进行上述试验以作对比,就可协助推断,当没有高柠檬酸根配体的 γ-羧酸根介人支架质子的传递、[Mo]-位上不能放氢时,N≡N 是否能进到[Mo-3Fe,3Fe]-位? 还是只能像乙炔那样进到[6Fe]-位?最近,我们和李季伦教授及其同事进行协作研究,通过 $nifV^-$ 突变种固氮酶(柠檬酸盐固氮酶)与克氏杆菌固氮酶 $Kp$(一种 $nifV^+$ 高柠檬酸盐固氮酶)对乙炔和 $N_2$ 在 $D_2O$ 中以及在 $H_2O$ 中酶促还原的对比实验,证实了柠檬酸盐固氮酶的固氮活力很低,并发现 $N_2$ 对乙炔还原的抑制也很小,这结果将另文发表。

表 1　氮气氛对乙炔酶促还原活性和顺式加氢(氘)选择性的影响

Tab. 1　Inhibition of $C_2H_2$ by $N_2$ reduction in $D_2O$ catalyzed by native nitrogenase, and corresponding effects on trans/cis $d_2$-ethene ratios

| 反应系统 | | | 相对活性 | 反/顺比 |
| --- | --- | --- | --- | --- |
| 气相组成 | 体积比 | 固氮酶 | | 反式(%) |
| $C_2H_2$/Ar | 1:13 | 有 | 100 | 0.7(估算值* 0.60) |
| $C_2H_2$/Ar | 1:108 | 无 | 0 | |
| $C_2H_2$/Ar | 1:108 | 有 | 26 | 1.6 |
| $C_2H_2$/Ar | 1:217 | 无 | 0 | |
| $C_2H_2$/Ar | 1:217 | 有 | 13 | 1.9 |
| $C_2H_2$/$N_2$ | 1:108 | 有 | 19(~81%抑制) | 3.2(估算值▲3.2) |
| $C_2H_2$/$N_2$ | 1:217 | 有 | 11(~89%抑制) | 5.0(估算值▲5.5) |

\* 估算方法见本文 1,2 节。

▲按氮和放氢只影响笼内的乙炔还原,而基本都不影响笼外的乙炔还原估算。

本课题的工作得到国家科委攀登计划项目(共生固氮体系中最佳固氮结瘤控制模型的研究)的经费资助,本工作是"优化固氮模型信息研究"子课题 1993、1994 和 1996 年年度进展报告的主要内容,参加本项目的刘爱民、周朝晖博士先后获得国家自然科学基金的资助,在本项目及其前期工作中我们多次得到卢嘉锡教授、唐敖庆教授、和过兴先教授的启发和鼓励,在此一并致谢。参加本工作的还有杨茹、黄河清、邱雪慧等同志。

## 参考文献

[1]Kim J，Rees D C. Nature，1992，360：553～560；cf. also Biochemistry 1994，**33**：389～397.

[2]Chan M K，Kim J，Rees D C. Science，1993，260：792～794；. cf. also Rees D C. et al. Advati. Inorg. Chem.，1993，**40**：89～119.

[3]Kim J ，Woo D，Rees D C. Biochemistry，1993，**32**：7 104～7 115.

[4]Bolin J T et al. in Molybdenum Enzymes，Cofactors，& Model Systems，（eds. E. I. stiefel etal）Washington D C：Am. Chem. Soc.，1993：186～195.

[5]Georgiadis M M et al. Science，1992，**257**：1 653～1 659.

[6]Carnahan J E，Mortenson L E et al. Biochhn. Biophys. Acta，1960，**38**：188.

[7]Shah V K，Brill W J. Proc. NatL Acad. Sci. USA. ，1977，**74**：3 249.

[8]Liang J H et al. Proc. Natl. Acad. Sci. U S A. ，1988，**85**：9 446～9 450.

[9]Nitrgen Fixation Research Group，Department of Chemistry，Jirin University（Tang，A. Q.），Scientia Sinica，1974，**17**：193，cf. Tang AQ，Xu JQ. in The Nitrogen Fixaticni and Its Research in China（ed. G. F. Hong），Springer Verlag，1992：31～62.

[10]福建物质结构研究所固氮研究组（卢嘉锡）. 化学模拟生物固氮进展（第二辑）. 北京：科学出版社，1974：1～53.

[11]厦门大学固氮研究组（蔡启瑞）. 厦门大学学报（自然科学版），1974，13（1）：111；厦门大学固氮研究组（蔡启瑞）. 化学模拟生物固氮进展（第二辑）. 北京：科学出版社，1976：460～209；Scientia Sinica，1976：460.

[12]福建物质结构研究所固氮研究组（卢嘉锡）. 科学通报，1975，**12**：540.

[13]Lu J X. in Nitrogen Fixation，Vot. I（eds. W. E. Newton & W. H. Orme-Johnson），Baltimore，USA：U-niversity Park Press，1980：343～371. cf. Liu C W et al. in The Nitrogen Fixation and Its Research in China（ed. G-F. Hong），Springer Verlag，1992：193～222. cf. Kang B S et al. ibid，1992：151～174.

[14]Tsai K R. in Nitrogen Fixation，Vol. I，Baltimore，USA：University Park Press，1980：373～387.

[15]Holm R H，Simhon E D. in Molybdenum Enzymes，N. Y. ：Wiley-Interscience，1985：2～87.

[16]Tsai K R et al. in The Nitrogen Fixation and Its Research in China，Springer Verlag，1992：87～117.

[17]Orme-Johnson W H. Science，1992，**257**：1 639.

[18]Shriver D F，Atkins P，Langford C H. Inorganic. Chemistry，2nd Ed. W. H. Freeman & Co.，1994：812.

[19]Orme-Johnson W H. In Molybdenum Enzymes，Cofactors，and Model Systems（eds. E. I. stiefel et al. ），Washington，D. C：Amer. Chem. Soc.，1993：257～270.

[20]Plas W. J. Mol. Struct. （Theochem. ），1994，**315**：53.

[21]吴新涛，卢嘉锡. 科学通报，1995，**40**：577.

[22]Tsai K R，Wan H-L. "New Perspectives on the Structures and Functions of Nitrogenase M-cluster and P-cluster Pair". Presented at the Int. Symp. on Molecular Structure，Fuzhou，China，1993；共生固氮攀登项目子课题"优化固氮模型信息研究"1993 年年度进展报告.

[23]Tsai KR,Wan H L. J. Cluster Sci. ,1995,**6**:485;万惠霖等. 共生固氮攀登项目子课题"优化固氮模型 信息研究"1995 年年度进展报告.

[24]Ertl G. in The Surface Chemical Bonds, North-Holland,1979:Chapter 5.

[25](a)Burgess B K. In Molybdenum Enzymes,Cofactors,and Model Systems(eds. E. I. stiefel et al. ),Wa shington,D. C:Amer. Chem. Soc. ,1993:144～169. (b) Burgess B K. Chem. Rev. , 1990,**90**:1 377. (c)Burgess B K. in Molybdenum Enzytnes, Wilev Interscience, 1985:221～281.

[26]McKenna C E et al. J. Am. Chem. Soc. ,1976,**98**:4 657. McKenna C E. in Nitrogen Fixation, Vol. I(eds. W. E. Newton & W. H. Orme-Johnson),Baltimore, USA:University Park Press, 1980:223～235.

[27]Seefeld L C et al. In Nitrgen Fixation:Fundamentals and Applications(eds. I. A. Tikhonovich et al. ), Kluwer Academic Publ.,1995:162.

[28]McKenna C E,McKenna M C,Huang C W. Proc. Natl. Acad. Sci. USA,1976,**76**:4773.

[29]Hardy R W F. in A Trealise on Dinitrogen Fixation(eds. R. W. H. Hardy et al. ),John Wiley & Sons,Inc.,1979:515～568.

[30]Imperial J et al. Biochemistry,1989,**28**:7796.

[31]Madden MS et al. Proc. Natl. Acad. Sci. USA,1990,**78**:6517.

[32]Kubas G J. Acc. Chem. Res.,1988,**21**:120.

[33]Li J L. Burris R H. Biochemistry,1983,**22**:4472.

[34] Newton W E et al. in Nitrogen Fixation:Fundamentals and Applications ( eds. I. A. Tikhonovich et al. Klawer Acad. Publ.,1995:91～96.

[35]Zhou C H,Wan H L et al. J. Chem,Crysrallogr. ,1995,**25**:807.

## Chemical-Probe Approach in the Structure and Function Relationship of Nitrogenase M-Cluster and P-cluster Pair

Hui-Lin Wan[1*] , Jing-Wei Huang[1] , Feng-Zhang Zhang[2] , Zhao-Hui Zhou[1] ,

Hong-Tu Zhang[1] , Liang-Shu Xu[2] ,K. R. Tsai[1*]

([1] Department of Chemistry & State Key Laboratory for Physical Chemistry of

Solid Surfaces,Xiamen University. * Corresponding authors.

[2] Department of Biology, Xiamen University)

**Abstract**　Based upon the principles of coordination catalysis and chemical approach,it has been inferred that in active,native nitrogenase,[Mo] directly takes part in binding dinitrogen,$N{\equiv}N$,and that the M-cluster of Kim-Rees-Chan Model must be a labile-mouthed cluster cage with molecular recognition in the bindings of substrates and inhibitors;the [Mo]-site of the [Mo-3Fe,3Fe]active center being available only to $N{\equiv}N$ $\mu_7$-coordination and only when there is a proton (and electron) relay pathway from the P-cluster pair to the [Mo]-site;and a second proton relay pathway is required for the reduction of the $\mu_7$-($\eta^2$, $\epsilon_4$, $\epsilon_3'$) coordinatively bonded $N{\equiv}N$; and that other substrates, according to molecular sizes and shapes,are bound at the [6Fe]-site inside the cage,or at the [2Fe]-site of the gape, while CO is most probably bound at the [4Fe]-site inside the cage and at the [2Fe]-site of the gape, thus not inhibiting $H_2$ evolution from the [Mo-Fe7]-site. The nature of the two proton-re lay pathways and their supporting groups is discussed. A chemical-probe method has been designed, based upon inhibition of HC≡CH reductive-deuteration by other substrates or inhibitors and the effects on trans-/cis-$d_2$-ethene ratios, for examining the binding sites of these substrates or inhibitors (whether inside the cage, or outside the cage), and the strength of binding; this method has been successfully used in obtaining strong support for the strong competitive bonding of N≡N vs. HC≡CH inside the cage, but practically no inhibition of HC≡CH by N≡N at the [2Fe]-site of the gape.

**Key words** Nitrogenase active-sites  Chemical-probe  Labiled-mouthed M-cluster cage  Molecular recognition  Nitrogen fixation & hydrogen evolution mechanisms  Proton-relay pathway.

■ 本文原载：Chinese Science Bulletin Vol. 42 No. 2, pp. 172～174, 1997.

# Biphasic Synergy Catalysis and Active Sites of VMgO Catalysts for Oxidative Dehydrogenation of Propane

Zhi-Min Fang, Wei-Zheng Weng, Hui-Lin Wan[①], Qi-Rui Cai(Khi-Rui Tsai)

(Department of Chemistry and State Key Laboratory for Physical Chemistry
of the Solid Surface, Xiamen University, Xiamen 361005, China)

**Acknowledgement**   This work was supported by the National Natural Science Foundation of China
(Grant No. 29392000)

The oxidative dehydrogenation of propane(ODP) to propene is one of the potentially important catalytic processes for the effective utilization of light alkanes. The VMgO catalysts which have better catalytic performances for the reaction have aroused much interest and argument. Kung et al.[1] proposed that the active phase was magnesium orthovanadate($Mg_3V_2O_8$), but Volta et al.[2] suggested that magnesium pyrovanadate ($\alpha$-$Mg_2V_2O_7$) was the active phase; in this phase, $V^{4+}$ ions which are associated to the formation of oxygen vacancies could stably exist, and $Mg_3V_2O_8$ is responsible for the total oxidation due to nonexistence of $V^{4+}$ ions. Based on the fact that the selectivity of $Mg_3V_2O_8$ could be promoted by the coexisting $\alpha$-$Mg_2V_2O_7$, Delmon et al.[3] suggested that the synergy catalysis probably existed between $Mg_3V_2O_8$ and $\alpha$-$Mg_2V_2O_7$ phases. In the present work, three series of VMgO catalysis were prepared by citrate method (C), impregnation method (I) and nitrate method (N), respectively, and the samples were characterized in detail with XRD, IR, LRS and EPR. Based on these results, in conjunction with the results of catalytic performance evaluation, the biphasic synergy catalysis and active sites of VMgO catalysts were investigated.

Three pure magnesium vanadates, i. e. $Mg_3V_2O_8$, $\alpha$-$Mg_2V_2O_7$ and $\beta$-$MgV_2O_6$, were prepared by the citrate method. Under the same propane conversion, the propene selectivities follow the order: $\alpha$-$Mg_2V_2O_7 > Mg_3V_2O_8 > \beta$-$MgV_2O_6$, indicating that $\alpha$-$Mg_2V_2O_7$ is the active phase with high propene selectivity. Under the same reaction conditions, $Mg_3V_2O_8$ (conversion 25.1%, selectivity 47.6%) is more active but less selective than $\alpha$-$Mg_2V_2O_7$ (C 22.8%, S 55.4% ), while $\beta$-$MgV_2O_6$ (C 20.3%, S 49.6%) demonstrates the worst catalytic activity. These results suggest that $Mg_3V_2O_8$ is probably beneficial to the activation of propane and $\alpha$-$Mg_2V_2O_7$ to the selective formation of propene.

The 2－60 VMgO-I catalysts (containing 2%～60% (weight percentage) of $V_2O_5$ supported on heavy MgO) with low and medium vanadia content prepared by the impregnation method present higher propane conversions and propene yields. At 500 ℃, the propene yield of 14.7% over 20 VMgO-I is 40%～50% higher than those reported in refs.[1,2] over 20 VMgO prepared by impregnation method, which, in addition to MgO, contained either $Mg_3V_2O_8$ or $\alpha$-$Mg_2V_2O_7$. The results of XRD, IR, LRS and EPR characterizations show that $Mg_3V_2O_8$ and $\alpha$-$Mg_2V_2O_7$ coexist in 2－60VMgO-I catalysts.

---

① To whom correspondence should be addressed.

Therefore, a compensative synergy catalysis between the two phases may exist.

The 62, 64, 67 VMgO-C catalysts also contained both $Mg_3V_2O_8$ and $\alpha$-$Mg_2V_2O_7$. Their propane conversions and propene selectivities are also higher than those obtained over pure $Mg_3V_2O_8$ and $\alpha$-$Mg_2V_2O_7$. At 500℃, the propene yield of up to 15.5%, which is the highest among the reported VMgO catalysts under the comparable conditions, is obtained over 62 VMgO-C.

Mechanical mixtures(M) of $Mg_3V_2O_8$ and $\alpha$-$Mg_2V_2O_7$ with the mole ratios of 3/1, 1/1 and 1/3(i. e. 62, 64, 67 VMgO-M) have been made. Under the same propane conversion, the propene selectivities of 62, 64, 67 VMgO-M catalysts are higher than those of pure $Mg_3V_2O_8$ and $\alpha$-$Mg_2V_2O_7$. This result provides further experimental evidence for the possible exisience of the synergy catalysis between $Mg_3V_2O_8$ and $\alpha$-$Mg_2V_2O_7$.

The EPR spectra of VMgO-I catalysts with low and medium vanadia content[2%~60% (weight percentage) $V_2O_5$] consist of two sets of octet hyperfine structure in parallel and perpendicular directions, respectively, indicating that $V^{4+}$ ions exist in two coordination environments, i. e. $V^{4+}$ in isolated $VO_4$ tetrahedra on $Mg_3V_2O_8$ and $V^{4+}$ in corner-sharing $VO_4$ tetrahedral on $\alpha$-$Mg_2V_2O_7$. The VMgO-C catalysts with low and medium vanadia content show only one set of octet hfs for the $V^{4+}$ ions in the isolated $VO_4$ tetrahedra on $Mg_3V_2O_8$. The results of catalytic performance evaluation show the parallel relationship between the spin concentrations of $V^{4+}$ and the catalytic activities for VMgO-I and VMgO-C catalysts, implying that a microsystem containing $V^{4+}$ ion is the active sites of VMgO catalysts for the oxidative dehydrogenation of propane. Since the $V^{4+}$ signals over fresh and used VMgO-N catalysts are very weak, the catalytic activities of catalysts are fairly low.

The catalytic performances of 2—60 VMgO-C are slightly lower than those of the corresponding 2—60 VMgO-I. This is attributable to the fact that, over the 2—60 VMgO-C, only $Mg_3V_2O_8$ phase was formed in addition to MgO; therefore the biphasic synergy catalysis does not exist in this system. However, the catalysts still show satisfactory performances since $V^{4+}$ ions can stably exist in $Mg_3V_2O_8$ phase.

Although both $Mg_3V_2O_8$ and $\alpha$-$Mg_2V_2O_7$ were detected in the VMgO-N catalysts with the vanadia content higher than 10%(weight percentage), the catalytic activities of 2—60 VMgO-N are much lower than in the other two systems, which is probably due to the fact that the microenvironment of vanadium ions in VMgO-N catalysts is different from that in VMgO-I and VMgO-C and therefore will be unfavorable to the stabilization of $V^{4+}$ ions.

Based on the above results, the following conclusions can be drawn: (i) VMgO catalysts prepared with different methods demonstrate different catalytic performances for oxidative dehydrogenation of propane to propene; (ii) only the Mg vanadates containing proper amount of $V^{4+}$ ions present better catalytic activities, so that Mg vanadates with proper amount of $V^{4+}$ may be the active phases of the catalysts, among which $\alpha$-$Mg_2V_2O_7$ shows the highest propene selectivity; (iii) biphasic synergy catalysis probably exists between $Mg_3V_2O_8$ and $\alpha$-$Mg_2V_2O_7$ phases.

# References

[1]Chaar, M. A., Patel, D., Kung, H. H., Selective oxidative dehydrogenation of propane over V-Mg-O catalysts, J. Catal. ,1988, **109**:463.

[2]Sam, D. S. H., Soenen, V., Volta, J. C., Oxidative dehydrogenation of propane over V-Mg-O catalysts, J. Catal, 1990, **123**:417.

[3]Gao, X. T., Ruiz, P., Xin, Q. et al. , Effect of coexistence of magnesium vanadate phases in the selective oxidation of propane to propene, J. Catal. ,1994, **148**:56.

■ 本文原载：Journal of Molecular Catalysis A：Chemical 122(1997)，pp. 147～157.

# Characterization and Catalysis of a SiO₂-Supported [Au₆Pt] Cluster [(AuPPh₃)₆Pt(PPh₃)]²⁺/SiO₂

You-Zhu Yuan[a,b], Kiyotaka Asakura[c], Hui-Lin Wan[b],

Khi-Rui Tsai[b], Yasuhiro Iwasawa[a]①

(ᵃ Department of Chemistry, Graduate School of Science,

The University of Tokyo, Hongo, Bunkyo-ku, Tokyo 113, Japan

ᵇ Department of Chemistry and the State Key Laboratory of Physical

Chemistry for Solid Surface, Xiamen University, Xiamen 361005, China

ᶜ Research Center for Spectrochemistry, Faculty of Science,

The University of Tokyo, Hongo, Bunkyo-ku, Tokyo 113, Japan)

Received 18 September 1996; accepted 28 November 1996

**Abstract**  Structures and catalytic properties of a SiO₂-supported [Au₆Pt] cluster, [(AuPPh₃)₆Pt(PPh₃)](NO₃)₂ **(1)**, were characterized by FT-IR, in-situ EXAFS, and TPR, and also by the reactions such as H₂-D₂ equilibration, ethene hydrogenation, and CO oxidation to get insight into the key issues of metal catalysis. The cluster **1** was supported on SiO₂ without fragmentation of the cluster framework at room temperature under Ar atmosphere. The cluster framework of **1**/SiO₂ was stable up to 400 K under vacuum. EXAFS analysis revealed that after heat-treatment of **1**/SiO₂ at 473 K under vacuum the coordination numbers of Pt－Au and Au－Au(Pt) decreased compared to those for the original clusters due to cluster deformation and after treatment of **1**/SiO₂ at 773 K the Au particles and Pt－P# species (P# : phosphine species) were produced. The combination of TPR and EXAFS showed that the reduction at 603 K caused complete cleavage of Pt－Au and Au－P# bonds accompanied with the formation of Au particles, while Pt－P# bonds remained at 773 K. The incipient **1**/SiO₂ showed catalytic activity in H₂－D₂ and ethene hydrogenation at 303 K without change of the cluster framework. The catalysis of **1**/SiO₂ was suggested to be referred to the platinum atom which was embedded in the six gold cluster by the results of CO-adsorption, EXAFS, and pulse reaction analysis. This is contrasted to the fact that both Pt(PPh₃)₄/SiO₂ and [Au₉(PPh₃)₈](NO₃)₃/SiO₂ showed no catalytic activity for these reactions.

**Key words**  Supported bimetallic [Au₆Pt] cluster  H₂－D₂ equilibration  Characterization by FT-IR  In-situ EXAFS and TPR

① Corresponding author. Fax：＋81-3-58006892；e-mail：iwasawa@utsc.s.u-tokyo.ac.jp.

# 1 Introduction

There has been currently significant interests in gold from both fundamental and industrial points of view, because Au is known as a good promoter for transition metal in catalysis and small Au particles on oxides such as $\alpha$-$Fe_2O_3$, $Co_3O_4$ and $TiO_2$ also exhibit high catalytic activity for low-temperature oxidation of CO(for examples of Pt—Au catalysts, see Ref. [1], for examples of supported Au catalysts, see Ref. [2]). Evaporating gold on platinum single-crystal surface shows different activity and selectivity from those of the monometallic Pt catalyst for conversion of $n$-hexene[3,4]. It is believed that alloys are formed between Group VIII metal such as Pt and IB element such as Au to alter the catalytic properties of platinum. Improved selectivity as well as suppressed deactivation are also beneficial aspects of these alloy catalysts. The 'alkali-metal-like' $d^{10}s^1$ electronic configuration of Au atom leads to relatively a tractable electronic structure compared to those for clusters of transition-metals with open d shells, which provides the theoretically interesting object particularly relevant to catalysis of the precious-metal-gold clusters. These studies can provide useful insight into reactivity of heteronuclear gold clusters and also provide a better understanding of metal-metal bonding and of synergistic effects in bimetallic catalysis. Recent advances in metal cluster syntheses may provide many opportunities to probe such modification at a molecular level in both homogeneous and heterogeneous systems[5—12]. Irrespective of large range of heterometallic cluster compounds containing gold synthesized to date, catalysis of only a few samples has been reported[13,14].

Recently, Pignolet et al. reported feasible application of cationic, phosphine-ligand stabilized AuPt clusters to heterogeneous catalysis by supporting them on $SiO_2$ and $Al_2O_3$ from surface organometallic interest[15—17]. They demonstrated that phosphine-ligated cationic clusters can be immobilized intactly on the oxide supports from the results of UV-VIS spectra and MAS $^{31}P$ NMR, and that the immobilization causes significant changes in catalytic activity. We independently reported characterization and performance of the catalyst obtained by supporting a bimetallic cluster$(AuPPh_3)_6Pt(PPh_3)](NO_3)_2$ (1) on $SiO_2$[18]. The EXAFS data revealed that the cluster was stable enough to adsorb essentially intact on $SiO_2$ without destruction of the framework, The catalyst 1/$SiO_2$ is reported to have a catalytic activity (turnover frequency)of 29. 8 s$^{-1}$ for $H_2$—$D_2$ equilibration at 303 K. Here we will present the surface structure characterization and catalytic performance of 1/$SiO_2$ by means of EXAFS, FT-IR, and TPR and kinetic measurements to examine the properties generated by heterogenization $Au_6$Pt cluster with the structure of a Pt atom embedded in $Au_6$ ensemble.

# 2 Experimental

## 2.1 Materials and catalyst preparation

A dark-yellow microcrystalline $[(AuPPh_3)_6Pt(PPh_3)](NO_3)_2$ (1) was synthesized according to the literature[19]. The framework of 1 is shown in Fig. 1. Silica(Aerosil 300) was heated at 673 K for 1 h under vacuum before using as a support. The treated silica was impregnated with a carefully dried ethanol solution of 1, followed by evacuation to remove the solvent. The loading of 1 was controlled to be 0. 5~1. 0 wt% based on Pt for the convenience of EXAFS measurements and CO and $H_2$ adsorption. All procedures were conducted in Ar atmosphere(99. 9999%)to avoid contact with air.

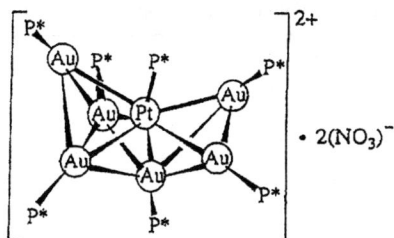

**Fig. 1    Framework of the cluster 1; P* = triphenylphosphine.**

## 2.2   Infrared spectroscopy

A pressed $SiO_2$ disk was placed in an IR cell with two NaCl windows, combined with a closed circulating system. The disk was heated at 673 K for 1 h in the cell, and impregnated dropwise with an ethanol solution of **1** in an Ar(99.9999%) atmosphere without contacting air. The supported cluster samples thus obtained were treated under vacuum at room temperature for 4 h. The FT-IR spectra were taken at 298 K after treatment at different temperatures under vacuum on a JEOL JIR-7000 spectrometer. A $SiO_2$ background spectrum was subtracted from the observed ones of the samples.

## 2.3   Temperature-programmed reduction(TPR)

TPR spectra were recorded in a fixed-bed flow reaction system equipped with a gas chromatograph. A dry-ice/acetone trap was used to eliminate the influences of water and hydrocarbon. The samples were treated at room temperature for 2 h under flowing argon and then switched to a reduction gas of 5% $H_2$/Ar with a flow-rate of 30 ml/min, and heated up to 773 K at a heating rate of 10 K/min for TPR measurements.

## 2.4   Extended X-ray absorption fine structure(EXAFS)spectroscopy

Au $L_3$-edge and Pt $L_3$-edge EXAFS spectra were measured in a transmission mode at room temperature at BL-10B of the Photon Factory in the National Laboratory for High Energy Physics (KEK-PF)(proposal No.:95G200). The measurements were carried out with a beam current of 250 — 350 mA and a storage-ring energy of 2.5 GeV in a transmission mode. The incident and transmitted x-rays were monitored by ionization chambers filled with $N_2$ and Ar(15%)−$N_2$(85%)gases, respectively. The samples were treated in a U-shape glass-tube connected to a closed circulating system, where the samples were treated at 400 K, or 473 K, or 773 K. The samples were also treated in the U-shape glass tube connected to a TPR measurement system, where the samples were reduced at given temperatures under a 5% $H_2$/Ar atmosphere. The treated samples were transferred to EXAFS cells, directly connected to the U-shape glass-tubes, without contacting air.

The analysis of EXAFS was performed by a curve-fitting method, using theoretically derived phase-shift and amplitude functions. The interactions of Pt−Au, Au−Au, Pt−P, Au−P, and Pt−Au−P were calculated by using the FEFF software[20]. Parameters used for FEFF calculations were listed in Table 1. The synthesized cluster **1**, Pt foil, Au foil, Pt(PH₃)₄ and Au(PH₃)₄ were used to determine the amplitude reduction factor S and to check the validity of these theoretically derived parameters. On the present analysis we took into account the error estimation recommended by international XAFS workshop on standards and criteria.

**Table 1  Structural data and Fourier transform ranges used for FEFF calculations**

| Parameters used for FEFF calculations | | | Fourier transform range | |
| --- | --- | --- | --- | --- |
| Shell | $N$ | $r(\text{Å})$ | $\Delta k(\text{Å}^{-1})$ | $\Delta r(\text{Å})$ |
| Pt—P for Pt(PPh$_3$) | 1.0 | 2.28 | 3.0~9.0 | 1.42~2.30 |
| Pt—Au for Pt(Au) | 1.0 | 2.68 | 3.0~9.0 | 2.40~3.20 |
| Pt—Pt for Pt foil | 1.0 | 2.78 | 3.0~9.0 | 2.10~3.15 |
| Au—P for Au(PPh$_3$) | 1.0 | 2.30 | 3.0~12.0 | 1.42~2.30 |
| Au—Au for Au(Au)or Au(Pt) | 1.0 | 2.87 | 3.0~12.0 | 2.40~3.20 |
| Au—Au for Au foil | 1.0 | 2.87 | 3.0~16.0 | 2.10~3.15 |

$N$: coordination number for absorber-backscatterer pair; $r$: interatomic distance; $\Delta k$: range used for Fourier transformation ($k$ is the wave vector); $\Delta r$: range used for inverse Fourier transformation ($r$ is the distance).

## 2.5  H$_2$-D$_2$ equilibration, ethene hydrogenation and CO oxidation

H$_2$/D$_2$ equilibration was tested in a fixed-bed flow reaction system equipped with a mass spectrometer. Ar was used as a diluent gas in a flow rate of 50 mL/min(H$_2$=D$_2$=2.0 mL/min). The calibrations for mass spectrometer signals of H$_2$, D$_2$ and HD were conducted under the same condition as in the reaction. Pulse experiments were carried out in the same system by using Ar as a carrier gas in a flow rate of 100 mL/min. At this flow rate the mean residence time in the reactor was less than 2 s. Purchased Ar, D$_2$, H$_2$, CO, and ethene of 99.999% purity were used without further purification.

Ethene Hydrogenation and CO oxidation were carried out under the condition of C$_2$H$_4$ : H$_2$=13 : 13 kPa and CO : O$_2$=13 : 13 kPa in a closed circulating system. The products in ethene hydrogenation and CO oxidation were analyzed by a gas chromatograph using VZ—10 and Unibeads C columns, respectively.

# 3  Results

## 3.1  Chemisorption of cluster 1 on SiO$_2$

When an ethanol solution of cluster **1** was brought into contact with a silica gel pretreated at 673 K under a high purity of Ar, the initial deep yellowish color became pale within several minutes. After the evacuation the supported cluster **1** was characterized by skeleton vibration of aryl group and the characteristic band of nitrate group by FT-IR. Fig. 2(a), (b), (c) are the spectra for PPh$_3$ on SiO$_2$, cluster **1** on SiO$_2$, and cluster **1** mixed with KBr, respectively. The bands attributed to PPh$_3$ ligands are almost the same among these samples, consistent with the hypothesis that the cluster framework did not decompose on the SiO$_2$ surface. However, the spectrum for **1**/SiO$_2$ did not show $v_{NO}$ peak at 1385 cm$^{-1}$ characterizing the nitrate ionic counterpart of cluster **1**. The spectra in the $v_{OH}$ region for OH groups of SiO$_2$ before and after supporting **1** are shown in Fig. 3(a), (b), respectively. The intensity of the $v_{OH}$ peak at ca. 3740 cm$^{-1}$ decreased by supporting **1** as proved in the difference spectrum shown in Fig. 3 (c). These results indicate that the OH groups on SiO$_2$ surface reacted with the NO$_3^-$ anions in the cluster probably through the ionic exchange reaction to form HNO$_3$ which can be eliminated during the evaporation of the sample.

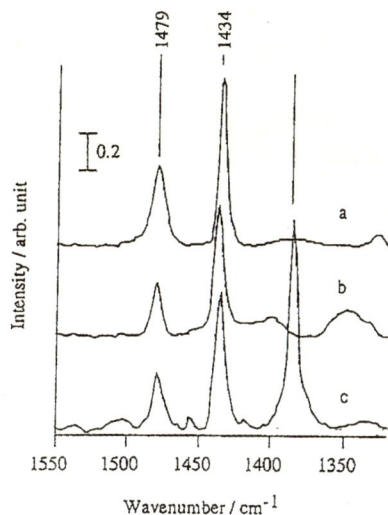

Fig. 2  FT-IR of $PPh_3/SiO_2$, $1/SiO_2$, and 1 mixed with KBr. (a)$PPh_3/SiO_2$, (b)1/$SiO_2$, (c)1+KBr.

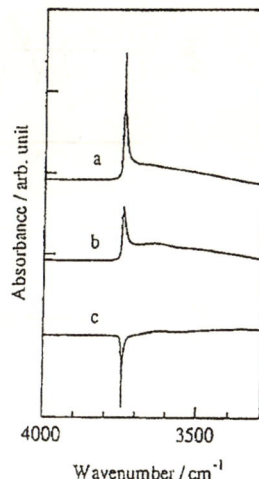

Fig. 3  Changes in the intensity of the $\nu(OH)$ peak by supporting the cluster 1 on $SiO_2$ at room temperatures; (a)$SiO_2$ pretreated at 673 K, (b) after supporting 1 on $SiO_2$ followed by evacuation, (c)the difference spectrum(a)−(b).

## 3.2  TPR performance of 1/SiO₂

TPR spectra of the incipient $1/SiO_2$ and the $1/SiO_2$ treated at 473 K under vacuum are shown in Fig. 4(a), (b). Reduction of the intact $1/SiO_2$ started at about 373 K and the TPR spectrum showed three peaks around 380, 520 and 680 K. In the TPR of the $1/SiO_2$ treated at 473 K, the first peak was not observed, but other two peaks remained essentially unchanged. The results imply that the cluster structure of the supported [Au₆Pt] species a little changed by the heat-treatment at 473 K, but still remains without destruction.

## 3.3  EXAFS observations during heat-treatment

EXAFS spectra of $1/SiO_2$ treated at 400, 473, and 773 K are shown in Fig. 5. Fig. 5(A) shows the EXAFS oscillation of Au L₃-edge for the incipient $1/SiO_2$ and its Fourier transform with three peaks around 2.0, 2.4 and 2.8 Å(phase shift uncorrected), which are assignable to Au−P, Au−Pt and Au−Au bonds, respectively.

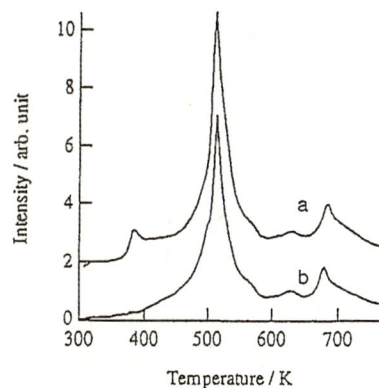

Fig. 4  TPR spectra of 1/SiO₂; (a) the incipient 1/SiO₂; (b) 1/SiO₂ treated 473 K.

Fig. 5(B) shows the EXAFS oscillation and its Fourier transform for the $1/SiO_2$ treated at 400 K, which are similar to those for $1/SiO_2$ (Fig. 5(A)). After treatment of $1/SiO_2$ at 473 K under vacuum, a decrease in intensity of the EXAFS Fourier transform for the peak around 2.7 Å was observed in Fig. 5(C). When $1/SiO_2$ was treated at 773 K under vacuum, the EXAFS data dramatically changed as shown in Fig. 5(D) which is due to the contribution of Au particles. The distance and the coordination number of the Au−Au bond were determined to be 2.87 Å and 10.0, respectively, by curve-fitting analysis(Fig. 6). The best fitting results are listed in Table 2. It is suggested that the cluster framework was decomposed to form metallic gold particles on $SiO_2$ surface at 773 K under vacuum.

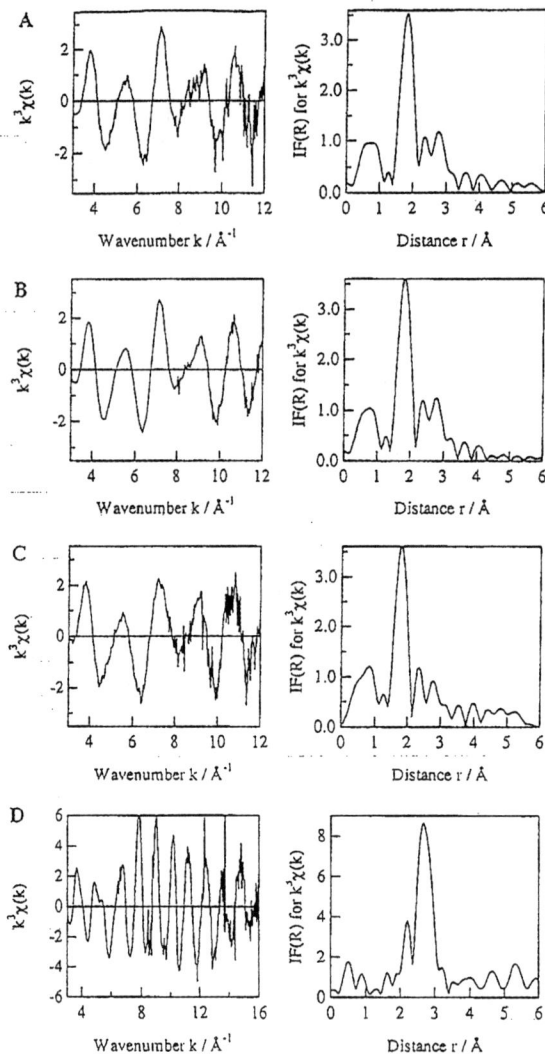

**Fig. 5** EXAFS oscillations and Fourier transforms for the Au $L_3$-edge data; (A) the incipient $1/SiO_2$, (B) $1/SiO_2$ treated at 400 K, (C) $1/SiO_2$ treated at 473 K, (D) $1/SiO_2$ treated at 773 K.

Fig. 7(A) shows the EXAFS oscillation of Pt $L_3$-edge for the incipient $1/SiO_2$ and its Fourier transform with two peaks around 2.0 and 2.6 Å(phase shift uncorrected), which are assigned to Pt—P and Pt—Au bonds. The EXAFS oscillation of Pt $L_3$-edge and its associated Fourier transform for the $1/SiO_2$ treated at 400 K under vacuum(Fig. 7(B)) were almost the same as those of the incipient $1/SiO_2$. By heating the $1/SiO_2$ at 473 K the intensity of the peak around 2.6 decreased a little(Fig,7(C)). No Pt—Au(Pt) bond was observed for the $1/SiO_2$ treated at 773 K under vacuum as shown in Fig. 7(D), where only a peak around 2.0 was observed. The curve-fitting results for the EXAFS data are shown in Fig. 8 and the obtained bond distance and coordination number are listed in Table 3.

### 3.4 EXAFS observations during TPR

EXAFS spectra of $1/SiO_2$ during TPR were observed for the samples heated to 423 K, 603 K, and 773 K under the TPR condition. Fig. 9 shows the Fourier transforms of the EXAFS data at Au $L_3$-edge and Pt $L_3$-edge for $1/SiO_2$ reduced at 423 K, 603 K and 773 K. By comparison with the EXAFS Fourier transforms for the incipient $1/SiO_2$ in Fig. 5(A) and Fig. 7(A), the peaks in Fig. 9(A) are

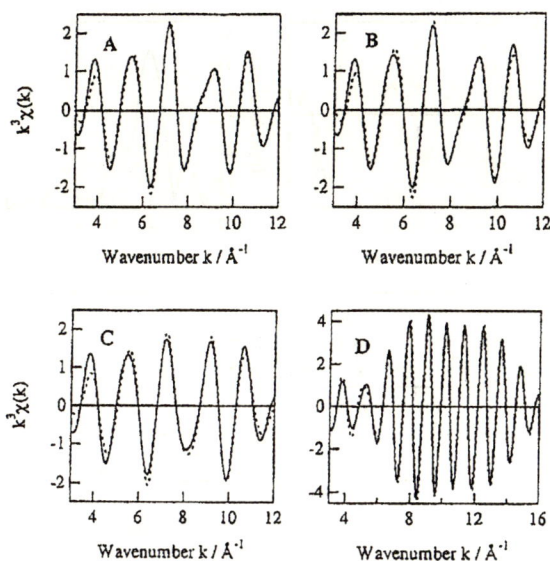

**Fig. 6** The curve-fitting analysis based on the two-shell model of Au − P and Au − Au(Pt) for 1/SiO₂ :
———, observed, ⋯calculated. (A) the incipient 1/SiO₂ , (B) 1/SiO₂ treated at 400 K, (C) 1/SiO₂ treated at 473 K, (D) 1/SiO₂ treated at 773 K.

straightforwardly assigned to the bonds of Au − P and Au − Pt(Au), and Pt − P and Pt − Au, respectively for the Au $L_3$-edge Fourier transform and the Pt $L_3$-edge Fourier transform. The best-fit results are listed in Table 4. The data demonstrate no change in the cluster framework of 1/SiO₂ by TPR up to 423 K within the limitation of EXAFS analysis. After raising the temperature to 603 K, however, the Au $L_3$-edge EXAFS Fourier transform exhibited only one peak which was assigned to Au − Au bond The curve fitting analysis for this peak revealed a bond distance of 2. 87 Å and a coordination number of 11. 1 as shown in Table 4, indicating the formation of metallic Au particles on SiO₂ surface at 603 K. After heating the sample of Au−Au bond to 773 K, the Au particles still have the distance and coordination number of Au−Au bond similar to those for the sample at 603 K. On the other hand the Fourier transforms for the Pt $L_3$ edge-EXAFS showed that the Pt − P# contribution(P# ; phosphine species) increased gradually throughout the TPR measurement. The coordination number of Pt − P# increased from 1. 0 at 423 K to 2. 8 at 603 K or to 2. 5 at 773 K in the TPR(Table 5).

**Table 2** Curve-fitting results for Au $L_3$-edge EXAFS data of 1/SiO₂ under vacuum

| | Au−P | | | | Au−Au(Pt) | | | | $R_f$(%) |
|---|---|---|---|---|---|---|---|---|---|
| | N | r(Å) | Δσ²(Å²) | ΔE(eV) | N | r(Å) | Δσ²(Å²) | ΔE(eV) | |
| 1/SiO₂ | 1. 0±0. 1 | 2. 30±0. 01 | 0. 0±0. 001 | 4±4 | 4. 0±0. 2 | 2. 83±0. 02 | 0. 0±0. 001 | 2±4 | 1. 8 |
| 1/SiO₂ᵃ | 1. 0±0. 1 | 2. 30±0. 01 | 0. 0±0. 001 | 3±4 | 4. 0±0. 2 | 2. 83±0. 02 | 0. 002±0. 001 | 4±4 | 2. 2 |
| 1/SiO₂ᵇ | 0. 6±0. 1 | 2. 28±0. 02 | 0. 0017+0. 0009 | 4±4 | 3. 0±0. 3 | 2. 83±0. 02 | 0. 004±0. 001 | 2±3 | 3. 1 |
| 1/SiO₂ᶜ | | | | | 10. 0+1. 0 | 2. 87±0. 02 | 0. 004±0. 001 | 0. 6±4 | 1. 0 |
| 1ᵈ | (1. 0) | (2. 30) | | | (4. 0) | (2. 83) | | | |

N: coordination number, r: interatomic distance, σ: Debye-Waller factor, ΔE: difference in the origin of photoelectron energy between reference and sample, $R_f$: residual factor in the curve fitting.

ᵃ Treated at 400 K.

ᵇ Treated at 473 K.

ᶜ Treated at 773 K.

ᵈ Crystallographic data of cluster 1. Au−Au(Au foil) : 2. 87 Å.

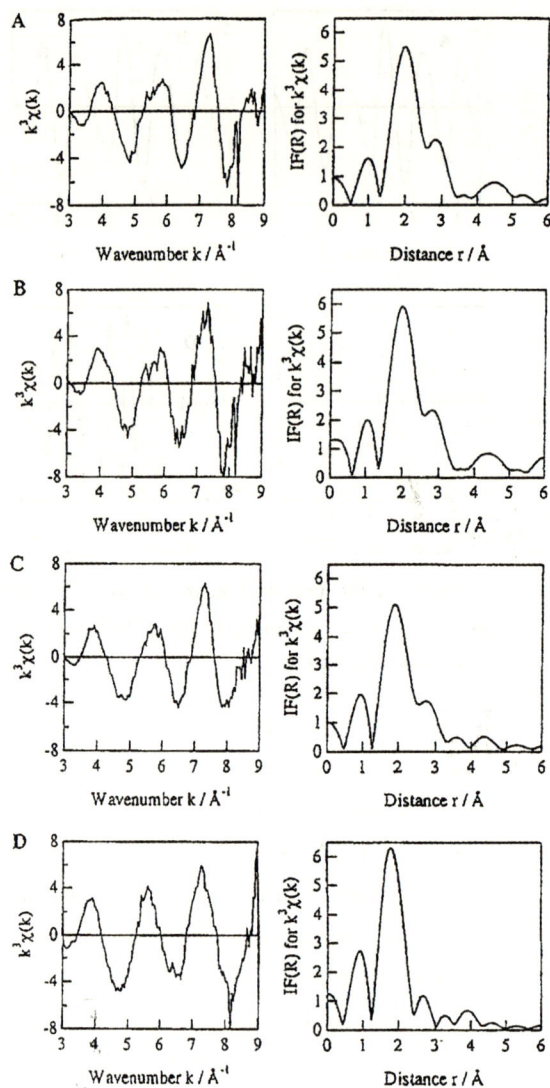

**Fig. 7** EXAFS oscillations and Fourier transforms for the Pt L$_3$-edge data; (A) the incipient 1/SiO$_2$, (B) 1/SiO$_2$ treated at 400 K, (C) 1/SiO$_2$ treated at 473 K, (D) 1/SiO$_2$ treated at 773 K.

**Table 3** Curve-fitting results for Pt L$_3$-edge EXAFS data of 1/SiO$_2$ under vacuum

| Sample | Pt−P | | | | Pt−Au | | | | $R_f(\%)$ |
|---|---|---|---|---|---|---|---|---|---|
| | $N$ | $r(\text{Å})$ | $\Delta\sigma^2(\text{Å}^2)$ | $\Delta E(\text{eV})$ | $N$ | $r(\text{Å})$ | $\Delta\sigma^2(\text{Å}^2)$ | $\Delta E(\text{eV})$ | |
| 1/SiO$_2$ | 1±0.2 | 2.28±0.02 | 0.0±0.001 | 6±4 | 6±0.2 | 2.69±0.02 | 0.0±0.001 | 0.3±4 | 1.1 |
| 1/SiO$_2$ [a] | 1±0.2 | 2.28±0.02 | 0±0.001 | 8±4 | 6±0.2 | 2.69±0.02 | 0.001±0.001 | 0.1±5 | 1.1 |
| 1/SiO$_2$ [b] | 1.3±0.3 | 2.27±0.02 | 0.001±0.001 | 5±5 | 4.1±0.2 | 2.69±0.02 | 0.003±0.001 | 3±4 | 1.2 |
| 1/SiO$_2$ [c] | 3.3±0.6 | 2.28±0.02 | 0.002±0.001 | 2±5 | | | | | 1.3 |
| 1 [d] | (1.0) | (2.28) | | | (4.0) | (6.0) | (2.69) | | |

[a] Treated at 400 K.

[b] Treated at 473 K. [c] Treated at 773 K. [d] Cluster 1.

**Table 4　Curve-fitting results for Au L₃-edge EXAFS data of 1/SiO₂ during TPR**

| $T(K)$ | Au−P | | | | Au−Au(Pt) | | | | $R_f(\%)$ |
|---|---|---|---|---|---|---|---|---|---|
| | $N$ | $r(Å)$ | $\Delta\sigma^2(Å^2)$ | $\Delta E(eV)$ | $N$ | $r(Å)$ | $\Delta\sigma^2(Å^2)$ | $\Delta E(eV)$ | |
| 423 | 1±0.1 | 2.28±0.02 | 0.001±0.001 | 4±4 | 3.6±0.8 | 2.8±0.02 | 0.000±0.001 | 3±4 | 1.6 |
| 603 | | | | | 11.1±1.0 | 2.87±0.02 | 0.001±0.001 | 1±4 | 2.7 |
| 773 | | | | | 11.2±1.0 | 2.87±0.02 | 0.001±0.001 | 1±4 | 1.6 |

**Table 5　Curve-fitting results for Pt L₃-edge EXAFS data of 1/SiO₂ during TPR**

| $T(K)$ | Pt−P | | | | Pt−Au | | | | $R_f(\%)$ |
|---|---|---|---|---|---|---|---|---|---|
| | $N$ | $r(Å)$ | $\Delta\sigma^2(Å^2)$ | $\Delta E(eV)$ | $N$ | $r(Å)$ | $\Delta\sigma^2(Å^2)$ | $\Delta E(eV)$ | |
| 423 | 1±1.0 | 2.3±0.02 | 0.001±0.001 | 4±4 | 4.9±0.8 | 2.8±0.02 | 0.001±0.001 | 3±4 | 0.1 |
| 603 | 2.8±0.6 | 2.28±0.02 | 0.001±0.001 | 0.4±4 | | | | | 2.6 |
| 773 | 3.2±0.7 | 2.27±0.02 | 0.003±0.001 | 0.7±4 | | | | | 2.7 |

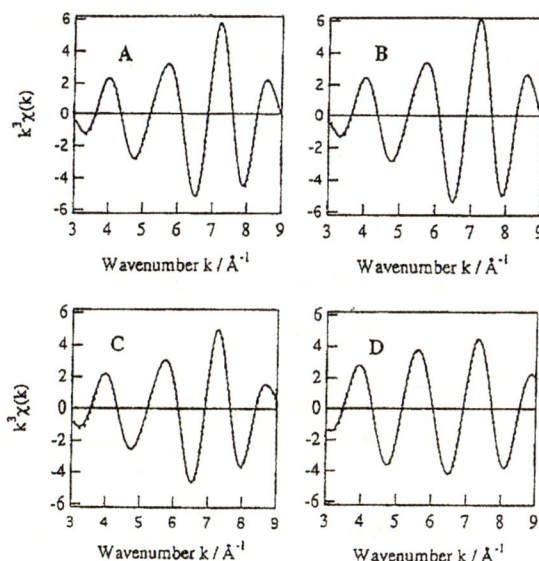

**Fig. 8　The curve-fitting analysis based on the two-shell model of Pt−P and Pt−Au for 1/SiO₂ :————, observed, ⋯cakculated. (A)the incipient 1/SiO₂. (B)1/SiO₂ treated at 400 K. (C)1/SiO₂ treated at 473 K. (D)1/SiO₂ treated at 773 K.**

**Table 6　H₂−D₂ equilibration(a),ethene hydrogenation(b)and CO oxidation(c)over several catalysts at 303 K**

| Catalysts | TOF/s⁻¹ | | |
|---|---|---|---|
| | (a) | (b) | (c) |
| 1 | 2.0 | | |
| 1/SiO₂ | 29.8 | 8×10⁻⁴ | 7×10⁻⁵ |
| 1/SiO₂ treated at 473 K | 11.1 | 5×10⁻⁴ | 3×10⁻⁵ |
| Pt(PPh₃)₄/SiO₂ | 0.0 | 0.0 | 0.0 |
| Pt(PPh₃)₄/SiO₂ treated at 473 K | 0.0 | 0.0 | 0.0 |
| [Au₉(PPh₃)₈](NO₃)₃/SiO₂ | 0.0 | 0.0 | 0.0 |

## 3.5 Catalysis of 1/SiO₂

$H_2-D_2$ equilibration on $\mathbf{1}/SiO_2$ at 303 K was carried out in a fixed-bed flow reactor. The catalytic reaction rates(turnover frequency; TOF)are shown in Table 6, where TOF is defined as(mol of HD) (mol of cluster)$^{-1}$(s)$^{-1}$. The activity of $\mathbf{1}$ in the solid state was promoted about 15 times by supporting $\mathbf{1}$ on $SiO_2$. Pignolet and coworkers also observed similar enhancement of the reaction rate by supporting[16,17]. The increase in activity may be attributed mainly to a high surface area of the support. It is to be noted that the catalytic activity of the incipient $\mathbf{1}/SiO_2$ was higher than that of the sample treated at 473 K. $[Au_9(PPh_3)_8](NO_3)_3/SiO_2$ was inactive for $H_2-D_2$ equilibration and $Pt(PPh_3)_4/SiO_2$ was also inactive as shown in Table 6.

The results of catalytic ethene hydrogenation and CO oxidation in a closed circulating system are also shown in Table 6. It was found that the hydrogenation of ethene to ethane over $\mathbf{1}/SiO_2$ proceeded at an initial rate of $8\times10^{-4}$ s$^{-1}$ at 303 K. The incipient $\mathbf{1}/SiO_2$ also catalyzed a very low reaction rate of CO oxidation(TOF of $7\times10^{-5}$ s$^{-1}$). TPR spectra and EXAFS data for $\mathbf{1}/SiO_2$ before and after ethene hydrogenation were the same, suggesting that no change occurred in the cluster framework of $\mathbf{1}/SiO_2$ during the ethene hydrogenation.

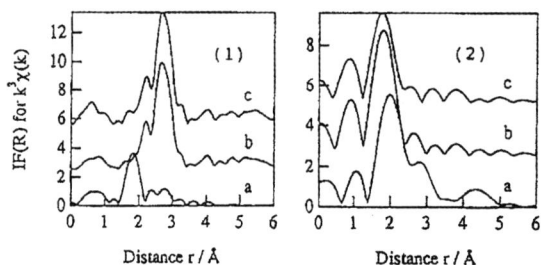

Fig. 9 Fourier transforms of the EXAFS oscillation at Au L₃-edge (1) and Pt L₃-edge(2) for 1/SiO₂ during TPR; (a)1/SiO₂ reduced by TPR to 423 K, (b)1/SiO₂ reduced by TPR to 603 K, (c)1/SiO₂, reduced by TPR to 773 K.

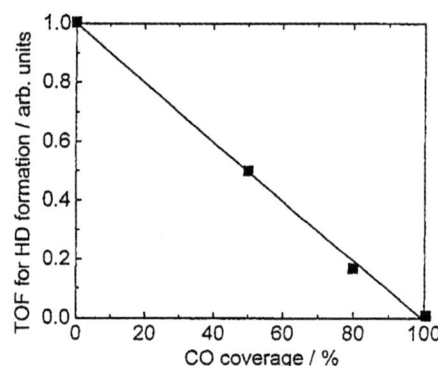

Fig. 10 CO inhibition: linear decrease of the TOF for $H_2-D_2$ equilibration by CO coverage.

## 3.6 Pulse reaction

In order to obtain information on active sites, the 1 ml pulse of a mixture of $H_2$, $D_2$, and $C_2H_4$ (1∶1∶1) was admitted to a 100 mL/min Ar flow and the products were monitored by mass spectroscopy. About 80% decrease in the rate of $H_2-D_2$ equilibration by the presence of $C_2H_4$ was observed. No ethane was formed, while a very small amount of $C_2H_3D$ was detected. When a mixture of $C_2H_4$ and $D_2$ was added as a pulse to the Ar flow, the HD formation was 1/20 as large as that in a $H_2-D_2$ pulse reaction and the amount of $C_2H_3D$ formed was the similar amount to that in the pulse reaction of $H_2-D_2-C_2H_4$. No $H_2-D_2$ equilibration proceeded when CO was pulsed to a gas flow of $H_2-D_2$.

## 3.7 CO poisoning effect on $H_2-D_2$ equilibration

The addition of molar excess of CO pulse to an Ar flow through $\mathbf{1}/SiO_2$, followed by Ar purge, resulted in the complete poisoning of H$_2$ — D$_2$ equilibration. By Ar flowing CO gradually desorbed accompanying by partial recovery of H$_2$ — D$_2$ equilibration. The catalytic activity was completely recovered after CO desorption. We monitored the amount of desorbed CO by an on-line GC and calculated the remained CO coverage on the cluster. The TOF for H$_2$ — D$_2$ equilibration was plotted against CO coverage in Fig. 10.

# 4 Discussion

## 4.1 Adsorption of 1 on SiO$_2$ and structure behavior of 1/SiO$_2$

The FT-IR spectra in Fig. 2 and the color of the solid which is sensitive to the coordination sphere in the cluster demonstrate that the cluster **1** is supported on SiO$_2$ with retention of the original cluster framework structure. The comparison of the IR spectrum of the cluster in KBr with that of the supported cluster, along with the decrease of the surface OH peak intensity by supporting the cluster **1** on SiO$_2$ suggests that the cluster—surface interaction occurred with the nitrate anions which reacted with the surface OH groups to form the surface bimetallic cluster, probably evolving HNO$_3$, as shown in Fig. 11. Thus it is proposed that the cluster was supported on SiO$_2$ by an ion exchange reaction, without loss of the strong interactions between Au and Pt metals.

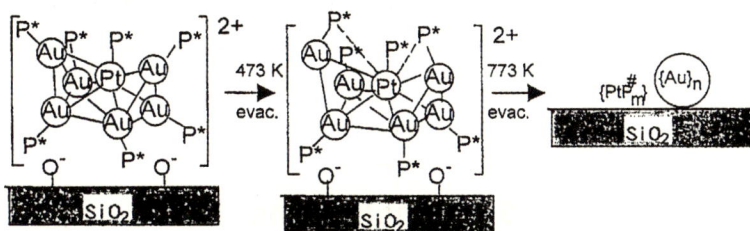

Fig. 11　Cluster framework transformation in 1/SiO$_2$ by heating in vacuum. Dotted lines are tentatively represented, P$^{\#}$ = phosphine species.

The retention of the cluster framework was also characterized by EXAFS, The data in Tables 2 and 3 demonstrate the cluster framework in **1**/SiO$_2$ retains its original structure after deposition on the SiO$_2$ surface, keeping the Pt—Au and Au—Au distances unchanged. The cluster framework is stable up to 400 K under vacuum. However, part of metal-metal bonds of **1** on SiO$_2$ at 470 K under vacuum is cleaved, judging from the results of the bond numbers of Au—Au(Pt) and Pt—Au which decreased to about 75% (Table 2) and 68% (Table 3) of the original ones, respectively, and also from a small increase in the coordination number of Pt—P$^{\#}$ (P$^{\#}$ : phosphine species). The TPR spectrum of the sample treated at 473 K was essentially the same as that of the incipient cluster in the temperature range above 400 K. It is likely that a local rearrangement of the cluster framework occurred at 473 K as proposed in Fig. 11. When the **1**/SiO$_2$ was heated at 773 K, the cluster of **1**/SiO$_2$ was completely destroyed to form metallic gold particles (Table 2) and Pt—(P$^{\#}$)$_n$ (n=3, Table 3) species when heated to 773 K (Fig. 11), No Pt—Au bonding was detected with the sample treated at 773 K.

The curve-fitting results of the EXAFS data during TPR of **1**/SiO$_2$ demonstrate the complete cleavage of Au—Pt and Au—P bonds to form Au particles at 603 K (Table 4 and Fig. 9). The TPR spectrum of Fig. 4 showed a principle peak around 420 K, which may be related to the destruction of the cluster framework characterized by EXAFS. The remaining Phosphine species coordinate to Pt atom even if the TPR temperature was reached to 773 K[①]. The formation of Au particles on $SiO_2$ surface by heat-treatment of **1**/$SiO_2$ under vacuum or in a flow of 5% $H_2$/Ar atmosphere indicates that the Au—P bond and the Au—Pt bond were much weaker than the Pt—P bond. Thus Pt—Au bimetallic particles may not be prepared by using a phosphine-stabilized Pt—Au cluster as precursor. However, the results also may imply that phosphine-stabilized gold clusters as precursors provide a method for preparation of supported ultra fine gold particles on appropriate oxides.

## 4.2  Catalytic site and synergism in 1/$SiO_2$

It was found that the adsorption of an equimolar CO on the [$Au_6Pt$] cluster supported on $SiO_2$ completely suppressed the $H_2$—$D_2$ equilibration as shown in Fig. 10. The CO poisoning was proportional to the CO coverage(coverage: amount of adsorbed CO/Pt atoms). By exposing **1**/$SiO_2$ to CO, yellowish color of **1**/$SiO_2$ immediately turned to orange-red; the CO-adsorbed **1**/$SiO_2$ may resemble the CO adduct of **1** in solution [15,16]. The CO adsorption on Pt atom in **1**/$SiO_2$ is also characterized by EXAFS which revealed the distortion of the metal phosphine bond angle[21]. Thus it is concluded that the Pt atom in the cluster **1** supported on $SiO_2$ contributes to the $H_2$—$D_2$ equilibration as active site. This is contrasted to the catalysis of Pt particles in which multi-site of Pt atoms are believed to be demanded for activation of hydrogen and $H_2$—$D_2$ equilibration.

Lower activity of **1**/$SiO_2$ treated at 473 K compared to **1**/$SiO_2$ is possibly referred to the increase in the coordination number of Pt—P bonds and to the decrease in the coordination number of Au—Pt bonds. The monometallic catalysts of Pt($PPh_3$)$_4$/$SiO_2$ and that pretreated at 473 K were inactive, which suggests that Au—Pt bonds may participate or play a synergistic role in the activation of $H_2$.

## 5  Conclusions

The obtained results show that the cluster framework of **1** can be supported intactly on $SiO_2$ surface. The $NO_3^-$ ionic counterpart in **1** is suggested to interact with the OH groups of $SiO_2$ to form [$Au_6(PPh_3)_6Pt(PPh_3)OSi \leqslant$]. The cluster framework in **1**/$SiO_2$ is stable up to 400 K without breaking of Au—Pt, Au—Au, Au—P* and Pt—P* bonds under vacuum. Local rearrangement in the cluster framework structure occurred when **1**/$SiO_2$ was heated at 473 K. By heating to 773 K Au particles and Pt(P#)species are produced on $SiO_2$ surface. The destruction of the cluster framework may begin from a cleavage of Pt—Au bond, accompanied with migration of P* ligand from Au atom to Pt atoms.

The intact **1**/$SiO_2$ shows a catalytic activity(TOF of 29.8 $s^{-1}$)for $H_2$—$D_2$ equilibration. The **1**/$SiO_2$ also catalyzes ethene hydrogenation and CO oxidation at 303 K with much lower activity. The catalysis of **1**/$SiO_2$ is due not to a platinum impurity but to the platinum atom which is embedded in the gold cluster framework. The catalytic reaction mechanism for $H_2$—$D_2$ equilibration on **1**/$SiO_2$ may be different from that on Pt metallic particles.

① By comparison, we conducted the EXAFS measurements for Pt(PPh$_3$)$_4$/$SiO_2$ after TPR at 773 K under the condition for **1**/$SiO_2$. The results showed that the P# species would bond to Pt atom and the coordination number is 2.8. The data is available upon request.

# References

[1] J. R. H. van Schaik, R. P. Dessing, V. Ponec, J. Catal. 1975, **38**: 273; J. K. A. Clarke, L. Manninger, T. Baird, J. Catal. 1978, **54**: 230; 1984, **9**: 85; J. H. Sinfelt, Bimetallic Catalysts (Wiley, New York, 1985); R. C. Yates, G. A. Somorjai, J. Catal. 1987, **103**: 208; K. Bakakrishnan, A. Sachdev, J. Sachwank, J. Catal. 1990, **121**: 441; P. A. Sermon, J. M. Thomas, K. Keryou, G. R. Millward, Angew. Chem. Int. Ed. Engl. 1987, **26**: 918; A. Sachdev, J. Schwank, J. Catal. 1989, **120**: 353; K. Balakrishnan, J. Sachwank, J. Catal. 1991, **132**: 451; D. Rouabah, J. Fraissard, J. Catal. 1993, **144**: 30; J. Sachtler, K. Balakrishnan, A. Sachdev, in: L. Guczi, F. Solymosi and P. Tetenyi(Eds. ), New Frontiers in Catalysis(Elsevier, Amsterdam, 1993)p. 905.

[2] N. W. Cant and W. K. Hall, J. Phys. Chem. 1971, **75**: 2914; S. Galvagno and G. Parravano, J. Catal. 1978, **55**: 178; J. Schwank, Gold Bull. 1983, **16**: 103; M. Haruta, S. Tsubota, N. Yamada, T. Kobayashi and S. Iijima, J. Catal. 1989, **115**: 301; M. Haruta, S. Tsubota, T. Kobayashi, H. Kageyama, M. J. Genet and B. Delmon, J. Catal. 1993, **144**: 174; S. Tsubota, A. Ueda, H. Sakurai, T. Kabayashi and M. Haruta, ACS Symp. Ser. 1993, **552**: 420.

[3] J. Sachtler and G. Somorjai, J. Catal. 1983, **81**: 77.

[4] J. Sachtler, J. Biberian and G. Somoijai, Surf. Sci. 1981, **110**: 43.

[5] P. Braunstein and J. Rose, in: I. Bemal(Ed. ), Stereochemistry of Oganometallic and Inorganic Compounds, Vol. 3(Elsevier, Amsterdam, 1988)ch. 1, p. 1.

[6] P. Braunstein and J. Rose, Gold Bull. 1985, **18**: 17.

[7] G. Suss-fink, G. Meister, Adv. Organomet. Chem. 1993, **35**: 41.

[8] B. C. Gates, L. Guczi, H, Knozinger(Eds. ), Stud. Surf. Sci. Catal. 1986, 29.

[9] M. Ichikawa, Adv. Catal. 1992, **38**: 283.

[10] Y. Iwasawa(Ed. ), Tailored Metal Catalysis(Reidel, The Netherlands, 1986).

[11] Y. Iwasawa, Catal. Today 1993, **18**: 21.

[12] B. C. Gates, Chem. Rev. 1995, **95**: 511.

[13] K. P. Hall and D. M. P. Mingoss, Prog. Inorg. Chem. 1981, **32**: 237.

[14] E. John and J. Gao, J. Chem. Soc, Chem. Commun. 1985, 39.

[15] I. Gubkina, L. Rubinstein and L. Pignolet, Abstr. ACS Meeting 1994, **208**: 405.

[16] I. Graf, J. Bacon, M. Consugar, M. Curley, L. Ito and L. Pignolet, Inorg. Chem. 1996, **35**: 689.

[17] M. A. Aubart, B. D, Chandler, R. A. T. Gould, D. A. Krogstad, M. J. Schoondergang and L. H. Pignolet, Inorg. Chem. 1994, **33**: 3724.

[18] Y. Yuan, K. Asakura, H. Wan, K. Tsai and Y. Iwasawa, Chem. Lett. 1996, 129.

[19] L. N. Ito, J. D. Sweet, A. M. Mueting, L. H. Pignolet, M. F. J. Schoondergang and J. J. Steggerda, Inorg. Chem. 1989, **28**: 3696.

[20] J. J. Rehr de Leon. J. Mustre, S. I. Zabinsky, R. C. Albers, J. Am. Chem. Soc, 1991, **113**: 5235.

[21] K. Asakura, Y, Yuan and Y. Iwasawa, Proc. of XAFS-IX, Grenoble 1996, in press.

■ 本文原载:Chemical Research in Chinese Universities Vol. 13 No. 1,pp. 83~85,1997.

# Coke-Resistant Ni-based Catalyst for Partial Oxidation and $CO_2$-Reforming of Methane to Syngas*

Ping Chen, Hong-Bin Zhang[①], Guo-Dong Lin, Zhong-Ping Guo, Khi-Rui Tsai

(Department of Chemistry & State Key Laboratory for Physical Chemistry of the Solid Surface, Xiamen University, Xiamen, 361005)

Received Oct. 12, 1996

**Key words**  Coke-resistant Ni catalyst  Methane  Partial oxidation  $CO_2$ reforming  Syngas

The catalytic partial oxidation and the $CO_2$-reforming of methane to syngas(POM and MCR)have become the two fashionable reactions recently[1,2] because the former is considered as a potential alternative of the conventional steam reforming of methane to syngas process,and the latter consumes $CH_4$ and $CO_2$,both of which are greenhouse gases. So far,the catalysts reported in literatures,especially Ni-based catalysts,are likely to coke severely[3-5]. One of the major challenges in the commercialization of both POM and MCR react ions is development of highly efficient and coke-esistant catalysts. In the present work,a modified Ni-Mg-Cr-M-$O_x$ catalyst for both POM and MCR reactions has been developed and studied. The experimental results demonstrated that besides the high catalytic activity being obtained,coke may be largely reduced in the case of MCR,and even eliminated in the case of POM. The performance stability of this newly developed catalyst is also remarkable.

The catalyst was prepared according to the procedure described as follows. The calculated amounts of $Ni(NO_3)_2 \cdot 6H_2O$,$Mg(NO_3)_2 \cdot 6H_2O$,$Cr(NO_3)_3 \cdot 9H_2O$ and the nitrate of a fourth metal(M) as promoter were mixed together with a certain amount of citric acid solution,and then dried at 373 K,and calcined at 1023 K for 4 h. The evaluation of the catalyst sample was performed in a fixed-bed continuous flow reactor-GC combination system. A catalyst sample of 40 mg was used each time of testing. A PCT−1 TG-DT A analyzer was used to measure the amount of coke formed on the catalyst.

The results of activity evaluation of the Ni-Mg-Cr-M-$O_x$ catalyst for the POM showed that 89% of methane conversion(ConvCH$_4$) and 95C% of CO selec-tivity(SelecCO)have been obtained with a $H_2$/CO molar ratio being 2. 08/1. 0 in the products under the reaction conditions of atmospheric pressure,1050 K,$CH_4/O_2$=2. 1/1. 0(mol/ mol),and GHSV=1. 0×10$^5$ mL/h-g catal. ;for the MCR reaction,the ConvCH$_4$ and the ConvCO$_2$ amounted stably to 88% and 86%,respectively,with a $H_2$/CO molar ratio being 0. 95/1. 0 in the products under the reaction conditions of atmospheric pressure,1053K,$CH_4/CO_2$ =1. 0/1. 08(mol/mol),and GHSV=1. 0×10$^4$ mL/h-g catal. .

The effects of temperature and GHSV on the catalytic performance of Ni-Mg-Cr-M-$O_x$ were

---

* Supported by the National Natural Science Foundation of China.

① To whom correspondence should be addressed.

investigated for both POM and MCR reactions, respectively. The experimental results (as shown in Figs. 1 and 2) showed that both conversion of reactants and the yield of syngas increased with elevating temperature, but dropped slightly with increasing GHSV; as GHSV increased from 34000 to 610000 mL/h-g catal., only 3.0% and 2.0% decreases in ConvCH$_4$ were observed for POM and MCR reactions, respectively, indicating the high efficiency of Ni-Mg-Cr-M-O$_x$ catalyst for activation and conversion of these reactant molecules.

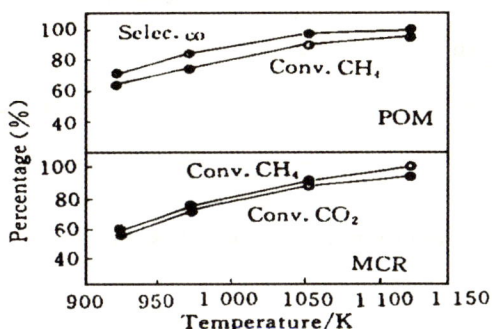

Fig. 1   Effect of temperature on the performance of Ni-Mg-Cr-M-O$_x$ catalyst for POM and MCR reactions, respectively.

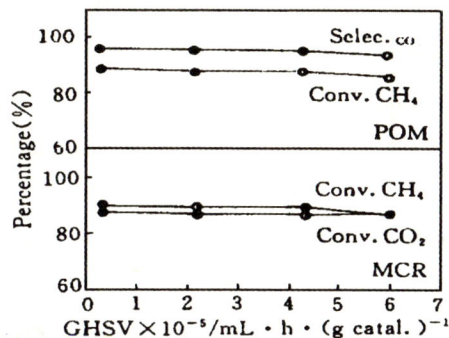

Fig. 2   Effect of GHSV on the catalytic performance of Ni-Mg-Cr-M-O$_x$ catalyst for POM and MCR reactions, respectively.

The amount of carbon deposited on Ni-Mg-Cr-M-O$_x$ and Ni-MgO as well as Ni-La$_2$O$_3$ catalysts, for a comparison, was measured after POM and MCR reactions, respectively, for a certain period of time (see Table 1). It can be found that after 50 h of POM operation, the total amount of carbon deposited on the Ni-Mg-Cr-M-O$_x$ catalyst was below the detectable limit of the PCT-1 TG-DTA analyzer; whereas, at a quite short time, much more coke was detected on the home-made Ni-MgO and Ni-La-O$_x$ catalysts, as well as on the Ni/La$_2$O$_3$ catalyst reported by Choudhary[4]. For MCR reaction, an average coke rate of 0.85 mass%/h for Ni-Mg-Cr-M-O$_x$ catalyst during 8 h of operation was observed, which was much lower than that for Ni-MgO and the Ni-La$_2$O$_3$ catalysts under the same reaction conditions; when temperature was elevated to 1100K, no carbon was detected on this newly developed catalyst after 6 h of the MCR operation (see Table 1).

Table 1   Coke formation on Ni-Cr-Mg-M-O$_x$, Ni-MgO and Ni-La$_2$O$_3$ catalysts after the POM and MCR reaction operations, respectively

| Reaction | Catalyst | Temp. (K) | Time(h) | Coke amount wt% | Coking rate wt%/h |
|---|---|---|---|---|---|
| MCS * | Ni-MgO | 1053 | 1.0 | 12.0 | 12.0 |
| | Ni-La$_2$O$_3$ | 1053 | 1.0 | 10.2 | 10.2 |
| | Ni-Cr-Mg-M-O$_x$ | 1053 | 8.0 | 7.0 | 0.85 |
| | | 1100 | 6.0 | — | — |
| POM * | Ni-MgO | 1053 | 8.0 | 30.0 | 3.77 |
| | Ni-La$_2$O$_3$ | 1053 | 15 | 4.5 | 0.3 |
| | Ni-Cr-Mg-M-O$_x$ | 1053 | 50.0 | — | — |
| | Ni/La$_2$O$_3$[4] | 773 | 6 | 9.1 | 1.5 |

* The catalytic activities of Ni-MgO, Ni-La$_2$O$_3$ and Ni-Mg-Cr-M-O$_x$ catalysts for POM and MCR are similar under the same reaction conditions.

The stability of Ni-Mg-Cr-M-O$_x$ catalyst is quite remarkable. POM reaction for 50 h or MCR for 10 h did not lead to any decrease in activity, as shown in Fig. 3. The results of the present work show that this catalyst developed newly has potential practically because of its high and stable activity and the excellent coke-resistant performance.

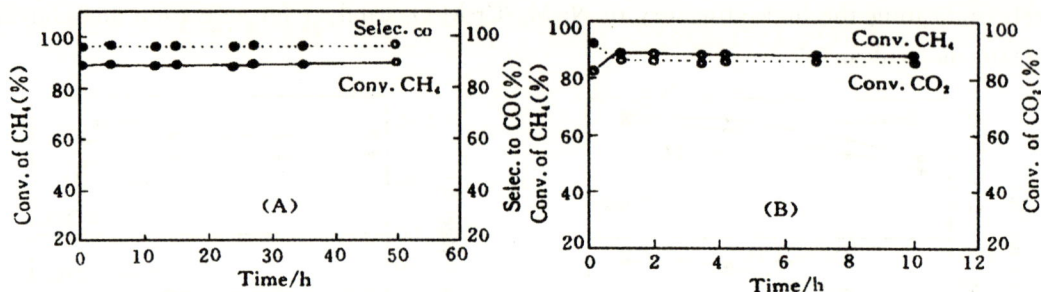

Fig. 3　Stability, at 1053 K, of Ni-Cr-Mg-M-O$_x$ catalyst for POM reaction for 50 h(A) and MCR reaction for 10 h(B).

# References

[1]A shcroft, A. T., Cheethan, A. K., Foord, J. S., Green, M. L. H., Grey, C. P., Murrel, A. J. and Vernon, P. D. T., Nature, 1990, **344**: 319.

[2]Hickmann, D. A., Haupfear, E. A. and Schmidt, L. P., Catal. Lett., 1993, **17**: 2223.

[3]Erdohelyi, A., Cserenyi, J. and Solymosi, J. Catal., 1993, **141**: 287.

[4]Choudhary, V. R., Rane, V. H. and Rajput, A. M., Catal. Lett., 1993, **22**: 289.

[5] Fleming, P. J., cossutta, W. and Jackson, P. J., Stud. Surf. Sci. Catal., p. 81: Natural Gas Conversion II(Eds. Curry-Hyde, H. E. and Howe, R. F. ), Elsevier, 1994, 321.

■ 本文原载：Carbon Vol. 35，No. 10~11，pp. 1495~1501，1997.

# Growth of Carbon Nanotubes by Catalytic Decomposition of CH₄ or CO on a Ni-MgO Catalyst*

P. Chen, H.-B. Zhang①, G.-D. Lin, Q. Hong, K. R. Tsai

(*Department of Chemistry and State Key Laboratory of Physical Chemistry for the Solid Surface, Xiamen University, Xiamen 361005, People's Republic of China*)

Received 20 May 1996; accepted in revised form 30 April 1997

**Abstract** By using a Ni-MgO catalyst, carbon nanotubes with small and even diameter could be prepared from catalytic decomposition of CH₄ or CO. These carbon nanotubes prepared by this method are more or less twisted, with the outer diameter at 15−20 nm, and the tube length up to 10 $\mu$m. The results of XRD measurements and pulse reaction testing indicated that the NiO and MgO components in this catalyst precursor formed, due to their highly mutual solubility, a $Ni_xMg_{1-x}O$ solid solution. The high dispersion of Ni-species in this solid solution and the effect of valence-stabilization by the MgO crystal field would be in favor of inhibiting deep reduction of $Ni^{2+}$ to $Ni^0$ and aggregation of the $Ni^0$ to form large metal particles at the surface of catalyst, making the carbon nanotubes grown on this catalyst relatively small and even in size of diameter. The experimental results also indicated that, in the growing process of carbon nanotubes, the rate-determining step was dependent upon the conditions of preparation(i. e. feedgas used, reaction temperature, flow-rate of the feedgas, etc.). The growth mechanism of the carbon nanotubes on the Ni-MgO catalyst is discussed together with the experimental results.© 1997 Elsevier Science Ltd

**Key Words** A. Catalytically grown carbon  A. nanotubes  B. pyrolytic deposition  C. TEM  C. X-ray diffraction.

## 1. INTRODUCTION

The preparation of carbon nanotubes(or carbon nanofibers)and the attempt to utilize this kind of new material have generated intense research activity in recent years[1−7]. There have been two ways for preparation carbon nanotubes on the laboratory scale, i. e. the carbon-arc process[2] and the catalytic decomposition of certain hydrocarbons or other organics(e. g. 2-methyl-1, 2′-naphthyl ketone)in the presence of various supported transition metal catalysts[3−6]. The former could produce relatively thin and straight carbon nanotubes, but suffered from low productivity and difficulty in separation of the carbon nanotubes from their original body. The latter has been widely studied in recent years. This

---

* Supported by the National Natural Science Foundation of China.

① Corresponding author.

process can produce a large amount of carbon nanofibers, with good facilities in purification of the products; however, the carbon nanofibers produced are relatively poor in crystallinity in comparison with those produced by the carbon-arc process, and the diameter of the nanofibers formed varies from 2 nm to 100 nm with the sizes of the crystallites of the catalyst used, and as a result, the carbon nanofibers produced are even smaller in diameter. In order to get relatively even carbon nanofibers, selection of feedgas and preparation of catalyst with ultrafine particles(about 10 nm of diameter), as well as reaction conditions used, all need to be taken into consideration.

A mechanistic interpretation on the growth of carbon nanofibers was first proposed by Baker[8]. The diffusion of carbon through the catalyst particle is generally considered to be the rate-determining step(RDS)in the growth of carbon nanofibers[9], and the driving force which pushed the carbon diffusion from the metal-gas interface(where adsorption and decomposition of feedgas molecules took place)to the metal-nanofiber interface ( where carbon precipitated to form carbon nanofibers ) was suggested to originate from the concentration gradient of dissolved carbon between the two interfaces described above[10]. This proposed model and the others refined later[11,12] provide a rational explanation for the formation of straight nanofibers, but they all suffer from obvious shortcomings, e. g. failing to predict the effect of feedgases and additives on the catalyst particles and on the morphological characteristics of the nanostructures.

In the present work, a new way for the preparation of carbon nanotubes small and even in size(15—20 nm of diameter) from decomposition of $CH_4$ or disproportionation of CO on a powder Ni-MgO catalyst is reported. By means of pulse reaction experiments, XRD and TEM methods, the performance of the catalyst and the texture of the carbon nanotubes produced were investigated. The growth mechanism of the carbon nanotubes on the Ni-MgO catalyst is discussed together with the experimental results.

# 2. EXPERIMENTAL

## 2.1　Preparation of the catalyst

1. 94 g of $Ni(NO_3)_2 \cdot 6H_2O$(the purity in AR grade)and 2. 56 g of $Mg(NO_3)_2 \cdot 6H_2O$(in AR grade)powders were mixed thoroughly, followed by addition of 2 g of citric acid and 20 ml of de-ionized water to form a solution, and then, the solution was evaporated. The solid obtained was dried at 373 K in air, and subsequently calcined at 973 K in air for 5 hours, and finally, a black and fluffy sample of catalyst precursor was obtained.

## 2.2　Evaluation system

A fixed-bed continuous flow reactor-GC combination system was used for the preparation of carbon nanotubes and the evaluation of the catalyst. The reactor was made of a quartz tube of 5 mm i. d. Gaseous reactants and products were analyzed by an on-line SQ-206 Model gas chromatograph(GC) equipped with a thermal conductivity detector and a 601# carbon molecular sieve column ( made by Beijing Analytic Instruments Company ), with $H_2$ as the carrier-gas. Pulse reaction experiments were conducted in the same reactor-GC combined system, with helium serving as carrier-gas for both the pulse feedgas and GC system;12 mg of the catalyst for each time of testing and 0. 1 ml of feedgas for each pulse were used, with the corresponding flow-rate of the carrier-gas He at 2000 ml h$^{-1}$; the exit-gases (including products, unconverted feedgas and carrier-gas ) were analyzed by the GC system;

conversions of $CH_4$ and CO were calculated by the reduction in area of the corresponding GC peak, respectively.

## 2.3　Preparation of carbon nanotubes

12 mg of the Ni-MgO catalyst precursor were packed into the reactor, followed by heating the sample in a flow of purified $H_2$ from room temperature to 873 K and keeping it at the same temperature for a few minutes, and then, introducing the feedgas, $CH_4$ or CO, with flow-rate at 2000 ml $h^{-1}$ for $CH_4$ and 600 ml $h^{-1}$ for CO, respectively; after 30 minutes of reaction, about 70 mg and 32 mg of the raw carbon nanotube products were obtained from $CH_4$ and CO, respectively. These raw products were further purified by means of immersion with a low concentration of nitric acid solution so as to dissolve the catalyst particles(containing metal $Ni^0$, NiO and MgO)attached at the extremities of the nanotubes, followed by washing by de-ionized water and drying at 473 K in a flow of $N_2$. About 53 mg and 18 mg of purified and only carbon-containing products were obtained from the $CH_4$ decomposition and the CO disproprotionation, respectively; and, from the TEM observation, it could be estimated that more than 90% of the deposited-carbon was in the form of nanotubes(including the central part and the tube-wall), and the rest was "encapsulating" carbon and carbon particles.

## 2.4　Spectroscopic characterization

X-ray diffraction measurements were carried out by using a Rigaku D/Max-C X-ray Diffractometer with Cu Kα radiation at a scanning rate of 8° $min^{-1}$. TEM observation was performed by a JEOL JEM—100CX transmission electron microscope. HRTEM observation was carried out by using a Hitachi H-9000 machine.

# 3.RESULTS AND DISCUSSIONS

## 3.1　Texture of the carbon nanotubes

Figure 1 gives the results of TEM observations on the carbon nanotubes prepared by decomposition of $CH_4$ and by disproportionation of CO, respectively, on the prereduced Ni-MgO catalyst at 873 K. It is obvious that these carbon nanotubes, with the outer diameter of 15—20 nm and the tube length up to 10 μm, are relatively small and even in diameter size in comparison with those produced by decomposition of $CH_4$ on a Ni-CaO catalyst(Fig. 2). The morphology of the carbon nanotubes prepared by this method is more or less twisted. Figure 3 shows the result of high resolution TEM observation of one of the carbon nanotubes prepared by disproportionation of CO, from which it can be seen that the carbon nanotube consists of a cylindrical arrangement, and the central part is less dense ( more electron transparent), even empty, whereas the outer region(i. e. tubewall)is constructed with many layers of carbon with graphite-like platelets.

## 3.2　Performance and characterization of the catalyst

It is well known that both NiO and MgO possess a rock-salt type crystal structure and that the ionic radius of $Ni^{2+}$ (0. 070 nm)is quite close to that of $Mg^{2+}$ (0. 065 nm), so that the dimensions of their crystal cells are very close to each other. It may be expected that the binary system consisting of NiO and MgO could form a solid solution due to very good mutual solubility between NiO and MgO. This can be brought out by the results of the XRD determination of the catalyst precursor NiO-MgO as shown in Fig. 4(a), in which, the XRD pattern observed does not have a significant difference in comparison with that of the pure MgO or NiO phase(see Fig. 4(b) and(c)), but the dimension of the crystal cell of this new formed phase is between those of NiO and MgO, perhaps implying the formation of $Ni_xMg_{1-x}O$ solid-solution.

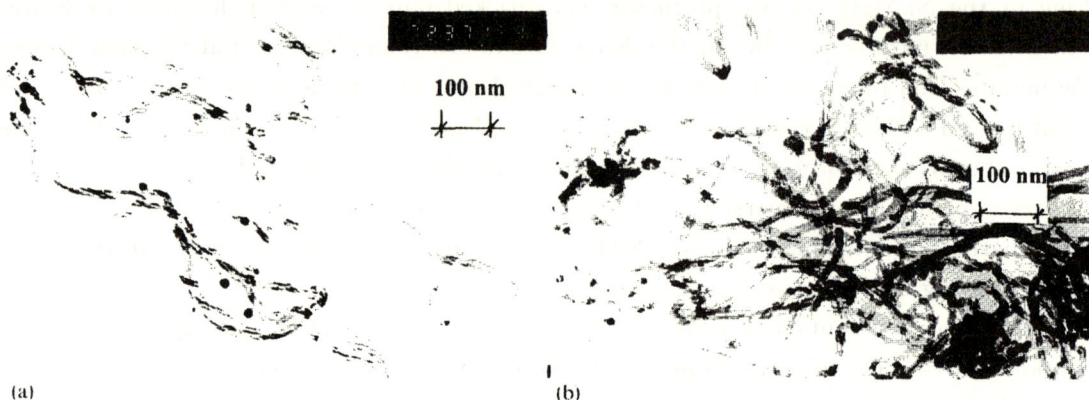

**Fig. 1** TEM images of nanotubes produced according to the procedure of preparation described in Section 2.3, by decomposition of: (a)$CH_4$ and (b)CO at 873 K on the prereduced Ni-MgO catalyst.

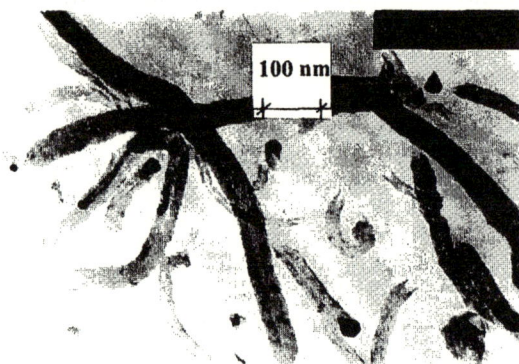

**Fig. 2** TEM image of carbon nanofibers prepared by decomposition of $CH_4$ at 873 K on a prereduced Ni-CaO catalyst.

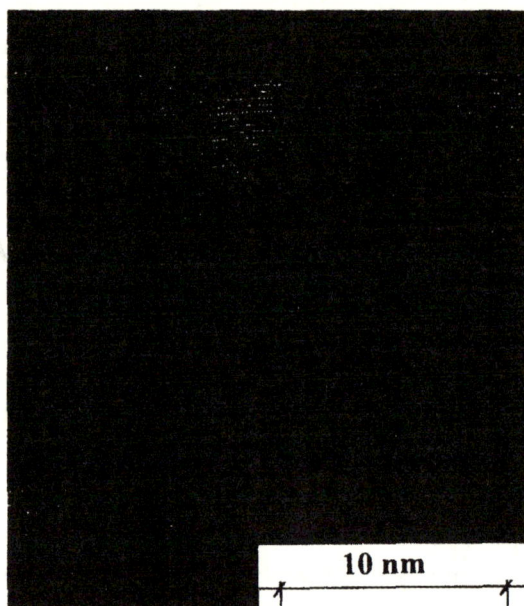

**Fig. 3** HRTEM image of one of the carbon nanotubes prepared by disproportionation of CO at 873 K on the Ni-MgO catalyst.

Being different from the other binary Ni-containing catalyst systems such as NiO-CaO, the $Ni^{2+}$ ions in the NiO-MgO system may be highly dispersed and evenly distributed in the lattice of MgO due to the mutual solubility between NiO and MgO, so that the Ni-component in the metal oxide phase of host-dopant type, $Ni_xMg_{1-x}O$, would be considerably difficult to reduce completely. In the XRD pattern of the hydrogen-reduced Ni-MgO catalyst(shown in Fig. 5(b)), the weak peaks of the metallic $Ni^0$-phase and strong peaks of the $Ni_xMg_{1-x}O$ phase were observed, indicating that only a quite small proportion of the $Ni^{2+}$ was reduced, most probably due to the effect of valence-stabilization by the MgO crystalfield, and the remaining large proportion of the $Ni^{2+}$ was left in the solid-solution. This is obviously different from the case of the hydrogen-reduced Ni-CaO system (see Fig. 5(a)), in which the dominant peak at $2\theta=43.24°$ due to the NiO-phase completely disappeared, simultaneously, a new dominant peak at $2\theta=44.30°$ due to the metallic $Ni^0$-phase appeared, indicating that the $Ni^{2+}$ had been almost completely reduced to $Ni^0$.

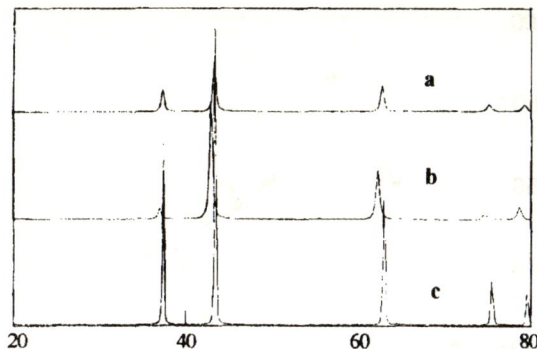

Fig. 4    XRD patterns of: (a) the NiO-MgO
catalyst precursor, (b) MgO, (c) NiO.

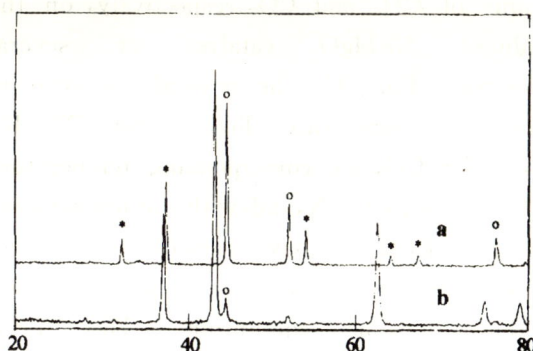

Fig. 5    XRD patterns of: (a) prereduced Ni-CaO catalyst, (b)
prereduced Ni-MgO catalyst; (○) represents the diffraction
peaks of metallic Ni, and ( * ) represents the diffraction
peaks of CaO.

The amount of the reducible $Ni^{2+}$ could be estimated from the results of the following pulse reaction of methane with the NiO-MgO and NiO-CaO catalyst precursors, respectively. A helium-carried pulsed-CH₄ feed-stream was continuously introduced into the reactor with 12 mg of the catalyst at 1053 K, until no formation of the oxygenated products, CO, $CO_2$ and $H_2O$, was detected. The amount of the lattice oxygen reacted was calculated from the measurement of the total amount of the oxygenated products. The results showed that the amount of oxygen being able to react with methane(to form CO, $CO_2$ and $H_2O$) was less than 15% of the total amount of oxygen corresponding to the NiO content in the NiO-MgO system, whereas it almost reached 100% in the case of NiO-CaO. It is conceivable that the high dispersion of Ni-species in the $Ni_xMg_{1-x}O$ solid solution and the effect of valence-stabilization by the MgO crystal-field would be very much in favor of inhibiting the deep reduction of the $Ni^{2+}$ to $Ni^0$ and aggregation of the $Ni^0$ to form large particles of metallic nickel at the surface, making the carbon nanotubes grown on this catalyst relatively small and even in cross section.

## 3.3    Reactivities of CH₄ and CO on the Ni-MgO catalyst and mechanism of the growth of carbon nanotubes

It has been accepted that the growth of carbon nanofibers by catalytic decomposition of hydrocarbons includes the following steps: at first, the feedhydrocarbon molecules are adsorbed and decomposed on certain surface active sites of the metal particles of the catalyst to form carbon species; and then, some of the surface carbon species dissolve into the bulk and diffuse through the metal particle from the front face(i. e. the metal-gas interface) to the rear face(i. e. the metal-nanofiber interface), where carbon is deposited in the form of the nanofiber[8]. It is generally agreed that the diffusion of carbon through the bulk of the catalyst particles is the rate-determining step(RDS) of the growth process of carbon nanofibers. The experimental finding in support of this contention is mainly that the activation energies for the growth of these nanofibers exhibit a remarkable correlation with those for diffusion of carbon through the corresponding metals. However, this seems not always to be true, because, based upon this viewpoint, it might be expected that the growing rate and the morphological

989

structure of carbon nanofibers would be mainly dependent on the metal used as catalyst, whereas independent of the feedgas used; but, from the results of our experimental observations, this is not the case.

Figure 6 shows the results of continual pulse reactions of $CH_4$ and CO, respectively, on the prereduced Ni-MgO catalyst at several temperatures. For $CH_4$, the detectable conversion occurred at temperatures higher than 723 K; whereas for CO, the corresponding temperature may be as low as 473 K, and, with the temperature elevated to 673 K, the conversions of CO-pulses, in reverse, decreased probably due to the restriction of thermodynamic equilibrium. The TEM observation of the catalyst samples after undergoing a treatment by $CH_4$ or CO feed-streams at 723 or 473 K, respectively, demonstrated the formation of carbon nanofibers, although they were short and winding in morphology.

From the results of monitoring the growing process of carbon nanofibers at various temperatures, it was found that, under the condition of temperate flow-rate of $CH_4$ (e. g. 1200

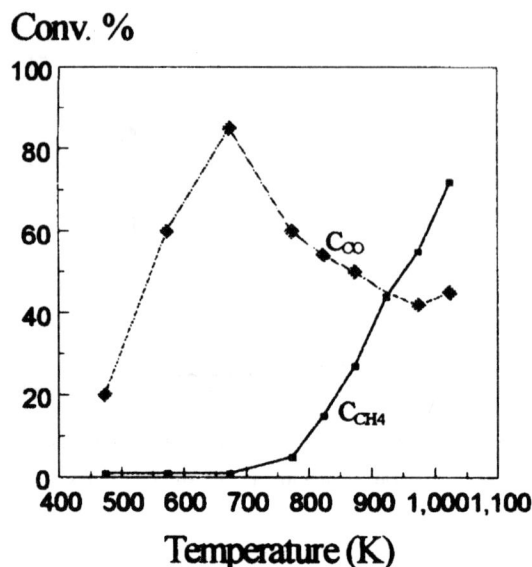

Fig. 6  Temperature dependence of $CH_4$ and CO conversions for $CH_4$ and CO pulses ( 0. 1 ml per pulse ), respectively, over the prereduced Ni-MgO catalyst, with the flow-rate of He carrier-gas at 2000 ml $h^{-1}$.

ml $h^{-1}$), the conversion of $CH_4$ increased with increasing reaction temperature and, at a certain temperature, could steadily keep up at a corresponding level during 30 minutes of testing(see Fig. 7(a—d)). However, the experiments also showed that, at a high temperature and with the flow-rate of $CH_4$ increasing up to some value(e. g. at 1073 K and with flow-rate up to 12000ml $h^{-1}$), a rapid decrease in $CH_4$ conversion would occur; the TEM observation ( see Fig. 8) revealed that the particles of the deactivated catalyst were almost all covered by a certain thickness of deposited carbon, and that, in the carbon-deposited products obtained in this case, the majority is "encapsulating" carbon, and the number of carbon nanotubes was dramatically decreased. This indicated that, in order to avoid the catalyst deactivation caused by deposition of carbon, the rate of $CH_4$ decomposition ( i. e. the rate of carbon formation)must be well matched with the rate of transfer/diffusion of the carbon species formed. At low temperature and in a certain flow-rate range, the rate of carbon formation may be lower than that of carbon transfer/diffusion, so that methane decomposition on the metal-gas interface would become the RDS of the growth of carbon nanofibers. At high temperature and with the flow-rate increasing, the rate of carbon formation would increase and probably exceed the rate of carbon transfer/diffusion, so that the carbon transfer/diffusion became the RDS of the growth of carbon nanotubes. In the latter case, the carbon species not being able to be transferred away in time would more or less cover the surface of the catalyst, leading to a decrease in the activity, and even deactivation of the catalyst(see Fig. 7(e)).

On the other hand, $H_2$ as one of the primary products of $CH_4$ decomposition may also play an adjusting role to a certain extent in the elimination and regeneration of the surface carbon-containing ad-species, $CH_x(x=0,1,2$ or $3)$, via the reversible secondary reaction:

$(4-x)/2H_2(g)+CH_x(a) \leftrightarrows CH_4(g)(x=0,1,2$ or $3)$ which would be in favor of self-cleaning of the catalyst surface and inhibiting the deposition and development of the "encapsulating" carbon.

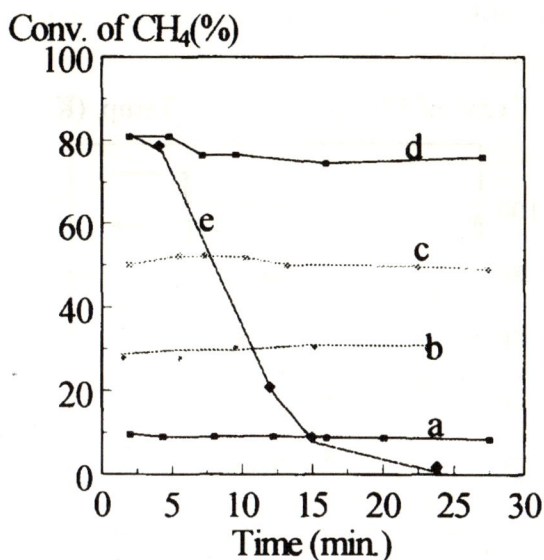

Fig. 7    Time dependence of CH₄ conversion at different flow-rates and temperatures over the prereduced Ni-MgO catalyst: (a) 773 K, (b) 873 K, (c) 973 K, (d) 1073 K, all with the flow-rate at 1200 ml h⁻¹ (e) 1073 K and flow-rate at 12000 ml h⁻¹.

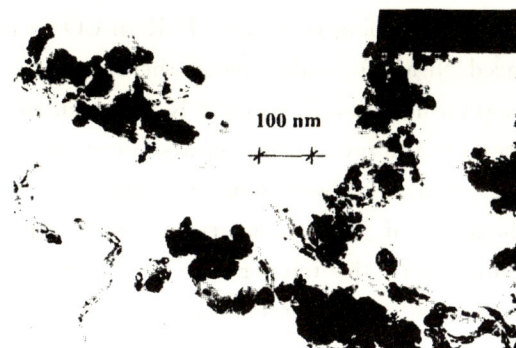

Fig. 8    TEM image of the morphology of deposited carbon produced by decomposition of CH₄ at 1073 K and with flow-rate at 12000 ml h⁻¹ on the Ni-MgO catalyst.

For the growth of carbon nanotubes by disproportionation of CO, the behavior is somewhat different. As revealed by the results described above (see Fig. 6), the disproportionation of CO, i. e. the Boudouard reaction, $2CO \rightarrow C + CO_2$, can occur at a temperature as low as 473 K. Figure 9(a—c) show the changes of the rate of CO disproportionation on the prereduced Ni-MgO catalyst with time at an unchanged flow-rate(600 ml h⁻¹)and at three different temperatures, respectively. It can be seen that, at 723 K(Fig. 9 (a)), the initial conversion of CO reached 82% and was so fast as to lead to the formation of plenty of carbon on the catalyst surface. Part of the carbon was unable to be transferred away in time due to the relatively low rate of the carbon diffusion in the bulk of the metal particle at this temperature and, thus, blocked the active catalyst surface, leading to a decrease in the reactivity; this was also brought out by the result of a TEM observation of the corresponding

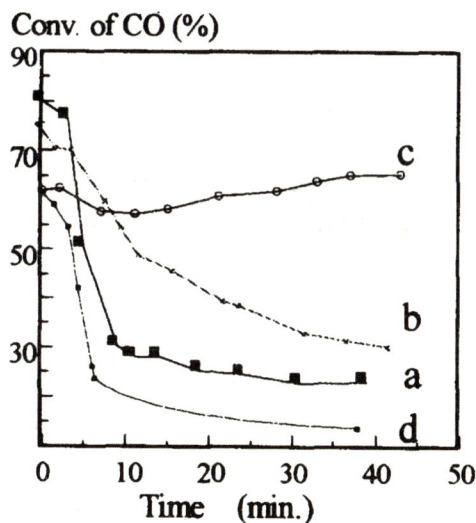

Fig. 9    Time dependence of CO conversion at different flow-rates and temperatures over the prereduced Ni-MgO catalyst: (a) 723 K, (b) 773 K, (c) 873 K, all with the flow-rate at 600 ml h⁻¹; (d) 873 K and flow-rate at 1200 ml h⁻¹.

sample, in which most of the catalyst particles were covered by "encapsulating" carbon. With the temperature elevated to 873 K(Fig. 9(c)), the rate of carbon diffusion in the bulk of the metallic particle was increased, and "catalyst self-cleaning" could occur to a certain extent, so that the rates of the carbon formation at the surface may well match with the rate of the carbon diffusion in the bulk, thus, the CO conversion observed could keep at a steady level; whereas, when the temperature was kept constant(i. e.

at 873 K) and with the flow-rate of CO increased to 1200 ml h$^{-1}$ (see Fig. 9(d)), the conversion of CO dropped again due to enhancement in the rate of formation of carbon.

Figure 10 shows the results of a temperature-pro-grammed-stepping-reaction (TPSR) of CO on the prereduced Ni-MgO catalyst with the flow-rate of CO at 600 ml h$^{-1}$ from 523 to 823 K, in which curve (b) represents the process of stepwise-programmed-elevating temperature with time and curve (a) shows the changes of CO conversion with time, equivalently, with the temperature. It can be seen that, in the first time interval from 0 to 5 minutes, with the corresponding temperature at 623 K, conversion of CO was rapidly decreased from 95% at the beginning down to 10% at the 5th minute due to the catalyst deactivation by carbon deposition, indicating that the rate of carbon formation exceeded greatly over the rate of the subsequent carbon transfer/diffusion, leading to the active catalyst surface being blocked by "encapsulating"

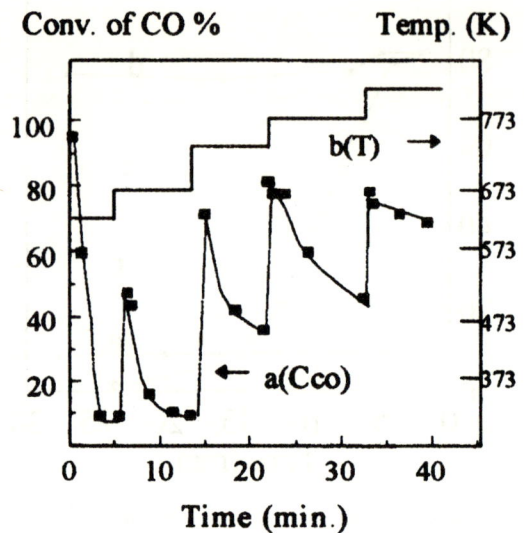

Fig. 10 Conversion of CO(a) with time on the prereduced Ni-MgO catalyst in the temperature-programmed-stepping(b)-reaction(TPSR) process.

carbon. With the temperature jumpily elevated stepwise from 623—673 to 723—773, and finally to 823 K, the rate of carbon diffusion in the bulk of the metallic particle was stepwise sped up, so that the "catalyst self-cleaning" could occur to a certain extent, thus, the CO-conversion was recovered to a level of about 68% in a general way. The above results indicated that the CO conversion was restricted by the extent of "catalyst self-cleaning", in other words, controlled by the transfer/diffusion of the carbon formed, and that the transfer/diffusion of the carbon formed on the metal-gas interface has been the RDS of the TPSR process. This once again demonstrated that the growth of the carbon nanotubes is affected by well-/ill-match between decomposition rate of the feedgas(i. e. the rate of carbon formation) and the transfer/diffusion rate of the carbon formed.

By comparison of the TEM images(see Fig. 1), it can also be seen that the average diameter of the carbon nanotubes derived from CO disproportionation was smaller than that from CH$_4$-decomposition (average diameter: $\sim$ 15 nm for CO and $\sim$ 20 nm for CH$_4$). The experiments also showed that, simultaneously with the growing process of the carbon nanotubes, fragmentation of the catalyst particles occurred to a certain extent, especially in the case where CO was used as feedgas, as brought out by SEM images obtained on the prereduced Ni-MgO catalysts treated with CH$_4$ and CO, respectively, at 873 K [13]. With CO as feedgas, the catalyst particles attached at the extremities of nanotubes are smaller than those with CH$_4$ as feedgas; correspondingly, the average diameter of the nanotubes formed is also smaller. The cause of the difference is under investigation.

# 4. CONCLUSIONS

By using the Ni-MgO catalyst developed by us, carbon nanotubes with small and even diameter can be prepared from catalytic decomposition of CH$_4$ or CO. The results of XRD measurements and pulse reaction testing indicated that the NiO and MgO components in this catalyst precursor formed a Ni$_x$Mg$_{1-x}$O solid solution, in which only a quite small proportion of the Ni$^{2+}$ could be reduced to Ni$^0$.

This would be in favor of inhibiting the formation of large metal particles at the surface of the catalyst, making the carbon nanotubes grown on this catalyst relatively small and even in diameter size. The experimental results also indicated that, in the growing process of carbon nanotubes, the rate-determining step was dependent upon the conditions of preparation (i. e. feedgas used, reaction temperature, flow-rate of the feedgas, etc.), and that the growth of the carbon nanotubes was strongly affected by well-/ill-match between decomposition rate of the feedgas(i. e. the rate of carbon formation) and the transfer/diffusion rate of the carbon formed.

# REFERENCES

[1]Rodriguez,N. M.,J. Mater. Res. ,1993,**8**(12),3233.

[2]Iijima,S.,Nature, 1991,**354**,56.

[3]Baker,R. T. K.,Carbon, 1989,**27**,315.

[4]Motojima,S.,Kawaguchi,M.,Nozaki,K. and Iwa-gana,H.,Appl. Phys. Lett. , 1990,**56**,321.

[5]Yudasaka,M.,Kikuchi,R.,Matsui,T.,Ohki,Y., Yos-himura,S. and Ota,E., Appl. Phys. Lett. , 1995,**67**(17),2477.

[6]Ivanov, V., Fonseca, A., Nagy, J. B., Lucas, A., Lambin, P., Bernaerts, D. and Zhang, X. -B., Carbon, 1995,**33**(12),1727.

[7]Endo,M.,Takeuchi,K.,Kobori,K.,Takahashi,K.,Kroto, H. W. and Sarkar,A.,Carbon, 1995, **33**(7),873.

[8]Baker,R. T. K., Barber, M. A., Harris, P. S., Feates, F. S. and Waite, R. J.,J. Catal. , 1972,**26**, 51.

[9]Alstrup,I.,J. Catal. , 1988,**94**,468.

[10]Baker,R. T. K., Harris,P. S., Thomas,R. B. and Waite. R. J.,J. Catal. , 1973,**30**,86.

[11]Sacco,A.,Thacker,P.,Chang,T. N. and Chiang,A. T. S.,J. Catal, 1984,**85**,224.

[12]Alstrup,I.,J. Catal, 1988,**109**,241.

[13]Chen,P.,Zhang, H. -B.,Lin,G. -D.,Tsai,K. R. and Zai,H. -S.,J. Xiamen Univ. (Nat. Sci. ), 1996,**35**(1),61.

■ 本文原载:Polyhedron Vol. 16,No. 1,pp. 75～79,1997.

# Molybdenum(VI) Complex with Citric Acid: Synthesis and Structural Characterization of 1:1 Ratio Citrato Molybdate $K_2Na_4[(MoO_2)_2O(cit)_2]$ · $5H_2O$

Zhao-Hui Zhou[①], Hui-Lin Wan, Khi-Rui Tsai

(Department of Chemistry and State Key Laboratory for Physical Chemistry of Solid Surface, Xiamen University, Xiamen 361005, P. R. China)
Received 2 April 1996; accepted 17 May 1996

**Abstract**　Sodium potassium citrato molybdate $K_2Na_4[(MoO_2)_2O(cit)_2]$ · $5H_2O$ has been prepared by the reaction of potassium trihydrogen citrate and sodium molybdate. Analysis of the crystal structure reveals that the anion of the complex contains a bent $(MoO_2)O(MoO_2)$ core with an Mo—O—Mo angle 142°. Each molybdenum has a distorted octahedral coordination and citrato ligands are tridentate to the two molybdenum atoms via the deprotonated hydroxy-, α-and β-carboxyl groups. Principal dimensions are:$[Mo=O(t)]_{av}$,1.706(4);$[Mo—O(b)]_{av}$,1.899(3);$[Mo—O(hydroxy)]_{av}$,1.944(3);$[Mo—O(\alpha\text{-carboxy})]_{av}$,2.207(3),and $[Mo—O(\beta\text{-carboxy})]_{av}$,2.264(3)Å. The IR,¹H and ¹³C NMR spectra are in agreement with this structure. Copyright © 1996 Elsevier Science Ltd

　**Key words**　oxomolybdenum(VI)　molybdate(VI)　citric acid citrate　nitrogenase　crystal structure.

Molybdenum is a trace element which plays an important role in different living plants and animal organisms[1]. It is known that some types of small biomolecules such as hydroxycarboxylic acids are strong molybdenum binders. Recent studies on the nitrogenase enzyme revealed its active site as a cagelike $MoFe_7S_9$ homocitrate cluster,in which the octahedral coordination sphere of the Mo within the cluster consists of three sulfides,a nitrogen atom from the imidazole group of a histidine residue,and two oxygen atoms from the hydroxy group and α-carboxyl moiety of homocitric acid[2], alternative polycarboxylic acids including citric,malic and citramalic by *in vitro* syntheses of the FeMo-co displayed $N_2$ reduction activity well above the background limits[3]. This prompted an investigation of the coordination chemistry of oxomolybdenum citrato complexes.

Citric acid($H_4$cit) has been found to act usually as a polydentate ligand,generally involving the central hydroxy and carboxyl group[4]. Of the four ionizable protons,three or four usually dissociate upon coordination,and a citrate with the charge—2 is less common. Its complexes with oxomolybdenum have been studied in the solid and in solution[5-13]. The first well characterized citrato molybdate was reported as $[Me_3N(CH_2)_6NMe_3]_2[Mo_4O_{11}(Hcit)_2]$ · $12H_2O$[8],and later a similar Mo:cit ratio 2:1 compound was prepared as $K_4[Mo_4O_{11}(Hcit)_2]$ · $6H_2O$[10]. Formation of 1:1 ratio complex was

---

①　Author to whom correspondence should be addressed.

obtained as early as $K_2[MoO_3(OH)(H_3cit)] \cdot 2H_2O$ and $K_3[MoO_4(H_3cit)] \cdot 2H_2O^{[9]}$, later $K_4[(MoO_2)_2O(Hcit)_2] \cdot 5H_2O$, $K_6[(MoO_2)_2O(cit)_2] \cdot 7H_2O$ and $K_4[MoO_3(cit)H_2O]^{[12]}$ were obtained by precipitation from aqueous solution at pH $4-8$. These complexes have been characterized by a variety of methods, however, the isolation of molybdenum(VI)citrate complexes in crystalline form suitable for X-ray structural analysis has been quite difficult[13]. The present X-ray analysis of the title compound shows that potassium trihydrogen citrate reacts with sodium molybdate to give the sodium potassium oxocitrato molybdate(VI)dimer, which shows a bent($MoO_2$)O($MoO_2$)core configuration, and the molybdenum (VI) atoms are coordinated tridentately by the oxygen atoms of hydroxy-and carboxylate of quadridentate citrate anion. The $^1H$ and $^{13}C$ NMR spectroscopy suggest this dimeric structure also exists in solution.

# 1 EXPERIMENTAL

All reagents are commercially available. Potassium trihydrogen citrate was prepared from the reaction of equimolar citric acid and sodium hydroxide. Infrared spectra were recorded as Nujol mulls between KBr plates using a Nicolet 740 FT-IR spectrometer. The $^1H$ and $^{13}C$ NMR spectra were recorded on a Varian UNITY 500 NMR spectrometer.

## Preparation of sodium potassium citrato molybdate $K_2Na_4[(MoO_2)_2O(cit)_2] \cdot 5H_2O$

Sodium molybdate(20 mmol)was reacted with an excess potassium trihydrogen citrate in a water bath. The resulting mixture was filtered,cooled and crystallized. The solid was collected and washed to give a white solid(6. 0 g,66%). Found:C,15. 4;H,1. 9. Calc. for $C_{12}H_{18}O_{24}Na_4K_2Mo_2$:C,15. 9;H, 2. 0%;IR(KBr):$v_{asym}$(C=O)1644 vs,1588 s;$v_{sym}$(C=O)1398 vs,$v$(Mo=O)950 s,902 s,$v$(Mo—O$_b$) 715 s,cm$^{-1}$,$^1H$ NMR $\sigma_H$(500 MHz,D$_2$O)ppm:2. 714(4H,$J_{AB}$ 16. 0 Hz,CH$_{2B}$),2. 516(4H,$J_{AB}$ 16. 5 Hz, CH$_{2A}$);$^{13}C$ NMR(500 MHz,D$_2$O)ppm:186. 111,185. 490 (CO$_2$)$\alpha$, 178. 372,178. 323,178. 275,178. 111 (CO$_2$)$_\beta$ 85. 738,84. 843(≡C—O),46. 131,45. 834,45. 119,44. 791,44. 085,43. 788 ppm(=CH$_2$).

Crystals of suitable quality for the subsequent X-ray diffraction studies were obtained as transparent plates by slow evaporation of a related solution of the title compound at room temperature. The resulting crystals were sealed to prevent loss of water molecules.

## X-ray data collection,structure solution and refinement

Crystallographic data for the citratomolybdate are summarized in Table 1. Diffraction data were collected on an Enraf-Nonius CAD-4 diffractometer with graphite monochromated Mo-K$_\alpha$ radiation at 296 K. Lp factor,anisotropic decay and empirical absorption corrections were applied. The structure was solved by heavy atom methods and refined by full-matrix least-squares procedures with anisotropic thermal parameters for all the non-hydrogen atoms. All H atoms were located on a difference Fourier map and refined isotropically. All calculations were performed on a 486 DX2/66 microcomputer using MoLEN software package.

# 2 RESULTS AND DISCUSSION

The title compound was prepared by the reactions of sodium molybdate and excess potassium trihydrogen citrate(KH$_3$cit). Note a similar reaction of sodium tungstate or ammonium metavanadate with citric acid has been reported[14,15]. However, the pH values in the reactions are controlled easily by citrate anions acting as both reactant and buffer agent. The loss of protons suggest the reaction(1), which corresponds to $[(MoO_4)_p(Hcit)_q(H^+)_r]^{(2p+3q-r)}$. stoichiometry of 2 : 2 : 4 showed in potentiometry, spectrophotometry, differential pulse polarography and calorimetry[13].

**Table 1　Crystal data and summaries of intensity data collection and structure refinement**

| | |
|---|---|
| Compound | $K_2Na_4[(MoO_2)_2O(cit)_2] \cdot 5H_2O$ |
| Color/shape | colorless, plate |
| Formula weight | 908.31 |
| Space group | $P2_1/n$ |
| Temp. (℃) | 23 |
| Cell constants[a] | |
| $a(\text{Å})$ | 10.036(2) |
| $b(\text{Å})$ | 13.408(2) |
| $c(\text{Å})$ | 19.816(3) |
| $\alpha(°)$ | |
| $\beta(°)$ | 90.41(1) |
| $\gamma(°)$ | |
| Cell volume($\text{Å}^3$) | 2666(1) |
| Formula units/unit cell | 4 |
| $D_{calc}$ (g cm$^3$) | 2.263 |
| $\mu_{calc}$ (cm$^{-1}$) | 13.948 |
| Diffractometer/scan | Enraf-Nonius CAD-4/ω-2θ |
| Radiation, graphite | Mo-K$_a$, ($\lambda$=0.71073 Å) |
| monochromator | |
| Max. crystal dimension, mm | 0.15×0.20×0.25 |
| Scan width | 0.45+0.35 tanθ |
| Standard reflections | 221；438 |
| Decay of standards | ± 2% |
| Reflections measured | 5367 |
| 2θ range(°) | 2≤2θ≤52 |
| Range of $h,k,I$ | ±12,−16,−24 |
| Reflections observed | 4672 |
| $[F_o \leqslant 3\sigma(F_o)]^b$ | |
| Computer programs[c] | MoLEN |
| Structure solution | MoLEN |
| No. of parameters varied | 470 |
| Weight | $[\sigma(F_v)^2+0.0001F_o^2+1]^{-1}$ |
| GOF | 0.81 |
| $R=\sum||F_o|-|F_c||/\sum|F_o|$ | 0.033 |
| $R_w$ | 0.037 |
| Largest feature final diff. map | 1.0 e$^-$ Å$^3$ |

[a]Least-squares refinement of $[(\sin\theta)/\lambda]^2$ values for 25 reflections θ>20.

[b]Corrections: Lorentz-polarization.

[c]Neutral scattering factors and anomalous dispersion corrections.

$$2MoO_4^{2-} + 2H_3cit^- \rightarrow [(MoO_2)_2O(cit)_2]^{6-} + 3H_2O \qquad (1)$$

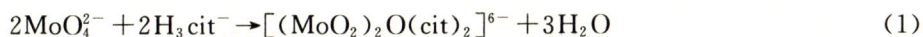

Similarly,good yield of citrato tungstate could be obtained by the reaction of sodium tungstate with sodium trihydrogen citrate as reaction(2).

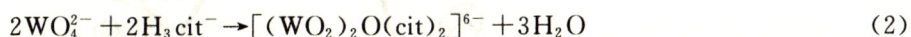

$$2WO_4^{2-} + 2H_3cit^- \rightarrow [(WO_2)_2O(cit)_2]^{6-} + 3H_2O \qquad (2)$$

This is also true for the reaction of ammonium metavanadate with potassium dihydrogen citrate as (3).

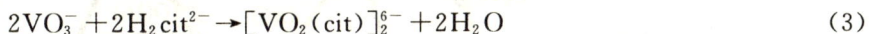

$$2VO_3^- + 2H_2cit^{2-} \rightarrow [VO_2(cit)]_2^{6-} + 2H_2O \qquad (3)$$

The complex crystallizes in the space group $P2_1/n$. Its structure is depicted in Fig. 1. Selected atomic distances and bond angles are given in Table 2.

Table 2　Selected bond lengths(Å)and angles(°)of $Na_4K_2[(MoO_2)_2O(cit)_2] \cdot 5H_2O$

| Mo(1)—O(1) | 1.916(3) | MO(2)—O(14) | 1.958(3) | O(10)—C(2) | 1.226(5) | C(1)—C(3) | 1.533(6) |
|---|---|---|---|---|---|---|---|
| Mo(1)—O(2) | 1.706(4) | Mo(2)—O(15) | 2.196(4) | O(14)—C(11) | 1.425(4) | C(1)—C(5) | 1.516(5) |
| Mo(1)—O(3) | 1.709(3) | Mo(2)—O(16) | 2.278(3) | O(15)—C(12) | 1.281(5) | C(3)—C(4) | 1.511(6) |
| Mo(1)—O(4) | 1.930(3) | O(4)—C(1) | 1.413(5) | O(16)—C(14) | 1.282(5) | C(5)—C(6) | 1.526(6) |
| Mo(1)—O(5) | 2.217(3) | O(5)—C(2) | 1.273(5) | O(17)—C(14) | 1.234(5) | C(11)—C(12) | 1.533(5) |
| Mo(1)—O(6) | 2.250(3) | O(6)—C(4) | 1.274(5) | O(18)—C(16) | 1.251(5) | C(11)—C(13) | 1.527(5) |
| Mo(2)—O(1) | 1.882(3) | O(7)—C(4) | 1.228(5) | O(19)—C(16) | 1.250(6) | C(11)—C(15) | 1.530(5) |
| Mo(2)—O(12) | 1.713(4) | O(8)—C(6) | 1.258(5) | O(20)—C(12) | 1.228(5) | C(13)—C(14) | 1.515(6) |
| Mo(2)—O(13) | 1.694(4) | O(9)—C(6) | 1.241(5) | C(1)—C(2) | 1.532(5) | C(15)—C(16) | 1.526(5) |
| O(1)—Mo(1)—O(2) | 102.1(1) | O(1)—Mo(1)—O(3) | 96.4(2) | O(1)—Mo(1)—O(4) | 148.9(1) |  |  |
| O(1)—Mo(1)—O(5) | 80.7(1) | O(1)—Mo(1)—O(6) | 76.6(1) | O(2)—Mo(1)—O(3) | 104.2(2) |  |  |
| O(2)—Mo(1)—O(4) | 97.8(1) | O(2)—Mo(1)—O(5) | 168.5(1) | O(2)—Mo(1)—O(6) | 91.9(1) |  |  |
| O(3)—Mo(1)—O(4) | 101.7(1) | O(3)—Mo(1)—O(5) | 86.6(1) | O(3)—Mo(1)—O(6) | 163.6(1) |  |  |
| O(4)—Mo(1)—O(5) | 75.3(2) | O(4)—Mo(1)—O(6) | 79.0(2) | O(5)—Mo(1)—O(6) | 77.7(2) |  |  |
| O(1)—Mo(2)—O(12) | 101.8(2) | O(1)—Mo(2)—O(13) | 101.1(2) | O(1)—Mo(2)—O(14) | 148.6(1) |  |  |
| O(1)—Mo(2)—O(15) | 79.4(1) | O(1)—Mo(2)—O(16) | 79.2(1) | O(12)—Mo(2)—O(13) | 102.7(2) |  |  |
| O(12)—Mo(2)—O(14) | 98.6(2) | O(12)—Mo(2)—O(15) | 163.4(1) | O(12)—MO(2)—O(16) | 85.9(2) |  |  |
| O(13)—Mo(2)—O(14) | 97.4(1) | O(13)—Mo(2)—O(15) | 93.2(2) | O(13)—Mo(2)—O(16) | 171.0(2) |  |  |
| O(14)—Mo(2)—O(15) | 74.4(2) | O(14)—Mo(2)—O(16) | 78.7(2) | O(15)—Mo(2)—O(16) | 78.0(1) |  |  |
| Mo(1)—O(1)—Mo(2) | 144.7(2) |  |  |  |  |  |  |

The dimeric anion consists of the common oxobridged $[Mo_2O_5]^{2+}$ entity which is noncentrosymmetric. The angle of Mo—O—Mo is 144.7(2)°,which is intermediate between the values of strictly 180° in $[(MoO_2)_2O(C_2O_4)_2(H_2O)_2]^{2-}$ and 136° in $[(MoO)_2O(O_2)_4(H_2O)_4]^{2-}$ [16,17]. The two terminal oxo groups are in a *cis*-configuration. Each molybdenum atom is six-coordinate with approximately octahedral geometry. The terminal and bridging oxogroups adopt a *fac*-stereochemistry. The *trans* positions are occupied by a fully deprotonated tridentate citrate. The coordination mode is similar to the proposed structure of citrato cubane $MoFe_3S_4$ cluster $(Et_4N)_2[MoFe_3S_4Cl_3(H_2cit)]$,and probably the hydroxy group in the cubane should be deprotonated[18].

Although no symmetry is imposed upon the molecule,there is a pseudo-twofold axis that passes through the bridging oxygen, which is perpendicular to the Mo—Mo vector. The lists of chemically equivalent bond distances and angles (Table 2)illustrate the approximately twofold symmetry. The

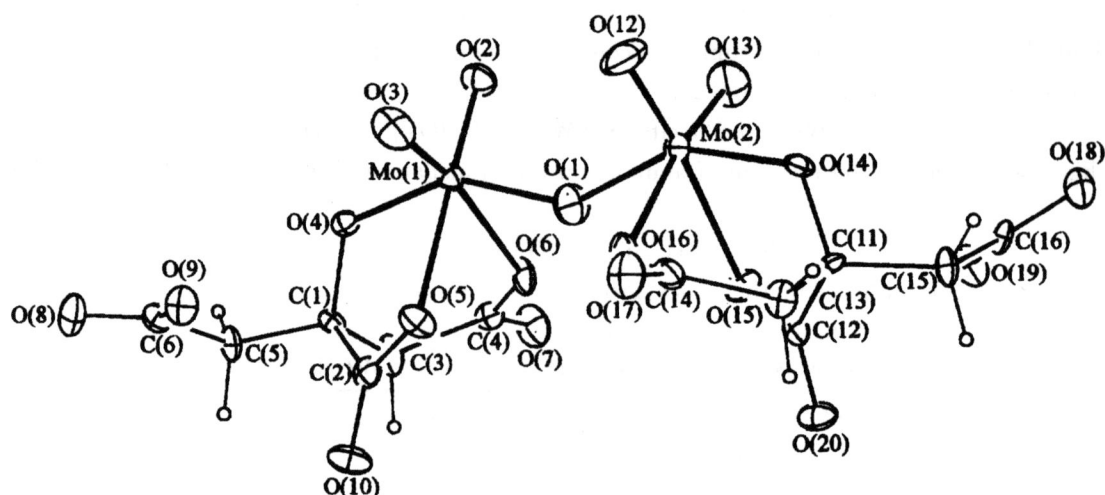

**Fig. 1  Perspective view of the anion structure of citratomolybdate.**

configuration is consistent with one of the proposed structures of 1 : 1 dinuclear complex of molybdenum(VI) and citrate[7,12,13]. The Mo—O distances can be classified into three types with the order Mo $=$ O(unshared oxygen)(1. 694~1. 713Å)< Mo—O(bridging oxo group)(1. 882~1. 916 Å)$\approx$ Mo—O(hydroxy)(1. 930 ~ 1. 958 Å)< Mo—O(α-carboxy)(2. 196 ~ 2. 217Å)$\approx$ Mo—O(β-carboxy) (2. 250~2. 278 Å). The Mo(1,2)—O(1)distances(av. 1. 899 Å)are very close to those in $[(MoO_2)_2O$ $(C_2O_4)_2(H_2O)_2]^{2-}$(1. 876 Å)[16] and $[(MoO_2)_2O(Hnta)_2]^{2-}$ [1. 880 Å, $H_3nta = N(CH_2CO_2H)_3]^{[19]}$. The Mo(VI)—O(hydroxy)distances(av. 1. 944 Å)are similar to those in $[MoO_2(Hmal)_2]^{2-}$ ($H_3mal =$ malic acid)(1. 939 Å)[20] and $[MoO(O_2)_2(H_2cit)]^{2-}$ (2. 011 Å)[21]. The Mo—O(α-carboxy)distances (av. 2. 207 Å)and the longest Mo—O(β-carboxy)(av. 2. 264 Å)show weak coordination of the carboxyl group with Mo$^{VI}$. The Mo—O(β-carboxy)is shorter than those already reported tetramers $[(MoO_2)_4O_3$ $(Hcit)_2]^{4-}$ (2. 323 Å)[14] and $[(MoO_2)_4O_3(R-mal)_2]^{4-}$ (2. 296 Å)[22] (Table 3).

**Table 3  Comparisons of the M—O bonds(Å)in malato or citrato complexes**

| Complex | Mo—O(hydroxy) | M—O(α-carboxy) | M—O(β-carboxy) | Ref. |
|---|---|---|---|---|
| $Cs_2[MoO_2(Hmal)_2]\cdot H_2O$, $K_2[MoO_2(Hmal)_2]\cdot H_2O$ | 1. 939(8) | 2. 243(9) | | [20] |
| $(NH_4)_4[(MoO_2)_2O_3(R-mal)_2]\cdot 6H_2O$ | 1. 925(5) | 2. 235(5) | 2. 296(5) | [22] |
| $K_2[MoO(O_2)_2(H_2cit)]\cdot 1/2H_2O_2\cdot 6H_2O$ | 2. 011(7) | 2. 220(8) | | [21] |
| $[Me_3N(CH_2)_6NMe_3]_2[(MoO_2)_4O_3(Hcit)_2]\cdot 12H_2O$, $K_4[(MoO_2)_4O_3(Hcit)_2]\cdot 6H_2O$ | 1. 976(5),1. 968(5) | 2. 185(5),2. 211(5) | 2. 318(5)av | [8] |
| $K_2Na_4[(MoO_2)_2O(cit)_2]\cdot 5H_2O$ | 1. 929(3),1. 958(3) | 2. 218(3),2. 197(4) | 2. 264(3)av | This work |
| $Na_6[(WO_2)_2O(cit)_2]\cdot 10H_2O$ | 1. 958(2) | 2. 195(3) | 2. 289(2) | [14] |
| $K_2(NH_4)_4[VO_2(cit)]_2\cdot 6H_2O$ | 1. 961(2),2. 005(2) | 1. 981(3) | | [15] |
| $K_2[VO_2(H_2cit)]_2\cdot 4H_2O$ | 1. 961(3),2. 010(4) | 1. 980(3) | | [23] |

Due to the Mo—O single bond which exists in the dimer and the existence of the monomer in narrow pH range[12],the equilibrium for the dimerization reaction might proceed easily as follows:

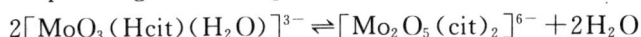

$$2[MoO_3(Hcit)(H_2O)]^{3-} \rightleftharpoons [Mo_2O_5(cit)_2]^{6-} + 2H_2O$$

Moreover,like the citratovanadates,the protonation and deprotonation of citratomolybdate might have less influence on the dimeric structure[23]. This means the different $p,q,r$ in $[(MoO_4)_p(Hcit)_q (H^+)_r]^{(2p+3q-r)}$ like $[2,2,4],[2,2,5],[2,2,6]$ can be transformed in solution without changing the dimeric structure of citratomolybdate as following:

$$[Mo_2O_5(cit)_2]^{6-} \rightleftharpoons [Mo_2O_5(Hcit)(cit)]^{5-} \rightleftharpoons [Mo_2O_5(Hcit)_2]^{4-}$$

The crystal structure comprises discrete $[(MoO_2)_2 O(cit)_2]^{6-}$ anions,sodium or potassium cations and water molecules. Waters of crystallization are distributed throughout the cell and together with various of the anion oxygens,make the sodium or potassium cations six-coordinate.

Moreover,the complex exhibits the following infrared absorption bands: $v_{asym}$(COO):1644 vs,1588 s, $v_{sym}$(COO):1398 vs, $v$(Mo=O):950 s,902 s cm$^{-1}$ and $v$(Mo—O$_b$—Mo):715 s. The separation($\Delta$) between $v_{asym}$ and $v_{sym}$ is 246 and 190 cm$^{-1}$,which is greater than the value of 150 cm$^{-1}$ for uncompleted sodium citrate salt,implying the presence of free or unidentately coordinated carboxyl groups. The strong band at 715 cm$^{-1}$ is assigned to the symmetric Mo—O—Mo stretch in agreement with previous studies. There is no other band at higher energy that can be assigned to asym stretching consistent with a bent disposition of the bridge. The Mo=O bands give rise to two major bands around 900 cm$^{-1}$, characteristic of the symmetric and asymmetric stretches of the *cis*-dioxo groups.

In the $^1$H NMR spectra the methylene protons of citric acid give rise to an AB quartet which shifts with pH[24]. Complexation with molybdenum(Ⅵ)leads to a slightly distorted AB quartet with $\delta q =$ 2. 714,2. 516 ppm respectively,indicating that both $CH_2CO_2$ arms are almost equivalent[12]. The shifts of the $^{13}$C resonances(in comparison with those for citrate)[12,24] of the alcoholic and α-carboxylic function show that the citrate is coordinated through these two groups,while the β-carboxylic region only a small shift is observed,indicating that the bonding to molybdenum is weak.

Acknowledgements—Financial support by the National Science Foundation of China ( No. 29503021)and the State Science and Technology Committee is gratefully acknowledged.

# REFERENCES

[1]E. I. Stiefel, Molybdenum Enzymes,Cofactors and Model Systems, p. 1,American Chemical Society,Washington,DC(1993).

[2](a)J. Kim and D. C. Rees,Science 1992,**257**,2677; (b)M. K. Chan,J. Kim and D. C. Rees, Science 1993,**260**,792.

[3]M. S. Madden,T. D. Paustian,P. W. Ludden and V. K. Shah,J. Bacteriol 1991,**173**,5403.

[4]J. P. Glusker,Acc. Chem. Res. 1980,**13**,345.

[5]J. D. Jesus and M. D. Farropas,Trans. Met. Chem. 1983,**8**,193.

[6]J. J. Cruywagen and R. F. Van de Water,Polyhedron 1986,**5**,521.

[7]M. Bartusek,J. Havel and D. Matula,Collect. Czech. Chem. Commun. 1986,**51**,2702.

[8]L. R. Nassimbeni,M. L. Niven,J. J. Cruywagen and J. B. B. Heyns,J. Crystalloqr. Spectrosc. Res. 1987,**17**,373.

[9]M. Dudek,E. Hodorowicz,A. Kanas,A. Samo-tus,B. Sieklucka,J. Szklarzewicz and E. Beltow-ska-Lehman,Proc. 11th Conf. Coord. Chem.,p. 57. Smolenice,Czechoslovakia.

[10]N. W. Alcock,M. Dudek,R. Grybos,E. Hodorowicz,A. Kanas and A. Samotus,J. Chem. Soc., Dalton Trans. 1990,707.

[11] D. O. Martire, M. R. Feliz and A. L. Capparelli, Phys. Chem. Leipzig. 1989, **270**, 951.

[12] A. Samotus, A. Kanas, M. Dudek, R. Grybos and E. Hodorowicz, Trans. Met. Chem. 1991, **16**, 495.

[13] J. J. Cruywagen, E, A. Rohwer and G. F. S. Wessels, Polyhedron 1995, **14**, 3481.

[14] (a) J. J. Cruywagen, L. J. Saayman and M. L. Niven, J. Crystallogr. Spectrosc. Res. 1992, **22**, 737; (b) E. Llopis, J. A. Ramirez, A. Domenech and A. Cervilla, J. Chem. Soc., Dalton Trans. 1993, 1121.

[15] (a) Z, H. Zhou, H. L. Wan and K. R. Tsai, Chinese Sci. Bull. 1995, **40**, 749; (b) Z. H. Zhou, H. L. Wan, S. Z. Hu and K. R. Tsai, Inorg. Chim. Acta 1995, **237**, 193.

[16] F. A. cotton, S. M. Morehause and S. J. Wood, Inorg. Chem. 1964, **3**, 1603.

[17] R. Stomberg, Acta Chem. Scand. 1968, **22**, 1076.

[18] K. D. Demadis and D. Coucouvanis, Inorg. Chem. 1995, **34**, 436.

[19] C. Knobler, B. R. Penfold, W. T. Robinson, C. J. Wilkins and S. H. Yong, J. Chem. Soc., Dalton Trans, 1980, 248.

[20] (a) C. B. Knobler, A. J. Wilson, R. N. Hider, I. W. Jensen, B. R. Penfold, W. T. Robinson and C. J. Wilkins, J. Chem. Soc., Dalton Trans. 1983, 1299; (b) R. Stomberg and S. Olson, Acta Chem. Scand., Ser. A 1985, **39**, 79.

[21] J. Flanagan, W. P. Griffith, A. C. Skapski and R. W. Wiggins, Inorg. Chim. Acta 1985, **96**, L23.

[22] (a) J. E. Berg, S. Brandange, L. Lindbolm and P. E. Werner, Acta Chem. Scand., Ser. A 1977, **31**, 325; (b) M. A. Porai-Koshits, L. A. Aslanov, G. V. Ivanova and T. N. Polinova, J. Struc. Chem. (Eng. Transl.) 1968, **9**, 401; (c) Z. H. Zhou, W. B. Yan, H. L. Wan and K. R. Tsai, Chinese J. Struc. Chem. 1995, **14**, 255.

[23] Z. H. Zhou, W. B. Yan, H. L. Wan, K. R. Tsai, J. Z. Wang and S. Z. Hu, J. Chem. Crystallogr. 1995, **25**, 807; (b) D. W. Wright, P. A. Humiston, W. H. Orme-Johnson and W. H. Davis, Inorg. Chem, 1995, **34**, 4194.

[24] (a) A. Loewenstein and J. D. Roberts, J. Chem. Soc. 1960, **82**, 2705; (b) A. Cervilla, J. A. Ramirez and E. Llopis, Trans. Met. Chem. 1986, **11**, 186.

■ 本文原载:JOURNAL OF CATALYSIS **170**,1997,pp. 191~199.

# Supported Au Catalysts Prepared from Au Phosphine Complexes and As-Precipitated Metal Hydroxides: Characterization and Low-Temperature CO Oxidation

You-Zhu Yuan[a,b], Anguelina P. Kozlova[a], Kiyotaka Asakura[c],
Hui-Lin Wan[b], Khi-Rui Tsai[b], Yasuhiro Iwasawa[a,①]

(*[a]Department of Chemistry, Graduate School of Science, The University of Tokyo,
Hongo, Bunkyo-ku, Tokyo 113, Japan; [b]Department of Chemistry and the State Key
Laboratory of Physical Chemistry for Solid Surface, Xiamen University,
Xiamen 361005, China; [c]Research Center for Spectrochemistry,
Faculty of Science, The University of Tokyo, Hongo, Bunkyo-ku, Tokyo 113, Japan)*

Received January 23, 1997; revised May 13, 1997; accepted May 14, 1997

**Abstract**    Supported Au catalysts were prepared by attaching Au phosphine complexes, $Au(PPh_3)(NO_3)(1)$ and $[Au_9(PPh_3)_8](NO_3)_3(2)$, on as-precipitated metal hydroxides $M(OH)_x^*$ ( *, as-precipitated; $M = Mn^{2+}$, $Co^{2+}$, $Fe^{3+}$, $Ni^{2+}$, $Zn^{2+}$, $Mg^{2+}$, $Cu^{2+}$, $Ti^{4+}$, $Ce^{4+}$, and $La^{3+}$ ), followed by temperature-programmed calcination in a flow of dry air. The obtained Au catalysts showed high catalytic activities in low-temperature CO oxidation. Among the obtained Au catalysts $1/Mn(OH)_2^*$ and $1/Co(OH)_2^*$ were most highly active even at 203 K. $1/Fe(OH)_3^*$ and $1/Ti(OH)_4^*$ also catalyzed CO oxidation at low temperatures 203—273 K, whereas $1/Fe_2O_3$ and $1/TiO_2$ prepared by supporting 1 on conventional $Fe_2O_3$ and $TiO_2$ showed negligible activity under the similar reaction conditions. It was estimated by TEM and XRD that the mean diameter of Au particles in $1/Fe(OH)_3^*$ was about 2.9 nm, which was about 10 times smaller than that for $1/Fe_2O_3$. EXAFS for $1/Ti(OH)_4^*$ revealed that the coordination number of Au—Au bond was 8—10, while that for $1/TiO2$ was 11.0, which also indicates that Au particle size for $1/Ti(OH)_4^*$ is smaller than that for $1/TiO_2$. The catalysts obtained by attaching the Au complexes on commercially available metal hydroxides also showed negligible activity for the low-temperature CO oxidation under identical conditions. These results demonstrate that supported Au catalysts with small Au particles, tremendously active for the low-temperature CO oxidation, can be prepared by attaching the Au phosphine complexes on the as-precipitated metal hydroxides. Sodium cations exhibited positive effect on the Au catalysis, whereas chloride anions drastically decreased the CO oxidation activity. © 1997 Academic Press

---

①   To whom all correspondence should be addressed. Fax:81-3-5800-6892. E-mail:iwasawa@utsc. s. u-tokyo. ac. jp.

# 1  INTRODUCTION

Low-temperature CO oxidation has been extensively studied as the key issue relevant to gas purification in $CO_2$ lasers, CO gas sensors, air-purification devices for respiratory protection, and pollution control devices for reducing industrial and environmental emission and removing a trace quantity of CO from the ambient air in enclosed atmospheres such as submarines and space crafts on long-duration missions[1-27]. Although $Pt/SnO_x$ and $Pd/SnO_x$ are effective catalysts for this reaction[1-3], there are complications of pretreatments for catalyst reduction and induction periods before reaching maximum catalytic activities[4]. It has been reported that Au particles dispersed on oxide supports are active for catalytic CO oxidation[5-7]. Haruta and co-workers screened a number of reducible-oxide-supported metals and found that small Au particles supported on $\alpha$-$Fe_2O_3$, $Co_3O_4$, $TiO_2$, and NiO were efficient CO oxidation catalysts at low temperatures[8-13]. Hoflund and co-workers also demonstrated that Au on $MnO_x$ was active for this reaction[18,19]. Numerous experimental results have demonstrated that supported Au particles work as CO oxidation catalysts[20-25].

Support effect and particle size effect on the Au catalysis have been substantiated with a number of Au catalysts. $Au/Fe_2O_3$, $Au/Co_3O_4$, $Au/NiO$, $Au/Al_2O_3$, and $Au/SiO_2$ showed an overall trend that the CO oxidation activity increased with decreasing size of the Au particles[9]. Ultrafine Au particles may be prerequisite for observation of a Au-support interaction[26]. Given this assumption, the reactivity of supported Au catalysts may be determined by the method of catalyst preparation which controls Au particle size and Au-support interaction.

Use of molecular metal complexes is a promising way to prepare tailored metal catalysts with more efficient interaction between metal and support to generate unique catalysis[27-32]. However, it is scarcely reported that supported Au catalysts with small particles are obtained by decomposition of Au complexes or Au clusters. Our earlier work[33,34] showed that phosphine-stabilized Au complexes and clusters supported on inorganic oxides were readily decomposed to form gold metallic particles by heat treatments in vacuum and the obtained catalysts were active for CO oxidation under mild conditions, typically at 300-400 K. XRD and EXAFS revealed that Au particles with sizes>12 nm were formed on the oxide surfaces by the heat treatment.

Recently, we developed a new way to prepare supported Au catalysts with highly dispersed Au particles by attaching a Au phosphine complex $Au(PPh_3)(NO_3)$ (1) and a Au phosphine cluster $[Au_9(PPh_3)_8](NO_3)_3$ (2), on as-precipitated metal hydroxides, followed by temperature-programmed calcination in a flow of air[35]. The highly dispersed Au particles on the specially obtained supports showed extremely high activity for catalytic CO oxidation at low temperatures. Here we report the characterization by EXAFS, XRD, and TEM and performance for low-temperature CO oxidation of the Au catalysts prepared from the Au complexes 1 or 2 and the as-precipitated metal hydroxides.

# 2  EXPERIMENTAL

## Preparation of Catalyst

Au phosphine complex, $Au(PPh_3)(NO_3)$ (1), and Au phosphine cluster, $[Au_9(PPh_3)_8](NO_3)_3$ (2) were synthesized according to the literature[36,37]. As-precipitated metal hydroxides were prepared by precipitation of metal nitrates(99.9% purity)with an aqueous solution of $Na_2CO_3$(99.9%)in a similar manner to the coprecipitation method[9-14]. As-precipitated vanadium hydroxide was obtained by hydrolysis of vanadium chloride(99.9%) with an aqueous solution of $Na_2CO_3$. The precipitates were repeatedly washed until no Cl was detected in the filtrates. Three kinds of as-precipitated $Ti(OH)_4^*$ were obtained by hydrolysis of $Ti(i\text{-}OC_3H_7)_4$ (titanium-tetra-*iso*-propoxide, 99.999%) with deionized water(denoted as $W\text{-}Ti(OH)_4^*$), by hydrolysis with an aqueous solution of $Na_2CO_3$ (denoted as $S\text{-}Ti(OH)_4^*$), and by hydrolysis with deionized water, followed by impregnation of $Na_2CO_3$ (Na/Au= 1.0 in an atomic ratio)(denoted as $D\text{-}Ti(OH)_4^*$). The precipitates were repeatedly washed with deionized water until the pH became 7.0. The as-precipitated $M(OH)_x^*$ were filtered and immediately impregnated with an acetone(99.9%)solution of **1** or a methanol(99.8%)solution of **2**, with vigorous stirring for at least 12 h, followed by evacuation for 5 h to remove the solvent at room temperature. The obtained samples were calcined in a glass tube at a heating rate of 4 K/min to 673 K and at 673 K for 4 h in a flow of air(30 mL/min). The catalysts thus prepared are denoted as $\mathbf{1}/M(OH)_x^*$ or $\mathbf{2}/M(OH)_x^*$. Note that the samples are always used as catalysts after the temperature-programmed calcination. For comparison, $Au/Fe_2O_3$ catalysts were prepared by a coprecipitation method by using $HAuCl_4$ and Fe $(NO_3)_3$ as reported in literature[9,10] and by impregnating **1** or **2** on commercially available $Fe_2O_3$ and $TiO_2$(Desussa P-25). These catalysts are denoted as coprec-$Au/Fe_2O_3$ and **1** or $\mathbf{2}/Fe_2O_3$ and **1** or $\mathbf{2}/TiO_2$. The impregnated samples were calcined in the similar way to that for **1** or $\mathbf{2}/M(OH)_x^*$. The Au loading on each support was controlled to be 3.0 wt% except for the coprecipitated catalyst(10 wt%).

## Characterization

The samples were characterized by X-ray diffraction(XRD), Transmission electron microscopy (TEM), and extended X-ray absorption fine structure(EXAFS). XRD patterns were recorded on a Rigaku powder X-ray diffractometer with $CuK\alpha$ radiation over the $2\theta$ range $20°\sim80°$. The TEM image for the Au catalysts was observed using a Hitachi H-1500 electron microscope operated at 800 kV. At least 300 particles were used to determine the mean diameter of Au particles. Au $L_3$-edge EXAFS spectra were measured in a transmission mode at BL-10B of the Photon Factory in the National Laboratory for High Energy Physics(Proposal No. 95G200). The measurements were carried out at a beam current of 250-350 mA and a storage-ring energy of 2.5 GeV. The samples were calcined in a U-shaped glass tube combined in a fixed-bed flow reaction system in a flow of air(30 mL/min) and transferred to EXAFS cells connected to the U-shaped glass tube. Data were analyzed by a curve-fitting method, using empirically derived phase-shift and amplitude functions. The interactions of Au-Au and Au-P were calculated by using the FEFF 6.0 software[38,39]. The parameters used for the analysis are summarized in Table 1.

## Catalytic CO Oxidation

CO oxidation reactions were carried out in a fixed-bed flow reactor equipped with a computer-controlled autosampling system by using 200 mg catalyst powder. The reaction gas containing 1.0% CO balanced with air purified through a molecular sieve column was passed through the catalyst bed at a flowrate of 67 mL/min (SV = 20,000 mL/h/g). Reaction products were analyzed by a gas chromatograph using a column of Unibeads C for $CO_2$ and a column of 5A molecular sieve for CO and $O_2$. The material balance in the catalytic reactions was checked from the concentration of $CO_2$ and CO, showing good balance under all the reaction conditions examined.

| FEFF calculation | | | Fourier range | |
|---|---|---|---|---|
| Shell | $N$ | $r(\text{Å})$ | $\Delta k$ range($\text{Å}^{-1}$) | $\Delta r$ range($\text{Å}$) |
| Au-P for Au(PPh$_3$) | 1.0 | 2.28 | 3.0—14.50 | 1.40—2.30 |
| Au-Au for Au(Au) | 1.0 | 2.87 | 3.0—14.50 | 2.00—3.20 |
| or Au(Pt) | | | | |
| Au-Au for Au foil | 1.0 | 2.87 | 3.0—16.0 | 2.10—3.15 |

*Note*. $N$, coordination number of backscatterer; $r$, distance; $\Delta k$, range used for Fourier transformation ($k$, wave number); $\Delta r$, range used for Fourier filtering ($r$, distance).

# 3  RESULTS

## Characterization of EXAFS, XRD, and TEM

EXAFS measurements in a transmission mode have been carried out at room temperature to characterize 1/TiO$_2$ and 1/Ti(OH)$_4^*$. The Fourier transforms of EXAFS oscillations at Au $L_3$-edge for 1/TiO$_2$ and 1/Ti (OH)$_4^*$, before and after calcination are shown in Fig. 1. The peaks around 1.8 Å in Figs. 1a,1b,and 1c are assignable to Au-P bond, which indicates that upon supporting the Au phosphine complex **1** on W-Ti (OH)$_4^*$, S-Ti(OH)$_4^*$, and TiO$_2$ the complex **1** is not decomposed. However, after calcination at 673 K in air, the EXFAS Fourier transform peak shifted to longer distance. The EXAFS curve-fitting analysis revealed that the new peak is due to Au-Au bonding. The coordination number and distance of Au-Au bond were determined to be 8.3 and 2.86 Å for 1/W-Ti(OH)$_4^*$ and 9.8 and 2.86 Å for 1/S-Ti (OH)$_4^*$, while those for 1/TiO$_2$ were 11.0 and 2.87 Å, as shown in Table 2. No Au-P bond was observed with the calcined samples. The results demonstrate that the Au complexes on the supports were decomposed to form metallic Au particles by the calcination at 673 K. The coordination numbers for 1/Ti(OH)$_4^*$ and 1/TiO$_2$ were different from each other, suggesting the formation of Au particles with different sizes.

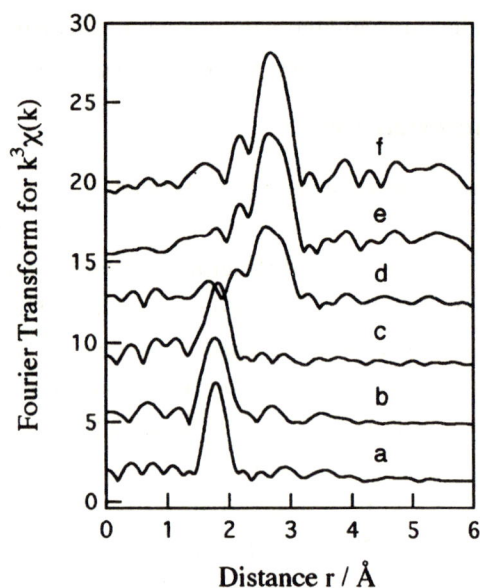

**FIG. 1  Fourier transforms of EXAFS oscillation at Au $L_3$-edge for 1/TiO$_2$ and 1/Ti(OH)$_4^*$, before and after calcination. (a) 1/W-Ti(OH)$_4^*$ (before calcination); (b) 1/S-Ti(OH)$_4^*$ (before calcination); (c) 1/TiO$_2$ (before calcination); (d) 1/W-Ti(OH)$_4^*$ (after the temperature-programmed calcination); (e) 1/S-Ti (OH)$_4^*$ (after the temperature-programmed calcination); (f) 1/TiO$_2$ (after the temperature-programmed calcination).**

The change in the Au species from the phosphine complex to metallic particles was accompanied by the development of the XRD peaks for crystalline TiO$_2$ as shown in Fig. 2. The peaks for TiO$_2$ derived from Ti(OH)$_4^*$ are much broader than those for TiO$_2$(P-25). When S-Ti(OH)$_4^*$ obtained by hydrolysis

of Ti($i$-OC$_3$H$_7$)$_4$ with an aqueous solution of Na$_2$CO$_3$ was used as a support for **1**, the calcined sample 1/
S-Ti(OH)$_4^*$ constituted an amorphous type of TiO$_2$ (Fig. 2d). Contrary to this, when Ti($i$-OC$_3$H$_7$)$_4$ was
hydrolyzed by H$_2$O, the TiO$_2$ structure in 1/W-Ti(OH)$_4^*$ was mainly of anatase type(Fig. 2b). The 1/
TiO$_2$ catalyst showed a Au(200)peak at $2\theta = 44.4°$, whereas the XRD peak for crystalline Au particles
was hardly observed with 1/W—Ti(OH)$_4^*$, 1/D-Ti(OH)$_4^*$, and 1/S-Ti(OH)$_4^*$ as shown in Figs. 2b—
2d.

**TABLE 2    Curve-Fitting Results for the Au L$_3$-Edge EXAFS Data of 1/(TiOH)$_4^*$ and 1/TiO$_2$**

| $T$(K) | Au−P | | | | Au−Au | | | | $R_f$(%) |
|---|---|---|---|---|---|---|---|---|---|
| | $N$ | $r$(Å) | $\Delta\sigma^2$(Å$^2$) | $\Delta E$(eV) | $N$ | $r$(Å) | $\Delta\sigma^2$(Å$^2$) | $\Delta E$(eV) | |
| | | | | 1/W-(TiOH)$_4^*$ | | | | | |
| 293 | 1.1±0.2 | 2.20±0.01 | 0.0021 | −5.42 | − | − | − | − | 1.7 |
| 673 | − | − | − | − | 8.3±1.0 | 2.86±0.01 | 0.0014 | 1.45 | 1.2 |
| | | | | 1/S-(TiOH)$_4^*$ | | | | | |
| 293 | 1.4±0.3 | 2.20±0.01 | 0.0006 | −3.46 | − | − | − | − | 1.3 |
| 673 | − | − | − | − | 9.8±1.0 | 2.86±0.01 | 0.0001 | 2.55 | 0.9 |
| | | | | 1/TiO$_2$ | | | | | |
| 293 | 1.3±0.3 | 2.22±0.01 | 0.0000 | −4.05 | − | − | − | − | 0.9 |
| 673 | − | − | − | − | 11.0±1.0 | 2.87±0.01 | 0.0001 | 2.12 | 0.7 |

Figure 3 shows the XRD patterns for the Au/Fe oxide catalysts, coprec-Au/Fe$_2$O$_3$, 1/Fe$_2$O$_3$, 1/Fe
(OH)$_3^*$, 2/Fe(OH)$_3^3$, Au(PPh$_3$)Cl/Fe(OH)$_3^*$, HAuCl$_4$/Fe(OH)$_3^*$, and calcined Fe(OH)$_3^*$. The
intensity of the diffraction peak at $2\theta = 38.2°$ for Au(111)depended on the kind of samples. The order of
the sharpness of the Au(111)diffraction peak for the samples was as follows: Au(PPh$_3$)Cl/Fe(OH)$_3^*$ >
HAuCl$_4$/Fe(OH)$_3^*$ ≥ 2/Fe(OH)$_3^*$ ≥ 1/Fe$_2$O$_3$ ≫ coprec-Au/Fe$_2$O$_3$ ≥ 1/Fe(OH)$_3^*$.

The Au particle sizes for **1**/Fe(OH)$_3^*$ and **1**/Fe$_2$O$_3$ were estimated by transmission electron
micrograph in Fig. 4. The mean diameter of Au particles in **1**/Fe(OH)$_3$ was about 2.9 nm, where most
of the particles ranged about 1—5 nm. In contrast, the mean diameter of Au particles in **1**/Fe$_2$O$_3$ was as
large as 30 nm, ranging 10—50 nm. Thus much smaller Au particles are produced on the as-precipitated
Fe(OH)$_3^*$ than on Fe$_2$O$_3$, which coincides with the results of EXAFS and XRD.

## Catalytic Performance

As a typical example, Fig. 5 depicts the catalytic activities(CO conversion) of the supported Au
catalysts, coprec-Au/Fe$_2$O$_3$, **1**/Fe(OH)$_3^*$, **1**/Fe(OH)$_3^*$(dried), Fe(OH)$_3^*$ alone, and **1**/Fe$_2$O$_3$, for CO
oxidation in a fixed-bed flow reactor as a function of reaction temperature. Under the present reaction
condition(SV=20,000 mL/h/g-cat;CO=1%),100% conversion corresponds to a reaction rate of 8.93
$\times 10^{-3}$ mol/h/g-cat. It is to be noted that **1**/Fe(OH)$_3^*$ is tremendously active, compared with **1**/Fe$_2$O$_3$.
The CO oxidation reaction on **1**/Fe$_2$O$_3$ proceeded only above 300 K, whereas the reaction on
**1**/Fe(OH)$_3^*$ occurred even at 203 K. Under the present conditions the catalytic activity of the coprec-
Au/Fe$_2$O$_3$ catalyst prepared by the coprecipitation method in the literature[9,10] was much lower than
that of the **1**/Fe(OH)$_3^*$ catalyst.

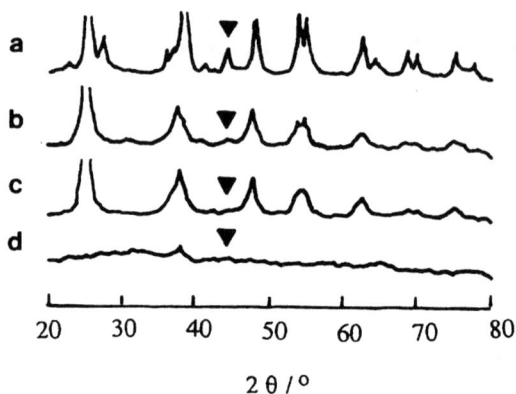

FIG. 2 XRD patterns of 1/Ti(OH)$_4^*$ and 1/TiO$_2$. ▼ indicates Au(200) peak. All the catalysts (hereinafter otherwise noted) are the ones after the temperature-programmed calcination. Catalysts: (a)1/TiO$_2$; (b)1/W—Ti(OH)$_4^*$; (c) 1/D—Ti(OH)$_4^*$; (d)1/S—Ti(OH)$_4^*$.

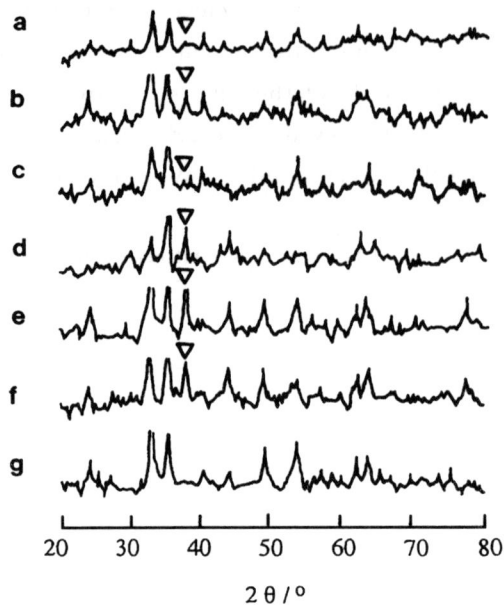

FIG. 3 XRD patterns of various Au/Fe oxide catalysts after the temperature-programmed calcination. ▽ indicates Au(111) peak. Catalysts: (a) coprec-Au/Fe$_2$O$_3$; (b)1/Fe$_2$O$_3$; (c)1/Fe(OH)$_3^*$; (d)2/Fe(OH)$_3^*$; (e) Au(PPhs)Cl/Fe(OH)$_3^*$; (f) treated HAuCl$_4$/Fe(OH)$_3^*$; (g)Fe(OH)$_3^*$.

Table 3 shows comparison of the catalytic activities of various 1/$M$(OH)$_x^*$ for CO oxidation. Among all the catalysts, those prepared by using the as-precipitated metal hydroxides originated from Mn(NO$_3$)$_2$, Co(NO$_3$)$_3$, Fe(NO$_3$)$_3$, Ni(NO$_3$)$_2$, Mg(NO$_3$)$_2$, Zn(NO$_3$)$_2$, and Ti($i$-OC$_3$H$_7$)$_4$ were very active in CO oxidation at the low temperatures.

TABLE 3 Catalytic CO Oxidation Activities of Various 1/$M$(OH)$_x^*$ Catalysts

| Original catalyst | $-T_s$(K) | $T_{50\%}$(K) | $T_{100\%}$(K) |
|---|---|---|---|
| 1/Mn(OH)$_2^*$ | <203 | <203 | 273 |
| 1/Co(OH)$_2^*$ | <203 | <203 | 273 |
| 1/Ni(OH)$_2^*$ | <203 | 230 | 273 |
| 1/Fe(OH)$_3^*$ | <203 | 206 | 273 |
| 1/Zn(OH)$_2^*$ | <203 | 248 | 273 |
| 1/Ce(OH)$_4^*$ | 243 | 263 | 283 |
| 1/Mg(OH)$_2^*$ | 203 | 250 | >373 |
| 1/Ti(OH)$_4^*$ | 253 | 304 | 433 |
| 1/Cu(OH)$_2^*$ | 223 | 334 | 443 |
| 1/La(OH)$_3^*$ | 283 | 335 | 503 |
| 1/Al(OH)$_3^*$ | 373 | 606 | >633 |
| 1/V(OH)$_3^*$ | ~383 | 649 | >773 |
| 1/Cr(OH)$_3^*$ | ~473 | 735 | >773 |

*Note.* $T_s$, temperature for CO oxidation to start; $T_{50\%}$, temperature for 50% conversion; $T_{100\%}$, temperature for 100% conversion.

## Influence of Sodium Cation

The CO conversions on $1/W\text{-}Ti(OH)_4^*$ ($Na^+ - $ free), $1/S\text{-}Ti(OH)_4^*$ (containing $Na^+$), and $1/D\text{-}Ti(OH)_4^*$ ($Na^+$-doped) are plotted as a function of reaction temperature in Fig. 6. It was found that $1/S\text{-}Ti(OH)_4^*$ was much more efficient for the low-temperature CO oxidation than $1/W\text{-}Ti(OH)_4^*$. Addition of a small amount of $Na^+$ ions to the as-precipitated $W\text{-}Ti(OH)_4^*$ by impregnation of aqueous $Na_2CO_3$ solution remarkably promoted the $Au/Ti(OH)_4^*$ catalysis.

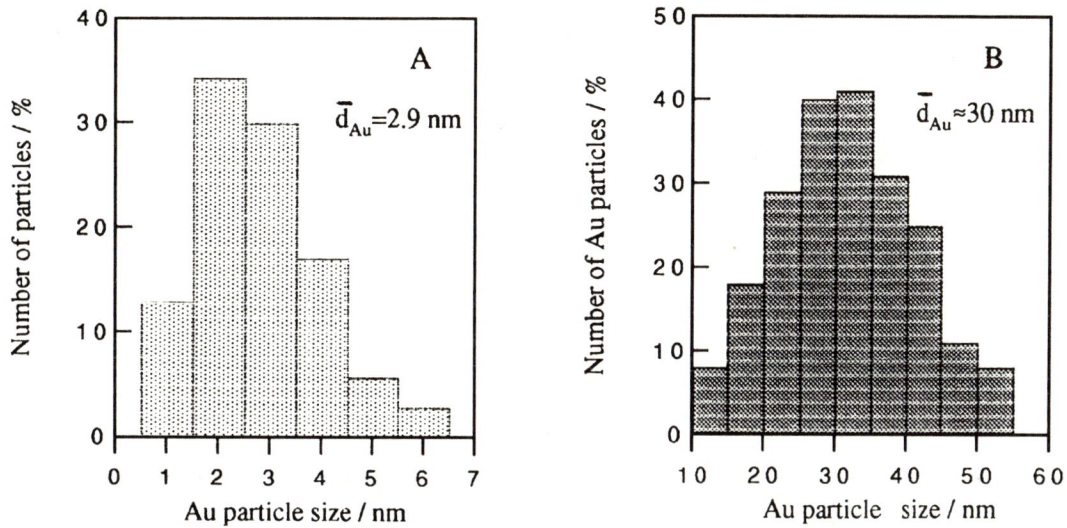

FIG. 4  Size distribution for Au particles estimated by TEM. Catalysts:(A)$1/Fe(OH)_3^*$;(B)$1/Fe_2O_3$.

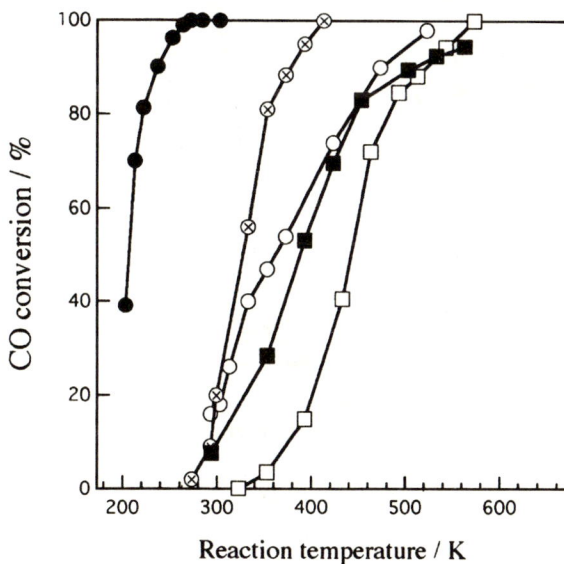

FIG. 5  Catalytic CO oxidation reactions on various Au/Fe oxide catalysts in a fixed-bed flow reactor(SV = 20, 000 ml/h/g) as a function of reaction temperature. ( ● ) $1/Fe(OH)_3^*$ ; ( ⊗ ) coprec-$Au/Fe_2O_3$; ( ○ ) $1/$ dried $Fe(OH)_3^*$ ($Fe(OH)_3$ dried before supporting the Au complex) ; ( ■ )$1/Fe_2O_3$ ;( □ )$Fe(OH)_3^*$ .

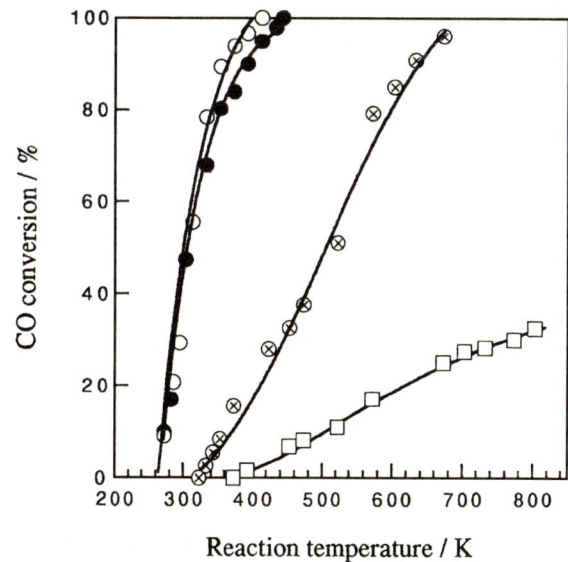

FIG. 6  Na effect on CO oxidation activity of the supported Au catalysts derived from different $Ti(OH)_3^*$ as a function of reaction temperature. (O)$1/D\text{-}Ti(OH)_4^*$ ∗ ;( ● )$1/S\text{-}Ti(OH)_4^*$ ;( ⊗ )$1/W\text{-}Ti(OH)_4^*$ ;( □ )$W\text{-}Ti(OH)_4^*$ .

**1007**

Figure 7 shows the catalytic activities of $1/S\text{-}Fe(OH)_3^*$ which was derived from the as-precipitated $Fe(OH)_3^*$ prepared by using an aqueous solution of $Na_2CO_3$ and $1/A\text{-}Fe(OH)_3^*$ which was derived from the as-precipitated $Fe(OH)_3^*$ prepared by using an aqueous $NH_3$ solution under the identical pH condition. Although the remaining amount of sodium cation in the sample of $1/S\text{-}Fe(OH)_3^*$ has not been determined, it is evident that the coexistence of Na ions on the $Fe(OH)_3^*$ also has a positive effect on the CO oxidation catalysis of $1/Fe(OH)_3^*$.

## Au Precursors

The CO conversions over the catalysts, $1/Fe(OH)_3^*$, $2/Fe(OH)_3^*$, $Au(PPhs)Cl/Fe(OH)_3^*$, $HAuCl_4/Fe(OH)_3^*$, $1/Mg(OH)_2^*$, and $2/Mg(OH)_2^*$, in a fixed-bed flow reaction system are plotted against the reaction temperature in Fig. 8. The catalysts derived from **1** were superior to those derived from **2** for both $Fe(OH)_3^*$ and $Mg(OH)_2^*$. Moreover, when $Au(PPh_3)Cl$ and $HAuCl_4$ were used as precursors, the obtained Au catalysts, $Au(PPh_3)Cl/Fe(OH)_3^*$ and $HAuCl_4/Fe(OH)_3^*$, showed much lower catalytic activities compared with **1** or $2/Fe(OH)_3^*$. The CO oxidation reaction on the Cl-containing precursor-derived catalysts proceeded only above 500 K (Fig. 8). These results may be attributed to poisoning by surface Cl ions which originate from the Au precursors and also to the larger Au particles in these catalysts as proved by XRD. The tremendous catalysis for low-temperature CO oxidation may be achieved by choosing the suitable Au precursors which is expected to interact with the as-precipitated hydroxide surfaces.

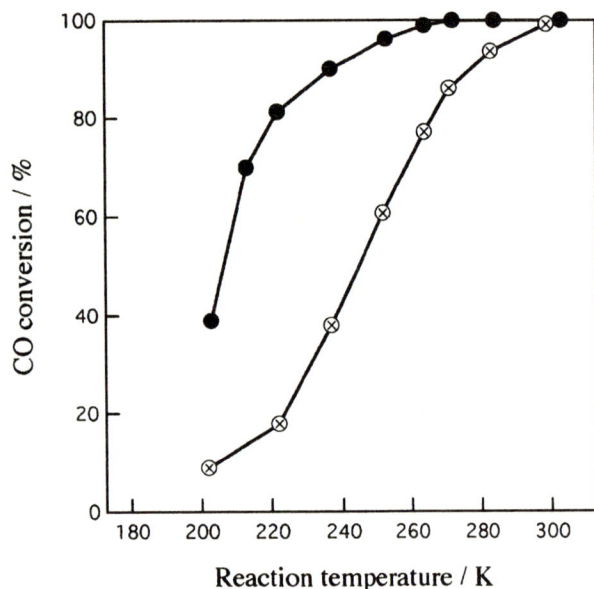

**FIG. 7** Comparison of catalytic CO oxidation activities of two kinds of $1/Fe(OH)_3^*$. (●) Catalyst derived from $Fe(OH)_3^*$ precipitated by aqueous $Na_2CO_3$; (⊗) catalyst derived from $Fe(OH)_3^*$ precipitated by aqueous $NH_3$.

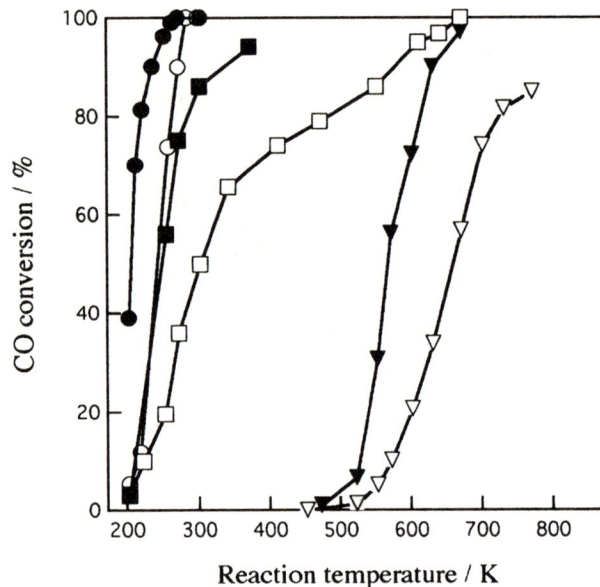

**FIG. 8** CO oxidation activities of supported Au catalysts derived from different Au precursors as a function of reaction temperature. (●) $1/Fe(OH)_3^*$; (○) $2/Fe(OH)_3^*$; (■) $1/Mg(OH)_2^*$; (□) $2/Mg(OH)_2^*$; (▼) $HAuCl_4Fe(OH)_3^*$; (▽) $Au(PPh_3)Cl/Fe(OH)_3^*$.

# 4　DISCUSSION

It has been demonstrated that use of suitable metal complexes and clusters as precursors provides a promising way to prepare highly dispersed metal particles on oxide supports as well as unique structures with chemical bonding or interaction with oxide surfaces[32]. To obtain highly dispersed Au particles on oxide surfaces, we chose phosphine-stabilized Au complex $Au(PPh_3)(NO_3)$ (1) and Au cluster $[Au_9(PPh_3)_8](NO_3)_3$ (2) as precursors which could be thermally decomposed on the oxide surfaces. When the Au complexes (1) and (2) were supported on the metal oxides such as $Fe_2O_3$ and $TiO_2$, however, the supported Au complexes aggregated to large Au particles(about 30 nm in diameter)during the calcination at 673 K as characterized by EXAFS(coordination number of Au-Au bond)in Fig. 1 and Table 2, XRD in Figs. 2 and 3, and TEM in Fig. 4. These Au complex-derived Au particles on the traditional oxides showed low catalytic activities for the CO oxidation as typically depicted in Fig. 5. As the Au aggregation and low activity were thought to be due to almost no or weak interaction of the Au precursors with the oxide surfaces, the oxides were exposed to water vapor to produce surface OH groups for anchoring the Au complexes. The resultant Au complexes supported on the water vapor-treated oxides were easily decomposed to aggregate to large Au particles by calcination at 673 K. Next, metal hydroxides commercially available were used as the supports for the Au complexes, but the obtained supported Au catalysts also showed low activities.

We newly prepared the as-precipitated metal hydroxides $M(OH)_x^*$ by hydrolysis of various metal nitrates, vanadium chloride and titanium-tetra-isopropoxide with an aqueous solution of $Na_2CO_3$ (5 wt%). Then the Au complexes were attached on the wet $M(OH)_x^*$ followed by temperature-programmed(4 K/min)calcination up to 673 K in a flow of air as shown in Scheme 1. It was found that the obtained Au catalysts showed remarkably high catalytic activities for CO oxidation at low temperatures 200—273 K as shown in Figs. 5—8. It is evident from the results of EXAFS(Fig. 1), XRD (Figs. 2 and 3), and TEM(Fig. 4)that the Au particles in 1 or $2/Ti(OH)_4^*$ or $Fe(OH)_3^*$ were several ten times smaller than those on the corresponding metal oxides, $TiO_2$ and $Fe_2O_3$. Negligible Au(200)and Au (111)peaks for $1/Ti(OH)_4^*$ (Fig. 2)and $1/Fe(OH)_3^*$ (Fig. 3), respectively, suggest that the diameter of the Au particles may be less than 3.0 nm, which coincides with the TEM data in Fig. 4. When the as-precipitated Fe hydroxide which was dried by evacuation at 473 K for 4 h was used as the support for 1, the obtained $1/dry-Fe(OH)_3^*$ catalyst was much less active than the $1/Fe(OH)_3^*$, as shown in Fig. 5.

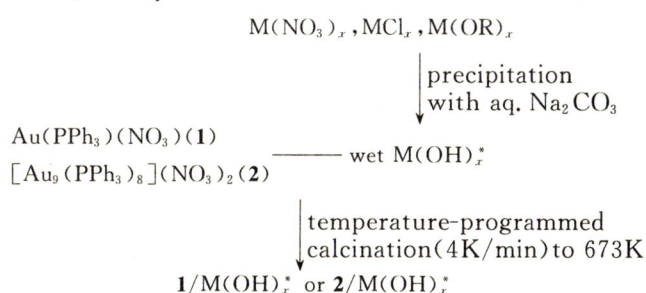

$$M(NO_3)_x, MCl_x, M(OR)_x$$

precipitation
with aq. $Na_2CO_3$

$Au(PPh_3)(NO_3)$ (1)
$[Au_9(PPh_3)_8](NO_3)_2$ (2)

wet $M(OH)_x^*$

temperature-programmed
calcination(4K/min)to 673K

$1/M(OH)_x^*$ or $2/M(OH)_x^*$

**SCHEME 1    Preparation of the $1/M(OH)_x^*$ or $2/M(OH)_x^*$ catalysts.**

Our early works[32—34] demonstrate that the Au complexes 1 and 2 on the oxide supports readily decompose to metallic gold at lower temperatures than 573 K. In the similar temperature range the as-precipitated metal hydroxides are dehydrated to form anhydrous metal hydroxides and mostly to partially dehydrated metal oxides. The supported Au complexes are expected to be stabilized on the surface of the as-precipitated metal hydroxides by chemical interaction between $[Au(PPh_3)]^+$ or $[Au_9 (PPh_3)_8]^{3+}$ and the hydroxyl groups of the as-precipitated metal hydroxide surfaces or water adsorbed on them. During calcination of the samples at 673 K the transformations of both Au complexes and as-precipitated hydroxides occur in parallel, which leads to the formation of highly dispersed Au particles on the *in situ* prepared oxide surfaces. On the other hand, in the case of Au complex/$M_xO_y$, there is only a limited number of OH groups on the oxide surface and the interaction with the Au complex may be insufficient and weak, hence leading to aggregation to large Au particles without significant stabilization of Au species during the calcination at 673 K.

It is also to be noted that the sharpness of Au(111) diffraction line at $2\theta = 38.2°$ for $1/Fe_2O_3$ was almost the same as that for $2/Fe(OH)_3^*$ (Figs. 3b and 3d), while the $2/Fe(OH)_3^*$ catalyst showed a much higher activity for the catalytic CO oxidation reaction than the $1/Fe_2O_3$ catalyst. The XRD patterns for the supports in the $1/Fe_2O_3$ and $2/Fe(OH)_3^*$ catalysts are different from each other. These results imply that the catalytic activity depends on not only the size of Au particles but also the nature of the support. It is evident that attaching the Au complexes on the as-precipitated metal hydroxides is superior to impregnating the Au complexes on the conventional metal oxides in preparation of small supported Au particles.

Thus it is suggested that the way to obtain highly dispersed Au particles active for the low-temperature CO oxidation is to choose suitable Au complexes as precursors which can be easily decomposed at mild temperatures and to prepare the as-precipitated metal hydroxides as precursors for oxide supports which have many surface OH groups reactive to the Au complexes, and to transform both precursors to Au particles and oxide supports under their chemical interaction by the temperature-programmed calcination in air.

Hoflund and co-workers claimed that some of Au is nonmetal and that this Au may be responsible for the low-temperature activity[18,40]. Haruta and co-workers reported that the sample having both metallic and nonmetallic Au species was not more active than the sample having only metallic Au species(10). Recently, it was suggested by XPS and ISS that the near-surface region of $Au/Fe_2O_3$ contains more Au as compared with that of $Au/Co_3O_4$, where Au is present as crystallites and small amounts of nonmetallic Au species are also present. Our XPS measurement provided no evidence on the presence of cationic Au species on the $1/Fe(OH)_3^*$ surface.

The structure of $TiO_2$ in $1/Ti(OH)_4^*$ was significantly affected by hydrolysis reagents used for preparation of the as-precipitated $Ti(OH)_4^*$ as shown in Fig. 2. Two types of $TiO_2$, anatase and amorphous $TiO_2$, were formed by using *deionized* water and an aqueous solution of $Na_2CO_3$ as hydrolysis reagents, respectively. The $1/S-Ti(OH)_4^*$ (amorphous $TiO_2$) catalyst exhibits much higher activity for CO oxidation than the $1/W-Ti(OH)_4^*$ (anatase $TiO_2$) catalyst as shown in Fig. 6. Thus the presence of $Na^+$ ions gives profound effects on both the morphology and catalytic property of the oxide. Figure 2 also shows the effect of Na doping to the as-precipitated $W-Ti(OH)_4^*$ on the morphology of the bulk and the size of Au particles. The anatase structure and the size of Au particles of $1/W-Ti(OH)_4^*$ were unchanged by the Na doping. On the other hand, the catalytic CO oxidation activity of $1/W-Ti(OH)_4^*$ was remarkably promoted by the Na doping to the $W-Ti(OH)_4^*$ ($1/D-Ti(OH)_4^*$), as shown in Fig. 6. These results assume that small Au particles on the titanium oxides prepared by the $Na_2CO_3$ treatments, independent of amorphous and anatase forms, were very active for the low-temperature CO oxidation. The mechanism for enhancement of CO oxidation by $Na^+$ dope is not clear at present.

The steady-state catalytic performance of the Au samples was recorded without deactivation over 8 h time-on-stream in the present study. The activities of the catalysts, particularly containing basic supports like MgO, can vary over a long induction period as reported in the literature[25]. In addition to the present results, further study on decay profile of the activity may be necessary for industrial application of the samples.

When the $Au(PPh_3)Cl/Fe(OH)_3^*$ catalyst was prepared and calcined in the similar manner to that for $1/Fe(OH)_3^*$, the XRD peak at 38. 2° for Au(111) in Fig. 3e was sharp and intense, indicating the formation of the larger Au particles and consequently the catalytic activity of the $Au(PPh_3)Cl/Fe(OH)_3^*$ for CO oxidation was very low so that it was active only above 500 K as shown in Fig. 8. $Au(PPh_3)(NO_3)$ was heat-sensitive and $[Au_9(PPh_3)_8](NO_3)_3$ decomposed at 503 K, while $Au(PPh_3)Cl$ (mp. ; 516 − 525 K) was stable around 570 K under air. The as-precipitated metal hydroxides are transformed to partially dehydrated metal oxides before the decomposition of $Au(PPh_3)Cl$. Under this situation the Cl-containing Au complex decomposes to Au metallic particles on metal oxide surface rather than metal hydroxide surface. The Au particle sizes in $2/Fe(OH)_3^*$ and $HAuCl_4/Fe(OH)_3^*$ are estimated to be similar and larger than 12 nm in Fig. 3. The former catalyst shows excellent catalysis for CO oxidation, whereas the latter catalyst shows negligible activity at the low temperatures. Thus the presence of Cl involved in the Au precursor not only increases the Au particle size but also suppresses catalytic CO oxidation at the surface, which agrees with the literature[17].

# 5  CONCLUSION

The supported Au catalysts with small Au particles were prepared by attaching the Au phosphine complexes, $Au(PPh_3)(NO_3)$ (1) and $[Au_9(PPh_3)_8](NO_3)_3$ (2), on the wet as precipitated metal hydroxides, followed by the temperature-programmed decomposition and calcination in a flow of air. The obtained 1 and $2/M(OH)_x^*$ catalysts ($M = Mn^{2+}$, $Co^{2+}$, $Fe^{3+}$, $Ni^{2+}$, $Zn^{2+}$, $Mg^{2+}$, $Cu^{2+}$, and $Ti^{4+}$) are highly active for CO oxidation at low temperatures below 273 K compared to the corresponding $Au/M_xO_y$ catalysts prepared by supporting 1 or 2 on conventional metal oxides. Such efficient catalysts were not obtained by supporting 1 or 2 on commercially available $M(OH)_x$ and water-exposed $M_xO_y$ followed by the similar calcination to that for the case of 1 or $2/M(OH)_4^*$. The Au particles in 1 or $2/M(OH)_x^*$ were estimated to be small, typically the size for $1/Fe(OH)_3^*$ being about 2. 9 nm by EXAFS, XRD, and TEM. The reactivity of the catalysts for CO oxidation is positively affected by sodium cation, whereas negatively affected by chloride ion. Sodium cations and chloride anions have significant effects on Au particle size and support morphology. The catalysts derived from 1 are more active for CO oxidation than those derived from 2.

# ACKNOWLEDGMENT

This work has been supported by CREST(Core Research for Evolutional Science and Technology) of Japan Science and Technology Corporation(JST).

# REFERENCES

[1]Stark, D. S., Crocker, A., and Steward, G. J., J. Phys. E: Sci. Instrum. **16**, 158(1983).

［2］Stark，D. S.，and Harris，M. R.，J. Phys. E：Sci. Instrum. **16**，492(1983).

［3］Stark，D. S.，and Harris，M. R.，J. Phys. E：Sci. Instrum. **21**，715(1988).

［4］Schryer，D. R.，Upchurch，B. T.，van Norman，J. D.，Brown，K. G.，and Schryer，J.，J. Catal. **122**，193(1990).

［5］Fuller，M. J.，and Warwick，M. E.，J. Catal. **34**，445(1974).

［6］Bond，G. C.，Molloy，L. R.，and Fuller，M. J.，J. Chem. Soc. Chem. Commun. ，796(1975).

［7］Croft，G.，and Fuller，M. J.，Nature **269**，585(1977).

［8］Haruta，M.，Kobayashi，T.，Sano，H.，and Yamada，N.，Chem. Lett. ，**405**(1987).

［9］Haruta，M.，Yamada，N.，Kobayashi，T.，and Iijima，S.，J. Catal. **115**，301(1989).

［10］Haruta，M.，Tsubota，S.，Kobayashi，T.，Kageyama，H.，Genet，M. J.，and Delmon，B.，J. Catal. **144**，175(1993).

［11］Tsubota，S.，Ueda，A.，Sakurai，H.，Kobayashi，T.，and Haruta，M.，ACS. Symp. Ser. **552**，420 (1994).

［12］Boccuzi，T.，Tsubota，S.，and Haruta，M.，J. Elec. Spectrosc. Relat. Phenom. **64/65**，241(1993).

［13］Haruta，M.，Tsubota，S.，Kobayashi，T.，Ueda，A.，Sakurai，H.，and Ando，M.，Stud. Surf. Sci. Catal. **75**，2657(1993).

［14］Gardner，S. D.，Hoflund，G. B.，Davidson，M. R.，and Schryer，D. R.，J. Catal. **115**，132(1989).

［15］Gardner，S. D.，Hoflund，G. B.，Schryer，D. R.，and Upchurch，B. T.，J. Phys. Chem. **95**，835 (1991).

［16］Schryer，D. R.，Upchurch，B. T.，Sidney，B. D.，Brown，K. G.，Hoflund，G. B.，and Herz，R. K.，J. Catal. **130**，314(1991).

［17］Gardner，S. D.，Hoflund，G. B.，Schyer，D. R.，Schryer，J.，Upchurch，B. T.，and Kielin，E. J.，Langmuir **7**，2135(1991).

［18］Gardner，S. D.，Hoflund，G. B.，Davidson，M. R.，Laitinen，H. A.，Schryer，D. R.，and Upchurch，B. T.，Langmuir **7**，2140(1991).

［19］Hoflund，C. B.，Gardner，S. D.，Schryer，D. R.，Upchurch，B. T.，and Kielin，E. J.，Appl. Catal. B：Environmental **6**，117(1995).

［20］Imamura，S.，Sawada，H.，Uemura，K.，and Ishida，S.，J. Catal. **109**，198(1988).

［21］Imamura，S.，Yoshie，S.，and Ono，Y.，J. Catal. **115**，258(1989).

［22］Lin，S. D.，Bollinger，M.，and Vannice，M. A.，Catal. Lett. **17**，245(1993).

［23］Knell，A.，Barnickel，P.，Baiker，A.，and Wokaun，A.，J. Catal. **137**，306(1992).

［24］Baiker，A.，Maciejewski，M.，Tagliaferri，S.，and Hug，P.，J. Catal. **151**，407(1995).

［25］Srinivas，G.，Wright，J.，Bai，C. -S.，and Cook，R.，Stud. Surf. Sci. Catal. **101**，427(1996).

［26］Bond，G. C.，and Sermon，P. A.，Gold Bull. **6**，102(1973).

［27］Yermakov，Y. I.，Kuznetsov，B. N.，and Zakharov，V. A.，"Catalysis by Supported complexes. " Elsevier，Amsterdam，1981.

［28］Basset，J. M.，J. Mol. Catal. **21**，95(1983).

［29］Iwasawa，Y.，Ed.，"Tailored Metal Catalysts，" Reidel，Amsterdam，1986.

［30］Maire，G.，Stud. Surf. Sci.，Catal. **29**，509(1986).

［31］Gates，B. C.，Chem. Rev. **95**，511(1995).

[32]Iwasawa,Y.,in "Proceedings,11th International Congress on Catalysis,Baltimore,1996)," p. 21. Studies in Surface Science and Catalysis,Vol. 101.

[33]Yuan,Y.,Asakura,K.,Wan,H.,Tsai,K.,and Iwasawa,Y.,Chem. Lett. ,129(1996).

[34]Yuan,Y.,Asakura,K.,Wan,H.,Tsai,K.,and Iwasawa,Y.,Catal. Lett. ,in press.

[35]Yuan,Y.,Asakura,K.,Wan,H.,Tsai,K.,and Iwasawa,Y.,Chem. Lett. ,755(1996).

[36]Mueting,A. M.,Alexander,B. D.,Boyle,P. D.,Casalnuovo,A. L.,Ito,L. N.,Johnson,B. J.,and Pignolet,L. H.,6 "Inorganic Syntheses"(R. N. Grimes, Ed. ),Vol. 29, p. 280. Wiley, New York,1992.

[37]Cariati,F.,and Naldini,L.,J. C. S. Dalton, 2286(1972).

[38]Rehr,J. J.,Mustre, de L. J.,Zabinsky,S. I.,and Albers,R. C.,J. Am. Chem. Soc. **113**,5235 (1991).

[39]Iwasawa, Y., Ed., "X-Ray Absorption Fine Structure for Catalysts and Surfaces." World Scientific,Singapore,1996.

[40]Epling,W. S.,Hoflund,G. B.,Weaver,J. F.,Tsubota,S.,and Haruta,M.,J. Phys. Chem. **100**, 9929(1996).

■ 本文原载:J. Coord. Chem., Vol. 42, pp. 131～141,1997.

# Syntheses, Structures and Spectroscopic Properties of Nickel(II) Citrato Complexes, $(NH_4)_2[Ni(Hcit)(H_2O)_2]_2 \cdot 2H_2O$ and $(NH_4)_4[Ni(Hcit)_2] \cdot 2H_2O$

Zhao-Hui Zhou ①, Yi-Ji Lin, Hong-Bin Zhang, Guo-Dong Lin, Khi-Rui Tsai

(Department of Chemistry, State Key Laboratory for Physical Chemistry, Xiamen University, 361005 Xiamen, China)

Received 30 August 1996

**Abstract** Dimeric ammonium diaquocitratonickelate(II) dihydrate $(NH_4)_2[Ni(Hcit)(H_2O)_2]_2 \cdot 2H_2O$, **1**, and its sodium and potassium salts, as well as ammonium dicitratonickelate(II) dihydrate $(NH_4)_4[Ni(Hcit)_2] \cdot 2H_2O$, **2**, ($H_4cit$ = citric acid) have been synthesized and characterized by spectroscopic methods. The crystal structures of **1** and **2** were determined by X-ray methods. Compound **1** is triclinic, space group $P\bar{1}$ with $a = 6.4071(7)$, $b = 9.4710(7)$, $c = 9.6904(5)$ Å, $\alpha = 105.064(5)$, $\beta = 91.992(7)$, $\gamma = 89.334(8)°$, $V = 567.5(1)$ Å$^3$, $Z = 1$, $R = 0.037$ for 1714 observed reflections. The structure consists of centrosymmetric dimers, $[Ni(Hcit)(H_2O)_2]_2^{2-}$. The principal Ni—O dimensions are Ni—O(hydroxy), 2.074(2) Å, Ni—O($\alpha$-carboxy), 2.020(3) Å, Ni—O($\beta$-carboxy), 2.031(2), 2.037(2) Å, Ni—O(water), 2.065(2), 2.072(3) Å. Compound **2** crystallizes in the monoclinic space group $P2_1/a$ with $a = 9.361(1)$, $b = 13.496(1)$, $c = 9.4238(7)$ Å, $\beta = 115.475(6)°$, $V = 1074.9(3)$ Å$^3$, $Z = 2$, $R = 0.052$ for 1507 observed reflections. The two terdentate citrato ligands coordinate symmetrically to one nickel via hydroxyl, $\alpha$-and $\beta$-carboxylato oxygens and the remaining acetato groups are bonded through strong hydrogen bonds with the hydroxy group of another dicitrato nickel anion [2.526(6) Å]. The principal Ni—O dimensions are: Ni—O(hydroxy), 2.021(3) Å, Ni—O($\alpha$-carboxy), 2.038(3) Å, Ni—O($\beta$-carboxy), 2.072(3) Å. Nickel atoms in the two compounds have octahedral geometry. The two compounds can be transformed easily by reaction with ammonium citrate or nickel chloride.

**Key words** nickel  citric acid  citrate  X-ray structure

## 1  INTRODUCTION

Citric acid($C_6H_8O_7 = H_4cit$)is of widespread importance in biological systems and has a number of key physiological functions[1]. Most of these depend on the chelating ability of citrate anions. Above pH 8.0, the $Hcit^{3-}$ is the predominant form, and deprotonation of the hydroxyl group is accelerated by its bonding with metal ions such as V(V), Mo(VI) and W(VI)[2-4]. In these metal complexes, the

---

① Author for correspondence.

tetraionized species is apparently formed at pH 5—7, while for Ni(II) complex, it is not formed below pH 9[5,6]. For several New Caledonian species of *Sebertia*, *Hybanthus* and *Homalium* containing nickel, citrate is involved in complex formation, where electrophoretic studies indicate that the complex is negatively charged[7]. In fact, nickel is recognized as an essential trace element for bacteria, plants, animals and humans[8], while its role remains largely undefined. In an entirely different mode, citric acid is used as a chelating agent in electroplating baths for various Ni alloys[9], and biodegredation of waste soluble nickel citrates generated from this process has been reported recently[10,11]. Moreover, the citrate method has important potential in the preparation of well-defined and high surface area particles[12], such as for Ni-based catalysts for hydrogenation or the partial oxidation of methane, in which $MNiO_2$ (M $= Mg^{2+}$, $Ca^{2+}$) or rare earth mixed oxides can be prepared from their corresponding nickel citrate precursors[13,14].

To now, there are only two well-characterized crystal structures of nickel(II) citrate complexes reported in literature. One is $\{[N(CH_3)_4]_5[Ni(II)_4(cit)_3(OH)(H_2O)] \cdot 18H_2O\}_2$ formed at pH 9.2[5], and the other is $K_2[Ni(Hcit)(H_2O)_2]_2 \cdot 4H_2O$ formed at pH 5[15]. Hedwig et al.[16], reported potentiometric and visible spectroscopic studies of various complexes formed between nickel(II) ions and citric acid in dilute solution and proposed four species $NiHL$, $NiH_2L$, $NiH_3L^+$ and $Ni(HL)_2^{4-}$ (L = cit). In the meantime, there was a report on solution studies of polynuclear species in the nickel(II) citrate system, including $Ni_4(OH)_4(HL)_3^{5-}$ $Ni_4(OH)(H_2L)_3^{5-}$, $Ni(HL)_2^{4-}$ and $NiH_2L$.[17] Bickley et al.[18], performed a comprehensive Raman spectroscopic study of nickel(II) citrate and its aqueous solutions, and concluded that nickel(II) citrate existed as $Ni(Hcit)^-$ species in aqueous solution in confirmation of electrochemical studies on this system. In the present work, the complexation of different mol ratio of nickel to citrate was investigated in concentrated solution($\geqslant 1M$) at pH 2—8, and several citrato nickel (II) complexes resulting from the reactions of nickel chloride or acetate and citrate were characterized. The X-ray crystal structures of $(NH_4)_2[Ni(Hcit)(H_2O)_2]_2 \cdot 2H_2O$, **1**, and $(NH_4)_4[Ni(Hcit)_2] \cdot 2H_2O$, **2** are reported.

# 2 EXPERIMENTAL

## Synthesis of $(NH_4)_2[Ni(Hcit)(H_2O)_2]_2 \cdot 2H_2O$, 1

Complex **1** was prepared in a similar way to that reported for its potassium salt with some improvement[15]. An equimolar solution(2.0 M, 50 cm³) of nickel chloride and ammonium citrate was reacted at room temperature overnight. The resulting mixture was placed in a refrigerator. After 2—3 days, green, platy crystals were obtained with a yield of 70%. *Anal* Calcd. for $C_{12}H_{30}O_{20}N_2Ni_2$(%): C, 22.5; H, 4.7. Found: C, 22.1; H, 4.8. Electronic spectrum $\lambda_{max}$(nm, $\varepsilon$, $M^{-1}$ $cm^{-1}$): 655(2.4), 392(5.9). IR bands observed using a Nicolet 740(KBr disc): $v_{as}$(COO)$1599_{vs.w}$; $v_s$(COO)$1419_s$; $v_{as}$(Ni—O)$891_s$; $v_s$(Ni—O)$834_m$ $cm^{-1}$. In a similar procedure, green sodium and potassium salts, $Na_2[Ni(Hcit)(H_2O)_2]_2 \cdot 2H_2O$, **3**, and $K_2[Ni(Hcit)(H_2O)_2]_2 \cdot 2H_2O$, **4**, were obtained with IR bands(KBr disc): $v_{as}$(COO)$1625_s$, $1592_{vs}$, $1562_{vs}$; $v_s$(COO)$1415_s$, $1385_s$; $v_{as}$(Ni—O)$921_m$, $896_m$; $v_s$(Ni—O)$842_m$ $cm^{-1}$ for complex **3**, and $v_{as}$(COO)$1610_{vs}$, $1566_{vs}$; $v_s$(COO)$1410_{vs}$, $1386_s$; $v_{as}$(Ni—O)$916_m$, $885_s$; $v_s$(Ni—O)$831_m$ $cm^{-1}$ for complex **4**.

## Synthesis of $(NH_4)_4[Ni(Hcit)_2] \cdot 2H_2O$, 2

$NiCl_2 \cdot 6H_2O$(0.02 mol) added slowly with stirring to a solution of ammonium citrate($NH_4)_3$Hcit

(1M, 40 cm³). The resulting mixture was kept on a water bath at 85 ℃ for two hours, then filtered and cooled. After 1—2 days, blue crystals of the complex were obtained with a yield of 60%. *Anal.* Calcd. for $C_{12}H_{30}O_{16}N_4Ni_1$ (%): C, 26.4; H, 5.5. Found: C, 26.1; H, 5.3. Electronic spectrum $\lambda_{max}$ (nm, ε, $M^{-1} \cdot cm^{-1}$): 638(2.0), 382(4.3). IR bands (KBr disc): $v_{as}$(COO) $1605_s$, $1566_{vs}$; $v_s$(COO) $1400_s$; $v_{as}$(Ni—O) $905_m$; $v_s$(Ni—O) $847_m$ $cm^{-1}$.

## Transformation of $(NH_4)_2[Ni(Hcit)(H_2O)_2]_2 \cdot 2H_2O$, 1, and $(NH_4)_4[Ni(Hcit)_2] \cdot 2H_2O$, 2

To a solution of $(NH_4)_2[Ni(Hcit)(H_2O)_2]_2 \cdot 2H_2O$, **1** (1M, 10 cm³), an equimolar amount of ammonium citrate was added. The mixture was kept on a water bath at 85 ℃ for two hours, then the solution was cooled slowly and placed in a refrigerator. After 1—2 days, blue crystals of **2** were obtained in 75% yield.

In another experiment, a solution of $(NH_4)_4[Ni(Hcit)_2] \cdot 2H_2O$, **2** (1M, 10 cm³) was reacted with $NiCl_2 \cdot 6H_2O$ (0.01 mol) on a water bath at 85 ℃ for two hours, and then the solution was cooled slowly and placed in a refrigerator. After 1—2 days, green crystals of **1** were obtained in 70% yield.

## X-ray data collection, structure solution and refinement

Crystals of suitable quality for X-ray diffraction were obtained as transparent green plates or blue prisms by slow evaporation of solutions of the title compounds at room temperature. The resulting crystals were sealed in capillaries to prevent loss of water molecules.

Crystallographic data for the two ammonium nickel citrates **1** and **2** are summarized in Table I. Diffraction data were collected on an Enraf-Nonius CAD-4 diffractometer with graphite-monochromated CuKα radiation at 296 K. Lp factor, anisotropic decay and empirical absorption corrections were applied. The structures were solved by the heavy atom method and refined by full-matrix least-squares procedures with anisotropic thermal parameters for all the nonhydrogen atoms. H atoms were located from difference maps. The hydrogen atoms were assigned appropriate isotropic temperature factors and included in structure factor calculations. All calculations were performed on a 586/100 microcomputer using the MoLEN software package. Positional parameters, selected atomic distances and bond angles are given in Tables II to V, respectively.

**TABLE I  Crystal data and summaries of intensity data collection and structure refinements**

| Compound | $(NH_4)_2[Ni(Hcit)(H_2O)_2]_2 \cdot 2H_2O$, 1 | $(NH_4)_4[Ni(Hct)_2] \cdot 2H_2O$, 2 |
|---|---|---|
| Colour/shape | green, plates | blue, prisms |
| Formula weight | 639.79 | 545.10 |
| Space group | $P\bar{1}$ | $P2_1/a$ |
| Temp., ℃ | 22 | 22 |
| Cell constants[a] | | |
| $a$, Å | 6.4071(7) | 9.361(1) |
| $b$, Å | 9.4710(7) | 13.496(1) |
| $c$, Å | 9.6904(5) | 9.4238(7) |
| $\alpha$, deg | 105.064(5) | |
| $\beta$ deg | 91.992(7) | 115.475(6) |
| $\gamma$, deg | 89.334(8) | |
| Cell volume, Å³ | 567.5(1) | 1074.9(3) |
| Formula units/unit cell | 1 | 2 |
| $D_{calc}$, g cm⁻³ | 1.872 | 1.684 |

续表

| Compound | $(NH_4)_2[Ni(Hcit)(H_2O)_2]_2 \cdot 2H_2O$, 1 | $(NH_4)_4[Ni(Hct)_2] \cdot 2H_2O$, 2 |
|---|---|---|
| $\mu_{kcalc}$, cm⁻¹ | 29.502 | 20.732 |
| Diffractometer/scan | Enraf-Nonius CAD-4/2θ | Enraf-Nonius CAD-4/2θ |
| Radiation | CuKα(λ=1.5418 Å) | CuKα(λ=1.5418 Å) |
| Max. crystal dimension,mm | 0.10×0.12×0.15 | 0.10×0.18×0.28 |
| Scan width | 0.55+0.35 tan θ | 0.65+0.35 tan θ |
| Standard reflections | 2 −3 −5；2 6 1 | −6 −2 6；−5 −3 −2 |
| Decay of standards | 1.1% | −1.5% |
| Reflections measured | 2145 | 2134 |
| 2θ range,deg | 3≤θ≤70 | 3≤θ≤70 |
| Range of $h,k,I$ | 7,±11,±11 | −11,−16,±11 |
| Reflections observed $I \geqslant 3\sigma(I)^b$ | 1714 | 1507 |
| Computer programs$^c$ | MoLEN | MoLEN |
| Structure solution | MoLEN | MoLEN |
| No. of parameters varied | 224 | 197 |
| Weights | unit weight | $[\sigma\|F_0\|^2+0.0004\|F_0\|^2+1]^{-1}$ |
| GOF | 0.791 | 0.831 |
| $R=\sum(\|F_v\|-\|F_c\|)/\sum\|F_v\|$ | 0.037 | 0.052 |
| $R_w$ | 0.038 | 0.058 |
| Largest feature final diff. map | 0.46e⁻ Å³ | 0.81 e⁻ Å³ |

$^a$Least-squares refinement of $[(\sin\theta)\lambda]^2$ values for 25 reflection θ>35° $^b$ Corrrections：Lorentzpolarization $^c$Neutral scattering factors and anomalous corrections.

TABLE Ⅱ  Atomic coordinates and thermal parameters for $(NH_4)_2[Ni(Hcit)(H_2O)_2]_2 \cdot 2H_2O$

| Atom | x/a | y/b | z/c | $B_{eq}$(Å²)* |
|---|---|---|---|---|
| Ni(1) | 0.94174(9) | 0.24197(6) | 0.21041(6) | 1.63(1) |
| N(1) | 0.7767(5) | 0.3691(3) | 0.6156(3) | 2.32(6) |
| O(1) | 1.2158(3) | 0.1247(2) | 0.1573(2) | 1.72(4) |
| O(2) | 0.8696(4) | 0.0637(2) | 0.2768(3) | 2.32(5) |
| O(3) | 1.0875(4) | 0.3332(2) | 0.4019(2) | 1.96(5) |
| O(4) | 1.3456(4) | 0.3668(3) | 0.5633(3) | 2.58(5) |
| O(5) | 1.1831(4) | −0.1401(2) | −0.0140(3) | 2.40(5) |
| O(6) | 1.3991(4) | −0.3143(3) | 0.0216(3) | 3.33(6) |
| O(7) | 1.0132(4) | −0.1347(2) | 0.3226(3) | 2.52(5) |
| C(1) | 1.2417(5) | 0.0384(3) | 0.2607(3) | 1.47(6) |
| C(2) | 1.0241(5) | −0.0168(3) | 0.2886(4) | 1.76(6) |
| C(3) | 1.3384(5) | 0.1367(3) | 0.3978(4) | 1.83(6) |
| C(4) | 1.2505(5) | 0.2883(3) | 0.4571(4) | 1.66(6) |
| C(5) | 1.3914(5) | −0.0876(3) | 0.2006(4) | 1.80(6) |
| C(6) | 1.3206(5) | −0.1907(4) | 0.0608(4) | 2.05(7) |
| O(w1) | 0.6698(4) | 0.3566(3) | 0.2724(3) | 2.40(5) |
| O(w2) | 1.0455(4) | 0.4159(2) | 0.1385(3) | 2.21(5) |
| O(w3) | 1.2110(4) | 0.3544(3) | −0.1219(3) | 2.55(5) |

* $B_{eq}=1/3\sum_i\sum_j B_{ij}a_ia_j$.

1017

**TABLE Ⅲ   Atomic coordinates and thermal parameters for$(NH_4)_4[Ni(Hcit)_2] \cdot 2H_2O$**

| Atom | $x/a$ | $y/b$ | $z/c$ | $B_{eq}(\text{Å}^2)^*$ |
|---|---|---|---|---|
| Ni(1) | 1.000 | 0.000 | 1.000 | 2.16(2) |
| N(1) | 1.0724(5) | 0.2741(3) | 0.9093(4) | 3.60(9) |
| N(2) | 0.7391(4) | 0.0230(4) | 1.2389(4) | 3.75(9) |
| O(1) | 0.8776(3) | −0.1127(2) | 0.8592(3) | 2.27(6) |
| O(2) | 0.7738(3) | 0.0407(2) | 0.9502(3) | 2.51(6) |
| O(3) | 0.9819(3) | 0.0769(2) | 0.8026(3) | 2.79(6) |
| O(4) | 0.8852(5) | 0.1197(3) | 0.5531(4) | 6.8(1) |
| O(5) | 0.5508(5) | −0.1660(3) | 0.8798(4) | 6.25(9) |
| O(6) | 0.5413(5) | −0.2974(3) | 0.7454(4) | 7.2(1) |
| O(7) | 0.5424(3) | 0.0431(3) | 0.7396(3) | 3.37(7) |
| C(1) | 0.7370(4) | −0.0707(3) | 0.7371(4) | 2.23(8) |
| C(2) | 0.6778(4) | 0.0090(3) | 0.8161(4) | 2.32(8) |
| C(3) | 0.7733(5) | −0.0220(3) | 0.6098(4) | 2.41(8) |
| C(4) | 0.8879(5) | 0.0644(4) | 0.6593(4) | 2.95(9) |
| C(5) | 0.6147(5) | −0.1526(3) | 0.6608(4) | 2.58(9) |
| C(6) | 0.5632(5) | −0.2072(3) | 0.7708(4) | 2.54(9) |
| O(w1) | 0.8041(4) | −0.1991(3) | 1.3378(4) | 5.3(1) |

$* \ B_{eq} = 1/3 \sum_i \sum_j B_{ij} a_i a_j.$

**TABLE Ⅳ   Selected bond lengths(Å)and angles(°)for$(NH_4)_2[Ni(Hcit)(H_2O)_2]_2 \cdot 2H_2O$**

| | | | | | | | |
|---|---|---|---|---|---|---|---|
| Ni(1)—O(1) | 2.074(2) | Ni(1)—O(2) | 2.020(3) | Ni(1)—O(3) | 2.031(2) | Ni(1)—O(5a) | 2.037(2) |
| Ni(1)—O(wl) | 2.065(2) | Ni(1)—O(w) | 2.072(3) | O(1)—C(1) | 1.452(4) | O(2)—C(2) | 1.262(4) |
| O(3)—C(4) | 1.276(5) | O(4)—C(4) | 1.245(4) | O(5)—C(6) | 1.287(4) | O(6)—C(6) | 1.239(5) |
| O(7)—C(2) | 1.248(4) | C(1)—C(2) | 1.551(5) | C(1)—C(3) | 1.526(4) | C(1)—C(5) | 1.528(4) |
| C(3)—C(4) | 1.512(4) | C(5)—C(6) | 1.508(4) | O(1)—H(1) | 1.00(4) | | |
| O(1)—Ni(1)—O(2) | 81.53(9) | 0(1)—Ni(1)—O(3) | 86.12(9) | O(1)—Ni(1)—O(5a) | 89.19(9) | | |
| O(1)—Ni(1)—O(w1) | 177.4(2) | O(1)—Ni(1)—O(w2) | 92.74(9) | O(2)—Ni(1)—O(3) | 90.6(1) | | |
| O(2)—Ni(1)—O(5a) | 88.5(2) | O(2)—Ni(1)—O(w1) | 96.7(2) | O(2)—Ni(1)—O(w2) | 174.2(1) | | |
| O(3)—Ni(1)—O(5a) | 175.30(9) | O(3)—Ni(1)—O(w1) | 92.0(1) | O(3)—Ni(1)—O(w2) | 88.4(1) | | |
| O(5a)—Ni(1)—O(w1) | 92.69(9) | O(5a)—Ni(1)—O(w2) | 92.1(1) | O(w1)—Ni(1)—O(w2) | 89.0(2) | | |
| Ni(1)—O(1)—C(1) | 105.1(2) | Ni(1)—O(2)—C(2) | 114.2(2) | Ni(1)—O(3)—C(4) | 128.3(2) | | |
| Ni(1)—O(5)—C(6) | 127.7(2) | C(1)—O(1)—H(1) | 102(3) | | | | |

Environment of $NH_4^+$ ion

| | | | | | | | |
|---|---|---|---|---|---|---|---|
| N(1)⋯O(3) | 2.877(4) | N(1)⋯O(4b) | 2.790(4) | N(1)⋯O(7c) | 2.768(4) | N(1)⋯O(w2d) | 2.910(3) |

Hydrogen bonds

| | Dist. | O—H⋯O | | Dist. | O—H⋯O |
|---|---|---|---|---|---|
| O(1)⋯O(5) | 2.629(3) | 145(4) | O(w1)⋯O(4d) | 2.691(3) | 164(4) |
| O(w1)⋯O(6e) | 2.792(4) | 155(4) | O(w2)⋯O(w3) | 2.691(4) | 174(4) |
| O(w2)⋯O(w3f) | 2.746(3) | 165(5) | | | |

Symmetry transformations $a,(2-x,-y,2-z)$; $b,(-1+x,y,z)$; $c,(2-x,-y,1-z)$; $d,(2-x,1-y,1-z)$; $e,(2-x,-y,-z)$; $f,(2-x,1-y,-z)$.

**TABLE Ⅴ  Selected bond lengths(Å)and angles(°)for(NH₄)₄[Ni(Hcit)₂] · 2H₂O**

| Ni(1)—O(1) | 2.021(3) | Ni(1)—O(2) | 2.038(3) | Ni(1)—O(3) | 2.072(3) | O(1)—C(1) | 1.441(4) |
|---|---|---|---|---|---|---|---|
| O(2)—C(2) | 1.269(4) | O(3)—C(4) | 1.267(5) | O(4)—C(4) | 1.240(6) | O(5)—C(6) | 1.217(6) |
| O(6)—C(6) | 1.240(6) | O(7)—C(2) | 1.247(4) | C(1)—C(2) | 1.5406(6) | C(1)—C(3) | 1.530(6) |
| C(1)—C(5) | 1.530(6) | C(3)—C(4) | 1.516(6) | C(5)—C(6) | 1.510(7) | O(1)—H(1) | 0.93(6) |

| O(1)—Ni(1)—O(1a) | 180.01(7) | O(1)—Ni(1)—O(2) | 79.2(1) | O(1)—Ni(1)—O(2a) | 100.8(1) |
|---|---|---|---|---|---|
| O(1)—Ni(1)—O(3) | 89.0(2) | O(1)—Ni(1)—O(3a) | 91.0(2) | O(2)—Ni(1)—O(2a) | 180.01(4) |
| O(2)—Ni(1)—O(3) | 89.0(1) | O(2)—Ni(1)—O(3a) | 91.1(1) | O(3)—Ni(1)—O(3a) | 180.01(3) |
| Ni(1)—O(1)—C(1) | 107.1(2) | Ni(1)—O(2)—C(2) | 111.1(3) | Ni(1)—O(3)—C(4) | 129.9(4) |
| C(1)—O(1)—H(1) | 111(3) | | | | |

Environment of NH₄⁺ ion

| N(1)⋯O(1a) | 2.973(5) | N(1)⋯O(3) | 2.843(5) | N(1)⋯O(5b) | 2.811(7) | N(1)⋯O(7c) | 2.887(5) |
|---|---|---|---|---|---|---|---|
| N(2)⋯O(2) | 2.885(5) | N(2)⋯O(3a) | 3.1105(5) | N(2)⋯O(4d) | 2.976(5) | N(2)⋯O(7e) | 2.871(5) |

| Hydrogen bonds | Dist. | O—H⋯O | | Dist. | O—H⋯O |
|---|---|---|---|---|---|
| O(1)⋯O(6f) | 2.526(6) | 177(5) | O(w1)⋯O(4a) | 2.845(6) | 154(5) |

Symmetry transformations $a$,$(2-x,-y,2-z)$;$b$,$(3/2-x,1/2+y,2-z)$;$c$,$(-1/2+x,-1/2-y,z)$;$d$,$(x,y,1+z)$;$e$,$(1-x,-y,2-z)$;$f$,$(-1/2+x,-3/2-y,z)$.

# 3  RESULTS AND DISCUSSION

Preparation of the title compounds depends on pH control and the mol ratio of the reactants. Hedwig[16] reported a potetiometric and visible spectrophotometric study of the nickel(Ⅱ)-citric acid system $(H^+)_p(Ni^{2+})_q(cit^{4-})_r$ in the pH range 3—6.

The dominant species is(1,1,1)for pH > 5 at 1 : 1 nickel/citrate. This is consistent with the preparation of sodium, potassium or ammonium dimeric nickel(Ⅱ)citrate complexes in concentrated solution. The preparation of **1** was suggested via the following reaction.

$$2NiCl_2 + 2(NH_4)_3Hcit + 2H_2O \longrightarrow (NH_4)_2[Ni(Hcit)(H_2O)_2]_2 \cdot 2H_2O + 4NH_4Cl \qquad (1)$$

Moreover, the preparation of dicitratonickelate(Ⅱ)can be fulfilled *via* the following reaction.

$$NiCl_2 + 2(NH_4)_3Hcit + 2H_2O \longrightarrow (NH_4)_4[Ni(Hcit)_2]_2 \cdot 2H_2O + 2NH_4Cl \qquad (2)$$

Transformations between **1** and **2** can be accomplished by the reaction with ammonium citrate or nickel chloride.

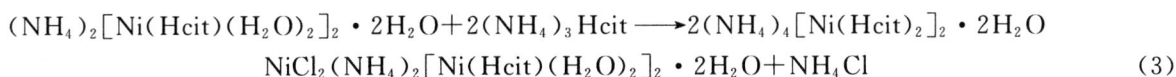

$$(NH_4)_2[Ni(Hcit)(H_2O)_2]_2 \cdot 2H_2O + 2(NH_4)_3Hcit \longrightarrow 2(NH_4)_4[Ni(Hcit)_2]_2 \cdot 2H_2O$$
$$NiCl_2(NH_4)_2[Ni(Hcit)(H_2O)_2]_2 \cdot 2H_2O + NH_4Cl \qquad (3)$$

Attempts to protonate **1** and **2** below pH 3 were unsuccessful, and led to decomposition of the complexes. This kind of reaction provide a good explantion for the removal and recovery of nickel ion in sodium citrate solution with a chelating resin containing a triethylenetetramine side chain[19].

In the electronic spectra of both nickel citrates, shifts to higher frequencies of the two bands in dicitratonickelate( II ), 2(638, 382 nm), compared to those of 1(655, 392 nm) are the result of the much stronger ligand field of the citrato ligand; there is also a marked shift compared to the $Ni^{2+}$ ion(720, 394 nm) consistent with the spectrochemical series of ligands[11,17,20]. Table VI shows comparisons of IR bands of the complexes. It can be seen that no IR band between $1740 - 1700$ $cm^{-1}$ is observed, indicateing that there is no associated carboxylic group. Typical of the spectra of 3 and 4 and dicitratonickelate( II ), 2, are well-resolved strong bands between 1625 and 1566 $cm^{-1}$, assignable to $v_{as}$(COO) and $v_s$(COO) bands between 1419 and 1385 $cm^{-1}$. They are all shifted with regard to citric acid or sodium citrate[21] and are not as well resolved in 1, implying coodination of carboxylato groups, as confirmed by X-ray structural analyses.

TABLE VI   Relevenl infrared data( $cm^{-1}$ )for the complexes

| citric acid | sodium citrate | 1 | 2 | 3 | 4 | assignt. |
|---|---|---|---|---|---|---|
| $3495_{vs}, 3448_w$ | | $3487_{vs}$ | $3550_{vs}$ | $3437_{vs}$ | $3511_{vs}, 3447_{vs}$ | v(OH) |
| $3298_{s,br}$ | | | | | | |
| | | $3115_{s,br}$ | $3172_{s,br}$ | | | v(NH) |
| $1750_{vs}$, | $1603_m$, | $1599_{vs,w}$ | $1605_{vs}$, | $1625_s, 1592_{vs}$, | | $v_{as}$(COO) |
| $1704_{vs}$ | $1587_s$ | | $1566_{vs}$ | $1562_{vs}$ | $1610_{vs}, 1566_{vs}$, | |
| $1430_s, 1390_s$ | $1439_s$, | $1419_s$, | $1400_{vs}$, | $1415_{vs}, 1385_s$ | $1410_{vs}, 1386_s$ | $v_s$(COO) |
| $1359_s$ | $1418_w$ | | | | | |
| | | $891_s$ | $905_m$ | $921_m, 896_m$ | $916_m, 885_s$ | $v_{as}$(NiO) |
| | | $834_m$ | $847_m$ | $842_m$ | $831_m$ | $v_s$(NiO) |

The complex $(NH_4)_2[Ni(Hcit)(H_2O)_2]_2 \cdot 2H_2O$, 1 exists as centrosymmetric dimers, as does its potassium salt[15], in which each citrate ion acts as a tetradentate ligand with the hydroxy, α-carboxyl and β-carboxyl oxygens coordinated to one nickel atom, and the other β-carboxylato group to a second nickel atom via a bridging bond ( forming an intramolecular hydrogen bond [O(1)—H(1)···O(5), 2.629(3) Å]). Octahedral coordination about the nickel atom is completed by the coordination of two water molecules. Figure 1 shows a perspective view of the dimeric anion.

The crystal structure of $(NH_4)_4[Ni(Hcit)_2]$ $\cdot 2H_2O$ 2 comprises discrete ammonium cations, water molecules and centrosymmetric dicitratonickelate anions. As shown in Figure 2, each citrate ion acts a as terdentate ligand coordinated to the same nickel atom via its hydroxyl, α- and β-carboxyl groups, while the

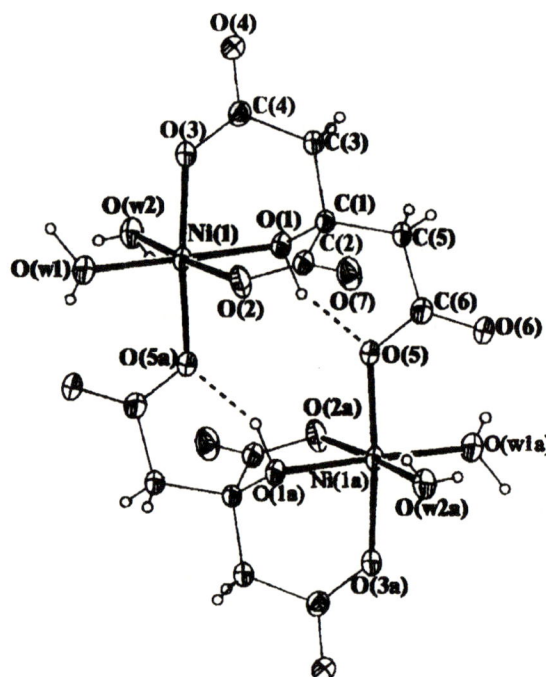

FIGURE 1   Perspective view of the anionic dimer $[Ni(Hcit)(H_2O)_2]_2^{2-}$ of 1.

other β-carboxyl group remains uncomplexed. The free β-carboxyl group is bonded with the hydroxy group of the other dicitrato nickel anion through a strong linear hydrogen bond [O(1)—H(1)···O(6f),

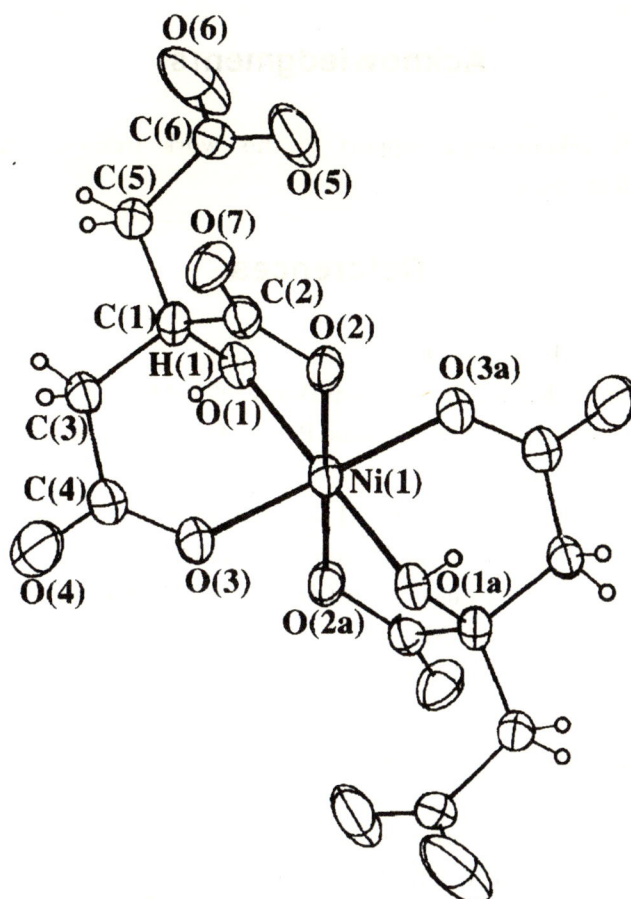

FIGURE 2   Perspective view of the anion $[Ni(Hcit)_2]^{4-}$ of 2.

2.526(6)Å].

Terdentate or tetradentate coordination of citrate through its hydroxyl, α-carboxyl and β-carboxyl group is a basic feature of mono- or dimeric nickel complexes. The structure of the Ni(II) complex of cit⁴⁻, where bridging is through the ionized hydroxyl oxygens and carboxyl groups at higher pH, is polymeric[5]. Thus, the pH of the solution and the ratio of reactants are crucial factors for the formation of these species. It is undoubted that the hydroxyl group of the citrate ion carries a proton in the structure of 1 and 2, since they were clearly visible in difference maps. The angle O(1)—Ni(1)—O(2) is 81.53(9)° for 1 and 79.2(1)° for 2, reflecting the small distortion of the five-membered chelate ring formed by the hydroxyl oxygen and α-carboxyl group. The Ni—O(α-carboxyl) bond lies between 2.020 (3) and 2.038(3) Å, for complex 2 and the Ni—O(hydroxyl) bond (2.021(3) Å) is shorter than that of Ni—O(β-carboxyl) bonds (2.072(3) Å), whereas for 1 and its potassium salt,[15] the Ni—O(hydroxyl) bond [2.074(2) or 2.125(3) Å] is much longer than the Ni—O(β-carboxyl) bonds [2.031(2), 2.037(2), 2.036, 2.075(3) Å]; this indicates the effects of different intramolecular and intermolecular hydrogen bonding.

## SUPPLEMENTARY MATERIAL

Complete lists of thermal parameters, bond distances, bond angles, hydrogen positions and observed and calculated structure factors for 1(10 pages) 2(9 pages) are available from the authors upon request.

# Acknowledgments

The authors gratefully acknowledge support to this work from the National Natural Science Foundation of China(No 29503021).

# References

[1]J P Glusker. Acc Chem Res,1980,**13**:345.

[2]Z H Zhou,H L Wan,K R Tsai. Inorg Chim Acta,1995,**237**:193.

[3]J J Cruywagen,E A Rohwer G F S Wessels. Polyhedron,1995,**14**:3481;Z H Zhou,H L Wan,K R Tsai. Polyhedron,1997,**16**:15.

[4]J J Cruywagen,L J Saayman,M L Niven,J Crystallogr Spectrosc Res,1992,**22**:737;E Llopis,J A Ramirez,A Domenech. J Chem Soc Dalton Trans,1993:1121.

[5]J Strouse,S W Layten,C E Strouse. J Am Chem Soc,1977,**99**:562.

[6]J Strouse. J Am Chem Soc,1977,**99**:572.

[7]J Lee,R D Reeves,R R Brooks. Phytochem,1977,**16**:1503.

[8]M A Halcrow,G Christou. Chem Rev,1994,**94**:2421.

[9]A Brenner. Electrodeposition of alloys:Principles and Practice,(Academic Press,New York,1963),**2**:219.

[10]G A Joshi-Tope,A J Francis. J Bacterioi,1995,**177**:1989.

[11]A J Francis,G A Joshi-Tope,C J Dodge. Environ Sci Technol,1996,**30**:562.

[12]X T Gao,D Ruitz,X X Guo. Catal Lett,1994,**23**:321.

[13]S Sato,F Nozaki,T Nakayama. Appl Catal A-Gen,1996:**139**:L1.

[14]Y Yoshimura,T Sato,H Shimada. Catal Today,1996,**29**:221.

[15]E N Bsker,H M Baker,B F Anderson. Inorg Chim Acta,1983,**78**:281.

[16]R Hedwig,J R Liddle,R D Reeves. Aus J Chem,1980,**33**:1685,and references therein.

[17]E R Still,P Wikberg. Inorg Chim Acta,1980,**46**:153.

[18]R I Bickley,H G M Edwards,R Gustar. J Mol Struct,1991,**246**:217.

[19]H Maeda,H Egawa. J Appi Poly Sci,1992,**45**:173.

[20]F A Cotton,G Wilkinson. Advanced Inorganic Chemistry,(John Wiley and Sons,New York,1988):744.

[21]K M Rao. Indian J Phys,1971,**45**:455.

■ **本文原载**:《厦门大学学报》(自然科学版)第 36 卷第 3 期(1997 年 5 月),第 381～387 页。

# 促进型甲酸甲酯氢解制甲醇铜基催化剂的研究[*]

李海燕　张鸿斌　林国栋　杨意泉　蔡启瑞

（厦门大学化学系　固体表面物理化学国家重点实验室　厦门　361005）

**摘　要**　研制开发了一种耐 CO 低温高效四组分(Cu-Cr-Mn-Ni)改进铜铬催化剂,其原料合成气中 CO 允许含量可高达 13V％;在常压,413 K,原料气 $H_2$/MF=4/1(v/v),流速为 1 800mL/h·g cat al 的反应条件下,MF 转化率高达 99.8％,甲醇选择性为 99.9％(在相同反应条件下二组分的 Cu—Cr 催化剂上 MF 转化率仅为 73.1％,甲醇选择性为 98.0％),当向原料气中添加 13.1V％ CO 时,MF 转化率仍可以稳定达到 95.9％,甲醇选择性为 96.3％。实验结果表明,CO 导致催化剂失活的重要原因之一是 CO 导致 $Cu^+$ 深度还原为 $Cu^0$,减少催化剂的活性中心。

**关键词**　甲酸甲酯氢解　甲醇　促进型铜铬催化剂

**中国图书分类号**　O 643.36

合成气经甲醇(MeOH)羰化-甲酸甲酯(MF)氢解两步串联催化一体化制甲醇是十分诱人的低温低压、高转化率的甲醇合成的催化过程[1-4]。其中,"羰化"和"氢解"两反应的反应热均约为现行 Cu—Zn—Al 催化剂上合成气直接合成甲醇反应热的一半,较易于移去,从而有利于单程转化率的提高及生产工艺的大型化;且产物只含甲醇和甲酸甲酯,无水甲醇的生成使能耗很高的甲醇-水分离得以省去。但自从 Christiansen 首次提出二步法[5]以来,目前仅有 Palekar 等[6,7]报道的将两个分立的液相反应合二为一,在浆态床反应体系中进行,催化剂体系尚无实质性改变,且因反应以浆态床方式进行,不可避免地存在产物与催化剂的分离以及时空收率低等问题。

本工作试图开发一条气体连续流动固定床串联催化一体化由合成气经甲酸甲酯制甲醇的新技术路线;即在同一反应器内上下两段固定床分别装填两种对反应条件(压力、反应气组成等)能相互兼容,在生产能力上能相互匹配的催化剂,分别催化甲醇羰化和甲酸甲酯氢解两步反应,以实现由合成气经甲酸甲酯到甲醇的串联合成。为实现这一目标,一种能适用于气体连续流动固定床操作的负载型强碱液膜固体催化剂已成功的研制出[4],率先实现了甲醇羰化制甲酸甲酯的多相化但一体化原料气中的高含量的 CO 却是导致目前现有甲酸甲酯氢解铜铬催化剂失活的主要因素,文献[8,9]曾报道,原料气中,即使 1V％～2V％ 的 CO 亦导致 Cu—Cr,RanyCu,Cu/$SiO_2$ 以及其他铜基催化剂的失活。因此研制能在较高 CO 含量氛围下工作的甲酸甲酯氢解催化剂将是实现气相甲醇羰化和甲酸甲酯氢解串联一体化由合成气制甲醇的关键。为此,本工作开发了一种四组分的改进铜铬催化剂,其耐 CO 能力达到原料气 CO 含量～13V％ 的水平。我们考察了甲酸甲酯在这种改进型铜铬催化剂上的氢解行为,以及 CO 对催化剂活性的影响,探讨其作用本质,为谱学表征研究以及进一步提高催化剂抗 CO 的能力提供动力学实验依据。

---

* 本文 1996 年 1 月 25 日收到;国家教委博士点基金和国家自然科学基金联合资助项目。

# 1　实验

采用文献[10]报道的共沉淀法制备包括助剂的系列铜铬氧化物催化剂,制好的粉末经压片、过筛,取60~80目样品备用。

称取上述催化剂 1 g 装入 Φ8 mm 不锈钢管反应器中,通入 H$_2$(10 mL/min),然后以 0.5 ℃/min 速度升温至 170 ℃,预还原 12 h 后,降至反应温度。通过转换阀使净化的 H$_2$ 通过鼓泡器带走纯化后的甲酸甲酯蒸气,甲酸甲酯的蒸发量可由鼓泡器温度的高低来调控,以调节原料气的 MF/H$_2$ 之比在~1/4(摩尔比)范围内,反应气的流速可通过尾气排空处的皂沫流量计来准确计量尾气通过加热带保温(~120℃)的 Φ3 mm 不锈钢管,经由六通阀采样,直接进入热导池检测器进行现场分析。所用色谱仪为上海分析仪器厂产品(103 型),色谱柱填料为 GDX401,柱长 2 m,柱温:100 ℃;载气(H$_2$)流速:20 mL/min。

# 2　结果与讨论

## 2.1　助剂对催化剂活性的影响

添加第四组分 Ni 前后 Cu—Cr 基系列催化剂对甲酸甲酯氢解反应的活性比较结果列于表 1。实验结果表明,添加 Mn 助剂有助于催化剂活性和甲醇选择性的提高。而在添加 Mn 的基础上再添加少量的助剂 Ni,催化剂的活性骤增(MF 转化率由 Cu-Cr-Mn 体系的 76.2％增至 Cu-Cr-Mn-Ni(1wt％)体系的 94.3％),甲醇选择性非但不下降,还略有上升(由原来的 99.6％升至 99.9％)。然而,Ni 助剂的添加量有一最佳值,约 1wt％;超过这一最佳值,催化剂活性骤降;当 Ni 添加量为 1.5wt％时,MF 转化率降至 44.1％,2wt％的 Ni 添加量导致催化剂完全失活。

表 1　助剂对催化剂催化甲酸甲酯氢解反应活性的影响

Tab 1　Effect of promoter on activity of catalysts for hydrogenolysis of methyl formate to methanol

| 催化剂 | MF 转化率(％) | MeOH 选择性(％) | 时空收率 (mmol/h · g catal) |
| --- | --- | --- | --- |
| Cu—Cr | 73.14 | 98.02 | 23.04 |
| Cu—Cr—Mn | 76.15 | 99.55 | 24.64 |
| Cu—Cr—Mn—Ni(1wt％) | 94.34 | 99.87 | 30.28 |
| Cu—Cr—Mn—Ni(1.5wt％) | 44.09 | 97.77 | 13.85 |
| Cu—Cr—Mn—Ni(2wt％) | 0 | 0 | 0 |

反应条件:压力＝0.1 MPa;温度＝413 K;GHSV＝1800 mL/h · g catal 原料气组成:H$_2$/MF＝4/1(mol/mol)

## 2.2　反应条件对催化剂活性的影响

(1)温度的影响　图 1 示出温度对 Cu—Cr—Mn—Ni(1wt％)催化剂上 MF 氢解反应活性的影响。为便于对照,相应反应条件下的热力学平衡转化率也作了计算。亦示于图 1 中。由实验得出的 MF 转化率随 T 的变化曲线 1-a 可看出,开始随着反应温度的上升,MF 的转化率单调上升,当温度升至 413 K 时,达到极大;而后,随着反应温度的继续上升,MF 的转化率反而下降。由热力学平衡计算所得的理论线 1-b 可知,由于该反应是一放热反应故 MF 的平衡转化率将随反应温度的上升而下降。比较 a、b 两曲线可知,在反应温度未达到 413 K 时,催化剂未达到其最佳活化状态,MF 的转化率远低于热力学平衡值。因此,在低于 413 K 的温度范围内,MF 的转化率未受热力学平衡的限制,随着反应温度的升高,反应速度加快。当温度升至 413 K 时,MF 的实验转化率(94.3%)接近理论计算值(95.50%),表明催化剂此时已基本达到其最高活性水平。于是,当温度继续升高时,反应受热力学平衡的限制,其 MF 转化率开始下降;当反应温度从 413 K 升至 433 K 时,MF 转化率的下降幅度同热力学平衡计算值的下降幅度相近,表明在此温度范围内的反应收率的降低是受热力学平衡所限,而非催化剂的活性下降。然而,当反应温度由 433 K 继续上升时,由图 1 中,曲线 a、b 的逼近程度之差别可见,MF 的转化率下降幅度大于理论计算值,这似乎表明,在 433 K 以上,温度升高引起 MF 转化率的下降不仅系热力学平衡所限,亦可能系催化剂活性降低所致。在另一方面,由选择性曲线 c 可知,当反应温度超过 423 K 时,甲醇的选择性亦明显下降。这可能是由于反应温度升高,导致 MF 更容易脱羰生成副产物 CO 所致。

(2)压力的影响　表 2 示出压力对 MF 氢解反应的影响。增加反应系统的压力,固然一方面有利于热力学平衡点的右移(该反应是体积缩小的反应),提高 MF 的转化率。但是,由表中括号内的理论计算值可知,这种压力的变化所引起的 MF 转化率的上升幅度并不大(当压力从 0.1 MPa 升至 1.2 MPa 时,MF 的转化率相应由 95.50% 升至 99.53%),但实际催化反应所达到的转化率的升高却显示出较大的幅度(当压力由 0.1 MPa 升至 1.2 MPa,其相应的转化率从 76.30% 增至 99.30%)其原因可能是增大压力,有利于增加 $H_2$ 和 MF 在催化剂表面的吸附,增大反应速率,从而提高了 MF 的转化率。但是,实际反应压力也不能升得过高。因为热力学计算表明,在此反应条件下,当压力升至 20 MPa 时,将会出现 MF 和 MeOH 的凝聚,浸湿催化剂,阻塞催化剂床层。在另一方面,根据文献[11]报道,由于 $H_2$ 与 MF 在 Cu 上的吸附能力是 MF>$H_2$,故增大压力,更有利于 MF 的吸附活化,使所生成的甲醛中间体因氢供应不上而来不及加氢就自身发生聚合,覆盖催化剂活性表面,从而导致催化剂活性下降。

表 2　压力对反应活性的影响

Tab. 2　Effect of pressure on activity of catalyst for hydrogenolysis of methyl formate to methanol

| 压力 | MF 转化率(%) | | MeOH 选择性(%) | 时空收率(mmol/h·g catal) |
|---|---|---|---|---|
| 0.1 | 76.30 | (95.50)[a] | 99.00 | 97.11 |
| 0.2 | 79.78 | | 99.02 | 101.54 |
| 0.4 | 95.91 | | 99.21 | 122.33 |
| 0.6 | 96.45 | (99.14)[a] | 99.99 | 123.98 |
| 0.8 | 97.22 | | 99.89 | 124.85 |
| 1.0 | 98.66 | (99.48)[a] | 99.90 | 126.71 |
| 1.2 | 99.30 | (99.53)[a] | 99.94 | 127.58 |
| 2.0 | | (99.74+)[a] | | |

反应条件:温度=413 K;GHSV=7200 mL/h·g catal 原料气组成:$H_2$/MF=4/1(mol/mol)

a 代表理论计算值;+代表气液共存态

(3)原料气组成的影响　图 2 示出原料气组成($H_2$/MF)对 MF 转化率的影响。随着 $H_2$/MF 比的上升,MF 的转化率相应上升,但 MeOH 的时空收率却逐渐下降。这一规律与热力学理论计算值所表现出的趋势完全一致。但比较 MF 转化率的实验值(曲线 a)和理论值(曲线 b),可以发现,随着 $H_2$/MF 比例的提高,a,b 两曲线逐渐靠近至 $H_2$/MF 达 12/1 时,两线重叠,实验 MF 转化率达到最大值。这多半暗示 $H_2$ 分压的提高,有利于 $H_2$ 的吸附和活化,提高了加氢反应步骤的推动力。

(4)空速的影响　1800 表 3 示出了空速对 MF 氢解的反应速率的影响。随着空速的提高,MF 的转化率有所下降,但 MeOH 的时空收率却递增。这是由于提高空速,一方面势将缩短反应分子与催化剂的接触时间,减少反应分子的吸附活化几率,从而导致 MF 转化率的下降。但在另一方面,空速的提高意味着单位时间内进入催化剂床层的反应物分子总数的增加,可以抵消由接触时间缩短带来的不利影响,因而表现出甲醇的时空收率有所提高。

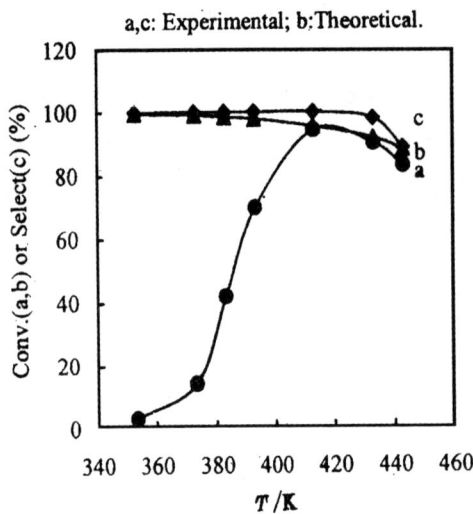

图 1　温度对反应活性的影响

反应压力＝0.1 MPa;GHSV＝1 800 mL/h·g catal;

原料气组成:H₂/MF＝4/1(mol/nol)

$a.$ MF 转化率与温度的关系曲线;

$b.$ MF 的平衡转化率与温度的关系曲线;

$c.$ 甲醇的选择性与温度的关系曲线

**Fig 1　Effect of Temperature on Activity for Hydrogenolysis of Methyl Formate to Methanol over the Cu—Cr—Mn—Ni catalyst**

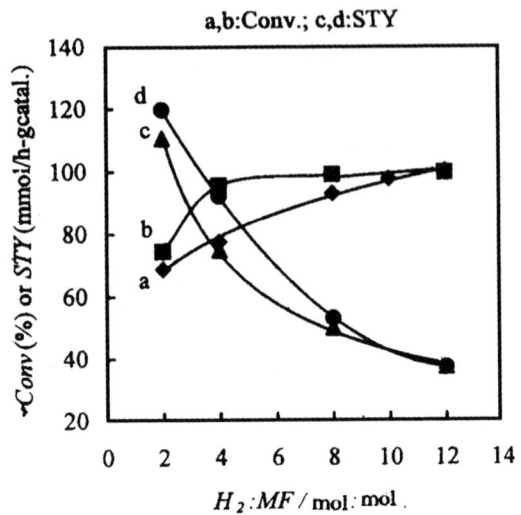

图 2　原料气组成对反应活性的影响

反应压力＝0.1 MPa;温度＝413 K GHSV＝5 400 mL/h·g catal

$a.$ MF 转化率与温度的关系曲线;

$b.$ MF 的平衡转化率与温度的关系曲线;

$c.$ 甲醇的时空收率与温度的关系曲线;

$d.$ 平衡时甲醇的时空收率与温度的关系曲线

**Fig 2　Effect of Feed composion on Activity for Hydrogenolysis of Methyl Formate to Methanol over the Cu—Cr—Mn—Ni catalyst**

表 3　空速对反应活性的影响

**Tab 3　Effect of GHSV of feed gases on activity of catalyst for hydrogenolysis of methyl formate to Methanol**

| 空速(mL/h·g catal) | MF 转化率(%) | MeOH 选择性(%) | 时空收率(mmol/h·g catal) |
| --- | --- | --- | --- |
| 1 800 | 94.3 | ~100 | 30.31 |
| 3 600 | 82.8 | ~100 | 53.23 |
| 4 800 | 78.6 | ~100 | 67.37 |
| 5 400 | 77.5 | ~100 | 74.73 |
| 7 200 | 76.3 | ~100 | 98.01 |

反应条件:温度＝413 K;压力＝0.1MPa;原料气组成:H₂/MF＝4/1(mol/mol)

## 2.3　CO 对催化剂活性的影响

文献[8,9]曾报道,原料气中,即使 1V%～2V%的 CO 亦导致 Cu—Cr,RanyCu,Cu/SiO₂ 以及其他铜基催化剂的失活。然而,对改进后的四组分 Cu—Cr—Mn—Ni(1wt%)催化剂,由表4示出 CO 对其催化活性影响的考察结果可知:当原料气中 CO 的含量由 0 逐渐升至 13.1%时,相应之 MF 转化率基本可维持在 93%以上,尚无明显下降趋势;当 CO 的含量升至 17.1%时,MF 的转化率出现明显下降;而当 CO 含量升至 21%时,MF 的转化率降至只有 8.75%。由此看来,添加助剂 Mn 和 Ni 后的四组分 Cu 基催化剂耐 CO 浓度可达 13.1%。其活性未见衰减。我们认为改进后的四组分催化剂抗 CO 能力之所以有所提高,可能同添加 Mn 和 Ni 后增加了氢的吸附活化位,从而在一定程度上补偿了较高浓度 CO 与 H₂ 争夺吸附位所带来的负面效应。但是,如果 CO 导致催化剂活性降低的原因仅仅是可逆的竞争吸附,那么可以期望当用惰性气体(Ar)彻底吹扫以除去含 CO 的气体及催化剂表面的吸附物后,导入不含 CO 的原料气时,催化剂活性应回升至未通 CO 前的水平。从实验结果来看,此时催化剂的活性回升远未能达到原有水平。这就意味着 CO 的阻化作用看来不仅仅是 CO 与 H₂ 的可逆竞争吸附所致,应该还有另一非可逆的导致催化剂失活的因素。根据活性催化剂是黑色,而失活后的催化剂呈红色,且颗粒度增大,我们推测高浓度 CO 深度还原 Cu⁺ 至 Cu⁰,大量的 Cu⁰ 在催化剂表面聚结,因而减少了催化剂的活性中心,是导致催化剂失活的主要原因。这一观点将由下文 XPS 实验结果得到进一步支持。

**表 4 CO 对反应活性的影响**

**Tab 4　Effect of CO in feed gases on activity of catalyst for hydrogenolysis of methyl formate to methanol**

| CO 含量(v%) | MF 转化率(%) | MeOH 选择性(%) | 时空收率(mmol/h·g catal） |
|---|---|---|---|
| 0 | 94.34 | 97.35 | 29.5 |
| 4 | 94.30 | 97.45 | 29.2 |
| 83 | 93.10 | 97.54 | 29.2 |
| 13.1 | 93.71 | 96.77 | 29.4 |
| 17.1 | 63.65 | 96.34 | 19.7 |
| 21.0 | 8.75 | 91.67 | 2.3 |
| 0ª | 51.82 | 98.93 | 16.2 |

反应条件:温度=413 K;压力=0.1 MPa;GHSV=1800 mL/h·gcatal 原料气组成:H₂/MF=4/1(mol/mol)

ª 上述反应后,用惰性气体(Ar)彻底吹扫(0.5 h)以除去含 CO 的气体及催化剂表面的吸附物后,导入不含 CO 的原料气继续实验

# 参考文献

[1]Evans J W,Cant N W,Trimm DL et al. Hydrogenolysis of ethyl formate over copper based catalysts,Appl Catal,1983,**6**(3):355~362.

[2]MontiDM,KohlerM A,Wainuright M S et al. Liquid phase hydrogeno lysis of methyl fomate in a semi batch reactor. Appl Catal,1986,**22**(1):123~136.

[3]Liu Z,Tierney J W,Shah Y T et al. Kinetics of two-Step methanol synthesis in the slurry phase. Fuel Process Tech. ,1988,**18**(2):185~199.

[4]Zhang H B,LiH Y,Lin GD et al. Study on catalytic synthesis ofmethanol from syngas via methylfor-mate in heterogeneous "One-Pot"reactions proc 11th ICC-40th anniversary. Studies in Surf. Sci and Catal,1996,**101**:1 369~1 378.

[5]Christiansen J A. U. S. Patent,1 302 011,April 29,1919.

[6]Palekar V M,Jung H,Tierney J W et al. Slurry phase synthesis of methanol with a potassium methoxide/copper chromite catalytic system. Appl Catal A:General,1993,**102**(1):13~34.

[7]Palekar V M,Tierney JW,Wender I. A lkali compounds and copper chromite as low-temperature slurry phase methanol catalysts. Appl Catal A:General,1993,**103**(1):105~122.

[8]Monti D M,Wainuright M S,Treimm D L et al. Kinetics of the vapor-Phase hydrogenolysis of methyl fomate over copper on sillica catalysts. Ind. Eng. Chem. Prod. Res Dev. ,1985,**24**(3):397~401.

[9]Evans J W,Casey P S,Wainuright M S et al. Hydrogenolysis of alkyl form ates over a copper

chromite catalyst. Appl Catal,1983,**7**(1):31～41.

[10]Adkins H,Burks R E. Equilibria in hydrogenation of eesters and of Indoles. J. A m. Chem. S oc.,1948,**70**(4):4 174～4 177.

[11]Monti D M,Cant N W,Trimm D L et al. Hydrogenolysis of methyl formate over copper on sillica:study of the mechanism using labeled compounds. J. Catal,1986,**100**(1):28～38.

# Study of Promoted Cu-based Catalysts for Hydrogenolysis of Methyl Formate to Methanol

Hai-Yan Li，Hong-Bin Zhang，Guo-Dong Lin，Yi-Quan Yang，Khi－Rui Tsai

(Dept of Chem. and State Key Lab for Phy. Chem.

of the Solid Surf.，Xiamen Univ.，Xiamen 361005)

**Abstract**    A promoted Cu-based catalyst composed of 4-components (Cu—Cr—Mn—Ni) for Hydrogenolysis of Methyl Formate to Methanol has been studied and developed. Under reaction condition of 413 K,0. 1MPa,$H_2$/MF=4/1(mol/mol),and GHSV=1800 mL(STP)/h・g catal,the conversion of MF reached to 99. 8% and selectivity to $CH_3OH$ up to 99. 9%；whereas under the same conditions,with the bi-components Cu—Cr catalyst,only 73. 1% MF was converted. Furthemore,even in the case that CO content was increased to 13. 1% in the feed stream,the MF-conversion and $CH_3OH$-selectivity still kept as high as 95. 5% and 96. 3%,respectively. The experimental results indicated that one of the important reasons for loss of the activity of catalyst caused by CO seemed that CO would lead to $Cu^+$ being deeply reduced to $Cu^0$,and as a consequence,the active sites on the surface of the functioning catalyst was going down.

**Key words**    Hydrogenolysis of methyl formate    Methanol    Promoted Cu—Cr based catalyst

■ **本文原载**:《高等学校化学学报》(1997 年 7 月)，第 1185～1193 页。

# 过渡金属配合物催化剂及其分子设计构思的发展与相互作用*

万惠霖[①]　袁友珠　高景星　张鸿斌　蔡启瑞

（厦门大学化学系　物理化学研究所　固体表面物理化学国家重点实验室，厦门　361005）

**摘　要**　讨论了某些不对称合成和 α-烯烃定向配位聚合及 α-烯烃氢甲酰化的过渡金属配合物定向络合催化剂的发展与催化剂分子设计构思的发展及相互促进作用，并藉以说明过渡金属配合物定向络合催化剂的分子设计已具备向计算机辅助设计发展的科学基础。

**关键词**　不对称合成　定向配位聚合　氢甲酰化　催化剂分子设计

催化剂设计可粗分为工程设计和分子设计。催化剂分子设计的目的在于为特定的化学转化设计出具有高催化功能的化学物质，为新催化剂的选择和研制指引方向。催化剂分子设计是建立在分子催化基础上的，即建立在对催化剂的结构与功能关系有着深入到分子、原子水平的认识基础上的。催化剂分子设计既可为催化剂体系的选择和研制指引方向，反过来又需根据催化剂研制、表征和评价结果与预期目标的对比来调整设计方案，乃至补充某些必要的分子催化信息，并通过分子设计与反复调试而趋近自洽优化结果。分子催化研究的发展与催化剂分子设计及其用于催化剂研制实践的发展是相互促进的。目前仅有少数类型的催化反应和催化剂的结构与功能关系了解得比较清楚，催化剂分子设计构思的发展也比较深入。如具有择形催化功能的分子筛催化剂和具有定向配位催化功能的过渡金属配合物催化剂等。但这些发展至今仍然依靠分子设计的创新构思与大量研制、调试的实验工作相结合来获得。关于有机金属催化剂（更广泛些就是金属配合物催化剂）的研制，直至最近才有计算机辅助催化剂分子设计软件开发的报道[1]。与此对比，药物合成已初步开发出不少的计算机辅助分子设计软件，并取得一定的应用效果[2]。药物合成的分子设计主要建立在药理和药物分子与蛋白质等生物高分子受体的结构契合基础之上；蛋白质结构测定的精度（分辨率一般低于 0.1～0.2 nm）比过渡金属配合物催化剂的低得多，而且生物高分子还有塑性和诱导变构等问题。因此，蛋白质受体与药物分子的结构契合（所谓的锁-钥契合），比起化学反应物种与过渡金属配合物催化剂的定向配位络合要复杂得多。可见过渡金属配合物定向络合催化剂的分子设计早应发展计算机辅助设计（CAD）的专家系统。

本文讨论了 3 种类型的过渡金属配合物定向络合催化剂的发展与催化剂分子设计构思发展的相互促进作用，并藉以说明过渡金属配合物定向络合催化剂的分子设计早已具备向计算机辅助设计发展的基础。这 3 种典型的过渡金属配合物定向络合催化剂为：某些类型的不对称合成催化剂，多次更新换代的 α-烯烃定向配位聚合催化剂和 α-烯烃氢甲酰化催化剂。前两类催化剂的定向络合催化选择性是对映体选择性和等规/间规配位聚合选择性，皆属于立体异构选择性范畴/而后一类催化剂的定向选择性是正构/异构（$n/i$）选择性，属于区位选择性范畴。这些反应相对于涉及反应分子中其他官能团的副反应则属

＊ 国家自然科学基金和中国石油化工总公司联合资助课题。

① 本文 1997 年 2 月 14 日收到；联系人及第一作者：万惠霖，男，58 岁，教授，博士导师。

于化学选择性范畴。后两类反应有时也可采用特殊设计的手性催化剂,并获得同时具有对映体选择性的产物。

# 1 某些反应类型的不对称合成催化剂的发展

不对称催化是手性增殖过程,即用少量催化剂(其手性源为不对称催化剂的手性配体)为模板,控制反应物的对映体选择性,通过催化循环产生大量的光学活性物质,因而是实现不对称合成最有效的手段。近 20 年来,不对称催化在不对称加氢、不对称环氧化、不对称环丙烷化、不对称异构化、不对称氢氰化等反应中已取得了重要成果,其对映体过量百分数已超过 90%,有的甚至接近 100%;包括左旋-多巴、薄荷醇和拟除虫菊酯等近十个不对称催化合成反应已实现工业化。不对称催化在药物、农药、香料合成等方面的用途正日益扩大。迄今,所研究的不对称催化反应绝大多数为均相催化反应,所用催化剂为过渡金属手性配合物。

不对称催化是"四维化学"。高效的不对称催化过程要求催化体系有理想的三维结构及与之适应的反应动力学特征,包括对催化剂(中心金属和配体)、底物分子和反应条件的合适选择。其中,手性配体的设计和合成至关重要,必须使之具有洽当的功能、构型及构型的刚性与柔韧性。在适宜的反应条件下,通过催化剂的分子识别过程能够对反应物基态和过渡态结构的稳定性及反应活性进行巧妙的控制,即对反应物分子及其过渡态的空间取向和反应通道进行控制,从而达到提高立体选择性并有足够活性的目的。

Noyori 等[3] 合成了被誉为超手性配体的联萘膦 BINAP,这是一种旋转受阻异构体,且能与中心金属形成环状结构,因此具有一定的刚性;但其 C(1)—C(1′) 轴可适当旋转和调整,双膦与中心金属原子螯合生成的是七元环,故而又有一定的柔韧性和结构可调整性。此外,与 CHIRAPHOS 和 DIPAMP 等手性配体一样,BINAP 具有低阶次的 $C_2$ 对称轴,能减少反应中非对映异构体过渡态可能出现的数目,使活性物种比较单纯,产物较易提纯。

Halpern 及 Brown 等[4,5] 基于对反应中间体的 NMR、X 射线结晶学研究及对反应动力学的详细分析,阐明了 Rh-膦配合物催化的(Z)-α-乙酰胺基肉桂酸酯不对称加氢的反应机理,即底物分子以其烯键和酰胺基中羰基与 Rh$^I$ 配位形成螯形配合物;$H_2$ 对金属的氧化加成及 Rh$^{III}$ 二氢基化合物中间体的形成,2 个氢基配体对底物烯键的连续邻位转移(还原消去)从而完成这种催化循环(图 1)。如二膦手性配体具有 $C_2$ 对称性,一方面由于烯键能以其 re 面或 si 面与 Rh$^I$ 配位络合,故可形成两种烯酰胺配合物的非对映异构体;另一方面,动力学研究结果表明,室温下涉及金属中心和非对映异构体

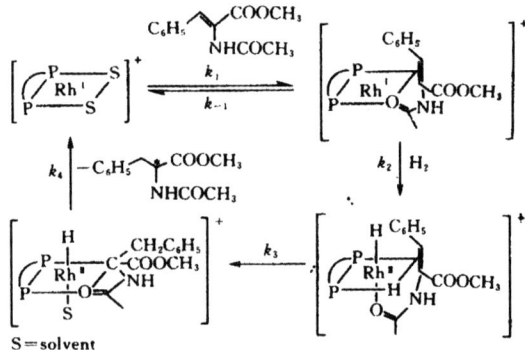

Fig. 1　Mechanism of Rh-based asymmetric hydrogenation

配键的 $H_2$ 的氧化加成,是诸基元反应中第一个不可逆的立体控制步骤。因而两种非对映体配合物的相对浓度(即相对热力学稳定性)和反应活性决定了产物的 对映体选择性。

该类反应所用的催化剂是(S,S)-CHIRAPHOS-Rh 或(R,R)-DIPAMP-Rh,底物为(Z)-α-乙酰胺基肉桂酸甲酯(MAC)或乙酯(EAC)。研究结果表明,大过量的主要非对映体 MAC-Rh 配合物并未转化成相应的主要对映体产物,而较不稳定的次要$[MAC-Rh-(CHIRAPHOS)]_{minor}^+$配合物却导致了主要对映体(R)-苯丙氨酸酯的生成。这是因为$[MAC-Rh-(CHIRAPHOS)]_{minor}^+$与 $H_2$ 反应的过渡态能垒比其主要异构体的过渡态能垒要低,生成的 Rh$^{III}$ 二氢基络合物也比较稳定。此外,Halpern 等[6]用 XRD 测定了主要

非对映体配合物$[(S,S)\text{-(CHIRAPHOS)-Rh-(EAC)}]^+_{major}$的结构(图2),由此加氢应产生 $S$-对映体,但由于次要的非对映体配合物与 $H_2$ 反应的活性至少比前者高上百倍(催化剂较有效地结合反应过渡态),所以得到的主要产物是 $R$-异构体。类似的例子还有$(R,R)$-DIPAMP-Rh 催化的上述反应及$[(R)-$BINAP-Rh$]^+$催化的$(Z)$-$\alpha$-苯酰胺基肉桂酸加氢等。这表明在不对称催化中,用谱学(XRD、NMR 等)方法能够检测到的热力学稳定的始态或中间体配合物很可能不直接与主要的催化循环有关(或很少参与);为获得高的立体选择性,手性金属配合物催化剂不仅要加速反应的进程,还要能以大约 10 kJ/mol 活化自由能差异的精确性区别非对映体过渡态。

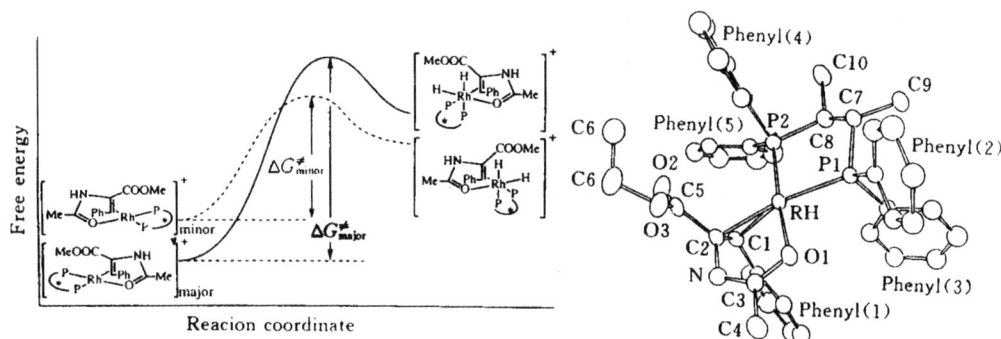

Fig. 2　Reactivity of diastereometric substrate-Rh complexes and crystal structure of the major $[\text{Rh}(S,S)\text{-chiraphos(EAC)}]^+\text{ClO}_4$ complex

上述重要研究结果将为有关不对称催化剂设计构思的形成和发展及其在结构、机理和动力学信息的运用方面提供重要参考。

## 2　$\alpha$-烯烃配位聚合 Ziegler-Natta 型催化剂的多次更新换代

用 Ziegler-Natta 型配位聚合方法生产的世界三大合成材料(合成塑料、合成纤维、合成橡胶,它们的产量吨位比约为 8:1.2:0.8),近年来已占 40% 左右。Ziegler-Natta 型催化剂已开始发展到了第三代[7]。这三代催化剂体系的"主催化剂"是ⅣB-Ⅷ族过渡金属卤化物、烷氧化合物或金属茂氯化物,"共催化剂"是ⅢA-ⅠA 族金属烷基化合物,或含卤素、氧的ⅢA 金属烷基化合物。第一代催化剂体系用于 $\alpha$-烯烃聚合时,典型的体系是由 Ti、V 或 Cr 等的过渡金属卤化物(或烷氧基化合物等)和烷基铝(或烷基铝氯化物)组成,均不用载体。Natta[8]从夹层型 $TiCl_3$ 的晶体结构出发,指出 $\alpha$-烯烃的螺旋性等规定向聚合与夹层状 $TiCl_3$ 晶体活性中心具有不对称结构有关(60 年代初我们发现,不对称性结构也从 $\alpha$-$TiCl_3$ 晶体中钛离子相邻氯离子的诱导偶极矩及其取向的计算结果得到反映)。关于 $\alpha$-烯烃定向聚合机理,60 年代初 Cossee[9]曾提出了著名的单金属活性中心模型,它是夹层型 $\alpha$-$TiCl_3$ 晶体的侧面上含 1 个空配位(单个 $Ti^{Ⅲ}$ 上)和 1 个烷基配体 R 的八面体结构的中心 $Ti^{Ⅲ}$ 离子,其他 4 个配体是氯离子。Cossee 机理的要点是:烯烃分子在空位上与活性中心的过渡金属配位络合,R 基配体按邻位转移插入机理,通过 1 个四中心过渡态转移到 $\alpha$-烯烃配位的 $\beta$-碳上,形成增长 2 个碳链节(并多一个 $CH_3$ 侧基)的链(R' 或 P),并与空配位互换了位置,接着 P 再转移到原来的位置上腾出空配位,恢复原来的活性中心构型以进行下一轮的催化循环。60 年代中期,Arlman、Cossee(基于单金属活性中心模型)及 Boor(基于双金属活性中心模型)[9-19]在深入分析已知的 $TiCl_3$ 夹层状晶体侧面结构特点的基础上,较好地说明了 $\alpha$-烯烃等规配位聚合活性中心微环境的立体化学控制因素。Boor[13-19]对 Lewis 碱添加剂的作用也进行了较系统的考察,说明了其主要作用是堵塞那些空间位阻太小、定向配位功能不够好的活性位,如边角的一些较大的空位;但 Lewis 碱添加剂也起着一定的电子因素作用。Natta、Cossee、Arlman 和 Boor 等的工作[8-19]都隐含着催化剂体系的分子设计构思。Cossee 机理虽被广泛引用,但还有些问题不容易用其来说明[7]。最近,Ystenes[20]提出了所谓的"扳机机理",其设想大意可理解为:1 个候补的单体分子紧跟在配位络合着的单体旁边,有随时乘机挤进配位界之势,当配位单体的 π-键刚开键准备插入到 M—R 键之间时,旁边的候补单体分子 π-键就开始与金属中心部分络合,及时补偿键能的损失,以降低过渡态能垒、协助促进配位单体分子的邻位插入。值得注意的是,这里并未明显提出是邻位转移插入。

70 年代中期,以 $MgCl_2$ 为载体的 Lewis 碱-金属烷基物-$TiCl_3/MgCl_2$ 负载型催化剂在 α-烯烃聚合方面取得了很大成功。这类负载型催化剂体系被称为第二代 Ziegler-Natta 催化剂[7]。它能大幅度地提高钛的利用率和定向聚合效率,并可防止暴聚;对丙烯定向聚合,由于载体 $MgCl_2$ 与 α-(δ-)$TiCl_3$ 同属夹层状晶型,晶胞大小相近,所以有利于活性组分 $TiCl_3$ 被诱导形成夹层状同晶薄层并均匀地负载在载体晶体表面,与金属烷基物和 Lewis 碱紧密接触,使得这类催化剂不但具有很高的比活性,而且有很高的等规选择性。可见第二代 Ziegler-Natta 催化剂的研制成功,包括载体及 Lewis 碱添加剂的合理选择,主要是建立在上述丙烯聚合立体选择性控制因素与催化剂晶体结构、夹层晶体侧面活性中心的形成及作用模式的关系等分子催化研究基础上的。但这种多相催化剂表面很不均匀,存在着多种微环境不同、反应动力学参数也不同的活性位,因此,聚合物分子量分布很宽,动力学分析困难。

第三代 Ziegler-Natta 催化剂是均相催化剂体系。Breslow[21]和 Natta[22]发现金属茂类均相 Ziegler-Natta 催化剂体系($Cp_2TiCl_2$-$AlR_2Cl$)乙烯聚合活性低,对丙烯聚合无活性。1980 年,Kaminsky 等[23]用甲基铝氧齐聚物(MAO)与 ⅣB 族金属茂制成活性极高的乙烯聚合催化剂,如 $Cp_2ZrCl_2$-MAO 催化剂在 20 ℃时乙烯生产能力高达 $9 \times 10^2$ g PE/(mol Zr·h·Pa),但这种催化剂活性中心对称性不够低,只能生产无规聚丙烯。可以设想,若在均相配位催化剂的活性中心周围构建一种类似于层状 $TiCl_3$ 活性中心那样的具有空间位阻的不对称微环境,则有可能用于丙烯的定向聚合。80 年代初,Brintzinger 等[24,25]合成了属于金属二茂基配合物类型的外消旋乙烯桥连(本文以斜体 *Et* 表示)二茚基锆二氯化物 *Et*(Ind)$_2$ZrCl$_2$、四氢化物 *Et*(H$_4$Ind)$_2$ZrCl$_2$ 及内、外消旋的钛同系物,并与 MAO 配成均相 Ziegler-Natta 催化剂。其中,外消旋的 *rac-Et*(Ind)$_2$ZrCl$_2$-MAO 和 *rac-Et*(H$_4$Ind)$_2$ZrCl$_2$-MAO 均相体系能高活性地催化丙烯的等规定向聚合,这是首次成功地采用均相 Ziegler-Natta 催化剂聚合得到的等规聚丙烯;反之,内消旋的钛同系物 *meso-Et*(Ind)$_2$TiCl$_2$-MAO 只能产生全无规聚丙稀。根据 Pino[26]提出的立体选择性控制模型(此外还有其他模型),外消旋的 2 个金属茂催化剂的茚基呈柄型排列(图 3),六元环(环己烯)向上的茚基配体像个柄,能迫使(σπ-配位络合的丙烯单体的甲基指向空间位阻小的一边,即指向六元环朝下的茚基配体(柄朝下)的一侧[图 3(A)],而不是指向六元环朝上的茚基配体[图 3(B)]。对于内消旋型同系物体系,与钛配位的 2 个茚基配体都是六元环朝下(上面两侧皆无柄),σπ-配位络合的丙烯单体的甲基可指向任何一侧[图 3(C)、(D)],因而这类内消旋型同系物体系对 α-烯烃的配位聚合无定向配位控制功能,这犹如线型的 β-$TiCl_3$ 晶体与烷基铝所构成的催化剂体系,只能生产无规聚丙烯。

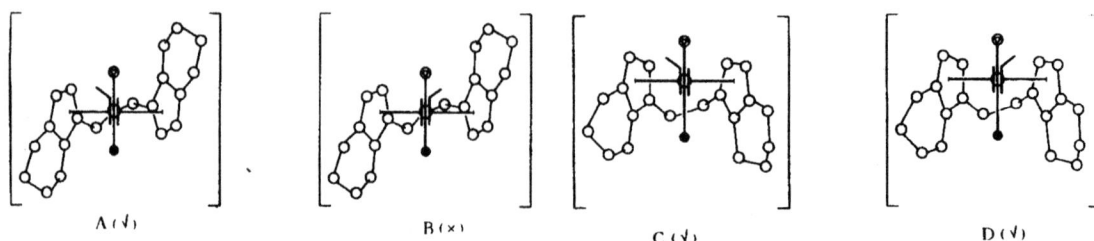

**Fig. 3 Possible states in the polymerization of propylene with *Et*(H$_4$Ind)$_2$TiCl$_2$-MAO(Pino's model)**

$\heartsuit$ : Isotactic(or atactic) growing chain;  $\rangle$ : Propene(σπ-ligand);  ● : O-Al(CH$_3$)[O-Al(CH$_3$)]$_n$-CH$_3$.

上述外消旋金属茂均相催化剂体系还有两种非常特殊的定向催化功能:(1)都能使环烯烃(如环戊烯、降冰片烯或二亚甲桥基八氢萘)定向聚合成不溶于普通烃类溶剂且耐 400℃ 以上高温的结晶态等规高聚物(不像多相 Ziegler-Natta 催化剂那样导致开环聚合),并能使环烯与乙烯共聚生成机械物理性能优越的高度透明高聚物(有可能用作光盘和光纤材料);(2)将外消旋混合物中的一种旋光异构体分离,可用于催化丙烯高等规度地聚合为具有旋光活性的纯左旋或纯右旋的高聚物或齐聚物。

烷基金属茂催化剂体系的活性物种本质也许都是含有 1 个空配位的烷基金属茂阳离子(金属中心为 $d^0$ 组态,配合物中心有 14 个价电子),一种金属茂催化剂体系也许只含一种阳离子活性物种,所以聚合产物分子量分布很窄($M_w/M_\eta \approx 2$),因而有"单一活性位"催化剂之称。但在 MAO 大量存在下,即使只有一种阳离子活性物种,也很可能被 MAO 溶媒化并与 MAO 形成多种松散的缔合物。MAO 齐聚物的本身结构尚不清楚,使得聚合反应动力学分析十分困难。虽然已提出了多种动力学模型,有的还使用计算机拟定出动力学参数[27,28],但所验证的都是表观动力学模型。Chien 等[29,30]巧妙地利用 $Ph_3C^+ B(C_6F_5)_4^-$ 与 $Et(Ind)_2Zr^+ \cdot (CH_3)_2$ 反应制得不含 MAO 的 $Et(Ind)_2Zr^+(CH_3) + B(C_6F_5)_4^-$-TEA(后加)催化剂体系,在 $-20℃$ 时丙烯定向聚合活性高达 $2 \times 10^9$ g PP/(mol Zr · mol $C_3H_6$ · h),PP 等规度为 93.8%。在 $-55℃$ 时活性降到 $3 \times 10^8$ g PP/(mol Zr · mol $C_3H_6$ · h);PP 等规度上升到 99.4%。可见 $T_p$ 很低时,缔合物解离度很小,而且这种金属茂均相催化剂体系对聚合物的立体选择性随 $T_p$ 上升迅速下降,说明这种均相配合物催化剂活性中心结构显得较"软",较易变形,不象固体催化剂活性中心结构那样坚实。

有关金属茂二氯化物与 MAO 通过氯离子与烷基交换和转移并形成阳离子活性物种与 MAO 的缔合物的机理比较容易理解。Kaminsky[31]提出了含金属中心的四元环中间态模型和配位丙烯直接插入 M—P 键的机理。有关含金属四元环中间态在烯烃催化歧化和歧化聚合机理方面已有不少先例,Corradini[32]则提出了含金属中心和烷基上的 1 个 H 基的五元环中间态模型和转移插入机理。可见烯烃单体究竟是直接插入,还是邻位转移插入尚需进一步研究。设法推断丙烯定向聚合时活性中心的配位结构(如有无弱配位 MAO 协助结合 R 链)及单体邻位插入的可能模式,对于这类螯形配体和催化剂的分子设计将会有重要的导向作用。

用金属茂均相催化剂体系进行烯烃配位聚合时,链转移机理包括 β-氢转移给金属中心、β-甲基转移给金属中心、β-氢转移给单体分子及 β-氢转移给 MAO 等反应方式。在温度不很低时,前 2 个反应进行得相当快,聚合物的分子量一般偏低。提高茂基配体的电子给予能力可削弱金属中心夺取 P 的 β-氢或 β-甲基的能力,从而提高分子量,并可提高活性。降低聚合温度也能提高分子量,但活性也降低。通过配体分子设计和试验,已研制出几种即将进行工业化试验的金属茂催化剂[33];其中有以甲基硅烯桥联的 2 个带有甲基和异丙基的茚基配体,每个茂芳环的 3-位上各带 1 个甲基,每个苯环的 2-位上各带 1 个异丙基(图4)。这些推电子基团的位置安排恰到好处,不致造成太大的空间位阻而影响活性。用此配体制得的甲基锆茂-MAO 催化剂体系丙烯定向聚合活性达到 105 kg PP/(mol Zr · h),$M_w$ 为 46 万,$t_m$ 为 152 ℃,等规度为 98%(用 $^{13}$C NMR 测定)。

金属茂配合物已被纳入有机金属化合物的数据库,通过选择和研制,并借助于催化剂分子设计构思,已获得具有多种催化聚合功能的金属茂催化剂。Ewen[34,35]研制出(Cp-i-Pr-fluorenyl)ZrCl₂-MAO 和(3-MeCp-i-Pr-fluorenyl)ZrCl₂-MAO 催化剂体系[图 4(B)、(C)],分别能使丙烯定向聚合为熔点达 185 ℃、透明度比 iPP 好的间规聚丙烯(sPP)和所谓半等规(hemiisotactic)聚丙烯,实际上可能是一种等规聚丙烯与较少量无规聚丙烯的镶嵌聚合物,如同使用非键联的 2 个苯-茚配体的金属茂催化剂时得到的等规、无规镶嵌聚丙烯[36]。最近,有人研制出一种不对称的外消旋的锆茂-MAO 催化剂(能使丙烯聚合为聚丙烯弹性体[37])及一种能催化 1,5-己二烯类单体进行对映体选择性环化聚合的手性金属茂-MAO 催化剂[38]。

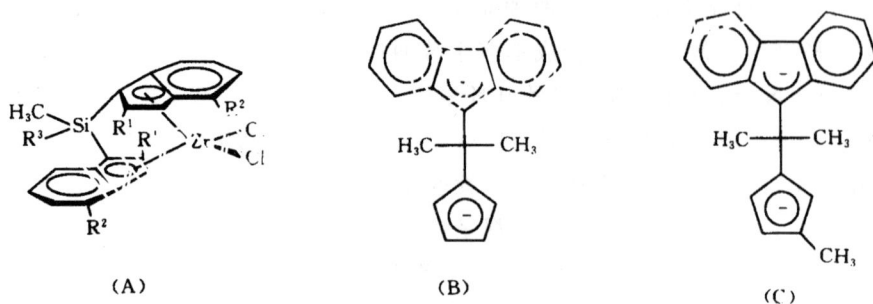

**Fig. 4  Metallocene catalysts, or chelate-ligands (planar projection) for metallocene catalysts with different stereo-selectivities**

(A) Isotactic catalyst; (B) Chelate-ligand for syndiotactic PP; (C) Chelate-ligand for hemiisotactic PP.

上述烯烃聚合物多数是新型的,有的很可能开发成为有重要用途的新材料,如环烯-乙烯共聚物。分子量分布窄且大小适当的 iPP 及 PE 可用于制备超细旦纤维,sPP 可能也有重要用途,全无规聚丙烯有可能大量用作充油橡胶的油料,而乙烯、丙烯等共聚体可能代替有环保问题的聚氯乙烯及某些含苯乙烯的共聚体品种,乃至某些高附加值的工程塑料等。其中最重要的显然是已工业化的所谓"单一活性位"烯烃聚合催化剂,用其生产的低压低密度聚乙烯(乙烯加少量长链 α-烯的均匀共聚体,L-LDPE)的性能比使用第二代 Ziegler-Natta 催化剂生产的优越得多(如高光洁度、高抗撕裂强度及易加工、低热封温度和低灰分等),L-LDPE 无疑将成为产量最大的塑料品种。最近,国外大力宣传的"单一活性位"烯烃聚合催化剂大概已通过分子设计构思和研制,实现了烯烃单体分子只能在单一活性物种的单一活性位插入到生长中的高分子链,不容它有第二个活性位的选择。如果这个设计目标已经达到,也可减少均相催化聚合 iPP 的结构位错缺陷,从而可进一步改进 iPP 及 EP 共聚物的性能。我国的丙烯主要来源为:石油烃催化裂化深加工新技术及碳一化学路线[尤其是合成气-甲醇(或二甲醚)-乙烯(丙烯,丁烯)路线副产丙烯]。因此,必须多方面开辟丙烯的新用途。

# 3  α-烯类氢甲酰化配位催化剂的发展

最近,Beller 等[39]详尽综述了 15 年来 α-烯类氢甲酰化配位催化剂的发展。典型的催化剂是含氢基、羰基及(或)其他配体(L)的Ⅷ族过渡金属(M)单核或双核配合物 $H_xM_y(CO)_zL_n$。除非使用多核配体,否则多核及双金属原子簇羰基配合物在氢甲酰化反应条件下都析出单核活性物种(起主要催化作用的是单核配合物)。所以近年来这类单核配合物成为研究重点,尤其是含单 P 的单齿配体及含双 P 的双齿配体的合成和配位化学的研究。大量筛选结果表明,中心原子 M 的活性次序为:Rh≫Co≫Ir,Ru>Os>Pt>Pd>Fe>Ni,催化活性与这些金属原子 $M^o$ 对 σπ-型配体反馈电子的能力大致成反平行关系。与配体 L 的碱性成平行关系,即 $Ph_3P≫Ph_3N≫Ph_3As$(膦也可用亚磷酸酯代替)。在区位选择性方面,直链产物选择性也是以 Rh-P 系最佳,但与配体结构在金属配位上产生的空间位阻大小密切相关(Tolman 的锥角规律[40]);锥角相近时,带亚磷酸酯配体的铑催化剂的产物直链选择性($n/i$)与带膦配体的铑催化剂的 $n/i$ 也相近,但前者催化活性略高。如果是双膦配体,它与 M 中心原子的结合力与配体结构也有密切关系(Casey[41]根据分子力学计算结果提出了"自然咬角"概念)。使用某些恰当空间位阻的庞大单亚磷酸酯配体(邻-叔丁基对-甲基苯酚亚磷酸酯)的铑催化剂时,$n/i$ 值可达 96/4;使用更复杂的双膦或双亚磷酸酯配体,$n/i$ 值可高达 98/2 及 90.6/0.4。工业上用得最多的是铑-膦系催化剂[如 $HRh(CO)(PR_3)_3$,约占 80%,主要用于低压法制正丁醛、丁辛醇]及钴系催化剂[如 $HCo(CO)_3PR_3$ 等约占 20%,主要用于制长链醛等]。

近十多年来,氢甲酰化催化剂的突破性发展是水溶性高效催化剂,主要是含磺酸化(钠盐)三苯基膦类单齿配体(如 TPPTS)的铑-膦系催化剂,并于 1984 年实现工业化[42]。使用二相连续搅拌高塔反应器,陆续清倾油相以蒸馏分离出丁醛,其丙烯转化率为 95%,$n/i=95/5$,丁醛选择性高达的 99%,铑损失少于 $10^{-9}$ gRh/kg-正丁醛(RCH 公司现已发展到年产 30 万吨以上正丁醛)。近来也在发展效率更高的水溶性(磺酸盐化的)双齿膦配体。其他公司则使用了油溶性的 Rh-TPP 系催化剂,$n/i$ 值仅约 84/16,丙烯转化率 84%~86%,时空产率和化学选择性都较低,催化剂分离等工艺也较复杂。带水溶性单膦或双膦配体的氢甲酰化铑系催化剂的发展可认为是催化剂分子设计构思与工程设计构思相结合的成功例子。

进一步发展出的担载水相(液膜)催化(SAPC)[43]和更广泛的担载液相催化剂(SLPC)[44]使二相分离工艺趋近简化,更便于连续操作,这主要应归功于催化剂的工程设计。介质情况不同引起的区位选择性($n/i$ 值)差异可由 R 基配体构象的差异所引起的界面能的不同来解析,但 SAPC 也存在着如何控制水相中水的最佳含量及使用简单配体(如 TPPTS)时 SAPC 对短链的 α-烯(丙烯)产物丁醛 $n/i$ 值偏低等问题。这些问题看来都不是不能解决的。因此,现在来判断 SAPC 和 SLPC 是否有工业化前景还为时尚早。对于乙烯的氢甲酰化,可使用担载的油溶性铑-膦催化剂,且不存在区位选择性及铑流失的问题,但丙醛年产量仅为长链醛的 1/16,丁醛的 1/45(正丁醛约 70%用于制丁、辛醇)。

另一个新发展是氢甲酰化不对称合成。对于 α-烯烃,要求甲酰基加在链端第二位才能形成不对称中心,所以 $n/i$ 值要小。因此,这里同时存在着对映体选择性和区位选择性问题,使用油溶性的铑-膦催化剂,$n/i$ 值可能趋小。

有关该类反应的催化剂作用机理比较简单,大致如 Heck 和 Breslow[45]于 1961 年提出的。在工业反应条件下,产物 RCHO 生成的最后一步是通过酰基-金属(M=CO,Rh)键的加氢,如 R(O)CCo(H-H)(CO)₃氢解生成,其中 Co(H-H)表示 H₂ 配位过渡态(Kubas complex,见图 5 虚框左边)。无论是氧化加成还是直接插入,大概都要经过这个过渡态。这一步往往是氢甲酰化反应的速率控制步骤(rds),即催化活性高低的决定性步骤。氢解成醛快慢既涉及电子因素,也涉及空间因素,问题较复杂,虽有大量的实验知识积累,但规律性不够明显。如果对过渡态结构有较深入的了解,应该利用计算机软件来协助解决问题。

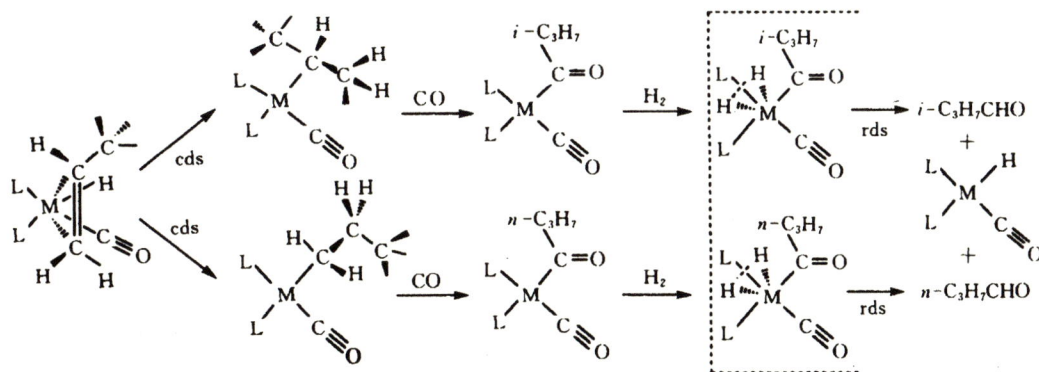

**Fig. 5** Configuration-determining step(cds) and rate-determining step(rds)
in hydroformylation of propene to *n*-and *i*-bntaldehydes

产物 $n/i$ 选择性却并不决定于最后一步(图 5),而主要决定于 R 基配体的形成,即 H 配体对 α-烯键配体的邻位转移插入步骤。其区位选择性是 H 配体插在链端第一个 C 上(端位)与第二个 C 上(侧位)之比;接下去的一步是 CO 插入 M—Cσ 键,形成酰基配体,它不能改变已经形成的 R 基 $n/i$ 比。由于对决定 R 基 $n/i$ 比这一步的中间态物种与催化剂结合的结构模式已有一定程度的了解,高区位选择性的配体选择和合成应该可仿照药物合成的计算机辅助设计软件,以开发 CAD 专家系统。区位选择性的专家系统可成为催化活性专家系统的一个子系统,催化剂分子设计必须能适当兼顾催化活性与选择性,才有实际意义。

## 参考文献

[1]Kranenburg M.,Van der Burgt Y. E. M.,Kamer C. P. J. et al.. Organometallics,1995,**14**:3 081.

[2]Kuntz I. D.,Meng E. C.,Shoichet B. K.. Acc. Chem. Res.,1994,**27**:117.

[3]Noyori R., Asymmetric Catalysis in Organic Synthesis. New York:John Wiley & Sons, Inc., 1994,Chapter 1 and 2.

[4]Halpern J.,Acc. Chem. Res. 1982,**15**:332.

[5]Brown J. M.,Maddox P. J.. J. Chem. Soc. Chem. Commun.,1987:1 276.

[6]Takaya H.,Ohta T.,Noyori R. ;Ojima I. Ed.. Catalytic Asymmetric Synthesis. VCH,1993: Chapter 1.

[7]Huang J.,Rempel G. L.. Prog. Polym. Sci.,1995,**20**:459.

[8]Natta G. J. Inorg. Nucl. Chem.,1958,**8**:589.

[9]Arlman E. J.,Cossee P.. J. Catal.,1964,**3**:99.

[10]Ariman E. J.. J. Catal.,1964,**3**:89.

[11]Ariman E. J.. J. Catal.,1966,**5**:178.

[12]Ariman E. J.. Recl. Trav. Chim. Pays-Bas,1968,**87**:1 217.

[13]Boor J.. J. Polym. Sci.,1962,**625**:45.

[14]Boor J.. J. Polym. Sci. Part C,1963,**1**:237.

[15]Boor J. Youngman E. A.. J. Polym. Sci.,Part B,1964,**2**:265.

[16]Boor J.. J. Polym. Sci., Part B,1965,**3**:7.

[17]Boor J.. J. Polym. Sci.,Part A,1965,**3**:995.

[18]Youngman E. A.,Boor J.. J. Polym. Sci.,Part B,1965,**3**:577.

[19]Boor J.. J. Polym. Sci.,Part A-1,1971,**9**:617.

[20]Ystenes M.,MakromoL Chem.,Macromol. Symp.,1993,**66**:71.

[21]Breslow D. S.,Newburg N. R.. J. Am. Chem. Soc.,1957,**79**:5 072.

[22]Natta G.,Pino P.,Mazzanti G. et al.. Chim. Ind. (Milan),1957,**39**:1 032.

[23]Sinn H.,Kaminsky W.. Ad. Organomet. Chem.,1980,**18**:99.

[24]Wild F. R. W. P.,Wasincionek M.,Hattner G. et al.. J. Organomet. Chem.,1985,**288**:63.

[25]Wild F. R,W. P.,Zsolnai L.,Huttner G. et al.. J. Organomet. Chem.,1982,**232**:233.

[26]Pino P.,Rotzinger B.,von Achenbach E. ;Keii T.,Soga K. Eds.. Catalytic Ploymerization of Olefins. Elsevier, Tokyo,1986:461.

[27]Jordan R. F.,Bajger C. S.,Willet R. et al.. J. Am. Chem. Soc.,1986,**108**:7 410.

[28]Herfert N.,Fink G.,Makromol. Chem.. Rapid Commun.,1993,**14**:91.

[29]Chien J. C. W.,Tsai W. -M.,Rausch M. D.. J. Am. Chem. Soc.,1991,**113**:8 570.

[30]Chien J. C. W.,Tsai W. -M.. Makromol. Chem.. Macromol Symp.,1993,**66**:141.

[31]Kaminsky W.,Steiger R.,Polyhedron,1988,**7**(22/23):2 375.

[32]Corradini P.,Guerra G.. Prog. Polym. Sci.,1991,**16**:239.

[33]Spaleck W., Antberg M., Rohrmann J. et al., Angew. Chem. Int. Ed. Engl.,1992,**31**(10):1 347.

[34] Ewen J. A., Elder M. J., Jones R. I. et al., Keii T., Soga K. Eds.. Catalytic Olefin Polymerization, Elsevier,Tokyo,1990:439.

[35]Ewen J. A.,Elder M. J.,Jones R. I. et al.. Makromol. Chem.,Macromol. Symp.,1991,**48/49**:253.

[36]Stevens J. E.. Stud. Surf. Sci. Catal. (Proc. 11th Int. Cingr. Catal.,Ealtimore),1996,**101**:10.

[37]Babu G. N.,Newmark R. A.,Cheng H. N. et al.. Macromoleaules,1992,**25**:7 400.

[38]Coates G. W.,Waymouth R. M.. J. Am. Chem. Soc.,1993,**115**:91.

[39]Beller M.,Cornils B.,Frohnmg C. D. et al.. J. Mol. Catal. A,1995,**104**:17.

[40]Tolman C. A.. J. Am Chem. Soc.,1970,**92**:2 956.

[41]Casey C. P.,Whiteke. G. T.,Melville M. G. et al.. J. Am. Chem. Soc.,1992,**114**:5 535.

[42]Wiebus E.,Cornils B.. Chem. Ing. Techn.,1994,**66**:916.

[43]Arhancet J. P.,Davis M. E.,Merola J. S. et al.. Nature,1989,**339**:454.

[44]Hjortkjaer J.,Scurrell M. S.,Simonsen P. et al. J. Mol. Catal.,1981,**12**:179.

[45]Heck R. F.,Breslow D. S.. J. Am. Chem. Soc.,1961,**38**:4 023.

## Evolution and Significance of Molecular Design Conception in the Development of Stereo-Selective Transition-Metal Complex Catalysts

Hui-Lin Wan*, You-Zhu Yuan, Jing-Xing Gao, Hong-Bin Zhang, Khi-Rui Tsai

(Department of Chemistry, Institute of Physical Chemistry, State Key Laboratory for Physical Chemistry of the Solid Surface, Xiamen University, Xiamen, 361005)

**Abstract** The mutual promotion of the development of stereo-selective coordination catalysts involving transition-metal coordination compounds and the development of the conception of molecular design of such catalysts was discussed, as illustrated by three typical types of stereo-selective coordination catalysis, namely, asymmetric synthesis, stereo-selective coordination polymerization of α-olefins, and hydroformylation of α-olefins. With this discussion, it is shown that for the molecular design of certain types of stereo-selective coordination catalysts, there has already been sound scientific basis for the development of computer-aided design(CAD).

**Key words** Asymmetric synthesis Stereo-selective coordination polymerization Hydroformylation Catalyst molecule design(Ed. :Y,A)

■ 本文原载:《物理化学学报》第 13 卷第 12 期(1997 年 12 月)，第 1057～1060 页。

# 碳纳米管负载铑催化剂上丙烯氢甲酰化[*]

张　宇　吴汜昕　张鸿斌　林国栋　袁友珠　蔡启瑞[①]

（厦门大学化学系　固体表面物理化学国家重点实验室，厦门　361005）

**关键词**:碳纳米管载体材料　氢甲酰化　丙烯　负载型铑催化剂

氢甲酰化反应迄今已有七十多年历史,但有关这一催化领域的研究一直未停止过。目前有关氢甲酰化催化剂的研究主要集中在新型膦配体的合成方面[1],对催化剂载体的研究却较少;然而载体对催化剂性能的影响有时也十分显著。碳纳米管是一种碳素新材料,具有纳米级的管状结构,以及尺寸小,机械强度高,比表面大,电导率高,界面效应强等特点,被认为在包括催化等诸多领域具有重要应用前景[2]。本文以自行制备的碳纳米管[3,4]作为新型载体材料,首次用于负载型烯烃氢甲酰化铑膦配合物催化剂的制备;结果显示,与现有的几种常规载体材料相比,碳纳米管负载的铑膦配合物催化剂对丙烯氢甲酰化制丁醛显示出高得多的催化活性和产物分子区位选择性。

## 1　实验

碳纳米管按前文[3,4]所描述的方法制备。JEM-100CX 电子透射显微镜被用于自行制备的碳纳米管的形貌观测。负载型铑膦配合物催化剂由等容浸渍法制备,即将计量的 $HRh(CO)(PPh_3)_3$ 和 $PPh_3$ 分别溶于苯制成溶液,浸渍负载于一定量的载体上,放置一定时间,接着真空干燥,而后在 $N_2$ 气氛下保存备用。丙烯氢甲酰化反应在固定床加压连续流动反应器-气相色谱仪组合系统上进行。反应产物由 102G-D 气相色谱仪氢焰检测器作在线分析。色谱分离柱填料为 Porapak Q,柱长 2 米。色谱分析数据按修正面积归一化法计算。在碳纳米管负载的铑膦配合物催化剂上进行丙烯氢甲酰化反应时,实验发现,在总产物中加氢副产物不超过 1%。

## 2　结果与讨论

表 1 示出碳纳米管及几种其他载体负载的铑膦配合物催化剂上丙烯氢甲酰化反应活性的评价结果。从表 1 可见,碳纳米管负载的催化剂显示出大大优于其他几种载体负载催化剂的催化性能;在 393 K,1.0 MPa,原料气组成为 $C_3H_6/CO/H_2 = 1/1/1(V/V)$,其流速为 9000 mL(STP)/h·g·catal. 的反应条件下,碳纳米管负载铑膦配合物催化剂上丙稀的转化率为 $SiO_2$ 负载催化剂的 2.1 倍,前者产物丁醛的正构/异构比$(n/i)$高达 11.6,而后者仅为 6.03。

---

[*] 国家自然科学基金和中国石化总公司专项经费联合资助项目。

[①] 1997 年 8 月 26 日收到初稿,1997 年 9 月 26 日收到修改稿。联系人:张鸿斌。

表 1　不同载体负载的铑膦配合物催化剂上丙烯氢甲酰化活性评价结果

Table 1　The results of activity assay of propene hydroformylation over HRh(CO)(PPh₃)₃ catalysts supported by different supports*

| | Carbon nanotubes | SiO₂ | Carbon molecular sieves | Active carbon | GDX-102** |
|---|---|---|---|---|---|
| Surf. area(m² · g⁻¹) | 100 | 500 | 800 | 1000 | 800 |
| Conv. of C₃H₆(%) | 21.3 | 10.1 | 8.9 | 0.50 | 23.5 |
| TOF (s⁻¹) | 0.079 | 0.038 | 0.033 | 0.002 | 0.087 |
| STY | 2771 | 1314 | 1158 | 65 | 3057 |
| (mmolC₃H₇CHO/h · g Rh)n/i | 11.6 | 6.03 | 14.7 | 2.52 | 5.45 |

* Reaction conditions：393 K，1.0 MPa，C₃H₆/CO/H₂＝1/1/1(V/V)，GHSV＝9000 mL(STP)/h · g catal.；0.01 mmol HRh(CO)(PPh₃)₃ and 0.03 mmol PPh₃ were loaded on 0.1 g support，with the corresponding P/Rh ratio at 6 and rhodium-loading amount at 0.88% (mass fraction)；data taken after 2 h of reaction operation.

** A copoiymer of styrene with divinylbenzene.

表2示出两种铑负载量时,碳纳米管负载催化剂的催化活性随不同膦/铑比的变化情况。实验结果表明,当膦铑比由4.5逐步增高时,丙烯转化率及产物丁醛的n/i比均明显提高;当膦/铑比达到9～12时,两者均趋于一最高值;此后,随着膦/铑比进一步增至18～21,丙烯转化率及产物丁醛的n/i比变化均不明显。为获得高的丁醛产率及其高的正/异比,对于碳纳米管负载的铑膦配合物催化剂,其膦/铑比看来以9～12为宜。

表 2　碳纳米苷负载铑膦配合物催化剂中膦铑比对其催化性能的彩响

Table 2　Effect of P/Rh ratio in the carbon nanotubes-supported Rh-catalysts on the catalytic performance*

| Rh-loading amount(mmol/0.1 g support) | P/Rh(mol/mol) | Conversion of C₃H₆(%) | TOF(s⁻¹) | n/i |
|---|---|---|---|---|
| 0.01 | 4.5 | 14.1 | 0.053 | 4.3 |
| | 6 | 21.8 | 0.081 | 11.2 |
| | 9 | 32.1 | 0.119 | 12.5 |
| | 15 | 36.2 | 0.135 | 14.4 |
| | 18 | 36.8 | 0.137 | 14.3 |
| 0.005 | 6 | 9.4 | 0.070 | 7.7 |
| | 9 | 9.7 | 0.072 | 11.8 |
| | 12 | 13.7 | 0.102 | 13.4 |
| | 15 | 11.2 | 0.083 | 10.2 |
| | 18 | 12.3 | 0.092 | 8.0 |
| | 21 | 11.8 | 0.088 | 10.7 |

* Reaction conditions：393 K，1.0 MPa，C₃H₆/CO/H₂＝1/1/1(V/V)，GHSV＝9000 mL(STP)/h · g catal.

碳纳米管负载铑膦配合物催化剂对丙烯氢甲酰化反应所表现出的如此优异的催化性能显然与这种新型载体材料的特殊结构和性质密切相关。本文所用自行制备的碳纳米管载体材料的 TEM 图象示如图1。由 TEM 观察可知,这类碳纳米管为圆柱状并或多或少有些弯曲的碳纤维,其断面直径约15～20 nm,长度可达微米甚至毫米级;它们的中心部分电子透射率较高,表明中心区碳的密度较低,甚至是中空的。

由我们对这类碳纳米管所作一系列谱学表征结果[4,5]可知,它们的 XRD 图与石墨相近,但相应的特征衍射峰峰形稍为宽化,表明其长程有序度较石墨低;HRTEM 观察揭示这类碳纳米管系由许多层具有类石墨片状结构的同心、等径、中空圆锥形面叠合而成;TPO 试验则显示,在所制备的这类碳纳米管产物中无定形碳的含量相当低,它们的整体结构石墨化程度较高。由相关几种谱学表征结果相互佐证可知,这类碳纳米管确系由类石墨碳六元环面一层层叠合成管状结构的碳(大)分子。

图 1　由 CH₄ 催化分解制备的碳纳米管的 TEM 图像

Fig. 1　TEM image of carbon nanotubes prepared by catalytic decomposition of CH₄

由表 1 所示几种载体的比表面测定结果可知,本文所用碳纳米管的比表面仅为～100 m² · g⁻¹,这比其他几种载体(SiO₂,碳分子筛,活性炭,或苯乙稀—二乙烯苯共聚物 GDX-102)低得多。可见,碳纳米管负载的铑膦配合物催化剂高的活性和高的区位选择性并非缘于其比表面因素。当三苯基膦配体大大过量时,在氢甲酰化反应温度(393 K)下,过量的三苯基膦固然可能局部液相化,在碳纳米管管壁内外表面上出现一些局部的微液膜区,使该催化剂一定程度上接近于负载液相催化剂。然而,这种过量三苯基膦配体在反应条件下可能的液膜化在其他载体上也可以发生,这无法说明象 SiO₂ 这样的载体(其孔径分布集中在 3～5 nm 范围)负载的催化剂,其活性和产物区位选择性因何低得多?因此,在我们看来,碳纳米管负载催化剂所表现出的优异催化性能更可能与碳纳米管的纳米级管腔结构,以及其内、外表面所具有的类石墨碳六元环结构密切相关。断面直径约 2～3 nm 的管腔适于容纳纳米级大小的铑膦催化活性络合物并留下供反应分子扩散、反应的适度空间,在保持高的转化频率的同时又有利于提高反应中间态及产物丁醛分子的空间选择性;而其表面类石墨碳六元环与配体三苯基膦苯基的碳六元环存在着结构及其电子性质的相似性,这可能有助于铑膦催化活性络合物及膦配体在碳纳米管管壁上的均匀分散与稳定负载,并有利于提高具有催化活性和区位选择性的铑膦络合物的稳定性,避免膦配体的解络与流失。很可能正是这些因素,使碳纳米管负载的铑膦配合物催化剂对丙烯氢甲酰化表现出优异的催化性能。

其他几种载体负载催化剂的评价结果提供了一系列鲜明的对比:同是由碳组成的活性炭载体,其孔径只有几个埃,无法容纳纳米级大小的铑膦催化活性络合物,使之几乎无催化活性;碳分子筛孔径稍大,但要容纳纳米级大小的铑膦催化活性络合物,并同时提供反应分子的进出通道看来仍不合适,其结果是产物丁醛正异比较高,但活性仍较低;苯乙烯-二乙烯苯共聚物载体 GDX-102 比表面虽大并富含芳香环结构,活性较高,但可能由于不具备管腔结构,产物丁醛的区位选择性并不高;一般的无机载体如硅胶(SiO₂),其比表面是碳纳米管的数倍,并有孔径在 3～5 nm 范围的大量微孔,但可能由于其表面不具备类石墨碳六元环结构,相应催化剂的活性及区位选择性只及碳纳米管负载催化剂的一半左右。有关碳纳米管载体的作用本质在进一步研究之中。

# 参考文献

[1]Yuan Youzhu(袁友珠),Yang Yiquan(杨意泉),Lin Guodong(林国栋)et al. Chemistry of Oxo-

synthesis(羰基合成化学). Yan Yuanqi el.（殷元骐主编），Chemical Indnstry Press（化学工业出版社），1995. p. 64.

[2]Rodriguez N M. J. Mater. Res.,1993,**8**(12):3 233.

[3]Chen Ping(陈萍),Zhang Hongbin(张鸿斌),Lin Guodong(林国栋)et al. Gaodeng Xuexiao Huaxue Xuebao(高等学校化学学报),1995,**16**(11):1 783.

[4]Chen Ping(陈萍),Zhang Hongbin(张鸿斌),Lin Guodong(林国栋)et al. Xiamen Daxue Xuebao, Ziran Kexueban(厦门大学学报自然科学版),1996,**35**(1):61.

[5]Chen P,Zhang H B,Lin G D,et al. CARBON, 1997,in press.

## Hydroformylation of Propene over Carbon Nanotubes-Supported Rh-Catalyst

Yu Zhang，Fan-Xin Wu，Hong-Bin Zhang，Guo-Dong Lin，You-Zhu Yuan，Khi-Rui Tsai

(*Department of Chemistry* & *State Key Laboratory for Physical Chemistry of the Solid Surface*, *Xiamen University*, *Xiamen* 361005)

**Abstract**   Effect of carbon nanotubes, as a novel support material, on the performance of Rh-catalyst supported by them was studied. Catalysts based on carbon nanotubes, $SiO_2$, carbon molecular sieves, active carbon, and GDX-102 (a copolymer of styrene with divinylbenzene), were prepared, and their catalytic behaviors for propene hydroformylation were investigated and compared. The results showed that, over the carbon nanotubes-supported Rh-catalyst, $C_3H_6$ conversion and regioselectivity of butyric aldehyde (represented by $n/i$, a ratio of n-butyric aldehyde to its isomer, $i$-butyric aldehyde, in the products) were pronouncedly improved; the average turnover frequency (TOF) for the catalytic hydroformylation of propene was 0.079 $s^{-1}$ at 393K, which was 2.1 times faster than that over the Rh catalyst based on $SiO_2$, and the $n/i$ ratio of the aldehyde products reached to 11.6, which was 1.9 times higher than that over the catalyst based on $SiO_2$. The roles of six-membered C-ring at the surface of the carbon-nanotubes on the stability of the catalytically active Rh-complexes and of the tubular nano-channel on the spatiospecific seletivity of reaction intermediate state and butyric aldehyde produced were discussed.

**Key words**   Carbon-nanotubes support material   Hydroformylation   Propene   Supported Rh-catalyst

■ **本文原载**:《化学通报》第 9 期(1997 年),第 1～6 页。

# 同位素方法在催化反应机理研究中的应用

汪海有[①]　蔡启瑞

(厦门大学化学系　固体表面物理化学国家重点实验室　361005)

**摘　要**　介绍了同位素方法在催化反应机理研究中的几种典型应用。

**关键词**　同位素　催化　反应机理

在多相催化体系中,确定反应中间体是建立反应网络、说明反应机理的重要方法。表征反应中间体的常用方法有化学捕获法和红外光谱法。而在上述方法中,同位素方法的应用对获得有关反应中间体的明确结论是十分有益的,有时甚至是不可缺少的。此外,同位素效应也是研究化学反应历程的一种有效方法。当反应键原子被其同位素取代后,产生的同位素效应对于反应历程尤其是速率控制步骤可提供重要的信息。作者在合成气制乙醇、甲烷部分氧化制合成气等课题的研究中,广泛使用了各种同位素方法,取得了许多关于上述催化过程的反应机理的重要结论。本文以上述工作为 基础,介绍同位素方法在催化反应机理研究中的几种典型应用。

## 1　同位素方法在确定反应物活性吸附态中的应用

以铑基催化剂上的合成气制乙醇反应体系为例,该反应中,反应物 CO 有多种吸附态如线式、桥式等。关于 CO 的活性吸附型式,文献中尚存争论。作者[1]首次应用 CO 吸附—程序升温表面反应—红外光谱动态技术结合同位素方法考察了 CO 吸附物种对氢(氘)的反应性能,对该问题进行了深入研究。图 1 是 Rh-Mn $(1:1)/SiO_2$ 上 $H_2$ 气流中进行 CO 吸附-TPSR 过程中不同温度下记录下来的红外光谱图。(a)中峰2026 $cm^{-1}$、1854 $cm^{-1}$ 分别归属于线式 CO、桥式 CO 吸附物种。随着温度的升高,线式 CO 峰强度的下降速率比桥式 CO 要快得多。类似的变化趋势也在 $Rh/SiO_2$ 上观察到。CO 吸附物种峰强度随温度升高而逐渐下降有两种可能的原因:一种是发生脱附;另一 种是发生反应。若只由脱附引起峰强度减弱,那么在 $D_2$ 气流中进行 CO 吸附-TPSR 动态过程应将观察到与图 1 相类似的结果。然而,如图 2 所示,$D_2$ 气流中线式 CO (2020 $cm^{-1}$)峰强度的下降速率比在 $H_2$ 气流中更快;与图 1 中线式 CO 峰强度升至 493 K 时仍高于桥式 CO 不同,在图

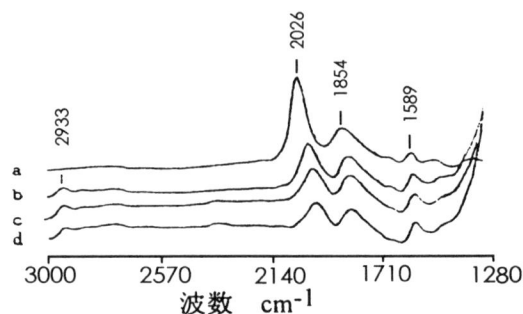

**图 1**　**CO 吸附-TPSR($H_2$ 气流中)动态过程记录得的 IR 光谱图**

催化剂:Rh-Mn$(1:1)/SiO_2$；a:328 K,b:428 K,c:463 K,d:493 K

---

①　汪海有,男,博士,副教授。从事催化研究。
　1996 年 10 月 15 日收稿,1997 年 1 月 22 日修回。

2 中,当温度升至 463 K 时,线式 CO 峰强度就已低于桥式 CO。可见,在升温过程中,CO 吸附物种与氢(氘)发生了反应。由于线式 CO 在 $H_2(D_2)$ 气流中比桥式 CO 对温度更敏感,表明线式 CO 对氢(氘)具有更活泼的反应性,即线式 CO 更有可能是反应物 CO 的主要活性吸附态。

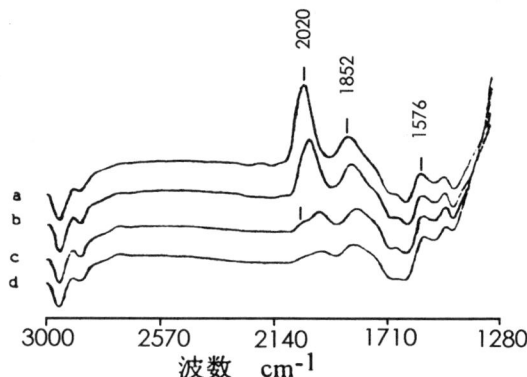

**图 2　CO 吸附-TPSR($D_2$ 气流中)动态过程记录得的 IR 光谱图**
催化剂:Rh-Mn(1∶1)/SiO₂;a:328 K,b:428 K,c:463 K,d:493 K

# 2　二级同位素效应在反应中间体红外光谱表征中的应用

在反应中间体的红外光谱表征中,利用一级同位素效应所产生的振动频率的位移以进一步证实某一归属是经常采用的做法。作者[1]在下面的例子中利用了二级同位素效应以确认 CO 加氢反应中的甲酰基中间体。由图 1 可以看到,在 Rh-Mn(1∶1)/SiO₂ 催化剂上,在 CO 吸附-TPSR($H_2$ 气流中)过程中,随温度升高,低频区峰 1589 $cm^{-1}$ 强度有所增强,该峰位与福冈淳等人[2]在 Ru-Co/SiO₂ 催化剂上 CO+$H_2$ 反应中观察到的 1584 $cm^{-1}$ 特征峰(被指论为 HC=O 的 $\nu_{C=O}$)相近。由图 2 可见,与 $H_2$ 气流中 1589 $cm^{-1}$ 相对应的峰,在 $D_2$ 气流中出现在 1576 $cm^{-1}$ 处,向低位移了 13 $cm^{-1}$,这说明 1589 $cm^{-1}$ 物种含有氢。由于仅有 13 $cm^{-1}$ 的位移,故 1589 $cm^{-1}$ 不能指论为 $\delta_{C-H}$ 的振动峰。与福冈淳等人相同,作者把 1589 $cm^{-1}$ 峰指认为甲酰基的 $\nu_{C=O}$ 特征振动峰。在 $D_2$ 气流中,由于生成的是 DC=O,D 的二级同位素效应使甲酰基中 $\nu_{C=O}$ 略向低波数位移,这就合理地解释了 1589 $cm^{-1}$ 峰在 $D_2$ 气流中发生红移的实验现象。由于有二级氘同位素效应作佐证,把 1589 $cm^{-1}$ 峰指论为甲酰基的 $\nu_{C=O}$ 特征振动峰就比较可靠了。

# 3　同位素方法在化学捕获反应中的应用

化学捕获是表征反应中间体常用的方法之一。所谓化学捕获方法,就是在现场反应中或反应后,加入一种能与预期存在的中间体反应的物质(捕获剂)进行化学捕获反应,然后根据捕获反应产物的组成、结构推断反应中间物种的组成、结构。同位素标志反应物或同位素标志捕获剂的使用有助于获得关于反应中间体的全面的、可靠的结论,以下的例子充分说明了这一点。

### 3.1　合成气制乙醇反应中甲酰基的化学捕获

对铑基催化剂上合成气制乙醇反应中预期存在的甲酰基中间体,以烷基化试剂如 CH₃I 作捕获剂进行捕获,捕获反应可表示为:HC=O+CH₃I⟶CH₃CHO,预期捕获反应产物是乙醛。由于乙醛是合成气转化反应的产物之一,为分辨出捕获反应中生成的乙醛,作者[3,4]采用 $D_2$ 代替 $H_2$ 进行合成气转化反应,这样,由捕获反应生成的乙醛是 CH₃CDO 而由合成气本身反应生成的乙醛是 CD₃CDO,借助质谱可区分这两种不同同位素构造的乙醛。

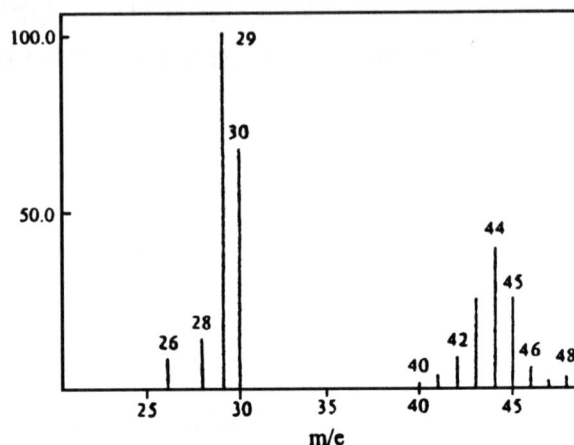

图 3 是 CO + $D_2$ 反应后用 $CH_3I$ 进行捕获反应生成的乙醛的质谱图。图中峰 45、44 的峰强度均显著大于峰 48(峰 48 归属于 CO+$D_2$ 反应中生成的全氘代乙醛 $CD_3CDO$)。由于在标样乙醛的质谱图中,乙酰基碎片峰强度比相应乙醛分子峰强度小,说明在捕获反应中生成了质量数为 45、44 的两种乙醛。此外,图 3 中,峰强度比 $I_{45}/I_{30}$、$I_{44}/I_{29}$ 分别为 0.38、0.39,均接近标样乙醛中乙醛分子峰与甲酰基碎片峰强度比($I_{44(CH_3CHO)}/I_{29(HCO)}=0.36$)。因此,捕获反应中生成的质量数分别为 45、44 的两种乙醛可分别主要归属于 $CH_3CDO$、$CH_3CHO$。$CH_3CDO$ 是甲酰

图 3　CO + $D_2$ 反应后用 $CH_3I$ 捕获生成的乙醛的质谱图

基存在的预期捕获反应产物,而 $CH_3CHO$ 的大量生成却是意料之外的。空白实验表明,在活化的催化剂上,只要存在 CO、$CH_3I$ 就会发生生成乙醛的反应。说明 $CH_3I$ 中的甲基在催化剂表面上发生了脱氢反应(生成 $\underline{H}$)、CO 插入反应(生成 $CH_3\underline{CO}$),且生成的 $\underline{H}$ 进一步将 $CH_3\underline{CO}$ 氢化为 $CH_3CHO$;或者生成的 $\underline{H}$ 先与 CO 反应生成 $H\underline{CO}$,而后进一步甲基化生成 $CH_3CHO$。捕获剂 $CH_3I$ 的这种副反应,说明文献中普遍使用的对甲酰基的化学捕获方法[5]存在明显的不合理之处。通过改进的捕获实验方法[3,4],作者证实了 $D\underline{CO}$ 物种的甲基化反应是捕获反应产物 $CH_3CDO$ 的主要来源,由此说明甲酰基是铑基催化剂合成气制乙醇反应中的 $C_1$ 含氧中间体。

### 3.2　合成气制乙醇反应中乙烯酮、乙酰基的化学捕获

对铑基催化剂上合成气制乙醇反应中预期存在的乙烯酮、乙酰基两个 $C_2$ 中间体,作者[6~8]分别采用 $CH_3OD$、$D_2^{18}O$ 作为捕获剂进行化学捕获,预期的捕获反应式分别为:

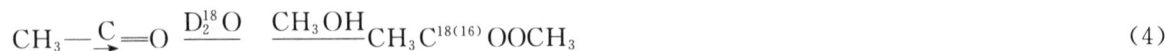

$$CH_2=C=O \xrightarrow{CH_3OD} CH_2=C\overset{OD}{\underset{OCH_3}{\Big|}}CH_2DCOOCH_3 \qquad (1)$$

$$CH_3-\underset{\downarrow}{C}=O \xrightarrow{CH_3OD} CH_3-C\overset{OD}{\underset{OCH_3}{\Big|}}CH_3COOCH_3 \qquad (2)$$

$$CH_2=C=O \xrightarrow[\text{}]{D_2^{18}O} \xrightarrow{CH_3OH} CH_2DC^{18(16)}OOCH_3 \qquad (3)$$

$$CH_3-\underset{\downarrow}{C}=O \xrightarrow[\text{}]{D_2^{18}O} \xrightarrow{CH_3OH} CH_3C^{18(16)}OOCH_3 \qquad (4)$$

以 $CH_3OD$ 为捕获剂,生成的是 α-氘代乙酸甲酯($CH_2DCOOCH_3$)和非氘代的乙酸甲酯($CH_3COOCH_3$),借助质谱可区分这两种具有不同同位素构造的乙酸甲酯。在以 $CH_3OD$ 为捕获剂的原位化学捕获反应中产物乙酸甲酯的质谱图中[7],峰 43、44 分别归属于碎片 $CH_3CO$、$CH_2DCO$;峰 74、75 分别归属于 $CH_3COOCH_3$、$CH_2DCOOCH_3$,表明捕获反应中生成了两种分子式为 $CH_3COOCH_3$、$CH_2DCOOCH_3$ 的乙酸甲酯。根据反应式(1)、(2),可以推断在合成气制乙醇反应中同时存在乙烯酮、乙酰基两个 $C_2$ 中间体。以 $D_2^{18}O$ 为捕获剂进行化学捕获反应,接着用含甲醇的 $N_2$ 气流吹扫,也检测到了按反应式(3)、(4)生成的四种分子式为 $CH_3COOCH_3$、$CH_2DCOOCH_3$、$CH_3C^{18}OOCH_3$、$CH_2DC^{18}OOCH_3$ 的乙酸甲酯[8],进一步证实了乙烯酮、乙酰基两个 $C_2$ 中间体的同时存在。福贵岛和等人[9]用 $^{13}CH_3OH$ 作为捕获剂进行

化学捕获反应,得到的捕获产物为 $CH_3COO^{13}CH_3$,由此认为反应中存在乙酰基中间体。必需指出的是即使合成气反应中同时存在乙烯酮中间体,用 $^{13}CH_3OH$ 作为捕获剂也是无法捕获到的。因为它与 $^{13}CH_3OH$ 反应生成的也是 $CH_3COO^{13}CH_3$。这个例子说明,捕获剂中同位素标志原子的选择是非常重要的。

# 4  同位素效应方法在反应机理研究中的应用

## 4.1  合成气制乙醇反应中的 $H_2/D_2$ 同位素效应

作者[10, 11]通过交替进行 $CO+H_2$、$CO+D_2$ 反应考察了铑基催化剂上合成气转化反应中的 $H_2/D_2$ 同位素效应。在 $Rh-Mn/SiO_2$ 催化剂上,每次由 $CO+H_2$ 切换为 $CO+D_2$ 时,CO 转化率、甲醇、乙醇时空得率均显著上升;反之,则显著下降。$CO+D_2$ 反应中的上述三个参数值分别是 $CO+H_2$ 反应中的 1.3、1.9、1.5 倍。上述结果表明,以 $D_2$ 代替 $H_2$ 进行反应时,CO 总转化反应、甲醇生成反应、乙醇生成反应均表现出氘反同位素效应。

根据文献[12],铑基催化剂上甲醇是由分子吸附态的 CO 逐步加氢生成的。键级守恒的计算结果表明,$COs+ Hs \longrightarrow HCOs$ 是甲醇生成过程中活化能量高的基元步骤,是可能的反应速控步骤。当用 $D_2$ 代替 $H_2$ 反应时,甲醇生成反应表现出氘反同位素效应与该反应的速控步骤为一步加氢反应是相符的。就甲醇生成反应而言,氢同位素效应的产生有两个来源:一个是 $H_2(D_2)$ 吸附的热力学同位素效应;另一个是速控步骤的动力学同位素效应。Bell 等人[13]的计算表明在 $453\sim543$ K 范围内,金属上氘、氢吸附平衡常数之比($K_D/K_H$)在 $0.79\sim0.62$ 之间。由此可以认为甲醇生成反应的氘反同位素效应是由反应速控步骤的动力学氘反同位素效应引起的。根据动力学同位素效应主要由活化络合物与反应物之间的零点能变化引起的观点[14],对 $COs+Hs \longrightarrow HCOs$ 基元反应表现出氘反同位素效应可作出合理的解释。

根据文献[3,6,7,8],乙醇的生成机理可示意为:

$COg \longrightarrow COs \xrightarrow{Hs} HCOs \xrightarrow{Hs} H_2COs(HCOH_s) \xrightarrow{Hs} CH_{2,s} \xrightarrow{COs} CH_2COs \xrightarrow{Hs} CH_3COs \xrightarrow{Hs} CH_3CH_2OHs \longrightarrow CH_3CH_2OHg$。键级守恒的计算结果表明,在金属铑表面上,$H_2COs(HCOHs) + Hs \longrightarrow CH_{2,s} + OHs$ 反应是乙醇生成反应中活化能量高的步骤,是可能的速控步骤。乙醇生成反应与甲醇生成反应同时表现出氘反同位素效应,表明与甲醇生成反应相同,乙醇生成反应的速控步骤也是一步加氢反应,这与键级守恒的计算结果相符。本例说明利用考察同位素效应的方法并结合其他催化研究方法,可获得反应速控步骤的有关信息。

## 4.2  甲烷部分氧化制合成气反应中的 $CH_4/CD_4$ 同位素效应

文献[15,16]通过交替进行 $CH_4+O_2$、$CD_4+O_2$ 反应考察了甲烷部分氧化制合成气反应中的 $CH_4/CD_4$ 同位素效应。如图 4a 所示,在 $0.5\%Rh/SiO_2$ 上,700 ℃下,当 $CH_4+O_2$ 反应切换至 $CD_4+O_2$ 反应时,甲烷转化率显著下降;反过来,则显著上升。交替反应中,CO、$CO_2$ 得率的变化规律与甲烷转化率一致。表明甲烷氧化总反应、CO、$CO_2$ 生成反应均存在正氘同位素效应。计算还表明,$CH_4+O_2$ 反应中的甲烷转化率、CO、$CO_2$ 得率值均是 $CD_4+O_2$ 反应中的 1.08 倍。如图 4b 所示,不考虑未达到稳定状态的第一个 $CH_4/O_2$ 脉冲反应点,在 $CH_4+O_2$、$CD_4+O_2$ 反应交替进行期间,CO 选择性几乎保持不变,意味着产物分布未发生重要变化。

一般说来,当一个 X—H 或 X—D 键在一个反应的速控步骤中断裂时,氢化合物的反应速率常数将超过相应的氘化合物,即该反应存在正氘同位素效应[17]。既然当 $CD_4$ 代替 $CH_4$ 进行反应时,甲烷氧化总反应、CO 和 $CO_2$ 生成反应均表现出相同大小的正氘同位素效应,由此可以推断甲烷中 C—H 键的解

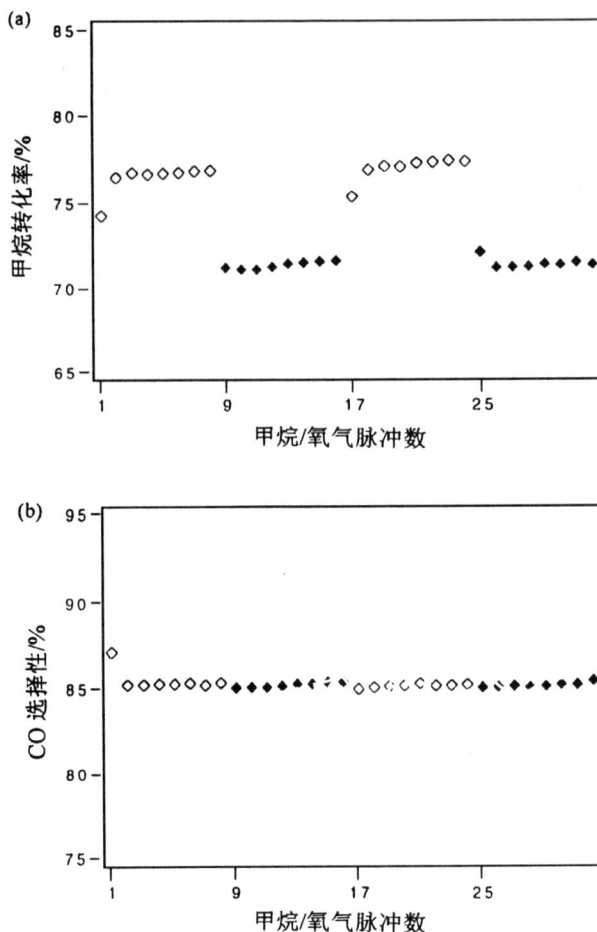

图 4　甲烷转化率(a)和 CO 选择性(b)与甲烷/氧气脉冲数的关系

空心、实心符号分别代表 $CH_4/O_2$、$CD_4/O_2$

离是甲烷氧化反应的关键步骤;CO 和 $CO_2$ 生成经历了某些共同的中间物种,而这些中间物种的生成速率受氘同位素效应的影响。很明显,这些共同的中间物种是由 $CH_4$ 分解而来的 $CH_x$($x=0$—3)物种。以上推论与 Schmidt 及其合作者提出的甲烷热解机理[18] 是一致的。根据下述甲烷热解机理

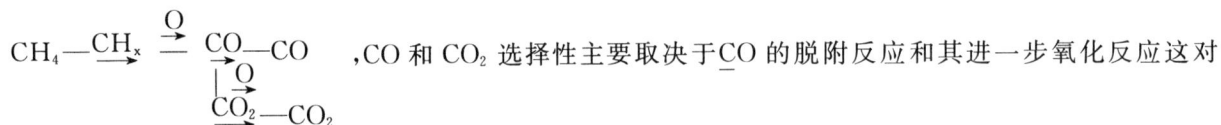

$$CH_4 - CH_x \underset{O}{\overset{O}{\rightleftarrows}} \begin{array}{c} CO - CO \\ \downarrow O \\ CO_2 - CO_2 \end{array}$$

,CO 和 $CO_2$ 选择性主要取决于 CO 的脱附反应和其进一步氧化反应这对竞争反应的相对速率。当 $CD_4$ 代替 $CH_4$ 进行反应时,由于正氘同位素效应,甲烷解离速率和 $CH_x$ 物种生成速率同时减小,导致 CO 和 $CO_2$ 生成速率均随着减小,CO 和 $CO_2$ 生成反应因此表现出正氘同位素效应;与此同时,CO 脱附反应对其进一步氧化反应的相对反应速率与氘同位素效应无关,因而 $CH_4+O_2$、$CD_4+O_2$ 两反应中 CO、$CO_2$ 选择性基本相同。在这个例子中,不仅获得了关于反应速控步骤的信息,也获得了关于反应历程的重要信息。

# 参考文献

[1]汪海有,刘金波,蔡启瑞等.分子催化,1994,8(2):111.

[2]福冈淳,肖丰收,市川胜等.触媒,1990,32(6):368.

［3］汪海有,刘金波,蔡启瑞等. 分子催化,1992,6(5):346.

［4］Wang Haiyou,Liu J inpo,Tsai Khirui et al Research on Chemical Intermediates,1992,17:233.

［5］Deluzarche A,Hinderm ann J P,Kiennem ann A et al J. Mol Catal,1985,31:225.

［6］Liu Jinpo,Wang Haiyou,Tsai Khirui et al Proc 9 ICC (Calgary,Canada),1988:735.

［7］汪海有,刘金波,蔡启瑞等. 物理化学学报,1991,7(6):681.

［8］Wang Haiyou,Liu J inpo,Tsai Khirui et al Catal Lett;,1992,12:87.

［9］福岛贵和,荒川裕则,市川胜等. 触媒,1986,28(2):60.

［10］汪海有,刘金波,蔡启瑞等. 分子催化,1992,6(2):156.

［11］汪海有,刘金波,蔡启瑞等. 高等学校化学学报,1993,14(8):1 157.

［12］Takeuchi A,Katzer J R. J. Phys. chem.,1981,85:937.

［13］Keller C S,Bell A T. J. Catal,1981,67:175.

［14］Ozaki A. in "Isotoic Studies of Heterogeneous Catalysis" (Academic Press),1977.

［15］汪海有,区泽棠·科学通报,1996,41 (22):2 055.

［16］Au C T,Wang H Y. J. Catal,1997,in press.

［17］Westheimer F H. Chem. Rev ,1961,61:265.

［18］Hickman D A,Schmidt L D. Science,1993,259:343.

■ 本文原载：Applied Catalysis A：General 166(1998)，pp. 343～350.

# Development of Coking-Resistant Ni-Based Catalyst for Partial Oxidation and CO$_2$-Reforming of Methane to Syngas

Ping Chen，Hong-Bin Zhang[①]，Guo-Dong Lin，Khi-Rui Tsai

(Department of Chemistry and State Key Laboratory for Physical Chemistry of the Solid Surfaces. Xiamen University, Xiamen 361005. China)

Received 27 March 1997; received in revised form 23 July 1997; accepted 22 August 1997

**Abstract**    Addition of small amount of trivalent-metal oxides, Cr$_2$O$_3$ and La$_2$O$_3$], to a Ni—Mg—O (Ni/Mg = 1/1, mol/mol) catalyst for partial oxidation of methane (POM) and CO$_2$-reforming of methane (MCR) reactions has been found to improve the performance of the catalyst for coking-resistance. The POM operation at 1053 K for 50 h, or the MCR operation at 1100 K for 6 h, did not leave any detectable amount of carbon deposit on the surface of the catalyst. Studies of XRD, XPS, and H$_2$-TPR spectroscopies showed that the doping of small amounts of Cr$^{3+}$ and La$^{3+}$ to the Ni—Mg—O system led to the formation of a host-dopant-type Ni—Mg—Cr—La—O solid solution, with a considerable number of Schottky defects in the form of cationic vacancies. An increase in the degree of disorder in the solid solution due to Cr$_2$O$_3$ and La$_2$O$_3$ dissolved in Ni$_x$Mg$_{1-x}$O lattice would be expected to enhance the mobility of the lattice oxygen anions. This would be in favor of speeding up the reaction between the carbon-containing species and reactive oxygen species via migration of the lattice O$^{2-}$ so as to inhibit the deposition of carbon on the surface of the catalyst. On the other hand, part of the Schottky defects in the form of cationic vacancies may diffuse to the surface, where Ni-species can be well accommodated and stabilized, thus, forming a rich-in-Ni (with mixed valence states) surface layer. As a result, the proportion of the reducible Ni-species was pronouncedly increased, but the temperature for their reduction was considerably raised, so that the surface Ni-species were maintained with higher possibility in positive valence states under POM and MCR reaction conditions. This would, to some extent, lead to the reduction of the rate of deep dehydrogenation of methane to carbon, therefore tending to reduce, if not avoid, coking caused by an excess of carbon on the surface. © 1998 Elsevier Science B. V.

**Key words**    Coking-resistance    Ni-based catalyst    Methane    Partial oxidation    CO$_2$-reforming

①    Corresponding author. Fax：+86 592 2086116；e-mail：hbzhang@xmu. edu. cn.

# 1  Introduction

The catalytic partial oxidation of methane (POM) and the $CO_2$-reforming of methane (MCR) to syngas have drawn much attention in recent years[1], because the former has potential significance as an alternative to the conventional steam reforming of methane, while the latter consumes $CH_4$ and $CO_2$, both of which are greenhouse gases. Utilization of these two processes is of economic and environmental benefit. Studies on the mechanisms of these two reactions and development of highly efficient catalysts are among the hot topics in the literature[1—10].

British researchers first reported that several noble metal catalysts in the form of supported or mixed oxides were very effective catalysts for POM and MCR reactions[2]; later, Choudhary et al. [4] and Hatakawa et al. [7] reported that some Ni- and co-based catalysts can also carry out POM or MCR reaction successfully. In practice, Ni-based catalysts are believed to be more promising than noble metal-based catalysts. Later experiments, however, showed[4,11,12] that almost all kinds of Ni catalysts, e. g. $Ni—Ln_2O_3$, $Ni/MOx$ (M = Al, Si, Zr, etc.), Ni—Mg—O, etc., suffered more or less from the disadvantage of serious deactivation caused by fouling of coke formed in the processes of POM and MCR, especially under rich-in-$CH_4$ and poor-in-oxidant conditions. Therefore, the development of coking-resistant non-noble metal catalysts has potential significance to commercial utilization of POM and MCR processes.

In the present work, a 4-component Ni-based catalyst, Ni—Mg—Cr—La—O, was developed and studied. In addition to the high activity and stability of both POM and MCR reactions, this catalyst displayed excellent performance in coking-resistance, even under the reaction conditions of $CH_4/O_2 =$ 2. 1/1 (v/v) for POM, and of $CH_4/CO_2 = 1/1. 08$ (v/v) for MCR. XRD, XPS and TPR methods were employed in the characteristic studies on the structure and properties of the catalyst. The results provide useful implications about the nature of the promoter behavior.

# 2  Experimental

## 2. 1  Preparation of catalyst

The catalyst was prepared by thoroughly mixing $Ni(NO_3)_2 \cdot 6H_2O$, $Mg(NO_3)_2 \cdot 6H_2O$, $Cr(NO_3)_3$ $\cdot 10H_2O$ and $La(NO_3)_3 \cdot nH_2O (n < 6)$ (all in A. R. grade of purity) with a molar ratio of Ni/Mg/Cr/ La = 9/9/1/1, and then by adding a certain amount of de-ionized water to form a homogeneous solution. The solution was dried in vacuum at 373K, and then calcined in air at 1073 K for 5 h. A porous catalyst precursor with a BER area of 80 $m^2$/g was obtained. The home-made Ni—Mg—O, Ni—Ca—O catalysts were prepared by the same procedure as described above, with the molar ratios of Ni/Mg, Ni/Ca and Ni/ La all at 1/1.

## 2. 2  Evaluation of catalyst activity

Evaluation of the catalyst activity was carried out in a fixed-hed continuous flow reactor-GC combination system operating under atmospheric pressure. The reactor is made of a quartz tube of 5. 0 mm I. D.. The temperature was controlled by an XCC-1000 programmed-temperature controller. A 24 mg of catalyst sample was used each time for testing. The sample of catalyst was first pre-reduced by using a flow of purified $H_2$ following a temperature programmed procedure of 40 min from room temperature to 973 K, and then a pre-mixed feed-gas was introduced into the reactor at the desired temperature. The catalytic POM and MCR reactions over the catalysts were carried out at a stationary state and under the following reaction conditions: 1053 K, a feedstream of mixture of 2. 1$CH_4$/1$O_2$(v/v) with a gas hourly space velocity (GHSV) at 1. 7×$10^5$ ml(STP)/h-g catal. for POM reaction; and 1053 K, 1$CH_4$/1. 08$CO_2$, GHSV at 6×$10^4$ ml(STP)/h-g catal. for MCR reaction. Activity data were taken after 30 min of operation, when a stationary state activity appeared. The gaseous reactants and products were analyzed by using an on-line SQ-206 GC equipped with a thermal conductivity detector (TC) and a 601$^{\#}$ carbon molecular sieve column, with helium as a carrier gas.

## 2. 3　Characterization of catalyst

Temperature programmed reduction (TPR) of catalyst was conducted on a fixed-bed continuous flow reactor-GC combination system. A $N_2$-carried 5v% $H_2$ gaseous mixture was used as reducing gas; the rate of elevating temperature was 10 K/min. ; an amount of catalyst equivalent to the content of 0. 1 mmol Ni was used each time for testing. X-ray diffraction (XRD) patterns were taken by using a Rigaku D/Max-C X-ray diffractometer with Cu$K\alpha$ radiation at a scanning rate of 6 ℃/min. XPS measurements were done on a VG ESCA LAB MK-2 machine with Al$K_a$ radiation (1487 eV) and UHV (1×$10^{-7}$Pa), calibrated internally by the carbon deposit C(1s) (B. E. ) at 284. 6 eV. The amount of coke was measured by a PCT-1 TG-DTA analyzer, with the temperature elevated at a rate of 20 K/min.

# 3　Results and discussion

## 3. 1　Performance of catalyst for POM and MCR

The activities of the Ni—Mg—Cr—La—O catalyst and the two binary Ni-based catalysts, Ni—Mg—O and Ni—La—O, for POM and MCR reactions, were tested at 1053 K. The results (shown in Table 1 and Fig. 1) indicated that this 4-component catalyst displayed high activities as compared to those of the Ni—Mg—O and Ni—La—O catalysts[9], with conversion of methane ($X_{CH_4}$) of 89% and selectivity of 96C% to CO ($S_{CO}$, calculated on the carbon number basis) for POM reaction, and $X_{CH_4}$ of 88% and conversion of $CO_2$ ($X_{CO_2}$) of 86% for MCR, respectively. Fig. 2(a and b) illustrate the results of stability testing of the Ni—Mg—Cr—La—O catalyst during 50 h of POM and 8 h of MCR operations, respectively. They indicate that this catalyst has rather high stability of operation for both POM and MCR reactions.

The amounts of coke formed on the Ni—Mg—Cr—La—O catalyst and, for comparison, on the Ni—Mg—O and Ni—La—O catalysts, after conducting POM or MCR reaction for a certain period of time, were measured by a TG-DTA analyzer. The results are summarized in Table 2. For the MCR reaction, the amounts of coke detected on the Ni—Mg—O and Ni—La—O catalysts reached up to 12 and 10. 2 wt% of the catalyst used, respectively, after 1 h of operation at 1053 K. On the Ni—Mg—Cr—La—O catalyst, only an amount of coke equivalent to 7 wt% of the catalyst used was detected after 8 h of operation at 1053 K; on elevating the temperature to 1100 K, no coke was detected after 6 h of operation. For the POM reaction, the coking-resistant performance of the Ni—Mg—Cr—La—O catalyst is more pronounced; the operation in 50 h under the reaction conditions of 1053 K and 2. 1 $CH_4$/1$O_2$ did not leave any detectable amount of carbon deposit on the catalyst, whereas, as a sharp contrast, on the Ni—Mg—O and Ni—La—O, considerable amounts of coke had been formed in a much shorter period of time for testing under the same conditions. These results demonstrate that the 4-component Ni—Mg—Cr—La—O catalyst not only retains high activities and selectivities for both POM and MCR reactions, but also possesses remarkable coking-resistance.

Table 1    The results of activity assays of the Ni—Mg—Cr—La—O, Ni—Mg—O and Ni—La—O
catalysts for the POM and MCR reactions, respectively, at 1053 K[a]

| Catalyst | Temperature (K) | POM | | MCR | |
|---|---|---|---|---|---|
| | | $X_{CH_4}$ (%) | $S_{CO}$ (C%) | $X_{CH_4}$ (%) | $X_{CO_2}$ (%) |
| Ni—Mg—O | 1053 | 89 | 96 | 88 | 86 |
| Ni—La—O | 1053 | 88 | 95 | 87 | 85 |
| Ni—Mg—Cr—La—O | 1053 | 89 | 96 | 88 | 86 |

[a] Reaction conditions: 0.1 MPa. CH$_4$/O$_2$ = 2.1/1(v/v). GHSV ~ 1.7 × 10$^5$ ml(STP)/h-g catal. for POM; and 0.1 MPa, CH$_4$/CO$_2$ —1/1.08(v/v). GHSV 6 × 10$^4$ ml(STP)h-g catal. for MCR.

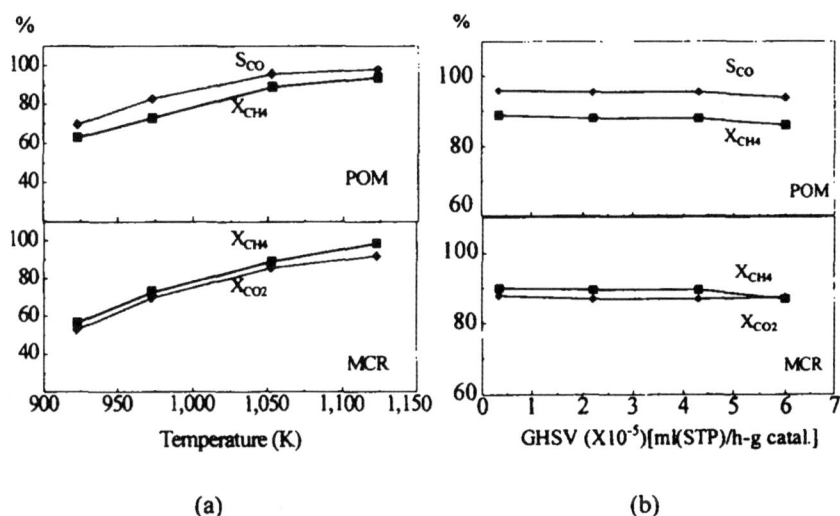

Fig. 1    Effect of temperature (a) and of feedgas hourly space velocity (GHSV) (b) on the performance of
the Ni—Mg—Cr—La—O catalyst for POM or MCR. Reaction conditions: (a) 0.1 MPa; CH$_4$/O$_2$
= 2.1/1(v/v), GHSV ~ 1.7 × 10$^5$ mL(STP)/h-g catal. for POM and CH$_4$/CO$_2$—1/1.08 (v/v),
GHSV ~ 6 × 10$^4$ mL (STP)/h-g catal. for MCR, respectively; (b) 1053 K, 0.1 MPa; CH$_4$/O$_2$—
2.1/1 (v/v) for POM and CH$_4$/CO$_2$ = 1/1.08(v/v) for MCR, respectively.

Table 2    coke formation on the Ni—Mg—Cr—La—O. Ni—Mg—O and Ni—La—O
catalysts after conducting POM or MCR reaction[*]

| Reaction | Catalyst | Temp. (K) | Reaction time (h) | Percentage of coking (wt%) | Coking rate (wt%/h) |
|---|---|---|---|---|---|
| | Ni—Mg—O | 1053 | 1.0 | 12.0 | 12.0 |
| MCR | Ni—La—O | 1053 | 1.0 | 10.2 | 10.2 |
| | Ni—Mg—Cr—La—O | 1053 | 8.0 | 7.0 | 0.85 |
| | | 1100 | 6.0 | — | — |
| | Ni—Mg—O | 1053 | 8.0 | 30.0 | 3.77 |
| POM | Ni—La—O | 1053 | 15.0 | 4.5 | 0.3 |
| | Ni—Mg—Cr—La—O | 1053 | 50.0 | — | — |
| | Ni/La$_2$O$_3$[11] | 773 | 6.0 | 9.1 | 1.5 |

[*] Reaction conditions: 0.1 MPa, CH$_4$/O$_2$ = 2.1/1 (v/v), and GHSV ~ 1.7 × 10$^5$ (STP) ml/h-g catal. for POM; and 0.1 MPa, CH$_4$/CO$_2$—1/1.08(v/v), GHSV ~ 6 × 10$^4$ (STP)ml/h-g catal. for MCR.

P. Chen et al./Applied Catalysis A: General 166 (1998) 343–350

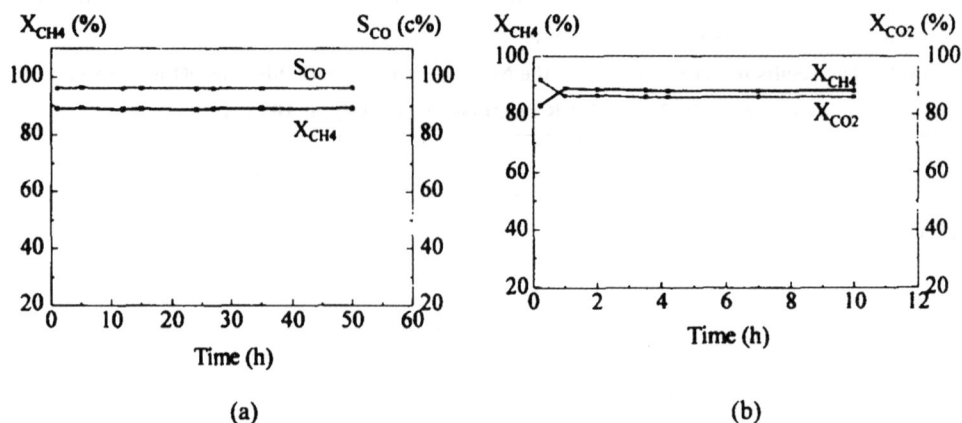

(a)                                          (b)

Fig. 2 Stability testing of the Ni—Mg—Cr—La—O catalyst for POM (a) and MCR (b). Reaction conditions: 0.1 MPa, 1053 K; CH$_4$/O$_2$ = 2.1/1(v/ v), GHSV ~ 1.7 × 10$^5$ ml(STP)/h-g catal. for POM and CH$_4$/CO$_2$ —1/1.08(v/v), GHSV ~ 6 × 10$^4$ ml(STP)/h-g catal. for MCR, respectively.

## 3.2 Spectroscopic characterization of catalyst

Fig. 3 shows the results of XRD measurements of several catalysts. It is well known that crystal structures of both NiO and MgO belong to the "NaCl" lattice-type. Since Ni$^{2+}$ and Mg$^{2+}$ have the same valence and quite close ionic radius values [0.070 nm for r(Ni$^{2+}$) and 0.065 nm for r(Mg$^{2+}$)], the dimensions of their crystal cells are very close to each other. Thus, the NiO and MgO components in the Ni—Mg—O system would easily form a Ni$_x$Mg$_{1-x}$O solid solution due to excellent mutual solubility between them, as evidenced by the XRD pattern of the Ni—Mg—O system. In Fig. 3(a), only the diffraction peaks assigned to a Ni$_x$Mg$_{1-x}$O solid-solution phase appeared; no peak ascribed to either discrete NiO or discrete MgO phase was observed. In the XRD pattern for the Ni—Mg—O catalyst in the functioning state (shown in Fig. 3(e)), the weak features of metallic Ni-phase and strong peaks of the Ni$_x$Mg$_{1-x}$O phase were observed, indicating that only a small proportion of the Ni$^{2+}$ component was reduced to Ni$^0$ and the remaining large proportion of the Ni$^{2+}$ remained in the phase of the Ni$_x$Mg$_{1-x}$O solid solution.

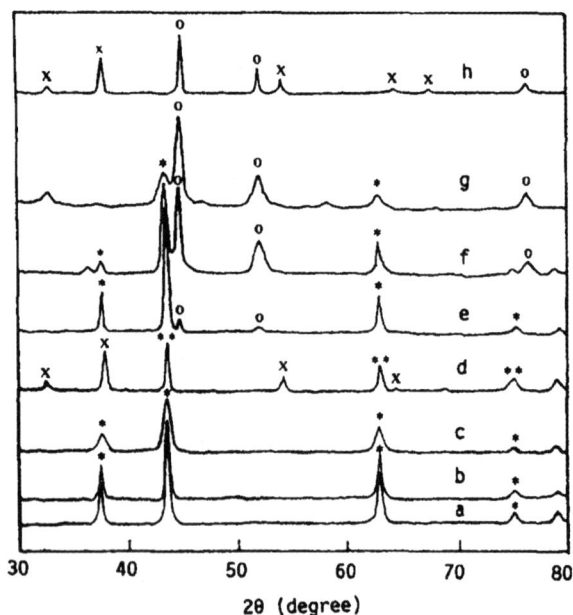

Fig. 3 XRD patterns of the catalyst precursor of: (a) Ni—Mg—O. (b) Ni—Mg—Cr—O. (c) Ni—Mg—Cr—La—O. and (d) Ni—Ca—O, and of the functioning catalyst of: (e) Ni—Mg—O, (f) Ni—Mg—Cr—O. (g) Ni—Mg—Cr—La—O. and (h) Ni—Ca—O. (o) represents the diffraction peaks of metallic Ni. (˙) represents the diffraction peaks of the Ni$_x$Mg$_{1-x}$O. or Ni—Mg—Cr—O, or Ni—Mg—Cr—La—O solid solutions. (˙˙) represents ihe diffraction peaks of NiO, (x) represents the diffraction peaks of CaO.

The high stability of the solid-solution phase of the $Ni_xMg_{1-x}O$ could also be demonstrated through the following comparative testing. On a functioning Ni—Ca—O ( Ni/Ca = 1/1, mol/mol ) catalyst undergoing POM operation at 1053 K for 30 min, only the diffraction peaks ascribed to metallic Ni-phase and CaO phase were observed, whereas the intensities of the features for NiO-phase were below the detectable limit ( Fig. 3( h ) vs. Fig. 3( d )). This indicated that almost all $Ni^{2+}$ component had been reduced to $Ni^0$, with a developed metallic Ni-phase. The reducible behavior of the Ni—La—O system (Ni/La=1/1,mol/mol) is similar. After undergoing the same pre-reduction treatment and reaction, the original $LaNiO_3$ composite oxide phase with $ABO_3$-type structure in the oxidation state was disintegrated, and the two discrete phases, metallic Ni and $La_2O_3$, were formed. However, for the Ni—Mg—O (Ni/Mg=1/1,mol/mol) system undergoing the same treatment, only a small proportion of the $Ni^{2+}$, perhaps located at the surface layer, was reduced to $Ni^0$. The remaining large proportion of the $Ni^{2+}$ could not be reduced, probably due to the strong valence stabilization of the crystal field of $Ni_xMg_{1-x}O$ with 'NaCl'-type structure, and remained in the phase of $Ni_xMg_{1-x}O$ solid solution.

Addition of a small amount of trivalent-metal oxides, $Cr_2O_3$ and $La_2O_3$, to the Ni—Mg—O catalyst did not lead to disintegration of the $Ni_xMg_{1-x}O$ solid-solution phase, but rather formed a host-dopant solid solution of the Ni—Mg—Cr—La—O, as evidenced by the XRD patterns ( Fig. 3( b and c )), where no discrete phases, i. e. NiO, MgO, $Cr_2O_3$, or $La_2O_3$, was observed. Solution of an amount of $Cr^{3+}$ [$r$ ($Cr^{3+}$) = 0. 064 nm] in the $Ni_xMg_{1-x}O$ lattice would result in the formation of cationic vacancies. Solution of $La^{3+}$ in the $Ni_xMg_{1-x}O$ lattice is expected to be relatively difficult in comparison with that of $Cr^{3-}$ due to its larger ionic radius [$r(La^{3+})$=0. 106 nm], and, thus, most of the $La^{3+}$ would be dispersed on the surface of the $Ni_xMg_{1-x}O$ crystallites. As a result, the doping of both $Cr^{3-}$ and $La^{3+}$ in the $Ni_xMg_{1-x}O$ system would lead to the decrease in the degree of long-range order in the solid-solution lattice, as supported by the broadening of the XRD peaks observed ( Fig. 3( b and c )).

Conceivably, an increase in the degree of disorder in the solid solution due to $Cr_2O_3$ and $La_2O_3$ dissolved in $Ni_xMg_{1-x}O$ lattice would enhance the mobility of the lattice oxygen anions. This would be in favor of improving the transport of the reactive oxygen-species via the migration of lattice $O^{2-}$, and, thus, favor the speeding up of the surface reaction between the carbon-containing species and reactive oxygen species so as to inhibit the deposition of carbon on the surface of catalyst. On the other hand, part of Schottky defects in the form of cationic vacancies will diffuse to the surface of $Ni_xMg_{1-x}O$ phase, where $Ni^+$ species can be well accommodated and stabilized, thus forming a rich-in-Ni (with mixed valence states) surface layer of $—(Ni_x)^+——O^2—Mg^{2-}—O^{2-}—$ to achieve valence and charge compensation for subsurface layers of $—M^{3+}—O^{2-}—Mg^{2+}—O^{2-}—$ As a result, the proportion of the reducible Ni-species was pronouncedly increased, but their deep reduction to $Ni^0$ seemed to be somewhat difficult due to the strong valence stabilization of the crystal field of $Ni_xMg_{1-x}O$. This can gain some experimental support from the XRD and XPS

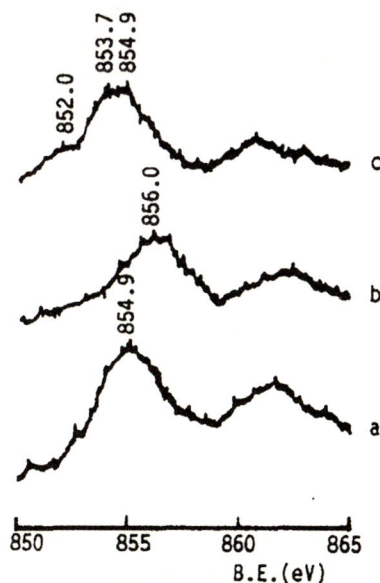

Fig. 4　XPS spectra of Ni(2p) on: (a) the precursor of the Ni—Mg—O catalyst, ( b ) the precursor of the Ni—Mg—Cr—La—O catalyst, and ( c ) the functioning Ni—Mg—Cr—La—O catalyst.

observations. As shown in Fig. 3(g), the diffraction peaks of metallic Ni in the XRD pattern of the functioning Ni—Mg—Cr—La—O catalyst were pronouncedly enhanced in their intensities, but broadened in the shapes, in comparison with those of the undoped Ni—Mg—O catalyst. The results of XPS measurements on the Ni—Mg—Cr—La—O catalyst indicated that the addition of a small amount of $Cr_2O_3$ and $La_2O_3$ to the Ni—Mg—O catalyst precursor shifted the Ni(2p) electron binding-energy from 854.9 to 856.0 eV (Fig. 4(a and b)), and that there coexisted some Ni-species with mixed valence states ($Ni^0$, $Ni^+$ and $Ni^{2+}$) on the surface of the functioning Ni—Mg—Cr—La—O catalyst, with a considerable proportion of Ni in $Ni^+$ (Fig. 4(c)).

$H_2$-temperature programmed reduction (TPR) of the catalyst precursors provided useful information about reducibility of the Ni-component in these catalysts. As shown in Fig. 5(b), two peaks (673, 900 K) were observed on the Ni—Mg—O catalyst, which may be reasonably assigned to the reduction of the reducible $Ni^{2+}$-species; tentatively, the 673 K peak was due to $Ni^{2+} \rightarrow Ni^+$, and the 900 K due to $Ni^+ \rightarrow Ni^0$). On the Ni—Mg—Cr—La—O, a strong peak at 718 K with two shoulder peaks at 618 and 683 K, respectively, and another strong peak at 1123 K and another distinct shoulder peak at 1000 K were observed (Fig. 5 (c)). It is known that $La^{3+}$ and $Mg^{2+}$ are nonreducible under the reaction conditions for POM and MCR. The $H_2$-TPR spectrum taken on a related system, Cr—Mg—O (Cr/Mg = 1/1, mol/mol) (Fig. 5(a)) demonstrated that reduction of a small amount of the $Cr^{3+}$ to its lower valence state may occur. The Cr(2p)-XPS spectra (Fig. 6) also indicated that there were Cr-species in lower valence state (perhaps in $Cr^2$) on the surface of the functioning Ni—Mg—La—O catalyst. Thus, the

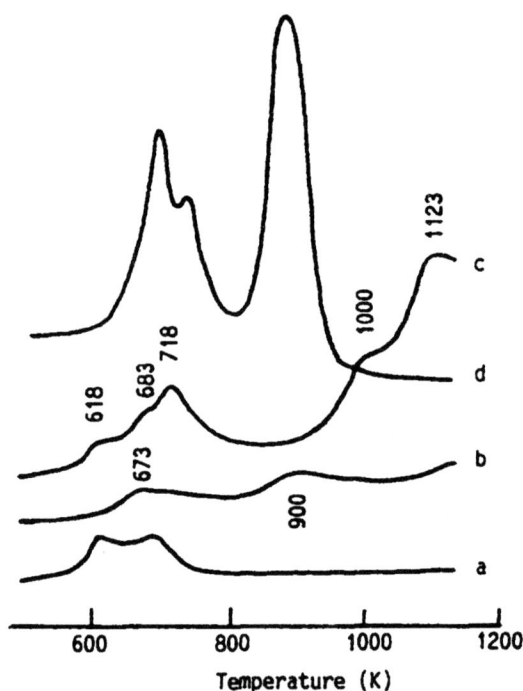

**Fig. 5** TPR spectra of the catalyst precursor of: (a) Cr—Mg—O, (b) Ni—Mg—O, (c) Ni—Mg—Cr—La—O, and (d) Ni—La—O, all at 10 K/min. of temperature-elevating rate and in a reductant stream of 5v%$H_2$—95v%He with a flow rate of 2000 ml/h.

two shoulder-peaks at 618 and 683 K may be assigned to the reduction of $Cr^{3+}$-species; and the 718 K peak and the 1000 K shoulder-peak may be due to $Ni^{2+} \rightarrow Ni^+$ and $Ni^+ \rightarrow Ni^0$, respectively; and the 1123 K peak might mainly originate from the reduction of the $Ni^{n+}$-species, newly separated from the Ni—Mg—Cr—La—O solid-solution phase under the high temperature. Therefore, by comparison of the $H_2$-TPR spectra of these systems, it could be concluded that on one hand, the doping of a small amount of $Cr^{3+}$ and $La^{3+}$ brought about an increase in the amount of reducible $Ni^{2+}$-species in comparison with the undoped Ni—Mg—O system, and on the other hand, it resulted in a pronounced raise in the reduction temperature for the $Ni^{2+}$-species, in agreement with an increase in the binding energy of electrons of the $Ni^{2+}$, revealed by the XPS spectra of Ni(2p) (Fig. 4(a and b)). As a resutt, the reduction of these reducible $Ni^{2+}$-species would be relatively difficult and somewhat large proportion of the Ni-species could be stablized at a positive valence state, as evidenced by the XPS result shown in Fig. 4(c). This is obviously different from the case of the Ni—La—O, in which all the $Ni^{2+}$ could be reduced to metallic Ni-state at much lower temperatures, with three $H_2$-TPR peaks observed correspondingly(Fig. 5(d)).

Our previous comparative study on the catalytic behavior of the first series of transition metals (including Cr, Mn, Fe, CO, Ni, Cu, etc. ) for the POM reaction has led to the following conclusions[10] : the POM activity is related to the valence-changeable transition-metal sites. $M^0/M^{n+}$ (e. g. $Ni^0/Ni^{2+}$ ), on the surface of the functioning catalysts; the transition-metal sites in zero-valence state seem to be responsible for the activation and dehydrogenation of methane by homolytic splitting of its C—H bonds on these sites, and a changeable valence of the active sites is required for the activation and rapid conversion of dioxygen. Therefore, it seems to us that one of roles that the doping of $Cr^{3+}$ and $La^{3+}$ plays is to generate a considerable number of Schottky defects in the form of cationic vacancies, which will diffuse to the surface of the catalyst, and. thus, provide favorable micro-environments for accommodating and

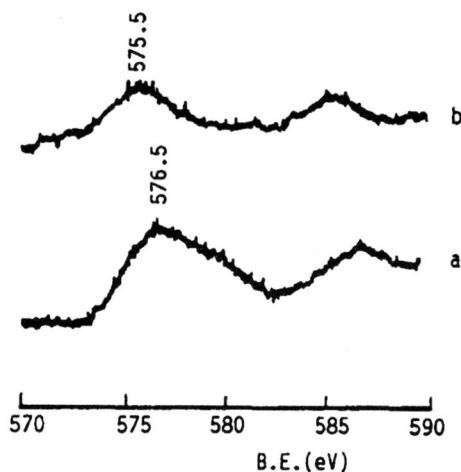

Fig. 6　XPS spectra of Cr(2p) on the Ni—Mg—Cr—La—O catalyst in: (a) the precursor state. and (b) the functioning state.

maintaining the surface Ni-species with higher possibility in positive valence states under POM and MCR reaction conditions. This would, to a certain extent, lead to reducing the rate of deep dehydrogenation of methane to carbon and therefore would tend to reduce, if not avoid, coking caused by an excess of carbon on the surface.

# 4. Conclusions

Addition of a small amount of trivalent-metal oxides, $Cr_2O_3$ and $La_2O_3$ to the Ni—Mg—O (Ni/Mg=1/1, mol/mol) catalyst for POM and MCR reactions has been found to remarkably improve the performance of the catalyst for coking-resistance.

The XRD measurements indicated that the doping of the small amount of $Cr^{3+}$ and $La^{3+}$ to the Ni—Mg—O system led to the formation of a host-dopant-type Ni—Mg—Cr—La—O solid solution, with a considerable number of Schottky defects in the form of cationic vacancies. The XPS observation provided evidence that there coexisted the Ni-species with mixed valence states ($Ni^0$, $Ni^+$ and $Ni^{2+}$, with a somewhat large proportion in $Ni^+$) on the surface of the functioning Ni—Mg—Cr—La—O catalyst. The $H_2$-TPR testing showed that the doping of a small amount of $Cr^{3+}$ and $La^{3+}$ brought about an increase in the amount of reducible $Ni^{2+}$-species in comparison with the undoped Ni—Mg—O system, and on the other hand, it resulted in a pronounced rise in their reduction temperature.

The above experimental results may lead to the following inferences: the doping of a small amount of $Cr^{3+}$ and $La^{3+}$, on one hand, resulted in an increase in the degree of disorder in the solid solution, which would be in favor of enhancing the mobility of the lattice $O^{2-}$, and thus speeding up the surface oxidation reactions; on the other hand, it also led to the formation of a rich-in-Ni (with mixed valence states) surface layer, where the surface Ni-species were maintained with higher possibility in positive valence states so that deep dehydrogenation of methane to carbon was inhibited to some extent. It is these factors that are closely related to the excellent performance of the catalyst for coking-resistance.

## Acknowledgements

The authors are grateful for financial support from the National Natural Science Foundation of China.

## References

[1]S. C. Tsang, J. B. Claridge, M. L. H. Green, Catal. Today **23**(1995)3.

[2]A. T. Ashcroft, A. K. Cheetham, J. S. Foord. M. L. H. Green, C. P. Grey, A. J. Murrell. P. D. F. Vernon, Nature **344**(1990)319.

[3]D A. Hickman. L. D. Schmidt, J. Catal. **138**(1992)267.

[4]V. R. Choudhary, A. M. Rajput, V. H. Rane, J. Phys. Chem. **96**(1992)8686; Catal. Lett. **16**(1992) 269.

[5]D. A. Hickman, E. A. Haupfear, L. D. Schmidt, Catal. Lett. **17**(1993)223.

[6]D. Dissanayak, M. P. Rosynek. K. C. C. Kharas, J. H. Lunsford, J. Catal. **132**(1991)117.

[7]J T. Hayakawa, A. G. Andersen, M. Shimizu, K. Suzuki. K. Takehira. Catal. Lett. **22**(1993)307.

[8] J. A. Lapszewicz, X. Z. Jiang, Prepr. Am. Chem. Soc. . Div. Pet. Chem. **37**(1992)252.

[9] P. Chen, H. B. Zhang, G. D. Lin. K. R. Tsai, Chem. Research in Chinese Univ. **12**(1996)70.

[10] P. Chen, H. B. Zhang, G. D. Lin. K. R. Tsai, Chem. Research in Chinese Univ. **11**(1995)323.

[11]J. B. Claridge, M. L. H. Green. S. C. Tsang. A. P. E York, A. T. Ashcroft, P. D. Battle, Catal. Lett. **22**(1993)299.

[12]A. K. Bhattacharya. J. A. Breach, S. Chand, et al., Appl. Calal. A; General **80**(1992)L1.

■ 本文原载:Chemical Research in Chinese Universities Vol. 14 No. 3,pp. 340~343,1998.

# Enantioselective Transfer Hydrogenation of Aromatic Ketones Catalyzed by New Diaminodiphosphine Ru(II)Complexes*

Pian-Pian Xu, Jing-Xing Gao①, Xiao-Dong Yi, You-Qing Huang,
Hui Zhang, Hui-Lin Wan, Khi-Rui Tsai, Ikariya Takao^a
(Department of Chemistry, State Key Laboratory for Physical Chemistry of
Solid Surface and Institute of Physical Chemistry, Xiamen University,
Xiamen, 361005; Department of Chemical Engineering^a,
Faculty of Engineering, Tokyo Institute of Technology,
2—12—1 O-okayama, Meguro-ku, Toyoto, Japan)
Received April 17, 1996

**Key words** Chiral ligand Ruthenium complex Asymmetric transfer hydrogenation

Chiral biphosphine ligands provide a useful tool for preparing optically active secondary alcohols and have been the interesting subject of numerous investigations[1]. However, it is noted that in the field of enantioselective transfer hydrogenation, the mostly used chiral auxiliary ligands should contain nitrogen as the donor atom[2]. Recently, the importance of nitrogen donors has been reviewed[3] and some chiral ruthenium complexes bearing nitrogen donors have been developed[4]. In the past few years, we have been interested in the synthesis of well-designed polydentate ligands possessing two "soft" phosphorus atoms and two "hard" nitrogen atoms as chelating ligands[5-8]. These ligands can act as bi-, tri-and tetra-dentate ligands, depending on the reaction conditions and display some interesting structures, chemical and catalytic properties[9]. This communication reports the synthesis and characterization of some new chiral Ru(II) complexes with a similar structure $N,N'$-bis[$o$-(diphenylphosphino)benzylidene] propane-1,2-diamine($P_2N_2$Me-Ru(II)$Cl_2$ for abbriviation) and $N,N'$-bis[$o$-(diphenylphosphino)benzyl] propane-1,2-diamine($P_2N_2H_4$Me-Ru(II)$Cl_2$ for abbriviation), their application in the enantioselective transfer hydrogenation of aromatic ketones as well.

When a mixture of $o$-(diphenylphosphino)benzylaldehyde and($R$)-propane-1,2-diamine in a molar ratio of 2 : 1 was stirring in dichloromethane with an excess of $Na_2SO_4$ as a dehy-drating agent, a tetradentate($R$)—$N,N'$-bis[$o$-(diphenylphosphino)benzylidene] propane-1,2-diamine[($R$)—1] was produced in a 83%~88% yield (Scheme 1). $^1$H NMR($CDCl_3$):$\delta$—8.75 for Ph—CH=N;$^{31}$P NMR ($CDCl_3$),$\delta$—11.81,—12.44. Furthermore, the reduction of ($R$)—1 with an excess of $NaBH_4$ was carried

* Supported by the National Natural Science Foundation of China and Union Laboratory of Asymmetric Synthesis (Chengdu Institute Organic Chemistry;Hong Kong Polytechnic;kuo Qing Chemical Co.,Ltd.).

① To whom conespondence should be addressed.

out in refluxing ethanol to afford the corresponding $(R)-N, N'$ bis [$o$-(diphenylphosphino) benzyl] propane-1,2-diamine[$(R)-2$] in a $68\% \sim 73\%$ yield($^1$H NMR(CDCl$_3$): $\delta$ 4.14 for Ph_CH$_2$_; $^{31}$P NMR(CDCl$_3$), $\delta-15.41, -15.51$).

The interaction of an equimolar mixture of $trans$-RuCl$_2$(DMSO)$_4$ and $(R)-1$ in refluxing toluene gave a dark-red solution. This solution was concentrated under reduced pressure and the crude product was chromatographed on a silica gel column (2 cm$\times$12 cm) with CH$_2$Cl$_2$ as the eluent solvent, giving a red colour ruthenium complex containing $(R)-N, N'$-bis [$o$-(diphenylphosphino) benzylidene] propane-1,2-diamine [$(R)-3$]) yield $78\% \sim 82\%$, $^1$H NMR(CDCl$_3$): $\delta$8.76 for Ph_CH $=$N_; $^{31}$P NMR(CDCl$_3$), $\delta$48.12, 48.51).

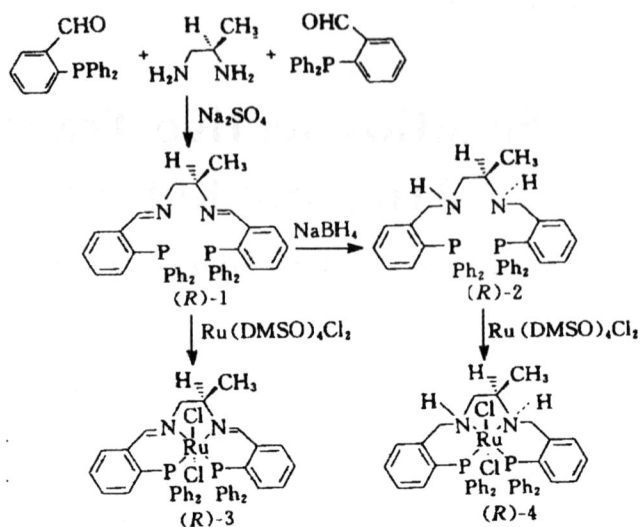

Scheme 1

A yellow Ru(II)complex($R$)-4 was prepared by the similar procedure and CH$_2$Cl$_2$/ace-tone (1:1) was used as eluent solvent when a silica gel column was used. The $^{31}$P NMR spectrum of($R$)-4 exhibits two singlets of equal intensity at $\delta$ 45.18 and 43.88, which suggest that the two phosphine groups are non-equivalent and coordinated to ruthenium atom. $^1$H NMR(500 MHz CDCl$_3$) for $(R)-4$: $\delta$ 0.91 (d, $J=5.8$ Hz, H$^3$, _CH$_3$), 3.28(t, H$^1$, _ CH$_2$), 3.01(d, 1H, _CH$_2$), 4.62(s, 1H, _NH_), 3.95(t, $J=12$ Hz, 1H, _NH_). 3.54(m, 1H, —HC $>$ ), 3.70(d, $J=12$ Hz, 1H, PhCH$_2$_), 4.06(d, $J=12$ Hz, 1H, PhCH$_2$ _), 4.75(d, $J=11$ Hz, 1H, PhCH$_2$_), 4.80(t, $J=11$ Hz, 1H, PhCH$_2$_), 6.82$\sim$7.34(m, 28 Hz, C$_6$H$_5$ _). $^{31}$P NM R(CDCl$_3$) for $(R)-4$: $\delta$ 45.18, 43.88. m. p. 226$\sim$228 ℃. Anal. Calcd. (%)for $(R)-4 \cdot 0.5$ C$_6$H$_{14}$(837.79): C 63.11; H 5.61; N 3.36; Found: C 63.04; H 5.46; N 3.35. IR(KBr), $\tilde{\nu}$/cm$^{-1}$: 3450 m, 3057 m, 2 867 m, 1 474 s, 1 431 vs, 1 089 s, 1027w, 950 s, 744 s, 692 vs, 516 vs. Ligands $(S)-1$, $(S)-2$ and Ru(II)complexes$(S)-3$, $(S)-4$ were easily prepared by means of the above similar procedures.

A suitable crystal of complex $(R)-3$ for X-ray diffraction was grown from a CH$_2$Cl$_2$/hexane mixture. The structure analysis of $(R)-3$ revealed a distorted octahedral trans-configuration for the complex (Fig. 1). The crystal data for $(R)-3$ C$_{41}$H$_{36}$N$_2$P$_2$Cl$_2$Ru are as follows: $M=790.67$, monoclinic, space group $P2_1$, $a=1.1569(1)$nm, $b=1.5079(1)$nm, $c=1.1972(1)$nm, $\beta=97.42(1)°$, $V=2.07113$ nm$^3$, $Z=2$, $D_c=1.540$ g/cm$^3$, $\mu=78.1$ cm$^{-1}$, F(000)=976. The two chloro-ligands in the axial position are mutually $trans$ to each other and the $(R)-1$ ligand acts as a tetradentate ligand around the Ru center with the two phosphino-groups $cis$ to each other. The attempt to get a suitable crystal of complex($R$)-4 for structure analysis is still unsuccessful. However, based on the spectroscopic data and the molecular structures of $trans$-RuCl$_2$(P$_2$N$_2$) and $trans$-RuCl$_2$ (P$_2$N$_2$H$_4$)$^{[9,10]}$, the structure of ruthenium complex $(R)$-4 is

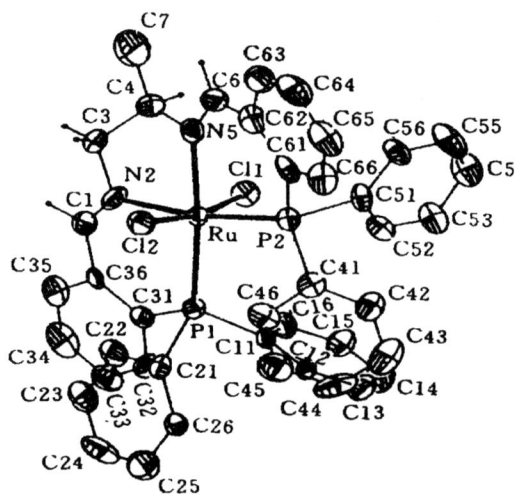

Fig. 1 Molecular structure of complex$(R)-3$.

assignable to an analogy of complex $(R)-3$.

complexes $(R)-3$, $(S)-3$, and $(R)-4$, $(S)-4$ have been tested as catalysts for the enantioseletive transfer hydrogenation of aromatic ketones in an iso-PrOH solution(Scheme2). The catalytic hydrogenation of acetophenone(1a) was conducted using some potassium 2-propoxide(1_3 equiv. with respect to Ru)as a promoter(Table 1).

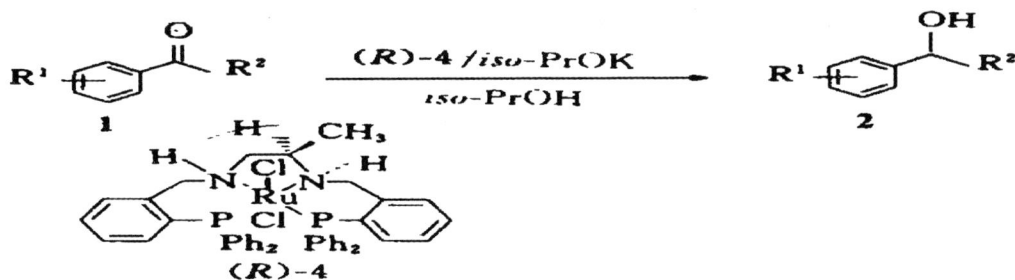

a. $R^1 = H, R^2 = Me; b. R^1 = H, R^2 = Et; c. R^1 = Cl, R^2 = Me; d. R^1 = OCH_3, R^2 = Me.$

**Scheme 2**

**Table 1    Asymmetric transfer hydrogenation of ketones catalyzed by chiral
$P_2N_2Me_2$-Ru( Ⅱ )Cl$_2$ and $P_2N_2H_4$Me-Ru( Ⅱ )Cl$_2$ complexes**[*]

| Ketone substrate | Catalyst | $n(S)n(C)n(iso$-PrOK)[a] | condition | | Alcohol product | | |
|---|---|---|---|---|---|---|---|
| | | | $t/℃$ | $t/h$ | Yield(%)[b] | e.e.(%)[c] | Cnfig.[d] |
| 1a | $(R)-3$ | 100·1·3 | 40 | 22 | 63 | 26 | S |
| 1a | $(S)-3$ | 100·1·3 | 40 | 22 | 65 | 14 | R |
| 1a | $(S)-3$ | 100·1·3 | 30 | 46 | 90 | 91 | S |
| 1b | $(S)-4$ | 100·1·2 | 45 | 48 | 55 | 88 | S |
| 1b | $(R)-4$ | 100·1·3 | 30 | 46 | 73 | 91 | R |
| m-1c | $(S)-4$ | 100·1·2 | 30 | 24 | 99 | 87 | S |
| p-1c | $(S)-4$ | 100·1·2 | 30 | 24 | 82 | 89 | S |
| m-1d | $(S)-4$ | 100·1·2 | 30 | 24 | 72 | 85e | S |
| p-1d | $(S)-4$ | 100·1·2 | 30 | 24 | 49 | 87e | S |

[*] conditions:catalysts 0.01 mmol;solvent iso-PrOH 20 mL;a. S/C/iso-PrOK=ketone/ Ru/iso-PrOK;b. GLC analysis;c. capillary GLC analysis using a chiral Chrompack CD-cyclidextrin-β-236 M—19 column unless otherwise specified;d. determined by comparison of the retention time of each of the enantiomers on the GLC traces with literature values;e. determined by HPLC analysis using a Daicel Chiralccl OB column([10·90] 2-propanol-hicxane.)

The concentration of iso-PrOK is an important factor for the catalytic activity and the catalytic system is inactive without a basic co-catalyst. The increase of reaction temperature accelerates the reaction rate with a slight loss of enantiomeric purity of the product. The ketones possessing an eletron-donating substituent such as methoxyl to the para position tend to lower the rate, but still show high stereoselectivity. It is noteworthy that the diimino complexes $(R)-3$ or $(S)-3$ and the diamino complexes $(R)-4$ or $(S)-4$ display the differences in reactivities and enantioseletivities. complex $(R)-4$ or $(S)-4$ with $sp^3$ hybridized nitrogens containing N_H bonds showed the higher reaction rate and enantioselectivity. The detailed reaction mechanism is now under investigating.

## Acknowledgement

*We would like to thank professor Ryoji Noyori for his very valuable discussion.*

## References

[1] Noyori R., Asymmetric Catalysis in Organic Synthesis, New York: John Wiley & Sons, Inc., 1994:16.

[2] Zassinvich G, Mestroni G., Chem. Rev., **92**, 1051(1992).

[3] Togni A. Venanzi L., Angew. Chem. Int. Ed. Engl., **33**, 497(1994).

[4] Hashiguchi S. Fujii A., Takehara J. et al., J. Am. Chem. Soc., **117**, 7562(1995).

[5] Wong W. K., Gao J. X., Zhou Z. Y. et al, Polyhedron **11**, 2965(1992).

[6] Wong W. K., Gao J. X., Wong W. T, Polyhedron, **12**, 1047(1993).

[7] Wong W. K., Gao J. X., Wong W. T. et al., Polyhedron, **12**, 2063(1993).

[8] Wong W. K., Gao J. X., Wong W. T. et al., J. Organoment. Chem., **471**, 277(1994).

[9] Gao J. X., Wan H. L., Wong W. K. et al., Poly hedr on, **15**, 1241(1996).

[10] Wong W. K., Gao J. X., Polyhedron, **12**, 1415(1993).

■ 本文原载：Catalysis Letters 53(1998)，pp. 119～124.

# Nonoxidative Dehydrogenation and Aromatization of Methane over W/HZSM-5-Based Catalysts*

Jin-Long Zeng，Zhi-Tao Xiong，Hong-Bin Zhang①，Guo-Dong Lin，K.R.Tsai

*(Department of Chemistry and State Key Lab of Physical Chemistry for the Solid Surfaces, Xiamen University, Xiamen 361005, P.R. China)*

Received 11 December 1997; accepted 25 April 1998

Highly active and heat-resisting W/HZSM-5-based catalysts for nonoxidative dehydroaromatization of methane (DHAM) have been developed and studied. It was found from the experiments that the $W-H_2SO_4$/HZSM-5 catalyst prepared from a $H_2SO_4$-acidified solution of ammonium tungstate (with a pH value at 2-3) displayed rather high DHAM activity at 973～1023 K,whereas the W/HZSM-5 catalyst prepared from an alkaline or neutral solution of $(NH_4)_2WO_4$ showed very little DHAM activity at the same temperatures. Laser Raman spectra provided evidence for existence of $(WO_6)^{n-}$ groups constructing polytungstate ions in the acidified solution of ammonium tungstate. The $H_2$-TPR results showed that the reduction of precursor of the 3％ $W-H_2SO_4$/HZSM-5 catalyst may occur at temperatures below 900 K,producing W species with mixed valence states, $W^{5+}$ and $W^{4+}$, whereas the reduction of the 3％ W/HZSM-5 occurred mainly at temperatures above 1023 K,producing only one type of dominant W species, $W^{5+}$. The results seem to imply that the observed high DHAM activity on the $W-H_2SO_4$/HZSM-5 catalyst was closely correlated with $(WO_6)^{n-}$ groups with octahedral coordination as the precursor of catalytically active species. Incorporation of Zn (or La) into the $W-H_2SO_4$/HZSM-5 catalyst has been found to pronouncedly improve the activity and stability of the catalyst for DHAM reaction. Over a 2.5％ $W-1.5％ Zn-H_2SO_4$/HZSM-5 catalyst and under reaction conditions of 1123 K,0.1 MPa,and GHSV=1500 ml/(hg-cat.),methane conversion$(X_{CH_4})$ reached 23％ with the selectivity to benzene at～96％ and an amount of coke for 3 h of operation at 0.02％ of the catalyst weight used.

**Key words** Methane Dehydro-aromatization W/HZSM-5 $W-H_2SO_4$/HZSM-5 $W-Zn$(or La)-$H_2SO_4$/HZSM-5 Polytungstates Promoting effect of Zn or La

## 1  Introduction

With the development of a series of new catalysts,nonoxidative dehydrogenation and aromatization of methane(DHAM) to aromatic hydrocarbons has drawn much attention in recent years[1-12]. This process has definite advantages,namely the less complicated technology and easy separation of the

---

* Supported by the National Natural Science Foundation of China.

① To whom correspondence should be addressed.

aromatic hydrocarbon products(benzene, toluene, etc.) from unconverted methane and the by-product hydrogen, which is a very valuable source of hydrogen for petroleum refining industry.

Bragin et al. [1] earlier reported 78% selectivity of benzene formation at 18% conversion of methane at 1023 K over a Pt-CrO$_3$/HZSM-5 catalyst in a pulse reactor. Since then, a few studies have been published on DHAM reaction over Mo/HZSM-5[3,4], MoO$_3$/ZSM-5 or MoO$_3$/SiO$_2$[4,6], Mo$_2$C/ZSM-5[6,7], Mo-W/HZSM-5[8], Mo-Pt/HZSM-5[9], Mo-Ru/HZSM-5[10], and Mo-Co(or Fe)/HZSM-5[11]. Our previous work[12] showed that benzene was formed from methane over a Mo-Zn-H$_2$SO$_4$/HZSM-5 catalyst with average conversion of methane of 19% for 12 h and selectivity to benzene of ~90% at 1018 K (which are slightly higher than the values estimated from thermodynamics due to disregarding part of methane converted to coke in the calculations of methane conversion and selectivity to benzene).

Most of the existing catalysts for DHAM reaction so far are operating at about 973 K, so that the methane conversion achieved is low due to the thermodynamic limitation. Thermodynamic calculations carried out by us indicate that equilibrium conversions of 11.3, 16, 21, 27 and 33% for methane to benzene(i. e., $6CH_4 \rightleftharpoons C_6H_6 + 9H_2$) are predicted at 973, 1023, 1073, 1123 and 1173 K, respectively. This means that the operation temperature as high as ~1073 K is required for methane conversion of ~20% practically. However, under such high temperature, Mo-based catalysts suffer from the disadvantage of serious deactivation by fouling by coke formed in the process of DHAM and by losing of Mo component by sublimation[12]. Therefore, development of DHAM catalysts with high activity and stability at higher temperatures has potential significance for commercial utilization of DHAM process.

In the present work, highly active and heat-resisting W-H$_2$SO$_4$/HZSM-5-based catalysts were developed and studied. Incorporation of Zn(or La) into the W-H$_2$SO$_4$/HZSM-5 catalyst was found to markedly improve the DHAM activity and stability of the catalyst operating under relatively high temperatures.

# 2  Experimental

The catalysts were prepared by the method of incipient wetness, with the HZSM-5 zeolite with a Si/Al molar ratio of 38(obtained from the Chemical Plant of Nankai University)as a carrier. A certain amount of the HZSM-5 zeolite was impregnated with ammonium tungstate aqueous solution containing a calculated amount of W, which was prepared by dissolving(NH$_4$)$_2$WO$_4$(in A. R. grade of purity)in deionized water and adding a little of H$_2$SO$_4$ to regulate the pH value of the solution to 2-3, followed by drying at 383 K for 2 h, and then calcining at 773 K for 4 h; thus, a precursor of the W-H$_2$SO$_4$/HZSM-5 catalyst was obtained. The precursor of the W-Zn-H$_2$SO$_4$/HZSM-5 catalyst was prepared by impregnating the precursor of the W-H$_2$SO$_4$/HZSM-5 catalyst with aqueous solution of ZnSO$_4$(in A. R. grade of purity), followed by drying at 383 K for 2 h, and then impregnating with a solution of NH$_4$OH, drying again at 383 K for 2 h, and finally calcining at 673 K for 4 h. With(NH$_4$)$_2$MoO$_4$ replacing (NH$_4$)$_2$WO$_4$, with La(NO$_3$)$_3$ replacing ZnSO$_4$, and an aqueous solution of(NH$_4$)$_2$SO$_4$ replacing the solution of NH$_4$OH, according to the same procedure of preparation as that for W-Zn-H$_2$SO$_4$/HZSM-5, the precursors of the Mo-Zn-H$_2$SO$_4$/HZSM-5 and the W-La-H$_2$SO$_4$/HZSM-5 catalysts were prepared, respectively. All these catalyst samples were pressed, crushed, and sieved to size of 40—60 mesh.

The evaluation experiment of catalyst activity was performed in a fixed-bed continuous-flow reactor-GC combination system operating under atmospheric pressure. The DHAM reaction over the catalysts was carried out at a stationary state and under the following reaction conditions:1023—1173 K, and feed gas(CH$_4$)hourly-space velocity(GHSV)of 1500 ml/(h g-cat.). 0.6 g of catalyst sample was used each time for testing. Feed gas methane was in 99.9% purity. The reactant and products were analyzed by an on-line 102GD GC equipped with a hydrogen flame ionization detector and a thermal conductivity detector and 2 m long dinonyl phthalate (DNP) column, with nitrogen or hydrogen as carrier gas. The data were taken all at 60 min after the reaction starting, unless otherwise noted. Methane conversion($X_{CH_4}$) and product selectivity($S_{product}$) were calculated upon the carbon number basis, with part of methane converted to coke neglected.

Raman spectra of the precursor solutions of tungstate to be supported were taken at room temperature by using a Spex Ramalog-6 laser Raman spectrometer, with the 514.5 nm line from a Spectra

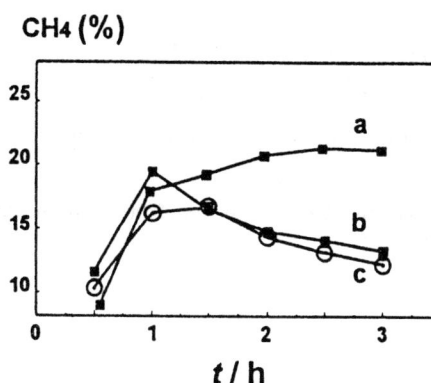

Figure 1.　Stability testing of the catalyst of: (a) 2% W-1.5% Zn-$H_2SO_4$/HZSM-5; (b) 2% Mo-1.5% Zn-$H_2SO_4$/HZSM-5; (c) 2% W-$H_2SO_4$/HZSM-5. Reaction conditions: 1073 K, 0.1 MPa, GHSV=1500 ml/(h g-cat.).

Physics model 164 argon ion laser used as the excitation source. Slit width settings correspond to a resolution of 4 cm$^{-1}$.

Temperature-programmed reduction(TPR)of catalyst in precursor(oxidation)state was conducted on a fixed-bed continuous-flow reactor-GC combination system. A $N_2$-carried 5 vol% $H_2$ gaseous mixture was used as reducing gas; the rate of elevating temperature was 10 K/min; 50 mg of catalyst was used each time for testing.

# 3　Results and discussion

## 3.1　Performance of catalyst for DHAM

The activity of W/HZSM-5 and Mo/HZSM-5(as a contrast)catalysts modified by Zn-$H_2SO_4$ for DHAM reaction was evaluated at 1073 K, respectively. The results are illustrated in figure 1. It can be seen that the 2%(percentage of mass, the same below)W-1.5% Zn-$H_2SO_4$/HZSM-5 catalyst displays rather high activity and stability for DHAM reaction in comparison with the 2% Mo~1.5% Zn-$H_2SO_4$/HZSM-5 catalyst. Under reaction conditions of 1073 K, 0.1 MPa, and GHSV=1500 ml/(h g-cat.), $X_{CH_4}$ reached stably 20.5%, and the total amount of coke for 3 h was 0.02% of the catalyst weight used. In addition to benzene and toluene, trace $C_{8+}$ aromatic hydrocarbons, including ethylbenzene, dimethylbenzene and naphthalene, can be found in the products(see sample #6 in table 1). Whereas over the Mo-Zn-$H_2SO_4$/HZSM-5 catalyst(sample #11 in table 1)and under the same conditions of reaction, $X_{CH_4}$ reduced gradually from 19% at the first hour down to 13% after 3 h of DHAM operation, predicting that deactivation of the catalyst had occurred to a certain extent, very likely due to fouling by coke formed in the reaction process(a total amount of coke for 3 h equal to 0.5% of the catalyst weight detected)and losing of Mo component by sublimation(a pale-yellow deposit of Mo component on the inner wall of exit tube of the reactor observed).

Figure 2 showed the effect of amount of W loading on DHAM performance of the W-based catalyst. It can be seen that both $X_{CH_4}$ and $S_{ben.}$ increased initially with increasing amounts of W loading, and tended toward stable levels at W loading of 3.5%. It seems that an amount of W loading no less than 2.5% would be appropriate for the HZSM-5 carrier.

**1063**

Figure 3 illustrates the effect of temperature on $X_{CH_4}$ and Sben. over the 2.5% W~1.5% Zn-$H_2SO_4$/HZSM-5 catalyst for DHAM reaction. $X_{CH_4}$ increased with elevating temperature from 10.7% at 973 K up to 20.2% at 1073 K and reached 23.2% at 1173 K, and the $S_{ben.}$ in this region of temperature is almost maintained at a level of about 95% (see sample #7 in table 1).

**Table 1**
**The results of activity assays of the catalysts with different compositions for dehydro-aromatization of methane.** [a]

| Sample No. | Catalyst | $T$ (K) | $X_{CH_4}$ (%) | Selectivity[b] (%) | | | | |
| --- | --- | --- | --- | --- | --- | --- | --- | --- |
| | | | | $C_2$ | $C_3$ | Ben. | Tol. | $C_8$ |
| 1 | HZSM-5 | 973 | 0.1 | 88.2 | — | 11.8 | — | — |
| 2 | $H_2SO_4$/HZSM-5 | 973 | 0 | — | — | — | — | — |
| 3 | Zn/HZSM-5 | 973 | 1.0 | 19.6 | 1.3 | 69.9 | 9.2 | — |
| 4 | 2.0% W/HZSM-5(pH[c]=7—9) | 1023 | 0 | — | — | — | — | — |
| | | 1073 | 5.7 | — | — | 94.3 | 4.7 | ~1 |
| 5 | 2.0% W-$H_2SO_4$/HZSM-5(pH=2—3) | 1023 | 7.1 | — | — | 94.2 | 4.8 | ~1 |
| | | 1073 | 16.7 | — | — | 94.7 | 4.3 | ~1 |
| | | 1123 | 21.4 | — | — | 95.5 | 3.5 | ~1 |
| 6 | 2.0% W-1.5% Zn-$H_2SO_4$/HZSM-5(pH=2—3) | 1073 | 20.5 | — | — | 96.0 | 3.0 | ~1 |
| 7 | 2.5% W-1.5% Zn-$H_2SO_4$/HZSM-5(pH=2—3) | 973 | 10.7 | — | — | 92.3 | 6.7 | ~1 |
| | | 1073 | 20.2 | — | — | 94.2 | 4.8 | ~1 |
| | | 1123 | 23.0 | — | — | 96.7 | 2.3 | ~1 |
| | | 1173 | 23.2 | — | — | 97.0 | 2.0 | ~1 |
| 8 | 2.5% W-1.5% La-$H_2SO_4$/HZSM-5(pH=2—3) | 1123 | 22.3 | — | — | 94.6 | 4.4 | ~1 |
| 9 | 4.0% Mo-$H_2SO_4$/HZSM-5 | 1023 | 12.5 | — | — | 91.8 | 7.2 | ~1 |
| 10 | 4.0% Mo-1.0% Zn-$H_2SO_4$/HZSM-5 | 1023 | 15.8 | — | — | 92.6 | 6.4 | ~1 |
| 11 | 2.0% Mo-1.5% Zn-$H_2SO_4$/HZSM-5 | 1073 | 19.0 | — | — | 93.8 | 5.2 | ~1 |

[a] Reaction conditions: 0.1 MPa, GHSV = 1500 ml($CH_4$)/(h g-cat.). All data were taken at 60 min after reaction starting.

[b] Trace $C_{8+}$ aromatic hydrocarbons(ethylbenzene, dimethylbenzene, naphthalene, etc.)can be found in the products; the selectivity data listed here are slightly higher than realities due to disregarding part of methane converted to coke in the calculations of $X_{CH_4}$ and selectivities.

[c] The pH value of the preparative solution of ammonium tungstate.

## 3.2 Effect of preparation condition on the catalyst activity

It has been found experimentally that selecting an appropriate pH value of the precursor solution of tungstate to be supported is one of crucial factors for preparation of highly active W/HZSM-5-based catalysts for DHAM. The result of a comparative testing demonstrated that a W/HZSM-5 catalyst prepared by impregnating the HZSM-5 carrier with the aqueous solution of ammonium tungstate with pH = 7—9, which was prepared by dissolving ammonium tungstate in de-ionized water or mixing tungstic acid and aqua ammonia, showed very little activity for DHAM reaction at the temperature of 973—1023 K, which is consistent with the result reported by Wang et al.[8]. Whereas over a W-$H_2SO_4$/HZSM-5 catalyst, prepared from a $H_2SO_4$-acidified aqueous solution of ammonium tungstate with the

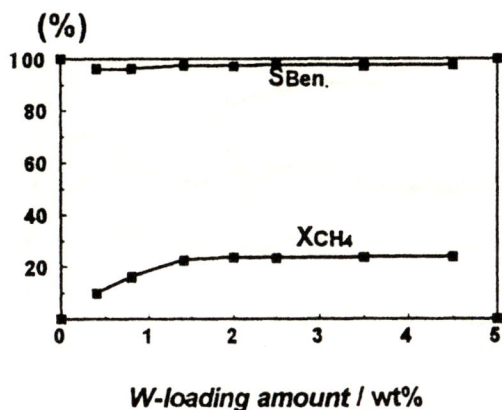

**Figure 2.** Effect of W loading in W-1.5% $Zn$-$H_2SO_4$/ HZSM-5 on the catalyst performance. Reaction conditions: 1123 K, 0.1 MPa, GHSV=1500 ml/(hg-cat.).

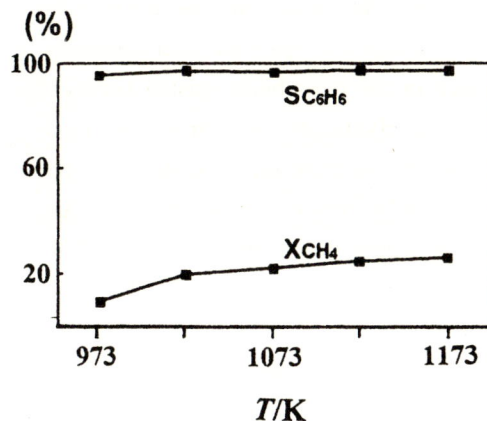

**Figure 3.** Effect of temperature on the performance of 2.5% W~1.5% $Zn$-$H_2SO_4$/HZSM-5 catalyst. Reaction conditions: 0.1 MPa, GHSV=1500 ml/(hg-cat.).

pH=2—7, both $X_{CH_4}$ and $S_{ben}$. were dramatically enhanced(see table 1, sample #4 vs. sample #5). It could be inferred that the roles that addition of a small amount of $H_2SO_4$ plays were probably not only in enhancing the solubility of $(NH_4)_2WO_4$ in water so as to favor an increase in W loading amount in single-pass operation of impregnation, more importantly, also in resulting in formation of the polytungstates in the precursor solution via the reactions as follows:

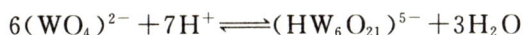

$$6(WO_4)^{2-}+7H^+ \rightleftharpoons (HW_6O_{21})^{5-}+3H_2O$$

or/and

$$12(WO_4)^{2-}+14H^+ \rightleftharpoons (H_2W_{12}O_{42})^{10-}+6H_2O$$

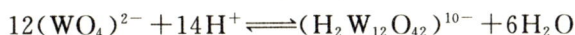

Lowering the pH value of the precursor solution would shift the equilibria of the above reactions to the right. It has been known that these polytungstate ions are built of octahedral groups $(WO_6)^{n-}$ sharing edges, which is obviously different from $(WO_4)^{2-}$ ions with tetrahedral coordination[13]. The above result seems to imply that there may exist some correlation between the $(WO_6)^{n-}$ species with octahedral coordination derived from these polytungstate ions and the high DHAM activity of the catalyst.

Laser Raman spectra of the precursor solution provide further evidence for the existence of the polytungstate ions. In figure 4(a) is shown the Raman spectrum of an $NH_4OH$ alkalified solution of $(NH_4)_2WO_4$ with pH=8—9, in which a Raman peak assignable to the symmetric stretching mode of tungstic radical ions $WO^{2-}$ with tetrahedral coordination($T_d$) was observed at 931 $cm^{-1}$[14]. For the neutral solution of $(NH_4)_2WO_4$ with pH=7, as shown in figure 4(b), the peak at 931 $cm^{-1}$ was obviously weakened, and simultaneously, a new peak appeared at 945 $cm^{-1}$, which may be due to the W=O stretching mode of tungstic acid molecules $H_2WO_4$ with $C_{2v}$ symmetry. This means that tungstic radical ions coexisted with tungstic acid molecules:

$$2NH_4^+ +(WO_4)^{2-}+2H_2O \rightleftharpoons 2NH_4OH+H_2WO_4$$

With pH value of the solution lowering to 2—3 by acidification of $H_2SO_4$, the peaks ascribed to discrete monotungstate species disappeared, and meanwhile, two new strong peaks at 983 and 971 $cm^{-1}$ were observed(see figure 4(c)). The peak at 983 $cm^{-1}$ is evidently due to the symmetric stretching mode of $SO_4^{2-}$ ion with $T_d$ symmetry[15]. The peak at 971 $cm^{-1}$ may originate from the stretching vibration of terminal W = O of polytungstate species such as $(HW_6O_{21})^{5-}$ and $(H_2W_{12}O_{42})^{10-}$, which is quite analogous to the case of polymolybdates[16].

On the other hand, the results of $H_2$-TPR investigation showed that obvious difference in the reducibility existed between the two catalysts prepared from the two solutions of ammonium tungstates with different pH values, respectively. As shown in figure 5(a), the main $H_2$-TPR peak observed on the 3% W/HZSM-5 sample was present at 1073 K. In view of the result of EPR measurement carried out by us[17] that the $H_2$ reduction of this catalyst at 1073 K for 30 min produced only one type of $W^{5+}$ species with a strong EPR signal ($g_\parallel = 1.82/g_\perp =$

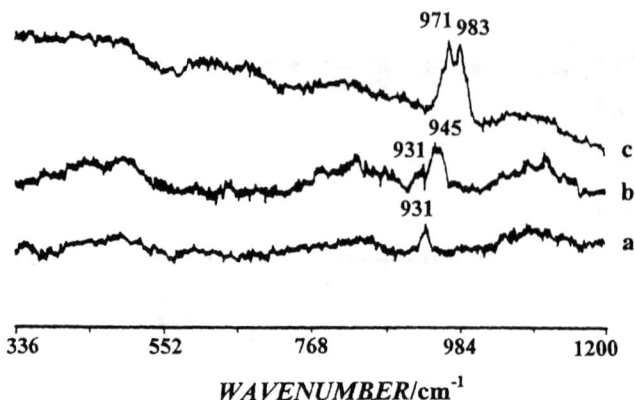

**Figure 4.** Raman spectra of the precursor solutions of: (a)$(NH_4)_2WO_4$ alkalified by $NH_4OH$ with $pH = 8-9$; (b)$(NH_4)2WO_4$ with $pH = 7$; (c)$(NH_4)_2WO_4$ acidified by $H_2SO_4$ with $pH = 2-3$.

1.95)and the intensity of this signal did not reduce with prolonging time for reduction(indicating that the further reduction of this type of $W^{5+}$ species was quite difficult), the $H_2$-TPR peak at 1073 K may be mainly ascribed to single-electron reduction of the $W^{6+}$ species derived from the$(WO_4)^{2-}$ ions, $W^{6+} + e^- \longrightarrow W^{5+}$. On the 3% $W$-$H_2SO_4$/HZSM-5, the observed three distinct peaks were present at 819, 891 and 1073 K, respectively. The peak at 1073 K is quite similar to the peak at 1073 K for the W/

HZSM-5 and, evidently, ascribable to the single-electron reduction of the same type of $W^{6+}$ species derived from $(WO_4)^{2-}$ ions, whereas the peaks at 819 and 891 K are very probably due to the sequent two steps of single-electron reduction of another type of $W^{6+}$ species, most probably derived from $(WO_6)^{n-}$ groups constructing polytungstate ions, $W^{6+} + e^- \longrightarrow W^{5+}$ and $W^{5+} + e^- \longrightarrow W^{4+}$. The area of the second peak is about 70% of that of the first one, implying that only about 70% of these $W^{5+}$ species produced from the first step of reduction can be further reduced to $W^{4+}$. This is consistent with the results of our EPR observation[17], which showed that $H_2$ reduction of the W-$H_2SO_4$/HZSM-5 sample at 1073 K for 15 min produced two types of $W^{5+}$ species with two strong EPR signals, corresponding to $g_\parallel^{(1)} =$

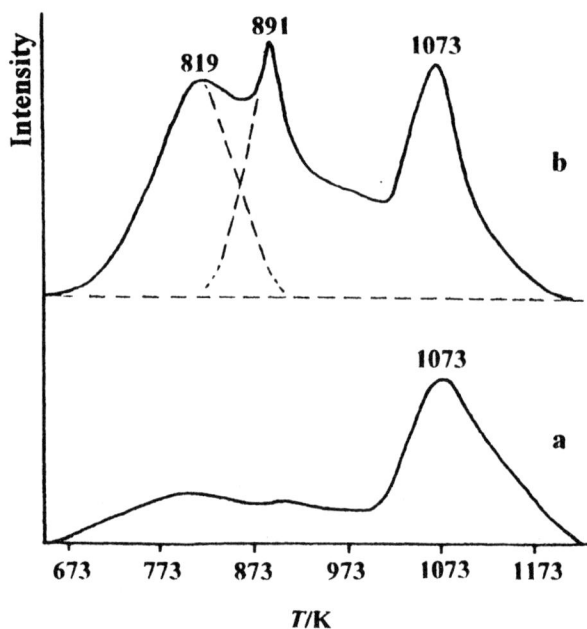

**Figure 5.** $H_2$-TPR spectra of the catalysts(in oxidation state) of: (a)3% W/HZSM-5; (b)3% W-$H_2SO_4$/HZSM-5.

$1.82/g_\perp^{(1)} = 1.95$ and $g_\parallel^{(2)} = 1.88/g_\perp^{(2)} = 1.97$, respectively, and the intensity of the latter EPR signal was greatly reduced with the time for reduction prolonged to 60 min, implying that a considerable proportion of the second type of $W^{5+}$ species was further reduced to $W^{4+}$ valence state of EPR silence.

Thus, it seems to us that, on the W-$H_2SO_4$/HZSM-5 catalyst, the observed high DHAM activity may originate from the pronounced reducibility of the supported W species in the catalyst precursor, of which a large proportion was derived from$(WO_6)^{n-}$ groups rather than$(WO_4)^{2-}$ ions; these $W^{6+}$ species derived from $(WO_6)^{n-}$ groups may be reduced at relatively low temperature, and a considerable proportion of them may be reduced to the valence state of $W^{4+}$ under the reaction conditions for DHAM.

## 3.3 The promoting effect of Zn, or La on DHAM activity of the *W-H₂SO₄/HZSM*-5 catalyst

It is experimentally found that the DHAM activity of W-H$_2$SO$_4$/HZSM-5 catalyst can be improved by addition of Zn, or La, as an additive. The results of activity assays of several catalysts with different compositions for dehydro-aromatization of methane are summarized in table 1. The HZSM-5 zeolite acidified by H$_2$SO$_4$ was completely inactive for methane conversion. The W-H$_2$SO$_4$/HZSM-5 (sample # 5) catalyst prepared from a solution of ammonium tungstate acidified by H$_2$SO$_4$ displayed relatively high activity for the methane conversion, but deactivated gradually, as shown in figure 1(c). Whereas over a Zn-promoted catalyst, 2.5% W ~ 1.5% Zn-H$_2$SO$_4$/HZSM-5 (sample # 7), methane conversion was pronouncedly enhanced, reaching 23% with a selectivity of 96.7% to benzene under reaction conditions of 1123 K, 0.1 MPa, and GHSV=1500 ml/(hg-cat.).

Addition of a small amount of Zn to the Mo-H$_2$SO$_4$/HZSM-5 catalyst can also improve the performance of the catalyst for DHAM reaction to a certain extent, as evidenced by the result of the activity assay of samples # 9 and # 10 at 1023 K shown in table 1. However, the losing of Mo component by sublimation would inevitably lead to deactivation of the catalyst under the reaction temperature as high as 1073 K and above, as shown in figure 1(b).

Incorporating La as an additive was also found to significantly enhance DHAM activity of the W-H$_2$SO$_4$/HZSM-5 catalyst, but its promoting effect was somewhat not so remarkable as that of Zn. The experimental results also showed that, in order to gain the promoting effect as large as possible, an optimum match of W with the promoter, Zn or La, in their loading amounts was requisite.

By comparison of the methane conversions and the selectivities over the singly and doubly acid-promoted W/HZSM-5 catalysts, it is easy to find that, with the addition of Zn(or La) promoter, not only is the DHAM activity of the catalyst improved significantly, but the catalyst stability is also prolonged. As shown in figure 1, after 3 h of the operation, the Zn-promoted W-H$_2$SO$_4$/HZSM-5 catalyst (sample # 6) maintained the methane conversion still at ~20%, whereas the singly acid-promoted catalyst W-H$_2$SO$_4$/HZSM-5 (sample # 5) deactivated gradually. It seems to us that the improvement of the activity and stability of the catalyst by addition of Zn(or La) promoter originates probably from the following factors: on the one hand, the presence of a Zn or La component in tungsten oxide matrix would be favorable to alleviate coke deposition and to inhibit aggregation of W-containing active species and formation of WO$_3$ crystallites at the surface of catalyst, and, on the other hand, the presence of a sulfate radical in the form of Zn(or La) sulfate would be in favor of stabilization of the B-acid sites by formation of hydrogen-sulfate radical($HSO_4^-$) and, therefore, would tend to reduce, if not avoid, losing of B-acid at the surface of the catalyst. For better understanding of the mechanism of the promoter action, more works, especially more detailed knowledge about the nature of catalytically active sites, are needed.

# 4  Conclusions

(1) The W-H$_2$SO$_4$/HZSM-5 catalyst prepared by impregnating the HZSM-5 zeolite (Si/Al = 38) carrier with an aqueous solution with pH=2—3 of ammonium tungstate acidified by H$_2$SO$_4$ has been demonstrated to be an highly effective catalyst for methane nonoxida-tive dehydro-aromatization. The

high DHAM activity of the W-H$_2$SO$_4$/HZSM-5 catalyst for methane nonoxidative conversion seems to be closely associated with (WO$_6$)$^{n-}$ groups with octahedral coordination as the precursor of cat-alytically active species.

(2) The incorporation of Zn (or La) into the W-H$_2$SO$_4$/ HZSM-5 catalyst is found to markedly enhance the activity and stability of the catalyst for methane nonoxidative dehydro-aromatization. The W-Zn-H$_2$SO$_4$/HZSM-5 catalyst is demonstrated to be a more promising catalyst for nonoxidative dehydro-aromatization of methane with its high activity and good performance of heat- and coking-resistance.

# References

[1] O. V. Bragin, T. V. Vasina, A. V Preobrazhenskii and K. M. Minachev, Izv. Akad. Nauk SSSR, Ser. Khim. (1982)954;(1989)750.

[2] T. Koerts, M. J. A. G. Deelen and R. A. van Santen, J. Catal. **138**(1992)101.

[3] L. Wang, L. Tao, M. Xie, G. Xu, J. Huang and Y. Xu, Catal. Lett. **21**(1993)35.

[4] D. J. Wang, J. H. Lunsford and M. P. Rosynek, J. Catal. **169**(1997)347.

[5] F. Solymosi, A. Erdohelyi and A. Szoke, Catal. Lett. **32**(1995)43.

[6] F. Solymosi, J. Cserenyi, A. Szoke, T. Bansagi and A. Oszko, J. Catal. **165**(1997)150.

[7] F. Solymosi, A. Szoke and J. Cserenyi, Catal. Lett. **39**(1996)157.

[8] S. Wong, Y. Xu, L. Wang, S. Liu, G. Li, M. Xie and X. Guo, Catal. Lett. **38**(1996)39.

[9] L. Chen, L. Lin, Z. Xu, T. Zhang and X. Li, Catal. Lett. **39**(1996)169.

[10] Y. Shu, Y. Xu, S. Wang, L. Wang and X. Guo, J. Catal. **170**(1997)11.

[11] S. Liu, Q. Dong, R. Ohnishi and M. Ichikawa, Chem. commun. (1997)1455.

[12] J. L. Zeng, Z. T. Xiong, G. D. Lin, H. B. Zhang and K. R. Tsai, in: 11th ICC, Baltimore, 1996, Po −158; J. Xiamen Univ. **35**(1996)900.

[13] A. F. Well, Structural Inorganic Chemistry, 5th Ed. (Clarendon Press, Oxford, 1984) p. 519, and references therein.

[14] N. Weinstock, H. Schulze and A. Muller, J. Chem. Phys. **59**(1973)5063.

[15] Landolt-Bornstein, Physikalisch-Chemische Tabellen, **2**(1951).

[16] H. Knozinger and H. Jeziorwski, J. Phys. Chem. **82**(1978)2002; **83**(1979)1166.

[17] J. L. Zeng, Z. T. Xiong, L. J. Yu, G. D. Lin and H. B. Zhang, J. Xiamen Univ. **37**(1998), in press.

■ 本文原载:Chemical Physics Letters 286(1998),pp. 163~170.

# Structural Properties of $[(AuPH_3)_6Pt(H_2)(PH_3)]^{2+}$: Theoretical Study of Dihydrogen Activation

X. Xu [a], Y. Z. Yuan [a], K. Asakura [b], Y. Iwasawa [c], H. L. Wan [a,①], K. R. Tsai [a]

*a State Key Lab for Physical Chemistry of Solid Surfaces,*
*(Department of Chemistry, Xiamen University, Xiamen 361005, China*
*b Research Center for Spetrochemistry, Faculty of Science,*
*Universiry of Tokyo. Hongo, Bunkyo-ku, Tokyo 113, Japan*
*c Department of Chemistry, Graduate School of Scince,*
*University of Tokyo, Hongo, Bunkyo-ku, Tokyo 113, Japan)*

Received 13 October 1997; in final form 9 January 1998

**Abstract** B3LYP studies have been performed on the phosphine-stabilized gold-platinum clusters by using model clusters such as $[(AuPH_3)_6 Pt(PH_3)]^{2+}$ and $[(AuPH_3)_6 Pt(H_2)(PH_3)]^{2+}$. Two stationary points have been located, one dihydrogen complex and the other dihydrido complex, which are the possible intermediates in the $H_2$—$D_2$ equilibration. Pt is characterized as being the active site, but Au atoms also play an important role in the activation of $H_2$. The electron transfer from the metal core to the $\sigma^*$ anti-bonding orbital of $H_2$ not only activates the hydrogen molecule, but also induces large metal core movement which provides open Au sites for bonding of the second $H_2$ or $D_2$. ⓒ 1998 Elsevier Science B. V.

## 1　Introduction

A wide variety of phosphine-stabilized, gold-platinum cluster compounds have been synthesized(see for example Ref. [1]). These clusters are unusual in that they are effective catalysts for the $H_2$—$D_2$ equilibration reaction($H_2 + D_2 = HD$) and they catalyze $H_2$—$D_2$ equilibration without any H/D exchange with solvent or ligand hydrogen atoms. Among this class of cluster compounds, $[(AuPPh_3)_6Pt(PPh_3)]$ $(NO_3)_2^{[2,3]}$ and $[(AuPPh_3)_8Pt](NO_3)_2^{[4]}$ with the structure of a single Pt atom embedded in the$(Au)_n$ ensemble have been most extensively studied. The Pt atom has been referred to as the catalytic site. This is contrary to the results reported for Pt metallic catalysts where multi-sites of Pt atoms are believed to be demanded for dihydrogen activation and $H_2$—$D_2$ equilibration. The mono-nuclear Pt complex Pt $(PPh_3)_4$ and pure Au cluster $[Au_9 (PPh_3)_8](NO_3)_2$ have been found to be inactive for $H_2$—$D_2$ equilibration[3]. Thus it is of interest to examine theoretically the catalytic feature of those clusters.

Both $[(AuPPh_3)_6Pt(PPh_3)](NO_3)_2$ and $[(AuPPh_3)_8 Pt](NO_3)_2$ have been shown to rapidly and

---

① Corresponding author. E-mail:hlwan @ xmu. edu. cn.

reversibly react with H₂ according to Eq. 1[3-5].

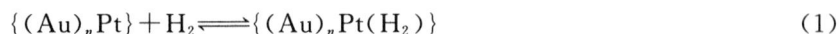

$$\{(Au)_nPt\} + H_2 \rightleftharpoons \{(Au)_nPt(H_2)\} \qquad (1)$$

Although $\{(Au)_nPt(H_2)\}$ is of obvious importance in the $H_2$—$D_2$ equilibration as it is likely to be an intermediate in the catalytic process, no structure information of this complex has yet been reported. On the $SiO_2$ supported $\{(Au)_6$—$Pt\}$ catalyst[6], an in situ EXAFS analysis has been performed. Unfortunately, the $H_2$ contribution to the EXAFS spectra was in an error level of EXAFS analysis. However, detailed NMR studies have been carried out with $\{(Au)_nPt\}$ ($n = 6, 7, 8$) clusters in solution[5]. In the case of $[(AuPPh_3)_8Pt](NO_3)_2$, reaction (I) was directly observed by [31]P{[1]H} NMR[5]. The triple Pt—H coupling pattern observed in the phosphorus decoupled [195]Pt NMR spectrum clearly indicated the presence of two hydride ligands. Since no terminal Pt—H stretching vibrations were detected by an IR spectrum, a bridging bonding mode to the hydride ligands has been assigned[7]. In the case of $[(AuPPh_3)_6Pt(PPh_3)](NO_3)_2$ in $CD_2Cl_2$ solution, however, no changes in the NMR spectra have been observed[5], In a basic solution like pyridine, deprotonation from $\{(Au)_n$—$Pt(H_2)\}$ to the monohydrido cluster $\{(Au)_n$—$Pt(H)\}$ was directly observed by[31] P{[1]H} NMR. This deprotonation reaction provided evidence that $H_2$ does react and become activated with $[(AuPPh_3)_6Pt(PPh_3)]$ $(NO_3)_2$[4,8].

In this Letter, we have performed B3LYP calculations on model clusters of $[(AuPH_3)_6Pt(PH_3)]^{2+}$ and $[(AuPH_3)_6Pt(H_2)(PH_3)]^{2+}$, trying to address the interesting catalytic features of these gold-platinum cluster compounds. Pt is found to be the active site with important assistance from Au atoms. The metal core movement should play an important role in the $H_2$—$D_2$ equilibration.

## 2  Computational details

A relativistic effective core potential (RECP) has been employed to replace the 1s to 4d core electrons of gold or platinum while the electrons arising from the $5s^2, 5p^6, 5d^{9/10}$ and $6s^1$ shells are treated explicitly[9]. The Gaussian-type orbital(GTO) basis set for the RECP of Au or Pt includes 5s, 6p, and 3d primitive functions contracted to [3s3p2d]. This basis set is referred to as LanL2DZ in GAUSSIAN-94[10]. The is to 2p core of phosphorus has been replaced with the ECP of Stevens et al. [11]. The GTO basis set for the ECP of P is CEP-31G*[11], with one d polarization function being added since the phosphorus d orbital is believed to be important for the bonding between the phosphine and the transition metal[12]. As is common practice[13,14], the $PH_3$ ligand has been used to model the effect of $PPh_3$. The basis set for hydrogen in $PH_3$ is STO-3G[10]. For the $H_2$ molecule, we use the(4s)/ (2s) set of Huzinaga-Dunning[15,16], plus p-type functions which are the first derivatives of the (2s) set[17]. Then, the Hellmann-Feynman theorem is essentially satisfied for the forces acting on the hydrogen nuclei.

In the density functional theory(DFT) we use the B3LYP functional[18], which is a hybrid functional including a mixture of Hartree-Fock exchange with DFT exchange-correlation. Full geometry optimizations have been carried out with GAUSSIAN-94[10] on the model clusters of $[(AuPH_3)_6Pt(PH_3)]^{2+}$ and $[(AuPH_3)_6Pt(H_2)(PH_3)]^{2+}$. No symmetry restriction is applied in all calculations. All model compounds are assumed to be singlet, since preliminary calculations indicate that the ground states are singlets.

# 3　Results and discussion

## 3. 1　Geometry of $[(AuPH_3)_6Pt(PH_3)]^{2+}$

It is well-known that phosphine ligands, like triphenylphosphine(PPh₃), play a principal role in the stabilization of gold clusters. Due to their size, however, these cluster compounds pose quite a challenge to quantum chemical methods and are commonly described by model compounds with simplified phosphine ligands[13,14]. PH₃ differs from PPh₃. in both steric and electronic effects. Naked gold clusters can hardly he stabilized unless phosphine ligands are present and no gold cluster with PH₃ ligands really exists. Therefore it would be anticipated that clusters with PPh₃ should be more stable than those with PH₃ The replacement of PPh₃ ligands with PH₃ would lead to a weaker metal-metal bonding and thus a longer metal-metal distance.

Table 1　Selected bond distances(Å)for $[(AuPH_3)_6Pt(PH_3)]^{2+}$ (1)

| | |
|---|---|
| Pt—Au1 | 2. 759(2. 693) |
| Pt—Au2 | 2. 797(2. 673) |
| Pt—Au3 | 2. 800(2. 699) |
| Pt—Au4 | 2. 725(2. 667) |
| Pt—Au5 | 2. 769(2. 665) |
| Pt—Au6 | 2. 769(2. 672) |
| Pt—P | 2. 331(2. 28) |
| Au1—Au2 | 3. 034(2. 912) |
| An1—Au3 | 3. 035(2. 893) |
| Au1—Au5 | 3. 023(2. 927) |
| Au1—Au6 | 3. 021(2. 824) |
| Au2—Au3 | 2. 841(2. 786) |
| Au2—Au4 | 3. 012(2. 938) |
| Au3—Au4 | 3. 003(2. 792) |
| Au5—Au6 | 2. 815(2. 723) |
| Au1—P1 | 2. 417(2. 32) |
| Au2—P2 | 2. 379(2. 30) |
| Au3—P3 | 2. 379(2. 28) |
| Au4—P4 | 2. 401(2. 30) |
| Au5—P5 | 2. 3S3(2. 29) |
| Au6—P6 | 2. 383(2. 29) |

Distances longer than 4. 21 Å are not included. Crystal data are shown in parentheses. The total energy is— 989. 994841 au.

Fig. 1 presents the optimized structure of $[(AuPH_3)_6Pt(PH_3)]^{2+}$ (1). A summary of the geometric data for selected distances is given in Table 1. As is shown in Fig. 1, the structure of the metal core is made up of a trigonal bipyramid(Pt, Aul, Au2, Au3, Au4)and a tetrahedron(Pt, Aul. Au5, Au6), which are fused together on the Pt—Au I bond. Qualitatively, Fig. 1 is in good agreement with the experimental findings from X-ray diffraction analysis of the $[(AuPPh_3)_6Pt(PPh_3)](NO_3)_2 \cdot Et_2O$

crystal[2] and EXAFS analysis of $[(AuPPh_3)_6 Pt(PPh_3)]$ $(NO_3)_2$ supported on $SiO_2$[6], The optimized Au—Pt bond distances range from 2.725 to 2.800 Å, with an average Au—Pt bond distance of 2.770 Å. These can be compared with the crystal data (average 2.678 Å, range 2.665 — 2.699 Å)[2] and the EXAFS results (average $2.69 \pm 0.03$ Å)[6]. All these are within the range of values 2.600 — 3.028 Å, generally observed in other Au—Pt clusters mainly with PPh$_3$ ligands[1]. The calculated Au—Au bond distances range from 2.841 to 3.035 Å, with an average Au—Au bond distance of 2.973 Å. These can be compared with the crystal data (average 2.849 Å, range 2.723— 2.938 Å)[2] and the EXAFS results (average $2.83 \pm 0.03$

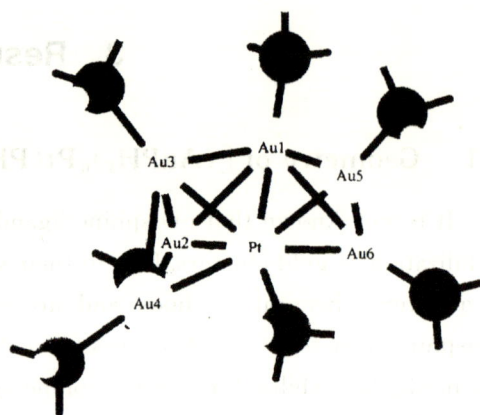

**Fig. 1   Optimized geometry for**
$[(AuPH_3)_6 Pt(PH_3)]^{2+}$ **(1).**

Å)[6]. All these are within the range of values 2.593—3.222 Å, generally observed in other Au—Pt clusters mainly with PPh$_3$ ligands[1]. The optimized Pt—P bond distance (2.331 Å) and Au—P bond distances (average 2.391 Å, range 2.379—2.417 Å) are within the range of values observed in gold (platinum)-phosphine complexes[1]. Au1 and Au4 are on the axial positions of the trigonal bipyramid. The Pt—Au1-P1 angle and the Pt—Au1—P1 angle are found to be the largest among all the Pt—Au—P angles, reproducing the observed data. Experimentally, the overall Pt—Au—P angle is 162.5°(range 153.1~175.6°)[2], while the optimized one is 160.6°(range 153.0—178.9°).

Although it has been concluded that for structural properties the PH$_3$ ligand provides a satisfactory model of the full PPh$_3$ ligand[19], our calculation results demonstrate an overall expansion of the bond distances in the model clusters. The correlation effects are found to play an important role in stabilizing the complex. The bond distances in the RHF optimized structure are generally 0.1 Å longer than those with B3LYP method. In general, we would like to conclude that the model compound with pH$_3$, provides a reasonable modeling of Au—Pt complexes with PPh$_3$ ligands in the present methodology of calculation.

## 3.2   Geometry of $[(AuPH_3)_6 Pt(H_2)(PH_3)]^{2+}$

Searching for the equilibrium geometry of the model cluster $[(AuPH_3)_6 Pt(H_2)(PH_3)]^{2+}$ proved to be nontrivial. We started our optimization by assuming that the H$_2$ adduct might be in an end-on form, similar to the CO adduct found in the literature[2]. However, the optimization leads to a free H$_2$ plus **1**, indicating that H$_2$ can hardly bind to a metal in an end-on form. Fig. 2 presents the optimized structure of $[(AuPH_3)_6 Pt(H_2)(PH_3)]^{2+}$ **(2)**. The H$_2$ is bonded sideways to the Pt atom with Pt-H bonds of 1.949 and 1.933 Å. Since the bond distance

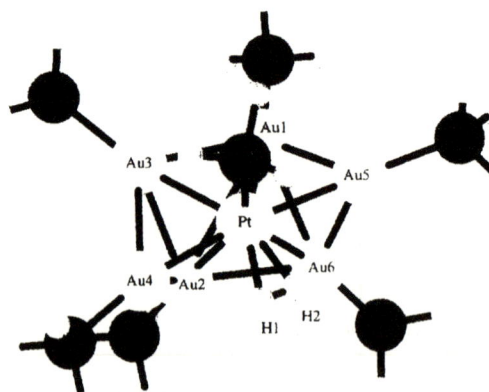

**Fig. 2   Optimized geometry for** $[(AuPH_3)_6 Pt(H_2)$ **$(PH_3)]^{2+}$ (2).**

between the two hydrogens is 0.810 Å (vs. 0.741 Å in free H$_2$[20]), **2** is plausible as a dihydrogen complex. Detailed geometric data for selected distances are summarized in Table 2. As is shown in Fig. 2, the structure of the metal core is grossly similar to the crystal structure of the CO adduct[2]. compared with the geometry of **1**, the bond distance of Au2—Au6 changes from 4.207 to 2.980 Å, indicating the formation of one additional Au2—Au6 bonding interaction; while the PPh$_3$ bonded to the Pt atom rotates up towards the Au1 atom(the calculated Au1—Pt—P bond angle changes from 156. 0 to 103. 3°), leaving space for the H$_2$ ligand. The optimized Au—Pt bond distances range from 2. 725 to 2. 796 Å, with an average Au—Pt bond distance of 2. 759 Å. The calculated Au—Au bond distances range from 2. 944 to 3. 538 Å, with an average Au—Au bond distance of 3. 062 Å. The optimized Pt—P bond distance is 2. 405 Å and the average Au—P bond distance is 2. 390 Å(range 2. 375—2. 400 Å). All these are within the range of values generally observed in other Au—Pt clusters mainly with PPh$_3$ ligands[1]. Interestingly, the Au1—Au2 bond lengthens from 3. 034 to 3. 538 Å, showing the weakening of Au1— Au2 bond.

**Table 2  Selected bond distance(Å)for $[(AuPH_3)_6Pt(H_2)(PH_3)]^{2+}$ (2)**

| | |
|---|---|
| Pt—Au1 | 2. 725 |
| Pt—Au2 | 2. 751 |
| Pt—Au3 | 2. 748 |
| Pt—Au4 | 2. 769 |
| Pl—Au5 | 2. 762 |
| Pt—Au6 | 2. 796 |
| Pt—P | 2. 405 |
| Pt—H1 | 1. 949 |
| Pt—H2 | 1. 933 |
| Au1—Au2 | 3. 538 |
| Au1—Au3 | 3. 113 |
| Au1—Au5 | 2. 998 |
| Au1—Au6 | 2. 991 |
| Au2—Au3 | 3. 025 |
| Au2—Au4 | 2. 999 |
| Au2—Au6 | 2. 980 |
| Au3—Au4 | 2. 944 |
| Au5—Au6 | 2. 967 |
| Au1—P1 | 2. 375 |
| Au2—P2 | 2. 400 |
| Au3—P3 | 2. 377 |
| Au4—P4 | 2. 390 |
| Au5—P5 | 2. 392 |
| Au6—P6 | 2. 392 |
| Au2—H1 | 3. 3% |
| Au2—H2 | 3. 706 |
| Au4—H1 | 2. 672 |
| Au4—H2 | 3. 388 |
| Au5—H1 | 3. 654 |
| Au5—H2 | 2. 974 |
| Au6—H1 | 3. 325 |
| Au6—H2 | 2. 938 |
| H1—H2 | 0. 810 |

Distances longer than 4. 21 Å are not included. The total energy is—991. 167586 au.

Fig. 3 presents another optimized structure of $[(AuPH_3)_6 Pt(H_2)(PH_3)]^{2+}$ (**3**). In this stationary point, the two hydrogens are separated by 2.468 Å. Thereby, **3** is a dihydrido complex where $H_2$ has been highly activated. Detailed geometric data for selected distances are summarized in Table 3. The optimized Au—Pt bond distances range from 2.713 to 2.793 Å, with an average Au—Pt bond distance of 2.780 Å. These values are normal, within the range of values generally observed experimentally[1]. The calculated Au—Au bond distances change quite strongly. The Au1—Au2 bond is greatly weakened (Au1—Au2, 3.753 Å); while the Au1—Au6 distance is too long (5.097 Å) to make any bond. The other calculated

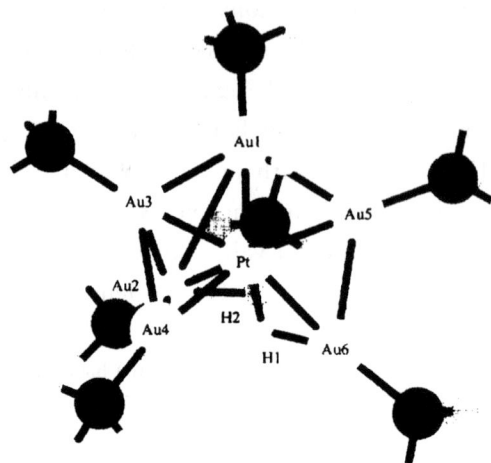

Fig. 3 Optimized geometry for $[(AuPH_3)_6 Pt(H_2)$ $(PH_3)]^{2+}$ (**3**).

Au—Au bond distances range from 2.960 to 3.133 Å, which are similar to the values 2.593~3.222 Å, generally observed in other Au—Pt clusters[1]. From the comparison between the optimized geometry and the crystal data of $[(AuPPh_3)_6 Pt(PPh_3)]^{2+[2]}$, we may conclude that the present methodology is acceptable for reliable prediction of geometry changes. Thereby we believe that in this dihydrido complex, the Au1—Au2 and Au1—Au6 bonds are at least partly broken. It is clear from the bond distances listed in Table 3 that H1 is bonded both to Pt and Au6 with bond distances of 1.742 Å for Pt—H1 and 1.791 Å for Au2—H1. H2 also bridges over Pt and Au2 with bond distances of 1.635 Å for Pt—H2 and 1.977 Å for Au2—H2. This shows the synergetic interaction of gold and platinum in activating the hydrogen molecule.

Table 3　Selected bond distances(Å)for $[(AuPH_3)_6 Pt(H_2)(PH_3)]^{2+}$ (**3**)

| | |
|---|---|
| Pt—Au1 | 2.713 |
| Pt—Au2 | 2.793 |
| Pt—Au3 | 2.786 |
| Pt—Au4 | 2.792 |
| Pt—Au5 | 2.725 |
| Pt—Au6 | 2.869 |
| Pt—P | 2.382 |
| Pt—H1 | 1.742 |
| Pt—H2 | 1.635 |
| Au1—Au2 | 3.753 |
| Au1—Au3 | 3.037 |
| Au1—Au5 | 2.969 |
| Au2—Au3 | 3.029 |
| Au2—Au4 | 3.030 |
| Au3—Au4 | 2.973 |
| Au5—Au6 | 3.133 |
| Au1—P1 | 2.389 |
| Au2—P2 | 2.359 |
| Au3—P3 | 2.396 |

续表

| | |
|---|---|
| Pt—Au1 | 2.713 |
| Au4—P4 | 2.395 |
| Au5—P5 | 2.396 |
| Au6—P6 | 2.375 |
| Au1—H2 | 2.821 |
| Au2—H1 | 3.364 |
| Au2—H2 | 1.977 |
| Au3—H2 | 3.501 |
| Au4—H1 | 2.428 |
| Au4—H2 | 3.554 |
| Au5—H1 | 3.565 |
| Au5—H2 | 2.819 |
| Au6—H1 | 1.791 |
| Au6—H2 | 3.019 |
| H1—H2 | 2.468 |

Distances longer than 4.21 Å are not included. The total energy is-991.175953 au.

Experimentally, no H/D exchange with water, ethanol or solvent has been observed. It is therefore concluded that mechanisms which lead to HD production do not involve heterolytic cleavage of $H_2$[1,4,21]. In addition, the production of HD should involve the addition of two molecules of reactant($H_2$ and $D_2$) to the same cluster. Based on a combination of kinetic and NMR measurements, following mechanism has been proposed[1,4,21]:

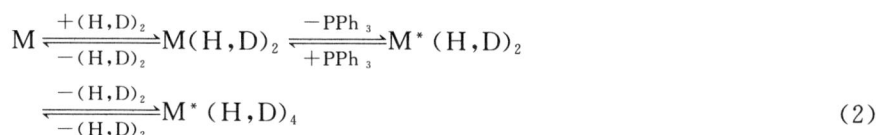

$$M \underset{-(H,D)_2}{\overset{+(H,D)_2}{\rightleftharpoons}} M(H,D)_2 \underset{+PPh_3}{\overset{-PPh_3}{\rightleftharpoons}} M^*(H,D)_2$$

$$\underset{-(H,D)_2}{\overset{-(H,D)_2}{\rightleftharpoons}} M^*(H,D)_4 \qquad (2)$$

Here M refers to the $\{(Au)_nPt\}$ cluster, $M^*$ refers to M with one less $PPh_3$ ligand and $(H,D)_2$ refers to the three isotopic possibilities $H_2$, $D_2$ and HD. The key step in this mechanism is assumed to be the release of a $PPh_3$ ligand to provide an open Au site for the bonding of the second $H_2$ or $D_2$ to form the $M^*(H,D)_4$ species. This was based on qualitative observations that $PPh_3$ inhibited the rate of HD production with $[(AuPPh_3)_8Pt](NO_3)_2$ as the catalyst[4].

In our calculation, the optimized Pt—P bond distance for 3 is 2.382 Å and the average Au—P bond distance is 2.385 Å(range 2.359—2.398 Å). The overall Au—P bond distance in 3 is similar to that in 1(average 2.391 Å). Our calculation results suggest that it is the movement of the metal core that gives open Au sites(e.g. Au1 site)for the bonding of the second $H_2$ or $D_2$. EXAFS analysis for the supported catalyst under CO showed an obvious decrease of the number of Au—Au(Pt)bonds with respect to the intact catalyst[6], which lends support to the calculation results.

## 3.3 Electron distribution in $[(AuPH_3)_6 Pt(PH_3)]^{2+}$ and $(AuPH_3)_6 Pt(H_2)(PH_3)]^{2+}$

We have performed natural atomic orbital and natural bond orbital analysis for the optimized structures of clusters 1—3. Table 4 summarizes the natural charges. The 5d, 6s and 6p orbitals are found to be involved in bonding. These orbitals should undergo complicated hybridization to facilitate the bond formation. Two kinds of electronic interactions are possible for the activation of $H_2$: one is the electron donation from the occupied metal orbital to the antibonding $\sigma^*$ of $H_2$ and the other is that from the bonding $\sigma$ orbital of $H_2$ to the unoccupied metal orbital. As shown in Table 4, $H_2$ looses some electrons in the dihydrogen compound; while $H_2$ gains some electrons in the dihydrido compound. The mechanism of filling the antibonding $\delta$ orbital of $H_2$ is more effective for the activation of the hydrogen molecule. The natural charge in the $AuPH_3$ unit is within the range of 0.30 to 0.50; while that *on* Pt atom is between $-0.35$ and $-0.58$ in all the Au—Pt clusters discussed here. This is in accordance with the theory of the isolobal analogy, where $[AuPPh_3]^+$ is isolobal to $H^+$[22]. Taking $[AuPPh_3]^{\delta+}$ as a building block, the Pt atom has to be negatively charged and 6p orbital of Pt has to be involved to accommodate the excess electrons. This feature makes Pt unique in the activation of the hydrogen molecule. It is to be noted that although Pt is the active center on which the main interaction with $H_2$ occurs, Au atoms also play an important role as indicated by the changes of the natural charges on the Au atoms upon interaction.

Table 4   comparison of natural charges

| | $[(AuPH_3)_6Pt(PH_3)]^{2-}$ (1) | $[(AuPH_3)_6Pt(H_2)(PH_3)]^{2+}$ (2) | $[(AuPH_3)_6Pt(H_2)(PH_3)]^{2+}$ (3) |
|---|---|---|---|
| Pt | $-0.580$ | $-0.513$ | $-0.377$ |
| Au1 | 0.007 | 0.172 | 0.134 |
| Au2 | 0.167 | 0.111 | 0.277 |
| Au3 | 0.166 | 0.170 | 0.104 |
| Au4 | 0.122 | 0.123 | 0.110 |
| Au5 | 0.145 | 0.1 16 | 0.080 |
| Au6 | 0.146 | 0.132 | 0.245 |
| $PH_3$ | 0.255 | 0.199 | 0.313 |
| $PH_3(1)$ | 0.214 | 0.248 | 0.243 |
| $PH_3(2)$ | 0.257 | 0.226 | 0.271 |
| $PH_3(3)$ | 0.258 | 0.255 | 0.232 |
| $PH_3(4)$ | 0.244 | 0.254 | 0.240 |
| $PH_3(5)$ | 0.258 | 0.250 | 0.232 |
| $PH_3(6)$ | 0.259 | 0.241 | 0.275 |
| H1 | — | 0.006 | $-0.268$ |
| H2 | — | 0.009 | $-0.113$ |

$PH_3$ is connected to Pt; while $PH_3(n)$ to the $n$th Au.

## 3.4   Energetics in the dihydrogen activation by $[Au(PH_3)_6Pt(PH_3)]^{2+}$

The reaction between $[Au(PPh_3)_6Pt(PPh_3)](NO_3)_2$, and $H_2$ has been shown to be a rapid reversible process[3-5]. When the equilibrium of Eq. 1 is reached, not only should the forward and reverse reactions take place on average at the same rate, but also the forward and reverse reactions should proceed via the same transition states. This is demanded by the principle of detailed balancing from statistical mechanics[23]. Since the reactions of Eq. 1 are fast, this suggests that the activation energies both for the forward and reverse reactions should be low. Therefore we should have a flat potential energy profile along the reaction path, where the reactants, activated complex and products possess similar energies. The calculations show that the dihydrogen compound **2** lies slightly (2.576 kcal) above the free reactants ($1 + H_2$), while the total energy of the dihydrido compound **3** is 2.674 kcal more stable than those of the free reactants. Generally speaking[24], the activation barrier in the oxidative addition of $H_2$ to a transition metal oxide arises primarily from the repulsive interaction between the doubly occupied $\delta$ orbital of $H_2$ and a doubly occupied metal d orbital. Thereby, the dihydrogen compound **2** is a possible candidate for a transition state. Though it is practically formidable to perform the vibrational frequency calculations so as to confirm that **2** is indeed a transition state, a similar 'early transition state' with a low activation energy (2.34 kcal/mol) and a H—H bond distance of 0.75 Å has been found in the literature[25] for the oxidative addition of $H_2$ to the smaller cluster $Pt(PH_3)_2$. The exothermicity for the formation of $Pt(H)_2(PH_3)_2$ was found to be 15.9 kcal/mol[24], much larger than that for the formation of **3**. The movements of the metal core in **3** should be responsible for this small exothermicity. A large part of the evolution of heat resulting from Pt—H bond formation has been absorbed by the metal core movements, while the tendency for the metal core to return to its optimal geometry would repel the hydrogen so as to facilitate the reverse reaction of eq. 1. It is interesting to recall that $H_2$ dissociation is non-activated on bulk Pt with strong Pt-H bonds formed, while bulk Au stands out as having both the highest activation barrier for $H_2$ dissociation and the least stable chemisorption state among the nearby metals[26]. It would be the Pt—Au bonding that plays an important role in the unique catalytic property of this gold-platinum cluster for the $H_2$—$D_2$ equilibration reaction.

# 4　Concluding remarks

　　B3LYP studies have been performed on the phosphine-stabilized gold-platinum clusters, trying to address the interesting question as to why these clusters are unique for a rapid and clean $H_2$—$D_2$ equilibration. Relativistic effective core potentials have been used to replace the core electrons of the heavy atoms and the $PPh_3$ ligands have been replaced with simple $PH_3$ ligands. The fully-optimized structure of the model cluster can be compared favorably to that of the crystal data for the intact cluster, which justifies the methodology employed here. Two stationary points have been located for the $H_2$ adducts which are the likely intermediates in the $H_2$—$D_2$ equilibration. The potential energy profile along the reaction path is found to be flat. This is consistent with the fact that the reaction between [Au $(PPh_3)_6Pt(PPh_3)](NO_3)_2$ and $H_2$ is a rapid reversible process. Pt is characterized as being the active site with important assistance from Au atoms. The dihydrogen complex would be the active intermediate in which $H_2$ is greatly activated and the induced metal core movements present, open Au sites for the bonding of the second $H_2$ or $D_2$. The charge in the $AuPH_3$ unit is found to be positive, which is in accordance with the theory of isolobal analogy, where $[AuPPh_3]^+$ is isolobal to $H^+$. The Pt is negatively charged, which facilitates the mechanism of electron donation from Pt to the $\sigma^*$ antibonding orbital of $H_2$. A detailed mechanistic study is in progress.

# Acknowledgements

　　This work is supported by the National Natural Science Foundation of China, the specific doctoral project foundation sponsored by the State Education commission of China.

# References

[1] L. H. Pignolct, M. A. Aubart. L. K. Craighead, R. A. T. Gould. D. A. Krogstad. J. S. Wiley, coord. Chem. Rev. **143**(1995)219.

[2] L. N. Ito, J. D. Sweet, A. M. Mueting, L. H, Pignolet, M. F. J. Schoondergang, J. J. Steggerda, Inorg. Chem. **28**(1989)3696.

[3] Y Z. Yuan. K. Asakura, H. L. Wan, K. R. Tsai, Y. Iwasawa. Chem. Lett. **290**(1996)129.

[4] M. A. Aubart, B. D. Chandler. R. A. T. Gould. D. A. Krogstad. M. F. J. Schoondergang, L. H. Pignoiet. Inorg. Chem. **33**(1994)3724.

[5] D C. Roe, J. Magn. Res. **63**(1985)388.

[6] Y. Z. Yuan, K. Asakura, H. L. Wan, K. R. Tsai, Y. Iwasawa, J. Mol. Catal. A **122**(1997)147.

[7] T. G. M. M. Kappen, J. J. Bour, P. P. J. Schlebos, A. M. Roelosen, J. G. M. van der Linden. J. J. Steggerda, M. A. Aubart, D. A. Krogstad, M. F. J. Schoondergang, L. H. Pignolet. Inorg. Chem. **32**(1993)1074.

[8] J. J. Bour, P. P. J. Schlebos, R. P. F. Kanters, M. F. J. Schoondergang, H. Addens, A. Overweg, J. J. Steggerda. Inorg. Chem. Acta **181**(1991)195.

[9] P. J. Hay, W. R. Wadt, J. Chem. Phys. **82**(1985)299.

[10] GAUSSIAN 94(Revision D. 2), M. J. Frisch, G. W. Trucks, H. B. Schlegel, P. M. W. Gill, B. G. Johnson, M. A. Robb, J R. Cheeseman, T. A. Keith, G. A. Petersson. J. A. Montgomery, K. Raghavachari, M. A. Al-Laham. V. G. Zakrzewski, J. V. Ortiz, J. B. Foresman, J. Cioslowski, B. B. Stefanov, A. Nanayakkara, M. Challacombe, C. Y. Peng, P. Y. Ayala, W. Chen, M. W. Wong, J. L. Andres, E. S. Replogle, R. Gomperts, R. L. Martin, D. J. Fox, J. S. Binkley, D. J. Defrees, J. Baker, J. P. Stewart, M. Head-Gordon, C. Gonzalez. J. A. Pople (Gaussian, Pittsburgh, PA, 1995).

[11] W. Stevens, H. Basch, J. Krauss, J. Chem. Phys. **81**(1984)6026.

[12] G. Pacchioni, P. S. Bagus, Inorg. Chem. **31**(1992)4391.

[13] P. Pyykko, N. Runeberg. J. Chem. Soc. Chem. commun., 1993, 1812.

[14] P. Fantucci, S. Polezzo, M. Sironi, A. Bencini, J. Chem. Soc. Dalton Trans.. 1995, 4121.

[15] S. Huzinaga, J. Chem. Phys. **42**(1965)1293.

[16] T. H. Dunning, J. Chem. Phys. **53**(1970)2823.

[17] H. Nakatsuji, K. Kanda. T. Ynezawa, Chem. Phys. Lett. **75**(1980)340.

[18] A. D. Becke, J. Chem. Phys. **98**(1993)5648.

[19] O. D. Haberlen. N. Rosch, J. Phys. Chem. **97**(1993)4970.

[20] K. P. Huber, G. Herzberg, Molecular Spectra and Molecular Structure, constants of Diatomic Molecules. Vol. IV(Van Nostrand, Princeton, NJ, 1979).

[21] L. I. Rubinstein, L. H. Pignolet, Inorg. Chem. **35**(1996)6755.

[22] G. Lavigne, F. Papageorgious. J. J. Bonnet, Inorg. Chem. **23**(1984)609.

[23] W. J. Moore, Physical Chemistry, 5th edn. (Longman Group, London) p. 342.

[24] F. Abu-Hasanayn, A. S. Goldman, K. Krogh-Jespersen, J. Phys. Chem. **97**(1993)5890.

[25] J. J. Low, W. A. Goddard Ⅲ, J. Am. Chem. Soc. **106**(1984)6928.

[26] B. Hammer, J. K. Norskov, Nature **376**(1995)238.

■ 本文原载：Catalysis Today 44(1998)，pp. 333~342.

# Supported Gold Catalysis Derived from the Interaction of a Au-Phosphine Complex with As-Precipitated Titanium Hydroxide and Titanium Oxide

You-Zhu Yuan[a,b], Kiyotaka Asakura[c], Anguelina P.Kozlova[a], Hui-Lin Wan[b],
Khi-Rui Tsai[b], Yasuhiro Iwasawa[a①]

(ᵃDepartment of Chemistry, Graduate School of Science, The University of
Tokyo, Hongo, Bunkyo-ku, Tokyo 113~0033, Japan
ᵇDepartment of Chemistry and State Key Laboratory of Physical Chemistry for
the Solid Surface, Xiamen University, Xiamen 361005, China
ᶜResearch Center for Spectrochemistry, Graduate School of Science,
The University of Tokyo, Hongo, Bunkyo-ku, Tokyo 113~0033, Japan)

**Abstract**    Supported gold catalysts derived from interaction of a Au-phosphine complex Au $(PPh_3)(NO_3)$ ( **1** ) with conventional titanium oxide $TiO_2$ and as-precipitated titanium hydroxide $Ti(OH)_4^*$ ( * , as-precipitated) have been characterized by means of XRD, XPS, EXAFS, and $^{31}P$ CP/MAS-NMR. The Au complex **1** was supported on $TiO_2$ and $Ti(OH)_4^*$ without loss of Au-P bonding at room temperature. The Au complex **1** on $TiO_2$ was readily and completely decomposed to form metallic gold particles. By calcination at 473 K, whereas only a small part of the complex **1** on $Ti(OH)_4^*$ was transformed to metallic gold particles. By calcination of $1/Ti(OH)_4^*$ at 573 K the formation of both metallic gold particles and crystalline titanium oxides became notable as evidenced by XRD, XPS and $^{31}P$ CP/MAS-NMR. The mean diameter of Au particles in $1/Ti(OH)_4^*$ calcined at 673 K was less than 30 Å as estimated from Au(2 0 0) diffraction, which was about one-tenth of that for the corresponding $1/TiO_2$. Thus the as-precipitated titanium hydroxide $Ti(OH)_4^*$ was able to stabilize the Au complex **1** to lead to the simultaneous decomposition of Au complex and $Ti(OH)_4^*$. The catalyst $1/Ti(OH)_4^*$ calcined at 673 K afforded remarkably high catalytic activity for low-temperature CO oxidation at 273 - 373 K as compared to the catalyst $1/TiO_2$. © 1998 Elsevier Science B.V. All rights reserved.

**Key words**：Supported Au phosphine complex; Supported gold catalyst; Characterization; CO oxidation

## 1  Introduction

It has been demonstrated that highly dispersed gold catalysts are tremendously active for low-temperature CO oxidation[1-21] and other catalytic reactions[22-31]. $Au/Fe_2O_3$ , $Au/Co_3O_4$ , $Au/NiO$, $Au/$

---

①  Corresponding author. Fax：81 3 5800 6892；e-mail：iwasawa@chem. s. u-tokyo. ac. jp.

$Al_2O_3$, and $Au/SiO_2$ showed an overall trend indicating that CO oxidation activity increased with decreasing size of the Au particles[5], where the size of Au particles may depend on the kind of supports due to the different Au-support interaction. Superfine Au particles may be a prerequisite for the observation of a Au-support interaction[32]. Thus to obtain highly efficient Au catalysts it is necessary to develop a new method for catalyst preparation which can control Au particle sizes and Au-support interaction during catalyst preparation.

Use of suitable metal complexes as precursors is a promising way to prepare tailored metal catalysts with more efficient interaction between metal and support to generate unique catalysis[33-38]. To obtain highly dispersed Au particles on oxide surfaces, we chose Au-phosphine complexes $Au(PPh_3)(NO_3)$ (1) and $[Au_9(PPh_3)_8](NO_3)_3$ (2) as precursors which could be thermally decomposed on oxide surfaces[39]. When the Au complexes were supported on conventional metal oxides such as $Fe_2O_3$ and $TiO_2$, however, the supported Au complexes aggregated to large Au particles(about 300 Å in diameter) during calcination at 673 K as characterized by EXAFS, XRD, and TEM[40]. These Au complex-derived Au particles on the traditional oxides showed low catalytic activity for CO oxidation. As the Au aggregation and low activity were thought to be due to almost no or weak interaction of the Au precursors with the oxide surfaces, the oxides were exposed to water vapor to produce surface OH groups which can interact with the Au complexes. The Au complexes supported on the water-treated oxides also aggregated to large Au particles by calcination at 673 K. Next, metal hydroxides commercially available were used as the support for the Au complexes, but the obtained supported Au catalysts also showed low activities for CO oxidation.

Very recently, we[40,41] successfully developed a new way to prepare supported Au catalysts with dispersed Au particles by supporting 1 and 2 on as-precipitated wet metal hydroxides, followed by temperature-programmed calcination in a flow of air. The obtained highly dispersed Au particles on the specially obtained supports showed extremely high activity for catalytic CO oxidation at low temperatures typically at 203—273 K. On the other hand, the catalysts obtained by supporting the Au-phosphine complexes 1 and 2 on conventional metal oxides showed CO oxidation activities only at 300—500 K under the identical reaction conditions. The Cl-containing precursors such as $Au(PPh_3)Cl$ and $HAuCl_4$ were unsuitable for the preparation of highly dispersed Au particles on the supports. The characterization studies using XRD, EXAFS and TEM revealed that the as-precipitated metal hydroxides are much superior to the conventional metal oxides as supports for small Au particles. Thus it is suggested that a way to obtain highly dispersed Au particles active for the low-temperature CO oxidation is to choose a suitable Au complex as a precursor and a suitable oxide precursor with many surface OH groups which can be transformed simultaneously to Au particles and oxide support by calcination in air.

Here we report the characterization and performance of the supported gold catalysts derived from the interaction of a Au-phosphine complex 1 with conventional $TiO_2$ and as-precipitated $Ti(OH)_4^*$ by means of EXAFS, XRD, XPS and $^{31}P$ CP/MAS(cross-polarization/magic-angle spinning)-NMR.

# 2. Experimental

## 2.1. Preparation of catalyst

A Au-phosphine complex, $Au(PPh_3)(NO_3)$ (1), was synthesized according to the literature[42]. As-precipitated titanium hydroxide $Ti(OH)_4^*$ was prepared by hydrolysis of $Ti(i\text{-}OC_3H_7)_4$ (titanium-tetra-iso-propoxide, 99.999% purity) with an aqueous solution containing 5.0% of $Na_2CO_3$ (99.9% purity). The precipitate was repeatedly washed several times with deionized water until the pH reached 7.0. The as-precipitated $Ti(OH)_4^*$ was filtered, and immediately impregnated with an acetone (99.9%) solution of **1** under vigorous stirring for at least 12 h, followed by evacuation for 5 h in vacuum to remove the solvent at room temperature. The obtained samples were calcined in a glass tube at a heating rate of 4 K/min to given temperatures and the temperatures were held for 4 h under a flow of air (30 mL/min). The catalysts thus prepared are denoted as $1/Ti(OH)_4^*$. Note that the samples are always used as catalysts after the temperature-programmed calcination. For comparison, catalysts $1/TiO_2$ were prepared by impregnating **1** on commercially available $TiO_2$ (Degussa P-25), followed by temperature-programmed calcination in a similar way to that for $1/Ti(OH)_4^*$. The Au loading on both supports was controlled to be 3.0 wt%.

## 2.2. Characterization

The samples were characterized by EXAFS, XRD, XPS, and $^{31}P$ CP/MAS-NMR.

XRD patterns were measured on a Rigaku powder X-ray diffractometer with Cu $K_a$ radiation over the $2\theta$ range $20 - 80°$. Mean Au particle sizes for $1/Ti(OH)_4^*$ and $1/TiO_2$ were estimated using the Sherrer equation and the width of a selected diffraction line of Au(2 0 0). XPS spectra were recorded on a Rigaku XPS-7000 spectrometer by using Mg$K_a$ radiation with an energy of 1253.6 eV. The binding energies were referred to C(1s) at $E_b = 284.6$ eV.

Au $L_3$-edge EXAFS spectra were measured in a transmission mode at BL-10B of the Photon Factory in the National Laboratory for High Energy Physics (KEK) (Proposal No.: 95G200). The measurements were carried out with a beam current of $250 - 350$ mA and a storage-ring energy of 2.5 GeV. The samples were calcined in a U-shaped glass tube combined in a fixed-bed flow reaction system in a flow of air (30 mL/min) and transferred to EXAFS cells connected to the U-shaped glass tube. Data were analyzed by a curve-fitting method, using empirically derived phase-shift and amplitude functions. The interactions of Au-Au and Au-P were calculated by using the FEFF6.0 software[43,44]. The parameters used for the analysis are summarized in Table 1.

**Table 1　Crystallographic data for FEFF calculation and Fourier transform ranges used in the EXAFS analysis**

| FEFF calculation | | | Fourier transform | |
|---|---|---|---|---|
| Shell | $N$ | $r(Å)$ | $\Delta k$ range($Å^{-1}$) | $\Delta r$ range($Å$) |
| Au-P for Au(PPh$_3$) | 1.0 | 2.28 | 3.0—14.50 | 1.40—2.30 |
| Au-Au for Au foil | 1.0 | 2.87 | 3.0—16.0 | 2.10—3.15 |

$N$: coordination number for absorber-backscatterer pair; $r$: distance; $\Delta k$: range for Fourier transformation ($k$: wave vector); $\Delta r$: range for shell isolation ($r$: distance).

$^{31}P$ CP/MAS-NMR measurements were performed at a resonance frequency of 121.616 MHz on a Chemagnetics Model-300 at room temperature. Chemical shifts were referred to $(NH_4)_2PO_4$ (1.33 ppm). Rotors were spun at 4 kHz and 4.5 $\mu$s pulses (90° pulses) were applied to obtain $^{31}P$ MAS spectra with proton high power decoupling. At least 1000 scans (a scan: 10 s) were acquired for each spectrum of **1**/supports. The chemical shift of triphenylphosphine in the solid state was measured as $-4.9$ ppm with respect to $(NH_4)_2PO_4$.

## 2.3. Catalytic CO oxidation

CO oxidation reactions were carried out in a fixed-bed flow reactor equipped with a computer-controlled auto-sampling system by using 200 mg of catalyst powder. The reaction gas containing 1.0% CO balanced with air purified through a molecular sieve column was passed through the catalyst bed at a flow rate of 67 mL/min (SV = 20 000 mL/h/g). The reaction products were analyzed by a gas chromatograph using a column of Unibeads C for $CO_2$ and a column of 5A molecular sieve for CO and $O_2$. The material balance in the catalytic reactions was checked from the concentration of $CO_2$ and CO, showing good balance under all the reaction conditions tested.

# 3. Results

## 3.1. Characterization by EXAFS and XRD

EXAFS measurements in a transmission mode were carried out at room temperature to characterize $1/TiO_2$ and $1/Ti(OH)_4^*$. The Fourier transforms of EXAFS oscillations at Au $L_3$-edge for $1/Ti(OH)_4^*$ and $1/TiO_2$, before and after calcination at 673 K are shown in Fig. 1. The peak around 1.8 Å in spectra a and b of Fig. 1 is assignable to Au-P bond, which indicates that upon supporting the Au-phosphine complex 1 on $Ti(OH)_4^*$ and $TiO_2$ the Au complex 1 was not decomposed. After calcination at 673 K, the EXAFS Fourier transform peak shifted to a longer distance as shown in Fig. 1. The EXAFS curve-fitting analysis revealed that the new peak is due to Au-Au bonding. The coordination number and distance of Au-Au bond were determined to be 9.8 and 2.86 Å for $1/Ti(OH)_4^*$, and 11.0 and 2.87 Å for $1/TiO_2$, respectively, as shown in Table 2. No Au-P bonding was observed with the calcined samples. The results demonstrate that the Au complex on the supports was decomposed to form metallic Au particles by calcination at 673 K. The coordination numbers for $1/Ti(OH)_4^*$ and $1/TiO_2$ were different from each other, suggesting the formation of Au particles with different sizes.

Table 2　Curve-fitting results for the Au $L_3$-edge EXAFS data of $1/Ti(OH)_4^*$ and $1/TiO_2$

| Pretreated temperature $T$(K) | Au-P | | | | Au-Au | | | | $R_f$ (%) |
|---|---|---|---|---|---|---|---|---|---|
| | $N$ | $r$(Å) | $\Delta\sigma^2$(Å$^2$) | $\Delta E$(eV) | $N$ | $r$(Å) | $\Delta\sigma^2$(Å$^2$) | $\Delta E$(eV) | |
| $1/Ti(OH)_4^*$ | | | | | | | | | |
| 293 | 1.4±0.3 | 2.20±0.01 | 0.0006 | −3.46 | — | — | — | — | 1.3 |
| 673 | — | — | — | — | 9.8±1.0 | 2.86±0.01 | 0.0001 | 2.55 | 0.9 |
| $1/TiO_2$ | | | | | | | | | |
| 293 | 1.3±0.3 | 2.22±0.01 | 0.0000 | −4.05 | — | — | — | — | 0.9 |
| 673 | — | — | — | — | 11.0±1.0 | 2.87±0.01 | 0.0001 | 2.12 | 0.7 |

EXAFS data of complex 1. Au-P: $N=1$, $r=2.20$ Å.

The transformation from the Au-phosphine complex to metallic Au particles was accompanied by the development of the XRD peaks for crystalline $TiO_2$ as shown in Fig. 2. The formation of metallic gold particles and crystalline $TiO_2$ were not detected after calcination at 473 K, but notable after calcination at 573 K. The XRD peak for crystalline Au particles was hardly observed with the $1/Ti(OH)_4^*$ calcined at 573 and 673 K, where the mean diameter of Au particles in $1/Ti(OH)_4^*$ was estimated to be less than 30 Å. In contrast, the $1/TiO_2$ catalyst showed the Au(200) peak at $2\theta=44.4°$ after calcination even at 473 K. The mean diameter of Au particles in $1/TiO_2$ calcined at 673 K was as large as 300 Å. Moreover, the peaks for crystalline $TiO_2$ produced from $Ti(OH)_4^*$ were much broader than those for $TiO_2$(P-25) as shown in Figs. 2 and 3.

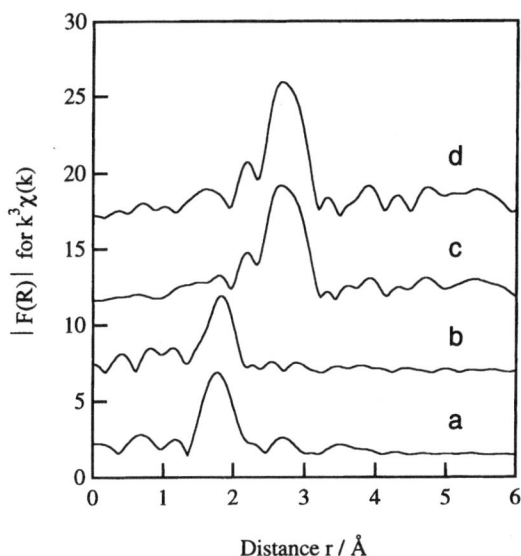

Fig. 1. The Fourier transforms of EXAFS oscillations at Au $L_3$-edge for $1/TiO_2$ and $1/Ti(OH)_4^*$ before and after calcination: (a) as-prepared $1/Ti(OH)_4^*$; (b) as-prepared $1/TiO_2$; (c) $1/Ti(OH)_4^*$ calcined at 673 K (temperature-programmed); (d) $1/TiO_2$ calcined at 673 K (temperature-programmed).

Fig. 2. The XRD patterns of $1/Ti(OH)_4^*$ before and after temperature-programmed calcination. (♦) Au(2 0 0): (a) as-prepared $1/Ti(OH)_4^*$; (b) $1/Ti(OH)_4^*$ after temperature-programmed calcination at 473 K; (c) $1/Ti(OH)_4^*$ after temperature-programmed calcination at 573 K; (d) $1/Ti(OH)_4^*$ after temperature-programmed calcination at 673 K; (e) $1/Ti(OH)_4^*$ after temperature-programmed calcination at 773 K.

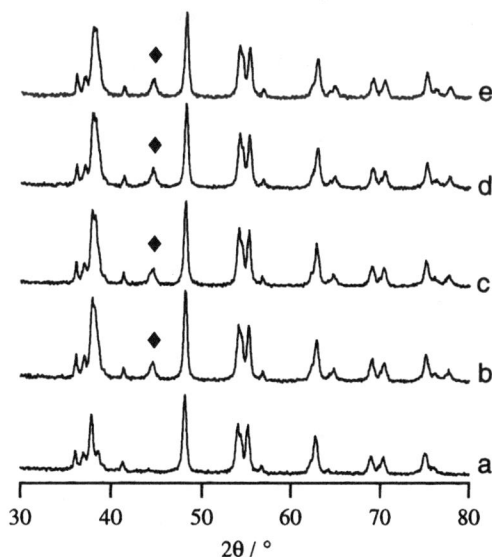

## 3. 2. Characterization by XPS

Fig. 4 shows the XPS spectra of P 2p and Au 4f levels in $1/Ti(OH)_4^*$ before and after calcination. The Au 4f peak in the intact $1/Ti(OH)_4^*$ was observed at 85. 5 eV, indicating that the Au atoms were still situated as monocationic ions probably bound to the phosphine ligand. After calcination at 473 K, the Au 4f peak intensity became weak and a new peak at 83. 8 eV appeared. This may be due to the coexistence of metallic gold particles and undecomposed Au complex in the sample. When $1/Ti(OH)_4^*$ was calcined at 573 K, no peak at 85. 4 eV was observed and a peak at 83. 8 eV due to metallic gold developed. The increase in the peak intensity at 83. 8 eV for metallic gold species by increasing calcination temperature from 573 to 673 K may be due to the growth and dispersion of

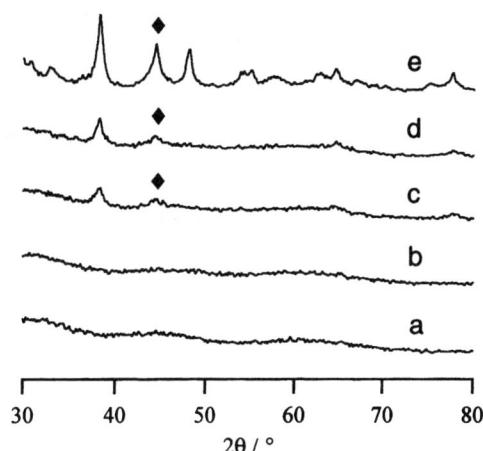

Fig. 3. The XRD patterns of $1/TiO_2$ before and after calcination. (♦) Au (200): (a) as-prepared $1/TiO_2$; (b) $1/TiO_2$ after temperature-programmed calcination at 473 K; (c) $1/TiO_2$ after temperature-programmed calcination at 573 K; (d) $1/TiO_2$ after temperature-programmed calcination at 673 K; (e) $1/TiO_2$ after temperature-programmed calcination at 773 K.

the small metallic particles on the catalyst surfaces. For the P 2p XPS spectra for phosphine species in the intact **1**/Ti(OH)$_4^*$, a peak at 131.9 eV was observed, which is assignable to the PPh$_3$ moiety bonding to Au. This species was also observed after calcination at 473 K. The results indicate that little change occurred significantly in the phosphine species. The peak at 131.9 eV shifted to higher binding energies(133.6 — 133.8 eV) by calcining the samples at 573 — 673 K, indicating the oxidation of phosphine species by calcination.

Fig. 5 depicts the XPS spectra of P 2p and Au 4f for **1**/TiO$_2$ before and after calcination. The Au 4f peak at 84.8 eV for the intact **1**/TiO$_2$ shifted to a lower binding energy of 83.7 eV by calcination at 473 K. No peak for the Au complex was observed with the sample calcined at 473 K. The P 2p XPS spectra

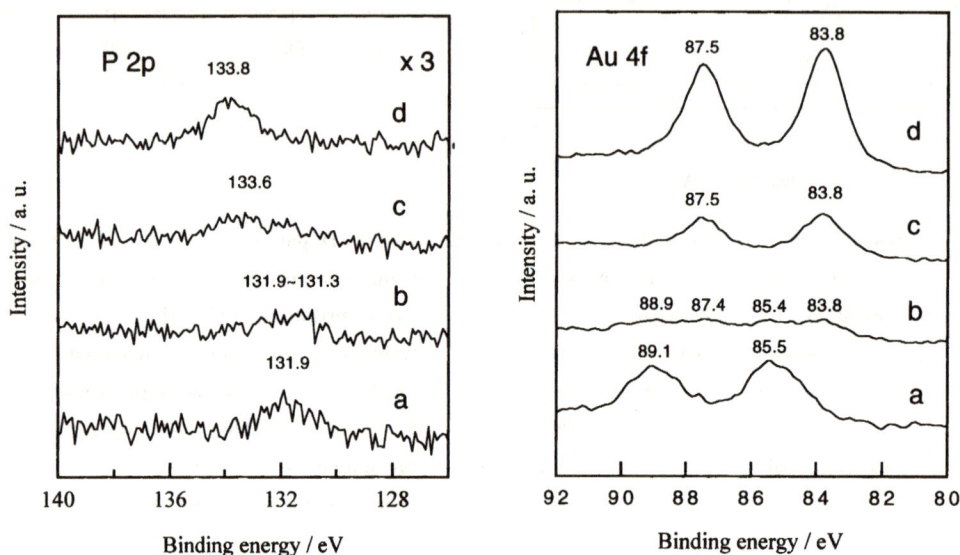

Fig. 4. P 2p and Au 4f XPS spectra for **1**/Ti(OH)$_4^*$ : (a) as-prepared sample; (b) after temperature-programmed calcination at 473 K; (c) after temperature-programmed calcination at 573 K; (d) after temperature-programmed calcination at 673 K.

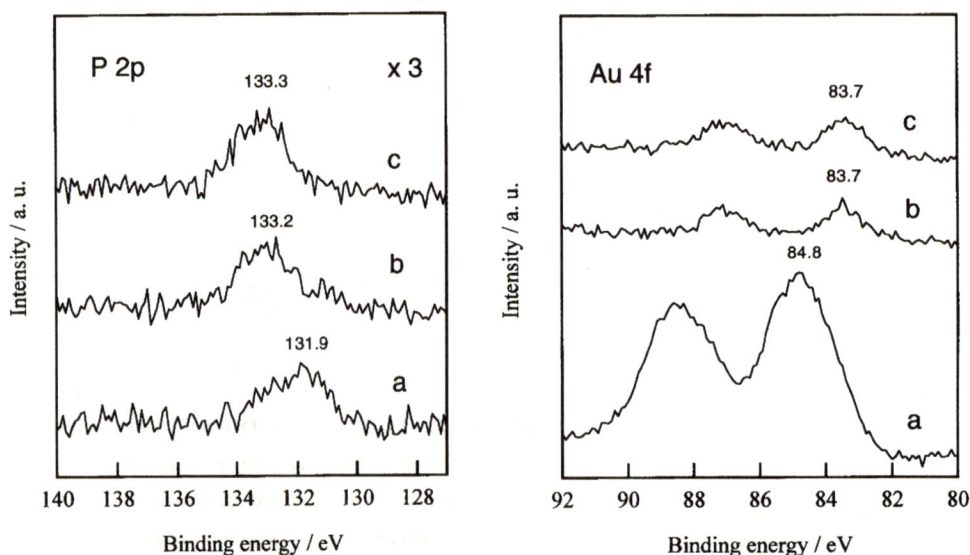

Fig. 5. P 2p and Au 4f XPS spectra for **1**/TiO$_2$ : (a) as-prepared sample; (b) after temperature-programmed calcination at 473 K; (c) after temperature-programmed calcination at 573 K.

showed that the peak at 131. 9 eV in the intact $1/TiO_2$ shifted to a higher binding energy of 133. 2 eV by calcination at 473 K, suggesting that the oxidation of phosphine species occurred in parallel to the formation of metallic gold particles.

## 3. 3. Characterization by $^{31}$P CP/MAS-NMR

The effects of calcination temperatures on the $^{31}$P CP/MAS-NMR spectra for $1/Ti(OH)_4^*$ and $1/$

TiO$_2$ are shown in Figs. 6 and 7, respectively. The spectra for both $1/Ti$ $(OH)_4^*$ and $1/TiO_2$ calcined at 473 K are similar to that for complex 1, though the chemical shift slightly moved from 24. 7 ppm(complex) to 30. 6 — 31. 6 ppm($1/Ti$ $(OH)_4^*$). The change in the chemical shift may be due to the chemical interaction between Au complex and Ti(OH)$_4^*$. No other resonances such as adsorbed PPh$_3$ (PPh$_3$/Ti(OH)$_4^*$, $\delta$ — 18. 8 ppm) and free PPh$_3$ ($\delta$ — 4. 9 ppm) liberated from the complex 1 were observed. After calcination at 673 K, a new resonance at 5. 9 ppm was observed as shown in Fig. 6. For $1/TiO_2$, the chemical shift of $^{31}$P in the intact sample was similar to that of $1/Ti(OH)_4^*$ (Fig. 7 (a)), but new resonances with chemical shifts at 75. 1, 42. 5 and 10. 0 ppm

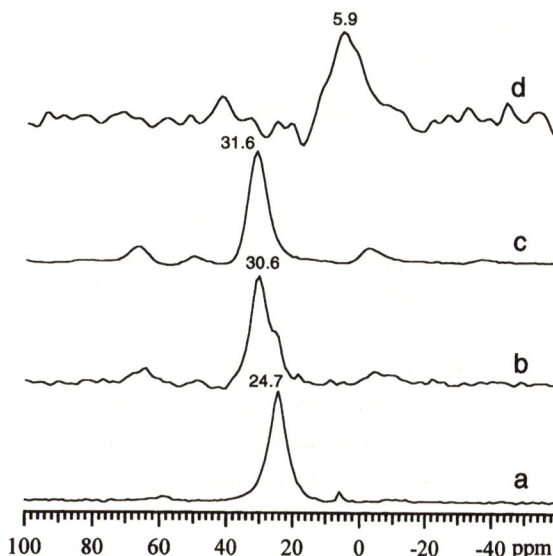

Fig. 6. $^{31}$P CP/MAS-NMR spectra of 1 /Ti(OH)$_4^*$ : (a) Au phosphine complex 1; (b) as-prepared 1/Ti(OH)$_4^*$ ; (c) after temperature-programmed calcination at 473 K; ( d) after temperature-programmed calcination at 673 K.

appeared in the NMR spectrum of $1/TiO_2$ calcined at 473 K(Fig. 7(b)). By calcination at 673 K, another new signal at 3. 0 ppm was observed. We assign the peaks at 3 — 6 ppm(Fig. 6 (d) and Fig. 7(c)) to surface$(PO_4)^{3-}$ from comparison with the spectrum of $H_3PO_4/TiO_2$ (Fig. 7(d)). The difference in the chemical shifts may be due to the difference in the concentrations of moisture in the samples[45].

## 3. 4. Catalytic activities

Fig. 8 depicts the catalytic activities(CO conversion) of supported Au catalysts, $1/Ti(OH)_4^*$ and $1/TiO_2$, for CO oxidation in a fixed-bed flow reactor as a function of reaction temperature. Under the present reaction conditions(SV = 20 000 mL/h/g-cat; CO = 1%), 100% conversion corresponds to a reaction rate of $8. 93 \times 10^{-3}$ mol/h/g-cat. It was found that the catalytic performances of $1/Ti(OH)_4^*$ and $1/TiO_2$ were remarkably affected by catalyst calcination temperatures. The CO oxidation on $1/Ti(OH)_4^*$ calcined at 673 and 773 K occurred at temperatures lower than 273 K. On the other hand, the catalysts $1/TiO_2$ calcined at 673 and 773 K were active for the CO oxidation only above 350 K. It was found that the catalytic activity of $1/Ti(OH)_4^*$ was much higher than that of $1/TiO_2$. When the catalyst $1/Ti(OH)_4^*$ calcined at 473 K was used, the CO conversion increased steeply with reaction temperature above 400 K. This might be due to an increase in the amount of small gold particles in the catalyst during the catalytic reaction by decomposition of the remaining Au complex in $1/Ti(OH)_4^*$ calcined at 473 K. It coincides with the observations of EXAFS, XRD and XPS for $1/Ti(OH)_4^*$ calcined at 473 K.

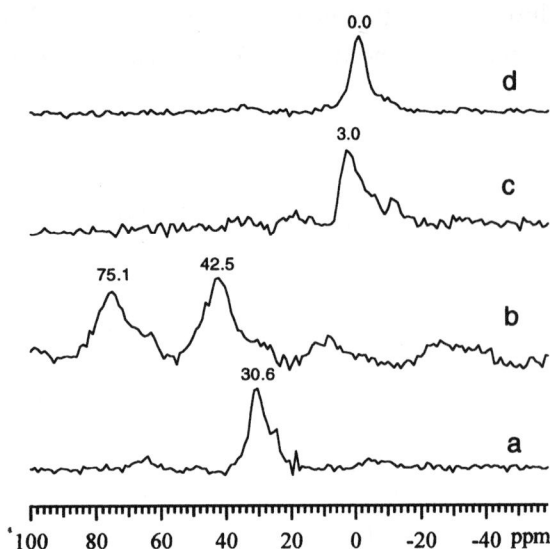

Fig. 7. $^{31}$P CP/MAS-NMR spectra of 1/TiO$_2$: (a) as-prepared 1/TiO$_2$; (b) after temperature-programmed calcination at 473 K; (c) after temperature-programmed calcination at 673 K; (d) as-prepared H$_3$PO$_4$/TiO$_2$.

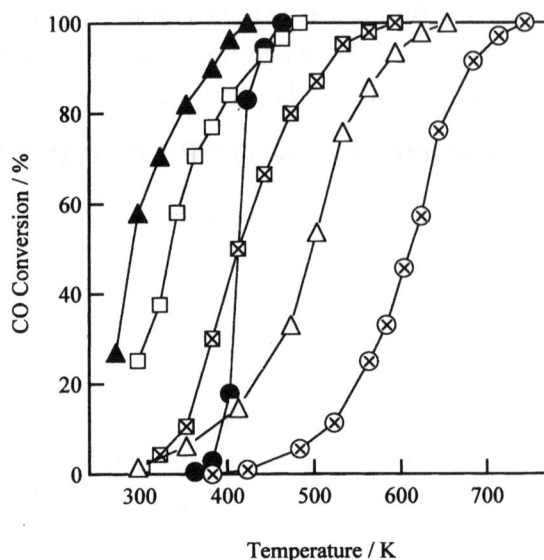

Fig. 8. CO oxidation reactions on various Au/Ti-oxide catalysts in a fixed-bed reactor (SV = 20 000 ml/h/g) as a function of reaction temperature: (▲) 1/Ti(OH)$_4^*$ (temperature-programmed calcination at 673 K); (□) 1/Ti(OH)$_4^*$ (temperature-programmed calcination at 773 K); (●) 1/Ti(OH)$_4^*$ (temperature-programmed calcination at 473 K); (⬤) 1/TiO$_2$ (temperature-programmed calcination at 773 K); (△) 1/TiO$_2$ (temperature-programmed calcination at 673 K); (⊗) 1/TiO$_2$ (temperature-programmed calcination at 473 K).

# 4. Discussion

The curve-fit results of the EXAFS data in Table 2 reveal that the Au-P bond lengths for 1/Ti(OH)$_4^*$ and 1/TiO$_2$ before calcination range between 2. 20 and 2. 22 Å. The Au-P distances are similar to that for crystalline Au(PPh$_3$)(NO$_3$)[46]. The results of XRD(Fig. 2(a) and Fig. 3(a)), XPS(Fig. 4(a) and Fig. 5(a)) and $^{31}$P CP/MAS-NMR(Fig. 6(a) and Fig. 7(a)) demonstrate that the Au phosphine complex Au(PPh$_3$)(NO$_3$)(1) is supported on the surfaces of as-precipitated titanium hydroxides and titanium oxides without significant loss of the Au-P bond at room temperature. The complex-surface interaction might occur through the nitrate anions which reacted with the surface OH groups to form surface Au-phosphine complexes [Au(PPh$_3$)]$^+$ as suggested from the comparison of the IR spectrum of Au phosphine complex 1 in KBr with that of the supported complex. ① Thus, the downfield chemical shift observed in the $^{31}$P CP/MAS-NMR spectra(Fig. 6(a) and Fig. 7(a)) of 1/Ti(OH)$_4^*$ and 1/TiO$_2$ may be due to the interaction between [Au(PPh$_3$)]$^+$ and the supports. When 1/Ti(OH)$_4^*$ was calcined at 473 K, only a partial cleavage of the Au-P bond occurred to form a small amount of metallic gold

---

① The data are obtainable upon request.

species, but no oxidation of the phosphine species was observed, judging from the results of XRD(Fig. 2
(b)), XPS(Fig. 4(b)) and $^{31}$P CP/MAS-NMR(Fig. 6(b)). The results indicate that the amorphous as-
precipitated Ti(OH)$_4^*$ interacts with the Au phosphine complex to stabilize the supported complex. It
was found that the formation of metallic gold particles and the transformation of the amorphous
Ti(OH)$_4^*$ support to crystalline titanium oxide took place at 573 K in the case of **1**/Ti(OH)$_4^*$ as
evidenced by XRD(Fig. 2(c)) and XPS(Fig. 4(c)). On the other hand, when **1**/TiO$_2$ was calcined at 473 K,
the interpretation of the newborn chemical shifts at 75.1, 42.5 and 10.0 ppm in $^{31}$P CP/MAS-NMR
(Fig. 7(b)) is not clearly done at present. The Au clusters such as [Au$_9$(PPh$_3$)$_8$]$^{3+}$ and
[Au$_8$(PPh$_3$)$_7$]$^+$ provide similar $^{31}$P-NMR shifts in the solid state[47-49]. Hence, a possible explanation
for the chemical shifts at 75.1 and 42.5 is that phosphine species were bound to Au clusters/particles
formed by the decomposition of the Au complex **1** at 473 K. Another possible explanation is that partial
oxidation of the phosphine species on the support occurred as suggested by XPS(Fig. 5(b)).

It may be of interest to note that the intensity of the Au 4f and P 2p XPS signals for **1**/Ti(OH)$_4^*$
significantly decreased after temperature-programmed calcination at 473 K as shown in Fig. 4(b). The
reason for the intensity reduction by the 473 K calcination is not clear at present, but it may not be
artificial because the intensity of the O1s XPS signal(not shown) was similar in all the samples treated
in Fig. 4. We propose that the reduction in the Au and P XPS signals is due to the change of the location
of the Au phosphine species from the catalyst surface to the subsurface/the bulk near the surface upon
calcination at 473 K. The occlude Au species are decomposed to metallic Au particles after calcination at
573 K as shown in Fig. 4(c), and the intensity increased, indicating that a part of them is moved to and
dispersed at the surface. After calcination at 673 K, Au particles are suggested to be grown and
dispersed at the support surface judging from the increase in the XPS signals(Fig. 4(d)).

Our previous studies[39-41,50,51] demonstrated that the Au complex on the oxide supports
decomposes to metallic Au particles at lower temperatures than 573 K. In the similar temperature range
the as-precipitated wet metal hydroxides are dehydrated to form anhydrous metal hydroxides, and
mostly to partially dehydrated metal oxides. The supported Au complex is expected to be stabilized on
the surface of the as-precipitated metal hydroxides like Ti(OH)$_4^*$ by chemical interaction between [Au
(PPh$_3$)]$^+$ and hydroxyl groups or adsorbed water. During the calcination of the samples the
decompositions of both the Au complex and the as-precipitated Ti(OH)$_4^*$ occur in parallel, which leads
to the formation of highly dispersed Au particles on the in situ prepared oxide surfaces. On the other
hand, in the case of **1**/TiO$_2$, there is only a limited number of OH groups on the oxide surface and the
property of the OH groups seems to be different from the OH groups of Ti(OH)$_4^*$, and the interaction
of the TiO$_2$ surface with the Au complex may be insufficient and weak, hence leading to aggregation to
large Au particles without significant stabilization of Au species during the calcination.

When the as-prepared **1**/Ti(OH)$_4^*$ was calcined at a temperature-programmed rate of 4 K/min up
to 673 K in a flow of air, it was found that the obtained Au catalysts showed high catalytic activities for
CO oxidation at low temperatures like 273 K as shown in Fig. 8. The diameter of the Au particles in
**1**/Ti(OH)$_4^*$ is likely to be 30 Å, judging from negligible Au(2 0 0) peak. EXAFS(Fig. 1 and Table 2)
and XRD(Figs. 2 and 3) estimate the size of Au particles in **1**/Ti(OH)$_4^*$ to be about one-tenth of that in
**1**/TiO$_2$.

Calcination of **1**/Ti(OH)$_4^*$ at 673 K oxidized the phosphine ligands to phosphoric species as
suggested by XPS(Fig. 4) and $^{31}$P CP/MAS-NMR(Fig. 6). There was no correlation between the
catalytic activity and the amount of the phosphoric species in the catalysts, but limited information in the
present study indicates the negative effect of the remaining phosphoric species on the CO oxidation

catalysis.

Hoflund and co-workers[13,52] claimed that nonmetallic Au species is responsible for the low-temperature activity. Haruta et al. [6] reported that the sample having both metallic and nonmetallic Au species was not more active than the sample having only metallic Au species. Recently, it was suggested by XPS and ISS that the near-surface region of Au/ $Fe_2O_3$ contains more Au as compared with that of Au/$Co_3O_4$, where Au is present as crystallites and small amounts of nonmetallic Au species are also present[52]. The present XPS date provided no evidence on the presence of cationic Au species on the surface of $1$/Ti(OH)$_4^*$, although XPS cannot detect a small amount of cationic Au species which coexists with metallic species in the 3.0 wt% Au catalysts.

# 5. Conclusions

The Au-phosphine complex **1** was supported intact on the surfaces of as-precipitated Ti(OH)$_4^*$ and $TiO_2$ at room temperature. The amorphous as-precipitated wet Ti(OH)$_4^*$ which contains large amounts of surface OH groups and physisorbed water, interacts with the Au complex **1** more efficiently than the conventional $TiO_2$. The transformation of the as-prepared **1**/ Ti(OH)$_4^*$ to crystalline titanium oxide was observed at 573 K, whereas the decomposition of **1**/$TiO_2$ accompanying the oxidation of phosphine species occurred by calcination at 473 K. The difference of the decomposition behavior results in the difference of the sizes of metallic Au particles formed on **1**/Ti(OH)$_4^*$ and **1**/$TiO_2$ by calcination at 673 K. The mean diameter of Au particles in **1**/Ti(OH)$_4^*$ was ten times smaller than that in **1**/$TiO_2$. The supported Au catalysts prepared from **1**/$TiO_2$ catalyzed CO oxidation above 350 K, while the catalysts derived from **1**/as-precipitated Ti(OH)$_4^*$ showed a high activity for CO oxidation at low temperatures.

# References

[1] M. J. Fuller, M. E. Warwich, J. Catal. **34**(1974) 445.

[2] G. C. Bond, L. R. Molloy, M. J. Fuller, J. Chem. Soc., Chem. commun. (1975) 796.

[3] G. Croft, M. J. Fuller, Nature **269**(1977) 585.

[4] M. Haruta, T. Kobayashi, H. Sano, N. Yamada, Chem. Lett. (1987)405.

[5] M. Haruta, N. Yamada, T. Kobayashi, S. Iijima, J. Catal. **115**(1989) 301.

[6] M. Haruta, S. Tsubota, T. Kobayashi, H. Kageyama, M. J. Genet, B. Delmon, J. Catal. **144**(1993) 175.

[7] S. Tsubota, A. Ueda, H. Sakurai, T. Kobayashi, M. Haruta, ACS Symp. Ser. **552**(1994) 420.

[8] T. Boccuzi, S. Tsubota, M. Haruta, J. Elec. Spectrosc. Relat. Phenom. **64/65**(1993) 241.

[9] M. Haruta, S. Tsubota, T. Kobayashi, A. Ueda, H. Sakurai, M. Ando, Stud. Surf. Sci. Catal. **75** (1993) 2657.

[10] S. D. Gardner, G. B. Hoflund, M. R. Davidson, D. R. Schryer, J. Catal. **115**(1989) 132.

[11] S. D. Gardner, G. B. Hoflund, D. R. Schryer, B. T. Upchurch, J. Phys. Chem. **95**(1991) 835.

[12] D. R. Schryer, B. T. Upchurch, B. D. Sidney, K. G. Brown, G. B. Hoflund, K. R. Herz, J. Catal. **130**(1991) 314.

[13] S. D. Gardner, G. B. Hoflund, D. R. Schryer, J. Schryer, B. T. Upchurch, E. J. Kielin, Langmuir **7**(1991) 2135.

[14] S. D. Gardner, G. B. Hoflund, M. R. Davidson, H. A. Laitinen, D. R. Schryer, B. T. Upchurch,

Langmuir **7**(1991) 2140.

[15]G. B. Hoflund, S. D. Gardner, D. R. Schryer, B. T. Upchurch, E. J. Kielin, Appl. Catal. B **6** (1995) 117.

[16]S. Imamura, H. Sawada, K. Uemura, S. Ishida, J. Catal. **109**(1988)198.

[17]S. Imamura, S. Yoshie, Y. Ono, J. Catal. **115**(1989) 258.

[18]S. D. Lin, M. Bollinger, M. A. Vannice, Catal. Lett. **17**(1993) 245.

[19]A. Knell, P. Barnickel, A. Baiker, A. Wokaun, J. Catal. **137**(1992) 306.

[20]A. Baiker, M. Maciejewski, S. Tagliaferri, P. Hug, J. Catal. **151**(1995) 407.

[21]G. Srinivas, J. Wright, C. -S. Bai, R. cook, Stud. Surf. Sci. Catal. **101**(1996) 427.

[22]M. Haruta, A. Ueda, A. Tsubota, R. M. Torres Sanchez, Catal. Today **29**(1996) 443.

[23]S. Naito, M. Tanimoto, J. Chem. Soc., Chem. commun. (1988) 832.

[24]D. Andreeva, V. Idakiev, T. Tabakova, A. Andreev, J. Catal. **158**(1996) 354.

[25]D. Andreeva, V. Idakiev, T. Tabakova, A. Andreev, R. Giovanoli, Appl. Catal. **A 134**(1996) 275.

[26]H. Sakurai, A. Ueda, T. Kobayashi, M. Haruta, J. Chem. Soc., Chem. commun. (1997) 271.

[27]H. Sakurai, S. Tsubota, M. Haruta, Appl. Catal. **A 102**(1993) 125.

[28]H. Sakurai, M. Harura, Catal. Today **29**(1996) 361.

[29]H. Sakurai, M. Haruta, Appl. Catal. **A 127**(1995) 93.

[30]M. Shibata, N. Kawata, T. Masumoto, H. Kimura, J. Chem. Soc., Chem. commun. (1988) 154.

[31]S. Qiu, R. Onishi, M. Ichikawa, J. Phys. Chem. 98 (1994) 2719; Y. Takita, T. Imamura, Y. Mizuhara, Y. Abe, T. Ishihara, Appl. Catal. **B 1**(1992) 79.

[32]G. C. Bond, P. A. Sermon, Gold Bull. **6**(1973) 102.

[33]Y. I. Yermakov, B. N. Kuznetsov, V. A. Zakharov, Catalysis by Supported Complexes, Elsevier, Amsterdam, 1981.

[34]J. M. Basset, J. Mol. Catal. **21**(1983) 95.

[35]Y. Iwasawa(Ed.), Tailored Metal Catalysts, Reidel, Netherlands, 1986.

[36]G. Maire, Stud. Surf. Catal. **29**(1986) 509.

[37]B. C. Gates, Chem. Rev. **95**(1995) 511.

[38]Y. Iwasawa, Proceedings of the 11th International Congress on Catalysis, Baltimore, 1996, Studies on Surface Science and Catalysis.

[39]Y. Yuan, K. Asakura, H. Wan, K. Tsai, Y. Iwasawa, Catal. Lett. **42**(1996) 15.

[40]Y. Yuan, A. P. Kozlova, K. Asakura, H. Wan, K. Tsai, Y. Iwasawa, J. Catal. **170**(1997) 191.

[41]Y. Yuan, K. Asakura, H. Wan, K. Tsai, Y. Iwasawa, Chem. Lett. **9**(1996) 756.

[42]A. M. Mueting, B. D. Alexander, P. D. Boyle, A. L. Casalnuovo, L. N. Ito, B. J. Johnson, L. H. Pignolet, in: R. N. Grimes(Ed.), Inorganic Synthesis, vol. 29, Wiley, New York, 1992, p. 280.

[43]J. J. Rehr, L. J. de Mustre, S. I. Zabinsky, R. C. Albers, J. Am. Chem. Soc. **113**(1991) 5235.

[44]Y. Iwasawa (Ed.), X-Ray Absorption Fine Structure for Catalysts and Surfaces, World Scientific, Singapore, 1996.

[45]J. H. Lunsford, P. R. William, S. Wenxia, J. Am. Chem. Soc. **107**(1985) 1540.

[46]P. F. Barron, L. M. Engelhardt, P. C. Healy, J. Oddy, A. H. White, Aust. J. Chem. **40**(1987) 1545.

[47]J. W. Diesveld, E. M. Menger, H. T. Edzes, W. S. Veeman, J. Am. Chem. Soc. **102**(1980) 7935.

[48]J. W. A. van der Velden, P. T. Beurskens, J. J. Bour, W. P. Bosman, J. H. Noordik, M. Kolenbrander, J. A. K. A. Buskes, Inorg. Chem. **23**(1984) 146.

[49]N. J. Clayden, C. M. Dobson, K. P. Hall, D. Michael, P. Mingos, D. J. Smith, J. Chem. Soc., Dalton Trans. (1985) 1811.

[50]Y. Yuan, K. Asakura, H. Wan, K. Tsai, Y. Iwasawa, Chem. Lett. **2**(1996) 129.

[51]Y. Yuan, K. Asakura, H. Wan, K. Tsai, Y. Iwasawa, J. Mol. Catal. **A 122**(1997) 147.

[52]W. S. Epling, G. B. Hoflund, J. F. Weaver, S. Tsubota, M. Haruta, J. Phys. Chem. **100**(1996) 9929.

■ 本文原载:Journal of the Chinese Chemical Society,45(1998),pp. 673~678.

# The Effects of $M_2O_3$ on Stabilizing Monocopper over the Surface of Cu-ZnO-$M_2O_3$ Catalysts for Mathanol Synthesis

Hong-Bo Chen[a,b], Dai-Wei Liao[a,b,c], La-Jia Yu[a,b],
Jun Yi[a], Hong-Bin Zhang[a,b,c], Khi-Rui Tsai[a,b,c]

([a]Institute of Physical Chemistry, Xiamen University, Xiamen 361005, China
[b]Department of Chemistry, Xiamen University, Xiamen 361005, China
[c]The State Key Laboratory of Physical Chemistry on Solid Surfaces,
Xiamen 361005, China)

**Abstract**　The effects of $M_2O_3$ (M＝Al, Sc etc.) in Cu-ZnO-$M_2O_3$ catalysts on methanol synthesis at low pressure were studied with ESR, XPS and TPR spectroscopy. The results of ESR showed that the generation of monovalent cationic defects was because the valence state and electronic charge on the ZnO lattice lost their balance as $M^{3+}$ doped into ZnO. The induced effect by $Sc^{3+}$ is stronger than that by $Al^{3+}$. The results of XPS and TPR indicated that the amount and stabilization of $Cu^+$ on the surface of reduced copper-based catalyst and its catalytic activity were affected by the monovalent cationic defects on the surface of ZnO.

## 1　INTRODUCTION

The effects of each component of the copper-based catalyst on methanol synthesis at low pressure have been extensively investigated[1-4] since the catalysts and their excellent catalysis are known as an outstanding example of synergistic catalysis. It is generally accepted that the coordination chemisorption and activation of carbon monoxide and the homogeneous splitting of hydrogen take place on $Cu^0$ or $Cu^+$, but the heterogeneous splitting of hydrogen take place on ZnO which provides $H^{\delta+}$ and $H^{\delta-}$ in the catalytic process[5]. However, many important aspects of methanol synthesis remain controversial. Herman and coworkers[1] suggested that the active species was $Cu^+$ dissolved in ZnO. Chinchen and Waugh[6] proposed that the activity of copper-based catalyst was directly proportional to the total copper surface area, including both $Cu^0$ and $Cu^+$, rather than depending on the special copper sites. The principal argument focused on whether $Cu^0$ or $Cu^+$ was the active phase for methanol synthesis. The appropriate answer is related to the formation of $Cu^+$ and its stability in reaction as well as the effect of carbon dioxide on methanol synthesis.

We have discussed the mechanism of methanol synthesis on Cu-ZnO-$M_2O_3$ before[7]. Now, in this paper, we describe the effect of $M_2O_3$ (M＝Al, Sc etc.) in Cu-ZnO-$M_2O_3$ catalysts on both the formation and stabilization of $Cu^+$ sites on the catalyst surfaces.

# 2　EXPERIMENTAL SECTION

## Preparation of Catalysts

The catalysts of $ZnO-M_2O_3$ and $Cu-ZnO-M_2O_3$ ( M = Al, Sc etc,) were prepared by modified coprecipitation method[1]. To avoid precipitation of carbonates with different solubility product, in turn, we simultaneously added drop by drop the mixing solution of nitrates and $Na_2CO_3$ at the desired ratio into a little deionized water, and kept the temperature of the mixing solution at 85 ℃ and the pH at 6. 5 —7. 0 while stirring strongly. Then the precipitate was slaked, filtered, washed, dried at 110 ℃ overnight, and calcined at 350 ℃ for 3—4 h.

## Spectroscopic Characterization ESR Spectroscopy

The ESR studies were carried out with a Bruker ER-200D-SRC type paramagnetic resonance spectrometer, using a powder sample of 80 mg in $\phi 3$ high-quality glass tubes at room temperature. The detective range was the X wave range. The output power was 20. 4 mW and the modulation frequency was 100 KHz, The modulation amplitude was 0. 8 $G_{pp}$ and the time constant was 1 ms.

## XPS Spectroscopy

The XPS measurements were carried out with a VG Escalab Mark-II type spectrometer, using Al-$K_a$(1486. 6 eV, 10. 0 kV) as radiation source and taking the bonding energy of $Zn(2p_{3/2})$(1022. 2 eV) as a reference energy for ZnO, $ZnO-M_2O_3$ and $Cu-ZnO-M_2O_3$ and taking the bonding energy of pollution carbon on the surface(284. 6 eV) as an internal reference energy for samples. The samples were reduced with mixing gas of $H_2$ and $N_2$($H_2/N_2=5/95$, v/v) before measurements.

## Temperature Program Reduction(TPR)

The experimental system for TPR includes the gas chromatograph(GC), microreactor with 40—60 mesh catalyst of 30 mg and temperature program controller. The temperature program reduction of catalyst was carried out with a mixing gas of $H_2$ and $N_2$($H_2/N_2=7/95$, v/v) at rate of 3 ℃/min, and the results were recorded by X-Y recorder.

## Measurement of Catalytic Activity for Cu-ZnO-M₂O₃

The catalytic activity measurements were carried out in the flow reaction system with a fixed-bed ($\phi 8$) microreactor made of stainless steel. The catalyst with 40—60 mesh was reduced by $H_2/N_2$(5/95, v/v) with flow velocity of 30 mL/min at atmospheric pressure and the temperature program was carried out. The determinations of catalytic activity for methanol synthesis were carried out with $CO/H_2$(1/2, v/v) of flow velocity of 30 mL/min at reaction pressure of 1. 0 Mpa and reaction temperature of 230±0. 5 ℃.

# 3　RESULTS AND DISCUSSIONS

## Induced effect of M₂O₃ on defects of ZnO-M₂O₃

The ESR spectra of ZnO, $ZnO-Al_2O_3$ and $Al_2O_3$ were detected under the same conditions of both preparation and measurement. No ESR signal was observed for either ZnO or $Al_2O_3$, but a strong and sharp ESR signal at g $=$ 1.9606$\pm$0.0002 with $\Delta H_{pp}=(2.8\pm0.1)\times10^{-4}$ T was observed.

The variation of ESR signal intensity vs. the atomic ratio of Zn/Al is shown in Fig. 1. The relative signal intensity is proportional to the $Y_m'(\Delta H_{PP})^2$ where the $Y_m'$ is amplitude of peak and the $\Delta H_{PP}$ is the distance of the peak to peak after first-order differential quotient. Therefore, the signal intensity is proportional to the amplitude of the peak since $\Delta H_{pp}$ is a constant. According to a standard sample for ESR, the intensities of signals in spin numbers per gram are calculated. As you can see from Fig. 1, the signal intensity increases from zero to $1.67\times10^{16}$ as the Al/Zn ratio increases from zero to 1/100. Then, as the Al/Zn ratio increases from 1/100 to 2/100, the enhancement of the signal intensity is slower than before. However, when the ratio of Al/Zn is larger than 2/100, the variation of the intensity is very gentle. This variation of ESR signal intensity, obviously, is based on the amount of the paramagnetic

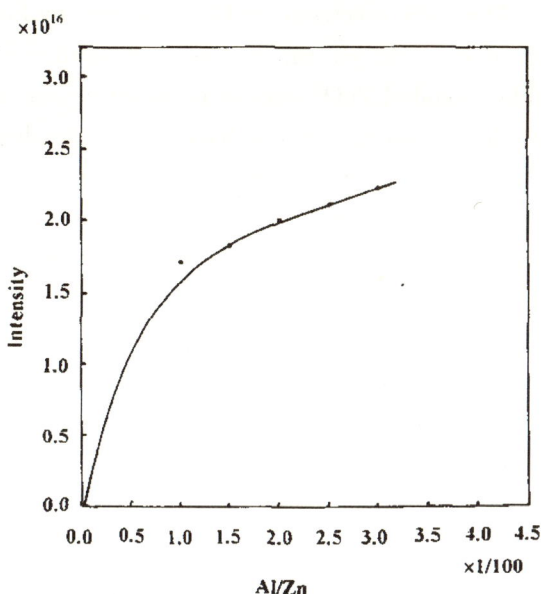

Fig. 1. The relationship between the ESR intensity and atomic ratio of Zn/Al.

species. A few outside objects, such as $Al_2O_3$, doped into ZnO crystal can produce a considerable amount of the paramagnetic species. However, as the amount of doping object continues to increase, the amount of paramagnetic species does not proportionally increase because the latter may be controlled by the thermodynamic equation conditions inside of the crystal.

Above ESR signals may be due to three possible factors: 1) The defects of monovalent cations were induced because of the doping of $Al_2O_3$ in the crystal of ZnO. 2) The precursors of ZnO, $ZnCO_3$, were decomposed at high temperature in the presence of $Al_2O_3$ and produced $Zn^+$ ions. 3) The paramagnetic species of oxygen was produced inside the crystal of ZnO. Now, we further discuss the later two possibilities based on the experimental results of both XPS and ESR spectroscopy.

The XPS spectra of Zn($L_3M_{4.5}M_{45}$) for ZnO and ZnO-$Al_2O_3$ are shown in Fig. 2. The band at $E_b=499.6$ eV(Fig. 2a) can be assigned to $Zn^{2+}$, but no bands can be assigned to $Zn^0$ with binding energy $E_b$ of 494.8 eV[8]. However, in both samples, there is an obvious shoulder peak at 496.6 eV(Fig. 2a and 2b) which is lower 3.0 eV than the peak of $Zn^{2+}$. This shoulder peak is not due to $Zn^+$ because the binding energy $E_b$ only changes about 1 eV as the oxidation

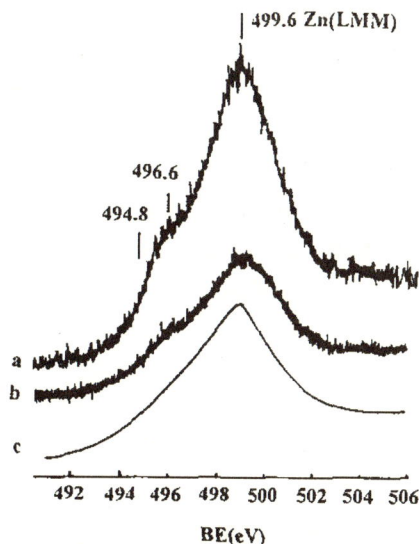

Fig. 2. XPS spectra of ZnO and ZnO-$Al_2O_3$. (a) ZnO; (b) ZnO-$Al_2O_3$; (c) Differential spectra of ZnO and ZnO-$Al_2O_3$.

state of an element changes one unit. Also, the shoulder peak is not due to the $Zn^+$ induced by $Al_2O_3$ since the shoulder peak disappeared as the analysis of differential spectra of both ZnO-$Al_2O_3$ and $Al_2O_3$ was carry out as shown in Fig. 2c. We think that the shoulder peak may be due to an unknown impurity.

**1093**

Therefore, according to the above analyses, the possibility that the ESR signal of the $ZnO-Al_2O_3$ sample is due to $Zn^+$ can be ruled out.

The XPS spectrum of $O_{ls}$ is shown in Fig. 3. The peak at $E_b = 532$ eV can be assigned to the $O^{2-}$ species inside crystal ol $ZnO^9$ since other negative ions of oxygen with lower valence states have higher binding energy than the $O^{2-}$ species. Meriaudean[10] and Gonzalez-Elipe[11] and their coworkers reported that the ESR signals assigned to $O_2^-$ 、 $O_3^-$ and $O_3^{3-}$ were observed for $TiO_2$, which was degassed under high vacuum, reduced with hydrogen at high temperature, again degassed under high vacuum and then oxidized, sequentially. These oxygen species, except of $O_2^-$, are all unstable at room temperature. There are three values of g for $O_2^-$ species, the highest being 2.0001. However, in this paper, the $ZnO-Al_2O_3$ sample did not undergo any degassing, reduction, or oxidation, and we observed only one ESR signal with the $g$ value of 1.9606, very far from the

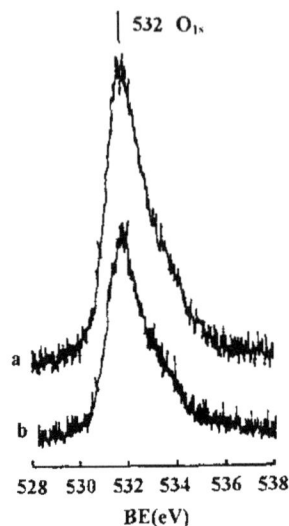

**Fig. 3. XPS spectra of $O_{ls}$ in ZnO and $ZnO-Al_2O_3$. (a) $ZnO-Al_2O_3$; (b) ZnO.**

characteristic $g$ value of 2.0001 for $O_2^-$. Therefore, according to the experimental results of both XPS and ESR, the ESR signal is not due to any paramagnetic species of oxygen.

We suggested that the ESR signal for the $ZnO-Al_2O_3$ sample was due to the doping effect of $Al_2O_3$. The $Al^{3+}$ replaces $Zn^{2+}$ in the crystal lattice of $-Zn^{2+}-O^{2-}-Zn^{2+}-O^{2-}$ and induces formation of the monovalent cationic defects, namely $2 Zn^{2+} \rightarrow Al^{3+} + \oplus$. The character of the ESR signal for this type of defect is a single, sharp peak[12]. These defects transfer so easily unto the surface that the concentration of defects there is larger than in the bulk[13]. The surface of lg ZnO has $Zn^{2+}$ numbers of approximately $7.33 \times 10^{19}$, about 6.57% of the total $Zn^{2+}$ numbers of both bulk and surfacc. As you can see from Fig. 2 the intensity of the XPS band of $Zn^{2+}$ for the $ZnO-Al_2O_3$ sample only is half that of the ZnO sample with the same total $Zn^{2+}$. In other words, as doping $Al_2O_3$ unto ZnO, the $Zn^{2+}$ on the surface is only about 3% of the total $Zn^{2+}$, and the other $Zn^{2+}$ of about 3% arc replaced by the monovalent cationic defects. The amount of these defects is about $2.1 \times 10^{18}$ per gram. However, the ESR results showed that the bulk spin number per gram of the $ZnO-Al_2O_3$ sample with an atomic ratio of $Zn/Al = 100/2$ is $2.0 \times 10^{16}$. Therefore, it is due to the enrichment of these defects on the surface that the surface concennation of $Zn^{2+}$ is 100 times as large as the bulk concentration[14].

Because the ion radius of $Al^{3+}$, 0.053 nm, is considerable different from the ion radius of $Zn^{2+}$, 0.074 nm, not much doping of $Al^{3+}$ into crystal of ZnO occurs, so the induced monovalent cationic defects in $ZuO-Al_2O_3$ arc also very few. According to the principle that ions of similar radii are more easily soluble with each other, the closer the ion radius between $M^{3+}$ and $M^{2+}$ is, the larger the induced effect is on defects. As the $Sc^{3+}$ with radius of 0.073 nm, as an outside ion, was doped into the system of $ZnO-Al_2O_3$, the results with ESR determonation are shown in Fig. 4, The $Sc^{3+}$ more easily induces the monovalent cationic detects as we predicted. In comparing the ESR results of the $ZnO-Al_2O_3-Sc_2O_3$ system (Fig. 4) with the $ZnO-Al_2O_3$ system (Fig. 1), the comprehensive effect of both $Al^{3+}$ and $Sc^{3+}$ on the formation of the monovalent cationic defects in ZnO, as the $Sc^{3+}$ is doped into the $ZnO-Al_2O_3$ sample, is more effective than $Al^{3+}$ at inducing the defects. Therefore, we can conclude that the formation of the monovalent cationic defects on the surface of $ZnO-Al_2O_3$ is due to the doping of the object, $M_2O_3$.

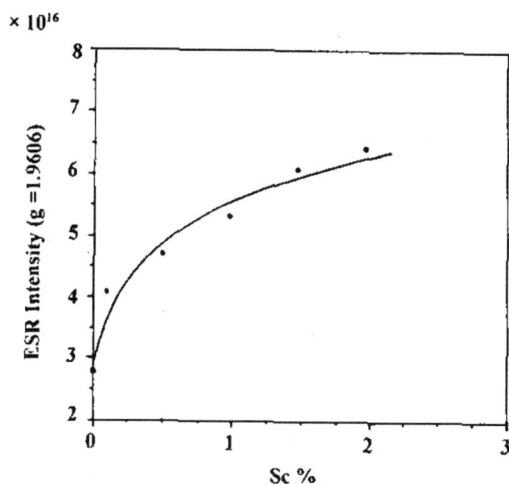

**Fig. 4.** Relationship of the intensities of ESR signal($g=1.9606$) with the content of trivalent metal ions in ZnO-Al$_2$O$_3$-Sc$_2$O$_3$. * Sc% = $[Sc^{3+}/(Zn^{2+}+Al^{3+}+Sc^{3+})] \times 100\%$.

## The effect of the monovalent cationic defects on stabilization of Cu$^+$ on catalyst surface

As mentioned above, the $M_2O_3$ in the ZnO-$M_2O_3$ sample caused the formation of the monovalent cationic defects and the enrichment of the defects on the ZnO surface. However, in Cu-ZnO-$M_2O_3$ catalyst, what effects on the valence states of copper on the surface of the catalyst do these monovalent cationic defects have? The X-ray induced Auger spectra and XPS spectra of the Cu-ZnO-Al$_2$O$_3$ with the atomic ratio of Cu : Zn : Al = 60 : 30 : 10 and Cu-ZnO-Al$_2$O$_3$-Sc$_2$O$_3$ with the atomic ratio of Cu : Zn : Al : Sc = 60 : 30 : 8 : 2, respectively, in reduction states are shown in Fig. 5 and Fig. 6. As shown in Fig. 5, both the peak at $E_b = 567.5$ eV assigned to Cu$^0$(2p) and the shoulder peak at $E_b = 569.5$ eV assigned to Cu$^{+15}$ were observed for the above two catalysts, As you can also see from Fig. 6, the XPS of the two catalysts showed a shoulder peak at 76.7 eV, which is in good agreement with the Cu$^+$(3p) XPS peak of CuCl and Cu$_2$O at 76.8 eV observed by Frost et al. [16] Obviously, there are a lot of Cu$^0$ and a few Cu$^+$ on the surfaces of reductive catalysts. The amount of Cu$^+$ on the surfaces is dependent on the composition of catalysts, but the amount of Cu$^+$ on the surfaces of catalysts with Sc is higher than those without Sc. The TPR spectra showed that there were two distinct reduction processes[17] (Fig. 7). As Sc$^{3+}$ content increased in catalysts, maximum peak assigned to Cu$^{2+} \rightarrow$ Cu$^+$ shifted to low temperature, but another maximum peak assigned to Cu$^+ \rightarrow$ Cu$^0$ moved to high temperature. These are due to that dispersion of Cu on the surfaces of catalysts increases with increase of special surface area from 23.6 m$^2$/g for Cu-ZnO-Al$_2$O$_3$ with the atomic ratio of Cu : Zn : Al = 60 : 30 : 10 to 40.2 m$^2$/g for Cu-ZnO-Al$_2$O$_3$-Sc$_2$O$_3$ with the atomic ratio of Cu : Zn : Al : Sc = 60 : 30 : 8 : 2 as the content of Sc$^{3+}$ increases, so reduction of Cu$^{2+}$ to Cu$^+$ becomes easier. However Cu$^+$ can not be reduced easily due to the exisience of monovalent cationic defects on the surface of catalysts, so the $M_2O_3$ in copper-based catalyst stabilizes Cu$^+$ on the surface in the reduction process.

## The relation of defect numbers in Cu-ZnO-Al$_2$O$_3$-Sc$_2$O$_3$ with catalytic activity

The relation of both the catalytic activity for methanol synthesis and the intensity of ESR with the content of scandium in Cu-ZnO- Al$_2$O$_3$-Sc$_2$O$_3$ is shown in Fig. 8. As can be seen from Fig. 8, as the ratio Sc$^{3+}$/(Cu$^{2+}$ + Zn$^{2+}$ + Al$^{3+}$ + Sc$^{3+}$) increases, both the monovalent cationic defects and the catalytic activity for methanol synthesis also increase. A small amount of $Sc_2O_3$ promoted methanol synthesis very ellectiveiy.

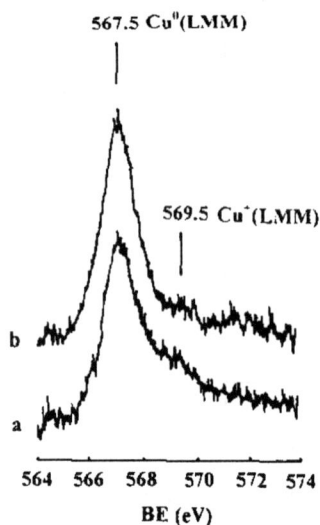

Fig. 5. The X-ray induced Auger spectra of the Cu-ZnO-Al$_2$O$_3$ and Cu-ZnO-Al$_2$O$_3$-Sc$_2$O$_3$ catalysts after reduction, (a) Cu-ZnO-Al$_2$O$_3$ (atomic ratio of Cu : Zn : Al=60 : 30 : 10). (b) Cu-ZnO-Al$_2$O$_3$-Sc$_2$O$_3$ (atomic ratio of Cu : Zn : Al : Sc=60 : 30 : 8 : 2).

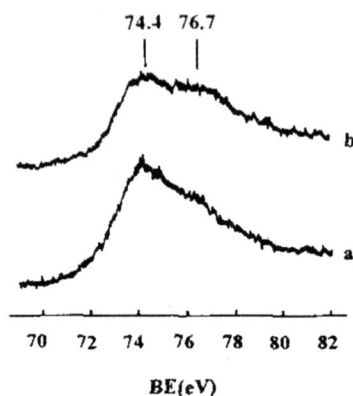

Fig. 6. XPS spectra of the Cu-ZnO-Al$_2$O$_3$ and Cu-ZnO-Al$_2$O$_3$-Sc$_2$O$_3$ catalysts after reduction. ( a ) Cu-ZnO-Al$_2$O$_3$ ( atomic ratio of Cu : Zn : Al = 60 : 30 : 10 ), (b) Cu-ZuO-Al$_2$O$_3$-Sc$_2$O$_3$ ( atomic ratio of Cu : Zn : Al : Sc=60 : 30 : 8 : 2).

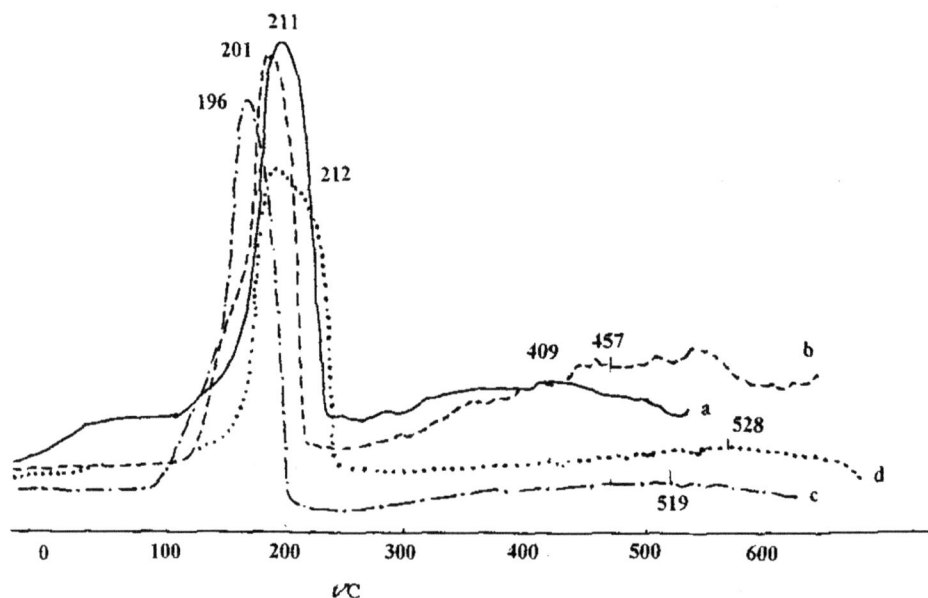

Fig. 7. TPR spcctra of Cu-ZnO-M$_2$O$_3$ : (a) Cu-ZnO-Al$_2$O$_3$ ( atomic ratio of Cu : Zn : Al=60 : 30 : 10). (b) Cu-ZnO-Al$_2$O$_3$-Sc$_2$O$_3$ ( atomic ratio of Cu : Zn : Al : Sc=60 : 30 : 8 : 1) , ( c ) Cu-ZnO-Al$_2$O$_3$-Sc$_2$O$_3$ ( atomic ratio of Cu : Zn : Al : Sc=60 : 30 : 8 : 2) , (d) Cu-ZuO-Al$_2$O$_3$-Sc$_2$O$_3$ ( atomic ratio of Cu : Zn : Al : Sc= 60 : 30 : 8 : 3).

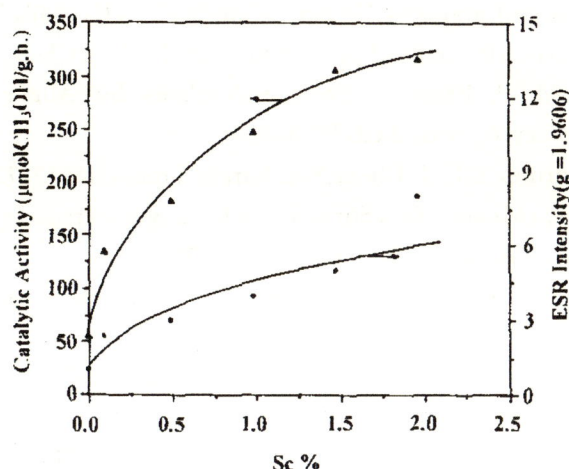

**Fig. 8.** The relation of both the catalytic activity for methanol synthesis and the intensity of ESR to the content of $Sc^{3+}$ in Cu-ZnO-$Al_2O_3$-$Sc_2O_3$ catalyst. * $Sc\% = [Sc^{3+}/(Cu^{2+} + Zn^{2+} + Al^{3+} + Sc^{3+})] \times 100\%$.

## 4  CONCLUSIONS

Based on the experiments with ESR, XPS, and the evaluation of catalytic activity, we can draw the following conclusion: 1) In the system with two components, ZnO-$M_2O_3$, doping of a few trivalent metal ions such as $Al^{3+}$ and $Sc^{3+}$, favors formation of the monovalent cationic defects on the crystal surface of ZnO. 2) In the system with three components, CuO-ZnO-$M_2O_3$, due to doping of a small amount of $M_2O_3$, the monovalent cationic defects formed on the surface of catalysts favor both enrichment and stabilization of $Cu^+$ on the surface in the reduction process. Therefore, these favor an increase in the catalytic activity for methanol synthesis of the copper-based catalysts and stabilize them under reduction conditions as well as extend their life. The above conclusions, obviously, are helpful to both modification of the copper-based catalysts and the development of new catalysts for methanol synthesis.

## ACKNOWLEDGMENT

This work has been supported by Natural Science Foundation of China.

Received February 24, 1998.

**Key Words**  Copper  Zinc  Aluminum  Scandium  Oxide  Catalysis  Defect  XPS  EPR TPR.

## REFERENCES

[1] Herman, R. G. ; Klier, K. ; Simmons, G. W. ; Finn, B. P; Bulko, H, B. J. Catal. 1978, **56**, 407.

[2] Klier, K. ; Chatikavanij, V. ; Herman, R. G. ; Simons, G. W. J. Catal 1982, **74**, 343.

[3] Chinchen, G. C. ; Denny, P. J. ; Jenning, J. R. ; Spencer, M. S. ; Waugh, K. Appl. Catal. 1988, **36**, 1.

[4] Tohji, K. ; Udagawa, Y. ; Mizushima, T. ; Ueno, A. J. Phys. Chem. 1985, **89**, 5671.

[5] Edwards, J. F. ; Schrader, G. L. J. Catal. 1985, **94**, 175.

[6] Chinchen, G. C. ; Waugh, K. J. Catal. 1986, **97**, 280.

[7] Chen, H. -B. ; Wang, S. -J. ; Liao、Y. -Y. ; Tsai, J. -X. ; Zhang, H. -B. ; Tsai, K. -R, in Proc. 9th

International congress on Catalysis, Calgary,Canada,Vol,II,1988,537.

[8]Okamoto,Y. ;Fukino,K. ;Imanaka,T. ; Teranishi,S. J. Phys. Chem. 1983,**87**, 3747.

[9] Briggs, D. ; Seah, M. P. Practical Surface Analysis by Auger and X-ray Photoelectron Spectroscopy;John Wiley & Sons Led,1983,330.

[10]Meriaudeau,P. ;Vedrine, J. C. J. Chem Soc. Farad. Trans,II1979,**72**,472.

[11]Gonzalez-EIipe,A. R. ;Munera,G. ;Soria,J. J. Chem. Soc. Farad. Trans. I1979,**75**,748.

[12]Ng,C. F. ;Leung,K. S. ;Chan,C. K. J. Catal. 1982,**78**,51.

[13]Kittel,C. in Introduction to Solid State Physics; 2nd ed. New York,Wiley,1956,Chapter 17,p 477.

[14]Chen,H. -B. Ph. D. Thesis,Xiamen University,China,1986,52.

[15]Flesch,T. H. ;Mieville,R. L. J. Catal. 1984,**90**,165.

[16]Frost,D. C. ;Ishitani,A. ;McDowell,C. A. Mol. Phys. 1972,**24**,861.

[17]Hurst,N. W. ;Gentry,S. J. ;Jones,A. Catal,Rev. 1982,**24**, 233.

■ **本文原载**:《高等学校化学学报》第 19 卷第 5 期(1998 年 5 月),第 765～769 页。

# 催化裂解 CH₄ 或 CO 制碳纳米管结构性能的谱学表征*

陈 萍 张鸿斌① 林国栋 蔡启瑞

(厦门大学物理化学研究所 固体表面物理化学国家重点实验室,厦门 361005)

**摘 要** 利用 TEM、HRTEM、XRD、XPS 和 TPO 等方法对 CH₄ 或 CO 催化分解生成的碳纳米管结构和性能进行了表征。结果表明,所得产物是管径 15～20 nm 的均匀碳纳米管。其 XRD 谱图与石墨的相近,但特征衍射峰稍宽化,表明其长程有序度较石墨的低。由 CH₄ 制备的碳纳米管系由多层具有类石墨片状结构的同心、等径及中空圆锥形面叠合而成,类石墨层面取向与管轴倾斜;而由 CO 制备的碳纳米管系由多层具有类石墨片状结构的圆柱形面围叠而成,类石墨层面取向与管轴平行。碳纳米管中 $C_{1s}$ 的电子结合能比石墨的下降约 0.5eV。TPO 试验结果显示所制备的两种产物中无定形碳含量都很低,其整体结构石墨化程度较高;由 CH₄ 制得的碳纳米管与 $O_2$ 反应的起燃温度比由 CO 制得的约高 100 K。

**关键词** 碳纳米管 碳纳米结构 催化裂解 CH₄ CO

**分类号** O643.35

碳纳米管在结构上与 $C_{60}$ 球烯同类,是近年来大力开发的纳米级碳素材料[1—11]。它们均由单个碳原子在一定条件下自然形成。典型的碳纳米管系由六元环组成的类石墨平面叠合而成的圆筒形管状结构,具有尺寸小、机械强度高、比表面大、电导率高、界面效应强等特点,在催化、吸附-分离、储能器件电极材料等诸多领域具有重要应用前景。最初发现的碳纳米管由碳电弧法产生,该法制得的碳纳米管管直,结晶度高,但产率低,且分离纯化困难。利用烃($C_2H_2$,$C_6H_6$ 等)在超细金属催化剂(N i 或 N i-Cu 合金)上分解长成碳纳米管是稍后发展起来的另一类方法[1],但该法制备条件比较苛刻(>973 K),产品结晶度不如碳电弧法高,且管径分布较宽(2～100 nm),粗细不均匀。

前文[9,11]曾报道在负载 Ni 催化剂上于较低温度(≤973 K)下制备管径均匀碳纳米管的方法。本文采用 TEM、HRTEM、XRD、XPS 及 TPO 等方法对由 CH₄ 或 CO 制备的管状碳纳米纤维材料的结构和性能进行了比较研究。

## 1 实验部分

### 1.1 催化剂的制备

催化剂制备按文献[11]方法进行:将一定量的硝酸镍与相应载体的硝酸盐依特定方法充分混合、烘

---

* 国家自然科学基金和国家教育委员会博士点基金资助课题。

① 联系人:张鸿斌。第一作者:陈萍,女,28 岁,博士研究生。

干,于 1 073 K 下空气气氛中灼烧 5 h,得粉末状催化剂前驱物。

### 1.2 碳纳米管的制备

碳纳米管的制备亦按文献[11]方法进行:取 20 mg 催化剂前驱物置于固定床常压连续流动反应器内,于 873 K、$H_2$ 气氛下预还原 0.5 h,迅速转换到反应所需温度,在一定流速下导入纯 $CH_4$ 或 CO 原料气,反应 0.5 h 后渐冷,收集样品,经稀硝酸溶液浸溶去附于碳纳米管末端的催化剂颗粒,再经水洗、烘干后,于 473 K 氮气流吹扫 1 h,得纯化过的碳纳米管。

### 1.3 谱学表征

分别在 JEOL JEM-100 CX 型电子透射显微镜上及 Hitachi H-9000 型电子透射显微镜上进行 TEM 及 HRTEM 观测;在 Rigaku D/Max-C 型 X 射线衍射仪上进行 XRD 测试,Cu $K\alpha$($\lambda = 0.154\,8$ nm)为辐射源,扫描速度 4°/min;在 VG ESCA LAB MK-II 型光电子能谱仪上进行 XPS 测试,以 $MoO_3$ 中 $Mo_{3d}$ 电子结合能为内标;用固定床常压连续流动反应器-色谱组合系统进行 TPO 试验,以 $He/O_2$(体积比为 95∶5)混合气为反应气(1 000 mL/h),升温速度 20 K/min。

## 2 结果与讨论

图 1 为 873 K 下 Ni 基催化剂上由 $CH_4$ 裂解和 CO 歧化分别制得的两种碳纳米纤维样品的 TEM 图象。由图 1 可见,所得产物为圆柱形碳纤维,长度可达微米甚至毫米级,直径小于 25 nm(以 $CH_4$ 及 CO 制备的碳纳米纤维断面直径分别为 ～20 nm 及 ～15 nm)。

从图 1 还可看出,圆柱纤维状产物的中心部分电子透射率较高,表明其中心区密度较低,甚至是中空的;可见该纤维状产物可归属为碳纳米管。未经纯化的粗产物末端附着有电子透射率较低的催化剂颗粒,其大小与碳纳米管的断面直径相当,形状似为较规则的多面体,并有部分表面裸露在反应气氛中供作反应活性表面。在碳纳米管粗产物的纯化处理过程中,该催化剂颗粒可被溶去。由 $CH_4$ 催化裂解或 CO 歧化制备的碳纳米管形状较为弯曲,这可能是由于制备过

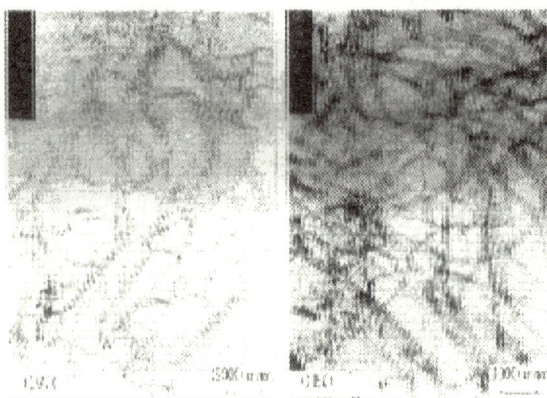

Fig. 1　TEM images of carbon-nanotubes

(A)Decomposition of $CH_4$;

(B)Disproportionation of CO.

程中,反应条件的波动造成碳原子在有序堆集排列过程中出现缺位或错位所致。由 $CH_4$ 制备的碳纳米管管径较由 CO 制备的略粗,管长也较短。

反应前后催化剂金属颗粒粒度的变化表明,在生成碳纳米管初期,伴随着 $CH_4$ 或 CO 分解以及 C 的生成、扩散和堆集,催化剂颗粒也不同程度地碎裂,并以 CO 为原料气时尤甚[10]。

两种碳纳米管的 XRD 分析结果(图 2)表明,其 XRD 特征主峰位置(2θ＝26.2°)与石墨的(2θ＝26.6°)相近,但峰形较宽,强度较低,表明碳纳米管结构的长程有序度较石墨的低;此外,碳纳米管相结构中还伴有亮石墨相(2θ＝21°和 23.9°)[12]出现。比较图 2(A)和 2(B)可见,由 $CH_4$ 制备的碳纳米管衍射峰相对较强、峰宽较窄,暗示其管壁中碳原子堆集有序度较高。

由纯化后碳纳米管样品的 XPS 分析(图 3)可见,位于 284.0 eV 附近的 $C_{1s}$ 峰略显不对称,可拟合为两个峰,其 $C_{1s}$ 结合能分别为 283.8 和 285.0 eV;结合能较高者可归属为污染碳及无定形碳的 $C_{1s}$ 峰,而位于 283.8 eV 的较强峰则可被指认为碳纳米管样品中类石墨状碳的 $C_{1s}$ 峰;两种碳纳米管类石墨状碳的

$C_{1s}$结合能均较一般石墨的(284.3 eV)低;可见在碳纳米管管状结构中的电子似变得更易于流动、逃逸。两种碳纳米管的$C_{1s}$结合能的微小差别可能与管壁中类石墨状碳的不同堆集方式(如类石墨层面相对于管轴的不同取向)有关。

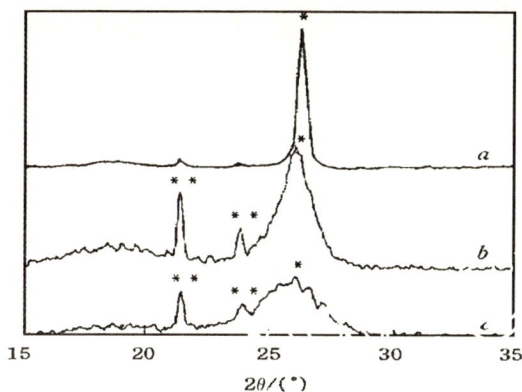

**Fig. 2  XRD patterns of graphite(a), carbon-nanotubes produced by decomposition of CH₄ (b) and disproportionation of CO(c)**

* Due to graphite; ** Due to chaoite

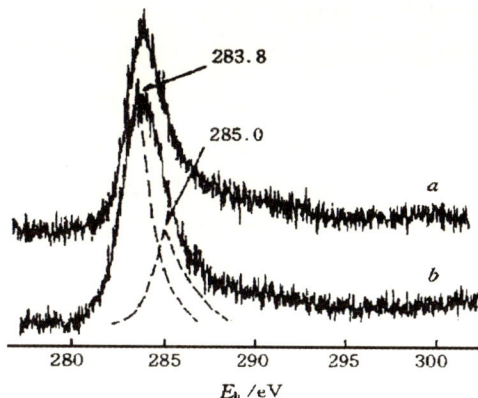

**Fig. 3  XPS spectra of carbon-nanotubes produced**

a. Decomposition of CH₄;

b. Disproportionation of CO.

利用 HRTEM 可在准原子水平上对碳纳米管中 C 原子的排列方式作更直接的观察,经 50 万倍放大后,碳纳米管的层状结构清晰可见(图 4)。

由 CO 制备的碳纳米管系由许多层具有类石墨片状结构的圆柱形面围叠而成,类石墨层面取向与管轴平行[图 4(A)];而由 CH₄ 制备的碳纳米管则由许多层具有类石墨片状结构的同心、等径、中空圆锥形面叠合而成,类石墨层面取向与管轴倾斜[图 4(B)]。

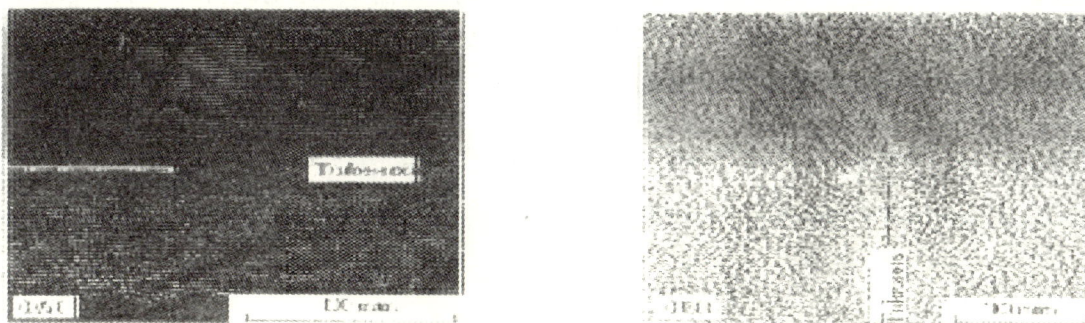

**Fig. 4  HRTEM images of carbon-nanotube**

(A)Disproportionation of CO;(B)Decomposition of CH₄.

由图 4(A)可见,由催化法制备的碳纳米管管壁存在明显缺陷:既有大范围的类石墨层面与层面间的错位,也存在点和线缺陷。碳纳米管的中心区透明度(即电子透过率)高,除内管壁可能沾粘少许无定形碳[间或也观察到个别斜插的碳束(须)]外,基本上是空的;而处于管壁内层的碳原子排列较为规整,即使出现长程的缺陷,其短程有序度也较高。管壁外层呈起伏状,表明碳原子排列无序度较高,这可能与反应条件下气相分子与成长中的碳纳米管外壁表面碳原子发生反应,导致表面遭受不同程度的"侵蚀"有关。在同一种 Ni 基催化剂上,两种碳纳米管的管壁类石墨层面相对于管中心轴线的取向存在明显差异(即 CH₄ 基的与管轴倾斜,而 CO 基的与管轴平行)的原因可能与两种原料气分子催化分解方式不同及其副产物(H₂ 或 CO₂)不同有关。不同的气相副产物可在催化剂上吸附,从而影响了催化剂金属颗粒表面碳纳米管生长区的微环境及碳纳米管管基界面区碳原子的聚集排列堆砌方式。

为获得有关碳纳米管中碳原子有序化排列、堆砌的规整度及整个管状结构抗氧化性能的信息,考察了 2 种碳纳米管的 TPO 反应活性(图 5)。由图 5(a)可见,由 CO 歧化制备的碳纳米管的氧化反应起始温度约为 650 K。随着温度升高,690 K 处出现一个弱的 $CO_2$ 响应峰,从 780 K 开始,$CO_2$ 的响应曲线急剧上升,960 K 时出现一个以 960 K 为峰温位置的不对称强带,表明有大量碳急剧氧化燃烧生成 $CO_2$;之后该峰迅速下降,直至 1 020 K TPO 过程基本结束。690 K 处的弱峰可归属于少量无定形碳(主要位于碳纳米管表面及中心区)的氧化燃烧峰,该峰面积小,表明碳纳米管中的无定形碳含量少,产物结晶度高;而峰温位置在 960 K 的不对称强带则可指认为碳纳米管管壁类石墨形态碳的氧化燃烧峰。随着氧

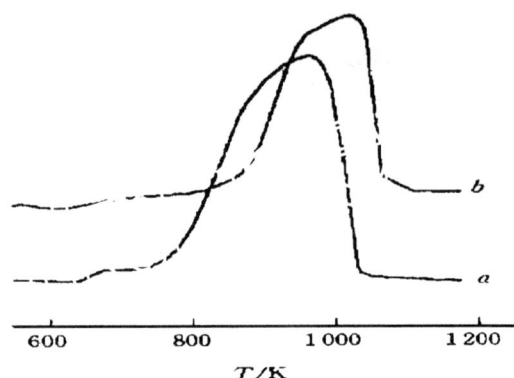

**Fig. 5  The TPO spectra of carbon-nano tubes**
  *a*. Disproportionation of CO;
  *b*. Decomposition of $CH_4$

化燃烧的进行,碳数量不断减少,碳氧化燃烧总反应表面积并非恒定;伴随着可观测量燃烧热的释放,反应床层温度可能局部急剧上升,因而整个反应过程并非处于稳态,这可能是造成 TPO 主峰峰形不对称,高温侧曲线急剧下降的主要原因。对于 $CH_4$ 分解制备的碳纳米管,TPO 过程也显示出类似的特点,只是 690 K 处的无定形碳氧化燃烧峰不明显,表明含无定形碳量很少,产物结晶度更高。880 K 处类石墨状碳的氧化燃烧峰急剧上升,至 1 020 K 达到峰顶,尔后迅速下降,至 1 080 K TPO 过程基本结束。该峰的起始温度(880 K)比由 CO 制备的碳纳米管约高 100 K,表明由 $CH_4$ 制备的碳纳米管比由 CO 制备的稳定性高,有较强的抗氧化能力。

为进一步考察 TPO 过程中碳纳米管表面碳的氧化燃烧状况,在另一组实验中,分别于 873 K 或 973 K 时中断 TPO 反应并急速冷却,收集相应试样作 TEM 观测。由图 6 所示的观测结果可以发现,由 $CH_4$ 分解制备的碳纳米管经 TPO 至 973 K 后,其外管壁表面已显得凹凸不平,但管端首尾部形貌则变化不大;看来碳的氧化燃烧并非从管的两端开始,而是沿管壁由表及里地进行。

对于以 CO 为原料气制备的碳纳米管在经由相同的 TPO 操作至 873 K 后,其形貌与由 $CH_4$ 制备的碳纳米管比较则有所不同:结晶度较低的碳纳米管的氧化燃烧可从管壁或从端面同时发生;结晶度较高的碳纳米管(管粗且直)的氧化燃烧则主要从管的端面开始(从图 6 可见碳纳米管的端面呈凹凸不平状)。

联系到两种碳纳米管产物中类石墨"碳六元环"面相对于管轴线的不同取向(即由 $CH_4$ 制得的碳纳米管,管的端面是"六元环"面,管壁圆柱体系"六元环"面的边线重叠堆砌而成;而由 CO 制得的碳纳米管,管壁的圆柱面就是"六元环"面)可推测,"六元环"面内

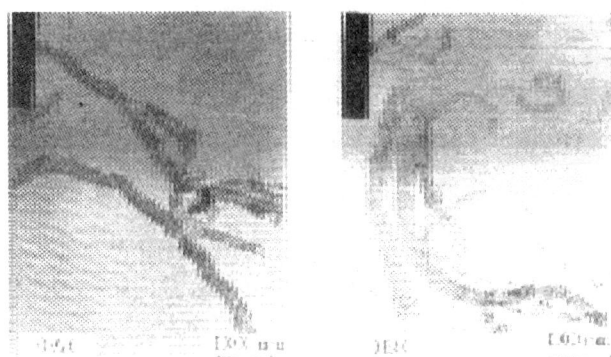

**Fig. 6**  **TEM images of carbon-nanotubes produced by (A) decomposition of $CH_4$, which underwent a TPO testing from room temperature to 973 K and (B) disproportionation of CO, which underwent a TPO testing from room temperature to 873 K**

的碳原子比"六元环"面周边的碳原子稳定,氧化燃烧反应应从"六元环"面周边不稳定的碳原子开始。

# 参考文献

〔1〕Rodriguez N. M.. J. Mater Res,1993,**8**(12):3 233.

〔2〕Iijima S. Nature,1991,**354**:56.

〔3〕Baker R. T. K . Carbon,1989,**27**:315.

〔4〕Motojima S,Kawaguchi M.,NozakiK. et al. Appl Phys. Lett,1990,**56**:321.

〔5〕Yudasaka M.,Kikuchi R.,M atsui T. et al. Appl Phys Lett,1995,**67**(17):2 477.

〔6〕Ivanov V.,Fonseca A.,N agy J. B. et al. Carbon,1995,**33**(12):1 727.

〔7〕Endo M.,Takeuchi K.,Kobori K. et al. Carbon,1995,**33**(7):873.

〔8〕A lstrup I. J. Catal 1988,**94**:468;1988,**109**:241.

〔9〕Chen Ping(陈萍),Wang Pei-Feng(王培峰),Lin Guo-Dong(林国栋)et al. Chem. J. Chinese Universities(高等学校化学学报),1995,**16**(11):1 783.

〔10〕Chen Ping(陈萍),Zhang Hong-Bin(张鸿斌),Lin Guo-Dong(林国栋) et a l.. J. Xiamen Univ. (Nat. Sci. )(厦门大学学报,自然科学版),1996,**35**(1):61.

〔11〕Chen Ping,Zhang Hong-Bin,Lin Guo-Dong et al. Carbon,1997,**35**(10 & 11):1495.

〔12〕Goresy,Donnay. Science,1968,**161**:363.

## Studies on Structure and Property of Carbon-nanotubes Formed Catalytically from Decomposition of CH₄ or CO

Ping Chen，Hong-Bin Zhang*，Guo-Dong Lin，Khi-Rui Tsai

(Institute of Physical Chemistry & State Key Lab of Physical Chemistry on the Solid Surfaces，Xiamen University，Xiamen，361005)

**Abstract**　By means of TEM,HRTEM,XRD,XPS and TPO methods,structure and properties of the carbon-nanofibers formed catalytically from decomposition of CH₄ or CO have been investigated. The TEM observation shows that these materials are even with nanotubes the outer diameters of $15\sim20$ nm. Their XRD patterns are very close to that of graphite,but the XRD features are somewhat broadened,indicating that the degree of long-range order of these nanostructures is relatively low in comparison with that of graphite HRTEM observation reveals that the wall of these nanotubes is constructed by many layers of carbon with graphite-like platelets in an extremely ordered arrangement, with the orientation of the conical graphite-like platelets inclined to the central axis of the tube for the material produced from CH₄ and the cylindrical graphite-like platelets parallel to the tube-axis for that from CO,respectively. XPS measurements show that $C_{1s}$ electron binding energy in these carbon-nanotubes is about 0.5 eV lower than that of graphite The results of the $O_2$-TPO testing demonstrate that content of amorphous carbon in these materials is very low,and then the nanostructures are entirely graphitic in nature; the TPO results also show that the temperature for gasification of the material produced from CH₄ is about 100 K higher than that from CO,indicating that the fomer has a better oxygen-resistibility than the latter.

**Key words**　Carbon-nanotube　Carbon-nanostructure　Catalytic decomposition　CH₄　CO

■ **本文原载**:《化学物理学报》第 11 卷第 5 期(1998 年 10 月),第 456～460 页。

# 铜基催化剂表面缺陷对催化性能的影响[*]

陈鸿博[①]　于腊佳　廖代伟　林国栋　张鸿斌　蔡启瑞

(厦门大学物理化学研究所　化学系　固体表面切理化学国家重点实验室,厦门 361005)

**摘　要**　用顺磁共振谱(ESR)研究外来离子含量与由这些正三价金属离子在 ZnO 表面诱导出来的正一价缺位数量之间的关系,结果表明,$Al^{3+}$ 和 $Sc^{3+}$ 对 ZnO 表面的正一价缺位都有诱导作用,其中 $Sc^{3+}$ 的掺杂对正一价缺位的诱导更有效。TPR 结果显示,随着样品中 $Sc^{3+}$ 含量的增加,归属于 $Cu^{2+} \rightarrow Cu^{+}$ 和 $Cu^{+} \rightarrow Cu^{0}$ TPR 谱最高峰温分别向低温和高温方向移动。在 Cu-ZnO-$M_2O_3$(M= Al、Sc)催化剂中,ESR 信号强度与催化活性两者随样品中 $Sc^{3+}$ 含量的增加而增加,这说明催化剂的甲醇合成催化活性与 $M^{3+}$ 掺杂后诱导出来的正一价缺位及还原态催化剂表面 $Cu^{+}$ 的稳定性之间存在着一定的关系。

**关键词**　铜基催化剂　甲醇合成　CO 氢化　缺位

## 1　前　言

晶体缺陷对晶体的物理化学性质起着重要的作用。在 ZnO 晶体中存在极少量的本征缺陷,当外来正三价离子掺入后,通过原子价诱导,在 ZnO 晶格中产生出一定数量的正一价缺位。本文较系统地研究在 ZnO、Cu-ZnO 中分别掺杂 $Al_2O_3$、$Al_2O_3$＋$Sc_2O_3$ 后产生的正一价缺位数量的变化,并且将掺杂量与缺陷数及催化剂的甲醇合成催化性能关联起来,以期说明正三价金属离子对 Cu-ZnO 催化剂的助催化作用本质及 Cu、ZnO,$Al_2O_3$ 在催化甲醇合成反应中的协同催化作用机理。

## 2　实验部分

### 2.1　催化剂制备

本文使用的催化剂均采用并流共沉淀法制备,即将按比例混合的硝酸盐溶液和 $Na_2CO_3$ 溶液同时滴加到少许的去离子水中,在剧烈搅拌的同时,保持混合液的温度为 85 ℃,pH=6.5～7.0。沉淀经熟化、过滤、洗涤,在 110 ℃干燥,并在 350 ℃灼烧 3～4 h。

### 2.2　ESR 波谱测定

ESR 测定是在 Bruker ER-200D-SRC 型顺磁共振仪上进行的,将粉末样品 80 mg 置于优质玻璃管($\phi$3)中,在 X 波段、输出功率 20.4 mW、调制频率 100 kHz、室温下录谱。

### 2.3　程序升温还原(TPR)实验

TPR(Temperature Program Reduction)实验装置是由气相色谱、反应炉和程序升温控制系统组成。

---

＊　国家自然科学基金和福建炼油厂资助项目。

①　通讯联系人。

实验时将 30 mg(0.25～0.45 mm)催化剂装入反应器中,通入 $H_2/N_2$(体积比为 5/95,按 3 ℃/min 速率程序升温还原,用 X-Y 记录仪记录其实验结果。

### 2.4 催化活性评价

将催化剂粉末成型、过筛,取 0.25～0.45 mm 颗粒作为催化活性测试样品。将 1 g 左右的催化剂样品装入不锈钢反应器($\phi$8)中,用 $H_2/N_2$(体积比为 5/95)还原气于常压、流速 30 mL/min 条件下还原。升温程序按文献[3]进行,甲醇合成催化活性评价条件:反应压力 1.0 MPa,反应温度(230±0.5)℃;合成气组成 $CO/H_2$(体积比为 1/2),合成气流速 30mL/min。产物用气相层析仪分析。

# 3 结果和讨论

## 3.1 $Sc_2O_3$ 对 $ZnO-Al_2O_3$ 缺位的诱导效应

我们曾对 ZnO、$ZnO-Al_2O_3$ 和 $Al_2O_3$ 样品进行了 ESR 测定[1],发现在相同的制备和测试条件下的 ESR 谱有很大的差异,$ZnO-Al_2O_3$ 样品存在一个尖锐的 $g=1.9606$ 的 ESR 信号(在我们的测试条件下 ZnO、$Al_2O_3$ 没有 ESR 信号),为了确定此信号的归属,我们用 XPS-Auger 和 XPS 谱对 $ZnO-Al_2O_3$ 进行表征,排除了此信号来源于 $Zn^+$ 和顺磁性氧物种的可能性。因此,我们将 $ZnO-Al_2O_3$ 样品中观测到的 ESR 信号归属于 $Al_2O_3$ 客体的掺杂,在 ZnO 晶体中产生的正一价阳离子缺位,这些缺位的形成是晶体中价态和电荷平衡的综合结果,即 1 个 $Al^{3+}$ 和 1 个正一价阳离子缺位代替了 ZnO 晶格中的两个 $Zn^{2+}$ 的平衡位置。但由于 $Al^{3+}$ 的离子半径(0.053nm)与 $Zn^{2+}$ 的离子半径(0.074nm)相差较大,根据相近相溶原理,$Al^{3+}$ 溶入 ZnO 晶格中的量不多,因此所诱导的正一价缺位也不多。如果选择正三价客体离子的离子半径接近 $Zn^{2+}$,理应产生较强的缺位诱导效应。我们选择了 $Sc^{3+}$(0.073 nm)作为外来离子,对 $ZnO-Al_2O_3$ 体系进行掺杂,经 ESR 测定,其结果如图 1b 所示。如我们所料,$Sc^{3+}$ 比 $Al^{3+}$ 容易诱导出正一价缺位,而且对于同一种掺杂离子,其掺杂量与正一价缺位的数量存在一定的关系。图 2 比较了 $ZnO-Al_2O_3$、$ZnO-Al_2O_3-Sc_2O_3$ 两个体系的 ESR 测定结果。从图 2a 可见,样品的 ESR 信号随 Al 的百分含量变化而呈现规律性变化,当 Al 含量从 0 增至 2.0%时,用以表示样品正一价缺位的 ESR 信

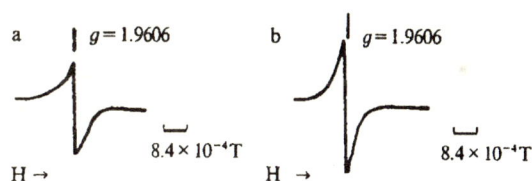

**图 1 $ZnO-Al_2O_3$ 和 $ZnO-Al_2O_3-Sc_2O_3$ 的 ESR 谱**
**Fig. 1 ESR spectra of $ZnO-Al_2O_3$ and $ZnO-Al_2O_3-Sc_2O_3$**
a. $ZnO-Al_2O_3$ (100/2); b. $ZnO-Al_2O_3-Sc_2O_3$ (100/2/0.1)

号强度(以谱峰振幅表示)相应地从 0 上升到 28 mm,以后 Al 含量从 2%增加到 5%,$g=1.9606$ 信号强度变化趋于平缓。这可能是受到 $Al^{3+}$ 在 ZnO 中的溶解度和 ZnO 晶体内部热力学平衡所控制。图 2b 示出 $ZnO-Al_2O_3$ 样品中掺入 $Sc^{3+}$ 后,$Al^{3+}$、$Sc^{3+}$ 两者对 ZnO 中正一价缺位形成的综合结果。显然,少量 $Sc^{3+}$ 的掺杂比 $Al^{3+}$ 对正一价缺位的诱导更有效。当 $ZnO-Al_2O_3$(100/2,原子比)样品用 $H_2/N_2$ 还原后,$g=1.9606$ 的 ESR 谱峰向高场移动 $2.3\times10^{-3}$ 高斯,谱线增宽,而且 ESR 信号强度略有增加(图 3),ESR 谱变化的原因可能是在还原后的 $ZnO-Al_2O_3$ 样品中产生了氧缺位($g=1.96$)[3],这些缺位与上述的正一价缺位的 ESR 信号叠加在一起。

## 3.2 还原前后 $Cu-ZnO-Al_2O_3$ 的 ESR 表征

为了进一步证实 2.1 节中 $g=1.9606$ ESR 信号的归属,我们比较了 $Cu-ZnO-Al_2O_3$ 催化剂还原前后的 ESR 谱。图 4 示出氧化态 $Cu-ZnO-Al_2O_3$ 催化剂存在 $Cu^{2+}$ 的 ESR 信号 $g=2.1$ 和 $g=1.9606$ 信号。ZnO、$Al_2O_3$ 在相同增益下无顺磁信号。用 $H_2/N_2$(体积比为 5%)还原 16 h 后,归属于 $Cu^{2+}$ 的顺磁

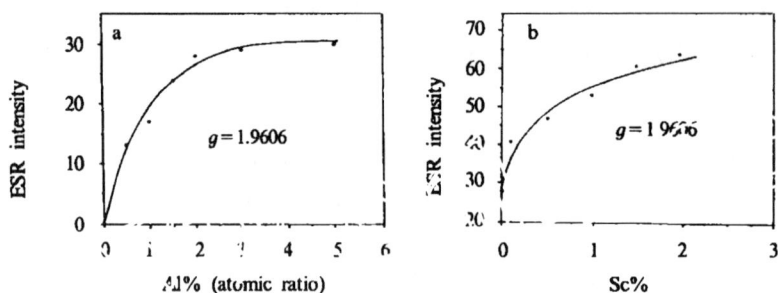

图 2    ZnO-Al$_2$O$_3$ 和 ZnO-Al$_2$O$_3$-Sc$_2$O$_3$ 样品的 ESR 信号强度与正三价金属离子含量的关系

Fig. 2    The relationship between ESR signal intensity and the content dopping trivalent cations on ZnO-Al$_2$O$_3$ and ZnO-Al$_2$O$_3$-Sc$_2$O$_3$

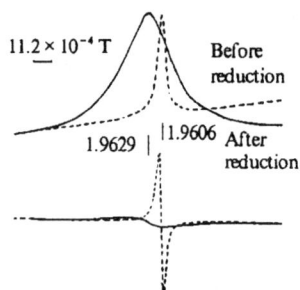

图 3    还原前后 ZnO-Al$_2$O$_3$ 的 ESR 谱

Fig. 3    ESR spectra of ZnO-Al$_2$O$_3$

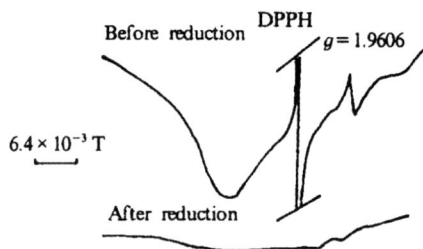

图 4    Cu-ZnO-Al$_2$O$_3$ 还原前后的 ESR 谱

Fig. 4    ESR characterization of Cu-ZnO-Al$_2$O$_3$

信号在相同增益下几乎消失,而 $g = 1.9606$ 的 ESR 信号也同时消失。这一现象可解释为催化剂经还原后,Cu$^{2+}$ 被还原为 Cu$^+$(Cu$^+$ 是满充壳层,其本身无顺磁特征),使在还原后的样品中归属于 Cu$^{2+}$($g = 2.1$)的 ESR 信号消失;而 ZnO 晶格中掺杂 Al$^{3+}$ 后产生的正一价缺位向 ZnO 表面迁移和富集[4],这些正一价缺位被还原过程中产生的 Cu$^+$ 所占据,因此使归属正一价缺位的 ESR 信号 $g = 1.9606$ 也在还原过程中消失。

### 3.3    Sc$_2$O$_3$ 在稳定 Cu-ZnO-M$_2$O$_3$ 催化剂中 Cu$^+$ 的作用

图 5 示出 Cu- ZnO-Al$_2$O$_3$-Sc$_2$O$_3$ 催化剂的 TPR 实验结果。由图可见,所有催化剂的 TPR 谱都存在两个分离的还原峰,这两个还原峰可能分别归属于 Cu$^{2+}$ → Cu$^+$ 和 Cu$^+$ → Cu$^0$ 两步还原过程[6]。当 Sc$^{3+}$ 的含量从 0.1% 到 2% 时,低温还原峰的最高峰温 $Tm_1$ 均比不含 Sc$^{3+}$ 的低,而高温还原峰的最高温 $Tm_2$ 则比不含 Sc$^{3+}$ 的高。由于 Sc$_2$O$_3$ 加入后使铜基催化剂的比表面增大(Cu-ZnO-Al$_2$O$_3$,60∶30∶8,23.6 m$^2$/g;Cu-ZnO-Al$_2$O$_3$-Sc$_2$O$_3$,60∶30∶8∶2,40.2 m$^2$/g),铜在催化剂表面上的分散度增加,因此,催化剂中的 Cu$^{2+}$ 比较容易被还原为 Cu$^+$。但是,由于催化剂表面

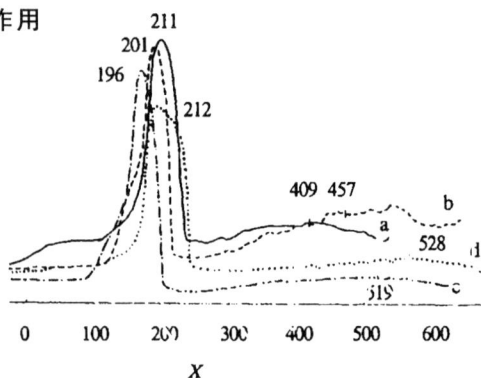

图 5    Cu-ZnO-Al$_2$O$_3$-Sc$_2$O$_3$ 的 TPR 谱

Fig. 5    TPR profiles for Cu-ZnO-Al$_2$O$_3$-Sc$_2$O$_3$

Cu∶Zn∶Al∶Sc    a. 60∶30∶10∶0, b. 60∶30∶9∶1,
c. 60∶30∶8∶2, d. 60∶30∶7∶3

存在 Cu$^+$ 易于坐落的正一价缺位,使 Cu$^+$ 处于稳定的位置,比较不容易被还原为 Cu$^0$,在 TPR 谱中体现为随着 Sc$^{3+}$ 含量的增加,$Tm_2$ 逐渐提高。说明 Sc$_2$O$_3$ 在铜基催化剂中的作用可能是通过形成正一价缺

位而使 $Cu^+$ 容易稳定在催化剂表面。

### 3.4  Cu-ZnO- $Al_2O_3$-$Sc_2O_3$ 催化剂中缺位数与催化活性的关联

图 6 示出 Cu-ZnO-$Al_2O_3$-$Sc_2O_3$ 催化剂的催化活性与 $Al^{3+}$、$Sc^{3+}$ 含量的关系,由图可见,随着 $Sc_2O_3$ 含量的增加,正一价缺位和甲醇合成催化活性也随着增加。少量 $Sc_2O_3$ 的存在对催化甲醇合成有好的助催化作用。图 5 将 Cu-ZnO-$Al_2O_3$-$Sc_2O_3$ 催化剂的表面缺位与催化剂中 $Cu^+$ 的稳定性联系起来。因此,与其说催化剂的催化活性与正一价缺位数有关,倒不如说催化剂的催化活性与催化剂表面 $Cu^+$ 浓度有关。本文及本研究组的另外一些实验结果[7]都从直接和间接方面证实了我们先前提出铜基甲醇合成催化剂上的 Cu、ZnO、$Al_2O_3$ 三者之间的协同作用机理及活性本质。

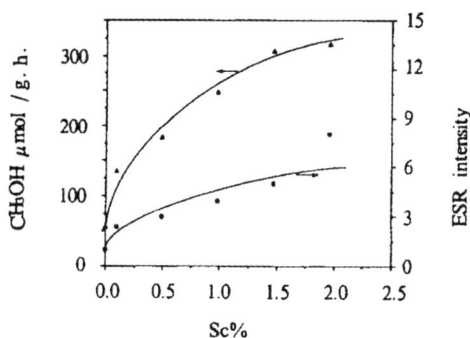

图 6   甲醇合成催化活性和 ESR 强度与催化剂中 Sc 含量的关系
**Fig. 6   Catalytic activity and ESR intensity as a function of $Sc^{3+}$ content in catalysis**

## 4   结   论

从 ESR、TPR 及催化剂的活性评价结果可以得到如下结论:在 ZnO 中掺杂少量的正三价金属离子（$Al^{3+}$ 和 $Sc^{3+}$）,有利于在 ZnO 晶体表面形成正一价缺位,而这些缺位有利于催化剂在还原过程中 $Cu^+$ 在其表面上的增浓及稳定,从而有利于提高铜基催化剂的甲醇合成催化活性、催化剂的稳定性和寿命。这一结果对改进铜基催化剂和开发新的甲醇合成催化剂具有实际的意义。

## 参考文献

[1]Tong Yourong(传友荣),Zhu Qiming(朱启明),et al.. Wuli Huaxue Xuebao(物理化学学报),1985,**1**:431.

[2]Chen Hongbo(陈鸿博),Yu Lajia(于腊佳),Yuan Youzhu(袁友珠),et al.. Fenzi Cuihua(分子催化),1994,**8**:58.

[3]Yu Lajia(于腊佳),Chen Hongbo(陈鸿博),Yuan Youzhu(袁友珠),et al.. Xiamen Daxue Xuebao,Ziran Kexueban(厦门大学学报,自然科学版),1995,**34**:57.

[4]Kasai P H. Phys. Rev.,1963,**130**:898.

[5]Kittel C. Introduction to Solid State Physics,2nd edition,New York,1956:477.

[6]Huret N W,et al.. Catal. Rev.,1982,**24**:233.

[7]Chen Hongbo(陈鸿博),Cai Junxiu(蔡俊修),Cai Qirui(蔡启瑞),et al.. Xumen. Daxue Xuebao,Ziran Kexueban(厦门大学学报,自然科学版),1990,**29**:417.

# Effect of Defects over the Surface of Copper-based Catalyst for Methanol Synthesis on the Catalytic Activities[*]

Hong-Bo Chen[**], La-Jia Yu, Dai-Wei Liao, Guo-Dong Lin, Hong-Bin Zhang, Qi-Rui Cai

(Institute of Phys. Chem., Dept, of Chem., State Key Lab. for Phys. Chem. of the Solid Surface, Xiamen University, Xiamen 361005)

**Abstract**　We studied that the relationship between quantity of extrinsic trivalent cations($Al^{3+}$ or $Sc^{3+}$) and the monovalent cationic vacancies induced by these trivalent cations using ESR spectroscopy. The results shown that the monovalent cationic vacancies over ZnO surface can be induced by doping either $Al^{3+}$ or $Sc^{3+}$. But later,$Sc^{3+}$,is more effective.

The TPR spectra shown that,as increasing contents of $Sc^{3+}$ in catalysts,maximum peak assigned to $Cu^{2+} \rightarrow Cu^+$ shifted to low temperature but the another maximum peak assiged to $Cu^+ \rightarrow Cu^0$ moved to high temperature. Both the catalytic activity for methanol synthesis and intensity of ESR increased as increasing the amount of $Sc^{3+}$ in $Cu\text{-}ZnO\text{-}M_2O_3$($M=Al$ or $Sc$). These indicated that catalytic activity for methanol synthesis considerably related with the monovalent cationic vacancies by doping $M^{3+}$ and stability of $Cu^+$ on the surface of reduced catalysts.

**Key words**　Copper-based catalyst　Methanol synthesis　Hydrogenation of CO　Defect

---

\* Project supported by the National Natural Science Foundation of China and Fujian Oil Refinery.

\*\* To whom correspondence should be addressed.

**1108**

本文原载:《中国科学院院刊》第 6 期(1998 年),第 413～415 页。

# 推进化工生产可持续发展的途径
## ——绿色化学与技术

闵恩泽等①

(中国科学院化学部 北京 100864)

**关键词** 绿色化学,绿色技术,化工生产,可持续发展,咨询报告

## 1 概 述

为了经济与社会的持续发展,必须对现有与化学有关的工业生产做重大的变革,否则在增加产量的同时,会增加废物的排放量。另一方面,人们已逐渐认识到某些化学品对生态环境、社区安全和人体健康的危害性,有的化工产品已不能再使用。因此,国际上对与化学有关的工业环境污染的治理已在治标(末端治理)的同时重视治本,加速研究和发展绿色化学与技术,从根本上减少或消除污染。

绿色化学(Green Chemistry)又称无害化学(Environmentally Benign Chemistry),在其基础上发展的技术称绿色技术(Green Technology)、环境友好技术(Environmentally Friendly Technology)等。理想的绿色技术应采用具有一定转化率的高选择性化学反应来生产目的产品,不生成或很少生成副产品或废物,实现或接近废物的"零排放";工艺过程使用无毒无害原料、溶剂和催化剂;生产环境友好产品。

近年来,绿色化学与技术的研究已是国际科研的前沿,其中以美国最为重视,发展最迅速。1996 年,绿色化学与技术在美国政府、企业和学术界进一步推动下,进入了一个新的加速发展时期。1996 年 7 月,第一届总统绿色化学挑战奖在华盛顿国家科学院举行,共有 67 个项目被提名,4 家化学公司和 1 位化学工程教授被授予总统绿色化学挑战奖,以奖励他们利用化学原理从根本上减少环境污染的成就,赞扬"受奖者在化学制造中开展了一个令人惊奇的革命,它将改变人类的生活"。1996 年,国际学术水平最高的戈登会议(Gordon Conference)也第一次以"环境无害的有机合成"为题,讨论了原子经济反应、环境无害溶剂等。同年,美国化学会主席 Ronald Brealow 发表了"化学的绿色化"的评论文章,指出了绿色化学的重要性和今后的发展方向。美国 1996 年还出版了第一部绿色化学文集《绿色化学——为环境而设计化学》。美国化学会等还召开了一系列和绿色化学有关的专题报告会。

1997 年,美国颁发了第二届总统绿色化学挑战奖;第二次讨论绿色化学的戈登会议在英国牛津举行。此外,还举行了一系列有关绿色化学与技术的会议,如美国化学会的"绿色化学/具有竞争的环境友好制造",美国化学会、美国化学工程师学会、美国环保局等联合举办的绿色化学与工程会议"2020 年环境的设想"等。美国在国家实验室、大学与企业之间联合成立了绿色化学院(The Green Chemistry Institute)。在英国出版了《绿色化学:理论与应用》专著。

我国在绿色化学与技术方面也进行了不少研究工作。中国科学院从 1995 年起组织开展了"工业生产中绿色化学与技术"的院士咨询活动;国家自然科学基金委员会与中国石化集团公司联合资助的"九

① 该文为闵恩泽等中国科学院院士提交的咨询报告。闵恩泽为该咨询课题组组长。课题组成员还有:陈家镛(副组长)、蔡启瑞、沈家骢、戴立信、胡 英。

五"重大研究项目"环境友好石油化工催化化学与化学反应工程"已于 1997 年启动,并已取得可喜的进展;中国科学技术大学于 1997 年初成立了"绿色科技研究与开发中心",并举行了专题讨论会;1997 年还举行了以"可持续发展问题对科学的挑战——绿色化学"为主题的香山科学会议。

# 2 建 议

## 2.1 加强宣传 扩大影响

鉴于我国目前对绿色化学与技术的意义和内容普遍认识不足,必须加强对绿色化学与技术的宣传,使广大企业家、科技人员、工人和各级领导都认识到绿色化学的重要性和意义,共同推动绿色化学与技术的发展。最近以来,《科技日报》、《中国科学报》等对绿色化学的意义和内容均有报道,一些刊物上也开始发表有关文章,还有一两本专著也在酝酿出版。这些活动均应得到积极推动和鼓励。此外,还可组织科普报告会、专题展览等活动来扩大影响。

## 2.2 制订法规和政策

环境保护,立法是关键。我国近年制订的防止空气污染、水污染等法规,大大推动了防止空气污染、水污染等末端治理技术的发展和应用。绿色技术的发展和推广应用,立法也是关键。

建议我国根据化学品对生态、健康、安全等的危害性,参考国外有关法规,结合我国有关生产及使用等实际情况,制订涉及化学品生产、使用的环境保护法规,全面推动绿色化学与技术的发展。国家还应制订对绿色化学与技术的奖励、支持政策,如对科研开发,可设绿色化学与技术专项奖;对于工厂采用绿色技术进行技术改造、建立新装置等,从资金安排、税收等方面给予政策上的支持。

## 2.3 加强开展学术活动

除在香山科学会议上继续组织有关活动外,建议由中国化学会组织一系列有关绿色化学的专题学术活动,如"原子经济的有机合成","代替氢氟酸、硫酸的固体酸"等系列专题研讨会;由中国化工学会组织"提高化学反应选择性的反应工程"等专题研讨会。

## 2.4 加快技术改造

建议加快对现有装置采用绿色技术的技术改造进程,以事实显示采用绿色技术比末端治理的优越性。如吉林化工集团的"H 酸生产装置和香兰素生产装置"的绿色技术改造。燕山石化公司的分子筛催化剂代替三氯化铝催化剂的合成异丙苯装置技术改造已经完成,应予宣传。

## 2.5 列人"九五"研究规划

绿色化学涉及化学的有机合成、催化、生物化学、分析化学等各学科,要开发新反应、新催化材料、新反应环境、新高分子、新酶催化剂和相应测试方法等。美国化学界已把"化学的绿色化"作为迈向 21 世纪化学进展的主要方向之一。

建议我国由科技部组织调研,列入"九五"基础研究规划。建议石化、制药、造纸、酿造、印染等行业在滚动修订"九五"发展规划时,逐步将绿色化学与技术的内容补充入规划中,同时安排科研工作。从国外发展看,一些已工业化的绿色技术的内容应列入有关行业的规划,如超临界二氧化碳代替有机挥发性溶剂用于涂料、塑料发泡剂,用无毒无害二氧化碳代替光气合成异氰酸酯等。我国国家自然科学基金委员会与中国石化集团公司联合支持开展了"环境友好石油化工催化化学与化学反应工程"重大基础研究项目,这是一种把导向性基础研究与技术创新相结合的较好组织形式,建议其他行业也与国家自然科学基金委员会联合,共同资助有关行业开展绿色化学与技术的研究。

对于我国引起严重环境污染的酿造、印染等行业,其有关的绿色化学与技术问题,这次未及调研,还需有关单位进行。

■ **本文原载**:《应用化学》第 15 卷(1998 年 6 月),第 76～78 页。

# 稀土氟氧化物在丙烷、异丁烷氧化脱氢中的催化作用*

张伟德　洪碧凤　古萍英　万惠霖　蔡启瑞

(厦门大学化学系,固体表面物理化学国家重点实验室,物理化学研究所　厦门 361005)

> **关键词**　丙烷　异丁烷　催化氧化脱氢　稀土氟氧化物

石油、天然气等化石原料的有效利用可以大大提高经济效益和社会效益,引起了越来越多的重视[1]。丙烷、异丁烷广泛存在于油田气和天然气中。经氧化成烯烃或含氧有机物可成为更有价值的基础有机化工原料[2]。我们曾经研究稀土含氟复合物在甲烷等低碳烃选择氧化中的作用[3]。本工作将讨论丙烷、异丁烷在稀土氟氧化物催化剂上氧化脱氢制相应烯烃。

稀土基氟氧化物催化剂用固体合成的方法,把稀土氧化物和氟化铈(摩尔比为 1∶4,分析纯,上海跃龙化工厂)粉末混合并充分研磨,加入一定量蒸馏水,调成糊状,120 ℃烘干,880 ℃焙烧 3h。压片、粉碎、筛分 0.22～0.5 mm 颗粒样品用以作催化剂活性评价。评价在固定床反应器中进行,反应器内径 6 mm。催化剂物相组成采用日本理学 D/Max-RC X 射线衍射仪分析,$CuK\alpha(\lambda=0.15406 \text{ nm})$为辐射源,管压 40 kV,管流 30 mA。原位拉曼光谱采用 JobinYven U-1000 谱仪测定。

## 结果与讨论

丙烷、异丁烷在 $Sm_2O_3/4CeF_3$、$Nd_2O_3/4CeF_3$ 和 $Y_2O_3/4CeF_3$ 上的氧化脱氢结果列于表 1。由表 1 可见,在 500 ℃反应温度下,丙烷转化率为 7.5%～9.0%,丙烯选择性为 92.8%～99.0%,没有裂解产物生成。反应温度提高到 520 ℃,丙烷转化率和丙烯收率大大提高,部分丙烷裂解产生甲烷和乙烯,丙烯选择性有所降低,如在 $Nd_2O_3/4CeF_3$ 催化剂上丙烷转化率达 32.7%,丙烯选择性为 71.3%,丙烯产率为 23.3%。异丁烷在 500 ℃时已有裂解产物甲烷和丙烯生成,而且异丁烷转化率比相应条件下丙烷转化率低。提高温度到 520 ℃,异丁烷转化率提高很少,如在 $Sm_2O_3/4CeF_3$ 催化剂上,500 ℃时为 5.61%,520 ℃时为 7.80%,相应异丁烯选择性分别为 84.7% 和 75.0%,异丁烯收率则为 4.75% 和 5.85%。

**表 1　丙烷、异丁烷在稀土基催化剂上氧化脱氢反应结果**

| 催化剂 | 丙烷转化率/% | 丙烯 | 甲烷 | 选择性/% 乙烷 | CO | $CO_2$ | 丙烯产率/% |
|---|---|---|---|---|---|---|---|
| $Sm_2O_3/4CeF_3$ | 7.5 | 92.8 | 0 | 0 | 0 | 7.2 | 6.96 |
| $Nd_2O_3/4CeF_3$ | 8.8 | 99.0 | 0 | 0 | 0 | 1.0 | 8.71 |
| $Nd_2O_3/4CeF_3$ * | 32.7 | 71.3 | 11.6 | 16.7 | 0 | 2.1 | 23.3 |
| $Y_2O_3/4CeF_3$ | 9.0 | 97.0 | 0 | 0 | 0 | 3.0 | 8.73 |
| $Y_2O_3/4CeF_3$ * | 33.3 | 65.4 | 14.2 | 16.3 | 0 | 4.0 | 21.8 |

* 国家自然科学基金资助项目。

续表

| 催化剂 | 异丁烷转化率/% | 异丁烯 | 甲烷 | 选择性/%<br>丙烯 | CO | $CO_2$ | 异丁烯产率/% |
|---|---|---|---|---|---|---|---|
| $Sm_2O_3/4CeF_3$ | 5.61 | 84.7 | 2.53 | 6.60 | 2.87 | 4.75 | 4.75 |
| $Sm_2O_3/4CeF_3$* | 7.80 | 75.0 | 4.78 | 10.30 | 4.62 | 5.31 | 5.85 |
| $Nd_2O_3/4CeF_3$ | 4.31 | 87.9 | 1.45 | 6.27 | 1.50 | 2.87 | 3.79 |
| $Nd_2O_3/4CeF_3$* | 4.94 | 75.7 | 2.15 | 17.1 | 1.93 | 3.22 | 3.74 |
| $Y_2O_3/4CeF_3$ | 6.41 | 78.6 | 3.96 | 14.4 | 1.50 | 1.54 | 5.04 |
| $Y_2O_3/4CeF_3$* | 11.9 | 72.6 | 5.03 | 17.8 | 2.10 | 2.49 | 8.64 |

反应温度:500 ℃(*520 ℃),原料气组成:丙烷(异丁烷):$O_2$:$N_2$=2:3:5,空速:6000/h.

催化剂活性评价结果表明,丙烷、异丁烷在稀土基含氟复合物催化剂上氧化脱氢反应时,异丁烷比丙烷更容易裂解,脱氢反应却更难,这是因为它们的分子结构的差异所致。丙烷分子有 2 个甲基(6 个 H,C-H 键能:410 kJ/mol)和 1 个亚甲基(2 个 H,C-H 键能:397.5 kJ/mol);异丁烷分子有 3 个甲基(9 个 H),1 个与叔碳相连的 H(C-H 键能:381 kJ/mol),从键能来考虑,这个 H 最容易解离,似乎异丁烷分子更容易进行脱氢反应。但因其周围有 3 个甲基,空间位阻很大,这个 H 很难接近催化剂表面,其反应最可能的途径只能是先解离甲基上的 1 个 H,这比丙烷分子首先解离亚甲基上的 1 个 H 要难。其宏观表现为异丁烷转化率不高,而且由于其更易裂解,因此,选择性也较低。

XRD 结果表明,$Sm_2O_3/4CeF_3$ 催化剂由六方 SmOF 和立方 $CeO_2$ 物相组成 $Nd_2O_3/4CeF_3$ 只有六方 $NdF_3$ 和立方 $CeO_2$ 物相。$Y_2O_3/4CeF_3$ 中有单斜 $YF_3$ 和 $CeO_2$ 及少量的六方 YOF 物相。表明在催化剂制备过程中,$CeF_3$ 中的 $F^-$ 与其他稀土氧化物中的 $O^{2-}$ 发生了阴离子的交换。在这些氟氧化物体系中,铈更易与氧结合生成氧化物,而其他稀土元素则更易于生成氟化物或氟氧化物。$CeO_2$ 本身具有氧缺陷,不同价态的 $O^{2-}$ 和 $F^-$ 离子交换也可以产生缺位。在这些催化剂体系中缺陷的存在有助于氧的活化,这有利于烃类的活化,因此,催化剂具有较高活性。并且,由于氟氧化物、氟化物的存在,表明 $F^-$ 离子的存在可以分隔催化剂表面的活性氧物种,避免深度氧化,有利于提高催化剂的选择性[4]。

图 1 为 $Nd_2O_3/4CeF_3$ 的现场拉曼光谱图。由图可见,在 700~1400 $cm^{-1}$ 范围内,在 500 ℃,$O_2$ 气氛条件下,没有任何拉曼峰(图 1a),当温度降低至 30 ℃时,可观测到在 1073 $cm^{-1}$ 有较强吸收峰(图 1b),此时通入 $O_2$ 和丙烷(1:1)混合气体,谱图并没有改变(图 1c);当温度升到 200 ℃时,这个峰显著减弱(图 1d),在 500 ℃时则完全消失,并在 1056 $cm^{-1}$ 产生一新吸收峰(图 1e).根据文献报道[5,6],可指认 1073 $cm^{-1}$ 峰为 $O_2^-$ 表面物种,1056 $cm^{-1}$ 处的峰则为 $CO_3^{2-}$ 的峰。表明在较高温度下,丙烷可与 $O_2^-$ 反应,并在催化剂表面产生碳酸盐。用异丁烷作现场反应也得到类似的结果。这一结果暗示,在烷烃氧化脱氢反应中,稀土含氟化合物催化剂表面 $O_2^-$ 是可能的活性物种。

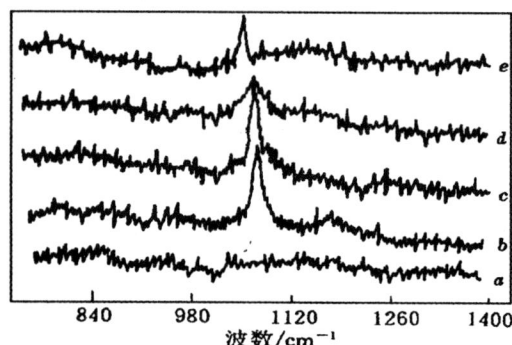

图 1 丙烷在 $Nd_2O_3/4CeF_3$ 催化剂上的原位拉曼光谱
a. 500 ℃,通氧;b. 冷却到 30 ℃;
c. 30 ℃,通入丙烷和氧(1:1);
d. 升温至 200 ℃;e. 升温至 500 ℃

# 参考文献

[1]蔡启瑞,张鸿斌,万惠霖等.第八届全国催化会议论文集.厦门:厦门大学出版社,1996:18.

[2]Bell A T,Boudard M,Ensley B D. Catalysis Looks to the Future Washington D C:National Academy Press,1992:21.

[3]Zhou X P,Zhang W D,Wan H L et al Catal Lett,1993,**21**:113.

[4]Khan A Z,Ruckenstein E. J Catal,1992,**138**:322.

[5]Eysel H H,Thym S. A norg A llg Chem,1975,**411**:97.

[6]Bo sch M,Kanzig W. H elv Phys A cta,1975,**48**:743.

## Catalytic Oxidative Dehydrogenation of Propane and Isobutane over Rare Earth Oxyfluorides

Wei-De Zhang*，Bi-Feng Hong，Ping-Ying Gu，Hui-Lin Wan，Khi-Rui Tsai

(Department of Chemistry，State Key Laboratory for Physical Chemistry of the Solid Surface and Institute of Physical Chemistry，Xiamen University，Xiamen 361005)

**Abstract**　　Three rare earth oxyfluorides $Sm_2O_3/4CeF_3$, $Nd_2O_3/4CeF_3$ and $Y_2O_3/4CeF_3$ were prepared. They were active in oxidative dehydrogenation of propane and isobutane with high selectivity to propene and isobutene. The propane is easier to dehydrogenate than isobtuane, and the isobutane is easier to crack. XRD measurements indicated the existence of new phase such as $CeO_2$ and $REF_3$ or REOF as a result of the exchange between $F^-$ and $O^{2-}$ during the preparation. The *in-situ* Raman spectra implied that $O_2^-$ was the active species.

**Key words**　Propane　Isobutane　Catalytic Dehydrogenation　Rare earth oxyfluoride

■ 本文原载:Journal of Molecular Catalysis A:Chemical 147(1999),pp. 105～111.

# Asymmetric Transfer Hydrogenation of Prochiral Ketones Catalyzed by Chiral Ruthenium Complexes with Aminophosphine Ligands

Jing-Xing Gao [a①], Pian-Pian Xu [a], Xiao-Dong Yi [a], Chuan-Bo Yang [a], Hui Zhang [a],
Shou-Heng Cheng [a], Hui-Lin Wan [a], Khi-Rui Tsai [a], Takao Ikariya [b]

( [a] Department of Chemistry, Institute of Physical Chemistry,
State Key Laboratory of Physical Chemistry of Solid Surface,
Xiamen University, Xiamen 361005, Fujian, China
[b] Department of Chemical Engineering, Faculty of Engineering,
Tokyo Institute of Technology, 2-12-1 O-okayama, Meguro-ku, Tokyo, Japan)

**Abstract**    The condensation of ( S )-propane-1, 2-diamine with two equivalents of o-(diphenylphosphino)benzaldehyde gives ( S )-N, N′-bis[o-(diphenylphosphino)benzylidene]propane-1,2-diamine [( S )-1] ligand. The reduction of( S )-1 with excess NaBH₄ is carried out in refluxing ethanol to afford corresponding ( S )-N, N′-bis[o-(diphenylphosphino)benzyl]propane-1, 2-diamine [( S )-2]. The interaction of trans-RuCl₂ (DMSO)₄ with one equivalent of ( S )-1 or ( S )-2 in refluxing toluene gives ( S )-3 or ( S )-4 in good yield, respectively. ( S )-1,( S )-2,( S )-3 and ( S )-4 have been fully characterized by analytical and spectroscopic methods. The structure of( R )-3 has been also established by an X-ray diffraction study. Catalytic studies showed that ( S )-4 as an excellent catalyst precursor for the asymmetric transfer hydrogenation of acetophenone with 90% yield and up to 91% enantiomeric excess. © 1999 Elsevier Science B. V. All rights reserved.

**Key words**   Chiral ligand   Ruthenium complex   Asymmetric transfer hydrogenation   Ketones

## 1   Introduction

Optically active secondary alcohols are versatile building blocks for synthesis of natural and unnatural biological active compounds as well as functional materials. Chiral biphosphine ligands provide a useful tool for preparing optically active secondary alcohols and have attracted considerable attention as chiral ligands for metal-catalyzed asymmetric reactions[1,2]. It should be noted,however,that in the field of enantioselective transfer hydrogenation the most used chiral auxiliary ligands contain nitrogen as the donor atom[3]. Recently, importance of nitrogen donor has been reviewed[4] and some chiral ruthenium complexes bearing nitrogen donors have been developed with great successes for asymmetric transfer hydrogenation of aromatic ketones[5,6]. For the past several years,we have been interested in the synthesis of well-designed ligands possessing two 'soft' phosphorus atoms and two 'hard' nitrogen

---

①   Corresponding author. Fax:+ 86-592-2188054;E-mail:jxgao@xmu. edu. cn.

■■■■■■■■■■■■■■■■■■■■■■■■■

atoms for preparation of polydentate-metal complexes. These ligands can serve as bi-, tri-and tetradentate ligands depending on the reaction conditions and display some interesting structural[7−11], chemical and catalytic properties[12,13]. In this paper we would like to describe the synthesis and characterization of new chiral ruthenium complexes with structurally similar (S)-1, (S)-2, (R)-1 and (R)-2 ligands, as well as their application to enantioselective transfer hydrogenation of aromatic ketones.

# 2  Experimental

## 2.1. General

All experiments were carried out under a nitrogen atmosphere using a vacuum line and standard Schlenk-tube techniques. IR spectra were recorded on a PE-Spectroy 2000 spectrophotometer. NMR spectra were recorded on a Varian Unity-500 spectrometer. $^{31}$P NMR spectra were measured with 85% phosphoric acid as an external standard. Elemental analyses were performed by the Fujian Institute of structure on the matter, Chinese Academy of Sciences. All the solvent were purified according to standard methods before use. *trans*-$RuCl_2(DMSO)_4$ was prepared as previously described[14].

The asymmetric hydrogen transfer reactions of ketones were performed according to the following procedure: The ruthenium complex (S)-3 or (S)-4 was dissolved in 20 mL of 2-propanol in a schlenk-tube under nitrogen atmosphere at room temperature and to the resulting solution appropriate amounts of ketone and *iso*-PrOK/*iso*-PrOH solution were added, respectively. The solution was stirred and heated to the desired temperature. The reaction mixture was measured by capillary GLC analysis using a chiral Chrompack CP-cyclodextrin-β-236-M-19 column.

## 2.2. Procedure for the preparation of the chiral ligands and the ruthenium complexes

### 2.2.1. (S)-N,N′-bis[o-(diphenylphosphino)benzylidene]propane-1,2-diamine,(S)-1

To a mixture of o-(diphenylphosphino)benzaldehyde(1.45 g,5.0 mmol) and anhydrous $Na_2SO_4$ (3.55 g,25.0 mmol) in $CH_2Cl_2$(15 ml) was added a solution of(S)-propane-1,2-diamine(0.24 g,2.5 mmol). The mixture was stirred at room temperature for 24 h. The resulting solution was filtrated and the solvent was removed under vacuum to leave a pale yellow solid(S)-1(1.45 g,88% yield). m.p. 60-63 ℃. Anal. calcd. for $C_{41}H_{36}N_2P_2$: C,79.63;H,5.82;N,4.53%. Found C,79.78;H,6.02;N,4.57%. IR (cm$^{-1}$): 3057m, 2838m, 1635s, 1582w, 1478w, 1431vs, 1347w, 1184w, 1089m, 1023m, 744vs, 696vs, 545m,497s. $^1$H NMR,δ: 8.75(m,2H,*Ph*CH=),6.82−7.90(m,28H,$C_6H_5$-),3.51(S,1H,-CH<), 3.37(S,2H,-CH$_2$-),0.89(m,3H,-CH$_3$). $^{31}$P NMR,δ: −11.81,12.44.

### 2.2.2. (S)-N,N′-bis[o-(diphenylphosphino)benzyl]propane-1,2-diamine,(S)-2

A mixture of(S)-1(1.24 g,2.0 mmol) and $NaBH_4$(0.46 g,12.0 mmol) in 40 ml of absolute $C_2H_5OH$ was heated at reflux with stirring for 36 h. Then,20 ml of $H_2O$ was added and the solvent was removed under reduced pressure. The white residue was extracted repeatedly with $CH_2Cl_2$(20 ml×3) and the combined extracts was neutralized by saturated $NH_4Cl$ solution. The solution was washed with $H_2O$ and organic layer was dried on anhydrous $MgSO_4$. Removal of the solvent afforded a white solid (S)-2(0.86 g,70% yield). m.p. 59−62 ℃. Anal. calcd. for $C_{41}H_{40}N_2P_2$:C,79.12;H,6.42;N,4.50%; Found C,78.62;H,6.50;N,4.20%. IR(cm$^{-1}$):3411m,3057m,1478m,1431vs,1347m,1156m,1089m, 1024w,744vs,696vs,545m,497s. $^1$H NMR,δ: 6.87−7.62(m,28H,$C_6H_5$-),4.14(m,2H,PhCH-), 3.99(m,2H,PhCH-),2.85(S,2H,-NH-),2.67(S,1H,-CH<),2.63(m,2H,-CH$_2$-),1.07(m,3H,

-CH$_3$). $^{31}$P NMR, $\delta$: $-15.41, -15.51$.

Ligands $(R)$-1 and $(R)$-2 are also prepared according to the above procedure.

### 2.2.3. $(R)$-RuCl$_2$(P$_2$N$_2$Me), $(R)$-3

A mixture of $(R)$-1(0.309 g, 0.5 mmol) and trans-RuCl$_2$(DMSO)$_4$(0.242 g, 0.5 mmol) in 15 ml of toluene was heated under reflux for 16 h. The resulting red solution was cooled to room temperature and the solvent was removed under vacuum to leave a red-brown residue. The solid was dissolved in a minimum amount of CH$_2$Cl$_2$ and chromatographed on a silica gel column (2 × 15 cm) using CH$_2$Cl$_2$/acetone(1 : 1) solution as an eluant. The solvent was removed to give a red solid $(R)$-3(0.32 g, 81% yield). m. p. 280−283 ℃. Anal. calcd. for C$_{41}$H$_{36}$N$_2$P$_2$Cl$_2$. 1.25CH$_2$Cl$_2$ : C, 56.60; H, 4.87; N, 3.12%; Found: C, 56.99; H, 4.55; N, 3.39%. IR(cm$^{-1}$): 3048m, 2914s, 1725w, 1626m, 1474m, 1426s, 1261s, 1090s, 1024w, 749s, 692vs, 526vs, 474s. $^1$H NMR, $\delta$: :8.76(m, 2H, PhCH=), 6.86−7.64(m, 28H, C$_6$H$_5$-), 4.70(S, 1H, -CH<), 4.29(m, 2H, -CH$_2$-), 1.72(m, 3H, -CH$_3$). $^{31}$P NMR, $\delta$: 48.12, 48.51.

A suitable crystal of $(R)$-3 for X-ray diffraction measurements was obtained from CH$_2$Cl$_2$/n-hexane. A dark-red crystal of dimensions 0.15 × 0.18 × 0.20 mm$^3$ was mounted on a CAD4 diffractometer. Reflection data were collected with graphitemonchromated Cu K$\alpha$($\lambda$=1.5418 Å) radiation by $\omega$/2$\theta$ scan mode. A total of 3679 independent reflections within 4°< 2$\theta$< 65° were collected of which 2692 reflections with $I$>3$\sigma$($I$) were used in the refinements. The structure analysis was completed on a PC computer with MOLEN program package. The intensity data were corrected by Lorentz-polarization factor and PSI empirical absorption. The structure was solved by Patterson method and $\Delta F$ syntheses. Full-matrix least-squares refinements were used with anisotropic temperature factors for all non-hydrogen atoms. All hydrogen atoms were not refined and extreme in the final difference map was 1.139 eÅ$^{-3}$. Final value is $R$ = 0.063, $R_w$ = 0.067, $\Delta/\delta$<0.01. A perspective drawing of

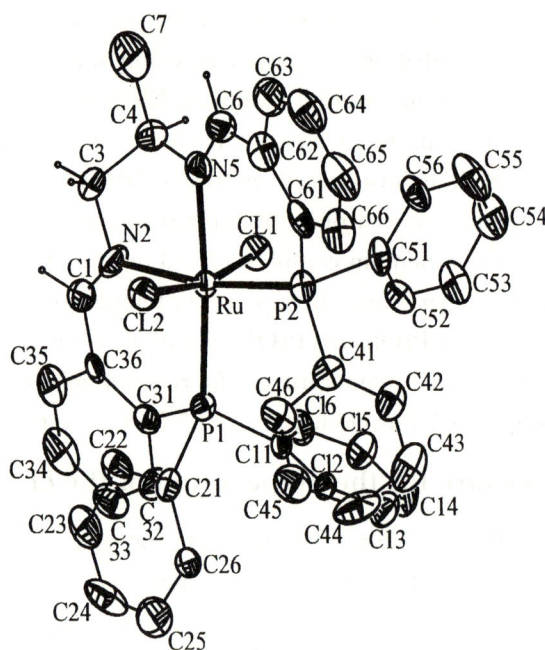

Fig. 1. A perspective view of the structure of trans- RuCl$_2$ (C$_{41}$H$_{36}$N$_2$P$_2$). Selected bond lengths(Å) and angles(°): Ru-CL(1) 2.399(4); Ru-CL(2), 2.440(4); Ru-P(1), 2.297(4); Ru-P(2), 2.277(4); Ru-N(2), 2.08(1); Ru-N(5), 2.07(1); N(2)-C(1), 1.30(2); N(2)-C(3), 1.51(2); N(5)-C(4), 1.45(2); N(5)-C(6), 1.35(2); CL(1)-Ru-CL(2), 170.2(1); P(1)-Ru-P(2), 99.8(1); P(1)-Ru-N(2), 91.0(4); P(1)-Ru-N(5); 170.0(5); P(2)-Ru-N(2), 169.1(4); P(2)-Ru-N(5), 95.2(1); N(2)-Ru-N(5), 80.3(6); CL(1)-Ru-P(2), 95.2(1).

$(R)$-3 is showing in Fig. 1. Crystal data for $(R)$-3: C$_{41}$H$_{36}$N$_2$P$_2$Cl$_2$Ru, $M$=790.67, monoclinic, space group P2$_1$, $a$=11.569(1) 2 Å, $b$=15.079(1) Å, $c$=11.972(1) Å, $\beta$=97.42(1)°, $V$=2071.13(1) Å$^3$, $Z$=2, $D_c$=1.540 g/cm$^3$, $\mu$=78.1 cm$^{-1}$, $F$(000)=976, $T$=23±1°.

### 2.2.4. $(S)$-RuCl$_2$(P$_2$N$_2$H$_4$Me), $(S)$-4

The procedure was similar to that of $(R)$-3, except $(S)$-2(0.311 g, 0.5 mmol) and trans-RuCl$_2$(DMSO)$_4$ (0.242 g, 0.5 mmol) were used. The resulting solid was recrystallized from CH$_2$Cl$_2$/hexane to afford pale-yellow crystals$(S)$-**4**(0.25 g,65% yield). m. p. 226—228℃. Anal. calcd. for $C_{41}H_{40}N_2P_2Cl_2Ru \cdot 0.5C_6H_{14}$: C,63.11;H,5.61;N,3.34%. Found C,63.04;H,5.46;N,3.35%. IR(cm$^{-1}$):3450m,3057m,2867m, 1474s,1431vs,1227w,1089s,950s,744s,692vs. $^1$H NMR,$\delta$: 6.82—7.34(m,28H,$C_6H_5$-),4.03(m,2H, PhCH$_2$-),3.98(m,2H,PhCH$_2$-),3.73(d,1H,-CH<),3.60(S,2H,-NH-),3.04(d,2H,-CH$_2$-),0.91 (m,3H,-CH$_3$). $^{31}$P NMR,$\delta$: 45.18,43.88.

# 3　Results and discussion

## 3.1. Synthesis of chiral ligands (S)-1 and (S)-2

When a mixture of $o$-(diphenylphosphino) benzylaldehyde and $(S)$-propane—1,2-diamine in molar ratio of 2 : 1 was stirring in dichloromethane with excess of Na$_2$SO$_4$ as dehydrating agent,a pale-yellow solid$(R)$-$N$,$N'$-bis[$o$-(diphenylphosphino) benzylidene]propane-1,2-diamine [$(S)$-**1**]was produced in 83%—88% yield(Scheme 1). The IR spectrum of $(S)$-**1** exhibits a strong C=N stretch at 1635 cm$^{-1}$. The $^1$H NMR spectrum exhibits a doublet($J_{(P-H)}$=4.5 Hz) at $\delta$ 8.75 for imino protons. The $^{31}$P NMR spectrum exhibits two singlets of equal intensities at $\delta$ −11.81 and −12.44,respectively. Elemental analysis and spectroscopic data indicate that $(S)$-**1** contains diimino and diphosphino groups.

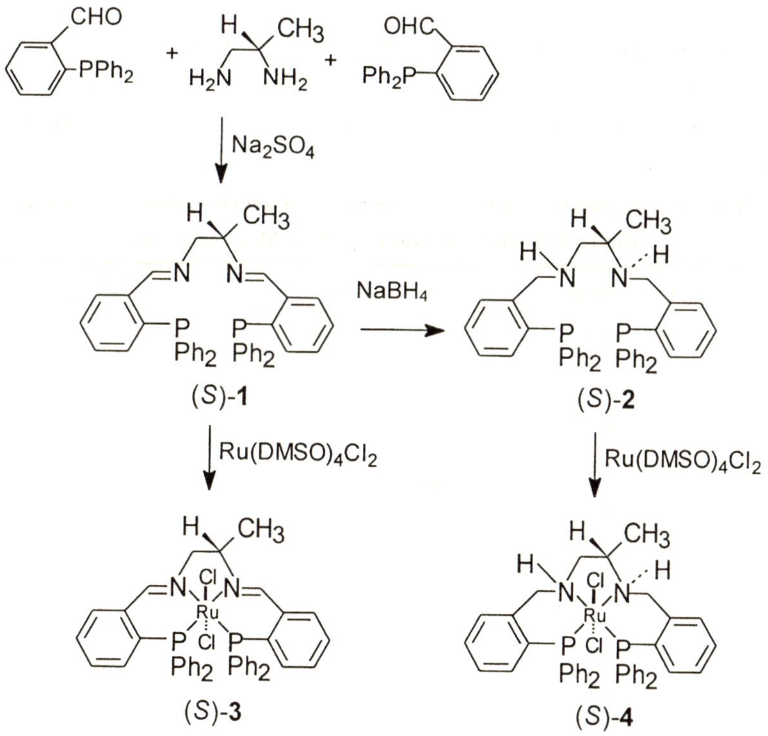

**Scheme 1.**

The reduction of ($S$)-**1** with excess NaBH$_4$ was carried out in refluxing ethanol to afford corresponding $(S)$-$N$,$N'$-bis[$o$-( diphenylphosphino) benzyl]propane-1,2-diamine [$(S)$-**2**] in 68%—73% yield. After reduction of $(S)$-**1**,disappearance of infrared band at 1635 cm$^{-1}$(for -C=N-) and its $^1$H NMR spectrum at 2.85 for the -NH- protons suggest that the two imino groups were reduced to corresponding diamino groups. $^{31}$P NMR spectrum exhibits two singlets of equal intensities at $\delta$ −15.14 and −15.51. Based on these spectroscopic data,the structure of $(S)$-**2** is similar to $(S)$-**1**.

Ligands $(R)$-**1** and $(R)$-**2** can be prepared by means of the above similar procedures.

### 3.2. Synthesis of ruthenium complexes (R)-3 and (R)-4

The interaction of *trans*-RuCl₂ (DMSO)₄ with one equivalent of ligand(R)-1 in refluxing toluene gave (R)-RuCl₂ (P₂N₂Me) [(R)-3] in good yield(81%). ³¹P NMR spectrum of(R)-3 exhibits two singlets of relative intensities 1/1 at δ 48.12 and 48.51, indicating that the two phosphino groups of (R)-1 are coordinated to the ruthenium center. The structure of(R)-3 was established by an X-ray diffraction study, which revealed a distorted octahedral *trans*-configuration for the complex(Fig. 1). The two chloro ligands in the axial position are mutually *trans* to each other and (R)-1 ligand behaves as a tetradentate ligand around the Ru center with the two phosphino group *cis* to each other.

Similar to that of (R)-3, yellow crystals of (R)-4 were obtained in moderate yield(65%). The ³¹P NMR spectrum exhibits two singlets of equal intensities at 45.18 and 43.88, indicating the two phosphino groups are coordinated to the Ru center. Attempt to get suitable crystals of complex(R)-4 for structure analysis was unsuccessful. However, based on the spectroscopic data and the molecular structures of *trans*-RuCl₂ (P₂N₂H₄)[9,12] and *trans*-RuCl₂ (cyclo-C₆P₂N₂H₄)[13], the structure of ruthenium complex(R)-4 is assignable to analogy with complex(R)-3. According to the above procedure ruthenium complexes (S)-3 and (S)-4 have been also prepared.

### 3.3. Asymmetric transfer hydrogenation of prochiral ketones

Complexes (S)-3, (R)-3, (S)-4 and (R)-4 have been tested as catalysts in enantioselective transfer hydrogenation of aromatic ketones in a *iso*-PrOH solution. The catalytic hydrogenation of acetophenone(1a) was conducted using some potassium 2-propoxide (1—3 equiv. with respect to Ru) as a promoter(Table 1 and Scheme 2).

**Table. 1  Asymmetric transfer hydrogenation of ketones catalyzed by chiral RuCl₂ (P₂N₂Me) and RuCl₂ (P₂N₂H₄Me) complexes[a]**

| Ketone substrate | Catalyst | S/C/*iso*-PrOK[b] (molar ratio) | conditions | | Alcohol product | | |
|---|---|---|---|---|---|---|---|
| | | | Temperature/℃ | Time/h | Yield/%[c] | ee/%[d] | configuration[e] |
| 1a | (R)-3 | 100 : 1 : 3 | 40 | 22 | 63 | 26 | S |
| 1a | (S)-3 | 100 : 1 : 3 | 40 | 22 | 65 | 14 | R |
| 1a | (S)-4 | 100 : 1 : 3 | 30 | 46 | 90 | 91 | S |
| 1b | (S)-4 | 100 : 1 : 2 | 45 | 48 | 55 | 88 | S |
| 1b | (R)-4 | 100 : 1 : 3 | 30 | 46 | 73 | 91 | R |
| *m*-1c | (S)-4 | 100 : 1 : 2 | 30 | 24 | 99 | 87 | S |
| *p*-1c | (S)-4 | 100 : 1 : 2 | 30 | 24 | 82 | 89 | S |
| *m*-1d | (S)-4 | 100 : 1 : 2 | 30 | 24 | 72 | 85[f] | S |
| *p*-1d | (S)-4 | 100 : 1 : 2 | 30 | 24 | 49 | 87[f] | S |

[a] conditions: catalyst, 0.01 mmol; solvent, *iso*-PrOH, 20 ml.

[b] S/C/*iso*-PrOK = Ketone/Ru/*iso*-PrOK.

[c] GLC analysis.

[d] Capillary GLC analysis using a chiral Chrompack CD-cyclodextrin-β-236 M-19 column unless otherwise specified.

[e] Determined by comparison of the retention times of the enantiomers on the GLC traces with literature values.

[f] Determined by HPLC analysis using a Daicel Chiralcel OB column(10 : 90 2-propanol-hexane).

The concentration of *iso*-PrOK is an important factor for catalytic activity and the catalytic system

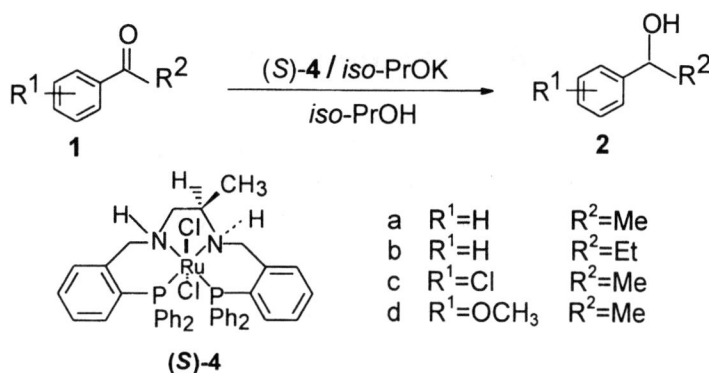

Scheme 2.

is inactive without a basic co-catalalyst. Increase of reaction temperature accelerates the reaction rate with a slight loss of enantiomeric purity of the product. The ketones possessing an electron-donating substituent such as methoxyl at the para position tend to lower the rate, but still show high stereoselectivity.

It is noteworthy that the diimino complexes (S)-**3** and the diamino complex (S)-**4** display the differences in reactivities and enantioselectivities. The reaction with the diimine complexes (R)-**3** or (S)-**3** proceeded in moderate yield with very low ee(14—26%). However, the diamine complexes (R)-**4** and(S)-**4** have proved to be an excellent catalyst precursor in asymmetric transfer hydrogenation of acetonphenone, leading to 2-phenylethanol in 90% yield with 91% ee. Complex (R)-**4** or (S)-**4** with sp³-hybridized nitrogens containing N-H bonds displayed higher reaction rate and enantioselectivity.

The reaction mechanism of hydrogen transfer hydrogenation can be envisaged by a 'hydridic route' or a 'direct hydrogen transfer'[15]. The later involves the process in which both the hydrogen donor and the substrate are bound on the catalytic active center. For the catalyst precursor (S)-**3** or (S)-**4**, a 'direct hydrogen transfer' pathway is difficult to be realized since the catalysts possess the configuration with saturated coordination. Therefore, a reaction mechanism by 'hydridic route' may be performed in hydrogen transfer hydrogenation of acetophenone in this work. Based on the fact that the catalyst precursor, (S)-**3** or (S)-**4**, has inactive in the absence of iso-PrOK and the conversion increases with the increase of iso-PrOK concentration, iso-PrOK may play a role for promoting the formation of catalytic active ruthenium hydride(Scheme 3)[15-19]. On the other hand, the presence of an NH groups in the ligands is possible to stabilize a six membered cyclic transition state by forming a hydrogen bond with oxygen atom of ketones[20,21]. The study on isolation and characterization of the catalytic active intermediate is now under investigating.

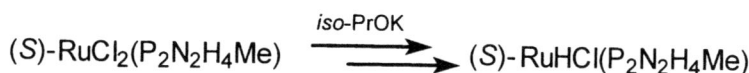

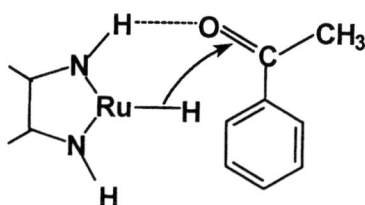

Scheme 3.

# Acknowledgements

We would like to thank the National Nature Science Foundation of China and Fujian Province, State Key Laboratory for Physical Chemistry of Solid Surface in Xiamen University for financial support and professor Ryoji Noyori (Nagoya University, Japan) for his very valuable assistance.

# References

[1] J. K. Whitesell, Chem. Rev. **89**(1989) 1581.

[2] R. Noyori, Asymmetric Catalysis in Organic Synthesis, Wiley, New York, 1994, p. 21.

[3] G. Zassinovich, G. Mestroni, S. Gladiali, Chem. Rev. **92**(1992) 1051.

[4] A. Togni, L. Venanzi, Angew. Chem., Int. Ed. Engl. **33**(1994) 497.

[5] S. Hashiguchi, A. Fujii, J. Takehara, T. Ikariya, R. Noyori, J. Am. Chem. Soc. **117**(1995) 7562.

[6] A. Fujii, S. Hashiguchi, N. Uematsu, T. Ikariya, R. Noyori, J. Am. Chem. Soc. **118**(1996) 2521.

[7] J. C. Jeffrey, T. B. Rauchfuss, P. A. Tucker, Inorg. Chem. **19**(1980) 3306.

[8] W. K. Wang, J. X. Gao, Z. Y. Zhou, T. C. W. Mak, Polyhedron **11**(1992) 2965.

[9] W. K. Wong, J. X. Gao, W. T. Wong, Polyhedron **12**(1993) 1415.

[10] W. K. Wong, J. X. Gao, W. T. Wong, C. M. Che, Polyhedron **12**(1993) 2063.

[11] W. K. Wong, J. X. Gao, W. T. Wong, W. C. Cheng, C. M. Che, J. Organomet. Chem. **471**(1994) 277.

[12] J. X. Gao, H. L. Wan, W. K. Wong, M. C. Tse, W. T. Wong, Polyhedron **15**(1996) 1241.

[13] J. X. Gao, T. Ikariya, R. Noyori, Organometallics **15**(1996) 1887.

[14] W. Mitchell, A. Spencer, G. Wilkinson, J. Chem. Soc. Dalton. Trans. (1973) 204.

[15] Y. Sasson, J. Blum, J. Org. Chem. **40**(1975) 1887.

[16] D. Morton, D. J. Cole-Hamilton, J. Chem. Soc. Chem. Commun. (1988) 1154.

[17] H. Chowdury, J. E. Backvall, J. Chem. Soc., Chem. Commun. (1991) 1063.

[18] C. Brown, P. V. Ramachandran, Acc. Chem. Res. **25**(1992) 16.

[19] J. Corey, R. K. Bakshi, S. Shibata, J. Am. Chem. Soc. **109**(1986) 5551.

[20] J. Haack, S. Hashiguchi, A. Fujii, T. Ikariya, R. Noyori, Angew. Chem., Int. Ed. Engl. **36**(1997) 285.

[21] R. Noyori, S. Hashiguchi, Acc. Chem. Res. **30**(1997) 97.

■ 本文原载:Bull. Chem. Soc. Jpn.,72(1999),pp. 2643~2653.

# Characterization of CO- and H$_2$-Adsorbed Au$_6$Pt-Phosphine Clusters Supported on SiO$_2$ by EXAFS, TPD, and FTIR

You-Zhu Yuan[a], Kiyotaka Asakura[b, ①], Hui-Lin Wan[a],

Khi-Rui Tsai[a], Yasuhiro Iwasawa*

(Department of Chemistry, Graduate School of Science,

The University of Tokyo, Hongo, Bunkyo-ku, Tokyo 113-0033

[a]Department of Chemistry and State Key Laboratory for Physical

Chemistry of Solid Surface, Xiamen University, Xiamen 361005, China

[b]Center of Spectrochemistry, Graduate School of Science,

The University of Tokyo, Hongo, Bunkyo-ku, Tokyo 113-0033)

Received June 11, 1999

**Abstract**    The adsorption of CO and H$_2$ on a SiO$_2$-supported Au$_6$Pt cluster [(AuPPh$_3$)$_6$Pt(PPh$_3$)] (NO$_3$)$_2$(**1**) has been studied by means of FTIR, TPD, and EXAFS. CO adsorbed on **1**/SiO$_2$ to form a CO-adduct complex, which exhibited an IR band at ca. 1972 cm$^{-1}$. The adsorbed CO was desorbed below 403 K, showing a peak maximum at 363 K in TPD. There was little change in the coordination number of Pt-Au in Pt L$_3$-edge EXAFS upon CO adsorption, but the coordination number of Au-Au in Au L$_3$-edge EXAFS slightly increased and the peak ascribed to Pt-(Au)-P in the Pt L$_3$-edge EXAFS Fourier transform remarkably increased. The increase in the intensity of the Pt-(Au)-P peak was interpreted by the multiple scattering effect owing to the change of Pt-Au-P bond angle. The original EXAFS oscillations at both Pt and Au L$_3$-edges were regenerated after evacuation of the CO-adsorbed sample for 2 h at 353 K, indicating the recovery of the original cluster structure. This is entirely different from the case of the cluster **1** in solution, where the cluster framework is fragmented by the CO adsorption-desorption process. The adsorption of H$_2$ on **1**/SiO$_2$ was totally reversible at room temperature; it provided no contribution to the EXAFS oscillation.

Supported gold and gold-noble metal catalysts have attracted considerable attention from chemical interests as well as from the viewpoint of industrial applications[1−3]. It has been reported that the addition of gold to Pt catalysts increases catalytic activity and selectivity and suppresses deactivation in hydrocarbon conversions[4−11]. Recently, a SiO$_2$-supported cluster [(AuPPh$_3$)$_6$Pt(PPh$_3$)](NO$_3$)$_2$ (**1**) with a unique structure of a Pt atom embedded in a six Au-atoms ensemble was found to show significant catalysis for H$_2$-D$_2$ equilibration at 303 K, whereas a supported mononuclear Pt complex

---

①   Present address:Catalysis Research Center,Hokkaido University,Kita-ku,Sapporo 060-0811,Japan.

Pt(PPh$_3$)$_4$/SiO$_2$ and a supported Au cluster [Au$_6$]$^{2+}$/SiO$_2$ showed almost no activity for the H$_2$-D$_2$ exchange reaction[12-15]. This observation may be of interest because multi-sites composed of several Pt atoms are believed to be active sites for dihydrogen activation and H$_2$-D$_2$ equilibration on Pt particle catalysts.

It has also been reported that H$_2$ coordination on both **1** and [(AuPPh$_3$)$_8$Pt(PPh$_3$)](NO$_3$)$_2$ (**2**) rapidly and reversibly occurs[16,17]: {(Au)$_n$Pt} + H$_2$ ⇌ {(Au)$_n$Pt(H$_2$)}. Although {(Au)$_n$Pt(H$_2$)} is likely to be an intermediate in the catalytic H$_2$-D$_2$ equilibration, no structural information for this complex has been reported. Detailed NMR studies have been carried out with {(Au)$_n$Pt} ($n$ = 6,7,8) clusters in solution[17]. In the case of **2**, the triple Pt-H coupling pattern was observed in the $^{31}$P decoupled $^{195}$Pt NMR spectrum, indicating the presence of two hydride ligands[17]. Since no terminal Pt-H stretching vibrations were detected by IR, the hydride ligands were assigned as bridge-bonding type[18]. In the case of **1** in CD$_2$Cl$_2$ solution, however, no changes in the NMR spectra have been observed[17]. In a basic solution like pyridine, deprotonation from {(Au)$_n$-Pt(H$_2$)} to monohydrido cluster {(Au)$_n$-Pt(H)} was directly observed by $^{31}$P{$^1$H} NMR and $^1$H NMR, which demonstrates that H$_2$ is activated to dissociate to H atoms on **1**[16,19,20].

The catalytic H$_2$-D$_2$ equilibration on **1**/SiO$_2$ is completely suppressed by CO adsorption on the Pt atom of **1**/SiO$_2$. The adsorbed CO creates a site blocking effect as well as an electronic effect on the catalysis of **1**/SiO$_2$. However, there is no information on any structural change of the supported cluster framework upon CO adsorption.

The adsorption and reactivity of metal organic and inorganic compounds on oxide surfaces have been extensively studied to gain a better understanding of the fundamental principles of surface catalysis and the relations among bonding, structure and reaction[21-24]. This class of catalysts prepared by well-defined precursors has a great advantage in accepting the characterization of catalytically active structures and compositions by physical techniques. Among the characterization techniques, extended X-ray absorption fine structure(EXAFS) spectroscopy is of great importance for the structural study of dispersed metal sites on oxide surfaces in the static state and during the course of catalysis[25,26]. In-situ EXAFS analysis provided information on the bonding feature of Pt-Au(Pt) and Pt(Au)-ligands(such as phosphine and CO) in **1**/SiO$_2$[13,15,27]. We found a reversible change of the cluster structure induced by CO adsorption, which is different from the cluster fragmentation in solution[27]. In this paper, we report the further investigation on the structural change of **1**/SiO$_2$ upon adsorption and desorption of CO and H$_2$ by means of in-situ EXAFS, TPD(temperature programmed desorption), and FTIR.

# 1　Experimental

Preparation of the SiO$_2$-supported [(AuPPh$_3$)$_6$Pt(PPh$_3$)](NO$_3$)$_2$ (**1**) catalyst has been reported elsewhere[13,15]. The loading of **1** on SiO$_2$ was controlled to be 0.5-1.0 wt% based on Pt. All the procedures for sample preparation were conducted in Ar atmosphere (purity: 99.999%) to avoid contacting air. Adsorptions of CO and H$_2$ were measured in a closed circulating system at 298 K. The uptake was monitored at an interval of every 15 min until the pressure did not change, typically for about 40 min. The uptake of H$_2$ on **1**/SiO$_2$ was somewhat difficult to measure due to the weak interaction between H$_2$ and **1**/SiO$_2$ at 298 K. The data for H$_2$ adsorption have larger errors than those for CO adsorption. The H$_2$ adsorption was reversible.

Temperature-programmed desorption(TPD) spectra were measured in a fixed-bed reactor system equipped with a gas chromatograph. A dry-ice/acetone trap was used to eliminate the influence of water and hydrocarbons. The samples were first treated at room temperature for 120 min under Ar and then switched to $CO(99.999\%)$ for 10 min in a flow of 30 ml min$^{-1}$. After the sample was purged by Ar at a flow rate of 30 ml min$^{-1}$ for 60 min, CO-TPD spectra were obtained with the carrier gas of Ar at a heating rate of 10 Kmin$^{-1}$.

IR spectra were measured on a JEOL JIR 7000 spectrometer with a resolution of 4 cm$^{-1}$. A pressed SiO₂ disk was placed in an IR cell with two NaCl windows, combined with a closed circulating system. The disk was pretreated at 673 K for 1 h in the cell and a dried ethanol solution of **1** was carefully added dropwise onto the disk in a flow of high-purity Ar. The supported sample was evacuated at 298 K for 4 h in-situ before CO adsorption. The spectra were recorded as difference spectra between before and after CO adsorption.

EXAFS measurements were carried out at BL-10B (for CO adsorption) and at BL-7C (for H₂ adsorption) of the Photon Factory in the Institute for Material Structure Science (PF-IMSS) (Proposal No. 95G200) operated at 2.5 GeV with a ring current 250 — 350 mA. The incident and transmitted X-rays were monitored by ionization chambers filled with N₂ and Ar(15%)-N₂(85%) gases, respectively. EXAFS data were measured at room temperature in a transmission mode. The samples was transferred to an in-situ EXAFS glass cell by a Schlenk technique and evacuated at room temperature for 4 h. After recording the EXAFS spectrum of the sample, CO of 300 Torr(1 Torr=133.322 Pa) was admitted into the EXAFS cell to measure the EXAFS spectrum for CO-adsorption sample. Then, the cell was evacuated at 353 K for 0.5 h and EXAFS was measured again. Similarly, the EXAFS measurement of H₂ adsorbed sample was carried out, but the measurement was conducted under 0.3 MPa of H₂.

The analysis of EXAFS was performed by removal of background using cubic smoothing and normalization with the edge height, whose energy dependency was taken into account using the McMaster equation[28]. The inverse Fourier-transformed EXAFS data were analyzed by a curve fitting method using phase-shifts and amplitudes calculated by the FEFF 6.01 software[29]. The parameters used for the FEFF calculations are listed in Table 1. The cluster **1**, Pt foil, Au foil, Pt(PH₃)₄ and Au (PH₃)₄ were used to determine the amplitude reduction factor S and to check the validity of these theoretically derived parameters. In the present analysis we took into account the error estimations recommended by the international XAFS workshop on standards and criteria.

**Table 1. Crystallographic Data for the FEFF Calculation and Fourier Transform Ranges Used in the EXAFS Analysis**

| Bond | FEFF Calculation | | Fourier transformation range | |
|---|---|---|---|---|
| | $N$ | $r$/nm | $\Delta k$/nm$^{-1}$ | $\Delta r$/nm |
| Pt-P for Pt(PPh₃) | 1.0 | 0.228 | 30 — 90 | 0.14 — 0.23 |
| Pt-Au for Pt(Au) | 1.0 | 0.267 | 30 — 90 | 0.24 — 0.32 |
| Pt-Pt for Pt foil | 1.0 | 0.278 | 30 — 90 | 0.24 — 0.32 |
| Pt-C for Pt(CO) | 1.0 | 0.198 | 30 — 90 | 0.14 — 0.20 |
| Pt-(C)-O for Pt(CO) | 1.0 | 0.304 | 30 — 90 | 0.14 — 0.32 |
| Pt-(Au)-P for Pt(Au-P) | 1.0 | 0.497 | 30 — 90 | 0.39 — 0.50 |
| Au-P for Au(PPh₃) | 1.0 | 0.230 | 30 — 130 | 0.14 — 0.23 |
| Au-Au for Au(Au) or Au(Pt) | 1.0 | 0.287 | 30 — 130 | 0.24 — 0.32 |
| Au-Au for Au foil | 1.0 | 0.287 | 30 — 160 | 0.21 — 0.32 |

$N$: coordination number for absorber-backscatterer pair; $r$: interatomic distance; $\Delta k$: wavenumber range for forward Fourier transformation; $\Delta r$: distance range for shell isolation.

# 2 Results

**FT IR Spectra.** When CO was admitted into $1/SiO_2$ in an in-situ IR cell, a single CO band at ca. 1972 cm$^{-1}$ was observed, as shown in Fig. 1A. This is assigned to linear CO adsorbed on Pt. This is similar to the IR band at ca. 1967 cm$^{-1}$ for $[(AuPPh_3)_6Pt(CO)(PPh_3)]^{2+}$ in solution[30]. Evacuation of the system caused a gradual decrease in the peak intensity, but the peak was still observable after evacuation for 4 h (Fig. 1A). The adsorbed CO can be completely removed by evacuation at 353 K (Fig. 1A-g).

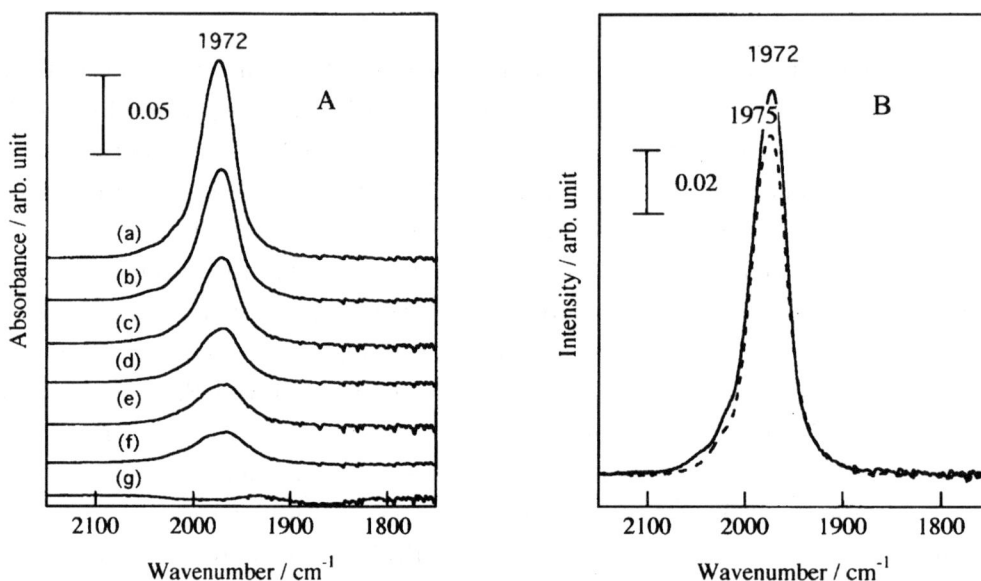

**Fig. 1.** FT-IR spectra for CO adsorbed on the intact $1/SiO_2$ (A) and the 473 K-treated $1/SiO_2$ (B); Evac. at RT for (a) 0 min; (b) 5 min; (c) 10 min; (d) 30 min; (e) 60 min; (f) 120 min; (g) Evac. at 353 K.

When the $1/SiO_2$ sample was treated at 473 K in vacuum, the peak of adsorbed CO was observed at 1975 cm$^{-1}$ (Fig. 1B). The previous EXAFS study indicated a small change in the cluster framework by the treatment at 473 K[15]. The IR peak frequency is slightly higher than that for the untreated sample; this may be related to a small structural change of the cluster **1** on $SiO_2$.

**Adsorption of CO and H$_2$.** Figure 2A shows irreversible CO adsorption on $1/SiO_2$ at 298 K as a function of CO pressure. The CO uptake seems to be explained by the Langmuir adsorption isotherm. The amount of the irreversible adsorption was calculated by subtracting the amount of CO adsorbed in the second run from that in the first run. The system was evacuated at 298 K for 20 min between the first and second runs. The CO adsorption saturated at CO pressure of ca. 250 Torr, in which the coverage was one CO per Pt.

The result of H$_2$ adsorption on $1/SiO_2$ at 298 K is shown in Fig. 2B. The amount of adsorbed H$_2$ was very low. The adsorption isotherms show that hydrogen adsorption is completely reversible at 298 K.

**CO-TPD on $1/SiO_2$.** CO-TPD spectra for $1/SiO_2$ untreated and treated at 473, 573, and 773 K are shown in Fig. 3. The TPD spectra for the intact $1/SiO_2$ and the 473 K-treated $1/SiO_2$ were similar to each other, whereas that for the 773 K-treated sample showed almost no peak in the same temperature range. In fact the CO uptake on the 773 K-treated sample was negligible. From the TPD data, we find

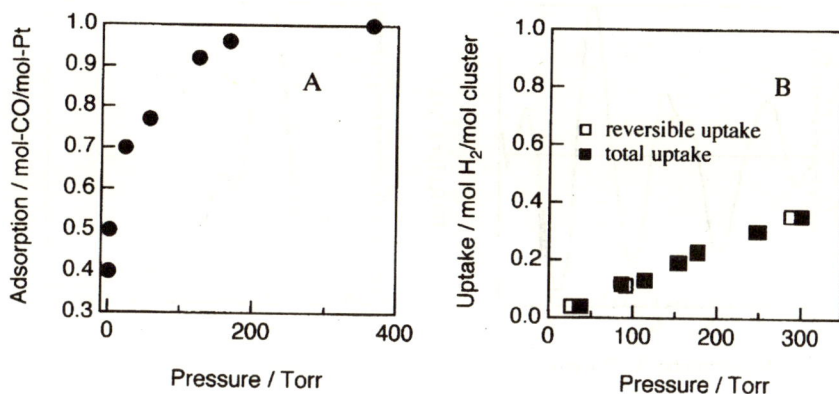

**Fig. 2. The adsorption of CO(A) and $H_2$(B) on $1/SiO_2$ at 298 K.**

that the CO uptakes on the intact and 473 K-treated samples were CO/Pt = 1. 0 and 0.8, respectively.

**EXAFS Measurements for CO-Adsorbed $1/SiO_2$.** We measured the Pt $L_3$-edge EXAFS spectra of $1/SiO_2$ before exposure to CO, after exposure to 300 Torr of CO and after evacuation at 353 K. The EXAFS oscillations and associated Fourier transforms are shown in Fig. 4. The peaks in the Fourier transforms that appeared in the range 0.1−0.35 nm are due to Pt-P and Pt-Au bondings. A small peak around 0.45 nm may be due to Pt-(Au)-P interaction. When $1/SiO_2$ was exposed to 300 Torr of CO at room temperature, the intensity of the peak at

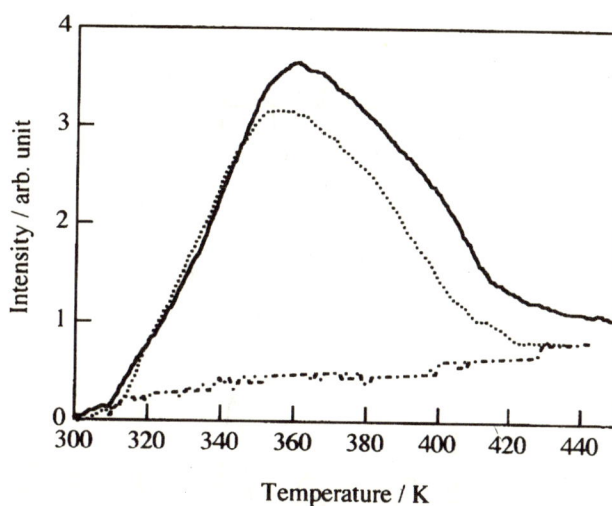

**Fig. 3. CO-TPD of $1/SiO_2$.**
— $1/SiO_2$ ; $\cdots 1/SiO_2$ pretreated at 473K;
-·- $1/SiO_2$ pretreated at 773K.

0.45 nm became more than twice that for $1/SiO_2$, indicating that some change occurred with the Pt-(Au)-P bonding. When the CO-exposed $1/SiO_2$ was subsequently evacuated at 353 K, no decomposition of the framework of the $Au_6$Pt cluster was observed, but the peak at 0.45 nm reduced significantly, as shown in Fig. 4C.

Since the Pt-P and Pt-Au peaks are well separated from each other, they can be analyzed by a curve fitting method. The curve fitting analysis by two shells(Pt-P and Pt-Au) in the Fourier transform range 0.14−0.35 nm reproduced the experimental results for the intact $1/SiO_2$ and the evacuated $1/SiO_2$, as shown in Figs. 5A and 5C. The best-fit results are listed in Table 2. However, the two-shell fitting analysis for the CO-adsorbed sample never reproduced the experimental data. Therefore, we performed 4-term curve fitting analysis(Pt-C, Pt-P, Pt-Au, and Pt-(C)-O) adding a CO molecule coordinated to Pt. This analysis requires 16 fitting parameters more than the maximum number of degrees of freedom $(2\Delta r\Delta k/\pi\approx10)$. To conduct the analysis we fixed $\Delta E$ and $\sigma$ values. The best-fit result is shown in Fig. 5B. The curvefitting analysis data in Table 2 demonstrate the retention of the Pt-P and Pt-Au bondings in all the samples.

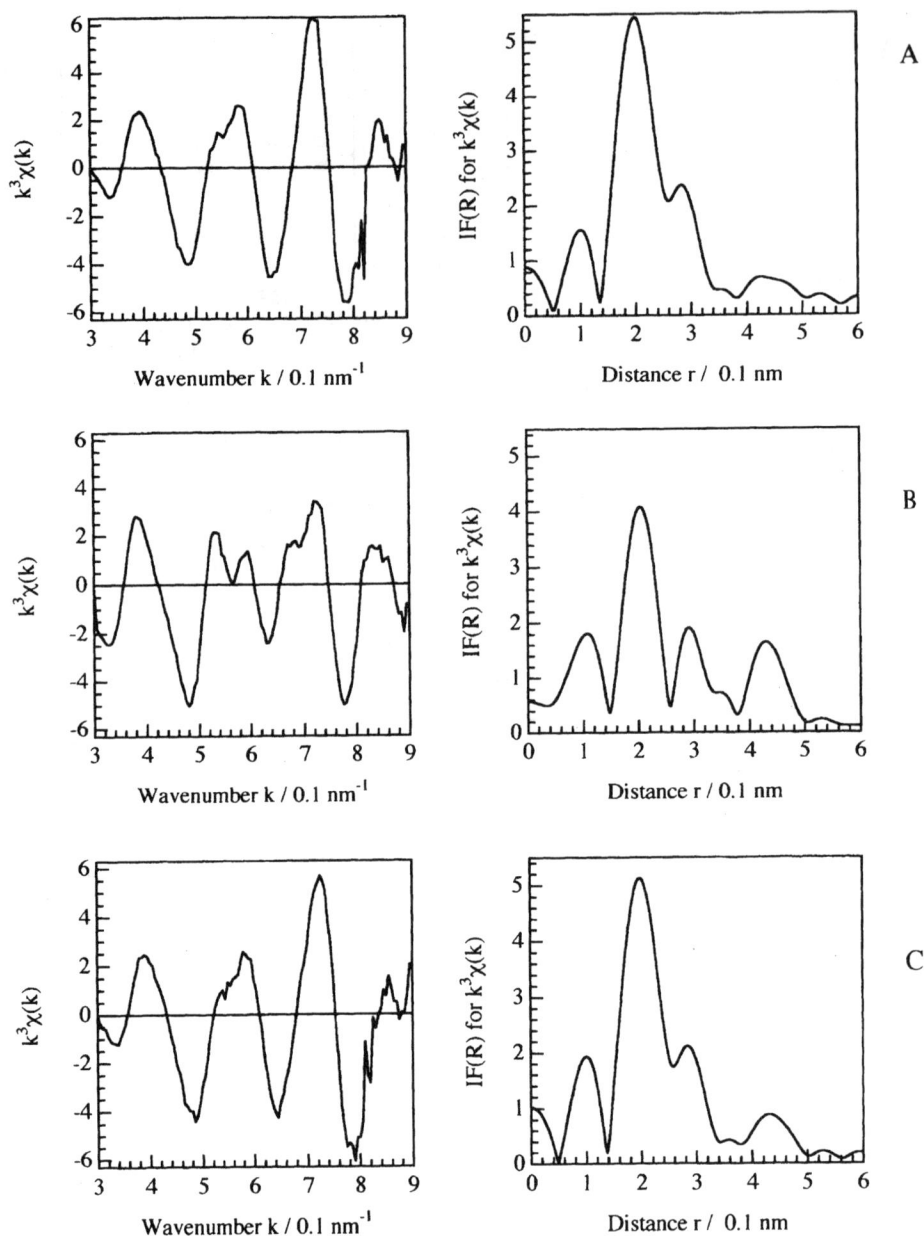

**Fig. 4.** Pt L$_3$-edge EXAFS oscillations and associated Fourier transforms for 1/SiO$_2$.

(A) 1/SiO$_2$; (B) 1/SiO$_2$ under 300 Torr of CO; (C) 1/SiO$_2$ after evacuation at 353 K.

**Table 2. Curve-Fitting Results for the Pt L$_3$-edge EXAFS Data of 1/SiO$_2$**

| Sample | Pt-P | | Pt-Au | |
|---|---|---|---|---|
| | $N$ | $r$/nm | $N$ | $r$/nm |
| **1** | 1[a] | 0.228[a] | 6[a] | 0.268[a] |
| **3** | 1[a] | 0.234[a] | 6[a] | 0.269[a] |
| **1**/SiO$_2$ | 1.1±0.1 | 0.228±0.001 | 6.0±0.5 | 0.268±0.001 |
| **1**/SiO$_2$ after adsorption of CO | 1.2±0.4 | 0.231±0.004 | 5.8±0.9 | 0.269±0.003 |
| 1/SiO2 after desorption of CO | 1.1±0.1 | 0.229±0.001 | 5.9±0.5 | 0.269±0.001 |

The $\Delta r$ range for the inverse Fourier transformation was 0.12—0.35 nm for both Pt and Au EXAFS.

The curve fitting range was 30—90 nm$^{-1}$ for Pt EXAFS. a) The averaged values derived from crystallographic data.[30]

b) The bond distances for Pt-C and Pt-(C)-O were 0.194 and 0.309 nm, respectively.

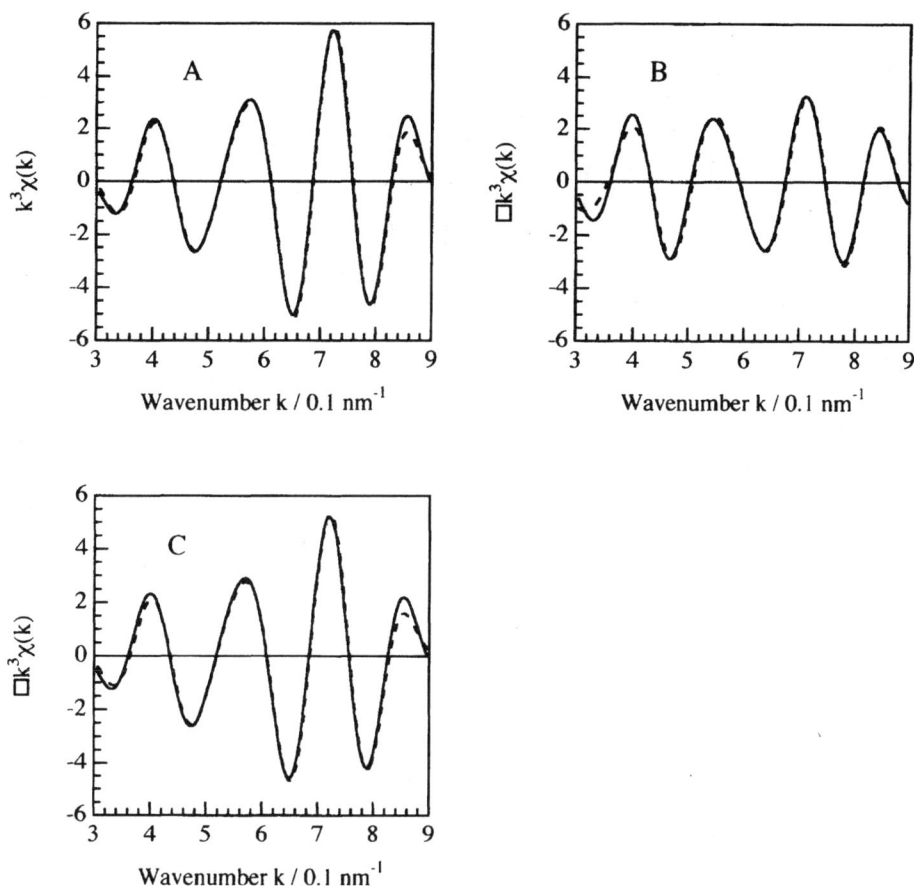

**Fig. 5.** The curve-fitting analysis for the Pt L$_3$-edge EXAFS data of 1/SiO$_2$ based on two-shells(Pt-P and Pt-Au) model(A and C) and four-shells(Pt-C, Pt-P, Pt-Au, and Pt-(C)-O) model(B): —, observed, ⋯, calculated. (A)1/SiO$_2$; (B) after CO adsorption; (C) after desorption of the adsorbed CO at 353 K.

Figure 6 shows the EXAFS data at Au L$_3$-edge. The curve-fitting result is shown in Fig. 7A. Fig. 6B shows the EXAFS oscillation and its Fourier transform for 1/SiO$_2$ exposed to 300 Torr of CO. No change in the intensity of the first peak attributable to Au-P was observed. The curve-fitting analysis revealed that Au-P and Au-Pt coordination numbers remained unchanged at nearly unity(Table 3). The Au$_6$Pt cluster framework was maintained with a small deformation after CO adsorption. Figure 6C shows the EXAFS oscillation and its associated Fourier transform for 1/SiO$_2$ after CO desorption by evacuation at 353 K. The coordination number of Au-Au decreased to the same value as that for the intact 1/SiO$_2$ (Table 3). It is concluded that the original structure of 1/SiO$_2$ was recovered after desorption of CO.

**Table 3. Curve-Fitting Results for the Au L$_3$-edge EXAFS Data of 1/SiO$_2$**

| Sample | Au-P | | Au-Pt | | Au-Au | |
|---|---|---|---|---|---|---|
| | N | r/nm | N | r/nm | N | r/nm |
| **1** | 1[a] | 0.229[a] | 1[a] | 0.268[a] | 2.67[a] | 0.285[a] |
| **3** | 1[a] | 0.229[a] | 1[a] | 0.269[a] | 3[a] | 0.289[a] |
| 1/SiO$_2$ | 1.0±0.1 | 0.228±0.001 | 1.1±0.1 | 0.268±0.001 | 2.6±0.3 | 0.285±0.001 |
| 1/SiO$_2$ after adsorption of CO | 1.0±0.1 | 0.228±0.004 | 1.1±0.1 | 0.269±0.001 | 3.0±0.3 | 0.290±0.001 |
| 1/SiO$_2$ after desorption of CO | 1.0±0.1 | 0.228±0.001 | 1.1±0.1 | 0.268±0.001 | 2.7±0.3 | 0.285±0.001 |

The $\Delta r$ range for the inverse Fourier transformation was 0.12—0.35 nm for both Pt and Au EXAFS. The curve fitting range was 30—120 nm$^{-1}$ for Au EXAFS. a) The averaged values derived from crystallographic data.[30]

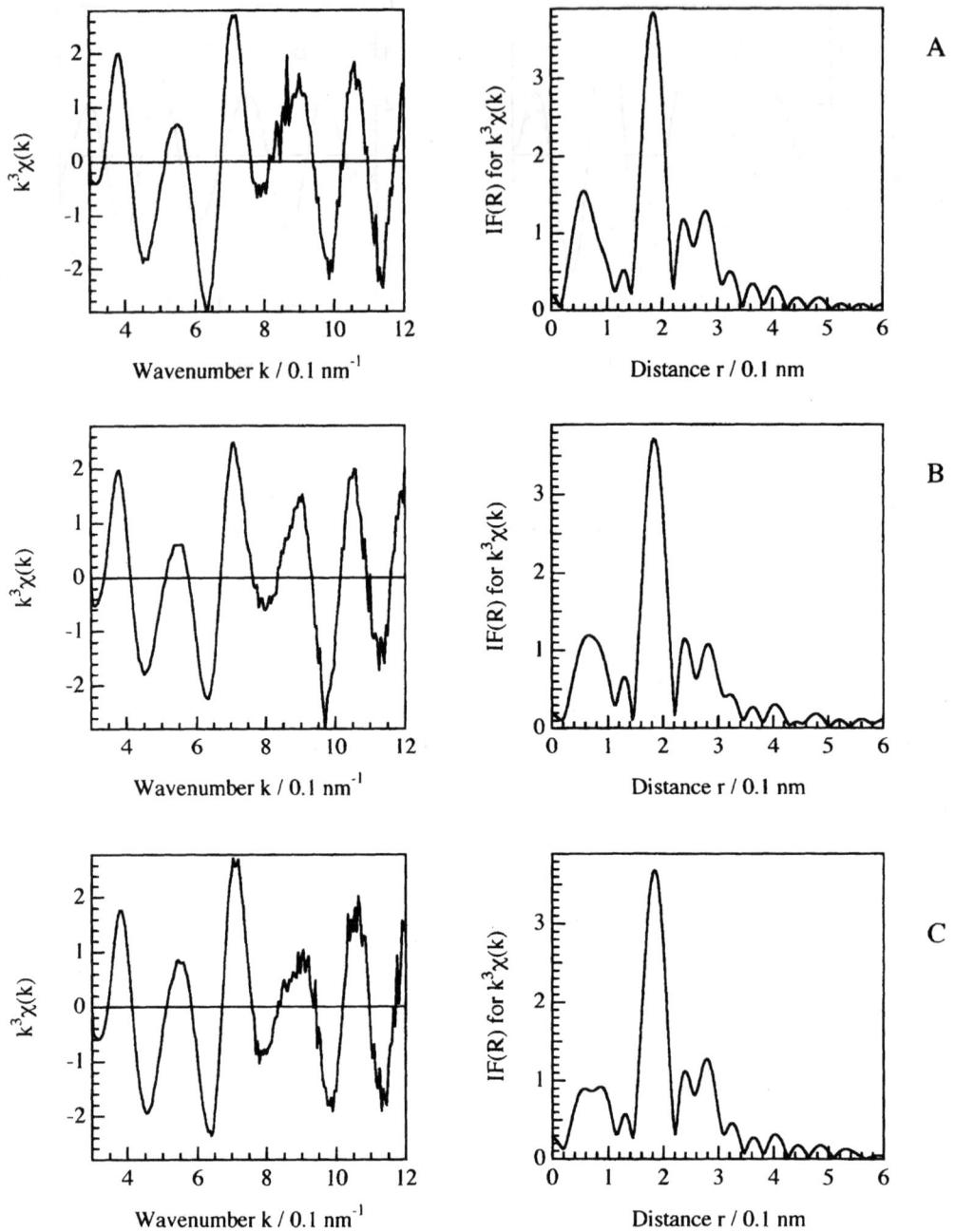

**Fig. 6.** Au $L_3$-edge EXAFS oscillations and associated Fourier transforms for $1/SiO_2$.
(A) $1/SiO_2$；(B) $1/SiO_2$ under 300 Torr of CO；(C) $1/SiO_2$ after evacuation at 353 K.

**EXAFS Measurements for $H_2$-Adsorbed $1/SiO_2$.** Since the adsorption of $H_2$ on $1/SiO_2$ was weak and small at room temperature, the EXAFS spectra for $1/SiO_2$ exposed to $H_2$ were measured under 0.3 MPa of $H_2$. Figures 8 and 9(A, B, and C) show the oscillations and associated Fourier transforms for the EXAFS data at Pt $L_3$-edge and Au $L_3$-edge for $1/SiO_2$ before exposure to $H_2$, under 0.3 MPa of $H_2$ and after evacuation, respectively. It was found that the EXASF spectra were quite similar to each other. There was little difference in the cluster structures among the three samples.

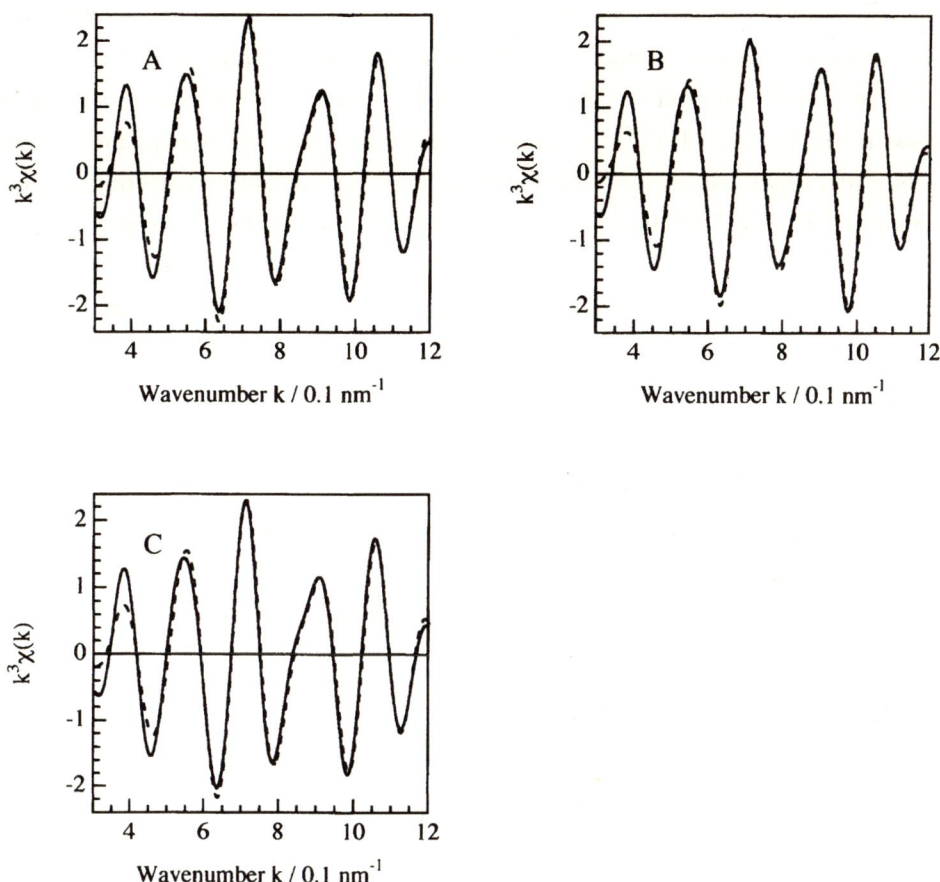

Fig. 7. The curve-fitting analysis for the Au L₃-edge EXAFS data of 1/SiO₂ based on two-shells model(Au-P and Au-Au
(Pt)):—,observed,…,calculated.
(A) 1/SiO₂ ;(B) 1/SiO₂ after adsorption of CO;(C) 1/SiO₂ after desorption of the adsorbed CO at 353 K.

# 3  Discussion

**CO Adsorption on 1/SiO₂.** From comparison with the crystal data and $v_{CO}$ peak of the cluster $[(AuPPh_3)_6Pt(CO)(PPh_3)]^{2+}$ synthesized by reaction of **1** with CO in solution, the peak at 1972 cm$^{-1}$ observed with the CO-adsorbed 1/SiO₂ is assigned to linear CO adsorbed on Pt. When CO was exposed to the 1/SiO₂ pretreated at 473 K, the $v_{CO}$ peak became slightly weaker in intensity and higher in wavenumber in Fig. 1B and the TPD-desorption amount of CO became smaller in Fig. 2A. These results support the indication of the previous EXFAS results that there is a small deformation in the cluster framework by heating the intact 1/SiO₂ at 473 K in vacuum, where the bond numbers of Au-Au(Pt) and Pt-Au decreased to about 75 and 68% of the original ones[13,15].

**Structure Transformation of 1/SiO₂ by CO Adsorption and Desorption.** We have observed a reversible structure transformation of 1/SiO₂ by the adsorption and desorption of CO without any fragmentation of the Au₆Pt framework. There are two features in this transformation found by EXAFS: (1) The Au-Au coordination number increases with the CO adsorption;(2) The peak of Pt-(Au)-P in the Fourier transform of Pt L₃-edge increases, though no increase in the coordination number of P around Au is observed. Thus the change occurs only on the Au atoms surrounding Pt.

A CO-adduct cluster $[(AuPPh_3)_6Pt(CO)(PPh_3)](PF_6)_2$ (**3**) can be easily derived by reaction of **1** with CO in $CH_2Cl_2$ solution and thereafter by metathesis of $NO_3^-$ salt from a methanol solution of $NH_4PF_6$[30]. The compound **3** has a longer Pt-P bond distance on average than the cluster **1**, which is compatible with the change found in the cluster **1** on $SiO_2$ after CO adsorption, as shown in Table 2.

The increase in the Pt-(Au)-P Fourier transform peak intensity can be explained by the bond angle of Pt-Au-P to be 180°, which causes a strong focusing effect in the multi scattering of photoelectrons. The average angles of Pt-Au-P in clusters **1** and **3** are 162. 5° and 164. 8°, respectively, based on the crystallographic data. We have calculated the Pt-(Au)-P EXAFS oscillation using the FEFF program on the assumption that the cluster structures on $SiO_2$ before and after CO adsorption are the same as the

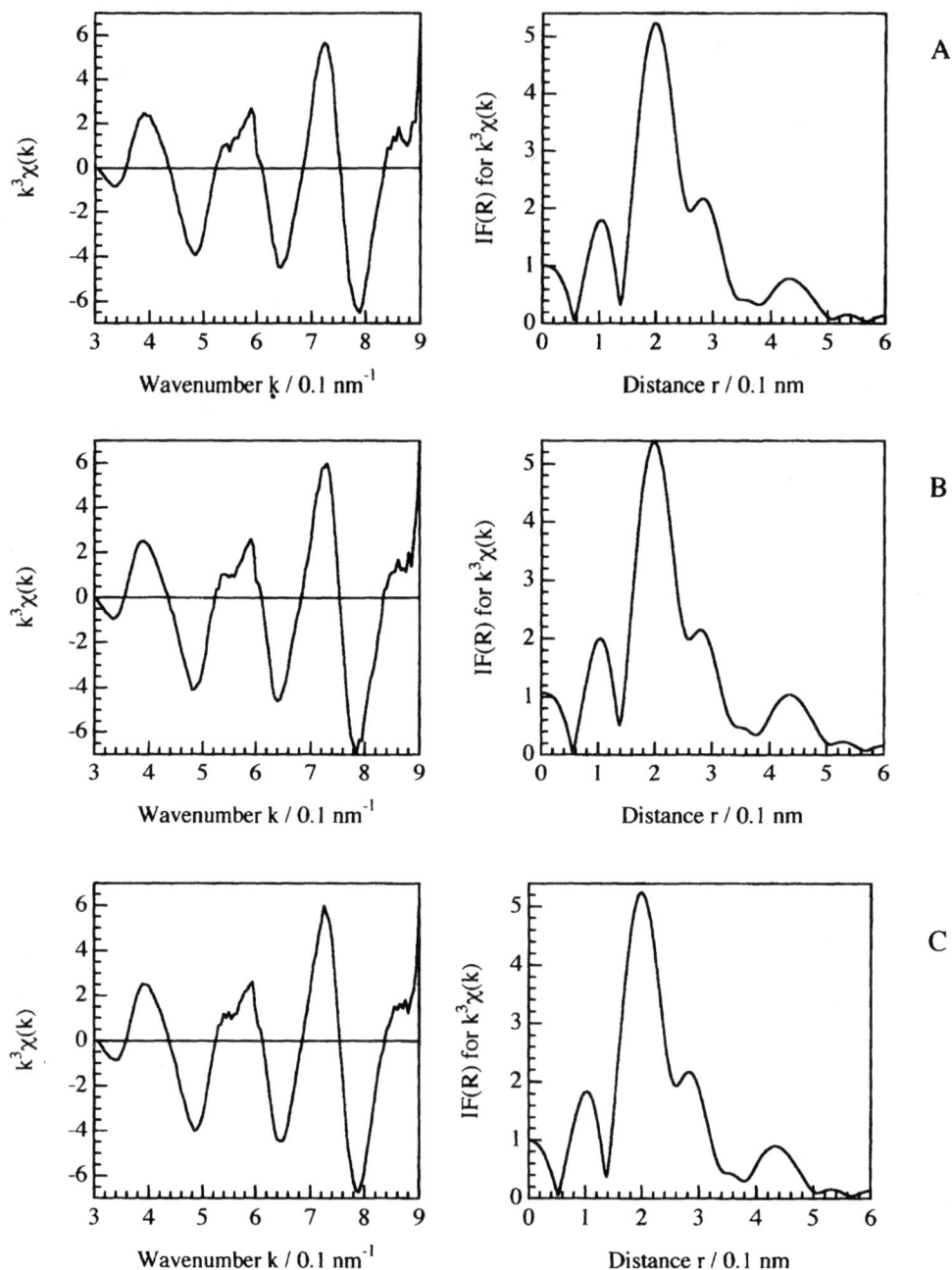

**Fig. 8.** Pt $L_3$-edge EXAFS oscillations and associated Fourier transforms for $1/SiO_2$.
(A) $1/SiO_2$; (B) $1/SiO_2$ under 0. 3 MPa of $H_2$; (C) $1/SiO_2$ after desorption of $H_2$.

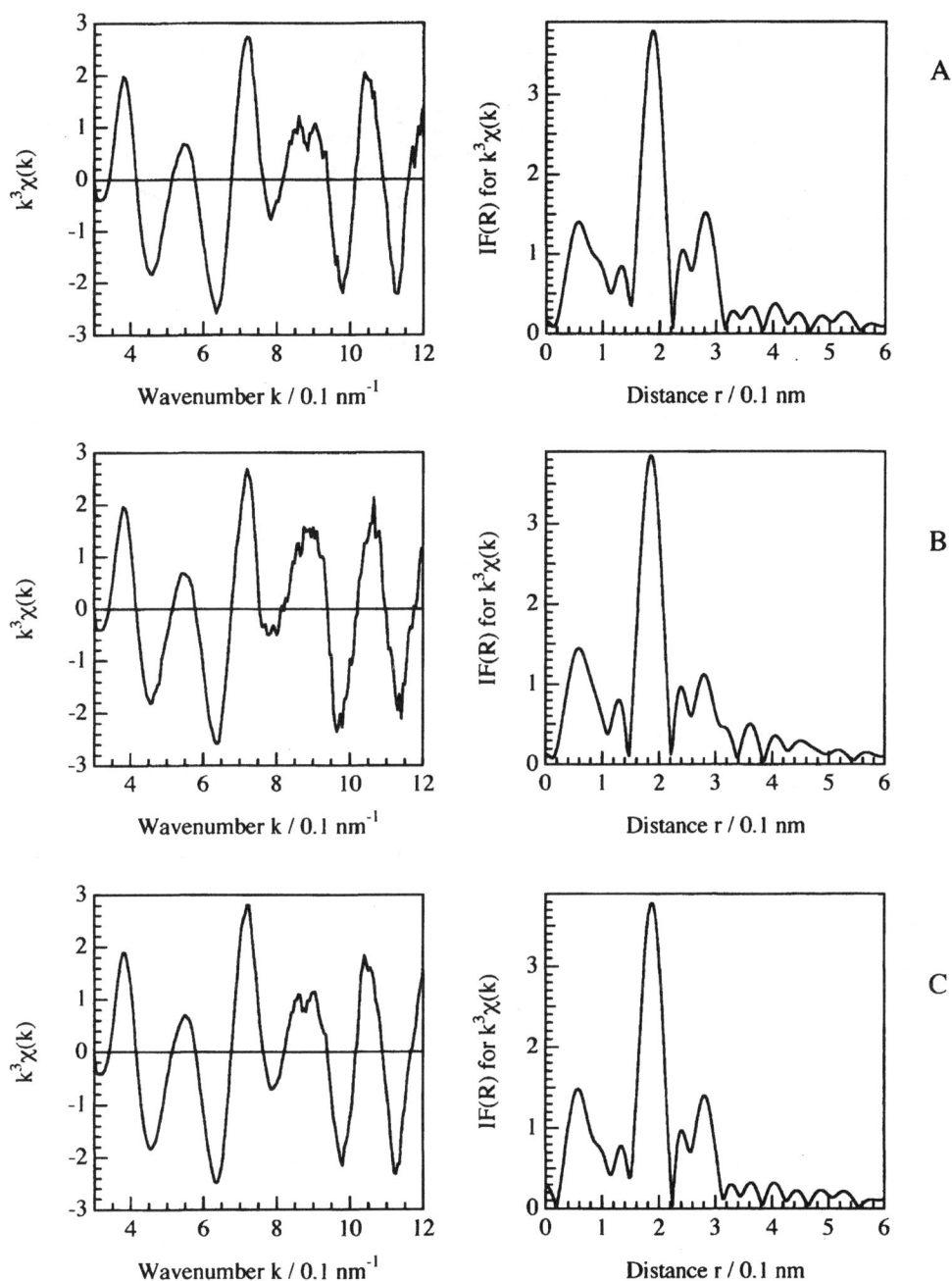

**Fig. 9.** Au L$_3$-edge EXAFS oscillations and associated Fourier transforms for 1/SiO$_2$.

(A) **1**/SiO$_2$ ; (B) **1**/SiO$_2$ under 0.3 MPa of H$_2$ ; (C) **1**/SiO$_2$ after desorption of H$_2$.

ones in compounds **1** and **3** given in the literature, respectively[30]. The results are depicted in Fig. 10. Because of the ambiguity in Debye-Waller factors, the relative height of the simulated Fourier transforms has some meaning. The height around 0.45 nm in **1** is 1.1 and it increases to 2.0 in **3**. Therefore, the increase in the peak of 0.45 nm in the Pt L$_3$-edge EXAFS Fourier transform (Fig. 4) can be explained by the transformation of the surface structure from **1** to **3** by the CO adsorption. Note that the difference in average Pt-Au-P angle between **1** and **3** is 2.3°, though the peak height for **3** is almost twice that for **1**. According to the FEFF calculation of the peak height of Pt-(Au)-P in the Fourier transform for several angles, the peak height sensitively varied with the angles at larger than 160°.

From the EXAFS results, we propose that the structure transformation occurs during CO adsorption, as shown in Fig. 11. It is to be noted that the adsorption of CO from Pt atom has never been observed with the cluster **3** in a solution system unless the framework structure is destroyed. Supporting the cluster **1** on $SiO_2$ has made it possible to undergo a reversible structural change without collapse of the cluster framework like $Ru_6C(CO)_{16}/MgO^{[31]}$. Such a reversible structural change may provide a molecular actuator for molecular mechanics and molecular devices.

**Adsorption of $H_2$ and Mechanism Consideration of $H_2$-$D_2$ Equilibration on $1/SiO_2$.** It has been reported that the cluster **1** showed a catalytic activity with TOFs of 0.6 $s^{-1}$ in solution and 2.0 $s^{-1}$ in gas-solid condition for $H_2$-$D_2$ equilibration at room

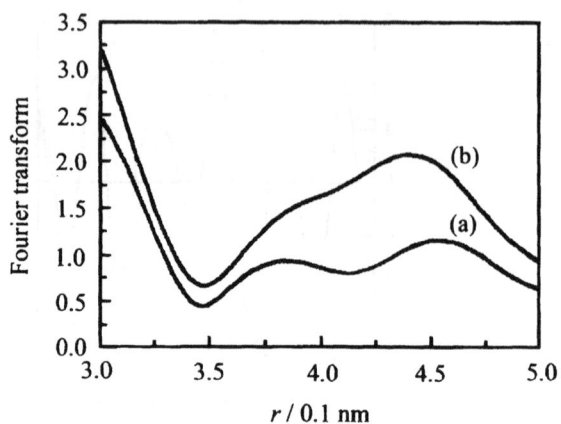

**Fig. 10.** Fourier transforms of the FEFF simulation based on the models **1** ( a ) and **3** ( b ). The Fourier transformation was carried out over $k = 30 - 90$ $nm^{-1}$.

temperature. It was also found that the TOF increased to 29.8 $s^{-1}$ by supporting **1** on $SiO_2$, mainly due to high dispersion of the active clusters where the reaction occurs on the Pt atom promoted by Au-Pt bonds$^{[12-15]}$. When $H_2$ was exposed to the intact $1/SiO_2$, however, only very weak adsorption of $H_2$ on $1/SiO_2$ occurred and almost no contribution of $H_2$ adsorption to the EXAFS observation was observed. It is not due to the weak scattering of hydrogen. Kubota et al. reported that a shape change 8 eV above the edge occurred when hydrogen was adsorbed on $Pt^{[32]}$. They attributed the shape change to a multiple scattering effect caused by H-Pt bonds$^{[33]}$. Koningsberger ascribed the change to an EXAFS structure based on their FEFF calculations$^{[34]}$. Thus if hydrogen was chemisorbed on Pt, one should find a change near the edge region. However, we could not find any change in the edge region of $1/SiO_2$ and no dissociative chemisorption occurred, but very weak adsorption of $H_2$ on $1/SiO_2$ did occurr.

The structure transformation of $1/SiO_2$ by CO adsorption and desorption encouraged us to examine the possibility that $H_2$ adsorption or $H_2$-$D_2$ equilibration on $1/SiO_2$ may result in core movements of the cluster framework through the angle of Pt-Au-P and the bond length of Au-Pt(Au). Recently, a B3LYP study was performed on the phosphine-ligated gold-platinum model clusters such as $[(AuPH_3)_6Pt(PH_3)]^{2+}$ and $[(AuPH_3)_6Pt(H_2)(PH_3)]^{2+[35]}$. The theoretical calculation revealed that there were two stable $H_2$ adducts ( dihydrogen complexes ), which might be possible intermediates for the $H_2$-$D_2$ equilibration. The potential profile along the reaction path was very flat. Pt was characterized as being the active site, but Au atoms also played an important role in $H_2$ activation. The $H_2$ activation may include the electron transfer from the metal core to the $\sigma^*$ anti-bonding orbital of $H_2$ and movements of the metal core$^{[35]}$. These findings support the experimental fact that the reaction between $1/SiO_2$ and $H_2$ is a rapid reversible reaction.

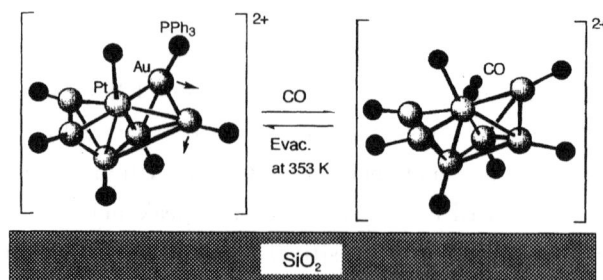

**Fig. 11.** Structural transformation of $1/SiO_2$ by CO adsorption and desorption.

# 4  Conclusion

The $SiO_2$-supported cluster $[(AuPPh_3)_6Pt(PPh_3)](NO_3)_2$ (**1**) reacted with CO on the Pt atom to give a CO-adduct **1**/$SiO_2$, which showed $\nu_{CO}$ at 1972 cm$^{-1}$. One CO molecule adsorbed on a Pt atom in **1**/$SiO_2$ when the CO pressure was higher than 250 Torr. The cluster framework of **1** on $SiO_2$ was deformed by the expansion in the bond angle of $\angle$Au-Pt-P when CO adsorbed on the Pt atom of **1**/$SiO_2$. The deformed structure was reconverted to the original structure without fragmentation by evacuation of CO at 353 K, unlike the case of the cluster in solution. The $SiO_2$ surface prevented the bimetal cluster structure from fragmenting during the adsorption-desorption process. The $H_2$ adsorption on **1**/$SiO_2$ was weak and reversible at 298 K. The contribution of $H_2$ to the EXAFS of **1**/$SiO_2$ was negligible.

This work has been supported by Core Research for Evolutional Science and Technology(CREST) of the Japan Science and Technology Corporation (JST). The EXAFS measurements have been performed with approval of the Photon Factory Advisory Committee(Proposal No. 95G200).

# 5  References

[1]For examples of supported Au catalysts: a) N. W. Cant and W. K. Hall, J. Phys. Chem. , **75**, 2914(1995). b) S. Galvagno and G. Parravano, J. Catal., **55**, 178(1978). c) J. Schwank, Gold Bull., **16**, 103 (1983). d) M. Haruta, S. Tsubota, N. Yamada, T. Kobayashi, and S. Iijima, J. Catal., **115**, 301(1989). e) S. D. Gardner and G. B. Hoflund, Langmuir, **7**, 2135(1991). f) S. D. Gardner, G. B. Hoflund, and M. R. Davidson, Langmuir, **7**, 2140 (1991). g) M. Haruta, S. Tsubota, T. Kobayashi, H. Kageyama, M. J. Genet, and B. Delmon, J. Catal., **144**, 174(1993). h) S. Tsubota, A. Ueda, H. Sakurai, T. Kabayashi, and M. Haruta, ACS Symp. Ser., **552**, 420 (1993). i) G. B. Hoflund, S. D. Gardner, D. R. Schryer, B. T. Upchurch, and E. J. Kielin, Appl. Catal., **6**, 117(1995). j) G. Srinivas, J. Wright, C. -S. Bai, and R. cook, Stud. Surf. Sci. Catal., **101**, 427(1996). k) Y. Yuan, K. Asakura, H. Wan, K. Tsai, and Y. Iwasawa, Catal. Lett. , **42**, 15 (1996). l) Y. Yuan, K. Asakura, H. Wan, K. Tsai, and Y. Iwasawa, Chem. Lett., **29**, 755(1996). m) Y. Yuan, A. P. Kozlova, K. Asakura, H. Wan, K. Tsai, and Y. Iwasawa, J. Catal, **170**, 191 (1997). n) Y. Yuan, K. Asakura, A. P. Kozlova, H. Wan, K. Tsai, and Y. Iwasawa, Catal. Today, **44**, 333(1998).

[2]For Examples of Pt-transition metal catalysts: a) Y. L. Lam and M. Boudart, J. Catal., **50**, 530 (1977). b) Y. L. Lam, J. Criado, and M. Boudart, Nouv. J. Chim., **1**, 461 (1977). c) G. A. Somorjai, Catal Lett. , **15**, 25(1992). d) W. D. Provine, P. L. Mills, and J. J. Lerou, Stud. Surf. Sci. Catal., **101**, 191(1996).

[3]For examples of Pt-Au catalysts: a) J. R. H. van Schaik, R. P. Dessing, and V. Ponec, J. Catal., **38**, 273(1975). b) J. K. A. Clarke, L. Manninger, and T. Baird, J. Catal., **54**, 230(1978). c) J. K. A. Clarke, L. Manninger, and T. Baird, J. Catal., **9**, 85 (1984). d) J. H. Sinfelt, "Bimetallic Catalysts," Wiley, New York(1985). e) R. C. Yates and G. A. Somorjai, J. Catal., **103**, 208 (1987). f) K. Balakrishnan, A. Sachdev, and J. Sachwank, J. Catal., **121**, 441(1990). g) P. A. Sermon, J. M. Thomas, K. Keryou, and G. R. Millward, Angew. Chem., Int. Ed. Engl., **26**, 918 (1987). h) A. Sachdev and J. Schwank, J. Catal., **120**, 353(1989). i) K. Balakrishnan and J. Sachwank, J. Catal, **132**, 451(1991). j) D. Rouabah and J. Fraissard, J. Catal., **144**, 30(1993). k)

J. Sachtler, K. Balakrishnan, and A. Sachdev, in "New Frontiers in Catalysis," ed by L. Guczi, F. Solymosi, and P. Tetenyi, Elservier, Amsterdam(1993), p. 905.

[4] J. W. A Sachtler and G. A. Somorjai, J. Catal., **81**, 77(1983).

[5] J. Sachlter, J. Biberian, and G. Somorjai, Surf. Sci., **110**, 43(1981).

[6] A. Sachdev and J. Schwank, J. Catal., **120**, 353(1989).

[7] J. R. H. van Shaik, R. P. Dessing, and V. Ponec, J. Catal., **38**, 273(1975).

[8] H. C. de Jongste, F. J. Kuijers, and V. Ponec, in "Proc. Int. Symp. Heterogeneous Catalysis," ed by B. Delmon, P. A. Jacobs, and G. Poncelet, Brussels(1975), Elsevier, Amsterdam(1976), p. 207.

[9] R. P. Dessing and V. Ponec, React. Kinet. Catal. Lett., **5**, 251(1976).

[10] A. F. Kane and J. K. A. Clarke, J. Chem. Soc., Faraday Trans. 1, **76**, 1640(1980).

[11] J. K. A. Clarke, I. Manninger, and T. Baird, J. Catal., **54**, 230(1978).

[12] I. Gubkina, L. Rubinstein, and L. Pignolet, Abstr. of ACS Meeting, **208**, 405(1994).

[13] Y. Yuan, K. Asakura, H. Wan, K. Tsai, and Y. Iwasawa, Chem. Lett., **1996**, 129.

[14] I. V. G. Graf, J. W. Bacon, M. B. consugar, M. E. Curley, L. N. Ito, and L. H. Pignolet, Inorg. Chem., **35**, 689(1996).

[15] Y. Yuan, K. Asakura, H. Wan, K. Tsai, and Y. Iwasawa, J. Mol. Catal. A: Chem., **122**, 147(1997).

[16] M. A. Aubart, B. D. Chandler, R. A. T. Gould, D. A. Krogstad, M. F. J. Schoondergang, and L. H. Pignolet, Inorg. Chem., **33**, 3724(1994).

[17] D. C. Roe, J. Magn. Res., **63**, 388(1985).

[18] T. G. M. M. Kappen, J. J. Bour, P. P. J. Schlebos, A. M. Reolofsen, J. G. M. van der Linden, J. J. Steggerda, M. A. Aubart, D. A. Krogstad, M. F. J. Schoonderhang, and L. H. Pignolet, Inorg. Chem., **32**, 1074(1993).

[19] J. J. Bour, P. P. J. Schlebos, R. P. F. Kanters, M. F. J. Schoondergang, H. Addens, A. Overweg, and J. J. Steggerda, Inorg. Chim. Acta, **181**, 195(1991).

[20] M. A. Aubart, J. F. D. Koch, and L. H. Pignolet, Inorg. Chem., **33**, 3852(1994).

[21] Y. Iwasawa, Stud. Surf. Sci. Catal. (Proc. 11th Int. congr. Catal., Baltimore), Vol. 101, Elsevier(1996), p. 21.

[22] Y. Iwasawa, "Tailored Metal Catalysts," Reidel, Holland(1986).

[23] Y. Iwasawa, Adv. Catal, **35**, 187(1987).

[24] Y. Iwasawa, Catal Today, **18**, 21(1993).

[25] Y. Iwasawa, "X-Ray Adsorption Fine Structure for Catalysts and Surfaces," World Scientific, Singapore(1996).

[26] Y. Iwasawa, Res. Chem. Intermed., **15**, 183(1991).

[27] K. Asakura, Y. Yuan, and Y. Iwasawa, J. Phys. IV, Fr., **7**, C2-863(1997).

[28] K. Asakura, in "X-Ray Adsorption Fine Structure for Catalysts and Surfaces," ed by Y. Iwasawa, World Scientific, Singapore(1996).

[29] J. J. Rehr, de Leon, J. Mustre, S. I. Zabinsky, and R. C. Albers, J. Am. Chem. Soc., **113**, 5235(1991).

[30] L. N. Ito, J. D. Sweet, A. M. Mueting, L. H. Pignolet, M. F. J. Schoodergang, and J. J. Steggerda, Inorg. Chem., **28**, 3696(1989).

[31] Y. Izumi, T. Chihara, H. Yamazaki, and Y. Iwasawa, J. Chem. Soc., Dalton Trans., **1993**, 3667.

[32] T. Kubota, K. Asakura, N. Ichikuni, and Y. Iwasawa, Chem. Phys. Lett., **256**, 445(1996).

[33] K. Ohtani, T. Fujikawa, T. Kubota, K. Asakura, and Y. Iwasawa, Jpn. J. Appl. Phys., **36**, 6504 (1997).

[34] D. C. Konigsberger, Proc. XAFS X, in press.

[35] X. Xu, Y. Yuan, K. Asakura, Y. Iwasawa, H. Wan, and K. Tsai, Chem. Phys. Lett., **286**, 163 (1998).

■ 本文原载：Transition Metal Chemistry **24**(1999)，pp. 605～609.

# Complexation Between Vanadium(V) and Citrate: Spectroscopic and Structural Characterization of a Dinuclear Vanadium(V) Complex

Zhao-Hui Zhou[①], Hui Zhang, Ya-Qi Jiang, Dong-Hai Lin, Hui-Lin Wan, Khi-Rui Tsai

(*Department of Chemistry and State Key Laboratory for the Physical Chemistry of Solid Surfaces, Xiamen University, Xiamen, 361005,P. R. China, e-mail: zhzhou@xmu. edu. cn*)

Received 18 November 1998; accepted 10 February 1999

**Abstract**    Investigation of the aqueous coordination chemistry for citrate and vanadium(V) resulted in the isolation and characterization of a dinuclear vanadium(V) citrato complex (**1**) $Na_2 K_2 [VO_2 (Hcit)]_2 \cdot 9H_2O$. complex (**1**) is an intermediate between the fully deprotonated and diprotonated citrate vanadate. It may represent an early mobilized precursor in the biosynthesis of FeV-co, as well as a relevant model in the proton transport relay process between P-cluster pair to M-cluster pair. The complex has been characterized by elemental analyses and i. r. spectroscopy. Its i. r. spectra are consistent with a oxo-bridged dinuclear structure as revealed by a single crystal X-ray diffraction study.

## 1  Introduction

Tricarboxylic acids and their metal complexes have attracted much attention in recent years, because of their involvement in the biosynthesis of the cofactor of nitrogenase and their ability to mobilize heterometals such as molybdenum or vanadium from the appropriate storage enzyme[1,2]. Molybdenum and vanadium are believed to be taken up firstly by organisms as $MoO_4^{2-}$ or $VO_4^{3-}$; this would be essential for the assembly of the final cofactor cluster from an oxomolybdenum-or oxovanadium-citrate precursor[3]. While the precise role of homocitrate in both the biosynthesis of FeMo-co and the mechanism of dinitrogen reduction is poorly understood, the discovery and elucidation of the key role played by this tricarboxylic acid in nitrogenase are of the great importance[4]. A more detailed understanding of the nature of vanadium and biologically important ligands such as citric acid interactions necessitates a detailed knowledge of the coordination environment and oxidation state of the metal(s) within the systems. Moreover, the variety of possible functions exhibited by citrate in its interactions with vanadium, a metal which plays an important role in different living plants and animal organisms[5], prompted our further investigation into the coordination chemistry of vanadium citrate complexes.

Complex formation between vanadate and citrate has been studied by u. v.,i. r.,e. p. r., potentiometric and $^{51}V$ n. m. r. measurements[6-16]. The first structurally characterized citrato vanadate

---

① Author for correspondence.

(V) was reported as $Na_2H_2[VO_2(cit)] \cdot H_2O$[17], and its potassium vanadate counterpart was prepared from the reaction of vanadium(V) oxide with critic acid and described as $K[VO_2(H_2cit)] \cdot H_2O$[18]; it was later confirmed structurally to be a dinuclear complex, $K_2[VO_2(H_2cit)] \cdot 2H_2O$[19,20]. Moreover, homocitrate complex $[K_2(H_2O)_5][(VO_2)_2(R, S\text{-}homo\text{-}H_2cit)_2] \cdot H_2O$ and the completely deprotonated citrato vanadate $Na_2(NH_4)_4[VO_2(cit)]_2 \cdot 6H_2O$ were reported[21,22]. Crystals of these two kinds of complexes were obtained from acidic or neutral solutions. Here we report the synthesis, spectroscopic properties, structure and transformation of a new dinuclear complex $Na_2K_2[VO_2(Hcit)]_2 \cdot 9H_2O$ (1) ($H_4cit$ = citric acid), which is the intermediate between the fully deprotonated and diprotonated citrate vanadates.

# 2　Experimental

## Preparation

A colourless solution of sodium metavanadate(V), prepared by dissolving $V_2O_5$ (30 mmol) in an aqueous solution of NaOH(60 mmol) at 40 ℃ overnight, was cooled in an ice bath, and aqueous potassium trihydrogen citrate($KH_3cit$, 60 mmol) was added dropwise. The solution was stirred at room temperature for 4 h and filtered. An excess of ethanol was then added until the solution turned cloudy. The mixture was kept in a refrigerator for several days and the deposited solid was collected and washed with EtOH to give a yellow solid(12. 2 g, 49%). Found:C,17. 1,H,3. 3 Calc. for $C_{12}H_{28}O_{27}K_2Na_2V_2$:C, 17. 4,H,3. 4%. IR(KBr): $v_{as}$(C=O) $1641_{vs}$, $v_s$(C=O) $1410_s$, $1377_m$, $v_s$(V=O) $929_s$, $v_{as}$(V=O) $871_m$.

## Transformation of $Na_2K_2[VO_2(Hcit)]_2 \cdot 9H_2O$(1)

To a solution of $Na_2K_2[VO_2(Hcit)]_2 \cdot 9H_2O$ (1) (5 mmol), an excess of citric acid was added and the solution was stirred for 4 h at room temperature. The mixture was then filtered and added to an excess of EtOH. The solid formed was collected and recrystallised to give a green yellow solid(68%). The i. r. spectrum of the product is similar to $K_2[VO_2(H_2cit)]_2 \cdot 4H_2O$ reported previously[18−20], IR (KBr): $v_{as}$(C=O) $1708_{vs}$, $v_s$(C=O) $1403_{vs}$, $1330_m$, $v_s$(V=O) $953_s$, $900_s$, $v_{as}$(V=O) $870_m$.

In another experiment, a solution of (1)(5 mmol) was added to an equivalent mixture of potassium dihydrogen citrate($K_2H_2cit$) and potassium hydrogen citrate($K_3Hcit$) and the solution was stirred for 4 h at room temperature. The mixture was filtered and added to an excess of EtOH. The solid formed was collected and recrystallised to give a green yellow solid(38%). The i. r. spectrum of the product is similar to $Na_2(NH_4)_4[VO_2(cit)]_2 \cdot 6H_2O$ reported previously[22], I. r. (KBr): $v_{as}$(C=O) $1634_s$, $1580_{vs}$, $v_s$(C=O) $1430_s$, $1387_s$, $1352_s$, $v_s$(V=O) $945_s$, $v_{as}$(V=O) $866_m$.

Crystals of suitable quality for the subsequent X-ray diffraction studies were obtained as transparent prism by cooling the saturated solution of compound (1) in refrigerator. The resulting yellow green crystals were sealed in a capillary (to prevent loss of water molecules) and kept in refrigerator.

## Physical measurements

Electronic spectra were recorded on a UV-240 spectrophotometer. I. r. spectra were recorded as Nujol mulls between KBr plates using a Nicolet 740 FT-IR spectrometer. [1]H n. m. r. spectra were recorded on a Varian UNITY 500 NMR spectrometer. Elemental analysis were performed using EA 1106 elemental analyzers.

## X-ray data collection, structure solution and refinement

Crystallographic data for the citratovanadate (**1**) are summarized in Table 1. Diffraction data were collected on an Enraf-Nonius CAD-4 diffractometer with graphite monochromated Cu-K$\alpha$ radiation at 296 K. Lp factor, anisotropic decay and empirical absorption corrections were applied. The structure was solved by the heavy atom method and refined by full-matrix least-squares procedures with anisotropic thermal parameters for all the non-hydrogen atoms. H atoms were located from difference Fourier map and refined isotropically. All calculations were performed on a 586 P/100 microcomputer using MoLEN software package. Selected atomic distances and bond angles are given in Table 2.

**Table 1. Crystal data summaries of intensity data collection and structure refinement for $Na_2 K_2 [VO_2 (Hcit)]_2 \cdot 9H_2O$ (1)**

| | |
|---|---|
| Formula | $C_{12} H_{28} O_{27} K_2 Na_2 V_2$ |
| Molecular weight | 830. 41 |
| Crystal colour, habit | Yellow green, prism |
| Crystal size(mm) | 0. 12×0. 17×0. 40 |
| Crystal system | Monoclinic |
| Cell constants: | $a=8. 8021(4)$ Å |
| | $b=19. 2186(9)$ Å |
| | $c=17. 935(2)$ Å |
| | $\beta=99. 569(5)°$ |
| | $V=2991. 6(6)$ Å$^3$ |
| Space group | $P2_1/n$ |
| Formula units/unit cell | 4 |
| $D_{calc}$ | 1. 844 |
| $F(000)$ | 1688 |
| $\mu(MoK\alpha)$ | 91. 17 |
| Diffractometer | Enraf-Nonius CAD-4 |
| Radiation | CuK$\alpha$ ($\lambda=1. 5418$ Å) |
| Temp. (°) | 23 |
| Scan Width | $0. 50+0. 34 \tan \theta$ |
| Decay of standards | $-6. 0\%$ |
| No. of reflections measured | 6044 |
| $2\theta$ Range(°) | 3—70 |
| Range of $h, k, l$ | ±10, 23, -21 |
| No. of reflections observed | 4072 |
| $[I>3\sigma(I)]^a$ | |
| No. of parameters varied | 495 |
| Weighting scheme(w = ) | $[\sigma(F_0)^2 + 0. 0004(F_0)^2 +]^{-1}$ |
| Goodness of fit | 1. 033 |
| $R=\sum(|F_0|-|F_c|)/\sum|F_0|$ | 0. 054 |
| $R_w$ | 0. 063 |
| Largest difference peak and hole(eÅ$^{-3}$) | 0. 69, -0. 07 |

$^a$ corrections: Lorentz-polarization.

Table 2.  Selected bond distances($\overset{\circ}{A}$) and angles(°)for Na$_2$K$_2$[VO$_2$(Hcit)]$_2 \cdot$ 9H$_2$O (1)

| Bond distances | | | |
|---|---|---|---|
| V(1)—O(1) | 2.005(3) | O(5)—C(2) | 1.232(7) |
| V(1)—O(2) | 1.976(4) | O(6)—C(4) | 1.222(7) |
| V(1)—O(3) | 1.628(4) | O(7)—C(4) | 1.281(6) |
| V(1)—O(4) | 1.611(5) | O(8)—C(6) | 1.214(7) |
| V(1)—O(11) | 1.973(4) | O(9)—C(6) | 1.306(7) |
| V(2)—O(1) | 1.966(3) | O(11)—C(11) | 1.436(6) |
| V(2)—O(11) | 1.993(3) | O(12)—C(12) | 1.287(7) |
| V(2)—O(12) | 1.982(4) | O(15)—C(12) | 1.239(7) |
| V(2)—O(13) | 1.621(4) | O(16)—C(14) | 1.235(7) |
| V(2)—O(14) | 1.629(4) | O(17)—C(14) | 1.291(7) |
| O(1)—C(1) | 1.439(6) | O(18)—C(16) | 1.226(8) |
| O(2)—C(2) | 1.298(7) | O(19)—C(16) | 1.284(7) |
| **Bond angles** | | | |
| O(1)—V(1)—O(2) | 78.1(1) | O(4)—V(1)—O(11) | 101.2(2) |
| O(1)—V(1)—O(3) | 130.6(2) | O(11)—V(2)—O(12) | 78.2(1) |
| O(1)—V(1)—O(4) | 122.7(2) | O(11)—V(2)—O(13) | 121.2(2) |
| O(1)—V(1)—O(11) | 71.7(1) | O(11)—V(2)—O(14) | 131.9(2) |
| O(2)—V(1)—O(3) | 95.3(2) | O(12)—V(2)—O(13) | 96.4(2) |
| O(2)—V(1)—O(4) | 98.6(2) | O(12)—V(2)—O(14) | 95.3(2) |
| O(2)—V(1)—O(11) | 149.4(2) | O(13)—V(2)—O(14) | 106.8(2) |
| O(3)—V(1)—O(4) | 106.6(2) | V(1)—O(1)—V(2) | 108.0(2) |
| O(3)—V(1)—O(11) | 101.1(2) | V(1)—O(11)—V(2) | 108.2(2) |
| **Hydrogen bonding** | | | |
| O(7)—H(7) | 1.04(7) | O(9)—H(9) | 0.99(6) |
| H(7)···O(17a) | 1.43(8) | H(9)···O(19b) | 1.51(6) |
| O(7)—H(7)···O(17a) | 177(7) | O(9)—H(9)···O(19b) | 164(6) |

Symmetry transformation:$a(-1/2+x,-1/2-y,-1/2+z)$,$b(-3/2+x,-1/2-y,-3/2+z)$.

# 3  Results and discussion

## Synthesis and general characterization

Preparation of the citrato vanadates depends mainly on pH control. Previous potentiometric and $^{51}$V n. m. r. measurements in the oxovanadium(V)-citrate system have proceeded in the 0—8. 0 pH range in solution[9]. The equilibria for the various possible complex formations may be represented by the general equation

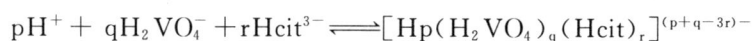

$$pH^+ + qH_2VO_4^- + rHcit^{3-} \rightleftharpoons [Hp(H_2VO_4)_q(Hcit)_r]^{(p+q-3r)-}$$

The major species are determined as 1 : 2 : 1,2 : 2 : 1 and 3 : 2 : 1 in a 2 : 1 ratio of vanadium and citrate in solution[9]. In our experiment only the 1 : 1(V:cit) ratio of citrato vanadates are separated in the acidic,weak acidic and neutral solution. The transformation and isolation of citrate complexes in different protonated form have proven that 1 : 1 ratio citrato:vanadate is the major species in solution, and the ratio of reactants is not such a crucial factor for the formation of these species. The 2 : 1 ratio is unfavorable in citrate vanadate(V) formation. In acidic solution, an excess of vanadium in the system may result in the formation of homopolyvanadium(V) acid. Although similar anions $[(MoO_2)_4O_3(cit)_2]^{4-}$ and $[(MoO_2)_4O_3(R \text{ or } S\text{-mal})_2]^{4-}$ are present in citrate and malate tetramolybdates[23-28].

Transformations between the dimeric oxocitrato vanadates can be accomplished by the controlling of pH, in which citric acid or citrate was added. The dimeric configuration and bidentate mode of citrate are retained.

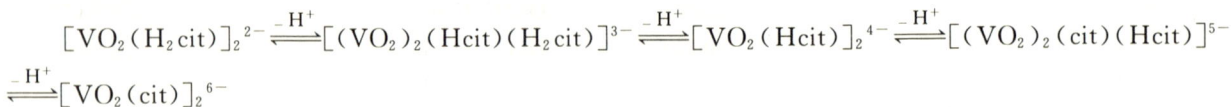

$$[VO_2(H_2cit)]_2^{2-} \underset{-H^+}{\rightleftharpoons} [(VO_2)_2(Hcit)(H_2cit)]^{3-} \underset{-H^+}{\rightleftharpoons} [VO_2(Hcit)]_2^{4-} \underset{-H^+}{\rightleftharpoons} [(VO_2)_2(cit)(Hcit)]^{5-}$$

$$\underset{-H^+}{\rightleftharpoons} [VO_2(cit)]_2^{6-}$$

The pH values in the reactions are controlled easily by citrate anions acting as both reactant and buffer agent. This is further supported by the preparation of dihydrocitrato vanadate or deprotonated citrato vanadate from ammonium vanadate and citric acid in acidic or neutral solution. The loss of protons suggests the following reactions for the products.

$$2VO_3^- + 2H_4cit \longrightarrow [VO_2(H_2cit)]_2^{2-} + 2H_2O$$

$$2VO_3^- + 2H_3cit^- \longrightarrow [VO_2(Hcit)]_2^{4-} + 2H_2O$$

$$2VO_3^- + 2H_2cit^{2-} \longrightarrow [VO_2(cit)]_2^{6-} + 2H_2O$$

Complex $Na_2K_2[VO_2(Hcit)]_2 \cdot 9H_2O$ (**1**) which separated at pH 4 is an intermediate between fully protonated and deprotonated dimeric 1 : 1 vanadium-citrate complex, in which one of the $\beta$-carboxyl groups of the citrate is protonated, while the other is deprotonated.

The vanadium(V) complex is featureless in the 350—600 nm visible range. The frequencies and assignments of selected i. r. absorption bands are shown in the experimental section. In 1800—1400 $cm^{-1}$ region, the bands at 1641, 1410 and 1377 $cm^{-1}$ correspond to the bound carboxyl group $v_{asym}$ and $v_{sym}(CO_2)$ respectively, this is in accord with the coordinated and free carboxyl of the citrato ligand. Loss or addition of the proton in compound (**1**) displaces the bands to lower or higher frequencies. In the ca. 900 $cm^{-1}$ region, the complexes show several bands which might result from the presence of *cis*-dioxo cores. The low frequency symmetric $VO_2$ stretch may be ascribed to intramolecular hydrogen bonding and the coordination of oxo-groups to the sodium or potassium cation.

In the $^1H$ n. m. r. spectra the methylene protons of citric acid give rise to an AB quartet which shifts with pH. Complexation with vanadium(V) leads to a slightly distorted AB quartet with $\delta q = 3.007, 2.932$ ppm respectively, indicating that both $CH_2CO_2$ arms are almost equivalent.

It is clear that one of the citrate ions carries two $\beta$-carboxyl group protons; this is not only because hydrogen atoms are clearly visible in difference maps, but also results from charge balance in the complex. The observed C—O distances of terminal $\beta$-carboxyl groups, with or without protons, are equivalent [O(6)—C(4), 1.222(7) Å and O(7)—C(4), 1.281(6) Å, O(8)—C(6), 1.214(7) Å and O(9)—C(6), 1.306 Å; O(16)—C(14), 1.235(7) Å and O(17)—C(4), 1.291(7) Å, O(18)—C(16), 1.226(8) Å and O(19)—C(16), 1.284(7) Å]. This is because of the existence of strong intermolecular hydrogen bonding between protonated and deprotonated free $\beta$-carboxyl groups [H(7)—O(17a), 1.43(8) Å, H(9)—O(19b), 1.51(6) Å], which makes the complex forming a network. This is further supported by i. r. bands at $1641_s$, and $1410_s$, $1377_m$ $cm^{-1}$ corresponding to bound $v_{as}$-and $v_s(CO_2)$, and the absence of i. r. bands between 1740 and 1700 $cm^{-1}$ for a typical undissociated carboxylic group, indicating the obvious shift to lower frequency caused by hydrogen bonding.

## Description of the structure

The crystal structure of (**1**) comprises discrete $Na^+$ and $K^+$ cations, water molecules and hydrocitrato dioxovanadate anions. As shown in Figure 1, the two vanadium(V) atoms are bridged by two alkoxide oxygen atoms of a fully deprotonated citrate ligand, O(11), and of one $H_2 cit^{2-}$ ligand, O(1), forming a common oxobridged $[V_2O_2]^{2+}$ entity which is quasi-centrosymmetric. Each citrate ion acts as a bidentate ligand with the hydroxyl and α-carboxyl coordinated to one vanadium atom, and the other two β-carboxyl groups remain uncomplexed, as does its deprotonated or dihydrocit-rato form[19-22], or the coordination mode of homocitrate in the FeMo cofactor[29]. The two terminal oxo groups are in a *cis*-configuration. Each vanadium atom is five-coordinate and exists in an intermediate geometry between an ideal square pyramid and trigonal bipyramid according to the dihedral data[30].

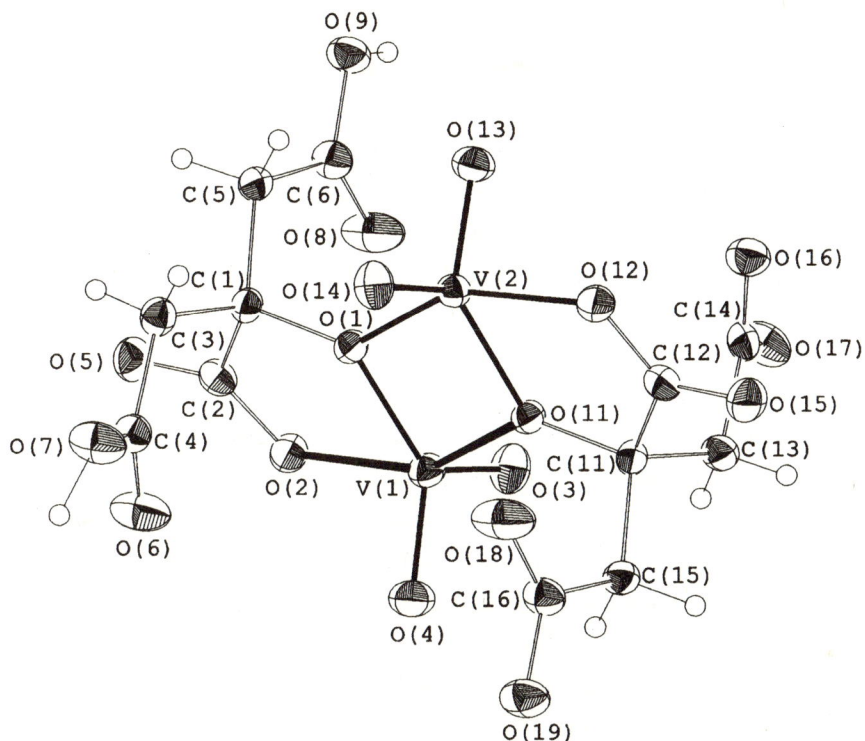

**Fig. 1. Perspective view of the anion structure of $Na_2 K_4 [VO_2 (Hcit)]_2 \cdot 9H_2O$ (1).**

As shown in Table 2, the V—O distances in citrato vanadate vary systematically. V = O is in the range 1.611(5) — 1.629(4) Å, indicating that they are double bonds. The resulting O = V = O angles, 106.6(2) and 106.8(2)°, are considerably larger than the 90° regular value for *cis* groups; this is expected from the greater O···O repulsions between oxygens with short bonds to the metal atom. The V—$O_b$(hydroxyl) bridging distance is 1.984(4)$_{av}$ Å. The V—O distances(α-carboxyl) are in a similar range [1.976(4), 1.982(4) Å]. Usually, vanadium is believed to be taken up by organisms as $VO_4^{2-}$, and a possible function of the tricarboxylic acid in the biosynthesis of the cofactor of nitrogenase is to mobilize vanadium from the appropriate storage enzyme and take part in mediating a P-cluster to Mo site $H^+$ relay process[3,4,19,21]. The two labile β-carboxylate and the bridged alkoxide in dimeric citratovanadate (1) indicate that it is easier to involve in this process, and substitute by histidine group to form bidentate citrate in FeV-co. Such structural changes would be essential for the assembly of the final cofactor cluster from an oxovanadium-citrate precursor. The transformation of citrate vanadates with pH range shows that bidentate tricarboxylic acid can take part in mediating a P-cluster to Mo site $H^+$ relay system specifically required for $N_2$ reduction without changing its configuration and skeleton[4]. Therefore, compound (1) represents a possible biomimetic precursor for the biosynthesis of

FeV-co,as well as a relevant form in the proton transport relay process from P-cluster pairing.

Additional material available from the Cambridge Crystallographic Data Centre comprises thermal parameters and remaining bond lengths and angles(Deposition number,113247)

# Acknowledgements

Financial support provided by the National Science Foundation of China ( No. 29503021 ) is gratefully acknowledged.

# References

[1]P. W. Ludden, V. K. Shah, G. P. Roberts, M. Homer, R. Allen, T. Paustian, J. Roll, R. Chatterjee, M. Madden and J. Allen, Molybdenum Enzymes, Cofactors and Model Systems, American Chemical Society, Washington DC, 1993, p. 196.

[2]E. M. Page and S. A. Wass, Coord. Chem. Rev., **164**, 203(1997).

[3]B. J. Hales, Adv. Inorg. Biochem., 165(1990).

[4]K. R. Tsai and H. L. Wan, J. Clust. Sci., **6**, 485(1995).

[5]A. Butler and C. J. Carrano, Coord. Chem. Rev., **109**, 61(1991).

[6]B. M. Nikolova and G. St. Nikolov, J. Inorg. Nucl. Chem., **29**, 1013(1967).

[7]R. H. Dunhill and T. D. Smith, J. Chem. Soc. A, 2189(1968).

[8]Yu. K. Tselinskii, L. V. Shevcheko and I. I. Kusel'man, Ukr. Khim. Zh., **46**, 656(1980).

[9]P. M. Ehde, I. Anderson and L. Pettersson, Acta Chem. Scand., **143**, 136(1989).

[10]S. G. Vul'fson, A. N. Glebov, O. Yu. Tarasov and Yu. I. Sal'nikov, Dokl. Akad. Nauk. SSSR, **314**, 386(1990).

[11]S. P. Arya and P. K. Sharma, J. Indian Chem. Soc., **69**, 793(1992).

[12]T. A. Dyachkova, R. S. Safin, A. N. Glebov and G. K. Budnikov, Zh. Neorg. Khim., **38**, 482 (1993).

[13]T. Kiss, P. Buglyo, D. Sanna, G. Micera, P. Decock and D. Dewaele, Inorg. Chim. Acta, **239**, 145 (1995).

[14]S. Burojevic, I. Shweky, A. Bino, D. A. Summers and R. C. Thompson, Inorg. Chim. Acta, **251**, 75(1996).

[15]Z. H. Zhou, H. L. Wan, S. Z. Hu and K. R. Tsai, Inorg. Chim. Acta, **237**, 193(1995).

[16]V. Murugesan, V. Babu, S. Sankaran, Inorg. Chem., **37**, 1336(1998).

[17]G. I. Fillin and V. N. Markin, Deposited Doc. VINITI 3475-3479, 162(1975); Chem. Abstr., **88**, 57252e(1976).

[18]C. Djordjevic, M. Lee and E. Sinn, Inorg. Chem., **28**, 719(1989).

[19]D. W. Wright, P. A. Humiston, W. H. Orme-Johnson and W. H. Davis, Inorg. Chem., **34**, 4194 (1995).

[20]Z. H. Zhou, W. B. Yan, H. L. Wan, K. R. Tsai, J. Z. Wang and S. Z. Hu, J. Chem. Crystallogr., **25**, 807(1995).

[21]D. W. Wright, R. T. Chang, S. K. Mandal, W. H. Armstrong, W. H. Z. H. Zhou, H. L. Wan and K. R. Tsai, Chinese Sci. Bull., **40**, 749(1995).

[23]J. D. Pedrosa de Jesus, M. de D. Farropas, P. O'Brien, R. D. Gillard and P. A. Williams,

Transition Met. Chem., **8**,193(1983).

[24]L. R. Assunbeni, M. L. Niven, J. J. Cruywagen and J. B. B Heyns, J. Crystallogr. Spectrosc. Res., **17**,373(1987).

[25]N. W. Alcock, M. Dudek, R. Grybos, E. Hodorowicz, A. Kanas and A. Samotus, J. Chem. Soc., Dalton Trans., 707(1990).

[26]M. A. Porai-Koshits, L. A. Aslanov, G. V. Ivanova and T. V. Polynova, J. Struct. Chem. (Engl. Transl. ), **9**,401(1968).

[27]J. E. Berg, S. Brandange, L. Lindblom and P. E. Werner, Acta. Chem. Scand. Ser. A., **31**,325 (1977).

[28]Z. H. Zhou, W. B. Yan, H. L. Wan and K. R. Tsai, Chin. J. Struc. Chem., **14**,255(1995).

[29]J. Kim and D. C. Rees, Biochem., **33**,389(1994).

[30]E. L. Muetterties and L. J. Guggenberger, J. Am. Chem. Soc., **96**,1748(1974).

TMCH 4397

■ 本文原载：Journal of Natural Gas Chemistry Vol. 8 No. 3，pp. 223～230，1999.

# Diluted Hydrogen Activation of Modified Copper-Based Catalyst
## —NC208 for Methanol Synthesis

Yi-Quan Yang， Shen-Jun Dai， You-Zhu Yuan，
Ren-Cun Lin， Hong-Bin Zhang， Khi-Rui Tsai

(Department of Chemistry, Institute of Physical Chemistry
State Key Laboratory of Physics Chemistry for the Solid Surface,
Xiamen University Xiamen, Fujian 361005)

**Abstract**   TPR, DTA and XRD methods were employed to study a modified copper-based catalyst, NC208, for methanol synthesis. A temperature-reduction program established on the basis of diluted hydrogen activation of the catalyst has been successfully used in the commercial production of methanol with a good yield.

**Key words**   Methanol synthesis   Copper-based catalyst   Activation   Temperature-reduction program

## 1   Introduction

A modified copper-based catalyst, NC208, for methanol synthesis, which was invented by Xiamen University and commercially tested by Nanjing Catalyst Factory, has passed qualified appraisal and has been put into production[1]. It has been known that the preactivation, namely the temperature-programmed reduction, of the catalyst is of great importance to a high yield of methanol[2]. The reduction of the catalyst is found to be a strong exothermic reaction, which may give rise to overheat and then sinter the catalyst if the heat formed is not removed in time. Over-fast rate of reduction to the catalyst may lead to an increase in particle sizes and a decrease in surface area. On the other hand, vigorous reduction condition may result in a higher pressure of water vapor during a relatively short period due to the considerable amount of water produced, hence may lead to smashing and fragmenting the catalyst grain. Thus it is a key issue to set up an efficient reduction program and to control strictly the reduction condition for obtaining the best catalytic performance. In the present work, TPR, DTA and XRD methods were employed to study the NC208 catalyst. The results revealed a close relation between the reduction process and the phase structure of the catalyst, thereby provided a sound experimental base to establish the temperature-reduction program commercially.

## 2　Experimental

### Pretreatment of the catalysts

Commercial methanol synthesis catalyst, C207, and its modified one, NC208, were obtained from Nanjing Catalyst Plant. The NC208 catalyst comprises the main components of CuO, ZnO and $Al_2O_3$, which are the components of C207 catalyst, and two promoters of metal oxides(ⅢB and ⅣB). Both of them are lustrous-dark cylinders with the diameter of $\phi 5$ mm and the length of 5.5 mm. Their density, surface area and pore volume are $1.4 \sim 1.6$ kg/L, $70 \sim 80$ $m^2$/g and $0.17$ mL/g, respectively. The catalysts with $20 \sim 40$ mesh were pretreated with a reduction gas of diluted hydrogen($5\%$ $H_2$ balanced by $N_2$) in a given temperature range, and then cooled to room temperature under Ar atmosphere.

### Characterization

TPR spectra were recorded in a fixed-bed flow reaction system equipped with a gas chromatograph. 50 mg of oxidized samples was treated at a heating rate of $10℃$/min with a reduction gas of $5\%$ $H_2/N_2$ at a flow rate of 10 mL/min for TPR measurement.

DTA spectra were obtained by using a LCT deviation-indicator with a measurement range of $\pm 50$ $\mu$V. 10 mg of samples was treated at a heating rate of $10℃$/min with a flowing diluted hydrogen of 30 mL/min. $Al_2O_3$ was taken as reference.

X-ray diffraction(XRD) analysis was carried out on a Rigaku Ru-200X diffractometer with Cu $K$ radiation at a scanning rate of $6°$/min in the $2\theta$ range of $20° \sim 80°$.

## 3　Results and discussion

The TPR spectrum of NC208 catalyst is shown in Fig. 1, in which that of industrial C207 catalyst is shown for the sake of contrast. It is obvious that there are two strong reduction peaks for the NC208 catalyst. The main peak at $240℃$ is possibly due to the reduction of $Cu^{2+} \rightarrow Cu^+$, while the shoulder peak at about $260℃$ may be caused by the reduction of $Cu^+ \rightarrow Cu^0$ according to literature[3]. As for the C207 catalyst, a strong peak at about $240℃$ is found to be obviously broadened, which may be probably due to the overlap of the two peaks corresponding to the reduction of $Cu^{2+} \rightarrow Cu^+$ and $Cu^+ \rightarrow Cu^0$. In addition, the weak shoulder peaks at about $174 \sim 190℃$ for both catalysts can be detected. It seems to indicate that a quantity of highly coordinately unsaturated $Cu^{2+}$ species is easily reduced at low temperature. Nevertheless, the reduction temperature at which the

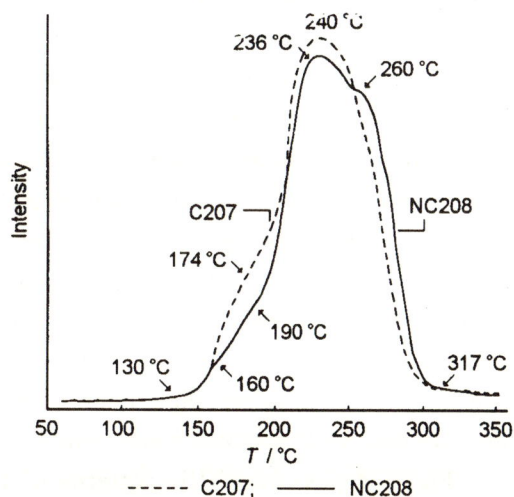

Fig. 1　TPR spectra of the catalysts

shoulder peak appeared for the NC208 catalyst is slightly higher than that of the C207 catalyst, implying that the NC208 catalyst may have the ability against deep reduction. It can be also observed from the TPR spectra that the two catalysts may be slightly reduced at about $100℃$, apparently reduced at about

130℃, and vigorously reduced at 160~190℃. It has been reported[4] that the active sites of the NC208 catalyst are found to be synergic species, namely $Cu_x^0$-$Cu^+$-$O$-$Zn^{2+}$, and that the key to enhance the catalytic activity would be to maintain a suitable ratio of $Cu^+$/$Cu^0$ on the catalyst surface. Therefore, from the above results, it may be reasonable to consider the temperature point of main reduction peak at 240℃ for the NC208 catalyst as its final reduction temperature. Hence, the reducing program can be reasonably designed as following stages: firstly, the temperature-raising stage from room temperature to about 100℃; secondly, the primary reduction stage from 130℃ to 160℃; thirdly, the predominant reduction stage from 160℃ to 190℃; finally, the reduction stage from 190℃ to 240℃. In commercial scale, because of the thickness of the catalyst bed and large temperature difference between up-and down-catalyst layers in synthetic column, the temperatures at both the predominant reducing stage and final reducing stage are permitted to be higher by 20~40℃ than the recommended ones, so as to ensure that the reduction of the catalysts conducts fully.

Fig. 2 shows the DTA curves of both catalysts under reducing conditions by using diluted hydrogen. It can be found from Fig. 2 that the temperature points at which the reducing peaks appeared are different from each other. For the NC208 catalyst, the exothermic peaks appear at 180℃, 236℃, 280℃ and disappear at about 310℃. Similarly, the exothermic peaks of the C207 catalyst appear at 184℃, 239℃, 265℃ and end at 304℃. From the results of TPR(Fig. 1) and DTA(Fig. 2), it is deducible that the DTA exothermic peak at

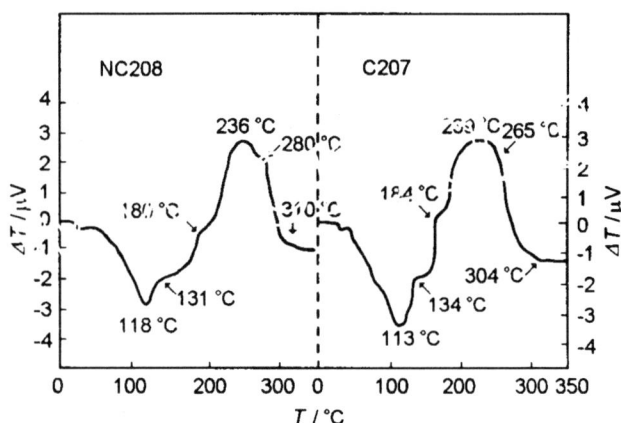

Fig. 2　DTA patterns of the C207 and NC208 catalysts

180℃ or 184℃ separately for the two catalysts may correspond to the reduction of unsaturetedlv coordinated surface $Cu^{2+}$ species; the main DTA peak at 236℃ or 239℃ may be due to the reduction of $Cu^{2+} \rightarrow Cu^+$; the shoulder DTA peak at 280℃ or 265℃ may correspond to the reduction of $Cu^+ \rightarrow Cu^0$. Fig. 2 also reveals that the intensity of the exothermic peaks of the C207 catalyst is evidently stronger than that of those of NC208. Moreover, the main peak at 236℃ for the NC208 catalyst is different from that at 239℃ for the C207 catalyst, the former is followed by a notable shoulder peak at 280℃. the later is both broad and blunt, no shoulder peak at 280℃ can be detected, indicating that a severe exothermic reaction may easily occur over the C207 catalyst. The results also demonstrate that the NC208 catalyst doped with promoters of metal oxides has the ability against deep reduction. Under the same reduction conditions, the NC208 catalyst is reduced at a slow rate to stabilize most Cu-species as $Cu^+$, a small amount of Cu-species on the NC208 catalyst is further reduced to $Cu^0$. It can be seen from the exothermic curve that a large quantity of heat is produced during the reduction, in order to avoid overheating the catalyst, high hour-space velocity of feed stock is required, the experiment results indicate that 3000~5000 $h^{-1}$ are suitable.

Fig. 3 depicts the XRD patterns of the NC208 catalyst at different reduction temperatures. By comparing with the patterns of the oxidized sample(Fig. 3a), it can be found that there are two new diffraction lines at $2\theta=43.2°$ and 36.4°, accompanying by the disappearance of CuO patterns at 35.4° and 38.8° in the reduction samples(Fig. 3, b~f)[4-7]. The new peak at $2\theta=43.2°$ may be assigned to the diffraction line of $Cu^0$, while the peak at 36.4° may be due to the diffraction of $Cu_2O$ or $Cu_2O$ dissolved in ZnO phase[6-9]. The intensity ratio $I_{36.4}/I_{43.2}$ decreases gradually with the increment of reduction temperature as shown in Table 1. At the initial reduction step at 130℃, the value of $I_{36.4}/I_{43.2}$ is 3.23. At reduction temperature of 240℃, at which the catalyst exhibits the best catalytic performance, the $I_{36.4}/I_{43.2}$ value decreases to 1.57. The increment of peak intensity at $2\theta=43.2°$ during the reduction is ascribed to the increase in the concentration of $Cu^0$ species. Whereas, the intensity of peak at $2\theta=36.4°$ increases in the early stage and reaches a maximum value when the reducing temperature is increased up to 160℃, indicating that most $Cu^{2+}$ species might be reduced to $Cu^+$ at 160℃. When the reduction temperature is increased up to 320℃, the value of $I_{36.4}/I_{43.2}$ decreases further to 1.15, implying that even more species $Cu^+$ would be reduced to $Cu^0$.

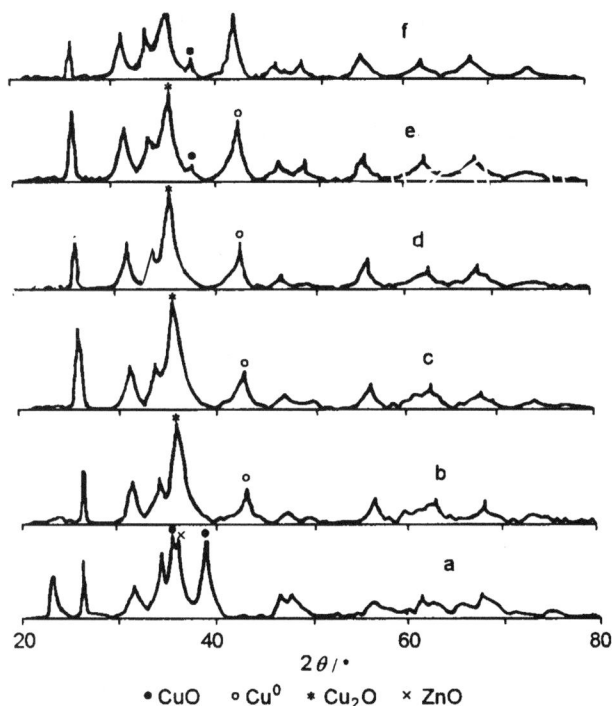

**Fig. 3** XRD patterns of the NC208 catalyst: (a) in oxide state, (b to f) reduced in diluted-$H_2$ at 130℃, 160℃, 200℃, 240℃ and 320℃, respectively

**Table 1** XRD data of the NC208 catalyst activated at different temperatures

| $T/℃$ | $2\theta/°$ | $I$ | WID | $d$/nm | $I_{36.4}/I_{43.2}$ |
|---|---|---|---|---|---|
| O* | 36.1 | 306 | 0.81 | 2.460 | – |
| 130 | 36.40 | 452 | 1.38 | 2.468 | 3.23 |
| | 43.20 | 140 | 1.05 | 2.092 | |
| 160 | 36.40 | 465 | 1.29 | 2.466 | 3.1 |
| | 43.40 | 150 | 1.44 | 2.089 | |
| 200 | 36.48 | 413 | 1.35 | 2.462 | 2.24 |
| | 43.42 | 184 | 1.05 | 2.082 | |
| 240 | 36.32 | 383 | 1.26 | 2.472 | 1.57 |
| | 43.34 | 244 | 1.14 | 2.086 | |
| 320 | 36.46 | 306 | 1.35 | 2.462 | 1.15 |
| | 43.36 | 266 | 1.14 | 2.085 | |

* O stands for the oxidized sample.

The parameters of temperature-reduction program and the temperature-reduced curve of the NC208 catalyst industrially operated in Tai-Zhou Fertilizer Factory are listed in Table 2 and shown in Fig. 4, respectively. It can be observed from Fig. 4 that the accumulative dehydrated weight is 245 kg during the reduction, which is about 20% higher than that by theoretical calculating. This may be probably caused by removing the physically adsorbed water on the catalysts and the chemically adsorbed hydrates. When the reducing temperature is elevated to 140℃, only 28.3% of the total quantity of water is produced. The first dehydrating peak appears at about 8 h of primary reduction in the temperature range of 60~80℃. When the reduction temperature reaches 160℃, another dewatering peak would appear. After 56 h of reduction(240℃), 93% of the total water would be removed, suggesting that the reduction process is basically completed.

**Table 2  Temperature-reduction program for the NC208 catalyst**

| Step | Time h | | Heating rate ℃/h | Temperature range ℃ | Pressure MPa | composition of inlet gas | | Outlet water kg/h | Vapor amount % |
|------|--------|-------|------|------|------|------|------|------|------|
| | Single step | Total | | | | $H_2$ | CO | | |
| Heating | | | | | | | | | |
| I | 3 | 3 | ~20 | ~20 | 5.0 | 0 | <1 | 0 | <0.3 |
| II | 18 | 21 | 2~5 | 70~130 | 5.0 | 0 | <2 | <5 | 0.3~0.5 |
| Reducing | | | | | | | | | |
| I | 10 | 31 | 2~3 | 135~160 | 6~7 | 0.5~2 | <2 | <5 | 0.3~8.5 |
| II | 16 | 47 | 3~5 | 160~220 | 7~9 | 2~5 | <2 | <5 | 0.3~0.5 |
| III | 6 | 53 | 5~7 | 220~250 | 9.5 | >5 | <2 | <5 | 0.3~0.5 |
| IV | 3 | 56 | ≤10 | 250~270 | 10 | >50 | <2 | <5 | 0.3~0.5 |
| Keeping | 3 | 59 | - | 270 | 10 | >50 | <2 | <5 | 0.3~0.5 |
| Gas switch | 4 | 63 | - | 265±5 | >10 | >50 | - | 0 | <0.3 |

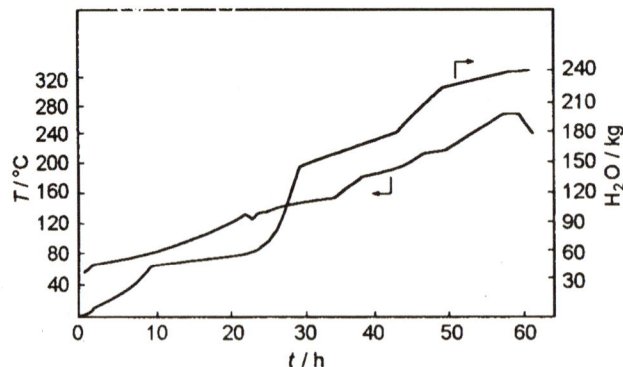

**Fig. 4  Temperature-reduction curve of the NC208 catalyst**

Table 3 shows the catalytic performances of both the C207 catalyst and the NC208 catalyst under the identical conditions of pretreatment and reaction. The data listed in Table 3 demonstrate that the activity life of the NC208 catalyst may be able to reach 186 days, which is 1.55 times as long as that of the C207 catalyst. Moreover, the capacity of methanol production per day by using the NC208 catalyst increases by 25.2% as compared with that by using the C207 catalyst, displaying that the NC208 catalyst has a higher efficiency for the reaction and that the reduction program proposed above for both catalysts is reasonable.

**Table 3　comparison in yields of methanol over NC208 and C207 catalysts**

| Item | NC208 | C207 | Increase ratio/% |
|---|---|---|---|
| Total yield in 120 days/m³ | 2420.9 | 1944 | 24.5 |
| Total yield in activity life* /m³ | 3801.9 | 2106 | 80.5 |
| Yield per day /m³/d | 20.44 | 16.33 | 25.2 |

\* 0.97 ton of NC208 catalyst was used, the catalytic life 186 days;

1.18 ton of C207 catalyst was used, the catalytic life 120 days.

## 4　Conclusions

The temperature-reduction program established and used in the present work to activate the modified copper-based catalyst NC208 for meihanol synthesis is testified to be practicable and effective in improving the catalytic performance of the catalyst. Under the same diluted hydrogen reduction conditions, the modified copper-based catalyst of the methanol synthesis, NC208, is found to be superior to industrial catalyst C207 both in the activity and in the stability. The TPR and DTA characterizations indicate that the modified catalyst NC208 has the ability against deep reduction, thereby maintaining the ratio of $Cu^+/Cu^0$ over the catalyst surface on a suitable level, which is found to be the key to enhance the catalytic activity.

The XRD patterns of NC208 show that the intensity ratio of the peak at $2\theta = 36.4°$, which can be ascribed to $Cu^+$ and that at $2\theta = 43.2°$, which belongs to $Cu^0$ decreasing with the increment of the reduction temperature. When the intensity ratio of $I_{36.4}/I_{43.2}$ reaches to 1.57, the catalyst exhibits the best catalytic performance, which corresponds to the state at the reduction temperature up to 240℃.

## 5　References

[1]Y. Q. Yang, C. Z. Che, and H. B. Zhang, J. Appl. Chem., **14**(3)(1997)62.

[2]Y. Q. Feng, Methanol Production, Beijing, Beijing Chemical Industry Press, (1989) 89.

[3]J. L. Li, Q. M. Zhu, and Z. H. Qian, 7th NCC, Dalian, (1994)283.

[4]Y. Q. Yang, Gaoxiao Huaxue Xuebao(Chem. J. Chinese University), **18**(1)(1997)103.

[5]Y. M. Li, Gongye Cuihua Yuanli(Principle of Industrial Catalysis), Tianjin University Press, Tianjin, (1992)222.

[6]Y. Okamoto, K. Fukino, T. Imanaka, and S. Teranishi, J. Phys. Chem., **87**(1983)3740.

[7]Y. Okamoto, K. Fukino, T. Imanaka, and S. Teranishi, J. Phys. Chem., **87**(1983)3747.

[8]P. B. Himelfarb, G. W. Simmons, K. Klier, and R. G. Herman, J. Catal, **93**(1985)442.

[9]J. B. Bulko, R. G. Herman, K. Klier, and G. W. Simmons, J. Phys. Chem., **83**(1979)3118.

■ 本文原载：Catalysis Today 51(1999)3～23.

# Forty Years of Applied Catalysis Research at Xiamen University and Its Interaction with Fundamental Catalysis Research

K. R. Tsai*, D. A. Chen, H. L. Wan, H. B. Zhang, G. D. Lin, P. X. Zhang

(Department of Chemistry and Institute of Physical Chemistry,
State Key Laboratory for Physical Chemistry of Solid Surfaces,
Xiamen University, Xiamen 361005, China)

**Abstract**　This paper gives a concise, chronological review of 40 years of applied catalysis research done at Xiamen University, in acetylene and olefin conversion, nitrogen fixation, and in more detail CO hydrogenation to methanol and ethanol, oxidative coupling of methane, and oxidative dehydrogenation of light alkanes; some references of more recent work in other areas related to applied catalysis are also given. © 1999 Elsevier Science B.Y. All rights reserved.

**Key words**　CO hydrogenation to methanol and ethanol　Formyl intermediate　Dipole-charge interaction with cation　Metal-carboxylic acid　Water-gas shift mechanism　Methane oxidative coupling　Rare-earth catalysts　composite oxides and oxyfluorides　Active oxygen species　Superoxide ion and active precursor　In situ Raman spectra

## 1. Introduction

Applied catalysis is among the few branches of applied science that can make tremendous contribution to the international competitiveness of a country. In many industrialised countries, as well as in some developing countries, applied catalysis research and application-oriented fundamental catalysis research are often well integrated, and their intimate interactions have been leading to rapid progress of catalysis science and technology. This paper gives a concise, chronological review of the 40 years of applied catalysis research, along with its interaction with fundamental catalysis research, done at Xiamen University, since 1957 when post-graduate education and research in catalysis at this university were started.

## 2. Applied catalysis in acetylene and olefin conversion

Before the successful exploitation of Daqing oil field, China was in dire need of basic organic chemicals for the production of synthetic rubber, plastics, and synthetic fibres. In the late 1950s, Xiamen University was in charge of an applied catalysis research project funded by the Ministry of Chemical

---

* Corresponding author.

Industry of China, and in the mid-1960s, in charge of a subproject of the State Key Project no. 29 on application-oriented fundamental catalysis research supported by the State Science and Technology Commission(SSTCC). Our first task was to find a practical catalyst to be used as a substitute for the highly toxic $HgSO_4$-$H_2SO_4$ catalyst used in acetaldehyde production from carbide-based acetylene. At first, we tried to improve on the co-precipitation-type$(Cd,Ca)_3(PO_4)_2$ catalyst, invented by Gorin[1,2] in USSR, for acetaldehyde production via vapour-phase hydration of acetylene. By cation exchange of $Ca_3(PO_4)_2$ pellets with $Cd(NO_3)_2 \cdot aq$, cation exchange-type $Cd_3(PO_4)_2/Ca_3(PO_4)_2$ catalyst was prepared. Laboratory test and pilot-plant test(fluidised-bed reactor) showed that the $Cd_3(PO_4)_2/Ca_3(PO_4)_2$ catalyst and the similarly prepared $Zn_3(PO_4)_2/Ca_3(PO_4)_2$ catalyst were more active than the corresponding catalyst prepared by co-precipitation method. Chemical analysis and thermometric titration with $n$-butylamine showed that the cation exchange was fast, but practically stopped when $Ca^{2+}$ ions of the surface layer of $Ca_3(PO_4)_2$ pellets were completely exchanged by $Cd^{2+}$ (or $Zn^{2+}$) ions. Thus much less amount of cadmium was required. But with fluidised-bed reactor, attrition loss of the phosphate catalysts was found to be too serious. Since it was inferred from mechanistic consideration[3] that surface $Cd^{2+}$, or $Zn^{2+}$ ions were mainly responsible for the catalyst activity, a more robust, more active, and less easily deactivated $ZnO/SiO_2$ catalyst was prepared from impregnated $Zn(NO_3)_2$ precursor and successfully used as a substitute for the phosphate catalysts. A 400 ton/a pilot-plant test ran smoothly at $643-673$ K($370-400$℃) for three months in 1973(with re-activation of the catalyst every five days, or continuously). Space-time yield with the catalyst was ca. 180 g $CH_3CHO$/l/h($\approx81\%$ selectivity) plus ca. 20 g crotonaldehyde/l/h($\approx10\%$ selectivity) as a by-product[3]. Subsequently, a small plant was set up to supply the $ZnO/SiO_2$ catalyst to more than half a dozen small acetic acid plants in China for about five years.

Our second task was to improve on the $CrO_x/SiO_2$ catalyst developed by the Shanghai Research Institute of Chemical Technology for benzene production via cyclotrimerisation of carbide-based acetylene. The $CrO_x/SiO_2$ catalyst was found to be highly active and selective, but rapidly deactivated above 413 K, thus excessive amount of cool water was required with practically no way to recover the enormous heat of reaction to produce superheated steam. From mechanistic reasoning and trial experiments, we soon found that $Nb_2O_5/SiO_2$ catalyst prepared from impregnated oxalate precursor of $Nb(V, d^0)$ was an excellent substitute for the $CrO_x/SiO_2$ catalyst[4]. A 100 ton/a pilot-plant test of the $Nb_2O_5/SiO_2$ catalyst at the third Chemical Factory of Xiamen was run very smoothly at $473-523$ K for two months, with continuous withdrawal of part of the catalyst for re-activation by carefully burning off very small amounts of nonvolatile carbonaceous deposits. With undiluted acetylene, ultra-pure benzene was produced ($\sim$ 1 kg/kg h). From the observed products distribution in experimental study of $HC\equiv CH$ and $CH_3C\equiv CH$ co-cyclotri-merisation, as well as from the high selectivity and activity, a Diels-Alder type mechanism for the benzene formation was proposed[4], based upon the formation of metallo-cyclopentadiene intermediate, $M(\underline{H}C=CHCH=C\underline{H})$ on coordination-unsaturated $Nb(V, d^0)$ active-site. Note that in this paper, underlined atomic symbols denote surface-bound atoms of adspecies.

The pilot-plant test of $Nb_2O_5/SiO_2$ catalyst and the small-scaled production of $ZnO/SiO_2$ catalyst were discontinued(though the two catalytic processes may be documented as two successful reserve technologies for the production of acetaldehyde and benzene from acetylene in case of petroleum shortage) after the successful large-scale exploitation of Daqing oil field. We then re-oriented our catalysis research direction in basic organic synthesis, first to catalysis in ethylene, propylene polymerisation and selective oxidation[5,6] for a few years, then focusing our attention to CO hydrogenation and light alkanes activation and selective oxidative-conversion. In the meantime, chemical modelling of biological nitrogen fixation was also started(including ammonia synthesis over iron-based catalysts).

## 3. Applied catalysis related to chemical modelling of biological nitrogen fixation

In 1972, a multi-disciplinary research project, "Chemical modelling of biological nitrogen fixation" was organised as a state key project, involving many research institutes and universities, and supported by the SSTCC for 10 years. The front line of the project was actually integrated with applied catalysis research on industrial ammonia synthesis catalysts, aimed at finding practical catalysts which could be used under lower temperature and pressure. An electron donor-acceptor(EDA)-type catalyst[7], Fe(II)-phthalocya-nin-K-graphite system, was tried by the joint efforts of two universities and two research institutes, as well as a factory. The EDA-type catalyst proved to be very active even at 573 K. But laboratory and pilot-plant tests failed to produce a catalyst with sufficient stability, conceivably due to high volatility and poison-sensitivity of metallic potassium. However, ammonia synthesis by the EDA-type catalyst conceivably involves $N_2$ coordination, dissociative chemisorption of $H_2$, H-spillover via concerted electron and proton transports, and electron storage afforded by the large conjugated system. This is in principle analogous to our concept about the mechanism of nitrogenase catalysis[8], this mechanistic study has led us to further development of the concept of catalysis by coordination activation[9] and to the study of the probable mechanistic relationship between ammonia synthesis catalysed by promoted iron catalysts and by nitrogenase[10,11]. In the experimental study of ammonia synthesis over promoted iron catalysts in our laboratory[12,13], in situ Raman spectroscopy was found to be very useful for the characterisation of chemisorbed species with sufficient polarisability, e. g., $N_2$, $N$, $NH$, and somewhat unexpectedly, $H(H^{\delta-}$ on $Fe^0)$ on the functioning catalyst.

Recently, in rationalisation of chemical modelling of nitrogenase catalysis, a labile-mouthed Mcluster-cage model of the substrate-binding FeMo-cofactor with two proton-transport pathways and mechanisms of shape-selective reduction of nitrogenase substrates have been proposed by Tsai and Wan[14,15]. Besides the theoretical interest of increasing our understanding about the functioning of this key enzyme with its multi-nuclear substrate-binding sites and proton-transport pathways, fundamental study of nitrogenase catalysis has led us to some ideas about catalysis by cluster complexes, and may have practical significance in guiding the design and preparation of catalysts for certain types of reactions; e. g., highly selective hydrogenation of acetylene to ethylene.

## 4. Metal-metal oxide catalysts for CO hydrogenation to methanol and ethanol

Since the two oil crises in the 1970s, catalysis in $C_1$ chemistry has become an area of extensive research internationally. In China, a State Key Project of this research subject was organised and supported by the National Natural Science Foundation of China(NNSFC) for five years since 1987. At Xiamen University, one of the research topics was molecular catalysis in CO hydrogenation to methanol (COHTM) over Cu-ZnO-based catalysts and to ethanol(COHTE) over Rh-metal-oxide-based catalysts, aimed at comparing the functions of metal oxide promoters(or cocatalysts) in these two systems, and getting some ideas for catalyst improvement from molecular catalysis approach.

## 4.1. CO hydrogenation to methanol over Cu-ZnO-based catalysts

For COHTM over Cu-ZnO-based catalysts, it has been generally accepted since the mid-1980s that formyl adspecies is a key intermediate, and that there is pronounced synergy between the copper component and ZnO[16,17]; the nature of this synergy has been a subject of extensive investigation. It is known that the energy barrier for partial hydrogenation of CO to formyl intermediate (HC=O) is about 20 kcal/mol[18] (1 kcal = 2.1868 kJ). Note that from the handbook[19], the dipole moment of aldehydic carbonyls ($2.3 - 2.7 \times 10^{-18}$ e.s.u.) is one order of magnitude larger than that of CO ($0.10 \times 10^{-18}$ e.s.u.). It has been postulated by Tsai and co-workers[20,21] that the formation of HCO in COHTM over Cu-ZnO-based catalysts may be greatly promoted by dipole-charge interaction of the $Cu^{\delta+}$-bound HC=O carbonyl with $Zn^{2+}$. The magnitude of this interaction energy roughly estimated from classical model, $N\mu(ze)\cos\theta/r^2 \approx N(2.3 \times 10^{-18})(2 \times 4.8 \times 10^{-10})\cos110°/(3.0 \times 10^{-8})^2$, is about 12 kcal/mol. This also implies that in every subsequent step of the surface reaction, the CO-derived O-end of the adsorbed intermediate will keep interacting or linked with $Zn^{2+}$, and that the HCO may be readily and successively hydrogenated to $CH_3$ O-bound to $Zn^{2+}$, and finally converted to $CH_3OH$ via protonation (with protonic hydrogen taken from a neighbouring -OH), or hydrogenation. This has led to the suggestion of $[Cu^0 H]_x Cu^{\delta+}$ OZn (H) OH active-centre for COHTM[20,21]. In view of the known experimental fact[16,17] that in COHTM over Cu-ZnO-based catalyst, $H_2O(g)$ in small amount acts as a promoter, and in larger amount as an inhibitor; while water-gas shift reaction (WGS) is not inhibited at all by $H_2O$[16,17], a virtually one-sited active-centre, $Cu^{\delta+}$ at the surface lattice of ZnO and linked with the epitaxy copper, has been suggested by Tsai and co-workers[20,21] for WGS, along with the suggestion that the WGS may involve the usual *cis*-migratory insertion of CO into Cu-OH to form Cu C(O)OH. This is *not* a Cu-ligating formate (HCOOCu or HCOOCu via *unusual* insertion of CO into Cu O- H), but a metal-carboxylic acid, of which the formation, structure and properties have been well reviewed by Bennett[22], who has mentioned that metal-carboxylic acid might be an intermediate of WGS in homogeneous catalysis. The CuC(O)OH may readily liberate $CO_2$ to give CuH, followed by $H_2$ liberation with a neighbouring H. As more information became available in the literature (e.g., formation of bidentate formate intermediate (HCOO −) in COHTM observed from in situ FTIR spectra[23], formation of dioxymethylene ($H_2COO$ −) in COHTM[24], as well as in HCHO adsorption on $Cu-ZnAl_2O_4$[25], and on $Cu/ZrO_2$[26,27], and H-spillover in either direction between copper surface and ZnO[28,29]), it has become possible for Cai et al.[30] to add to the COHTM scheme more details taken from relevant literature. The whole scheme is shown in Scheme 1, the metal-carboxylic acid mechanism for WGS[20,21] is assumed to be still applicable (see Scheme 2).

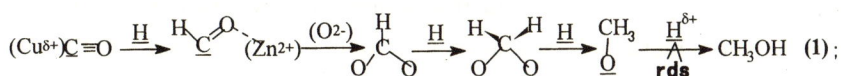

**Scheme 1.**

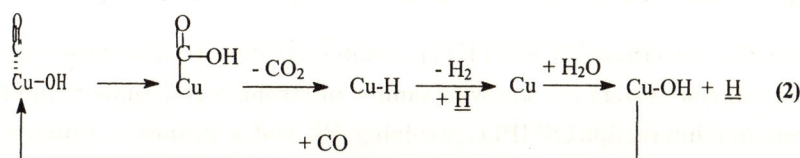

**Scheme 2.**

The role of $CO_2$ is a key issue in methanol synthesis from syngas over Cu-based catalysts. It is known[16,17] that the rate of COHTM is greatly promoted by a small proportion of $CO_2$ in the syngas. However, it has since been well illustrated, mainly by the ICI scientists[31,32] that, with more than a few percents of $CO_2$ in the syngas, $CO_2$ hydrogenation to methanol($CO_2$HTM) with Cu-bound bidentate formate(H$\underline{C}$OO) as a key intermediate becomes predominant, with practically complete suppression of COHTM; that under the same operating conditions, catalyst activities for $CO_2$HTM and WGS both depend only on the surface area of copper, practically irrespective of the supports (ZnO-$Al_2O_3$, $SiO_2$, $Al_2O_3$, MgO, MnO, ZnO); and that during the $CO_2$HTM reaction in the presence of very little CO, about 40% of the copper surface is covered by oxygen(i. e., about 80% saturation of available sites by $N_2O$ titration[33]). This amount of surface oxygen is determined by titration with CO to give $CO_2$, followed by liberation of $H_2$ in an amount corresponding to about one monolayer of $\underline{H}$, which is assumed to come mostly from sub-surface $\underline{H}$[33]. Though the kinetics and mechanism of $CO_2$HTM are beyond the scope of this report, it is pertinent to note here that the predominance of $CO_2$HTM(with bidentate formate as a key intermediate) and the absence of carbon-containing, common intermediate in $CO_2$HTM and WGS[32,33] may be taken as direct evidence against the formate mechanism for WGS(suggested by many researchers, as seen in some comprehensive reviews[34,35]), but not against the metal-carboxylic acid mechanism and the redox mechanism for WGS, the former mechanism may even be more tenable than the latter, or at least as an alternative of the latter. A considerable fraction of the copper surface may actually be covered by hydroxyls (as also mentioned by Ghiotti and Boccuzzi[35]), rather than by [Cu]-bound oxygen. *Cis*-insertion of C$\underline{O}$ into $Cu^{\delta+}$-$(OH)^{\delta-}$ anywhere on the copper surface to form Cu-C(O)OH should be able to take place readily, followed by liberation of $CO_2$ and $H_2$, as shown in Scheme 2. Thus CO titration cannot differentiate surface $\underline{O}$ from -$\underline{O}$H. However, with metal-carboxylic acid mechanism for WGS, it is not necessary to assume that most of the $H_2$ liberated may have come from sub-surface $\underline{H}$(which may be able to combine readily with surface $\underline{O}$ to form surface hydroxyls). Moreover, copper with much sub-surface $\underline{H}$ may not be able to displace $H_2$ from $H_2O$ fast enough, Cu being less electropositive than $H_2$. Waugh[33] has remarked that the molecular description of the reaction pathways for $CO_2$HTM and WGS are still crude. It is desirable to obtain a clearer picture of molecular catalysis in WGS and $CO_2$HTM on copper-based catalysts. This is also a requisite for setting up reliable kinetic model in process development[36].

More recently, a Cu-ZnO-based catalyst promoted with cerium oxide has been worked out by Yang et al.[37] for methanol synthesis with $H_2$-rich syngas. The catalyst has been commercialised by the Nanjing Catalyst Factory for use in coproduction of methanol in ammonia synthesis, with improved methanol yield and catalyst stability.

Incidentally, COHTM over Pd-$Mg^{2+}$/$SiO_2$ has also been studied by Shen et al.[38-40] for a period of time. Research is now also in progress here in the design and preparation of compatible dual-catalysts system for high conversion methanol synthesis via "one pot" reactions of the known, consecutive methyl-formate formation and hydrogenation.

## 4.2. CO hydrogenation to ethanol(COHTE) over promoted rhodium catalysts

In CO hydrogenation to ethanol(COHTE), promoter effects are also observed with metal oxide promoters[41,42]. The catalyst system is a known example of strong metal-support interaction(SMSI)[43], or strong metal-promoter interaction(SMPI), involving Rh and a promoter cation of variable valency and/or strong oxyphilicity, such as $Mn^{(2,3,4)+}$, $Ti^{(3,4)+}$, or $Zr^{4+}$, $Nb^{5+}$. Some researchers[44,45] have proposed that the function of the metal oxide promoter is to assist in weakening the $\underline{C}\equiv\underline{O}$ bond for direct bond rupture via formation of bridging or tilted $\underline{C}$O(a large lowering of $v_{C-O}$ to 1715 cm$^{-1}$ being taken as an argument), followed by hydrogenation of $\underline{C}$ to form $\underline{C}H_2$, or $\underline{C}H_3$ and reaction with $\underline{C}O$ by *cis*-coupling or *cis*-insertion to form ketene or acetyl as $C_2$-adspecies. This is called the dissociative mechanism for the formation of $CH_3$ in the $CH_3CH_2OH$. However, in the study of CO hydrogenation over $Rh/SiO_2$ catalyst, Orita et al. [46] have observed an IR band at 1587 cm$^{-1}$, ascribable to symbol $v_{C-O}$ of $H\underline{C}=O$ and observable only in the presence of both CO and $H_2$ (though the two characteristic $v_{C-H}$ bands were not observed, probably due to too low concentration of $H\underline{C}=O$ on $Rh/SiO_2$). They have also shown by isotopic tracings that the $CH_3$ groups of the light alkanes side-products and that of $CH_3CH_2OH$ are from the same $C_1$ intermediate, thus extending the significance of searching for this $C_1$ species to alkane formation over rhodium catalyst. Many investigators have observed that, over rhodium catalysts, hydrogen-assisted CO dissociation is faster than direct dissociation of CO in the absence of $H_2$, so they have favoured some form of associative mechanism in which C-O bond rupture in the presence of hydrogen is preceded by partial hydrogenation of CO. Thus Takeuchi and Katzer[47] have proposed a CO-formyl-carbene-ketene mechanism for the major pathway of $C_2$ formation from CO; they have speculated ketene rearrangement to epoxyethene cyclic intermediate before further hydrogenation to ethanol to account for the observed isotopic scrambling in the ethanol formed, starting with $H_2$ and $C^{18}O/^{13}CO$. But Deluzarche et al. [48] have pointed out that the observed isotopic scrambling might be accounted for by rapid isotopic exchange of the acetaldehyde intermediate with the $H_2O$ formed in CO hydrogenolysis, without recourse to the speculation of epoxyethene cyclic intermediate. For CO hydrogenation to $CH_4$ and other alkanes. Riecke and Bell[49,50] have suggested that C-O bond breaking is preceded by partial hydrogenation of CO to $H_2\underline{C}O$, and promoted by oxyphilic interaction of the promoter cation with the oxygen-end of $H_2CO$. Even in CO hydrogenation to $CH_4$ over $Ni/SiO_2$, Mori et al. [51] have observed hydrogen-assisted CO dissociation faster than direct dissociation of CO, and deuterium inverse kinetic isotope effect ($k_H/k_D = 0.75$) on the methanation reaction rate. Incidentally, hydrogen-assisted CO dissociation in metal-catalysed CO hydrogenation might be even more general; it has been remarked[52] that, under certain conditions, deuterium inverse kinetic isotope effects of similar magnitude for methanation reaction have been observed with many other group VIII metals, including Fe and Ru.

Based on the literature knowledge about the compositions and properties of the catalyst systems and the product and side products distribution, and by similar reasoning as in the study of promoter action in COHTM, it has been proposed by Tsai and co-workers[53-55] that in COHTE over promoted rhodium catalysts the oxyphilic promoter-cation may also promote $H\underline{C}O$ formation through dipole-charge interaction, or even through formation of metalloxy-carbene (similar to Bercaw's zirconoxy-carbene[56]); as well as hydrogenolysis of $H\underline{C}O$ to form Rh-bound $\underline{C}H_2$ and promoter-cation bound-$\underline{O}H$, and that this may be followed by the formation of $H_2\underline{C}=\underline{C}=O$(a $\pi$-olefinic-Rh complex[57]) promoted by dipole-charge interaction with $Li^+$ and successive hydrogenation to ethanol. ($Li^+$, better than $Na^+$ and $K^+$, can also effectively catalyse COHTM over rhodium, most probably by dipole-charge interaction of the $H\underline{C}=O$ with $Li^+$, and probably also of $H_2\underline{C}=O$ carbonyls with $Li^+$.) This may occur before the removal of the -$\underline{O}H$ from the promoter cation (e. g., $Mn^{3+}$) by hydrogenation, to account for the high $C_2$ selectivity. Thus a scheme of the main pathway of COHTE have been suggested, as shown in the central part of Scheme 3, i. e., a scheme of CO-formyl-carbene-ketene-acetyl-(acetaldehyde, or $CH_3\underline{C}OH$)-ethanol pathway, and the systematic examination and establishment of this scheme have been set out by a combination of various methods. The framed blocks in Scheme 3 show the chemical trapping of ketene with $CH_3OD$(likewise, for MeAc side-product formation), and side-reactions for methanol and methane formation.

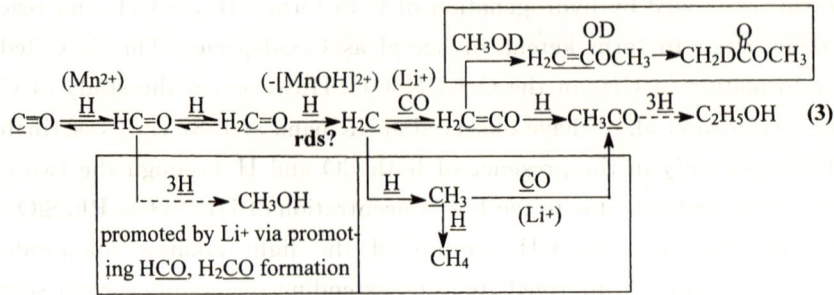

$$CH_3OD \qquad OD \qquad O$$
$$\rightarrow H_2C=COCH_3 \rightarrow CH_2DCOCH_3$$

$$(Mn^{2+}) \qquad (-[MnOH]^{2+}) (Li^+) $$
$$\underline{C}=O \underset{H}{\overset{H}{\rightleftharpoons}} H\underline{C}=O \overset{H}{\rightleftharpoons} H_2\underline{C}=O \underset{rds?}{\overset{H}{\rightarrow}} H_2\underline{C} \overset{CO}{\underset{H}{\rightleftharpoons}} H_2C=\underline{C}O \overset{H}{\rightarrow} CH_3\underline{C}O \overset{3H}{\dashrightarrow} C_2H_5OH \quad (3)$$

$$\dashrightarrow CH_3OH$$

promoted by Li+ via promoting H$\underline{C}$O, H$_2\underline{C}$O formation

$$\underset{H}{\overset{H}{\rightarrow}} \underline{C}H_3 \overset{CO}{\underset{(Li^+)}{\longrightarrow}}$$
$$CH_4$$

**Scheme 3.**

First of all it is essential to detect and determine the formyl adspecies. Following the method of Delu-zarche et al. [58], Wang et al. [59,60] used CH$_3$I in excess as chemical trapping agent for trapping the formyl intermediate, D$\underline{C}=\underline{O}$, in ethanol synthesis from CO-2D$_2$ (isotope purity $\geq$ 99. 9%) over Rh-MnO/SiO$_2$ catalyst, and found quite unexpectedly from the GC-MS mass spectrogram of the acetaldehyde fraction slightly stronger signals of mass 29(HCO) and mass 44(CH$_3$CHO) than that of

the expected mass 30 (DCO) and mass 45 (CH$_3$CDO)(Fig. 1A). It appeared that some of the trapping agent had dissociated at the CH$_3$ group on rhodium surface to give $\underline{H}$, which was trapped in some way (either by CH$_3\underline{C}$O, or successively by $\underline{C}$O and $\underline{C}$H$_3$) to give CH$_3$CHO. Since there were considerable amounts of $\underline{C}$O and $\underline{D}$ besides D$\underline{C}$O on the working rhodium catalyst before the excess CH$_3$I was injected onto it, the possibility that CH$_3$I, besides trapping D$\underline{C}$O, might also trap considerable amount of the $\underline{C}$O and $\underline{D}$ successively to give CH$_3$CDO, must be taken into consideration. Considering that D$\underline{C}$O is directly bound to Rh and also held by the promoter cation Mn$^{2+}$ through strong dipole-charge interaction, while $\underline{D}$ and $\underline{C}$O (the dipole moment of which is probably one order of magnitude smaller than that of D$\underline{C}$O) are more weakly bound to the surface,a slight modification of the chemical trapping method has been made by Wang et al. [59,60]: by sweeping the catalyst bed with a stream of N$_2$ for varying periods of time(0,3,5,13 min) after the reaction and before the injection of excess CH$_3$I, it was found that after 3 min of sweeping with N$_2$,the amount of D$\underline{C}$O(probably along with small amounts of $\underline{C}$O

**Fig. 1** A: GC-MS pattern of acetaldehyde fraction from chemical trapping with CH$_3$I of D$\underline{C}$O in CO+2D$_2$ reaction over Rh/MnO$_x$/ SiO$_2$ at 493 K,0. 1 MPa. B:GC-MS pattern of methylacetate fraction from in situ chemical-trapping of H$_2\underline{C}=\underline{C}$=O with CH$_3$OD added into CO-H$_2$ (2∶1) over Rh/MnO$_x$/Li$^+$/SiO$_2$ at 493 K,0. 1 MPa.

and $\underline{D}$) trapped by CH$_3$I to form CH$_3$CDO decreased by only 50% while the amount of surface $\underline{D}$ decreased by about 96%. The results showed that D$\underline{C}$O intermediate was actually formed in COHTE, and more tightly bound at the catalyst surface than $\underline{D}$(and $\underline{C}$O).

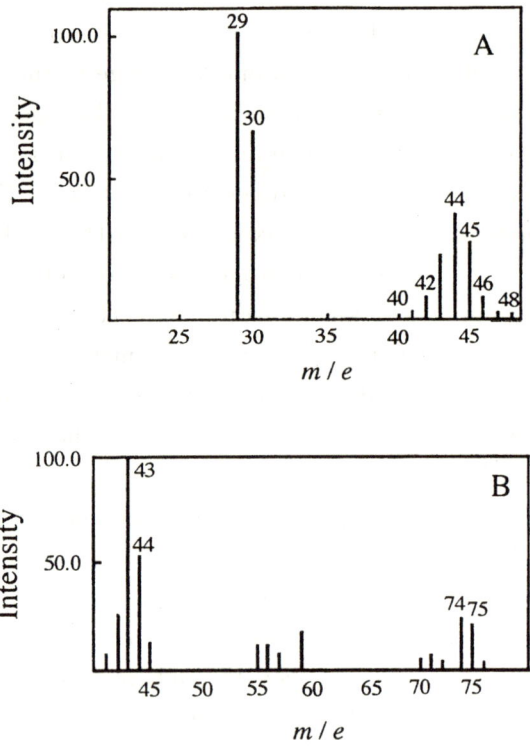

Conceivably, the active centre of COHTE on promoted rhodium catalysts must be situated at the metal-promoter boundary, as suggested by many investigators. Du et al.[61] have substantiated this point by their studies of silica-supported rhodium catalysts promoted separately with various rare-earth oxides. They have demonstrated that the metal-promoter interaction and promoter effect are both more pronounced with rare-earth oxides of variable cationic valency than with those of stable cationic valency. The strong metal-promoter interaction has been illustrated by Hu and Wang[62] with quantum-chemical calculations (DV-X$\alpha$-SCC method), using a Rh$_4$OMnO cluster model suggested by Wan. A section in a review by Hindermann et al.[63] provides valuable reference for CO adsorption.

More information about formyl intermediate in COHTE over Rh-MnO/SiO$_2$ catalyst, together with relative activities of the linear and bridging CO, or tilted CO, has been obtained by Wang et al.[59,64] from temperature-programmed surface reaction (TPSR)-FTIR study of pre-adsorbed CO on Rh-MnO/SiO$_2$

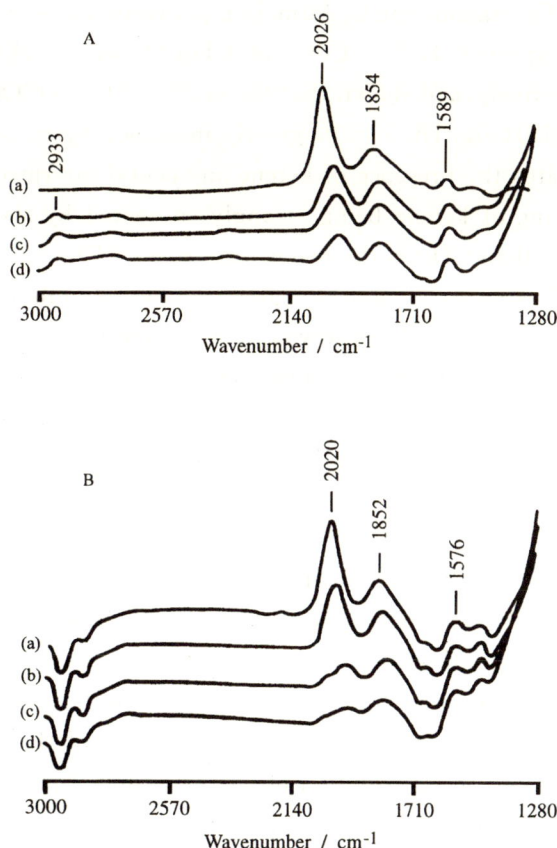

Fig. 2  TPSR-FTIR of pre-adsorbed CO on Rh/MnO$_x$/SiO$_2$ with flowing A: H$_2$, and B: D$_2$, at (a) 328, (b) 428, (c) 463, and (d) 493 K.

with H$_2$ flow and with D$_2$ flow. The characteristic $v_{C-O}$ IR bands of HCO(1589 cm$^{-1}$, very close to that observed in this laboratory by Fu et al.[65] at 1591 cm$^{-1}$, by slow displacement and reaction of pre-adsorbed H on Rh-MnO/SiO$_2$ with CO at 503 K, the HCO characteristic $v_{C-H}$ at 2708 and 2659 cm$^{-1}$ also being observed) and DCO(1576 cm$^{-1}$) were clearly observed and identified by the frequency red-shift ($\sim$13 cm$^{-1}$) of $v_{C-O}$ due to the secondary isotope effect of D. This is comparable with the $v_{C-O}$ red-shift (9 cm$^{-1}$) from $v_{C-O}$ of HCO at 1520 cm$^{-1}$ to $v_{C-O}$ of DCO at 1512 cm$^{-1}$ on Cu-ZnAl$_2$O$_4$ observed by Lavalley et al.[24]. Note that, as seen from Fig. 2, the intensity of the $v_{C-O}$ IR band (at 2020 cm$^{-1}$) of linear CO decreased conspicuously faster with temperature programming than that (at 1852 cm$^{-1}$) of bridging CO, especially from Fig. 2B CO + D2, due to the deuterium inverse kinetic isotope effects on COHTE and COHTM (these two reactions have slightly different deuterium inverse kinetic isotope effects[66] and different activation energies[67]). This may be taken as direct evidence[68] showing that linear CO is in fact more reactive in COHTE and COHTM than the bridging or tilted CO, in spite of the considerably lower $v_{C-O}$ of the latter than that of linear CO. Direct evidence has also been reported by Chuang[69,70] for much faster hydrogenation and *cis*-insertion of linear CO than bridging or tilted CO. This is in line with the proposal[9,21,53-55] that the promoter acts to lower the activation energy of the reaction by partial stabilisation of the highly unstable intermediate, rather than by binding the reactant molecule more tightly to weaken the bond to be broken (unless direct bond rupture happens to be the rds). This is reminiscent of Pauling's famous remark that enzymes bind the transition states of reactions, while antibodies bind the ground states of the substrates.

To examine the $C_2$ formation pathway, Liu et al. [54,55] have carried out in situ chemical trapping of the expected $H_2\underline{C}=\underline{C}O$ and $CH_3\underline{C}O$ with $CH_3OD$ to form $CH_2DCOOCH_3$ and $CH_3COOCH_3$, respectively, and shown that the mol% of $CH_2DCOOCH_3$ in the methyl acetate fraction (determined by GC-MS) (Fig. 1B) can be greatly increased by increasing the $CH_3OD/H_2$ ratios in the feed, the result indicates the presence of ketene and acetyl intermediates in comparable quantities; and in situ chemical trapping of ketene being strongly competed by its further hydrogenation. It is inferred that ketene, rather than $\underline{C}H_3$, is the major precursor of $CH_3\underline{C}O$.

Lastly, to account for the isotopic scrambling in the ethanol obtained from hydrogenation of $C^{18}O-^{13}CO(1:1)$ observed by Takeuchi and Katzer[47], in situ chemical trapping of ketene and acetyl intermediates with oxygen isotopic-exchange has been carried out by Wang et al. [59,71] by adding varying amounts (ca. $5-15$ mol% of the feed) of $D_2^{18}O$ into the feed $CO/nH_2$ ($n=1,2,3$) over $Rh/TiO_2/SiO_2$ catalyst at 493 K, 0.1 MPa, and collected the organic compounds (along with the co-produced water) in the reactor effluent in liquid-nitrogen cold trap, and also recovered the ketene, acetyl, and acetate adspecies on the catalyst surface in the forms of four methyl esters (one unlabelled, one doubly labelled, and two singly labelled with $^{18}O/^2H$) by thoroughly purging the organic adspecies from the catalyst bed with a stream of $CH_3OH$-containing $N_2$. It was found that the mol% of $^{18}O$-labelled esters was $2-3$ times larger than that of mono-deuterated acetate in each case of the $CO/H_2/D_2^{18}O$ ratios, indicating that the reversible hydration and dehydration of $H_2\underline{C}=C=O$ with nonligating carbonyl-O were considerably faster than the rearrangement of the hydrated ketene, $H_2\underline{C}=C(^{18}OD)(^{16}OD)$, to mono-deuterated acetic acid. In a blank test, ethanol showed practically no $^{18}O$ isotopic-exchange with $H_2^{18}O$. Statistical calculations[59,71] demonstrated that two rounds of ketene hydration-dehydration reactions with the $H_2O$ co-produced in COHTE are nearly enough to account for the isotopic scrambling observed by Takeuchi and Katzer[47]. Some contribution to the observed isotopic scrambling must have come from reversible hydration and dehydration of the acetyl and acetaldehyde intermediates.

Thus the proposed scheme for the main reaction pathway of COHTE over promoted rhodium catalysts has been verified in some detail[59,60,64,72]. The results show that partial hydrogenation of $\underline{C}O$ to $H\underline{C}O$ promoted by strong dipole-charge interaction with oxyphilic cation followed by hydrogenolysis of $H\underline{C}O$ to $\underline{C}H_2$ is energetically more effective than direct cleavage of the very strong CO bond promoted by multi-nuclear coordination of $\underline{C}O$ (which in fact seriously retards cis-insertion of $\underline{C}O$[69,70]). This is mainly due to the much larger carbonyl dipole moment and O-end nucleophilicity of $H\underline{C}=O$ than that of $\underline{C}O$. This may have more general significance in fundamental and applied catalysis in CO hydrogenation.

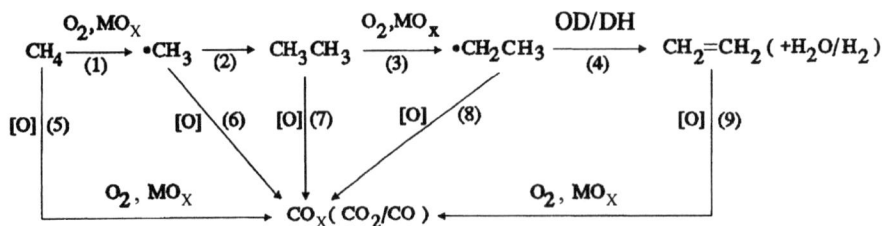

Fig. 3　Schematic diagram of OCM reactions and major side-reactions (over irreducible rare-earth-based composite oxides).

## 5. Rare-earth-based catalysts for OCM and oxidative dehydrogenation of light alkanes

Oxidative coupling of methane(OCM) to ethane and ethylene[73] over reducible or irreducible metal oxides-containing catalysts with cyclic (redox)-feed[73] or cofeed[74] mode of operation has attracted great, world-wide interest for its potentiality to be developed into a "smart process" for utilising the enormous, global natural-gas resources. It is generally accepted, as seen from many review articles(e. g. ,[75-78]), that OCM involves a series of complex heterogeneous-homogeneous free radical reactions, initiated by H-abstraction from $CH_4$ by surface active-oxygen species to liberate $CH_3$ radicals, which couple in the gas phase to give $C_2H_6$ as the primary product, along with coproduced $H_2O$. Ethylene appears to be a secondary product formed via oxidative dehydrogenation of ethane at the catalyst surface, and via free radical hydrogen-transfer reaction in the gas phase, and probably in minor amount by thermal dehydrogenation at the high temperature of OCM. All of the hydrocarbons may undergo deep oxidation to $CO_2$ and CO at the catalyst surface, or in the gas phase in the presence of $O_2$. A diagrammatic sketch of the reaction intermediates, products, and major side-pro-ducts formation is shown in Fig. 3.

It is also known that OCM catalysts with stable cationic valency can only be used in cofeed operation and show practically no activity with methane in the absence of $O_2$[75-79], while those with variable cationic valency can be used in both cofeed and cyclic-feed operations. With irreducible OCM catalysts, lattice $O^{2-}$ ions associated with cations of stable valency are obviously inactive towards methane, and certain active oxygen-adspecies must be there to initiate the OCM reaction via H-abstraction from $CH_4$ to form $CH_3$. Adsorption of $O_2$ on metal oxides with stable cationic valency may lead to the formation of a variety of oxygen adspecies, such as $O_2$, $O_2^-$, $O_2^{2-}$, and $O^-$; by stepwise transfer of $1-2$ electrons to $O_2$ from $1-2$ lattice $O^{2-}$ ions; with concomitant reduction of $1-2$ $O^{2-}$ to $1-2$ $O^-$, which may also be formed from dissociation of $O_2^{2-}$. Further transfer of an electron from $O^{2-}$ to $O^-$ produces no new species, only an exchange of sites; namely, hole migration. This is further complicated by the possibility that $O_2$(or $O_2^-$) may incorporate $O^-$(or $O^{2-}$) to form $O_3$(or $O_3^{2-}$). In the literature, each of the anionic-oxygen adspecies $O^-$, $O_2^-$, $O_2^{2-}$, and $O_3^-$ has been suggested[75-78] to be the principal oxygen-adspecies for OCM, and no unified view has been reached.

With supported or composite metal oxides containing cations of variable valency, cations in the reduced state can readily donate electrons to each $O_2$, most probably one by one into the antibonding orbitals of $O_2$ to form similar sequence of the diatomic precursors as intermediates, finally, reduce it into two $O^{2-}$, rather than via direct rupture of the strong $O=O$ double-bond followed by reduction to two $O^{2-}$. Each of the lattice $O^{2-}$ associated with cations in the oxidised state might have partial radical-anion character(especially those $O^{2-}$ at the surface lattice), and should be able to abstract a H from $CH_4$, forming $CH_3$ and $OH^-$, aided by concomitant abstraction of an electron from $O^{2-}$ by a neighbouring cation in the oxidised state. However, under certain conditions, the transient precursors $O_2^{2-}$, $O^-$, and even $O_2^-$ (as to be shown later) might also have some chance to react with $CH_4$ and other hydrocarbons.

In China, research on catalysis in OCM was a part of the State Key Project "Catalysis in $C_1$ chemistry" supported by the NNSFC for five years and carried over in 1992, along with oxidative dehydrogenation of light alkanes, into the State Key Project "Catalysis fundamentals in rational utilisation of fossil fuels resources". Research related to OCM has also been carried out in some laboratories supported by other sources of funding.

At Xiamen University and SKLPCSS, studies of irreducible OCM catalysts ($La_2O_3/BaCO_3$ and $K^+/BaCO_3$) were started in 1987[80]. Since 1990, attention has been focused on REO-AEO-based irreducible composite oxides of the host-dopant type as OCM catalysts[81-87], especially those with defective fluoride structure(DFS)[88,89], because of their very high activity, high selectivity, wide range of anionic-vacancy adjustability, outstanding thermal stability and long life. Novel fluoride-containing REO(F)-AEO(F) catalysts for OCM, as well as oxidative dehydrogenation of ethane and propane(ODE and ODP), have been systematically studied by Wan et al.[90-98] since 1992. In OCM and ODE studies, laser Raman spectroscopy(complemented with FTIR spectroscopy) was found to be very useful for in situ and ex situ characterisation of oxygen-containing surface species.

## 5.1. Study of REO-AEO composite oxide catalysts with stable cationic valency

In the study of OCM catalysts, Liu et al.[81,82] have found that with $La_2O_3$-based catalysts promoted separately with alkali and alkaline earth cations as dopants, the $C_2$ selectivity appeared to increase with increasing tendency of the dopant cations to form superoxides and peroxides. However, the promoter actions of $Na^+$, $K^+$, $Rb^+$, and $Cs^+$ rapidly declined, probably due to formation of stable carbonates(of $Na^+$ or $K^+$) or stable hydroxides.

As a prelude to in situ spectroscopic characterisation, Fig. 4 illustrates some information obtainable from ex situ spectroscopic characterisation of adsorbed species on metal oxides.

From Fig. 4A and B it can be seen that, in the room-temperature(r. t.), ex situ laser Raman spectra of oxygen-containing surface-species on samples of metal oxides after various treatments, the Raman bands of the $Ba^{2+}$-bound $\underline{O_2^{2-}}$, and $CO_3^{2-}$, and of the $La^{3+}$-bound $CO_3^{2-}$ surface species are all well-resolved and characteristic[81,82]. The Raman bands of the surface $CO_3^{2-}$ resemble that of unidentate-carbonato ligands in metal complexes[99]. In fact, the Raman spectroscopic features foretell the feasibility of characterising oxygen-containing surface species on metal oxide catalysts by in situ Raman spectroscopy.

It may be inferred from the relative intensities of the Raman signals of $\underline{O_2^{2-}}$ and $\underline{O_2}$ in Fig. 4A(a) versus in A(b) that $\underline{O_2^{2-}}$ is considerably more reactive than $\underline{O_2^-}$ towards $CO_2$[82].

Fig. 4C(a) shows the r. t. laser Raman spectrum obtained by Liu et al.[81] of a 10 mol% CsOH/$La_2O_3$ catalyst freshly activated in air at 973 K, then quenched to r. t. Besides a strong, $La^{3+}$-bound surface-carbonate band at 1086 $cm^{-1}$(s) and some un-identified, overlapping weak bands at 1030−1060 $cm^{-1}$, the Raman signals(r. t.) at 1118(w) and 800 $cm^{-1}$(w) were ascribable, respectively, to $\underline{O_2^-}$ and $O_2^{2-}$. The disappearance of these two signals from partially deactivated sample C(b) after 3 h on stream of $CH_4$-$O_2$-$N_2$ feed at 973 K, then quenched to r. t. gave some indication that both $\underline{O_2^-}$ and $\underline{O_2^{2-}}$ might exist at 973 K and both might be active-oxygen species towards $CH_4$.

By the high-temperature-low-pressure-interdiffu-sion-rapid-quenching-EPR method of Osada et al.[100] for the characterisation of $\underline{O_2^-}$ on an oxide catalyst, Liu et al.[81,82] have shown that $\underline{O_2^-}$ might exist at 1023 K and 16 kPa $O_2$ on $La_2O_3$-$ThO_2$(La:Th = 3 : 7, single-phase with DFS confirmed by XRD) since inter-diffusion of $CH_4$(16 kPa) into the cell with 16 kPa $O_2$, for 18 min caused the characteristic EPR signal(taken with the quenched sample at 77 K) to disappear, while similar inter-diffusion of $N_2$(16 kPa) into the cell with 16 kPa of $O_2$ at 1023 K for 18 min produced not much change in the intensity of the $\underline{O_2^-}$ EPR signal(at 77 K)(Fig. 4D(a)−(c)). In this way, Osada et al.[100] have previously shown that $\underline{O_2^-}$ might exist on $Y_2O_3$/CaO catalyst and might be reactive towards $CH_4$ at 1023

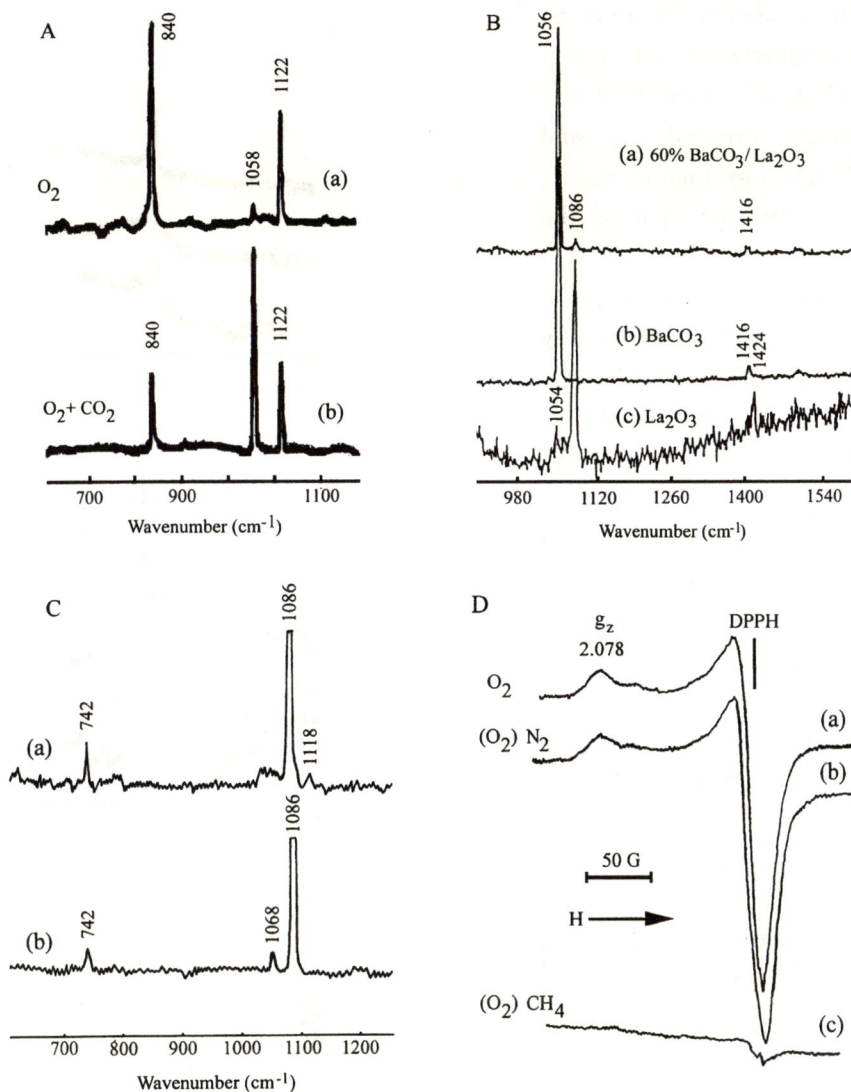

Fig. 4 Ex situ spectra of oxygen-containing species on/in various metal oxides. A: Raman spectra(r. t. ) of $BaO_2$/Ba $(O_2)_2$ quenched to r. t. after heating from 303 to 973 K for 40 min, (a) in $O_2$ stream versus(b) in $O_2$-$CO_2$(74 : 26) stream. B: Raman spectra(r. t. ) of(a) coprecipitated $La_2O_3$/$BaCO_3$ (Ba: La = 6 : 4, ignited in air at 1123 K for 5 h, then quenched to r. t. );(b) and(c) reference samples of $BaCO_3$ and $La_2O_3$. C: Raman spectra (r. t. ) of 10 mol% CsOH/$La_2O_3$ catalyst(a) freshly activated by heating in air at 973 K;(b) same catalyst after 3 h on stream of $CH_4$-$O_2$-$N_2$ feed at 973 K. D: EPR spectra(77 K) of uni-phasic $La_2O_3$-$ThO_2$(La: Th = 3 : 7, ignited) quenched to 77 K after separate treatments at 1023 K:(a) heating in $O_2$(0. 1 MPa, 0. 5 h, in EPR cell), evacuation($<$0. 013 kPa, 1 h), heating in $O_2$(16 kPa, 18 min), showing EPR signal of $\underline{O_2^-}$;(b) diffusion of $N_2$(16 kPa, 18 min) into(a);(c) diffusion of $CH_4$(16 kPa, 18 min) into(a).

K. Similarly, Yang et al. [101] have shown that $\underline{O_2^-}$ might exist at 1053 K and 13. 3 kPa $O_2$ on a $La_2O_3$/CaO(10 wt%) catalyst even in the presence of some $CO_2$, and might be reactive towards $CH_4$ at 1053 K. However, with similar ex situ EPR method, Louis et al. [102] have found no such reactivity of $\underline{O_2^-}$ on undoped $La_2O_3$, or(LaO)$_2$ (O, $CO_3$) catalyst with $CH_4$. The $\underline{O_2^-}$ adspecies was found to be easily displaced from the sample by $CO_2$. The different behaviours of doped and undoped(LaO)$_2$(O, $CO_3$) is a matter for further investigation.

Much more valuable information about active-oxy-gen species under OCM reaction conditions can be obtained from in situ Raman spectroscopy. For this purpose, a quartz tubular (4 mm i. d.) Raman cell (and micro-reactor) capable of oscillating vertically up and down has been designed by Liao et al.[103] for scanning around the hot spot of the sample.

With a single-phase Th-La-$O_x$ (Th:La = 70:30) OCM catalyst(DFS, confirmed by powders XRD) on stream of $CH_4$-$O_2$ (4:1) feed at(a) 953,(b) 1013,(c) 1073, and(d) 1133 K, in situ Raman spectra have been obtained by Liu et al.[84] with bands at 1140 cm$^{-1}$ (mw to w, decreased with increasing temperature) ascribable to $v_{O-O}$ of $\underline{O_2^-}$ (Fig. 5B(a) — (d)), along with surface carbonate band around 1060 cm$^{-1}$ (vs). When the $CH_4$-$O_2$(4:1) flow was switched to pure $O_2$ flow over the same Th-La-$O_x$ catalyst at 1013 K, no Raman signal for $\underline{O_2^-}$ was observed(Fig. 5A(a)).

It has been found by Cai et al.[86] that, with a sample of $(LaO)_2(O, CO_3)$/$BaCO_3$ (La:Ba = 2:8, prepared by coprecipitation and ignition) OCM catalyst after 10 h on stream of pure $O_2$ at 973 K, Raman signals of surface-carbonate at 1052 cm$^{-1}$ (vs) and of $\underline{O_2^{2-}}$ around 810 cm$^{-1}$ (w and broad) were observed, but no Raman signal of $\underline{O_2^-}$ (Fig. 6A(a)). Upon switching the $O_2$ flow to a flow of $CH_4$, the Raman signal around 810 cm$^{-1}$ (w) disappeared right away, only the surface-carbonate signal at 1052 cm$^{-1}$ (vs) remained (Fig. 6A(b)). One hour after switching the $CH_4$ flow to a stream of $CH_4$-$O_2$-He (24:6:70, v/v) feed, two more surface-carbonate Raman bands at 1080(s) and 740 cm$^{-1}$ (wm) appeared, along with a very weak band around 1120 cm$^{-1}$, which might be $\underline{O_2^-}$ signal, but too weak to be sure (Fig. 6A(c)). With a similarly prepared OCM catalyst La-Ca-O/CaO/(Ba, Ca)$CO_3$ (La:Ca:Ba = 2:4:4)

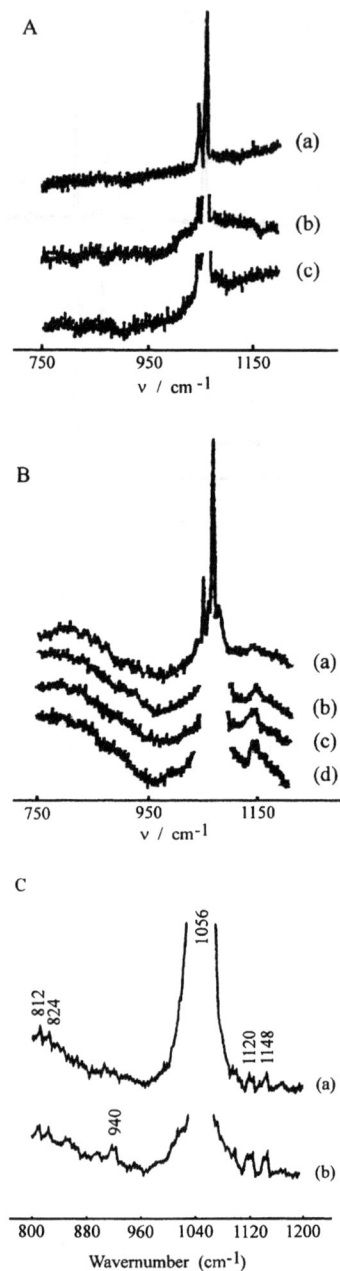

Fig. 5 In situ Raman spectra of A: uni-phasic $La_2O_3$-$ThO_2$, 1 h after switching gas flow from $CH_4$-$O_2$(4:1) feed to flowing(a) $O_2$, (b) $CO_2$, and (c) $CO_2$-$H_2O$(92:8), all at 1013 K; B: functioning $La_2O_3$-$ThO_2$ catalyst in flowing $CH_4$-$O_2$(4:1) at (a) 1133, (b) 1073, (c) 1013, and(d) 953 K[84]. C: functioning Th-La-O/BaCO$_3$ catalyst on stream of $CH_4$-$O_2$(4:1) feed at(a) 1013 K, showing $\underline{O_2^-}$ signals at 1120(w to wm) and 1148 cm$^{-1}$(w), besides carbonate signal at 1056 cm$^{-1}$(vs) and two $\underline{O_2^{2-}}$ signals at 812 and 824 cm$^{-1}$(w);and(b) 773 K, showing a new signal at 940 cm$^{-1}$(wm) besides the signals in (a)[85].

after 10 h in a flow of pure $O_2$ at 973 K, 0. 1 MPa, only the Raman signal at 1052 $cm^{-1}$ (s) of surface-carbonate was observed (Fig. 6B(a)). After switching the $O_2$ flow to a stream of the $CH_4$-$O_2$-He feed and reacting for 20 min at 973 K, a Raman signal at 1140 $cm^{-1}$ (w) ascribed to $\underline{O_2^-}$ was clearly observable, along with a second surface-carbonate signal at 1076 $cm^{-1}$ (m), besides the first one at 1052 $cm^{-1}$ (s) (Fig. 6B(b)). After 20 min more on stream of the feed, the $\underline{O_2^-}$ signal shifted to 1136 $cm^{-1}$ (w), still clearly observable; and the surface-carbonate signal at 1076 $cm^{-1}$ (s) grew in intensity at the expense of the carbonate signal at 1052$cm^{-1}$ (ms) (Fig. 6B(c)).

In situ Raman spectra of Th-La-$O_x$/$BaCO_3$ catalyst (Fig. 5C) in flowing $CH_4$-$O_2$(4 : 1) at 1013 K and 2. 0 × $10^4$ $h^{-1}$ GHSV have been obtained by Zhang et al.[85], with Raman band at 1056 $cm^{-1}$ (vs) due to surface carbonate, two bands at 1120 (mw) and 1148 $cm^{-1}$ (mw), and two more bands at 812(w) and 820 $cm^{-1}$ (w), ascribable to $v_{O\text{-}O}$ of two $\underline{O_2^-}$ and two $\underline{O_2^{2-}}$ adspecies, respectively, in different microenvironments. Note that the existence of sub-surface $\underline{O^-}$ and $\underline{O_2^-}$ have been shown by Che and coworkers[102,104] by means of EPR method. At lower temperature, the Raman bands of the two superoxide adspecies conspicuously increased in intensity, and at 773 K, an additional Raman band at 940 $cm^{-1}$ (mw) appeared. This band disappeared at temperature only slightly above 773 K. Note that the unidentified species with Raman signal at 940 $cm^{-1}$ (mw) is in the roughly estimated frequency range of $v_{1(O\text{-}O)}$ stretching of an expected angular species $\underline{O_3^{2-}}$, as active precursor of $\underline{O_2^-}$, as discussed in Section 5. 2. A species with Raman signal at 948

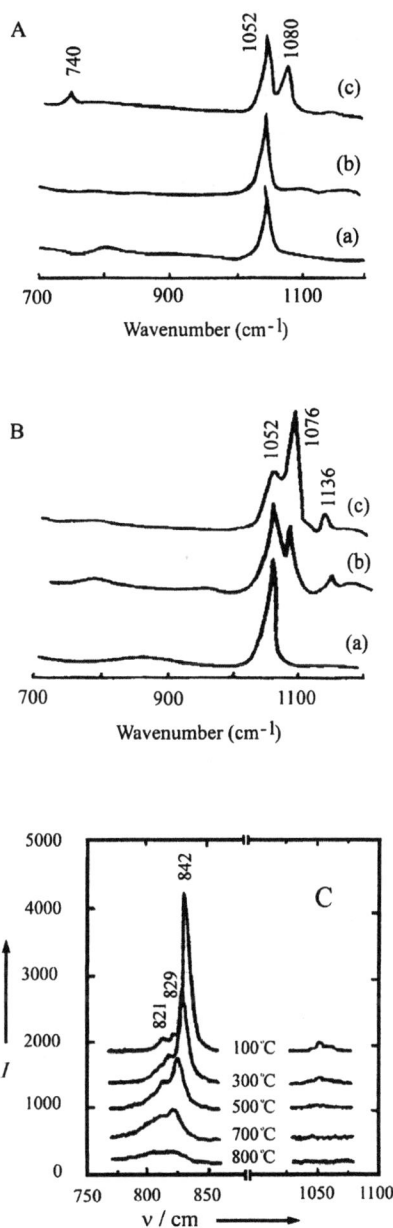

Fig. 6 In situ Raman spectra of OCM catalysts at 0. 1 MPa under various reaction conditions. A: $(LaO)_2(O,CO_3)$/$BaCO_3$ (La : Ba＝2 : 8, ignited) after(a) 10 h on stream of $O_2$ at 973 K; (b) ca. 10 s after exposing(a) to flowing $CH_4$ at 973 K; (c) 1 h after swtching the $CH_4$ flow to flowing $CH_4$-$O_2$-He feed at 973 K. B: $(LaO)_2(O,CO_3)$/CaO/$(Ca,Ba)CO_3$ (La : Ba : Ca＝2 : 4 : 4) catalyst after(a) 10 h on stream of $O_2$ at 973 K; (b) after switching the $O_2$ flow to the feed stream at 973 K and reacting for 20 min; and(c) after reacting for 40 min at 973 K[86]. C(reproduced from [106]): 0. 5 mol% BaO/MgO, in $O_2$ flow at 373 — 1073 K, showing $\underline{O_2^{2-}}$ Raman signal, and carbonate signal.

cm$^{-1}$(m-w) on fluoride-containing sample observable in helium flow and in C$_2$H$_6$ flow up to 673 and 573 K, respectively, will be shown in Section 5. 2.

Note that in situ Raman spectra of La$_2$O$_3$, Na$^+$/La$_2$O$_3$, and Sr$^{2+}$/La$_2$O$_3$ in O$_2$ atmosphere at 973 K have been reported by Mestl et al.[105]. Over La$_2$O$_3$, a band at 863 cm$^{-1}$ was observed and ascribed to $v_{O-O}$ of O$_2^{2-}$. After the O$_2$ flow was switched to a flow of CH$_4$, the Raman band gradually decreased in intensity, shifted to 813 cm$^{-1}$, and finally disappeared. After switching the CH$_4$ flow to a flow of CH$_4$-O$_2$, the 863 cm$^{-1}$ band reappeared. Unfortunately, the observation of the 863 cm$^{-1}$ Raman signal was not very reproducible according to Lunsford[75]. In situ Raman spectra of 0. 5 mol% BaO/MgO in a flow of O$_2$ at 373—1073 K have been reported by Lunsford et al. [106], with a major Raman band at 842—829 cm$^{-1}$, ascribable to O$_2^{2-}$, and a band at 1050 cm$^{-1}$(w) due to surface carbonate(Fig. 6C), but no O$_2^-$ signal in CH$_4$ and O$_2$ flow under OCM working conditions, only a strong carbonate signal was observed. Note that the absence of O$_2^-$ Raman signal in O$_2$ atmosphere is no longer a surprise. However, the Ba$^{2+}$/MgO is quite different from the alien-valence host-dopant compo-site-oxides system with comparable cationic radii, so the behaviours might be different. It would be interesting to examine the presence or absence of O$_2^-$ Raman signal around 1115—1165 cm$^{-1}$ with the Sr$^{2+}$/La$_2$O$_3$ catalyst under OCM cofeed-operation conditions.

The in situ Raman spectra obtained with La$_2$O$_3$-ThO$_2$ uniphasic catalyst and La-Ca-O/CaO/(Ba, Ca)CO$_3$ multi-phasic catalyst under OCM cofeed conditions versus in pure O$_2$ stream reviewed above definitely confirm that O$_2^-$ adspecies can exist on the rare-earth-based composite oxides under OCM cofeed reaction conditions in the presence of large amounts of surface-carbonates and considerable concentration of CO$_2$ in the gas phase(contrary to the suggestion by some authors[107,108] that the formation of O$_2^-$ adspecies might be sensitive to the inhibition by surface carbonates and CO$_2$), and up to very high temperature(e. g., 1133 K), much higher than the decomposition temperatures of Ba(O$_2$)$_2$, KO$_2$, and CsO$_2$(contrary to the suggestion by some authors[109,110] that O$_2^-$ might be too unstable to exist at the high temperature of OCM). Evidently, the high-tem-perature stability of O$_2^-$ is due to additional stabilisation of the O$_2^-$ adspecies by the surface crystal-field, with reference to the energy diagram of various oxygen species(O$^{2-}$, O$_2^{2-}$, O$^-$, O$_2^-$, O$_2$) in the gas phase, in the solid phase, and in the surface of a binary oxide given in a review by Bielanski and Haber[111]. Conceivably, the crystal-field at a surface-lattice anionic site of rare-earth dioxide(REO$_2$) or sesquioxide(Ln$_2$O$_3$) is considerably stronger than that of the alkali oxides and AEO.

With the existence and reactivity of O$_2^-$ observed at high temperature, Liu et al. [82,83] have been able to advance several arguments in favour of O$_2^-$ being the active-oxygen adspecies with higher OCM selectivity than O$_2^{2-}$: (1) The OCM selectivity of many catalyst systems is known to increase with increasing temperature in the range 923—1073 K, while O$_2^{2-}$ may be reactive towards CH$_4$ at much lower temperature as indicated by the reactivity of metal peroxides[109,110]. (2) From the relevant bond energies listed as an appendix under Fig. 3, H-abstraction from CH$_4$ by O$_2^-$ may be endothermic by about 18 kcal/mol(while that by O$^-$ may be exothermic by several kcal/mol); this is areaction step of breaking a stronger bond and forming a weaker bond, so deuterium normal kinetic isotope effect, $k_H/k_D>1$, for this step may be expected from zero-point energy consideration; and deuterium normal kinetic isotope effect has been observed for OCM under usual OCM reaction conditions[77], though the rate-determining step of OCM may change under other conditions[75]. (3) It is known that CO$_2$ can exert significant promotional effect on OCM selectivity, probably by virtue of inhibiting the less selective active-sites[75]; and that CO$_2$ can react very fast with O$_2^{2-}$, in fact, considerably faster than O$_2^-$, according to the observation of Liu et al. [82]; as shown in Fig. 4A(a) versus(b); this again indicates that $O_2^{2-}$ is less likely to be an oxygen-adspecies of high OCM selectivity.

In any metal oxide catalyst with stable cationic valency, electrons for the reduction of adsorbates can only come from the $O^{2-}$ ions. The reduction of each $O_2$ adspecies to $O_2^-$ or $O_2^{2-}$ adspecies is accompanied by the formation of, respectively, one or two $O^-$ in the oxide lattice. Though some of $O^-$ ions might tug away in the bulk[104], still there would be quite a large number of $O^-$ in the surface lattice in the absence of some way to tie them up. The $O^-$ ions are known to be highly reactive, capable of attacking $C_2H_4$ even at a very low temperature[104]. In the absence of some way to tie the $O^-$ up, how could high $C_2^+$ selectivity at moderate level of methane conversion be actually obtained with many of the OCM catalysts with stable cationic valency? But the most interesting puzzle is the absence of Raman signal of $O_2^-$ when the rare-earth-oxide-based catalyst is in contact with pure $O_2$ at 0.1 MPa, in contrast with the presence of clearly observable Raman signal of $O_2^-$ when the catalyst is in contact with $CH_4$-$O_2$ in the feed with $P_{O_2}$ only about 0.02—0.002 MPa at the reactor entry.

In an attempt to solve these puzzles, it has been suggested by Tsai et al. [83] that there might be certain elusive precursor of $O_2^-$, which might be present in very low concentration in OCM, but much more reactive than $O_2^-$ towards $CH_4$, and that it might react with $CH_4$ with the formation of $O_2^-$ besides hydrocarbon products. It has been speculated by these authors[83] that $O_3^{2-}$ might be formed by incorporation of $O_2$ adspecies with a neighbouring $O^{2-}$ ion at about 2.7—2.9 Å nucleus-to-centre distance from the anionic vacancy occupied by $O_2^-$. The process of this incorporation may be envisaged as follows. An $O_2$ molecule adsorbs reversibly at the anion-vacant site, the diatomic adspecies $O_2$ abstracts an electron from a neighbouring $O^{2-}$ ion, resulting in the formation of two radicals($O_2^-$ and $O^-$) at neighbouring sites. $O_2^-$ is known to be both Raman and IR active[98], so it is not likely to be a flat-lying adspecies. In the absence of steric hindrance by neighbouring adsorbate, the two radicals may then unite to form the triatomic, angular anionic adspecies, $O_3^{2-}$, thus tying up the damaging $O^-$ species. This process is shown as the dotted rectangular box in Scheme 4.

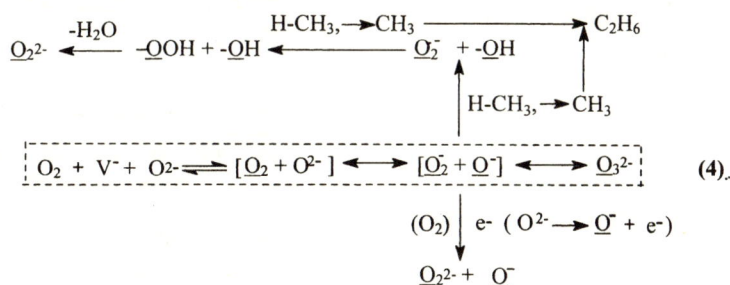

$$O_2^{2-} \xleftarrow{-H_2O} -OOH + \cdot OH \longleftarrow \begin{array}{c} H\text{-}CH_3, \rightarrow CH_3 \longrightarrow C_2H_6 \\ O_2^- + \cdot OH \\ \uparrow \\ H\text{-}CH_3, \rightarrow CH_3 \end{array}$$

$$[\,O_2 + V^- + O_2^- \rightleftharpoons [O_2 + O^{2-}] \longleftrightarrow [O_2^- + O^-] \longleftrightarrow O_3^{2-}\,] \qquad (4).$$

$$(O_2) \downarrow e^- \; (O^{2-} \rightarrow O^- + e^-)$$

$$O_2^{2-} + O^-$$

**Scheme 4.**

The internuclear distance between the two terminal-$O$ of the angular $O_3^{2-}$ is estimated to be about 2.52 Å, with the bond angle taken to be ca. 116.8°, as that of $O_3$[112] and the O-O single-bond length about 1.48 Å[19]. On the surface of a host rare-earth oxide(e.g., on the non-polar(110) surface of $La^{3+}$-doped or $Sr^{2+}$-doped $ThO_2$ (DFS), or the non-polar (110) minor surface of $Ca^{2+}$-doped, hexagonal $La_2O_3$) the closest anionic-anionic internuclear distance of two $O^{2-}$ ions, or an $O^{2-}$ and an anionic vacancy is about 2.8 Å So, in the absence of steric hindrance imposed by neighbouring adsorbate, the angular $O_3^{2-}$ should be able to stretch the distance of its two terminal-O by ca. 0.3 Å, and adapt itself (with approximate $C_{2v}$ symmetry) to the twin-sites of the surface lattice(Fig. 7).

The reactivity of this active precursor species in OCM may be envisaged as follows. In H-abstraction from $CH_4$, this active precursor might react like a biradical, $[O_2^-, O^-]$ to form $CH_3$, $-OH$, and $O_2^-$ since H-abstraction activity of $\underline{O}^-$ is known to be much higher than that of $\underline{O}_2^-$. On the other hand, in a flow of $O_2$ with no hydrocarbons, $O_3^{2-}$ might capture an electron from the oxide lattice and be reduced to form $\underline{O}_2^{2-}$ plus $\underline{O}^-$, rather than $\underline{O}_2^-$ plus $\underline{O}_2^-$, since the binuclear $\underline{O}_2^-$ is more electrophilic than $\underline{O}^-$. These two different pathways are shown in the central part of Scheme 4 above and below the dotted frame. (In the dotted frame of Scheme 4, resonance hybrid structures are connected by double-headed arrows, and anionic vacancy denoted by $V^-$.) The $\underline{O}^-$ is likely to react with $CH_4$ considerably more slowly than its active precursor species $O_3^{2-}$, of which the reactivity might be between that of $\underline{O}_2^{2-}$ and $\underline{O}_2^-$ in H-abstraction from $CH_4$. So a dynamic concentration of $\underline{O}_2^-$ may be built up to Raman spectroscopically detectable level, though it is likely that some of the $\underline{O}_2^-$ ions may be further reduced to $\underline{O}_2^{2-}$. Collision of the newly produced $CH_3$ with $\underline{O}H^-$ will be harmless, resulting in no further H-abstraction. Collision of $CH_3$ with $\underline{O}_2^-$ will also be relatively harmless since H-abstraction from $CH_3$ by $\underline{O}_2^-$ is weak (as can be inferred from the low BE of 87 kcal/mol for H-OOH, and the relatively high BE of $H-CH_2$, 110 kcal/mol, ca. 5 kcal/mol higher than that of $H-CH_3$)[19]. Moreover, at moderate level of methane conversion, there are more $CH_4$ molecules than $CH_3$ radicals around each site. So the chance of H-abstraction from $CH_4$ by $\underline{O}_2^-$ is much greater than that from $CH_3$. Thus at OCM temperature when $\underline{O}_2^-$ is sufficiently active, there is a good chance for the production and coupling of two $CH_3$ radicals formed successively from two $CH_4$ molecules via H-abstraction by the resonance hybrid of $O_3^{2-}$,

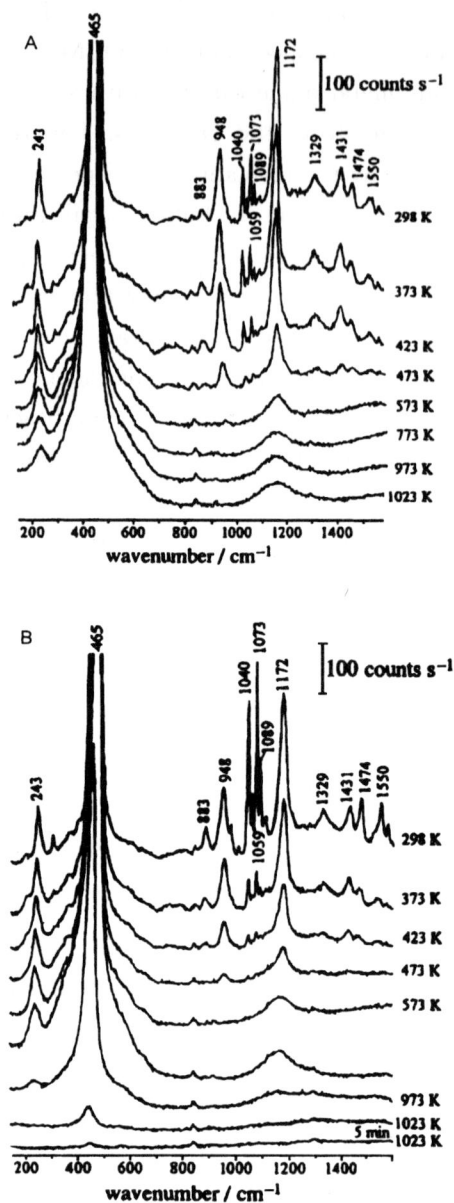

Fig. 7 Reproduced from [98]: Microprobe Raman spectra of $O_2$-pretreated $BaF_2/CeO_2$ ( Ba : Ce = 4 : 1 ) catalyst at the indicated temperature in A: Helium, and B: $C_2H_6$ atmosphere.

concomitant with the successive formation of $\underline{O}H^-$ and $^-\underline{O}OH$ followed by dehydration to give $\underline{O}_2^{2-}$. The sequence of reactions may take place so fast that most of the first $CH_3$ and the -OH formed cannot have time to diffuse or migrate away. The sequence of reactions is shown at the right and left corners above the dotted frame of Scheme 4. Thus the attainable high $C_2^+$ selectivity can be explained.

The angular $O_3^{2-}$ species may dissociate from either one of the two O-O single bonds for $O_2$ to desorb reversibly from either side; thus the rapid oxygen isotope exchange with labelled $O_2$ and a doped $REO_2$ or $Ln_2O_3$ can be explained.

From the known O-O stretching frequencies of $O_2$ ($v_{O-O}=1550$ cm$^{-1}$), $Ba^{2+}$-bound $\underline{O}_2^-$ and $\underline{O}_2^{2-}$ (cf.

Fig. 4A, $v_{0,0} = 1122$ and 840 cm$^{-1}$, respectively), and $O_3$ ($v_1 = 1136$ cm$^{-1}$), $\underline{O_3^-}$ ($v_1 = 1016$ cm$^{-1}$[113]), the Raman-active $v_1$ of $\underline{O_3^{2-}}$ may be roughly estimated to be around $900 - 950$ cm$^{-1}$, depending on the micro-environment. A check of the frequency range and H-abstraction reactivity of $\underline{O_3^{2-}}$ by quantum-chemical calculations with embedded cluster model analogous to that used by Borce and Patterson[114] is in progress.

The species with Raman signal at 940 cm$^{-1}$ (mw) in the in situ Raman spectrum (Fig. 5C) of Th-La-O$_x$/BaCO$_3$ catalyst in flowing CH$_4$-O$_2$ (4 : 1) at 773 K (along with the two $\underline{O_2^-}$ Raman signals at 1120 (mw) and 1148 cm$^{-1}$ (mw)) observed by Zhang et al. [85] may be the elusive $\underline{O_3^{2-}}$ species they have been looking for as the active precursor of $\underline{O_2^-}$ and $\underline{O_2^{2-}}$.

In some systems, O$_2$ may adsorb at an F-centre to form $\underline{O_2^-}$. There may also be some sites where $\underline{O_3^{2-}}$ may not be formed due to steric hindrance, or lack of neighbouring O$^{2-}$ at suitable distance, then $\underline{O_2^-}$ may be formed from $\underline{O_2}$ by electron transfer from the oxide lattice.

Biphasic REO-AEO OCM catalysts with DFS have been prepared by Liu et al. [82,87] by co-precipitation method. The Th-La-Ca-Ba-O/(Ba, Ca)CO$_3$ catalyst thus prepared showed very high stability (no change in performance after 1000 h micro-reactor evaluation at 1053K plus intensified sintering test by heating at 1273 K for 1 h), high CH$_4$ conversion (ca. 33% at $6 \times 10^4$ h$^{-1}$ GHSV), C$_2$ selectivity ($\sim$62%; 65.7% in terms of C$_2^+$) and C$_2$ yield ($\sim$20.5%, $\sim$3% higher than that of La$_2$O$_3$/BaCO$_3$ catalysts under the same conditions). Thus $x + S \approx 99\%$. Further improvement in micro-reactor performance via composition optimisation may be expected. But the catalyst was found to deactivate rapidly due to hot spots overheating in an adiabatic stationary-bed reactor with 50 ml sample. So there are also engineering design problems to be solved.

## 5.2. Systematic studies of novel REO(F)-AEO(F) catalysts for OCM, ODE, and ODP

Since there is a review by Wan et al. [90] of this work, here only a brief account is given.

The study of fluoride doping of REO-AEO-based catalyst systems was started by one of our research groups in 1992, drawing inspiration from the work of Lunsford et al. [75,78] showing the marked enhancement of C$_2$H$_4$ selectivity by doping some OCM and ODE catalysts with chlorides. Further elaboration of the idea of fluoride-doping was guided by structure-function relationship consideration and experimental development. The anionic radii of F$^-$ (1.33 Å) and O$^{2-}$ (1.32 Å) are almost equal, but the ionisation potential of F$^-$ is much higher than that of O$^{2-}$. The crystal habits of most of the alkaline-earth fluorides (MF$_2$), lanthanide oxides (LnO$_2$), sesquioxides (Ln$_2$O$_3$), and oxyfluorides (LnOF) are quite similar[19,115−118] being the cubic fluorite structure and defective fluorite structure, or slightly distorted fluorite structures with tetragonal, or rhombohedral settings. So wide ranges of mutual solubility of the MF$_2$ and the REO$_2$, Ln$_2$O$_3$, or LnOF may be expected, with wide ranges of anionic vacancy concentrations and anionic mobility, and wide ranges of modification of work functions, surface basicity and resistance to CO$_2$ inhibition. The fluoride ions may also serve to disperse the various active-oxygen species to decrease deep oxidation of the hydrocarbons.

The results from the studies of fluoride-containing catalysts are summarised below.

With lanthanide ions of stable valency (e. g., La Nd$^{3+}$, Sm$^{3+}$, Gd$^{3+}$, and Y$^{3+}$), the $20 - 50$ mol% SrF$_2$-Ln$_2$O$_3$ catalyst systems showed higher C$_2$-selectivity (57.3 − 54.6%), and on the average slightly higher ethylene selectivity than that of the corresponding REO-AEO catalysts. However, the possible effect of "hot spot" temperature on the precision of the temperature measurements must be taken into consideration. Results of catalyst evaluation of these SrF$_2$-Ln$_2$O$_3$ catalysts with stable cationic valency show that the higher the anionic conductivity of the Ln$_2$O$_3$ the slightly higher the C$_2$ selectivity. This is also the order of increasing $Ln^{3+}$ ionic sizes, with the exception of $Y_2O_3$ with comparatively smaller size of $Y^{3+}$, but somewhat higher oxide-ion conductivity and higher $C_2$ selectivity[93,94]. With lanthanide ions of variable valency(e. g., $Ce^{3+/4+}$), replacement of the AEO in the REO-AEO composite-oxides by $BaF_2$ (or $SrF_2$) has the marvellous effect of changing the deep-oxidation catalysts(e. g., $CeO_2$-AEO), into good OCM catalysts(e. g., $CeO_2/BaF_2$, 54. 6% $C_2$ selectivity and 17. 6% $C_2$ yield at 1073 K) and excellent ODE catalysts[90-95]. As shown by in situ microprobe Raman spectra of $O_2$-pretreated $CeO_2$-$BaF_2$[90-93,98], the $\underline{O_2^-}$ with Raman band at 1172 $cm^{-1}$ (s-m-w, under helium atmosphere at 298-373-973 K) and lattice $O^{2-}$ (probably only those associated with $Ce^{4+}$) with lattice-vibration band at 465 $cm^{-1}$(very strong, even at 1023 K under helium atmosphere) were observed, along with a medium-to-weak band at 243 $cm^{-1}$. These have been assigned, respectively, to $F_{2g}$ mode and TO (transverse optical) mode of lattice vibrations of slightly distorted $CeO_2$ with DFS. So they are both related to $Ce^{4+}$-$O^{2-}$. Each of these two bands appeared to be affected separately to different extent by $CH_4$, $C_2H_6$, and $C_2H_4$ in the order of increasing rates, and disappeared in the order of markedly decreasing times at 1023 K. Thus $\underline{O_2^-}$ and the lattice $O^{2-}$ backed up by $Ce^{4+}$ must be regarded here as significant active-oxygen species[90,98]. There was an unidentified species with Raman band at 948 $cm^{-1}$ (ms at 298 K, m at 373 K, wm at 423 K, and w at 473 K, w at 573 K); the signal fell below detection limit above 573 K under He, or above 473 K under $C_2H_6$ atmosphere[90,98]. This may be compared with the species with Raman signal at 940 $cm^{-1}$ on Th-La-$O_x$/$BaCO_3$ catalyst(Fig. 5C(b)) in flowing $CH_4$-$O_2$(4∶1) at 773 K observed by Zhang et al.[85](Fig. 8). There was a Raman band at 1550 $cm^{-1}$ (w at 298 K, vw at 373 K), ascribable to $O_2$ adspecies. There were also Raman bands at 1040(m), 1073(m), 1089(w), 1329(wm), 1431(w), and 1474 $cm^{-1}$(w), all of these signals fell below detection limit above 473 K under either He or $C_2H_6$ atmosphere. Compared with the known Raman spectra of $O_2$-pretreated $La_2O_3$/$BaCO_3$ catalysts taken in argon at room temperature, these may be assigned to three carbonate species at different micro-environments(although the possibility of over-lapping with some other $\underline{O_2^-}$ species and some other $O_2$ species cannot be ruled out). These surface carbonate species might have come from combustion of carbonaceous contaminants during the $O_2$-pretreatment of the catalyst, and they were easily removed by heating the catalyst sample in helium to 473 K due to the much lower basicity of 20 mol% $BaF_2$/$CeO_2$

than that of $La_2O_3$/$BaCO_3$. It is interesting to note that the $O_2$-pretreated catalyst sample under $CH_4$ atmosphere at 573-773-973-1023 K gave microprobe Raman band at 1165 $cm^{-1}$ (s-s-s-ms, the signal fell below detection limit after heating the system at 1023 K for 10 min), and another band at ca. 840 $cm^{-1}$ (w-w-w-w, persisting after heating the system at 1023 K for 10 min and becoming very weak (vw) after heating the system for 40 min); no other bands being observed in the range 800—1400 $cm^{-1}$. The first Raman band may be ascribed to $\underline{O_2^-}$ and the second to $\underline{O_2^{2-}}$, as a downstream adspecies in further reduction of

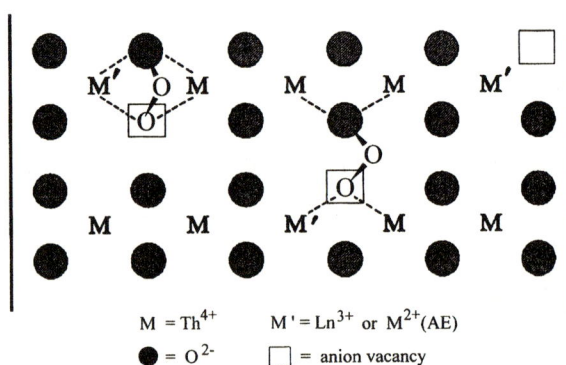

**Fig. 8 A model of probable structure of $\underline{O_3^{2-}}$ ( ~$C_{2v}$, internuclear distance O(1)—O(3) of angular $\underline{O_3^{2-}} \geq 2.5$ Å). on $Ln_2O_3$-$ThO_2$ (uniphasic, DFS, closest O-O ≈ 2. 7—2. 8 Å(1 10) surface.**

M = $Th^{4+}$    M′ = $Ln^{3+}$ or $M^{2+}$(AE)

● = $O^{2-}$    □ = anion vacancy

$\underline{O_2^-}$. Note also that with $O_2$-pretreated catalyst sample in a flow of $CH_4$-$O_2$(3. 4∶1) at 923, 973, and 1023 K, only a strong Raman band of $\underline{O_2^-}$ at 1153 $cm^{-1}$ was observed in the wave number range 780—

$1500 \ cm^{-1}$, indicating that this superoxide adspecies was steadily produced at these temperatures, to practically saturated, dynamic concentration, from a much more reactive precursor oxygen-species. The downstream peroxide adspecies at these temperature was too reactive to give detectable Raman signal, as in the case of OCM reaction over $Ba^{2+}/MgO$[106].

Conceivably, due to the higher work function of a fluoride-containing REO-AEO-based catalyst than that of the corresponding REO-AEO catalyst, $O_2^-$ adspecies on the fluoride-containing catalyst may not be so easily reduced further to form the downstream adspecies, $O_2^{2-}$ and $2 \ O^-$, thus making it possible to detect the $O_2^-$ on $O_2$-pretreated sample by in situ FTIR and Raman spectroscopic techniques, and to study its reaction with methane, or other reactants. $O_2^-$ adspecies on $O_2$-pretreated $SrF_2/La_2O_3$ with Raman bands at $1113 \ cm^{-1}$ (923 K) have been found to react with $CH_4$ with the formation of $C_2H_4$, $CO_2$, and surface carbonate; and likewise $O_2^-$ adspecies with Raman band at $1114 \ cm^{-1}$ on $SrF_2/Nd_2O_3$ at 973 K[90-93]. Thus for the first time, direct evidence for $O_2^-$ being a principal OCM-active oxygen-adspecies was obtained. It has also been demonstrated that there is no simple dependency of catalyst activity and selectivity for OCM and ODE on catalyst surface basicity-acid-ity, and that p-type conductivity is not always required for good OCM performance[90-93].

Based upon the principle of structure-directed constituent selection[95-97], a series of $BaF_2$-doped, tetragonal LaOF catalysts showing excellent performance for both OCM and ODE were prepared. With a 10 mol% $BaF_2$/LaOF catalyst for OCM, under the conditions of 1043 K, $CH_4/O_2 = 6$, and GHSV = $15 \ 000 \ h^{-1}$, methane conversion($x$) of 19.5% and $C_2$ selectivity($S$) of 81.2% were obtained[90,93], $x + S$ exceeding 100%. With a 30 mol% $BaF_2$/LaOF catalyst for ODE, under the conditions of 913 K, $C_2H_6/O_2 = 2$, GHSV = $11600h^{-1}$, $C_2H_6$ conversion of 80.8% and $C_2H_4$ selectivity of 70.8% were obtained[96,97]. Although, the graduate, hydrolytic loss of fluoride(HF) at high reaction temperature is still quite a problem for these fluoride-containing REO-AEO-based catalyst systems, they may find application in some other oxidative-coupling, or oxi-dative-dehydrogenation reactions that can take place at lower temperatures. More importantly, the systematic studies of these halide-containing REO-AEO-based catalyst systems have already furnished important information for clarifying some controversial issues, establishing some fundamental principles for host and dopant constituents selection, and opening the way for further research. For example, by proper design of experiments, it may be possible to reveal the mechanistic mystery of the remarkable promotion of ethylene selectivity by halide-doping of the OCM and ODE catalysts.

# 6. Concluding remarks

Most of the research work reviewed in this paper was done in the last 10 years or so(picking up speed with the development of postgraduate research programme and reconstruction of the SKLPCSS), with the main objective of co-ordinating with the expected development of power industry in China based on coal, as well as on natural gas, by the integrated gasification and combined cycles(IGCCs) technology, for co-production of liquid fuels and basic organic chemicals with the generation of electric power. Over the past 40 years, 19 Chinese patents were obtained, most of them in the last few years. There has been more recent research work done by our team in other areas related to applied catalysis; this is not covered in this review, though there have been some very interesting results, such as new catalytic materials[118] and new catalytic processes[119], exhaust pollution control catalysts for methanol-fuelled automobiles[120], high enantio-selective catalysts for asymmetric transfer-hydrogenation of carbonyl functions[121,122], supported liquid-phase catalysts for olefin hydroformylation[123], K$_2$O-promoted iron oxides-based catalysts for ethyl-benzene dehydrogenation[124,125] (which has been commercialised by the chemical plant at Xiamen University by supplying to about a dozen small styrene plants in China for more than 10 years), structural basis for catalysts preparation by the citrate-complex method[126,127], and computer-aided catalyst design[128]. It may be expected that these recent achievements in catalysis will be reviewed in the near future by the young, principal investigators themselves.

# Acknowledgements

The support of this work by the State Science and Technology Commission and National Natural Science Foundation of China is gratefully acknowledged.

# References

[1]Yu. A. Gorin, Chem. Ind. (Russ.)(1959) 1944.

[2]Yu. A. Gorin et al., Chem. Ind. (Russ.)(1964) 265.

[3] Xiamen Acetic Acid Factory, Xiamen Synth. Rubber Processing Factory, Chemistry Department, Xiamen University, Acta Chim. Sinica(Chinese) 3(2)(1975) 114.

[4]Catal. Div., Chemistry Department, Xiamen University, Scientia Sinica(Chinese)(1973) 373.

[5]J. L. Zeng et al., CN Patent No. 85102996. 5.

[6]W. Z. Weng et al., Chem. J. Chinese Univ. B(4)(1990) 346.

[7]K. Tamaru, Catal. Rev. 4(1970) 161.

[8]K. R. Tsai, in: W. E. Newton, W. H. Orme-Johnson(Eds.), Nitrogen Fixation, vol. I, University Park Press, Baltimore, USA, 1980 p. 373; Scientia Sinica(Engl. Ed.)(1976) 460.

[9]K. R. Tsai, H. L. Wan, in: M. Tsutsui, Y. Ishi, Y. -Z. Huang(Eds.), Fundamental Research in Organometallic Chemistry, University Park Press, Baltimore, USA, 1982, p. 1.

[10]K. R. Tsai, H. -B. Zhang, G. D. Lin, Adv. Sci. China, Chem. 2(1987) 125.

[11]K. H. Huang, in: A. T. Seiyama, D. K. Tanabe(Eds), Proceedings of the Seventh International Congress on Catalysis, New Horizons in Catalysis, Studies in Surface Science and Catalysis, vol. 7, Kodansha, Tokyo, 1981, p. 554.

[12]D. W. Liao, H. B. Zhang, Z. Q. Wang, K. R. Tsai, Scientia Sinica, Ser. B(Engl. Ed.)(1987) 246.

[13]H. B. Zhang, K. R. Tsai, Catal. Lett. 3(1989) 129.

[14]K. R. Tsai, H. L. Wan, J. Cluster Sci. 6(1895) 485.

[15]H. L. Wan, J. W. Huang, F. Z. Zhang, Y. Wu, L. S. Xu, J. L. Li, K. R. Tsai, in: C. Elmerich et al. (Eds.), Biological Nitrogen Fixation for the 21st Century, Kluwer Academic Publishers, Dordrecht, 1998, p. 78.

[16]K. Klier, Adv. Catal. 31(1982) 243.

[17]G. A. Vedage, R. Pitchai, R. G. Herman, K. Klier, in: Proceedings of the Eighth International Congress on Catalysis, Berlin, vol. 2, 1984, p. 47.

[18]J. Halpern, Acc. Chem. Res. 15(1982) 238.

[19]CRC Handbook of Chemistry and Physics, 66th ed., 1985.

[20]H. Chen, S. Wang, Y. Liao, J. Cai, H. Zhang, K. Tsai, in: Proceedings of the Third China-Japan-USA Symposium on Catalysis, Xiamen, 1987, p. 97.

[21]H. Chen,S. Wang,Y. Liao,J. Cai,H. Zhang,K. Tsai,in:Proceedings of the Ninth International Congress on Catalysis,Calgary,1988,p. 537.

[22]M. A. Bennett,J. Mol. Catal. **41**(1987) 1.

[23]J. F. Edwards,G. L. Schrader,J. Phys. Chem. **89**(1985) 782.

[24]J. C. Lavalley,J. Saussey,T. Rais,A. Chakor-Alami,J. P. Hindermann,A. Kiennemann,J. Mol. Catal. **26**(1984) 159.

[25]H. Idress,J. P. Hindermann,A. Kiennemann et al.,J. Mol. Catal. **42**(1987) 205 and relevant references cited.

[26]M. -Y. He,J. G. Ekerdt,J. Catal. **87**(1984) 237.

[27]M. -Y. He,J. G. Ekerdt,J. Catal. **87**(1984) 381.

[28]R. Burch,R. J. Chappell,S. E. Golunski,J. Chem. Soc.,Faraday Trans. I **85**(1989) 4569.

[29]D. Duprez,Z. Ferhat-Hamida,M. M. Bettanar,J. Catal. **124**(1990)1.

[30]J. Cai, Y. Liao, H. Chen, K. Tsai, in: Proceedings of the 10th International Congress on Catalysis,Budapest,1992,p. 2769.

[31]G. C. Chinchen,M. S. Spenser,K. C. Waugh,D. A. Whan,Faraday Symp. Chem. Soc. 21(1986) **21**(Paper 18).

[32]G. C. Chinchen,P. J. Denny,J. R. Jennings,M. S. Spenser,K. C. Waugh,Appl. Catal. **36**(1988) 1—65.

[33]K. C. Waugh,Catal. Today **18**(1993) 147.

[34]J. C. J. Bart,R. P. A. Sneeden,Catal. Today **2**(1987) 1.

[35]G. Ghiotti,F. Boccuzzi,Catal. Rev. -Sci. Eng. **29**(1987) 151.

[36]A. Cybulski,Catal. Rev. -Sci. Eng. **36**(1994) 557.

[37]Y. Q. Yang,H. B. Zhang,G. D. Lin,H. Z. Chen,Y. Z. Yuan,K. R. Tsai,J. Xiamen Univ. **33** (1994) 477.

[38]Y. F. Shen,L. F. Cai,J. T. Li,S. J. Wang,K. H. Huang,Catal. Today **6**(1989) 47.

[39]Y. F. Shen,S. J. Wang,K. H. Huang,Appl. Catal. **57**(1990) 55.

[40]Y. F. Shen et al.,Appl. Catal. **59**(1990) 61.

[41]T. P. Wilson,P. H. Kasai,P. C. Ellgen,J. Catal. **69**(1981) 193.

[42]M. Ichikawa,T. Fukushima,K. Sikakura,in:Proceedings of the Eighth International Congress on Catalysis,Berlin,vol. 2,1984,p. 69.

[43]S. J. Tauster et al.,J. Am. Chem. Soc. **100**(1978) 170.

[44]W. M. H. Sachtler,in:Proceedings of the Eighth International Congress on Catalysis,Berlin, vol. 1,1984,p. 151.

[45]W. M. H. Sachtler,M. Ichikawa,J. Phys. Chem. **90**(1986) 4752.

[46]S. Orita,S. Naito,K. Tamaru,J. Catal. **90**(1984) 183.

[47]J. S. Rieck,A. T. Bell,J. Catal. **113**(1985) 341.

[48]A. T. Bell,J. Mol. Catal. **100**(1995) 1—11.

[49]A. Takeuchi,J. R. Katzer,J. Phys. Chem. **86**(1982) 2438.

[50]A. Deluzarche,J. P. Hindermann,R. Kieffer et al.,J. Phys. Chem. **88**(1984) 4493.

[51]T. Mori,H. Masuda,H. Imal,J. Phys. Chem. **86**(1972) 2753.

[52]M. A. Logan,G. A. Somorjai,J. Catal. **85**(1985) 317.

[53]G. S. Gu,J. P. Liu,J. K. Fu,K. R. Tsai,Acta Phys. Chim. Sinica **1**(1985) 177.

[54]J. P. Liu,H. Y. Wang,J. K. Fu,Y. G. Li,K. R. Tsai,in:Proceedings of the Ninth International Congress on Catalysis,Calgary,1988,p. 735.

[55]J. P. Liu, H. Y. Wang, J. K. Fu, Y. G. Li, K. R. Tsai, Proceedings of the Third China-Japan-USA Symposium on Catalysis, Xiamen, 1987, p. 99.

[56]P. T. Wolczanski, J. E. Bercaw, Acc. Chem. Res. **13**(1980) 121.

[57]P. H. McBreen, W. Erley, H. Ibach, Surf. Sci. **148**(1984) 292.

[58]A. Deluzarche, J. P. Hindermann, A. Kiennemann, J. Mol. Catal. **31**(1985) 225.

[59]H. Y. Wang, J. P. Liu, J. K. Fu, H. B. Zhang, K. R. Tsai, Catal. Lett. **12**(1992) 87.

[60]H. Y. Wang, J. P. Liu, J. K. Fu, H. L. Wan, K. R. Tsai, Res. Chem. Interm. **17**(1992) 233.

[61]Y. H. Du, D. A. Chen, K. R. Tsai, Appl. Catal. **35**(1987) 77.

[62]Y. M. Hu, H. Y. Wang, J. Mol. Catal. (China) **7**(1993) 119.

[63]J. P. Hindermann, G. J. Hutchings, A. Kienemann, Catal. Rev. **39**(1993) 1~127.

[64]H. Y. Wang, J. P. Liu, K. R. Tsai, in: K. R. Tsai, S. Y. Peng et al. (Eds.), Catalysis in $C_1$ Chemistry, Chem. Ind. Press, Beijing, 1995, pp. 167~186.

[65]J. K. Fu, J. L. Xu, J. P. Liu, H. Y. Wang, K. R. Tsai, J. Xiamen Univ. (Nat. Sci.) **32**(1993) 604.

[66]H. Y. Wang, J. P. Liu, K. R. Tsai, J. Mol. Catal. (China) **5**(1991)16.

[67]M. Ichikawa, in: Y. Iwasawa(Ed.), Tailored Metal Catalysis, Reidel, Dordrecht, 1986, p. 183.

[68]H. Y. Wang, K. R. Tsai, Hua Xue Tong Bao **9**(1997) 1.

[69]S. S. C. Chuang, S. I. Pien, J. Catal. **138**(1992) 536.

[70]S. S. C. Chuang et al., Appl. Catal. A **151**(1997) 334.

[71]H. Y. Wang, J. P. Liu, J. K. Fu, K. R. Tsai, Acta Phys. Chim. Sinica **7**(1991) 681.

[72]H. Y. Wang, J. P. Liu, K. R. Tsai, J. Mol. Catal. (China) **8**(1994) 472~486.

[73]G. E. Keller, M. M. Bhasin, J. Catal. **73**(1982) 9.

[74]W. Hinsen, M. Baerns, Chem. Ztg. **107**(1983) 229.

[75]J. H. Lunsford, Angew. Chem., Int. Ed. Engl. **34**(1995) 970~980.

[76]G. J. Hutchings, M. S. Scurrell, in: E. E. Wolf (Ed.), Methane Conversion by Oxidative Processes, Van Nostrand Reinhold, New York, 1992, pp. 201~258.

[77]M. Baerns, J. R. H. Ross, in: J. M. Thomas, K. I. Zamaraev (Eds.), IUPAC Monograph Perspective in Catalysis, 1990, pp. 315~334.

[78]J. H. Lunsford, Catal. Today **6**(1990) 235.

[79]Z. Kalenik, E. Wolf, Catal. Today **13**(1992) 255~264.

[80]Z. L. Zhang, C. T. Au, K. R. Tsai, Appl. Catal. **62**(1990) L29.

[81]Y. D. Liu, G. D. Lin, H. B. Zhang, J. X. Cai, H. L. Wan, K. R. Tsai, in: Preprints Fuel. Chem. Div., ACS National Meeting, San Francisco, vol. 37, no. 1, 1992, p. 356.

[82]Y. D. Liu, H. B. Zhang, G. D. Lin, K. R. Tsai, in: K. R. Tsai, S. Y. Peng et al. (Eds.), Catalysis in $C_1$ Chemistry, Chem. Ind. Press, Beijing, 1995, p. 46.

[83]Y. D. Liu, G. D. Lin, H. B. Zhang, K. R. Tsai, in: H. E. Curry-Hyde, R. F. Howe(Eds.), Natural Gas conversion II, Elsevier, Amsterdam, 1994, p. 131.

[84]Y. D. Liu, H. B. Zhang, G. D. Lin, Y. Y. Liao, K. R. Tsai, J. Chem. Soc., Chem. commun. (1994) 1871.

[85]H. B. Zhang, Y. D. Liu, G. D. Liu, Y. Y. Liao, K. R. Tsai, 207th ACS National Meeting, 1994, Abstracts of Papers(2), PHYS 0277, submitted for publication.

[86]J. X. Cai, A. M. Huang, Y. Y. Liao, H. L. Wan, in: N. T. Yu, X. Y. Li(Eds.), Proceedings of the Fourth International conference on Raman, Wiley, New York, 1994, p. 526.

[87]K. R. Tsai, H. L. Wan, H. B. Zhang, G. D. Lin, invited paper, Sect. 5, 34th IUPAC congress, Beijing, 1993, in: Book of Abstracts, p. 671, submitted for publication.

[88]J. L. Dubois, C. J. Cameron, Appl. Catal. **67**(1990) 49.

[89]A. G. Anshits, E. N. Voskesenkaya, L. I. Kurteeva, Catal. Lett. **6**(1990) 49.

[90]H. L. Wan, W. Z. Weng, Y. Iwasawa, Hyomen(Surface) **36**(1998) 53.

[91]X. P. Zhou, S. Q. Zhou, W. D. Zhang, Z. S. Chao, W. Z. Weng, R. Q. Long, D. L. Tang, H. Y. Wang, J. X. Cai, H. L. Wan, K. R. Tsai, 207th ACS National Meeting, Preprints Div. Petro. Chem. Inc. **39**(1994) 222.

[92]M. M. Bhasin, D. W. Slocum(Eds. ), Methane and Alkane conversion Chemistry, Plenum, New York, 1995, p. 19.

[93]R. Q. Long, S. Q. Zhou, Y. P. Huang, W. Z. Weng, H. L. Wan, K. R. Tsai, Appl. Catal. A **133** (1995) 269.

[94]R. Q. Long, Y. P. Huang, W. Z. Weng, H. L. Wan, K. R. Tsai, Catal. Today **30**(1996) 59.

[95]Z. S. Chao, X. P. Zhou, H. L. Wan, K. R. Tsai, Appl. Catal. A **130**(1995) 127.

[96]X. P. Zhou, Z. S. Chao, J. Z. Luo, H. L. Wan, K. R. Tsai, Appl. Catal. A **133**(1995) 263.

[97]H. L. Wan, Z. S. Chao, W. Z. Weng, X. P. Zhou, J. X. Cai, K. R. Tsai, Catal. Today **30**(1996) 67.

[98]R. Q. Long, H. L. Wan, J. Chem. Soc., Faraday Trans. **93**(1997) 355.

[99]K. Nakamoto, Infrared and Raman Spectra of Inorganic and Coordination Compounds, 4th ed., Wiley, New York, 1986.

[100]Y. Osada, S. Kolke, T. Fukushima, S. Ogasawara, Appl. Catal. **59**(1990) 59.

[101]T. -L. Yang, L. -B. Feng, S. K. Shen, J. Catal. **145**(1994) 384.

[102]C. Louis, T. L. Chang, M. Kermarec, T. L. Van, J. M. Tatibouet, M. Che, Catal. Today **13** (1992) 283.

[103]Y. Y. Liao, P. H. Hong, J. X. Cai, in: N. T. You, X. Y. Li(Eds. ), Proceedings of the Fourth International Conference on Raman Spectroscopy, Wiley, New York, 1994, p. 1086.

[104]M. Che, G. C. Bond(Eds. ), Adsorption and Catalysis on Oxide Surfaces, Studies in Surface Science and Catalysis, vol. 21, Chapters 1 and 2, Elsevier, Amsterdam, 1985, pp. 1 and 11.

[105] D. Mestl, H. Knoginger, J. H. Lunsford, Ber. Bunsenges. Phys. Chem. **97**(1993) 319.

[106] J. H. Lunsford, X. Yang, K. Haller, J. Laane, G. Mestl, H. Knozinger, J. Phys. Chem. **97** (1993) 13810.

[107] L. Dubois, M. Bisiaux, M. Mimoun, C. J. Cameron, Chem. Lett. (1990) 967.

[108] L. Dubois, M. Bisiaux, M. Mimoun, C. J. Cameron, Chem. Lett. (1991) 1089.

[109]K. Otsuka, A. A. Said, K. Jimmo, Komatsu, Chem. Lett. (1987) 77.

[110]Otsuka et al., Chem. Lett. (1986) 967.

[111]A. Bielanski, J. Haber, Catal. Rev. -Sci. Eng. **19**(1979) 1.

[112]N. W. Greenwood, A. Earushaw, Chemistry of the Elements, Pergamon Press, Oxford, 1984, pp. 707~711.

[113]L. Andrews, P. C. Spiker Jr., , J. Chem. Phys. **59**(1973) 863.

[114]K. J. Borve, Pettersson, J. Phys. Chem. **95**(1991) 3214.

[115]A. F. Wells, Structural Inorganic Chemistry, 5th ed., Oxford University Press, New York, 1984, pp. 545~546.

[116]W. H. Zachariasen, Acta Cryst. **4**(1951) 231.

[117]D. B. Shinn, H. A. Eick, Inorg. Chem. **8**(1969) 232~235.

[118]P. Chen, H. B. Zhang, G. D. Lin, Q. Hong, K. R. Tsai, CARBON **35**(10)(11)(1997) 1495.

[119]H. B. Zhang, H. Y. Li, G. D. Lin, K. R. Tsai, in: J. W. Hightower, W. N. Delgass, E. Iglesia, A. T. Bell(Eds. ), Proceedings of the 11th ICC, Stud. Surf. Sci. Catal. **101**(1996) 1369.

[120]L. F. Yang, D. H. Chen, J. X. Cai, H. L. Wan, in: Proceedings of the APCAT'97, South Korea, 1997, p. 183.

[121]J. X. Gao, P. P. Xu, P. Q. Huang, H. L. Wan, K. R. Tsai, J. Mol. Catal. (China) **11**(1997) 413.

[122]J. X. Gao, T. Ikariya, R. Noyori, Organometallics **15**(1996) 1087.

[123]Y. Z. Yuan, J. L. Xu, H. B. Zhang, K. R. Tsai, Catal. Lett. **29**(1994) 387.

[124]H. Z. Chen, D. Y. He, Z. L. Xiao, K. R. Tsai, Chem. J. Chinese Univ. **6**(1985) 433.

[125]J. P. Chen, D. Y. He, S. J. CA0, Chem. J. Chinese Univ. **7**(1986) 1020.

[126]Z. H. Zhou, Y. J. Lin, H. B. Zhang, G. D. Lin, K. R. Tsai, J. Coord. Chem. **42**(1997) 131~141.

[127]Z. H. Zhou, H. L. Wan, S. Z. Hu, K. R. Tsai, Inorg. Chim. Acta **237**(1995) 193~197.

[128]D. W. Liao, Z. N. Huang, Y. Z. Lin, H. L. Wan, H. B. Zhang, K. R. Tsai, J. Chem. Inf. Comput. Sci. **36**(1996) 1178.

■ 本文原载:Journal of Molecular Catalysis A:Chemical 147(1999)pp. 99~104.

# Hydrogenation and Hydroformylation of Olefins with Water-Soluble $Ru_3(CO)_9(TPPMS)_3$ Catalyst

Jing-Xing Gao*, Pian-Pian Xu, Xiao-Dong Yi, Hui-Lin Wan, Khi-Rui Tsai

(Department of Chemistry, Institute of Physical Chemistry,

State Key Laboratory of Physical Chemistry of Solid Surface,

Xiamen University, Xiamen 361005, Fujian, China)

**Abstract**   Water-soluble and air-stable triruthenium carbonyl cluster $Ru_3(CO)_9(TPPMS)_3$ (TPPMS = sodium diphenylphosphinobenzene-*m*-sulphonate) was used as catalyst precursor to hydrogenated acrylic acid in good yield. The cluster also catalyzed the hydroformylation of propylene with syngas in water, the main product is *n*-butyraldehyde; side-products are *iso*-butyraldehyde and a small amount of 1-butyl and isobutyl alcohols. At 120 ℃ and propylene, CO and $H_2$ partial pressures of 0.7 MPa, 2.0 MPa and 2.0 MPa, respectively, catalytic turnover of 61.2 mol products/mol cluster h and product *n* : *i* ratio of 15.9 were obtained. For ethylene hydroformylation, the main product is propanal; side-products are 3-pentanone, 1-propanol and 2-methyl-pent-2-en-1-al. The catalyst was characterized before and after the reaction by IR and. X-ray photoelectron spectroscopy and the results were discussed as related to the possible catalytic active species. © 1999 Elsevier Science B.V. All rights reserved.

**Key words**   Hydrogenation   Hydroformylation   Propylene   Ethylene   Water-soluble cluster catalyst

## 1. Introduction

Ruthenium carbonyl cluster complexes, $Ru_3$-$(CO)_{12}$, $H_4Ru_4(CO)_{12}$, $[PPN][HRu_3(CO)_{11}]$ and $[PPN][H_3Ru_4(CO)_{12}]$ have been reported to catalyse a variety of CO transformation, including syngas conversion[1], the water gas shift reaction[2] and also alkene hydroformylation[3,4]. The latter reaction has been thought to involve the anion $[HRu_3(CO)_{11}]^-$, rather than $[H_3Ru_4(CO)_{12}]^-$, which has been observed in some water gas shift systems[5].

In our previous work, FTIR spectra in the C-O stretching region indicated anionic $[HRu_3(CO)_{11}]^-$ as the catalytic active component on a ruthenium carbonyl hydroformylation system and a possible cycle mechanism of catalysis by triruthenium cluster has been proposed[6,7].

However, ruthenium anions $[HRu_3(CO)_{11}]^-$ and $[H_3Ru_4(CO)_{12}]^-$ are air-sensitive and unstable in water. Thereby, we have been interested in the catalytic properties of water-soluble and air-stable ruthenium cluster $Ru_3(CO)_9(TPPMS)_3$[8]. In this paper, we would like to report the use of $Ru_3(CO)_9(TPPMS)_3$ as a water-soluble catalyst precursor for the hydrogenation of several functional olefins and the hydroformylation of propylene and ethylene.

---

* Corresponding author. Fax:+86-592-2188054;E-mail:jxgao@xmu.edu.cn

# 2. Experimental

The water-soluble ligand TPPMS and cluster $Ru_3(CO)_9(TPPMS)_3$ was prepared according to literature methods[8,9]. IR spectra were recorded on a PE-Spectroy 2000 spectrophotometer. XPS spectra were measured on VG Escalab MKII photoelectron spectrometer. All catalytic experiments were carried out in a 160 mL stainless steel autoclave with a glass liner. After introduction of the catalyst components, the autoclave was flushed with $H_2$ to remove oxygen, and then pressurized to the desired pressure and $CO/H_2$ ratio. The autoclave was heated to the required temperature and the solution was stirred vigorously during the reaction. At the end of the experiment, the autoclave was cooled, depressurized and organic products were analyzed by conventional GC techniques.

# 3. Results and discussion

## 3.1. Selective hydrogenation of functional olefins by $Ru_3(CO)_9(TPPMS)_3$

Some olefins with various functional groups have been reduced by using water-soluble $Ru_3(CO)_9(TPPMS)_3$. These are listed in Table 1 which indicates that acrylic acid and phenylethylene have been hydrogenated in good to excellent yields, while 2-methylacrylic acid having an methyl substituent group, only lower rate is observed. Acetophenone was rather difficult to be reduced even at 100 ℃ and hydrogen pressure of 5.0 MPa.

**Table 1  Hydrogenation of various functional olefins with $Ru_3(CO)_9(TPPMS)_3^a$**

| Substrate | Product | Temperature (℃) | Hydrogen pressure(MPa) | Yield (%) | Turnover (h$^{-1}$) |
|---|---|---|---|---|---|
| Acrylic acid | Propionic acid | 40 | 3.0 | 93.6 | 780 |
| 2-Methylacrylic acid | Isobutyric acid | 40 | 3.0 | 26.4 | 220 |
| Phenylethylene | Phenylethane | 40 | 3.0 | 85.4 | 712 |
| Acetophenone | 1-Phenylethanol | 100 | 5.0 | 7.1 | 59 |
| Benzaldehyde | Benzylalcohol | 100 | 5.0 | 38.2 | 318 |
| 2-Hydroxybenzaldehye | 2-Hydroxybenzyl alcohol | 100 | 5.0 | 45.7 | 381 |
| Cyclohexanone | Cyclohexanol | 100 | 5.0 | 28.1 | 234 |

a $Ru_3(CO)_9(TPPMS)_3$, 0.012 mmol.
Substrate:catalyst = 2500:1(molar ratio).
MeOH, 35 mL; 3 h.

## 3.2. Hydroformylation of propylene with $CO/H_2$ in water catalyzed by $Ru_3(CO)_9(TPPMS)_3$

In the attempt of searching for non-rhodium and water-soluble transition metal complexes as catalysts, several catalytic systems were investigated(Table 2). The $Fe(CO)_5/TPPMS$ system showed only negligible activity. When $Co_2-(CO)_8/TPPMS$ or $Ru_3(CO)_{12}/TPPMS$ were used, the activity improved slightly but still low. When $Ru_3(CO)_9(TPPMS)_3$ was employed as catalyst at 120 ℃ and a total gas pressure of 4.7 MPa, high activity(turnover = 61.2 $h^{-1}$) was observed. Such turnover value is a magnitude higher than the value when $[Et_4N][HRu_3(CO)_{11}]$ was used as catalyst. The main product was $n$-butyraldehyde; side products were $iso$-butyraldehyde and a small amount of 1-butyl and isobutyl alcohols due to further hydrogenation of the corresponding butanals.

**Table 2　Hydroformylation of propylene with $CO/H_2$ catalyzed by $Ru_3(CO)_9(TPPMS)_3^a$**

| Catalyst | Solvent | Temperature (℃) | Composition of products(%)[b] | | | | Turnover[c] | $n:i$[d] |
| | | | A | B | C | D | | |
|---|---|---|---|---|---|---|---|---|
| $Fe(CO)_5/(TPPMS)$ | 0.05 M $NaOH/H_2O$ | 80 | 3.2 | 45.4 | 51.4 | — | <1.0 | 0.8 |
| $Co_2(CO)_8/(TPPMS)$ | 0.05 M $Na_3PO_4/H_2O$ | 100 | 8.2 | 75.4 | 14.9 | 1.5 | 24.0 | 3.3 |
| $Ru_3(CO)_{12}/(TPPMS)$ | $H_2O$ | 100 | 5.5 | 57.2 | 35.2 | 2.1 | 17.6 | 1.5 |
| $Ru_3(CO)_9(TPPMS)_3$ | $H_2O$ | 80 | 5.5 | 73.5 | 19.5 | 1.5 | 34.4 | 3.0 |
| $Ru_3(CO)_9(TPPMS)_3$ | $H_2O$ | 100 | 2.4 | 88.1 | 4.9 | 4.6 | 148.8 | 12.7 |
| $Ru_3(CO)_9(TPPMS)_3$ | $H_2O$ | 120 | 3.0 | 89.5 | 2.9 | 4.6 | 489.6 | 15.9 |
| $Ru_3(CO)_9(TPPMS)_3$ | 0.1 M $HOAc/0.1$ M $NaOAc$ | 100 | 8.6 | 85.2 | 5.4 | 0.8 | 82.4 | 6.1 |
| $Ru_3(CO)_9(TPPMS)_3$ | 0.05 M $Na_3PO_4/H_2O$ | 100 | 6.8 | 76.9 | 15.2 | 1.1 | 32.8 | 3.5 |

[a] Catalyst, 0.05 mmol; solvent, 25 mL; $PC_3H_6/PCO/PH_2 = 0.7/2.0/2.0(MPa)$; 8 h.

[b] A = isobutyraldehyde, B = 1-butanal, C = isobutyl alcohol, D = 1-butanol.

[c] Turnover = overall products mol/cluster mol.

[d] $n/i = B+D/A+C$(molar ratios).

The effect of catalyst concentration variation is shown in Fig. 1. The increase of reaction temperature and partial pressure of the gaseous reagents enhanced the catalytic activity. The effect of temperature was remarkable; a change of temperature from 100 ℃ to 120 ℃ would double the turnover value(Fig. 2). In the cases of high activity, high $n:i$ molar ratios were also observed. Increase of reaction temperature seems to favour the formation of $n$-butyraldehyde. At 120 ℃, the highest $n:i$ ratio of 15.9 was obtained.

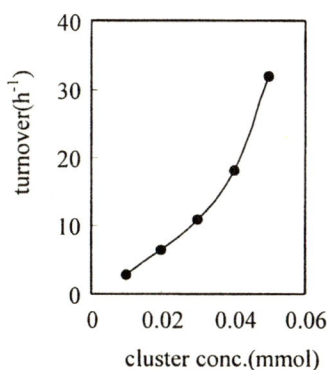

Fig. 1　Dependence of reaction rate upon the concentration of $Ru_3(CO)_9(TPPMS)_3$. Reaction conditions: $H_2O$, 25 mL; 100 ℃; partial pressure: 0.7 MPa propylene/2.0 MPa CO/2.0 MPa $H_2$.

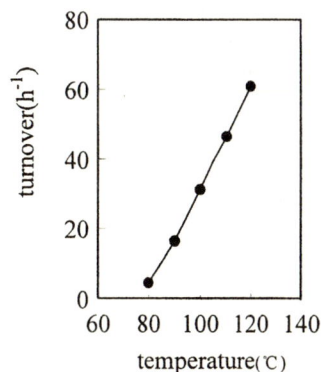

Fig. 2　Dependence of reaction rate on temperature. Reaction conditions: $Ru_3(CO)_9(TPPMS)_3$, 0.05 mmol; $H_2O$, 25 mL; others are the same as shown in Fig. 1.

The pH value of the reaction solution also affects the activity. In our earlier work in which $Ru_3(CO)_{12}$ was used as catalyst to hydroformylate ethylene, basic media were needed to form the catalytic active anion $[HRu_3(CO)_{11}]^-$ [1,2]. In the present studies, neutral water was found to be the most suitable for propylene hydroformylation.

## 3.3. Hydroformylation of ethylene with CO / H$_2$ in water catalyzed by Ru$_3$(CO)$_9$(TPPMS)$_3$

The results for the hydroformylation of ethylene with CO/H$_2$ are shown in Table 3. With Ru$_3$(CO)$_9$(TPPMS)$_3$ alone, the main product is propanal, side-products are 3-pentanone, 1-propanal and 2-methyl-pent-2-en-1-al. Interestingly, the addition of halide or alkali metal cation promoters increased the catalytic activity and led to the formation of 3-pentanone. In the case of KBr as promoter, the highest yield(52. 7%) of 3-pentanone was obtained. The organic halide promotes have little effect on the formation of 3-pentanone.

**Table 3　Hydroformylation of ethylene with CO/H$_2$ in water catalyzed by Ru$_3$(CO)$_9$(TPPMS)$_3$**

| Halide-additive | Ru$_3$(CO)$_9$(TPPMS)$_3$/ halide(molar ratios) | Composition of products(%)[b] | | | | Turnover[c] |
|---|---|---|---|---|---|---|
| | | A | B | C | D | |
| — | — | 55. 3 | 37. 0 | 6. 2 | 1. 5 | 113. 4 |
| NaI | 1∶3 | 55. 1 | 36. 0 | 6. 9 | 2. 0 | 121. 5 |
| LiBr · H$_2$O | 1∶3 | 44. 5 | 44. 6 | 8. 4 | 2. 5 | 137. 6 |
| CsCl | 1∶3 | 44. 8 | 41. 8 | 11. 7 | 1. 7 | 214. 6 |
| KBr | 1∶3 | 40. 3 | 44. 6 | 13. 8 | 13. 3 | 255. 4 |
| KBr | 1∶9 | 48. 0 | 40. 8 | 6. 4 | 4. 8 | 180. 8 |
| KBr | 1∶15 | 39. 9 | 52. 7 | 5. 7 | 1. 7 | 138. 9 |
| Bu$_4$NI | 1∶3 | 58. 7 | 29. 7 | 3. 9 | 7. 7 | 64. 3 |
| Bu$_4$NBr | 1∶3 | 72. 0 | 15. 7 | 4. 3 | 8. 0 | 61. 2 |
| I$_2$ | 1∶1 | 55. 5 | 35. 7 | 4. 4 | 4. 4 | 89. 9 |

[a] Ru$_3$(CO)$_9$(TPPMS)$_3$, 0. 05 mmol; solvent, H$_2$O, 30 mL; PC$_2$H$_4$/PCO/PH$_2$=3. 5/0. 75/0. 75(MPa); 100 ℃, 2. 5 h.

[b] A=Propanal, B=3-Pentanone, C=1-Propanol, D=2-Methyl-pent-2-en-1-al.

[c] Turnover = products mol/cluster mol.

**Table 4　Effect of alkali metal halide promoters on carbonylation of ethylene with CO[a]**

| Additive | Temperature (℃) | Composition of products(%) | | Turnover (h$^{-1}$) |
|---|---|---|---|---|
| | | Methyl propionate | 3-Pentanone | |
| — | 150 | 73. 5 | 26. 5 | 5 |
| — | 190 | 90. 8 | 9. 2 | 17 |
| LiBr | 150 | 89. 1 | 10. 9 | 36 |
| NaI | 150 | 99. 0 | 1. 0 | 50 |
| NaI | 190 | 94. 7 | 5. 3 | 315 |
| I$_2$ | 150 | 40. 6 | 59. 4 | 58 |
| Bu$_4$NI | 150 | 98. 9 | 1. 1 | 28 |

[a] Ru$_3$(CO)$_9$(TPPMS)$_3$, 0. 05 mmol; PC$_2$H$_4$/PCO=2. 0∶4. 0(MPa); CH$_3$OH, 30 mL; 5 h.

## 3.4. Carbonylation of ethylene with $CO/H_2$ in mathanol catalyzed by $Ru_3(CO)_9$ $(TPPMS)_3$

$Ru_3(CO)_9(TPPMS)_3$-halide systems have also been used to catalyzed the carbonylation of ethylene with CO in methanol. The results are giving in Table 4. With $Ru_3(CO)_9(TPPMS)_3$ alone, the catalytic activity is very low. The addition of alkali metal halide such as NaI, remarkably enhanced the activity and led to produce a main product of methyl propionate, however the addition of $I_2$ led to the selective formation of 3-pentanone.

### 3.5. Study on catalytic active species

$Ru_3(CO)_9(TPPMS)_3$ is violet-red in colour. The colour changed to yellow after the reaction. Such colour change could also be observed if the violet-red solution was treated with CO(2.0 MPa) and $H_2$ (2.0 MPa) mixture at 100 ℃ for 8 h. After the organic products were removed by distillation, the yellow solution still exhibited good catalytic activity and could be used repeatedly. Similar to $Ru_3(CO)_9(TPPMS)_3$, this yellow complex was soluble in water and methanol but insoluble in dichloromethane and diethyl ether. It is still stable in air as judged by IR spectroscopy(Figs. 3 and 4) and its catalytic performance after exposed to air for several days.

A yellowish-brown solid was obtained by drying the yellow solution. The solid was examined by X-ray photoelectron spectroscopy (Table 5). Comparing with the results of $Ru_3(CO)_{12}$ and $Ru_3(CO)_9(TPPMS)_3$ complexes, the Ru $3d_{5/2}$ peak appeared at around 281.0 eV in binding energy in all the three cases; that is roughly 1 eV higher than that of metallic ruthenium. For $Ru_3(CO)_{12}$, the C 1s and O 1s peaks were at 286.8 eV and 533.3 eV respectively, corresponding to the presence of carbonyl groups.

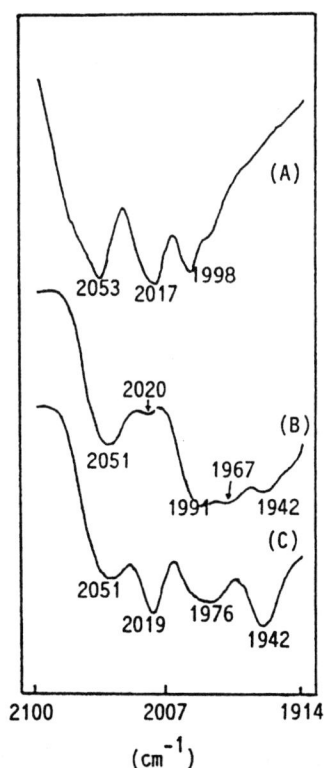

Fig. 3 IR ($v_{CO}$) of sample in KBr disk. (A) $Ru_3(CO)_{12}$; (B) $Ru_3(CO)_9(TPPMS)_3$ (before reaction); (C) $Ru_3(CO)_9(TPPMS)_3$ (after reaction).

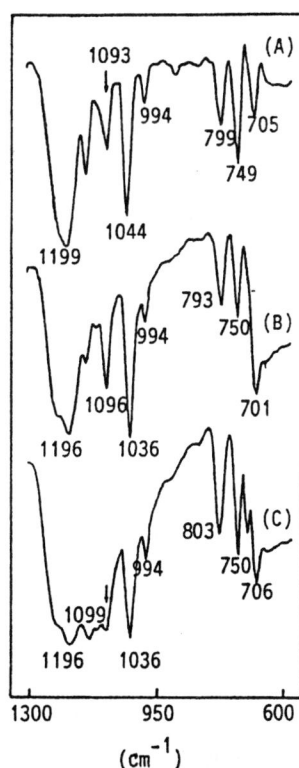

Fig. 4 IR ($v_{SO}$) of sample in KBr disk. (A) free ligand TPPMS; (B) same as Fig. 3(B); (C) same as Fig. 3(C).

**Table 5 X-ray photoelectron spectroscopic studies of the complexes**

| Complexes | Peak positions; in binding energy(eV)[a] | | | | | | In kinetic energy(eV) NaKL$_{23}$L$_{23}$ |
|---|---|---|---|---|---|---|---|
| | Ru 3d$_{5/2}$ | C 1s | O 1s | S 2p | P 2p | Na1s | |
| Ru$_3$(CO)$_{12}$ | 281.0 | 285.8 | 533.3 | — | — | — | — |
| Ru$_3$(CO)$_9$(TPPMS)$_3$ | 280.6 | 284.0 | 531.0 | 167.4 | 130.7 | 1071.5 | 990.3 |
| Yellow complex | 280.7 | 284.0 | 531.0 | 167.4 | 131.1 | 1071.6 | 990.2 |

[a]Calibrated against the Au 4f levels(833.8 eV,87.45 eV).

When Ru$_3$(CO)$_9$(TPPMS)$_3$ was formed, the Ru 3d$_{5/2}$ peak was greatly attenuated and the C 1s and O 1s peak shifted to 284.0 eV and 531.0 eV respectively, indicating the ligand TPPMS is very much on the surface with the ruthenium atoms and CO ligands buried underneath. The S 2p, P 2p and Na 1s speak positions were at 167.4 eV, 130.7 eV and 1071.5 eV respectively before and after the reaction, an implication that the ruthenium cluster remained essentially intact through out the reaction. IR studies led to similar conclusion (Figs. 3 and 4). By direct comparisons, one can see that both Ru$_3$(CO)$_9$(TPPMS)$_3$ and the yellow complex are very similar in IR pattern, especially in the IR band position within the $v_{CO}$ and $v_{SO}$ regions. The result clearly indicate the presence of the soluble ligand TPPMS in the yellow complex and the possible analogous structure between the two compounds. Whether this yellow ruthenium carbonyl complex represents the actual active species is under further investigation.

# 4. Conclusion

Water-soluble and air-stable cluster Ru$_3$(CO)$_9$(TPPMS)$_3$ was used as effective catalyst for hydrogenation of acrylic acid and phenyleth-ylene. For hydroformylation of propylene, catalytic turnover of 61.2 h$^{-1}$ and product $n : i$ ratio of 15.9 were obtained and neutral water was the most suitable for the reaction. Ru$_3$(CO)$_9$(TPPMS)$_3$ have also been used to catalyze the hydroformylation and carbonylation of ethylene. The addition of halide promoters enhanced the catalytic activity and led to the selective formation of 3-pentanone. According to the results from IR and XPS studies of catalyst before and after reaction, the triruthenium cluster remained essentially intact through out the reaction and represented the possible active species.

# Acknowledgements

Financial support from the National Nature Foundation of China and Fujian Province, State Key Laboratory for Physical Chemistry of Solid Surface in Xiamen University is gratefully acknowledged.

# References

[1]B. D. Dombek, J. Am. Chem. Soc. **103**(1981) 6508.

[2]P. C. Ford, R. G. Rindker, C. Ungermann, R. M. Laine, V. Landis, S. A. Moya, J. Am. Chem. Soc. **100**(1978) 4595.

[3]G. Süss-Fink, J. Organomet. Chem. **193**(1980) C20.

[4]G. Süss-Fink, J. Reiner, J. Mol. Catal. **16**(1982) 231.

[5] J. C. Bricker, C. C. Nagel, S. G. Shore, J. Am. Chem. Soc. **104**(1982) 1444.

[6] J. X. Gao, J. Evans, J. Catal. (Chinese) **8**(1987) 384.

[7] J. Evans, J. X. Gao, H. Leach, A. C. Street, J. Organomet. Chem. **372**(1989) 61.

[8] B. Fontal, J. Orlewski, C. C. Santini, J. M. Basset, Inorg. Chem. **25**(1986) 4320.

[9] S. Ahrland, J. Chatt, N. R. Davies, A. A. William, J. Chem. Soc. (1958) 276.

■ 本文原载：Applied Surface Science 147(1999)85～93.

# Influence of Trivalent Metalions on the Surface Structure of a Copper-Based Catalyst for Methanol Synthesis

Hong-Bo Chen [a,b], Dai-Wei Liao [a,b,c] *, La-Jia Yu [a,b], Yi-Ji Lin [b], Jun Yi [a], Hong-Bin Zhang [a,b,c], Khi-Rui Tsai [a,b,c]

[a] Institute of Physical Chemistry, Xiamen University, Xiamen 361005, China
[b] Department of Chemistry, Xiamen University, Xiamen 361005, China
[c] The State Key Laboratory of Physical Chemistry on Solid Surfaces, Xiamen 361005, China

Received 6 November 1998; accepted 21 December 1998

**Abstract** The method of doping trivalent metal ions into a copper-based catalyst for methanol synthesis is effective in modifying the surface structure of the catalyst. The promotion effect and its relation to catalytic activity for hydrogenation of CO to methanol after doping with trivalent metal ions such as $Al^{3+}$, $Sc^{3+}$, and $Cr^{3+}$ into Cu-ZnO have been investigated by XRD, ESR, XPS, TPR, and the evaluation of catalytic activity. The results show that doping trivalent metal ions into ZnO assists in the formation of monovalent cationic defects on the surface of ZnO. These monovalent cationic defects both enrich and stabilize monovalent copper on the surface of copper-based catalysts for methanol synthesis during reduction and reaction. They increase catalytic activity for methanol synthesis and extend the life of catalysts. © 1999 Elsevier Science B.V. All rights reserved.

**Key words** Trivalent metal ion  Catalyst  Methanol synthesis

## 1. Introduction

The amount and stability of monovalent copper on the surface of copper-based catalysts for methanol synthesis are important for both catalytic activity and catalyst life. It is generally accepted that the coordination chemisorption and activation of carbon monoxide and the homogeneous splitting of hydrogen takes place on $Cu°$ or $Cu^+$, and that the heterogeneous splitting of hydrogen, which provides $H^{\delta+}$ and $H^{\delta-}$ in the catalytic process, takes place on $ZnO$[1]. Therefore, the nature of the valence states of copper on the surface of catalysts for methanol synthesis is an important problem. However, there still exist many controversies. Herman et al.[2] suggested, for example, that $Cu^+$ dissolved in ZnO was the active species. The $Cu^+$ sites adsorbed and activated carbon monoxide and hydrogen. Chinchen et al.[3] proposed that the activity of copper-based catalyst was directly proportional to the total copper,

---

* Corresponding author. Institute of Physical Chemistry, Department of Chemistry, Xiamen University, Xiamen 361005, China. Tel.:+ 86-592-2183045; Fax:+ 86-592-2183043; E-mail:dwliao@xmu. edu. cn

both $Cu^{\circ}$ and $Cu^{+}$, and that the activity does not depend on special copper sites. The arguments focus on the active phase of copper for methanol synthesis. The answer relates to both the formation and stability of $Cu^{+}$ in reduction and reaction process. In this paper, we will discuss the influence of trivalent metal ion dopants in the preparation of catalysts regarding the formation and stability of $Cu^{+}$ on the surface of the catalyst.

# 2. Experiments

## 2.1. Preparation of catalysts

The catalysts $ZnO$, $ZnO$-$Al_2O_3$, $ZnO$-$Al_2O_3$-$Sc_2O_3$, and $Cu$-$ZnO$-$Al_2O_3$, $Cu$-$ZnO$-$Al_2O_3$-$Sc_2O_3$, and $Cu$-$ZnO$-$Al_2O_3$-$Cr_2O_3$, were prepared by the coprecipitation method[2]. To avoid precipitation of carbonates with different solubility products, we modified this method. First, we simultaneously added, drop by drop, the mixing solution of nitrates and $Na_2CO_3$ with the desired ratio into a small amount of deionized water, and while stirring strongly, kept the mixing solution at 85 ℃ and a pH of 6.5~7.0. Then the precipitation was filtered and washed after slaking for 1 h. Finally, the precursors were dried at 110 ℃ overnight, and calcined at 350 ℃ for 3~4 h.

## 2.2. Characterization of the catalysts

### 2.2.1. ESR spectroscopy

A powdered sample of 80 mg in 3-mm-diameter high-quality glass tubes was characterized with a Bruder ER-200D-SRC type paramagnetic resonance spectrometer. The detective range is in X-wave range. The output power is 20.4 mW and the modulation frequency is 100 kHz. The modulation amplitude is 0.8 $G_{pp}$ and the time constant is 1 ms.

### 2.2.2. XPS and X-ray-induced Auger spectroscopy

The XPS and X-ray-induced Auger measurements were carried out with a VG Escalab Mark-II type spectrometer. The Al $K_a$ line(1486.6 eV, 10.1 kV) was used as the radiation source. The binding energy of $Zn(2 P_{3/2})$(1022.2 eV) was used as a reference energy for $ZnO$, $ZnO$-$M_2O_3$ and $Cu$-$ZnO$-$M_2O_3$(M= Al, Sc, and Cr). The binding energy of carbon contamination on the surface(284.6 eV) was used as an internal reference energy for samples. The samples were reduced with mixing gas of $H_2$ and $N_2$($H_2$/$N_2$ = 5/95, v/v) before characterization.

### 2.2.3. XRD for structure analysis

XRD for structure analysis of catalyst was carried out with Rigaku Rotafles D/mas-C XRD type X-ray diffractometer, using Cu $K_a$ radiation with a tube voltage of 40 kV, a tube current of 30 mA, and scan rate of 8°rmin.

### 2.2.4. Temperature Program Reduction(TPR)

The experimental apparatus for TPR included the gas chromatograph(GC), a microreactor with 40~60 mesh catalyst of 30 mg and a temperature program controller. The temperature program reduction of catalyst was carried out with a mixing gas of $H_2$ and $N_2$($H_2$/$N_2$ = 7/93, v/v) at a rate of 3 ℃/min and the results were recorded by an X-Y recorder.

### 2.2.5. Measurement of catalytic activity for $Cu$-$ZnO$-$M_2O_3$

The catalytic activity measurements were carried out in a flow reaction system with a fixed-bed 8-mm-diameter microreactor made of stainless steel. Using a temperature program, the catalyst with 40~60 meshes of about 1 g was reduced by $H_2$/$N_2$(5/95, v/v) with a flow velocity of 30 mL/min at atmospheric pressure. The catalytic activity for methanol synthesis was evaluated with $CO/H_2$ (1/2) with a flow velocity of 30 mL/min at a reaction pressure of 1.0 MPa and a reaction temperature at 230 $\pm 0.5$ ℃.

# 3. Results and discussions

## 3.1. Effect of $M_2O_3$ on the formation of monovalent defects in $ZnO-M_2O_3$

### 3.1.1. Formation of monovalent defects after doping $Al^{3+}$ and $Sc^{3+}$ into the ZnO lattice

The ESR spectra of ZnO, $ZnO-Al_2O_3$ and $Al_2O_3$ were detected under the same conditions regarding both preparation and measurement. No ESR signal was observed for either ZnO or $Al_2O_3$, but a strong and sharp ESR signal at $g = 1.9606 \pm 0.0002$ with $\Delta H_{pp} = (2.8 \pm 0.1) \times 10^{-4}$ T was observed for $ZnO-Al_2O_3$. The relative signal intensity is proportional to $Y'_m (\Delta H_{pp})^2$ where $Y'_m$ is the amplitude of the peak and $\Delta H_{pp}$ is the distance between peaks of the first-order differential quotient. Therefore, the signal intensity is proportional to the amplitude of the peak since $\Delta H_{pp}$ is a constant. The signal intensities in spin numbers per gram were calculated according to a standard sample for ESR. The variation of ESR signal intensity vs. the atomic ratio of Al/Zn was shown in Fig. 1. As can be seen from Fig. 1, the intensity of the ESR signal at $g = 1.9606$ increases from zero to $1.67 \times 10^{16}$ as the Al/Zn radio increases from zero to 1/100. Then as the Al/Zn ratio increases from 1/100 to 2/100, the enhancement of the signal intensity is less. When the ratio of Al/Zn increased beyond 2/100 the variation of the intensity was very slow. It is obvious that the variation of the ESR signal is connected with the amount of the paramagnetic species. As the doping amount continued to increase, however, the amount

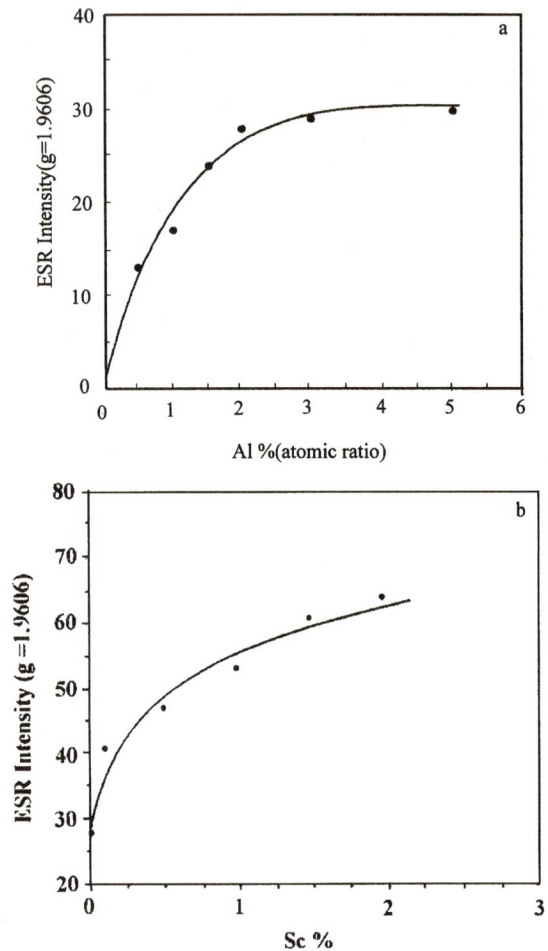

Fig. 1 The relationship between the ESR intensity and the $M^{3+}$ content in $Zn-M_2O_3$ (M=Al, Sc). (a) The changes of ESR intensities at $g = 1.9606$ as the atomic ratio of Al/Zn increases; (b) relationship of the intensities of the ESR signal at $g = 1.9606$ with the content of $Al^{3+}$ and $Sc^{3+}$ ions in $Zn-Al_2O_3-Sc_2O_3$. * % $Sc = [Sc^{3+}/Zn^{2+} + Al^{3+} + Sc^{3+}] \times 100\%$.

of paramagnetic species did not increase proportionally because the latter may be controlled by the thermodynamic equation conditions inside of the crystal.

The above ESR signals may be produced by three possible mechanisms: (1) Monovalent cationic defects were induced due to the doping of $Al_2O_3$ into ZnO. (2) $Zn^+$ ions were produced when the precursors of ZnO and $ZnCO_3$ were decomposed at high temperature. (3) The paramagnetic species of oxygen was produced inside the ZnO lattice. Our experimental results showed that the ESR signal at $g$ = 1.9606 did not come from $Zn^+$ ions and the paramagnetic species of oxygen[4]. We suggest that the ESR signal for the $ZnO-Al_2O_3$ sample is due to the doping effect of $Al_2O_3$. The $Al^{3+}$ or $Sc^{3+}$ displaced $Zn^{2+}$ in the crystal lattice of-$Zn^{2+}$-$O^{2-}$-$Zn^{2+}$-$O^{2-}$-and formed monovalent cationic defects. That means one aluminum ion and one monovalent cationic defect were located at the position of two zinc ions. The character of the ESR signal for this type of monovalent defect is a single sharp peak[5]. These defects are enriched easily on the surface[6]. Therefore, the X-ray-induced Auger spectra showed that the concentration of defects on the surface was larger than in the bulk(Fig. 2).

Calculations show that the surface of 1 g ZnO has $Zn^{2+}$ numbers of approximately $7.33 \times 10^{18}$, about 6.57% of the total $Zn^{2+}$ numbers. The intensity of the X-ray-induced Auger intensity of $Zn^{2+}$ for the $ZnO-Al_2O_3$ sample was only half that of the ZnO sample with the same total $Zn^{2+}$. In

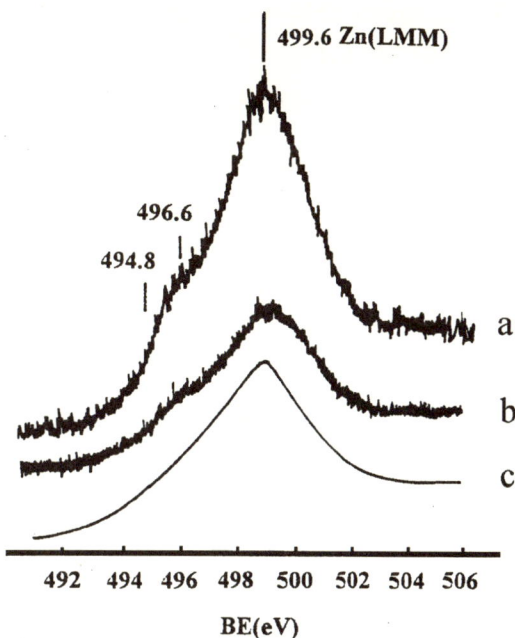

Fig. 2  X-ray-induced Auger spectra of ZnO and $Zn-Al_2O_3$. (a) ZnO; (b) $Zn-Al_2O_3$; (c) differential spectra of ZnO and $Zn-Al_2O_3$.

other words, the surface $Zn^{2+}$ is only about 3% of the total $Zn^{2+}$ as doping $Al_2O_3$ into ZnO, and the other $Zn^{2+}$ of about 3% were replaced by the monovalent cationic defects. The amount of these defects was about $2.1 \times 10^{18}$ per gram. The ESR result showed, however, that the bulk spin number per gram of the $ZnO-Al_2O_3$ sample with an atomic ratio of Al/Zn = 2/100 was $2.0 \times 10^{16}$. Therefore, the surface concentration of defects on ZnO was 100 times as large as the bulk concentration on account of the enrichment of these defects on the surface[7].

According to the principle that ions are more readily soluble in oxides having ions of similar radii, $Sc^{3+}$ ions are more readily doped into ZnO than $Al^{3+}$ ions because the ionic radius of $Sc^{3+}$, 0.073 nm, is closer to the radius of $Zn^{2+}$ ion, 0.074 nm, than that of $Al^{3+}$. Comparing the ESR results of the $ZnO-Al_2O_3-Sc_2O_3$ system(Fig. 1b) with the $ZnO-Al_2O_3$ system(Fig. 1a), we find that the influence of both $Al^{3+}$ and $Sc^{3+}$ on the formation of monovalent cationic defects in ZnO, $Sc^{3+}$ doped into the $ZnO-Al_2O_3$ sample is more effective than $Al^{3+}$ alone in inducing defects, as we predicted. Therefore, it is obvious that the formation of monovalent cationic defects on the surface of $ZnO-Al_2O_3$ is due to the doping $M_2O_3$.

### 3.1.2. The effect of $Cr_2O_3$ on the formation of defects in $ZnO-Al_2O_3-Cr_2O_3$

When a small quantity of $Cr_2O_3$ was doped into the $ZnO-Al_2O_3$ sample an effect similar to that mentioned above was observed. The ESR spectra of the $ZnO-Al_2O_3-Cr_2O_3$ sample, however, was more complicated than that of $ZnO-Al_2O_3-Sc_2O_3$. As shown in Fig. 3, the ESR signals presented an interesting change as the doping of $Cr_2O_3$ into $ZnO-Al_2O_3$ increases gradually. There was only an ESR signal at $g$ = 1.9606 when the ratio of Zn : Al : Cr was equal to 100 : 2 : 0.5. Another ESR signal at $g$ = 1.9833 was hardly observed. The area of the ESR signal at $g$ = 1.9833 increased but the signal at $g$ = 1.9606 decreased gradually as the quantity of $Cr_2O_3$ increased. When the ratio of $Zn:Al:Cr$ was equal to $100:2:3$ a shoulder peak at $g=1.9833$ only appeared. The result may imply that there are three paramagnetic species in the sample. In comparison to the ESR spectrum of $ZnO\text{-}Al_2O_3$, we suggested that the change of ESR spectra in $ZnO\text{-}Al_2O_3\text{-}Cr_2O_3$ samples was produced as a result of doping $Cr_2O_3$ into the $ZnO\text{-}Al_2O_3$.

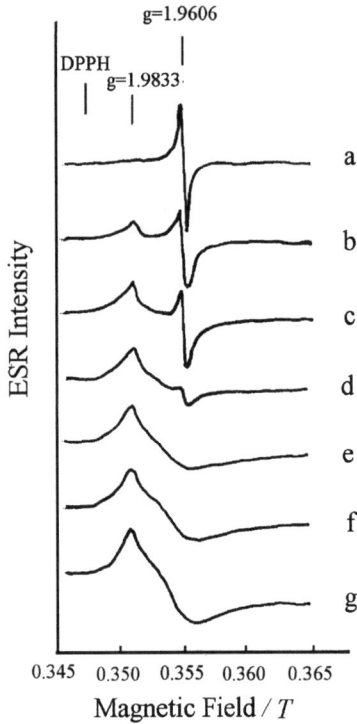

Fig. 3　ESR spectra of $Zn\text{-}Al_2O_3\text{-}Cr_2O_3$.
(a) $Zn\text{-}Al_2O_3\text{-}Cr_2O_3$ ($100:2:0.1$);
(b) $Zn\text{-}Al_2O_3\text{-}Cr_2O_3$ ($100:2:0.5$);
(c) $Zn\text{-}Al_2O_3\text{-}Cr_2O_3$ ($100:2:1.0$);
(d) $Zn\text{-}Al_2O_3\text{-}Cr_2O_3$ ($100:2:1.5$);
(e) $Zn\text{-}Al_2O_3\text{-}Cr_2O_3$ ($100:2:2.0$);
(f) $Zn\text{-}Al_2O_3\text{-}Cr_2O_3$ ($100:2:3.0$);
(g) $Zn\text{-}Al_2O_3\text{-}Cr_2O_3$ ($100:2:5.0$).

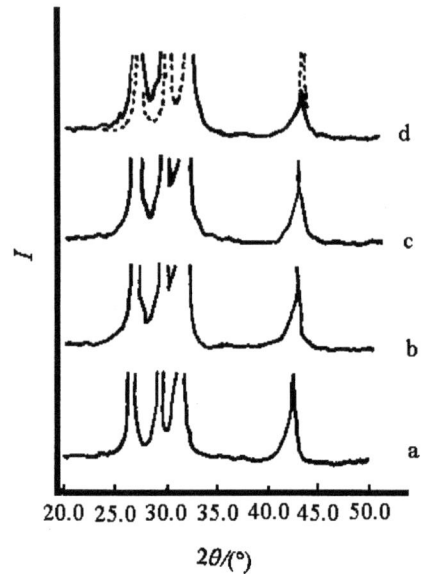

Fig. 4　XRD spectra of $Zn\text{-}Al_2O_3\text{-}Cr_2O_3$.
(a) $ZnO$; (b) $Zn\text{-}Al_2O_3\text{-}Cr_2O_3$ ($100:2:0.5$); (c) $Zn\text{-}Al_2O_3\text{-}Cr_2O_3$ ($100:2:1.0$); (d) $Zn\text{-}Al_2O_3\text{-}Cr_2O_3$ ($100:2:2.0$).

In order to study the existent forms of chromium in the $ZnO\text{-}Al_2O_3\text{-}Cr_2O_3$, surface analysis was carried out by XRD. As can be seen from Fig. 4 the XRD spectrum of $ZnO\text{-}Al_2O_3$ sample was sharp. However, the XRD peak lowered and broadened in diffraction range from $d=2.9$ to $2.4$, i.e., diffraction angles $2\theta$ from $30°$ to $40°$ as the quantity of Cr gradually increased. There are three diffraction peaks of XRD in $ZnCr_2O_4$ sample, i.e. $2.94$, $2.51$ and $1.47$[8]. The peaks at $2.94$ and $2.51$ are

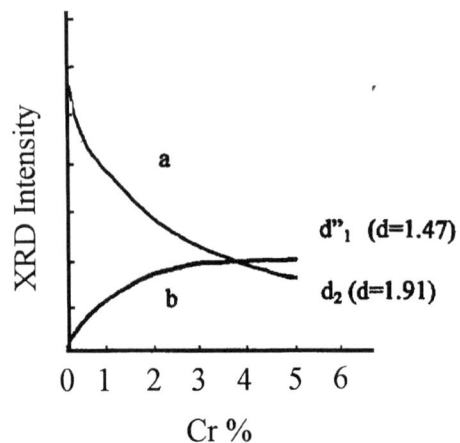

Fig. 5　The changes of intensity of XRD Peak d2 for ZnO and d1' for $ZnCr_2O_4$ with increasing Cr content in $Zn\text{-}Al_2O_3\text{-}Cr_2O_3$.

near those at 2. 82,2. 60,and 2. 47,which are the main diffraction peaks of pure ZnO. Hence,it became difficult to observe the quantity of $ZnCr_2O_4'$ (or other compounds) at the surface of the $ZnO-Al_2O_3-Cr_2O_3$ sample due to overlap of the XRD peaks. However,the XRD diffraction peak of pure ZnO at 1. 91 does not overlap that of $ZnCr_2O_4$,and can be used as a calculation standard for ZnO in the overlapped XRD peak. Curve $d_2$ in Fig. 5 indicates that the intensity of the diffraction peak of ZnO at $d=1.91$ in $ZnO-Al_2O_3-Cr_2O_3$ decreased as the quantity of chromium increased. Curve $d_1'$ in Fig. 5 shows that the intensity of the $ZnCr_2O_4$ diffraction peak at $d=1.47$,after deducting the intensity of the ZnO XRD peak in the overlapping peaks,increased with increasing chromium content in $ZnO-Al_2O_3-Cr_2O_3$.

The XRD and ESR experiments showed that the ESR spectrum at $g=1.9833$ came from the paramagnetic species of $ZnCr_2O_4$ or other compound produced by the reaction between ZnO and $Cr_2O_3$.

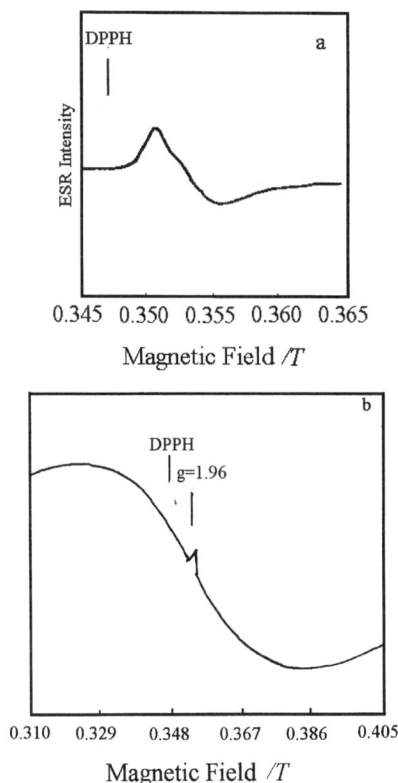

Fig. 6   ESR spectra of Zn-Al₂O₃-Cr₂O₃ (Zn ∶ Al ∶ Cr = 100 ∶ 2 ∶ 1. 5). (a) Before reduction;(b) after reduction.

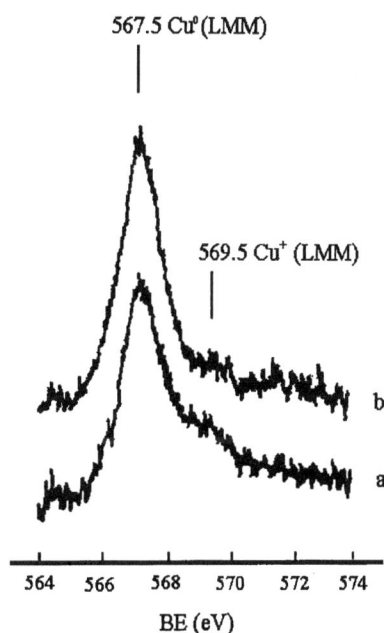

Fig. 7   The X-ray-induced Auger spectra of Cu-Zn-Al₂O₃ and Cu-Zn-Al₂O₃-Sc₂O₃ catalysts after reduction. (a) Cu-Zn-Al₂O₃(atomic ratio of Cu ∶ Zn ∶ Al= 60 ∶ 30 ∶ 10);(b) Cu-Zn-Al₂O₃-Sc₂O₃ (atomic ratio of Cu ∶ Zn ∶ Al ∶ Sc=60 ∶ 30 ∶ 8 ∶ 2).

The ESR spectra in Fig. 6 shows that the ESR peak at $g=1.9833$ disappeared after the sample of $ZnO-Al_2O_3-Cr_2O_3$(100 ∶ 2 ∶ 1. 5) was reduced by a mixing gas of $H_2/N_2$(5/95,v/v) for 7 h.,but that an ESR peak with high symmetry at $g=1.98$ appeared. The linear broadness of several hundreds,or even several thousand, gauss was introduced due to the effect of zero field[9]. Although this spectrum overlapped the ESR spectrum at $g=1.9606$,it can be identified. This indicated that the paramagnetic species can transfer into $Cr^{3+}$ and induce monovalent cationic defects as $Al^{3+}$ and $Sc^{3+}$. They are different in the induced effect of monovalent defects. The effect of $Cr^{3+}$ is larger than $Al^{3+}$.

## 4. The effects of the monovalent cationic defect on enrichment and stabilization of monocopper over the surface of catalyst

As mentioned above, the $M_2O_3$ in the $ZnO\text{-}M_2O_3$ sample caused the formation of monovalent cationic defects and the enrichment of the defects on the surface of ZnO. Recently, Chen et al.[10] identified, with XPS, TPD, and the measurement of the adsorption isotherm, that monovalent copper was a constituent of the active center for hydrogenation of CO to methanol. However, in $Cu\text{-}ZnO\text{-}M_2O_3$ catalyst, what effects on the amount of monovalent copper on the surface of the catalyst do these monovalent cationic defects have? X-ray-induced Auger spectra of $Cu\text{-}ZnO\text{-}A_2O_3$ with the atomic ratio of $Cu : Zn : Al = 60 : 30 : 10$, $Cu\text{-}ZnO\text{-}Al_2O_3\text{-}Sc_2O_3$ with the atomic ratio of $Cu : Zn : Al : Sc = 60 : 30 : 8 : 2$ and $Cu\text{-}ZnO\text{-}Al_2O_3\text{-}Cr_2O_3$ with the atomic ratio of $Cu : Zn : Al : Cr = 60 : 30 : 8 : 2$, respectively, in reduction states are shown in Figs. 7 and 8. Both the peak of $E_b = 567.5$ eV assigned to $Cu^o$(LMM) and the shoulder peak at $E_b = 569.5$ eV assigned to $Cu^+$ (LMM)[11] were observed for the above three catalysts. Obviously, there was much $Cu^o$ but little $Cu^+$ on the surfaces. Their content was dependent on the composition of catalysts. The amount of $Cu^+$ on the surface of catalysts with Sc and Cr was higher than those without Sc and Cr. As shown in Fig. 6, this indicates that the more monovalent defects present on the surface of $ZnO\text{-}M_2O_3$, the more monovalent copper on the surface of the reduced catalyst. Fig. 8 shows a similar result. This appears to form the fine environment for enriching and stabilizing the monovalent copper on the surface that results from the existence of monovalent defects.

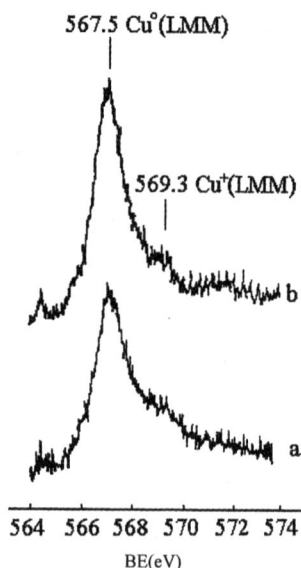

Fig. 8   The X-ray-induced Auger spectra of Cu-Zn-Al$_2$O$_3$ and Cu-Zn-Al$_2$O$_3$-Cr$_2$O$_3$ catalysts after reduction. ( a ) Cu-Zn-Al$_2$O$_3$ ( atomic ratio of Cu : Zn : Al = 60 : 30 : 10); (b) Cu-Zn-Al$_2$O$_3$-Cr$_2$O$_3$ (atomic ratio of Cu : Zn : Al : Cr = 60 : 30 : 8 : 2).

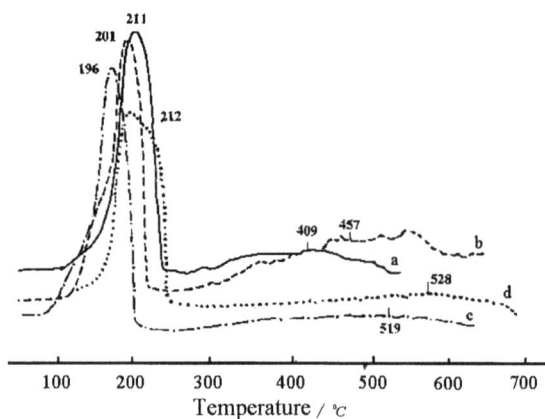

Fig. 9   TPR spectra of Cu-Zn-M$_2$O$_3$. ( a ) Cu-Zn-Al$_2$O$_3$ (atomic ratio of Cu : Zn : Al = 60 : 30 : 10); (b) Cu-Zn-Al$_2$O$_3$-Sc$_2$O$_3$ (atomic ratio of Cu : Zn : Al : Sc = 60 : 30 : 8 : 1); ( c ) Cu-Zn-Al$_2$O$_3$-Sc$_2$O$_3$ (atomic ratio of Cu : Zn : Al : Sc = 60 : 30 : 8 : 2); (d) Cu-Zn-Al$_2$O$_3$-Sc$_2$O$_3$ (atomic ratio of Cu : Zn : Al : Sc = 60 : 30 : 8 : 3).

The TPR spectra showed two distinct reduction processes[12] (Fig. 9). As the $Sc^{3+}$ content increased the maximum peak assigned to $Cu^{2+} \rightarrow Cu^+$ shifted to lower temperature, but another maximum peak assigned to $Cu^+ \rightarrow Cu°$ moved to higher temperature. These are due to the increment of dispersion of Cu on the surfaces of catalyst. The special surface area(23.6 $m^2/g$) for Cu-ZnO-$Al_2O_3$ with the atomic ratio of Cu : Zn : Al=60 : 30 : 10 increased to 40.2 $m^2/g$ for Cu-ZnO-$Al_2O_3$-$Sc_2O_3$ with the atomic ratio of Cu : Zn : Al : Sc=60 : 30 : 8 : 2 as the $Sc^{3+}$ content increased. Therefore, the reduction of $Cu^{2+}$ to $Cu^+$ became easier. However, $Cu^+$ cannot be reduced easily due to the existence of monovalent cationic defects on the surface of catalysts. The $M_2O_3$ in the copper-based catalysts stabilized $Cu^+$ on the surface in both reduction and reaction processes. The schematic diagram is as follows:

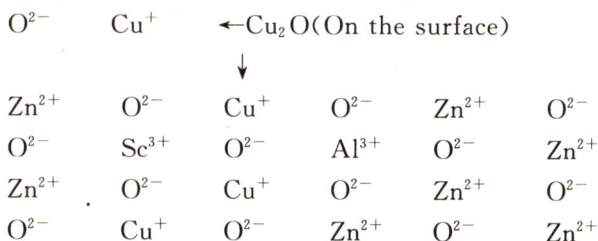

$$O^{2-} \quad Cu^+ \quad \leftarrow Cu_2O(\text{On the surface})$$
$$\downarrow$$

| | | | | | |
|---|---|---|---|---|---|
| $Zn^{2+}$ | $O^{2-}$ | $Cu^+$ | $O^{2-}$ | $Zn^{2+}$ | $O^{2-}$ |
| $O^{2-}$ | $Sc^{3+}$ | $O^{2-}$ | $Al^{3+}$ | $O^{2-}$ | $Zn^{2+}$ |
| $Zn^{2+}$ | $O^{2-}$ | $Cu^+$ | $O^{2-}$ | $Zn^{2+}$ | $O^{2-}$ |
| $O^{2-}$ | $Cu^+$ | $O^{2-}$ | $Zn^{2+}$ | $O^{2-}$ | $Zn^{2+}$ |

# 5. The effects of $M_2O_3$ on the catalytic activity of Cu-ZnO-$M_2O_3$

## 5.1. The relation of ESR intensity with catalytic activity in Cu-ZnO-$Al_2O_3$-$Sc_2O_3$

Fig. 10 shows the relation between the catalytic activity for methanol synthesis and the intensity of ESR with the content of scandium in Cu-ZnO-$Al_2O_3$-$Sc_2O_3$. As the ratio $Sc^{3+}/(Cu^{2+} + Zn^{2+} + Al^{2+} + Sc^{3+})$ increases, both the monovalent cationic defects and catalytic activity for methanol synthesis also increases. Catalytic activity is very effectively promoted by a small amount of $Sc_2O_3$.

## 5.2. The effect of $Cr_2O_3$ on the catalytic activity of Cu-ZnO-$Al_2O_3$-$Cr_2O_3$

The relation between the catalytic activity for methanol synthesis in Cu-ZnO-$Al_2O_3$-$Cr_2O_3$ and the ESR intensity of ZnO-$Al_2O_3$-$Cr_2O_3$ with chromium content is shown in Fig. 11. We can see from the figure that, as the content of Cr in sample increased, the intensity of ESR at 1.9833 increased(curve a) and that at 1.9606 decreased(curve b). Curve c in Fig. 11 shows the comprehensive results of the ESR signals at 1.9833 and 1.9606. Curve d indicates the relationship between catalytic activities for methanol synthesis and the content of Cr in samples. The change of curve c and curve d is the same. This implies that the paramagnetic species corresponding to the ESR signals at $g=1.9606$ and 1.9833 ($Cr^{3+}$ and $ZnCr_2O_4$, etc. ) related to the catalytic activity. In other words, $Cr_2O_3$ may induce the monovalent cationic defects in the lattice of ZnO by $Cr^{3+}$ (or transfer $Cr_2O_4^-$ to $Cr^{3+}$) and increase the concentration of $Cu^+$ on the surface of the catalyst. Therefore, $Cr_2O_3$ in Cu-ZnO-$Al_2O_3$-$Cr_2O_3$ catalyst is a better promotor for methanol synthesis. At a Cr content of 1% the catalyst has its highest ESR signal and catalytic activity.

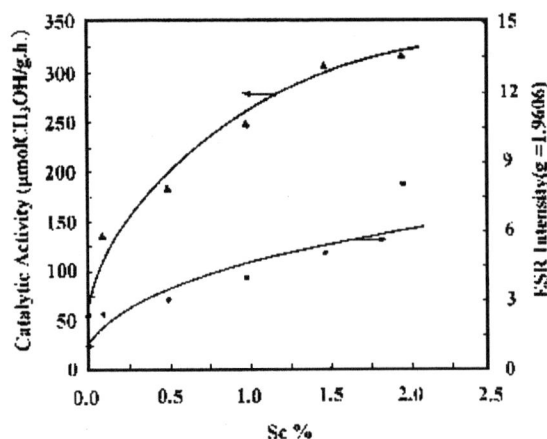

Fig. 10  The relation of both the catalytic activity for methanol synthesis and the intensity of ESR to the content of $Sc^{3+}$ in $Cu-Zn-Al_2O_3-Sc_2O_3$ catalyst. $^*$ % $Sc = [Sc^{3+}/Zn^{2+} + Al^{3+} + Sc^{3+}] \times 100\%$.

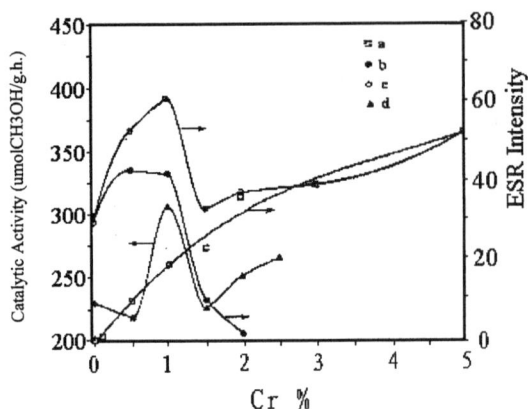

Fig. 11  The relationship of the catalytic activity for methanol synthesis and ESR intensity of $Zn-Al_2O_3-Cr_2O_3$ with the content of Cr in the samples: (a) and (b) ESR intensity of $g = 1.9833$ and $g = 1.9606$, respectively; (c) ESR intensity of(a) plus(b); (d) catalytic activity.

# 6. Conclusions

Based on the experiments with ESR, XPS, TPR, and the evaluation of catalytic activity, we can obtain the following conclusions: (1) The promotion effect is produced by trivalent metal ions. The use of $Sc^{3+}$ in particular increased the specific surface area and greatly enhanced the methanol yield. (2) The increase and decrease in the catalytic activity are totally correlated to the ESR intensity. (3) In the system with the two components $ZnO-M_2O_3$, doping of trivalent metal ions such as $Al^{3+}$, $Sc^{3+}$ and $Cr^{3+}$ enhances the formation of monovalent cationic defects on the crystal surface of ZnO. (4) In the system with three components $Cu-ZnO-M_2O_3$, due to doping of a small amount of $M_2O_3$, the monovalent cationic defects on the surface of catalysts enhance both enrichment and stabilization of $Cu^+$ on the surface in the reduction and reaction processes. Therefore, these are favorable regarding both catalytic activity and lifetime for methanol synthesis of copper-based catalysts. (5) The correlation between TPR, ESR, specific surface area, and catalytic activity is a useful tool to explore the promotion effect. The above conclusions, obviously, are helpful regarding both modification and development of catalysts for methanol synthesis.

# Acknowledgements

This work was supported by Natural Science Foundation of China.

# References

[1]J. F. Edwards, G. L. Schrader, J. Catal. **94**(1985) 175.

[2]R. G. Herman, K. Klier, G. W. Simmons, B. P. Fim, H. B. Buiko, J. Catal. **56**(1978) 407.

[3]G. C. Chinchen, P. J. Denny, J. R. Jenning, M. S. Spencer, K. Waugh, Appl. Catal. **36**(1988) 1.

［4］H. B. Chen, L. J. Yu, Y. Z. Yuan et al., Mol. Catal. (Chinese) **8**(1)(1994) 58.

［5］C. F. Ng, K. S. Leung, C. K. Chan, J. Catal. **78**(1982) 51.

［6］Kittel, C., Introduction to Solid State Physics, 2nd edn., Chap. 17, Wiley, New York, 1956, p. 477.

［7］Chen, H. B., PhD thesis, Xiamen University, China, 1986, p. 52.

［8］Powder Diffraction File, Inorganic Volume, S-22iR sets 21 to 22.

［9］M. M. Huang, Y. H. Duan, Q. H. Wang, J. Spectrosc. (Chinese) **7**(1)(1990) 98.

［10］H. B. Chen, J. X. Cai, H. B. Zhang et al., J. Xiamen University **29**(4)(1990) 411.

［11］Y. Okamoto, K. Fuklno, T. Imanaka et al., J. Phys. Chem. **87**(19)(1983) 3740.

［12］N. W. Hurst, S. J. Gentry, A. Jones, Catal. Rev. **24**(1982) 233.

■ **本文原载**：Applied Catalysis A：General 187（1999）pp. 213～224.

# Preparation, Characterization and Catalytic Hydroformylation Properties of Carbon Nanotubes-Supported Rh-Phosphine Catalyst

Yu Zhang，Hong-Bin Zhang*，Guo-Dong Lin，Ping Chen，You-Zhu Yuan，K. R. Tsai

(*Department of Chemistry & State Key Lab of Phys. Chem. for the Solid Surfaces，Xiamen University，Xiamen 361005，China*)

Received 9 April 1999; received in revised form 31 May 1999; accepted 31 May 1999

**Abstract**   Two kinds of carbon nanotubes grown catalytically, as a novel material for catalyst carrier, were prepared and characterized. Propene hydroformylation catalyzed by the Rh-phosphine complex catalysts supported by carbon nanotubes was investigated, and compared to that catalyzed by the Rh-phosphine complex catalysts supported by $SiO_2$ (a silica gel), TDX-601 (a carbon molecular sieve), AC (an active carbon), and GDX-102 (a polymer carrier). Activity assay of the catalysts showed that the carbon nanotubes-supported Rh-phosphine complex catalysts displayed not only high activity of propene conversion but also excellent regioselectivity to the product butylaldehyde. Under the reaction conditions of 393 K, 1.0 MPa, $C_3H_6/CO/H_2 = 1/1/1$ (v/v), GHSV = 9 000 mL (STP) $h^{-1}$ (g catal.)$^{-1}$, P/Rh = 9～12 (molar ratio), and Rh-loading at 0.1 mmol Rh (g carrier)$^{-1}$, the molar ratio of normal/branched ($n/i$) aldehydes reached 12～13 at a turnover frequency (TOF) of 0.12 $s^{-1}$, corresponding to propene conversion of ～32%. The characterization by using TEM, HRTEM, XRD, Raman, XPS, BET and temperature programmed desorption (TPD) methods indicated that the carbon nanofibers prepared were quite even nanotubes with the outer diameters at 15～20 nm and the inner diameters (i.e., pore diameters) at ～3 nm. Each tube wall was constructed of many layers of carbon with graphite-like platelets in a cross-section orientation of 'parallel type' or 'fishbone type'; their C (1 s) electron binding energy was about 0.5 eV lower than that of graphite. These results, together with the results of comparative studies of the Rh-phosphine complex catalysts supported by several other carriers, implied strongly that the tubular channels with the inner diameter of ～3 nm in the carbon nanostructures and its hydrophobic surface consisting of six-membered C-rings played important roles in enhancing the activity of propene hydroformylation, especially the regioselectivity of butylaldehyde on the Rh-phosphine complex catalysts supported by them. © 1999 Elsevier Science B. V. All rights reserved.

**Key words**   Carbon nanotubes grown catalytically   Carbon nanotubes carrier   Propene   Supported Rh-phosphine complex catalyst   Hydroformylation   Spatio-selective catalysis

* Corresponding author. Tel.:+86-592-2086580;fax:+86-592-2086116

E-mail address: hbzhang@xmu.edu.cn(H.-B. Zhang)

# 1 Introduction

Hydroformylation of olefins in the presence of a homogeneous catalyst to form aldehydes containing an additional carbon atom has been applied commercially for producing higher aldehydes from olefins and syngas for years[1]. However, a homogeneous catalyst system has disadvantages, such as discontinuous operation and difficulties in the separation of catalyst from the reaction mixture. In recent years, a considerable effort has been directed towards the development of immobilization of homogeneous catalysts, and special attention has been given to the supported aqueous phase catalysts (SAPCs)[2,3] and the supported liquid phase catalysts (SLPCs)[4]. These supported catalyst systems allow the separation of catalyst from the reaction mixture to be greatly simplified, thus facilitating successive operation, but are still faced with the problem of how to improve the low regioselectivity to the product aldehydes(i. e., molar ratio of normal/branched($n/i$) aldehydes).

Carbon nanotubes, one type of tubular carbon nanofibers, as a novel material for catalyst carrier, are attracting increasing attention recently[5]. Scientists are surprised by the tubular nanochannel($2\sim3$ nm for inner diameter) together with the tube wall with graphite-like structure and the surface constructed of six-membered carbon rings. Noticeable also is its large specific surface area($100\sim700$ $m^2 g^{-1}$); the hydrophobic or hydrophilic character of the surface can be controlled by chemical treatment and/or modification. All these peculiarities make carbon nanotubes a novel material for catalyst carrier, with a range of properties suitable for supporting many types and amounts of metal or active complex species.

In the present work, two kinds of carbon nanotubes grown catalytically and the Rh-phosphine complex catalysts supported by them were prepared and characterized. Their property of catalytic propene hydroformylation was investigated and compared to those of the Rh-phosphine complex catalysts supported by $SiO_2$(a silica gel), TDX-601(a carbon molecular sieve), AC(an active carbon), and GDX-102(a copolymer of styrene with divinylbenzene). The results have potential significance for better understanding of the peculiarities of the carbon nanotubes as a novel material for catalyst carrier and for the development of new supported catalyst systems based on the carbon nanotubes.

# 2 Experimental

## 2.1 Preparation of the carbon nanotubes grown catalytically

A Ni-MgO catalyst used for catalytic growth of the carbon nanotubes was prepared by our previously reported method[6]; i. e., 2.91 g of Ni($NO_3$)$_2$ · $6H_2O$(purity in AR grade) and 2.56 g of Mg ($NO_3$)$_2$ · $6H_2O$(in AR grade) powders were mixed thoroughly, followed by addition of 2 g of citric acid (in AR grade) and 20 mL of deionized water to form a solution; subsequently, the solution was evaporated by evacuation from room temperature to 353 K in a programmed way. The solid material obtained was dried at 373 K in air, and subsequently, calcined at 973 K in air for 5 h, and finally, a black and fluffy sample of the catalyst precursor was obtained. The result of XRD measurements indicated that the NiO and MgO components formed a $Ni_xMg_{1-x}O$ solid solution in the catalyst precursor after calcination.

The growth of carbon nanotubes was performed according to our method described previously[6] by catalytic decomposition of $CH_4$ or CO on the Ni-MgO catalyst. The catalyst precursor of 40 mg was reduced in a flow of purified hydrogen at 973 K for 30 min, followed by introducing the feedgas of $CH_4$ or CO at 873K, with a flow rate of 2 400 mL h$^{-1}$ for $CH_4$ and 1 200 mL h$^{-1}$ for CO. After 60 min of the reaction operation, about 150 and 90 mg of the raw products were obtained from $CH_4$ and CO, respectively. The raw products were further purified by means of immersion in a certain concentration ($\sim$4 M) of nitric acid solution so as to dissolve the catalyst particles (containing metal $Ni^0$, NiO, and MgO) attached at the extremities of the nanotubes, followed by washing with deionized water, then drying at 473 K in a flow of nitrogen, and finally, hydrogen-treating at 673 K for 1 h. Thus, the carbon nanotubes products with hydrophobic surface were obtained. From the TEM observation, it could be roughly estimated that more than 90% of the carbon deposit products were in the form of nanotubes.

## 2.2 Preparation of supported Rh-phosphine complex catalysts

The complex $HRh(CO)(PPh_3)_3$ was prepared by the known method[7], with the observed infrared absorption bands at 2010 cm$^{-1}$ for Rh-H stretching and 1920 cm$^{-1}$ for C-O stretching of carbonyl, and with $^{31}P(^1H)$-NMR double peaks at 43$\sim$41 ppm, which were in good agreement with values reported earlier[8]. $^{31}P(^1H)$-NMR measurement showed that the purity of the Rh-phosphine complex synthesized reached above 98%.

The supported Rh-phosphine complex catalysts were prepared by an incipient wetness technique. A solution containing desired amounts of the $HRh(CO)(PPh_3)_3$ complex in benzene was impregnated onto the purified and hydrogen-treated carbon nanotubes carrier; after 30 min, the solvent benzene was evaporated by evacuation at room temperature, and if necessary, a solution containing additional amounts of $PPh_3$ in benzene was impregnated again onto the above sample, followed by another 30 min wait, and then evacuation at room temperature to remove the solvent benzene from the solid material. The sample of catalyst precursor obtained was further dried and preserved in an atmosphere of purified nitrogen.

With $SiO_2$ (80$\sim$100 mesh, product from the Ocean-chemical Plant of Qingdao), or TDX-601(80$\sim$100 mesh, produced by the Institute of Inorg. Chem. of Shanghai), or AC(80$\sim$100 mesh, supplied by the Active Carbon Plant of Shanghai), or GDX-102(40$\sim$60 mesh, produced by the Chem. Reagent Plant of Tianjin), to replace the carbon nanotubes, the Rh-phosphine complex catalysts supported by each material were prepared, following the same procedure as that described above.

## 2.3 Evaluation of catalyst activity for propene hydroformylation

The evaluation of the catalyst activity was performed in a fixed-bed continuous flow reactor-GC combination system, operating under a pressure of 1.0 MPa. The hydroformylation reaction of propene over the catalysts was carried out at a stationary state and under the following reaction conditions: 393 K, 1.0 MPa, feedgas composition $C_3H_6/CO/H_2 = 1/1/1(v/v)$, GHSV = 9 000 mL(STP)h$^{-1}$(g catal.)$^{-1}$. A catalyst sample of 100 mg was used each time for testing. The reactants and products were analyzed by an on-line SQ-206 Model gas chromatograph (GC, made by Beijing Analytic Instruments Co.), equipped with a thermal conductivity detector and a 6 m long polyethyleneglycol(PEG)/#102 white support column, with hydrogen as carrier gas. All data were taken at 2 h after the start of the reaction, unless otherwise noted.

## 2.4 Spectroscopic measurements

TEM and HRTEM observations were performed by using a JEOL JEM-100CX transmission electron microscope and a Hitachi H-9000 machine, respectively. BET surface area, pore volume and pore diameter of the carriers were determined by using a Sorptomatic-1900 Surface Area Analyzer(Carlo

Erba Instruments, Italy). X-ray diffraction measurements were carried out by using a Rigaku D/Max-C X-ray Diffractometer with Cu K$\alpha$ radiation at a scanning rate of 8° min$^{-1}$. Laser Raman spectra were taken by using a highly sensitive confocal microprobe Raman system (LabRam I from Dilor, France) with an air-cooled 1024×256 pixels CCD (Wright, England). The 514.5 nm line from a Coherent-Innova Model 200 argon ion laser was used for excitation. $^{31}$P($^{1}$H)-NMR spectra of solution and supported liquid-film were taken by using a Varian Unity +500 NMR spectrometer at room temperature and 200 MHz, calibrated externally by 85% phosphoric acid in aqueous solution. Temperature-programmed desorption (TPD) of ammonia and benzene on the carbon nanotubes and SiO$_2$ was conducted on a fixed-bed continuous-flow reactor-GC combination system. Purified gaseous helium was used as carrier gas; the rate of elevation of temperature was 10 K min$^{-1}$; 50 mg of adsorbent sample was used each time for testing. NH$_3$-pre-adsorption was carried out by treating the adsorbent samples by passing a stream of purified gaseous ammonia with the flow rate of 30 m min$^{-1}$ at room temperature (RT) for 10 min; and pre-adsorption of benzene was performed by treating the adsorbent samples by passing a stream of purified benzene-saturated helium at RT and the flow rate of 10 mL min$^{-1}$ for 10 min, followed by elevating temperature at a rate of 10 K min$^{-1}$ from RT to 373 K in a stream of purified helium as carrier gas at the flow rate of 30 mL min$^{-1}$ and keeping at 373 K in the stream of helium for 30 min. This would remove the liquid benzene formed due to coagulation condensation in the micropores of the adsorbents. Then the temperature was lowered to RT for TPD measurement.

## 3　Results and discussion

### 3.1　Catalytic propene hydroformylation behavior of the supported Rh-phosphine complex catalysts

Activities of propene hydroformylation over the supported Rh-phosphine complex catalysts based on the carbon nanotubes and the other four carriers selected were evaluated; parts of the results are shown in Tables 1 and 2. The Rh-phosphine complex catalysts supported by the two kinds of carbon nanotubes displayed not only high activity of propene conversion but also excellent regioselectivity (represented by $n/i$, a molar ratio of $n$-butylaldehyde to its isomer, $i$-butylaldehyde, in the products), in comparison with those for the Rh-phosphine complex catalysts supported by the other carriers, such as SiO$_2$, TDX-601, AC, and GDX-102. Under the reaction conditions of 393 K, 1.0 MPa, C$_3$H$_6$/CO/H$_2$=1/1/1(v/v), GHSV = 9 000 mL(STP) h$^{-1}$ (g catal.)$^{-1}$ and the molar ratio of P/Rh at 12, the average turnover frequencies (TOFs) of propene at each Rh site were 0.14 and 0.19 s$^{-1}$, respectively, over the two carbon nanotubes-supported Rh-phosphine complex catalysts with the Rh-loading at 0.05 mmol Rh (g carrier)$^{-1}$. Fig. 1(a) and(b) illustrate the results of the stability test of the two carbon nanotubes-supported Rh-phosphine complex catalysts with the Rh-loading at 0.1 mmol Rh(g carrier)$^{-1}$ and P/Rh =9(molar ratio) during 30 h of hydroformylation operation; under the reaction conditions mentioned above, the conversion of propene reached ~30% and ~36%(the corresponding TOF at 0.11 and 0.14 s$^{-1}$, respectively) at the initial stage of the reaction, and subsequently descended slowly and came down to 19% and 24%(the corresponding TOF at 0.07 and 0.09 s$^{-1}$, respectively) after 30 h of the reaction operation; however, the descent tendencies in the regioselectivity were much more gentle, with the corresponding $n/i$ ratios still maintained at a level of 12~10.

**Table 1　Results of the activity assays of the Rh-catalysts supported by the carbon nanotubes for hydroformylation of propene[a]**

| Carrier | Rh-loading (mmol Rh(g carrier$^{-1}$)) | P/Rh (mol/mol) | Conversion of $C_3H_6$ (%) | TOF ($s^{-1}$) | STY (mol h$^{-1}$(g Rh$^{-1}$)) | Selectivity ($n/i$) |
|---|---|---|---|---|---|---|
| Carbon nanotubes (derived from $CH_4$) | 0.025 | 12 | 10.7 | 0.16 | 5.57 | 10.5 |
| | 0.05 | 6 | 17.0 | 0.13 | 4.42 | 6.2 |
| | | 9 | 20.8 | 0.15 | 5.41 | 10.5 |
| | | 12 | 18.8 | 0.14 | 4.91 | 11.4 |
| | | 15 | 16.8 | 0.12 | 4.37 | 12.0 |
| | 0.1 | 6 | 26.1 | 0.10 | 3.40 | 9.0 |
| | | 9 | 30.8 | 0.12 | 4.01 | 12.0 |
| | | 12 | 30.6 | 0.12 | 3.98 | 13.5 |
| | | 15 | 28.9 | 0.11 | 3.76 | 13.5 |
| Ends-unopened carbon nanotubes(derived from $CH_4$)[b] | 0.1 | 6 | 17.2 | 0.06 | 2.10 | 6.0 |
| Carbon nanotubes (derived from CO) | 0.025 | 12 | 16.1 | 0.24 | 8.38 | 6.9 |
| | 0.05 | 6 | 21.6 | 0.16 | 5.62 | 5.1 |
| | | 9 | 24.8 | 0.18 | 6.45 | 9.0 |
| | | 12 | 25.4 | 0.19 | 6.61 | 11.2 |
| | | 15 | 22.4 | 0.17 | 5.83 | 12.4 |
| | 0.1 | 6 | 31.0 | 0.12 | 4.03 | 7.2 |
| | | 9 | 34.1 | 0.13 | 4.44 | 10.3 |
| | | 12 | 32.5 | 0.12 | 4.23 | 11.2 |
| | | 15 | 32.0 | 0.12 | 4.16 | 13.7 |

[a] Reaction conditions:393 K,1.0 MPa,feedgas composition $C_3H_6/CO/H_2 = 1/1/1$(v/v),GHSV$=9\,000$ mL(STP)h$^{-1}$(g catal.)$^{-1}$;the by-products formed from side reactions of hydrogenation(i. e.,propane,$n$-butanol and $i$-butanol) were below 1%(molar fraction) of the total carbon-based products.

[b] The catalyst particles attached at the extremities of those nanotubes were not removed due to their being free from an acid-immersion treatment;thus,their ends were not opened.

**Table 2　Results of the activity assays of the Rh-catalysts supported by the other four carriers for hydroformylation of propene[a]**

| Carrier | Rh-loading (mmol Rh(g carrier$^{-1}$)) | P/Rh (molar ratio) | Conversion of $C_3H_6$ (%) | TOF ($s^{-1}$) | STY (mol h$^{-1}$(g Rh$^{-1}$)) | Selectivity ($n/i$) |
|---|---|---|---|---|---|---|
| $SiO_2$ | 0.05 | 6 | 19.3 | 0.14 | 5.02 | 6.4 |
| | | 9 | 23.9 | 0.18 | 6.22 | 7.3 |
| | | 12 | 22.5 | 0.17 | 5.85 | 7.5 |

续表

| Carrier | Rh-loading (mmol Rh(g carrier$^{-1}$)) | P/Rh (molar ratio) | Conversion of $C_3H_6$ (%) | TOF ($s^{-1}$) | STY (mol h$^{-1}$(g Rh$^{-1}$)) | Selectivity ($n/i$) |
|---|---|---|---|---|---|---|
| | | 15 | 19.6 | 0.15 | 5.10 | 9.1 |
| | 0.1 | 6 | 30.3 | 0.11 | 3.94 | 7.0 |
| | | 9 | 32.8 | 0.12 | 4.27 | 7.9 |
| | | 12 | 31.7 | 0.12 | 4.12 | 8.2 |
| | | 15 | 30.5 | 0.11 | 3.97 | 9.9 |
| | 0.15 | 6 | 28.5 | 0.07 | 2.47 | 9.1 |
| TDX-601(carbon molecular sieve) | 0.05 | 6 | 11.3 | 0.08 | 2.94 | 15.3 |
| | | 9 | 11.4 | 0.08 | 2.97 | 19.4 |
| | | 12 | 10.2 | 0.08 | 2.65 | 21.2 |
| | | 15 | 10.4 | 0.08 | 2.71 | 23.8 |
| | 0.1 | 6 | 9.1 | 0.034 | 1.18 | 15.3 |
| | 0.15 | 6 | 10.9 | 0.027 | 0.95 | 17.8 |
| AC(active carbon) | 0.05 | 6 | 2.9 | 0.02 | 0.76 | 2.8 |
| | 0.1 | 6 | 5.3 | 0.02 | 0.69 | 2.5 |
| GDX-102$^b$ | 0.05 | 6 | 22.3 | 0.17 | 5.80 | 3.4 |
| | 0.1 | 6 | 38.5 | 0.14 | 5.01 | 5.6 |

[a] Reaction conditions: 393 K, 1.0 MPa, feedgas composition $C_3H_6/CO/H_2=1/1/1$(v/v), GHSV=9 000 mL(STP)h$^{-1}$ (g catal.)$^{-1}$; the by-products formed from side reactions of hydrogenation (i.e., propane, $n$-butanol and $i$-butanol) were below 1%(molar fraction) of the total carbon-based products.

[b] A carrier of co-polymer of styrene with divinylbenzene.

Over the Rh-phosphine complex catalysts supported by the other two carriers with large surface area and large pore diameters, $SiO_2$ and GDX-102, the conversions of propene were close to those of the carbon nanotube-supported catalysts, but their regioselectivities were relatively low, even in the case of high P/Rh molar ratios. It is worth noting that the TDX-601-supported system displayed the highest regioselectivity among these catalyst systems investigated in the present work, but the corresponding conversion activity of propene reached only about a half of those for the carbon nanotubes-based systems. As for the AC-supported catalyst system, its activity and regioselectivity were both extraordinarily low.

It should be mentioned that the same deactivation phenomenon was also observed on other Rh-phosphine complex catalysts in liquid phase or supported aqueous phase or supported liquid phase. The $SiO_2$-supported HRh(CO)(PPh$_3$)$_3$ catalyst also displayed the deactivation characteristic analogous to that of the carbon nanotubes-supported systems. The stability test of the $SiO_2$-supported catalyst for 30 h of reaction under comparable conditions(i.e., 1.0 MPa, $C_3H_6/CO/H_2=1/1/1$(v/v), GHSV=9 000 mL(STP) h$^{-1}$(g catal.)$^{-1}$, Rh-loading at 0.1 mmol(g carrier)$^{-1}$ and P/Rh=9(molar ratio)) showed that propene conversion descended slowly from the initial 32.8%(corresponding to a TOF of 0.12 s$^{-1}$) down to 20.3%(the TOF at 0.07 s$^{-1}$) after 30 h of reaction; correspondingly, the $n/i$ ratio descended from the initial 7.9 down to 6.4 after 30 h of reaction. As is generally accepted now, the loss of activity with time may be due to degradation reactions of the ligand[1] and/or to the oxidation of the PPh$_3$-ligand to triphenylphosphine oxide(OPPh$_3$)[9].

In order to get better information about the location of the supported Rh-phosphine complex species on the carbon nanotubes carrier,an ends-unopened carbon nanotube sample ( i. e. , the catalyst particles attached at the extremities of those nanotubes were not removed due to being free from an acid-immersion treatment; thus their ends were not opened) was used as the carrier. The Rh-phosphine complex species were not able to get into the inner channel of the carbon nanotubes in the catalyst supported by it,and were only dispersed on the outer surface. Under the comparable conditions described in Table 1,propene conversion and the $n/i$ ratio on the ends-unopened carbon nanotubes-supported catalyst reached only 17. 2%(corresponding to a TOF of 0. 06 s$^{-1}$) and 6. 0,respectively,which were obviously lower than those of the catalyst supported by the ends-opened carbon nanotubes (i. e. ,26. 1% of propene conversion and 9. 0 of the $n/i$ ratio). This result provides an important indication that the high propene conversion and the excellent regioselectivity ($n/i$ ratio) on the ends-opened carbon nanotubes-supported catalysts were mainly due to the contributions

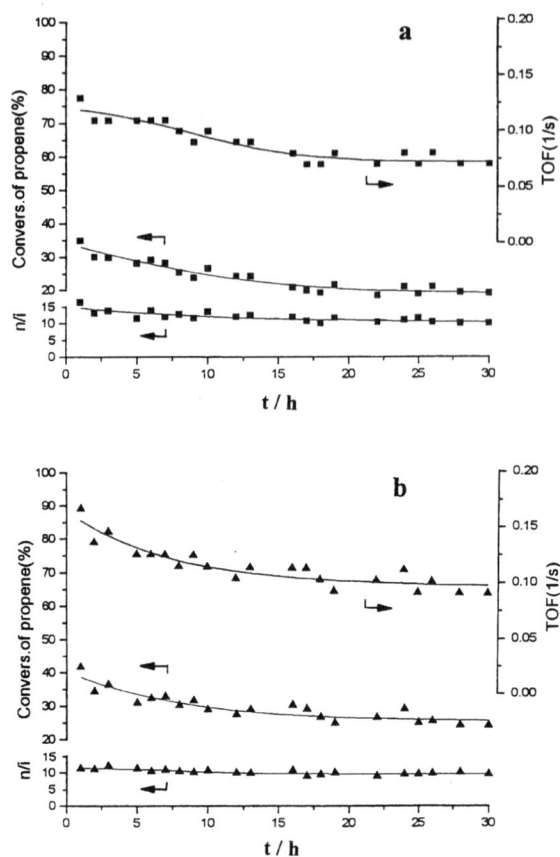

Fig. 1  Stability testing for 30 h of the supported Rh-catalysts based on the carbon nanotubes produced from CH$_4$ ( a ) and CO ( b ). Reaction conditions: 393 K, 1. 0 MPa, C$_3$H$_6$/CO/H$_2$ = 1/1/1 ( v/v ) , GHSV = 9000 ml ( STP ) h$^{-1}$ ( g catal. )$^{-1}$, Rh-loading at 0. 1 mmol Rh ( g carrier )$^{-1}$, and P/Rh =9( molar ratio ).

made by the catalytically active Rh-phosphine complex species located on the inner surface of the tubular nanochannel.

The molar ratio P/Rh has a pronounced effect on the catalytic hydroformylation behavior of the supported Rh-phosphine complex catalysts. It is shown by the data in Tables 1 and 2 that the TOF of propene and the molar ratio of normal/branched aldehydes ($n/i$) both increased initially as addition of PPh$_3$ was increased,and tended towards a stable level with the P/Rh molar ratio reaching 12～15. It seems that a P/Rh molar ratio of 9～12 would be appropriate for the Rh-phosphine complex catalysts supported by the two kinds of carbon nanotubes.

The results of the investigation about effects of Rh-loading on the hydroformylation performance of the carbon nanotubes-supported Rh-phosphine complex catalysts are also shown in Table 1. The experimental results showed that the TOF of propene and the selectivity to $n$-butylaldehyde($n/i$) both tended towards stable levels at a Rh-loading of 0. 1 mmol Rh(g carrier)$^{-1}$. It seems that the Rh-loading of ～0. 1 mmol Rh(g carrier)$^{-1}$ was proper for the two kinds of carbon nanotube carriers.

Table 3 shows the effect of temperature on the hydroformylation catalyzed by the Rh-phosphine complex catalysts supported by the carbon nanotubes. The TOF of propene and the STY of butylaldehyde were both enhanced, but the $n/i$ ( molar ratio ) descended somewhat, with elevating reaction temperature. The experimental results indicated that the by-products formed from side reactions of hydrogenation(i. e., propane, n-butanol and i-butanol) were below 1% (molar fraction) of the total carbon-based products for reaction temperatures not over 393 K. It was also found that a reaction temperature higher than that which is enough(e. g., above 403 K) would easily bring about an increase in the by-products of hydrogenation and condensation, and easily lead to the stability of catalytically active complex species descending, thus speeding up deactivation of the catalysts, which is consistent with the report by Pelt et al. [9].

**Table 3　Effect of temperature on propene hydroformylation over the carbon nanotubes-supported Rh-catalysts[a]**

| Carrier | Temperature (K) | Conversion of $C_3H_6$ (%) | TOF ($s^{-1}$) | STY(mol $C_3H_7CHO$ $h^{-1}$(g $Rh^{-1}$)) | Selectivity ($n/i$) |
|---|---|---|---|---|---|
| Carbon nanotubes derived from $CH_4$ | 383 | 24.6 | 0.09 | 3.20 | 14.1 |
| | 393 | 30.6 | 0.12 | 3.98 | 13.5 |
| | 403 | 39.1 | 0.15 | 5.09 | 13.5 |
| Carbon nanotubes derived from CO | 383 | 27.7 | 0.10 | 3.60 | 13.9 |
| | 393 | 32.5 | 0.12 | 4.23 | 11.2 |
| | 403 | 39.3 | 0.15 | 5.11 | 11.2 |

[a] Reaction conditions: 1.0 MPa, feedgas composition $C_3H_6/CO/H_2 = 1/1/1$(v/v), GHSV = 9 000 mL(STP)$h^{-1}$(g catal. )$^{-1}$, Rh-loading at 0.1 mmol(g carrier)$^{-1}$, and P/Rh = 12(molar ratio).

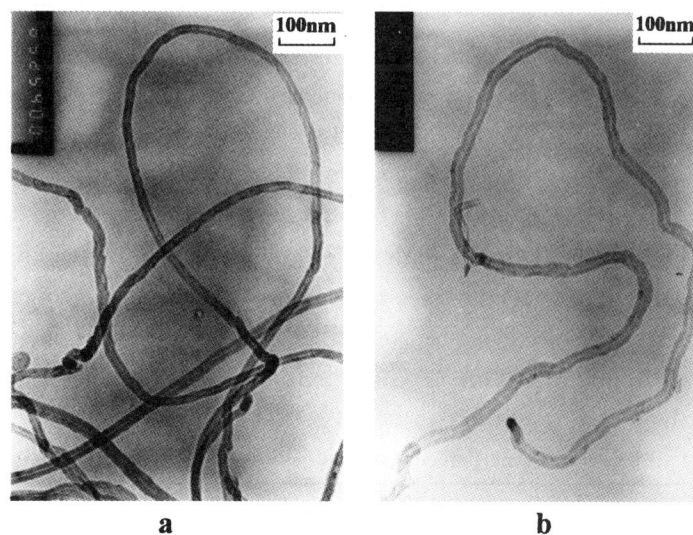

**Fig. 2　TEM images of the carbon nanotubes prepared by catalytic decomposition of CO(a) and $CH_4$(b).**

## 3.2　Characterization of the carbon nanotubes carriers

It is evident that the exceedingly good performance that the carbon nanotubes-supported Rh-phosphine complex catalysts displayed in catalytic hydroformylation is closely related to the peculiar structure and properties of the carbon nanotubes. Fig. 2 gives the TEM images of the raw products of the two kinds of carbon nanotubes prepared from the decomposition of methane and the disproportionation of carbon monoxide, respectively, on the Ni-MgO catalyst. These carbon nanotubes are small and even in diameter size, and their morphology is more or less twisted. The catalyst particles attached at the extremities of the nanotubes can be eliminated by dissolution by the nitric acid solution, leading to the end of the tubes to be open. From the TEM images, it could be roughly estimated that their outer diameters were 15~20 nm and the inner diameters were ca. 3 nm, and the tube lengths were probably as long as 10 $\mu$m. In a general way, the carbon nanotubes derived from $CH_4$ have thicker trunks than those derived from CO. The HRTEM observation further reveals that the wall of the nanotube is constructed by many layers of carbon with graphite-like platelets in an extremely ordered arrangement: the orientation of the cylindrical graphite-like platelets is parallel to the tube axis (so-called 'parallel type') for the nanotube produced from CO and the conical graphite-like platelets are inclined to the central axis (so-called 'fishbone type'[10]) for that derived from $CH_4$, respectively, as shown in Fig. 3.

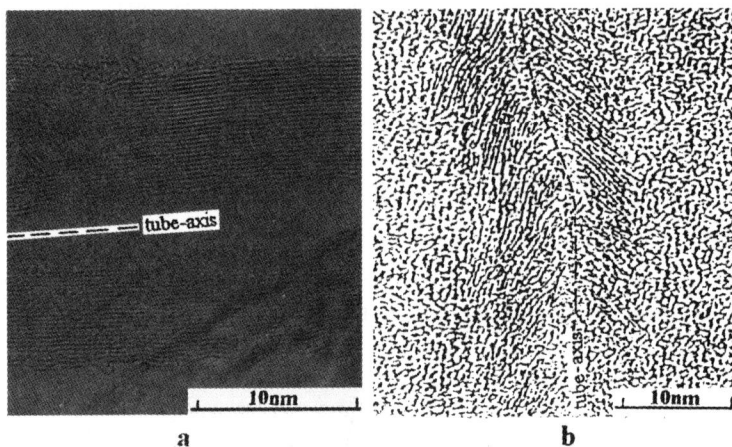

**Fig. 3　HRTEM images of one of the carbon nanotubes produced from catalytic decomposition of CO(a) and $CH_4$(b).**

The surface area, pore volume, and pore diameters of the carbon nanotubes and the other four carriers were measured by the BET method; the results are shown in Table 4. The surface area of the carbon nanotubes produced from CO was as high as ~237 $m^2$ $g^{-1}$, and higher than that (~155 $m^2$ $g^{-1}$) of those produced from $CH_4$. However, the sizes of the pore diameters (i. e., inner diameter) of the two kinds of carbon nanotubes were quite close to each other: 2.4~3.2 nm for those derived from CO and 3.2~3.6 nm for those from $CH_4$. This is consistent with the results estimated from the TEM observation mentioned above.

**Table 4　BET surface area, pore volume, and pore diameter of the various carriers**

| Carrier | BET surface area($m^2 g^{-1}$) | Pore volume($cm^3 g^{-1}$) | Pore diameters(nm) |
|---|---|---|---|
| Carbon nanotubes(derived from $CH_4$) | 155 | 0.46 | 3.2—3.6 |
| Carbon nanotubes(derived from CO) | 237 | 1.33 | 2.4—3.2 |
| $SiO_2$ | 380 | 1.03 | 6—12 |
| TDX-601(carbon molecular sieve) | 1210 | 0.43 | 1.6—2.4 |
| AC(active carbon) | 592 | 0.21 | 1.4—2.0 |
| GDX-102[a] | 501 | 1.66 | 20—100 |

[a] A carrier of co-polymer of styrene with divinylbenzene.

Fig. 4 shows the XRD patterns of the carbon nanotubes and graphite. The results show that the main XRD feature of the carbon nanotubes at $2\theta = 26.2°$ is close to that of graphite at $2\theta = 26.6°$, but somewhat broadened, indicating that the degree of long-range order of these nanostructures is relatively low in comparison with that of graphite. In addition, the other two weak peaks at $2\theta = 21.0°$ and $2\theta = 23.9°$ are also observed; these are ascribed to chaoite[11]. The Raman spectra shown in Fig. 5 indicate that the Raman spectral features of the carbon nanotubes are rather different from those of graphite, but much closer to those of a low-order carbon, indicating that the long-range-order degree of the arrangement of carbon atoms on the surface of the carbon nanotubes is not as high as that of graphite. The results of the XPS measurement shown in Fig. 6 reveal that the binding energy of C(1 s) electron in these carbon nanotubes was $283.8 \sim 284.0$ eV, and $0.6 \sim 0.4$ eV lower than that (284.4 eV) of graphite, implying that the valence electrons on the carbon atom, especially on C-rings, in the carbon nanotubes could escape somewhat more easily or become delocalized in comparison with those in graphite.

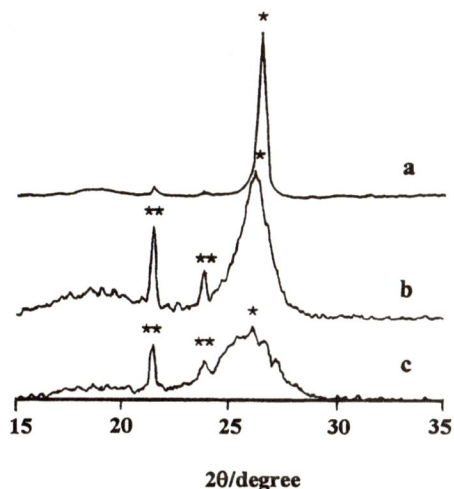

Fig. 4    XRD patterns of graphite (a), and the carbon nanotubes produced from catalytic decomposition of CH$_4$ (b) and CO(c).
* peaks due to graphite phase; ** peaks due to chaoite phase.

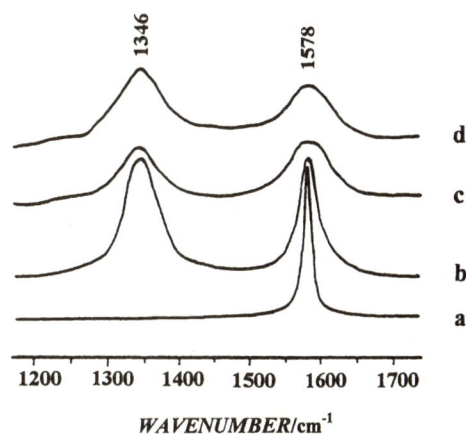

Fig. 5    Laser Raman spectra of graphite (a), a low-ordering carbon (b), the carbon nanotubes produced from catalytic decomposition of CO(c) and CH$_4$ (d).

The comparative investigation of adsorption of NH$_3$ and C$_6$H$_6$ on the carbon nanotubes and SiO$_2$ may be expected to provide useful information about the nature of the surface of the carbon nanotubes. The TPD spectra of NH$_3$ and C$_6$H$_6$ shown in Fig. 7 revealed that an obvious difference in the adsorption/desorption behavior towards NH$_3$ and C$_6$H$_6$ existed between the carbon nanotubes and SiO$_2$. For ammonia adsorption on SiO$_2$, a strong NH$_3$-TPD peak at 364 K(Fig. 7Ic) was observed; while, on the carbon nanotubes, almost all NH$_3$-TPD peaks were quite weak and ambiguous beyond recognition (Fig. 7Ia and b), indicating that the interaction of the surface of the carbon nanotubes with basic and/or hydrophilic molecules such as NH$_3$ was weak. But for benzene adsorption and desorption on these carriers, the situation was completely different: strong C$_6$H$_6$-TPD peaks at 431 and 410 K were observed on the two kinds of carbon nanotubes, respectively(Fig. 7IIa and b); while there was not any C$_6$H$_6$-TPD signal detected on the silica-gel(Fig. 7IIc). These results demonstrate distinctly that the surface of the carbon nanotubes is indeed markedly hydrophobic; it is worth noting that their interaction with an aromatic compound molecule such as benzene is strong.

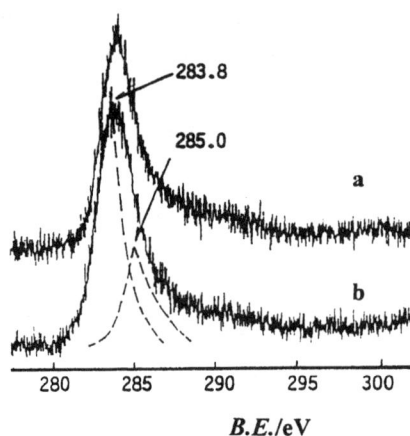

**Fig. 6** C ( 1s )-XPS spectra of the carbon nanotubes produced from the catalytic decomposition of $CH_4$ (a) and CO(b).

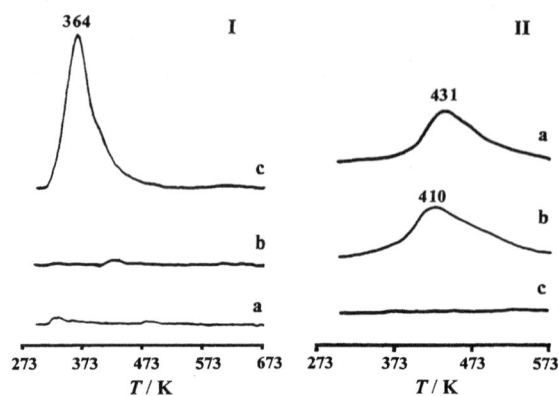

**Fig. 7** TPD spectra of $NH_3$ ( I ) and $C_6H_6$ ( II ) adsorbed on the carbon nanotubes derived from $CH_4$ (a), CO(b) and $SiO_2$ (c).

## 3.3 Nature of the promoting action of the carbon nanotubes carrier

The BET measurements have shown that the surface areas of the two kinds of carbon nanotubes were ～237 and ～155 $m^2 g^{-1}$, respectively; such values were much lower than those of the other four carriers(i. e., $SiO_2$, TDX-601, AC, and GDX-102). It is thus evident that, in comparison with the catalysts supported by the other four carriers, the high propene conversion activity, especially the excellent regioselectivity($n/i$) on the carbon nanotubes-supported Rh-phosphine complex catalysts were not due to the difference in their surface areas. On the other hand, it can be estimated according to the well-known Kelvin equation based upon the capillary coagulation theory that, under the actual reaction condition(i. e.,1. 0 MPa,393 K,$C_3H_6$/CO/$H_2$＝1/1/1(v/v)) and even in the case of propene conversion attaining to 45% (corresponding to a $C_3H_7$CHO partial pressure of 0. 21 MPa in the reaction exit gas), the capillary coagulation of $C_3H_7$CHO occurred only in the hydrophobic micropores with the pore diameters of ≤2. 0 nm. According to the pressure boundary of gas-phase for the capillary condensation of $n$-butylaldehyde shown in Table 5 and the pore diameter data in Table 4, as well as the actual propene conversion levels attained in the present work, one can conclude that the capillary coagulation of $n$-butylaldehyde could not occur on the carbon nanotubes carriers with pore diameters of ca. 3. 0 nm as well as on the other four carriers. The possibility that the 3 nm pores act merely to capillary-condense a carbon nanotubes-supported liquid phase catalyst may be ruled out.

**Table 5** Relation of the pressure boundary of $n$-butylaldehyde in gas-phase for capillary condensation in the carbon nanotubes with their pore diameters at different temperatures

| $T$(K) $T$(K) | $P^0$($C_3H_7$CHO) (atm) | $P$($C_3H_7$CHO) (atm) | Pore diameter (nm) |
| --- | --- | --- | --- |
| 383 | 2. 796 | 2. 1 | 3. 9 |
|  |  | 1. 8 | 2. 6 |
|  |  | 1. 5 | 1. 8 |
| 393 | 3. 606 | 2. 5 | 3. 0 |
|  |  | 2. 1 | 2. 0 |
|  |  | 1. 8 | 1. 6 |

续表

| $T(K)$ $T(K)$ | $P^0(C_3H_7CHO)$ (atm) | $P(C_3H_7CHO)$ (atm) | Pore diameter (nm) |
|---|---|---|---|
| | | 1.5 | 1.2 |
| | | 1.2 | 1.0 |
| | | 0.50 | 0.56 |
| | | 0.30 | 0.44 |
| 403 | 4.586 | 2.5 | 1.8 |
| | | 1.8 | 1.2 |
| | | 1.2 | 0.8 |

As a matter of fact, the $n/i$ ratio on the liquid phase Rh-phosphine complex catalyst system was not as high as that on the corresponding complex catalyst supported by carbon nanotubes under comparable conditions. It was earlier reported by Wilkinson[12] that, when $RhH(CO)(PPh_3)_3$ was used as catalyst precursor, a $n/i$ ratio of 2.33 was obtained under conditions of 373 K, $\sim 3.5$ MPa, $CO/H_2 = 1/1$ (v/v) and P/Rh = 10 (mole ratio); when $PPh_3$ was used as solvent for the catalytic reaction system, corresponding to P/Rh $\cong$ 600 (mole ratio), the $n/i$ ratio attained to 15.3 under the conditions of 398 K, 1.25 MPa, $CO/H_2 = 1/1$.

Thus, it seems to us that the excellent catalytic performance of the carbon nanotubes-supported Rh-phosphine complex catalysts is most probably associated closely with the nanochannels of the carbon nanotubes carrier and its surface constructed of the graphite-like six-membered C-rings. The tubular nanochannels with pore diameters of $\sim 3$ nm are quite suited to accommodating the $\sim 1.8$ nm $RhH(CO)(PPh_3)_3$ complex (the catalyst precursor)[13] and/or its fragment $RhH(CO)(PPh_3)_2$ (the functioning catalytically active species), leaving an appropriate space for diffusion and reactions of the reactant molecules. This would be in favor of enhancing the regioselectivity to reaction intermediates and the product butylaldehyde molecules by means of spatiospecific selective catalysis by rigorous spatio-restraint. Moreover, as shown by the above TPD measurements, a strong interaction may exist between the phenyl of the phosphine ligands and the hydrophobic carbon nanotubes surface constructed of the graphite-like six-membered C-rings, due to their similarities in structural and electronic properties. This, on one hand, would be beneficial to even dispersion of the Rh-phosphine complex species on the surface; but, on the other hand, it would also be easy to bring about decomplexation of coordinated $PPh_3$ ligands, thus leading to lowering of the activity and the regioselectivity, unless it is compensated by adding a proper excess of the $PPh_3$ ligand. The experimental results indicated that a P/Rh molar ratio of $9 \sim 12$ would be appropriate. The mobility of the excess phosphine ligands on the $PPh_3$-modified surface of the carbon nanotubes was probably also in favor of maintaining the Rh-phosphine complex species with higher probability in their catalytically active configurations via complexation-decomplexation-recomplexation of mobile $PPh_3$ ligands. Moreover, one does not rule out the possibility that plenty of delocalizable valence electrons on the graphite-like six-membered C-rings on the surface of the carbon nanotubes might delocalize partially onto the phenyl-ring of coordinated $PPh_3$ ligands, thus favoring the donation of electron from coordinated $PPh_3$ ligands to the Rh central atom. The two factors would conduce to enhancing TOF of the reactant molecules on the Rh active sites. Most probably, these factors make the carbon nanotubes-supported Rh-phosphine complex catalysts display the excellent performance of catalytic hydroformylation.

The results of activity assay of the Rh-phosphine complex catalysts supported by the other carriers provide a set of distinct contrasts. As mentioned in Section 3. 1,propene conversion and the $n/i$ ratio on the ends-unopened carbon nanotubes(surface area: $\sim 115$ m$^2$ g$^{-1}$)-supported catalyst were obviously low due to the inner surface and the nanochannels of the nanotube carrier being unable to get utilized. The surface of the inorganic oxide carrier SiO$_2$ is hydrophilic;its interaction with the phenyl-ring of the phosphine ligands can be expected to be weak,probably due to a lack of similarities in their structural and electronic properties. This can get support from the contrastive experiments described above of adsorption/desorption of ammonia and benzene, respectively, on the carbon nanotubes and SiO$_2$. Moreover,the pore diameters of the SiO$_2$ carrier is $2\sim 4$ times as large as those for the carbon nanotubes,and its spatio-restraint is smaller than that of the carbon nanotubes. These are probably the reasons why the observed $n/i$ ratio of the product butylaldehyde on the SiO$_2$-supported Rh-phosphine complex catalyst was lower than that on the carbon nanotubes-supported catalysts. Both pore diameter (1. 4~2. 0 nm) and pore volume(0. 21 cm$^3$ g$^{-1}$) of the active carbon carrier are comparatively small,and probably too small to accommodate the catalytically active Rh-phosphine complex,so that most of the inner surface could not be utilized,resulting in quite a low TOF and poor regioselectivity. The pore diameter of TDX-601 carbon molecular sieve is somewhat larger than that of active carbon. Though not large enough to accommodate the RhH(CO)(PPh$_3$)$_3$ complex of $\sim 1.8$ nm size,it may accommodate its catalytically active fragment, such as RhH(CO)(PPh$_3$)$_2$, and leave a certain space for diffusion and reaction of the reaction molecules. As a result,the $n/i$ ratio of butylaldehyde reached was extraordinarily high due to the pronounced spatioselective catalysis by the more rigorous spatio-restraint, but the conversion of propene was still quite low. Over the GDX-102 supported Rh-phosphine complex catalyst, the TOF of propene was high,but the corresponding regioselectivity of butylaldehyde was relatively low,probably due to the pore diameter being too large to give rise to the micro-environments needed for the spatioselective catalysis.

# 4　Conclusions

1. The structure and properties of the carbon nanotubes,as a novel material of catalyst carrier, produced catalytically from the decomposition of CH$_4$ or CO,have been characterized by means of TEM, HRTEM, XRD, Raman, XPS, BET and TPD methods. The results demonstrated that the prepared carbon nanofibers were even nanotubes with the outer diameters of 15~20 nm and an inner diameter of ca. 3 nm,and that their tube walls were constructed of many layers of carbon with graphite-like platelets in an extremely ordered arrangement:with the orientation of the cylindrical graphite-like platelets parallel to the tube axis for those derived from CO,and the conical graphite-like platelets inclined to the central axis of the tube for those produced from CH$_4$, respectively. The degree of long-range order of these carbon nanostructures was relatively low in comparison with that of graphite. Their surface possesses strong hydrophobicity,especially strong adsorption ability towards benzene.

2. Such carbon nanotubes grown catalytically have been first employed as a novel carrier material for preparation of the supported Rh-phosphine complex catalysts. The carbon nanotubes-supported Rh-phosphine complex catalysts prepared for propene hydroformylation displayed not only a high activity for propene conversion but also excellent regioselectivity to the product butylaldehyde. These,together with the results of comparative study of the Rh-phosphine complex catalysts supported by the ends-unopened carbon-nanotubes,SiO$_2$,TDX-601,AC,and GDX-102,are in favor of the viewpoint that the tubular nanochannels of the carbon nanostructures and their hydrophobic surface consisting of six-

membered C-rings play important roles in enhancing the propene hydroformylation activity, and especially regioselectivity to the product butylaldehyde. But this is a preliminary suggestion. For better understanding of the promoting action of the carbon nanotube carriers, more detailed knowledge about the interaction of the supported Rh-phosphine complex species with the carrier surface and the reaction chemistry of the reactant molecules in the tubular nanochannels is needed.

3. The present work also provides a successful example for the application of the carbon nanotubes as an alternative to middle-pore molecular sieves under certain circumstances.

## Acknowledgements

The authors gratefully acknowledge the financial supports from the Fujian Provincial Natural Science Foundation and the National Natural Science Foundation of China.

## References

[1] M. Beller, B. Cornils, C. D. Frohning, C. W. Kohlpaintner, J. Mol. Catal. A **104**(1995) 17, and the related references therein.

[2] J. P. Arhancet, M. E. Davis, J. S. Merola, B. E. Hanson, Nature **339**(1989) 454.

[3] J. P. Arhancet, M. E. Davis, J. S. Merola, B. E. Hanson, J. Catal. **121**(1990) 327.

[4] J. Hjortkjaer, M. S. Scurrell, P. Simonsen, J. Mol. Catal. **10**(1981) 127.

[5] N. M. Rodriguez, J. Mater. Res. **8**(12)(1993) 3233.

[6] P. Chen, H. B. Zhang, G. D. Lin, Q. Hong, K. R. Tsai, Carbon **35**(10−11)(1997) 1495.

[7] Shou-shan Chen, Zheng-zhi Zhang, Xu-kun Wang(Eds.), Handbook of Synthesis of Organometallic Compounds, Chem. Industry Press, Beijing, 1986, p. 291.

[8] D. Evans, G. Yagupsky, G. Wilkinson, J. Chem. Soc. A. 1968, 2660.

[9] H. L. Pelt, P. J. Gijsmao, R. P. J. Verburg, J. J. F. Scholten, J. Mol. Catal. **33**(1985) 119.

[10] M. S. Hoogenraad, M. F. Onwezen, A. J. van Dillen, J. W. Geus, in: J. W. Hightower, W. N. Delgass, E. Iglesia, A. T. Bell(Eds.), Proc. 11th Int. congr. on Catalysis-40th Anniversary, Stud. Surf. Sci. Catal., vol. 101, Elsevier, Amsterdam, 1996, p. 1331.

[11] D. Goresy, Science **161**(1968) 363.

[12] G. Wilkinson, US Patent 4 108 905, 1978.

[13] M. E. Davis, J. P. Arhancet, B. E. Hanson, US Patent 4 947 003, 1990, US Patent 4 994 427, 1991.

■ 本文原载:《催化学报》第 20 卷第 1 期(1999 年 1 月),第 35～40 页。

# 甲烷脱氢芳构化 W/ HZSM-5 基催化剂的制备<sup>*</sup>

熊智涛　曾金龙　张鸿斌　林国栋　蔡启瑞[①]

(厦门大学化学化工学院　固体表面物理化学国家重点实验室,厦门　361005)

**摘　要**　由 $H_2SO_4$ 酸化的 $(NH_4)_2WO_4$ 溶液(pH=2～3)制备的 W-$H_2SO_4$/HZSM-5 催化剂对甲烷脱氢芳构化(DHAM)反应的催化活性比不经酸化的 $(NH_4)_2WO_4$ 溶液(pH=8～9)制备的 W/HZSM-5 催化剂高得多。氧化前驱态催化剂的 $H_2$-TPR 研究表明,W-$H_2SO_4$/HZSM-5 试样的还原所需温度比 W/HZSM-5 低得多;前者可还原至较低价态的 W 物种在负载总 W 量中所占比例也比后者高得多。$H_2$-TPR 及相关体系的 EPR 和 Raman 表征结果相互佐证表明,用 $H_2SO_4$ 酸化导致负载钨酸铵前驱溶液中一些由共边正八面体配位结构单元 $(WO_6)^{n-}$ 构成的聚钨酸根物种的生成;W-$H_2SO_4$/HZSM-5 催化剂对 DHAM 反应催化活性之所以高,主要由于所负载的 W 前驱物种较大部分是 $(WO_6)^{n-}$ 基的 W 物种,而非 $(WO_4)^{2-}$-基 W 物种;在较低温度下就可被还原,且在 DHAM 反应条件下有一部分可还原至较低价态 $W^{4+}$ 的 $(WO_6)^{n-}$ 物种很可能是催化活性位的前驱组成部分。

**关键词**　甲烷　非氧化脱氢芳构化　钨　硫酸　HZSM-5 沸石　苯

无氧条件下甲烷脱氢芳构化(DHAM)制芳烃是近年来甲烷优化利用的新方向之一[1~9]。迄今已报道的催化剂多为负载的过渡金属(如 Pt,Cr,Mo,Re 等)的氧化物体系。Bragin 等[1]曾率先报道用脉冲反应器在 Pt-$CrO_3$/HZSM-5 上于 1 023 K 得到甲烷转化率和苯选择性分别为 18% 和 78% 的结果。此后陆续发表的 DHAM 催化剂体系主要有 Mo/HZSM-5[3],$MoO_3$/ZSM-5 或 $MoO_3$/$SiO_2$[4],Mo-W/HZSM-5[5],$Mo_2$C/ZSM-5[6],Pt-Mo/ZSM-5[7]和 Mo-Zn-$H_2SO_4$/HZSM-5[8,9],其中以改进的 Mo/HZSM-5 基催化剂为佳。现有的大多数 DHAM 催化剂的操作温度约为 973 K,由于热力学平衡转化的限制,所达到的甲烷转化率比较低;而在稍高温度下操作时 Mo 组分容易流失,且积炭明显增加,导致催化剂快速失活。因此,寻求耐高温抗结炭的新催化剂体系对 DHAM 反应的实用化有重要意义。

本研究组新近发现,由碱性或中性钨酸铵溶液(pH=7~9)制备的 W/HZSM-5 催化剂即使在反应温度高达 1 023 K 时也几乎无 DHAM 催化活性;而由经 $H_2SO_4$ 酸化的钨酸铵溶液(pH=2~3)制备的 W-$H_2SO_4$/HZSM-5 催化剂在相同反应条件下对 DHAM 反应的催化活性却相当高;在 W-$H_2SO_4$/HZSM-5 中添加一定量的 Zn 或 La 还能进一步地改进该催化剂的 DHAM 性能[10]。

本文利用 $H_2$-TPR 和 EPR 等谱学方法跟踪考察了不同方法制备的催化剂氧化前驱 W 物种的还原行为,表征了工作状态下催化剂中 W 组分的催化活性价态,其结果对于深入了解该类催化剂活性位的本质有重要意义,并可为高活性和高选择性 W/HZSM-5 基催化剂的制备提供科学根据。

---

* 国家自然科学基金(批准号 29392002)和教育部博士点科研基金资助项目。

① 联系人:张鸿斌。第一作者:熊智涛,男,1972 年生,博士生。

## 1 实验部分

前驱态催化剂采用等容逐步浸渍法制备,即分别用 pH＝8～9 或经 $H_2SO_4$ 酸化至 pH＝2～3 的钨酸铵(纯度为 AR 级)水溶液浸渍 HZSM-5 分子筛(南开大学化工厂产品,$n(Si)/n(Al)＝38$)载体,经 383 K 烘干 2 h,673 K 焙烧 4 h,即得 W/HZSM-5 或 W-$H_2SO_4$/HZSM-5 氧化态前驱物;Zn 或 La 促进的催化剂系用计量的促进剂 Zn 或 La 的相应盐溶液和计量的 $H_2SO_4$ 先后浸渍 W-$H_2SO_4$/HZSM-5 前驱物而得。本文所制备的催化剂中 W 的负载量均为 2％(质量分数,下同);促进型 W/HZSM-5 催化剂中 Zn 或 La 的添加量为 1.5％。

催化剂活性评价试验在固定床连续流动反应器-GC(102 GD 型)组合系统上进行。原料气甲烷纯度为 99.9％。反应在 0.1 MPa,973～1 173 K,GHSV＝1 500 mL/(h·g)的反应条件下进行。产物由在线气相色谱仪氢焰和热导双检测器联合作现场分析。甲烷转化率采用"工作曲线法"测定,苯选择性由氢焰检测器分析数据按"碳基归一法"(不考虑结焦反应)计算。

TPR 实验装置由一套气体净化系统、微型反应器、在线 GC(102 G 型)、台式双笔自动平衡记录仪等仪器和单元组件组成。每次试验催化剂用量 50 mg,粒度为 40～60 目。TPR 测试系将氧化态催化剂在 673 K 下用 Ar 气吹扫 0.5 h 以净化其表面,降至室温后直接用 $H_2$-$N_2$ 混合气($V(H_2)/V(N_2)＝5/95$)作原料气-载气进行程序升温还原;原料气-载气流速为 30 mL/min,吹扫气流速为 50 mL/min。

负载前驱溶液的 Raman 光谱在 Spex Ramalog-6 型 Raman 光谱仪上进行测试。催化剂的 EPR 谱由 ER-200D 型 X-波段电子顺磁共振谱仪在室温下记录。

## 2 结果与讨论

### 2.1 钨酸铵浸渍液 pH 值对催化剂性能的影响

表 1 的实验结果表明,由碱性或中性钨酸铵溶液(pH＝7～9)制备的 W/HZSM-5 催化剂,在 973～1 023 K 温度范围内,对甲烷转化无催化活性;当反应温度升至 1 073 K 时遂显示出一定的 DHAM 催化活性,相应的甲烷转化率为 5.7％,苯选择性为 95.3％。而由经 $H_2SO_4$ 酸化的钨酸铵溶液(pH＝2～3)制备的 W-$H_2SO_4$/HZSM-5 催化剂,在 1 023 K 反应条件下对 DHAM 反应的催化活性已相当明显,相应的甲烷转化率为 7.1％,苯选择性为 95.2％;当反应温度升至 1 073 K 时,甲烷转化率达到 16.7％,苯选择性为 95.7％。由此可见,钨酸铵浸渍液的 pH 值对所制得催化剂的 DHAM 催化性能有显著影响。初步推断,这与不同 pH 值钨酸铵溶液中钨组分前驱物种的组成和结构的差别有关。为对这些差别有进一步了解,本文利用 $H_2$-TPR 并辅以 EPR 方法考察了经烘干、焙烧后的氧化前驱态催化剂的还原行为。

表 1 甲烷非氧化脱氢芳构化催化剂的活性评价结果

Table 1 The results of activity assays of the catalysts with different compositions for non-oxidative dehydro-aromatization of methane at different reaction temperatures

| Sample | Catalyst | T/K | Conversion of $CH_4$(％) | Selectivity(％)* | | | |
| --- | --- | --- | --- | --- | --- | --- | --- |
| | | | | $C_2$ | $C_3$ | Benzene | Toluene |
| 1 | HZSM-5 | 973 | 0.1 | 88.2 | — | 11.8 | — |
| 2 | $H_2SO_4$/HZSM-5 | 973 | 0 | | | | |
| 3 | Zn/HZSM-5 | 973 | 1.0 | 19.6 | 1.3 | 69.9 | 9.2 |

续表

| Sample | Catalyst | $T/K$ | Conversion of $CH_4$(%) | Selectivity(%)* | | | |
|---|---|---|---|---|---|---|---|
| | | | | $C_2$ | $C_3$ | Benzene | Toluene |
| 4 | W/HZSM-5 | 1023 | 0 | — | — | — | — |
| | | 1073 | 5.7 | — | — | 95.3 | 4.7 |
| 5 | W-$H_2SO_4$/HZSM-5 | 1023 | 7.1 | — | — | 95.2 | 4.8 |
| | | 1073 | 16.7 | — | — | 95.7 | 4.3 |
| | | 1123 | 21.4 | — | — | 96.5 | 3.5 |
| 6 | W-Zn-$H_2SO_4$/HZSM-5 | 1073 | 20.5 | — | — | 97.0 | 3.0 |
| | | 1123 | 23.0 | — | — | 97.7 | 2.3 |
| 7 | W-La-$H_2SO_4$/HZSM-5 | 1123 | 22.3 | — | — | 95.6 | 4.4 |

Reaction conditions :0.1 MPa,GHSV=1 500 mL/(h • g)

* Trace $C_{8+}$ aromatic hydrocarbons(ethyl-benzene,dimethyl-benzene,naphthalene etc. ) can be found in the products ; the selectivity data listed here are slightly higher than realities due to disregarding part of methane converted to coke and $C_{8+}$ aromatic hydrocarbons in the calculations of selectivity

### 2.2 氧化态催化剂前驱物中 W 物种的还原行为

图 1 示出两种氧化态催化剂前驱物及其相关基底试样的 $H_2$-TPR 谱。在该系列催化剂中,载体 HZSM-5 分子筛及 $H_2SO_4$ 组分在 $H_2$-TPR 测试或 DHAM 反应条件下均为不可能被还原的无机酸盐或氧化物组分;这可以从图 1(c)所示基底试样 $H_2SO_4$/HZSM-5 未显示任何 $H_2$-TPR 峰的实验结果得到证实。因此,图 1(a)和(b)中出现的若干 $H_2$-TPR 谱峰显然都应归属于 W 组分被 $H_2$ 还原的贡献。

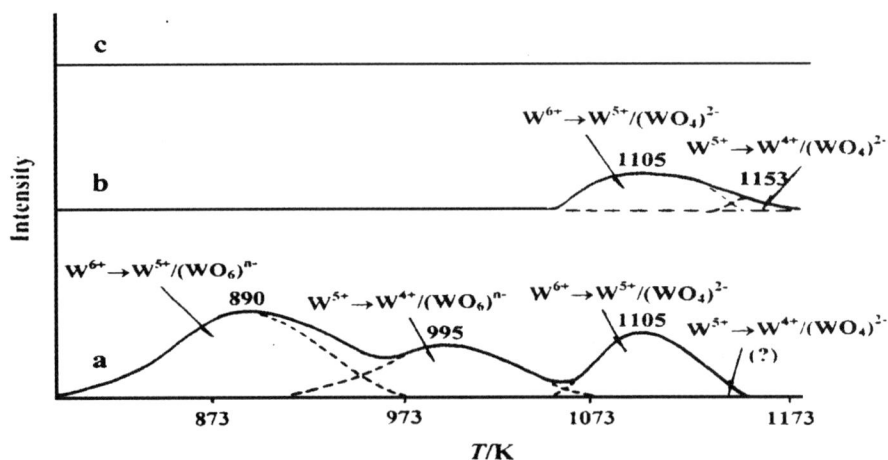

图 1　催化剂及相关体系的 $H_2$-TPR 谱

**Fig. 1　$H_2$-TPR spectra of(a) W-$H_2SO_4$/HZSM-5,(b) W/HZSM-5 and(c) $H_2SO_4$/HZSM-5**

图 1(b)中试样 b(W/HZSM-5)仅在 1 105 K 处出现一个 $H_2$ 还原峰。该峰略显不对称,在其高温侧似包含一弱的肩峰;经分析拟合,可分解出一个峰温为 1 153 K 的弱峰。前后两峰面积的相对比例可粗略估算为 $A$(1 153 K)/$A$(1 105 K)=15/100。EPR 测试结果显示,该试样在 1 073 K 经 $H_2$ 还原 30 min 只产生一类 $W^{5+}$ 物种,相应的 EPR 信号特征为 $g\parallel$=1.82,$g\perp$=1.95;随着 $H_2$ 还原时间延长至 1 h,该信号的强度并无减弱,表明该类 $W^{5+}$ 物种难以进一步还原至没有 EPR 信号的 $W^{4+}$ 价态。鉴于此,本文倾向于将 1 105 K 处的 $H_2$-TPR 峰指认为一类氧化态 W 前驱物种的单电子还原:$W^{6+}\xrightarrow{+e}W^{5+}$。在

1 153 K处强度甚弱的肩峰则很可能源于少量 $W^{5+}$ 物种(约占 15％)在更高温度下被进一步还原至 $W^{4+}$ 价态。

由图 1(a)可见,与试样 b 不同,由酸性钨酸铵溶液制备的试样 a($W\text{-}H_2SO_4$/HZSM-5)上有三个 $H_2$-TPR 峰,其峰温分别为 890,995 和 1 105 K。不同还原阶段 $W\text{-}H_2SO_4$/HZSM-5 试样的 EPR 谱(见图 2)显示,该试样在 1 073 K 用 $H_2$ 还原 10 min 产生两类具有不同微环境的 $W^{5+}$ 物种,相应的 $g$ 值分别为 $g_{\parallel}^{(1)}=1.82$,$g_{\perp}^{(1)}=1.95$ 和 $g_{\parallel}^{(2)}=1.88$,$g_{\perp}^{(2)}=1.97$。第(1)类 $W^{5+}$ 的 EPR 信号强度并不随还原时间的延长而减弱,表明相应的 $W^{6+}$ 物种只能被还原至 $W^{5+}$ 价态。而第(2)类 $W^{5+}$ 的 EPR 信号强度却经历一个由弱→强→弱的变化过程:在还原 20 min 时强度达到最大;随着还原时间的延长,该信号强度减弱,暗示处于第(2)类微环境的 $W^{6+}$ 物种在被还原至 $W^{5+}$ 价态后,其中有相当一部分可进一步还原至没有顺磁信号的 $W^{4+}$ 价态。参照这些 EPR 结果并鉴于试样 a 在 1 105 K 处 TPR 峰的峰温及峰形与试样 b 的相当接近,可以推测该处的 TPR 峰很可能同属于一类较难被还原的 W 物种的单电子还原:$W^{6+}\xrightarrow{+e^-}W^{5+}$;而 890 和 995 K 两个 TPR 峰则很可能源于另一类较易被还原的 W 物种的连续两步单电子还原:$W^{6+}\xrightarrow{+e^-}W^{5+}$ 和 $W^{5+}\xrightarrow{+e^-}W^{4+}$。图 1(a)中 995 K 和 890 K 两峰面积的相对比例可粗略估算为:$A(995\ K)/A(890\ K)=50/100$,意味着该类较易被还原的 W 物种在还原至 $W^{5+}$ 价态之后,随着 $H_2$-TPR 过程的继续,其中约 50％的 $W^{5+}$ 可进一步还原至 $W^{4+}$ 价态。于是,在还原态的 $W\text{-}H_2SO_4$/HZSM-5 催化剂上共存着价态不同和数量可观的 $W^{5+}$ 和 $W^{4+}$ 物种。

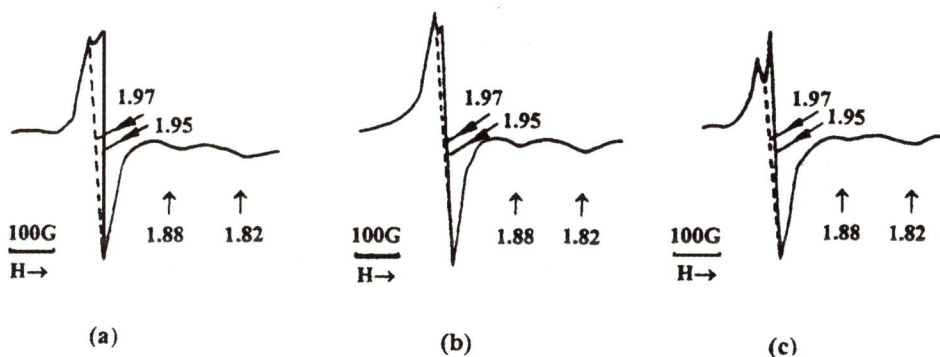

图 2　氢还原不同阶段 $W\text{-}H_2SO_4$/HZSM-5 催化剂的 EPR 谱

Fig. 2　EPR spectra of the $W\text{-}H_2SO_4$/HZSM-5 catalyst reduced by $H_2$ at 1 073 K for(a) 10 min,(b) 20 min and(c) 30 min,respectively

从图 1(a)和(b)所示 $H_2$-TPR 谱峰面积的相对大小还可估算出,直到温度升至 1 200 K(即整个 $H_2$-TPR 过程结束),试样 a($W\text{-}H_2SO_4$/HZSM-5)中可被还原至 $W^{5+}$ 价态的 W 量约为有相同 W 总负载量的试样 b(W/HZSM-5)的 3.2 倍,试样 a 中可被还原至 $W^{4+}$ 价态的 W 量约为试样 b 的 10 倍。

### 2.3　不同 W 前驱物种与催化活性的相关性

已知在钨酸铵水溶液中存在着钨酸根离子与一系列聚钨酸根离子之间的平衡:

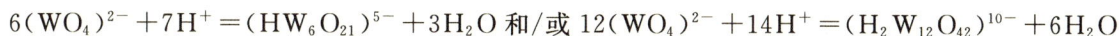

$$6(WO_4)^{2-}+7H^+=(HW_6O_{21})^{5-}+3H_2O \text{ 和/或 } 12(WO_4)^{2-}+14H^+=(H_2W_{12}O_{42})^{10-}+6H_2O$$

我们认为,钨酸铵浸渍液的酸化不仅有助于增大钨酸盐组分的溶解度,从而可提高单程浸渍的负载量;更重要的是酸化将通过移动上述可逆反应的平衡点,促进聚钨酸根离子的形成,这和聚钼酸盐体系十分相似[11,12]。已知这些聚钨酸根离子系由一些共边或共角的正八面体配位结构单元 $(WO_6)^{n-}$ 构成[13]。Raman 光谱实验结果表明,不同 pH 值的钨酸铵溶液中分别主要地存在聚钨酸根离子(诸如 $(HW_6O_{21})^{5-}$,$(H_2W_{12}O_{42})^{10-}$ 等),钨酸分子($H_2WO_4$)和钨酸根离子($WO_4$)$^{2-}$:在酸性(pH=2～3)、中性

(pH＝7)和碱性(pH＝8～9)三种溶液中,与钨酸根物种相关的 Raman 谱峰分别出现在 971,945 和 931 cm$^{-1}$ 处,它们可分别归属于聚钨酸根离子端基 W＝O 键的伸缩振动、$H_2WO_4$ 分子($C_{2v}$)的 W＝O 伸缩振动,以及($WO_4$)$^{2-}$ 离子($T_d$)的对称伸缩振动[14]。

2.1 节中有关催化剂活性的评价结果已表明,在相同反应条件下,W-$H_2SO_4$/HZSM-5 催化剂对甲烷转化的催化活性比 W/HZSM-5 催化剂高得多。结合上述 $H_2$-TPR 等谱学表征结果不难看出,催化剂的 DHAM 活性与相应氧化态 W 前驱物种的可还原性有着密切的联系。由碱性浸渍液负载制备的 W/HZSM-5 催化剂(即试样 b)主要含有一类 W 物种,该类物种显然系由溶液中单核的钨酸根离子($WO_4$)$^{2-}$ 负载衍生而来。在温度≤1 023 K 时,由于其不可能被 $H_2$ 所还原,以致该催化剂对 DHAM 反应几乎无活性;只有在更高的操作温度下部分该类 W 物种被还原后才显示出一定的 DHAM 催化活性,但仍比相同条件下 W-$H_2SO_4$/HZSM-5 的活性水平低得多(见表 1)。而由经酸化处理的钨酸铵溶液制备的负载 W-$H_2SO_4$/HZSM-5 催化剂(即试样 a)则含有两类 W 物种,其中较难还原的一类与试样 b 的十分相似;另一类则很可能系由溶液中($WO_6$)$^{n-}$-基的聚钨酸根离子负载衍生而来,比较容易被还原(在～995 K 温度下就已有相当大的一部分 W 物种可还原至 $W^{4+}$ 价态),因而,在 1 023 K,该催化剂上实际观测到的甲烷转化活性和芳烃选择性已相当高。由此可见,高的 DHAM 反应活性与催化剂上较高价态 W 物种($W^{5+}$)的浓度无明显的顺变关系,而与 $W^{4+}$ 物种的浓度密切相关。由酸性钨酸铵浸渍液负载制备的 W-$H_2SO_4$/HZSM-5 催化剂对 DHAM 反应的催化活性之所以高,主要是由于其所负载的 W 前驱物种较大部分是($WO_6$)$^{n-}$-基 W 物种(而非($WO_4$)$^{2-}$-基 W 物种)在较低温度下就可被还原,且在 DHAM 反应条件下相当一部分可还原至较低价态 $W^{4+}$ 的($WO_6$)$^{n-}$ 物种很可能是催化活性位的前驱组成部分。

### 2.4 Zn 或 La 的促进作用

实验发现,在 W-$H_2SO_4$/HZSM-5 中添加一定量的 Zn 或 La 能进一步改进催化剂的 DHAM 性能。W-$H_2SO_4$/HZSM-5 催化剂的活性虽高,但不够稳定,在 1 073 K 反应 3 h 后甲烷转化率已由第 1～2 h 的 16.7％降至 12.5％。当添加 Zn 之后,在相同反应温度下连续操作 3 h,甲烷转化率一直保持在 20.5％左右而未见下降,苯选择性为 97％,3 h 总积炭量约为催化剂质量的 0.02％。La 的促进作用与 Zn 相仿,仅甲烷转化率的提高幅度不如 Zn 的大(约少一个百分点)。由此可见,Zn 或 La 的加入不仅提高了催化剂的 DHAM 活性,还使其在高温下操作较稳定,不易失活。初步推测,这可能与下列因素有关:一方面 Zn 或 La 在 W 物种间的掺杂有利于抑制 W 物种的聚集及 $WO_3$ 微晶的形成,并减轻炭的沉积;另一方面,以 Zn 或 La 的硫酸盐形态存在的 $SO_4^{2-}$ 通过形成 $HSO_4^-$ 离子以稳定 B 酸位,有利于避免或减少表面 B 酸位(这类表面活性位在 $CH_4$ 第 1,2 步分步脱氢生成卡宾中间物种($CH_2$)的过程中可能起重要作用[8,9])的流失。这两方面因素都有利于催化剂 DHAM 活性和操作稳定性的提高。为深入了解 Zn 或 La 的促进作用机理,尚需做更多的工作,特别是获取有关催化活性位本质的信息。

## 参考文献

[1]Bragin O V,Vasina T V,Preobrazhenskii A V et al. Izv Akad Nauk SSSR,Ser Khim ,1989,**3**: 750.

[2]Koerts T,Deelen M J A G,van Santen R A. J Catal,1992,**138**:101.

[3]Wang L S,Tao L X,Xie M S et al. Catal Lett,1993,**21**:35.

[4]Solymosi F,Erdohelyi A,Szoke A. Catal Lett,1995,**32**:43.

[5]Wong S T,Xu Y D,Wang L S et al. Catal Lett,1996,**38**:39.

[6]Solymosi F,Szoke A,Cserenyi J. Catal Lett,1996,**39**:157.

[7]Chen L,Lin L W,Xu Z S et al. Catal Lett,1996,**39**:169.

[8]Zeng J L,Xiong Z T,Lin G D et al. 11th International Congress on Catalysis. Baltimore,1996. 158.

[9]曾金龙,熊智涛,林国栋等. 厦门大学学报(自然科学版),1996,**35**(6):900.

[10]曾金龙,熊智涛,林国栋等. 高等学校化学学报,1998,**19**(1):123.

[11]Knozinger H,Jeziorowski H. J Phys Chem,1978,**82**:2002.

[12]Jeziorowski H,Knozinger H. J Phys Chem,1979,**83**(9):1166.

[13]Well A F. Structural Inorganic Chemistry. 5th Ed. Oxford:Clarendon Press,1984. 519.

[14]Weinstock N,Schulze H,Muller A. J Chem Phys,1973,**59**:5063.

# Study of Preparation of W/Hzsm-5-Based Catalysts for
# Non-Oxid Ative Dehyd Ro-Aromatization of Methane

Zhi-Tao Xiong, Jin-Long Zeng, Hong-Bin Zhang, Guo-Dong Lin, Khi-Rui Tsai

(Institute of Chemistry and Chemical Engineering, State Key Laboratory of Physical Chemistry for the Solid Surfaces, Xiamen University, Xiamen 361005)

**Abstract** It has been found experimentally that the acidification of the solution of $(NH_4)_2WO_4$ used for impregnation by $H_2SO_4$ enhances markedly the activity of the W/HZSM-5-based catalysts for non-oxidative dehydro-aromatization of methane(DHAM). The $H_2$-TPR results indicated that the temperature required for the $H_2$-reduction of the catalyst precursor of the W-$H_2SO_4$/HZSM-5 derived from acidified solution of $(NH_4)_2WO_4$(with pH$=2\sim3$) was much lower than that of the W/HZSM-5 derived from non-acidified solution of $(NH_4)_2WO_4$(with pH$=8\sim9$). It was estimated that,in the temperature region of $298\sim1\,200$ K,the proportions of the W-species reducible to $W^{5+}$ and $W^{4+}$ in the total amount of loaded W for the former were $\sim3.2$ and $\sim10$ times as high as that for the latter, respectively. This along with the results of the EPR and Raman investigations of the related systems would be in favor of the following viewpoint,*i.e.*,that the acidification by $H_2SO_4$ resulted in the formation of polytungstates species being built of octahedral group $(WO_6)^{n-}$ sharing edges, via the reaction such as $6(WO_4)^{2-} + 7H^+ = (HW_6O_{21})^{5-} + 3H_2O$, in the precursor solution, and that $(WO_6)^{n-}$-derived W-species were more easily reduced by $H_2$ than those derived from $(WO_4)^{2-}$,and that the high DHAM activity was closely correlated with the polytungstate group $(WO_6)^{n-}$ as precursor of catalytically active species.

**Key words** Methane Non-oxidative dehydro-aromatization Tungsten Sulfuric acid HZSM-5 zeolite Benzene

■ **本文原载**：《高等学校化学学报》第 20 卷第 10 期（1999 年 10 月），第 1589～1594 页。

# 酸碱存在下水溶性铑膦配合物的 NMR 表征及氢甲酰化性能研究[*]

张　宇　袁友珠　廖新丽　叶剑良　姚春香　蔡启瑞[①]

（厦门大学化学系　物理化学研究所　固体表面物理化学国家重点实验室，厦门　361005）

**摘　要**　$^{31}P(^1H)$ NMR 和 $^1H$ NMR 研究表明，当 NaOH 加入到水溶性铑膦配合物 $HRh(CO)(TPPTS)_3$ [TPPTS：$P(m\text{-}C_6H_4SO_3Na)_3$] 后，可观察到有少量的 OTPPTS[OTPPTS：$O=P(m\text{-}C_6H_4SO_3Na)_3$] 出现，但配合物的特征谱峰即使在高浓度的 NaOH 存在下也基本保持不变，表明 NaOH 对配合物分子结构的影响较小；当吡啶加入到 $HRh(CO)(TPPTS)_3$ 中，$^{31}P(^1H)$ NMR 谱图中出现游离配体 TPPTS 的 $^{31}P$ 谱峰及若干结构未知的新水溶性配合物的 $^{31}P$ 谱峰，表明吡啶分子将与配合物分子中的配体 TPPTS 发生配体交换反应。在 $HRh(CO)(TPPTS)_3$ 中分别加入一定量的 HCl，$HNO_3$，$H_2SO_4$ 和 $H_3PO_4$ 等无机酸时，随着酸量的增加，配合物的 $^{31}P$ 物种含量逐渐下降，而 OTPPTS 量明显上升，直至配合物 $^{31}P$ 物种完全消失；高浓度乙酸对配合物结构的影响与上述无机酸类似 $HRh(CO)(TPPTS)_3$ 的 1-己烯氢甲酰化催化反应结果表明，碱存在下可获得较高的正异构醛比值，但催化活性降低；酸存在下所得产物正异构醛比值相对较低且呈淡黄色。可见，酸对 $HRh(CO)(TPPTS)_3$ 的结构和性能的影响比碱大

**关键词**　$^{31}P(^1H)$ NMR　$^1H$ NMR　$HRh(CO)(TPPTS)_3$　酸碱性　氢甲酰化

有关水溶性铑膦配合物 $HRh(CO)(TPPTS)_3$ [TPPTS：$P(m\text{-}C_6H_4SO_3Na)_3$（间-三苯基膦三磺酸钠盐）] 用于烯烃氢甲酰化催化反应的文献讨论了添加物对催化剂性能的影响[1-8]。所研究的添加物包括表面活性剂、有机溶剂、无（有）机盐类等。添加物的作用主要有两点：增加油溶性反应底物在水溶性催化剂中的溶解性和调节反应体系的 pH 值，以此提高催化剂活性和选择性。文献[6,9]采用原位、动态谱学方法（如动态 $^{31}P$ NMR）研究某些添加物的作用机理和配合物的稳定性，把该领域的研究逐步推向分子水平。已有的研究结果表明，配合物催化剂水层的 pH 值直接影响着催化剂活性、选择性和稳定性，以中性至偏碱性的催化剂水层 pH 值为佳[8,10]。在对水溶性铑膦配合物体系和负载型水溶性铑膦配合物体系等烯烃氢甲酰化催化剂的研究过程中，外界或反应体系内的酸性对其结构的影响远比碱性大[11,12]。

为从分子水平探讨关于酸碱对水溶性铑膦配合物 $HRh(CO)(TPPTS)_3$ 结构及其氢甲酰化催化性能的影响，本文在高分辨 NMR 谱仪上，对水溶性铑膦配合物 $HRh(CO)(TPPTS)_3$ 分别在酸和碱存在下的 $^{31}P$ 和 $^1H$ NMR 谱图进行了较系统的研究，并考察了酸碱对该铑膦配合物 1-己烯氢甲酰化催化反应的影响，以期为该催化剂体系的烯烃氢甲酰化工艺条件的确定和催化剂失活机理的研究提供基础。

---

\* 国家自然科学基金（批准号：29873037）和中国石油化学总公司联合资助"九五"重大项目（批准号：29792075）。

① 联系人：袁友珠。第一作者：张宇，男，27 岁，博士。

# 1 实验部分

## 1.1 样品制备

间-三苯基膦三磺酸钠盐(TPPTS)按文献[1,13]方法制备,纯度 85%,其余为 $H_2O$ 和 OTPPTS(间-三苯基氧化膦三磺酸钠盐),使用时经丙酮-甲醇-水体系进一步提纯[14]。OTPPTS 为本实验室分离所得,纯度>99% [$^{31}P(^1H)$ NMR 结果]。$HRh(CO)(TPPTS)_3$ 按文献[13]方法制备和提纯,纯度>98% [$^{31}P(^1H)$ NMR 结果]。所制备样品均置于干燥器中保存备用。

## 1.2 NMR 表征实验

$^{31}P(^1H)$ NMR 和 $^1H$ NMR 谱图在 Varian Unity+500 超导核磁共振波谱仪上于室温下得到。以 $D_2O$ 为溶剂,磷谱采用 85% $H_3PO_4$ 作外标,氢谱采用 DSS 作内标。

## 1.3 氢甲酰化催化反应

1-己烯氢甲酰化反应在 GCF-025 型体积为 250 mL 的强磁力搅拌高压釜中进行,将 10 mL 催化剂[0.003 2 mmol $HRh(CO)(TPPTS)_3/H_2O$]和 10 mL 1-己烯/庚烷(质量比为 1:4)加入高压釜,用经除水除氧净化的低压(1.0 MPa)$CO/H_2$(体积比为 1:1)置换 3 次,然后引入 3.0 MPa 除水除氧净化的 $CO/H_2$,控制反应温度为 363 K,反应过程中剧烈搅拌(搅拌速度为 400 r/min)。反应 4 h 后,冷却、放空、开釜,由气相色谱分析产物,色谱仪为北京分析仪器分 SQ-206 气相层析仪,TCD 检测器,色谱柱:15%聚乙二醇癸二酸酯/102 白色担体(6 m)。按修正面积归一化法计算 1-己烯转化率和产物选择性。

# 2 结果与讨论

## 2.1 $HRh(CO)(TPPTS)_3$ 在碱存在下的 $^{31}P(^1H)$ 和 $^1H$ NMR 结果

图 1(A)和图 1(B)分别示出了 $HRh(CO)(TPPTS)_3-D_2O$ 溶液中加入不同量的 NaOH 溶液后测得的 $^{31}P(^1H)$ NMR 和 $^1H$ NMR 谱图。从图 1(A)可见,$^{31}P$ NMR 谱图中化学位移 $\delta=35.0$,归属为 OTPPTS 的谱峰强度随 NaOH 加入量增加而略有增加,但未见其特征谱峰有明显变化或产生其他新的物种。说明加入 NaOH 对 $HRh(CO)(TPPTS)_3$ 的结构影响较小;微量的 OTPPTS 增量可能缘于溶液中高浓度的电解质破坏少量 $HRh(CO)(TPPTS)_3$ 分子,并使解离下来的 TPPTS 进一步被氧化成 OTPPTS。上述 $^{31}P(^1H)$ NMR 谱中物种的变化情况在图 1(B)示出的 $^1H$ NMR 谱图中也得到很好的验证。随着 NaOH 量的增加,在 $^1H$ NMR 谱中 $\delta=-9.56$ 附近依然保留裂分清晰的四重峰,即 H-Rh 键未断裂且配合物的三角双锥构型未受影响。加入 NaOH 后,在芳环的氢峰区域,归属为配位的 TPPTS 芳环质子谱峰($\delta 7.1\sim7.7$)略有增宽,且在 $\delta 7.7\sim8.2$ 处归属为 OTPPTS 的谱峰强度略有增强,这可能是加入 NaOH 后,配合物分子结构中各质子的弛豫加快,但并未改变分子的结构排布,同时也说明,NaOH 的加入将使 OTPPTS 含量增加。

图 2(A)和图 2(B)分别示出了 $HRh(CO)(TPPTS)_3-D_2O$ 溶液中加入吡啶后测得的 $^{31}P(^1H)$ NMR 和 $^1H$ NMR 谱图。从图 2 可见,$HRh(CO)(TPPTS)_3$ 与吡啶作用后,谱图变化较复杂。随着吡啶量的增加,配合物的特征 $^{31}P$ 谱双峰(化学位移为 44.5,43.8)略向高场位移,且其强度逐渐减弱。同时,在 $\delta 33.1\sim34.8$ 处出现若干新谱峰,在 $\delta -5.0\sim-5.3$ 附近可观察到逐渐增加的游离配体 TPPTS 的 $^{31}P$ 谱峰强度,但未见有 OTPPTS 的 $^{31}P$ 谱峰出现;从图 2(B)可见,当吡啶加入量与配合物浓度比约为 1.0 时,归属为 H-Rh 键的质子峰将消失,但 OTPPTS 芳环质子谱峰强度没有增加趋势,仅在芳环质子峰区域观察

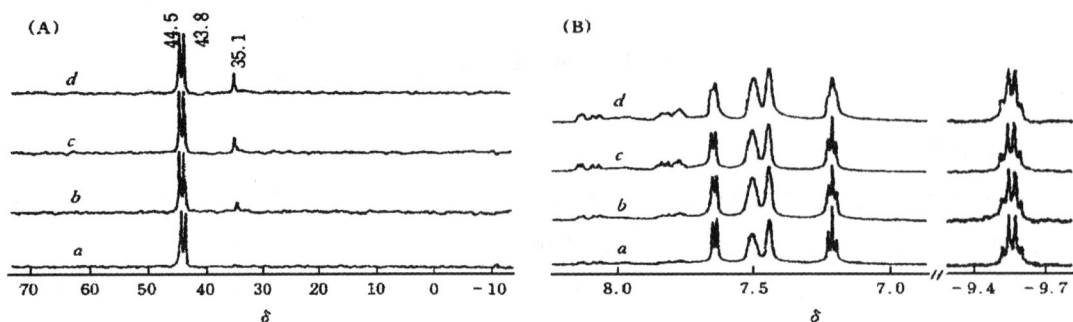

**Fig. 1** $^{31}P(^{1}H)(A)$ and $^{1}H$ NMR(B) spectra of HRh(CO)(TPPTS)$_3$ in the presence of NaOH

a. HRh(CO)(TPPTS)$_3$-D$_2$O;b. $n$(NaOH)$/n$(Rh)=0.4,pH=11.7;c. $n$(NaOH)$/n$(Rh)=4.0,pH=12.6;d. $n$(NaOH)$/n$(Rh)=40.0,pH=13.4.

到 HRh(CO)(TPPTS)$_3$ 的 TPPTS 芳环质子谱峰与吡啶环质子谱峰并存的$^1H$ NMR 谱图。上述结果说明,在 HRh(CO)(TPPTS)$_3$ 体系中加入吡啶后,吡啶分子将与 TPPTS 和/或铑上的 H,CO 等配体竞争和 Rh 配位络合,形成一系列含吡啶基团的铑膦配合物,由此使铑中心更加富电子,$^{31}P$ NMR 谱图中显示三角双锥配合物的特征$^{31}P$谱双峰向高场偏移;当吡啶大量存在时,Rh-H 键消失,且 Rh-TPPTS-吡啶配合物复合体已成为 Rh 配合物的主要物种。Rh-TPPTS-吡啶配合物物种的具体结构有待进一步阐明。

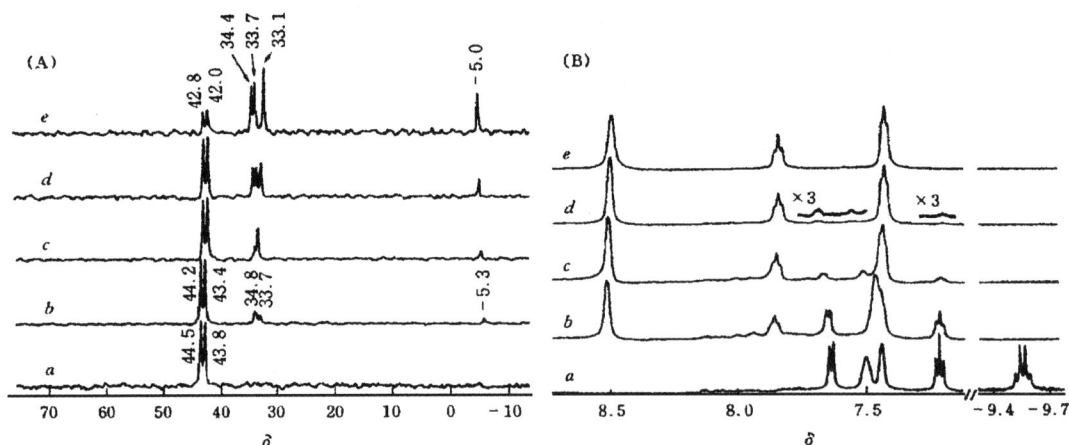

**Fig. 2** $^{31}P(^{1}H)$ NMR(A) and $^{1}H$ NMR(B) spectra of HRh(CO)(TPPTS)$_3$ in the presence of pyridine(Py)

(A)a. $n$(Py)$/n$(Rh)=0.36,pH=7.8;b. $n$(Py)$/n$(Rh)=9.4,pH=8.5;c. $n$(Py)$/n$(Rh)=45.4,pH=9.1;d. $n$(Py)$/n$(Rh)=99.4,pH=9.4;e. $n$(Py)$/n$(Rh)=189.4,pH=9.5(B) a. HRh(CO)·(TPPTS)$_3$;b. $n$(Py)$/n$(Rh)=1.0;c. $n$(Py)$/n$(Rh)=2.7;d. $n$(Py)$/n$(Rh)=5.4;e. Py.

### 2.2 HRh(CO)(TPPTS)$_3$ 在酸存在下的$^{31}P(^{1}H)$ NMR 的结果

与加入 NaOH 不同,在 HRh(CO)(TPPTS)$_3$-D$_2$O 溶液中加入一定量的无机酸(如 HCl,H$_2$SO$_4$,HNO$_3$,H$_3$PO$_4$)后,HRh(CO)(TPPTS)$_3$ 的结构将发生较大变化。图3示出了在 HRh(CO)(TPPTS)$_3$ 中上述 4 种无机酸浓度逐步增加,所体现出的$^{31}P$ NMR 谱图变化。4 种无机酸逐步加入后,配合物特征的$^{31}P$ 物种谱峰强度逐渐减弱,而 OTPPTS 谱峰逐渐增强,在较高浓度的无机酸存在下,配合物特征的$^{31}P$物种均无一例外地被降至零,此时溶液中主要含 OTPPTS 和一类化学位移为 27~30 的$^{31}P$ 物种。根据实验现象并结合下面的氢甲酰化催化反应结果,认为后者很可能为一类含有高价磷物种的 Rh-膦复合体 H$_3$PO$_4$ 对 HRh(CO)(TPPTS)$_3$ 的结构影响程度相对弱于其他 3 种无机酸。从上述结果可以预料,铑膦配合物催化剂水溶液若偏酸性将比偏碱性对催化剂性能和稳定性危害更大些。

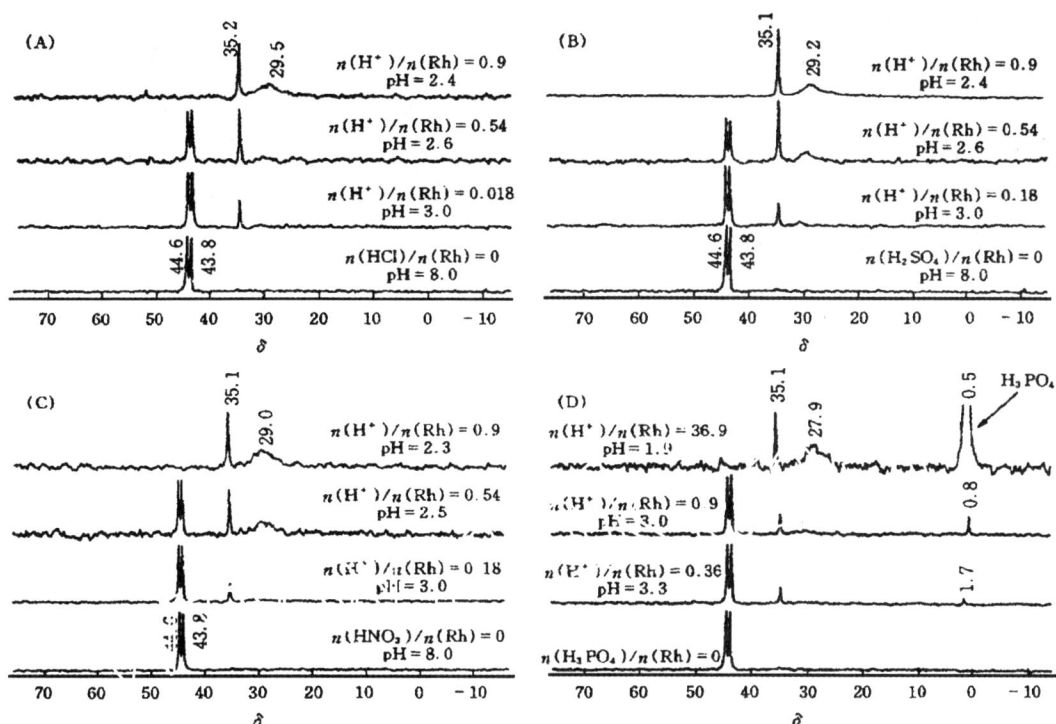

Fig. 3   $^{31}P(^1H)$ NMR spectra of HRh(CO)(TPPTS)$_3$ in the presence of HCl(A),
H$_2$SO$_4$(B), HNO$_3$(C) and H$_3$PO$_4$(D)

图 4 示出了在乙酸存在下,HRh(CO)(TPPTS)$_3$的$^{31}P(^1H)$ NMR 表征结果。从$^{31}P(^1H)$ NMR 谱图来看,随着乙酸量的增加,配合物的特征$^{31}$P 谱峰相对强度减弱,OTPPTS 的含量随之上升,并与上面无机酸存在下的物种变化情形很类似,但可能由于乙酸的酸性较弱,只有在很高浓度的乙酸存在下才能使配合物完全分解。此外,由$^{31}$PNMR 谱图可见,当加入乙酸后,将出现一系列化学位移为 32~29 的$^{31}$P 物种,这可能系 Rh-膦-乙酸复合配合物的贡献。

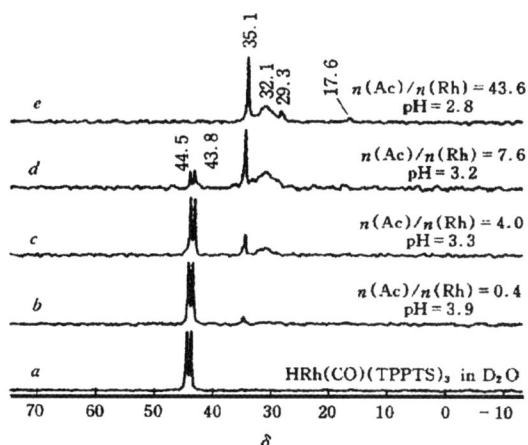

Fig. 4   $^{31}P(^1H)$ NMR spectra of HRh(CO)(TPPTS)$_3$ in the presence of CH$_3$COOH(Ac)

## 2.3   在酸碱存在下 HRh(CO)(TPPTS)$_3$ 的 1-己烯氢甲酰化催化性能

在酸碱存在下,HRh(CO)(TPPTS)$_3$ 的 1-己烯氢甲酰化结果示于表 1。需要指出的是,本文采用化学计量式的膦铑比配合物[即 HRh(CO)(TPPTS)$_3$]作催化剂,未加入过量的配体 TPPTS,因此,表 1 中获得的催化活性和产物醛正异构($n/i$)比值均较低。由表 1 可见,碱性对催化剂活性和选择性影响相对较小,产物清澈透明;吡啶存在下,催化剂表现较好的 $n/i$ 比,可能应归结为吡啶分子参与铑的配位,产生新的水溶性配合物。在硫酸存在下,HRh(CO)(TPPTS)₃ 催化剂虽表现出相对较高的活性,但反应产物颜色淡黄,产物中醛的正异构($n/i$)值也较低,可能暗示硫酸对配合物结构有较大的破坏作用,使 Rh 周围的 TPPTS 膦配体减少,虽提高了催化剂活性,但对醛的选择性大为降低;在加入等物质量的乙酸量与配合物等摩尔的情况下,对反应活性和产物选择性均无明显影响,这可能说明乙酸酸性较弱,尚不足以对配合物的结构产生大的破坏作用。

**Table 1　1-Hexene hydrofoimylation catalyzed by HRh(CO)(TPPTS)₃ in the presence of base or acid***

| Catalyst system | Conversion (%) | Aldehyde Selectivity | $n/i$ | TOF/ $h^{-1}$ | Color of product |
|---|---|---|---|---|---|
| HRh(CO)(TPPTS)₃,pH=8.0 | 55.8 | 100 | 1.32 | 1.53 | Clear |
| $n$[HRh(CO)(TPPTS)₃]/$n$(NaOH)=1/1,pH=11.9 | 54.5 | 100 | 1.42 | 1.51 | Clear |
| $n$[HRh(CO)(TPPTS)₃]/$n$(Pyridine)=1/1,pH=8.1 | 49.6 | 99 | 1.56 | 1.34 | Clear |
| $n$[HRh(CO)(TPPTS)₃]/$n$(H₂SO₄)=1/1,pH=2.3 | 65.1 | 97 | 1.06 | 1.75 | Yellow |
| $n$[HRh(CO)(TPPTS)₃]/$n$(CH₃COOH)=1/1,pH=3.3 | 54.8 | 98 | 1.36 | 1.52 | Clear |

* Reaction conditions:$V$(CO)/$V$(H₂)=1/1,$p$=3.0 MPa,$T$=363 K,$t$=4 h;Catalyst phase:HRh(CO)(TPPTS)₃,0.0032 mmol in 10 mL H₂O;Organic phase:$w$(1-hexene)/$w$(heptane)=1/4,10 mL. TOF:$n$(aldehydes)/[$n$(Rh)·h].

## 3　结　论

$^{31}$P($^{1}$H)和$^{1}$H NMR 表征结果表明,NaOH 对水溶性铑膦配合物 HRh(CO)(TPPTS)₃ 的分子结构破坏较小,含有配位原子 N 的有机碱吡啶加入到 HRh(CO)(TPPTS)₃ 后,配合物分子中部分 TPPTS 配体被交换取代下来,产生新水溶性配合物。无机酸和有机酸均对 HRh(CO)(TPPTS)₃ 的有较明显的结构破坏作用,直至使配合物完全解离,并使解离下来的 TPPTS 氧化成无配位能力的 OTPPTS。配合物在碱存在下可提高其 1-己烯氢甲酰化正异醛比值,但使催化活性下降;而在酸存在下,特别是在无机酸存在下,将获得较差的反应选择性。NMR 结果和 1-己烯氢甲酰化反应活性均表明,酸性对配合物、配体等化合物结构的负面影响较大。因此,以水溶性铑膦配合物为催化剂的水/油两相催化体系的反应条件控制中,应随时监测溶液的 pH 值,避免呈酸性。

## 参考文献

[1]Kuntz E. U. S. Patent,4 248 802,1981.

[2]Morel D.,Jenck J.. Fr. Patent,2 550 202,1982.

[3]Cornila B.,Bahrmann H.,Lipps W. et al.. Eur. Patent,173 219,1986.

[4]Varre C.,Desbois M.,Nouvel J.. Fr. Patent,2 561 650,1985.

[5]Ding H.,Hanson B. E.. J. Chem Soc,Chen. Comm.,1994:2 747.

[6]Ding H.,Hanson B. E.,Glass T. E.. Inorg. Chem. Acta,1995,**229**:329.

[7]Ding H.,Hanson B. E.. J. Mol. Catal.,1995,**99**:131.

[8]Chen H.,Li Y. Z,Chen F. M. et al.. Chinese J. Mol. Catal.,1994,**8**:347.

[9]Horvath I T.,Kastrup R. V.,Oswald A. A. et al.. Catal. Lett.,1989,**2**:85.

[10]Chen Hua（陈华），Hu Jia-Yuan（胡家元），Li Yao-Zhong（黎耀忠）et al.. Abst. of the 7th Chinese National Conf. on Catal.（第七届全国催化学术会议论文摘要集），Dalian,1994:776.

[11]Chen Hua（陈华），Hu Jia-Yuan（胡家元），Li Yao-Zhong（黎耀忠）et al.. Abst. of the 9th Chinese National Conf. on Catal.（第九届全国催化学术会议论文摘要集），Beijng,1998:E-J-12.

[12]Yuan Y. Z., Zhang Y., Fu Q. J. et al.. Abst. of the 9th Intern. Symp. on Relations Between Homogeneoas and Heterogeneous Catal.,Southampton,1998:162.

[13]Arhancet J. P.,Davis M. E.,Merola J. S. et al.. J. Catal.,1990,**121**:327.

[14]Bartik T.,Bartik B.,Hanson B. E. et al.. Inorg. Chem.,1992,**31**:2 667.

# NMR Characterization and Catalytic Hydroformylation of Water-Soluble Rhodium-Phosphine Complex in the Presence of Acid and Base

Yu Zhang，You-Zhu Yuan* ，Xin-Li Liao，Jian-Liang Ye，Chun-Xiang Yao，Khi-Rui Tsai

(Department of Chemistry, Institute of Physical Chemistry, State Key Laboratory of Physical Chemistry for the Solid Surface, Xiamen University, Xiamen 361005, China)

**Abstract**　When NaOH was added into $D_2O$-solution of $HRh(CO)(TPPTS)_3$ [TPPTS:$P(m-C_6H_4SO_3Na)_3$, trisodium salt of tri-($m$-sulfophenyl)-phosphine],there were no changes in the characteristic peaks for the water-soluble complex although a small peak at $\delta$ 35.1 for OTPPTS [OTPPTS: $O=P(m-C_6H_4SO_3Na)_3$, trisodium salt of tri-($m$-sulfophenyl)-phosphine oxide] was formed under a high concentration of NaOH, as evidenced by the spectra of $^{31}P(^1H)$ NMR and $^1H$ NMR, indicating that the influence on the molecular structure of the complex by NaOH may be limited Several new signals at $\delta$ 29~34 accompanied by the appearance of free ligand TPPTS at $\delta$ -5.0~-5.3 appeared in the $^{31}P(^1H)$ NMR spectra when pyridine was introduced into $HRh(CO)(TPPTS)_3$, probably due to the reaction of ligand exchange among the coordinated ligands (such as TPPTS, hydrogen, and CO ) in the complex $HRh(CO)(TPPTS)_3$ and pyridine molecule. The water-soluble complex can be readily decomposed, however, when inorganic acids such as HCl, $H_2SO_4$, $HNO_3$ and $H_3PO_4$ were introduced into the $D_2O$-solution of $HRh(CO)(TPPTS)_3$, as shown in $^{31}P(^1H)$ NMR spectroscopic data. The decomposition of the complex was completed by the formation of OTPPTS at $\delta$ 35.1 and some new phosphate species at $\delta$ 27~29 in the $^{31}P(^1H)$ NMR spectra in the presence of above inorganic acids. Analogous results to those by addition of inorganic acid were obtained when acetate acid was exceeded in mole ratio to $HRh(CO)(TPPTS)_3$. An increment in $n/i$ ratio of heptyl aldehydes and a depression in TOF were obtained in case of the addition of base, in contrary, a lower $n/i$ ratio of aldehydes in yellowish product was obtained in case of the addition of acid in 1-hexene hydrofomylation catalyzed by $HRh(CO)(TPPTS)_3$. The results obtained showed that the molecular structure and catalytic perfomance of $HRh(CO)(TPPTS)_3$ may be affected by acid more disserviceably than by base.

**Key words**　$^{31}P(^1H)$ NMR　$^1H$ NMR　$HRh(CO)(TPPTS)_3$　Acidity and basicity　Hydrofoimy lation

(Ed. :U,X)

■ **本文原载:**《化学学报》第 57 期(1999 年),第 907～913 页。

# 重水中固氮酶催化还原乙炔产物的1H NMR 研究[*]

陈　忠[①]　林国兴　蔡淑惠[a]　徐　昕　黄静伟　万惠霖　蔡启瑞

(厦门大学化学系　[a]物理系　固体表面物理化学国家重点实验室　厦门 361005)

**摘　要**　用1H NMR 研究了固氮酶在重水中催化还原乙炔的反应产物氘代乙烯。利用群对称性对1H NMR 谱图进行了归属,计算了几种可能的 $C_2H_2D_2$ 结构以及 $C_2H_3D$ 结构的 NMR 谱线频率和强度,得出了理论谱。通过理论谱与实验谱的比较,表明固氮酶在重水中催化还原乙炔的产物主要以顺式结构 $C_2H_2D_2$ 为主,并含有较多的单氘代乙烯。单氘代乙烯相对乙烯的化学位移往高场移动约 4.0Hz,而双氘代乙烯向高场的位移大约是单氘代乙烯的 2 倍左右。

**关键词**　氘代乙烯　1HNMR　高自旋体系　谱分析

90 年代初,Kim 和 Rees[1]提出了 K-R 固氮模型,在国际上再次引起固氮研究的热潮。为探讨固氮酶催化反应中底物的配合位以及配合的可能情况,黄静伟等[2]以乙炔为探针,研究了固氮酶在重水中催化还原乙炔为乙烯的反应。由于还原反应是在重水中进行,双氘代乙烯 $C_2H_2D_2$ 应是主要反应产物。$C_2H_2D_2$ 具有图式 1 所示的三种可能结构。理论研究表明,该反应产物是图式 1(a)的顺式双氘代乙烯还是图式 1(b)的反式双氘代乙烯对于揭示固氮酶催化反应的机理具有十分重要的意义[2],为此需对反应产物的结构进行分析表征。然而,图式 1 所示的三种 $C_2H_2D_2$ 的物理和化学性质差异非常小,而且生成产物的量很少,难以用一般的分析测试方法给予定性定量的表征。考虑到1H NMR 对结构变化比较敏感,且具有较高的灵敏度,本文对该反应产物进行了1H NMR 研究。由于2D 的核自旋量子数 $I=1$,而 $I=1$ 的偶合体系比 $I=1/2$ 的偶合体系表述明显复杂得多,即使是弱偶合体系,其基函数、本征值和本征函数的计算过程也很复杂。虽然 Pegg 等[3,4]提出了用 Heissenberg 矢量模型方法和 Heissenberg 表示理论来描述高自旋体系,但这两种方法都不能很好地解析谱线分裂和谱线强度等问题。Nakashima 采用密度矩阵的方法在计算机辅助下计算 NMR 信号[5],这不仅计算量相当大,而且物理意义不太明确。本文利用群论方法[6,7]计算了三种可能的 $C_2H_2D_2$ 以及单氘代 $C_2H_3D$ 的1H NMR 谱线频率和强度。群对称性的利用可以减少久期方程中不为零非对角元的数目和所需求解的子行列式的维数,并且谱线跃迁的计

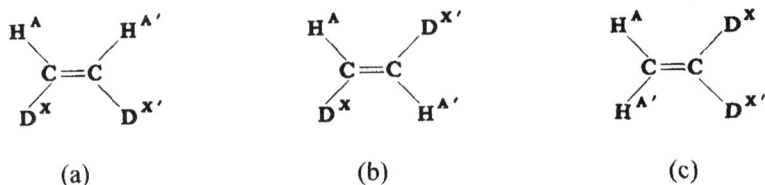

图式 1　$C_2H_2D_2$ 的三种结构

* 国家攀登(29892166)预选项目、国家自然科学基金(19605004)项目和固体表面物理化学国家重点实验室基金项目。

① 陈忠,男,34 岁,博士,副教授。

算只需在同种群对称性间进行,从而大大降低了计算量,有利于对复杂谱图的完整合理归属。

# 1 实 验

NMR 实验是在 Varian 公司生产的 Unity plus 500 超导核磁共振谱仪上进行的,探头为 5mmSW 探头,温度 293K。常压下将固氮酶在重水中催化还原乙炔的产物气 250mL 溶解于装有 0.7mL$C_6D_6$ 的核磁管中,并用万能胶将核磁管封口。为便于比较,将相同量的纯 $C_2H_4$ 溶于含 $C_6D_6$ 的另一核磁管中并密封。由于样品浓度较低,每一实验采样 1 000 次。在匀场较好的条件下,将采样时间取为 5s 以获得较高的分辨率。实验 $^1$H 谱用溶剂 $C_6D_6$ 中残留的 $^1$H 峰(相对于 TMS 峰的位移为 7.4)定标。为了分析方便,把 $C_2H_3D$ 左起第三个谱峰定为 0 Hz.

# 2 实验结果和谱分析

产物气的 $^1$H NMR 谱如图 1(e)所示。该谱图不仅复杂而且谱峰重叠严重,难以对谱峰进行合理的归属。虽然从原理上说,利用 $^{13}$C NMR 能给出更多的信息,但由于产物气的量及其在 $C_6D_6$ 中的溶解度有限,因此用天然丰度的 $^{13}$C 观察不到 $^{13}$C NMR 信号。而纯的已知构型的氘代乙烯样品又难以得到,为此只能凭借理论分析来帮助谱图解析。本文将双氘代乙烯 $C_2H_2D_2$ 作为 AA'XX' 体系,该结构的对称操作属于 $C_2$ 群。利用 $C_2$ 群的对称性,可以简化 36 阶久期方程的求解,从而得到本征值和对应的本征函数。根据跃迁只能在同种不可约表示的子空间进行,再结合选择定则和跃迁强度公式,可得到理论谱如图 1(c)~1(c″)所示。对于单氘代乙烯 $C_2H_3D$,我们用类似方法拟合出其 $^1$H NMR 谱,如图 1(b)所示。

对图式(1)所示的几种可能的 $C_2H_2D_2$ 结构,由对称性有:

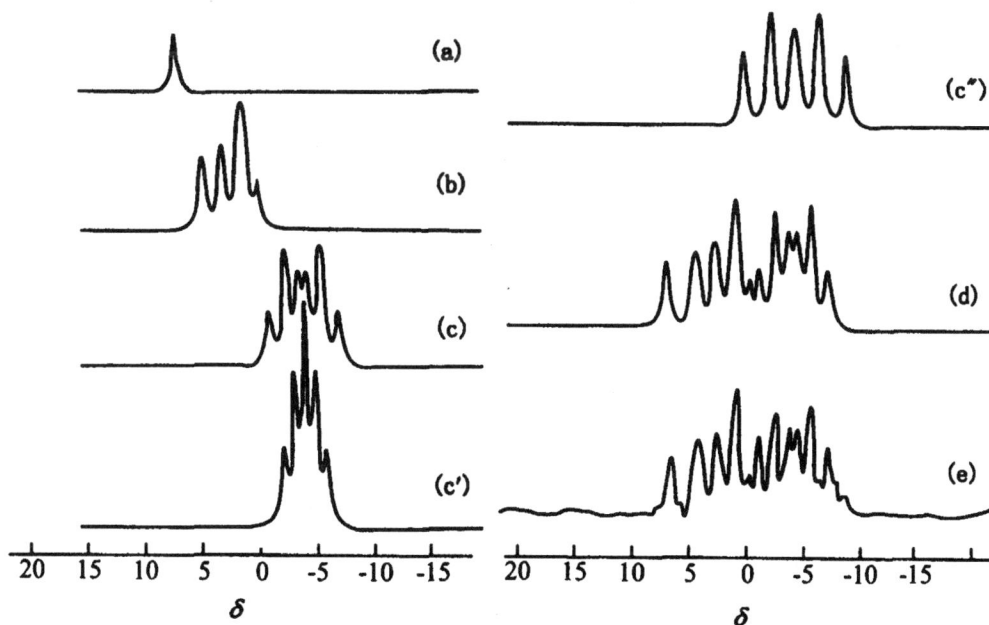

图 1 $^1$H NMR 谱

a—$C_2H_4$ 理论谱;b—$C_2H_3D$ 理论谱;c—图式 1(a)结构 $C_2H_2D_2$ 理论谱;

c′—图式 1(b)结构 $C_2H_2D_2$ 理论谱;c″—图式 1(C)结构 $C_2H_2D_2$ 理论谱;

d—3%$C_2H_4$＋30%$C_2H_3D$＋67%顺式 $C_2H_2D_2$ 理论谱;e—实验谱

$$\begin{cases} \omega_A = \omega_{A'} \\ \omega_X = \omega_{X'} \\ J_{AX} = J_{A'X'} \equiv J \\ J_{A'X} = J_{AX'} \equiv J' \end{cases} \tag{1}$$

其中 $\omega_A = \omega_{A'}$ 为两个 H 原子的化学位移,在固定坐标系中在 500MHz 附近(在 500MHz 谱仪上);$\omega_X = \omega_{X'}$ 为两个 D 原子的化学位移,在 76.8MHz 附近。由于 $\omega_A$ 与 $\omega_X$,$\omega_A - \omega_X$,$\omega_X - \omega_A$ 的频率相差很远,观测 $^1$H 谱时,只能观测到 500MHz 附近只含 $\omega_A$ 的各条谱线,其他含 $\omega_X$,$\omega_A - \omega_X$,$\omega_X - \omega_A$ 的跃迁虽然存在,但不在 $^1$H 谱的观测范围内。计算时,由选择定则 $\Delta m = -1$ 出发,挑选出频率只含 $\omega_A$ 而不含 $\omega_X$ 的各条谱线计算其相对强度,即可得出理论 $^1$H 谱。由于此时 $\omega_A$ 只代表相对的偏移,其具体数值对结果并无影响。可不必引入 $J_{ik}$ 是核 $i$ 和核 $k$ 之间的偶合常数。因此,体系的哈密顿算符为:

$$H = -\{\omega_A(I_{Az} + I_{A'z}) + \omega_X(I_{Xz} + I_{X'z}) + J(I_{Az}I_{Xz} + I_{A'z}I_{X'z})$$
$$+ J'(I_{Az}I_{X'z} + I_{A'z}I_{Xz}) + J_{AA'}I_{Az}I_{A'z} + \frac{1}{2}J_{AA'}(I_A^+ I_{A'}^- + I_A^- I_{A'}^+) \tag{2}$$
$$+ J_{XX'}I_{Xz}I_{X'z} + \frac{1}{2}J_{XX'}(I_X^+ I_{X'}^- + I_X^- I_{X'}^+)\}$$

这里 $I_{iz}$ 是核 $i$ 的核自旋在 $z$ 方向的分量,$I_i^+$ 和 $I_i^-$ 分别是核 $i$ 的核自旋的升降算符。$C_2$ 群的对称元素 $E$ 和 $C_2$ 对 36 个乘积自旋函数作用结果见表 1。

表 1 乘积自旋函数及对称元素对它们的作用

| No. | $m$ | 乘积自旋函数 | $E$ 作用后 No. | $C_2$ 作用后 No. | No. | $m$ | 乘积自旋函数 | $E$ 作用后 No. | $C_2$ 作用后 No. |
| --- | --- | --- | --- | --- | --- | --- | --- | --- | --- |
| 1 | 3 | $\alpha\alpha\chi(1)\chi(1)$ | 1 | 1* | 19 | 0 | $\beta\alpha\chi(1)\chi(-1)$ | 19 | 18 |
| 2 | 2 | $\alpha\alpha\chi(1)\chi(0)$ | 2 | 3 | 20 | 0 | $\beta\alpha\chi(0)\chi(0)$ | 20 | 17 |
| 3 | 2 | $\alpha\alpha\chi(0)\chi(1)$ | 3 | 2 | 21 | 0 | $\beta\alpha\chi(-1)\chi(1)$ | 21 | 16 |
| 4 | 2 | $\alpha\beta\chi(1)\chi(1)$ | 4 | 5 | 22 | 0 | $\beta\beta\chi(1)\chi(0)$ | 22 | 23 |
| 5 | 2 | $\beta\alpha\chi(1)\chi(1)$ | 5 | 4 | 23 | 0 | $\beta\beta\chi(0)\chi(1)$ | 23 | 22 |
| 6 | 1 | $\alpha\alpha\chi(1)\chi(-1)$ | 6 | 8 | 24 | -1 | $\alpha\alpha\chi(-1)\chi(-1)$ | 24 | 24* |
| 7 | 1 | $\alpha\alpha\chi(0)\chi(0)$ | 7 | 7* | 25 | -1 | $\alpha\beta\chi(0)\chi(-1)$ | 25 | 28 |
| 8 | 1 | $\alpha\alpha\chi(-1)\chi(1)$ | 8 | 6 | 26 | -1 | $\alpha\beta\chi(-1)\chi(0)$ | 26 | 27 |
| 9 | 1 | $\alpha\beta\chi(1)\chi(0)$ | 9 | 12 | 27 | -1 | $\beta\alpha\chi(0)\chi(-1)$ | 27 | 26 |
| 10 | 1 | $\alpha\beta\chi(0)\chi(1)$ | 10 | 11 | 28 | -1 | $\beta\alpha\chi(-1)\chi(0)$ | 28 | 25 |
| 11 | 1 | $\beta\alpha\chi(1)\chi(0)$ | 11 | 10 | 29 | -1 | $\beta\beta\chi(1)\chi(-1)$ | 29 | 31 |
| 12 | 1 | $\beta\alpha\chi(0)\chi(1)$ | 12 | 9 | 30 | -1 | $\beta\beta\chi(0)\chi(0)$ | 30 | 30* |
| 13 | 1 | $\beta\beta\chi(1)\chi(1)$ | 13 | 13* | 31 | -1 | $\beta\beta\chi(-1)\chi(1)$ | 31 | 29 |
| 14 | 0 | $\alpha\alpha\chi(0)\chi(-1)$ | 14 | 15 | 32 | -2 | $\alpha\beta\chi(-1)\chi(-1)$ | 32 | 33 |
| 15 | 0 | $\alpha\alpha\chi(-1)\chi(0)$ | 15 | 14 | 33 | -2 | $\beta\alpha\chi(-1)\chi(-1)$ | 33 | 32 |
| 16 | 0 | $\alpha\beta\chi(1)\chi(-1)$ | 16 | 21 | 34 | -2 | $\beta\beta\chi(0)\chi(-1)$ | 34 | 35 |
| 17 | 0 | $\alpha\beta\chi(0)\chi(0)$ | 17 | 20 | 35 | -2 | $\beta\beta\chi(-1)\chi(0)$ | 35 | 34 |
| 18 | 0 | $\alpha\beta\chi(-1)\chi(1)$ | 18 | 19 | 36 | -3 | $\beta\beta\chi(-1)\chi(-1)$ | 36 | 36* |

\* 代表在对称元素 $C_2$ 作用后得到的基函数没有改变

由表 1,以乘积自旋函数为基函数的 $C_2$ 对称群可约表示中 $E$ 和 $C_2$ 的表示矩阵的迹分别为 $\mathrm{tr}E = 36$,$\mathrm{tr}C_2 = 6$. $C_2$ 对称群的矩阵表示约化公式为

$$n_i = \frac{1}{g} \sum_{\text{g}} X(Rg) X_i^*(Rg) \tag{3}$$

其中 $g$ 为群的阶，$X(Rg)$ 为要约化的矩阵表示中对称元素 $Rg$ 的特征标，$X_i^*(Rg)$ 为第 $i$ 个不可约矩阵表示中对称元素 $Rg$ 的特征标的复共轭。根据(3)式，以 36 个乘积自旋函数为基函数的 $C_2$ 对称群可约表示可以约化为 $1/2 \times (36 \times 1 + 6 \times 1) = 21$ 个不可约表示 $A$ 和 $1/2 \times (36 \times 1 - 6 \times 1) = 15$ 个不可约表示 $B$。用投影算符 $P_A = (E + C_2)$ 和 $P_B = 1/2(E - C_2)$ 作用在 36 个乘积自旋函数上，可以得到 $A$，$B$ 对称化的基函数，见表 2。

### 表 2  对称化基函敏和对角元

| No. | $m$ | 对称化基函数 | 对角元 | 顺式结构的对角元数值 |
|---|---|---|---|---|
| \multicolumn{5}{c}{A 对称} | | | | |
| 1 | 3 | $\alpha\alpha\chi(1)\chi(1)$ | $-\omega_A - 2\omega_X - J - J' - 1/4 J_{AA'} - J_{XX'}$ | $-\omega_A - 2\omega_X - 6.3$ |
| 2 | 2 | $1/\sqrt{2}\{\alpha\alpha\chi(1)\chi(0) + \alpha\alpha\chi(0)\chi(1)\}$ | $-\omega_A - \omega_X - 1/2 J - 1/2 J' - 1/4 J_{AA'} - J_{XX'}$ | $-\omega_A - \omega_X - 4.7$ |
| 3 | 2 | $1/\sqrt{2}\{\alpha\beta\chi(1)\chi(1) + \beta\alpha\chi(1)\chi(1)\}$ | $-2\omega_X - 1/4 J_{AA'} - J_{XX'}$ | $-2\omega_X - 3.1$ |
| 4 | 1 | $1/\sqrt{2}\{\alpha\alpha\chi(1)\chi(-1) + \alpha\alpha\chi(-1)\chi(1)\}$ | $-\omega_A - 1/4 J_{AA'} + J_{XX'}$ | $-\omega_A - 2.6$ |
| 5 | 1 | $\alpha\alpha\chi(0)\chi(0)$ | $-\omega_A - 1/4 J_{AA'}$ | $-\omega_A - 2.85$ |
| 6 | 1 | $1/\sqrt{2}\{\alpha\beta\chi(1)\chi(0) + \beta\alpha\chi(0)\chi(1)\}$ | $-\omega_X - 1/2 J + 1/2 J' + 1/4 J_{AA'}$ | $-\omega_X + 4.15$ |
| 7 | 1 | $1/\sqrt{2}\{\alpha\beta\chi(0)\chi(1) + \beta\alpha\chi(1)\chi(0)\}$ | $-\omega_X - 1/2 J - 1/2 J' + 1/4 J_{AA'}$ | $-\omega_X + 1.15$ |
| 8 | 1 | $\beta\beta\chi(1)\chi(1)$ | $\omega_A - 2\omega_X + J + J' - 1/4 J_{AA'} - J_{XX'}$ | $\omega_A - 2\omega_X + 0.1$ |
| 9 | 0 | $1/\sqrt{2}\{\alpha\alpha\chi(0)\chi(-1) + \alpha\alpha\chi(-1)\chi(0)\}$ | $-\omega_A + \omega_X + 1/2 J + 1/2 J' - 1/4 J_{AA'} - J_{XX'}$ | $-\omega_A + \omega_X - 1.5$ |
| 10 | 0 | $1/\sqrt{2}\{\alpha\beta\chi(1)\chi(-1) + \beta\alpha\chi(-1)\chi(1)\}$ | $-J + J' + 1/4 J_{AA'} + J_{XX'}$ | $5.7$ |
| 11 | 0 | $1/\sqrt{2}\{\alpha\beta\chi(0)\chi(0) + \beta\alpha\chi(0)\chi(0)\}$ | $-1/4 J_{AA'}$ | $-2.85$ |
| 12 | 0 | $1/\sqrt{2}\{\alpha\beta\chi(-1)\chi(1) + \beta\alpha\chi(1)\chi(-1)\}$ | $-J + J' + 1/4 J_{AA'} + J_{XX'}$ | $0.5$ |
| 13 | 0 | $1/\sqrt{2}\{\beta\beta\chi(1)\chi(0) + \beta\beta\chi(0)\chi(1)\}$ | $\omega_A - \omega_X + 1/2 J + 1/2 J' - 1/4 J_{AA'} - J_{XX'}$ | $\omega_A - \omega_X - 1.5$ |
| 14 | -1 | $\alpha\alpha\chi(-1)\chi(-1)$ | $-\omega_A + 2\omega_X + J + J' - 1/4 J_{AA'} - J_{XX'}$ | $-\omega_A + 2\omega_X + 0.1$ |
| 15 | -1 | $1/\sqrt{2}\{\alpha\beta\chi(0)\chi(-1) + \beta\alpha\chi(-1)\chi(0)\}$ | $\omega_X - 1/2 J + 1/2 J' + 1/4 J_{AA'}$ | $\omega_X + 4.15$ |
| 16 | -1 | $1/\sqrt{2}\{\alpha\beta\chi(-1)\chi(0) + \beta\alpha\chi(0)\chi(-1)\}$ | $\omega_X + 1/2 J - 1/2 J' + 1/4 J_{AA'}$ | $\omega_X + 1.55$ |
| 17 | -1 | $1/\sqrt{2}\{\beta\beta\chi(1)\chi(-1) + \beta\beta\chi(-1)\chi(1)\}$ | $\omega_A - 1/4 J_{AA'} + J_{XX'}$ | $\omega_A - 2.6$ |
| 18 | -1 | $\beta\beta\chi(0)\chi(0)$ | $\omega_A - 1/4 J_{AA'}$ | $\omega_A - 2.85$ |
| 19 | -2 | $1/\sqrt{2}\{\alpha\beta\chi(-1)\chi(-1) + \beta\alpha\chi(-1)\chi(-1)\}$ | $2\omega_X - 1/4 J_{AA'} - J_{XX'}$ | $2\omega_X - 3.1$ |
| 20 | -2 | $1/\sqrt{2}\{\beta\beta\chi(0)\chi(-1) + \beta\beta\chi(-1)\chi(0)\}$ | $\omega_A + \omega_X - 1/2 J - 1/2 J' - 1/4 J_{AA'} - J_{XX'}$ | $\omega_A + \omega_X - 4.7$ |
| 21 | -3 | $\beta\beta\chi(-1)\chi(-1)$ | $\omega_A + 2\omega_X - J - J' - 1/4 J_{AA'} - J_{XX'}$ | $\omega_A + 2\omega_X - 6.3$ |
| \multicolumn{5}{c}{B 对称} | | | | |
| 1 | 2 | $1/\sqrt{2}\{\alpha\alpha\chi(1)\chi(0) - \alpha\alpha\chi(0)\chi(1)\}$ | $-\omega_A - \omega_X - 1/2 J - 1/2 J' - 1/4 J_{AA'} - J_{XX'}$ | $-\omega_A - \omega_X - 4.2$ |
| 2 | 2 | $1/\sqrt{2}\{\alpha\beta\chi(1)\chi(1) - \beta\alpha\chi(1)\chi(1)\}$ | $-2\omega_X + 1/4 J_{AA'} - J_{XX'}$ | $-2\omega_X + 8.3$ |
| 3 | 1 | $1/\sqrt{2}\{\alpha\alpha\chi(1)\chi(-1) - \alpha\alpha\chi(-1)\chi(1)\}$ | $-\omega_A - 1/4 J_{AA'} + J_{XX'}$ | $-\omega_A - 2.6$ |
| 4 | 1 | $1/\sqrt{2}\{\alpha\beta\chi(1)\chi(0) - \beta\alpha\chi(0)\chi(1)\}$ | $-\omega_X - 1/2 J + 1/2 J' + 1/4 J_{AA'}$ | $-\omega_X + 4.15$ |
| 5 | 1 | $1/\sqrt{2}\{\alpha\beta\chi(0)\chi(1) - \beta\alpha\chi(1)\chi(0)\}$ | $-\omega_X - 1/2 J - 1/2 J' + 1/4 J_{AA'}$ | $-\omega_X + 1.55$ |
| 6 | 0 | $1/\sqrt{2}\{\alpha\alpha\chi(0)\chi(-1) - \alpha\alpha\chi(-1)\chi(0)\}$ | $-\omega_A + \omega_X + 1/2 J + 1/2 J' - 1/4 J_{AA'} + J_{XX'}$ | $-\omega_A + \omega_X - 1.0$ |
| 7 | 0 | $1/\sqrt{2}\{\alpha\beta\chi(1)\chi(-1) - \beta\alpha\chi(-1)\chi(1)\}$ | $-J + J' + 1/4 J_{AA'} + J_{XX'}$ | $5.7$ |

续表

| | | | | |
|---|---|---|---|---|
| 8 | 0 | $1/\sqrt{2}\{\alpha\beta\chi(0)\chi(0)-\beta\alpha\chi(0)\chi(0)\}$ | $1/4J_{AA'}$ | 8.55 |
| 9 | 0 | $1/\sqrt{2}\{\alpha\beta\chi(-1)\chi(1)-\beta\alpha\chi(1)\chi(-1)\}$ | $J-J'+1/4J_{AA'}+J_{XX'}$ | 0.5 |
| 10 | 0 | $1/\sqrt{2}\{\beta\beta\chi(1)\chi(0)-\beta\beta\chi(0)\chi(1)\}$ | $\omega_A-\omega_X+1/2J+1/2J'-1/4J_{AA'}+J_{XX'}$ | $\omega_A-\omega_X-1.0$ |
| 11 | −1 | $1/\sqrt{2}\{\alpha\beta\chi(0)\chi(-1)-\beta\alpha\chi(-1)\chi(0)\}$ | $\omega_X-1/2J+1/2J'+1/4J_{AA'}$ | $\omega_X+4.15$ |
| 12 | −1 | $1/\sqrt{2}\{\alpha\beta\chi(-1)\chi(0)-\beta\alpha\chi(0)\chi(-1)\}$ | $\omega_X+1/2J-1/2J'+1/4J_{AA'}$ | $\omega_X+1.55$ |
| 13 | −1 | $1/\sqrt{2}\{\beta\beta\chi(1)\chi(-1)-\beta\beta\chi(-1)\chi(1)\}$ | $\omega_A-1/4J_{AA'}+J_{XX'}$ | $\omega_A-2.6$ |
| 14 | −2 | $1/\sqrt{2}\{\alpha\beta\chi(-1)\chi(-1)-\beta\alpha\chi(-1)\chi(-1)\}$ | $2\omega_X+1/4J_{AA'}-J_{XX'}$ | $2\omega_X+8.3$ |
| 15 | −2 | $1/\sqrt{2}\{\beta\beta\chi(0)\chi(-1)-\beta\beta\chi(-1)\chi(0)\}$ | $\omega_A+\omega_X-1/2J-1/2J'-1/4J_{AA'}+J_{XX'}$ | $\omega_A+\omega_X-4.2$ |

表中只给出顺式结构数据

表中顺式结构的数据由文献[7]中的实验值结合关系式 $J_{AB}=K\gamma_A\gamma_B$ 和 $\gamma_H/\gamma_D=6.5$ 得到：$J=0.3$ Hz，$J'=1.8$ Hz，$J_{AA'}=11.4$ Hz，$J_{XX'}=0.3$ Hz。从表2看，21个不可约表示A对称化的基函数构成21维子空间。根据相同 $m$ 值的态才能混合的原则，可以把 $21\times21$ 的子行列式进一步分解成 $1\times1,2\times2,5\times5,5\times5,5\times5,2\times2,1\times1$ 等7个子行列式。同样，15个不可约表示B对称化的基函数构成15维子空间，而 $15\times15$ 的子行列式可进一步分解成 $2\times2,3\times3,5\times5,3\times3,2\times2$ 等5个子行列式。

表3 非对角元，$H_{ij}=H_{ji}$

| A 对称性 | | B 对称性 | | |
|---|---|---|---|---|
| 4 | 5 | $1\sqrt{2}J_{XX'}$ | 4 | 5 | $1/2J_{AA'}-J_{XX'}$ |
| 6 | 7 | $-(1/2J_{AA'}+J_{XX'})$ | 7 | 8 | $-J_{XX'}$ |
| 10 | 11 | $-J_{XX'}$ | 7 | 9 | $1/2J_{AA}$ |
| 10 | 12 | $-1/2J_{AA'}$ | 8 | 9 | $-J_{XX'}$ |
| 11 | 12 | $-J_{XX'}$ | 11 | 12 | $1/2J_{AA'}-J_{XX'}$ |
| 15 | 16 | $-(1/2J_{AA'}+J_XX')$ | | | |
| 17 | 18 | $-\sqrt{2}J_{XX'}$ | | | |

这些行列式的对角元和不等零的非对角元分别列于表2和表3。

从表中可以看出，在利用群对称性的对称化基函数的情况下，不等零的非对角元很少。限于篇幅，我们只给出顺式结构的数据。用消元法解这些久期方程，可得到体系的本征值与本征函数见表4。由于跃迁只能在同种不可约表示的对称性间进行，再根据选择定则 $\Delta m=-1$ 及相对强度正比于 $|\langle\phi_{m-1}|I^-|\phi_m\rangle|^2$，可得表5所列质子共振频率及相对强度，从而画出理论谱如图1(c)。基于谱的对称性，我们只给出谱的一半。

# 3  结果与讨论

在同一分子内，各条谱线的相对强度与 $|\langle\phi_{m-1}|I^-|\phi_m\rangle|^2$ 成正比。在同一样品管中，某种构型分子 $^1$H 谱线的总强度与每个分子所含 H 原子的数目和分子摩尔浓度的乘积成正比。将实验谱中的谱线用理论谱归属后，由于每种构型分子所含 H 原子的数目是已知的，只要利用各种构型分子谱线的积分强度，再算出每种构型分子中 $^1$H 谱线的相对强度，就可以求出各种构型分子的相对摩尔浓度（可进一步换算成百分比浓度），从而得到不同构型分子间谱线强度对比。将各种构型分子的理论谱按强度对比进行叠加，即得图1(d)。由图1(d)与图1(e)中谱线位置与强度的对比可以认为，固氮酶在重水中催化还原乙炔的产物——氘代乙烯中主要以顺式结构 $C_2H_2D_2$ 存在[图式1(a)]，另外还含有较多的单氘代乙烯，这可能是由于重水氘代不完全的缘故。为什么产物中单氘代乙烯含量比重水中氘代不完全的程度明显多，更确切的证据，还在进一步研究中。图1(e)中第一个谱峰，处于低场，是没有被氘代的乙烯的谱峰。图1(d)左起第六条谱线强度小于图1(e)实验线，这可能是由于(b)，(c)两种构型的氘代乙烯很少，理论谱中略去了它们的贡献，从而引起强度的偏差。实验发现，单氘代乙烯相对乙烯的化学位移往高场移动

约 4.0 Hz,双氘代乙烯往高场的位移大约是单氘代乙烯的 2 倍。这可能是由于 $C_2H_4$ 上的氢被较重的同位素氘取代,使分子的位能降低,从而[1]H 的化学位移略向高场移动。此结果与文献[7]中单氘代乙烯相对乙烯的化学位移往高场移动是一致的。

**表 4  本函数及本征值**

| $\psi$ | $m$ | 本征函数 (A 对称) | 本征值 (A 对称) | $m$ | 本征函数 (B 对称) | 本征值 (B 对称) |
|---|---|---|---|---|---|---|
| 1 | 3 | $\varphi_a(1)$ | $-\omega_A-2\omega_X-6.3$ | 2 | $\varphi_\beta(1)$ | $-\omega_A-\omega_X-4.2$ |
| 2 | 2 | $\varphi_a(2)$ | $-\omega_A-\omega_X-4.7$ | | $\varphi_\beta(2)$ | $-2\omega_X+8.3$ |
| 3 | | $\varphi_a(3)$ | $-2\omega_X-3.1$ | 1 | $\varphi_\beta(3)$ | $-\omega_A-2.6$ |
| 4 | | $\sqrt{6}/3\varphi_a(4)-1/\sqrt{3}\varphi_a(5)$ | $-\omega_A-2.35$ | | $0.785\varphi_\beta(4)+0.619\varphi_\beta(5)$ | $-\omega_X+8.45$ |
| 5 | | $1/\sqrt{3}/3\varphi_a(4)+2/\sqrt{6}\varphi_a(5)$ | $-\omega_A-3.1$ | | $0.619\varphi_\beta(4)-0.785\varphi_\beta(5)$ | $-\omega_X-2.75$ |
| 6 | 1 | $0.630\varphi_a(6)+0.778\varphi_a(7)$ | $-\omega_X-3.19$ | | $\varphi_\beta(6)$ | $-\omega_A+\omega_X-1.0$ |
| 7 | | $-0.778\varphi_a(6)+0.630\varphi_a(7)$ | $-\omega_X+8.89$ | | $0.789\varphi_\beta(7)-0.345\varphi_\beta(8)+0.509\varphi_\beta(9)$ | 9.49 |
| 8 | | $\varphi_a(8)$ | $-\omega_A-2\omega_X+0.1$ | 0 | $0.295\varphi_\beta(7)+0.938\varphi_\beta(8)+0.182\varphi_\beta(9)$ | 8.42 |
| 9 | | $\varphi_a(9)$ | $-\omega_A+\omega_X-1.5$ | | $-0.541\varphi_\beta(7)+0.006\varphi_\beta(8)+0.841\varphi_\beta(9)$ | $-3.17$ |
| 10 | | $-0.841\varphi_a(10)+0.006\varphi_a(11)+0.541\varphi_a(12)$ | 9.365 | | $\varphi_\beta(10)$ | $\omega_A-\omega_X-1.0$ |
| 11 | 0 | $0.286\varphi_a(10)-0.841\varphi_a(11)+0.455\varphi_a(12)$ | $-2.63$ | | $0.785\varphi_\beta(11)+0.619\varphi_(12)$ | $\omega_X+8.45$ |
| 12 | | $0.459\varphi_a(10)+0.535\varphi_a(11)+0.709\varphi_a(12)$ | $-3.39$ | $-1$ | $0.619\varphi_\beta(11)-0.785\varphi_(12)$ | $\omega_X-2.753$ |
| 13 | | $\varphi_a(13)$ | $\omega_A-\omega_X-1.5$ | | $\varphi_\beta(13)$ | $\omega_A-2.6$ |
| 14 | | $\varphi_a(14)$ | $-\omega_A+2\omega_X+0.1$ | | $\varphi_\beta(14)$ | $2\omega_X+8.3$ |
| 15 | | $0.630\varphi_a(15)+0.778\varphi_a(16)$ | $\omega_X-3.19$ | $-2$ | $\varphi_\beta(15)$ | $\omega_A+\omega_X-4.2$ |
| 16 | $-1$ | $-0.778\varphi_a(15)+0.630\varphi_a(16)$ | $\omega_X+8.89$ | | | |
| 17 | | $\sqrt{6}/3\varphi_a(17)-1/\sqrt{3}\varphi_a(18)$ | $\omega_A-2.35$ | | | |
| 18 | | $1/\sqrt{3}\varphi_a(17)+2/\sqrt{6}\varphi_a(18)$ | $\omega_A-3.1$ | | | |
| 19 | $-2$ | $\varphi_a(19)$ | $2\omega_X-3.1$ | | | |
| 20 | | $\varphi_a(20)$ | $\omega_A+\omega_X-4.7$ | | | |
| 21 | $-3$ | $\varphi_a(21)$ | $\omega_A+2\omega_X-6.3$ | | | |

表中只给出顺式结构数据,$\varphi_a$ 为 A 对称性对称化基函数;$\varphi_\beta$ 为 B 对称性对称化基函数,见表 2

表 5　顺式结构 $C_2H_2D_2$ 中质子的共振频率及相对强度

| A 对称性 | | | | | | | | B 对称性 | | | |
|---|---|---|---|---|---|---|---|---|---|---|---|
| No. | 跃迁 | 频率/Hz | 相对强度 | No. | 跃迁 | 频率/Hz | 相对强度 | No. | 跃迁 | 频率/Hz | 相对强度 |
| 1 | 1→3 | $-\omega_A-3.2$ | 2 | 7 | 4→12 | $-\omega_A+1.04$ | 0.067 | 1 | 1→4 | $-\omega_A-12.65$ | 0.027 |
| 2 | 2→6 | $-\omega_A-1.51$ | 1.978 | 8 | 5→10 | $-\omega_A-12.47$ | 0.027 | 2 | 1→5 | $-\omega_A-1.447$ | 1.97 |
| 3 | 2→7 | $-\omega_A-13.59$ | 0.016 | 9 | 5→11 | $-\omega_A-0.47$ | 0.30 | 3 | 3→7 | $-\omega_A-12.09$ | 0.0784 |
| 4 | 3→8 | $-\omega_A-3.2$ | 2 | 10 | 5→12 | $-\omega_A+0.29$ | 1.67 | 4 | 3→8 | $-\omega_A-11.02$ | 0.0127 |
| 5 | 4→10 | $-\omega_A-11.72$ | 0.062 | 11 | 6→13 | $-\omega_A-1.69$ | 1.978 | 5 | 3→9 | $-\omega_A+0.566$ | 1.91 |
| 6 | 4→11 | $-\omega_A+0.28$ | 1.67 | 12 | 7→13 | $-\omega_A+10.39$ | 0.0216 | 6 | 4→10 | $-\omega_A+9.45$ | 0.027 |
| | | | | | | | | 7 | 5→10 | $-\omega_A-1.753$ | 1.97 |

# References

[1] J. Kim, D. C. Rees, Science, 1992, **257**, 1677.

[2] Huang Jing-Wei, Ph. D. Thesis of Xiamen Univ., Xiamen, 1994.

[3] D. T. Pegg, M. R. Bendall, D. M. Doddrell, J, Magn. Reson., 1984, **58**, 14.

[4] D. T. Pegg, M. R. Bendall, J. Magn. Reson., 1983, **53**, 229.

[5] T. T. Nakashima, R. E. D. McClung, B. K. John, J. Magn. Reson., 1981, **58**, 27.

[6] P. L. Corio, "Structure of High-Resolution NMR Spectra", Academic Press, New York, 1966.

[7] G. S. Reddy, J. H. Geldstein, J. Mol. Spectrose., 1962, **8**, 485.

# [1]H NMR Study on the Products of the Catalytic Reduction of Ethyne by Nitrogenase in $D_2O$

Zhong Chen[*], Guo-Xing Lin, Shu-Hui Cai[a], Xin Xu,

Jing-Wei Huang, Hui-Lin Wan, Qi-Rui Cai

(Department of Chemistry, [a] Department of Physics, Xiamen University, Xiamen, 361005)

**Abstrbct**　The products of the catalytic reduction ethyne by nitrogenase in $D_2O$ were studied by [1]H NMR. In order to assign the [1]H NMR spectrum, the theoretical spectra of $C_2H_2D_2$ were calculated using group symmetry. The final eigenfunctions, eigenvalues, allowed transitions and intensities of $C_2H_2D_2$ and $C_2H_3D$ were obtained. It can be concluded that the products comprised mainly of $cis$-$C_2H_2D_2$. There was also some $C_2H_3D$. [1]H chemical shift in $C_2H_3D$ moved by ca. 4.0Hz to the upfield compared with that in $C_2H_4$. The upfield shift in $C_2H_2D_2$ was about twice as large as that in $C_2H_3D$. The theoretical results are in good agreement with experimental ones.

**Key words**　Ethylene -$d$　[1]H NMR　High spin system　Spectra analysis

■ 本文原载：J. Chem. Soc., Dalton Trans., 1999, pp. 4289～4290.

# Bidentate Citrate with Free Terminal Carboxyl Groups, Syntheses and Characterization of Citrato Oxomolybdate(VI) and Oxotungstate(VI), $\Delta/\Lambda$-Na$_2$[MO$_2$(H$_2$cit)$_2$]·3H$_2$O(M=Mo or W)

Zhao-Hui Zhou, Hui-Lin Wan, Khi-Rui Tsai

*Department of Chemistry and State Key Laboratory for Physical Chemistry of Solid Surface, Xiamen University, Xiamen, 361005, P.R. China)*

E-mail: zhzhou@xmu.edu.cn

Received 2nd November 1999, accepted 8th November 1999

**Abstract** The first example of monomeric bidentate citrato oxomolybdate(VI) and -tungstate (VI) have been prepared by cleavage of dimeric citrato molybdate and tungstate.

The FeMo-cofactor in nitrogenase, where dinitrogen is bound and transformed, is a MoFe$_7$S$_9$ cluster, and homocitrate is an integral part in the coordination sphere of molybdenum through vicinal carboxylate and alkoxide[1]. Alternatively polycarboxylic acid (i.e., MoFe$_7$S$_9$: citrate) results in low nitrogen fixing ability, but is able to reduce C$_2$H$_2$ to C$_2$H$_4$[2]. It has been suggested that a possible function of the tricarboxylic acid in the biosynthesis of the cofactor of the nitrogenase is to mobilize molybdenum from the appropriate storage enzyme. Molybdenum is believed to be taken up by organisms as MoO$_4^{2-}$, this would be essential for the assembly of the final cofactor cluster from an oxomolybdenum-homocitrate precursor[3]. While the precise role of homocitrate is poorly understood, the discovery and elucidation of the interactions between molybdenum and the tricarboxylic acid are of great importance[4]. Moreover, high valent molybdenum oxotransferases are enzymes that catalyze oxygen atom transfer from a substrate. In order to understand the chemistry and oxo-transfer properites of these enzymes, numerous dioxo-MO(VI) complexes having a wide range of ligands have been prepared and structurally charactersed[5]. Although a few molybdenum and tungsten complexes of citrate are known[6,7], none of these meets the requirements for a satisfactory structural model, since they are binuclear or tetranuclear complexes and contain the citrate in a tridentate mode. Here we report closer structural models of the mononuclear precursors: Na$_2$[MO$_2$(H$_2$cit)$_2$]·3H$_2$O(M=Mo **3** or W **4**) prepared from the cleavages of dimeric tridentate citrato molybdate(VI) and tungstate, which contain the correctly coordinated carboxylate-alkoxide moiety.

Dimeric citrato complexes Na$_4$[Mo$_2$O$_5$(cit)$_2$]·4H$_2$O **1** and Na$_4$[W$_2$O$_5$(cit)$_2$]·8H$_2$O **2** were prepared from the reactions of sodium molybdate or tungstate with citric acid (H$_4$cit) as described previously[8]. When **1**(4.8 g, 5 mmol) was added with an excess of citric acid(2.5 g, 12 mmol) and the solution heated at 60 ℃ for 4 h, Na$_2$[MoO$_2$(H$_2$cit)$_2$]·3H$_2$O **3**† was obtained in 43% yield(2.6 g). Similarly, when **2** was reacted with an excess of citric acid, a 35% yield of Na$_2$[WO$_2$(H$_2$cit)$_2$]·3H$_2$O

4† was separated. Both complexes were characterised by single X-ray structure determination and were shown to be isomorphous. ‡

Direct cleavage of dinuclear citrato complexes with excess citric acid results in the preparation of the title complexes, in which the pH values remain in the range of 3. 5[8]. The crystal structures of 3 and 4 comprise discrete sodium cations, water molecules and dihydrocitrato dioxo molybdate or tungstate anions, Fig. 1 and Fig. 2 Show plots of the anion structures of 3 and 4 respectively, which show one of the stereoisomers($\Delta$ configuration) of each complex. Selected bond parameters are given in the captions. Both molybdenum and tungsten atoms are quasi-octahedrally coordinated by two *cis*-oxo groups and two citrate ligands. Each citrate ion acts as a bidentate ligand *via* its alkoxy and $\alpha$-carboxyl groups, while the other two $\beta$-carboxylic acid groups remain uncomplexed. Bidentate coordinaon of citrate through its alkoxy and $\alpha$-carboxyl group is rare. Complex $Na_2[MoO_2(H_2cit)_2] \cdot 3H_2O$ 3 represents the first example of a structurally characterised monomeric oxomolybdenum-citrate complex, which exhibits the bidentate coordination mode of the tricarboxylic acid to molybdenum. Such is the case in the related dinuclear citrato vanadate $K_2[VO_2(H_2cit)]_2 \cdot 4H_2O$[9], in which the citrate ligand is bidentate. There are two enantiomers of the two complexes, which results from the asymmetric octahedral coordination environment around molybdenum or tungsten. This is similar to that of FeMo-co, in which the octahedral coordination geometry for Mo is typically asymmetric.

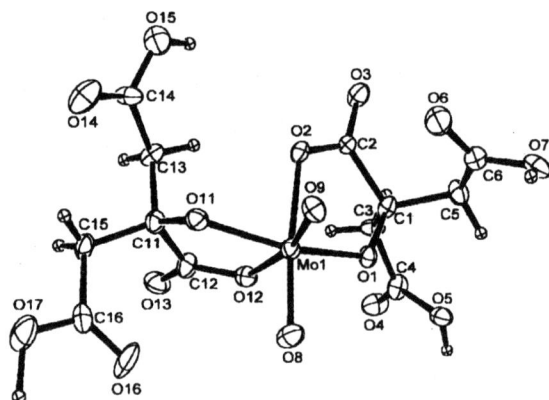

Fig. 1 Perspective View of the anion structure of $Na_2[MoO_2(H_2cit)_2] \cdot 3H_2O$ 3, showing the 50% thermal ellipsoids. Bond lengths: Mo(1)—O(1) 1.953(6), Mo(1)—O(2) 2.190(7), Mo(1)—O(11) 1.960(7), Mo(1)—O(12) 2.247(6), Mo(1)—O(8) 1.703(8), Mo(1)—O(9) 1.704(6) A. Bond angles: O(1)—Mo(1)—O(2) 75.1(3), O(11)—Mo(1)—O(12) 74.2(2), O(8)—Mo(1)—O(9) 103.3(3)°.

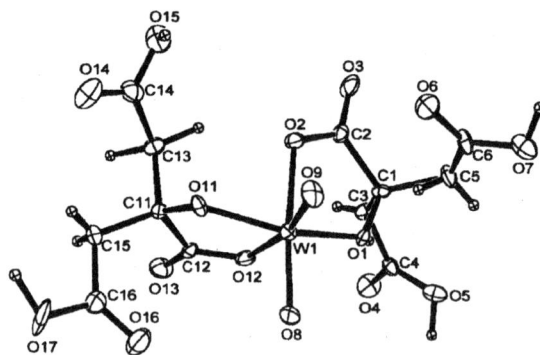

Fig. 2 Perspective view of the anion structure of $Na_2[WO_2(H_2cit)_2] \cdot 3H_2O$ 4, showing the 50% thermal ellipsoids. Bond lengths: W(1)—O(1) 1.945(6), W(1)—O(2) 2.189(8), W(1)—O(11) 1.968(7), W(1)—O(12) 2.227(7), W(1)—O(8) 1.733(8), W(1)—O(9) 1.727(8) Å. Bond angles: O(1)—W(1)—O(2) 74.9(3), O(11)—W(1)—O(12) 74.3(3), O(8)—W(1)—O(9) 102.4(4)°.

As shown in the Figure captions, the Mo—O and W—O distances in citrato molybdate 3 and tungstate 4 vary systematically M=O bonds are 1.703(8) and 1.704(6) Å for Mo and 1.727(8) and 1.733(8) Å for W, indicating that they are double bonds. The resulting O=M=O angles, 103.3(3) and 102.4(4)° for 3 and 4, respectively are considerably larger than the 90° regular octahedron value for *cis* groups. This is expected from the greater O···O repulsions between oxygens with short bonds to the metal atom. The Mo—O and W—O(alkoxy) bonds are slightly longer [1. 953(6) and 1. 960(7)Å for Mo and 1. 945(6) and 1. 968(7)Å for W], indicating the deprotonation of the hydroxyl group in the citrate anion, and those to the $\alpha$-carboxyl are longer [2. 190(7) and 2. 247(6) Å for Mo and 2. 189(8) and 2. 227(7) Å for W]. This is compatible with the Mo—O (alkoxy) [1. 996 and 2. 035 Å] and Mo—O ($\alpha$-carboxyl) bonds [2. 167 and 2. 206 Å] of coordinated homocitrate ligand in MoFe protein and its putaive transition-state complex[10]. Therefore, compound **3** may be relevant to the assembly of NifV FeMo-co, in which molybdenum is coordinated by the citrate ion.

# Acknowledgements

Financial support by the National Science Foundation of China (NO. 29503021 and 29933040) is gratefully acknowledged.

# Notes

† Selected spectroscopic data. For **3**. IR(KBr); $\nu_{as}$(CO)1733s, 1671s, 1636s, 1596s, $\nu_s$(CO)1406s, 1391s, 1360s, $\nu$(MoO)927s, 912s. $\delta_H$(500 MHz, D$_2$O): 3. 01(d, 4H, $J$ 14. 1 Hz, CH$_2$), 2. 98(d, 4H, $J$ 14. 3 Hz, CH$_2$); $\delta_C$(75 MHz, D$_2$O): 183. 7(CO$_2$)$_\alpha$, 174. 0(CO$_2$)$_\beta$, 85. 8($\equiv$C—O); 43. 5(=CH$_2$). For **4**. IR (KBr); $\nu_{as}$(CO) 1734s, 1671s, 1631s, 1607s, $\nu_s$(CO) 1398s, 1391s, 1362s, $\nu$(MoO)914s, 904s. $\delta_H$(500 MHz, D$_2$O): 3. 01(d, 4H, $J$ 15. 8 Hz, CH$_2$), 2. 84(d, 4H, $J$ 15. 8 Hz, CH$_2$); $\delta_C$(75 MHz, D$_2$O): 184. 5 (CO$_2$)$_\alpha$, 174. 0, 173. 9(CO$_2$)$_\beta$, 85. 7($\equiv$C—O); 43. 5, 43. 2(=CH$_2$).

‡Crystal data: for **3**: colorless crystal, Na$_2$[MoO$_2$(H$_2$cit)$_2$] · 3H$_2$O, C$_{12}$H$_{18}$O$_{19}$Mo$_1$Na$_2$, $M$=608. 19, monoclinic, space group $P2_1/n$, $a$=7. 6326(8), $b$=18. 351(1), $c$=14. 913(2) Å, $\beta$=103. 15(1)°, $V$= 2031. 5(7)Å$^3$, $Z$=4, $D_c$=1. 99 g cm$^{-3}$, $F$(000)=1224. Crystal dimensions: 0. 05×0. 05×0. 08 mm, $\mu$ (Mo-K$\alpha$)=7. 64 cm$^{-1}$. $N$=4004, $N_o$[$I$>3$\sigma$($I$)]=2439, $R$=0. 058, $R_w$=0. 072.

For **4**: colorless crystal, Na$_2$[WO$_2$(H$_2$cit)$_2$] · 3H$_2$O, C$_{12}$H$_{18}$O$_{19}$W$_1$Na$_2$, $M$=696. 10, monoclinic, space group $P2_1/n$, $a$=7. 6329(8), $b$=18. 318(3), $c$=14. 860(2) Å, $\beta$=102. 85(1)°, $V$=2023. 2(9)Å$^3$, $Z$ =4, $D_c$=2. 29 g cm$^{-3}$, $F$(000)=1352. Crystal dimensions: 0. 03×0. 10×0. 13mm, $\mu$(Mo-K$\alpha$)=59. 7 cm$^{-1}$. $N$=4918, $N_o$[$I$>3$\sigma$($I$)]=3512, $R$=0. 046, $R_w$=0. 058.

Unique diffractometer data sets were measured at *ca*. 296 K to 2$\theta_{max}$=52°(CAD4 diffractometer, 2$\theta$-$\theta$ scan mode, monochromatic Mo-K$\alpha$ radiation, $\lambda$=0. 71073 Å), $N$ independent reflections were obtained $N_o$ being considered 'observed' and used in the full-matrix least squares refinements after Gaussian absorption correction, Anisotropic thermal parameters were included constrained at estimated values for **3** and **4**. CCDC reference number 186/1725.

# References

[1]J. B. Howard and D. C. Rees, Chem. Rev., 1996, **96**, 2965.

[2]B. K. Burgess, Chem. Rev., 1990, **90**, 1377.

[3]A. Müller and E. Krahn, Angew. Chem., Int. Ed. Engl., 1995, **34**, 1071; D. W. Wright, R. T. Chang, S. K. Mandal, W. H. Armstrong, W. H. Orme-Johnson and W. H. Davis, J. Bioinorg. Chem., 1996, **1**, 143.

[4]K. L. C. Grönberg, C. A. Gormal, M. C. Durrant, B. E. Smith and R. A. Henderson, J. Am. Chem.

Soc.,1998,**120**,10613;K. R. Tsai and H. L. Wan,J. Clust. Sci.,1995,**6**,485.

[5]R. Hille,Chem. Rev.,1996,**96**,2757;M. K. Johnson,D. C. Rees and M. W. W. Adams,Chem. Rev.,1996,**96**,2817.

[6]L. R. Nassunbeni,M. L. Niven,J. J. Cruywagen and J. B. B. Heyns,J. Crystallogr. Spectrosc, Res.,1987,**17**,373;N. W. Alcock,M. Dudek,R. Grybos,E. Hodorowicz,A. kanas and A. Samotus,J. Chem. Soc.,Dalton Trans.,1990,707;Z. H. Zhou,H. L. Wan and K. R. Tsai, Polyhedron,1997,**16**,75;Y. H. Xing,J. Q. Xu,H. R. Sun,D. M. Li.,R. Z. Wang,T. G. Wang,W. M. Bu,L. Ye,G. D. Yang and Y. G. Fan,Acta Crystallogr.,Sect. C,1998,**54**,1615;Y. H. Xing,J. Q. Xu,H. R. Sun,D. M. Li,Y. Xing,Y. H. Lin and H. Q. Jia,Eur. J. Solid State Inorg. Chem., 1998,**35**,745;J. Q. Xu,X. H. Zhou,L. M. Zhou,T. G. Wang,X. Y. Huang and B. A. Averill, Inorg. Chim. Acta,1999,**285**,152.

[7]J. J. Cruywagen,L. J. Saayman and M. L. Niven,J. Crystallogr. Spectrosc. Res.,1992,**22**,737;E. Llopis,J. A. Ramírez,A. Doménech and A. Cervilla,J. Chem. Soc.,Dalton Trans.,1993,1121;Y. H. Xing,J. Q. Xu,H. R. Sun,Z. Wang,R. Z. Wang,W. M. Bu,L. Ye,G. D. Yang and Y. G. Fan, Chem. Res. Appl.,1999,**11**,27.

[8]Z. H. Zhou,H. L. Wan and K. R. Tsai,Inorg. Chem.,in press.

[9]Z. H. Zhou,W. B. Yan,H. L. Wan,K. R. Tsai,J. Z. Wang and S. Z. Hu,J. Chem. Crystallogr., 1995,**25**,807;D. W. Wright,P. A. Humiston,W. H. Orme-Johnson and W. H. Davis,Inorg. Chem.,1995,**34**,4194.

[10]J. Kim and D. C. Rees,Nature,1992,**360**,553;H. Schinderlin,H. C. Kisker,J. L. Schlessman,J. B. Howard and D. C. Rees,Nature,1997,**387**,370.

■ 本文原载：CHIRALITY 12(2000)，pp. 383~388.

# New Chiral Catalysts for Reduction of Ketones[1] *

Jing-Xing Gao[1] ①, Hui Zhang[1], Xiao-Dong Yi[1], Pian-Pian Xu[1],
Chun-Liang Tang[1], Hui-Lin Wan[1], Khi-Rui Tsai[1], Takao Ikariya[2]
(*1Department of Chemistry, Institute of Physical Chemistry,
State Key Laboratory for Physical Chemistry of Solid Surfaces,
Xiamen, University, Xiamen, Fujian, P. R. China
*2Department of Chemical Engineering, Faculty of
Engineering, Tokyo Institute of Technology, Tokyo, Japan )*

**Abstract**    The condensation of *o*-(diphenylphosphino)benzaldehyde and various chiral diamine gives a series of diimino-diphosphine tetradentate ligands, which are reduced with excess $NaBH_4$ in refluxing ethanol to afford the corresponding diamino-diphosphine ligands in good yield. The reactivity of these ligands toward *trans*-$RuCl_2(DMSO)_4$ and $[Rh(COD)Cl]_2$ had been investigated and a number of chiral Ru(Ⅱ)and Rh(Ⅰ)complexes with the PNNP-type ligands were synthesized and characterized by microanalysis and IR, NMR spectroscopic methods. The chiral Ru(Ⅱ)and Rh(Ⅰ)complexes have proved to be excellent catalyst precursors for the asymmetric transfer hydrogenation of aromatic ketones, leading to optically active alcohols in up to 97% ee.
*Chirality* 12:383~388,2000. © 2000 Wiley-Liss,Inc.
    **Key words**:Ruthenium  Rhodium  Chiral amino/phosphine ligands  Asymmetric catalysis Ketones  Alcohols  Enantioselective hydrogenations

Enantioselective reduction of prochiral ketones to optically active secondary alcohols is an important subject in synthetic organic chemistry because the resulting chiral alcohols are extremely useful, biologically active compounds. Catalytic asymmetric synthesis using chiral metal complexes as catalyst precursors offers an ideal method for reducing ketones to chiral alcohols[1,2]. For the past 10 years, a developed efficient method is the catalytic transfer hydrogenation of ketones using chiral Rh(Ⅰ),Ir(Ⅰ),Ru(Ⅱ), or lanthanoid complexes as catalyst precursors and *iso*-PrOH/KOH or HCOOH/Et₃N as hydride source[3-5]. The chiral ligands employed in asymmetric transfer reactions are very important for obtaining high enantioselectivity. Generally, the chiral biphosphines are the most widely used auxiliaries; however, in the field of asymmetric transfer reactions the most-used chiral

* Contract grant sponsor: the National Natural Science Foundation of China; Contract grant number: 29873038; Contract grant sponsor: the Fujian Provincial Science and Technology Commission; Contract grant number: B9810002; Contract grant sponsor:the Ministry of Education of China;Contract grant number:1998-121.

① Correspondence to: Jing-Xing Gao, Department of Chemistry, Xiamen University, Xiamen 361005, Fujian, P. R. China. E-mail:jxgao@xmu. edu. cn

Received for publication 3 December 1999; Accepted 5 January 2000.

auxiliaries contain nitrogen as a donor atom[3]. Recently, optically active nitrogen compounds have been reviewed in detail[6,7] and developed with great success[8-12]. Furthermore, the importance and application of nitrogeneous auxiliaries have been further extended by a combination phosphorus centers and nitrogen donor atoms in PN[13], NPN[14], PNP[15,16], or PNNP[17-19]-type ligands.

In earlier studies, we reported the synthesis of some Ru, Fe, Cu, Ag, Cr, and Mo complexes with racemic PNNP-type ligands[20-23]. The PNNP tetradentate systems possess $C_2$-symmetry and contain two "soft" phosphorus atoms and two "hard" nitrogen atoms, which should easily modify the steric and electronic properties of the resulting complexes and their catalytic reactivites[16,24-27]. In a previous communication, we briefly reported the synthesis of Ru(Ⅱ) complex with a chiral $C_2$-symmetric tetradentate ligand and its use in asymmetric transfer hydrogenation of ketones[28]. As an extension of our recent work, we report here the synthesis and characterization of some well-designed Ru(Ⅱ) and Rh(Ⅰ) complexes bearing the chiral PNNP tetradentate ligands, and the character of asymmetric transfer hydrogenation of ketones catalyzed by these chiral complexes.

## MATERIALS AND METHODS

Prior to use, 2-propanol was refluxed from calcium hydride followed by distillation. All the other solvents were purified according to standard methods. o-(Diphenylphos-phino) benzaldehyde (PCHO) and various chiral diamines from Aldrich (Milwaukee, WI) were used as received. All other chemicals were purchased from commercial sources.

IR spectra were recorded on a PE-Spectroy 2000 spectrometer. NMR spectra were recorded on a Varian Unity-500 spectrometer. [1]H NMR chemical shifts are reported in ppm relative to TMS. [31]P NMR spectra were referenced to 85% $H_3PO_4$ as external standard. The elemental analyses were carried out on a Fisons EA 1110. All melting points were measured in sealed tubes and are not corrected.

### General Procedure for Preparation of the Chiral Diiminodiphosphine Ligands 2,5,8, and 11, Depicted for 2

A solution of (S,S)-1,2-diaminocyclohexane (0.35 g, 3.0 mmol), o-(diphenylphosphino) benzaldehyde(1.74 g, 6.0 mmol) and anhydrous $Na_2SO_4$ (2.56 g, 18.0 mmol) in $CH_2Cl_2$ (20 mL) was stirred for 24 h. A pale-orange solution was obtained. The solution was filtered, then concentrated under reduced pressure to ca. 5 mL. To the solution was added 20 mL of ethanol and the resulting solution was cooled to −18 ℃ to give (S,S)-2 as a yellow solid (mp 60~62; 1.78 g, 90% yield). IR(KBr): 3049 m, 2928 s, 2857 s, 1636 s, 1433 s, 1087 w, 748 vs. 697 vs. 547 w, 503 s cm$^{-1}$. [1]H NMR(CDCl$_3$): δ 8.69(d, 2H, $J=4.0$ Hz), 6.79-7.74(m, 28H), 3.11(m, 2H), 1.65(d, 2H, $J=6.0$ Hz), 1.45(m, 2H), 1.38(d, 2H, $J=7.8$ Hz), 1.26(m, 2H); [31]P NMR(CDCl$_3$): δ-12.98. Anal. Calcd for $C_{44}H_{40}N_2P_2 \cdot 0.5C_2H_5OH$: C, 79.30; H, 6.31; N, 4.11. Found: C, 79.10; H, 6.39; N, 4.23. The chiral ligands 5, 8, and 11 were prepared by the same procedure.

### General Procedure for Preparation of the Chiral Diaminodiphosphine Ligands 3,6,9, and 12, Depicted for 3

A solution of compound (S,S)-2 (1.65 g, 2.5 mmol) and NaBH$_4$ (0.5 g, 15 mmol) in absolute ethanol(30 mL) was refluxed with stirring for 24 h. The solution was cooled to room temperature and H$_2$O(15 mL) was added to destroy excess NaBH$_4$. The mixture solution was extracted with CHCl$_3$(30 mL×3). The combined extracts were washed with saturated NH$_4$Cl solution(10 mL×3), H$_2$O(10 mL×3) and the organic layer was dried over anhydrous $MgSO_4$, followed by filtration, concentrated to ca. 5 mL, 15 mL of ethanol was added and cooled to $-18$ ℃ to give cream-white crystals $(S,S)$-3 (mp 54~56; 1.34 g, 80% yield). IR(KBr): 3051 m, 2924 s, 2852 m, 1565 w, 1433 s, 1183 m, 1114 m, 746 vs. 696 vs. 544 m, 502 m cm$^{-1}$. $^1$H NMR(CDCl$_3$: δ 6.81~7.52(m, 28H), 4.00(d, 2H, $J=13.6$ Hz), 3.82(d, 2H, $J=13.6$ Hz), 2.11(d, 2H, $J=8.5$ Hz), 1.98(d, 2H, $J=12.8$ Hz), 1.87(br, 2H), 1.58(d, 2H, $J=6.4$ Hz), 1.08(t, 2H), 0.87(m, 2H); $^{31}$P NMR(CDCl$_3$): δ$-15.21$. Anal. Calcd for $C_{44}H_{44}N_2P_2 \cdot 0.5 C_2H_5OH$: C, 78.83; H, 6.86; N, 4.23. Found: C, 78.39; H, 6.78; N, 4.27. The chiral ligands 6, 9, and 12 were also prepared by the same procedure.

## General Procedure for Preparation of the Ruthenium(Ⅱ)Complexes 13, 14, 15, and 16, Depicted for 14

A mixture of $(S,S)$-3 (0.20 g, 0.3 mmol) and *trans*-RuCl$_2$(DMSO)$_4$, (0.15 g, 0.3 mmol) in toluene (15 mL) was refluxed with stirring for 16 h. The resulting orange-red solution was cooled to room temperature and the solvent was removed under vacuum to leave an orange-red residue. The solid was purified by column chromatography on silica with CH$_2$Cl$_2$/acetone(1 : 1) as an eluant to give orange-yellow crystals $(S,S)$-14 (mp 306~310; 0.21 g, 84% yield). IR(KBr): 3412 m, 3052 m, 2930 m, 1590 w, 1479 m, 1434 s, 1185 m, 1091 m, 1046 m, 751 s, 698 vs. 520 vs. cm$^{-1}$. $^1$H NMR(CDCl$_3$): δ 6.88~7.25 (m, 28H), 4.66(m, 2H), 4.03(m, 2H), 3.88(d, 2H,), 2.94(m, 2H), 2.71(d, 2H), 1.78(d, 2H), 1.17 (m, 4H); $^{31}$P NMR(CDCl$_3$): δ 43.28. Anal. Calcd for $C_{44}H_{44}Cl_2N_2P_2Ru$: C, 63.32; H, 5.28; N, 3.36. Found: C, 63.07; H, 5.45; N, 3.30. Similarly, the chiral ruthenium complexes 13, 15, and 16 were also prepared by the above procedure.

## General Procedure for Preparation of the Rhodium(Ⅰ)Complexes 17, 18, 19, and 20, Depicted for 18a

To a mixture of $(S,S)$-3 (0.20 g, 0.3 mmol) and [Rh-(COD)Cl]$_2$ (0.074 g, 0.15 mmol) were added benzene(6 mL) and methanol(6 mL). The mixture was stirred at room temperature for 12 h. After removal of solvent, the residue was dissolved in a minimum of methanol and precipitated by addition of a solution of NH$_4$PF$_6$ (0.082 g, 0.5 mmol) in H$_2$O(3 mL). The precipitate was collected and washed successively with H$_2$O(3 mL×3) and diethyl ether(3 mL), then dried in vacuo to afford 18a as a yellow solid(mp 242~245, dec. ; 0.20g, 74% yield). IR(KBr): 3414 m, 3056 m, 2857 w, 1591 w, 1482 w, 1436 s, 1167 m, 1098 s, 750 m, 723 m, 697 vs. 541 vs. 464 w cm$^{-1}$. $^1$H NMR(CDCl$_3$: δ 6.52~7.82(m, 28H), 4.86(m, 2H), 4.36(m, 2H), 4.01(d, 2H, $J=10$ Hz), 3.19(s, 2H), 2.17(br, 2H), 1.77(d, 2H, $J=28.5$Hz), 1.12(m, 4H); $^{31}$P NMR(CDCl$_3$: δ 33.03. Anal. Calcd for $C_{44}H_{44}N_2F_6P_3Rh \cdot 2H_2O$: C, 55.84; H, 5.28; N, 2.91. Found: C, 55.83; H, 5.11; N, 2.96.

## General Procedure for Asymmetric Hydrogenation of Ketones Catalyzed by Ruthenium or Rhodium Complexes

The catalyst precursor(0.01 mmol) was added to a Schlenk tube and 2-propanol(20 mL), *iso*-PrOK/*iso*-PrOH solution(0.1 M, 0.1 mL) were introduced under nitrogen. The mixture was stirred for 10 min, acetophenone was added, and the solution was stirred at the desired temperature for the required reaction time. At the end of the experiment, the reaction products were determined by GLC using a chiral chrompack CP-cyclodextrin-p-236-M-19 column.

# RESULTS AND DISCUSSION

## Synthesis of the Ligands

The chiral diiminodiphosphine ligands, such as **2**, **5**, **8**, and **11**, were prepared by the condensation of PCHO and an appropriate chiral diamine in dichloromethane using anhydrous $Na_2SO_4$ as a dehydrating agent. The IR spectrum of 2 exhibits a strong C=N stretch at 1636 cm$^{-1}$; and the $^1$H NMR spectrum presents a doublet $J_{P-H}$=4. 0 Hz) at $\delta$ 8. 69 for the imino protons. The $^{31}$P NMR spectrum of **2** exhibits a singlet at $\delta$−12. 98 in accord with two equivalent phosphino groups (Scheme 1).

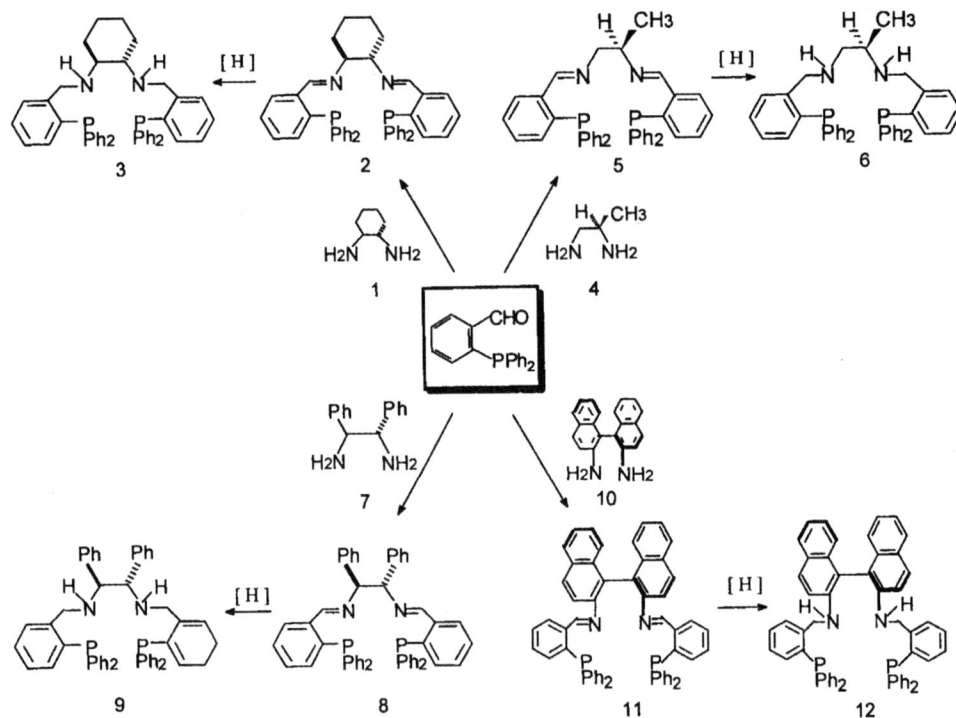

Scheme 1. Synthesis of the ligands

Reduction of the diiminodiphosphine ligands **2**, **5**, **8**, and **11** with excess NaBH$_4$ was carried out in refluxing ethanol to afford the corresponding diaminodiphosphine ligands in 75％ ∼ 80％ yield, respectively. For the ligands **2** and **3**, the disappearance of infrared band at 1636 cm$^{-1}$ of **2** and the appearance of $^1$H NMR spectrum at $\delta$ 1. 87 of **3** for the —NH— protons indicate that the two imino groups were reduced to the corresponding diamino groups.

## Synthesis of the Chiral Ru(Ⅱ)and Rh(Ⅰ)Complexes

The coordination chemistry of these PNNP ligands has been investigated and shows that both the phosphorus and nitrogen donor atoms are coordinated to the Ru(Ⅱ)or Rh(Ⅰ)center. The interaction of *trans*-RuCl$_2$(DMSO)$_4$ with one equivalent of ligand (S,S)-**3** gave (S,S)-**14** in good yield(84％). $^{31}$P NMR spectrum of (S,S)-**14** displayed a singlet at $\delta$ 43. 28, indicating that the two phosphino groups are coordinated and equivalent (Fig. 1). In an analogous manner, using (R, R)-**3** instead of (S, S)-**3**, ruthenium complex (R,R)-**14** was also prepared.

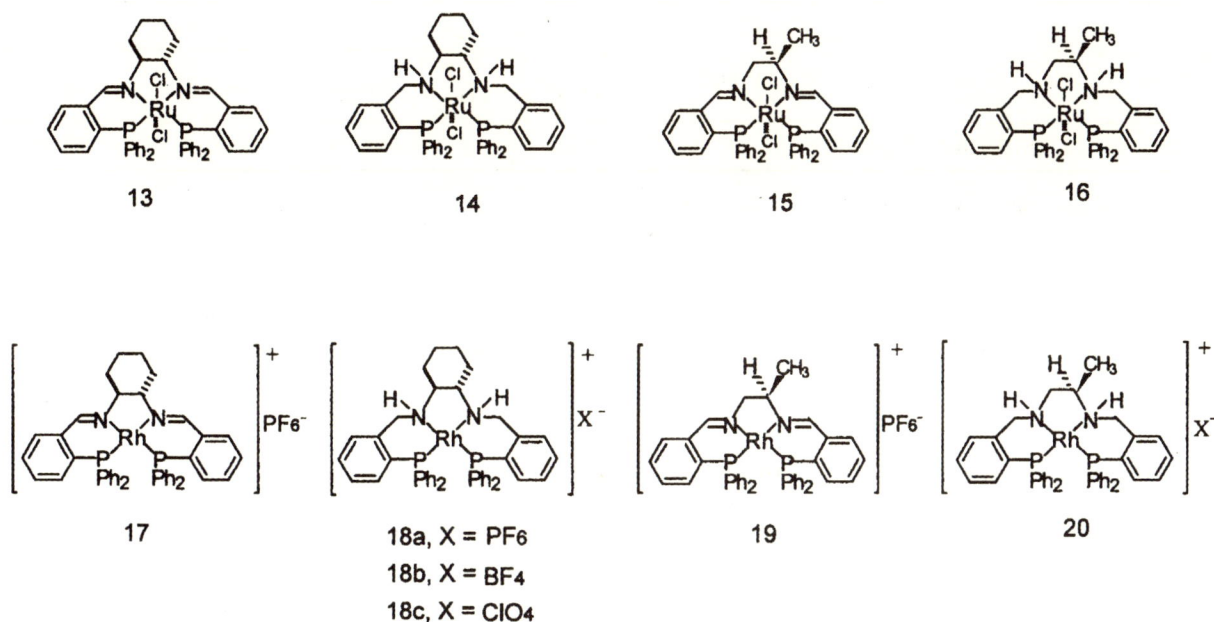

Figure 1. Ruthenium(II) and rhodium(I) complexes with chiral tetradentate diamine/diphosphine ligands.

The treatment of [Rh(COD)Cl]$_2$ with two equivalents of ligand $(S,S)$-3 in a 1 : 1 mixture of methanol-benzene and then precipitated by the addition of a solution of NH$_4$PF$_6$ in water, affording a yellow solid of cationic rhodium complex $(S,S)$-18a in moderate yield(74%). The $^{31}$P NMR spectrum of $(S,S)$-18a exhibited a singlet at δ 39.6, attributed to the two coordinated and equivalent phosphorus atoms. In a similar fashion as described for $(S,S)$-18a, using NaBF$_4$ or NaClO$_4$ instead of NH$_4$PF$_6$, complexes $(S,S)$-18b and $(S,S)$-18c were also obtained, respectively.

Table 1.　Asymmetric transfer hydrogenation of acetophenone catalyzed by chiral Ru(II) and Rh(I) complexes[a]

| Entry | Catalyst | S/C/$^i$PrOK[b] | Temp. (℃) | Time (h) | Alcohol | | |
|---|---|---|---|---|---|---|---|
| | | | | | yield, %[c] | ee, %[d] | config. [e] |
| 1 | $(S,S)$-13 | 200 : 1 : 0.5 | 23 | 48 | 3 | 18 | $R$ |
| 2 | $(S,S)$-13 | 200 : 1 : 0.5 | 83 | 4 | 7 | 5 | $R$ |
| 3 | $(S)$-15 | 100 : 1 : 3 | 40 | 22 | 65 | 14 | $R$ |
| 4 | $(S,S)$-19 | 100 : 1 : 1 | 83 | 7 | 40 | 40 | $R$ |
| 5 | $(R,R)$-14 | 200 : 1 : 0.5 | 23 | 25 | 1 | 97 | $S$ |
| 6 | $(S,S)$-14 | 200 : 1 : 0.5 | 45 | 7 | 3 | 97 | $R$ |
| 7 | $(S)$-16 | 100 : 1 : 3 | 30 | 46 | 0 | 91 | $S$ |
| 8 | $(S,S)$-18a | 100 : 1 : 1 | 83 | 7 | 7 | 91 | $R$ |
| 9 | $(S,S)$-18a | 400 : 1 : 1 | 83 | 24 | 5 | 89 | $R$ |
| 10 | $(R,R)$-18a[f] | 100 : 1 : 1 | 83 | 24 | 58 | 96 | $S$ |
| 11 | $(S,S)$-18b | 100 : 1 : 1 | 83 | 7 | 98 | 80 | $R$ |
| 12 | $(R,R)$-18c | 100 : 1 : 1 | 83 | 9 | 87 | 86 | $S$ |

Con. Table 1

| | | | | | Alcohol | | |
|---|---|---|---|---|---|---|---|
| Entry | Catalyst | S/C/$^i$PrOK$^b$ | Temp. (℃) | Time (h) | yield,%$^c$ | ee,%$^d$ | config.$^e$ |
| 13 | (S,S)-18c$^g$ | 100∶1∶1 | 83 | 24 | 73 | 92 | S |

$^a$Conditions:catalyst,0.01 mmol;solvent,iso-PrOH,20 mL.

$^b$S/C/$^i$PrOK=[ketone]/[M](M=Ru,Rh)/[iso-PrOK].

$^c$GLC analysis.

$^d$Capillary GLC analysis using a chiral Chrompack CD-cyclodextrin-β-236-M-19 column unless otherwise specified.

$^e$Determined by comparison of the retention times of the enantiomers on the GLC traces with literature values.

$^{f,g}$Ligand (R,R)-3(0.01 mmol)was added.

## Asymmetric Transfer Hydrogenation of Acetophenone

The chiral Ru(Ⅱ) and Rh(Ⅰ)complexes bearing chiral diimino- or diamino-diphosphine ligands have been tested as catalyst precursors for transfer hydrogenation of aceto-phenone in 2-propanol solution (Scheme 2);the results are summarized in Table 1. The reaction with the diimino complexes, such 13,15, and 19, proceeded in low yield(3%～65%) and ee(5%～40%)(Table 1,entries 1—4 ). However,diamino complexes,such as 14,16,and 18 have proved to be an excellent catalyst precursor in asymmetric transfer hydrogenation of acetophenone,leading to 2-phenylethanol in 93% yield with 97% ee(entries 5—13). The difference in reactivity and enantioselectivity indicates that the NH moiety in ligand is an important factor for obtaining high catalytic efficiency. The NH linkage possibly can stabilize a six-membered cyclic transition state by forming a hydrogen bond with oxygen atoms of ketones(Fig 2)[4,29]. The chiral Ru(Ⅱ)complex (S,S)-14 and the cationic Rh (Ⅰ) complex (S,S)-18a have proved to be excellent catalysts,as can be seen from Table 1(entries 5,6,8). Although the yields gradually decreased on increasing the mole ratios of [ace-tophenone]/[Rh] from 100∶1 to 400∶1,the yield and enantioselectivity are still high (entries 8,9).

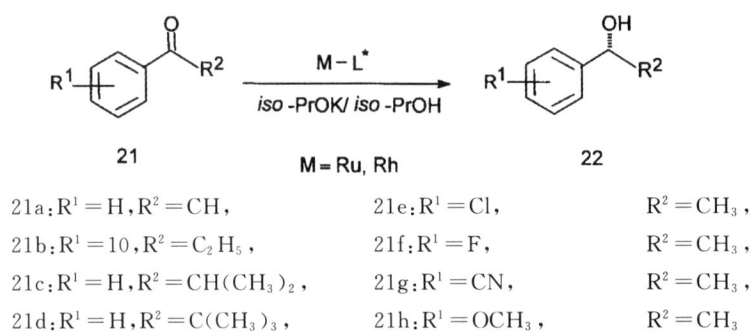

21a:R$^1$=H,R$^2$=CH,     21e:R$^1$=Cl,     R$^2$=CH$_3$,

21b:R$^1$=10,R$^2$=C$_2$H$_5$,     21f:R$^1$=F,     R$^2$=CH$_3$,

21c:R$^1$=H,R$^2$=CH(CH$_3$)$_2$,     21g:R$^1$=CN,     R$^2$=CH$_3$,

21d:R$^1$=H,R$^2$=C(CH$_3$)$_3$,     21h:R$^1$=OCH$_3$,     R$^2$=CH$_3$

**Scheme 2.**

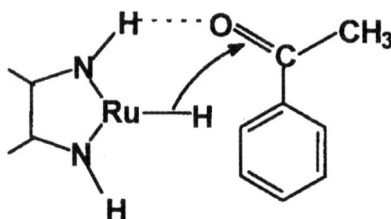

**Figure 2. A possible six-membered cyclic transition state for activation and reduction of acetophenone.**

## Asymmetric Transfer Hydrogenation of Various Ketones

Asymmetric transfer hydrogenation of various aromatic ketones catalyzed by chiral Ru(Ⅱ)and Rh (Ⅰ)complexes was examined; the results are summarized in Table 2. A variety of simple aryl alkyl ketones can be transformed to the corresponding secondary alcohols with high enantiomeric purity. The reaction rate and enantioselectivity are delicately affected by the steric and electronic properties of the ketones. Although the reactivity gradually decreased by increasing the bulkiness of the alkyl groups (Table 2,entries 1—5),good enantioselectivity was also observed.

**Table 2.** Asymmetric transfer hydrogenation of aromatic ketones catalyzed by chiral Ru(Ⅱ)and Rh(Ⅰ) complexes[a]

| Entry | Ketone | Catalyst | S/C/$^i$PrOK | Temp. (℃) | Time (h) | Alcohol | | |
|---|---|---|---|---|---|---|---|---|
| | | | | | | yield,% | ee,% | config. |
| 1 | 21b | (S,S)-14 | 200 : 1 : 0.5 | 45 | 7 | 78 | 96 | R |
| 2 | 21b | (R,R)-18a | 100 : 1 : 1 | 83 | 22 | 91 | 80 | S |
| 3 | 21b | (R)-20 | 200 : 1 : 3 | 30 | 46 | 73 | 91 | R |
| 4 | 21c | (R,R)-18a | 200 : 1 : 1 | 83 | 22 | 18 | 95 | S |
| 5 | 21d | (R,R)-18a | 100 : 1 : 1 | 83 | 22 | 13 | 76 | S |
| 6 | o-21e | (S,S)-14 | 200 : 1 : 0.5 | 45 | 5 | 15 | 91 | R |
| 7 | m-21e | (S,S)-14 | 200 : 1 : 0.5 | 45 | 6 | 99 | 95 | R |
| 8 | m-21e | (S)-16 | 100 : 1 : 2 | 30 | 24 | 99 | 87 | S |
| 9 | m-21e | (R,R)-18a | 100 : 1 : 1 | 83 | 22 | 97 | 90 | S |
| 10 | p-21e | (S,S)-14 | 200 : 1 : 0.5 | 45 | 5 | 95 | 94 | R |
| 11 | p-21e | (S)-16 | 100 : 1 : 2 | 30 | 24 | 82 | 89 | S |
| 12 | p-21f | (S,S)-14 | 200 : 1 : 0.5 | 45 | 6 | 99 | 89[b] | R |
| 13 | p-21g | (S,S)-14 | 200 : 1 : 0.5 | 45 | 6 | 97 | 80[c] | R |
| 14 | m-21h | (S,S)-14 | 200 : 1 : 0.5 | 45 | 6 | 74 | 95 | R |
| 15 | m-21h | (S)-16 | 100 : 1 : 2 | 30 | 24 | 72 | 85 | S |
| 16 | m-21h | (R,R)-18a | 100 : 1 : 1 | 83 | 22 | 99 | 94 | S |
| 17 | p-21h | (S)-16 | 100 : 1 : 2 | 30 | 24 | 49 | 87 | S |
| 18 | p-21h | (R,R)-18a | 100 : 1 : 1 | 83 | 22 | 49 | 71 | S |

[a] Reaction was carried out in 2-propanol(20 mL); catalyst, 0.01 mmol; S/C/$^i$PrOK =[ketone]/[M](M=Ru,RH)/[iso-PrOK];
the others were the same as shown in Table 1.

[b] Chiralcel OB column(5 : 95 2-propanol-hexane).

[c] Chiralcel OB column(10 : 90 2-propanol-hexane).

The introduction of an electron-withdrawing group,such as chloro,accelerated the reaction with high enantioselectivity(entries 6—13),but o-chloroacetophenone reacted very slowly with high ee(entry 6). The ketone having an electron-releasing group,such as methoxy to the meta position,was reduced smoothly with up to 99% yield and 94% ee (entries 14—16),while the methoxy to the para position tended to lower the rate and stereoselectivity (entries 17,18).

## Consideration of the Reaction Mechanisms

The discovery of these chiral Ru(Ⅱ) and Rh(Ⅰ) complex catalysts provides an efficient method for the catalytic transfer hydrogenation of ketones using 2-propanol as hydride source. Generally, the reaction mechanism of the transfer hydrogenation has been envisaged to a "hydridic route" or a "direct hydrogen transfer"[3]. The latter involves a process in which both the hydrogen donor and the substrate ketone are bound on the catalytic active center. For complexes **14** and **16**, used in this work, a direct hydrogen transfer pathway is difficult to realize since these complexes possess a distorted octahedron structure with the coordinate saturation[20,29]. Therefore, a "hydridic route" may be performed by the chiral Ru(Ⅱ) complexes with the PNNP-type ligand and a possible catalytic cycle can be proposed, as shown in Figure 3.

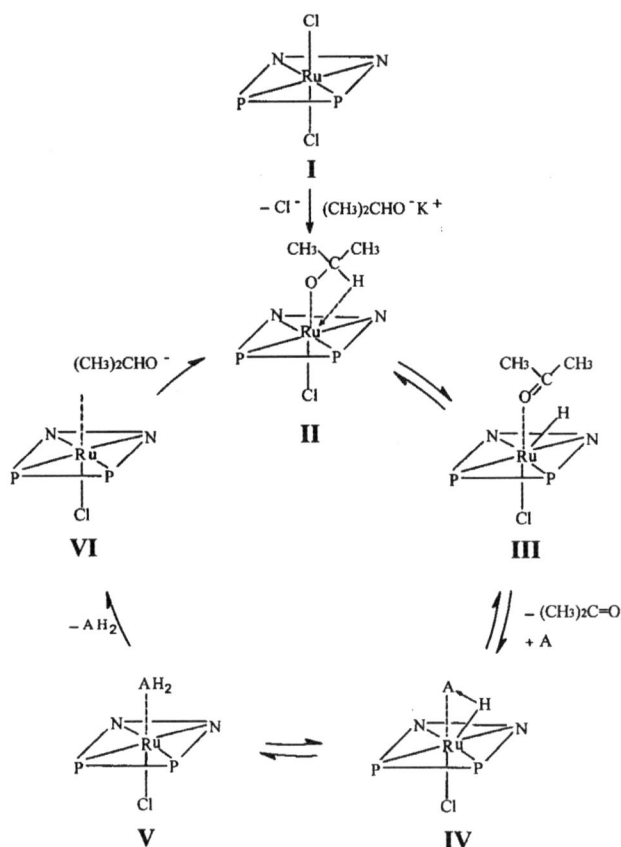

**Figure 3. Proposed mechanism for the hydride transfer hydrogenation of acetophenone.**

The addition of *iso*-PrOK is an important factor for transfer hydrogenation; the catalytic system is inactive without a basic cocatalyst. First, *iso*-PrO$^-$ attacks the central Ru atom and replaces the ligand Cl$^-$ to form a *iso*-PrO$^-$-Ru intermediate Ⅱ, and then transfers its active H atom to the central Ru atom to form a catalytic active ruthenium hydride species Ⅲ, which gives a $^1$H NMR signal at $\delta = -9.98$ in CDCl$_3$. In the following step, the substrate acetophenone is coordinated and the intramolecular hydrogen transfer takes place, leading to the formation of 1-phenylethanol (intermediate Ⅳ and Ⅴ; for a more detailed structure of Ⅳ, see Fig. 1). After releasing the product 1-phenylethanol, the *iso*-PrO$^-$ group with complex Ⅵ would restore the starting intermediate Ⅱ, completing the catalytic cycle.

# ACKNOWLEDGMENTS

We thank Professor Ryoji Noyori(Nagoya University,Japan)for valuable help.

# REFERENCES

[1]Noryri R. Asymmetric catalysis in organic synthesis. New York:John Wiley;1994. pp. 1~82.

[2]Takaya H,Ohta T,Noyori R. Catalytic asymmetric synthesis. In:Ojima I,editor. Asymmetric hydrogenation. Berlin:VCH;1993. pp. 20~31.

[3]Zassinovich G,Mestroni G,Gladiali S. Asymmetric hydrogen transfer reactions promoted by homogeneous transition metal catalysts. Chem Rev 1992;**10**:1051~1069.

[4]Noyori R,Hashiguchi S. Asymmetric transfer hydrogenation catalyzed by chiral ruthenium complexes. Acc Chem Res 1997;**30**:97~102.

[5] Palmer MJ, Wills M. Asymmetric transfer hydrogenation of C=O and C=N bonds. Tetrahedron:Asymmetry 1999;**10**:2045~2061.

[6]Togni A,Venanzi LM. Nitrogen donors in organometallic chemistry and homogeneous catalysis. Angew Chem Int Ed Engl 1994;**33**:497~526.

[7]Lucet D,Gall TL,Mioskowski C. The chemistry of vicinal diamines. Angew Chem Int Ed Engl 1998;**37**:2581~2627.

[8]Hashiguchi S,Fujii A,Takehara J,Ikariya T,Noyori R. Asymmetric transfer hydrogenation of aromatic ketones catalyzed by chiral ruthenium(Ⅱ)complexes. J Am Chem Soc 1995;**117**:7562 ~7563.

[9]Fujii A,Hashiguchi S,Uematsu N,Ikariya T,Noyori R. Ruthenium(Ⅱ)-catalyzed asymmetric transfer hydrogenation of ketones using a formic acid-triethylamine mixture. J Am Chem Soc 1996;**118**:2521~2522.

[10]Murata K,Ikariya T,Noyori R. New chiral rhodium and iridium complexes with chiral diamine ligands for asymmetric transfer hydrogenation of aromatic ketones. J Org Chem 1999;**64**:2186 ~2187.

[11]Jiang Y,Jiang Q,Zhang X. A new chiral bis(oxazolinylmethyl)amine ligand for Ru-catalyzed asymmetric transfer hydrogenation of ketones. J Am Chem Soc 1998;**120**:3817~3818.

[12]Touchard F,Bernard M,Fache F,Delbecq F,Guiral V,Sautet P,Lemaire M. Optically active nitrogen ligands in asymmetric catalysis. Effect of nitrogen substitution on the enantioselective hydride transfer reduction of acetophenone. J Organomet Chem 1998;**567**:133~136.

[13]Newkome GR. Pyridylphosphines. Chem Rev 1993;**93**:2067~2089.

[14]Jiang Y,Zhu Q,Zhang X. Highly effective NPN-type tridentate ligands for asymmetric transfer hydrogenation of ketones. Tetrahedron Lett 1997;**38**:215~224.

[15] Sablong R, Osborn JA. The asymmetric hydrogenation of imines using tridentate $C_2$ diphosphins complexes of iridium(Ⅰ)and rhodium(Ⅰ). Tetrahedron Lett 1996;**37**:4937~ 4940.

[16]Rahmouni N,Osborn JA,Cian AD,Fischer J,Ezzamarty A. Ruthenium(Ⅱ)hydrido complexes of 2,6-(diphenylphosphinomethyl)-pyridine. Organometallics 1998;2470~2476.

[17]Trost BM,Vranken DLV,Bingle C. A modular approach for ligand design for asymmetric

allylic alkylations via enantioselective palladium-catalyzed ionizations. J Am Chem Soc 1992; **114**:9327~9343.

[18] Trost BM, Patterson DE. Enhanced enantioselectivity in the desymmetrization of meso-biscarbanates. J Org Chem 1998;**63**:1339~1341.

[19] Kless A, Kadyrov R, Börner A, Holz J, Kagan HB. A new chiral multidentate ligand for asymmetric catalysis. Tetrahedron Lett 1995;**36**:4601~4606.

[20] Gao JX, Wan HL, Wong WK, Tse MC, Wong WT. Synthesis and characterization of Iron( II ) and Ru( II ) diimino-, diamido-diphosphine complexes. X-ray crystal structure of trans-RuCl$_2$ (P$_2$N$_2$C$_2$H$_4$)CHCl$_3$. Polyhedron 1996;**15**:1241~1251.

[21] Wong WK, Gao JX, Wong WT, Che CM. Preparation of Cu( I ) BINAP-P$_2$N$_2$ complexes; crystal and molecular structure of [Cu(BINAP-P$_2$N$_2$)Br]. Polyhedron 1993;**12**:2063~2066.

[22] Wong WK, Gao JX, Wong WT, Cheng WC, Che CM. Synthesis and reactivity of N, N'-bis[o-(diphenylphosphino) benzylidene-2, 2'-diimino-1, 1'-binaphthylene] ( Binap-P$_2$N$_2$ ). Crystal structure of [Ag(Binap-P$_2$N$_2$)[BF$_4$]. J Organomet Chem 1994;**471**:277~282.

[23] Wong WK, Gao JX, Wong WT. Reactivity of P$_2$N$_2$ and P$_2$N$_2$H$_4$ towards C$_7$H$_8$M(CO)$_4$ (M= Cr, Mo). X-ray structure of fac-(CO)$_3$Mo[PN$_2$P(O)]. Polyhedron 1993;**12**:1047~1053.

[24] Burckhardt U, Hintermann L, Schnyder A, Togni A. Synthesis and structure of pyrazole-containing ferrocenyl ligands for asymmetric catalysis. Organometallics 1995;**14**:5415~5425.

[25] Amurrio D, Khan K, Kündig EP. Asymmetric addition of organolithium reagents to prochiral arene tricarbonylchromium complexes. J Org Chem 1996;**61**:2258~2259.

[26] Siwek MJ, Green JR. (R)-and (S) enantioselective lithiation of (arene) tricarbonylchromium acetal complexes with chiral alkyllithiums. J Chem Soc Chem Commun 1996;2359~2360.

[27] Prétôt R, Pfaltz A. New ligands for regio- and enantiocontrol in Pd-catalyzed allylic alkylations. Angew Chem Int Ed Engl 1998;**37**:323~325.

[28] Gao JX, Ikariya T, Noyori R. A Ruthenium( II )complex with a C$_2$-symmetric diphosphine/diamine tetradentate ligand for asymmetric transfer hydrogenation of aromatic ketones. Organometallics 1996;**15**:1087~1089.

[29] Haack K-J, Hashiguchi S, Fujii A, Ikariya T, Noyori R. The catalyst precursor, catalyst and intermediate in the Ru$^{II}$-promoted asymmetric hydrogen transfer between alcohol and ketones. Angew Chem Int Ed Engl 1997;**36**:285~287.

■ **本文原载**:《高等学校化学学报》(2000 年 9 月),第 1445~1447 页。

# 高稳定度 CH$_4$/CO$_2$ 重整 Ni/MgO 催化剂的研究*

李基涛[①]   陈明旦   严前古   万惠霖   蔡启瑞

(厦门大学化学系   物理化学研究所   固体表面物理化学国家重点实验室,厦门   361005)

**摘 要** 用 TPR,TPD,TPO,TPMC(程序升温 CH$_4$ 解离积炭)和活性评价等手段研究了普通浸渍法与载体盐助分散浸渍法制得的 CH$_4$/CO$_2$ 重整制合成气 Ni 基催化剂的性能。结果表明,用载体盐助分散浸渍制备的催化剂 Ni—O—Mg 间作用较强,吸附 CO$_2$ 能力较大,CH$_4$ 解离积炭量少,因此其稳定性及寿命较好。

**关键词** CH$_4$/CO$_2$ 重整   载体盐助分散浸渍   Ni 基催化剂

**中图分类号** O643.1   **文献标识码** A   **文章编号** 0251-0790(2000)09-1445-03

CH$_4$ 蒸气重整制得的合成气中 $n($H$_2)/n($CO$)\geqslant3$,适于合成氨,但不适于羰基合成和 F-T 合成[1],而以 CH$_4$/CO$_2$ 重整制得的合成气中 $n($H$_2)/n($CO$)\leqslant1$,是富含碳资源物质,可用于羰基合成和 F-T 合成[2,3]。CH$_4$/CO$_2$ 重整制合成气催化剂以 Rh,Ru,Pd,Ir 等贵金属为佳[1]。Ni 基催化剂与贵金属的活性相近,但易积炭失活[4~7]。本文以载体盐助分散浸渍法制得了 Ni 催化剂,其寿命比普通浸渍法制备的长 9 倍,分散度高 30%,积炭率减少 48 倍,这为实际应用提供一种新的催化剂制备方法。

## 1 实验部分

### 1.1 催化剂制备及活性评价

载体 MgO(国营上海化学原料厂)经压片、筛分,取 0.4 mm~0.9 mm 颗粒度,以 $m[$Ni(NO$_3)_2$·6H$_2$O$]/m[$MgO$]$=5/95 等容浸渍 4 h 后得催化剂 A;以 $m[$Ni(NO$_3)_2$·6H$_2$O$]/m[$Mg(NO$_3)_2$·6H$_2$O$]/n[$MgO$]$=5/5/90,Ni(NO$_3)_2$·6H$_2$O 和 Mg(NO$_3)_2$·6H$_2$O 混溶等容浸渍 4 h 后得催化剂 B。催化剂 A,B 均经 110 ℃烘干过夜,600 ℃灼烧 4 h。Ni(NO$_3)_2$·6H$_2$O(上海试剂二厂)及 Mg(NO$_3)_2$·6H$_2$O(北京化工厂)均为 A.R. 级试剂。

采用 3 气路、双石英反应器同时进行对比评价(其中 1 路作原料气分析)。原料气按 $n($CO$_2)/n($CH$_4)$=1.1/1 预配于高压钢瓶中,平放 10 d 后起用。反应后催化剂的积炭量用 EA 1110 CHN S-O 元素分析仪测定。分散度参照文献[8]方法测定。

### 1.2 TPR,TPO 及 TPD 实验

TPR 及 TPO 实验参照文献[9]方法进行,CO$_2$-TPD 实验取 40 mg 催化剂,经 750 ℃,H$_2$ 气还原 0.5

---

* 收稿日期:1999 年 11 月 4 日。

  基金项目:国家重点基础研究发展规划项目(批准号:G1999-022408)资助。

① 联系人简介:李基涛(1940 年出生),男,副教授,从事 C$_1$ 化学的催化研究。

h 后切换成 He 气,并快速升温至 800 ℃,He 气吹扫 1 h 后切换成 CO₂ 吸附 30 min,在 He 气氛下风冷降至室温,然后以 20 ℃/min 进行 TPD 实验。

### 1.3 TPMC-程序升温 CH₄ 解离积炭

10 mg 催化剂经同法还原,于 800 ℃ He 气吹扫 1 h 后风冷快速降至室温,改用 $V(CH_4)/V(N_2)=$ 10/90 的混合气吹扫,待基线平稳后,以 10 ℃/min 进行 TPMC 实验$(CH_4 \rightarrow C + 2H_2)$。

## 2 结果与讨论

### 2.1 催化剂性能测试

表 1 为催化剂 A 及 B 的分散度、催化活性和积炭量的测试结果。由表 1 可见,催化剂 B 的催化活性较高。其分散度比催化剂 A 高近 30%,积炭量比催化剂 A 减少 48.3 倍,说明催化剂 B 的制备方法较好。

Table 1  Results of dispersion, catalytic activity and carbon deposition*

| Catalysts | Dispersion of Ni(%) | $S_{BET}$/ (m² · g⁻¹) | $CH_4$ conv. (%) | $CO_2$ conv. (%) | Comp. of dry outlet gases(%) | | | | Carbon deposition/ (mg · g⁻¹ cat · h⁻¹) |
|---|---|---|---|---|---|---|---|---|---|
| | | | | | $CH_4$ | $CO_2$ | CO | $H_2$ | |
| A | 10.4 | 19.1 | 90.0 | 87.1 | 4.7 | 5.4 | 46.0 | 43.9 | 2.9 |
| B | 13.5 | 20.1 | 92.2 | 87.2 | 4.0 | 4.4 | 46.3 | 45.3 | 0.06 |

* Reaction condition:0.03 MPa,800 ℃,14 400 mL · g⁻¹cat · h⁻¹,$n(CO_2)/n(CH_4)=$ 1.1/1.

等物质量的载体盐与活性组分盐在浸渍前均匀混合,浸渍时能较均匀地分散在载体表面,烘干灼烧时,此 2 种盐基本上能同时分解,并可能进行复杂反应(复分解、准固相或固相反应),致使催化剂 B 表面的 Ni—O—Mg 间产生较强的相互作用,使其分散度较大,可吸附较多的 CO₂,消炭能力增加。

图 1 是催化剂稳定性的评价结果由图 1 可见,催化剂 B 经 105 h 测试后,其 CH₄ 转化率未见明显下降,而催化剂 A 评价 10 h 后,CH₄ 转化率明显降低说明催化剂 B 的寿命较长。

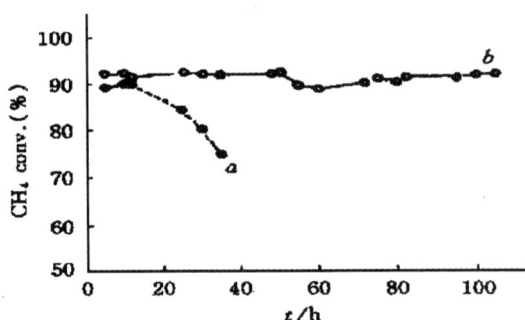

Fig. 1  Relationship between the CH₄ conversion and reaction time
a. Catalyst A; b. Catalyst B.

### 2.2 TPR 及 TPO 谱

由催化剂的 TPR 结果(图 2)可见,两种催化剂的 TPR 峰面积相近,但催化剂 B 的峰温比 A 的高 50 ℃,说明催化剂 B 中的 Ni 原子与 MgO 间的作用较强,使之较难还原由 CH₄ 解离积炭和 CO 歧化积炭的 TPO 谱(图 3)可见,催化剂 B 的 CH₄-TPO 和 CO-TPO 的峰面积均较催化剂 A 的小,说明其抗积炭能力较强,这与其稳定性高、积炭量少相关联。

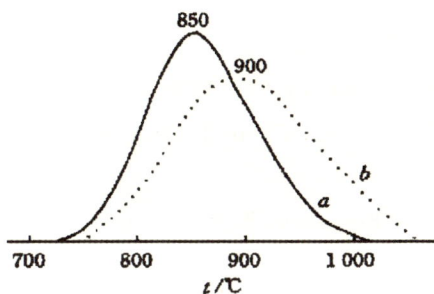

Fig. 2　TPR spectra of catalysts
*a*. Cat. A;*b*. Cat. B.

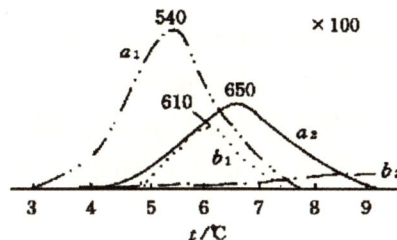

Fig. 3　TPO spectra of catalysts after CH₄ deccmposition and CO disproportionation

*a*₁,*b*₁. CO-TPO for cat. A and B;

*a*₂,*b*₂. CH₄-TPO for cat. A and B.

### 2.3　TPD 及 TPMC 谱

由催化剂 CO₂-TPD 的结果(图 4)可见,催化剂 B 脱附的 CO₂ 峰面积较大且峰温较高,说明催化剂 B 对 CO₂ 吸附能力较强,这与催化剂 B 积炭量较少、寿命较长有关。

图 5 是催化剂的 TPMC 谱。由图 5 曲线 *c* 可见,CH₄ 约 900 ℃ 开始裂解成 C 和 H₂,960 ℃ 以上时,CH₄ 裂解速度非常快。当空管中进行的 CH₄-TPMC 实验至 1 000 ℃ 时,切断 CH₄ 和加热电源,通 CO₂ 并让其自然降至室温后,发现管壁非常干净,无炭沉积现象,说明 CO₂ 有消炭作用。

比较图 5 曲线 *a*,*b* 可见,两种催化剂均有两个 TPMC 峰,高温峰因与 CH₄ 自裂解峰相重叠,故较难判定。对于低温峰,催化剂 B 的峰温比 A 的高 120 ℃,而峰面积却小得多,说明催化剂 A 的 CH₄ 解离积炭能力较强,这与性能测试和 TPO 的结果相符。

综上所述,载体盐助分散浸渍法能制备出分散度高、抗积炭能力强、寿命长的 Ni 基催化剂。

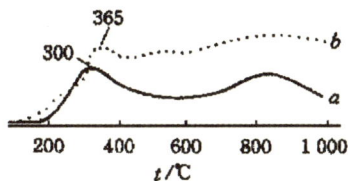

Fig. 4　TPD spectra of catalysts after adsorption CO₂

*a*. Cat. A;*b*. Cat. B;

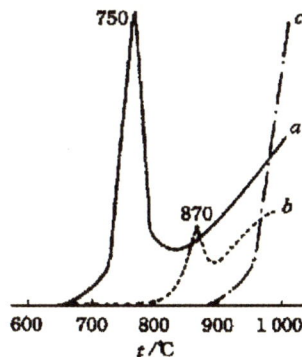

Fig. 5　TPMC spectra of catalysts
*a*. Cat. A;*b*. Cat. B;*c*. No Cat.

# 参考文献

[1]Bhat R. N.,Sachtler W. M. H.. Appl. Catal. A:General[J],1997,**150**:279～296.

[2]Ashcroft A. T.,Cheecham A. K.,Green M. L. H. et al.. Nature[J],1991,**352**:225～226.

[3]SUN Xi-Xian(孙希贤),LI Xin-Min(李新民),LI Ting-Hua(李廷化)et al.. Chem. J. Chinese

Universities(高等学校化学学报)[J],1992,**13**(10):1302~1306.

[4]Querini C. A.,Fung S. C.,Catal. Today[J],1997,**37**(3):277~283.

[5]Tomishige K.,Chen Y. G.,Fujimoto K . J.,Catal. [J],1999,**181**(1):91~103.

[6]JI Min(纪敏),BI Ying-Li(毕颖丽),ZHEN Kai-Ji(甄开吉) et al.. Chem. J. Chinese Universities(高等学校化学学报)[J], 1997,**18**(10): 1698~1699.

[7]JI Tao(姬涛),LIN Wei-Ming(林维明). J. Natural Gas Chemistry(天然气化工)[J],1998,**23**(3): 24-28.

[8]JI Min(纪敏),BI Ying-Li(毕颖丽),ZHEN Kai-Ji(甄开吉) et al.. J. Mol. Catal. (China)(分子催化)[J],1998,**12**(3): 199~205.

[9]Tang S. ,Lin J., Tan K. L..Catal. Lett. [J],1998,**51**: 169~175.

# Development of Higher Stable Ni/MgO Catalyst for $CO_2$ Reforming of Methane

Ji-Tao Li[*]，Ming-Dan Chen，Qian-Gu Yan，Hui-Lin Wan，K.R.Tsai

(*Department of Chemistry*，*Institute of Physical Chemistry*，

*State Key Laboratory for Physical Chemistry of Solid Surface*，

*Xiamen University*，*Xiamen* 361005，*China*)

**Abstract**　Methane reforming by carbon dioxide has been studied over Ni/MgO catalysts prepared by two impregnation methods which were common and promoted with support salt(magnesia salt). The catalysts were characterized by TPR, TPD, TPO and TPMC (temperature programmed methane decomposition to carbon deposition) techniques. The promoted catalyst exhibited a higher ability to adsorb $CO_2$ and lower ability of $CH_4$ decomposition. Therefore the carbon fomation was inhibited and the catalyst stability was increased more.

**Key words**　$CH_4$ refoming by $CO_2$　Promoted impregnation by support salt　Ni-based catalysts

■ 本文原载:《厦门大学学报》(自然科学版)第1期(2000年1月),第128～131页。

# 介孔分子筛负载型铑膦配合物催化剂的制备*

蔡 阳[①] 袁友珠 傅琪佳 杨意泉 蔡启瑞

(厦门大学化学系 物理化学研究所 固体表面物理化学国家重点实验室,福建厦门 361005)

摘 要 研究介孔分子筛MCM-41及其负载铑膦配合物催化剂的制备,并用XRD和FT IR对催化剂进行表征,活性结果表明,该催化剂在丙烯氢甲酰化反应中的催化活性比其他分子筛负载的催化剂高。

关键词 介孔分子筛MCM-41 氢甲酰化 铑膦配合物

在分子筛孔腔中制备配(簇)合物(如"瓶中造船")或利用分子筛作载体担载配(簇)合物,并对其结构和催化性能等方面的研究已相当多见。人们期望利用分子筛孔道的规整性,得到比均相配合物催化剂性能更优异或相当的多相化催化剂。在以往的国内外文献中,所采用的分子筛孔径多数在纳米级以下,研究的分子筛孔腔金属配(簇)合物种类受到一定的限制,所得的催化剂多数不具备空间多维性。随着分子筛合成和开发上的飞速发展,新型分子筛不断问世,使分子筛固载化的主体有了相当大的变化。美国Mobil公司首先报道合成的介孔分子筛如MCM-41[1],其孔径可在1.5～10 nm范围内变化,具有很大的比表面积和吸附量,及规整的一维孔道和很高的热稳定性。MCM-41介孔分子筛的出现引起了人们极大兴趣,因其是一类理想的大分子反应催化剂载体,也为化学分离、吸附剂及高新材料研制提供了新机会。但是,对利用诸如MCM-41等介孔分子筛的纳米级孔腔空间,设计并制备、负载分子尺寸足够大且催化选择性很高的配合物催化剂等方面的研究,在国内外均有待系统开展。

利用MCM-41这种孔径较大且大小均一并具大表面积的新型介孔分子筛作为载体,负载均相配合物而实现均相催化多相化的研究,近年已有不少报道[2,3],但迄今未见有用于烯烃氢甲酰化催化反应中的研究结果。本文研究了MCM-41负载铑膦配合物催化剂的制备及其丙烯氢甲酰化反应的催化性能,对所制备催化剂进行了XRD和FT-IR表征,以开拓均相催化剂多相化新途径、扩展MCM-41介孔分子筛的用途。

# 1 实验部分

纯硅型M CM-41介孔分子筛参考文献[4]的水热法合成。按类似步骤合成了铝硅介孔分子筛Al—M CM—41,将不同比例的硫酸铝溶液$Al_2(SO_4)_3$和硅源TEOS(原硅酸四乙酯)混合搅拌均匀,用NaOH调节pH值,加入CTMABr(十六烷基三甲基溴化铵)溶液,搅拌直至得到溶胶。其余操作同

* 收稿日期1999年4月12日。

基金项目:国家自然科学基金(29873037)和国家教育优秀年轻教师基金资助项目。

① 作者简介:蔡阳(1975— ),女,硕士研究生。

文献[3]。所得样品在 Rigaku D/MAX-C 型旋转靶 X 射线衍射仪上进行小角度 XRD 测定,以高灵敏度 CuKₐ为激发光源,扫描速度为 2°/min,狭缝参数为 DS= 1/6,SS= 1/6,RS= 0.15。红外光谱表征在 Nicolet 740 FT-IR 型红外光谱仪上进行,室温录谱,分辨率为 4 cm⁻¹采用浸渍法制备催化剂,分别将 Rh(acac)(CO)₂(乙酰丙酮二羰基铑)和 PPh₃ 配成一定浓度的 CH₂Cl₂ 溶液,依次负载于 MCM-41 载体上,抽除溶剂后遂制成 MCM-41 负载铑膦络合物催化剂。

丙烯氢甲酰化反应在固定床加压连续流动微型反应器-102GD 型气相色谱仪组合装置上进行,气体产物由氢火焰离子检测器作在线分析,色谱柱填料为聚乙二醇癸二酸酯/102-白色担体,柱长 3 m。丙烯转化率及产物丁醛区位 选择性用修正面积归一化法计算。

## 2 结果与讨论

### 2.1 XRD 粉末衍射谱图

图 1 示出了制备的 MCM-41 介孔分子筛、负载 Rh(acac)(CO)₂-PPh₃ 后和具氢甲酰化催化活性的工作态催化剂之 XRD 谱图。典型的 MCM-41 其 XRD 谱图在 2θ 为 6°以上基本无衍射峰,最强峰通常出现在 2θ=2.2°处,对应[100]晶面。图 1 所示出各样品的衍射谱相似,并与文献[3]报道的一致,说明 MCM-41 的结构在负载铑膦配合物和氢甲酰化反应后,均未发生变化。

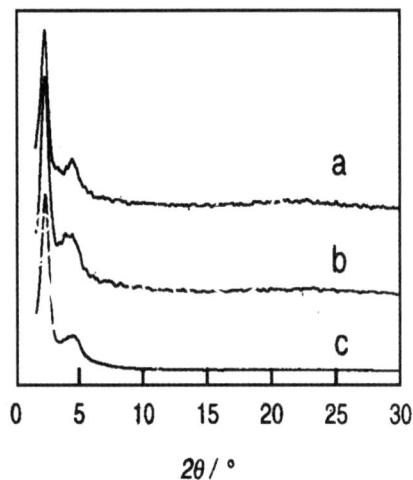

图 1　XRD 谱图

**Fig. 1　XRD Patterns of(a)MAC-41(Si/Al=∞);(b)Rh(acac)(CO)₂-PPh₃/MCM-41 prepared freshly;(c) after 3 h of operation for propene hydroformlation**

### 2.2 FTIR 表征结果

不同硅铝比(38,50,100 和∞)叫的介孔分子筛 MCM-41,其 IR 谱图如图 2 所示。从图 2 可见,在 1 085,798 与 460 cm⁻¹附近有分子筛骨架中的 Si—O—Si 振动,在 961 cm⁻¹附近有 Si 吸附了羟基后形成 Si—O—H 的 Si—O 基团伸缩振动。图 3 示出了 MCM-41(Si/Al=∞)负载铑膦配合物前后和丙烯氢甲酰化反应前后的 IR 谱图。由图 3c 可见,在具反应活性的工作态催化剂中,于 2039 cm⁻¹和 1922 cm⁻¹处有新谱峰生成,可分别归属为铑膦配合物的 Rh—H 键和配位羰基的伸缩振动吸收峰,说明在氢甲酰化反应条件下进行适当时间的催化反应,将在 MCM-41 表面生成具反应活性的铑膦配合物物种。

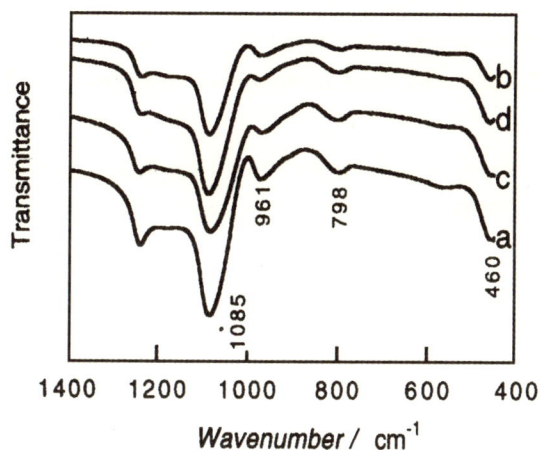

图2　介孔分干筛 MCM-41R FT-IR 谱图

**Fig. 2　FT-IR spectra of MCM-41：(a)Si/Al＝∞；(b)Si/Al＝100；(c)Si/Al＝50；(d)Si/Al＝38**

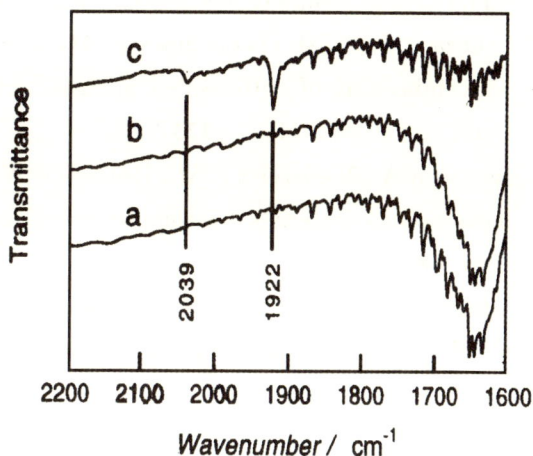

图3　介孔分子筛 MCM-41R 负载型铑膦配合物 的 FT-IR 光谱图

**Fig. 3　FT-IR spectra of(a)MCM-41：(a)Si/Al＝∞；(b)Rh (acac)(CO)₂-PPh₃/MCM-41 prepared freshly(c) after 3 h of operation propene hydroformylation**

## 2.3　丙烯氢甲酰化反应结果

　　表1示出了 MCM-41 介孔分子筛及几种其他载体负载的铑膦配合物催化剂上丙烯氢甲酰化反应活性的评价结果。从表1可以看出以 MCM-41 为载体的催化剂在反应中有较高的转化率和区位选择性。而同是硅铝结构的分子筛 ZSM-5 和 NaY 载体,其产物正、异构醛的比值较高,但其催化活性低,这可能是因为这两种分子筛的孔径太小,无法同时提供铑膦配合物和反应分子的进出孔道,而 MCM-41 具有较大的有效表面积和纳米级的孔径结构,这可能导致其负载型铑膦催化剂有较好的氢甲酰化性能。

表1　不同载体负载的铑膦配合物催化剂上丙烯氢甲酰化活性评价结果

**Tab. 1　Propene hydroformylation on different carriers supported Rh-complex catalysts**

| 载体 | 丙烯转化率/(%) | 转化数/s⁻¹ | 时空产率/mmol/h·g-Rh | 正异构醛比值/n/i |
|---|---|---|---|---|
| SMCM-41 | 34.0 | 0.126 | 4425 | 7.0 |
| SiAl MCM-41(Si/Al＝ 38) | 18.2 | 0.068 | 2369 | 9.5 |
| SiAl MCM-(Si/Al＝ 50) | 23.9 | 0.089 | 3111 | 9.7 |
| SiAl MCM-(Si/Al＝ 100) | 22.6 | 0.084 | 2941 | 8.6 |
| NaY | 13.9 | 0.052 | 1809 | 11.0 |
| HZXM-5 | 16.1 | 0.060 | 2296 | 10.7 |

反应条件:393 K,1.0MPa,$C_3H_6$/CO/$H_2$＝1/1/1(V/V),GHSV＝ 9 000 mL(STP)/h(g-catal);

催化剂:0.01mmolRh(acac)(CO)₂负载于 0.1 g 的载体上,P/Rh＝ 9/1(n/n),铑担载量＝0.08 wt%;数据取自反应 3 h 后。

# 参考文献

[1]Kresge C T,Leonowicz M E,Roth,et al. Ordered mesoporous molecular sieves synthesized by a

liquid-crystal template mechanism [J]. Nature,1992:710~712.

[2]Frunza L,Kosslish H,Landmesser H,et al. Host/guest interations in nanoporous materials I. The embedding of chiral salen manganese(Ⅲ) complex into mesoporous silicates [J]. J. Mol Catal A:Chemical,1997 ,**123**:179.

[3]Jaanssen A ,Niederer J P M ,Holderich W F,Investigation of rhodium complexes in micro-and meso-porous materials by computermodeling,FT-IR,and 31 PMA SNM R [J]. Catal. Lett., 1997,**48**(3-4):165~171.

[4]Ulagappan N ,Rao C N R,Synthesis and characterization of the mesoporous chromium silicates, Cr MCM-41 [J]. Chem. Commun ,1997:1047~1048.

## Preparation of Mesoporous-Molecular-Sieve Supported Rhodium-Phosphine Complex

Yang Cai，You-Zhu Yuan，Qi-Jia Fu，Yi-Quan Yang，K.R.Tsai

(Dept. of Chem., Inst. of Phys Chem., State Key Lab. of Phys Chem.

for Solid Surf. Xiamen Univ., Xiamen 361005，China)

**Abstract**    The mesoporous molecular sieve (MCM-41) and the MCM-41 supported Rh-PPh$_3$ complex(PPh$_3$：triphenylphosphine) were prepared and characterized by XRD and FT-IR. The results showed that the MCM-41 supported Rh-PPh$_3$ complex performed higher catalytic activity for hydroformylation of propene than some other molecular sieves supported catalysts under the same reaction conditions.

**Key words**    Mesoporous molecular sieve(MCM-41)    Hydroformylation    Propene    Rh-phopshine complex

■ 本文原载:《应用化学》第 17 卷第 5 期(2000 年 10 月),第 530~532 页。

# 抗积炭 $CH_4/CO_2$ 重整用 $Ni/Al_2O_3$ 催化剂*

李基涛　严前古　陈明旦　万惠霖　蔡启瑞

（厦门大学化学系　物理化学研究所　固体表面物理化学国家重点实验室,厦门　361005）

**关键词**　$Ni/Al_2O_3$ 催化剂　积炭　$CH_4/CO_2$ 重整

甲烷转化制备的合成气是合成液体燃料和含氧有机化合物的原料。甲烷转化制合成气的方法有甲烷蒸汽重整、甲烷部分氧化和甲烷、二氧化碳重整 3 种[1-3]。对于 $CH_4/CO_2$ 重整反应,调节进料比可制备出 $H_2/CO \leqslant 1$、富含 CO 的合成气,它适于羰基合成和 F-T 合成。这种方法一方面充分利用碳资源,缓解能源危机;一方面可减少温室气体的排放,改善人类的居住环境,目前倍受关注。

$CH_4/CO_2$ 重整制合成气,Rh、Ru、Pd、Ir 等贵金属有很高的活性和稳定性[4],但其价格昂贵,高温易流失,商业化困难。Ni 基催化剂的活性与贵金属相当,但它易积炭而失活。故研究 Ni 基催化剂的关键是抗积炭[5]。Ni 基催化剂的积炭与载体的组成、助剂的选择以及制备方法等有密切关系。本文研究载体盐助分散浸渍法,用此法能制备出高分散度,低积炭率的 Ni 基催化剂,这为理论研究和实际应用提供有益的讯息。

催化剂制备:1# 催化剂用摩尔分数为 3% 的醋酸镍(AR,中国泗联化工厂)等容浸渍在摩尔分数为 97% 的 $\gamma$-$Al_2O_3$(上海试剂五厂)载体上;2# 催化剂取摩尔分数为 3% 的醋酸镍 + 摩尔分数为 6% 的硝酸铝(AR,上海振兴试剂厂)共溶于水后,等容浸渍在摩尔分数为 91% 的 $\gamma$-$Al_2O_3$ 载体上。浸渍后均经 80℃ 真空烘干过夜,600℃ 灼烧 4 h 而成。

活性评价:催化剂用 3 气路,双石英管反应器进行对比评价(其中 1 气路作原料气分析用)。原料气按 $n(CO_2):n(CH_4) = 1:1$,预先配于高压钢瓶中,转化率计算参考文献[6],催化剂在 750 ℃ $H_2$ 还原 1 h 后,通入原料气反应,产物用在线色谱分析。反应后催化剂的积炭量用 EA 1110 CHN S—O 元素分析仪(CE Instruments 意大利)测定。

TPR 谱测定见文献[7]。TPD 谱:取 40 mg 催化剂在 750℃、$H_2$ 还原 0.5 h,改用 He 气并升温至 800℃ 吹扫 1 h,然后通 $CH_4$ 或 $CO_2$ 吸附 0.5 h,在各自吸附气氛下电扇风冷却降至室温,改用 He 气吹扫,待色谱基线平稳后,以 20℃/min 作 TPD 谱。程序升温甲烷解离积炭(TPMC)谱:取 10 mg 催化剂在 750℃、$H_2$ 还原 0.5 h 后,He 气气氛下升温至 800℃,吹扫 1 h,电扇风冷却降至室温,改用 $CH_4$ 和 $N_2$ 摩尔分数分别为 10% 和 90% 的混合气吹扫,待基线平稳后,以 10℃/min 作 TPMC 谱,分散度测定详见文献[8]。

## 结果与讨论

催化剂性能:催化剂活性和积炭量测定结果如表 1 所示。由表 1 可见,2# 催化剂的分散度比 1# 增加

---

* 2000 年 2 月 23 日收稿,2000 年 5 月 8 日修回。

72.7％,而积炭率却比 1# 减少 84％,说明载体盐助分散浸渍法有一定的优点。2# 催化剂活性组分 Ni 盐在浸渍前与少量的载体盐液相混合均匀,等容浸渍时,它能较均匀地分散在载体 γ-Al₂O₃ 的表面上,经 80℃真空干燥,600℃高温快速分解和灼烧时,活性组分盐与载体盐在载体 γ-Al₂O₃ 表面可能进行较复杂的反应(复分解反应、准固相或固相反应),从而 Ni,Al 和 O 间作用较强,这有利于 $CO_2$ 吸附和消炭能力的增加。

**Tab. 1  Activity and carbon deposition of catalysts**

| Catalysts | Dispersion /% | BET /(m²·g⁻¹) | CH₄ conv. /% | CO₂ conv. /% | CH₄ | CO₂ | CO | H₂ | Carbon deposition /(mg·g⁻¹·h⁻¹) |
|---|---|---|---|---|---|---|---|---|---|
| 1# | 8.8 | 89.0 | 86.5 | 83.4 | 5.4 | 6.5 | 43.7 | 44.4 | 4.4 |
| 2# | 15.2 | 85.4 | 87.5 | 84.6 | 5.0 | 6.3 | 44.0 | 44.7 | 0.7 |

Reaction condition：0.03 MPa,750℃,18000 h⁻¹：$n(CO_2)$：$n(CH_4)$ = 1:1.

TPR 谱:催化剂的 TPR 谱如图 1 所示,图中可见,2 种催化剂均有 3 个峰,峰面积较小的低温峰可理解为是催化剂表面少量的 NiO 还原产生,2 个高温峰则可能是铝酸镍和熔进载体晶格的 NiO 还原所至,比较图 1 的 a、b 曲线可见,2# 催化剂的 TPR 峰温较高,说明 2# 催化剂表面的 Ni,Al 和 O 间作用较强,从而使其抗积炭能力较大。

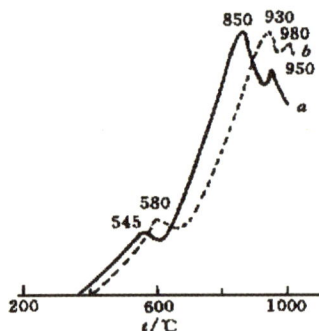

Fig. 1  TPD spectra of catalysts 1# (a) and 2# (b)

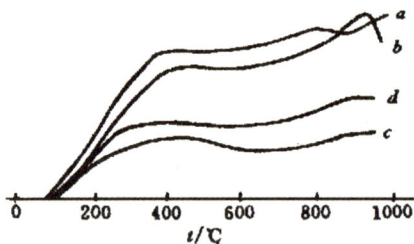

Fig. 2  TPD spectra of catalysts after adsorption of CH₄ and CO₂
a. 1# cat CH₄-TPD;b. 2# cat CH₄-TPD; c. 1# cat CO₂-TPD;d. 2# cat CO₂-TPD

TPD 谱:图 2 是催化剂吸附 CH₄ 和 CO₂ 后的 TPD 谱。比较图 2 可见,1# 催化剂吸附 CH₄ 的能力较强,而 2# 催化剂吸附 CO₂ 的能力较强。参考表 1 的数据,说明 1# 催化剂较强吸附的 CH₄ 不利其转化而利其积炭,而 2# 催化剂吸附较多的 CO₂,这有利于 CH₄ 转化和消炭,这与性能测试结果相符。

TPMC 谱:图 3 是催化剂的 TPMC 谱,图中曲线 c 是石英管没有装任何催化剂时所测得的结果,它说明 CH₄ 在 880℃时开始自行裂解积炭,950℃ 以上 CH₄ 裂解速度很快,1000℃时,曲线 c 仍未下降。空管 CH₄-TPMC 至 1000℃,切断加热电源和 CH₄,把 CO₂ 通到反应管中,同时让其自然降温,室温下取出反应管,可看到管壁非常干净,无炭沉积现象,说明 CO₂ 有消炭作用。

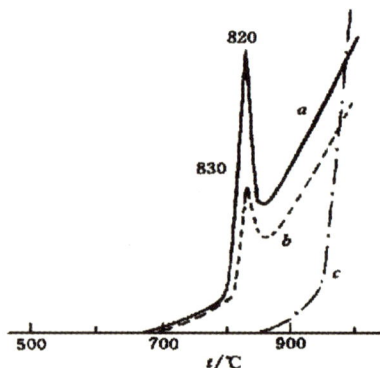

Fig. 3  Temperature programmed methane decomposition carbon deposition spectra of catalysts
a. 1# cat.; b. 2# cat.; c. no catalyst

比较图 3 中的 a,b 曲线可知,2 种催化剂均有 2 个 TPMC

峰,高温的 TPMC 峰因与 CH$_4$ 自行裂解峰相重叠,故无法判定。比较它们低温的 TPM C 峰可见,2$^\#$ 催化剂的 CH$_4$-TPMC 峰 面积较小,这与其积炭量较少相关联,说明载体盐助分散浸渍法制备催化剂确有一定的优点。

# 参考文献

[1]Bhat R N,SachtlerW M. Appl Catal A General,1997,**150**:279.

[2]严前古(YAN Qian-Gu),于军胜(YU Jun-Sheng),于作龙(YU Zuo-Long),et al 应用化学(Yingyong Huaxue),1997,**14**(3):19.

[3]许峥(XU Zheng),张继炎(ZHANG Ji-Yan),张鎏(ZHANG Liu). 石油化工(shi you Hua gong),1997,**26**(6):402.

[4]Ashcroft A T,Cheecham A K,GreenM L H,et al. Lett. to Nature,1991,**352**:225.

[5]严前古(YAN Qian -Gu),吴廷华 WU Ting-Hua),李基涛(LI Ji-Tao),et al. 应用化学(Ying yong Huaxue),1999,**16**(4):20.

[6]姬涛(JI Tao),林维明(LIN Wei-Ming). 天然气化工(Tian ranqi Huagong),1998,**23**(3):24.

[7]李基涛(LI Ji-Tao),严前古(YAN Qian-Gu),张伟德(ZHANG Wei-De),et al. 分子催化(Fenzi Cuihua),1999,**13**(3):205.

[8]纪敏(JI Min),毕颖丽(BI Ying-Li),甄开吉(ZHEN Kai-Ji),et al. 分子催化(Fenzi Cuihua),1998,**12**(3):199.

## Ni/Al$_2$O$_3$ Catalyst with Repressed Coke Formation in CH$_4$/O$_2$ Reforming

Ji-Tao Li$^*$, Qian-Gu Yan, Ming-Dan Chen, Hui-Lin Wan, K. R. Tsai

(Department of Chemistry, Institute of Physical Chemistry, State Key Laboratory for Physical Chemistry of Solid Surface, Xiamen University, Xiamen 361005)

**Abstract**　The Ni/Al$_2$O$_3$ catalysts were prepared by two impregnation methods: common and promoted with support salt, for CH$_4$/CO$_2$ reforming and studied by using TPR, TPD, TPMC (Temperature programmed methane decomposition carbon deposition) and activity test techniques. The results showed that the coke formation of the catalysts promoted with support salt(aluminium nitrate) was reduced to one sixth and the dispersion increased by 72.7% in comparison with catalyst prepared by common impregnation method. The interactions between Ni and Al with O on the surface of the support salt promoted catalyst appeared stronger.

**Key words**　Ni/A l$_2$O$_3$ catalyst　Coke formation　CH$_4$/CO$_2$ reforming

■ 本文原载:《分子催化》第 14 卷第 1 期(2000 年 2 月),第 20~24 页。

# 载体对负载型水溶性铑膦配合物结构和氢甲酰化性能的影响*

袁友珠[①]  张宇  蔡阳  杨意泉  张鸿斌  蔡启瑞

(厦门大学化学系  物理化学研究所  固体表面物理化学国家重点实验室,厦门  361005)

**摘 要** 以丙烯、1-己烯氢甲酰化为目标反应,研究了载体表面酸碱性、比表面积及孔径等对负载型水溶性铑膦配合物结构和催化性能的影响。$^{31}P(^1H)$-NMR 结果显示,水溶性铑膦配合物 $HRh(CO)(TPPTS)_3$ 负载到未经处理的 $SiO_2$ 表面后,配合物主要以 $\{Rh(CO)(TPPTS)_2\}$ 物种形式存在。负载型催化剂的活性和选择性与载体表面的酸碱性、比表面积、孔径密切相关。载体孔径应足够大,以利于活性物种的负载及反应物/产物的进出。

**关键词** 负载型催化剂 水溶性铑膦配合物 氢甲酰化 NMR

采用负载型水溶性铑膦配合物(SAPC)催化烯烃的氢甲酰化反应,突出的优点是催化剂制备简单、催化活性较高、易于连续操作。国内外分别对 SAPC 体系进行了一系列研究,得出了不少有意义的结果[1~15]。但迄今未见系统研究载体对 SAPC 催化剂结构和催化性能影响的文献报道[2,14]。我们考察了 SAPC 制备过程中,载体的表面酸碱性、孔结构特性(如孔径及比表面积等)对 SAPC 催化烯烃氢甲酰化反应性能的影响,并结合 $^{31}P(^1H)$NMR 表征、TPD 等实验方法,推测负载配合物的组成和结构,以加深对 $SiO_2$ 负载水溶性铑膦配合物催化剂的认识。

# 1 实验部分

催化剂采用两种方法制备:(a)配合物 $HRh(CO)(TPPTS)_3$ 按文献[2]方法合成后,将其配成一定浓度的溶液,采用等容浸渍法负载于硅胶载体上,遂制成相应的 $SiO_2$ 负载铑膦配合物催化剂;(b)采用等容浸渍法,将一定浓度的 $Rh(acac)(CO)_2$ 的环己烷溶液浸渍于 0.8 mL 载体上,摇动浸渍 30 min 后,真空抽干(10 min)。再将一定浓度的 TPPTS 水溶液浸渍其上,摇动浸渍 30 min 后真空抽干(120 min),硅胶先变为血红色,之后逐渐变黄,即制得 SAPC 催化剂以固定的抽真空时间和温度来控制催化剂的含水量。

采用固定床加压流动反应器—气相色谱组合装置,进行催化剂活性评价。反应前先用合成气(4 0 MPa)于室温下吹扫 2 h,后升温至反应温度。用 SY-02A 双柱塞微量计量泵压入反应物液体,产物用 102-GD 气相色谱仪定量分析,氢焰检测,用 CDMC-1CX 型色谱数据处理机计算峰面积,转化率用修正面积归一化法计算。色谱柱为 10%邻苯二甲酸二壬酯,柱长 2 m。

$^{31}P(^1H)$NMR 测试在 Varian FT Unity+500 核磁共振谱仪上进行。氢气气氛下,将适量固体催化

---

* 收稿日期:1999 年 3 月 1 日;修回日期:1999 年 8 月 17 日。

基金项目:国家自然科学基金和中国石化总公司联合资助课题(29792075,29873037)。

① 作者简介:袁友珠,男,36 岁,博士,教授。

剂装入核磁管,滴入适量 $D_2O$(恰好润湿固体表面),盖好核磁管待测于室温下 200 MHz 处录谱。液体测试以 85% 磷酸水溶液作外标。

比表面积及孔径分布用意大利 CARLO ERBA 仪器公司产 SORPTOMATIC-1900 吸附仪,以高纯氮为吸附质,于液氮温度下测试,利用 MLE-STONE 200 软件进行计算。

# 2 结果与讨论

## 2.1 载体表面酸碱性的影响

经 $Na_2CO_3$ 处理前后的 $SiO_2$ 载体负载水溶性配合物 $HRh(CO)(TPPTS)_3$ 的丙烯氢甲酰化催化性能比较列于表 1。可以发现,经 $Na_2CO_3$ 处理的 $SiO_2$ 载体负载的催化剂,其催化活性远低于不经 $Na_2CO_3$ 处理、表面酸性较强的 $SiO_2$ 载体负载的催化剂,前者活性只及后者的 1/8,其产物区位选择性略高于后者。

表 1  $HRh(CO)(TPPTS)_3/SO_2$ 催化剂催化丙烯的氢甲酰化反应

Table 1 Propene hydrofoimylation on $HRh(CO)(TPPTS)_3/SiO_2$ catalyst

| $SiO_2$ | Conv. (%) | $n/i$ | STY (mmol/(h·g Rh)) | TOF ($s^{-1}$) |
|---|---|---|---|---|
| No pretreatment | 12.8 | 5.0 | 1664 | 0.048 |
| $Na_2CO_3$ pretreated | 1.6 | 5.1 | 2208 | 0.006 |

Reaction conditions:$T=393$ K;$C_3H_6/CO/H_2=1/1/1$(volume ratio);1.0 MPa;GHSV(STP):1100 $h^{-1}$; Rh loading: 0.01 mmol/0.36 g $SiO_2$

图 1 为 TPPTS 及配合物 $HRh(CO)(TPPTS)_3$ 分别负载到 $SiO_2$ 载体表面后的 $^{31}P(^1H)$NMR 谱可以看出,TPPTS 负载到 $SiO_2$ 载体表面以后,其特征 $^{31}P(^1H)$NMR 谱峰的化学位移并没有发生明显的变化,仍与在 $D_2O$ 溶液中时的 5.1 基本一致[11]。图 1 中在 35.2 处的核磁共振峰可归属为 OTPPTS 之 P 原子的共振信号。其化学位移也与其在相应溶液中时的位置基本一致。当采用等容浸渍法将配合物 $HRh(CO)(TPPTS)_3$ 的重水溶液负载到 $SiO_2$ 载体表面后,其 $^{31}P(^1H)$MMR 谱图发生了明显的变化(见图 1c)。原配合物中 P 原子在化学位移 44.4 处的特征双峰已完全消失;在化学位移为 29.5 处出现一个新的宽峰,相应膦物种含 P 量约占体系总 P 量的 60%(摩尔分数)左右。这是配合物在 $SiO_2$ 载体表面形成的一种新含膦配合物物种,其 TPPTS 与 Rh 的比例约为 2:1(摩尔比)。且由于载体表面的非均一性,使该物种中 P 原子所处的化学环境各向异性较大,谱峰较宽。在 35.2 处属 OTPPTS 物种的 P 原子的 NMR 谱峰化学位移不变,但其在总 P 量中所占比例大为增加,从负载前的约 10% 增加到负载后的 30%(摩尔分数)左右。实验发现,对这一负载型催化体系而言,在丙烯氢甲酰化反应 4 h 后,催化剂表面各膦物种谱峰的化学位移及相对强度基本不变(图 1d)。

若先用 TPPTS 对 $SiO_2$ 载体表面进行修饰,而后再负载配合物,其 $^{31}P(^1H)$NMR 谱(图 1e)与前者明显不同。这时,属原配合物 P 原子的特征 NMR 双峰仍保留一定强度,氧化配体 OTPPTS 的 P 原子的共振峰明显增强,后者的 P 原子含量约占此时体系总 P 量的 50%。这说明,在 $SiO_2$ 载体表面先覆盖一层配体后,再后续负载的配合物所接触到的表面与直接负载时不同,铑膦配合物与载体的相互作用有所减弱,配合物较大程度上可维持其在溶液中的构型,从而证实了载体 $SiO_2$ 的表面性质是引起负载配合物结构发生变化的主要原因。

当以经 $Na_2CO_3$ 预处理的 $SiO_2$ 载体负载配合物时,相应体系($HRh(CO)(TPPTS)_3/Na_2CO_3/SiO_2$)

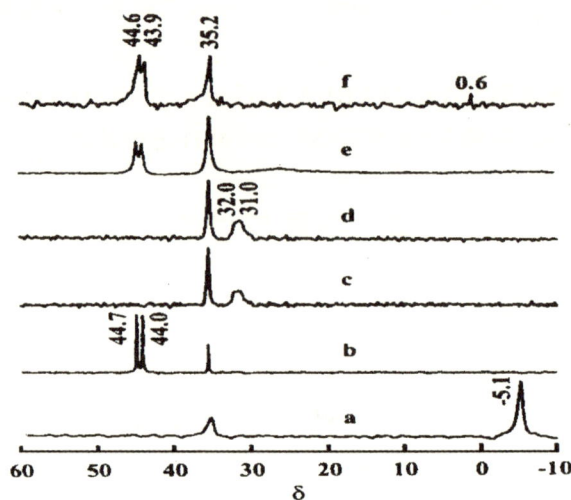

**图1 SiO₂负载HRh(CO)(TPPTS)₃的³¹P(¹H)-NMR谱图**

**Fig. 1 ³¹P(¹H)-NMR spectra of SiO₂-supported HRh(CO)(TPPTS)₃**

(a)TPPTS on SiO₂;(b)HRh(CO)(TPPTS)₃ in D₂O;(c)HRh(CO)(TPPTS)₃ on SiO₂;(d)HRh(CO)(TPPTS)₃ on SiO₂,after propene hydroformylation for 3 h;(e)HRh-(CO)(TPPTS)₃ on SiO₂ pre-impregnated with TPPTS;(f)HRh(CO)(TPPTS)₃ on SiO₂ modified with Na₂CO₃

液膜的³¹P(¹H)NMR谱示于图1f。可见,在$\delta=44.4$附近、属原配合物的 NMR 共振特征双峰依然存在,只是比在溶液中时有所宽化(这同样可归因于载体表面的各向异性);从相应谱峰之相对面积同样可估算出,以原配合物形态存在的膦物种中的 P 原子含量约占体系总 P 量的60%,其余 P 原子主要以 OTPPTS 形式存在;此外,尚有很小一部分是吸附在载体表面、属未配位的配体 TPPTS,其化学位移在 0.6 附近。这一结果表明,在经 Na₂CO₃ 处理的碱性 SiO₂ 载体表面,被负载的铑膦配合物与载体表面的相互作用相当弱,基本上保留其在 D₂O 溶液中的三膦配合物构型特征。

我们在用³¹P(¹H)MAS NMR 谱法表征 Rh-(acac)(CO)₂ 和 TPPTS 在 SiO₂ 载体表面原位生成的配合物物种时,并没有观察到本应在$\delta=44.4$处出现的、HRh(CO)(TPPTS)₃中配位膦物种的特征双峰;而在工作态催化剂中,观察到可归属为表面配合物{Rh(CO)(TPPTS)₂}的³¹P 物种[15]。SiO₂ 载体以亲水性的极性表面为主要特征,其表面具有强酸性,这可能是引起上述配合物负载以后,其结构发生变化的主要原因。为此,我们测定了在 SiO₂ 载体上吸附吡啶的程序升温脱附(Pyridine-TPD)谱,结果示于图2。实验表明,SiO₂ 载体对吡啶有很强的吸附能力,在 420 K 附近出现一个较大的脱附峰,证实了 SiO₂ 载体表面的强酸性用弱碱性的 Na₂CO₃ 水溶液对 SiO₂ 载体表面进行处理,然后用水洗至中性,在 773 K 下灼烧 2 h,再经水洗至

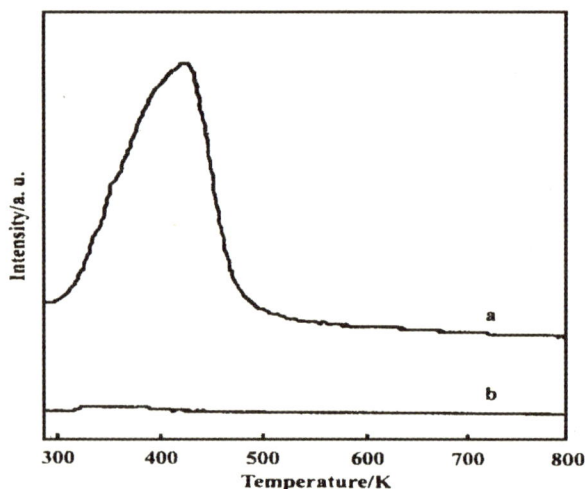

**图2 Na₂CO₃ 处理前后 SiO₂ 的吡啶-TPD 谱图**

**Fig. 2 Pyridine-TPD patterns of as-received SiO₂(a) and Na₂CO₃-treated SiO₂(b)**

中性,并以该 SiO₂ 载体作吡啶-TPD 实验时,发现此载体表面对吡啶的吸附能力大大降低,与处理前相比,表面酸性几乎完全消失。

以上实验结果表明，配合物 HRh(CO)(TPPTS)$_3$ 可与载体 SiO$_2$ 的酸性表面发生较强的相互作用，导致配合物的表面结构不同于在溶液中。进一步实验表明，只有那些与表面直接接触的配合物才发生这种结构上的变化。根据这一实验事实，可能通过测定不同负载量试样液膜的 $^{31}$P($^1$H)NMR 谱，从其中含 P 物种在表面存在形式的变化，来估算配合物在 SiO$_2$ 载体表面的单层负载量，结果示于图 3。从谱图中可以看出，在负载量较低（≤0.1 mmol Rh/g SiO$_2$）时，该负载体系只在 δ=35.2 处出现，属于氧化配体 OTPPTS 的 P 原子共振信号和 δ=29.5 处属于与表面有较强的相互作用的含膦配合物物种 P 原子的共振信号（见图 3a 和 b）。当负载量达到 0.2 mmol Rh/g SiO$_2$ 时，才出现属溶液中配合物 HRh(CO)(TPPTS)$_3$ 构型特征（δ=44.4 附近）的共振双峰（图 3c）。此时，保持原有配合物构型特征的表面配合物物种含 p 量约占体系总 P 量的 30%。由此可以估算出，配合物 HRh(CO)(TPPTS)$_3$ 在所用 SiO$_2$ 载体表面的单层负载量约为 0.14 mmol Rh/g SiO$_2$。这也为测算负载组分在载体上的单层负载量提供了一种新方法。

可以初步断定，配合物 HRh(CO)(TPPTS)$_3$ 负载到表面呈酸性的 SiO$_2$ 载体表面以后，由于与 SiO$_2$ 载体表面酸性位发生较强的相互作用，使配合物构型发生变化，在溶液中的原有构型消失，代之出现游离的氧化膦配体 OTPPTS 和化学位移在 δ=30 附近的、未知的表面含磷配合物物种，配合物物种的 TPPTS/Rh≈2∶1（摩尔比）。

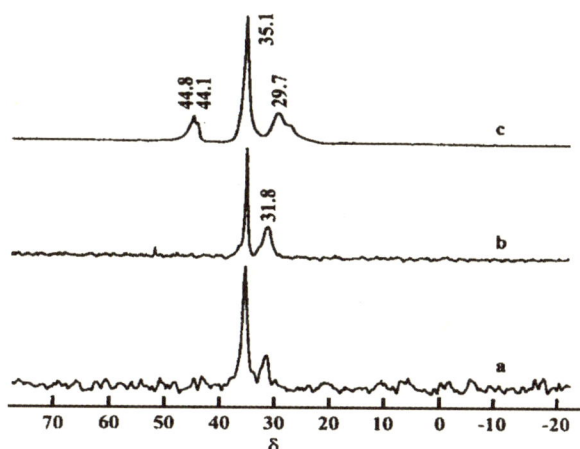

图 3　SiO$_2$ 负载 HRh(CO)(TPPTS)$_3$ 的 $^{31}$P($^1$H)NMR 谱图

Fig. 3　$^{31}$P($^1$H)NM R spectra of SiO$_2$-supported HRh(CO)(TPPTS)$_3$ on SiO$_2$

(a)0.0025 mmol HRh(CO)(TPPTS)$_3$/g-SiO$_2$；(b)0.0125 mmol HRh(CO)(TPPTS)$_3$/g-SiO$_2$；

(c)0.2 mmol HRh(CO)(TPPTS)$_3$/g-SiO$_2$

结合 $^{31}$P($^1$H)NMR 表征（见图 1）和催化活性评价结果，可以推断，在硅胶负载的铑膦配合物催化剂上，对丙烯氢甲酰化具催化活性的配合物物种是 ⎨Rh(CO)(TPPTS)$_2$⎬（P 原子的化学位移在 30 附近），而非 HRh(CO)(TPPTS)$_3$（P 原子化学位移为 44.2）。这是因为原配合物的配位数为 5，中心金属铑原子周围已达到 18 电子的饱和配位结构，反应物分子不能配位活化；当 SiO$_2$ 表面经 Na$_2$CO$_3$ 处理之后，表面呈碱性，与碱性的 TPPTS 配位基难以发生强的相互作用，配合物在负载以后仍保持原有结构，因此催化活性很低。而当配合物 HRh(CO)(TPPTS)$_3$ 负载于酸性的 SiO$_2$ 载体表面时，由于碱性的 TPPTS 配位基与酸性载体表面发生较强的相互作用，使部分配位膦配体解络，并生成了表面物种 ⎨Rh(CO)(TPPTS)$_2$⎬，其中心金属铑原子的配位数为 4，属配位不饱和结构，有空位供反应物分子进行配合活化，因而反应活性高。

## 2.2 载体孔结构特性(孔径及比表面积)的影响

载体对 SAPC 体系催化性能有显著影响。考察了几种不同载体负载的催化剂体系,实验结果列于表 2 和表 3。Davis 等[4]认为,在负载型水溶性催化体系中,催化剂与反应物是在水/有机物两相界面进行反应的。而从表 2 和表 3 可以看出,当 SAPC 使用不同比表面积的同类载体(SiO₂)时,其反应活性无明显差别,区位选择性也相差不大。这表明,当活性物种总负载量相同,即使因载体比表面积因素而造成在水/有机物两相界面活性物种的浓度有一定差别时,仍显示出基本相同的反应活性。由此可以推断,在这类体系上反应似乎不只局限在水—有机相界面上进行,处于负载水相中的活性物种也可发挥催化活性中心的作用。

表 2　载体对 1-己烯氢甲酰化反应结果的影响

Table 2　Effect of support on the SA PC perfonn ance for 1-hexene hydroformylation

| Support | Specific surface area($m^2$/g) | Pore-diameter(nm) | $X$(%) | $n/i$ |
|---|---|---|---|---|
| MgO | 42 | 24.8 | 17.1 | 2.30 |
| SiO₂ | 400 | 11.8 | 63.6 | 2.81 |
| SiO₂ | 113 | 18.6 | 63.7 | 2.99 |
| SiO₂ | 110 | 28.2 | 66.0 | 2.87 |

Reaction conditions：Rh-loading：0.02 mmol/0.36 g-support；TPPT S/Rh = 10(molar ratio)；$T$ = 373 K；liquid-feeding rate：1-hexene：1.0 mL/h；CO/H₂ = 1/1(volume ratio)；4.0 MPa；GHSV(STP) = 900 h⁻¹；feed-syngas being wetted by passing through a bubbler filled with water at 323 K,data taken during 2~4 h after reaction start

表 3　载体对 1-己烯氢甲酰化反应结果的影响

Table 3　Effect of support on the SAPC perfonn ance for 1-hexene hydroformylation

| Support | Specific surface area($m^2$/g) | Pore-diameter(nm) | $X$(%) | $n/i$ |
|---|---|---|---|---|
| SiO₂ | 526 | 8.4 | 75.4 | 2.86 |
| SiO₂ | 690 | 1.4 | 15.2 | 3.33 |
| SiO₂ | 400 | 11.8 | 78.5 | 2.91 |
| NaY | 912 | 1.42 | ~0 | >3 |

Reaction conditions：liquid-feeding rate：1-hexene/n-heptane(1/4,mass ratio)：2.0 mL/h；GHSV(STP) = 1800 h⁻¹. Others conditions are the same as in Table 2.

表 4 列出了表 3 中后两份催化剂试样在负载前后及反应后比表面积及平均孔径的测试结果。从表 4 可以看出,孔径较小的 NaY 型分子筛,无法有效地负载体积较大的 HRh(CO)(TPPTS)₃ 活性物种(直径约 20 nm),致使活性组分堆积在载体表面,堵塞微孔孔道。负载后比表面积远远低于负载前,催化剂活性组分难以发挥作用,催化剂活性很低,经 4 h 反应后,活性组分被液态反应物洗提带走,流失严重;比表面积虽有所恢复,但因活性组分大量流失而无活性。看来,要达到有效负载,载体孔径至少应不小于 20 nm。然而当孔径太大时,往往造成比表面积锐减(如 MgO),从而大大降低反应物与催化剂的接触面积,导致反应活性下降。

表 4　不同孔径载体负载配合物前后及反应前后的比表面积及平均孔径

Table 4　Specific surface area and pore-diameter of two SAPC before and after supporting and after 4 h reaction

| Support | Before supporting | | After supporting | | After 4 h reaction | |
|---|---|---|---|---|---|---|
| | Surf. area ($m^2$/g) | Pore-radius (nm) | Surf. area ($m^2$/g) | Pore-radius (nm) | Surf. area ($m^2$/g) | Pore-radius (nm) |
| NaY | 912 | 1.42 | 24 | 16.52 | 539.0 | 1.42 |
| SiO₂ | 400 | 11.80 | 411 | 9.42 | 392.7 | 13.94 |

## 3　结　论

3.1　在酸性的 $SiO_2$ 载体表面,水溶性配合物 $HRh(CO)(TPPTS)_3$ 负载以后,由于碱性的 TPPTS 配体与硅胶载体酸性表面发生较强的相互作用,使部分配位膦配体解络,负载后的配合物主要以 {Rh(CO)(TPPTS)$_2$} 形式存在。该物种是工作态催化剂上主要的催化活性物种。

3.2　只要载体孔径大到足够提供活性物种负载及反应物/产物进出的反应通道,以保证反应物/产物的扩散不会成为速率控制步骤,即使比表面积有一定差别,也不会引起催化活性的明显改变。

## 参考文献

[1] Arhancet J P, Davis M E, Merola J S, et al Hydroformylation by Supported A queous-Phase Catalysis:a New Class of Heterogeneous Catalysts [J]. Nature, 1989,**339**:454.

[2] Arhancet J P, Davis M E, Merola J S, et al. Supported Aqueous-Phase Catalysts [J] J Catal, 1990,**121**:327～339.

[3] Arhancet J P, Davis M E, Hanson B E, et al Supported Aqueous-Phase, Rhodium Hydroformylation Catalysts I:New Methods of Preparation [J]. J Catal,1991,**129**:94～99.

[4] Arhancet J P, Davis M E, Hanson B E, et al Supported A queous-Phase, Rhodium Hydroformylation Catalysts II: Hydroformylation of Linear, Terminal and Interal Olefins[J]. J Catal, 1991,**129**:100～105.

[5] Davis M E. Supported Aqueous-Phase Catalysis [J]. Chem tech,1992,**22**:489～502.

[6] 袁友珠,陈鸿博,蔡启瑞 $SiO_2$ 负载的磺化三苯基膦配合物催化高碳烯氢甲酰化[J]. 应用化学, 1993,**10**(4):13.

[7] 袁友珠,杨意泉,张鸿斌等.担载型水溶性膦铑配合物催化剂研究[J].高等学校化学学报,1993, **14**(6):863～865.

[8] 袁友珠,刘爱民,许金来等. $SiO_2$ 负载的磺化三苯基膦配合物催化高碳烯氢甲酰化及反应中的氘逆同位素效应[J].分子催化,1993,**7**(6):384～390.

[9] You zhu Yuan, Jin lai Xu, Hong bin Zhang, et al. The Beneficial Effect of Alkali Metal Chlorides on Supported A queous-Phase Catalysts for Hydroformylation [J]. Catal. Lett,1994,**29**:387～395.

[10] You zhu Yuan, Yi quan Yang, Jin lai Xu, et al. Rate-Deteimining Step in Olefin Hydroformylation over Supported Aqueous-Phase Catalysts[J]. Chinese Chem. Lett.,1994,**5** (4):291～294.

[11] 刘海超,陈华,黎耀忠等.负载水溶性铑-膦配合物催化 1-己烯氢甲酰化反应的研究[J].分子催化,1994,**8**(1):22～28.

[12] 袁友珠,张宇,杨意泉等.负载型水相膦铑配合物催化剂上丙烯氢甲酰化制丁醛[J].厦门大学学报(自然科学版),1996,**35**(2):220～225.

[13] 袁友珠,杨意泉,林国栋等.负载型水溶性膦铑配合物催化剂上气、液态烯烃氢甲酰化.殷元骐主编.羰基合成化学[M].北京:化学工业出版社,1996,64～91.

[14] 张宇,袁友珠,傅金印等.载体对 1-己烯在 SAPC 催化剂上 OXO 性能的影响[C].见:第九届全

国催化会议论文集. 北京:海潮出版社,1998. 1171～1172.

[15]袁友珠,张宇,陈忠等. 负载型水溶性铑膦配合物催化剂的结构和性能[J]. 物理化学学报,1998, **14**(11):1013～1019.

# The Influence of Supports on the Structure and Hydroformylation Performance of Supported Aqueous-Phase Catalysts

You-Zhu Yuan，Yu Zhang，Yang Cai，Yi-Quan Yang，Hong-Bin Zhang，Qi-Rui Cai

(*Department of Chemistry，Institute of Physical Chemistry，State Key Laboratory for Physical Chemistry of Solid Surface，Xiamen University，Xiamen* 361005)

**Abstract**　In this article, influences of supports on the structure and catalysis of supported aqueous-phase catalysts(SAPC) for hydroformylation of propene and 1-hexene were studied. According to the results of $^{31}P(^1H)$-NMR, it was found that when water-soluble phosphine-rhodium complex of $HRh(CO)(TPPTS)_3$ was brought to $SiO_2$, the decomplexation of part of phosphine ligands in the complex occurs due to the strong interaction of basic ligand TPPTS with acidic surface of silica-gel carrier, forming a surface complex containing two phosphine ligands, i e, $\{Rh(CO)(TPPTS)_2\}$. However, when using $Na_2CO_3$-pretreated $SiO_2$, the twin signals at $\delta=44.8$ for $HRh(CO)(TPPTS)_3$ in the $^{31}P(^1H)$-NMR was observed as dominant species in the supported catalyst under identical conditions. The hydrofomylation perfomance of SAPC is also affected by surface area and pore size of the supports.

**Key words**　Supported catalysts　Water-soluble phosphine rhodium complex　Hydrofomyaltion　NMR

■■■■■■■■■■■■■■■■■■■■■■■■■■■■

■ 本文原载：Inorganic Chemistry，Vol. 39，No. 1，pp. 59～64，2000.

# Syntheses and Spectroscopic and Structural Characterization of Molybdenum(VI) Citrato Monomeric Raceme and Dimer, $K_4[MoO_3(cit)] \cdot 2H_2O$ and $K_4[(MoO_2)_2O(Hcit)_2] \cdot 4H_2O$

Zhao-Hui Zhou[①], Hui-Lin Wan, Khi-Rui Tsai

(Department of Chemistry and State Key Laboratory for Physical Chemistry of Solid Surface, Xiamen University, Xiamen, 361005, People's Republic of China)

Received January 6, 1999

**Abstract**    Investigation of the aqueous coordination chemistry for citrate and molybdenum(VI) resulted in the isolation of molybdenum(VI) citrato monomeric raceme and dimer $K_4[MoO_3(cit)] \cdot 2H_2O(1)$ and $K_4[(MoO_2)_2O(Hcit)_2] \cdot 4H_2O(2)(H_4cit = $ citric acid). Complex 1 can serve as the first structurally characterized monomeric citrato molybdate and may represent an early mobilized precursor in the biosynthesis of FeMo-co (FeMo-cofactor). The two complexes have been characterized by elemental analyses and IR and NMR spectroscopies. The IR and NMR spectra are consistent with a monomeric species or a monooxo-bridged dinuclear structure, as revealed by a single crystal X-ray diffraction study. Compound 1 is monoclinic space group $P2_1/c$ with $a = 7.225$ (1) Å, $b = 9.151(2)$ Å, $c = 22.727(2)$ Å, $\beta = 94.93(1)°$, $V = 1497.1(7)$ Å$^3$, and $Z = 4$. Full-matrix least-squares refinement resulted in residuals of $R = 0.027$ and $R_w = 0.032$. The molybdenum atom forms an octahedral coordination with three oxo groups and one tridentate citrate, in which the latter is coordinated through the alkoxy and vicinal carboxyl and much more weakly by one of the two terminal groups [2.411(3) Å]. Compound 2 is triclinic space group $P\bar{1}$ with $a = 8.2728(8)$ Å, $b = 8.9514(8)$ Å, $c = 10.0605(9)$ Å, $\alpha = 101.673(8)°$, $\beta = 100.672(7)°$, $\gamma = 112.938(7)°$, $V = 642.5$ (3) Å$^3$, and $Z = 1$. Full-matrix least-squares refinement resulted in residuals of $R = 0.033$ and $R_w = 0.039$. The complex anion contains a linear $(O_2Mo)O(MoO_2)$ core with the bridging oxo group lying at the center of inversion symmetry $(Mo—O_b—Mo, 180°)$. Each citrate ligand is three-coordinated to one molybdenum atom through the deprotonated hydroxy, $\alpha$-carboxyl, and one $\beta$-carboxyl group, making each metal atom six-coordinate.

## 1    Introduction

Recent single-crystal X-ray structural analysis of the nitrogenase proteins has revealed the structure of FeMo-co (FeMo-cofactor) as a cagelike $MoFe_7S_9$ homocitrate cluster[1−9], in which the Mo is

---

①    To whom correspondence should be addressed. E-mail: zhzhou@ xmu. edu. cn.

essentially octahedrally coordinated by three $\mu_3$-S ligands, a histidine, and a bidentate homocitrate through the alkoxy and vicinal carboxyl groups, which may be termed an $\alpha$-carboxyl group with reference to the alkoxy carbon atom as $\alpha$-carbon atom. Early studies have shown that the mutant MoFe proteins lacking homocitrate, which may contain a Mo-bound citrate ligand to molybdenum as a replacement to homocitrate, prevented the enzyme from strong binding and efficient reduction of $N_2$ while acetylene and proton reduction remained at a high level[10,11]. Moreover, it has been suggested that a possible function of the tricarboxylic acid in the biosynthesis of the cofactor of nitrogenase is to mobilize molybdenum or vanadium from the appropriate storage enzyme. Molybdenum or vanadium is believed to be taken up by organisms as $MoO_4^{2-}$ or $VO_4^{3-}$; this would be essential for the assembly of the final cofactor cluster from an oxomolybdenum- or oxovanadium-citrate precursor[12—15]. Tricarboxylic acid may play an early and essential role in the mobilization of the heterometal during cofactor biosynthesis, and the mobilized oxoheterometal tricarboxylic acid fragment must then undergo reduction, exchange oxo ligands for sulfide ligands, and merge with nif B-co.

While the precise role of homocitrate in both the biosynthesis of FeMo-co and the mechanism of dinitrogen reduction is still regarded as poorly understood[16], the elucidation of the key role played by this tricarboxylic acid in nitrogenase catalysis has been pursued with great interest[17]. As part of our systematic study of the coordination chemistry of vanadium (V/IV) and molybdenum (VI) with hydroxycarboxylic acids, complexes formed from aqueous solutions of vanadate, vanadyl or molybdate, and citric acid ($H_4$cit) or its salt have been studied. We have first reported the preparations and structures of $K_2[VO_2(H_2cit)]_2 \cdot 4H_2O$, $Na_2K_2[VO_2(Hcit)]_2 \cdot 9H_2O$, $Na_2(NH_4)_4[VO_2(cit)]_2 \cdot 6H_2O$, and $Na_4[VO(cit)]_2 \cdot 6H_2O$[18—21]. The pH of the medium is the principal variable controlling complex formation and interconversion equilibria[22,23]. At high pH ($>6$) the anions of vanadium complex are vanadate(V), $[VO_2(cit)]_2^{6-}$, or vanadyl(IV) anion, $[VO(cit)]_2^{4-}$, while at lower pH three different dinuclear anions $[VO_2(H_2cit)]_2^{2-}$, $[VO_2(Hcit)]_2^{4-}$, or $[(VO)_2(cit)(Hcit)]^{3-}$ were observed[18—26].

Complex formation between molybdate and citrate has been reported in different pH ranges by potentiometry, spectrophotometry, difference pulse polargraphy, and calorimetry[27—32]. The first well structurally characterized citrato molybdates with 2:1 ratios (Mo:cit) were reported as $[Me_3N(CH_2)_6NMe_3]_2[Mo_4O_{11}(cit)_2] \cdot 12H_2O$ and $K_4[Mo_4O_{11}(cit)_2] \cdot 6H_2O$[33,34]. Formation of the 1:1 complex was first obtained as $K_2[MoO_3(OH)(H_3cit)] \cdot 2H_2O$ and $K_3[MoO_4(H_3cit)] \cdot 2H_2O$[35], later $K_4[(MoO_2)_2O(Hcit)_2] \cdot 5H_2O$, $K_6[(MoO_2)_2O(cit)_2] \cdot 7H_2O$, and $K_4[MoO_3(cit)H_2O]$ were obtained by precipitation from aqueous solution at pH 4—8[36]. In the neutral solution, the complex has been separated and structurally confirmed as a dimeric oxomolybdenum citrate as $K_2Na_4[(MoO_2)_2O(cit)_2] \cdot 5H_2O$ or $K_6[(MoO_2)_2O(cit)_2] \cdot 2H_2O$[37,38]. It is also shown that citric acid is the most effective eluent for the separation of W(VI) and Mo(VI) oxoanions[39]. In the use of eluents without alkoxy groups, W(VI) and Mo(VI) oxoanions were strongly retained, due to the formations of W(VI) and Mo(VI) polyanions. These complexes have been characterized by chemical analyses and various other methods and remained uncertain for their composition and the degree of aggregation. Moreover, the variety of possible functions exhibited by citrate in its interactions with molybdenum, a metal which plays an important role in different living plants and animal organisms[40], prompted our further investigation of the coordination chemistry of molybdenum citrate complexes.

## 2　Experimental Section

**Preparation of $K_4[MoO_3(cit)] \cdot 2H_2O$ (1).** Potassium molybdate (20 mmol) prepared from the reaction of molybdenum trioxide and potassium hydroxide (82%) was added with an excess potassium dihydrogen citrate (K$_2$H$_2$cit, 30 mmol) from the reaction of citric acid and potassium hydroxide. The solution was stirred in a water bath at 80 ℃ for 4 h and filtered. An excess amount of ethanol was added until the solution turned cloudy. The mixture was kept refrigerated for several days, and the solid was collected and recrystallized from H$_2$O-EtOH to give a white solid (5.1 g, 49%). Anal. Found: C,13.3; H,1.6. Calcd for C$_6$H$_8$K$_4$MoO$_{12}$: C,13.7; H,1.5. IR (KBr): $\nu_{asym}$(C=O) 1603$_{s,sh}$, 1575$_{vs,b}$, $\nu_{sym}$(C=O) 1425$_{s,sh}$, 1398$_{vs}$, $\nu$(Mo=O) 937$_w$, 897$_m$, 848$_s$, 826$_s$. $^1$H NMR (500 MHz,D$_2$O; ppm): $\delta_H$ 2.666 (d,J 16.7 Hz,CH$_2$), 2.646 (d,J 14.9 Hz,CH$_2$), 2.543 (d,J 15.1 Hz,CH$_2$), 2.524 (d,J 16.6 Hz,CH$_2$). $^{13}$C NMR (D$_2$O;ppm): $\delta_C$ 187.5 (CO$_2$)$_\alpha$, 179.6, 179.4 (CO$_2$)$_\beta$, 81.7 (≡CO), 46.4,46.3 (=CH$_2$).

**Preparation of K$_4$[(MoO$_2$)$_2$O(Hcit)$_2$]·4H$_2$O (2).** Potassium molybdate (20 mmol) prepared from the reaction of molybdenum trioxide and potassium hydroxide was added with an excess citric acid (22 mmol) and a small amount of potassium trihydrogen citrate (2 mmol). The solution was stirred in a water bath at 60 ℃ for 4 h. The mixture was filtered and added with an excess amount of ethanol. The solid formed was collected and recrystallized from EtOH-H$_2$O to give a white solid. (3.6 g, 41%). Anal. Found: C,16.0; H,2.0. Calcd for C$_{12}$H$_{18}$K$_4$Mo$_2$O$_{23}$: C,16.4; H,2.1. IR (KBr): $\nu_{asym}$(C=O) 1715$_s$, 1652$_{vs,b}$, 1604$_s$, 1553$_{s,b}$; $\nu_{sym}$(C=O) 1440$_m$, 1426$_m$, 1404$_s$, 1346$_m$; $\nu$(Mo=O) 933$_s$, 908$_{s,sh}$, 895$_{vs,b}$; $\nu_{as}$(MoO$_b$Mo), 785$_{vs}$; $\nu_s$(MoO$_b$Mo) 690$_s$. $^1$H NMR (500 MHz,D$_2$O; ppm): $\delta_H$ 2.796 (d,4H,J 16.9 Hz,CH$_2$), 2.590 (d,4H,J 17.0 Hz,CH$_2$). $^{13}$C NMR (D$_2$O; ppm): $\delta_C$ 185.3 (CO$_2$)$_\alpha$, 176.3 (CO$_2$)$_\beta$, 84.6 (≡CO), 43.6 (=CH$_2$).

Crystals of suitable quality for the subsequent X-ray diffraction studies were obtained as transparent prism or rhombhedral blocks by slow evaporation of the related solution of compounds **1** or **2** at room temperature. The resulting crystals were sealed in capillary to prevent loss of water molecules.

**Physical Measurements.** Infrared spectra were recorded as Nujol mulls between KBr plates using a Nicolet 740 FT-IR spectrometer. Elemental analyses were performed using EA 1106 elemental analyzers. $^1$H NMR and $^{13}$C NMR spectra were recorded on Varian UNITY 500 NMR and 300 NMR spectrometers, respectively, using DDS (sodium 2,2-dimethyl-2-silapentane-5-sulfonate) as internal reference.

**X-ray Data Collection, Structure Solution, and Refinement.** Crystallographic data for the citratomolybdates **1** and **2** are summarized in Table 1. Diffraction data were collected on an Enraf-Nonius CAD-4 diffractometer with graphite monochromated Mo Kα radiation at 296 K. A Lorentz-polarization factor, anisotropic decay, and empirical absorption corrections were applied. The structures were solved by heavy atom methods and refined by full-matrix least-squares procedures with anisotropic thermal parameters for all the non-hydrogen atoms. H atoms were located from different Fourier map and not refined. All calculations were performed on a 586 P/100 microcomputer using the MoLEN software package[41]. Selected atomic distances and bond angles are given in Table 2.

# 3  Results and Discussion

Preparation of the title compounds depends on pH control and the mole ratio of the reactants[30,31]. In this experiment the pH values in the reactions are controlled easily by citrate anions acting as both reactant and buffer agent. This is further supported by the preparation of deprotonated dimeric citrato molybdate[37]. The interconversion of the monomeric and dimeric oxocitrato molybdates is shown in

Scheme 1. Transformation of monomeric and dimeric oxocitrato molybdates can be accomplished by controlling the pH.

Table 1. Crystal data summaries of intensity data collection and structure refinement for
$K_4[MoO_3(cit)] \cdot 2H_2O$ (1) and $K_4[(MoO_2)_2O(Hcit)_2] \cdot 4H_2O$ (2)

| | 1 | 2 |
|---|---|---|
| emp formula | $C_6H_8K_4MoO_{12}$ | $C_{12}H_{18}K_4Mo_2O_{23}$ |
| fw | 524.47 | 878.55 |
| cryst color, habit | colorless, prism | colorless, rhombic |
| cryst dimers (mm) | $0.05 \times 0.05 \times 0.08$ | $0.10 \times 0.10 \times 0.12$ |
| cryst syst | monoclinic | triclinic |
| no. of refins used for unit cell determn ($2\theta$ range) | 25 (15.0—17.0°) | 25 (15.0—17.0°) |
| space group | $P2_1/c$ | $P\bar{i}$ |
| formula units/unit cell | 4 | 1 |
| cell constants: | | |
| $a$(Å) | 7.225(1) | 8.2728(8) |
| $b$(Å) | 9.151(2) | 8.9514(8) |
| $c$(Å) | 22.727(2) | 10.0605(9) |
| $\alpha$(deg) | | 101.673(8) |
| $\beta$(deg) | 94.93(1) | 100.672(7) |
| $\gamma$(deg) | | 112.938(7) |
| $V$(Å³) | 1497.1(7) | 642.5(3) |
| $D_{calc}$(g/cm³) | 2.327 | 2.271 |
| $F_{000}$ | 1032 | 434 |
| $\mu$(Mo K$\alpha$)(cm⁻¹) | 20.3 | 17.0 |
| diffractometer | Enraf-Nonius CAD-4 | Enraf-Nonius CAD-4 |
| radiation | Mo K$\alpha$ ($\lambda = 0.7107$ Å) | Mo K$\alpha$ ($\lambda = 0.7107$ Å) |
| temp | 23° | 23° |
| scan width | $0.37 + 0.35\ \tan\theta$ | $0.41 + 0.35\ \tan\theta$ |
| decay of standards (%) | $\pm 2$ | $-2.0$ |
| no. of refins measd | 3172 | 2695 |
| $2\theta$ range (deg) | $2 \leqslant 2\theta \leqslant 52$ | $2 \leqslant 2\theta \leqslant 52$ |
| Range of $h,k,l$ | 8, $-11$, $\pm 28$ | 10, $\pm 11$, $\pm 12$ |
| no. of refins obsd $[F_o \leqslant 3\sigma(F_o)]^a$ | 2564 | 2196 |
| computer programs[b] | MoLEN | MoLEN |
| structure solution | MoLEN | MoLEN |
| no. of params varied | 209 | 190 |
| weight | $[\sigma(F_o)^2 + 0.0001(F_o)^2 + 1]^{-1}$ | $[\sigma(F_o)^2 + 0.0001(F_0)^2 + 1]^{-1}$ |
| GOF | 0.86 | 0.75 |
| $R = \sum(|F_o| - |F_c|)/\sum|F_o|$ | 0.027 | 0.033 |
| $R_w$ | 0.032 | 0.039 |
| largest feature final diff. map (e⁻ Å⁻³) | 1.0 | 0.8 |

[a]Corrections: Lorentz-polarization. [b]Neutral scattering factors and anomalous dispersion corrections.

**Table 2. Selected bond distances (Å) and angles (deg) for $K_4[MoO_3(cit)] \cdot 2H_2O$ (1)
and $K_4[(MoO_2)_2O(Hcit)_2] \cdot 4H_2O$ (2)[a]**

| | | | |
|---|---|---|---|
| Mo(1)—O(1) | 2.052(2) | Mo(1)—O(1) | 1.958(3) |
| Mo(1)—O(2) | 2.237(7) | Mo(1)—O(2) | 2.210(3) |
| Mo(1)—O(3) | 2.411(3) | Mo(1)—O(3) | 2.276(3) |
| Mo(1)—O(8) | 1.740(3) | Mo(1)—O(8) | 1.703(3) |
| Mo(1)—O(9) | 1.731(3) | Mo(1)—O(9) | 1.714(5) |
| Mo(1)—O(10) | 1.759(3) | Mo(1)—O(10) | 1.8766(4) |
| O(1)—Mo(1)—O(2) | 72.90(9) | O(1)—Mo(1)—O(3) | 79.0(1) |
| O(1)—Mo(1)—O(3) | 75.5(1) | O(1)—Mo(1)—O(8) | 98.9(1) |
| O(1)—Mo(1)—O(8) | 96.6(1) | O(1)—Mo(1)—O(9) | 97.2(2) |
| O(1)—Mo(1)—O(9) | 91.8(1) | O(1)—Mo(1)—O(10) | 150.9(1) |
| O(1)—Mo(1)—O(10) | 150.8(2) | O(2)—Mo(1)—O(3) | 79.5(1) |
| O(2)—Mo(1)—O(3) | 77.65(9) | O(2)—Mo(1)—O(8) | 89.2(2) |
| O(2)—Mo(1)—O(8) | 90.7(1) | O(2)—Mo(1)—O(9) | 165.3(1) |
| O(2)—Mo(1)—O(9) | 160.1(1) | O(2)—Mo(1)—O(10) | 82.69(9) |
| O(2)—Mo(1)—O(10) | 85.6(2) | O(3)—Mo(1)—O(8) | 168.7(2) |
| O(3)—Mo(1)—O(8) | 167.4(2) | O(3)—Mo(1)—O(9) | 86.7(2) |
| O(3)—Mo(1)—O(9) | 86.3(1) | O(3)—Mo(1)—O(10) | 78.56(9) |
| O(3)—Mo(1)—O(10) | 80.9(1) | O(8)—Mo(1)—O(9) | 104.6(2) |
| O(8)—Mo(1)—O(9) | 104.0(1) | O(8)—Mo(1)—O(10) | 99.5(1) |
| O(8)—Mo(1)—O(10) | 103.3(1) | O(9)—Mo(1)—O(10) | 99.8(1) |
| O(9)—Mo(1)—O(10) | 103.6(1) | Mo(1)—O(10)—Mo(1a) | 180.000 |
| O(1)—Mo(1)—O(2) | 75.1(1) | | |

[a] $(-x, -y, -z)$.

Previously the dominant mononuclear oxomolybdenum-citrate species in solution were formulated as $[MoO_4(H_2cit)]^{4-}$, $[MoO_3(cit)]^{4-}$, $[MoO_3(H_2cit)_2]^{4-}$, $[MoO_3(Hcit)]^{3-}$, $[MoO_2(cit)]^{2-}$, $[MoO_2(OH)(cit)]^{3-}$, and their protonated forms[29-31], and the dinuclear oxomolybdenum-citrate species were in the compositions $[Mo_2O_5(cit)_2]^{6-}$, $[Mo_2O_5(cit)(H_2O)_3]^{2-}$, and their protonated forms. Furthermore, salts separated from the solution indicated monomeric forms, $K_2[MoO_3(OH)-(H_3cit)] \cdot 2H_2O$, $K_3[MoO_4(H_3cit)]2H_2O$, $K_4[MoO_3(cit)H_2O]$; their dinuclear forms, $K_4-[(MoO_2)_2O(Hcit)_2] \cdot 5H_2O$, $K_6[(MoO_2)_2O(cit)_2] \cdot 7H_2O$; or their tetramer[33-36]. The precise distribution of the products for this reaction system is expected to be complex. Only 2 : 1 (Mo : cit) and 1 : 1 citrato molybdates have been structurally characterized as tetramer and dimer[33,34,37]. Complex $K_4-[MoO_3(cit)] \cdot 2H_2O$ (1) represents a first example of a structurally characterized monomeric 1 : 1 molybdenum-citrate complex which exhibits the coordination of the polycarboxylic acid to the molybdenum, and the isolated hydroxy and water molecule are not involved in the coordination of the molybdenum site. Such is the case in the related peroxide adduct $K_2[MoO(O_2)_2(H_2cit)] \cdot 1/2H_2O_2 \cdot 3H_2O$[42], in which two peroxo groups and the citrate ligand are bidentate.

The crystal structure of **1** comprises discrete potassium cations, water molecules, and citrato trioxo molybdate anions. As shown in Figure 1, each citrate ion acts as a tridentate ligand coordinated to the molybdenum atom via its alkoxy, $\alpha$-carboxyl, and one $\beta$-carboxyl group, while the other $\beta$-carboxyl group remains uncomplexed. Tridentate coordination of citrate through its alkoxy or hydroxyl, $\alpha$-carboxyl, and $\beta$-carboxyl group is a basic feature of mono- or dimeric citrate complexes. A similar type of coordination has also been seen in the mononuclear complexes of $(NH_4)_5Fe(C_6H_4O_7)_2 \cdot 4H_2O$, $(NH_4)_5Al(C_6H_4O_7)_2 \cdot 2H_2O$, and $(NH_4)_4[Ni(C_6H_5O_7)_2] \cdot 2H_2O^{[43-45]}$. There are two enantiomers of this complex, which resulted from the asymmetric coordination environment around molybdenum. This is similar to that of FeMo-co (Chart 1), in which the octahedral coordination geometry for Mo is typically asymmetrical. Attempts to resolve the enantiomers of **1** were unsuccessful.

As shown in Figure 2, the complex $K_4[(MoO_2)_2O(Hcit)_2] \cdot 4H_2O(\mathbf{2})$ exists as a centrosymmetric

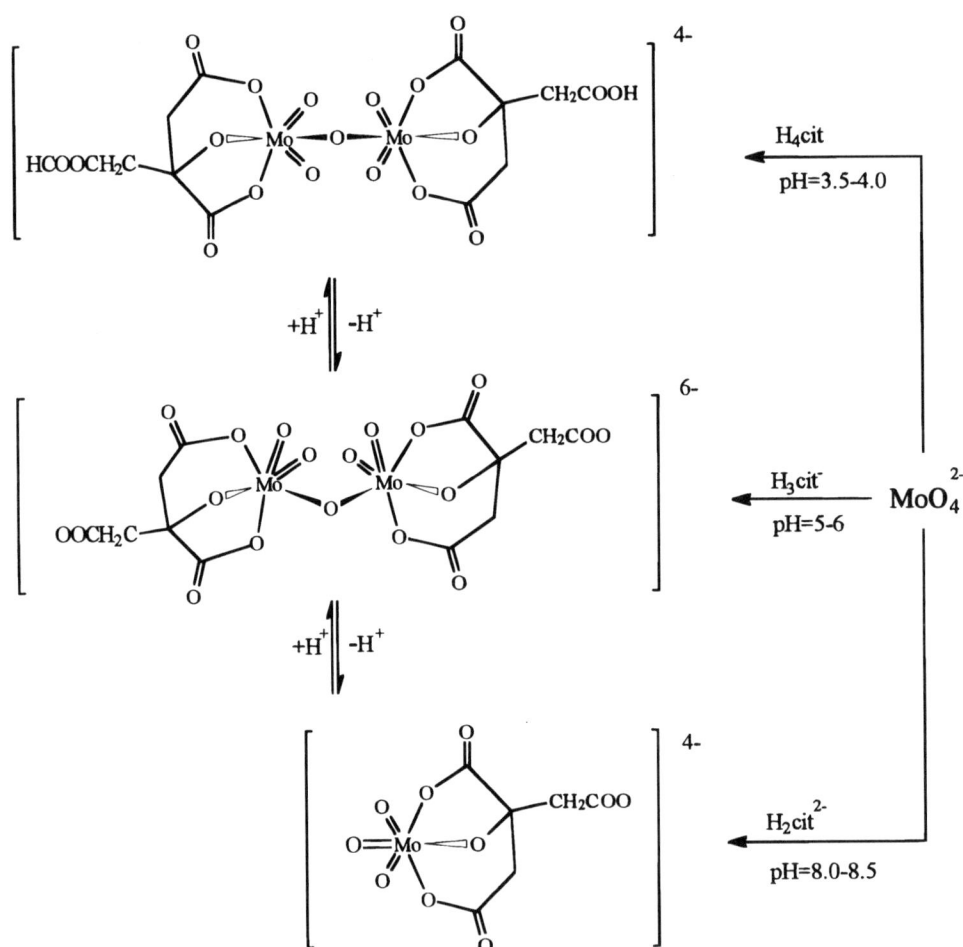

**Scheme 1.   Syntheses and Transformation of Monomeric and Dimeric Citratomolybdates**

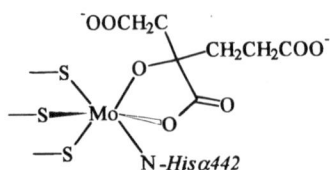

**Chart 1.   Schematic Representation of the First Coordination Sphere of Molybdenum in FeMo-cofactor[1]**

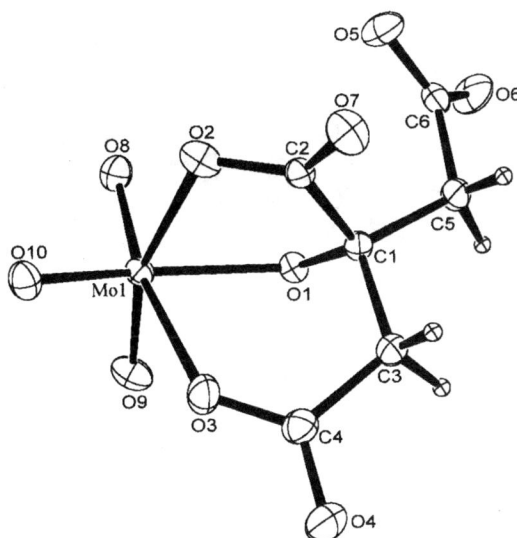

**Figure 1** **Perspective view of the anion structure of $K_4[MoO_3(cit)] \cdot 2H_2O$.**
**Thermal ellipsoids are drawn by ORTEP and represent 50% probability surfaces.**

dimer. Each citrate ion also acts as a tridentate ligand with the alkoxy, α-carboxyl, and one β-carboxyl oxygens coordinated to the molybdenum atom, and the other β-carboxyl group remains uncomplexed, as does its deprotonated form. The dimeric anion consists of a common oxobridged $[Mo_2O_5]^{2+}$ entity which is centrosymmetric. The angle of the Mo—O—Mo bridge is the same as that in $[(MoO_2)_2O(C_2O_4)_2(H_2O)_2]^{2-}$ (180°)[46], and is different from the angles of the Mo—O—Mo bridge [144.7(2) and 137.1(4)°] in deprotonated citratomolybdate[37,38]. The two terminal oxo groups are in a *cis*-configuration. Each molybdenum atom is six-coordinate with approximately octahedral geometry. The terminal and bridging oxo groups adopt a *fac*-stereochemistry. The *trans* positions are occupied by a tridentate citrate.

**Table 3. Relevant Infrared Data, $v/cm^{-1}$, for Citrato Molybdates 1-4, 1, $K_4[MoO_3(cit)] \cdot 2H_2O$;**
**2, $K_4[(MoO_2)_2O(Hcit)_2] \cdot 4H_2O$; 3, $K_2Na_4[(MoO_2)_2O(cit)_2] \cdot 5H_2O$; 4, $K_4[Mo_4O_{11}(cit)_2] \cdot 6H_2O$**

| | 1 | 2 | 3 | 4 |
|---|---|---|---|---|
| $v(OH)$ | $3417_{vs}$ | $3623_s, 3527_s$ $3423_s$ | 3421 | |
| $v(CH_2)$ | $2969_m$ | $2960_m, 2921_m$ | $2968_m, 2934_m$ | |
| $v(OH)$, carboxy | | $2717_m, 2604_m$ $2504_m$ | | |
| $v_{as}(C=O)$ | $1603_{s,sh}, 1575_{vs,b}$ | $1715_s,\quad 1652_{vs,b},$ $1604_s, 1553_{s,b}$ | $1644_{vs}, 1588_s$ | $1720_s, 1660_{vs}, 1620_{vs,sh}, 1595_{vs}, 1560_{vs}$ |
| $v_s(C=O)$ | $1425_{s,sh}, 1398_{vs}$ | $1440_m,\quad 1426_m,$ $1404_s, 1346_m$ | $1398_{vs}$ | $1430_s, 1410_{vs}$ |
| $v(Mo=O)$ | $937_w, 897_m, 848_s, 826_s$ | $933_s,\quad 908_{s,sh},$ $895_{vs,b}$ | $950_s, 902_s$ | $950_{vs}, 920_{vs}, 900_{vs}, 890_{vs,sh}, 870_m, 850_m,$ $820_m, 800_m$ |
| $v_{as}(Mo—O_b—Mo)$ | | $785_{vs}$ | $780_s$ | $740_{vs,sh}, 730_{vs,br}$ |
| $v_s(Mo—O_b—Mo)$ | | $690_s$ | $715_s$ | $690_{vs,sh}, 650_{vs}, 620_{vs}$ |
| ref | this work | this work | 37 | 34 |

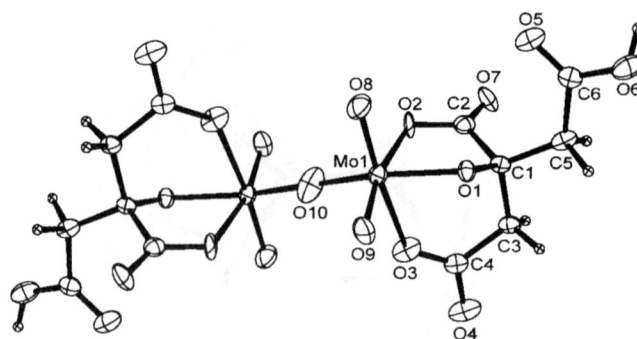

**Figure 2 Perspective view of the anion structure of K₄[(MoO₂)₂O(Hcit)₂]·4H₂O.**
**Thermal ellipsoids are drawn by ORTEP and represent 50% probability surfaces.**

As shown in Table 2, the Mo—O distances in citrato molybdates vary systematically. Mo=O is in the range 1.703(3)—1.759(3)Å, indicating that they are double bonds. The resulting O=Mo=O angles, 104.0(1), 103.3(1), 103.6(1), and 104.6(2)°, are considerably larger than the 90° regular octahedron value for cis groups; this is expected from the greater O···O repulsions between oxygens with short bonds to the metal atom. The Mo—O(10)—Mo bridging distance is 1.8766(4)Å. The Mo—O(alkoxy) bonds are slightly longer [2.052(2) and 1.958(3)Å], indicating the deprotonation of the hydroxyl group, and those to the α-carboxyl are longer [2.237(2) and 2.210(5)Å]. This is compatible with the Mo—O(alkoxy) [1.996 and 2.035Å] and Mo—O(α-carboxyl) bonds [2.167 and 2.206Å] of coordinated homocitrate ligand in MoFe protein and its putative transition-state complex[2,47].

The longest Mo—O(β-carboxyl) distances [2.411(3)Å] of monomer **1** show weak coordination of the β-carboxyl group to Mo(Ⅵ). This is much longer than those of dimeric **2** [2.276(3)Å], full deprotonated dimeric **3** [2.264(3) Å (a)], and tetrameric citratomolybdate **4** [2.318(5) Å (av)]. The significantly longer β-carboxylate—Mo distance is notable. In proteinbound FeMo-cofactor, only the alkoxy and α-carboxyl sites of homocitrate are coordinated to molybdenum. It suggests that coordinated β-carboxylate is probably much easier to replace by another ligand like histidine imidazole to form bidentate—citrato—Mo in FeMo-co biosynthesis.

It is believed molybdenum is taken up by organisms as $MoO_4^{2-}$, and a possible function of the tricarboxylic acid in the biosynthesis of the cofactor of nitrogenase is to mobilize molybdenum from the appropriate storage enzyme. Such structural changes would be essential for the assembly of the final cofactor cluster from an oxomolybdenum-homocitrate precursor[12—14]. Therefore, compound **1** may represent a close relevant form as a possible biomimetic precursor for the biosynthesis of FeMo-co, as well as a physiologically relevant form of metabolized molybdenum(Ⅵ) utilized in the assembly of FeMo-co.

The ¹H NMR spectrum of dimeric citrato molybdate **2** gives a sharp AB quartet for methylene protons of the coordinated citrate ligand, and the ¹H magnetic equivalence of the methylene groups gives only one unshifted ¹³C NMR signal compared with KH₃cit at the same pH (3.4). [KH₃cit. ¹³C NMR(D₂O;ppm)$\delta_C$ 177.6 $(CO_2)_\alpha$, 174.0 $(CO_2)_\beta$, 73.6 ($\equiv$CO), 43.0 ($=CH_2$). K₃Hcit. ¹³C NMR (D₂O; ppm): $\delta_C$ 182.2 $(CO_2)_\alpha$, 178.6 $(CO_2)\beta$, 75.6 ($\equiv$CO), 45.4 ($=CH_2$).] This is similar to the NMR spectra of its deprotonated form K₂Na₄[(MoO₂)₂O(cit)₂]·5H₂O and a dimeric citrato tungstate Na₆[(WO₂)₂O(cit)₂]·11H₂O[37,38]. The large low-field shift of some ¹³C resonances in comparison with KH₃cit ions shows that both alkoxy (about δ11) and α-carboxyl (about δ 8) groups are coordinated.

In the monomer form, the $^1$H NMR spectrum of **1** shows two groups of sharp AB quartets in a 1 :
1 ratio, and the $^{13}$C NMR signals of β-carboxyl and methylene groups are doubled. The large low-field
shift of some $^{13}$C resonances of **1** in comparison with K$_3$Hcit ions (see $^{13}$C NMR data given earlier and
ref 49) at the same pH (7.5) clearly shows that both alkoxy (about δ 6) and α-carboxyl (about δ 5)
group are coordinated, while β-carboxyl groups gives only a small shift(Δδ1 ppm) of $^{13}$C NMR signals,
indicating that the bonding to molybdeum is weak.

The frequencies and assignments of selected IR absorption bands are given in Table 3. In the region
between 1800 and 1400 cm$^{-1}$ compound **2** gives a typical band of a nonbonded and undissociated
carboxylic acid group at 1715 cm$^{-1}$. The bands between 1660 and 1540 cm$^{-1}$ and between 1440 and 1340
cm$^{-1}$ correspond to a bound carboxyl group $v_{asym}$ and $v_{sym}$(CO$_2$M), respectively; this is in accord with a
chelate ring and bridging by the citrato ligand. Loss of the proton in compounds **1** and **3** and the absence
of a citrate bridge as in **1**—**3** reduce the number of bands and displace them to lower frequencies.

In the region between 1000 and 600 cm$^{-1}$, the complexes show several bands which might result
from the presence of *cis*-dioxo cores in two different environment. The low-frequency symmetric MoO$_2$
stretching may be explained by intramolecular hydrogen bonding and the coordination with potassium
cation. The band positions 890 and 840 cm$^{-1}$ of an assumed *fac*-trioxo core were found to be like the
other complexes with the MoO$_3$ core[50]. The strong IR band around 700 cm$^{-1}$ observed only for the
dimers or tetramer is attibuted to the Mo—O$_b$—Mo bridges. Evidently, the β-carboxyl group of the
citrate ion carries a proton in the structure of **2**; this is not only shown by the visibility of hydrogen
atom in difference maps but also by the difference between the C—O distances of terminal carboxylate
[O(5)—C(6), 1.203(6) Å; O(6)—C(6), 1.322(6) Å], as well as from the consideration of charge
balance. The conclusion that full deprotonation of monomer **1** occurs can be drawn from the observed
carbon-oxygen bond distances of β-carboxyl groups, which are equivalent [1.259(5), 1.251(5)Å]. This
is further supported by IR bands found at 1603$_{s,sh}$, 1575$_{vs,b}$ and 1425$_{s,sh}$, 1398$_{vs}$ cm$^{-1}$ corresponding to $v_{as}$
and $v_s$(CO$_2$M) (bound carboxyl group), and the absence of IR bands between 1740 and 1700 cm$^{-1}$,
indicating the presence of fully deprotonated carboxyl groups.

## Acknowledgments

We thank the referee for his critical review. Financial support by the National Science Foundation
of China (Grants 29503021 and 29933040) is gratefully acknowledged.

**Supporting Information Available**: Tables of X-ray crystal structure refinement data, positional and
thermal parameters for K$_4$[MoO$_3$(cit)] · 2H$_2$O and K$_4$[(MoO$_2$)$_2$O(Hcit)$_2$] · 4H$_2$O. This material is
available free of charge via the Internet at http://pubs.acs.org.

## References

[1]Kim, J.; Rees, D. C. Science 1992, **257**, 1677.

[2]Kim, J.; Rees, D. C. Nature 1992, **360**, 553.

[3]Kim, J.; Wood, D.; Rees, D. C. Biochemistry 1993, **32**, 7104.

[4]Chan, M. K.; Kim, J.; Rees, D. C. Science 1993, **260**, 792.

[5]Rees, D. C.; Chan, M. K.; Kim, J. Adv. Inorg. Chem. 1993, **40**, 89.

[6]Kim, J.; Rees, D. C. Biochemistry 1994, **33**, 389.

[7]Bolin, J. T. ; Ronco, A. E. ; Mortenson, L. E. ; Morgan, T. V. ; Xuong, N. H. Proc. Natl. Acad. Sci. U. S. A. 1993, **90**, 1078.

[8] Bolin, J. T. ; Campobasso, N. ; Muchmore, S. W. ; Minor, W. ; Morgan, T. V. ; Mortenson, L. E. New Horizons in Nitrogen Fixation; Kluwer: Dordrecht, The Netherlands, 1993; 83.

[9]Dean, D. R. ; Bolin, J. T. ; Zheng, L. J. Bacteriol. 1993, **175**, 6737.

[10]Hoover, T. R. ; Imperial, J. ; Ludden, P. W. ; Shah, V. K. Biochemistry. 1989, **28**, 2768.

[11]Burgess, B. K. Chem. Rev. 1990, **90**, 1377.

[12]Shah, V. K. ; Allen, J. R. ; Spangler, N. J. ; Ludden, P. W. J. Biol. Chem. 1994, **269**, 1154.

[13]Hales, B. J. Adv. Inorg. Biochem. 1990, **165**.

[14]Wright, D. W. ; Chang, R. T. ; Mandal, S. K. ; Armstrong, W. H. ; OrmeJohnson, W. H. ; Davis, W. H. J. Bioinorg. Chem. 1996, **1**, 143.

[15]Müller, A. ; Krahn, E. Angew. Chem., Int. Ed. Engl. 1995, **34**, 1071.

[16]Grönberg K. L. C. ; Gormal, C. A. Durrant, M. C. ; Smith, B. E. ; Henderson, R. A. J. Am. Chem. Soc. 1998, **120**, 10163.

[17]Tsai, K. R. ; Wan, H. L. J. Cluster Sci. 1995, **6**, 485.

[18]Zhou, Z. H. ; Yan, W. B. ; Wan, H. L. ; Tsai, K. R. ; Wang, J. Z. ; Hu, S. Z. J. Chem. Crystallogr. 1995, **25**, 807.

[19]Zhou, Z. H. ; Wan, H. L. ; Tsai, K. R. Chin. Sci. Bull. 1995, **40**, 749.

[20]Zhou, Z. H. ; Wan, H. L. ; Hu, S. Z. ; Tsai, K. R. Inorg. Chim. Acta 1995, **237**, 193.

[21]Zhou, Z. H. ; Zhang, H. ; Jiang, Y. Q. ; Lin, D. H. ; Wan, H. L. ; Tsai, K. R. Trans. Met. Chem. 1999, **24**, 612.

[22]Ehde, P. M. ; Anderson, I. ; Petterson, L. Acta Chem. Scand. 1989, **43**, 136.

[23]Kiss, T. ; Buglyo, P. ; Sanna, D. ; Micera, G. ; Decock, P. ; Dewaele, D. Inorg. Chim. Acta 1995, **239**, 145.

[24]Wright, D. W. ; Humiston, P. A. ; Orme-Johnson, W. H. ; Davis, W. H. Inorg. Chem. 1995, **34**, 4194.

[25]Burojevic, S. ; Shweky, I. ; Bino, A. ; Summers, D. A. ; Thompson, R. C. Inorg. Chim. Acta 1996, **251**, 75.

[26]Murugesan, V. ; Babu, V. , Sankaran, S. Inorg. Chem. 1998, **37**, 1336.

[27]Creager, S. E. ; Aikens, D. A. ; Clark, H. M. Electrochim. Acta 1982, **27**, 1307.

[28]Pedrosa de Jesus, J. D. ; Farropas, M. de D. ; O'Brien, P. ; Gillard, R. D. ; Williams, P. A. Trans. Met. Chem. 1983, **8**, 193.

[29]Bartusek, M. ; Havel, J. ; Matula, D. Collect. Czech. Chem. Comm. 1986, **51**, 2702.

[30]Cruywagen, J. J. ; Van de Water, R. F. Polyhedron 1986, **5**, 521.

[31]Cruywagen, J. J. ; Rohwer, E. A. ; Wessels, G. F. S. Polyhedron 1995, **14**, 3481.

[32]Martie, D. O. ; Feliz, M. R. ; Capparelli, A. L. Z. Phys. Chem. (Leipzig) 1989, **270**, 951. ,

[33]Nassunbeni, L. R. ; Niven, M. L. ; Cruywagen, J. J. ; Heyns, J. B. B. J. Crytallogr. Spectrosc. Res. 1987, **17**, 373.

[34]Alcock, N. W. ; Dudek, M. ; Grybos, R. ; Hodorowicz, E. ; Kanas, A. ; Samotus, A. J. Chem. Soc., Dalton Trans. 1990, **707**.

[35] Dudek, M. ; Hodorowicz, E. ; Kanas, A. ; Samotus, A. ; Sieklucka, B. ; Szklarzewicz, J. ; Beltowska-Lehman, E. Proceedings of the 11th Conference on Coordination Chemistry;

Smolenice,Czechoslovakia,1987;57.

[36]Samotus,A.;Kanas,A.;Dudek,M.;Grybos,R.;Hodorowicz,E. Trans. Met. Chem. 1991,**16**, 495.

[37]Zhou,Z. H.;Wan,H. L.;Tsai,K. R. Polyhedron1997,**16**,75.

[38]Xing,Y. H.;Xu,J. Q.;Sun,H. R.;Li. D. M.;Wang,R. Z.;Wang,T. G.;Bu,W. M.;Ye,L.; Yang,G. D.;Fan,Y. G. Acta Crystallogr. 1998,**C54**,1615.

[39]Maruo,M.;Hirayama,N.;Shiota,A.;Kuwamoto,T. Anal. Sci. 1992,**8**,511.

[40] Molybdenum and Molybdemum-Containing Enzymes;Coughlan, M. P., Ed.;Pergmon:New York,1980.

[41] MoLen. An interactive Intelligent System for Crystal Analysis;Enraf-Nonius:Delft, The Netherlands,1990.

[42]Flangan,J.;Griffithe,W. P.;Skapski,A. C.;Wiggins,R. W. Inorg. Chim. Acta 1985,**96**,23.

[43]Matzapetakis,M.;Raptopoulou,C. P.;Terzis,A.;Lakatos,A.;Kiss,T.;Salifoglou,A. Inorg. Chem. 1999,**38**,618.

[44]Matzapetakis,M.;Raptopoulou,C. P.;Tsohos,A.;Papaefthymiou,V.;Moon,N.;Salifoglou, A.J. Am. Chem. Soc. 1998,**120**,13266.

[45]Zhou,Z. H.;Lin,Y. J.;Zhang,H. B.;Lin,G. D.;Tsai,K. R. J. Coord. Chem. 1997,**42**,131.

[46]Cotton,F. A.;Morehause,S. M.;Wood,S. J. Inorg. Chem. 1964,**3**,1603.

[47] Schinderlin, H.; Kisker, H. C.; Schlessman, J. L.; Howard, J. B.; Rees, D. C. Nature1997,**387**,370.

[48]Llopis, E.; Ramírez, J. A.; Doménech, A.; Cervilla, A. J. Chem. Soc., Dalton Trans. 1993, **1121**.

[49]Loewenstein, A.;Roberts, J. D. J. Chem. Soc. 1960, **82**, 2705.

[50] Stiefel, E. I. In Comprehensive Coordination Chemistry; Wilkinson, G., Gillard, R. D., McCleverty,J. A.,Eds.;Pergamon Press:London,1987,**3**,1380.

■ 本文原载：Catalysis Letters Vol. 73. No. 2~4, pp. 141~147, 2001.

# Active-Oxygen Species on Non-Reducible Rare-Earth-Oxide-Based Catalysts in Oxidative Coupling of Methane *

H. B. Zhang, G. D. Lin, H. L. Wan, Y. D. Liu, W. Z. Weng, J. X. Cai, Y. F. Shen, K. R. Tsai [①]

(Department of Chemistry and Institute of Physical Chemistry,
Xiamen University, and State Key Laboratory for Physical Chemistry of
the Solid Surface, Xiamen 361005, Fujian, P.R.China)
E-mail: krtsai@xmu.edu.cn

Received 14 November 2000; accepted 16 March 2001

**Abstrat** From supplementary *in situ* Raman spectroscopic studies of active-oxygen species on non-reducible rare-earth-oxide-based catalysts in the oxidative coupling of methane (OCM) and structural adaptability considerations, further support has been obtained for our proposal that there may be an active and elusive precursor (of $O_2^-$ and $O_2^{2-}$ adspecies), most probably $O_3^{2-}$ formed from reversible redox coupling of an $O_2$ adspecies at an anionic vacancy with a neighboring $O^{2-}$ in the surface lattice. This active precursor may initiate H abstraction from $CH_4$ and be itself converted to $OH^- + O_2^-$, or it may abstract an electron from the oxide lattice and be converted to $O_2^{2-} + O^-$. The prospect of developing this type of OCM catalysts is discussed.

**Key words** Methane oxidative coupling  Th—La—$O_x$/$BaCO_3$  Non-reducible rare-earth-oxide-based catalysts  Surface oxygen species  $O_3^{2-}$  $O_2^-$  $O_2^{2-}$  *In situ* Raman spectroscopy

## 1 Introduction

Certain supported or composite metal-oxides catalysts can catalyze the oxidative coupling of methane (OCM) to ethane and ethylene as well as to minor amounts of other $C_{2+}$ hydrocarbons from $CH_4$—$O_2$ fed alternatively (redox cyclic-feed operation for OCM catalysts containing multivalent cations[1]), or simultaneously (co-feed operation[2] for OCM catalysts containing multivalent cations, or only valence-stable cations) through the catalyst bed. On account of the potentially practical significance in naturalgas resource utilization and the fundamental significance of methane activation, catalysis research on OCM has been the focus of worldwide attention for about a decade since the 8th ICC (Berlin, 1984). As shown by a more recent review[3], it has been established that OCM and side reactions are heterogeneous-homogeneous reactions initiated by H abstraction from gaseous phase $CH_4$ by active-oxygen species on the catalyst surface, liberating ·$CH_3$ radicals (along with co-produced $H_2O$), which couple mainly in the gas phase to form ethane as the primary product, and that ethylene

---

* Work supported by the National Priority Fundamental Research Project (No. G1999022400) of PR China.

① To whom correspondence should be addressed.

are formed mainly by further reactions of $C_2H_6$ both heterogeneously and homogeneously, while by-products $CO_X$ (i. e., $CO_2$ and CO) are formed by deep oxidation of any of the hydrocarbons on the catalyst surface and in the gas phase. It has been found that $C_2$ selectivity $S$ (in%) decreases steadily with increasing $CH_4$ conversion, k (in %); the maximum $k+S$ that has been obtained so far is only about 100%. Obviously, as $\cdot CH_3$ and $C_2H_4$ in the reaction mixture increase in proportion relative to $CH_4$, their chances of further dehydrogenation and deep oxidation will also increase since the bond strength[4] of $H-CH_3$ ($\sim$440 kJ $mol^{-1}$) is only slightly smaller than that of $H-CH_2$ ($\sim$460 kJ $mol^{-1}$) or $H-C_2H_3$ ($\geqslant$452 kJ$mol^{-1}$). This is the main problem that has hampered the development of OCM to be a practical process for the direct production of ethylene from natural gas. However, with better understanding of the nature of active-oxygen species for OCM and the controlling factors for $C_2$ selectivity, the situation may be improved.

In the case of OCM catalysts containing multivalent cations, surface-lattice $O^{2-}$ associated with cations in the oxidized state may be the active-oxygen species. Some OCM catalysts of this category have been found to show good activity and selectivity, with $k+S$ around 100% [3,5]. However, OCM catalysts of this category generally require an alkali-ion promoter to modulate the oxidizing power of the multivalence metal oxide, and thus suffer from the disadvantage of gradual decline in activity and selectivity due to volatility loss of the alkali-ion promoter. OCM catalysts with valence-stable cations are practically inactive in $CH_4$ stream without $O_2$[3,6], indicating that lattice $O^{2-}$ associated with valence-stable cations is not an active-oxygen species.

The nature of active-oxygen species on non-reducible metal-oxide catalysts is still a matter of dispute[3]. Thus with alkali-doped alkaline-earth-oxide catalysts (e. g., the most thoroughly investigated $Li^+/MgO$), $\underline{O^-}$, $\underline{O_2^{2-}}$, and $\underline{O_2^-}$ have separately been considered to be the principal active-oxygen species for OCM [3,7,8]. However, $\underline{O^-}$ is known to be reactive toward $C_2H_4$ even at very low temperature[9], and the theoretical value of the energy barrier for H abstraction from $H_3C-H$ by $[\underline{O^-}-Li^+]/MgO$ has been shown by Borve and Pettersson[10] with the most refined model calculations to be only 25 kJ $mol^{-1}$, much smaller than the experimental value of 96 kJ $mol^{-1}$ from kinetic measurements[11]. Some investigators [8] have suggested that $\underline{O_2^-}$ is less likely than $\underline{O_2^{2-}}$ to be a principal active-oxygen species for OCM since $\underline{O_2^-}$ is inactive around 723 K and readily liberates $O_2$ at higher temperature. However, $\underline{O_2^-}$ is most probably the precursor of $\underline{O_2^{2-}}$, which is known to be reactive towards $CH_4$ and $C_2H_4$ above 773 K, while the $C_2$ yield is known to increase with increasing temperature in the usual temperature range of OCM above 923 K; and Liu et al. [12] have argued that $\underline{O_2^-}$ may become active-oxygen species at OCM reaction temperature. Some direct evidence for $O_2^-$ being reactive towards $CH_4$ at 973 K has been obtained by Wan et al. [13] with $SrF_2/Nd_2O_3$ and similar fluoride-containing rare-earth-oxide-based catalysts. For rare-earth-oxide-based catalysts, some investigators[14] have suggested that $\underline{O_2^{2-}}$ may be the principal active-oxygen species, and $\underline{O_2^-}$ formation may be sensitive to inhibition by surface carbonate. However, $\underline{O_2^-}$ formation on $La_2O_3/CaO$[15] has been found to be less sensitive to inhibition by $CO_2$ and surface carbonate than on undoped $La_2O_3$[16].

*Ex situ* Raman spectroscopy was first used by Liu et al. [17] to study the $\underline{O_2^-}$, $\underline{O_2^{2-}}$, and surface carbonate species on a $Cs^+/La_2O_3$ catalyst in relation to the OCM catalytic activity, as well as the selectivity-promoting effect of $CO_2$ in $O_2$ stream used for reactivation of the catalyst. More recently, interesting results have been obtained from *in situ* Raman spectroscopic studies of active-oxygen species in OCM [18-23]. Thus Raman signals of $\underline{O_2^{2-}}$ on functioning $La_2O_3$, $Na^+/La_2O_3$, and $Sr^{2+}/La_2O_3$ at 973 K in flowing $CH_4-O_2$ feed as well as in $O_2$ stream have been reported by Mestl et al. [18]; and Raman signals(confirmed with [18]$C_2$)of $O_2^{2-}$ at 842 and 821 cm$^{-1}$(vw)from 0.5 mol% BaO/MgO in $O_2$ stream at 373 and 583 K(along with a very weak carbonate signal),shifting to very broad and weak band at 973 and 1073 K,have been observed by Lunsford et al. [19],who found that the $O_2^{2-}$ adspecies reacted rapidly with $CO_2$ at OCM reaction temperature,especially in the presence of water vapor. With single-phase $ThO_2$—$La_2O_3$ of defective fluoride structure(DFS)[24] in $CH_4$—$O_2$(4:1,v/v)co-feed stream,the weak Raman signal of $O_2^-$ at 1140 cm$^{-1}$(w)(along with a very strong carbonate signal around 1060 cm$^{-1}$(vs)) has been observed by Liu et al. [20] at 1133 K with conspicuous increase in signal intensity with decreasing temperature;interestingly,the $O_2^-$ Raman signal disappeared after switching the cofeed stream to pure $O_2$ stream at 1013 K and 0.1 MPa. Similar results have been reported by Cai et al. [21] with(LaO)$_2$(O,CO$_3$)—CaO/(Ba,Ca)CO$_3$ catalysts. With functioning Th—La—O$_x$/BaCO$_3$ catalyst(Th:La:Ba=20:3:40,molar ratio)in $CH_4$—$O_2$(4:1,v/v)stream at 1013 K and $2\times10^4$ h$^{-1}$ GHSV, Raman signals at 1120(w)and 1148 cm$^{-1}$(w)and at 812(w)and 824 cm$^{-1}$(w)(assigned,respectively,to $\nu_{O-O}$ of two $O_2^-$ adspecies and two $O_2^{2-}$ adspecies in different microenvironments)have been observed by Zhang et al. [22,23],besides a very strong mono-dentate $CO_3^{2-}$ Raman band centering around 1056 cm$^{-1}$ (vs);after lowering the reaction temperature to 773 K,the two $O_2^-$ signals were found to increase conspicuously in intensity(though still weak),while the two $O_2^{2-}$ signals increased slightly in intensity, and a new band at 940 cm$^{-1}$(w)appeared,along with another weak $O_2^{2-}$ band around 850 cm$^{-1}$(w).

These interesting observations have led us [23,25] to postulate that there might be an active and elusive precursor(of $O_2^-$ and $O_2^{2-}$),most probably a bent $O_3^{2-}$(with approximate C$_{2v}$ symmetry and $v1$ (A)at 940 cm$^{-1}$),which might be formed from reversible redox coupling of $O_2$(adsorbed at an anionic vacancy)with a neighboring $O^{2-}$ in the surface lattice. However,more experimental and theoretical support for this assertion is required. For example,the possibility that the Raman signal at 940 cm$^{-1}$ might be due to a lattice vibration or to peroxide and the structural adaptability of such a bent $O_3^{2-}$ surface species must be carefully examined. Incidentally,the Th—La—O$_x$/BaCO$_3$ catalyst has been found in plug-flow microreactor evaluation to exhibit good $C_2$ yield and excellent thermal stability[26], but has failed in scale-up evaluation,probably due to hot spot over-heating and/or catalyst inactivation by $CO_2$ under oxygen-limiting conditions.

# 2 Experimental

## 2.1 Preparation of catalyst

A Th—La—O$_x$/BaCO$_3$(Th:La:Ba=10:0.2:10,molar ratio)catalyst was prepared from the aqueous nitrates by coprecipitation as carbonates and hydroxides,as reported previously[20];the thoroughly washed precipitate was dried in air at 378 K,and ignited in air at 1123 K,then cooled down to room temperature and carefully pulverized and sieved to the desired mesh in a gloved box to avoid any physical contact with the hazardous thorium compounds. The sample was stored in closed vessel before use. In this preparation,lower proportions of Ba and La were used to lessen the extent of carbonate formation.

## 2.2 In situ Raman spectroscopic experiments

A horizontal quartz-tubular sample cell(with very uniform diameter of ~2.5 mm i.d.)heated with a tubular copper-block heater(with a small opening to accommodate the incident laser beams and scattered

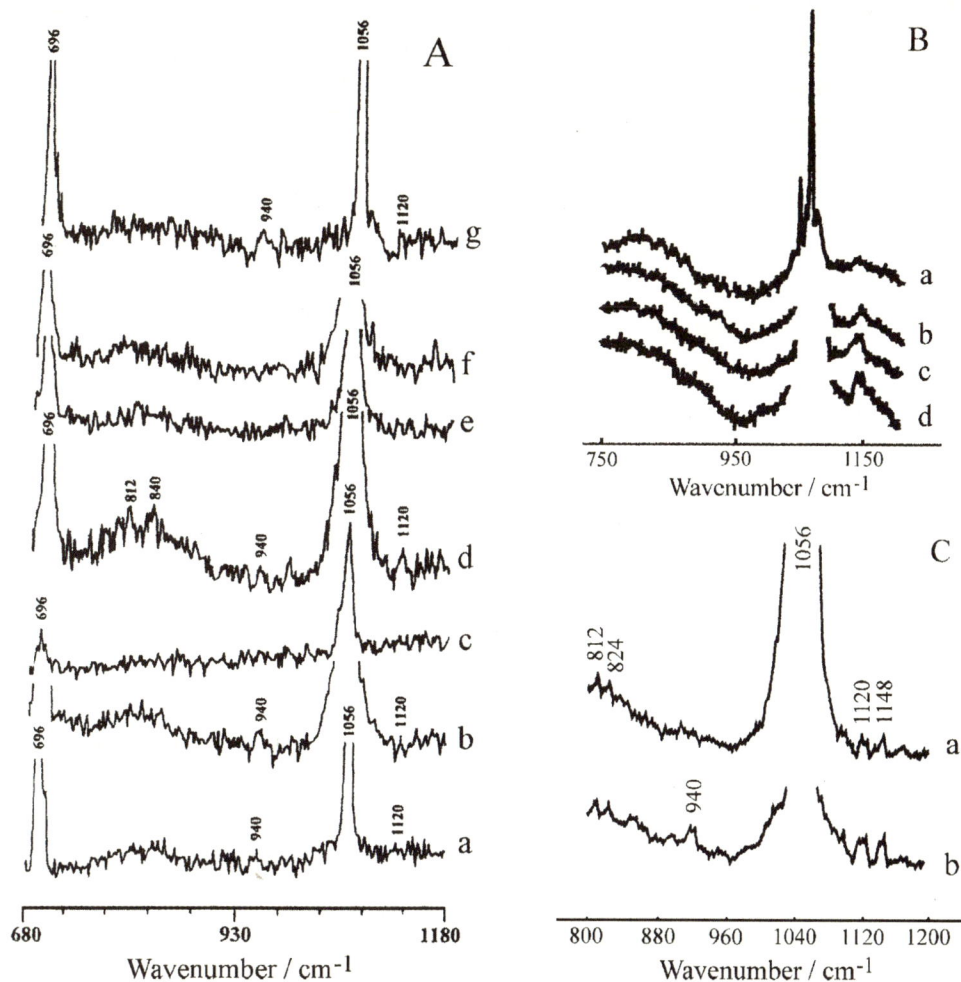

Figure 1    (A) Raman spectra of Th—La—$O_x$/$BaCO_3$ (Th : La : Ba＝10 : 0. 2 : 10, molar ratio) catalyst taken at 0. 1 MPa, $1\times10^4$ $h^{-1}$ GHSV and under the following conditions: (a) Taken after exposing the fresh catalyst to $O_2$ stream at 648 K for 15 min, followed by lowering temperature to 383 K and switching to $N_2$ stream, and then lowering temperature to 298 K. (b) Taken with the sample after (a) and in $CH_4$—$O_2$ (4 : 1, v/v) stream at 973 K (at the hot spot), with the incident laser beam focusing at the spot of the sample about 1—2 mm up-stream from the hot spot of the functioning sample. (c) Taken at the same spot with the sample after (b) and exposing to $CH_4$ stream at 973 K for 5 min. (d) Taken at the same spot with the sample after (c) and exposing to $CH_4$—$O_2$ (4 : 1, v/v) stream at 773 K. (e) Taken at the same spot with the sample after (d) and exposing to $CH_4$ stream at 773 K for 5 min. (f) Taken with the sample after (e) and exposing to $CH_4$—$O_2$ (4 : 1, v/v) stream at 973 K (hot-spot temperature), but with the incident laser beam focusing on the spot of the sample about 1—2 mm down-stream from the hot spot of the functioning sample. (g) Taken at the same spot as in (f) with the sample after (f) and cooling to 648 K and then switching to $O_2$ stream, followed by cooling to 383 K and then switching to $N_2$ stream, and finally cooling to 298 K. (B) In situ Raman spectra of functioning $ThO_2$—$La_2O_3$ (Th : La＝7:3, molar ratio; DFS, single-phase) in $CH_4$—$O_2$ (4 : 1, v/v) stream at 0. 1 MPa and (a) 1133, (b) 1073, (c) 1013 and (d) 953 K (reproduced from [20]). (C) In situ Raman spectra of functioning Th—La—$O_x$/$BaCO_3$ (Th : La : Ba＝20 : 3 : 40, molar ratio) in $CH_4$—$O_2$ (4 : 1, v/v) stream at $2. 0\times10^4$ GHSV, and (a) 1013 and (b) 773 K (reproduced from [22, 23]).

radiations and to facilitate visual observation of the hot spot) was used. The position of the sample cell (mounted together with the heater on a movable stand) could be forward-or-backward and upward-or-downward, as well as leftward-or-rightward, moved a few cm, and the incident laser beam was focused at the desired spot of the sample as indicated below. The quartz tubular sample cell also served as a plug-flow microreactor with 0.06 mL of 40—60 mesh catalyst sample filled to a length of about 1.2 cm and placed between two quartz-wool plugs, and the preheating zone was filled with 40—60 mesh quartz chips. The temperature of the catalyst sample was registered with a tiny NiCr-NiAl thermocouple (in thin, stainless-steel sheath ca. 1.0 mm o. d.) placed right at the hot spot of the sample during the OCM reaction. *In situ* Raman spectra of the sample in different gaseous streams were taken by using a Spex Ramalog 6 laser Raman spectrometer, with the 514.5 nm line from a Coherent-Innova model 200 argon ion laser used as the excitation source. Slit width settings correspond to a resolution of 4 cm$^{-1}$. The laser beam (30 mW) was focused on the desired spot of the stationary sample (except that the tubular reactor was manually rotated at intervals) under various treatments as indicated in the legend of figure 1(A). Since the temperature of the sample appeared to be very uneven along the feed-gas stream around the hot spot during the OCM reactions, the oscillation of the sample tube was not carried out in this supplementary experiment.

# 3  Results and discussion

## 3.1  In situ Raman spectroscopic studies

Figure 1(A, curve(a)) shows that, from the ignited sample after exposure to $O_2$ stream at 648—383 K and cooling in $N_2$ stream to 298 K, weak but distinct Raman signal at 940 cm$^{-1}$ (w) and weaker Raman signal around 1120 cm$^{-1}$ (w) due to $\underline{O_2^-}$ were observed, along with a weak and very broad band around 820~840 cm$^{-1}$ probably due to $\underline{O_2^{2-}}$ besides two strong bands at 1056 and 696 cm$^{-1}$ assigned to $\nu_2(A_1)$ and $P_{OCO}(B_2)$, of mono-dentate carbonate, $\underline{OCO_2^{2-}}$ [27], which were formed (in the preparation of sample) during the ignition in air usually containing 1 mol% $CO_2$. The features of *in situ* Raman spectra shown in curves (b) and (d) of figure 1(A) taken with the functioning $Th—La—O_x/BaCO_3$ (Th : La : Ba = 10 : 0.2 : 10, molar ratio) catalyst with the incident laser beam focused at the same spot slightly up-stream of the sample (thus at temperature slightly below the hot-spot temperature 973 or 773 K for figure 1(A(b)) or (A(d)), respectively) may be compared with that taken with functioning $Th—La—O_x/BaCO_3$ (Th : La : Ba = 20 : 3 : 40, molar ratio) catalyst at 1013 and 773 K (in slightly larger and vertically oscillating sample tube) shown in figure 1(C(a)) and (C(b)) (in the insertion reproduced from [23]). It can be seen that the general features of figure 1(A(d)) and (C(b)) appeared to be quite similar, in spite of the much lower level of doping with $La^{3+}$ and slightly lower temperature (probably 20°~30° lower than 773 K) in the former case; and that in figure 1(A(b)) with the Raman spectrum taken at the same spot of the sample as in figure 1(A(d)) (and thus at temperature at least 30° lower than the hot-spot temperature 973 K), the 940 cm$^{-1}$ signal was stronger, while the $\underline{O_2^{2-}}$ and $\underline{O_2^-}$; Raman signals both weaker than that shown in figure 1(A(d)) taken at temperature slightly lower than the hot-spot temperature of 773 K. In figure 1(C(a)) with the Raman spectrum taken at 1013 K, the Raman signal at 940 cm$^{-1}$ could hardly be seen; thus it appeared to be an elusive species, existing only at temperature below 1013K.

From the weaker Raman signal at 940 cm$^{-1}$ (w) compared with that of $\underline{O_2^{2-}}$ and $\underline{O_2^-}$ shown in figure 1 (A(d)) taken at ~773 K, and the stronger Raman signal at 940 cm$^{-1}$ compared with that of $\underline{O_2^{2-}}$ and $\underline{O_2^-}$

shown in figure 1(A(b))taken at $\sim$973 K, as well as from the disappearance of this signal after short exposure of the sample to $CH_4$ stream at 973 or 773 K, as shown in figure 1(A(c))and(A(e)), the possibility of this 940 $cm^{-1}$ signal being due to lattice vibration may be ruled out; it also appears very unlikely that this signal is due to a species of $O_2^{2-}$. On the other hand, these results are in line with the assertion [23] that the 940 $cm^{-1}$ signal might be due to an active precursor of $O_2^{2-}$ and $O_2^-$, most probably $O_3^{2-}$ formed(with a gain in electron delocalization energy)by reversible redox coupling of $O_2$ at an anionic vacancy with a neighboring $O^{2-}$ in the surface lattice with adequate inter-nuclear distance between neighboring anions. This active precursor $O_3^{2-}$ might react like a resonance hybrid $[O_3^{2-}\leftrightarrow O_2^- + O^-]$ in abstracting an electron from the oxide lattice and be itself converted to $O_2^{2-} + O^-$ since the electron affinity of $O_2^-$ is greater than that of $O^-$; or it might reversibly dissociate into $O_2^- + O^-$; or it might be active at the OCM reaction temperature in H abstraction from $CH_4$ and be itself converted to $O_2^-$ plus $OH^-$ since H abstraction by $O^-$ is much stronger than by $O_2^-$.

Since the surface site for the formation of this postulated active precursor $O_3^{2-}$ conceivably overlaps with that of its down-stream species $O_2^{2-}$ and $O_2^-$, and since this active precursor appears to be thermally less stable than $O_2^{2-}$ and, in H abstraction from $CH_4$, slightly less active than $O_2^{2-}$, but more active than $O_2^-$, the relative intensities of the observed Raman signals of these oxygen adspecies shown in figure 1 (A(d))and(A(b))may be interpreted as follows. In $CH_4$—$O_2$(4 : 1, v/v)co-feed stream at the lower temperature slightly below 773 K, $O_2^{2-}$ would not react very fast with $CH_4$, and $O_2^-$ would be inactive at this low temperature in H abstraction from $CH_4$, so most of the anionic vacancies would be occupied at the steady state by these down-stream adspecies, $O_2^{2-}$ and $O_2^-$ and a relatively smaller number of anionic vacancies would be left for the adsorption of $O_2$ and the formation of $O_3^{2-}$, resulting in weaker Raman signal of $O_3^{2-}$ at 940 $cm^{-1}$(w)compared with that at 840—812(w-m)and 1120 $cm^{-1}$(w-m)due to $O_2^{2-}$ and $O_2^-$, respectively, as shown in figure 1(A(d)). On the other hand, in the same co-feed stream at higher temperature(slightly below 973 K), most of the $O_2^{2-}$ adspecies would be scavenged by the fast reaction with $CH_4$, and the $O_2^-$ adspecies would also react with $CH_4$ to some extent, besides some extent of further reduction to $O_2^{2-}$ by the oxide lattice, and some extent of thermal decomposition(slightly aided by the 30 mW laser-beam irradiation)with liberation of $O_2$; thus there would be more anionic vacancies at this higher temperature for the adsorption of $O_2$ and formation of $O_3^{2-}$ as shown in figure 1(A(b))by the stronger Raman signal at 940 $cm^{-1}$(w-m).

From figure 1(A(f)), it can be seen that under the OCM reaction conditions and at the spot of the catalyst sample about 1—2 mm down-stream from the hot spot at 973 K, there was no distinct *in situ* Raman signal for any of the oxygen adspecies mentioned above, indicating that $O_2$ appeared to be almost exhausted at a short distance down-stream from the hot spot. That means most of the catalyst sample downstream from the hot spot was exposed to the reaction mixture rich in unconverted $CH_4$ and the $C_{2+}$ products, as well as $CO_X$ and $H_2O$(g), but almost exhausted in $O_2$. Note that over such highly active Th—La—$O_x$/$BaCO_3$-based catalysts under similar OCM reaction conditions, 94~98% conversion of $O_2$ in the $CH_4$—$O_2$(4 : 1, v/v) co-feed at high GHSV has been observed from GC analysis of the microreactor effluents [26].

As shown in figure 1(A(g)), after switching the above sample to $O_2$ stream at 648~383 K and cooling to 298 K in $N_2$ stream, the Raman signal at 940 $cm^{-1}$(w—m)and that around 1120 $cm^{-1}$(w) could still be observed(along with the strong carbonate signal). Note that these two signals shown in figure 1(A(g))were both of higher intensity than the corresponding signals shown in figure 1(A(a)) since in(a)exposure of the sample to $O_2$ stream at 648~383 K was started with a fresh sample ignited and cooled in air, while in(g) exposure of the sample to $O_2$ stream at $648-383$ K was started after(f) in which almost all the active-oxygen adspecies at this part of the sample(down-stream past the hot spot) could be well scavenged by the OCM reaction mixture rich in $CH_4$, $C_{2+}$, $H_2O(g)$, and $CO_x$, but almost exhausted in $O_2$.

Incidentally, with the $Th—La—O_x/BaCO_3$-based catalysts on prolonged exposure of the catalyst sample in $O_2$ stream around or above 973 K, neither the 940 $cm^{-1}$ Raman signal, nor the Raman signal of $O_2^-$; around $1120\sim1150$ $cm^{-1}$, could be observed. This is reminiscent of the disappearance of the Raman signal of $O_2^-$ from either the $La—Ba—Ca—O_x/BaCO_3$ catalyst at 973 K [21], or the single-phase $ThO_2—La_2O_3$ catalyst at 1013 K [20] on prolonged exposure to $O_2$ stream, whereas with either of the last two catalysts in $CH_4—O_2$ co-feed stream at the corresponding temperature, distinct Raman signal of $O_2^-$; has been observed even in the presence of very strong carbonate signals.

The disappearance of the Raman signal of $O_2^-$ (in contrast to the presence of the $O_2^{2-}$ signal) when a non-reducible metal-oxide catalyst for OCM is exposed to $O_2$ stream around 973 K is most probably explained as follows. The adspecies $O_2^{2-}$ is formed via interaction of $O_2$ with the oxide lattice in two steps with the formation of $O_2^-$ as a precursor or intermediate, which is known to be thermally less stable than $O_2^{2-}$ (though $O_2^{2-}$ would be unstable as a free anion [28]). Though the formation of $O_2^-$ is most probably faster than its further conversion to $O_2^{2-}$, and an initial transient concentration of $O_2^-$ might be built up, but it would soon decline, due to the gradual conversion of $O_2^-$ to the thermally more stable $O_2^{2-}$. At the steady state, most of the available anionic vacancies would be occupied by $O_2^{2-}$, and the surface concentration of $O_2^-$ might fall below the detection limit of the Raman spectrometer if the temperature is sufficiently high. However, in $CH_4—O_2$ co-feed stream at 973 K or higher OCM temperature, $O_2^-$ might react with $CH_4$ almost as fast as its formation from $O_2^-$, thus most of the $O_2^{2-}$ would be scavenged, leaving more anionic vacancies for the fast adsorption of $O_2$ and the formation of $O_2^-$, which at sufficiently high OCM temperature(973 K or higher) may have a chance to react with $CH_4$ before its further conversion to $O_2^{2-}$. By similar reasoning, the absence of the 940 $cm^{-1}$ signal from the $Th—La—O_x/BaCO_3$ catalyst in $O_2$ stream around 973 K may be taken as an indication that this signal is due to an OCM-active(though apparently not as active as $O_2^{2-}$) but thermally less stable precursor of $O_2^-$ and $O_2^{2-}$, and that the surface site for this precursor species overlaps with that for $O_2^-$ and $O_2^{2-}$. This again is in line with the suggestion that this active precursor is most probably $O_3^{2-}$ formed via reversible redox coupling of $O_2$ with a nearest $O^{2-}$ in the surface lattice.

Conceivably, the higher thermal stability of $O_2^{2-}$ is mainly due to the larger Madelung stabilization energy for the doubly charged $O_2^{2-}$ compared with that for the singly charged $O_2^-$ at the same anionic site of the surface lattice. Similarly, $O_3^{2-}$ may be thermally more stable than an ozonide anion, $O_3^-$, mainly due to the larger Madelung stabilization energy for $O_3^{2-}$ compared with that for $O_3^-$ at the same surface site.

It is desirable to examine the possibility of detecting the changes in the concentrations of $O_3^{2-}$ and/or $O_2^-$ as the transient precursors, and of $O_2^{2-}$ as the thermally more stable adspecies, with time of exposure of a thoroughly degassed $Th—La—O_x/BaCO_3$ or a similar catalyst to $O_2$ stream around 973 K. The use of a fast spectrometer, e. g., a microfocus Raman spectrometer or a fast FT-IR spectrometer, will be tried.

It is also desirable to carry out isotopic tracing experiments with $^{18}O_2$ to confirm the assignment of the 940 $cm^{-1}$ signal to $\nu_1$ (A) of $O_3^{2-}$ from the characteristic frequency redshifts, and to study the mechanism of isotopic exchange on the $Th—La—O_x/BaCO_3$-based catalysts. The reversible redox coupling of $^{18}O_2$ with lattice $O^{2-}$ to form $^{18}O^{18}O^{16}O$ followed by decoupling on this type of catalysts may lead to very fast isotopic exchange, liberating $^{18}O^{16}O$ as a primary product, as pointed out previously [23]. Note that fast isotopic exchange of $^{18}O_2$ with $Ca^{2+}/ThO_2$, or $La_2O_3$, or $Sr^{2+}/La_2O_3$, starting at about 773, 723 and 650 K, respectively, has been observed by Kalenik and Wolf [6], but no isotopic exchange at all with undoped $ThO_2$.

## 3.2  Structural adaptability for the formation of bent $\underline{O}_3^{2-}$

The frequency of the postulated $\underline{O}_3^{2-}$ species was estimated to be around $900 \sim 950$ cm$^{-1}$ from the known O—O stretching frequencies of $O_2$ ($1580.4$ cm$^{-1}$[27]), $\underline{O}_2^-$; ($\sim 1122$ cm$^{-1}$ for $Ba(O_2)_2$[26]), $O_2^{2-}$ ($\sim 840$ cm$^{-1}$ for $BaO_2$[26]), and the known stretching frequencies $\nu_1$ (A) and structural parameters [29,30] of $O_3$ ($\nu_1 = 1134.9$ cm$^{-1}$[27]; $O_1$—$O_2 = 0.1278$ nm, $O_1$—$O_3 = 0.218$ nm, bond angle $= 116.7°$[29]), and $\underline{O}_3^-$ ($\nu_1 = 1016$ cm$^{-1}$[27,29]; $O_1$—$O_2 = 0.135$ nm, $O_1$—$O_3 = 0.226$ nm, bond angle $= 113.5°$[30]). Thus from $O_3$ to $\underline{O}_3^-$, a frequency red-shift of about 119 cm$^{-1}$ has been observed from Raman spectroscopy. This is much smaller than the frequency red-shift of about 460 cm$^{-1}$ from $O_2$ to $\underline{O}_2^-$, as expected. Nevertheless, this is in accord with the observed structural parameters and indicates that there are a small increase in bond length and decrease in bond order from $O_3$ to $\underline{O}_3^-$. From $\underline{O}_2^-$ to $O_2^{2-}$, there is a considerably smaller frequency red-shift than from $O_2$ to $\underline{O}_2^-$ (ca. 280 cm$^{-1}$ vs. ca. 460 cm$^{-1}$). It may be expected that from $\underline{O}_3^-$ to the postulated $\underline{O}_3^{2-}$ there would also be a frequency red-shift considerably smaller than that from $O_3$ to $\underline{O}_3^-$, though there would still be some partial bonding between the two terminal oxygens, $O_1$ and $O_3$.

The $\nu_1$ (A) of $O_3$ and $\underline{O}_3^{2-}$ might be compared with the stretching frequencies of the corresponding isoelectronic bidentate carboxylate ligands, e. g., $HCOO^-$ ($\nu_{OCO} \approx 1366$ cm$^{-1}$[27], 1355 cm$^{-1}$[30]) or $CH_3COO^-$ ($\nu_{OCO} \approx 1414$ cm$^{-1}$[27], bond angle $\sim 121°$, inter-nuclear distance between the two terminal oxygens 0.218 nm [30]), and a dioxymethylene bidentate ligand, $\mu$-$H_2CO_2^{2-}$ ($\nu_{OCO} \approx 1160$ and 1130 cm$^{-1}$[31], ca. 200 cm$^{-1}$ lower than that of $HCOO^-$). Incidentally, $\mu$-$H_2CO_2^{2-}$ has been proposed to be a surface reaction intermediate in the hydrogenation of CO to methanol over Cu/ZnO-based catalyst [23,31]; $\mu$-$H_2CO_2^{2-}$ might also be formed in the adsorption of $H_2CO$ on $ZnAl_2O_4$[31], conceivably via coupling of $H_2CO$ with a lattice $O^{2-}$ (analogous to the redox coupling of $\underline{O}_2$ with $O^{2-}$).

From the above considerations, the frequency red-shift for $\underline{O}_3^-$ to $\underline{O}_3^{2-}$ may be only about $50\% \sim 70\%$ of that for $O_3$ to $\underline{O}_3^-$; i. e., $\nu_1$ of $\underline{O}_3^{2-}$ may be estimated to be about $930 \sim 960$ cm$^{-1}$; and the internuclear distance about $0.25 - 0.26$ nm, assuming that the bond angle may be about $116° \sim 120°$. This might be stretched to about $0.27 - 0.28$ nm at the surface anionic site, from the consideration of surface Madelung stabilization energy (especially with multiply charged cations). The two electronic units of negative charge on $\underline{O}_3^{2-}$ would be more or less concentrated at the two terminal oxygens. Thus in the case of the doped $ThO_2$ (DFS) with the closest inter-anionic distance of about 0.279 nm in the non-polar (110) surface lattice, formation of $\underline{O}_3^{2-}$ via reversible redox coupling of $\underline{O}_2$ with a neighboring $O^{2-}$ is highly probable, as shown in figure 2(a) in two different microenvironments, $a$ and $b$. In the case of doped $La_2O_3$ shown in figure 2(b), the closest inter-nuclear distance between two neighboring anions in the non-polar (110) surface lattice is 0.285 nm, slightly larger than the sum of V. d. W. radii of the two terminal O; this would be marginal for the formation of $\underline{O}_3^{2-}$, so here such an active-oxygen species (if it could be formed transiently at all) might be too elusive to show any observable Raman signal. For $Li^+/MgO$ and $Na^+/CaO$, both with rock salt structure for the host phase, formation of $\underline{O}_3^{2-}$ is improbable because of too large inter-nuclear distances, respectively, 0.297 and 0.340 nm, for two neighboring anions in the non-polar (100) surface lattice.

It is of fundamental interest to make refined quantumchemical calculations with an appropriate

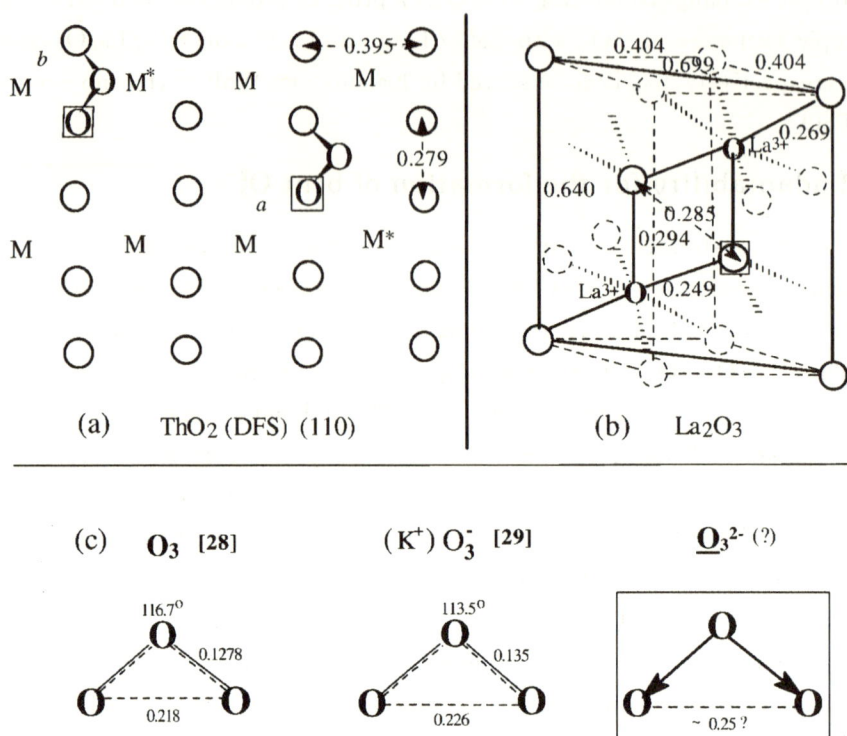

Figure 2. (a)(110)surface of ThO$_2$(DFS by doping with La$_2$O$_3$ or BaO,SrO,CaO);M,M$^*$,O,and square-bracketed O denote,respectively,Th(Ⅳ),La(Ⅲ)or M(Ⅱ),lattice oxide ion,and anionic vacancy. Formation of postulated bent $\underline{O_3^{2-}}$ from O$_2$ adspecies(at an anionic vacmicy)and a lattice O$^{2-}$ are shown at two types of sites,a and b. (b)Unit-cell of hexagonal La$_2$O$_3$(doped)with each La(Ⅲ)ligated by 7 O$^{2-}$ in mono-capped octahedral configuration. Solid lines and circles denote the non-polar(110)surface lattice, (c)Bent structure of O$_3$ and $\underline{O_3^{2-}}$. Unit of length:nm.

cluster model for the energy of formation and the vibration frequencies of $\underline{O_3^{2-}}$ at the defective ThO$_2$ surface lattice shown in figure 2(a),as well as the energy barrier for the H abstraction from H—CH$_3$ by $\underline{O_3^{2-}}$.

## 3.3　Rational design of non-reducible rare-earth-oxide-based OCM catalysts

With the Th—La—O$_x$/BaCO$_3$-based OCM catalysts, the formation of $\underline{O_3^{2-}}$ from O$_2$ via redox coupling with a neighboring O$^{2-}$ is not accompanied with the formation of the undesirable O$^-$ species in the surface lattice,in contrast with the formation of $\underline{O_2^-}$ from $\underline{O_2}$ via redox reaction with $\underline{O^{2-}}$ in the surface lattice with co-production of O$^-$.

The $\underline{O_3^{2-}}$ species may react neatly with CH$_4$ to produce a ·CH$_3$,and $\underline{O_3^{2-}}$ converted to $\underline{OH}^-$ plus $\underline{O_2^-}$,which is harmless to the ·CH$_3$ just formed,if collision occurs,since $\underline{O_2^-}$ may not be able to abstract an H from ·CH$_3$ because of the large difference in bond strengths [4] of H—OO(ca. 364. 8 kJ mol$^{-1}$ for $D^0$ of H—OOH)vs. H—CH$_2$($D^0$ 460 kJ mol$^{-1}$). At high enough OCM temperature $\underline{O_2^-}$ may be able to abstract an H from another CH$_4$ molecule(439. 7 kJ mol$^{-1}$ for $D^0$ H—CH$_3$)to produce one more ·CH$_3$. The two ·CH$_3$ radicals may have a high probability to couple and form C$_2$H$_6$. Thus in a series of comparative assays with stationary-bed microreactor, the Th—La—O$_x$/BaCO$_3$-based OCM catalysts have consistently been found by Liu et al. [26] to give 2～3%(absolute)higher C$_{2+}$ yield than the La$_2$O$_3$/BaCO$_3$-based catalysts,which have been found to form $\underline{O_2^-}$,but no $\underline{O_3^{2-}}$ species,under OCM reaction conditions.

For further improvement of the non-reducible rare-earth-oxide-based catalysts, the active sites of the catalyst should be highly dispersed and methane conversion should be adequate, but not be too high, in order to obtain high $C_2$ selectivity at the industrially acceptable level of $C_2$ yield and to decrease the extent of the highly exothermic $CO_2$ formation so as to avoid excessive hot-spot overheating. A catalyst with DFS host phase and adequate internuclear distance between an anionic vacancy and its nearest anionic neighbor for the formation of the active precursor, $O_3^{2-}$, and well-dispersed active sites would be desirable. But the use of hazardous thorium compound is undesirable. Probably, certain valence-stable $M_2O_3$-type lanthanide-oxide-based solid solution with cubic, defective fluorite structure may be used as the active phase. This must be well dispersed in a thermally stable carrier(e.g., $BaCO_3$), which may also serve as effective quencher of free-radical reactions in the gaseous phase [26]. This type of catalysts would not be robust enough for use in a fluidized-bed reactor. So a certain type of stationary-bed reactor must be designed, that would have the advantage of avoiding attrition loss of the catalyst. But then heat dissipation would become a formidable engineering problem.

# 4  Concluding remarks

It is now beyond doubt that on non-reducible rare-earth-oxide-based OCM catalysts, $O_2^-$ may be observed by its Raman signal in the presence of a large amount of surface carbonate, and that this adspecies is most probably an active-oxygen species under OCM reaction conditions. The supplementary *in situ* laser Raman spectroscopic study of active-oxygen species on $Th$—$La$—$O_x$/$BaCO_3$ OCM catalyst and structural adaptability considerations have provided further support for the suggestion [23,25] that there might be an active and elusive precursor of $O_2^-$ and $O_2^{2-}$, most probably $O_3^{2-}$ formed by reversible redox coupling of $O_2$ with a neighboring $O^{2-}$. However, further studies of the transient adspecies with fast *in situ* Raman and FT-IR spectroscopic methods and of the mechanism of $^{18}O_2$ exchange with this type of non-reducible rare-earth-oxide-based catalysts, as well as refined quantum-chemical calculations of $O_3^{2-}$ formation, vibration frequencies, and H-abstraction reactivity using appropriate surface-cluster models, are highly desirable. Such fundamental studies will be very helpful in rational design of this type of OCM catalysts.

# References

[1] G. E. Keller and M. M. Bhasin, J. Catal. **73**(1982)9.

[2] W. Hinsen, W. Bytyn and M. Baerns, in: Proc. 8th Int. Congr. on Catalysis, Vol. **3**(1984)p. 581.

[3] J. H. Lunsford, Angew. Chem. Int. Ed. Engl. **34**(1995)970.

[4] CRC Handbook of Chemistry and Physics, 66th Ed. (CRC Press, Boca Raton, 1985).

[5] Z. C. Jiang, C. J. Yu, X. P. Fung, S. B. Li and H. L. Wang, J. Phys. Chem. **97**(1993)12870.

[6] Z. Kalenik and E. E. Wolf, Catal. Today **13**(1992)255.

[7] K. D. Campbell, E. Morales and J. H. Lunsford, J. Am. Chem. Soc. **109**(1987)7900.

[8] M. Hatano and K. Otsuka, J. Chem. Soc. Faraday Trans. **85**(1989)199.

[9] M. Che and G. C. Bond, eds., Adsorption on Oxide Surfaces, Stud. Surf. Sci. Catal., Vol. **21** (Elsevier, Amsterdam, 1985)p. 1, 11.

[10] K. J. Borve and L. G. M. Pettersson, J. Phys. Chem. **95**(1991)3214.

[11]M. Xu,C. Shi,X. Yang,M. P. Rosynek and J. H. Lunsford,J. Phys. Chem. **96**(1992)6395.

[12]Y. D. Liu,G. D. Lin,H. B. Zhang and K. R. Tsai,in:Natural Gas Conversion II,Stud. Surf. Sci. Catal.,Vol. 81,eds. H. E. Curry-Hyde and R. F. Howe(Elsevier,Amsterdam,1994)p. 131.

[13]H. L. Wan,X. P. Zhou,W. Z. Weng,R. Q. Long,Z. S. Chao,W. D. Zhang,M. S. Chen,J. Z. Luo and S. Q. Zhou,Catal. Today **51**(1999)161.

[14]J. L. Dubois and C. J. Camron,Appl. Catal. **67**(1990)49.

[15]T. L. Yang,L. B. Feng and S. K. Shen,J. Catal. **145**(1994)384.

[16]C. Louis,T. L. Chang,M. Kermarec,T. L. Van,J. M. Taibouet and M. Che,Catal. Today **13**(1992)283.

[17]Y. D. Liu,G. D. Lin,H. B. Zhang,J. X. Cai,H. L. Wan and K. R. Tsai,in:Prepr. Fuel Chem. Div.,ACS Nat. Mtg.,San Francisco,1992,Vol. **37**(1),p. 356.

[18]G. Mestl,H. Knozinger and J. H. Lunsford,Ber. Burnsenges Phys. Chem. **97**(1993)319.

[19]J. H. Lunsford,X. Yung,K. Haller,J. Laane,G. Mestl and H. Knozinger,J. Phys. Chem. **97**(1993)13810.

[20]Y. D. Liu,H. B. Zhang,G. D. Lin,Y. Y. Liao and K. R. Tsai,J. Chem. Soc. Chem. Commun. (1994)1871.

[21]J. X. Cai,A. M. Huang,Y. Y. Liao and H. L. Wan,in:Proc. 4th Int. Conf. Raman Spectroscopy, eds. N. T. Yu and X. Y. Li(Wiley,New York,1994)p. 526.

[22]H. B. Zhang,Y. D. Liu,G. D. Lin,Y. Y. Liao and K. R. Tsai,Abstracts of Papers PHYS 0277, ACS Nat. Mtg.,San Diego,1994.

[23]K. R. Tsai,D. A. Chen,H. L. Wan,H. B. Zhang,G. D. Lin and P. X. Zhang,Catal. Today **51**(1999)3.

[24]A. G. Anshits,E. N. Voskesenkaya and L. I. Kurteeva,Catal. Lett. **6**(1990)49.

[25]K. R. Tsai,H. B. Zhang,G. D. Lin,H. L. Wan,Y. D. Liu,W. Z. Weng,J. X. Cai and Y. F. Shen, Abstracts of Papers CATL 028,ACS Nat. Mtg.,Boston,1998.

[26]Y. D. Liu,H. B. Zhang,G. D. Lin and K. R. Tsai,in:Catalysis in C1 Chemistry,eds. K. R. Tsai, S. Y. Peng et al. (Chem. Ind. Press,Beijing,1995)p. 46.

[27]K. Nakamoto,Infrared and Raman Spectra ofInorganic and Coordination Compounds,4th Ed. (Wiley,New York,1986)pp. 104,112,232,255.

[28]A. Bielansky and J. Habor,Catal. Rev. Sci. Eng. **19**(1979)1.

[29]N. N. Greenwood and A. Earnshaw,Chemistry ofthe Elements(Pergamon,Oxford,1984)p. 707.

[30]T. C. W. Mak and G. D. Zhou,Crystallography in Modern Chemistry (Wiley,New York,1992) p. 392.

[31]A. Vallet,C. Chauvin,J. C. Lavalley and P. Chaumette,J. Mol. Catal. **42**(1987)205.

■ 本文原载：Inorganica Chimica Acta 314(2001) pp. 184～188.

# Note Tungsten-Malate Interaction. Synthesis, Spectroscopic and Structural Studies of Homochiral $S$-Malato Tungstate( VI ), $\Lambda$-Na$_3$[WO$_2$H(S-mal)$_2$]

Zhao-Hui Zhou[①], Guo-Fu Wang, Shu-Ya Hou, Hui-Lin Wan, Khi-Rui Tsai

(Department of Chemistry and State Key Laboratory for Physical Chemistry of Solid Surface, Xiamen University, Xiamen 361005, People's Republic of China )

Received 26 August 2000; accepted 22 November 2000

**Abstract**  Reaction of the sodium tungstate and malic acid at 1 to 2 ratio in weak acidic solution (pH 3. 0—6. 0)resulted in the first isolation and structural characterization of a tungsten( VI )malato complex $\Lambda$-Na$_3$[WO$_2$H(S-mal)$_2$](1)(malic acid＝H$_3$mal). The complex has been characterized by elemental analysis, UV and IR spectroscopy. The IR spectra are consistent with an oxo-coordinated mononuclear structure as revealed by single crystal X-ray diffraction study. The tungsten atoms are quasi-octa-hedrally coordinated by two $cis$-oxo groups and two bidentate malate ligands via its alkoxy and $\alpha$-carboxyl groups, while the $\beta$-carboxylic and carboxylate groups remain uncomplexed. The latter are bonded each other through strong hydrogen bonding, forming a catenarian chain, and the anions are connected in homochiral form. It is proposed that the chiral configuration of metal center in wild-type FeMo-co biosynthesis might be induced by the coordination of the chiral $R$-homocitric acid, while a mixture of raceme might be obtained in the biosynthesis of $NifV$—FeMo-cofactor. © 2001 Elsevier Science B. V. All rights reserved.

**Key words**  Tungstate ( VI )  Malic acid  Malate  Nitrogenase  Crystal structures  Biosynthesis

## 1. Introduction

The $S$-malic acid is a natural constituent and common metabolite of plants and animals, being involved in the Krebs cycle and in the glyoxylic acid cycle. Formations of tungsten malate in solutions have been studied extensively by UV-Vis, potentiometry, polarimetry, cryoscopy and NMR spectroscopy [1—13]. As very few structures of malato complexes are known [14,15], there is also no direct evidence about the structure of the malato tungstate( VI ). Part of the current interest in this system is stimulated from the structural report of the FeMo cofactor in nitrogenase, in which homocitrate is chelated bidentately to

---

① Corresponding author. Tel. ：＋ 86-592-2184531；fax：＋ 86-5922183047.

*E-mail address*：zhzhou@xmu. edu. cn(Z.-H. Zhou).

**1279**

a molybdenum atom [16]. Tungsten and molybdenum are chemically analogous elements. Because of their great similarities in the properties, it was reasoned that replacing Mo with W might provide insight into the catalytic role of Mo in various enzymes. In fact, for almost all of the known tungstoenzyme, there is an analogous molybdoenzyme that is present within the same organism or in a very closely related species, while of the vast array of life forms that utilize Mo, a very small subset are also able to use W [17]. It is also shown that malic acid and citric acid are effective eluent for the separation of W(Ⅵ) and Mo(Ⅵ) oxoanion [18]. In the use of eluents without alkoxy groups, W(Ⅵ) and Mo(Ⅵ) oxoanions were strongly retained, due to the formations of W(Ⅵ) and Mo(Ⅵ) polyanions. In this work, we report on the synthesis, spectroscopic and structural characterization of the first tungsten(Ⅵ)-malate complex isolated from aqueous solutions.

# 2. Experimental

## 2.1 Preparation

### 2.1.1 Λ-Na₃[WO₂H(S-mal)₂](1)

Sodium tungstate(15 mmol) was added to a solution of S-malic acid(15 mmol, Sigma product) and sodium dihydrogen malate(15 mmol), which is prepared from the reaction of S-malic acid and sodium hydroxide. The mixture(pH 3.5) was heated in a water bath at 75℃ for 5 h, and kept in refrige for several days. The resulted solid was collected and washed with ethanol to give a white solid(6.0g, 73%). Found: C, 17.1, H, 1.7. Calc. for $C_8H_7O_{13}Na_3W$, C, 17.5, H, 1.3%. IR(KBr): $\nu_{as}$(C=O)1666$_{vs}$, $\nu_s$ (C=O)1409$_s$, 1381$_s$, $\nu_s$(W=O)929$_s$, 895$_m$, 865$_s$ cm$^{-1}$.

Crystals of suitable quality for the subsequent X-ray diffraction studies were obtained as transparent prism by cooling the saturated solution of compound **1** in refrige. The resulting colorless crystals were mounted in capillary for X-ray analysis.

**Table 1  Crystal data summaries of intensity data collection and structure refinement for Λ-Na₃[WO₂H(S-mal)₂](1)**

| Formula | $C_8H_7O_{12}Na_3W$ |
|---|---|
| Molecular weight | 547.96 |
| Crystal colour, habit | colorless, prism |
| Crystal size(mm) | 0.22×0.22×0.38 |
| Crystal system | monoclinic |
| Unit cell dimensions | |
| $a$(Å) | 11.6154(8) |
| $b$(Å) | 10.6150(6) |
| $c$(Å) | 5.7164(3) |
| $\beta$(°) | 92.995(5) |
| $v$(Å³) | 703.9(1) |
| Space group | C2 |
| Formula units/unit cell | 2 |
| $D_{calc}$ | 2.585 |
| $F$(000) | 516 |

Note Tungsten-Malate Interaction. Synthesis, Spectroscopic and Structural Studies of

Homochiral *S*-Malato Tungstate( $\text{VI}$ ) , $\Lambda$-Na$_3$[WO$_2$H(S-mal)$_2$]

Con. Table 1

| | |
|---|---|
| $\mu$(Mo K$\alpha$) | 85. 34 |
| Diffractometer | Enraf-Nonius CAD-4 |
| Radiation Mo K$\alpha$(Å) | $\lambda=0.7107$ |
| Temperature(°) | 23 |
| Scan width | $0.90+0.35 \tan \theta$ |
| Decay of standards | $\pm 1.0\%$ |
| No. of reflections measured | 692 |
| $2\theta$ Range(°) | 2—50 |
| No. of reflections observed | 658($R_{int}=0.040$) |
| $[I>3\sigma(I)]^a$ | |
| No. of parameters varied | 124 |
| Weighting scheme | $w=[\sigma(F_o)^2+0.0001(F_o)^2+1]^{-1}$ |
| Goodness-of-fit | 0. 71 |
| $R = \sum(|F_o|-|F_c|)/\sum|F_O|$ | 0. 011 |
| Rw | 0. 015 |
| Largest difference peak and | 0. 87 and $-0.08$ |
| hole(e Å$^{-3}$) | |

$^a$Corrections: Lorentz-polarization.

**Table 2　Selected bond lengths(Å)and bond angles(°)for $\Lambda$-Na$_3$[WO$_2$H(S-mal)$_2$]$^a$**

| Bond lengths | | | |
|---|---|---|---|
| W(1)—O(1) | 1. 973(4) | O(2)—C(2) | 1. 278(8) |
| W(1)—O(2) | 2. 170(4) | O(5)—C(2) | 1. 224(8) |
| W(1)—O(6) | 1. 748(4) | O(3)—C(4) | 1. 228(6) |
| O(1)—C(1) | 1. 407(7) | O(4)—C(4) | 1. 273(8) |
| *Bond angles* | | | |
| O(1)—W(1)—O(1') | 151. 6(2) | O(2)—W(1)—O(2') | 80. 4(2) |
| O(1)—W(1)—O(2) | 75. 0(2) | O(2)—W(1)—O(6) | 164. 6(2) |
| O(1)—W(1)—O(2') | 83. 3(2) | O(2)—W(1)—O(6') | 90. 2(2) |
| O(1)—W(1)—O(6) | 91. 8(2) | O(6)—W(1)—O(6') | 101. 4(2) |
| O(1)—W(1)—O(6') | 106. 2(2) | | |
| *Hydrogen bonding* | | | |
| O(4)—O(4') | 2. 429(6) | | |

$^a$Symmetry transformation$(2-x, y, 2-z)$.

## 2.2 Physical measurements

Electronic spectra were recorded on UV-240 spectrophotometer. Infrared spectra were recorded as Nujol mulls between KBr plates using a Nicolet 740 FT-IR spectrometer. Elemental analysis was performed using EA 1110 elemental analyzers.

## 2.3 X-ray data collection, structure solution and refinement

Crystallographic data for the malato tungstate 1 are summarized in Table 1. Diffraction data were collected on an Enraf-Nonius CAD-4 diffractometer with graphite monochromated Mo K$\alpha$ radiation at 296 K. Lp factor, anisotropic decay and empirical absorption corrections were applied. The structure was solved by heavy atom method and refined by full-matrix least-squares procedures with anisotropic thermal parameters for all the non-hydrogen atoms. H atoms were located from difference Fourier map and not refined. All calculations were performed on an AST Premium P/100 microcomputer using MOLEN software package [19] and published scattering factors [20]. Selected atomic distances and bond angles are given in Table 2.

# 3. Results and discussion

## 3.1 Synthesis and general characterization

Preparation of the malato tungstate depends mainly on the requisite proportions of tungstate and malate. The effect of pH variation seems no so crucial for the product. Instead of the protonated $Na_2[WO_2(S\text{-}Hmal)_2]$ or deprotonated malato tungstate( VI ) $Na_4[WO_2(S\text{-}mal)_2]$, solid separation from the reaction of sodium tungstate with malic acid and sodium dihydrogen malate resulted only the main product of $Na_3[WO_2H(S\text{-}mal)_2]$, which represents an intermediate between the $Na_2[WO_2(S\text{-}Hmal)_2]$ and malato tungstate $Na_4[WO_2(S\text{-}mal)_2]$. The title compound is the most stable species over the pH range 3.0—6.0(Scheme 1). This implies the special stability of the suprastructure. The tendency of the transformations between the oxomalato tungstate is smaller than that of malato molybdates [14].

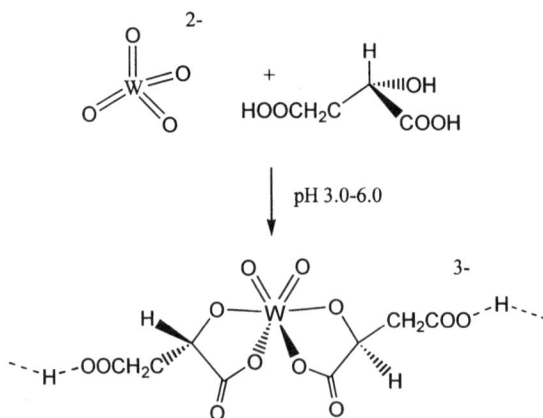

**Scheme 1.**

Note Tungsten-Malate Interaction. Synthesis, Spectroscopic and Structural Studies of

Homochiral *S*-Malato Tungstate( Ⅵ ), Λ-Na$_3$[WO$_2$H(S-mal)$_2$]

$$Na_2[WO_2(S\text{-}Hmal)_2] \underset{+H^+}{\overset{-H^+}{\rightleftharpoons}} Na_3[WO_2H(S\text{-}mal)_2] \underset{+H^+}{\overset{-H^+}{\rightleftharpoons}} Na_4[WO_2(S\text{-}mal)_2]$$

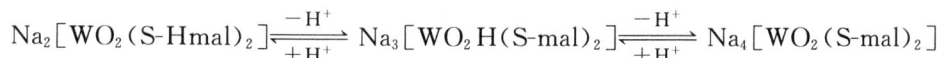

There was no direct evidence of the other crystalline phase of malato tungstate in the pH 3.0 to 6.0, although solution studies have suggested the existences of dinuclear species $[W_2O_6(mal)_2]^{6-}$, $[W_2O_5(OH)(mal)_2]^{5-}$, $[W_2O_5(mal)_2]^{4-}$, $[W_2O_6(mal)(H_2O)_2]^{3-}$, tetranuclear species $[W_4O_{10}(OH)(Hmal)_4]^{5-}$ and the other ratios of W : ligand complexes [13]. Attempts to separate the products only result in the isolation of isopoly-tungstates. The preferred stoichiometry of the product appears to be 1 : 2 for metal-to-ligand ratios, which is much stable than 1 : 1, 2 : 2 and 2 : 1 complexes. This is similar to those complexes forming by α-hydroxycar-boxylate ligands having only one hydroxy and one carboxylic group [21−25].

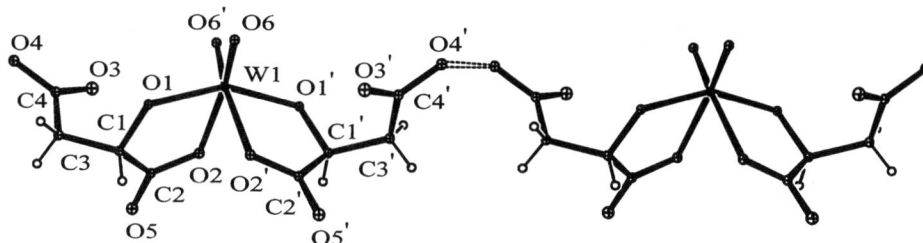

Fig. 1　Perspective view of the polymeric anion structure of Λ-Na$_3$[WO$_2$H(S-mal)$_2$].

The tungsten( Ⅵ )complex exhibits featureless in the visible range 350−600 nm. The frequencies and assignments of selected i. r. absorption bands are shown in experimental section. In the region between 1800 and 1400 cm$^{-1}$, the bands of 1666, 1409 and 1381 cm$^{-1}$ are corresponding to bound carboxyl group $\nu_{as}$ and $\nu_s(CO_2)$, respectively, this is in accord with the coordinated and free carboxyl of the malato ligand. In the region around 900 cm$^{-1}$, the complexes show several bands that might result from the presence of *cis*-dioxo cores. The low frequency of symmetric $WO_2$ stretching may be explained by the coordination of oxogroups to sodium cation.

It is undoubted that one of the malate ions in the complex carries one proton in the β-carboxyl group, this is not only the result from charge balance in the complex, but also the result by the measurement of the molar electrical conductivity($2.8 \times 10^{-2}$ S m$^2$ mol$^{-1}$)of the title compound. The value closes to the trivalent coordinated anion($2.4 \times 10^{-2}$ ca. $2.7 \times 10^{-2}$ S m$^2$ mol$^{-1}$)[26]. The existence of very strong intermolecular hydrogen bonding is between protonated and deprotonated free β-carboxyl groups [O(4)—O(4a), 2.430(6)Å], which make the complex forming a polymeric network. This is further supported by IR bands found at 1666$_{vs}$, 1409$_s$ and 1381$_s$ cm$^{-1}$ corresponding to bound $\nu_{as}$ and $\nu_s$ ($CO_2$), and the absence of IR bands between 1740 and 1700 cm$^{-1}$ for a typical undissociated carboxylic group, indicating the obvious shift to lower frequency caused by hydrogen bonding.

## 3.2　Description of the structure

The crystal structure of **1** comprises discrete sodium cations and hydrogen dimalato dioxotungstate anions. As shown in Fig. 1[27], the tungsten ( Ⅵ ) atoms are coordinated by two oxo groups, one protonated and one fully deprotonated malate ligands. Each malate ion acts as a bidentate ligand with the alkoxy and α-carboxyl coordinated to one tungsten atom, and the other β-carboxyl groups remain uncomplexed. This is similar to that of protonated malato molybdate( Ⅵ )Cs$_2$[MoO$_2$(Hmal)$_2$] • H$_2$O [14], or the coordination mode of homocitrate to molybdenum in the FeMo cofactor [16]. The two terminal oxo groups are in a *cis*-configuration. Each tungsten atom is six-coordinate and exists in quasi-octahedral geometry. The chiral Λ-configuration is achieved through enantiopure chiral *S*-malate ligands. An enantioselective aggregation with homochirality-$\Lambda\Lambda\Lambda$-is obtained within one dimensional catenarian chain through strong hydrogen bonding, which results in the formation of homochiral supramolecular entities [28]. Thus, the chirality of the whole complex is induced by the chirality of the chiral ligand.

**Table 3  Comparisons of M—O distances(Å)(M＝Mo,W)and absolute configuration in mononuclear citrato and malato complexes(H$_4$cit＝citric acid)**

| Complex | M—O(alkoxy) | M—O($\alpha$-carboxy) | Configuration | Ref. |
|---|---|---|---|---|
| Na$_2$[MoO$_2$(H$_2$cit)$_2$]·3H$_2$O | 1.953(6),1.960(7) | 2.190(7),2.247(6) | $\Delta/\Lambda$ | [29] |
| Na$_2$[WO$_2$(H$_2$cit)$_2$]·3H$_2$O | 1.945(6),1.968(7) | 2.189(8),2.227(7) | $\Delta/\Lambda$ | [29] |
| Cs$_2$[MoO$_2$(Hmal)$_2$]·H$_2$O | 1.939(8) | 2.243(9) | $\Delta$-S | [14] |
| Na$_3$[WO$_2$H(S-mal)$_2$] | 1.973(4) | 2.170(4) | $\Lambda$-S | this work |

As shown in Table 2, the W—O distances in malato tungstate vary systematically. W＝O bonds are 1.748(4) Å, indicating that they are double bonds. The resulting O＝W＝O angles, 101.4(2)° is considerably larger than the 90° regular value for *cis* groups, this is expected from the greater O···O repulsions between oxygens with short bonds to the metal atom. The W—O(alkoxy)distance is longer [1.973(4)Å], which shows the deprotonation of the hydroxyl group in the malate anion, and those to the $\alpha$-carboxyl is longest [2.170(4)Å].

Table 3 shows the comparisons of the M—O distances and absolute configuration. The malato tungstate possesses the longest M—O(alkoxy) and the shortest M—O($\alpha$-carboxy) bonds in the four citrato and malato complexes [14,29]. Moreover, achiral citrate ligand only results in the formation of a mixture of $\Lambda$-and $\Delta$-complex, and the racemic ligand forms complex coordinated by both $R$ and $S$-ligands, like [K$_2$(H$_2$O)$_5$][V$_2$O$_4$($R$-homocitrate)($S$-homocitrate)]·H$_2$O or M$_2$[V$_2$O$_2$(O$_2$)$_2$($R$-lact)($S$-lact)](M＝K$^+$,NBu$_4$)[30-32]. While the chiral $S$-malate ligand resulted in the separation of title compound $\Lambda$-Na$_3$[WO$_2$H(S-mal)$_2$] in our experiment. The $\Lambda$-configuration of the chiral complex is determined by the chiral ligand and the non-centrosymmetric space group $C2$ of the crystal. It is deduce that a racemic complex might be obtained in the biosynthesis of $Nif$V—FeMo-cofactor from achiral citrate [33,34], while the chiral configuration of metal center in wild type FeMo-co might be induced by the coordination of the chiral $R$-homocitrate.

# 4. Supplementary material

Crystallographic data for the structural analysis have been deposited with the Cambridge Crystallographic Data Centre, CCDC no. 156644. Copies of this information may be obtained free of charge from The Director, CCDC, 12 Union Road, Cambridge, CB2 1EZ, UK(fax:＋44-1223-336-033;e-mail:deposit@ccdc.cam.ac.uk or www:http://www.ccdc.cam.ac.uk). Computed and observed structure factor moduli, along with infrared and electron absorption data are available from the authors on request.

# Acknowledgements

The present work was financially support by the National Science Foundation of China(no. 29933040,29973032) and the Foundation for University Key Teacher by the Chinese Ministry of Education.

# References

[1]E. Richardson,J. Inorg. Nucl. Chem. **13**(1960)84.

[2]S. Prasad,K. S. R. Krishnaiah,J. Indian Chem. Soc. **38**(1961)153.

[3]M. J. Baillie,D. H. Brown,J. Chem. Soc. (1961)3691.

[4]D. H. Brown,D. Neumann,J. Inorg. Nucl. Chem. **37**(1975)330.

[5]M. K. Rastogi,R. K. Multani,Indian J. Chem. Sect. A **15**(1977)912.

[6]Y. K. Tselinskii,I. I. Kusel'man,T. M. Ponomareva,Zh. Neorg. Khim. **21**(1976)715.

[7]Y. K. Tselinskii,B. V. Pestryakov,A. I. Olivson,T. M. Pekhtereva,Zh. Neorg. Khim. **24**(1979) 1243.

[8]A. Cervilla,A. Beltran,J. Beltran,Can. J. Chem. **57**(1979)773.

[9]V. M. S. Gil,M. E. T. L. Saraiva,M. M. Caldeira,A. M. D. Pereira,J. Inorg. Nucl. Chem. **42** (1980)389.

[10](a) Y. V. Kokunov,V. A. Bochkareva,Y. A. Guslaev,Koord. Khim. **7**(1981)1853. (a) Y. V. Kokunov,V. A. Bochkareva,Y. A. Guslaev,Chem. Abstr. **96**(1983)114870w.

[11]A. Cervilla,J. A. Ramirez,E. Elopis,Can. J. Chem. **63**(1985)1041.

[12]M. Hlaibi,S. Chapelle,M. Benaïssa,J. F. Verchère,Inorg. Chem. **34**(1995)4434.

[13]J. J. Cruywagen,L. Krüger,E. A. Rohwer,J. Chem. Soc.,Dalton Trans. (1997)1925.

[14]C. B. Knobler,A. J. Wilson,R. N. Hider,I. W. Jensen,B. R. Penfold,W. T. Robinson,C. J. Wilkins,J. Chem. Soc.,Dalton Trans. (1983)1299.

[15]C. Djordjevic,M. Lee-Renslo,E. Sinn,Inorg. Chim. Acta **233**(1995)97.

[16]J. B. Howard,D. C. Rees,Chem. Rev. **96**(1996)2965.

[17]M. K. Johnson,D. C. Rees,M. W. W. Adams,Chem. Rev. **96**(1996)2817.

[18]M. Maruo,N. Hirayama,A. Shiota,T. Kuwamoto,Anal. Sci. **8**(1992)511.

[19]C. K. Fair,MOLEN,An Interactive Intelligent System for Crystal Structure Analysis. Enraf-Nonius,Delft,The Netherlands,1990.

[20]D. T. Cromer,J. T. Waber,International Tables for X-ray Crystallography, vol. IV, Kynoch Press,Birmingham,1974.

[21]V. M. S. Gil,Pure. Appl. Chem. **61**(1989)841.

[22]J. J. Cruywagen,L. Kriiger,E. A. Rohwer,J. Chem. Soc.,Dalton Trans. (1993)105.

[23]M. M. Caldeira,M. L. Ramos,V. M. S. Gil,Can. J. Chem. **65**(1987)827.

[24]J. J. Cruwagen,E. A. Rohwer,J. Chem. Soc.,Dalton Trans. (1995)3433.

[25]Y. Shii,Y. Motoda,T. Matsuo,F. Kai,T. Nakashima,J. P. Tuchagues,N. Matsumoto,Inorg. Chem. **38**(1999)3513.

[26]R. J. Angelici,Synth. Tech. Inorg. Chem. (1977)13.

[27]L. J. Farrugia,J. Appl. Crystallogr. **30**(1997)565.

[28]U. Knof,A. von Zelewsky,Angew. Chem. Int. Ed. Engl. **38**(1999)302.

[29]Z. H. Zhou,H. L. Wan,K. R. Tsai,J. Chem. Soc.,Dalton Trans. (1999)4289.

[30]D. W. Wright,R. T. Chang,S. K. Mandal,W. H. Armstrong,W. H. Orme-Johnson,J. Bioinorg. Chem. **1**(1996)143.

[31]F. Demartin,M. Biaioli,L. Strinna-Erre,A. Panzanelli,G. Micera,Inorg. Chim. Acta **299**(2000) 123.

[32]P. Schwendt,P. Švancárek,I. Smatanová,J. Marek,J. Inorg. Biochem. **80**(2000)59.

［33］K. C. Grönberg，C. A. Gormal，M. C. Durrant，B. E. Smith，R. A. Henderson，J. Am. Chem. Soc. **120**(1998)10613.

［34］V. R. Almeida，C. A. Gormal，K. L. C. Grönberg，R. A. Henderson，K. E. Oglieve，B. E. Smith，Inorg. Chim. Acta **291**(1999)212.

■ **本文原载**:《厦门大学学报》(自然科学版)第 40 卷第 2 期(2001 年 3 月),第 320～329 页。

# 固氮酶催化作用机理及其化学模拟*

周朝晖[1]　颜文斌[1]　张凤章[2]　万惠霖[1]　蔡启瑞[1]①

(1. 厦门大学化学系,2. 厦门大学生物系,福建　厦门　361005)

**摘　要**　固氮酶是某些微生物在常温常压下固氮成氨的主要催化剂,其催化机理和化学模拟是国际上长期致力研究的对象。尽管人们已经揭示了固氮酶催化活性中心——铁钼辅基(FeMo-co) $MoFe_7S_9$ ($R$-高柠檬酸)的结构,然而,有关 FeMo-co 生物合成的详细途径和其中包含的钼铁转移和利用;固氮酶在哪些活性位和按什么模式结合 $N_2$,$C_2H_2$、CO 等多种底物或抑制剂;以及其中涉及的电子和质子传递过程等许多问题,至今尚有待解决。本文将综述这些问题的研究进展和厦门大学固氮组最近五年来的相关研究工作。

**关键词**　固氮酶　铁钼辅基　铁钒辅基　化学模拟　催化作用　高柠檬酸　柠檬酸　不对称

**中图分类号**:Q 946　　　**文献标识码**:A

固氮酶是某些微生物在常温常压下固氮成氨的主要催化剂,它能将生物体无法直接利用的分子氮($N_2$)转化成可利用的氨态氮($NH_3$),而且不需要如工业合成氨过程那样消耗大量的能源,不降低土壤活性,不污染环境。全球每年约有 24 亿吨的氨态氮是通过微生物的固氮过程实现的,约占全球氮资源的65％,而工业合成氨过程提供约 25％。因而固氮酶的催化作用机理和化学模拟一直是国际上长期致力研究的对象。

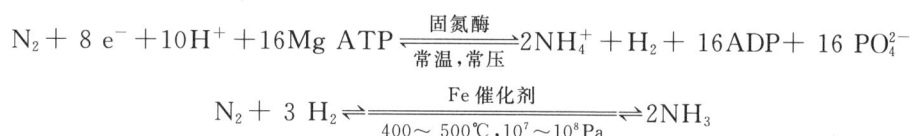

$$N_2 + 8\ e^- + 10H^+ + 16Mg\ ATP \underset{\text{常温,常压}}{\overset{\text{固氮酶}}{\rightleftharpoons}} 2NH_4^+ + H_2 + 16ADP + 16\ PO_4^{2-}$$

$$N_2 + 3\ H_2 \underset{400\sim500℃,10^7\sim10^8 Pa}{\overset{Fe\ 催化剂}{\rightleftharpoons}} 2NH_3$$

60 年代以来,固氮酶的研究已取得了多次突破和重要进展。最近一次突破是 1992 年 Rees 等发表了固氮酶钼铁蛋白和铁蛋白的 X 光晶体结构,揭示了 FeMo-co(铁钼辅基)$MoFe_7S_9$($R$-高柠檬酸)的本质[1,2]。在该结构中,Mo 原子处于一端的角落位置上,并和 3 个 $\mu_3$-硫配体、一个组氨酸和一个高柠檬酸配位,形成 A 面体的络合物。高柠檬酸以 $\alpha$-完氧基和 $\alpha$-羧基直接同钼形成双齿配位,而 $\alpha$-羧基和 $\alpha$-羧基与金属不成键,形成 1:1 型的高柠檬酸钼簇结构,如图 1。

## 1　固氮酶铁钼辅基对氮分子的活化方式和催化作用机理

固氮酶铁钼辅基结构的阐明,为 FeMo-co 对 $N_2$ 的活化方式和催化作用机理的研究提供了依据。对

---

*　收稿日期:2001 年 2 月 15 日。

基金项目国家自然科学基金(29933040);高等学校博士学科点基金;教育部骨干教师基金和厦门大学科学研究基金资助项目。

①　作者简介:周朝晖,(1964—),男,教授;蔡启瑞(1914—),男,院士。

通讯联系人;email:krtsai@xmu.edu.cn。

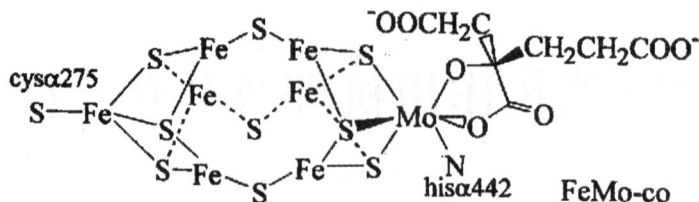

图1 钼和铁在固氮酶铁钼辅基中的配位结构[1,2]

Fig 1 Experimental coordination structure of iron molybdenum cofactor(Fe Mo-co)in nitrogenase[1,2]

FeMo-co 活化 $N_2$ 的探讨不仅有利于对固氮酶催化作用本质的认识,而且对化学模拟生物固氮的研究有重要的指导意义。因此,对这一问题的探讨一直引起了人们广泛的重视[3-9]。到目前为止,有关 FeMo-co 对 $N_2$ 的活化方式已提出几十种模型,图2示出其中有代表性的 4 种模型。就 $N_2$ 与 FeMo-co 的位置而言可分为两大类:一类是 $N_2$ 的活化在 FeMo-co 的外部;另一类是 $N_2$ 的活化在 FeMo-co 的内部($N_2$ 的活化在 FeMo-co 的内部与外部兼有的方式也归入此类);就配位的金属而言,涉及到铁位或钼位 2 种或协同活化方式。

## 2 铁钼(钒)辅基生物合成的化学模拟

从钒的 EXAFS 谱图中可发现钒固氮酶中的 FeV 辅基和 FeMo 辅基相似[10],柠檬酸($H_4$cit = $C_6H_8O_7$)和其他羟基多羧基酸有可能在如 Fe、V 和 Mo 这些金属的新陈代谢中起重要的作用野生型固氮酶含有 $R$ 构型的高柠檬酸,$R$-高柠檬酸有可以和钼配位的 $\alpha$-羟基、$\alpha$-羧酸、$\beta$ 和 $\gamma$-羧基;而固氮酶突变种(如:*Klebsiella pneumoniae*)仅含柠檬酸配位的铁钼辅基 MoFe$_7$S$_9$(柠檬酸),柠檬酸比高柠檬酸少一个亚甲基,没有光学活性。固氮酶突变种的固氮活性很低,但能还原乙炔和放氢[11]。目前的研究认为,在固氮酶辅基的生物合成中,三羧酸根的功能可能是转移和存储酶中的钼或钒,$MoO_4^{2-}$ 或 $VO_4^{3-}$ 与(高)柠檬酸的复合物经还原、硫交换和 $Nif$ B 辅基结合等过程形成 FeMo-co,FeMo-co 插入缺辅基的钼铁蛋白形成有活性的固氮酶[10-12](图3)。在固氮酶的固氮过程中,高柠檬酸发挥了重要的作用。研究钼(钒)酸盐同(高)柠檬酸间的反应,对于阐明铁钼(钒)辅基生物合成的第一步和其中包含的钼(钒)转移作用具有重要的意义。

### 2.1 钒酸盐和(高)柠檬酸体系的研究

对于钒酸盐和(高)柠檬酸的反应,我们最早分离和表征了柠檬酸钒(V)酰钾的结构(图4)[13],几乎是同时,MIT 的研究小组也独立地分离了该络合物[14]。络合物的典型结构特征是柠檬酸配体通过桥联的 $\alpha$-烷氧基和单齿的 $\alpha$-羧基双配位到钒原子上。这种独特的配位形式代表着第一个固氮酶辅基中三元羧酸配位到金属原子的模型,它的配位模式与 FeMo-co 中的高柠檬酸配位相似。

之后,我们分别从弱酸性和中性溶液中,分离了含柠檬酸钒(V)酰负四价和负六价阴离子络合物。两个阴离子同样含有心对称的四元环 $V_2O_2$ 二聚体,柠檬酸通过 $\alpha$-烷氧基、$\alpha$-羧酸与 V 进行配位,而另两个 $\beta$-羧酸基团未参与配位,钒原子保持五配位。其中负四价中间体可视为由强氢键组成的配位聚合物(图5)。对于这类柠檬酸钒(V)酰络合物,$\beta$-羧酸基团的质子分步加合与离解并不影响双核物种的结构和配位形式,柠檬酸钒(V)酰络合物呈现多元羧酸的基本性质[15-20]。这些性质为羟基多元羧酸在固氮酶反应体系中作为质子传递链提供了可能性[11,21],也为固氮酶中高柠檬酸搭桥的质子(电子)传递途径提供了依据。

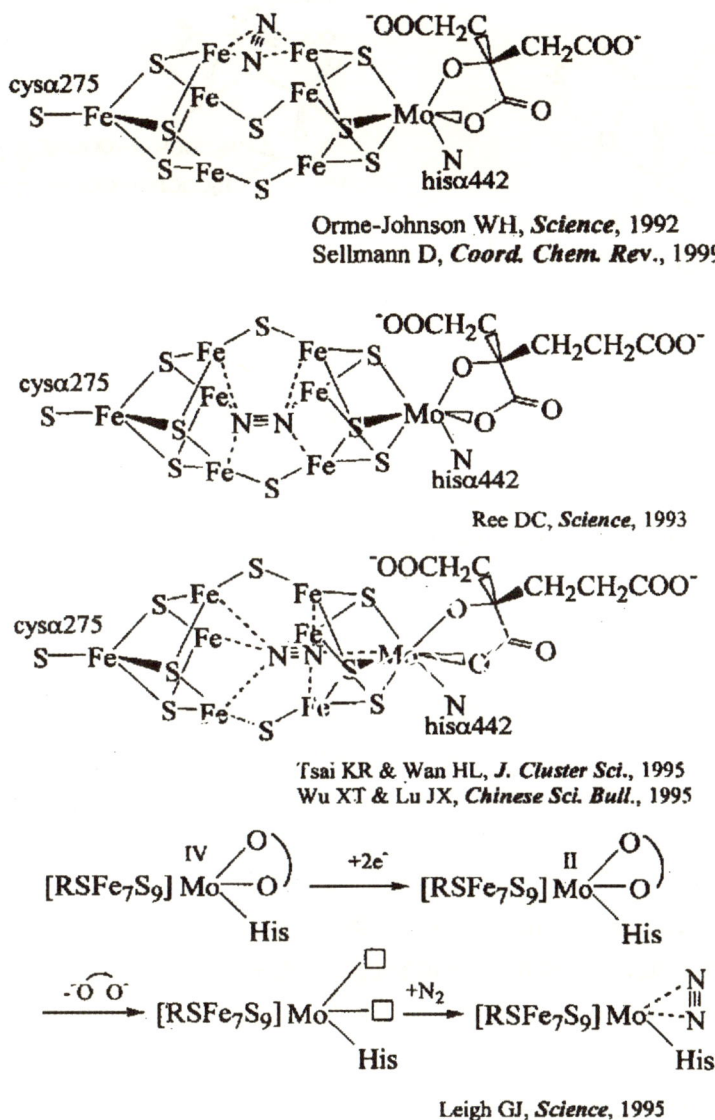

Orme-Johnson WH, *Science*, 1992
Sellmann D, *Coord. Chem. Rev.*, 1999

Ree DC, *Science*, 1993

Tsai KR & Wan HL, *J. Cluster Sci.*, 1995
Wu XT & Lu JX, *Chinese Sci. Bull.*, 1995

Leigh GJ, *Science*, 1995

图 2　氮分子在铁钼辅基上可能的活化方式

Fig 2　Different models of the active site of $N_2$ in iron molybdenum co. factor of nitrogenase

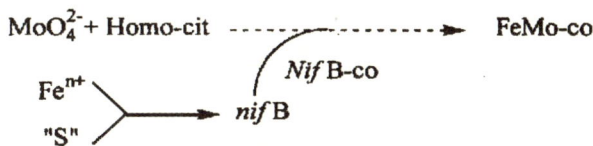

图 3　铁钼辅基的生物合成途径

Fig. 3　Biosynthesis of iron molybdenum cofactor

将上述柠檬酸钒（Ⅴ）酰络合物还原，可以得到对称的柠檬酸钒（Ⅳ）酰二聚体（图 6）。络合物中柠檬酸离子的 α-烷氧基、α-羧酸和两个 β-羧酸基团全部参与络合，α-烷氧基桥式配位在两个钒原子上形成 $V_2O_2$ 四元环[22−24]，钒的配位形式为畸变的八面体[15,25]。

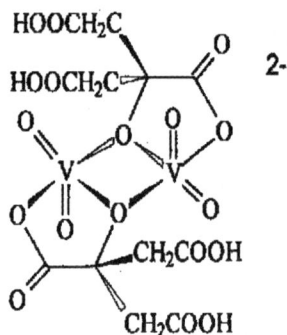

图 4　柠檬酸钒（Ⅴ）酰钾 $K_2[VO_2(C_6H_6O_7)]_2 \cdot 4H_2O$ 的络离子结构

Fig 4　Anion structure of potassium dioxocitrato vanadate(Ⅴ) $K_2[VO_2(C_6H_6O_7)]_2 \cdot 4H_2O$

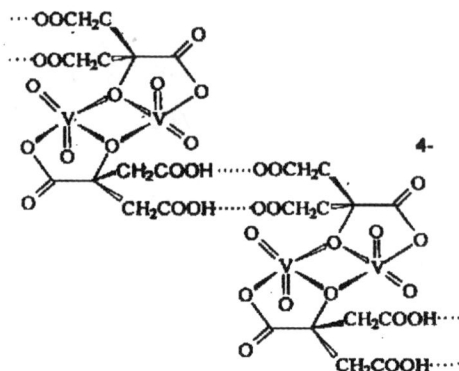

图 5　双质子柠檬酸钒（Ⅴ）酰酸钾钠 $Na_2K_2[VO_2(C_6H_5O_7)]_2 \cdot 9H_2O$ 的络离子聚合结构

Fig. 5　Polymeric anion structure of potassium sodium dioxohydrogencitrato vanadate (Ⅴ) $Na_2K_2[VO_2(C_6H_5O_7)]_2 \cdot 9H_2O$

　　Bino 等推测还原态的对称的柠檬酸钒（Ⅴ）酰二聚体与不对称的柠檬酸钒（Ⅴ）酰二聚体处在平衡中[26]，他们分离出不对称的柠檬酸钒（Ⅳ）酰二聚体 $(Hneo)_3[(VO)_2(cit)(Hcit)] \cdot 4H_2O$，络合物中的柠檬酸分别以三齿和四齿与钒（Ⅳ）络合，两个钒离子分别处于六配位畸变的八面体和五配位畸变的四方单锥构型中。其后，Orme-Johnson 等合成和结构表征了内消旋的高柠檬酸钒（Ⅴ）酰络合物 $[K_2(H_2O)_5][(VO_2)_2(R,S\text{-}homocitrate)_2]H_2O$[27]，络合物的配位形式与柠檬酸钒（Ⅴ）酰钾相似，高柠檬酸仍以独特的双齿配位与钒（Ⅴ）形成二聚体，分子中分别含有 $R$-和 $S$-构型配位的高柠檬酸组成的二聚体。

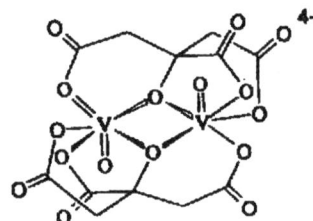

图 6　柠檬酸钒（Ⅳ）酰酸钠 $Na_4[VO(C_6H_4O_7)]_2 \cdot 6H_2O$ 的配离子结构

Fig. 6　Anion structure of sodium oxocitrato vanadate (Ⅳ) $Na_4[VO(C_6H_4O_7)]2 \cdot 6H_2O$

## 2.2　钼酸盐和（高）柠檬酸体系的研究

　　我们从自然界的钼源出发，以比 R-高柠檬酸少一个亚甲基的柠檬酸为模拟物，研究了钼酸盐与柠檬酸在 pH 为 3～9 之间的反应行为，化学模拟了固氮酶突变种生物合成中的钼源转移过程的第一步。柠檬酸既是反应物又是缓冲剂，通过控制反应物比例、溶液的 pH 值和反应温度，可以实现单体和二聚络合物的转化。结果发现，在接近固氮酶体系的 pH（7～8）环境下，柠檬酸配体以 $\alpha$-烷氧基、$\alpha$-羧基和 $\beta$-羧基同钼三齿配位，形成一比一的单核柠檬酸络合物，其中 $\beta$-羧基为弱配位，其钼氧单键键长较长 $[0.2411(3)nm]$。而且，三齿配位的钼络合物可进一步转化为仅含 $\alpha$-氧基和 $\alpha$-羧基的双齿配位柠檬酸钼络合物，这种配位形式与高柠檬酸在铁钼辅基中钼的配位类似进而说明羟基多羧酸钼（Ⅵ）络合物的形成可能是钼源转移的第一步。也就是说，柠檬酸钼（Ⅵ）络合物可能是突变种型铁钼辅基生物合成的前驱体，其突出的结构特征如下（图 7）[28]：

　　由此推测野生菌 FeMo-co 中的高柠檬酸配体以及生物体体外合成突变种中所含的高柠檬酸衍生物二元、三元羧酸也是与上图相似的螯型配体，其 $\alpha$-C 上的其他取代基与钼没有配位。

　　当 pH 值进一步降低到 5～6 后，单核柠檬酸钼络合物可转化为全脱质子的双核物种 $[(MoO_2)_2O\text{-}(cit)_2]^{6-}$，该物种是一个单氧桥联的对称的柠檬酸钼二聚体，柠檬酸仍然通过 $\alpha$-烷氧基、$\alpha$-羧基和一个

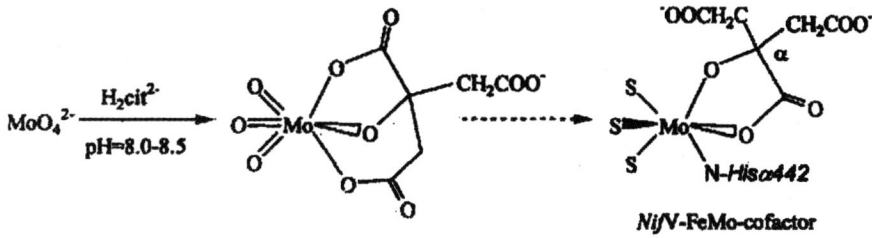

图 7  固氮酶生物合成中可能的钼源转移途径

Fig 7  Possible pathway of molybdenum transfer in nitrogenase

$\beta$-羧基三配位,两个钼都形成八面体构型。双核全脱质子柠檬酸钼可质子化,形成含质子的双核柠檬酸钼络离子$[(MoO_2)_2O(Hcit)_2]^{4-}$,含质子的双核络合物在过量配体的作用下,形成柠檬酸双齿配位的单核络离子$[MoO_2(H_2cit)_2]^{2-}$(图 8)[29-32]。

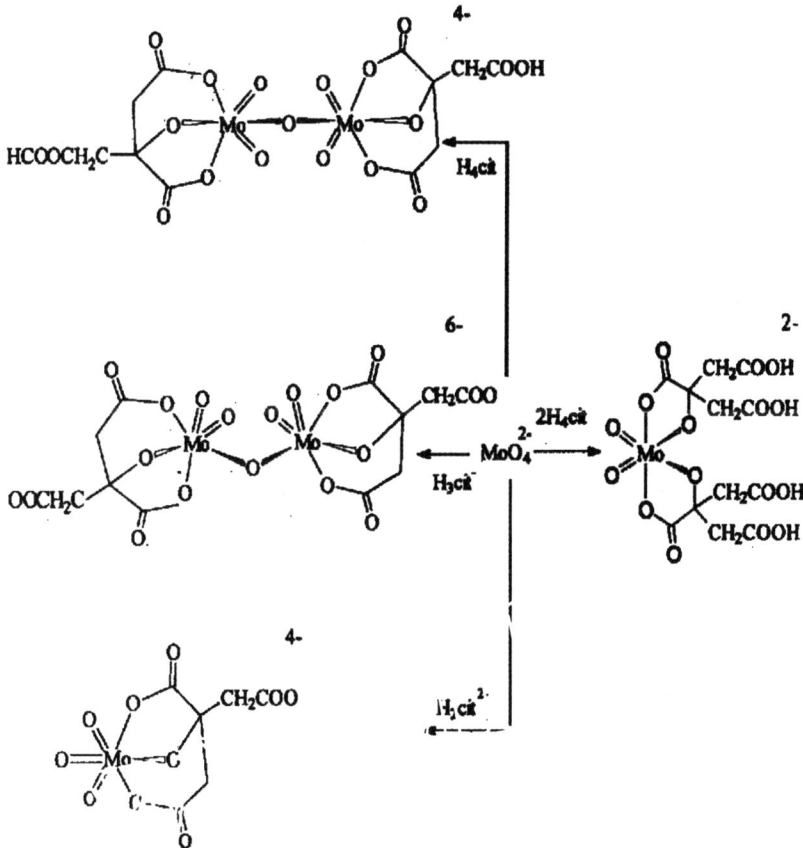

图 8  柠檬酸钼单核和双核物种的合成和转化

Fig 8  Syntheses and transformation of monomeric and dimeric citratomolybdates

## 2.3  野生型和突变种型固氮酶铁钼辅基前驱体的绝对构型问题

以同 $R$-高柠檬酸同样具有手性的 $S$-苹果酸为模拟物与钼酸盐反应,光学纯的 $S$-手性苹果酸诱导了中心金属不对称钼(Ⅵ)或钨(Ⅵ)络合物非对映异构体 $\Delta$-$S$ 和 $\Lambda$-$S$ 的分离,而对于非手性的柠檬酸配体,反应得到的钼(Ⅵ)或钨(Ⅵ)络合物 $\Lambda$ 为和 $\Delta$ 的一比一外消旋产物,表 1 示出柠檬酸和手性苹果酸的

Mo(W)—O键长的比较和我们对这些络合物绝对构型的指认。

　　从表1的比较可以看出,非手性的柠檬酸配体制备的络合物只能是 Λ 和 Δ 的外消旋体;而手性的苹果酸配体制备的络合物可导致手性分离。由此暗示了光学纯高柠檬酸在固氮酶铁钼辅基合成中的作用:对于非手性的柠檬酸配体参与的生物合成,得到的固氮酶铁钼辅基突变种可能是 Λ 和 Δ 的一比一外消旋产物;而在含 R-高柠檬酸的野生型固氮酶铁钼辅基生物合成中,光学纯的 R-高柠檬酸配体可能诱导含手性中心金属钼原子的 Λ-R 和 Δ-R 非对映异构体的分离。根据上述钼酸盐与柠檬酸的反应和铁钼辅基生物合成的研究结果,我们推测铁钼辅基的生物合成可能途径(图 9)包括:

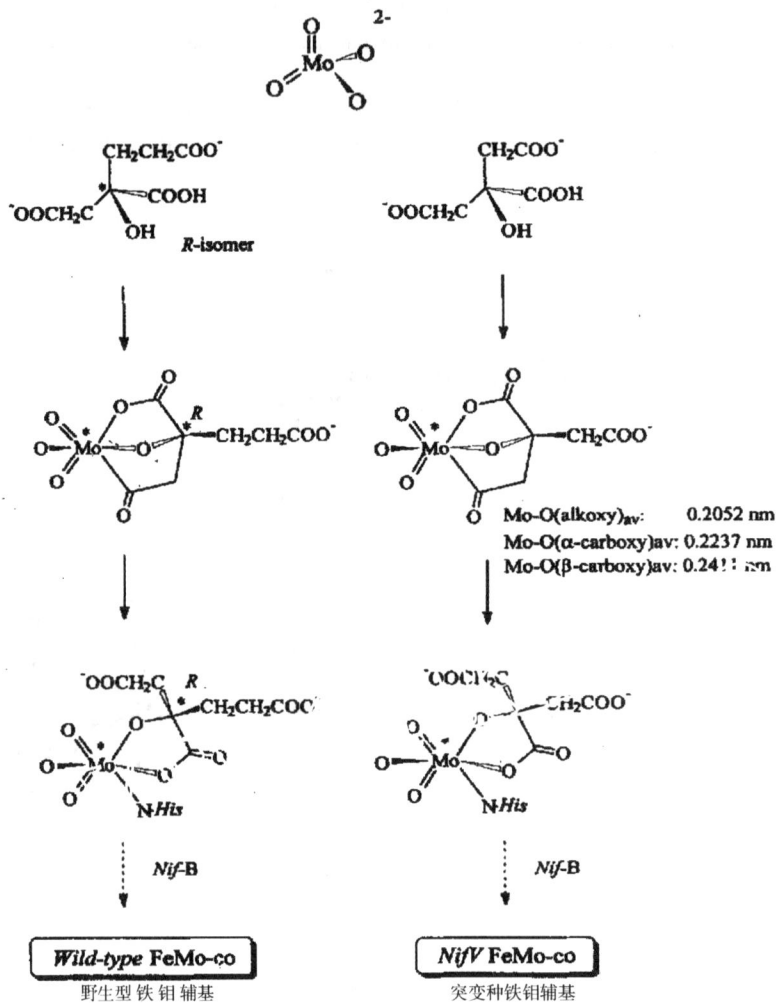

图 9　固氮酶铁钼辅基可能的生物合成途径

Fig 9　Possible biosynthesis pathway of iron molybdenum co factor in nitrogenase

表 1　柠檬酸和手性苹果酸的 Mo(W)—O 键长的比较和我们对这些络合物绝对构型的指认

Tab. 1　Comparisons of Mo(W)—O distances($\times 10^{-1}$ nm )and our assignm ent of ab solute configuration in citrato and malato complexes($H_4$ cit＝citric acid)

| Complex | M—O(alkoxy) | M—O($\alpha$-carboxy) | Configuration assignm ent | ref. |
|---|---|---|---|---|
| $Na_2P[MoO_2(H_2cit)_2]\cdot 3H_2O$ | 1 953(6),1.960(7) | 2.190(7),2.247(6) | Δ/Λ | 32 |
| $Na_2[WO_2(H_2cit)_2]\cdot 3H_2O$ | 1.945(6), 1.968(7) | 2.189(8),2.227(7) | Δ/Λ | 32 |

续表

| Complex | M—O(alkoxy) | M—O($\alpha$-Tcarboxy) | Configuration assignm ent | ref. |
|---|---|---|---|---|
| $(NH_4)_4[Mo_4O_{11}(mal)_2]6H_2O$ | 2.00 | 2.21 | $\Delta,\Delta,\Lambda,\Lambda,-S,S$ | 33 |
| $(NH_4)_4[Mo_4O_{11}(mal)_2]H_2O^*$ | 1.93(15) | 2.50(15) | $\Delta,\Delta,\Delta,\Delta-S,S$ | 34,35 |
| $(NH_4)_4[Mo_4O_{11}(mal)_2]6H_2O$ | 1.925(6) | 2.226(6) | $\Delta,\Delta,\Lambda,\Lambda-R,R$ | 36 |
| $Cs_2D[MoO_2(Hmal)_2]H_2O$ | 1.939(8) | 2.243(9) | $\Delta-S$ | 37 |
| $Na_3[WO_2H(S\,mal)_2]$ | 1.973(4) | 2.170(4) | $\Lambda-S$ | 38 |

固氮酶催化是典型的络合催化,70 多种底物及抑制剂是检验酶活性中心结构的最有效化学探针。我们曾根据配位催化原理和化学探针思路及 $N_2$ 对 $C_2H_2$ 在 $D_2O$ 中的竞争抑制降低 $trans/cis—C_2H_2D_2$ 比例的实验结果推断,在酶转化中 FeMo-co 原子簇笼应该是活口的,即中部 3 个 $\mu$-S 中之一能成 $H_2S$ 而移去,$N\equiv N$ 与 $HC\equiv CH$ 的三重健(或环丙烯、环丙偶氮的准三重健)能从笼口[2Fe]位进入笼内,分别结合在笼内[Mo,3Fe,3Fe]位和[3Fe,3Fe]位,而一端有烷基小尾巴的一些底物,如丙炔、甲基异晴,只能络合在笼口[2Fe]位。这样才能说明许多底物结合力的大小已知顺序,例如,$N\equiv N\approx CH_2(HC\!=\!=\!CH)$,$CH_2(N\!=\!=\!N)>HC\equiv CH>CH_3N\!=\!C>CH_3C\equiv CH>CH_3C\!=\!=\!N$,而且 $HC\equiv CH$ 只能还原到 $H_2C\!=\!=\!CH_2$,以及在 $D_2O$ 中和笼内无外源抑制剂时生成的 $C_2H_2D_2$ 99% 以上为顺式[6,7,39-43]。

高柠檬酸固氮酶催化放氢反应不受 CO 抑制,而柠檬酸固氮酶虽有中等程度的酶促放氢及还原氢化乙炔的活性,固氮催化活性却大大降低,$N_2$ 对乙炔还原氢化的竞争抑制也很弱,且放氢反应对 CO 的抑制敏感[11]从这些实验事实可以推测。从 P 簇到 M 簇有两条质子传递链质子传递可能是按同步位移接力传递方式进行的,才不致过分滞后于电子传递,可以设想,由 P 簇到 M 簇的两条质子传递链中,一条可用高柠檬酸的 $\gamma$-羧基的两个羧基 O 作为两个桥墩,各架接一个 HOH 所构成的氢键,$H^+$ 从 $Cys^{a62}$—HN—$Gly^{a61}$ 的酰胺基的一个羰基 O,再从另一个羰基 O 到 $His^{442}$ 的咪唑 N$\delta$ 进入笼内[Mo]位。同样,另一条可用 $\beta$-羧酸链的两个羧基 O 作为两个桥墩,各架接一个 HOH 所构成的氢键,从 $Cys^{a62}$—HN—$Gly^{a61}$ 的一个羰基 O;再从另一个羰基 O 到 $His^{a195}$ 的咪唑 N$\in$ 而进入笼内。这样的构架是否合理,正在用分子力学、量化计算和化学探针等方法进一步检验。

# 参考文献

[1]Kim J S,Rees D C. Structural models for the metal centers in the nitrogenase molybdenum-iron protein[J]. Science,1992,**257**:1 677~1 682.

[2]Howard J B,Rees D C. Structural basis of biological nitrogen fixation [J]. Chem. Rev.,1996,**96**:2 965~2 982.

[3]Orme-Johnson W H. Nitrogenase structure:Where to now? [J]. Science,1992,**257**:1 639~1 640.

[4]Sellmann D,U tz J,Blum N. On the function of nitrogenase FeMo cofactor and competitive catalysts:chemical principles,structural blue-prints,and the relevance of iron-sulfur complexes for $N_2$ fixation [J]. Coord Chem. Rev ,1999,190—192:607~627.

[5]Chan M K,Kim J S,Rees D C. The nitrogenase FeMo-cofactor and P-cluster pair-2 2Å resolution structures[J]. Science,1993,**260**:792~794.

[6]Tsai K R,W an H L. On the structure-function relationship of nitrogenase M-cluster and P-cluster pairs[J]. J. Cluster Sci ,1995,**6**:485～501.

[7]万惠霖,黄静伟,张凤章,等.化学探针方法研究固氮酶 M-簇和 P-簇对的结构与功能关系[J].厦门大学学报(自然版),1996,**35**:890～899.

[8]吴新涛,卢嘉锡.固氮酶活性中心网兜模型研究的回顾和前瞻[J].科学通报,1995,**40**:577～581.

[9]Leigh A J. A fixation with fixation [J]. Science,1995,**268**:827～828.

[10]Eady R R,Structure-function relationships of alternative nitrogenases[J]. Chem. Rev.,1996,**96**,3 013～3 030.

[11]Granberg K L C,Gorm al C A ,Durrant M C,et al. Why R-Homocitrate is essential to the reactivity of FeMo-cofactor of nitrogenase:studies on N if V⁻ extracted FeMo-cofactor [J]. J. Am. Chem. Soc ,1998,**120**:10 613～10 621.

[12]MÜller A,Krahn E. On the synthesis of the FeMo cofactor of nitrogenase:Gene-controlled in nature versus laboratory-produced by man [J]. A ngew. Chem. Ind Ed. Engl ,1995,**34**:1 071～1 078.

[13]Zhou Z H,Yan W B,Wan H L,et al. Meta-hydroxycarboxylate interactions syntheses and structures of $K_2[VO_2(C_6H_6O_7)]_2 \cdot 4H_2O$ and $(NH_4)_2[VO_2(C_6H_6O_7)]_2 \cdot 2H_2O$ [J]. J. Chem. Crystallogr ,1995,**25**:807～811.

[14]W right D W ,Humiston P A ,Orme-Johnson W H,et al. A unique coordination mode for citrate and a transition metal:$K_2[VO_2(C_6H_6O_7)]_2 \cdot 4H_2O$[J]. Inorg Chem.,1995,**34**:4 194～4 197.

[15]Zhou Z H,Wan H L,Hu S Z,et al. Syntheses and structures of potassium-ammonium dioxocitratovanadate(Ⅴ) and sodium oxocitratovanadate(Ⅳ) dimers[J]. Inorg. Chim. Acta,1995,**237**:193～197.

[16]Zhou Z H,Zhang H T,Huang T S. Syntheses and characterization of citrato complexes of vanadium [J]. J. Mol Sc.,1995,**11**:220～225.

[17]Zhou Z H,Wan H L,Tsai K R. Synthesis and crystal structure of sodium ammonium dimeric (citrato)dioxovanadium(Ⅴ)[J]. Chinese Sci Bull ,1995,**40**:749～752.

[18]Zhou Z H,Wan H L,Hu S Z,et al. Structural access to citrate process-synthesis and structure of sodium-potassium oxocitratovanadate(Ⅳ)dimer[J]. Chinese J. Struct Chem.,1995,**14**:337～341.

[19]周朝晖,缪建英,万惠霖.含质子柠檬酸氧钒(Ⅴ)配合物中间体的合成和晶体结构[J].高等学校化学学报,1997,**18**:11～14.

[20]Zhou Z H,Zhang H,Jiang Y Q,et al. Complexation between vanadium(Ⅴ) and citrate: spectroscopic and structural characterization of a dinuclear vanadium(Ⅴ)complex [J]. Trans Metal Chem.,1999,**24**:605～609.

[21]Szilagyi R K,Musaev D G,M orokuma K. Theoretical studies of biological nitrogen fixation. Part II Hydrogen bonded networks as possible reactant and product channels J. Mol Struct (Theor .),2000,**506**:131～146.

[22]Zhou Z H,Wang J Z,Wan H L,et al. Synthesis and structure of dimeric (glycollato) oxovanadium(Ⅴ)[J]. Chem. Res Chinese Univ.,1994,**10**:102～106.

[23]Svancárek P,Schwendt P,Tatiersky J,et al. Oxo peroxo glycolato complexes of vanadium(V) [J]. Crystal structure of(NBu$_4$)$_2$[V$_2$O$_2$(O$_2$)$_2$(C$_2$H$_2$O$_3$)$_2$]・H$_2$O. Monatsh. Chem.，2000,**131**: 145~154.

[24]BiagioliM, Strinna-Erre L,M icera G,et al. Molecular structure,characterization and reactivity of dioxo complexes formed by vanadium(V)with α-Thydroxycarboxylate ligands [J]. Inorg Chim. Acta，2000,**310**:1~9.

[25] Velayutham M, Varghese B, Subramanian S. Magneto-structural correlation studies of a ferrom agnetically coupled dinuclear vanadium(IV)complex. Single-crystal EPR study [J]. Inorg Chem.,1998,**37**:1 336~1 340.

[26]Burojevic S, Shweky I,B ino A, et al. Synthesis, structure and magnetic properties of an asymm etric dinuclear oxocitratovanadate(IV)complex [J]. Inorg Chim. Acta,1996, **251**:75~ 79.

[27]Wright D W,Chang R T,Mandal S K,et al. A novel vanadium(V)homocitrate complex: synthesis,structure,and biological relevance of [K$_2$(H$_2$O)$_5$][(VO$_2$)$_2$(R,S-homocitrate)$_2$]・ H$_2$O [J]. J. Biol Inorg Chem.，1996,**1**:143~151.

[28]Zhou Z H,Wan H L, Tsai K R. Syntheses and spectroscopic and structural characterization of molybdenum(VI)citrato monomeric raceme and dimmer [J]. Inorg Chem.，2000, **39**:59~64.

[29]Zhou Z H,Wan H L, Tsai K R. Molybdenum(VI)complex with citric acid:synthesis and structural characterization of 1:1 ratio citrato molybdate K$_2$Na$_4$[(MoO$_2$)$_2$O(cit)$_2$]・5H$_2$O [J]. Polyhedron, 1997,**16**:75~79.

[30]Xing Y H,Xu J Q, Sun H R,et al. Hexapotassium μ-oxo-bis[(citrato)dioxomolybdenum] dihydrate [J]. ActaCryst , 1998,C54:1 615~1 616.

[31]Zhou X H,Xing Y H,Li D M,et al. Molybdenum(VI)-oxygen complex containing citrate ligand:synthesis and characterization of K$_6$[(MoO$_2$)$_2$O(cit)$_2$]・5H$_2$O [J]. Solid State Sci , 1999,**1**:189~198.

[32]Zhou Z H,Wan H L, Tsai K R. Bidentate citrate with free terminal carboxyl groups,syntheses and characterization of the bidentate citrato oxomolybdate(VI)and oxotungstate(VI),Δ/Λ-Na$_2$ [MO$_2$(H$_2$cit)$_2$]・3H$_2$O(M＝Mo or W)[J]. J. Chem. Soc. Dalton Trans,1999,(24):4 289~ 4 290.

[33]Porai-Koshits M A, A slanov L A, Ivanova G V,Polynova T N. X-ray diffraction ammonium dimolybdomalate [J]. Zh. Strukt Khm.，1968,**9**:475~480.

[34]Berg J E, B randange S, L indblom L, W erner P E. Crystal and molecular structure of tetraammonium aa'-μ-oxobis { [ gied'-μ$_3$-(S)-malato-O$^1$, O $^2$, O$^4$, O$^4$']-di-μ-oxobis [dioxomolybdate(VI)]} monohydrate$^{13}$C NMR studies[J]. Acta Chem. Scand , 1977,**A 31**, 325~328.

[35]Berg J E,Werner P E. On the use of Guinier-Hagg film data for structure analysis,the crystal structure of tetraammonium aa'-μ-oxobis{ [gied'-μ$_3$-(S)-malato-O$^1$,O $^2$,O$^4$,O$^4$']-di-μ-oxobis [dioxomolybdate(VI)]} monohydrate[J]. Z Kristallogr ,1977,**145**,310~320.

[36]Zhou Z H,Yan W B,Wan H L ,et al. Complexation of molybdenum(VI)with R-and S -malic acid,the crystal structure of(NH$_4$)$_4$[(MoO$_2$)$_4$O$_3$(R -mal)$_2$]・6H$_2$O [J]. Chinese J. Struct

Chem. 1995,**14**,255～260.

[37]Knobler C B,Wilson A J,Hider R N,et al. Molybdenum(Ⅶ)complex with malic acid:their interrelationships,and the crystal structure of dicaesium bis [(S)-malato(2−)]-cis-dioxomolybdate(Ⅵ)-water(1/1)[J]. J. Chem. Soc Dalton Trans 1983,1 299～1 303.

[38]Zhou Z H,W ang G F,Hou S Y,et al. Tungsten-malate interaction. Synthesis,spectroscopic and structural studies of homochiral S-malato tungstate(Ⅵ),Λ—Na₃[WO₂H(S-mal)₂][J]. Inorg Chim. Acta,2001,**314**:184～188.

[39]Qiu X H,Dong E H,Zhou Z H,et al. The effects of citrato molybdate on the growth rate of A zotobacter V inelandii[J]. Chem. J. Chinese Univ. (Sppl),2000,**12**:152.

[40]Zhang F Z,Huang J W ,Huang H Q,et a. The binding site of N₂ and N₂O in nitrogenase indirectly detected by the change of trans/cis of DHC═CHD in products [J]. Acta Biophy Sinica,1999,**15**:18.

[41]张凤章,黄河清,龙敏南,等.固氮酶铁钼辅基在分离纯化中结构变化的新证据[J].中国生物化学与分子生物学报,1999,**15**:165～168.

[42]陈忠,林国兴,蔡淑惠.重水中固氮酶催化还原乙炔产物的¹HNMR研究[J].化学学报,1999,**57**:907～913.

[43]张凤章,黄静伟,邱雪慧,等.固氮酶中N₂和N₂O结合位的一种新的鉴定方法[J].厦门大学学报(自然科学版),1999,**38**(4):611～616.

## Catalytic Mechaniam of Nitrogen Reduction by Nitrogenase and its Chemical Modeling

Zhao-Hui Zhou[1] ， Wen-Bin Yan[1] ， Feng-Zhang Zhang[2] ， Hui-Lin Wan[1] ， Khi-Rui Tsai[1]

(1. Dept. of Chem.， 2. Dept. of Biol， Xiamen Univ.， Xiamen 361005，China)

**Abstract** Nitrogenase catalyzes the reduction of dinitrogen to ammonia coupled to the hydrolysis of ATP,which is central to the process of biological nitrogen fixation X-raycrystallographic structures has revealed its catalytic metal center FeMo-cofactor as a cage structure of MoFe₇S₉(R-homocitrate),in which the homocitrate coordinate bidentately to molybdenum. Recent efforts towards establishing the mechanism of activation of nitrogen around the metal center and the early biosynthesis process of FeMo-co are reviewed,which reflect a combination of structural,spectroscopic,synthetic,biochemical and theoretical approaches to this challenging problem pursued especially from the nitrogen fixation group of Xia men University.

**Key words** Nitrogenase Iron-molybdenum cofactor Iron-vanadium cofactor Chemical modeling Homocitric acid Citric acid Chiral Configuration assignment

■ **本文原载**:《厦门大学学报》(自然科学版)第 40 卷第 2 期(2001 年 3 月),第 407～417 页。

# 水溶性铑膦配合物催化剂的制备、结构和性能[*]

袁友珠[①]　杨意泉　林国栋　张鸿斌　蔡启瑞

(固体表面物理化学国家重点实验室　厦门大学化学系,福建厦门　361005)

**摘　要**　概述了烯烃氢甲酰化水溶性铑膦配合物催化剂分子设计的发展和水平,讨论有关水溶性铑膦配合物的结构、性能和在氢甲酰化催化反应中的应用,内容包括水溶性铑膦配合物催化剂和负载型水溶性铑膦配合物催化剂的设计、制备及其催化剂失活机理,并藉以说明水溶性配体和催化剂的合成、两相催化的应用开拓以及新型催化体系的设计乃将是今后研究的重点。

**关键词**　两相催化　水溶性膦配体　氢甲酰化　负载水相催化剂　催化剂结构

**中图分类号**　O 643　　**文献标识码**　A

烯烃氢甲酰化催化剂经过 60 多年的发展,有许多催化剂被研制出来,如烃溶性的 Rh-TPP(三苯基膦)系催化剂[1]、固载型烃溶性催化剂(SLPC)[2]、含磺酸化(钠盐)三苯基膦单齿配体(如 TPPTS:P($m$-$C_6H_4SO_3Na$)$_3$P)的铑膦系水溶性催化剂[3]、担载型水溶性液膜催化剂(SAPC)[4]等体系研究表明,采用均相 Rh-TPP 系催化剂进行丙烯氢甲酰化,时空产率和区位选择性都较低,为达到醛正异比为 10,膦铑比高达 250～300,不仅在配体上的耗资很大,尤其是还存在产物与催化剂分离困难而使铑严重流失等问题。虽然 SLPC 部分克服了均相反应的缺点,但未能彻底解决底物通过多相的传质问题使反应催化活性下降和铑从载体上脱落等问题[5]。

70 年代中期出现的以水溶性金属配合物催化剂在很大程度上简化了催化剂的分离回收过程[6,7]。1984 年,Ruhrchemie 公司和 Rhone-Poulenc 公司第一次成功地将 Kuntz 水溶性铑膦配合物,即 HRh(CO)(TPPTS)$_3$,用于两相催化体系催化丙烯氢甲酰化的工业化生产(简称 RCH/RP 过程)[3]。该过程与已有的均相催化过程相比,显示出的环境效益和经济效益,大大地促进了有关两相催化体系的基础研究和应用开发研究[8-11]。随后仅仅十余年地时间,就有多个以水溶性过渡金属—膦配合物为催化剂的两相催化体系推向工业应用[11],其发展之迅速,已成为目前最为活跃的研究领域之一。本文结合本课题组和国内外的研究结果,简要概述近年在水溶性铑膦配合物催化剂的设计、结构及其催化机理等方面的发展现状。

## 1　水溶性铑膦配合物催化剂的设计和制备

水溶性两相催化中,催化剂与产物的简便分离是通过使用具有良好水溶性的金属有机配合物作为催化剂而实现的,合成具有良好水溶性金属配合物催化的关键,是寻找合适的水溶性配体 Kuntz 经过大量的研究工作后提出,要使催化剂向有机相的流失减小到在经济上能够接受的程度,所使用的水溶性配体必须具有足够大的水溶性,更重要的是,它不能被有机物反萃取到有机相中去[12]。根据这一原则,Kuntz 首次合成了具有良好水溶性的三苯基膦三磺酸纳盐,即 TPPTS,它在水中的溶解度可达到 1 100 g/L(20

[*]　收稿日期:2001 年 2 月 15 日。

　基金项目:国家自然科学基金(29792075,29873037)和国家重点基础发展规划(G2000048008)资助项目。

[①]　作者简介:袁友珠(1963—　),男,研究员。

℃)[3]。事实表明,以 TPPTS 为配体的水溶性铑催化剂在 RCH/RP 过程中获得了极大的成功,而早在 1958 年报道的 TPPMS[Ph₂P(m-C₆H₄SO₃Na)][12],因在水中溶解度较差(80 g/L,20 ℃),未能有效地将贵金属离子"固定"在水相。

由于含磺酸钠盐基团的膦配体具有较高的水溶性,许多水溶性膦配体都是由相应的烃溶性膦配体直接磺化而制得[13-16]。如图 1 所示,除 TPPTS 外,在丙烯氢甲酰化中具有较高活性的水溶性的配体 **1**,**2**,**3** 和具表面活性的磺化芳烷基膦 **4**,**5** 和 **6**,以及水溶性的手性双膦配体 **4-6** 和 **9-12** 等。但用直接磺化的方法时,存在膦易被氧化及磺化程度难于控制的不足,不易得到纯的水溶性膦配体,往往需要比较复杂的分离和纯化步骤。

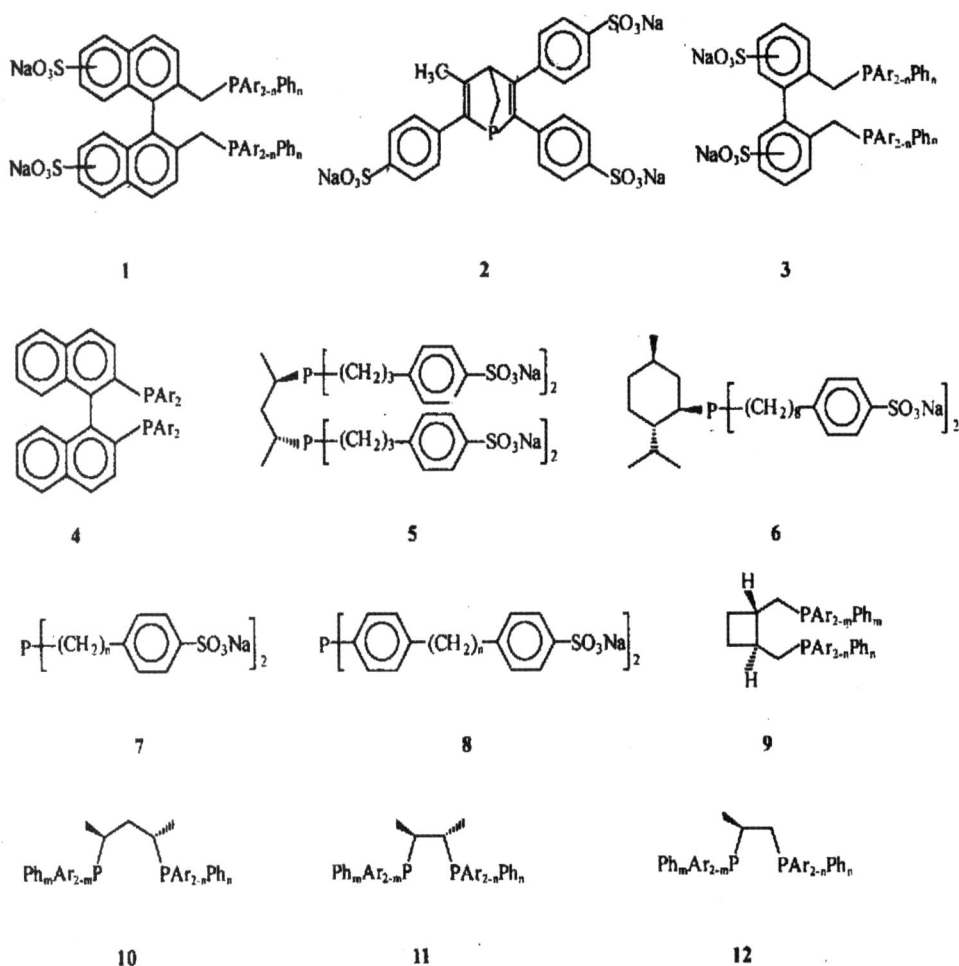

图 1　若干水溶性膦配体的分子结构

Fig. 1　Structures of several water-soluble phosphine ligands Ar＝m-C₆H₄SO₃Na;m＝0,1;n＝0,1

利用对-氟苯磺酸盐在超强碱溶液中(DMSO/KOH)与 PH₃、PhPH₂ 或 Ph₂PH 反应(图 2),可以制得具有不同取代程度的三苯基膦磺酸盐,且磺酸根基团位于磷原子的对位(配体 **13**,反应式 1)[17]。此外,PH₃ 与 4-氟-1,3-苯二磺酸钾反应时,由于空间位阻的原因,只能生成二取代的膦 **14a**(反应式 2)。**14a** 与氟代苯反应可得到叔膦 **14** 或其他取代的叔膦[18]。配体 **13** 和 **14** 与 TPPTS 在电子和空间结构上的差异,也许其相应的铑配合物在氢甲酰化性能方面可提供一些有意义的结果。

除磺酸盐外,含其他水溶性基团(如 PO₃⁻、CO₂⁻、⁺NR₄ 及 ⁅CH₂CH₂OCH₂CH₂O⁆ₙ 等)的水溶性膦配体的合成也得到广泛的研究[19,20]。各种水溶性膦配体的出现,大大地丰富了水溶性金属配合物的催化化学[19−21]。目前看来,由于三苯基膦已有工业产品提供,制备 TPPTS 的成本较低,且催化性能良好,采用 TPPTS 作为配体的两相体系的研究报道最为活跃。

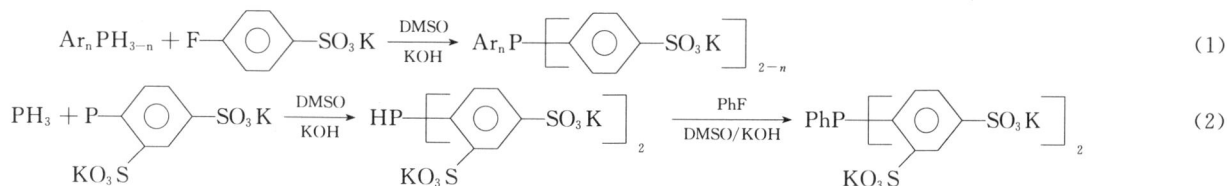

$$Ar_nPH_{3-n} + F\!-\!\!\bigcirc\!\!-SO_3K \xrightarrow[KOH]{DMSO} \left[Ar_nP\!-\!\!\bigcirc\!\!-SO_3K\right]_{2-n} \tag{1}$$

$$PH_3 + P\!-\!\!\bigcirc\!\!-SO_3K \xrightarrow[KOH]{DMSO} \left[HP\!-\!\!\bigcirc\!\!-SO_3K\right]_2 \xrightarrow[DMSO/KOH]{PhF} PhP\!-\!\!\left[\bigcirc\!\!-SO_3K\right]_2 \tag{2}$$

图 2  三苯基膦磺酸盐的合成路线

**Fig. 2  Synthesis pathways for sulfonated triphenyl phosphine**

## 2  水溶性铑膦配合物的结构稳定性

当三氯化铑溶于严格脱气的水中,得到水合三价铑物种并将在室温下与水溶性 TPPTS 配体反应,产生氧化的 TPPTS [即 OTPPTS:$O=P(m\text{-}C_6H_4SO_3Na)_3$]和一价的铑膦络合物 $Rh^ICl(TPPTS)_2$ 和 $[Rh^ICl(TPPTS)_2]_2$[22]。同位素标记实验结果证明了氧源系反应中水被分解的缘故这一结果虽表明了可采取牺牲一部分 TPPTS 的方法,得到低氧化态的催化活性铑膦物种,但这是 一个平衡过程,过量的 TPPTS 将得到低产率的一价铑膦络合物以及 OTPPTS。

随着烯烃碳链的增长,烯烃在水中的溶解性减小,烯烃和催化剂水溶液间的传质速度成为两相催化体系的速控因素,可因转化率太低而失去工业化价值。解决的办法是在体系中加入助剂,或者是制备成负载型水相催化(SAPC),以改善两相间的传质过程[1,3,4,23−32]。常用的助剂包括表面活性剂、有机溶剂、无(有)机盐类等。一些学者[24,32−35]采用原位、动态谱学方法(如$^{31}P$ NMR)来研究某些添加物的作用机理及其对配合物稳定性的影响,把该领域的研究逐步推向分子水平的认识高度。

助剂的作用主要包括增加油溶性反应底物在水溶性催化剂中的溶解度和调节反应体系的 pH 值。Hanson 等的研究结果表明[33],水溶液中的离子强度不仅影响反应速率,还影响产物选择性离子强度增加,选择性提高,但反应活性的增减则取决于膦配体是否能形成胶束———一 种新的表面活性膦。采用能形成胶束膦的配体,反应活性将随离子强度增加而增加实际上,配合物催化剂水层的 pH 值也直接影响着催化剂活性、选择性和稳定性。在碱存在下,水溶性铑膦配合物 $HRh(CO)(TPPTS)_3$ 可提高其 1-己烯氢甲酰化的正异醛比值,但催化活性略有下降(表 1);在酸存在下,特别是在无机酸存在下,将获得较差的反应选择性。因此,在实际操作中推荐采用中性至偏碱性的催化剂水层 pH 值[26,34]。NMR 研究结果表明,外加的酸性,无论是对水溶性铑膦配合物还是对负载型水溶性铑膦配合物结构,其影响程度远比碱性来得大[34,36,37]。NaOH 对水溶性铑膦配合物 $HRh(CO)(TPPTS)_3$ 的分子结构破坏很小,但含有配位原子 N 的有机碱吡啶加入到 $HRh(CO)(TPPTS)_3$ 后,配合物分子中部分 TPPTS 配体被交换取代下来,产生新水溶性配合物。无机酸和有机酸均对 $HRh(CO)(TPPTS)_3$ 有较明显的结构破坏作用,直至使配合物完全解离,并使解络下来的 TPPTS 氧化成弱配位能力的 OTPPTS[34](图 3)。阴离子对铑膦络合物的氢甲酰化催化活性也有相当的影响[38],在 TPPTS/Rh=30 和表面活性剂 CTAB 存在的情况下,表面活性剂阴离子和 $SO_4^{2-}$ 阴离子对反应几乎没有什么影响[39]。若干阴离子对催化性能影响的大小顺序为:$SO_3^{2-} > CO_3^{2-} > S_2O_3^{2-} > Cl^- > B_4O_7^{2-} > Br^- > HCO_3^- > I^- > EDTA^{2-} > H_2PO_4^- > SO_4^{2-}$。

我们曾用高分辨 NMR 研究了 NaCl、$NiSO_4$、$CuSO_4$、$Fe_2(SO_4)_3$ 和 $Cr_2(SO_4)_3$ 对水溶性铑膦配合物 $HRh(CO)(TPPTS)_3$ 分子结构的影响[35]。$^{31}P(^1H)$ 和 $^1H$ NMR 谱显示,于室温下在 $HRh(CO)(TPPTS)_3$ 中加入 NaCl 和 $NiSO_4$,对配合物的特征$^{31}P(^1H)$和$^1H$ NMR 谱峰无明显影响,仅可观察到膦

表1 在酸碱存在下 HRh(CO)(TPPTS)₃ 的 1-己烯氢甲酰化催化活性结果

Tab 1  1-Hexene hydroformylation catalyzed by HRh(CO)(TPPTS)₃ in the presence of base or acid

| Catalyst system | Conversion /% | Selectivity Aldehyde,% | n/i | TOF (h⁻¹) | Color of Product |
|---|---|---|---|---|---|
| HRh(CO)(TPPTS)₃,pH=8.0 | 55.8 | 100 | 1.32 | 1.53 | Clear |
| HRh(CO)(TPPTS)₃+NaOH(1/1 mol/mol),pH=11.9 | 54.5 | 100 | 1.42 | 1.51 | Clear |
| HRh(CO)(TPPTS)₃+ Pyridine(1/1 mol/mol),pH=8.1 | 49.6 | 99 | 1.56 | 1.34 | Clear |
| HRh(CO)(TPPTS)₃+ H₂SO₄(1/1 mol/mol),pH=2.3 | 65.1 | 97 | 1.06 | 1.75 | Yellow |
| HRh(CO)(TPPTS)₃+ CH₃COOH(1/1 mol/mol),pH=3.3 | 54.8 | 98 | 1.36 | 1.52 | Clear |

Reaction conditions:CO/H₂(1/1,V/V)=3.0 MPa,T=363 K,reaction time=4 h;

Catalyst phase:HRh(CO)(TPPTS)₃=0.003 2 mmol in 10 mL H₂O;

Organic phase:1-hexene/heptane (1/4,wt/wt) 10 mL;

TOF:aldehydes (mol)/[Rh](mol)·h.

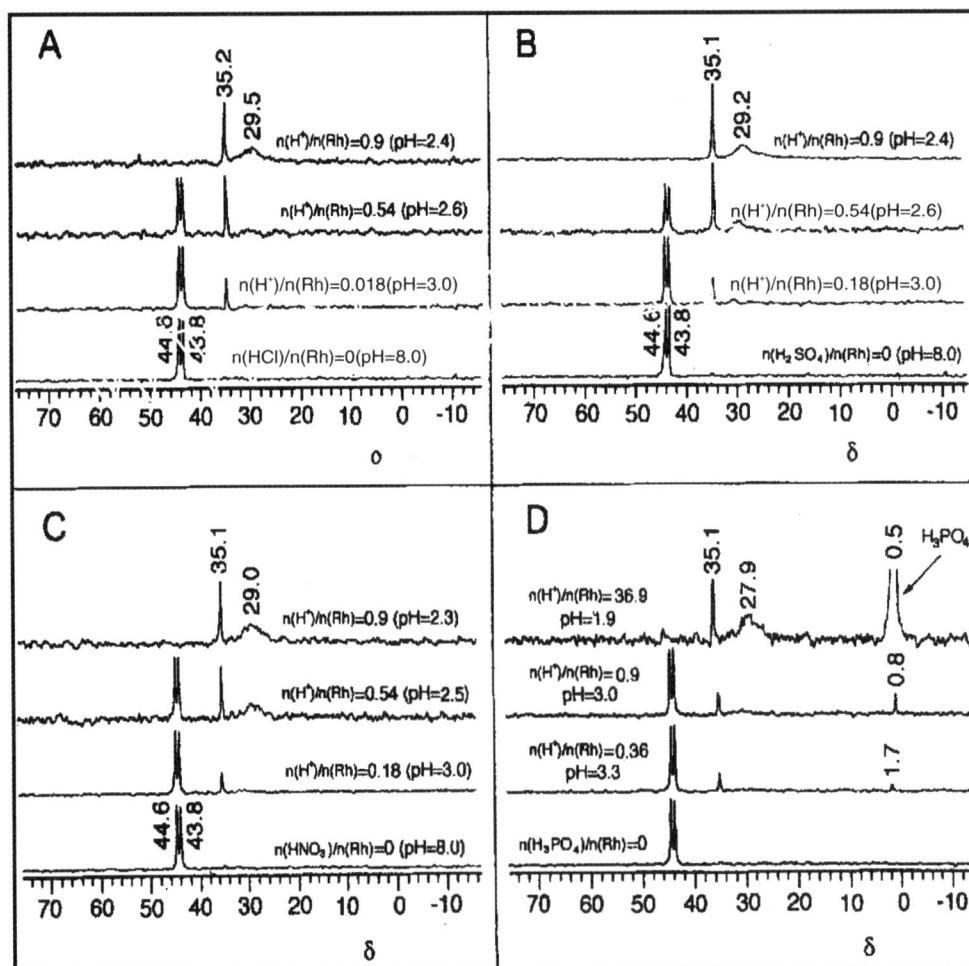

图3 若干无机酸存在下 HRh(CO)(TPPTS)₃ 的³¹P(¹H)核磁谱图

Fig. 3  ³¹P(¹H)NMR spectra for HRh(CO)(TPPTS)₃ in the presence of several kinds of inorganic acids

配体氧化物的谱峰强度有渐强趋势；当加入 $CuSO_4$ 后，配合物的 Rh-H 质子峰强度弱化明显，进而消失；与此相对应，原配合物的特征磷谱峰强度减弱，新生成的磷物种谱峰逐渐成为磷谱的主要物种。而当加入 $Fe_2(SO_4)_3$ 和 $Cr_2(SO_4)_3$ 后，三价金属离子的强顺磁性使 NMR 灵敏度下降，谱峰宽化，该 2 种盐均易与水溶性铑膦配合物产生强烈的相互作用，易使配合物特征谱峰消失(图 4)。实验结果表明，上述金属盐对配合物结构破坏性大小的顺序为：$Fe_2(SO_4)_3 > Cr_2(SO_4)_3 > CuSO_4 \gg NiSO_4 \sim NaCl$ 在 TPPTS/Rh = 30 (摩尔比)和表面活性剂 CTAB 存在的情况下，水溶性铑膦配合物如 $RhCl(CO)(TPPTS)_2$ 催化 1-己烯氢甲酰化反应的结果也表明了不同价态金属离子有不同程度的影响[39]。总体来看，加入金属离子均使反应活性呈下降趋势，三价金属离子的影响程度最大，二价次之，一价最小。

由研究 RCH/RP 氢甲酰化过程得到的结果，证实了水溶性铑膦络合物催化剂与烃溶性的铑膦催化剂有类似的失活机理[1]。普遍认为，氢甲酰化活性的油溶性铑膦络合物催化剂和 RCH/RP 水溶性铑膦络合物催化剂均遵循配体降级的催化机理。业已由若干位研究者阐明，铑膦络合物不发生邻位环金属化反应[40-42]。如图 5 所示，铑膦络合物的失活起源于铑原子对三苯基膦 P-C 键的氧化加成，所得芳基-铑物种紧接着进行的一系列反应。在油溶性铑膦络合物体系中，反应产物中分离出如 $C_3H_7PPh_2$ 和磷酸。对于水溶性铑膦配合物，由于磺酸根的存在，可不必考虑邻位环金属化，而直接认为铑原子对 TPPTS 中 P-C 的氧化加成反应，最后得到了间-苯甲醛磺酸钠盐 $m$-OHCC$_6$H$_4$SO$_3$Na。这一结果表明铑原子进攻了与磷原子键合的碳原子，随后进行 CO 插入、还原消除反应得到了苯甲醛磺酸钠。

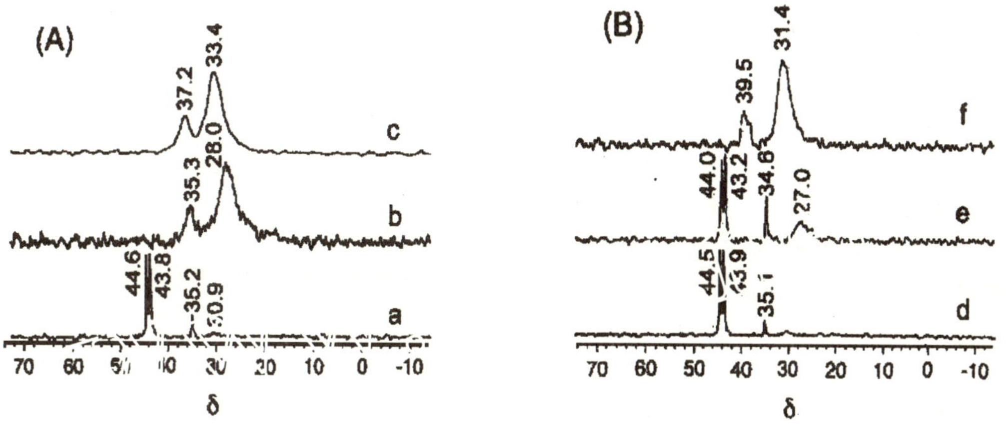

图 4  $Fe_2(SO_4)_3$(A)和 $Cr_2(SO_4)_3$(B)存在下 HRh(CO)(TPPTS)$_3$的$^{31}$P($^1$H)核磁谱图

$n_{[Fe^{3+}]}/n_{[Rh]}$：a. 0.2；b. 1.2；c. 4.5；$n_{[Cr^{3+}]}/n_{[Rh]}$：d = 0.2；e. 2.4；f. 8.4

Fig. 4  $^{31}$P($^1$H)NMR spectra of HRh(CO)(TPPTS)$_3$ in the presence of $Fe_2(SO_4)_3$(A)and $Cr_2(SO_4)_3$(B)

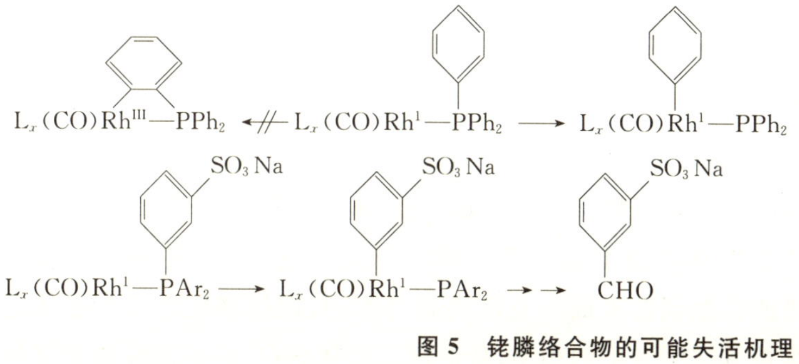

图 5  铑膦络合物的可能失活机理

Fig. 5  The possible deactivation mechanism of water-soluble phosphine rhodium complexes

# 3　SAPC 催化剂的结构与性能关联及其催化剂失活机理

Davis 等和我们的研究结果均指出[4,43,44]，SAPC 催化剂集中了均相和多相催化剂的优点，并克服了均相催化剂的缺点，适于更广泛碳数烯烃的氢甲酰化反应。近年对 SAPC 催化剂的研究侧重于催化剂结构与性能的关联，以及催化剂的失活机理研究[36,37,43,44]。

图 6 为 TPPTS 及配合物 HRh(CO)(TPPTS)$_3$ 分别负载到 SiO$_2$ 载体表面后的 $^{31}$P($^1$H) NMR 谱，可以看出，TPPTS 负载到 SiO$_2$ 载体表面以后，其特征 $^{31}$P($^1$H) NMR 谱峰的化学位移并没有发生明显的变化，仍与在 D$_2$O 溶液中时的-5.1 基本一致。图 6 中在 35.2 处的核磁共振峰可归属为 OTPPTS 之 P 原子的共振信号。其化学位移也与其在相应溶液中的位置基本一致当采用等容浸渍法将配合物 HRh(CO)(TPPTS)$_3$ 的重水溶液负载到 SiO$_2$ 载体表面后，其 $^{31}$P($^1$H) NMR 谱图发生了明显的变化（见图 6c）。原配合物中 P 原子在 44.4 处的特征双峰已完全消失，而在化学位移为 29.5 处出现一个新的宽峰，相应膦物种含 P 量约占体系总 P 量的 60% 摩尔分数左右。这是配合物在 SiO$_2$ 载体表面新形成的一种含膦配合物物种，相应之 TPPTS 与 Rh 的比例约为 2∶1（摩尔比），且由于载体表面的非均一性，使该物种中 P 原子所处的化学环境各向异性较大，谱峰较宽在 35.2 处属 OTPPTS 物种之 P 原子的 NMR 谱峰化学位移不变，但其在总 P 量中所占比例大为增加，从负载前的 10% 摩尔分数左右增加到负载后的 30% 摩尔分数左右。实验结果显示，对这一负载型催化体系而言，在 4 h 丙烯氢甲酰化反应之后，催化剂表面各磷物种谱峰的化学位移及相对强度基本不变（图 6d）[37]。

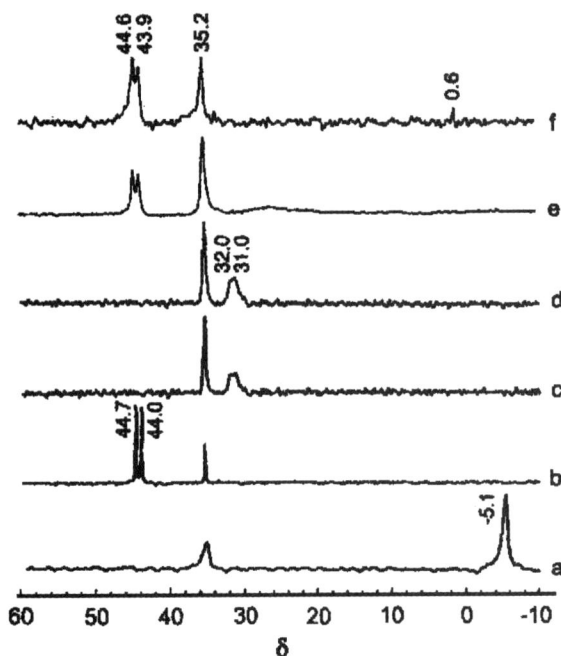

图 6　SiO$_2$ 负载 HRh(CO)(TPPTS)$_3$ 的 $^{31}$P($^1$H)NMR 谱图

(a)TPPTS on SiO$_2$；(b)HRh(CO)(TPPTS)$_3$ in D$_2$O；

(c)HRh(CO)(TPPTS)$_3$ on SiO$_2$；

(d)HRh(CO)(TPPTS)$_3$ on SiO$_2$, after propene hydrofomylation for 4 h；

(e)HRh(CO)(TPPTS)$_3$ on SiO$_2$ prempregnated with TPPTS；

(f)HRh(CO)(TPPTS)$_3$ on SiO$_2$ modified with Na$_2$CO$_3$

Fig. 6　$^{31}$P($^1$H)-NMR spectra for SiO$_2$-supported HRh(CO)(TPPTS)$_3$ catalyst

若先用 TPPTS 对 SiO$_2$ 载体表面进行修饰，而后再负载配合物，其 $^{31}$P($^1$H) NMR 谱（图 6e）与前者明显不同。这时，属原配合物 P 原子的特征 NMR 双峰仍保留一定强度，氧化配体 OTPPTS 之 P 原子的共振峰明显增强，后者之 P 原子含量约占此时体系总 P 量的 50% 摩尔分数。这说明，在 SiO$_2$ 载体表面先覆盖一层配体后，再后续负载的配合物所接触到的表面与直接负载时不同：即铑膦配合物与载体的相互作用有所减弱，配合物较大程度上可维持其在溶液中的构型，从而证实载体 SiO$_2$ 的表面性质是促使负载配合物结构发生变化的主要原因。

以 Na$_2$CO$_3$ 修饰的 SiO$_2$ 为载体，所制得负载配合物（即 HRh(CO)(TPPTS)$_3$/Na$_2$CO$_3$/SiO$_2$）的液

膜[31]P([1]H)NMR 谱示于图 6f。可见在 δ＝44.4 附近属原配合物的特征 NMR 共振双峰依然存在,只是比在溶液中时有所宽化(这同样可归因于载体表面的各向异性);从相应谱峰之相对面积同样可估算出,以原配合物形态存在的膦物种 P 原子含量约占体系总 P 量的 60% 摩尔分数,其余 P 原子主要以 OTPPTS 形式存在;此外,尚有很小一部分以吸附在载体表面属未配位的配体 TPPTS 形式存在,其化学位移在 δ＝0.6 附近。这一结果表明,在经 Na₂CO₃ 修饰的碱性 SiO₂ 载体表面,被负载的铑膦配合物与载体表面相互作用相当弱,基本上保留其在 D₂O 溶液中三膦配合物的构型特征。

SiO₂ 载体表面性质以亲水性的极性表面为主要特征,其表面具有强酸性。这可能是引起上述配合物负载以后,其结构发生变化的主因。为此,我们测定了在 SiO₂ 载体上吸附吡啶的程序升温脱附(Pyridine-TPD)谱,实验表明 SiO₂ 载体对吡啶有很强的吸附能力,在 420 K 附近出现一个较大的脱附峰,证实了 SiO₂ 载体表面的强酸性。当用弱碱性的 Na₂CO₃ 水溶液对 SiO₂ 载体表面进行处理,然后用水洗至中性,再经 773 K 下灼烧 2 h,而后再经水洗至中性。当以经 Na₂CO₃ 处理过的 SiO₂ 载体作吡啶-TPD 实验时,发现这时载体表面对吡啶的吸附能力已经大大降低与处理前相比,表面酸性几乎完全消失。

以上结果表明,配合物 HRh(CO)(TPPTS)₃ 可与载体 SiO₂ 的酸性表面发生较强的相互作用,导致配合物在表面上的结构与在溶液中时不同。进一步实验表明,只有那些与表面直接接触的配合物才发生这种结构上的变化。由于碱性的膦配体与酸性的载体表面强相互作用,在单层以下担载铑膦配合物时,含二个膦配体的铑膦配合物是该催化剂的主要活性物种。

用魔角旋转固体核磁共振磷谱([31]P([1]H)MASNMR)法表征 Rh(acac)(CO)₂ 和 TPPTS 在 SiO₂ 载体表面原位生成的配合物物种时,并没有观察到本应在 δ＝44.4 处出现的配合物 HRh(CO)(TPPTS)₃ 中配位膦物种的特征双峰,但在工作态催化剂中观察到可归属为表面配合物 {Rh(CO)(TPPTS)₂} 的[31]P 物种[36]。当 SiO₂ 担载-Rh(acac)(CO)₂ 制成的负载型水溶性催化剂进行 1-己烯氢甲酰化催化反应时,引入适量水蒸气可显著提高催化活性(表 2)。用[31]P([1]H)MAS NMR 谱表征得到,在新制备的催化剂中,吸附于 SiO₂ 表面但未参与配位的 TPPTS,约占总膦物种的 70 mol% 以上,而位于 δ＝32.4 处的表面配合物 {Rh(CO)(TPPTS)₂} 膦物种量约为 15 mol%,其他膦 10 mol% 左右催化剂经干燥合成气在 373 K 处理 2 h,或经湿合成气在较低温度(333 K)下处理 2 h 后,{Rh(CO)(TPPTS)₂} 的增加量仅约为 10～15 mol%,其他膦物种的变化量也较小;但催化剂经湿合成气于 373 K 处理 2 h 后,{Rh(CO)(TPPTS)₂} 的净增量大于 40 mol%;在工作态催化剂中,也观察到 {Rh(CO)(TPPTS)₂} 大量生成、未配位 TPPTS 量减小;经 43 h 反应运转后,催化剂活性下降,归属为 {Rh(CO)(TPPTS)₂} 的膦谱峰宽化,揭示有部份配合

**表 2  合成气中水蒸气量对 SAP 铑膦催化剂 1-己烯氢甲酰化活性的影响[a]**

**Tab 2  The influence of water vapor in syngas on hydroformylation activity of l-hexene on SAP Rh catalyst**

| Temperature of water container (K) | 296 | 313 | 323 | 333 | 343 | 323 |
|---|---|---|---|---|---|---|
| Water-feed in syngas (mmol/h) | 0.11[b] | 0.26[b] | 0.41[b] | 0.65[b] | 0.98[b] | 0.41[c] |
| Conversion of 1-hexene (%) | 75.6 | 84.1 | 87.9 | 91.3 | 91.9 | 79.9 |
| TOF of aldehyde (h$^{-1}$) | 244.8 | 272.3 | 284.6 | 295.6 | 297.6 | 258.7 |
| $n/i$ of aldehyde | 2.3 | 2.3 | 2.0 | 2.2 | 2.0 | 2.5 |

a)Reaction conditions:l-hexene/n-heptane=1/4($wt/wt$),feed-rate=20 mL/h;T=373 K;TPPTS/Rh=10 (mol//ol);CO/H₂=1/1 ($V/V$),4.0 MPa,GHSV=1 800 h$^{-1}$;data taken after 4 h reaction.

b)Pretreated under wet-syngas with the same amount of water-feed as in reaction. c)Pretreated under dry-syngas at room temperature

物解络、部分 TPPTS 被氧化成 OTTPTS。本研究结果证实,适量水可促进催化剂中具氢甲酰化催化活性的铑膦物种形成,提高活性;但随反应进行,配合物将逐渐解络、膦配体逐渐被氧化,从而使催化剂逐渐失活。

# 4　结语

两相催化体系的烯烃氢甲酰化经 20 多年的研究,已取代令人瞩目的进展,各种新型的水溶性配体,如手性双膦配体,离子、非离子及表面活性膦配体,以及其金属配合物的合成,为水溶性催化剂的选择、催化应用提供了广阔的空间,但在将两相体系推向工业化的进程中,尚有大量理论和实践上的问题有待解决。可以预料,水溶性配体和催化剂的合成、两相催化在各类反应中的应用、以及新型催化体系的设计仍将是两相催化研究的三个重点。

正如 1994 年召开的"水相有机金属化学和催化"北约远景研究专题研讨会中指出[10]:水作为催化反应中的一相,无论从工业应用还是从环境保护的角度来看,都是极富价值的。

# 参考文献

[1]Beller M,Cornils B,Frohning C D,et al. Progress in homogeneous and carbonylation[J]. J. Mol Catal A:Chemical,1995,**104**:17～85.

[2]Hjortkjaer J,ScurrelM S,Simonsen P. Heterogeneous hydroformylation catalysts produced by direct interaction between rhodium complexes and the support[J]. J. Mol. Catal.,1981,**10**:127～132.

[3]Kuntz G. Homogeneous catalysis in water[J]. *Chemtech*,1987,**15**:570～575.

[4]Arhancet J P,Davis M E,Merola J S,et al. Hydrofonnylation by supported aqueous-phase catalysis:a new class of heterogeneous catalysts[J]. Nature,1989,**339**:454～455.

[5]Pelt H L,Gijsmao P J,Verburg R P J,et al. The thermal and chemical stability limits of supported liquid phase rhodium catalysts in the hydroformylation of propene [J]. J. Mol. Catal. ,1985,**33**:119～128.

[6]Chatt J,Leigh G J,Slade R M. Rhodium(Ⅰ),rhodium(Ⅱ),palladium(Ⅰ),and platinum(Ⅱ) complexes Containing ligands of the type $PR_nQ_{3-n}$(n＝0,1 or 2;R＝Me,Et,Bu',or Ph;Q＝CH₃OCOMe or CH₂OH))[J]. J. Chem. Soc.,Dalton Trans,1973,2 021～2 028.

[7]Joo F,Toth Z,BeckM T. Homogeneous hydrogenations in aqueous solutions catalyzed by transition metal phosphine complexes[J]. Inorg Chem. Acta,1977,**25**:L 61－L 62.

[8]Cornils B,Wiebus E. Aqueous catalysts for organic reactions[J]. Chem tech,1995,**23**:33～38.

[9] Wiebus E,Cornils B. Water-soluble catalysts improve hydroformylation of olefins [J]. Hydrocarbon Processing,1996,63～66.

[10]Cornils B,Kuntz E G. Introducing TPPTS and related ligands for industrial biphasic processes [J]. J. Organomet Chem. ,1995,**502**:177～186.

[11]Cornils B,Hermann W F,Eckl R W. Industrial aspects of aqueous catalysis[J]. J. Mol. Catal A:Chemical,1997,**116**:27～33.

[12]Ahrland S,Chatt J,DaviesN R,et al. The Relative affinities of co-ordinating atoms for silver

ion. Part II nitrogen,phosphorus,and arsenic[J]. J. Chem. Soc,1958,276~279.

[13]Hemann W A,Albanese G P,Manetsberger R B,et al. New process for the sulfonation of phosphine ligands for catalysts[J]. Angew. Chem. Int Ed Engl,1995,**34**(7):811~813.

[14]陈华,黎耀忠,李东文,等.水溶性均相络合催化研究进展[J].化学世界,1998,**10**(2):146~156.

[15]郑晓来,王艳华,左焕培.水溶性膦配体的合成及进展[J].分子催化,1996,**10**(1):70~80.

[16]陈华,黎耀忠,陈骏如,等.水溶性膦配体的合成方法[J].分子催化,**13**(2):151~160.

[17]Herd O,Heβler A,Langhans K P,et al. Water soluble phosphine II[J]. J. Organomet Chem. ,1994,**475**:99~111.

[18]Bitterer F,Herd O,Hessler A,et al. Water-soluble phosphine 6. Tailor-made syntheses of chiral secondary and tertiary phosphines with sulfonated aromatic substituents:Structural and quantum chemical[J]. Inorg Chem. ,1996,**35**:4 103~4 113.

[19]Hermann W A,Kohlpaintner C W. Water-soluble ligands, metal complexes, and catalysts: synergism of homogeneous and heterogeneous catalysis [J]. Angew. Chem. Int Ed Engl,1993,**32**:1 524~1 544.

[20]郑晓来,蒋景阳,王兵,等.两相催化—均相催化多相化新进展[J].化学进展,1997,**9**(2):111~122.

[21]杨波,左焕培,金子林.水溶性膦铑络合物金属催化剂的进展[J].分子催化,1993,**7**(1):75~81.

[22]Larpent C,Dabard R,Patin H. Rhodium(Ⅰ) production during the oxidation by water of a hydrosoluble phosphine [J]. Inorg Chem. ,1987,**26**:2 922~2 924.

[23]Ding H,Hanson B E. Reaction activity and selectivity as a function of solution ionic strength in oct-1-ene hydrofomylation with sulfonated phosphines [J]. J. Chem. Soc,Chem. Commun,1994:2 747~2 748.

[24]DingH, Hanson B E, Glass T E. The effect of salt on selectivity in water soluble hydroformylation catalysts[J]. Inorg. Chem. Acta,1995,**229**:329~333.

[25]Ding H,Hanson B E. Spectator cations and catalysis with sulfonated phosphiines The role of cations in detemining reaction selectivity in the aqueous phase hydrofomylation of olefins[J]. J. Mol. Catal. A:Chemical,1995,**99**:131~137.

[26]陈华,黎耀忠,程溥明,等.水溶性铑-膦配合物催化 1-十二碳烯常压氢甲酰化反应研究[J].分子催化,1994,**8**:347~352.

[27]Arhancet J P,Davis M E,Merola J S,et al. Supported aqueous-phase catalysts [J]. J. Catal,1990,**121**:327~339.

[28]Horvath I T. Hydroformylation of olefins with the water soluble HRh(CO)[P(m-$C_6H_4SO_3Na$)$_3$]$_3$ in supported aqueous-phase. Is it really aqueous? [J]. Catal. Lett,1990,**6**:43~48.

[29]Davis M E. Supported Aqueous-Phase Catalysis[J]. Chem. tech,1992,**22**:498~502.

[30]袁友珠,杨意泉,张鸿斌,等.担载型水溶性膦铑配合物催化剂研究[J].高等学校化学学报,1993,**14**(6):863~865.

[31]袁友珠,陈鸿博,蔡启瑞.SiO₂负载的磺化三苯基膦铑配合物催化高碳烯氢甲酰化[J].应用化学,1993,**7**(6):384~390.

[32]Yuan Y Z,Xu J L,Zhang H B,et al. The beneficial effect of alkali metal chloride on supported

**1305**

aqueous-phase catalysts for hydrofomylation[J]. Catalysis Letters,1994,**29**:387~395.

[33]Bartik T,Bartik B,Hanson B E,et al. Comments on the synthesis of triphenylphosphine: reaction monitoring by NMR spectroscopy[J]. Inorg Chem. ,1992,**31**:2 667~2 672.

[34]张宇,袁友珠,廖新丽,等.酸碱存在下水溶性铑膦配合物的 NMR 表征及氢甲酰化性能研究[J].高等学校化学学报,1999,**20**(10):1 589~1 594.

[35]袁友珠,张宇,叶剑良等.高分辨 NMR 研究金属盐对水溶性铑膦配合物分子结构的影响[J].高等学校化学学报,1999,**20**(6):914~917.

[36]袁友珠,张宇,陈忠,等.负载型水溶性铑膦配合物催化剂的结构与性能[J].物理化学学报,1998,**14**(11):1 013~1 019.

[37]袁友珠,张宇,蔡阳,等.载体对负载型水溶性铑膦配合物结构和氢甲酰化性能的影响[J].分子催化,2000,**14**(1):20~24.

[38]黄裕林,黎耀忠,陈华,等.阴离子对 RhCl(CO)(TPPTS)$_2$ 催化 1-己烯氢甲酰化反应的影响[J].分子催化,1999,**13**(3):215~218.

[39]黎耀忠,黄裕林,陈华,等.金属离子对 RhCl(CO)(TPPTS)$_2$ 催化 1-己烯氢甲酰化反应的影响[J].分子催化,1999,**13**(3):212~214.

[40]Dubois R A,Garrou P E,Lavin K D,et al. Cobalt-mediated phosphorus-aryl bond cleavage during hydroformylation[J]. Organometallics,1984,**3**:649~650.

[41]Abatjoglou A G,Billig E,Bryant D R. Mechanism of rhodium-promoted triphenylphosphine reactions in hydroformylation processes[J]. Organometallics,1984,**3**:923~926.

[42]Deshpande R M,Divekar S S,Gholap R V,et. al. Deactivation of homogeneous HRh(CO)(PPh$_3$)$_3$ catalysts in hydroformylation of 1-hexene[J]. J. Mol. Catal. ,1991,**67**:333~338.

[43]袁友珠,张宇,杨意泉,等.负载型水溶性铑膦配合物催化剂上丙烯氢甲酰化制丁醛[J].厦门大学学报(自然科学版),1996,**35**(2):220~225.

[44]袁友珠,杨意泉,林国栋,等.负载型水溶性铑膦配合物催化剂上气、液态烯烃氢甲酰化[A];(殷元骐主编)羰基合成化学[C],北京:化学工业出版社,1996:64~91.

## Preparation,Structure and Performance of Water-Soluble Phosphine Rhodium Complexes

You-Zhu Yuan，Yi-Quan Yang，Guo-Dong Lin，Hong-Bin Zhang，Khi-Rui Tsai

(State Key Lab for Phys. Chem. of Solid Surf.，Dept. of Chem.，XiamenUniv.，Xiamen 361005，China)

**Abstract**　The development of biphasic-catalysis involving water-soluble phosphine rhodium complexes catalyzed olefin hydroformylation was discussed,as illustrated by the synthesis,structure stability and deactivation mechanism of water-soluble phosphine ligand coordinated rhodium complexes and supported aqueous-phase catalysts. With this discussion,it is shown that for the biphasic-catalysis system,the molecular design of novel ligand and water-soluble complexes associated with the exploration of new catalysis remains as key issues in the near future.

**Key words**　Biphasic catalysis　Water-soluble complex　Hydrofomylation　Supported aqueous-phase catalyst　Catalyst structure

■ **本文原载**:《厦门大学学报》(自然科学版)第 40 卷第 2 期(2001 年 3 月),第 387～397 页。

# 碳纳米管的催化合成、结构表征及应用研究*

张鸿斌① 林国栋 蔡启瑞

(厦门大学化学化工学院 固体表面物理化学国家重点实验室,福建厦门 361005)

**摘 要** 概述近 10 年来国内外在碳纳米管催化合成及其应用研究领域的发展动态,着重介绍本研究组在管径小而均匀碳纳米管的催化合成、结构性能表征、及其在作为新型催化剂载体材料及储氢等应用领域的研究进展

**关键词** 碳纳米管 多壁碳纳米管 催化合成 碳纳米管的表征 碳纳米管的应用 新型催化剂载体 氢的储存

**中图分类号** O 643 **文献标识码** A

碳纳米管(Carbon Nanotubes,简写为 CNTs,下同)是近年来引起高度兴趣的一类新奇纳米碳素材料。在结构上它与"碳富勒烯($C_{60}$等)"属同一类;它们都是单个碳原子在一定条件下聚集自然形成的。这些碳原子堆集结合时会组成各种几何图形。$C_{60}$是碳五元环和六元环混合排列而成的球面结构的碳分子,而典型的 CNTs 完全是由碳六元环组成的类似于石墨的平面、一片片按一定方式叠合而成的纳米级管状结构。迄今发现 CNTs 有单壁的(Single Walled)(简写为 SWCNTs,下同)和多壁的 Multiple Walled)(简写为 MWCNTs,下同)两大类:前者直径约一纳米,后者外径可达数十纳米。鉴于这类新奇管状纳米碳素材料在氢及其他某些气体的储存、复合材料、电子器件、场发射、催化剂载体材料与吸附分离介质等科技领域有许多现实或潜在的重要用途,自 1991 年 Iijma[1]在电弧放电法制备 $C_{60}$ 和其他富勒烯的碳沉积物中首次发现 CNTs(一种外径为 5.5 nm、内径 2.3 nm、仅仅由两层同轴类石墨园柱面叠合而成的多壁纳米管状结构)以来,有关 CNTs 的研究在国际上迅速形成热潮。大量实验和理论的研究工作致力于弄清它们的结构和电子性质以及它们的生成条件。

CNTs 可由不同方法制备[2,3]。电弧放电法制备的 CNTs 形直、壁薄(多壁甚至单壁的),但产率低,分离纯化比较困难。稍后发展由含碳底物催化分解长成 CNTs 的化学催化生长法[4-6],所制得 CNTs 管长可达数十甚至数百 μm,管壁较厚(多为多壁的),管径分布宽达 2～100 nm 范围,产率较高,产物分离纯化容易,但管径粗细较不均匀,管壁结构缺陷较多。近年来相继开发的还有激光烧蚀(蒸发)法和模板催化生长法。前者被用于制备 SWCNTs[7];后者曾用于阵列式 MWCNTs 的制备。Li et al[8]报道用负载于介孔硅胶中的 Fe 作模板催化剂,Ren et al[9]和 Li et al[10]分别用涂 Ni 的平面玻璃和负载于氧化铝阳极微孔内的 Co 作模板催化剂不同方法制备的 CNTs 产物,其形貌结构及物化性能可能有一定差别。新近的评论[11,12]认为,由于电弧放电法和激光烧蚀(蒸发)法所产生的往往是多种碳材料的混合物,为获得单一的 CNTs 产物,后续的纯化工序毕竟相当麻烦费事;从应用的角度(即适宜于大量制备,费用现实)考

* 收稿日期:2001 年 2 月 15 日。

基金项目:国家自然科学基金"八五"重大项目子课题(29392002)、面上项目(29773038 和 50072021),福建省自然科学基金重大项目(B9830001)和教育部科学技术重点项目(99069)。

① 作者简介:张鸿斌(1940— ),男,教授。

虑,催化生长法是最有希望的。

# 1 碳纳米管的催化合成

就催化生长法而言,早在 1889 年发表的一项专利就涉及利用含碳气体分子与某些灼热的金属表面发生相互作用长成碳纤维(也称碳须)[13],但是真正了解其结构却是在不久前先进电子显微镜技术问世之后。迄今已报道可用于催化长成 CNTs 和/或碳纳米纤维(Carbon Nanofibers,简写为 CNFs,下同)的金属主要有 Fe,Co,Ni,Pt,Ru,Cr,V,Mo 等以及它们的某些合金[4-6,14-17]。用作为催化剂的这些金属组分可以是块状的颗粒(典型粒度为 100 nm),也可以是负载于一定载体的颗粒(粒度为 10~50 nm)。重要的是,这些金属都能溶解碳和/或生成相应的金属碳化物。在 700~1 200 K 温度范围内,甲烷、一氧化碳、合成气($H_2$/CO)、乙炔、乙烯、和苯等均能被用以提供碳源。典型地,所长成 CNTs 一端呈半球面封口状,另一端(管基端)附着催化剂颗粒,利用某些溶剂容易将其溶解而除去使管基端得以开口。CNTs 的形态除一维须状外,还有双向生长的[18],螺旋状、麻花状、分枝状[17-20]、和环状的[21]。催化法制备 CNTs 的管径粗细及其分布依所用催化剂、碳源气体、以及合成反应条件(温度、原料气流速等)而异;业已见诸之文献报道所达到的管径分布往往较宽。如何优化选择各种制备参数,以实现对 CNTs 的形貌结构、管径大小及其均匀程度、晶体化程度、比表面积、电导率等物化性能的选择控制,有许多科学和技术上的问题有待解决;研究活动十分活跃。

一种可用于 873~973 K 温度下催化 $CH_4$ 或 CO 分解长成 CNTs 的 Ni-Mg-O 催化剂及合成管径小而均匀碳纳米管的相应技术在本实验室已研发成功[22-25]。所制得 CNTs 属多壁的一类,外径在 15~30 nm 范围,内径 3 nm,长度可达微米甚至毫米量级。最初的合成反应在固定床管状反应器(φ6 mm)进行,单程产率约 100 mg;后经放大,在 φ60 mm 固定床反应器单程产率可达 50 g 粗产物。纯化后产物中,约 90%碳具有纳米管状结构,非碳素杂质残留量<1 wt%。图 1 示出以 $CH_4$ 或 CO 为碳源分别制得的两类 CNTs 的 TEM 图。

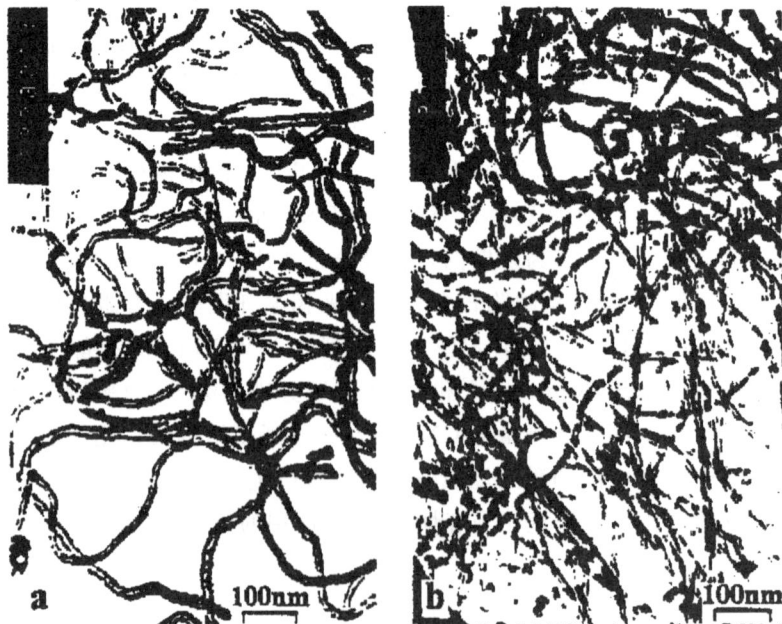

图 1 $CH_4$(a)和 CO(b)在 Ni-Mg-O 催化剂上分解长成 CNTs 的 TEM 图

Fig. 1 TEM images of the CNTs prepared by the Ni-Mg-O catalyzed decomposition of:$CH_4$(a)and CO(b)

XRD 测量和脉冲反应试验结果表明,所制得的氧化前驱态 Ni-Mg-O 催化剂由于其 NiO 和 MgO 两组分之间高度互溶性,生成一种 $Ni_xMg_{1-x}O$ 固体溶液。该固溶体中 $Ni^{2+}$ 物种高的分散度以及 MgO 晶体场的价态稳定化效应,使得仅约 15 mol% 的 $Ni^{2+}$ 可被还原,而大部分 $Ni^{2+}$ 仍存在于固溶体相中;这有利于还原后的催化剂表面 Ni 金属颗粒维持在纳米级尺寸,不易集结生成大尺寸金属 Ni 颗粒,遂使该催化剂上长成的 CNTs 管径较小且粗细比较均匀。

除催化剂外,反应条件对 CNTs 产物的形貌结构也有显著影响。考查结果表明,适当高的反应温度(不高于 973 K)和大的原料气空速,有利于制得管径小、管壁薄的 CNTs 产物。在几种 Ni 基催化剂之中,Ni-Mg-O 催化剂上长成的 CNTs 产物管径较小且较均匀,中空结构明显;而 Ni-Ca-O 催化剂上长成的产物则主要是断面直径较粗的碳纤维(见图 2)。

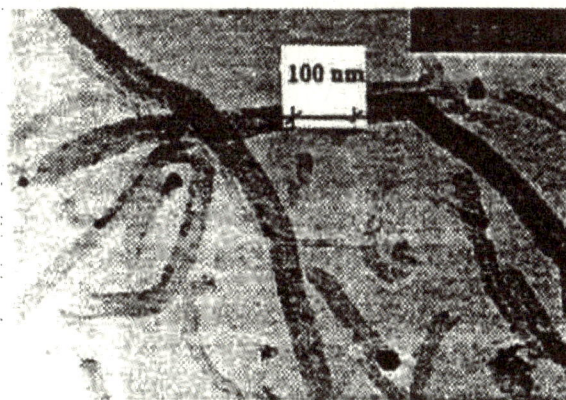

图 2　$CH_4$ 在 Ni-Ca-O 催化剂上分解长成 CNFs 的 TEM 图

Fig. 2　TEM image of CNFs grown by catalytic decomposition of $CH_4$ on a Ni-Ca-O catalyst

## 2　碳纳米管的催化生长机理

一般认为 CNTs 的催化生长过程包括以下几个步骤:首先,反应气体分子在催化剂金属颗粒一定晶面上吸附分解生成碳物种;接着,碳物种溶解入金属颗粒体相并经扩散迁移至金属颗粒背面(即金属-碳纳米管生长界面),进而长成 CNTs 促使初级反应生成的表面碳物种由气-固反应界面向 CNTs 生长界面扩散的推动力被认为是来自两界面之间溶解碳的浓度差,并认为碳原子在催化剂金属颗粒中的扩散是 CNTs 生长过程的速率控制步骤。以上模型可从以 $C_2H_2$ 为原料时碳纤维的生长速率与碳在相应催化剂金属颗粒中扩散的活化能存在相关性获得支持。它可对线形管状碳纤维的生长提供合理的解释,但看来仍存在不足;例如,按此模型则可预期,CNTs 的生长速率和形貌结构应主要依赖于所用的催化剂金属,而与原料气无关,但本研究组的实验结果并非如此。跟踪考察 Ni-Mg-O 催化剂上 CNTs 的生长过程发现,对于特定催化剂,CNTs 生长过程的速率控制步骤与其催化生长条件(反应温度、原料气组成及空速)也密切相关。在我们看来,原料气的分解(也即碳物种的生成)速率与所产生碳物种进一步扩散传输速率的匹配好坏对 CNTs 的生长影响显著。为获得高结晶度的 CNTs 产物并避免因积碳而致使催化剂失活,原料气 $CH_4$ 或 CO 的分解速率须保持不超过碳原子在催化剂金属颗粒中的扩散速率。原料气的组成不同,所制得 CNTs 的织构和结构以及管壁中类石墨平面相对于管中心轴线的取向可能有别。在 CNTs 生长过程中,还可能伴随发生催化剂颗粒一定程度的碎裂,这在以 CO 作为原料气时较为明显。甲烷分解的产物之一,氢,在影响管壁中类石墨平面相对于管轴的取向方面,可能起着不可忽视的作用。

# 3　碳纳米管结构和物化性能的表征

迄今有关 CNTs 结构的表征研究,主要地是利用 TEM、HRTEM、SEM 等技术对其结构/织构进行观测;利用多种谱学技术(如 XRD、XPS、Raman、TPO、TPD、TPR、以及 BET 等)从多种不同角度对 CNTs 产物进行协同表征仍鲜见报道。在本研究组,由于初步实现高纯度碳纳米管的可重复合成,解决了进行谱学观测所需批量试样的来源问题,遂使开展多种谱学的协同表征研究成为可能。下面简要介绍我们这方面工作的主要结果。

XRD 测试结果[27]表明,以 $CH_4$ 或 CO 为碳源制得的两类 CNTs 均具有类似石墨的体相结构,但 XRD 特征峰稍宽,表明这些纳米结构的长程有序度不如石墨的高(见图 3);较高的反应温度有利于产物体相石墨化程度的提高。XPS 测试结果[27]揭示,所制得两类 CNTs 产物的 C(1s)电子结合能为 284.0 eV,比石墨的(284.4 eV)和无定型碳的(284.6 eV)都低,暗示这些 CNTs 上的电子(尤其是价电子)比石墨或无定型碳的更容易离去。这两类 CNTs 的 Raman 光谱[28](见图 4)显示,它们的特征 Raman 谱峰稍有别于石墨而更接近于"低序碳"的,暗示它们的管壁外表面层碳原子排列的有序度较石墨的低;相对地说,CO-基 CNTs 管壁外表面碳原子排列的有序度比 $CH_4$-基的稍为高。

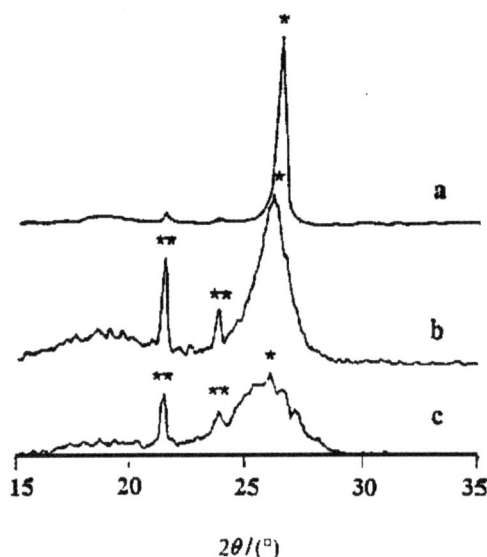

图 3　石墨(a),$CH_4$-基 CNTs(b)和 CO-基 CNTs(c),的 XRD 谱

**Fig. 3　XRD patterns of: graphite (a); $CH_4$-derived CNTs(b);CO-derived CNTs(c)**

* peaks due to graphite phase;

* * peaks due to chaoite phase

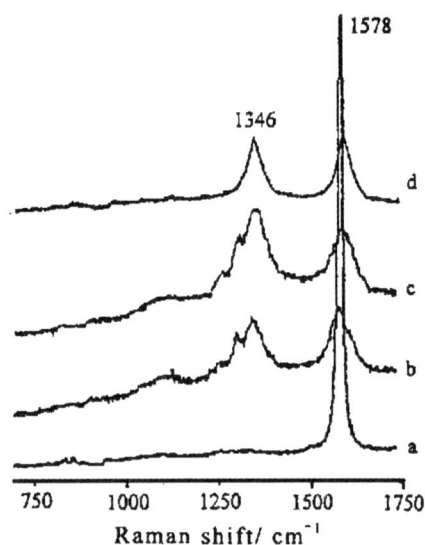

图 4　石墨(a),$CH_4$-基 CNTs(b),CO-基 CNTs(c)和低序碳(d)的喇曼光谱

**Fig. 4　Raman spectra of: graphite (a), $CH_4$-derived CNTs(b),CO-derived CNTs(c), and a low-ordering carbon(d)**

高分辨电镜(HRTEM)观测[27,28]揭示,$CH_4$-基 CNTs 系由一片片具有石墨片状结构的缺顶圆锥形面沿管中心轴线叠合而成,石墨层面取向与管轴倾斜(谓之"鱼骨型");而 CO-基 CNTs 系由一层层具有石墨片状结构的圆柱形面围绕管轴叠合而成,石墨层面取向与管轴平行(谓之"平行型")(见图 5)。在同一种 Ni 催化剂上所制得两类 CNTs 管壁类石墨层面相对于管中心轴线的取向存在明显差异的原因可能与两种原料气分子催化分解方式不同及其副产物($H_2$ 或 $CO_2$)不同有关。不同的气相副产物可在催化剂上吸附,从而影响催化剂金属颗粒表面 CNTs 生长区的微环境及 CNTs 管基界面区碳原子的聚集排列堆

砌方式。

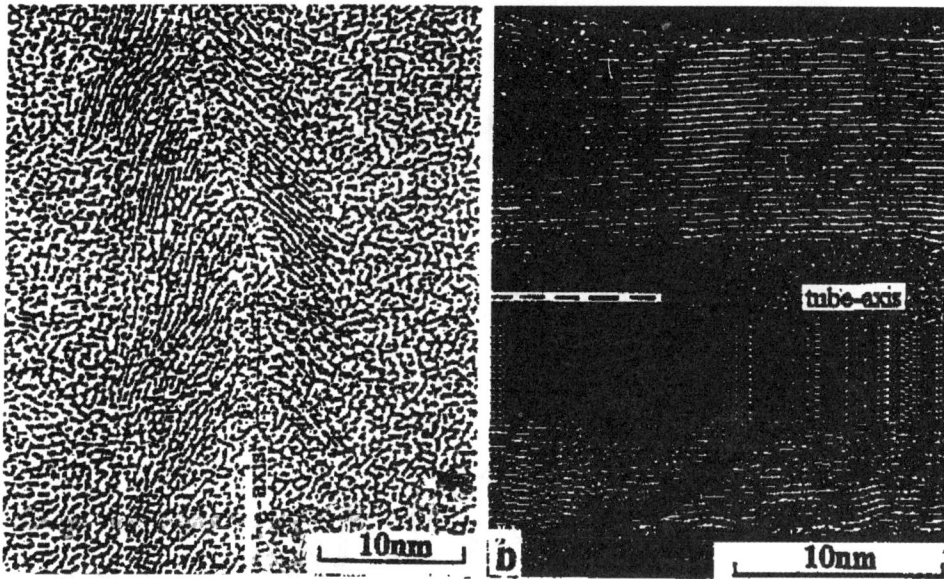

图 5　CH₄(a)和 CO(b)在 Ni-Mg-O 催化剂上分解长成 CNT 的 HRTEM 图

Fig. 5　HRTEM images of one of the CNTs produced from catalytic decomposition of：CH₄(a)and CO(b)

O₂-TPO(程序升温氧化)测试可以提供有关 CNTs 中含碳物种的种类、碳原子排列堆砌的规整度、以及整个管状结构抗氧化性能的信息。图 6 示出两类 CNTs 的 O₂-TPO 谱。CO-基 CNTs 的 O₂-TPO 谱在 690 K 处出现一弱的 CO₂ 响应峰,主峰在 960 K 处。前者系缘于少量无定形碳的氧化燃烧,小的峰面积表明其含量少;后者则归属于占主体量的石墨状碳的氧化燃烧 CH₄-基 CNTs 的 O₂-TPO 谱显示出相似的特点,但 690 K 处无定形碳燃烧峰不明显,暗示无定形碳量更少,石墨状碳含量更高;主峰的起始温度(880 K)比 CO-基的(780 K)约高 100 K,暗示所制得 CH₄-基 CNTs 比 CO-基的有较强的抗氧化能力。

为进一步考察 TPO 过程中 CNTs 表面碳的氧

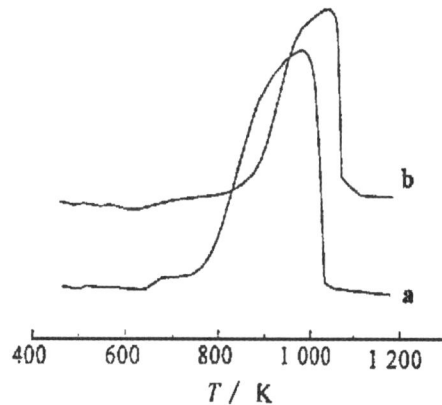

图 6　CO-基 CNTs(a)和 CH₄-基 CNTs(b)的 O₂-TPO 谱

Fig. 6　O₂-TPO spectra of：CO-derived CNTs(a)and CH₄-derived CNTs(b)

化燃烧状况,在另一组实验中,分别于 873 K 或 973 K 时中断 TPO 反应并急速冷却,收集相应试样作 TEM 观测。由图 7 可见,CH₄-基 CNTs 的氧化燃烧并非从管的两端开始,而是沿管壁由表及里地进行; CO-基 CNTs 的情形则有所不同,它的燃烧看来是先从管的两端开始。联系到这两类 CNTs 产物中类石墨"碳六元环"面相对于管中心轴线的不同取向,可以推测,"六元环"面内的碳原子比"六元环"周边碳原子来得稳定,氧化燃烧反应首先从较不稳定的"六元环"周边碳原子开始。

## 4　碳纳米管的应用研究

近 10 年来随着 CNTs 的发现及其制备技术的不断改进,CNTs 的应用研究也随之起步并形成热潮。

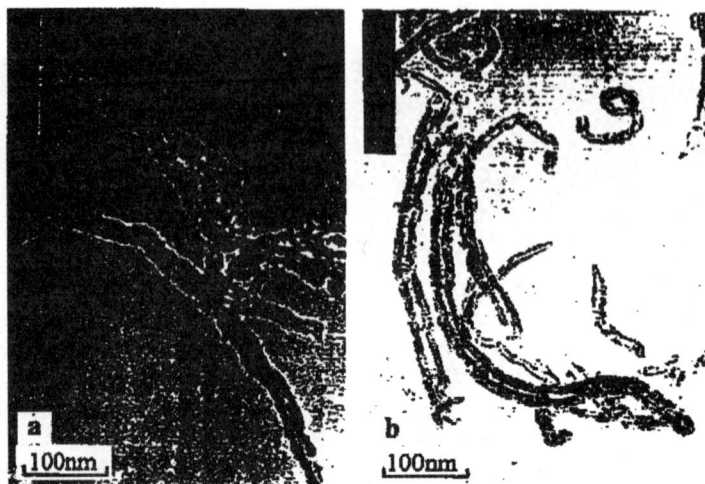

图 7 经受一定 $O_2$-TPO 试验之后 $CH_4$-基 CNTs(a)和 CO-derived CNTs(b)的 TEM 图

Fig. 7 TEM images of:$CH_4$-derived CNTs after undergoing a $O_2$-TPO test from R. T. to 973 K(a)and

CO-derived CNTs after undergoing a $O_2$-TPO test from R. T. to 873 K(b)

迄今已见诸于文献报道有关 CNTs 的应用研究主要有如下 4 个领域:气体(主要是 $H_2$)的储存、电子器件、作为聚合物添加剂、以及作为催化剂的载体材料。

CNTs 用于气体的储存近年来已受到科技界的极大关注。这从文献[29]称 CNTs 为"The World's Smallest Gas Cylinders"(世界上最小的气体钢瓶)可见一斑。该文将热均衡压缩的 Ar(923 K,170MPa)储存于 CNTs 中,观察到在冷却、减压之后 Ar 仍能保存在该"封闭"CNTs 中。从未来的应用角度着眼,氢的储存可能更加重要。Dillon 等[30]的初步研究似乎表明在近乎常温、低于大气压的压力下,在电弧放电法制备的 SWCNTs 上 $H_2$ 的吸附就可发生。Chambers 等[31]报道用催化生长法制备的石墨状 CNFs/CNTs 储氢,在常温、12MPa 压力下最大氢吸附容量达 20 NL($H_2$)/g-carbon,这与一项专利[32]提供的 CNFs 对氢的吸附容量,20 N mL($H_2$)/g,相比,提高 1 000 倍。稍后,Liu 等[33]报道他们由电弧法制备的 SWCNTs 在室温、10MPa 对氢的吸附容量为 4.2 wt%(相当于 0.47 NL($H_2$)/g-carbon);Chen 等[34]利用本研究组研发的技术[24]制备 MWCNTs,用 Li 对其进行修饰,后者在 473～673 K、大气压力下对氢的吸附容量据报可达 20 wt%(相当于 2.24NL($H_2$)/g-carbon)。不同研究者所采用的储氢材料及氢吸附实验条件不一,所报氢吸附容量数据差别悬殊,尤其是一些高水平的数据迄今未获重复;看来这方面的研究工作亟待深入。

本研究组新近利用 TPD 技术对 $H_2$ 在我们自行制备的 MWCNTs 及碱金属盐修饰的相应体系上的吸附-脱附行为进行了表征研究[35],结果显示,$H_2$ 在该种碳纳米管上的吸附在常温、常压下即可发生,碱金属盐的修饰使 CNTs 基质对 $H_2$ 的吸附容量获明显提高;被调查的 4 种体系对 $H_2$ 吸附容量的相对大小顺序为:$K^+$-CNTs>$Cs^+$-CNTs>$Na^+$-CNTs>CNTs;$K^+$-CNTs 上吸附氢的 $H_2$-脱附峰温度最高,达 373 K;$Na^+$-CNTs(353 K)居次;在 $Cs^+$-CNTs 上有 67% 吸附氢在室温下用常压 He 气流吹扫即行脱附为 $H_2$,另外 33% 吸附氢的 $H_2$-脱附温度为 345 K。接着进行的紫外 Raman 光谱研究[36]进一步表明,在常温常压下 $H_2$ 在 CNTs 上吸附可产生解离和非解离两大类氢吸附物种,$CHx(x=2,3)$(a)和 $H_2$(a),相应的 Raman 光谱特征峰出现在 2 875 和 2 987 $cm^{-1}$(分别属于表面 $CH_3$ 基的对称和不对称 C-H 伸缩振动),3 146 $cm^{-1}$(表面 $CH_2$ 基的不对称 C-H 伸缩振动),以及 3 923 $cm^{-1}$ 处(非解离吸附分子氢的 H-H 伸缩振动);TPD-MS 测试进一步揭示,在从室温至 973 K 的 TPD 过程中,并非所有氢吸附物种都以分子氢形式脱附,在 673 K 以上观测到烃分子(主要是 $CH_4$,$C_2H_2$)的脱附。

关于 CNTs 在电子器件方面的应用,迄今见诸于文献报道的典型实例有如:Niu 等[37]将其用于高能电化学电容器的研制,Che 等[38]研究了 CNTs/CNFs 在 Li-离子电池和燃料电池电极方面的应用;此外,还有将其用于制备高表面积电极,以及作为抗电磁辐射的防护材料。

CNTs 兼具独特的机械强度和电子学性质,在用作为高分子材料的添加剂以增强复合材料的机械强度和/或提高材料的导电率方面,具有诱人的应用前景。文献[12]报道,用 CNTs 或 CNFs 代替常规碳纤维作为尼龙或聚酯的添加剂以提高其电导率,其用量可大大下降——仅添加常规用量的 25%～30%就达到相同的添加效果。

CNTs 兼具独特的类石墨结构的管壁、纳米级的孔腔、大的比表面、以及高的机械强度和电导率,比常规的一些催化剂载体($Al_2O_3$,$SiO_2$ 等)更具特点,也为一些新催化应用领域的开拓带来新的机遇;CNTs 的管腔是多种气体快速吸附/脱附的理想介质,辅以适当的表面处理可引入一定功能团,这将大大增加其在选择吸附/脱附-分离技术领域的应用机会。在过去 5 年中,有关用 CNTs 作为催化剂载体的研究文献上已有 2～3 篇报道。Planeiz 等[39]用电弧放电法制备的碳纳米管材料作为载体制备一种 Ru/CNTs 催化剂,其对肉桂醛加氢制肉桂醇的催化活性比常规的 Ru/C 催化剂高得多(达到 80%转化率和 92%选择性);他们将高的选择性归因于金属-载体的相互作用。Hoogenraad 等[40]用由合成气/Fe 体系催化生长的所谓"鱼骨型"CNFs 作载体制备负载型 Pd/CNFs 催化剂,用于催化硝基苯液相加氢制苯胺,结果表明"鱼骨型"CNFs 负载催化剂的活性明显高于活性炭或"平行型"CNFs 负载的同类催化剂。遗憾的是,上述工作均欠缺进一步的表征数据,以致尚难对其结果作出较详细而合理的解释。

1997 年本研究组用 $CH_4$ 或 CO 为碳源由催化法制备的两类 CNTs 作为载体,成功地设计研制出一种用于催化烯烃氢甲酰化制醛的 CNTs 负载 Rh-膦配合物催化剂[26,28]。从示于表 1 的丙烯氢甲酰化活性评价结果可见,在表 1 所示反应条件下,CNTs 负载铑膦催化剂上不仅丙烯转化率高(达 32%),产物丁醛的区位选择性(即正/异构丁醛的摩尔比,n/i)也很高(达 12～13),相应每个 Rh 活性位转化频率(TOF)为 $0.12$ $s^{-1}$,这些指标均明显地优于管端未经开口的 CNTs、以及 $SiO_2$、TDX-601、AC、和 GDX-102 等常规载体负载的同类催化剂的水平。催化剂的活性评价和谱学表征结果,以及同几种常规载体负载催化剂进行比较研究的结果,有力地支持如下观点,即:CNTs 内径约 3 nm 的管腔以及比面积大并富含类石墨碳六元环结构的疏水性表面在促进丙烯氢甲酰化转化活性,尤其是产物丁醛的区位选择性的提高起着重要作用;断面直径约 3 nm 的内管腔相当适宜于容纳 1.8 nm 大小的催化剂前驱物 $HRh(CO)(PPh_3)_3$ 及催化活性物种 $HRh(CO)(PPh_3)_2$,同时留下供反应分子扩散及反应的适度空间以利于维持足够高的反应活性,而严格的空间约束则有利于发挥其空间选择催化作用,对反应底物分子及相应反应中间态的空间取向产生强的导向,于是促进产物丁醛的正异构比(n/i)大为提高。以上工作作为 CNTs 作为可以起空间选择催化作用的新型催化剂载体材料,在国际上提供了第一个成功实例[28,41,42],并表明在某种情形下它们可以作为中大孔沸石分子筛的替代物。

表 1　碳纳米管负载铑膦络合物催化剂及若干相关体系上丙烯氢甲酰化反应活性评价结果*

Tab 1　Results of the activity assays for hydroformylation of propene over the Rh-catalysts supported by the carbon nanotubes and related systems*

| Carrier | Surf. area ($m^2 g^{-1}$) | Pore diameter (nm) | Rh-loading (mmol Rh g-carrier$^{-1}$) | P/Rh (molar ratio) | Convers. of $C_3H_6$(%) | TOF ($s^{-1}$) | STY (mmol h$^{-1}$ g-catal$^{-1}$) | Select. (n/i) |
|---|---|---|---|---|---|---|---|---|
| | 155 | 3.2～3.6 | 0.1 | 6 | 26.1 | 0.10 | 29.9 | 9.0 |
| CNTs (CH$_4$-based) | | | | 9 | 30.8 | 0.12 | 35.3 | 12.0 |
| | | | | 12 | 30.6 | 0.12 | 35.0 | 13.5 |
| | | | 0.15 | 12 | 43.8 | 0.11 | 50.2 | 16.2 |

续表

| Carrier | Surf. area $(m^2 g^{-1})$ | Pore diameter (nm) | Rh-loading (mmol Rh g-carrier$^{-1}$) | ·P/Rh (molar ratio) | Convers. of $C_3 H_6$(%) | TOF ($s^{-1}$) | STY (mmol h$^{-1}$ g-catal$^{-1}$) | Select. (n/i) |
|---|---|---|---|---|---|---|---|---|
| CNTs (CO-based) | 237 | 2.4~3.2 | 0.1 | 6 | 31.0 | 0.12 | 35.5 | 7.2 |
| | | | | 9 | 34.1 | 0.13 | 39.1 | 10.3 |
| | | | | 12 | 32.5 | 0.12 | 37.2 | 11.2 |
| | | | 0.15 | 12 | 48.4 | 0.12 | 55.4 | 15.2 |
| Ends$^-$ unopened CNTs($CH_4-$ based) | 115 | — | 0.1 | 6 | 17.2 | 0.06 | 19.7 | 6.0 |
| SiO$_2$ | 380 | 6~12 | 0.1 | 9 | 32.8 | 0.12 | 37.6 | 7.9 |
| | | | | 12 | 31.7 | 0.12 | 36.3 | 8.2 |
| TDX-601 | 1210 | 1.6~2.4 | 0.1 | 6 | 9.1 | 0.034 | 10.4 | 15.3 |
| AC | 592 | 1.4~2.0 | 0.1 | 6 | 5.3 | 0.02 | 6.1 | 2.5 |
| GDX-102 | 501 | 20~100 | 0.1 | 6 | 38.5 | 0.14 | 44.1 | 5.6 |

* Reaction conditions:393 K,1.0 MPa,Feedgas $C_3 H_6$/CO/$H_2$=1/1/1(v/v),GHSV=9 000 mL(STP)h$^{-1}$(g catal)$^{-1}$; the by-products formed from side-reactions of hydrogenation, propane and butanol, were below 1%(molar fraction)of the total carbon-based products. The data of pore-diameter of the support materials were obtained from $N_2$-BET measurements.

# 5  结束语

化学催化法制碳纳米管及其应用开发研究,是介于新型碳素纳米材料和化学催化之间快速进展着的交叉研究领域CNTs这类新型碳素纳米材料正处于商品化前夕,目前需解决的问题首先是具有一定形貌结构和性质规范(包括:管径大小及分布、管壁结构、比表面、导电率、产品纯度等)的碳纳米管的可重复合成,实现价格合理的批量生产和供应,以便有效地推动其应用开发研究。自1991年发现碳纳米管至今历时不到10年,国内外科技界对碳纳米管的理论认识已大为深化,在其制备技术及应用开发方面也有长足进展。可以预期,未来10年内在碳纳米管研发领域,将会有更多振奋人心的研究成果问世,并为碳纳米管的商品化、产业化创造市场条件。

## 参考文献

[1]Iijma S. Nature,1991,**354**:56.

[2]Dresselhaus M S,Dresselhaus G,and Eklund P C. Science of fullerenes and carbon nanotubes [M]. Academic Press,1996.

[3]Ebbesen T W. Carbon nanotubes:preparation and properties[M]. New York:CRC Press,1997.

[4]Baker R T K and Harris P S. In Chemistry and Physics of Carbon [M]. (Eds Walker Jr P L and

Thrower P A)New York：MarcelDekker,1978,**14**：83.

[5]Audier M,Oberlin A,Oberlin M,et al. Carbon,1981,**19**：217.

[6]Hernadi K,Fonseca A,and Nagy J B,et al. Catalytic synthesis of carbon nanotubes using zeolite support[J]. Zeolites,1996,**17**：416.

[7]Thess A. Crystalline ropes of metallic carbon nanotubes[J]. Science,1996,**273**：483.

[8]Li W Z,Xie S S,Qian L X,et al. Large-scale synthesis of aligned carbon nanotubes[J]. Science,1996,**274**：1 701.

[9]Ren Z F,Huang Z P,Xu J W,et al. Synthesis of large arrays of well-aligned carbon nanotubes on glass[J]. Science,1998,**282**：1 105.

[10]Li J,Moskovits M,Haslett T L. Nanoscale electroless metal deposition in aligned carbon[J]. Chem. Mater,1998,**10**：1 963.

[11]Ajiayan P M. Nanotubes from carbon[J]. Chem. Rev.,1999,**99**：1 787.

[12]De Jong K P,Geus J W. Carbon nanofibers：catalytic synthesis and applications[J]. Catal. Rev. Sci. Eng.,2000,**42**：481～510.

[13]Hughes T V,Chambers C R. U S Patent,1889,405～480.

[14]Oberlin A,Endo M,Koyama T. J. Cryst. Growth,1976,**32**：335.

[15]Dresselhaus M S,Dresselhaus G,Sugihara K,et al. Graphite fibers and filaments [M]. Springer Series in Materials Science 5,New York：Springer-Verlag,1988.

[16]Figueiredo J L,Bernardo C A,Baker R T K,et al.（eds）. Carbon Fibers,Filaments and Composites [M]. NATO ASI Series Dordrecht,The Netherlands：Kluwer Academic Publishers,1989,Vol. 177,405 & 562.

[17]Rodriguez N M. A review of catalytically grown carbon nanofibers[J]. J. Mater. Rev.,1993,**8**：3 233～3 250.

[18]Kim M S,Rodriguez N M,Baker R T K. The interaction of hydrocarbons with copper-nickel and nickel in the formation of carbon filaments[J]. J. Catal,1991,**131**：60.

[19]Kim M S,Rodriguez N M,Baker R T K. The role of interfacial phenomena in the structure of carbon deposits[J]. J. Catal.,1991,**134**：253.

[20]Motojima S,Kawaguchi M,Nozaki K,et al. Carbon,1991,**29**：379.

[21]Ahlskog M,Seynaeve E,Vullers R J M,et al. Ring formation from catalytically synthesized carbon nanotubes[J]. Chem. Phys. Lett.,1999,**300**：202-206.

[22]陈萍,王培峰,林国栋,张鸿斌,蔡启瑞.低温催化裂解烷烃法制备碳纳米管[J].高等学校化学学报,1995,**16**：1 783～1 784.

[23]陈萍,林国栋,张鸿斌,蔡启瑞,翟和生.低温催化裂解烷烃法制备碳纳米管[J].厦门大学学报（自然科学版）,1996,**35**：61～66.

[24]Chen P,Zhang H B,Lin G D,et al. Growth of carbon nanotubes by catalytic decomposition of $CH_4$ or CO on a Ni-MgO catalyst[J]. Carbon,1997,**35**：1 495～1 501.

[25]陈萍,张鸿斌,林国栋,蔡启瑞.过渡金属催化剂及用于制备均匀管径碳纳米管的方法[P]. 中国专利：ZL 96 1 10252.7.

[26]张宇,吴范昕,张鸿斌,林国栋,袁友珠,蔡启瑞.碳纳米管负载铑催化剂上丙烯氢甲酰化[J].物理化学学报,1997,**13**：1 057～1 060.

[27]陈萍,张鸿斌,林国栋,蔡启瑞. 催化裂解 $CH_4$ 或 CO 制碳纳米管结构性能的谱学表征[J]. 高等学校化学学报,1998,**19**:765~769.

[28]Zhang Yu,Zhang Hong-Bin,Lin Guo-Dong,et al. Preparation,characterization and catalytic hydroformylation properties of carbon nanotubes-supported Rh-phosphine catalyst[J]. Appl. Catal. A:General,1999,**187**:213~224.

[29]Gadd G E,Blackford M,Moricca S,et al. The world's smallest gas cylinders[J]. Science, 1997,**277**:933.

[30]Dillon A C,Jones K M,Bekkedahl T A,et al. Storage of hydrogen in single-walled carbon nano-tubes [J]. Nature,1997,**386**:377.

[31]Chambers A,Park C,Baker R T K,et al. Hydrogen storage in graphite nanofibers[J]. J. Phys. Chem. B,1998,**102**:4 253.

[32]Rodriguez N M,Baker R T K. U S Patent:5 653 951,1997.

[33]Liu C,Fan Y Y,Liu M,et al. Hydrogen storage in single-walled carbon nanotubes at room temperature[J]. Scinece,1999,**286**:1 127~1 129.

[34]Chen P,Wu X,Lin J,et al. High $H_2$ uptake by alkali-doped carbon nanotubes under ambient pressure and moderate temperatures[J]. Science,1999,**285**:91~93.

[35]林国栋,周振华,董鑫,张鸿斌. $H_2$ 在碱金属修饰碳纳米管基材料上的吸-脱附特性[A]. 钟炳等:新世纪的催化科学与技术[M]. 太原:山西科学技术出版社,2000,955~956.

[36]周振华,林国栋,陈铜,等. 多壁碳纳米管及其对 $H_2$ 吸附体系的 Raman 光谱[J]. 厦门大学学报(自然科学版),2001,**40**(1):34~39.

[37]Niu C,Sichel E K,Hoch R,et al. High power electrochemical capacitors based on carbon nanotubes electrodes[J]. Appl. Phys. Lett.,1997,**70**:1 480.

[38]Che G,Lakshmi B B,Fisher E R,et al. Carbon nanotubule membranes for electrochemical energy storage and production [J]. Nature,1998,**393**:346.

[39]Planeix J M,Coustel N,Coq B,et al. Application of carbon nanotubes as supports in heterogeneous catalysis[J]. J. Am. Chem. Soc.,1994,**116**:7 935~7 936.

[40]Hoogenraad M S,Onwezen M F,van Dillen A J,et al. Supported catalysts based on carbon fibers[A]. Hightower J W,Delgass W N,Iglesia E,et al. (eds). Stud. Surf. Sci. Catal. [M]. Amsterdam:Elsevier,1996,**101**:1 331~1 339.

[41]Zhang H B,Zhang Y,Lin G D,et al. Carbon nanotubes-supported Rh-phosphine complex catalysts for propene hydroformylation[A]. Corma A,Melo F V,Mendioroz S,et al. (eds). Stud. Surf. Sci. Catal. [M]. Amsterdam,Elsevier,2000,**130**:3 885~3 890.

[42]张鸿斌,袁友珠,张宇,等. 碳纳米管基催化剂的空间选择催化作用[A]. 钟炳等:新世纪的催化科学与技术[M]. 山西:山西科学技术出版社,2000.425~426.

# Catalytic Synthesis and Applications of Carbon Nanotubes

Hong-Bin Zhang，Guo-Dong Lin，Khi-Rui Tsai

(Dept. of Chem. & State Key Lab of Phys. Chem. for the Solid Surfaces,

Xiamen Univ., Xiamen 361005，China)

**Abstract**   In this review, recent progresses in catalytic synthesis of carbon nanotubes(CNTs)and characterization of their structure and texture as well as the probable applications in the areas, such as hydrogen storage, electronic devices, polymer additive and catalyst support material, are summarized and briefly reviewed, with an emphasis on description of the research progress in our laboratory here.

**Key words**   Carbon nanotubes   Multi-walled carbon-nanotubes   Catalytic synthesis   Carbon-nanotubes characterization   Carbon-nanotubes application   Novel catalyst support   Hydrogen storage

■ **本文原载**:《高等学校化学学报》第 22 卷第 12 期(2001 年 12 月),第 1967～1970 页。

# 苹果酸钼外消旋体的合成、光谱性质和结构表征*

颜文斌[1]　　周朝晖[2]①　　章　慧[2]　　万惠霖[2]　　蔡启瑞[2]

(1.吉首大学化学系,吉首　416000;2.厦门大学化学系,厦门　361005)

**摘　要**　钼酸铵和外消旋苹果酸溶液反应得到外消旋苹果酸钼(Ⅵ)配合物,$(NH_4)_4[Mo_2O_2(S,S\text{-}Hmal)_2] \cdot [Mo_2O_2(R,R\text{-}Hmal)_2]$,对该配合物进行了元素分析、电导测定、旋光和红外光谱表征,并测定了晶体结构。该化合物晶体属单斜晶系,$P2_1/a$ 空间群,晶胞参数:$a=0.806\ 1(2)$ nm,$b=1.328\ 6(2)$ nm,$c=1.323\ 2(2)$ nm,$\beta=91.80(2)°$,$V=1.416\ 4(9)$ nm$^3$,$Z=2$,$D_c=2.008$ g·cm$^{-3}$,$F(000)=864$,$\mu=9.70$ cm$^{-1}$,一致性因子 $R=0.051$,$R_w=0.058$。在该单核配合物阴离子中,钼上的两个苹果酸配体具有相同手性,它以 $\alpha$ 烷氧基和 $\alpha$-羧基双齿配位形成畸变的八面体构型,而另一个 $\beta$-羧酸则保持自由状态。

**关键词**　苹果酸　苹果酸盐　钼(Ⅵ)配合物　固氮酶　晶体结构

**中图分类号**　O 614.61　**文献标识码**　A　**文章编号**　0251-0790(2001)12-1967-04

近年来,钼和钨的多羧酸配体的配合物研究比较活跃[1—5]。一方面,多羧酸配合物可作为催化剂等材料制备的前驱体[6,7],另一方面,钼和钨的多羧酸配合物与固氮酶铁钼辅基中高柠檬酸钼有着相类似的配位结构[8,9]。由于苹果酸与高柠檬酸的结构相似,两者同为生物体中存在的羟基多羧酸类配体,且具有光学活性,加之高柠檬酸的化学制备比较繁琐且售价昂贵[10—12],因此通过对苹果酸钼配合物的研究有望进一步了解高柠檬酸在固氮酶中的作用。有关苹果酸钼的溶液化学已有不少研究[13—15],现已分离并确定了晶体结构的苹果酸钼配合物有 $(NH_4)_4[Mo_4O_{11}(S\text{-}mal)_2] \cdot 6H_2O$[16—18],$(NH_4)_4 \cdot [Mo_4O_{11}(R\text{-}mal)_2] \cdot 6H_2O$[19] 和 $Cs_2[MoO_2(S\text{-}Hmal)_2] \cdot H_2O$[20],它们都是具有光学活性的配合物,配体中的苹果酸分别为 $R$ 或 $S$ 构型。本文在弱酸性条件下,利用钼酸铵与外消旋苹果酸($R,S\text{-}H_3mal$)反应,合成了外消旋苹果酸钼(Ⅵ)配合物 $(NH_4)_4[Mo_2O_2(S,S\text{-}Hmal)_2] \cdot [Mo_2O_2(R,R\text{-}Hmal)_2]$,确认配合物的阴离子中同时存在 $R$ 和 $S$ 构型苹果酸,并对该化合物进行了光谱表征及结构分析。

## 1　实验与晶体结构分析

### 1.1　仪器与试剂

$R,S$-苹果酸为南京国海公司生产,其他试剂均为国产分析纯试剂。

EA 1110 元素分析仪测定 C,H 和 N 的含量;DDS-11A 型电导率仪;以钠光灯为光源,采用 WZZ-1 自动指示旋光仪测定旋光度;Nicolet 740 FTIR 红外光谱仪,KBr 压片;Enraf-Nonius CAD-4 四圆衍

---

*　收稿日期:2000 年 11 月 28 日。
　　基金项目:国家自然科学基金(批准号:29933040)、高等学校骨干教师资助计划和博士学科点基金资助。
①　联系人简介:周朝晖(1964 年出生),男,教授,从事配位催化研究。　　E-mail:zhzhou@xmu.edu.cn。

射仪。

### 1.2 $(NH_4)_4[Mo_\Lambda aO_2(S,S\text{-}Hmal)_2]\cdot[Mo_\Delta O_2(R,R\text{-}Hmal)_2]$ 的合成

将等物质的量的钼酸铵(20 mmol)与 $R,S$-苹果酸(20 mmol)溶液混合,搅拌下逐滴加入氨水,控制溶液 pH=3,加热浓缩反应液体积至 20 mL 左右,然后将混合液于室温下缓慢挥发,几天后析出无色晶体,将晶体先后用 50%和 95%的乙醇洗涤,产量 24 g,产率 56%。元素分析实验值(%,$C_{16}H_{32}\cdot N_4O_{24}$ $Mo_2$ 计算值):C 22.2(22.4),H 3.7(3.8),N 6.6(6.5)。

### 1.3 $(NH_4)_4[Mo_\Lambda O_2(S,S\text{-}Hmal)_2]\cdot[Mo_\Delta O_2(R,R\text{-}Hmal)_2]$ 的光谱表征

配合物的红外光谱由 Nicolet 740 FTIR 光谱仪测定,数据归属如下,$\tilde{\nu}/cm^{-1}$:$\nu_{as}(C=O)$ 1 725, 1 620;$\nu_s(C=O)$ 1 380;$\nu(Mo=O)$ 920,885。旋光测定以钠光灯为光源,在 WZZ-1 自动指示旋光仪测定的旋光度为 0。

### 1.4 晶体结构分析

选取 0.08 mm×0.08 mm×0.23 mm 大小的晶体封于毛细管中用于衍射实验。使用 Enraf-Nonius CAD4 四圆衍射仪(Mo $K\alpha$,$\lambda=0.701\,7$ nm,石墨单色器),在 $2°\leqslant 2\theta\leqslant 52°$ 范围内收集 2 913 个独立反射点,其中 $I\geqslant 3\sigma(I)$ 的 1 902 个反射点用于结构解析和最小二乘修正。强度数据经 LP 因子和经验吸收校正。晶体属单斜晶系,$P2_1/a$ 空间群,晶胞参数:$a=0.806\,1(2)$ nm,$b=1.328\,6(2)$ nm,$c=1.323\,2(2)$ nm,$\beta=91.82(2)°$,$V=1.416\,4(9)$ nm³,$Z=2$,$D_c=2.008$ g·cm⁻³,$F(000)=864$,$\mu=9.700$ cm⁻¹。在 PC 微机上运行 WinGX 和 MoLEN 程序包,以直接法求解,从差分 Fourier 函数图求得全部非氢原子坐标和阴离子中的氢原子坐标,并采用各向异性热参数对非氢原子进行全矩阵最小二乘修正。最后偏离因子 $R=0.051$,$R_w=0.058$,最终差值电子密度图上最高峰为 $8.0\times10^{-4}$ e·nm³。

## 2 结果与讨论

### 2.1 配合物的合成

苹果酸和钼酸盐反应体系的产物的组成较复杂。Caldeira 等[13]利用 NMR 谱对苹果酸同钼酸根离子的反应体系进行了研究,推测 pH 在 4~7.5 范围内可能存在钼(VI)与苹果酸(根)比为 2∶1,1∶1 和 1∶2 的 3 个组分。Beltrán 等[14]利用电位分析,推测溶液在 pH=1~8 范围内存在 6 个不同物种。反应的最终产物与反应物的比例和溶液的 pH 值有关。

实验表明,将 $R,S$-苹果酸和钼酸盐按 1∶1 或 1∶2 比例混合,当 pH 值小于 3 时,析出的产物以 $(NH_4)_4[Mo_\Lambda O_2(S,S\text{-}Hmal)_2]\cdot[Mo_\Delta O_2(R,R\text{-}Hmal)_2]$ 为主。当溶液的 pH 值接近 6 时,析出的物种经红外光谱分析和旋光度测定,确定为先前报道过的 $(NH_4)_4[Mo_4O_{11}(mal)_2]\cdot6H_2O^{[16-19]}$,与钼酸铵和苹果酸按 2∶1 比例混合反应时析出的产物相同;而当溶液的 pH 值大于 6 时,产物为钼的同多酸盐。我们曾试图通过改变 $R,S$-苹果酸和钼酸铵的比例及控制温度和溶液的 pH 值,分离和表征钼(VI)和苹果酸比例为 1∶1 的配合物,但没有获得成功。

### 2.2 配合物的组成和表征

元素分析结果和结构表征数据表明,配合物的组成为 $(NH_4)_4[MoO_2(Hmal)_2]_2$。配合物在水溶液中的摩尔电导值为 $3.5\times10^2$ S·cm²·mol⁻¹,表明该配合物为 4∶2 型强电解质[21]。旋光度测定值为 0,说明是一外消旋的配合物。配合物的红外光谱数据表明,自由配体的羧基双键在 1 700 cm⁻¹处的伸缩振动吸收峰在配合物中发生了分裂,分别出现自由羧酸根 1 725 cm⁻¹ 的振动吸收峰和配位羧酸根的 1 620 cm⁻¹ 吸收,$\Delta(\nu_{as}-\nu_s)$ 差值为 240 cm⁻¹,可认定配位羧酸根为单齿配位[22],配合物在 920 和 885 nm

处呈现 Mo=O 特征的强振动吸收。

### 2.3 配合物的晶体结构

从溶液中分离出的固体经单晶衍射分析可知,产物中钼与苹果酸物质的量比为 1:2。图 1 为晶体中两种对映阴离子的结构,非氢原子坐标及热参数见表 1,表 2 为重要的键长和键角值。由于存在 $R$ 和 $S$ 苹果酸配位的阴离子,在晶体中,苹果酸钼(Ⅵ)酸铵由铵离子和两个单核苹果酸钼配阴离子的对映体组成。这两种对映阴离子的结构分别为 $[Mo_{\Delta}O_2(S,S\text{-Hmal})_2]^{2-}$ 和 $[Mo_{\Delta}O_2(R,R\text{-Hmal})_2]^{2-}$。

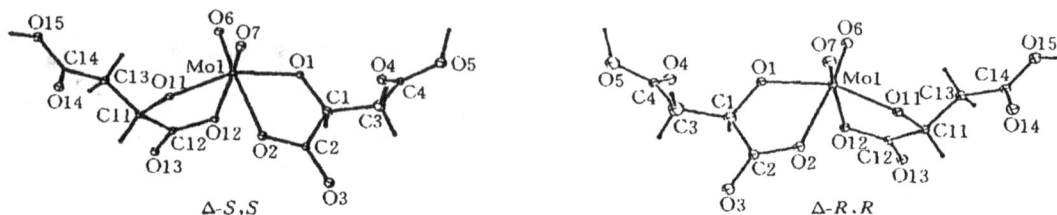

Fig. 1　Perspective view of the two enantio-anion structure for $(NH_4)_4[Mo_{\Delta}O_2(S,S\text{-Hmal})_2] \cdot [Mo_{\Delta}O_2(R,R\text{-Hmal})_2]$

Table 1　Atomic coordinates and thermal parameters($\times 10^4 \, nm^2$)

| Atom | $x$ | $y$ | $z$ | $U_eq$ | Atom | $x$ | $y$ | $z$ | $U_eq$ |
|---|---|---|---|---|---|---|---|---|---|
| Mo1 | 0.698 71(8) | 0.025 41(5) | 0.773 86(5) | 1.60(1) | O13 | 0.728 6(7) | −0.173 3(5) | 0.524 2(4) | 2.6(1) |
| N1 | 0.507 2(8) | −0.296 3(6) | 0.395 8(5) | 2.3(1) | O14 | 0.201 3(7) | −0.158 4(5) | 0.698 8(5) | 2.7(1) |
| N2 | 0.577 0(9) | 0.182 2(6) | 0.082 7(5) | 2.7(1) | O15 | 0.108 8(7) | −0.059 3(5) | 0.573 0(4) | 2.9(1) |
| O1 | 0.923 7(6) | 0.087 1(4) | 0.789 6(4) | 1.8(1) | C1 | 1.059(1) | 0.030 2(8) | 0.825 5(7) | 3.1(2) |
| O2 | 0.845 5(7) | −0.085 6(4) | 0.854 0(4) | 2.3(1) | C2 | 0.996(1) | −0.067 8(6) | 0.872 8(6) | 2.0(1) |
| O3 | 1.091 4(8) | −0.123 0(5) | 0.922 3(5) | 2.8(1) | C3 | 1.181(1) | 0.085 0(8) | 0.887 1(8) | 4.5(2) |
| O4 | 1.278 2(8) | 0.177 3(5) | 0.743 4(5) | 3.6(1) | C4 | 1.266(1) | 0.169 2(6) | 0.833 4(7) | 2.2(2) |
| O5 | 1.328 0(9) | 0.236 9(5) | 0.897 3(5) | 3.9(2) | C11 | 0.533 4(9) | −0.132 6(6) | 0.651 3(6) | 1.8(1) |
| O6 | 0.641 4(7) | 0.112 9(5) | 0.685 5(5) | 2.6(1) | C12 | 0.700 9(9) | −0.124 5(6) | 0.601 0(6) | 1.9(1) |
| O7 | 0.606 6(8) | 0.061 5(5) | 0.881 7(5) | 3.2(1) | C13 | 0.397 5(9) | −0.084 8(7) | 0.584 6(6) | 2.2(2) |
| O11 | 0.543 4(6) | −0.081 3(4) | 0.744 7(4) | 2.0(1) | C14 | 0.227(1) | −0.107 5(6) | 0.625 5(6) | 2.1(1) |
| O12 | 0.805 5(6) | −0.064 5(4) | 0.645 3(4) | 2.0(1) | | | | | |

Table 2　Selected bond lengths($\times 10 \, nm$) and bond angles($°$) of $(NH_4)_4[Mo_{\Delta}O_2(S,S\text{-Hmal})_2] \cdot [Mo_{\Delta}O_2(R,R\text{-Hmal})_2]$

| | | | | | | | |
|---|---|---|---|---|---|---|---|
| Mo1—O1 | 0.199 5(6) | Mo1—O2 | 0.215 0(6) | Mo1—O11 | 0.192 3(5) | Mo1—O12 | 0.227 0(5) |
| Mo1—O6 | 0.170 2(7) | Mo1—O7 | 0.169 8(6) | C1—O1 | 0.140(2) | C2—O2 | 0.126(1) |
| C2—O3 | 0.123(1) | C4—O4 | 0.120(2) | C4—C5 | 0.132(2) | C11—O11 | 0.141 1(9) |
| C12—O12 | 0.128 9(9) | C12—O13 | 0.123(1) | C14—O14 | 0.121(2) | C14—O15 | 0.133(1) |
| O1—Mo1—O2 | 75.3(2) | O1—Mo1—O6 | 91.0(2) | O1—Mo1—O7 | 102.5(3) | O1—Mo1—O11 | 155.2(2) |
| O1—Mo1—O12 | 85.9(3) | O2—Mo1—O6 | 160.2(2) | O2—Mo1—O7 | 91.6(3) | O2—Mo1—O11 | 86.4(2) |
| O2—Mo1—O12 | 78.0(3) | O6—Mo1—O7 | 105.6(3) | O6—Mo1—O11 | 101.8(3) | O6—Mo1—O12 | 86.9(2) |
| O7—Mo1—O11 | 94.5(3) | O7—Mo1—O12 | 164.6(3) | O11—Mo1—O12 | 73.8(3) | | |

在每个配阴离子中,苹果酸配体与钼形成畸变的八面体结构,中心钼(VI)和与之配位的两个苹果酸配体的手性构型分别为 $\Lambda$-$S,S$ 和 $\Delta$-$R,R$。配阴离子内的两个苹果酸以双齿形式同钼配位,每个苹果酸分别通过 $\alpha$-烷氧基和 $\alpha$-羧基同钼形成五元螯环,未配位的 $\beta$-羧基与另一分子的 $\alpha$-烷氧基的氧或 $\alpha$-羧基的氧形成分子间氢键,键长分别为 0.285 8(9) 和 0.265 4(8) nm。配阴离子中的 Mo—O 键键长有规律地变化:Mo=O 双键平均键长为 0.170 0(7) nm,钼与苹果酸 $\alpha$ 烷氧基配位键的平均键长为 0.195 9(6) nm,钼同 $\alpha$-羧基配位的平均键长为 0.221 0(6) nm,键长依次增大。与已报道的类似苹果酸钼配合物 $Cs_2[MoO_2(S\text{-}Hmal)_2] \cdot H_2O$ 相比[20],它们的结构相似,Mo—O 键的键长基本接近(见表3),配合物的两个五元螯环的 $\alpha$-烷氧基和 $\alpha$-羧基配位键键长之差 0.007 2 和 0.012 0 nm 略大,可能同外消旋苹果酸配体的不对称有关。

Table 3 Comparison of Mo=O and Mo—O length (nm) in the complexes of molybdenum(VI) and malic acid

| Complex | M—O(alkoxy) | Mo—O($\alpha$-carboxy) | Ref |
| --- | --- | --- | --- |
| $(NH_4)_4[Mo_4O_{11}(S\text{-}mal)_2] \cdot 6H_2O$ | 0.200 | 0.221 | 16 |
| $(NH_4)_4[Mo_4O_{11}(S\text{-}mal)_2] \cdot H_2O$ | 0.193(15) | 0.250(15) | 17,18 |
| $(NH_4)_4[Mo_4O_{11}(R\text{-}mal)_2] \cdot 6H_2O$ | 0.192 5(6) | 0.222 6(6) | 19 |
| $Cs_2[MoO_2(S\text{-}Hmal)_2] \cdot H_2O$ | 0.193 9(8) | 0.224 3(9) | 20 |
| $(NH_4)_4[MoO_2(R,S\text{-}Hmal)_2]_2$ | 0.199 5(6),0.192 3(5) | 0.215 0(6),0.227 0(5) | This work |

从配合物的结构图可以看出,分子中含有 3 个手性中心,可以形成 6 种非对映异构体。除了本文表征的 $(NH_4)_4[Mo_\Lambda O_2(S,S\text{-}Hmal)_2] \cdot [Mo_\Delta O_2(R,R\text{-}Hmal)_2]$ 复合物外,溶液中还可能存在下列 4 个物种: $(NH_4)_2[Mo_\Lambda O_2(R,S\text{-}Hmal)_2]$, $(NH_4)_2[Mo_\Delta O_2(R,S\text{-}Hmal)_2]$, $(NH_4)_2[Mo_\Lambda O_2(R,R\text{-}Hmal)_2]$ 和 $(NH_4)_2[Mo_\Delta O_2(S,S\text{-}Hmal)_2]$。实验析出产物为一对光学异构体混合形成的外消旋配合物,说明在本反应条件下没有自动拆分苹果酸对映体的能力。

# 参考文献

[1]Denadis K. D.,Malinak S. M.,Coucouvanis D.. Inorg. Chem. [J],1996,**35**:4 038~4 046.

[2]LI Dong-Mei(李冬梅),LI Ya-Feng(李亚丰),Xing Yong-Heng(邢永恒)et al. Chem. J. Chinese Universities(高等学校化学学报)[J],2000,**21**(10):1 464~1 467.

[3]XU Ji-Qing,ZHOU Xiao-Hua,ZHOU Li-Ming et al. Inorg Chim. Acta[J],1999,**285**:152~154.

[4]ZHOU Zhao-Hui,WAN Hui-Lin,TSAI Khi-Rui. J. Chem. Soc Dalton. Trans [J],1999:4 289~4 290.

[5]ZHOU Zhao-Hui,WAN Hui-Lin,TSAI Khi-Rui. Inorg. Chem. [J],2000,**39**:59~64.

[6]Yasuda H.,Higo M.,Yoshitomi S et al. Catal. Today[J],1997,**39**:77~87.

[7]ZHOU Zhao-Hui(周朝晖),MIAO Jian-Ying(缪建英),ZHANG Hui(章慧)et al. Chem. J. Chinese Universities(高等学校化学学报)[J],1999,**20**(8):1 168~1 171.

[8]Kim J.,Rees D. C. Biochemistry[J],1994,**33**(2):389~397.

[9]TSAI K. R.,WAN H. L.. J. Cluster Sci. [J],1995,**6**:485~501.

[10]LI Zhen-Chun,XU Ji-Qing Molecules[J],1998,**3**(2):31~34.

[11]LI Zhen-Chun,XU Ji-Qing Chem. Res Chinese Universities[J],1998,**14**(1):93～96.

[12]Rodriguez G. H.,Biellmann J. F.. J. Org Chem. [J],1996,**61**:1 822～1 824.

[13]Caldeira M. M.,Emilia M.,Saraiva T. L. et al. Inorg. Nucl. Chem. Lett. [J],1981,**17**:295～304.

[14]Beltrán A.,Avalos A. C,Beltrán J.. J. Inorg. Nucl. Chem. [J],1981,**43**:1 337～1 341.

[15]Cruywagen J. J.,Rohwer E. A.,van de Water R. F.. Polyhedron [J],1997,**16**:243～251.

[16]Porai-Koshits M. A.,Aslanov L. A.,Ivanova G. V. et al. Zh. Strukt. Khim. [J],1968,**9**:475～480.

[17]Berg J. E,Brandange S,Lindblom L. et al. Acta Chem. Scand A [J],1977,**31**:325～328.

[18]Berg J. E,Werner P. E.. Z Kristallogr. [J],1977,**145**:310～320.

[19]ZHOU Zhao-Hui,YAN Wen-Bin,WAN Hui-Lin et al. Chinese J. Struct. Chem. [J],1995,**14**:255～260.

[20]Knobler C. B.,Wilson A. J.,Hider R. N. et al. J. Chem. Soc. Dalton Trans [J],1983:1 299～1 303.

[21]GearyW. J.. Coord. Chem. Rev. [J],1971,**7**:81～122.

[22]Nakamoto Kazuo. Infrared and Raman Spectra of Inorganic and Coordination Compounds,Part B:Applications in Coordination,Organometallic and Bioinorganic Chemistry[M],New York:John Wiley & Sons Inc.,5th Ed,1997:59～69.

# Synthesis and Spectroscopic and Structural Characterization of Racemic Ammonium Dimeric(malato)dioxomolybdenum

Wen-Bin Yan[1] , Zhao-Hui Zhou[2] * , Hui Zhang[2] , Hui-Lin Wan[2] , Khi-Rui Tsai[2]

(1. *Department of Chemistry*, *Jishou University*, *Jishou* 416000, *China*;

2. *Department of Chemistry*, *Xiamen University*, *Xiamen* 361005, *China*)

**Abstract** The complex $(NH_4)_4[Mo_\Lambda O_2(S,S\text{-}Hmal)_2][Mo_\Delta O_2(R,R\text{-}Hmal)_2]$ was prepared from the reaction of ammonium molybdate and $R,S$-malic acid at pH $= 2-3$. It was characterized by elementary analysis,conductivity measurement,optical rotation,IR and X-ray diffraction. The complex crystallizes in monoclinic space group $P2_1/a$ with unit cell parameters:$a=0.806\ 1(2)$ nm,$b=1.328\ 6(2)$ nm,$c=1.323\ 2(2)$ nm,$\beta=91.82(2)°$,$V=1.416\ 4(9)$ nm$^3$,$Z=2$,$D_c=2.008$ g·cm$^{-3}$,$F(000)=864$,$\mu=9.700$ cm$^{-1}$ for 1 902 reflections with $I\geqslant3\sigma(I)$. The molybdenum atom is coordinated by two bidentate homochiralmalic acid in distorted octahedron,while the β-carboxyl groups remain uncomplexed. The crystal contains two isolated monomeric malatomolybdates (Ⅵ) anions in the configurations of $\Lambda$-$S,S$ and $\Delta$-$R,R$. The two enantiomeric anions result in the formation of racemic compound. There is no evidence for the other diastereoisomers like $(NH_4)_2[Mo_\Lambda O_2(R,S\text{-}Hmal)_2]$, $(NH_4)_2[Mo_\Lambda O_2(R,S\text{-}Hmal)_2]$,$(NH_4)_2[Mo_\Delta O_2(R,R\text{-}Hmal)_2]$ and $(NH_4)_2[Mo_\Delta O_2(S,S\text{-}Hmal)_2]$ in this experiment.

**Key words** Malic acid  Malate  Oxomolybdenum(Ⅵ)complex  Nitrogenase  Crystal structure

(Ed. :A,G)

■ 本文原载：Inorganic Chemistry Communications 5 （2002），pp. 388～390.

# A Hydrogen-Bonded Oxo-μ-Bis[ *Trans*-Nitrilotriacetato-Gs-Dioxotungstate( VI )] Synthesized Directly from an Alkali Tungstate and Nitrilotriacetic Acid

Zhao-Hui Zhou [a,①], Shu-Ya Hou [a], Zhi-Jie Ma [a],
Hui-Lin Wan [a], Khi-Rui Tsai [a], Seik Weng Ng [b]

( [a] *Department of Chemistry and State Key Laboratory for Physical Chemistry of Solid Surface, Xiamen University, Xiamen 361005, China*
[b] *Institute of Postgraduate Studies, University of Malaya, 50603 Kuala Lumpur, Malaysia* )
Received 20 December 2001; accepted 20 March 2002

**Abstract** Potassium tungstate( VI )reacts with nitrilotriacetic acid in water to form tripotassium hydrogen oxo-μ-bis [ *trans*-nitrilotriacetato-*cis*-dioxotungstate ( VI )] dihydrate. The octahedrally coordinated tungsten atom is $N, O, O$-chelated by the nitrilotriacetate entity, which is linked by its acid hydrogen [$O\cdots O=2.54(2)$ Å] to form a chain that propagates along a diagonal of the *b-c* face of the monoclinic unit cell. © 2002 Elsevier Science B. V. All rights reserved.
**Key words** Dioxotungstate( VI )  Tungsten  Nitrilotriacetic acid  Crystal structure

## 1　Introduction

The deprotonated trianion of nitrilotriacetic acid( $H_3$nta )forms water-soluble salts with a plethora of polyvalent metal ions for which the nitrilotriacetato ligand behaves as a tetradentate chelate that binds through the amino nitrogen and three carboxyl oxygen atoms[1-9]. Among these complexes, several oxomolybdates feature tridentate chelation instead of tetradentate chelation[10-12]. The corresponding oxotungstates have been investigated by largely by spectroscopic methods only[13-16], and to date, there is no crystallographic documentation of an $N, O, O$-chelated oxotungstate ( VI ). Molybdenum and tungsten oxotransferases are enzymes that catalyze oxygen atom transfer to and from a substrate. In order to understand the chemistry and oxo-transfer properties of these enzymes, numerous dioxo—Mo( VI )and W( VI )complexes having a wide range of ligands have been prepared and structurally characterized[17,18].

Curiously, a direct approach by merely reacting an alkali tungstate with nitrilotriacetic acid has not been reported in the literature. With the use of potassium tungstate, this afforded the title compound, whose crystal structure is detailed here.

---

①　Corresponding author. Tel. :+86-592-2184-531;fax:+86-592-218-3047.
*E-mail address*:zhzhou@xmu. edu. cn(Z. -H. Zhou).

## 2　Experimental

The compound was prepared by the reaction of equimolar quantities of potassium tungstate and nitrilotriacetic acid in about 50% yield. The reagents were heated in water until they dissolved completely; the compound was separated as colorless crystals, which were then collected and washed with ethanol. [①]

## 3　Results and discussion

The formation of nitrilotriacetate-tungstate complexes depends on the pH, cation and $W/H_3$nta molar ratio. Comparing with the pH dependence of citrato vanadates[19-25], at low pH and with potassium tungstate as the reagent, it is reasonable that the monoprotonated Hnta ion yields $K_2[W_2O_5(Hnta)_2]$ whereas at higher pH, $K_4[W_2O_5(nta)_2]$ is formed. With the present reaction, in the absence of other reagents, the resulting $K_3[W_2O_5H(nta)_2] \cdot 2H_2O$[②] represents a composition that is an intermediate between $K_2[W_2O_5(Hnta)_2]$ and $K_4[W_2O_5(nta)_2]$. Attempts to separate the other two products only result in the isolation of isopolytungstates. The title compound is the species that is stable within the pH 2.0—3.5 range. This implies the special stability of the suprastructure.

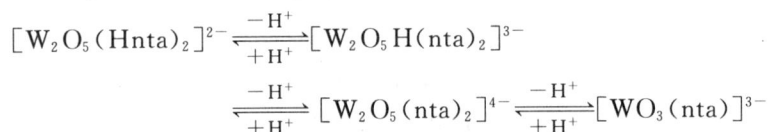

$$[W_2O_5(Hnta)_2]^{2-} \xrightleftharpoons[+H^+]{-H^+} [W_2O_5H(nta)_2]^{3-}$$
$$\xrightleftharpoons[+H^+]{-H^+} [W_2O_5(nta)_2]^{4-} \xrightleftharpoons[+H^+]{-H^+} [WO_3(nta)]^{3-}$$

In the infrared spectrum, the $v_{as}(CO_2)$ absorption bands appear as broad bands at approximately 1650 cm$^{-1}$. The absence of IR bands between 1740 and 1700 cm$^{-1}$ for a typical undissociated carboxylic acid group, indicating the obvious shift to lower frequency caused by very strong hydrogen bonding. In the region around 900 cm$^{-1}$, the complex shows several bands that might result from the presence of cis-dioxo tungstate.

The crystal structure comprises potassium cations, hydrogen oxo-$\mu$-bis[*trans*-nitrilotriacetato-*cis*-dioxo-tungstate] anions and lattice water molecules. The dimeric anion, which lies on the crystallographic twofold axis, consists of $WO_2$(Hnta) and $WO_2$(nta) units bridged by an oxygen linkage [W—O—W = 161.9(7)°]; the two oxo atoms are *cis* to each other in the octahedral geometry. The two cis-dioxo groups are in *trans*-configuration, which is in different configuration with that found in dipotassium tetrasodium oxobis[*cis*-citrato-*cis*-dioxomolybdate(Ⅵ)] [Mo—O—Mo = 144.7(2)°][26]. The other three sites are occupied by the atoms belonging to the nta entity. Adjacent $W_2O_5H(nta)_2$

---

①　C&H elemental analysis. Found: C, 14.3%, H, 1.7%. Calc. for $C_{12}H_{17}NO_{19}K_3W_2$, C, 14.7%, H, 1.8%. IR(KBr): $v_{as}$(C=O) 1675$_{vs}$, 1642$_{vs}$, $v_s$(C=O) 1443$_m$, 1371$_s$, 1340$_s$, $v_s$(W=O) 945$_s$, 922$_s$, 895$_s$, 878$_s$ cm$^{-1}$. $^1$H-NMR, $\delta$($D_2O$, ppm): 4.264(s, $CH_2$), 4.247(s, $CH_2$), 4.172(s, $CH_2$). $^{13}$C-NMR $\delta$($D_2O$, ppm): 180.9, 179.7, 178.3($CH_2$)$_\beta$, 66.2, 66.0, 65.3 ($CH_2$).

②　Crystallographic data for $K_3[W_2O_5H(nta)_2] \cdot 2H_2O$: $C_{12}H_{17}O_{19}K_3W_2$, $FW=978.28$, monodinic, C2/c, $a=14.981$(1)Å, $b=10.188$(1)Å, $c=16.895$(2)Å, $\beta=107.387$(6)°, $V=2460.8$(4)Å$^3$, $Z=4$, $D_c=2.641$ gcm$^{-3}$, Mo-$K\alpha$($\lambda=0.71073$ Å), $\mu=9.94$ mm$^{-1}$, $T=298$ K, $R=0.055$ for the 2045$I>2\sigma(I)$ of the 2418 absorption-corrected reflections measured on an Enraf-Nonius CAD-4 diffractometer. The calculations were carried out with the WinGX package, SHELXL97 and SHELXS97 programs[31,32].

units are linked through the acid hydrogen [O···O=2. 54(2)Å] into a chain(Fig. 1). The acid hydrogen is located at the center-of-symmetry in the crystal structure. The space between the chains are occupied by the potassium cations and the lattice water molecules; one potassium atom has seven-coordination number whereas the other has only six-coordination number.

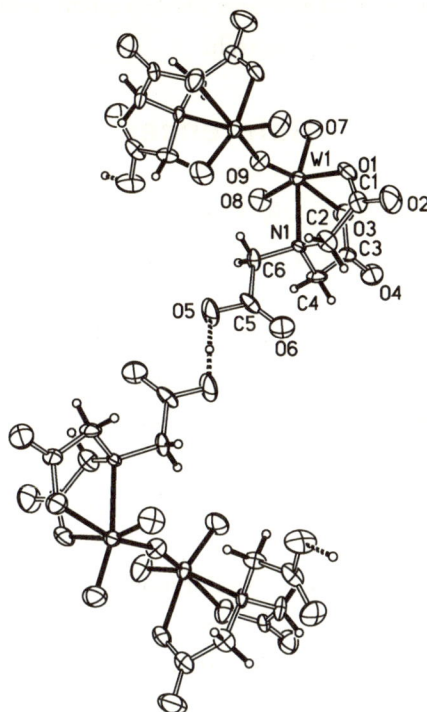

Fig. 1　ORTEP plot(50% probability)illustrating the geometry of the tungsten atom in the hydrogen-bonded oxo-μ-bis[*trans*-nitrilotriace-tato-*cis*-dioxotungstate(Ⅵ)] chain. Selected bond distances and angles: W1—O1＝2. 131(8)Å, W1—O3 ＝2. 024(9)Å, W1—O7＝1. 720(9)Å, W1—O8＝1. 729(8)Å, W1—O9＝1. 880(2)Å, W1—N1＝2. 412(9)Å; O1—W1—O3＝78. 7(4)°, O1—W1—O7＝91. 0(4)°, O1—W1—O8＝161. 9(4)°, O1—W1—O9＝84. 9(3)°, O1—W1—N1＝73. 6(3)°, O3— W1—O7＝97. 7(4)°, O3—W1—O8＝92. 3(4)°, O3—W1—O9＝154. 5(3)°, O3—W1—N1＝73. 8(3)°, O7—W1—O8＝105. 9(4)°, O7—W1— O9＝102. 1(4)°, O7—W1—N1＝163. 5(4)°, O8—W1—O9＝97. 6(4)°, O8—W1—N1＝88. 8(4)°, O9—W1—N1＝82. 9(3)°.

The W—O bond distance in nitrilotriacetato tungstate is consistent with bond order between W and O. W=O double bonds are 1. 720(9) and 1. 729(a)Å the resulting O=W=O angle, 105. 9(4)°is considerably larger than the 90° regular value for *cis* groups. This is expected from the greater O···O repulsions between oxygens with short bonds to the metal atom. The bridged W—O distance is longer [1. 880(2)Å] and the bent W—O—W is obtained. In the related protonated citrato oxomolybdate(Ⅵ) and deprotonated oxotungstates(Ⅵ), the Mo—O—Mo and W—O—W units are centrosymmetric[27−29]. The W(Ⅵ)—O$_{av}$(β-carboxyl)[2. 078(9)Å] distances are smaller than those in the protonated nitrilotriacetato oxomolybdates(Ⅵ)Na$_2$[Mo$_2$O$_5$(Hnta)$_2$]・8H$_2$O[2. 102(5)Å] and(pyH)$_2$[Mo$_2$O$_5$ (Hnta)$_2$][2. 112(4)Å][11,12], indicating a contraction. The W—N distance is the longest [2. 412(9)Å].

One of the nitrilotriacetate ions in the complex carries one proton in the β-carboxyl group. This is not only the result from charge balance in the complex, but also the result by the measurement of the molar electrical conductivity(2. 5×10$^{-2}$ S・m$^2$ mol$^{-1}$)of the title compound. The value is close to the trivalent coordinated anion(2. 4×10$^{-2}$−2. 7×10$^{-2}$ S・m$^2$ mol$^{-1}$)[30].

# Acknowledgements

We thank the National Natural Science Foundation of China(No. 29933040,20021002),the National Key Project Fundamental Research(001CB1089)and the Excellent Young Teachers Fund of the Ministry of Education for generously supporting our research.

# References

[1]K. Okamoto,J. Hidaka,M. Fukagawa,K. Kanamori. Acta Crystallogr. C,1992,**48**:1025.

[2]O. P. Gladkikh,A. F. Borina,A. K. Lyashchenko,B. A. Itin,N. D. Mitrofanova. Koord. Khim, 1994,**20**:215(in Russian).

[3]H. G. Visser,W. Purcell,S. S. Basson,Q. Claassen. Polyhedron,1997,**16**:2851.

[4]H. G. Visser,W. Purcell,S. S. Basson. Polyhedron,1999,**18**:2795.

[5]M. Sokolov,H. Imoto,T. Saito. Inorg. Chem. Commun,2000,**3**:96.

[6]Y. Chen,B. Q. Ma,Q. D. Liu,J. R. Li,S. Gao. Inorg. Chem. Commun,2000,**3**:319.

[7]B. W. Sun,Z. M. Wang,S. Gao. Inorg. Chem. Commun,2001,**4**:79.

[8]H. G. Visser,W. Purcell,S. S. Basson. Polyhedron,2001,**20**:85.

[9]L. A. Zasurskaya,I. N. Polyakova,T. N. Polynova,A. L. Poznyak,V. S. Sergienko. Russ. J. Coord. Chem,2001,**27**:270.

[10]R. J. Butcher,B. R. Penfold. J. Cryst. Mol. Struct,1976,**6**:13.

[11]C. B. Knobler,W. T. Robinson,C. J. Wilkins,A. J. Wilson. Acta Crystallogr. C,1983,**39**,443.

[12]K. Matsumoto,Y. Marutani,S. I. Ooi. Bull. Chem. Soc. Jpn,1984,**57**:2671.

[13]K. F. Miller,R. A. D. Wentworth. Inorg. Chem,1978,**17**:2769.

[14]K. Zare,P. Lagrange. J. Lagrange,J. Chem. Soc.,Dalton Trans. 1979:1372.

[15]M. A. Freeman,D. R. van der Vaart,F. A. Schultz,C. N. Reilley. J. Coord. Chem,1981,**11**:81.

[16]G. Anderegg. Pure Appl. Chem,1982,**54**:2693.

[17]R. Hille. Chem. Rev,1996,**96**:2757.

[18]M. K. Johnson,D. C. Rees,M. W. W. Adams. Chem. Rev,1996,**96**:2817.

[19]Z. H. Zhou,W. B. Yan,H. L. Wan,K. R. Tsai,J. Z. Wang,S. Z. Hu. J. Chem. Crystallogr,1995, **25**:807.

[20]D. W. Wright,P. A. Humiston,W. H. Orme-Johnson,W. H. Davis. Inorg. Chem,1995,**34**: 4194.

[21]Z. H. Zhou,H. L. Wan,S. Z. Hu,K. R. Tsai. Inorg. Chim. Acta,1995,**237**:193.

[22]S. Burojevis,I. Shweky,A. Bino,D. A. Summers,R. C. Thompson. Inorg. Chim. Acta,1996, **251**:75.

[23]M. Velayutham,B. Varghese,S. Subramanian. Inorg. Chem,1998,**37**:1336.

[24]M. Tsaramyrsi,D. Kavousanaki,C. P. Raptopoulou,A. Terzis,A. Salifoglou. Inorg. Chim. Acta, 2001,**320**:47.

[25]M. Tsaramyrsi,M. Kaliva,A. Salifoglou,C. P. Raptopoulou,A. Terzis,V. Tangoulis,J. Giapintzakis. Inorg. Chem. 2001,**40**:5772.

[26]Z. H. Zhou,H. L. Wan,K. R. Tsai. Polyhedron,1997,**16**:75.

[27]Z. H. Zhou,H. L. Wan,K. R. Tsai. Inorg. Chem,2000,**39**:59.

[28] J. J. Cruywagen, L. J. Saayman, M. L. Niven, J. Crystallogr. Spectrosc. Res, 1992, **22**: 737.

[29] E. Llopis, J. A. Ramirez, A. Doménech, A. Cervilla, J. Chem. Soc., Dalton Trans, 1993: 1121.

[30] R. J. Angelici, in: Synthesis and Technique in Inorganic Chemistry, second ed., Saunders, Philadelphia, 1977, p. 13.

[31] L. J. Farrugia. J. Appl. Cryst, 1999, **32**: 837.

[32] G. M. Sheldrick, SHELXL97 and SHELXS97, University of Göttingen, Germany, 1997.

■ 本文原载：Ctalysis Today 74(2002)，pp. 5～13.

# Structure Aspects and Hydroformylation Performance of Water-Soluble HRh(CO)[P(m-C$_6$H$_4$SO$_3$Na)$_3$]$_3$ Complex Supported on SiO$_2$

You-Zhu Yuan[①]，Hong-Bin Zhang，Yi-Quan Yang，Yu Zhang，Khi-Rui Tsai

(*State Key Laboratory for Physical Chemistry of Solid Surfaces, Department of Chemistry, Xiamen University, Xiamen 361005, P.R. China* )

**Abstract** The solution NMR($^{31}$P and $^1$H) and FTIR spectroscopies were employed to investigate the structure information of water-soluble complex HRh(CO)[P(m-C$_6$H$_4$SO$_3$Na)$_3$]$_3$ (**1**) [P(m-C$_6$H$_4$SO$_3$Na)$_3$：trisodium salt of tri-(m-sulfophenyl)-phosphine, TPPTS] supported on SiO$_2$ (**1**/SiO$_2$). The $^{31}$P($^1$H)NMR spectra showed that a pair of new twin-peak at about 31.5,32.1 ppm while no typical twin-peak at 44.0,44.7 ppm for the phosphorus species ascribed to the complex **1** were observed at **1**/SiO$_2$. However, the typical phosphorus peaks for the complex **1** appeared in the case of using TPPTS or Na$_2$CO$_3$-preimpregnated SiO$_2$ as supports. Moreover, the immobilization caused a considerable oxidation of the liberated TPPTS to OTPPTS(OTPPTS, i. e. OP(m-C$_6$H$_4$SO$_3$Na)$_3$：trisodium salt of tri-(m-sulfophenyl)-phosphine oxide)species as evidenced by $^{31}$P($^1$H) NMR spectroscopy. The phosphorus-31 peaks at 31.5,32.1 ppm at **1**/SiO$_2$ were found to be unchanged before and after the propene hydroformylation. The FTIR results revealed that the CO band appeared at about 1870 cm$^{-1}$ for the **1**/SiO$_2$ catalyst, which was lower than that for the precursor complex **1**. It is concluded that there exists a strong interaction between the complex **1** and the acidic support of SiO$_2$, resulting in the deformation of the Rh-phosphine complex containing less than three TPPTS ligands, which further transformed by dehydrogenation and dimerization under evacuation to form a species likely [Rh(CO)(TPPTS)$_2$]$_2$ as a main surface complex species at the catalyst **1**/SiO$_2$. © 2002 Elsevier Science B. V. All rights reserved.

**Key words** Water-soluble Rh-complex NMR Hydroformylation FTIR Supported aqueous-phase catalyst

## 1 Introduction

Research into heterogenization of HRh(CO)[P(C$_6$H$_4$)$_3$]$_3$ (P(C$_6$H$_4$)$_3$：triphenylphosphine) for hydroformylation and selective hydrogenation has received high attention in the past several decades from both academic and industrial interests[1]. Besides the milestone preparation of water-soluble TPPTS and corresponding Rh-phosphine complexes such as HRh(CO)(TPPTS)$_3$ (**1**)[2,3], supported

---

① Corresponding author. Fax：+86-592-2183047. *E-mail address*：yzyuan@xmu. edu. cn(Y. Yuan).

liquid-phase catalysts (SLPC)[4,5] and supported aqueous-phase catalysts (SAPC)[6] are known as attractive alternatives for heterogenized catalysts for olefin hydroformylation. The SAPC where the complex **1** or rhodium complex precursor and water-soluble ligand like TPPTS are immobilized in a thin water layer adhered within the pores of a high surface-area silicates[7] makes the separation of catalyst from reaction mixture to be greatly simplified, thus, facilitating successive operation. Using this system higher alkenes can be converted at a relatively high rate without metal leaching. It still faces the problems, however, how to improve the lower regioselectivity to the product aldehydes(i. e. molar ratio of normal : branched (*n* : *i*) aldehydes) and also the catalytic stability due to the oxidation of water-soluble phosphine ligands[8,9]. In literature, the results on SAPC for hydroformylation, thus, far were mostly based on rhodium, cobalt and platinum/nickel with TPPTS as the ligand[10,11] or sulfonated diphosphines including those with a large bite angle as ligands[12], limited attention has been paid on the structure information in the SAPC system[13].

It is well documented in the relationship of structure-performance in homogeneous and biphasic catalytic systems due to the easy characterization and well-defined structure as compared to the corresponding SAPC system[14]. Use of high resolution MAS NMR technique is one of the efficient ways to clarify the coordination chemistry in the SAPC system[8,13,15]. However, the measurements of MAS NMR spectra are time-consuming and usually MAS spectra with lower resolution are far from satisfactory for catalyst characterization. Alternatively, when the dried supported catalyst of SAPC is exposed to water it is possible to record a $^{31}$P NMR spectrum using conventional solution NMR techniques[15]. The objective of this work is to investigate the structural information at SiO₂-supported water-soluble TPPTS-Rh-complex catalysts in connection to the hydroformylation performance by means of solution NMR with the combination of pyridine-TPD and FTIR spectra.

# 2 Experimental

A commercial available SiO₂ was washed by distilled water and sieved to 80—100 mesh which has pore diameter of 6—12 nm, pore volume of 1.03 cm³g⁻¹ and BET surface-area of 380 m²g⁻¹. For comparison, SiO₂ was preimpregnated with 2% Na₂CO₃ follwed by drying in an oven at 393K for over night. TPPTS was synthesized by the known method[2,16-18]. $^{31}$P($^1$H) NMR spectroscopy showed it consisted of 15% OTPPTS(trisodium salt of tri-(*m*-sulfophenyl)-phosphine oxide). Thermogravometric analysis(TGA)on the TPPTS showed a 10% weight loss attributed to the residual water. The TPPTS had a solubility of 0.63 gml⁻¹ at room temperature and the same IR bands as reported in[16-18]. The complex **1** was synthesized according to the literature method[15]. $^{31}$P($^1$H) NMR: 44.7, 44.0ppm; IR (cm⁻¹):2010s,1928s.

A water solution of **1**(0.01 mmol ml⁻¹)was poured into a Schlenk flask containing 0.1g degassed SiO₂. After further degassing the mixture by vacuum boiling, argon was introduced and the slurry was kept at room temperature for 1 h while vibration. The water was removed under vacuum at room temperature. The final product was a dry, free-flowing yellow powder and denoted as **1/SiO₂**, which was stored under Ar at room temperature. The water content was analyzed by TGA to be 8—10%.

Solution $^{31}$P($^1$H)NMR spectra were recorded on a Varian FT Unity⁺ 500 spectrometer at 200 MHz at room temperature. Phosphorous-31 NMR chemical shifts are reported relative to 85% H₃PO₄. Samples of SAPC in the NMR tube were introduced about 40% D₂O before measurements. FTIR spectra were measured on a Nicolet-740 spectrometer with a resolution of 4 cm⁻¹. The surface-areas of the samples were measured by a conventional BET nitrogen adsorption method at 77 K using a

SORPTOMATIC-1900 machine.

The propene hydroformylation reaction was performed in a high-pressure fixed-bed flow reactor connecting an on-line gas chromatgraph(102-GD,Shanghai Analysis Factory,China)equipped with FID detector and a column of Porapak Q(2 m). The propene：CO：H$_2$ ratio used here was always 1：1：1. The results showed the total hydrogenation products were lower than 1% under present experiment condition.

# 3　Results

## 3.1　Pyridine—TPD

It is known that the hydrophilic SiO$_2$ support possesses acidity at the surface. Pyridine-TPD spectra for SiO$_2$ and Na$_2$CO$_3$/SiO$_2$ are shown in Fig. 1. The pyridine-TPD for the SiO$_2$ showed a peak at about 420 K,whereas that for Na$_2$CO$_3$/SiO$_2$ showed no peak in the same temperature range. In fact the pyridine adsorption on Na$_2$CO$_3$/SiO$_2$ sample was negligible.

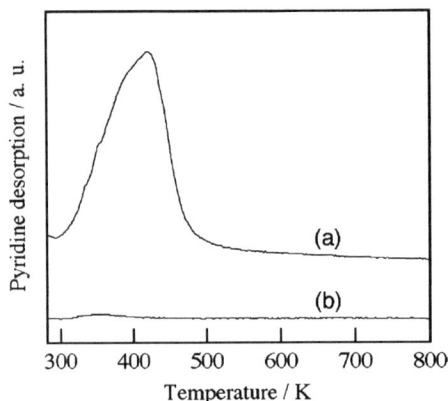

Fig. 1　TPD spectra of pyridine adsorbed on SiO$_2$(a) and Na$_2$CO$_3$/SiO$_2$(b).

## 3.2　Solution $^{31}$P($^1$H)NMR spectra of SiO$_2$-supported 1

The solution $^{31}$P($^1$H)NMR spectra for TPPTS,1,and a SiO$_2$-supported TPPTS and a SAPC sample containing about 40 wt% D$_2$O are shown in Fig. 2a-c,respectively. When TPPTS was supported on SiO$_2$,the $^{31}$P($^1$H)NMR showed two peaks at 5.1 and 35.2 ppm straightforwardly assignable to the free TPPTS and OTPPTS,respectively. However,no typical twin phosphorus-31 peaks at 44.7,44.0 ppm for the complex 1 but another pair of twin-peaks with broadness at about ～32.1 and 31.5 ppm for newly formed phosphorus species and an increased peak at 35.2ppm for the phosphorus in the OTPPTS were observed in the SAPC sample. The intensity for the peak at 35.2 ppm significantly increased from 10mol% in the precursor 1 to about 30 mol% in the SAPC sample,probably due to an oxidation of the phosphorous ligands liberated from the complex during the impregnation. The new peaks at 32.1 and 31.5 ppm contained 60 mol% amount of phosphorous species. Taking the weak or no coordination ability of OTPPTS to rhodium atom into account,we estimated the ratio of the phosphorous in the new peak to rhodium atom to be closely to 2：1(mol). Further NMR experimental results revealed that the newly formed Rh-TPPTS species at SiO$_2$ surface strongly interacted with SiO$_2$ so that it was impossible to be simply removed from the SiO$_2$ surface by washing with D$_2$O.

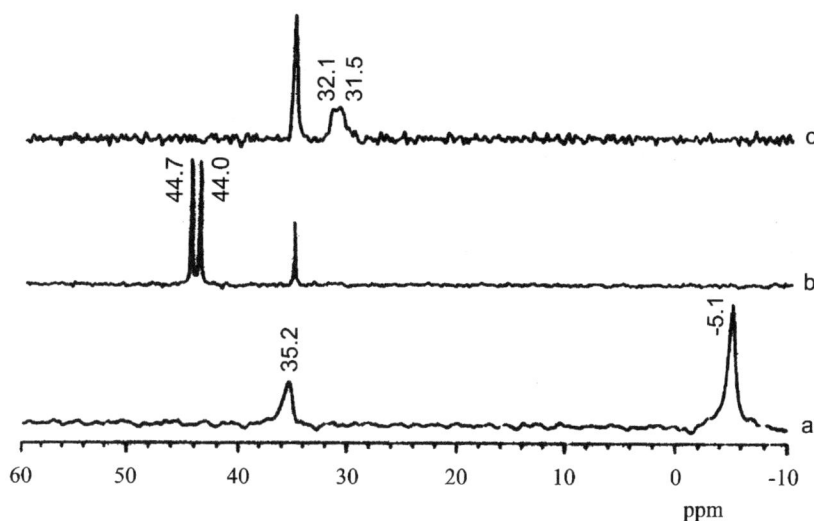

Fig. 2    $^{31}$P($^1$H)NMR spectra for: (a)TPPTS/$SiO_2$ containing 40% $D_2O$;
(b)1 in $D_2O$, pH=6.9; (c)1/$SiO_2$ containing 40% $D_2O$.

To investigate the property of the interaction between the Rh-complex **1** and $SiO_2$ support and clarify the reason why the interaction generated, we chose several supports, such as GDX-102 (a copolymer of ethylene benzen and xylene), $Na_2CO_3$-preimpregnated $SiO_2$ ($Na_2CO_3$/$SiO_2$), TPPTS-impregnated $SiO_2$ (TPPTS/$SiO_2$) to immobilize the Rh-complex **1** and characterize the structure information by the solution NMR. All the samples for the NMR measurements contained about 40% $D_2O$. The results are shown in Fig. 3. Clearly, the $^{31}$P($^1$H) NMR peaks at 44.7 and 43.9ppm for **1** appeared when it was supported on GDX-102, though there was an increase in the peak intensity for OTTPTS at 35.2ppm(Fig. 3e). The typical $^{31}$P($^1$H) NMR peaks for the phosphorus-31 species in the

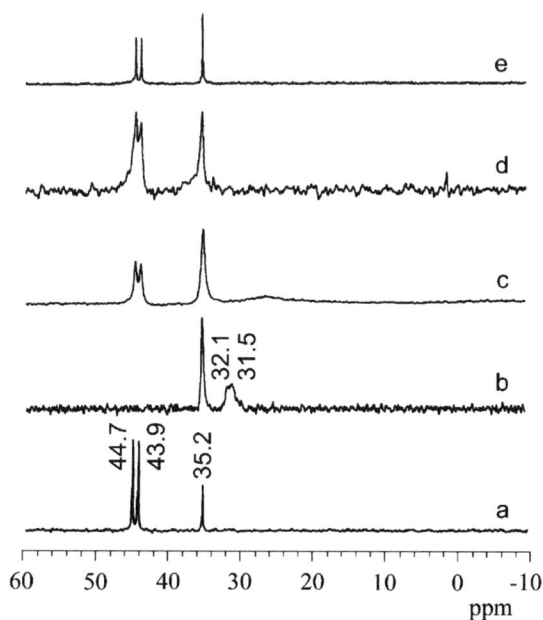

Fig. 3    $^{31}$P($^1$H)NMR spectra for: (a)1 in $D_2O$; (b)1/$SiO_2$ containing 40% $D_2O$; (c)1/TPPTS/$SiO_2$ containing 40% $D_2O$;
(d)1/$Na_2CO_3$/$SiO_2$ containing 40% $D_2O$; (e)1/GDX-102 containing 40% $D_2O$.

**1331**

precursor **1** were appeared when using TPPTS/SiO₂ or Na₂CO₃/SiO₂ as supports, though the peak intensity for the OTPPTS was increased considerably.

Before and after propene hydroformylation at 373 K, the $^{31}$P($^1$H)NMR spectra for the catalyst **1**/SiO₂ were no significant difference each other as shown in Fig. 4, indicating that the SiO₂ surface could stabilize the deformed structure of Rh-complex at **1**/SiO₂.

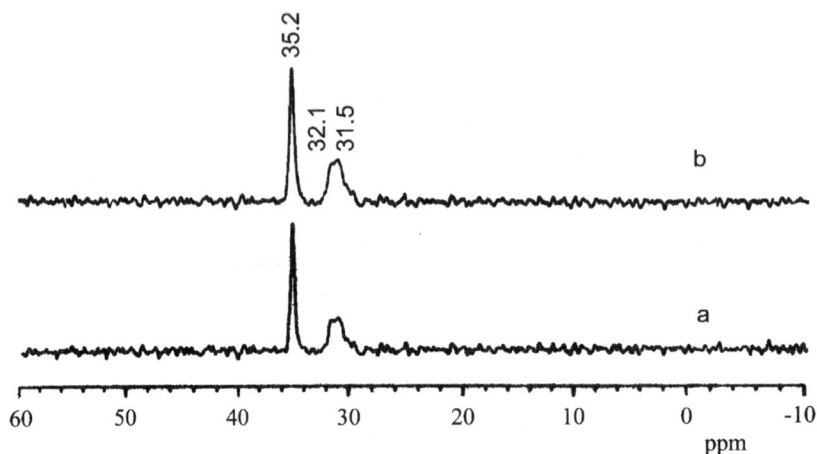

**Fig. 4** $^{31}$P($^1$H)NMR spectra for **1**/SiO₂ containing 40% D₂O; (a)before propene hydroformylation; (b)after propene hydroformylation for 4 h.

Fig. 5 shows that $^{31}$P($^1$H)NMR spectra of **1**/SiO₂ with different Rh loading weights. In the case of loading weigh less than 0.1 mmol Rh/γ-SiO₂, we found that the spectroscopy only showed a signal at δ=35.2ppm for OTPPTS and a signal at about δ=30 ppm for the species which strongly interacted with SiO₂. The signals at 44.7 and 43.9ppm for the complex **1** were appeared in the case of loading weight of 0.2 mmol Rh/γ-SiO₂, in which there were about 30 mol% of phosphorus species in the state of original complex **1**.

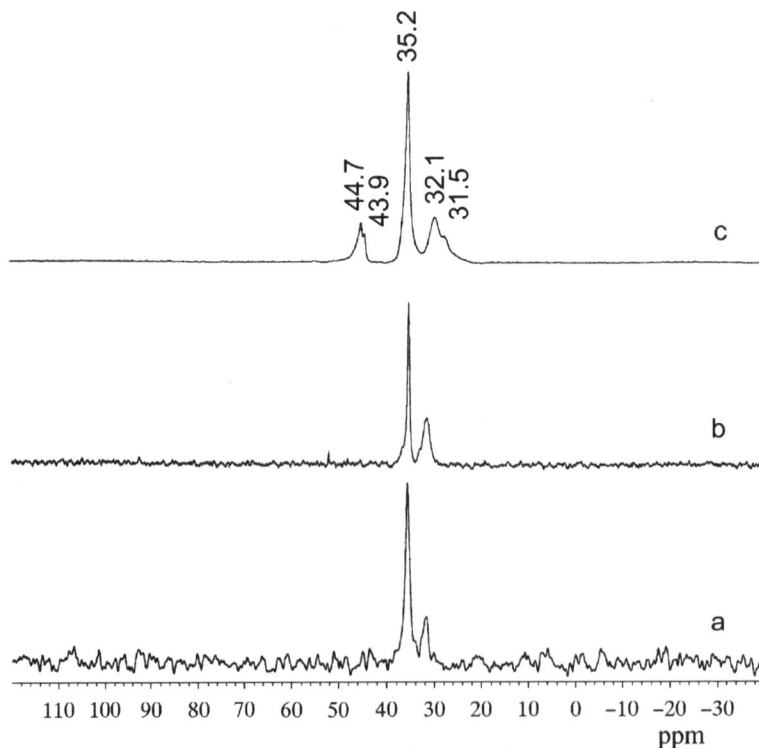

**Fig. 5** $^{31}$P($^1$H)NMR spectra for **1**/SiO₂ containing 40% D₂O; (a)0.0025 mmol **1**/γ-SiO₂; (b)0.0125 mmol **1**/γ-SiO₂; (c)0.2 mmol **1**/γ-SiO₂.

## 3.3 FTIR spectra for 1/SiO$_2$

The FTIR spectra for the complex **1** and **1**/SiO$_2$ are shown in Fig. 6. The complex **1** presented an IR band at 2010 cm$^{-1}$ for the stretch vibration of Rh-H and a peak at 1928 cm$^{-1}$ for the carbonyl species, which are coincided with the results in[13]. The peak at 1630 cm$^{-1}$ was assignable to the water existed in the sample. After immobilization on SiO$_2$, the IR peaks for the surface complex were differed from those for the precursor **1** as shown in Fig. 6b. The peaks at 2010 and 1928 cm$^{-1}$ for the complex **1** were replaced by those at 1992 and 1870 cm$^{-1}$ for the **1**/SiO$_2$, respectively. After evacuation at 373 K for 2h, the peak at 1992 cm$^{-1}$ became of broadness while that at 1870 cm$^{-1}$ increased in intensity as shown in Fig. 6c.

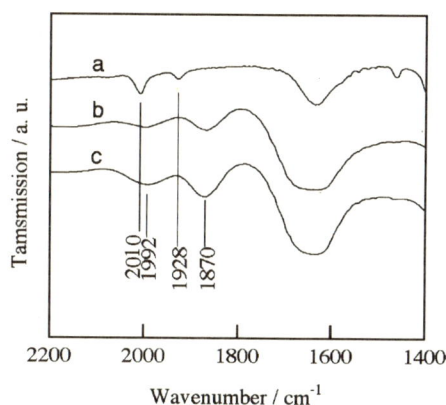

Fig. 6   FTIR spectra for; (a)**1**/KBr; (b)**1**/SiO$_2$; (c)**1**/SiO$_2$ after evacuation at 373 K for 2h.

## 3.4   $^{31}$P($^1$H)NMR spectra of 1 in the presence of solution base and acid

Fig. 7 A and B shows the spectra of $^{31}$P($^1$H)NMR and $^1$H NMR for the complex **1** in D$_2$O solution in the presence of NaOH. It was observed that the intensity of chemical shift at $\delta = 35.1$ ppm for the OTPPTS increased slightly with the addition of NaOH, but no other new peaks produced. The corresponding $^1$H NMR spectra revealed the peaks at $\delta = -9.56$ ppm keeping quartet splitting with the NaOH amount as shown in Fig. 6b, indicating that the bonding of Rh-H remained intact by adding NaOH up to high concentration. The broadness of the peaks at aryl-H range(7.1—7.7ppm) may be due to the fast spin-spin interaction in the presence of NaOH. The results suggested that there were no change in structure of **1** in the presence of NaOH.

However, the $^{31}$P($^1$H)NMR spectra for the **1** changed remarkably in the presence of inorganic acids as shown in Fig. 7. When introducing HCl, H$_2$SO$_4$, HNO$_3$ or H$_3$PO$_4$ different concentration, the peak intensities for the

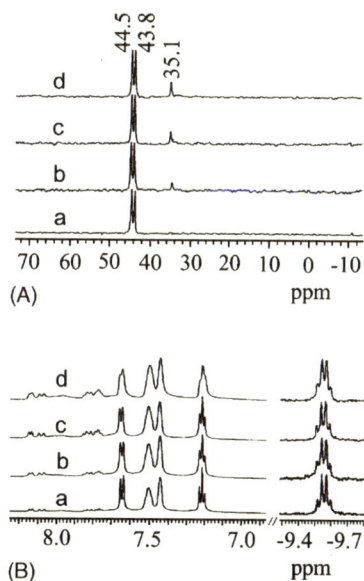

Fig. 7   (A and B)$^{31}$P($^1$H)NMR spectra for **1** in the presence of solution NaOH.

phosphorus of the complex **1** decreased drastically while the intensity for the OTPPTS increased with the acid amount. In the case of high concentration of acid, the phosphorus-31 peak due to the complex **1** became negligible, while the phosphorus-31 due to OTPPTS at 35.0ppm and the new species at $27-30$ ppm increased significantly. The results suggested that the structure of the complex **1** was unstable in the presence of acid media.

## 3.5 Hydroformylation performance on SAPC

Table 1 shows that propene hydroformylation performance on the catalysts of **1**/$SiO_2$, **1**/TPPTS/$SiO_2$ and **1**/$Na_2CO_3$/$SiO_2$ at 373 K, respectively. Lower activity but slightly higher regioselectivity to buty-laldehyde on **1**/TPPTS/$SiO_2$ and **1**/$Na_2CO_3$/$SiO_2$ were obtained as compared to that on **1**/$SiO_2$. When a catalyst with a loading weight of complex **1** on $SiO_2$ higher than monolayer, it presented much lower catalytic activity for the propene hydroformylation. From the $^{31}P(^{1}H)$ NMR results and the catalysis in Table 1, we considered that active species on **1**/$SiO_2$ likely to be $[Rh(CO)(TPPTS)_2]_n$ ($n=1,2$) (phosphorus-31 showed a peak at about 32.1, 31.5 ppm) rather than intact **1** on $SiO_2$ (phosphorus-31 showed the twin-peaks at 44.7, 43.9ppm). The coordination unsaturated species $[Rh(CO)(TPPTS)_2]$ which could be readily generated from $[Rh(CO)(TPPTS)_2]_n$ ($n=1,2$) at $SiO_2$ surface provided a possible coordinating position for activation of propene, leading to an increase in the reaction activity.

**Table 1 Propene hydroformylation on 1/$SiO_2$ catalyst[a]**

| Support | Rh loading (mmol $g_{SiO_2}^{-1}$) | L/Rh | Conversion (%) | $n:i$ | STY (mmolh$^{-1}$ $g_{Rh}^{-1}$) | TOF(s$^{-1}$) |
|---|---|---|---|---|---|---|
| $SiO_2$ | 0.3 | 3 | 31.4 | 1.6 | 1362 | 0.039 |
| | 0.03 | 3 | 18.7 | 1.7 | 8113 | 0.232 |
| TPPTS/$SiO_2$[b] | 0.03 | 30 | 3.3 | 2.3 | 1432 | 0.041 |
| $Na_2CO_3$/$SiO_2$ | 0.03 | 3 | 4.8 | 2.0 | 2082 | 0.059 |

[a] Reaction conditions: $T=373$ K, $C_3H_6$ : CO : $H_2 = 1:1:1$(v/v), total pressure $=1.0$MPa, GHSV(STP)$=9000$ml h$^{-1}$g$_{cat}^{-1}$. The data were achieved after 120 min when static state was reached.

[b] The complex **1** was immobilized on $SiO_2$ after preimpregnation of the excessive TPPTS ligand.

# 4 Discussion

We previously[8] reported that by the results of $^{31}P(^{1}H)$MAS NMR characterization a large part of phosphine ligands TPPTS were not in the states of coordination in the SAPC prepared freshly from Rh $(acac)(CO)_2$ complex and TPPTS, and that a considerable quantity of the complexes containing two phosphine ligands with the chemical shift of 32.4 ppm(assigned to $[Rh(CO)TPPTS_2]$ species) were formed in-situ after starting of the reaction, which are considered as the catalytically active complex species responsible for hydroformylation. When the water-soluble complex **1** was immobilized on $SiO_2$, no typical twin phosphorus-31 chemical shifts at $\delta=44.7, 43.9$ppm, but new twin-peaks at $\delta=32.1$, 31.5 ppm and significant increase in intensity at $\delta=35$ for OTPPTS species were observed by the conventional solution $^{31}P(^{1}H)$NMR(Fig. 2). The typical twin phosphorus-31 NMR chemical shifts could be detected while using $SiO_2$ preimpregnated with $Na_2CO_3$ or TPPTS($Na_2CO_3$/$SiO_2$ or TPPTS/$SiO_2$)as supports(Fig. 3). The peaks at 32.1 and 31.5 ppm were estimated to be a TPPTS-Rh-complex containing phosphorus:rhodium ratio closely to 2:1, likely the species of $[Rh(CO)(TPPTS)_2]n(n=1, 2)$at $SiO_2$ surface.

The acidic properties at the support surface were considered to be one of the main reasons that caused the structure change of the Rh-complex **1** while immobilization. Before loading the complex **1**, the

SiO$_2$ support was pre impregnated by TPPTS or Na$_2$CO$_3$, the $^{31}$P($^1$H)NMR spectra for the samples, thus, prepared, 1/TPPTS/SiO$_2$ and 1/Na$_2$CO$_3$/SiO$_2$, were found to be significant different from that for the 1/SiO$_2$ as shown in Fig. 2b and c. The results suggested that use of TPPTS/SiO$_2$ and Na$_2$CO$_3$/SiO$_2$ as supports prevent the complex 1 from an interaction with the acidity at SiO$_2$ surface, leading to that the surface complex at 1/TPPTS/SiO$_2$ and 1/Na$_2$CO$_3$/SiO$_2$ could mainly remain the structure feature of the intact complex 1. With the results of the pyridine-TPD for SiO$_2$ and Na$_2$CO$_3$/SiO$_2$, it is reasonable to deduce that the structure of 1 on support depends on the property of the support, particularly the acidity/polarity of the support.

We found that the Rh-TPPTS complex 1 could strongly interacted with unpretreated SiO$_2$ support, which resulted in the structure deformation while immobilization. The deformed complex at 1/SiO$_2$ was suggested to be exited at the interfacial layer between the SiO$_2$ surface and the surface complex. Therefore, it becomes possible to determine the monolayer loading weight of the complex by solution $^{31}$P ($^1$H)NMR technique by the observations of the structure changes. In the case of loading weigh less than 0. 1 mmol Rh/$\gamma$-SiO$_2$, we found that the $^{31}$P($^1$H)NMR spectroscopy presented a signal at $\delta=35.2$ppm for OTPPTS and a signal at about $\delta=30$ ppm for the species as shown in Fig. 5. The signals at 44. 7 and 43. 9 ppm for the complex 1 were appeared in the case of loading weight of 0. 2 mmol Rh/$\gamma$-SiO$_2$, in which there were about 30mol% of phosphorus species in the state of original complex 1. It was estimated from the NMR results the monolayer loading of the Rh-complex 1 to be 0. 14 mmol Rh/$\gamma$-SiO$_2$.

The coordination state of the Rh atom in the species of [Rh(CO)(TPPTS)$_2$]$_n$ was unsaturated in the case of $n=1$. To understand the complex structure at SiO$_2$ more clearly, we measured the IR spectra of 1/SiO$_2$ at different treatment conditions as shown in Fig. 6. The FTIR spectra for the 1/SiO$_2$ shown that there were two peaks at 1992 and 1870 cm$^{-1}$. After evacuation at 373 K for 2h, the relative intensity of the peak at 1987 cm$^{-1}$ increased a little. By comparison to the FTIR data of the complex 1, the peaks at 1992 and 1870 cm$^{-1}$ at the 1/SiO$_2$ might be due to the de-coordination of the ligands from the complex[19] and then the dimerization to a dinuclear complex as shown in the Scheme 1. The broad peak

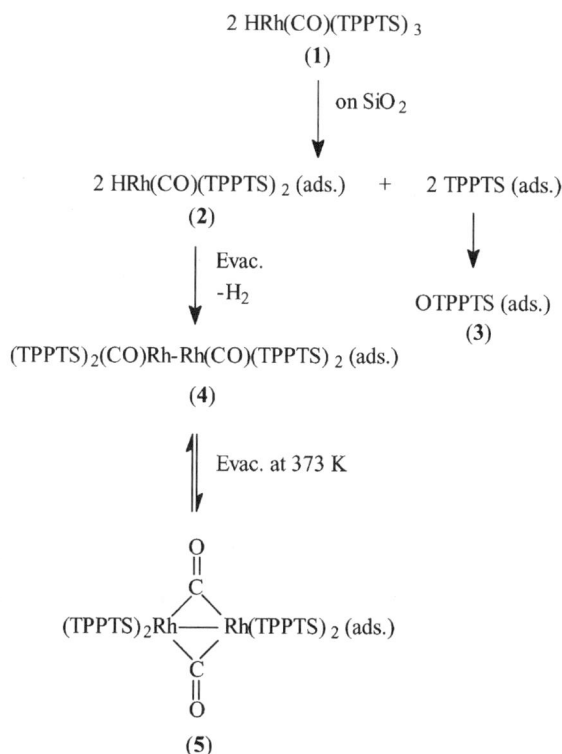

Scheme 1. A possible structure transformation of the complex 1 at SiO$_2$ surface.

at 1992 cm$^{-1}$ may be due to the contributions from following species: the Rh-H stretching at the mononuclear complex **2** and the linear carbonyl species, and also the linear carbonyl species at dinuclear complex **4**. The overlaps of above species caused the broadness of the peak at 1992 cm$^{-1}$. The peak at 1870 cm$^{-1}$ may be due to the contribution of bridge carbonyl in the dinuclear complex **5**. The treatment at 373 K caused further dehydrogenation to dimeriza-tion, thus, resulted in an increase in the peak intensity at 1870 cm$^{-1}$.

The results obtained indicate that there are de-coordination of the ligand TPPTS from **1** and oxidation of ligand TPPTS while deposition of the complex **1** onto SiO$_2$ due to the strong interaction of basic ligand TPPTS with acidic surface of the carrier silica-gel, forming a surface complex containing two phosphine ligands, which was further dehydrogenation and dimerization to dinuclear surface Rh-TPPTS complex most probably [Rh(CO)(TPPTS)$_2$]$_2$ (with the $^{31}$P($^1$H)NMR peak at ~32.1, 31.5 ppm). The dinuclear species were in-situ and readily activated to form a catalytic active species of [Rh(CO)(TPPTS)$_2$] at the SiO$_2$ surface during hydroformylation condition. The deformed complexes with a ratio of Rh:TPPTS=1:2 were stable enough during the propene hydroformylation(Fig. 5). The experiments also demonstrated that the de-coordination and oxidation could be partially/principally depressed when using Na$_2$CO$_3$/SiO$_2$ or TPPTS/SiO$_2$ as supports, which were supported by the NMR data for the complex **1** in the presence of NaOH and acids(Figs. 7 and 8). Higher catalytic activities for the hydroformylation of propene were achieved over **1**/SiO$_2$ than that on **1**/Na$_2$CO$_3$/SiO$_2$ due to the easier generation of the coordination unsaturated species [Rh(CO)(TPPTS)$_2$] at **1**/SiO$_2$.

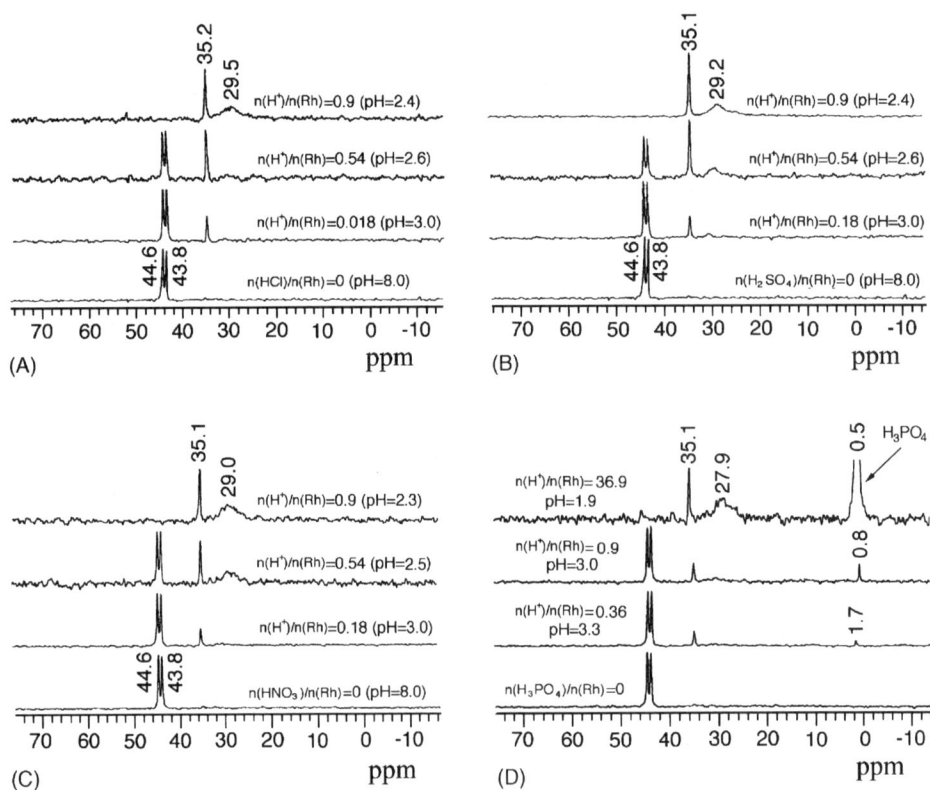

Fig. 8  (A—D)$^{31}$P($^1$H)NMR spectra for **1** in the presence of solution acids.

# 5  Conclusions

1. The strong interaction between the complex **1** and SiO$_2$ support surface was due to a interaction through the basic ligands TPPTS and the acidity at the SiO$_2$ surface, which resulted in the liberation of the coordinated TPPTS and the deformation of the complex **1** at SiO$_2$ by dehydrogenation and dimerization to form a new species of [Rh(CO)(TPPTS)$_2$]$_2$.

2. The de-coordination and oxidation could be partially/principally depressed when using Na$_2$CO$_3$/SiO$_2$ or TPPTS/SiO$_2$ as supports for immobilizing the complex **1**.

3. A higher catalytic activity in the propene hydro-formylation was obtained on **1**/SiO$_2$ than that on **1**/Na$_2$CO$_3$/SiO$_2$.

# Acknowledgements

The authors gratefully acknowledge the financial supports from the Sino-pec, the National Natural Science Foundation of China(Project Nos. 29792075, 29873037 and 20021002), the State "973" Research Project(G2000048008) and the Ministry of Education of China.

# References

[1]M. Beller, B. Cornils, C. D. Frohning, C. W. Kohlpaintner. J. Mol. Catal. A, 1995, **104**: 17.

[2]E. G. Kuntz. US Patent 4, 1981, **248**: 802.

[3]E. G. Kuntz. Chemtech, 1987, 570.

[4]J. Hjortkjaer, M. S. Scurrell, P. Simonsen, H. Svendsen. J. Mol. Catal, 1981, **12**: 179.

[5]H. L. Pelt, P. J. Gijsman, R. P. J. Verburg, J. J. F. Scholten. J. Mol. Catal, 1985, **33**: 119.

[6]J. P. Arhancet, M. E. Davis, J. S. Merola, B. E. Hanson. Nature, 1989: **339**, 454.

[7]M. E. Davis. Chemtech, 1992: 498.

[8]J. P. Arhancet, M. E. Davis M E, B. E. Hanson. J. Catal. 1991, **129**: 100.

[9]Y. Z Yuan, Y. Zhang, Z. Chen, H. B. Zhang, K. R. Tsai. Acta Phy. Chem. Sinica, 1998, **14**: 1013.

[10]W. A. Herrmann, C. W. Kohlpainter. Angew. Chem. Int. Ed. Engl, 1993, **32**: 1524.

[11]M. S. Anson, M. P. Leese, L. Tonks, J. M. J. Williams. J. Chem. Soc. Dalton Trans, 1998: 3529.

[12]A. J. Sandee, V. F. Slagt, J. N. H. Reek, P. C. J. Kamer, P. W. N. M. van Leeuwen. Chem. Commun, 1999: 1633.

[13]B. B. Bunn, T. Bartik, B. Bartik, W. R. Bebout, T. E. Glass, B. E. Hanson. J. Mol. Catal, 1994, **94**: 157.

[14]B. E. Hanson, H. Ding, C. W. Kohlpaintner. Catal. Today, 1998, **42**: 421.

[15]J. P. Arhancet, M. E. Davis, J. S. Merola, B. E. Hanson. J. Catal, 1990, **121**: 327.

[16]W. A. Herrmann, J. A. Kulpe, W. Konkol, H. Bahrmann. J. Organomet. Chem, 1990, **389**: 85.

[17]M. E. Davis, J. P. Arhancet, B. E. Hanson, EP 0372615A2(1990).

[18]T. Bartik, B. Bartik, B. E. Hanson, T. Glass, W. R. Bebout. Inorg. Chem, 1992, **31**: 2667.

[19]B. E. Hanson, M. E. Davis, D. Taylor, E. Rode. Inorg. Chem, 1984, **23**: 52.

■ **本文原载**：Journal of Inorganic Biochemistry 90(2002),pp. 137～143.

# Synthesis and Characterization of Homochiral Polymeric S-Malato Molybdate(VI): toward the Potentially Stereospecific Formation and Absolute Configuration of Iron-Molybdenum Cofactor in Nitrogenase

Zhao-Hui Zhou, Wen-Bin Yan, Hui-Lin Wan, Khi-Rui Tsai[①]

*(Department of Chemistry and State Key Laboratory for Physical Chemistry of Solid Surface, Xiamen University, Xiamen 361005, China )*

Received 20 August 2001; received in revised form 14 February 2002; accepted 15 February 2002

**Abstract**  Reaction of sodium or potassium molybdate and excess malic acid in a wide range of pH values(pH 4.0—7.0)resulted in the isolation of two *cis*-dioxo-bis(malato)-Mo(VI)complexes,viz. $Na_3[MoO_2H(S-mal)_2]$ and $K_3[MoO_2H(S-mal)_2] \cdot H_2O(H_3 mal = malic acid)$. The sodium complex is also characterized by an X-ray structure analysis,showing that the mononuclear Mo units are linked together via very strong symmetric $CO_2 \cdots H \cdots O_2 C$-hydrogen bond $[2.432(5)Å]$,forming a polymeric chain. The molybdenum atoms are *quasi*-octahedrally coordinated by two *cis*-oxo groups and two bidentate malate ligands via its alkoxy and α-carboxyl groups,while the β-carboxylic and carboxylate groups remain uncomplexed,as the coordination of vicinal carboxylate and alkoxide of homocitrate in FeMo cofactor of nitrogenase. The absolute configuration of the metal center in this S-malato complex is assigned as $\Lambda$ and the homochirality within the chain is established as a homochiral form $\cdots \Lambda_s$-$\Lambda_s$-$\Lambda_s$-$\Lambda_s \cdots$. It is proposed that the chiral configuration of the metal center in wild-type FeMo-co biosynthesis might be induced by the early coordination of the chiral *R*-homocitric acid,while a mixture of raceme might be obtained in the biosynthesis of $NifV^-$ FeMo-co factor. The absolute configuration of wild-type FeMo-co factor is assigned as $\Delta_R$. © 2002 Elsevier Science Inc. All rights reserved.

**Key words**  Stereospecific  Absolute configuration  Malic acid  Malate  Molybdate(VI)  Nitrogenase  Cofactor biosynthesis  Crystal structure

## 1  Introduction

Molybdenum is widely available to biological systems as an intergral component of the multinuclear metal center of nitrogenase and as the mononuclear active sites of oxygen transferases[1-10]. The

---

① Corresponding author. Tel.:186-592-218-4531;fax:186-592-2183047.

*E-mail addresses*: zhzhou@xmu.edu.cn(Z.-H. Zhou),krtsai@xmu.edu.cn(K.-R. Tsai).

interaction of molybdenum and $R$-homocitrate displays a unique bidentate coordination in the iron-molybdenum cofactor(FeMo-co)and was considered to be biologically essential in the mobilization and the early biosynthesis of FeMo-co [MoFe$_7$S$_9$ ($R$-homocitrate)] in nitrogenases[11,12]. Alternative nitrogenase containing citrate or malate as a replacement to homocitrate prevented the enzyme from strong binding and efficient reduction of N$_2$, while acetylene and proton reduction remained at a relatively high level[13-18].

$S$-Malic acid is a natural constituent and common metabolite of plants and animals, being involved in the Krebs cycle and in the glyoxylic acid cycle[19]. It is the principal acid in apples, but is also present in many other fruits[20]. Formations of molybdenum malate and its kinetics in solutions have been studied extensively by UV-Vis, potentiometry, polarimetry, cryoscopy, and nuclear magnetic resonance (NMR) spectroscopy[21-38]. The well structurally characterized malato molybdate with 2 : 1 and 1 : 2 ratios(Mo:mal) were reported as (NH$_4$)$_4$[Mo$_4$O$_{11}$(mal)$_2$] • H$_2$O[39,40], (NH$_4$)$_4$[Mo$_4$O$_{11}$-(mal)$_2$] • 6H$_2$O[41,42] and Cs$_2$[MoO$_2$(Hmal)$_2$] • H$_2$O[31]. However, there is no previous assignment about the metal configuration of the malato molybdate ( Ⅵ ). This promotes our further investigation of chiral malato molybdenum complex. In this work, we report the synthesis, spectroscopic data, structural characterization, assignment of absolute configuration of a homochiral polymeric molybdenum ( Ⅵ )-malate complex isolated from aqueous solution, and its relationships with the stereospecific formation and absolute configuration of iron-molybdenum cofactor in nitrogenase.

# 2　Experimental

## 2. 1　Materials, instruments and analytical methods

Sodium molybdate, potassium molybdate and malic acid were purchased from Sigma and Fluka. Electronic spectra were recorded on Shimadzu UV2501 spectrophotometer, and IR data were recorded as Nujol mulls between KBr plates on a Nicolet 360 Fourier transform(FT)IR spectrometer. Elemental analysis was performed using EA 1110 elemental analyzers. The optical rotation in water was determined on a WZZ-1 polarimeter. Conductometric measurements were performed on a DDS-11A conduc-tometer. $^1$H- and $^{13}$C-NMR spectra were recorded on a Varian UNITY 500 NMR spectrometer using DDS(sodium 2,2-dimethyl-2-silapentane-5-sulfonate)as internal reference.

## 2. 2　Preparation of Λ-(-)$_D$-Na$_{3n}$[MoO$_2$H(S-mal)$_2$]$_n$(1)

Sodium molybdate (10 mmol) was added to a solution of $S$-malic acid (10 mmol) and sodium dihydrogen malate(10 mmol)prepared from the reaction of $S$-malic acid and sodium hydroxide. The mixture(pH 5. 0)was heated in a water bath at 75 ℃ for 5 h, and kept in a refrigerator for several days. The resulting solid was collected and washed with ethanol to give a white solid(2. 5 g, 54%). $[\alpha]_D$ −44. 2°(water, $c$, 5) Found: C, 20. 8, H, 1. 7. Calculated for C$_8$H$_7$O$_{12}$Na$_3$Mo, C, 20. 9, H, 1. 5%. IR (KBr):$\nu_{as}$(C=O)1656$_{vs}$, $\nu_s$(C=O)1446$_m$,1410$_s$,1385$_s$,$\nu_s$(Mo=O)910$_s$,862$_s$ cm$^{-1}$. $^1$H-NMR(500 MHz, D$_2$O,ppm):$\delta_H$ 2. 695(m,3H),$^{13}$C-NMR(75 MHz,D$_2$O,ppm):$\delta_c$ 184. 5(CO$_2$)$_\alpha$, 178. 1,177. 0(CO$_2$)$_\beta$, 81. 4(≡C—O),41. 4,40. 0(=CH$_2$).

## 2. 3　Preparation of(-)$_D$-K$_{3n}$[MoO$_2$H(S-mal)$_2$]$_n$ • nH$_2$O(2)

Potassium molybdate(20 mmol)was added to a solution of $S$-malic acid(20 mmol)and potassium dihydrogen malate(20 mmol) prepared from the reaction of *S*-malic acid and potassium hydroxide. The mixture(pH 5. 0) was heated in a water bath at 75 ℃ for 5 h and evaporated to above 10 ml, and kept in a refrigerator for several days. The resulting solid was collected and washed with ethanol to give a white solid(3. 9 g, 37%). $[\alpha]_D$ $-28. 8°$(water, $c$, 5) Found: C, 18. 3, H, 1. 7. Calculated for $C_8H_9O_{13}K_3Mo$, C, 18. 3, H, 1. 7%. IR(KBr): $v_{as}$(C=O)$1680_{sh}$, $1647_{vs}$, $1627_{vs}$, $v_s$(C=O)$1403_m$, $1367_s$, $v_s$(Mo=O)$932_s$, $912_s$, $847_m$ cm$^{-1}$. $^1$H-NMR(500 MHz, $D_2O$, ppm): $\delta_H$ 2. 715(m, 3H), $^{13}$C-NMR(75 MHz, $D_2O$, ppm): $\delta_C$ 184. 5$(CO_2)_a$, 178. 0, 177. 0$(CO_2)_\beta$, 81. 4($\equiv$C—O), 41. 2, 39. 9($=CH_2$).

Crystals of suitable quality for the subsequent X-ray diffraction studies were obtained as transparent prism by cooling the saturated solution of compound **1** in a refrigerator. The resulting colorless crystals were mounted in a capillary for X-ray analysis.

## 2. 4　X-Ray data collection, structure solution and refinement

Crystallographic data for the malato molybdate **1** are summarized in Table 1. Diffraction data were collected on an Enraf-Nonius CAD-4 diffractometer with graphite monochromated Mo Kα radiation at 296 K. A Lorentz-polarization factor, anisotropic decay and empirical absorption corrections were applied. The structure was primary solved by WinGX package[43] and refined by full-matrix least-squares procedures with anisotropic thermal parameters for all the non-hydrogen atoms. H atoms were located from difference Fourier map and not refined. Final calculations were performed using Shelxl97 software package and published scattering factors[44,45]. Selected atomic distances and bond angles are given in Table 2.

**Table 1　Crystal data summaries of intensity data collection and structure refinement for $\Lambda$-(-)$_D$-Na$_{3n}$[MoO$_2$H(*S*-mal)$_2$]$_n$ 1**

| | |
|---|---|
| Empirical formula | $C_8H_7O_{12}Na_3Mo$ |
| Formula weight | 460. 05 |
| Crystal color, habit | Colorless, prism |
| Crystal dimers(mm) | 0. 11×0. 15×0. 34 |
| Crystal system | Monoclinic |
| Space group | C2 |
| Formula units /unit cell | 2 |
| Cell constants: | |
| $a$(Å) | 11. 6045(9) |
| $b$(Å) | 10. 6323(9) |
| $c$(Å) | 5. 7179(2) |
| $\beta$(°) | 93. 714(4) |
| $V$(Å$^3$) | 704. 0(3) |
| Density$_{calc}$ (g/cm$^3$) | 2. 170 |
| $F_{000}$ | 452 |
| Diffractometer | Enraf-Nonius CAD-4 |
| Radiation | Mo Kα(λ=0. 7107 Å) |
| Temperature(℃) | 25 |
| Scan width | 0. 45+0. 35 tan $\theta$ |
| Decay of standards(%) | −1. 3 |
| Number of reflections measured | 730 |
| 2$\theta$ range(°) | 2-52 |
| Number of reflections observed | 730 |
| [$I$>2$\sigma$( I )] | ($R_{int}$=0. 0081) |

续表

| Number of parameters varied | 112 |
| Flack parameter | 0.00(3) |
| Weight | $1/[\sigma^2(F_o^2)+(0.0274P)^2]$ |
| | where $P=(F_o^2+2F_c^2)/3$ |
| $S$ | 1.13 |
| $R[F^2>2\sigma(F^2)]$ | 0.014 |
| $wR(F^2)$ | 0.038 |
| Largest feature final diff. map(e$^-$ Å$^{-3}$) | 0.23,$-$0.47 |

**Table 2  Selected bond distances(Å)and angles(°)for $\Lambda$-(-)$_D$-Na$_{3n}$[MoO$_2$H(S-mal)$_2$]$_n$**

| | | | |
|---|---|---|---|
| Mo(1)—O(1) | 1.935(2) | O(2)—C(2) | 1.279(4) |
| Mo(1)—O(2) | 2.182(3) | O(3)—C(2) | 1.226(4) |
| Mo(1)—O(6) | 1.726(3) | O(4)—C(4) | 1.276(5) |
| O(1)—C(1) | 1.405(4) | O(5)—C(4) | 1.224(4) |
| | | | |
| O(1)—Mo(1)—O(1$'$) | 151.6(1) | O(2)—Mo(1)—O(2$'$) | 81.3(2) |
| O(1)—Mo(1)—O(2) | 75.41(9) | O(2)—Mo(1)—O(6) | 164.4(1) |
| O(1)—Mo(1)—O(2$'$) | 83.05(9) | O(2)—Mo(1)—O(6$'$) | 89.2(1) |
| O(1)—Mo(1)—O(6) | 91.3(1) | O(6)—Mo(1)—O(6$'$) | 102.7(2) |
| O(1)—Mo(1)—O(6$'$) | 106.6(1) | | |
| | | | |
| Hydrogen bonding | | | |
| O(4)—H(4) | 1.227(3) | O(4)—O(4$''$) | 2.432(5) |
| O(4)$\cdots$H(4)$\cdots$O(4$''$) | 165.3(2) | | |

Symmetry transformation $a'(-x,y,-z)$,$a''(-1-x,y,-z)$.

# 3  Results and discussion

## 3.1  Synthesis and spectroscopic characterization

Preparation of the malato molybdate depends mainly on the requisite proportions of molybdate and malate. The effect of pH variation between pH 4~7 seems not so crucial for the product. Instead of the deprotonated malato molybdate(Ⅵ), like Na$_4$[MoO$_2$(S-mal)$_2$], solid separation from the reaction of sodium or potassium molybdate with sodium or potassium dihydrogen malate resulted in only the main product of Na$_3$[MoO$_2$H(S-mal)$_2$] or K$_3$[MoO$_2$H(S-mal)$_2$]·H$_2$O, which represents an intermediate between the protonated M$_2$[MoO$_2$(S-Hmal)$_2$](M=Na$^+$ or K$^+$)and deprotonated malato molybdate M$_4$[MoO$_2$(S-mal)$_2$]. The title compound is the most stable species over a wide range of pH 4.0—7.0. This implies the special stability of the superstructure:

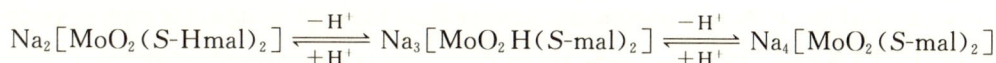

$$Na_2[MoO_2(S-Hmal)_2]\xrightleftharpoons[+H^+]{-H^+}Na_3[MoO_2H(S-mal)_2]\xrightleftharpoons[+H^+]{-H^+}Na_4[MoO_2(S-mal)_2]$$

There was no direct evidence of the other crystalline phase of malato molybdate in the pH 4.0—7.0,although solution studies have suggested the existences of mononuclear and dinuclear species

$[MoO_3(mal)]^{3-}$, $[Mo_2O_5(mal)_2]^{4-}$, and $[Mo_2O_6(mal)(H_2O)_2]^{3-}$ and their protonated forms[38]. Attempts to separate the products in high pH value only resulted in the isolation of homopolymolybdates. The preferred stoichiometry of the product appears to be 1 : 2 for metal-to-ligand ratios, which is more stable than 1 : 1 and 2 : 2 complexes. This is different from the complexes formed by citrate, which has one more carboxylic group[46].

The molybdenum(VI) complex exhibits featureless absorption spectrum in the visible range 350—600 nm. The frequencies and assignments of selected IR absorption bands are shown in the experimental section. In the region between 1800 and 1400 $cm^{-1}$, the bands of 1656, 1446, 1410 and 1385 $cm^{-1}$ are corresponding to bound and free carboxyl group $v_{as}$ and $v_s(CO_2)$, respectively. In the region around 900 $cm^{-1}$, the complexes show several bands that might result from the presence of *cis*-dioxo cores. The low frequency symmetric $MoO_2$ stretching may be explained by the coordination of the oxo groups to the sodium or potassium cation(Fig. 1).

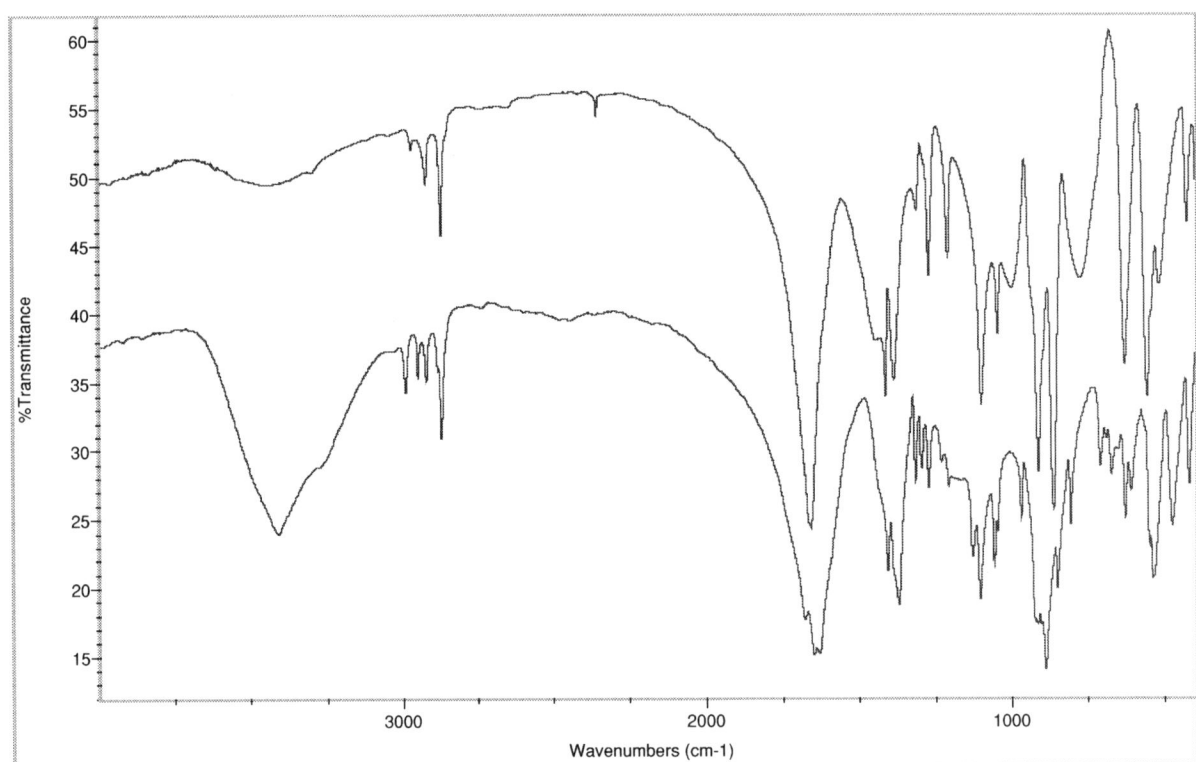

**Fig. 1** IR spectra of $Na_{3n}[MoO_2H(S\text{-}mal)_2]_n$(top)and $K_{3n}[MoO_2H(S\text{-}mal)_2]_n \cdot nH_2O$(bottom).

One of the malate ions carries one proton in the β-carboxyl group, this is seen from the experiment results that not only that hydrogen atom is clearly visible in difference maps and located in the symmetrical position, but also from charge balance in the complex. This is further supported by the measurements of the molar electrical conductivity(2. 8×10$^{-2}$ S m$^2$ mol$^{-1}$ for sodium malato molybdate and 3. 1×10$^{-2}$ S m$^2$ mol$^{-1}$ for potassium malato molybdate)of the title compounds[47]. The observed differences between C-O distances of terminal(β-carboxyl groups [O(4)-C(4), 1. 276(5)Å and O(5)-C(4), 1. 224(4)Å] with or without a proton are smaller than those of protonated malato molybdate(VI) [O(5)-C(4), 1. 19(2)Å and O(6)-C(4), 1. 31(2)Å][31]. This indicates the existence of very strong intermolecular hydrogen bonding between protonated and deprotonated free β-carboxyl groups [O(4)-O(4'), 2. 432(5)Å], which makes the complex form polymeric network. This is also supported by the

absence of IR bands between 1740 and 1700 cm$^{-1}$ for a typical undissociated carboxylic group,indicating
the obvious shift to lower frequency caused by strong hydrogen bonding.

## 3.2 Description of structure

The crystal structure of **1** comprises discrete sodium cations and hydrogen dimalato *cis*-dioxomolybdate anions. As shown in Scheme 1,the molybdenum(Ⅵ)atoms are coordinated by two oxo groups,one "mono" protonated and one fully deprotonated malate ligands. Each malate ion acts as a bidentate ligand with the alkoxy and α-carboxyl coordinated to one molybdenum atom,and the other β-carboxyl groups remain uncomplexed. The two terminal oxo groups are in a *cis*-configuration. Each molybdenum atom is six-coordinate and exists in *quasi*-octahedral geometry. The chiral Λ-configuration is achieved through enantiomerically pure chiral S-malate ligands. An enan-tioselective aggregation with homochirality-ΛΛΛ-is obtained within one dimensional catenarian chain through strong hydrogen bonding,which results in the formation of homochiral supermolecular entities. The chirality of the whole complex is induced by the chirality of the chiral ligand[48](Fig. 2).

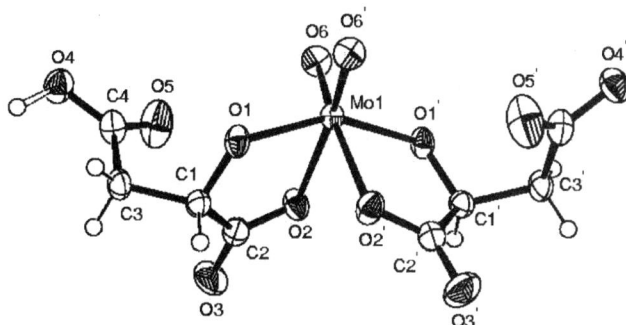

Fig. 2    Perspective view of the anion structure of Λ-Na$_{3n}$[MoO$_2$H(S-mal)$_2$]$_n$. Thermal ellipsoids
are drawn by ORTEP and represent 50% probability surfaces.

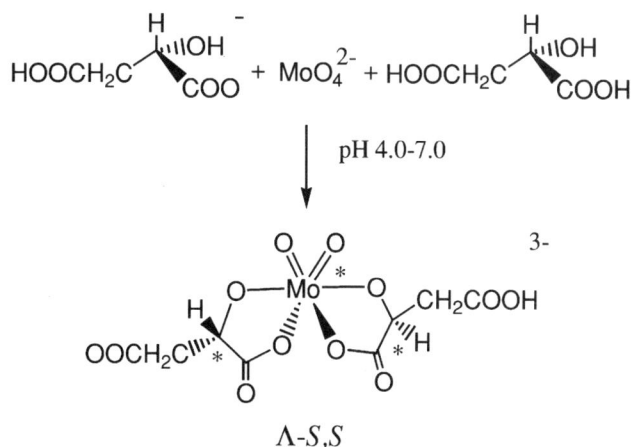

Scheme 1    Reaction of molybdate with malic acid.

As shown in Table 2,the Mo-O distances in malato molybdate vary systematically. Mo=O bonds are 1.726(3)Å,indicating that they are double bonds. The resulting O=Mo=O angles,102.7(2)°is considerably larger than the 90° regular value for *cis* groups,this is expected from the greater O···O repulsions between oxygens with short bonds to the metal atom. The Mo-O(alkoxy)distance is longer

[1. 935(2)Å], indicating the deprotonation of the hydroxyl group in a malate anion, and those to the α-carboxyl O is the longest [2. 182(3)Å], which may be the result of the strong *trans*-influence of the oxo group. Scheme 1 shows the reaction of malate ion with molyb-date(Ⅵ). The product is isostructural with tungstate(Ⅵ)congener complex of malate $Na_{3n}[WO_2H(mal)_2]_n$ and similar to that of protonated malato molybdate(Ⅵ)$Cs_2[MoO_2(Hmal)_2] \cdot H_2O^{[31,49]}$, or the coordination mode of homocitrate to molybdenum in the FeMo cofactor[2].

## 3.3 Biological relevance

Table 3 shows the comparisons of the M-O distances and absolute configuration. The polymeric malato molyb-date(Ⅵ)**1** possesses the shortest M-O [α-alkoxy, 1. 935(2)Å] and M-O [α-carboxy, 2. 182 (3)Å] bonds in the homocitrato, citrato and malato molybdenum complexes. They are compatible with the M-O(alkoxy) and M-O(α-carboxy) bonds of bidentately coordinated homocitrate ligand in MoFe protein[2,8,9], except the significantly longer α-alkoxy bond in the recent *Kp*1 models[9].

**Table 3** Comparisons of Mo-O distances(Å)and our assignment of absolute configuration in chiral metal center of FeMo-cofactor, citrato and malato complexes($H_3$mal＝malic acid, $H_4$cit＝citric acid, $H_4$homocit＝homocitric acid)

| Complex | M-O(alkoxy) | M-O(α-carboxy) | Configuration assignment | Ref. |
|---|---|---|---|---|
| $Na_2[MoO_2(H_2cit)_2] \cdot 3H_2O$ | 1. 953(6), 1. 960(7) | 2. 190(7), 2. 247(6) | Δ/Λ | [50] |
| $K_4[MoO_3(cit)] \cdot 2H_2O$ | 2. 052(2) | 2. 237(7) | Δ/Λ | [46] |
| $[MoFe_7S_9(R\text{-}homocit)(N\text{-}His)]$ | 1. 996 | 2. 167 | $Δ^b$-R | [2] |
| $[MoFe_7S_9(R\text{-}homocit)(N\text{-}His)]$ | 2. 035 | 2. 206 | $Δ^b$-R | [8] |
| $[MoFe_7S_9(R\text{-}homocit)(N\text{-}His)]$ | 2. 35(0)$_{re}$, 2. 35(2)$_{ox}$ | 2. 29(1)$_{re}$, 2. 29(2)$_{ox}$ | $Δ^b$-R | [9] |
| $(NH_4)_4[Mo_4O_{11}(mal)_2] \cdot 6H_2O$ | 2. 00 | 2. 21 | Δ, Δ, Λ, Λ-S, S | [39] |
| $(NH_4)_4[Mo_4O_{11}(mal)_2] \cdot H_2O^a$ | 1. 93(15) | 2. 50(15) | Δ, Δ, Δ, Δ-S, S | [40,41] |
| $(NH_4)_4[Mo_4O_{11}(mal)_2] \cdot 6H_2O$ | 1. 925(6) | 2. 226(6) | Δ, Δ, Λ, Λ-R, R | [42] |
| $Cs_2[MoO_2(Hmal)_2] \cdot H_2O$ | 1. 939(8) | 2. 243(9) | Δ-S | [31] |
| $Na_{3n}[WO_2H(S\text{-}mal)_2]_n$ | 1. 973(4) | 2. 170(4) | Λ-S | [49] |
| $Na_{3n}[MoO_2H(S\text{-}mal)_2]_n$ | 1. 935(2) | 2. 182(3) | Λ-S | This work |

a The complex should be a hexahydrate compound with the same configuration with Δ, Δ, Λ, Λ-$(NH_4)_4[Mo_4O_{11}(S\text{-}mal)_2] \cdot 6H_2O$ with misassignment of the absolute configuration.

b The absolute configuration of chiral Mo center is temporarily assigned as Δ(Fig. 3).

Moreover, achiral citrate ligand only results in the formation of a racemic mixture of Λ-and Δ-complexes[46,50]. The racemic homocitrate, malate or lactate ligands form mesomeric dinuclear complexes coordinated by R-and S-ligands, respectively, like $[K_2(H_2O)_5][V_2O_4(R\text{-}homocitrate)(S\text{-}homocitrate)] \cdot H_2O^{[51]}$, $Cs_2[V_2O_4(R\text{-}mal)(S\text{-}mal)^{[52]}$, and $(NBu_4)_2[V_2O_2(O_2)_2(R\text{-}lact)(S\text{-}lact)] \cdot 2H_2O^{[53]}$. In some cases, the racemic dinuclear complex $K_4[V_2O_2(O_2)_2(R\text{-}lact)_2] \cdot [V_2O_2(O_2)_2(S\text{-}lact)_2]$ is separated[54]. In our experiment, the chiral S-malate ligand resulted in the separation of polymeric homochiral malato molybdate(Ⅵ) Λ-$Na_{3n}[MoO_2H(S\text{-}mal)_2]_n$. The Λ-configuration of the chiral mononuclear complex is determined by the chiral ligand and the non-centrosymmetric space group C2 of the crystal. This is further supported by the recent separation of dinuclear S-lactato perox-ovanadate

( V )[53]. In the literature, the stereospecific formation of nonplanar $V_2O_2$ core in the anion of ($NBu_4$)$_2$ $[V_2O_2(O_2)_2(S\text{-lact})_2] \cdot 2H_2O$ is typical for the coordination of the chiral S-lactato ligand, which is one of the enantiomers in the racemic complex $K_4[V_2O_2(O_2)_2(R\text{-lact})_2] \cdot [V_2O_2(O_2)_2(S\text{-lact})_2]^{[54]}$, while the planar $V_2O_2$ core is characteristic of the $R,S$-combination of chiral lactato ligands.

Although it is undetermined whether the formation of homochiral polymeric malato molybdate is resulted from different solubility or spontaneous resolution, complex **1** may represent an example for the interpretation of the potentially stereospecific formation of iron-molybdenum cofactor in nitrogenase. It is proposed that α-hydroxy-carboxyte ligands take part in the incorporation of molybdenum in the cofactor of the active site of nitrogen-ases. Metal ions are adsorbed in the form of the oxoanions molybdate ( Ⅵ ) and then reduced to the oxidation states III or IV. The coordination of hydroxycarboxylic ligands seems essential to the mobilization and reduction processes[51,55,56]. Alternative substitution of organic acids for homocitrate in the in vitro FeMo-co synthesis shows the differences of the $R$-and $S$-malate with $R$-malate as the preferred configuration[16].

Based on our assignment of metal configuration of chiral malate and achiral citrate complexes, it is deduced that a racemic Λ/Δ complex might be obtained in the biosynthesis of $NifV$-FeMo-co factor from achiral citrate[13,17,18], while the chiral configuration of metal center in the wild type FeMo-co might be induced by the coordination of the chiral $R$-homocitrate, in which the chiral configuration of metal center is temporarily assigned as Δ-configuration(Fig. 3). It is noteworthy that only the Δ-configuration of the molybdenum center is achieved in FeMo-co from different sources of nitrogenases like *Azotobacter*

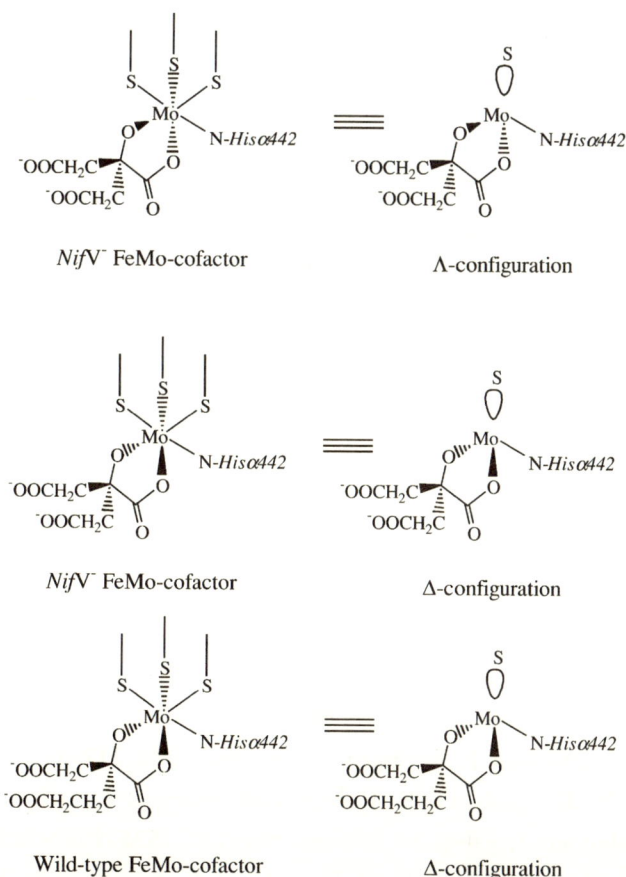

Fig. 3　Assignment of the absolute configuration of molybdenum center in wild-type(LHS) and *NifV*⁻ (RHS) cofactors.

*vinelandii*($A v1$), *Clostridium pasteurianum* ($C p1$), and *Klebsiella pneumoniae* ($K p1$), which implies the stereospecific formation of FeMo-cofactor in nitrogen-ase and the complexity in the biosynthesis of FeMo-cofactor.

## 4 Supplementary material

Tables S1-S4 containing tables of X-ray crystal structure positions, thermal parameters, detailed bond lengths and angles and a polymeric anion structure of $\Lambda$-(-)$_D$-Na$_{3n}$[MoO$_2$H(S-mal)$_2$]$_n$, are available in electronic form. (See the Elsevier website)

## Acknowledgements

Financial support by the Ministry of Science and Technology (001CB1089), National Natural Science Foundation of China(No. 29933040), the Doctoral Foundation, and the University Key Teacher Foundation from the Chinese Ministry of Education are gratefully acknowledged. We are grateful to Dr. Seik Weng Ng(Malaysia) for helpful discussion.

## References

[1] M. Coughlan(Ed.), Molybdenum and Molybdenum Containing Enzymes, Pergamon, 1980.

[2] J. Kim, D. C. Rees. Science, 1992, **257**: 1677~1682.

[3] J. T. Bolin, N. Campobasso, S. W. Muchmore, TV Morgan, L. E. Mortenson. ACS Symp. Ser, 1993, **535**: 186~195.

[4] R. Hille. Chem. Rev, 1996, **96**: 2757~2816.

[5] J. B. Howard, D. C. Rees. Chem. Rev, 1996, **96**: 2965~2982.

[6] B. K. Burgess, D. J. Lowe. Chem. Rev, 1996, **96**: 2983~3011.

[7] C. Kisker, H. Schindelin, D. C. Rees. Annu. Rev. Biochem, 1997, **66**: 233~267.

[8] H. Schindelin, C. Kisker, J. L. Schlessman, J. B. Howard, D. C. Rees. Nature, 1997, **387**: 370~376.

[9] S. M. Mayer, D. M. Lawson, C. A. Gormal, S. M. Roe, B. E. Smith. J. Mol. Biol, 1999, **292**: 871~891.

[10] D. C. Rees, J. B. Howard. Curr. Opin. Chem. Biol, 2000, **4**: 559~566.

[11] T. R. Hoover, A. D. Robertson, R. L. Cerny, R. N. Hayes, J. Imperial, V. K. Shah, P. W. Ludden. Nature, 1987, **329**: 855~857.

[12] T. R. Hoover, J. Imperial, P. W. Ludden, V. K. Shah. Biochemistry, 1989, **28**: 2768~2771.

[13] J. H. Liang, M. Madden, V. K. Shah, R. H. Burris. Biochemistry. 1990, **29**: 8577~8581.

[14] P. A. Mclean, R. A. Dixon. Nature, 1981, **292**: 655~656.

[15] T. R. Hawkes, P. A. Mclean, B. E. Smith. J. Biochem, 1984, **217**: 317~321.

[16] P. W. Ludden, V. K. Shah, G. P. Roberts, M. Homer, R. Allen, T. Paustian, J. Roll, R. Chatterjee, M. Madden, J. Allen. ACS Symp. Ser, 1993, **535**: 196~215.

[17] K. L. C. Grönberg, C. A. Gormal, M. C. Durrant, B. E. Smith, R. A. Henderson. J. Am. Chem. Soc, 1998, **120**: 10613~10621.

[18] K. L. C. Grönberg, C. A. Gormal, M. C. Durrant, B. E. Smith, R. A. Henderson. J. Inorg.

Biochem,1999,**74**:30.

[19]H. R. Mahler,E. H. Cordes. Biological Chemistry,Harper and Row,1966.

[20]S. R. Berger,Kirk-Othmer Encyclopedia of Chemical Technology,Wiley,1981.

[21]E. Richardson. J. Inorg. Nucl. Chem,1960,**13**:84~90.

[22]M. J. Baillie,D. H. Brown. J. Chem. Soc,1961:3691~3695.

[23]S. Prasad,L. P. Pandey. J. Indian Chem. Soc,1965,**42**:783~788.

[24]D. H. Brown,J. Macpherson. J. Inorg. Nucl. Chem,1971,**33**:4203~4207.

[25]M. D. Azevedo,R. G. Costa,M. T. Vilhena. Rev. Port. Quim,1973,**15**:35~38.

[26]K. Ogura,Y. Enaka,K. Morimoto. Electrochim. Acta,1978,**23**:289~292.

[27]Y. K. Tselinskii,B. V. Pestryakov,A. I. Olivson,T. M. Pekhtereva. Zh. Neorg. Khim,1979,**24**:
1243~1247.

[28]E. Mikanová,M. Bartušek. Scripta Fac. Sci. Nat. Univ. Purk. Brun,1981,**11**:439~449.

[29]A. Beltrán,A. C. Avalos,J. Beltrán. J. Inorg. Nucl. Chem,1981,**43**:1337~1341.

[30]M. M. Caldeira,M. Emilia,T. L. Saraiva,VM. S. Gil. Inorg. Nucl. Chem. Lett,1981,**17**:295~
304.

[31]C. B. Knobler,A. J. Wilson,R. N. Hider,I. W. Jensen,B. R. Penfold,W. T. Robinson,C. J.
Wilkins. J. Chem. Soc.,Dalton Trans,1983:1299~1303.

[32]M. Bartušek,J. Havel,D. Matula. Collect. Czech. Chem. Commun,1986,**51**:2702~2711.

[33]D. O. Mártire,M. R. Féliz,A. L. Capparelli. Polyhedron 7,1988:2709~2714.

[34]D. O. Mártire,M. R. Féliz,A. L. Capparelli. Polyhedron 8,1989:1387~1389.

[35]D. O. Mártire,M. R. Féliz,A. L. Capparelli. Ann. Assoc. Quim. Agent,1989,**77**:233~245.

[36]D. O. Mártire,M. R. Féliz,A. L. Capparelli. Ann. Quim,1990,**86**:117~121.

[37]VM. S. Gil. Pure Appl. Chem,1989,**61**:841~848.

[38]J. J. Cruywagen,E. A. Rohwer,R. F. van de Water. Polyhedron,1997,**16**:243~251.

[39]J. E. Berg,S. Brandange,L. Lindblom,P. E. Werner. Acta Chem. Scand. A,1977,**31**:325~328.

[40]J. E. Berg,P. E. Werner. Z. Kristallogr,1977,**145**:310~320.

[41]M. A. Porai-Koshits,L. A. Aslanov,G. V. Ivanova,T. N. Polynova. Zh. Strukt. Khim,1968,**9**:
475~480.

[42]Z. H. Zhou,W. B. Yan,H. L. Wan,K. R. Tsai. Chinese J. Struct. Chem,1995,**14**:255~260.

[43]L. J. Farrugia,WinGX"A Windows Program for Crystal Structure Analysis",University of
Glasgow,Glasgow,1998.

[44]G. M. Sheldrick,SHELXL97 and SHELXS97,University of Göttingen,Göttingen,1997.

[45]D. T. Cromer,J. T. Waber,International Tables for X-Ray Crystallography,Kynoch Press,
Birmingham,1974.

[46]Z. H. Zhou,H. L. Wan,K. R. Tsai. Inorg. Chem,2000,**39**:59~64.

[47]R. J. Angelici,Synthesis and Technique in Inorganic Chemistry,Saunders,Philadelphia,PA,
1977.

[48]U. Knof,A. von Zelewsky. Angew. Chem. Int. Ed,1999,**38**:302~322.

[49]Z. H. Zhou,G. F. Wang,S. Y. Hou,H. L. Wan,K. R. Tsai. Inorg. Chim. Acta,2001,**314**:184~
188.

[50]Z. H. Zhou,H. L. Wan,K. R. Tsai,J. Chem. Soc. . Dalton Trans,1999:4289~4290.

[51]D. W. Wright,R. T. Chang,S. K. Mandal,W. H. Armstrong,W. H. Orme-Johnson. . J.
Bioinorg. Chem,1996①:143~151.

[52]M. Biagioli,L. Strinna-Erre,G. Micera,A. Panzanelli,M. Zema. Inorg. Chim. Acta,2000,**310**:1~9.

［53］P. Schwendt, P. Švančâarek, I. Smatanová, J. Marek. J. Inorg. Biochem, 2000, **80**: 59～64.

［54］F. Demartin, M. Biagioli, L. Strinna-Erre, A. Panzanelli, G. Micera. Inorg. Chim. Acta, 2000, **299**: 123～127.

［55］R. R. Eady. Chem. Rev, 1996, **96**: 2965～2982.

［56］A. Müller, E. Krahn. Angew. Chem. Ind. Ed. Engl, 1995, **34**: 1071～1078.

■ **本文原载**:《厦门大学学报》(自然科学版)第41卷第2期(2002年3月),第135～140页。

# 碳纳米管促进Cu-基高效甲醇合成催化剂*

董 鑫① 张鸿斌② 林国栋 袁友珠 蔡启瑞

(厦门大学化学化工学院 固体表面物理化学国家重点实验室,福建厦门 361005)

**摘 要** 用自行制备的碳纳米管(CNTs)作为促进剂,研制出一类高效甲醇合成催化剂 $Cu_iZn_jAl_k-O_x-wt\%CNTs$,评价它们对 $CO/CO_2$ 加氢成甲醇的催化活性,并与非 CNTs 促进的相应体系作对比研究。实验发现,碳纳米管能显著地促进甲醇合成反应活性的提高。在 493 K, 5.0MPa, $H_2/CO/CO_2/N_2=62/30/5/3(V/V)$,GHSV$=8\,000\,h^{-1}$ 的反应条件下,在 $Cu_6Zn_3Al_1-O_x-$ 12.5wt%CNTs 催化剂上,甲醇的时空产率达 $1\,064\,mg\,h^{-1}(g\text{-catal})^{-1}$;产物中甲醇的选择性达 98%以上;而在相同的制备和反应条件下、在非促进相应催化剂 $Cu_6Zn_3Al_1-O_x$ 上,甲醇的时空产率 只达 $729\,mg\,h^{-1}(g\text{-catal})^{-1}$。$H_2$-TPD 观测揭示,常压下在 CNTs 材料、以及 CNTs 促进催化剂 $Cu_iZn_jAl_k-O_x-wt\%CNTs$ 上,可以吸附存储着数量相当可观、在 423～573 K 温度范围处于可逆吸、 脱附的吸附氢物种。这一特性将有助于在甲醇合成反应条件下,营造较高氢稳态浓度的表面氛围, 以有利于提高表面加氢反应的速率;与此同时,很可能由于加氢活性的提高,使得碳纳米管促进催化 剂上甲醇合成反应所需温度比非促进的相应体系下降15～25 K,这在相当大程度上将有利于提高 CO 的平衡转化率和甲醇的平衡产率。本文结果表明,碳纳米管对 $H_2$ 优异的吸附、活化及存储性能 对于促进其所改进催化剂上甲醇合成反应活性的显著提高,起着关键作用。

**关键词** 碳纳米管 碳纳米管促进剂 $Cu_iZn_jAl_k-O_x-wt\%CNTs$ 甲醇合成 $CO/CO_2$ 加氢

**中图分类号** O 643 **文献标识码** A

甲醇是用途最广泛的 $C_1$ 化学品。它已被证实可作为高辛烷值、低污染的车用燃料,或作为 PEM 燃料电池氢燃料载体,也是重要的基本化工原料,应用前景十分广阔。现有的甲醇合成工艺中,原料合成气只有一小部分(10%)转化为甲醇,大部分未反应合成气需经分离后,反复、多次进行循环反应以提高原料合成气的利用率,工艺流程及设备比较复杂,并需额外消耗用于分离、循环的能量。因而,高度活泼的甲醇合成催化剂的研制及相应低温高转化率甲醇合成过程的开发一直是许多研究工作追求之目标。

碳纳米管是一类纳米级新型碳素材料。典型的碳纳米管是由碳六元环组成的类石墨平面、按一定方式组合而成的纳米级管状结构。作为一种新型碳素催化剂载体,碳纳米管近年来已引起国内外催化学界的日益注意[1]。Planeiz 等[2]制备一种碳纳米管负载 Ru 催化剂,其对肉桂醛加氢制肉桂醇的催化活性比常规 Ru/C 催化剂高。Zhang 等[3]研制出一种碳纳米管负载铑膦配合物催化剂,其对丙烯氢甲酰化制丁醛显示出高活性和对正构产物分子优异的区位选择性,并提出碳纳米管纳米级的管腔可以起空间选择催化作用的新观点。从化学催化角度考虑,碳纳米管诱人的特性,除其高的机械强度、大而可修饰的表面、

* 收稿日期:2001 年 12 月 30 日。

基金项目:国家自然科学基金(50072021),教育部科技基金(99069)和福建省自然科学基金(2001H017)资助项目。

① 作者简介:董鑫(1974— ),男,博士研究生。

② 通讯作者:张鸿斌,E-mail:hbzhang@xmu.edu.cn。

类石墨的管壁结构、以及纳米级的管腔外,其优良的电子传递性能、对氢强的吸附能力并可期产生的氢溢流效应也很值得注意[4]。

本文报道一类碳纳米管促进 Cu-基高效甲醇合成催化剂。初步工作显示,其对 $CO/CO_2$ 加氢制甲醇显示出很高的低温催化活性;在 493 K 反应温度下,其 CO 转化率及甲醇时空产率是现有工业甲醇合成 $Cu-ZnO-Al_2O_3$ 催化剂的 ~1.46 倍。本文结果对于增进对碳纳米管促进作用本质的理解,以及促进低温高效甲醇合成催化剂的研究开发,均有重要意义。

# 1 实验

本文所用碳纳米管按前文[5,6]方法制备。所制得碳纳米管是一种多壁碳纳米管,其外管径为 10~45 nm,内管径 2.4~3.6 nm,含碳量≥99wt%,石墨状碳含量≥85wt%,比表面积 130 $m^2/g$.

碳纳米管促进催化剂(记为 $Cu_iZnAl_k-O_x-wt\%$ CNTs)由共沉淀法制备:将计算量的 $Cu(NO_3)_2 \cdot 3H_2O$,$Zn(NO_3)_2 \cdot 6H_2O$ 和 $Al(NO_3)_3 \cdot 9H_2O$(纯度均为 AR 级)三者混合,加入计量去离子水制成溶液 A,另将计量无水 $Na_2CO_3$(纯度为 AR 级)溶于计量去离子水,制成与溶液 A 等离子当量浓度的溶液 B,在 353 K 恒温并不断搅拌条件下,将溶液 A 和 B 等速、并流滴入一预置有计量 CNTs 的反应容器内进行共沉淀反应,保持料液 pH=6.8,连续搅拌 5 h,停止加热继续搅拌 3 h,后让其静置陈化过夜,料液经抽滤,滤饼经反复洗涤洗去 $Na^+$,于 393 K 烘干 4 h,氮气氛保护下 543~573 K 灼烧 23 h,冷至室温即得氧化前驱态催化剂;经压片、破碎,筛选出 40~80 目试样供活性评价之用。

催化剂对甲醇合成的催化活性评价在加压固定床连续流动反应器-GC 组合系统上,453~563 K,2.0~5.0 MPa,$H_2/CO/CO_2/N_2=62/30/5/3$(V/V),GHSV=1 000 8 000 $h^{-1}$ 的反应条件下进行。每次试验催化剂用量 500 mg(约 0.5 mL)。反应前,氧化前驱态催化剂先经低氢还原气(5% $H_2$+95% $N_2$)按一定升温程序进行原位预还原,后切换为原料合成气在一定温度、压力、空速条件下进行反应。产物由配备以 TCD 检测器和双色谱柱的 102GD 型气相色谱仪作在线分析。两支色谱柱分别充填 5A 分子筛和有机担体-401;前者用于分离 $N_2$ 和 CO,后者用于分离 $CH_3OH$,$CO/N_2$,$CO_2$,MF,DMC,CO 转化率由 $N_2$-内标法测算,甲醇时空产率直接由外标法(即工作曲线法)测量,并与内标法间接测算的结果相核校。$H_2$-TPD(程序升温脱附)测试系将待测试样 30 mg 置于石英质吸附-脱附管中,在流动 $H_2$(纯度 99.999%)气氛下从室温程序升温至 973 K 并保持 1 h,后降至室温并保持 2 h,切换为高纯 Ar 吹扫,直至色谱仪基线平稳,始以 10 K/min 升温速率进行程序升温脱附,氢信号变化由 102GD 型气相色谱仪 TCD 检测器进行在线跟踪监测。

# 2 结果与讨论

原料气空速优化试验结果表明,在 2.0 MPa、453~523 K 反应条件下,当空速(GHSV)低于 3 000 $h^{-1}$ 时,CO 转化率和甲醇时空产率均随空速增加而上升,并在 GHSV=3 000 $h^{-1}$ 附近达到极大,而后随空速继续增加而缓慢下降。据此可以认为,在上述温度、压力范围,当 GHSV=3 000 $h^{-1}$ 时反应已进入动力学控制区。本文催化剂活性评价试验均选定在 GHSV≥3 000 $h^{-1}$ 的条件下进行。已知原料气中的 $CO_2$ 是合成气持续快速转化为甲醇之不可或缺的参与组分,但评价实验显示反应尾气中 $CO_2$ 的含量与原料合成气相比变化很小,故本文所有活性评价结果均只示出 CO 转化率和甲醇时空产率。

图 1 示出在 453~523 K 温度范围,Cu/Zn/Al=6/3/1(摩尔比),添加以不同 CNTs 量的 4 种催化剂上,$CO/CO_2$ 加氢合成甲醇的活性评价结果。4 种催化剂上 CO 转化率及甲醇时空产率均先随温度升高而上升,在 493~513 K 附近达到极大值,而后随温度继续上升而下降。这很可能缘于:在极大值前的较

低温度下,反应受动力学控制;在极大值之后随着反应温度升高,反应转而受热力学平衡限制,CO 转化率及相应的甲醇时空产率均随反应温度升高而呈下降趋势。因此,在 CO 转化率及甲醇时空产率对温度的变化曲线上均出现一极大点。四种不同 CNTs 添加量的催化剂上甲醇合成活性的高低顺序为:a) $Cu_6Zn_3Al_1-O_x$-12.5wt%CNTs > b)$Cu_6Zn_3Al_1-O_x$-10.0wt%CNTs > c)$Cu_6Zn_3Al_1-O_x$-15.0wt%CNTs > d)$Cu_6Zn_3Al_1-O_x$-0.0wt%CNTs,以添加 12.5wt%CNTs 的为最佳。在该催化剂上 493 K 反应温度下,CO 转化率及甲醇时空产率分别达到 42.3% 和 537 mg-$CH_3OH\ h^{-1}$(g-catal.)$^{-1}$,而在相同条件下制备之不含 CNTs 的非促进催化剂(d)上,在其最佳活性温度 508 K(比碳纳米管促进的相应体系高约 15 K),CO 转化率及甲醇时空产率分别只达 30.2% 和 379 mg-$CH_3OH\ h^{-1}$(g-catal.)$^{-1}$,明显低于 CNTs 促进之催化剂。

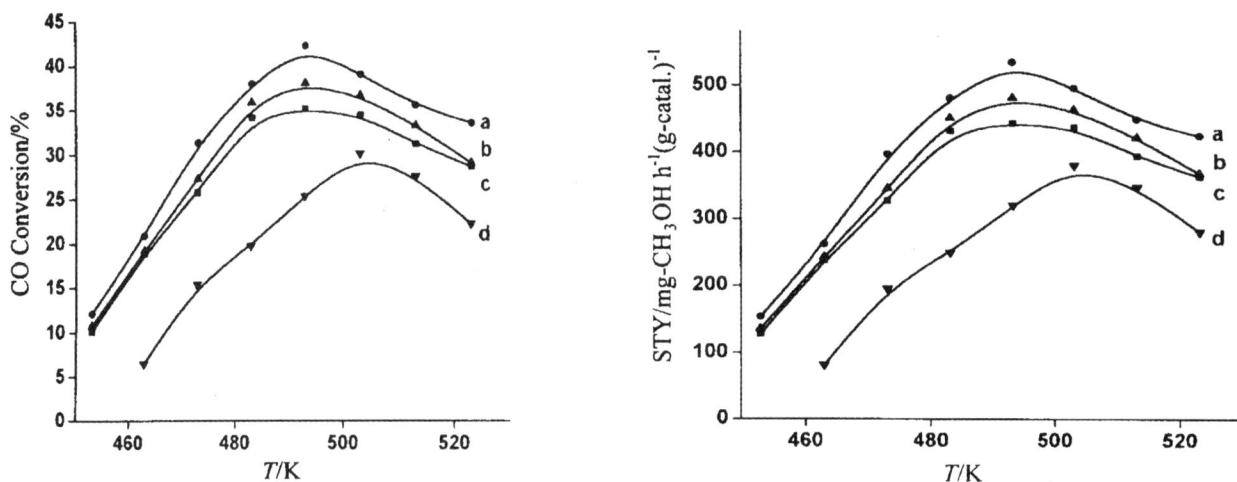

图 1  不同 CNTs 添加量的 $Cu_6-Zn_3-Al_1-O_x-$wt%CNTs 催化剂上 CO 转化率和甲醇时空产率随温度的变化

Fig. 1  Dependence of CO-conversion and $CH_3OH$-STY on temperature over a series of catalysts: (a) $Cu_6Zn_3Al_1-O_x$-12.5wt%CNTs; (b)$Cu_6Zn_3Al_1-O_x$-10.0wt%CNTs; (c)$Cu_6Zn_3Al_1-O_x$-15.0wt%CNTs; (d)$Cu_6Zn_3Al_1-O_x$-0.0wt%CNTS. Reaction conditions: 2.0 MPa, $H_2$/CO/$CO_2$/$N_2$=62/30/5/3(V/V), GHSV=3 000 $h^{-1}$

在优化 CNTs 添加量的基础上,保持 CNTs 添加量为 12.5wt% 不变,进而考察催化剂活性与金属元素组成的相关性,结果(图 2)表明,所考察的四种催化剂对合成气制甲醇催化活性的高低顺序为:a)$Cu_6Zn_3Al_1-O_x$-12.5wt% CNTs > b)$Cu_5Zn_{2.5}Al_1-O_x$-12.5wt% CNTs > c)$Cu_4Zn_2Al_1-O_x$-12.5wt% CNTs ≈ d)$Cu_7Zn_{3.5}Al_1-O_x$-12.5wt%CNTs,以 Cu/Zn/Al=6/3/1(摩尔比)为最佳:在 493 K 工作温度下,CO 转化率及甲醇时空产率分别达 42.1% 和 535 mg-$CH_3OH\ h^{-1}$(g-catal.)$^{-1}$,产物中甲醇的选择性≥98.0%。

为进一步考察该类催化剂在较高反应度工作状态下的性能,合成气制甲醇反应在较高的压力(5.0MPa)和空速(8 000 $h^{-1}$)条件下进行。结果表明,在 493 K,5.0 MPa,$H_2$/CO/$CO_2$/$N_2$=62/30/5/3(V/V),GHSV=8 000 $h^{-1}$ 的反应条件下,甲醇的时空产率达 1 064 mg $h^{-1}$(g-catal)$^{-1}$;产物中甲醇的选择性≥98%;而相同条件下制备之不含 CNTs 的非促进催化剂,在相同反应条件下甲醇的时空产率只达 729 mg $h^{-1}$(g-catal)$^{-1}$。若干碳纳米管促进催化剂 $Cu_i-Zn_j-Al_k-O_x-$wt%CNTs 及相应非碳纳米管促进体系上合成气制甲醇的对比活性评价结果示于表 1. 比较而言,CNTs 促进催化剂上甲醇合成的时空产率是非 CNTs 促进之相应体系的 ~1.46 倍。

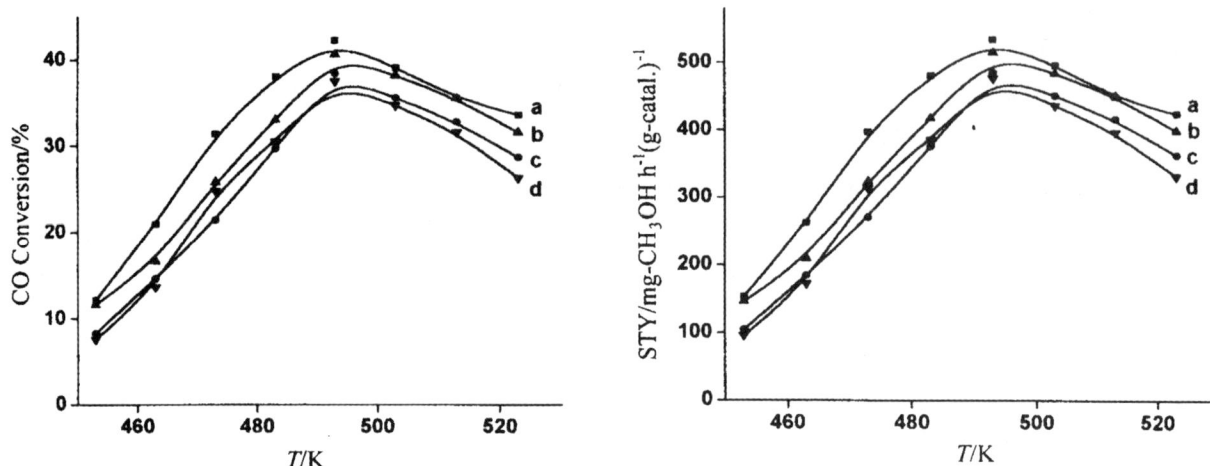

图 2　不同 Cu/Zn/Al 摩尔比的 $Cu_i\text{-}Zn_j\text{-}Al_k\text{-}O_x\text{-}12.5wt\%CNTs$ 催化剂上 CO 转化率和甲醇时空产率随温度的变化

Fig. 2　Dependence of CO-conversion and $CH_3OH$-STYon temperature over a series of catalysts: ( a ) $Cu_6Zn_3Al_1\text{-}O_x\text{-}12.5wt\%$ CNTs; ( b ) $Cu_5Zn_{2.5}Al_1\text{-}O_x\text{-}12.5wt\%$ CNTs; ( c ) $Cu_4Zn_2Al_1\text{-}O_x\text{-}12.5wt\%$ CNTs; ( d ) $Cu_7Zn_{3.5}Al_1\text{-}O_x\text{-}12.5wt\%$CNTs. Reaction conditions: 2.0 MPa, $H_2/CO/CO_2/N_2=62/30/5/3(V/V)$, GHSV=3 000 $h^{-1}$

表 1　碳纳米管促进催化剂 $Cu_i\text{-}Zn_j\text{-}Al_k\text{-}O_x\text{-}wt\%CNTs$ 上甲醇合成的反应活性

Tab. 1　Reactivity of $CH_3OH$ synthesis from syngas over a series of catalysts of $Cu_i\text{-}Zn_j\text{-}Al_k\text{-}O_x\text{-}wt\%CNTs$

| 催化剂试样 | 反应起动温度下 CO 转化率及甲醇时空产率 | | | 工作温度下 CO 转化率及甲醇时空产率 | | |
|---|---|---|---|---|---|---|
| | $T/K$ | $X_{co}/\%$ | STY | $T/K$ | $X_{co}/\%$ | STY |
| $Cu_6Zn_3Al_1\text{-}O_x\text{-}0.0wt\%CNTs$ | 463 | 6.5 | 82.1 | 503 | 30.2 | 379 |
| | | | | 493* | 20.3* | 729* |
| $Cu_6Zn_3Al_1\text{-}O_x\text{-}10.0wt\%CNTs$ | 453 | 11.9 | 149 | 493 | 39.1 | 493 |
| $Cu_6Zn_3Al_1\text{-}O_x\text{-}15.0wt\%CNTs$ | 453 | 10.7 | 135 | 493 | 38.1 | 492 |
| $Cu_6Zn_3Al_1\text{-}O_x\text{-}12.5wt\%CNTs$ | 453 | 14.1 | 178 | 493 | 42.3 | 537 |
| | | | | 493* | 31.1* | 1064* |
| $Cu_5Zn_{2.5}Al_1\text{-}O_x\text{-}12.5wt\%CNTs$ | 453 | 11.6 | 146 | 493 | 40.6 | 515 |
| $Cu_4Zn_2Al_1\text{-}O_x\text{-}12.5wt\%CNTs$ | 453 | 8.3 | 105 | 493 | 38.4 | 486 |
| $Cu_7Zn_{3.5}Al_1\text{-}O_x\text{-}12.5wt\%CNTs$ | 453 | 7.6 | 96.2 | 493 | 37.5 | 477 |

Reaction conditions: 2.0 MPa, $H_2/CO/CO_2/N_2=62/30/5/3(V/V)$, GHSV=3 000 $h^{-1}$; STY at mg-$CH_3OH$ $h^{-1}$ (g-catal.)$^{-1}$. * Data of assays at 5.0 MPa and GHSV=8 000 $h^{-1}$.

　　碳纳米管促进催化剂 $Cu_i\text{-}Zn_j\text{-}Al_k\text{-}O_x\text{-}wt\%$ CNTs 对 $CO/CO_2$ 加氢成甲醇所表现出的优异催化性能,显然同这种碳纳米材料特殊的结构和性质密切相关。图 3 示出该类 CNTs 的 TEM 图象。TEM, SEM 和 $N_2$-BET 联合观测证实,所制得 CNTs 为一类多壁碳纳米管(MWCNTs),外管径在 $15\sim45$ nm 范围,内管径为 $2.6\sim3.4$ nm,管长达数十至数百 $\mu$m,比表面积为 130 $m^2$ $g^{-1}$. 它们的 XRD 图(图 4)与石墨的相近,但相应特征衍射峰强度明显变弱,峰形较为宽化。HRTEM 观测进一步揭示,它们是由一片片具有类石墨片状结构的锥形面沿中空的管轴叠合而成,管壁纵截面呈所谓"鱼骨形",类石墨结构锥形层面与管空心轴倾斜。元素分析结果表明,经纯化的碳纳米管产物中,碳素含量$\geqslant99\%$;$O_2$-TPO(程序升温氧化)测试结果显示,总碳中石墨状碳含量$\geqslant85\%$,无定形碳含量在 $10\%$ 以下。

　　从化学催化的角度考虑,CNTs 诱人的特性,除显而易见高的机械强度,大而可修饰的表面,类石墨

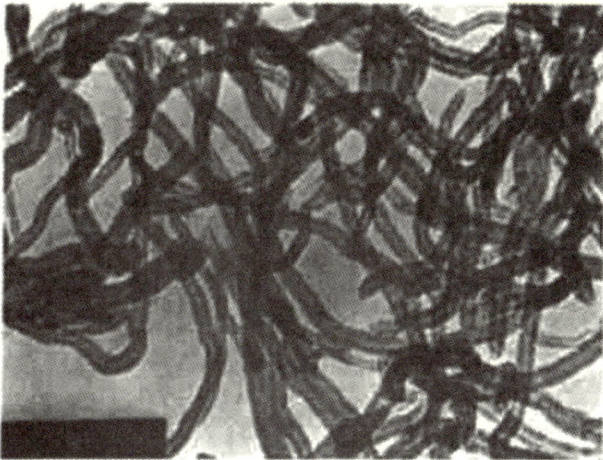

图 3　由 CH₄ 催化分解长成的碳纳米管的 TEM 图象

Fig. 3　TEM image of the CNTs grown catalytically from CH₄ decomposition

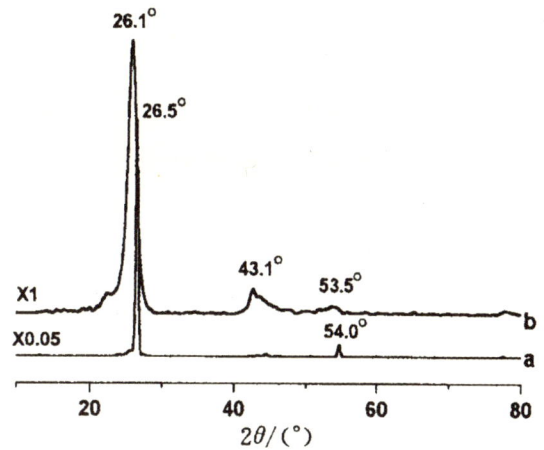

图 4　石墨和 CH₄ 催化分解制备碳纳米管的 XRD 图

Fig. 4　XRD patterns of: a) graphite; b) the CNTs grown catalytically from CH₄ decomposition

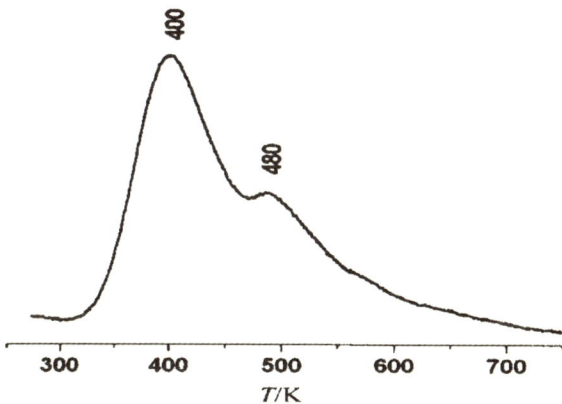

图 5　在碳纳米管材料上吸附氢的 TPD 谱

Fig. 5　TPD spectrum of hydrogen adsorbed on the CNTs material

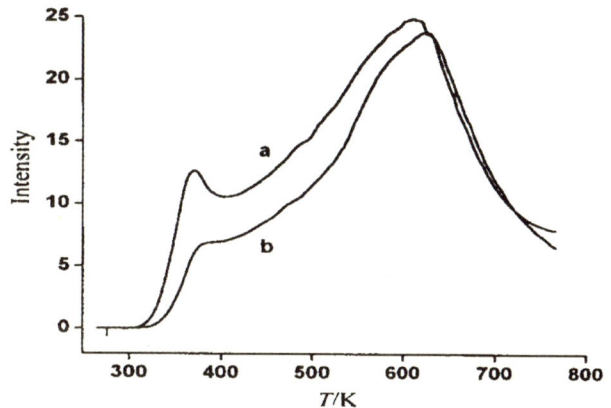

图 6　在碳纳米管促进 Cu-基催化剂上吸附氢的 TPD 谱

Fig. 6　TPD spectra of hydrogen adsorbed on the catalysts: (a) $Cu_6$-$Zn_3$-$Al_1$-$O_x$-12.5wt%CNTs; (b) $Cu_6$-$Zn_3$-$Al_1$-$O_x$-10.0wt%CNTs

结构的管壁,以及纳米级的内管腔之外,CNTs 高的电子传递能力,强的对氢的吸附、存储能力以及可期产生的氢溢流作用也很值得注意。图 5 示出常温常压下吸附 $H_2$ 的 CNTs 试样的 TPD 谱。低温主脱附峰峰温在 400 K,在 480 K 处出现一小肩峰,直到 723 K 才脱附完毕。图 6 示出 CNTs 促进 $Cu_i$-$Zn_j$-$Al_k$-$O_x$ 催化剂上吸附氢的 TPD 谱。十分显然,在本文 CO/$CO_2$ 加氢成甲醇反应温度范围(423～573 K),在 CNTs 促进 $Cu_i$-$Zn_j$-$Al_k$-$O_x$ 催化剂上存在着数量相当可观的可逆吸附氢物种。这一特性将有助于在甲醇合成反应条件下,营造较高氢稳态浓度的表面氛围,从而有利于提高 CO/$CO_2$ 加氢成醇过程中一系列加氢单元反应的速率。与此同时,由于加氢活性的提高,使得碳纳米管促进催化剂上甲醇合成反应所需温度比非促进的相应体系下降 15～25 K,这在相当大程度上将有利于提高 CO 的平衡转化率和甲醇的平衡产率。在我们看来,碳纳米管对 $H_2$ 优异的吸附、活化及存储性能对其所促进催化剂上甲醇合成反应活性的显著提高,起着关键作用。

# 参考文献

[1] De Jong K P, Geus J W. Carbon nanofibers: catalytic synthesis and applications [J]. Catal. Rev. - Sci. Eng., 2000, **42**: 481~510.

[2] Planeix J M, Coustel N, Coq B. Application of carbon narr otubes as supports in heterogeneous catalysis[J]. J. Am. Chem. Soc., 1994, **116**: 7935~7936.

[3] Zhang Yu, Zhang Hong-Bin, Lin Guo-Dong, et al. Preparation, characterization and catalytic hydroformylation properties of carbon nanotubes-supported Rh-phosphine catalyst[J]. Ap-pl. Catal. A: General, 1999, **187**: 213~224.

[4] 张鸿斌, 林国栋, 蔡启瑞. 碳纳米管的催化合成、结构 表征及应用研究[J]. 厦门大学学报(自然科学版), 2001, **40**(2): 387~397.

[5] Chen Ping, Zhang Hong-Bin, Lin Guo-Dong, et al. Growth of carbon nanotubes by catalytic decomposition of $CH_4$ or CO on a Ni-MgO catalyst[J]. Carbon, 1997, **35**(10—11): 1495~1501.

[6] 陈萍, 张鸿斌, 林国栋, 蔡启瑞. 过渡金属催化剂及用于制备均匀管径碳纳米管的方法[P]. 中国发明专利, ZL 96 1 10252.7.

## Highly Active CNTs-Promoted Cu-based Catalyst for Hydrogenation of $CO/CO_2$ to $CH_3OH$

Xin Dong, Hong-Bin Zhang, Guo-Dong Lin, You-Zhu Yuan, K R. Tsai

(Dept. of Chem. & State Key Lab of Phys. Chem. for the Solid Surfaces, Xiamen Univ., Xiamen 361005, China)

**Abstract**　With a kind of multi-walled carbon nanotubes(CNTs)synthesized catalytically in-house and the nitrates of the corresponding metallic components, highly active CNTs-piomoted catalysts, $Cu_iZn_jAl_k-O_x$-wt%CNTs, were prepared by coprecipitation method, and their catalytic performance for synthesis of methanol from $H_2/CO/CO_2$ was investigated and compared with the corresponding CNTs-fnse catalyst. It is experimentally found that incorporation of the CNTs into the $Cu_iZn_jAl_k-O_x$ can significantly advance the catalyst activity for methanol synthesis. Over a $Cu_6Zn_3Al_1-O_x$-12.5wt%CNTs catalyst and under reaction conditions of 493 K, 5.0 MPa, feed-gas $H_2/CO/CO_2/N_2=62/30/5/3(V/V)$, GHSV=8 000 $h^{-1}$, conversion of CO reached 31%, with the corresponding methanol-STY at 1 064 mg-$CH_3OH$ $h^{-1}$(g-catal. )$^{-1}$, which was about 1. 46 times as high as that(729 mg $h^{-1}$(g-catal. )$^{-1}$)over the corresponding CNTs-free catalyst $Cu_6Zn_3Al_1-O_x$. The $H_2$-TPD measurement demonstrated that the CNTs material and preduced $Cu_iZn_jAl_k-O_x$-wt%CNTs system could reversibly adsorb and storage a considerable quantity of hydrogen under atmospheric pressure in temperature region from room temperatuns to ca. 573 K. This peculiarity would be conducive to generating surface circumstances with higher stationary-state concentration of hydrogen-adspecies on the functioning catalyst, and in turn fawurable to enhancement of speed of the surface hydrogenation reactions. Meanwhile, the operation temperature for methanol synthesis over the CNTs-promoted catalyst can be 1525 K lower than that over the corresponding CNTs-free catalyst, which would considerably contribute to an increase in equilibrium CO-conversion and $CH_3OH$-yield. The results of the present work indicated that the carbon nanotubes could served as an excellent promoter, and that its peculiarity of adsorbing and storing $H_2$ played an important role in promoting enhancement of reactivity of methanol synthesis over the CNTs-promoted $Cu_iZn_jAl_k-O_X$ catalysts.

**Key words**　Carbon nanotubes　Carbon nanotubes-promoter　$Cu_i-Zn_jAl_k-O_X-wt\%$ CNTs　Methanol synthesis　$CO/CO_2$ hydrogenation

■ 本文原载:《高等学校化学学报》2002 年第 23 卷,第 902～905 页。

# 负载型铼催化剂体系与甲醇选择氧化性能的关系[*]

袁友珠[1,①],曹为[1],蔡启瑞[1],岩泽康裕[2]

(1.厦门大学化学系 物理化学研究所 固体表面物理化学国家重点实验室,厦门 361005;
2.东京大学化学系,东京 113－0033)

**摘 要** 以铼酸铵为前驱体,制备了氧化物负载型铼催化剂并研究其甲醇选择氧化反应的催化性能。结果表明,$Fe_2O_3$ 和 $V_2O_5$ 等氧化物负载型铼催化剂表现出很高的甲醇选择氧化制备二甲氧基甲烷的催化性能,选择性可达 90%～94%(摩尔分数)。选择氧化反应活性与铼担载量有关。在 $\alpha$-$Fe_2O_3$ 担载的铼催化剂中,以担载质量分数为 1%～3%铼的催化剂活性最高[450 mmol/(h·gRe)],而高于 3%的铼担载量,单位铼催化活性逐渐下降。XRD,XPS 和脉冲反应等结果表明,铼酸铵负载于 $\alpha$-$Fe_2O_3$ 载体上,并于 He 气氛中焙烧后,所得表面铼物种与担载量有关,当低于单层担载量时以 $Re^{6+}$ 占主导,而高于单层担载量时则 $Re^{6+}$ 与 $Re^{4+}$ 物种共存。

**关键词** 负载型铼催化剂 甲醇 选择氧化 二甲氧基甲烷

**中图分类号** O643.3 **文献标识码** A **文章编号** 0251-0790(2002)05-0902-04

由于铼的多样性化学氧化态和电子结构,使其具有许多独特的化学魅力。铼可与碱土金属、稀土金属及过渡金属形成稳定的复合氧化物[1];铼催化剂在石油化工工业中具有重要用途,表现出优异的催化活性和选择性[2,3]。以甲基三氧铼(MTO)为代表的有机金属铼化合物具有高效活化双氧水和分子氧的能力,是一类适应性广、选择性很高的新型均相选择氧化催化剂[4-6]。但在多相催化反应中,较高温度的氧化反应条件下,铼易于形成升华性的高价态铼 $Re^{7+}$ 物种,使铼催化剂性能不稳定[7]。因此,用含铼复合氧化物或负载型铼氧化物作为多相选择氧化反应催化剂的研究尚未引起人们重视[8-10]。本文制备了氧化物负载型铼催化剂,考察了其对甲醇选择氧化的催化性能,以期开拓铼催化剂的新催化作用。

# 1 实验部分

## 1.1 催化剂制备

负载型铼催化剂采用浸渍担载法制备。将计量的铼酸铵用适量去离子水溶解后,加入氧化物载体,搅拌 4 h,在水浴上蒸干后置于 393 K 烘箱中干燥 12 h,再在 He 气氛下程序升温(4 K/min)至 673 K,焙烧 6 h,所得催化剂记为 $ReO_x/M_yO_z$($M_yO_z$ 为氧化物载体),置于干燥器中备用。

## 1.2 催化剂活性评价

甲醇选择氧化活性评价在常压流动态微反应—在线色谱装置上进行。催化剂每次用量为 0.05 g。由

———————————

[*] 收稿日期:2000 年 12 月 14 日。
　　基金项目:国家自然科学基金(批准号:29873037 和 20023001)、教育部优秀青年教师基金和高等学校骨干教师计划课题资助。
　[①] 联系人简介:袁友珠(1963 年出生),男,博士,教授,博士生导师,主要从事配位催化和催化新材料研究。

气体质量流量计控制 He 和 $O_2$ 的流量;甲醇由 He 气鼓泡带入催化剂床层。He,$O_2$,MeOH 的摩尔分数分别为 86.3%,9.7%,4.0%,时空速率为 40 000 mL/(h·g cat)。反应器出口至色谱仪的管路用加热带保温(423 K 左右)装置以防产物冷凝。色谱柱:Poropak N,3 m,分析含氧化合物;Unibeads C,3 m,分析 CO 和 $CO_2$。

### 1.3 催化剂表征

催化剂的 XRD 谱在 Rigaku Miniflex 折射仪上测定,用 Cu $K\alpha$ 为辐射源($\lambda = 0.154\ 18$ nm),电压 30 kV,电流 15 mA,扫描速度 2°/min。XPS 谱在 Rigaku XPS-7000 能谱仪上测定,用 Mg $K\alpha$ 为辐射源(1 253.6 eV),加速电压 20 kV,电流 10 mA。样品于高纯氮气氛下转移至能谱仪预处理腔体中。样品污染碳($C_{1s} = 284.6$ eV)作内标。催化剂比表面在 BELSORP36 全自动比表面测定仪上测定。

# 2　结果与讨论

## 2.1　载体的影响

表 1 列出不同氧化物负载铼催化剂在 513 K 的催化活性和产物选择性。由表 1 可见,除 $Bi_2O_3$ 和 $Sb_2O_3$ 负载铼催化剂外,多数负载型铼催化剂表现出很高的甲醇选择氧化活性,主产物为二甲氧基甲烷,其中 $\alpha$-,$\gamma$-$Fe_2O_3$ 和 $V_2O_5$ 负载型铼催化剂上,二甲氧基甲烷选择性达到 90%～94%,以市售 $ReO_3$ 为催化剂时,二甲氧基甲烷的选择性高达 98%,而在 $ReO_2$ 上二甲氧基甲烷的选择性则相对较低。

**Table 1　The results of methanol selective oxidation on supported rehenium catalysts at 513 K**

| Catalyst | $S_{BET}$/ $(m^2 \cdot g^{-1})$ | MeOH conversion molar ratio (%) | Rate (mmol·h$^{-1}$ g$^{-1}$Re) | $CH_2$ $(OCH_3)_2$ | HCHO | $(CH_3)_2O$ | HCOOCH$_3$ | $CO_x^a$ |
|---|---|---|---|---|---|---|---|---|
| 10.0% $ReO_x$/$TiO_2$-R$^b$ | 5 | 53.7 | 351.2 | 83.1 | 1.9 | 0.7 | 9.1 | 5.2 |
| 10.0% $ReO_x$/$TiO_2$-A$^c$ | 50 | 59.5 | 389.1 | 78.5 | 4.1 | 1.1 | 11.7 | 4.6 |
| 10.0% Re/$V_2O_5$ | 6 | 21.5 | 140.6 | 93.7 | 0 | 4.3 | 0 | 2.0 |
| 10.0% Re/$ZrO_2$ | 9 | 35.8 | 234.1 | 89.4 | 2.0 | trace | 7.6 | 1.0 |
| 10% Re/$\alpha$-$Fe_2O_3$ | 3 | 15.5 | 101.4 | 90.5 | 2.0 | 1.0 | 7.0 | 0.5 |
| 10% Re/$\gamma$-$Fe_2O_3$ | 16 | 48.4 | 319.2 | 91.0 | 2.4 | 1.0 | 4.6 | 1.0 |
| 10% Re/$Fe_2O_3^d$ | 23 | 51.6 | 337.5 | 82.7 | 2.6 | 0.6 | 9.5 | 4.6 |
| 10% Re/$SiO_2$ | 36 | 15.1 | 98.8 | 60.7 | 1.3 | trace | 11.9 | 26.1 |
| 10% Re/$\alpha$-$Al_2O_3$ | 10 | 16.3 | 106.6 | 88.3 | 2.8 | trace | 5.9 | 2.9 |
| 10% Re/$Sb_2O_3$ | 1 | 0 | 0 | 0 | 0 | 0 | 0 | 0 |
| 10% Re/$Bi_2O_3$ | 1 | 0 | 0 | 0 | 0 | 0 | 0 | 0 |
| 10% Re/$MoO_3$ | 5 | 9.1 | 59.5 | 80.0 | 0 | 19.0 | 0 | 1.0 |
| $\alpha$-$Fe_2O_3$ | 3 | 0 | 0 | 0 | 0 | 0 | 0 | 0 |
| $V_2O_5$ | 6 | 9.3 | 10.8$^e$ | 1.0 | 91.5 | 7.4 | 0 | trace |
| $ReO_3$ | 1 | 12.4 | 10.2 | >98.0 | 0 | <1.0 | <1.0 | 0 |
| $ReO_2$ | 7 | 65.3 | 50.0 | 64.6 | 6.4 | 2.0 | 10.2 | 16.8 |
| $ReO_x^f$ | 1 | 10 | 7.7 | 84.2 | 0 | 8.9 | 5.9 | 1.0 |

a. $CO_x = CO_2 + CO$; b. $TiO_2$-R: rutile type; c. $TiO_2$-A: anatase type; d. $Fe_2O_3$ was prepared by adding an aqueous solution of $NH_4OH$ (Wako, purity 99.9%) to an aqueous solution of Fe($NO_3$)$_3$ followed by filtration, washing with deionized water, drying at 393 K, and calcining at 673 K for 4 h in air; e. mol·h$^{-1}$·g$_{-v}^{-1}$; f. $ReO_x$ was obtained by temperature-programmed calcination(4 K/min) of $NH_4ReO_4$ at 673 K for 6 h in He, and the crystalline phase thus obtained is mainly due to monoclinic $ReO_2$ by XRD.

## 2.2 反应温度的影响

图 1 为在 $\alpha$-Fe$_2$O$_3$ 负载铼催化剂上进行甲醇选择氧化时,反应温度对转化率和产物选择性的影响。结果表明,随着反应温度升高,甲醇转化率单调升高;而产物中二甲氧基甲烷的选择性则在反应温度约为 513 K 时达到最大值,随后逐渐下降。

## 2.3 负载量的影响

图 2 为 $\alpha$-Fe$_2$O$_3$ 负载铼催化剂上,铼担载量对甲醇选择氧化反应活性和对二甲氧基甲烷选择性的影响。由图 1 可知,当铼担载量从 0.5% 提高到 1.0% 时,单位铼的反应活性骤升;但当铼担载量为 1% ~ 3.0% 以上时,单位铼的活性为 450 mmol/(h·g Re)。高于 3.0% 的铼担载量,单位铼的反应活性下降,但对二甲氧基甲烷的选择性影响不大。根据测定的 $\alpha$-Fe$_2$O$_3$ 的比表面和 ReO$_3$ 的结构单元尺寸,估算出 2.0% 左右的铼担载量时,ReO$_3$ 将以近似单层形式分散在催化剂表面上。

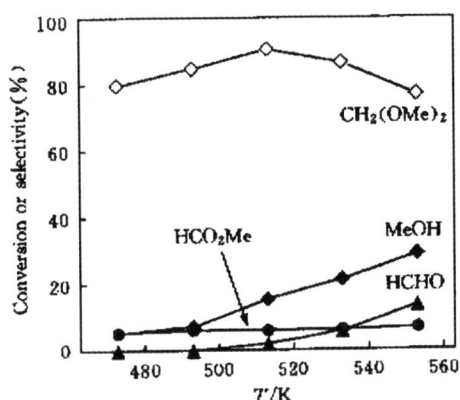

**Fig. 1 Catalytic methanol oxidation on 10% Re/$\alpha$-Fe$_2$O$_3$ as a function of reaction temperature**

GSHV $= 40\,000$ mL/(h·g cat); $n$(He)/$n$(O$_2$)/$n$(MeOH) $= 86.3/9.7/4.0$; $10^5$ Pa.

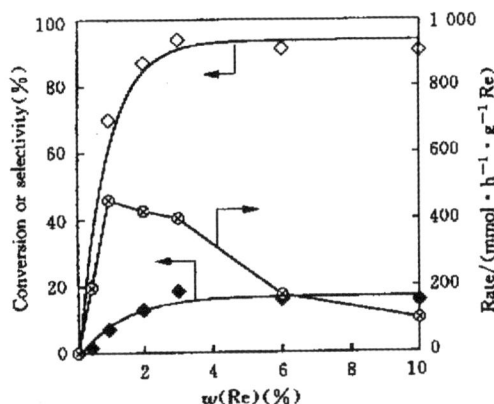

**Fig. 2 Catalytic methanol oxidation on Re/$\alpha$-Fe$_2$O$_3$ as a function of Re loading weight**

$T = 513$K; $\diamond$MeOH conversion; $\diamond$dimethoxymethane selectivity; $\otimes$reaction rate.

## 2.4 XRD 和 XPS 表征结果

铼酸铵在惰性气氛下于 673 K 焙烧,得到的主要为单斜晶系 ReO$_2$[10]。图 3 为不同氧化铁负载铼催化剂的 XRD 谱图。以 $\alpha$-Fe$_2$O$_3$ 为载体时,若铼担载量在 3% 以上,可观察到 ReO$_2$ 的 XRD 谱线随铼含量的增加而逐渐增强;而用 $\gamma$-Fe$_2$O$_3$ 为载体,即使铼担载量为 10%,仍未观察到 ReO$_2$ 的 XRD 谱线。考虑到 $\alpha$-Fe$_2$O$_3$ 载体的比表面比 $\gamma$-Fe$_2$O$_3$ 的小约 5 倍,推测所得负载型铼催化剂的表面铼物种与其在表面上的分散形式密切相关。铼担载量接近单层分散或低于单层分散时,所得的铼物种应与多层分散的表面铼物种不同。从 XRD 结果看出,当采用高于单层分散的担载量,所形成的表面氧化铼物种有一部分将与体相铼酸铵在 He 气氛下焙烧后的情况相似,即为 ReO$_2$。

图 4 为所制备的 $\alpha$-Fe$_2$O$_3$ 负载型铼催化剂中,不同铼负载量的 XPS 谱图。由图 4 可见,在催化剂焙烧前,Re $4f_{7/2}$ 的结合能为 46.1 eV,是典型的 Re$^{7+}$ 物种;He 气氛下经 673 K 焙烧,Re 物种的 XPS 结合能有所下降。当铼担载量为 2% 以下时,为单一铼物种,Re $4f_{7/2}$ 结合能为 42.3 eV;当铼担载量高于 3%,催化剂表面除存在 Re $4f_{7/2}$ 结合能在 42.3 eV 的铼物种外,还存在 45.3 eV 的铼物种。本文暂且把 Re $4f_{7/2}$ 结合能为 42.3 eV 的铼物种归属为 Re$^{4+}$,而结合能为 45.3 eV 的物种归属为 Re$^{6+}$。结果说明,担载量 3% 以上,催化剂表面铼物种为 Re$^{6+}$ 与 Re$^{4+}$ 共存。因此,利用某些载体与氧化铼之间的相互作用,可以避免铼酸铵在 He 气氛下焙烧时被还原成 Re$^{4+}$ 物种。

图 5$a$ 为焙烧后质量分数为 10% Re/$\alpha$-Fe$_2$O$_3$ 催化剂的 XPS 谱图,图 5$b$ 为样品与 10 个纯甲醇脉

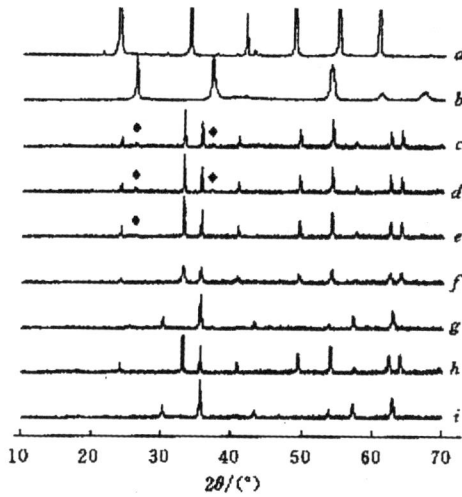

Fig. 3 XRD patterns for Re/α-, γ-Fe₂O₃ catalysts

◇ReO₂. a. ReO₃; b. ReO₂; Re/α-Fe₂O₃:c. 10%;
d. 6%; e. 3%; f. 10% Re/Fe₂O₃;
g. 10% Re/γ-Fe₂O₃; h. α-Fe₂O₃; i. γ-Fe₂O₃.

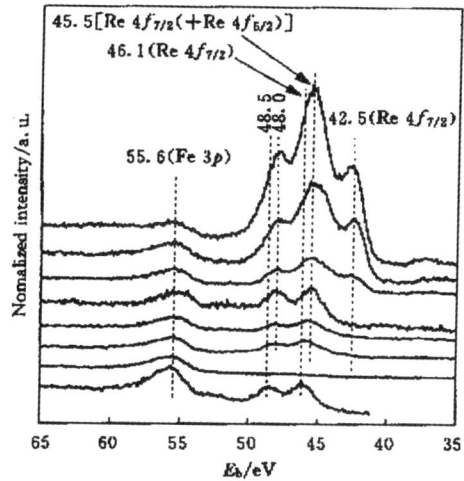

Fig. 4 XPS spectra for Re/α-Fe₂O₃

$w(Re)(\%)$:a. 10,before calcination;
b. 0;c . 0.5;d. 1.0;e. 2.0;f. 3.0;
g. 6.0;h. 10.

冲反应后的 XPS 谱图,图 5c 为样品再进行 2 h 的甲醇选择氧化反应后的 XPS 谱图。将所有样品转移至 XPS 谱仪样品室的过程均在 N₂ 气氛下完成,没有与空气接触。由图 5 可见,在焙烧后,表面存在着 $Re^{6+}$ 和 $Re^{4+}$ 两种铼物种,其 Re 4$f_{7/2}$ 结合能分别为 42.3 和 45.3 eV;在与无氧的甲醇脉冲反应后,$Re^{6+}$ 的 XPS 峰强度明显减少,$Re^{4+}$ 物种量增加;在氧存在下的甲醇氧化反应后,$Re^{6+}$ 和 $Re^{4+}$ 物种含量与新鲜样品接近。关联流动态的甲醇选择氧化反应结果,初步认为,甲醇在 Re/α-Fe₂O₃ 催化剂上以 90% 以上的选择性氧化生成二甲氧基甲烷的本质可能是由于氧化铼与某些载体的相互作用所致,在较低温度下的甲醇选择氧化反应中,伴随着表面铼物种 $Re^{4+}$ 和 $Re^{6+}$ 间的"氧化-还原"过程。

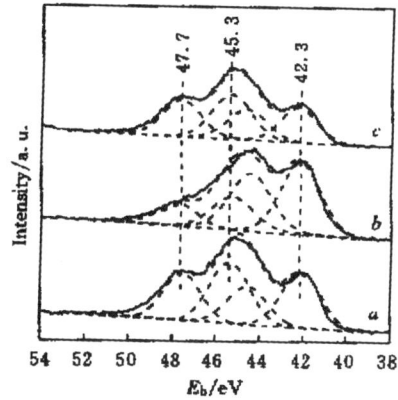

Fig. 5 XPS spectra for 10% Re/α-Fe₂O₃

a. before reaction;b. after 10th MeOH pulse at 513 K;
c. after MeOH oxidation at 513 K for 2 h.

# 参考文献

[1]Butz A.,Miehe G.,Paulus H. et al.. J. Solid State Chem. [J],1998,**138**:232～237.

[2]Okal J.,Kubicka H.. Appl. Catal. A:General[J],1998,**171**:351～359.

[3]Okal J.,Kepi n ski K.,Krajczyk L. et al.. J. Catal[J],1999,**188**:140～153.

[4]Roma o C. C.,Kühn F. E.,Herrmann W. A.. Chem. Rev. [J],1997,**97**:3 197～3 246.

[5]Gregory S. O. ,Joachin A.,Mahdi M. A. O.. Catal. Today[J],2000,**55**:317～363.

[6]Espenson J. H.. Chem. Commun[J],1999,479～488.

[7]Kim D. S., Wachs I. E.. J. Catal. [J], 1993, **141**: 419~429.

[8]Liu H. C., Gaigneaux E. C., Imoto H. et al.. J. Phys. Chem. [J], 2000, **104**: 2 033~2 043.

[9] Yuan Y. Z., Liu H. C., Imoto H. et al.. J. Catal. [J], 2000, **195**: 51~61.

[10]Abakumov A. M., Shpanchenko R. V., Antipov E. V.. Z. Anorg. All. Chem. [J], 1998, **624**: 750~753.

# Correlation of Performance and Structure of Supported Rhenium Catalysts for Methanol Selective Oxidation

You-Zhu Yuan[1*], Wei Cao[1], Khi-Rui Tsai[1], IWASAWA Yasuhiro[2]

(1. *Department of Chemistry, Institute of Physical Chemistry, State Key Laboratory for Physical Chemistry of Solid Surface, Xiamen University, Xiamen 361005, China;*

2. *Department of Chemistry, The University of Tokyo, Tokyo 113-0033, Japan*)

**Abstract**    Supported Re catalysts were prepared by impregnation of ammonium perrhenate precursor on oxide supports, followed by calcination at 673 K in He steam. Iron-oxides and $V_2O_5$ supported Re catalysts showed a higher activity and selectivity for the catalytic methanol oxidation to dimethoxymethane$[3CH_3OH + (1/2)O_2 \longrightarrow CH_2(OCH_3)_2 + 2H_2O]$. The selectivity was as high as $90\% - 94\%$ (molar fraction) at 513 K. The highest reaction rate of 450 mmol $\cdot$ h$^{-1}$ $\cdot$ g$^{-1}$ Re on $\alpha$-$Fe_2O_3$ supported Re catalysts was achieved with the Re loading weight of 2% (mass fraction). The reaction rate gradually decreased with the increase of Re loading weight. Calcination of ammonium perrhenate alone at 673 K in He gave monoclinic $ReO_2$ as a main product. However, the application of the iron-oxide may prevent $Re^{7+}$ from reducing to $Re^{4+}$ during the calcination at 673 K in He in the case of proper Re loading weight probably through the interaction of ammonium perrhenate with the supports, leading to the growth of $Re^{6+}$ on the catalyst surface. Above the monolayer loading weight of $ReO_3$, there was coexistence of $Re^{6+}$ and $Re^{4+}$ on the catalyst surface. A redox mechanism between $Re^{6+}$ and $Re^{4+}$ species on the catalyst surface was proposed to be responsible for the high performance in the methanol selective oxidation to dimethoxymethane.

**Key words**    Supported rhenium catalyst    Methanol    Selective oxidation    Dimethoxymethane

■ 本文原载：Catalysis Letters Vol. 85. Nos. 3~4. February 2003( © 2003)，pp. 237~246.

# Highly Active CNT-Promoted Cu-ZnO-Al₂O₃ Catalyst for Methanol Synthesis from H₂/CO/CO₂

Xin Dong，Hong-Bin Zhang①，Guo-Dong Lin，You-Zhu Yuan，K. R. Tsai
(Department of Chemistry & State Key Laboratory of Physical Chemistry for the Solid Surfaces, Xiamen University, Xiamen 361005, China )
Received 14 August 2002；accepted 7 November 2002

**Abstract**　With types of in-house-synthesized multi-walled carbon nanotubes(CNTs) and the nitrates of the corresponding metallic components, highly active CNT-promoted Cu-ZnO-Al₂O₃ catalysts, symbolized as $Cu_i Zn_j Al_k$-$x$% CNTs, were prepared by the co-precipitation method. Their catalytic performance for methanol synthesis from H₂/CO/CO₂ was studied and compared with the corresponding CNT-free coprecipitated catalyst, $Cu_i Zn_j Al_k$. It was shown experimentally that appropriate incorporation of a minor amount of the CNTs into the $Cu_i Zn_j Al_k$ could significantly increase the catalyst activity for methanol synthesis. Under the reaction conditions of 493 K, 5.0 MPa, H₂/CO/CO₂/N₂ = 62/30/5/3 (v/v), GHSV = 8000h⁻¹, the observed CO conversion and methanol formation rate over a co-precipitated catalyst of $Cu_6 Zn_3 Al_1$-12.5% CNTs reached 36.8% and 0.291 μmol CH₃OHs⁻¹(m²-surf. Cu)⁻¹, which was about 44 and 25% higher than those(25.5% and 0.233 μmol CH₃OHs⁻¹(m²-surf. Cu)⁻¹) over the corresponding CNT-free co-precipitated catalyst, $Cu_6 Zn_3 Al_1$. Addition of a minor amount (10 ~ 15wt%) of the CNTs to the $Cu_6 Zn_3 Al_1$ catalyst was found to considerably increase specific surface area, especially Cu surface area of the catalyst. H₂-TPD measurements revealed that the CNTs and the pre-reduced CNT-promoted catalyst systems could reversibly adsorb and store a considerably greater amount of hydrogen under atmospheric pressure at temperatures ranging from room temperature to ~573 K. This unique feature would be beneficial for generating microenvironments with higher stationary-state concentration of active hydrogen adspecies on the surface of the functioning catalyst, especially at the interphasial active sites since the highly conductive CNTs might promote hydrogen spillover from the Cu sites to the Cu/Zn interphasial active sites, and thus be favorable for increasing the rate of the CO hydrogenation reactions. Alternatively, the operation temperature for methanol synthesis over the CNT-promoted catalysts can be 15 ~ 20 degrees lower than that over the corresponding CNT-free contrast system. This would contribute considerably to an increase in equilibrium CO conversion and CH₃OH yield. The results of the present work indicated that the CNTs could serve as an excellent promoter.

**Key words**　Carbon nanotubes　$Cu_i Zn_j Al_k$-$x$% CNTs catalysts　Methanol synthesis　CO/CO₂ hydrogenation.

---

①　To whom correspondence should be addressed. E-mail：hbzhang@xmu.edu.cn.

**1361**

# 1    Introduction

Among the $C_1$ chemicals, methanol is the species most widely used in various chemical applications. Recently, it has been used as a clean synthetic fuel additive and considered as an alternative fuel source[1]. That methanol is a better and much cleaner automobile fuel with high octane number, and lower emissions of $NO_x$, ozone, CO, and aromatic vapors has been confirmed by a long-mileage road test carried out in the USA around 1990[2]. Methanol is also a convenient hydrogen carrier for PEM fuel cells. A recent breakthrough in the conversion of methanol to $H_2$ via oxidative steam reforming around 500 K is able to produce very pure $H_2$ with less than a few tens ppm level of CO. $H_2$ of such high purity can be used directly in $H_2$-air fuel cells[3].

In existing methanol synthesis technology, only a small portion($\sim 10\%$) of the syngas feed is converted to methanol, while a larger portion of the unreacted feed, after separation from methanol, must be recycled so as to enhance the utilization ratio of the syngas feed. The process and equipment involved are relatively complicated, and extra energy is consumed for the separation and recycling. Thus, finding more active catalysts and lower-temperature processes with high single-pass conversion of syngas has been one of the key objectives for research and development efforts.

The kinetics and mechanism of the methanol synthesis on $Cu$-$ZnO$-$Al_2O_3$ catalysts have been extensively studied since the late 1970s. A number of studies and excellent review articles have been published on this subject, including those by Klier[4], Kung[5], and Herman[6]. Progress in this field has considerably contributed to the growing understanding of the nature of this catalytic reaction system.

Carbon nanotubes(CNTs) have been drawing increasing attention recently[7]. This type of new carbon material possesses a series of unique features, such as its nanosize channel, the highly conductive graphite-like tube wall, the $sp^2$-carbon-constructed surface, and its excellent performance with regard to hydrogen adsorption. These features make the CNTs full of promise for being novel catalyst carriers[8-11] or even promoters. We recently reported a type of highly active CNT-supported Cu-based catalyst for hydrogenation of $CO/CO_2$ to methanol[12]. It was experimentally found that the carrier could significantly affect the activity for methanol synthesis. The space-time-yield(STY) of methanol over the CNT-supported catalyst at 503 K was 1.95 and 2.57 times as high as those of the corresponding catalysts supported by AC and $\gamma$-$Al_2O_3$ at their optimum operating temperatures, 523 and 543 K, respectively. In the present work, a series of CNT-promoted $Cu$-$ZnO$-$Al_2O_3$ catalysts was prepared by the co-precipitation method. Their catalytic performance for $CO/CO_2$ hydrogenation to methanol was studied and compared with that of the corresponding CNT-free conventional co-precipitated $Cu$-$ZnO$-$Al_2O_3$ catalyst. The results shed some light on understanding the nature of promoter action by the CNTs and the prospect of developing highly active catalysts.

# 2    Experimental

## 2.1    Catalyst preparation

The CNTs were synthesized by the catalytic method reported previously[13]. The prepared CNTs were a type of multi-walled carbon nanotubes, with O.D. of 10—50 nm and I.D. of ~3nm. The freshly prepared CNTs were treated with boiling nitric acid for 4 h, followed by rinsing with deionized water, then drying at 473 K under dry nitrogen. Open-end CNTs with somewhat hydrophilic surfaces were then obtained.

A series of CNT-promoted Cu-ZnO-Al₂O₃ catalysts, symbolized as $Cu_iZn_jAl_k$-$x\%$ ( mass percentage)CNTs, was prepared by the constant pH co-precipitation method[14]. An aqueous solution containing calculated amounts of Cu, Zn, and Al(total equivalent concentration of metallic cations at 4N),which was prepared by dissolving $Cu(NO_3)_2 \cdot 3H_2O, Zn(NO_3)_2 \cdot 6H_2O$, and $Al(NO_3)_3 \cdot 9H_2O$ (all of AR grade) in deionized water, and an aqueous $Na_2CO_3$ solution(4N) were simultaneously added dropwise under vigorous stirring into a Pyrex flask containing a calculated amount of the CNTs at constant temperature of 353K. The addition was adjusted to maintain a constant pH of ∼7. The precipitation procedure was completed in 1 h. The precipitate was then continuously stirred for 5 h at a temperature of 353K, followed by aging overnight at room temperature and then filtering. It was repeatedly rinsed with deionized water until the content of $Na^+$ ions in the eluant fell below 0.1 ppm as detected by the flame ion absorption method. The precipitate was then dried at 393 K for 4h and calcined at 543−573 K for 3 h to yield the precursor of the CNT-promoted catalysts. A CNT-free conventional co-precipitated Cu-ZnO-Al₂O₃ catalyst was prepared in a similar way. All samples of catalyst precursor were pressed, crushed, and sieved to a size of 40−80 mesh for the activity evaluation.

## 2.2 Catalyst evaluation

The catalyst was tested for methanol synthesis in a continuous-flow micro-reactor-GC combination system. An amount of 0.50g of catalyst(i.e.,equivalent to ∼0.5 ml of catalyst sample) was used for each test. Prior to the reaction, the catalyst sample was prereduced by 5% $H_2 + N_2$ for 16h. Methanol synthesis from $H_2/CO/CO_2$ over the catalysts was conducted at a stationary state with feed gas composition of $H_2/CO/CO_2/N_2 = 62/30/5/3$(v/v)under 453−563 K and 2.0 or 5.0 MPa. The reactants and products were determined by on-line GC(Model 102-GD)equipped with a TC detector and dual columns filled with 5A zeolite molecular sieve and 401 porous polymer,respectively. The former column was used for analysis of CO and $N_2$(as internal standard),and the latter for $CH_3OH, CO/N_2, CO_2$,MF, and DMC. CO conversion was calculated by an internal standard analysis method, and methanol formation rate was evaluated by an external standard(i.e.,working curve)method.

## 2.3 Catalyst characterization

BET specific surface area(SSA)of the CNT sample and the catalyst precursors was measured by $N_2$ adsorption using a Sorptomatic-1900 (Carlo Erba) system. TEM observations were performed using a JEOL JEM-100CX transmission electron microscope. X-ray diffraction measurements were carried out using a Rigaku Rotaflex D/Max-C X-ray diffractometer with CuKα radiation at a scanning rate of 8° min⁻¹. XPS measurements were done using a VG Esca Lab MK-2 system with MgKα radiation(10kV, 20 mA,$h\nu$=1253.6eV)under UHV($10^{-7}$Pa),calibrated internally by the Al(2p)binding energy(BE)at 73.6 eV and the carbon deposit C(1s)BE at 284.6eV.

Tests of $H_2$-TPR and $H_2$-TPD of the catalyst were conducted using a fixed-bed continuous-flow microreactor or adsorption-desorption system. A KOH column and a 3A zeolite molecular sieve column were installed in sequence at the reactor exit to remove water vapor formed by the reduction of metallic oxide components of the catalyst sample. The rate of temperature increase was 10Kmin⁻¹. Change of hydrogen signal was monitored using on-line GC (Shimadzu GC-8A) with a TC detector. For TPR measurements,20 mg of catalyst sample was used for each test. The sample was first flushed by an Ar (of 99.999% purity)stream at 673 K for 30 min to clean its surface, and then cooled down to room temperature, followed by switching to a $N_2$-carried 5 vol% $H_2$ gaseous mixture as reducing gas to start the TPR observation. For TPD tests, 200 mg of catalyst sample was used each time. Prior to TPD measurement, the catalyst sample was pre-reduced *in situ* in the TPD equipment by a $N_2$-carried 5 vol% $H_2$ gaseous mixture, with the highest reduction temperature reaching 483 K and lasting 16 h. Shortly after, the reduced sample was cooled to 433 K, followed by switching to a $H_2$ (of 99.999% purity) stream and maintaining at that temperature for 30 min, subsequently cooling down to room temperature and maintaining at room temperature for 1 h, and then flushing by an Ar (of 99.999% purity) stream at room temperature until the stable baseline of GC appeared.

Determination of the metal Cu surface area (Cu-SA) of the catalyst was carried out according to the improved method suggested by Bond and Namuo[15]. The above TPR apparatus was modified to allow flows of (1) 5% $H_2$ (of 99.999% purity) in $N_2$ (of 99.999% purity), (2) Ar (of 99.999% purity), and (3) $N_2O$ (of 99% purity, obtained from Aldrich Chemical Co.) to pass sequentially through the reactor and detector. A normal TPR was first performed on a calcined sample in the temperature region 293—593 K, which provided an estimate on the Cu content and produced a reduced catalyst in a suitable state for reaction with $N_2O$. The heating rate was 10 Kmin$^{-1}$ and the flow rate of the $N_2$-carried 5% $H_2$ gaseous mixture was 50 ml min$^{-1}$. The reduced sample was then cooled to the reaction temperature of 333 K in the Ar stream, and after 30 min the $N_2O$ was allowed to flow over the sample for 1 h at that temperature at a flow rate of 85 mlmin$^{-1}$. The $N_2O$ flow was then replaced by Ar and the sample cooled to room temperature. Finally, the 5% $H_2+N_2$ mixture was introduced and a second TPR was performed. A $H_2$-TPR peak due to the reduction of the adsorbed O atoms was observed. After the second TPR, a further calibration with a dose of $H_2$ (of 99.999% purity) was carried out, and the number of surface Cu atoms, $Cu_s$ (assuming $O/Cu_s=0.5$), and also the Cu surface area (assuming $1.47\times10^{19}$ atoms m$^{-2}$) was thus calculated.

# 3　Results and discussion

## 3.1　Reaction activity of methanol synthesis over the $Cu_iZn_jAl_k$-x% CNTs catalysts

Investigation of the effect of the feed gas GHSV on the reactivity of methanol synthesis over the CNT-promoted catalysts showed that at 2.0 MPa and a certain temperature (e.g. 493 K), CO conversion increased with increasing GHSV from zero to ~3000 h$^{-1}$, and reached a maximum at GHSV close to 3000 h$^{-1}$. It then started to decrease with further increase of GHSV. It was believed that, as the GHSV approached 3000 h$^{-1}$, the reaction of methanol formation was in the kinetics-controlled region. In the present work, reactivity tests of methanol synthesis were performed under the reaction condition of GHSV=3000 h$^{-1}$, unless otherwise specified.

It is well known that $CO_2$ in the feed gas is a participating component indispensable to continuously fast conversion of the syngas to methanol. However, the net conversion of $CO_2$ indicated by GC analysis of the inlet and exit gases of the reactor was quite low, and all the assay results of the catalyst activity in the present work are shown with CO conversion and methanol formation rate.

Figure 1 shows the assay results of reactivity of methanol synthesis at 453—523 K over the catalysts with different CNT percentages, $Cu_6Zn_3Al_1$-x% CNTs. On all four catalysts, both the CO conversion and the methanol formation rate first went up with increasing temperature, reached a maximum at their respective optimum operating temperature, and then went down as the temperature increased further. This strongly indicated that at the lower temperature range before the maximum, the methanol synthesis reaction was controlled by kinetics, and after the maximum, the reaction became equilibrium limited due to the high reaction temperature. The activity sequence observed on these catalysts for methanol synthesis is: (a) $Cu_6Zn_3Al_1$-12. 5% CNTs > (b) $Cu_6Zn_3Al_1$-10. 0% CNTs > (c) $Cu_6Zn_3Al_1$-15. 0%CNTs > (d) $Cu_6Zn_3Al_1$-0. 0%CNTs. The optimal value of the CNT percentage seemed to be ~ 12.5%. Over the $Cu_6Zn_3Al_1$-12. 5% CNTs catalyst under the reaction conditions of 493K, 2. 0MPa, feed gas $H_2/CO/CO_2/N_2$ = 62/30/5/3 (v/v), GHSV = 3000 $h^{-1}$, CO conversion reached 42. 4%, with the corresponding methanol formation rate at 0. 118 $\mu$mol $CH_3OHs^{-1}$(m²-surf. Cu)$^{-1}$. In comparison, the values only reached 29. 5% and 0. 095 $\mu$mol $CH_3OHs^{-1}$(m²-surf. Cu)$^{-1}$, respectively, over the CNT-free corresponding catalyst prepared in the same manner, $Cu_6Zn_3Al_1$, at its optimal operating temperature of 503 K.

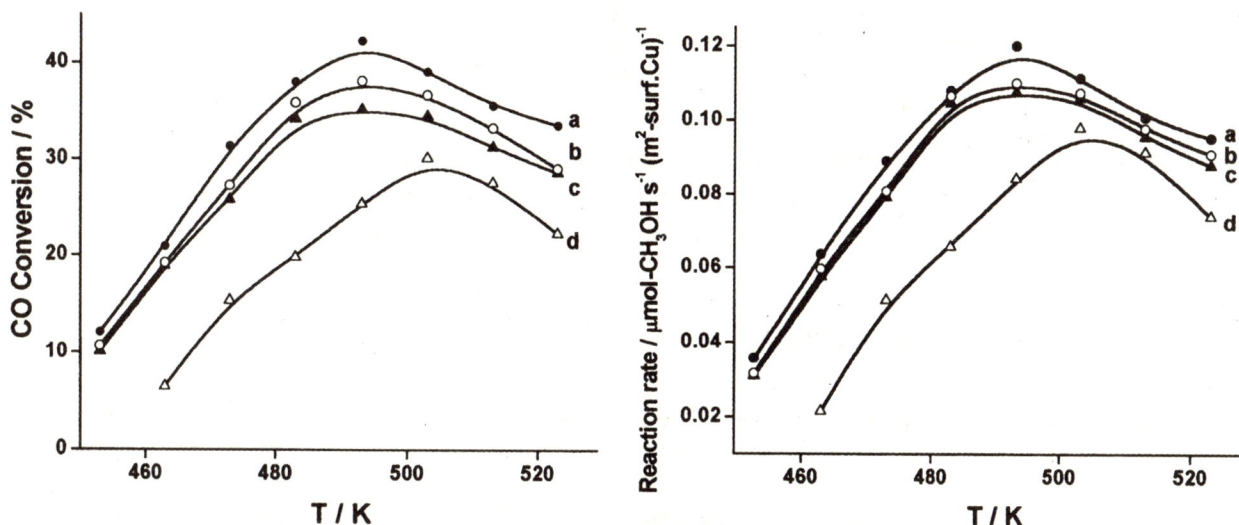

**Figure 1**    CO conversion and CH₃OH formation rate at different reaction temperatures over a series of catalysts: (a) Cu₆Zn₃Al₁-12. 5%CNTs; (b) Cu₆Zn₃Al₁-10. 0% CNTs; (c) Cu₆Zn₃Al₁-15. 0%CNTs; (d) Cu₆Zn₃Al₁-0% CNTs. Reaction conditions: 2. 0 MPa, H₂/CO/CO₂/N₂ = 62/30/5/3(v/v), GHSV=3000 h⁻¹.

Figure 2 shows the assay results of reactivity of methanol synthesis at 453—523 K over the 12. 5% CNT-promoted catalysts with different compositions of Cu/Zn/Al(molar ratio). The observed activity sequence of those catalysts for methanol synthesis was: (a) $Cu_6Zn_3Al_1$-12. 5%CNTs > (b) $Cu_5Zn_{2.5}Al_1$-12. 5%CNTs ≥ (c) $Cu_4Zn_2Al_1$-12. 5% CNTs > (d) $Cu_7Zn_{3.5}Al_1$-12. 5% CNTs. Over the catalyst with optimal molar ratio of Cu/Zn/Al=6/3/1, $Cu_6Zn_3Al_1$-12. 5%CNTs, CO conversion reached 42. 3%, with the methanol formation rate at 0. 118 $\mu$mol $CH_3OHs^{-1}$(m²-surf. Cu)$^{-1}$, under the reaction conditions of 2. 0 MPa, 493 K, $H_2/CO/CO_2/N_2$ = 62/30/5/3 (v/v), GHSV = 3000 $h^{-1}$. The corresponding $CH_3OH$ selectivity reached 98. 0% and above in the products.

In order to evaluate the performance of the catalysts under working conditions with a higher extent of reaction, the synthesis reaction of methanol from syngas was conducted at higher pressure and GHSV. The results are shown in figure 3. Under the reaction conditions of 493 K, 5. 0 MPa, $H_2/CO/CO_2/N_2$ = 62/30/5/3 (v/v) and GHSV = 8000 $h^{-1}$, methanol formation rate reached 0. 291 $\mu$mol $CH_3OHs^{-1}$(m²-surf. Cu)$^{-1}$ over the $Cu_6Zn_3Al_1$-12. 5% CNTs catalyst, which was about 25% higher than that (0. 233 $\mu$mol $CH_3OH$ $s^{-1}$ (m²-surf. Cu)$^{-1}$) over the corresponding CNT-free system, $Cu_6Zn_3Al_1$. Figure 4 shows the operation stability of this catalyst for methanol synthesis lasting 250 h under reaction conditions of 493 K, 5. 0 MPa, $H_2/CO/CO_2/N_2$ = 62/30/5/3(v/v) and GHSV=8000 $h^{-1}$,

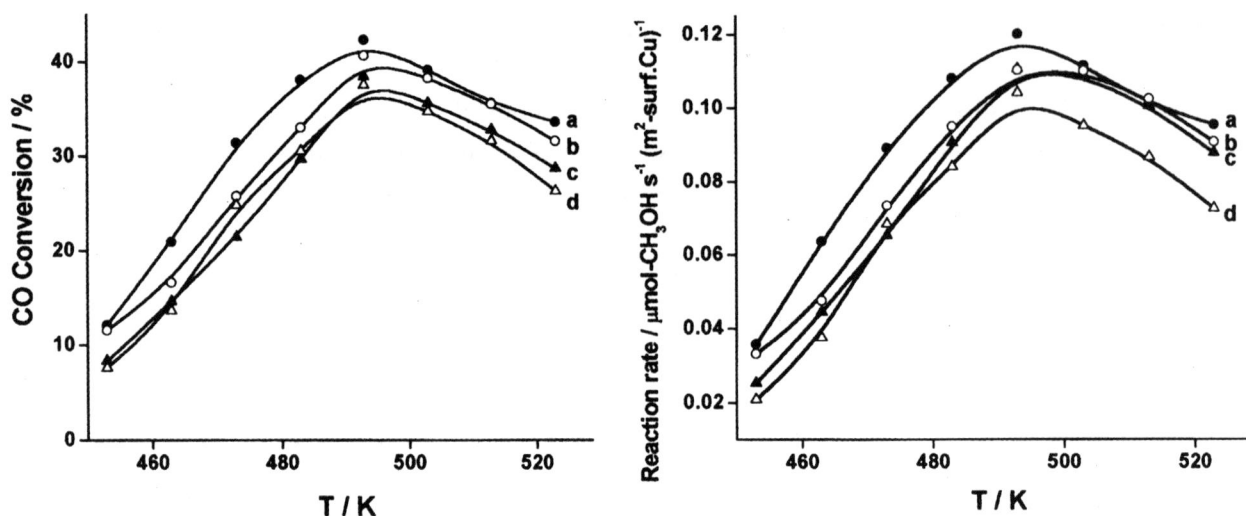

**Figure 2** CO conversion and CH$_3$OH formation rate at different reaction temperatures over a series of catalysts: (a) Cu$_6$Zn$_3$Al$_1$-12.5%CNTs; (b) Cu$_5$Zn$_{2.5}$Al$_1$-12.5%CNTs; (c) Cu$_4$Zn$_2$Al$_1$-12.5%CNTs; (d) Cu$_7$Zn$_3$.5Al$_1$-12.5% CNTs. Reaction conditions: 2.0 MPa, H$_2$/CO/CO$_2$/N$_2$=62/30/5/3(v/v), GHSV=3000h$^{-1}$.

without obvious deactivation of the catalyst observed. The results of the activity assay and the SSA/Cu-SA determinations of a series of Cu$_i$Zn$_j$Al$_k$-$x$%CNTs catalysts for methanol synthesis are summarized in table 1.

**Table 1** Reactivity of CH$_3$OH synthesis from syngas over a series of the CNTs-promoted catalysts Cu$_i$Zn$_j$Al$_k$-$x$%CNTs[a]

| Catalyst sample | SSA [m$^2$(g-catal.)$^{-1}$] | Cu-SA [m$^2$(g-catal.)$^{-1}$] | Reaction temperature (K) | CO conversion (%) | Methanol formation rate [$\mu$mol CH$_3$OHs$^{-1}$ (m$^2$-surf. Cu)$^{-1}$] |
|---|---|---|---|---|---|
| Cu$_6$Zn$_3$Al$_1$-15.0%CNTs | 60.3 | 35.5 | 453 | 10.7 | 0.031 |
|  |  |  | 493 | 35.1 | 0.109 |
| Cu$_6$Zn$_3$Al$_1$-10.0%CNTs | 60.2 | 36.8 | 453 | 11.9 | 0.031 |
|  |  |  | 493 | 38.1 | 0.113 |
| Cu$_6$Zn$_3$Al$_1$-12.5%CNTs | 61.4 | 39.4 | 453 | 14.1 | 0.036 |
|  |  |  | 493 | 42.4 | 0.118 |
|  |  |  | 493[b] | 36.8[b] | 0.291[b] |
| Cu$_6$Zn$_3$Al$_1$-0.0%CNTs | 50.3 | 33.9 | 463 | 6.5 | 0.021 |
|  |  |  | 493 | 25.3 | 0.082 |
|  |  |  | 503 | 29.5 | 0.095 |
|  |  |  | 493[b] | 25.5[b] | 0.233[b] |
| Cu$_5$Zn$_{2.5}$Al$_1$-12.5%CNTs | 55.7 | 38.3 | 453 | 11.6 | 0.033 |
|  |  |  | 493 | 39.8 | 0.114 |
| Cu$_4$Zn$_2$Al$_1$-12.5%CNTs | 54.2 | 36.0 | 453 | 8.3 | 0.025 |
|  |  |  | 493 | 37.5 | 0.114 |
| Cu$_7$Zn$_{3.5}$Al$_1$-12.5%CNTs | 62.6 | 39.9 | 453 | 7.6 | 0.021 |
|  |  |  | 493 | 36.5 | 0.100 |

[a] Reaction conditions: 2.0MPa, H$_2$/CO/CO$_2$/N$_2$=62/30/5/3(v/v), GHSV=3000h$^{-1}$.

[b] Data at 5.0 MPa and GHSV=8000 h$^{-1}$.

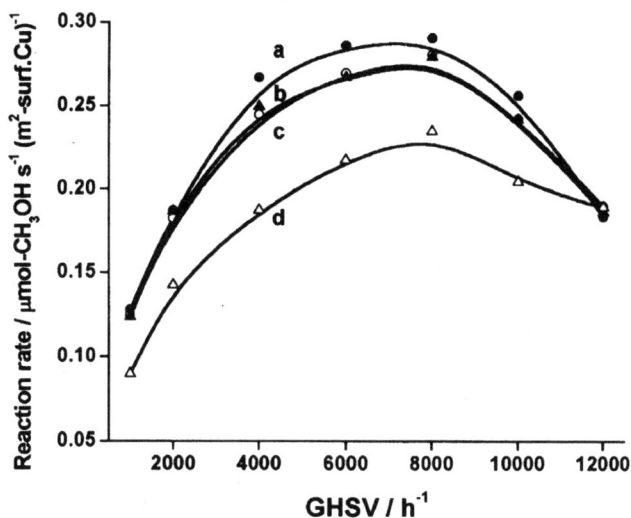

**Figure 3** Change of CH$_3$OH formation rate with GHSV of the feed gas over a series of catalysts: (a) Cu$_6$Zn$_3$Al$_1$-12.5% CNTs; (b) Cu$_6$Zn$_3$Al$_1$-10.0%CNTs; (c) Cu$_6$Zn$_3$Al$_1$-15.0%CNTs; (d) Cu$_6$Zn$_3$Al$_1$-0%CNTs. Reaction conditions: 493 K, 5.0MPa, H$_2$/CO/CO$_2$/N$_2$=62/30/5/3(v/v).

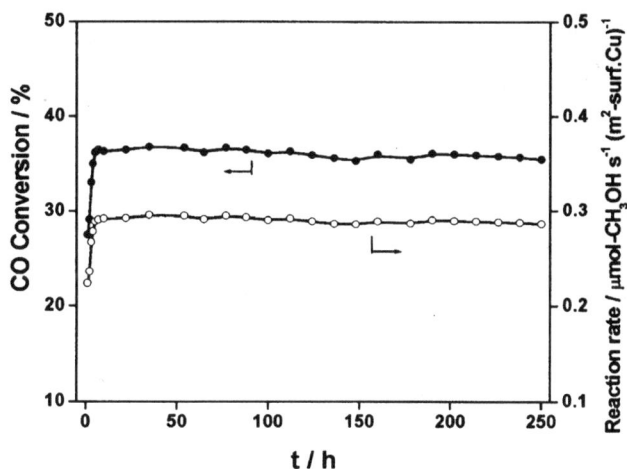

**Figure 4** Operation stability of CH$_3$OH synthesis over the catalyst of Cu$_6$Zn$_3$Al$_1$-12.5%CNTs lasting 250 h. Reaction conditions: 493 K, 5.0MPa, H$_2$/CO/CO$_2$/N$_2$=62/30/5/3(v/v), GHSV=8000h$^{-1}$.

## 3.2 Physicochemical properties of the CNTs as promoter

It is quite evident that the considerably better performance of the CNT-promoted catalysts for methanol synthesis from H$_2$/CO/CO$_2$ is closely related to the unique structures and properties of the CNTs as promoter. TEM and SEM observations showed that CNTs were almost the only species in the purified CNT products. Elemental analysis showed no other elements than carbon. Figure 5 shows the TEM and SEM images of the CNTs grown from catalytic decomposition of CH$_4$. These CNTs were uniform in diameter along the tube length with outer diameters in the range 10~50 nm. The HRTEM observation demonstrated that those CNTs derived from CH$_4$ were constructed by a superposition of many graphene layer facets, which were tilted at a certain angle with respect to the axis of the central hollow nanofiber, as if a number of cones were placed one on top of another[13,16].

The XRD pattern of the CNTs(figure 6(b)) was close to that of graphite(figure 6(a)), but the strongest 002 peak was weakened and somewhat broadened, and the peak position shifted to $2\theta = 26.1°$ from $2\theta = 26.5°$ for graphite, which corresponded to an increase in the spacing between the $sp^2$ carbon layers from 0.336 nm for graphite to 0.341 nm for the CNTs. The half-peak width was enlarged from $\sim 0.50°$ for graphite to $\sim 1.43°$ for the CNTs, indicating that the degree of long-range order of these nanostructures was lower than that of graphite. On the other hand, the 100 and 004 reflections of the CNTs were considerably enhanced in comparison with those of graphite, implying that their 100 and 004 faces were more exposed.

The spectra of $O_2$-TPO(temperature-programmed oxidation, with a He-carried 5 vol% $O_2$ gaseous mixture as oxidant) showed that the carbon nanostructures in the CNT sample were predominantly graphite-like(with the corresponding main oxidation peak at 960~1000 K). The content of amorphous carbon(the corresponding oxidation peak at around 700 K)was very low(less than 10% estimated).

It was known from the BET measurements with $N_2$ as adsorbate that the specific surface area of this type of CNTs was $\sim 140$ m$^2$ g$^{-1}$. The size of their inner diameter was in the range 2.6~3.4 nm, which is consistent with that estimated from the HRTEM image[13,16].

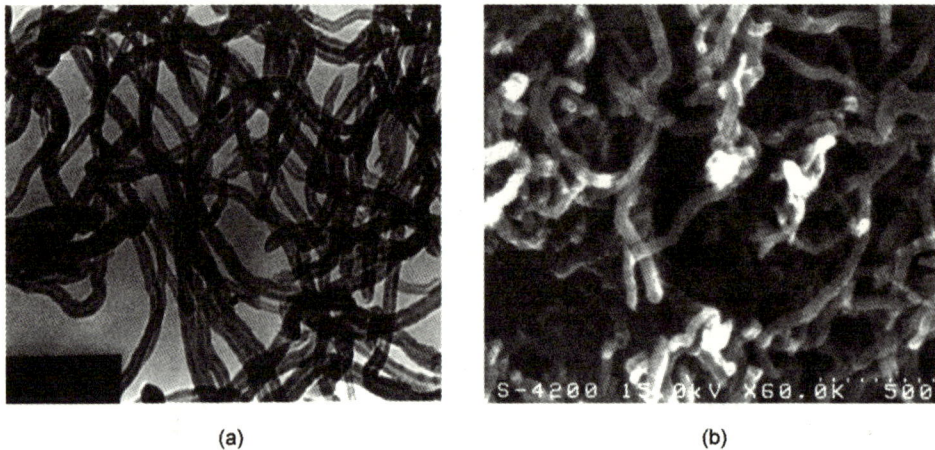

(a)　　　　　　　　(b)

Figure 5　(a)TEM and(b)SEM images of the CNTs grown catalytically by CH$_4$ decomposition.

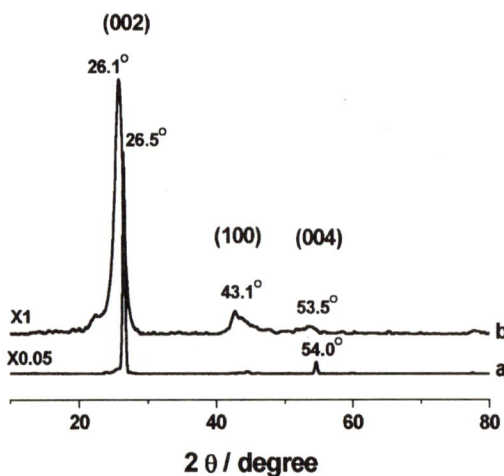

Figure 6　XRD patterns of(a)graphite and(b) the CNTs grown catalyti-cally by CH$_4$ decomposition.

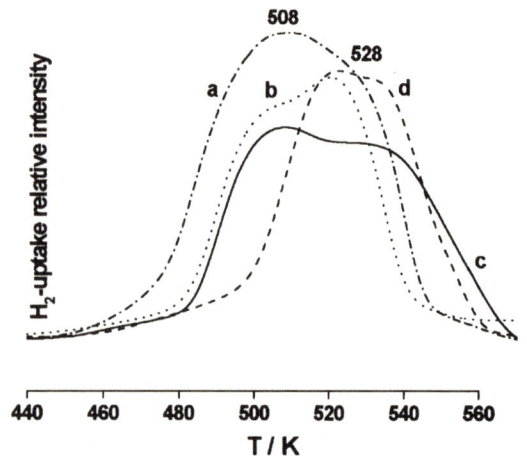

Figure 7　H$_2$-TPR spectra of the catalyst precursors：(a)Cu$_6$Zn$_3$Al$_1$- 12.5%CNTs；(b)Cu$_6$Zn$_3$Al$_1$-10.0%CNTs；(c)Cu$_6$Zn$_3$Al$_1$- 15.0%CNTs；(d)Cu$_6$Zn$_3$Al$_1$-0%CNTs.

## 3.3 TPR and TPD/TPSR characterization of the catalysts

The H$_2$-TPR spectra of the oxidation precursors of the catalysts are shown in figure 7. The H$_2$ reduction of Cu$_6$Zn$_3$Al$_1$-12.5%CNTs sample started at ~453K and the main TPR peak appeared at round 508 K. On the CNT-free Cu$_6$Zn$_3$Al$_1$ catalyst, apparent reduction by H$_2$ started at 478 K, and reached a peak value at ~528K, which was ~20K higher than that of the 12.5% CNTs-promoted system. The observed reducibility sequence of the four samples was: (a)Cu$_6$Zn$_3$Al$_1$-12.5%CNTs >(b) Cu$_6$Zn$_3$Al$_1$-10.0%CNTs > (c) Cu$_6$Zn$_3$Al$_1$-15.0% CNTs > (d) Cu$_6$Zn$_3$Al$_1$-0% CNTs. It is conceivable that a lower reduction temperature would be beneficial to inhibiting the growth of metal Cu crystallite produced by the H$_2$ reduction, thus favorable to increasing the Cu exposed area, which, in turn, enhances the catalyst activity. The above reducibility sequence was consistent with the sequence of the Cu exposed area of those catalysts and their catalytic activity for methanol synthesis(see table 1).

Figure 8 shows the TPD spectrum taken on the CNT sample adsorbing H$_2$(99.999% purity)at room temperature. With increasing temperature, the first main TPD peak was present at ~400 K,followed by a shoulder peak at ~485 K and the second main peak at 990 K. Further investigation by GC and MS indicated that the desorbed product was almost exclusively gaseous hydrogen at temperatures lower than ~693 K. At temperatures higher than ~693 K, the product contained a small amount of CH$_4$,C$_2$H$_4$, and C$_2$H$_2$,in addition to a considerable amount of H$_2$[17]. This result suggested that H$_2$ adsorption on this type of CNTs may be in two forms: associative (molecular state)and dissociative(atomic state), as demonstrated by the recent UV-vis Raman spectroscopic

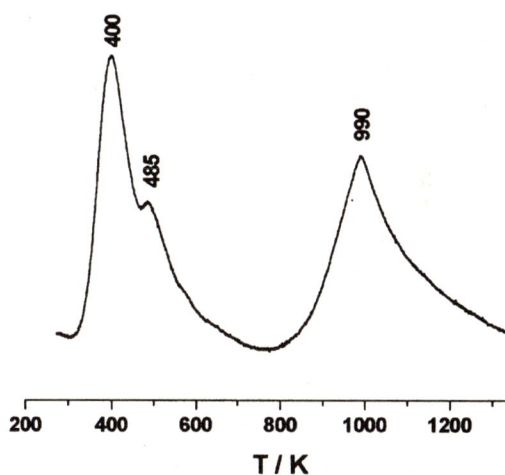

**Figure 8** TPD spectrum of the CNTs adsorbing H$_2$ at room temperature.

investigation of the H$_2$/CNTs adsorption system,for which Raman peaks at 3950,2856,and 2967 cm$^{-1}$ assignable to H-H stretch of molecularly adsorbed H$_2$(a), symmetric C-H stretch of surface CH$_2$, and asymmetric C-H stretch of surface CH$_3$(both originated from dissociative adsorption of H$_2$ on the CNTs),respectively,were observed[18].

Figure 9 shows the TPD spectra of H$_2$ adsorbed at 433 K followed by cooling down to room temperature on the pre-reduced catalysts. Overall,each spectrum contained a lower-temperature peak (peak I)centered round 383 K and a higher-temperature peak(peak II)spanning from 453 to 773 K or higher. The lower-temperature peaks resulted from the desorption of weakly adsorbed species,most probably molecularly adsorbed hydrogen H$_2$(a), and the higher-temperature peaks were attributed to the desorption of strongly adsorbed species,perhaps dissociatively adsorbed hydrogen H(a). The relative area-intensities of peak Is and peak IIs for these catalyst samples were estimated,and the results are shown in table 2. It is conceivable that at the temperatures for methanol synthesis(453-563 K for the present work),the surface concentration of hydrogen adspecies associated with peak I was expected to be very low,and most of hydrogen adspecies at the surface of the functioning catalysts corresponded to peak II. It was probably those strongly adsorbed H species that were closely associated with the reaction activity of methanol synthesis. The ratio of relative area-intensities of peak IIs for these catalysts was: (Cu$_6$Zn$_3$Al$_1$-12.5% CNTs )/(Cu$_6$Zn$_3$Al$_1$-10.0% CNTs )/(Cu$_6$Zn$_3$Al$_1$-15.0% CNTs )/

$(Cu_6Zn_3Al_1\text{-}0\%CNTs)=100/82/74/43$ (see table 2). This suggested that the sequence of increasing surface concentration of hydrogen adspecies at the functioning catalysts was: $Cu_6Zn_3Al_1\text{-}12.5\%CNTs>Cu_6Zn_3Al_1\text{-}10.0\%CNTs > Cu_6Zn_3Al_1\text{-}15.0\%CNTs > Cu_6Zn_3Al_1\text{-}0\%CNTs$, in line with the observed sequence of catalytic activity of these catalysts for methanol synthesis.

**Table 2　Relative area-intensity of $H_2$-TPD peaks I and II for the $Cu_6Zn_3Al_1$-$x\%$CNTs catalysts**

| Catalyst sample | Relative area-intensity [a] | |
| --- | --- | --- |
| | Peak I | Peak II |
| $Cu_6Zn_3Al_1\text{-}12.5\%CNTs$ | 42 | 100 |
| $Cu_6Zn_3Al_1\text{-}10.0\%CNTs$ | 33 | 82 |
| $Cu_6Zn_3Al_1\text{-}15.0\%CNTs$ | 32 | 74 |
| $Cu_6Zn_3Al_1\text{-}0\%CNTs$ | 14 | 43 |

[a] With area-intensity of the strongest peak II as 100.

With CO(99.999% purity) in place of $H_2$ as adsorbate in the above experiments, the obtained CO-TPD spectra are shown in figure 10. Each of these spectra has three peaks, i. e., peak I at ~393 K, peak II at 468—473 K, and peak III at 543—558 K. The relative area-intensities of these peaks were estimated and are listed in table 3. It is well known that there exist three kinds of sites for CO adsorption at the surface of copper-zinc oxide catalyst for methanol synthesis: $Cu^+$, $Cu^0$, and $Zn^{2+}$, and only $Cu^+$ species were known to adsorb CO strongly[19]. Thus, the observed low-temperature peaks(peak Is) could be reasonably attributed to desorption of weakly adsorbed CO species, most likely non-dissociatively adsorbed CO species on $Zn^{2+}$ sites. The higher-temperature peaks(peak IIs and IIIs) corresponded to desorption of medium-to-strongly and strongly adsorbed CO species, most likely non-dissociatively adsorbed CO species on $Cu^0$ and $Cu^+$ sites, respectively. Additional evidence in supporting these assignments also came from the experimental fact that the area-intensity sequence of peak IIs and IIIs of these catalyst samples(see table 3)was in keeping with the sequence of their Cu exposed area(see table 1), i. e., $Cu_6Zn_3Al_1\text{-}12.5\%CNTs > Cu_6Zn_3Al_1\text{-}10.0\%CNTs > Cu_6Zn_3Al_1\text{-}15.0\%CNTs > Cu_6Zn_3Al_1\text{-}0\%CNTs$, while the area-intensity sequence of peak Is was not.

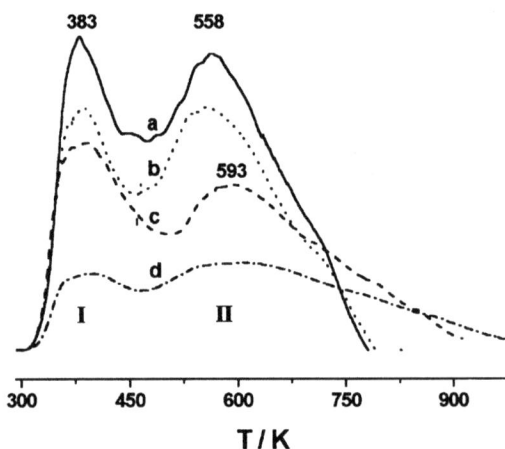

Figure 9　TPD spectra of hydrogen adsorbed on the pre-reduced catalysts: (a) $Cu_6Zn_3Al_1$-12.5% CNTs; (b) $Cu_6Zn_3Al_1$-10.0% CNTs; (c) $Cu_6Zn_3Al_1$-15.0%CNTs; (d)$Cu_6Zn_3Al_1$-0%CNTs.

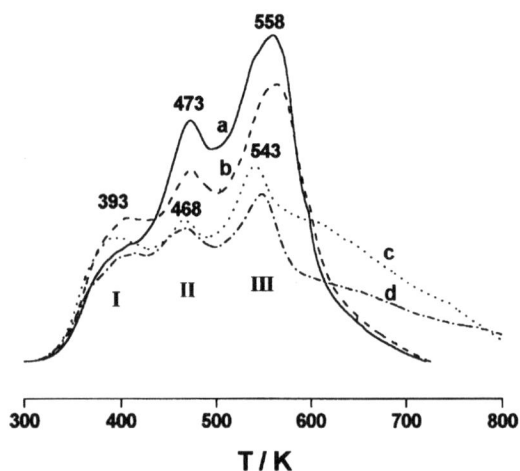

Figure 10　TPD spectra of CO adsorbed on the pre-reduced catalysts: (a) $Cu_6Zn_3Al_1$-12.5% CNTs; (b) $Cu_6Zn_3Al_1$-10.0%CNTs; (c) $Cu_6Zn_3Al_1$-15.0% CNTs; (d)$Cu_6Zn_3Al_1$-0%CNTs.

**Table 3** Relative area-intensity of CO-TPD peaks I, II, and III for the $Cu_6Zn_3Al_1$-$x\%$CNTs catalysts

| Catalyst sample | Relative area-intensity[a] | | |
|---|---|---|---|
| | Peak I | Peak II | Peak III |
| $Cu_6Zn_3Al_1$-12.5%CNTs | 27 | 46 | 100 |
| $Cu_6Zn_3Al_1$-10.0%CNTs | 31 | 41 | 90 |
| $Cu_6Zn_3Al_1$-15.0%CNTs | 27 | 28 | 85 |
| $Cu_6Zn_3Al_1$-0%CNTs | 23 | 25 | 60 |

[a] With area-intensity of the strongest peak III as 100.

Conceivably, at the temperatures for the methanol synthesis reaction (453 — 563 K for the present work), the surface concentration of weakly adsorbed CO species associated with peak I was expected to be very low, and most CO adspecies at the surface of the functioning catalyst were those associated with peak II, especially peak III. Most likely it was those strongly adsorbed CO species that were related to the reaction activity of methanol synthesis. It was evident that the incorporation of a minor amount of the CNTs into $Cu_6Zn_3Al_1$ led to a significant increase of the surface active sites for CO adsorption due to its increasing the Cu exposed area of the catalyst, which would be conducive to increasing the concentration of adsorbed CO species on the surface of the functioning catalyst, and thus to enhancing the reaction rate of methanol synthesis.

Using the feed gas of $H_2/CO/CO_2 = 62/30/5$ (v/v) in place of $H_2$ or CO as adsorbate in the above experiments, the observed TPD-TPSR (temperature-programmed surface reaction) spectra are shown in figure 11. On the sample of $Cu_6Zn_3Al_1$-12.5%CNTs pre-adsorbing the feedgas of $H_2/CO/CO_2$, a main peak at $\sim$ 451 K and a shoulder peak at $\sim$ 523 K were observed. The desorbed product was mainly $H_2$ and CO at temperatures below 453 K. It contained methanol in addition to $H_2$ and CO at temperatures of 453 K and above. With temperature increasing and the surface reaction speeding up, most of the adspecies of $H_2$, CO, and $CO_2$ remaining at the surface may be converted to methanol, which was easy to desorb, so that the whole process of TPD-TPSR came to an end with the temperature reaching $\sim$

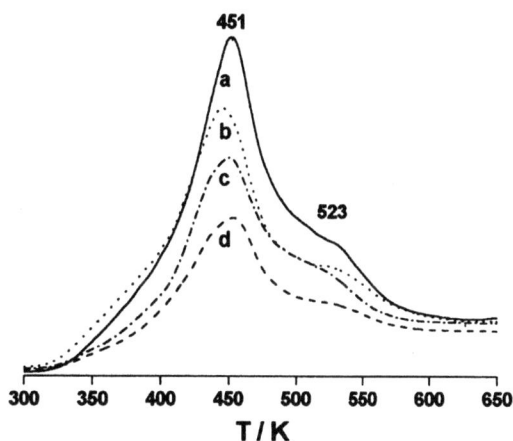

**Figure 11** TPD-TPSR spectra of syngas ($H_2/CO/CO_2$) adsorbed on the pre-reduced catalysts: (a) $Cu_6Zn_3Al_1$-12.5% CNTs; (b) $Cu_6Zn_3Al_1$-10.0%CNTs; (c) $Cu_6Zn_3Al_1$-15.0%CNTs; (d) $Cu_6Zn_3Al_1$-0%CNTs.

600 K. Similar TPD-TPSR behavior was also observed on the other three co-precipitated catalysts, but their adsorption capacities of the feed syngas of $H_2/CO/CO_2$ were relatively low, especially on the CNT-free catalyst, $Cu_6Zn_3Al_1$. The observed sequence of adsorption capacity of these catalysts toward the feed syngas of $H_2/CO/CO_2$ is: $Cu_6Zn_3Al_1$-12.5% CNTs $>$ $Cu_6Zn_3Al_1$-10.0% CNTs $>$ $Cu_6Zn_3Al_1$-15.0%CNTs$>$$Cu_6Zn_3Al_1$-0%CNTs, again in line with the observed sequence of their catalytic activity for methanol synthesis.

## 3.4 XRD and XPS characterization of the catalysts

Figure 12 shows the XRD patterns of the CNT-free $Cu_6Zn_3Al_1$ and $Cu_6Zn_3Al_1$-12.5% CNTs systems in the different chemical states. For the CNT-containing system in the oxidation state or in the functioning(reduction)state after 5 h of operating for the methanol synthesis, the graphite-like feature due to the CNTs appeared at $2\theta = 26.1°$(see figures 12(b) and (d)), in keeping with the CNT feature shown in figure 6(b). The features at $2\theta = 35.8°$ and $39.5°$ ascribed to CuO phase could be easily identified for the two oxidized samples(figures 12(a) and (b)). On the two samples in the functioning state, strong $Cu^0$ phase peaks($2\theta = 43.2°, 50.4°$, and $74.4°$)and a relatively weak $Cu_2O$ peak ($2\theta = 36.4°$) were simultaneously observed(figures 12(c) and (d)), indicating that major amounts of the Cu species were in the metal $Cu^0$ phase and minor amounts in the $Cu_2O$ phase in the functioning catalysts.

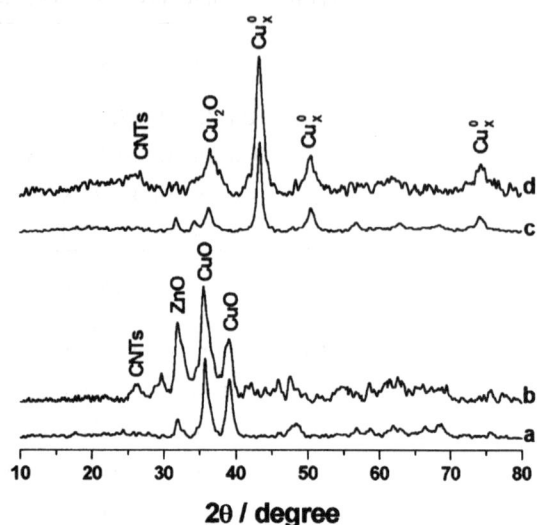

**Figure 12** XRD patterns of (a) $Cu_6 Zn_3 Al_1$ and (b) $Cu_6 Zn_3 Al_1$-12.5% CNTs both in the oxidation state; (c) $Cu_6 Zn_3 Al_1$ and (d) $Cu_6 Zn_3 Al_1$-12.5% CNTs both in the functioning state.

The XPS-Auger measurements provide direct experimental evidence about the valence states of the Cu species at the surface of the catalyst. On the oxidized samples of CNT-free $Cu_6 Zn_3 Al_1$ and $Cu_6 Zn_3 Al_1$-12.5%CNTs, $Cu^{2+}$ ($2p_{3/2}/2p_{1/2}$) XPS peaks appeared at 934.0/953.6 eV(BE), with the corresponding $Cu^{2+}$ satellite peak and Cu($L_3 M_{45} M_{45}$) Auger peak at 943.7 and 570.0 eV(BE), respectively(see figures 13(a) and (b)). At the surface of the functioning catalysts after 5h of operating for the methanol synthesis, $Cu^0$ was the dominant Cu species, with $Cu^0$($2p_{3/2}/2p_{1/2}$) at 932.6/952.5 eV (BE)and $Cu^0$($L_3 M_{45} M_{45}$) at 569.2eV(BE) (see figures 13(c) and(d))[20]. The surface concentration of $Cu^+$ was under the XPS detection limit, even though a minor amount of $Cu_2O$ crystallite was observed in the XRD measurements(see figures 12(c) and(d)).

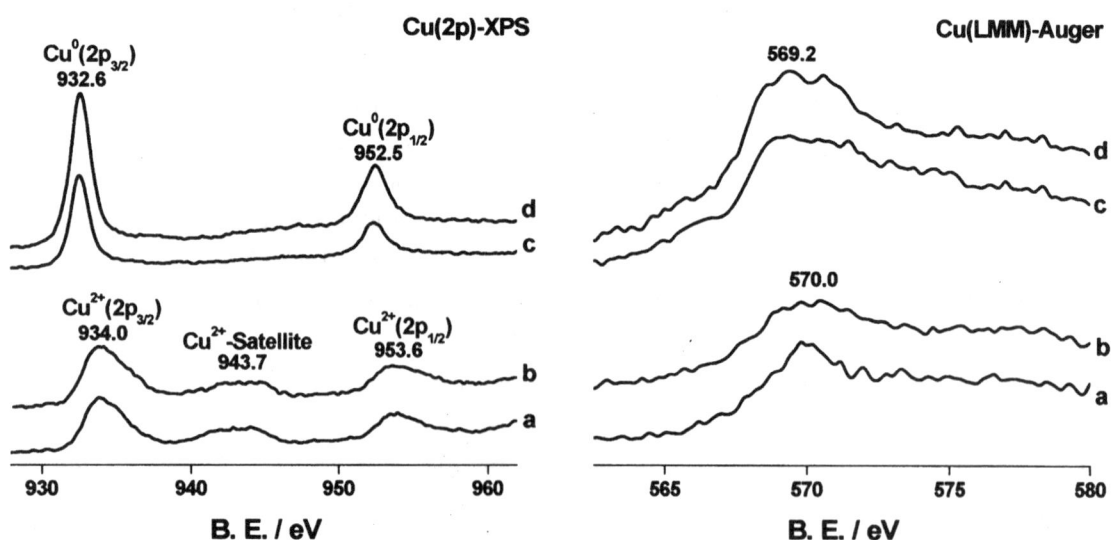

**Figure 13** Cu(2p)XPS-Auger spectra of(a)$Cu_6 Zn_3 Al_1$ and(b)$Cu_6 Zn_3 Al_1$-12.5%CNTs both in the oxidation state; (c)$Cu_6 Zn_3 Al_1$ and(d)$Cu_6 Zn_3 Al_1$-12.5%CNTs both in the functioning state.

From figures 12 and 13, it can be seen that the XRD pattern and Cu(2p)XPS-Auger spectra of the CNT-promoted system, Cu$_6$Zn$_3$Al$_1$-12.5%CNTs, were quite close to those of the CNT-free contrast system, Cu$_6$Zn$_3$Al$_1$, in the position and shape as well as relative intensity of the features associated with the Cu component. It appears that the doping of a minor amount of the CNTs into Cu$_6$Zn$_3$Al$_1$ did not lead to significant change in the phase composition of the Cu component in the functioning catalyst and in the relative concentrations of catalytically active Cu species with different valence states at the functioning catalyst surface.

## 3.5 Nature of the CNT promoter action

The results of the BET specific surface area and Cu surface area measurements shown in table 1 indicated that the doping of a minor amount of the CNTs led to a considerable increase in specific surface area(SSA)and Cu surface area(Cu-SA)of the catalyst. For the Cu$_6$Zn$_3$Al$_1$-12.5%CNTs catalyst, a 22% increase of specific surface area and a 16% increase of Cu surface area were observed as compared with those of the catalyst Cu$_6$Zn$_3$Al$_1$. The increment of specific surface area, especially Cu surface area, was undoubtedly in favor of enhancing the specific activity of the catalysts(i. e., activity of unit mass of catalyst). Nevertheless, it would be difficult to believe that the increase of as high as 44% of CO conversion(i. e., 36.8% vs. 25.5% for the Cu$_6$Zn$_3$Al$_1$-12.5% CNTs and the CNT-free Cu$_6$Zn$_3$Al$_1$, respectively, see table 1) was solely attributed to the difference in their specific surface area and Cu surface area. Besides, the difference in the surface areas could hardly justify the 25% increase of the intrinsic methanol formation rate(i. e., from 0.233 $\mu$mol CH$_3$OHs$^{-1}$(m$^2$-surf. Cu)$^{-1}$ for the CNT-free Cu$_6$Zn$_3$Al$_1$ to 0.291 $\mu$mol CH$_3$OH s$^{-1}$(m$^2$-surf. Cu)$^{-1}$ for the Cu$_6$Zn$_3$Al$_1$-12.5%CNTs).

Thus, it appears that the high reactivity, especially the high intrinsic reaction rate, for methanol synthesis over the CNT-promoted catalyst is closely related to the peculiar structure and properties of the carbon nanotubes as promoter. From a chemical catalysis point of view, the excellent performance of the CNTs in hydrogen adsorption/storage and electron transport is very attractive, in addition to its high mechanical strength, nanosize channel, $sp^2$-C-constructed surface, and graphite-like tube wall. It could be inferred from the above TPD/TPSR investigations that there would exist a considerable amount of reversibly adsorbed hydrogen species on the CNT-containing catalyst under the conditions of methanol synthesis used in the present study. This would lead to higher stationary-state concentration of active hydrogen adspecies on the surface of the functioning catalyst, especially at the interphasial active sites since the highly conductive CNTs might promote hydrogen spillover from the Cu sites to the Cu/Zn inter-phasial active sites, and thus be favorable to increasing the rate of a series of surface hydrogenation reactions in the process of CO/CO$_2$ hydrogenation to methanol. Alternatively, the operating temperature for methanol synthesis of the catalysts appropriately promoted with a minor amount of CNTs can be 15~20 K lower than that of the corresponding CNT-free contrast system. This would contribute considerably to an increase in equilibrium conversion of CO and yield of CH$_3$OH. The results of the present study indicated that CNTs could serve as an excellent promoter, and that its unique feature of adsorbing H$_2$ may play an important role in effectively promoting the methanol synthesis.

In summary, the origins of the promoter action by CNTs in enhancing the reaction activity of the methanol synthesis most likely involve the following aspects:

(a) The CNTs could serve as an excellent dispersant for Cu-ZnO-Al$_2$O$_3$ components. Proper incorporation of a minor amount of the CNTs into the Cu$_i$Zn$_j$Al$_k$ significantly increases the Cu exposed area, generating more catalytically active Cu sites at the catalyst surface for CO/CO$_2$

hydrogenation reaction.

(b) The CNTs could serve as an excellent adsorbent, activator, and reservoir of $H_2$, which would be beneficial to generating microenvironments with higher stationary-state concentration of active hydrogen adspecies on the surface of the functioning catalyst, and thus favorable to enhancing the rate of the $CO/CO_2$ hydrogenation reactions.

Further studies on the kinetics and the possible effect of promoting hydrogen spillover are currently under investigation.

# Acknowledgments

The authors are grateful for the financial support from National Natural Science Foundation (Project No. 50072021), Education-Ministerial Sci. Foundation (Project No. 99069), and Fujian Provincial Natural Science Foundation(Project No. 2001H017)of China.

# References

[1]L. V. MacDougall,Catal. Today **8**(1991)337.

[2]R. M. Bata,Reprints 207th ACS Meeting(San Diego,1994);Div. Fuel Chem. **39**(1994)299.

[3]R. F. Savinell,in:Proc. Int. Conf. on Applied Electrochemistry(University of Hong Kong,1995).

[4]K. Klier, Adv. Catal. **31**(1982)243.

[5]H. H. Kung,Stud. Surf. Sci. Catal. **45**(1989)228.

[6]R. H. Herman,Stud. Surf. Sci. Catal. **64**(1991)265.

[7]K. P. De Jong and J. W. Geus,Catal. Rev. Sci. Eng. **42**(2000)481.

[8]J. M. Planeix,N. Coustel,B. Coq,V. Brotons,P. S. Kumbhar,R. Dutartre,P. Geneste,P. Bernier and P. M. Ajiayan,J. Am. Chem. Soc. **116**(1994)7935.

[9]M. S. Hoogenraad, M. F. Onwezen, A. J. van Dillen and J. W. Geus,Stud. Surf. Sci. Catal. **101** (1996)1331.

[10]Y. Zhang,H. B. Zhang,G. D. Lin,P. Chen,Y. Z. Yuan and K. R. Tsai,Appl. Catal. A:General **187**(1999)213.

[11]H. B. Zhang,Y. Zhang,G. D. Lin,Y. Z. Yuan and K. R. Tsai,Stud. Surf. Sci. Catal. **130**(2000) 3885.

[12]H. B. Zhang, X. Dong, G. D. Lin, Y. Z. Yuan and K. R. Tsai,Preprints 223rd ACS Meeting (Orlando,2002);Div. Fuel Chem. **47**(2002)284.

[13]P. Chen,H. B. Zhang,G. D. Lin,Q. Hong and K. R. Tsai,Carbon **35**(1997)1495.

[14]P. G. Herman,K. Klier,G. W. Simmons,B. F. Finn,J. B. Bulko and T. P. Kobylinski,J. Catal. **56**(1979)407.

[15]G. C. Bond and S. N. Namuo,J. Catal. **118**(1989)507.

[16]P. Chen,H. B. Zhang,G. D. Lin and K. R. Tsai,Chem. J. Chinese Univ. **19**(1998)765.

[17]Z. H. Zhou,X. M. Wu,Y. Wang,G. D. Lin and H. B. Zhang,Acta Physico-Chemica Sinica (Chinese)**18**(2002)692.

[18]H. B. Zhang,G. D. Lin,Z. H. Zhou,X. Dong and T. Chen,Carbon **40**(2002)2429.

[19]Y. Y. Huang,J. Catal. **30**(1973)187.

[20]S. Marisa,Chem. Phys. Lett. **63**(1979)52.

■ 本文原载：Topics in Catalysis Vol. 22. Nos. 1/2，January 2003(© 2003)，pp. 9～15.

# Selective Methanol Conversion to Methylal on Re-Sb-O Crystalline Catalysts: Catalytic Properties and Structural Behavior

You-Zhu Yuan[a], Khi-Rui Tsai[a], Hai-Chao Liu[b], Yasuhiro Iwasawa[b],[①]

([a]State Key Laboratory for Physical Chemistry of Solid Surfaces,
Department of Chemistry, Xiamen University, Xiamen 361005, China
[b]Department of Chemistry, Graduate School of Science, University of Tokyo,
Hongo, Bunkyo-ku, Tokyo 113-0033, Japan)
E-mail: iwasawa@chem. s. u-tokyo. ac. jp

**Abstract**　Three crystalline compounds, $SbOReO_4 \cdot 2H_2O$, $Sb_4Re_2O_{13}$ and $SbRe_2O_6$, and several supported Re catalysts were employed as catalysts for the selective oxidation of methanol to methylal ($3CH_3OH + \frac{1}{2}O_2 \rightarrow CH_2(OCH_3)_2 + 2H_2O$). A high selectivity of 92. 5% to methylal at a conversion of 6. 5% under conditions of $GHSV = 10\ 000\ mLh^{-1}g^{-1}_{cat}$ and 573 K was obtained on the new $SbRe_2O_6$ catalyst, while no significant formation of methylal was observed with the other two catalysts. No structural change in the bulk and surface of $SbRe_2O_6$ and $Sb_4Re_2O_{13}$ occurred after methanol oxidation below 593 K, but $SbOReO_4 \cdot 2H_2O$ was transformed to $Sb_4Re_2O_{13}$, as characterized by XRD, Raman spectroscopy, XPS and SEM. The high performance of $SbRe_2O_6$ for the selective methylal synthesis was ascribed to Re oxide species stabilized by a specific connection with Sb oxides at the crystal surface.

**Key words**　Re-Sb-O crystalline catalysts　Re-based catalysts　Selective catalytic oxidation of methanol　Methylal synthesis　XRD　XPS　Raman　SEM

## 1　Introduction

The development of efficient catalysts for the selective oxidation if methanol to formaldehyde, methyl formate and dimethoxymethane(methylal)has received much attention from both academic and industrial interests. The oxidation of methanol to formaldehyde has been extensively studied and commercialized on silver and ferric molybdate catalysts[1]. Methyl formate has also been produced with high yields by the catalytic oxidation of methanol on V-Ti oxides[2], Sn-Mo oxides[3] and Bi-based mixed oxides[4]. However, methylal, which is used as a gasoline additive, a solvent in the perfume industry, a key intermediate for preparing high-concentration formaldehyde and a reagent in organic synthesis, has not successfully been produced by catalytic methanol oxidation. Catalytic methylal synthesis from

---

① To whom correspondence should be addressed.

methanol has been reported on V/TiO$_2$[2], V-Mo-O[5], PMoH-5.75/SiO$_2$[6], Mo/MCM-41[7], electrocatalysts[8], etc.[1], but the selectivities to methylal on those catalysts were low and practically insignificant. Hence, the discovery of a new selective oxidation catalyst is the key issue to realize the direct synthesis of methylal from methanol, where three methanol molecules are incorporated into a methylal molecule ($3CH_3OH + \frac{1}{2}O_2 \rightarrow CH_2(OCH_3)_2 + 2H_2O$).

Rhenium-based catalysts are active for petroleum reforming, metathesis, selective catalytic reduction of NO$_x$ with NH$_3$, methanol oxidation, hydrodesulfurization(HDS) of heavy fractions of crude oil, selective hydrogenation of organic compounds and dehydroaromarization of methane to hydrogen and benzene[9-19]. In solution, structurally well-defined methyltrioxorhenium (VII) (MTO) represents a rhenium catalyst for the selective expoxidization of alkenes at high catalytic turnovers at room temperature and even below[20,21]. Thus Re may constitute a key element to develop new promising catalysts for selective oxidation of alcohols and hydrocarbons when Re oxide species are stabilized under the reaction conditions. Also, Sb has been employed as a promoter element in several mixed oxides, such as V-Sb-O, Sn-Sb-O, Mo-Sb-O, Fe-V-O and U-Sb-O, for the selective oxidation/ammoxidation of hydrocarbons[22-26]. Recently, three crystalline Re-Sb-O mixed oxides, SbOReO$_4$ · 2H$_2$O, Sb$_4$Re$_2$O$_{13}$ and particularly SbRe$_2$O$_6$, have been reported to be capable of catalyzing the selective oxidation reactions of isobutane and isobutylene to methacrolein at 673 — 773 K[27-29] and the selective ammoxidation of isobutane, isobutylene and propene to methacrylonitrile and acrylonitrile, respectively[30-32]. SbOReO$_4$ · 2H$_2$O has also been applied for the selective oxidation of ethanol[33]. Methylal synthesis from methanol involves two different reaction steps: selective oxidation and dehydrative condensation. Rhenium oxides have both redox and acidic properties with suitable strength, which may be an active component for one-stage methylal synthesis. Very recently, we have discovered that the selective catalytic oxidation of methanol to methylal efficiently proceeded on SbRe$_2$O$_6$ with a selectivity as high as 92.5% at 573K, in which both redox and acid-base properties of Re oxides in a good balance are required[34,35]. In the SbRe$_2$O$_6$ catalysts, active Re oxides make a stable crystalline binary oxide by connecting with Sb oxides. In this paper, we examine and compare the performances for the selective catalytic oxidation of methanol to methylal and the surface/bulk structures of the three crystalline Re-Sb-O compounds by means of X-ray diffraction(XRD), X-ray photoelectron spectroscopy(XPS), scanning electron microscopy(SEM) and confocal laser Raman microscopy(Raman), to clarify the key issues of the catalytic phenomenon.

# 2　Experimental

## 2.1　Catalyst preparation

Three crystalline Re-Sb-O compounds, SbOReO$_4$ · 2H$_2$O, Sb$_4$Re$_2$O$_{13}$ and SbRe$_2$O$_6$, were prepared according to procedures reported previously[27-29,36,37]. For comparison, supported Re oxides were prepared by an incipient wetness impregnation method using aqueous NH$_4$ReO$_4$ (Soekawa, purity 99.9%) solutions. Sb$_2$O$_3$ (Soekawa, purity 99.99%) and SiO$_2$ (Aerosil 200) were used as supports for the Re oxides. A mechanical mixture catalyst of Sb$_2$O$_3$ with NH$_4$ReO$_4$ was also prepared by a known method[38]. The decomposition of the NH$_4$ReO$_4$ precursor to Re$_2$O$_7$ in the supported and mechanically mixed samples was performed by temperature programmed calcination(4 K min$^{-1}$) up to 573 K in a flow of He/O$_2$ = 90.0/10.0(mol%) at atmospheric pressure and the samples were further calcined at 573 K for 2h. They are denoted as Re$_2$O$_7$/Sb$_2$O$_3$, Re$_2$O$_7$/SiO$_2$ and Re$_2$O$_7$+Sb$_2$O$_3$. As for the samples involving

$Re_2O_7$ formed by $NH_4ReO_4$ decomposition, small amounts of sublimated $Re_2O_7$ were deposited at the exit of the reactor during the sample preparation. The Re loadings of those samples were regulated to be 10 wt% by taking into account the loss of $Re_2O_7$ by sublimation.

## 2.2 Catalytic methanol oxidation

Catalytic performances were examined in a conventional fixed-bed flow reactor using 200 mg of catalyst at 1 atm. Methanol(Wako, purity 99.8%) was introduced to the flow reactor by bubbling He gas through a glass saturator filled with methanol at 273 K. The reactant mixture of $He/O_2/MeOH$ was adjusted to 86.3/9.7/4.0(mol%) by mass flow controllers. Typical performances were investigated at $GHSV=10\ 000\ mLh^{-1}g_{cat}^{-1}$. The outlet stream line from the reactor to a gas chromatgraph was heated at about 423 K in order to avoid condensation of reaction products. The products were analyzed with an on-line gas chromatgraph(Shimadzu GC-8A) using two columns(3 m Porapak N and 3 m Unibeads C) at 423 K.

## 2.3 Catalyst characterization

The Re-Sb-O catalysts were characterized *ex situ* before and after the selective oxidation of methanol by BET surface area, powder XRD, XPS, SEM and Raman.

Surface areas of the samples were measured by a BET nitrogen adsorption method at 77 K using a BELSORP36 machine. XRD patterns were measured on a Rigaku Miniflex goniometer. The analysis was carried out in a continuous $\theta/2\theta$ scan reflection mode using CuK$\alpha$ radiation($\lambda=0.15418$ nm). The anode was operated at 30 kV and 15 mA. The $2\theta$ angles were scanned from 5 to 60 at a rate of 2 ° min$^{-1}$. Raman spectra were recorded under ambient atmosphere by using a confocal microprobe Raman system (LabRam I). A holographic notch filter was equipped to filter the excitation line and a 1800 gmm$^{-1}$ holographic grating was employed to disperse the scattered light. The excitation wavelength was 632.8 nm with a power of 12mW from an internal He-Ne laser. The size focused on the sample surface was ca. 5 $\mu$m. XPS spectra were obtained using a Rigaku XPS-7000 spectrometer using MgK$\alpha$ radiation(1253.6 eV)with an X-ray power of 200 W(accelerating voltage 20 kV, emission current 10 mA). Samples were pressed into thin disks, placed on holders and outgassed to less than $2.6\times10^{-5}$ Pa in a prechamber, and transferred to an analysis chamber. The binding energies of XPS peaks were referred to 284.6 eV for C 1s. The XPS peak intensities were normalized by the peak height of Sb 4d at 34.4 eV. Peak deconvolution and fitting were performed using SpXzeigR2.1 software running with Igor Pro and Gaussian-Lorentzian line shape, fixing both spin-orbit splitting and the relative intensity of spin-orbit components. SEM images were obtained using a Hitachi S-4500 microscope equipped with a field emission gun operated with an acceleration voltage of 5 kV and an emission current of 10 $\mu$A. The samples were imaged without any metallic coating. SEM images were obtained at many different places of the samples to obtain a common topographic feature of the samples.

# 3 Results and discussion

The performances of several Re-Sb-O catalysts at 573 K for the selective conversion of methanol to mathylal are listed in table 1. Sb oxides such as $Sb_2O_3$, $Sb_2O_4$ and $Sb_2O_5$, a mechanical mixture of $Sb_2O_3$ and $Re_2O_7$ ($Re_2O_7 + Sb_2O_3$), and Re oxides supported on $Sb_2O_3$ and $SiO_2$ ($Re_2O_7/Sb_2O_3$ and $Re_2O_7/SiO_2$, respectively) showed no or negligible activities for methylal formation. The crystalline Re-Sb oxides, $Sb_4Re_2O_{13}$ and $SbOReO_4 \cdot 2H_2O$, also produced almost no methylal. Only $SbRe_2O_6$ among the

Re-Sb-O compounds synthesized to date was active for the selective oxidation of methanol to methylal. The methylal selectivity was as high as 92.5% at a conversion of 6.5% at 573 K(table 1). The three crystalline Re-Sb-O compounds did not show any loss of Re species under the catalytic reaction conditions. However, for the supported $ReO_x$ catalysts small amounts of sublimation of $Re_2O_7$ occurred in the initial stage of the catalytic reactions, but the $Re_2O_7$ (at least its surface) was readily reduced to $ReO_3$ or species similar to $ReO_3$ showing a reddish color under the present reaction conditions. In this stage no significant sublimation of Re species was observed during the reaction. Thus the low and zero activity of the supported $ReO_x$ catalysts is due to their properties.

**Table 1    Methanol conversion on several Re-Sb-O catalysts and $SbO_x$ bulk at 573 K[a].**

| Catalyst | $S_{BET}$ $(m^2 g^1)$ | Conversion (%) | Selectivity(%) | | | | | | |
|---|---|---|---|---|---|---|---|---|---|
| | | | $CH_2(OCH_3)_2$ | $CH_3OCH_3$ | $HCOOCH_3$ | $HCOOH$ | $CO_2$ | $CO$ | $HCHO$ |
| $SbRe_2O_6$ | 1 | 6.5 | 92.5 | 6.3 | 1.2 | 0 | 0 | 0 | 0 |
| $Sb_4Re_2O_{13}$ | 1 | 0 | 0 | 0 | 0 | 0 | 0 | 0 | 0 |
| $SbOReO_4 \cdot 2H_2O$ | 1 | 4.7 | 1.0 | 99.0 | 0 | 0 | 0 | 0 | 0 |
| $Re_2O_7/Sb_2O_3$ | 1 | 2.5 | 7.4 | 42.7 | 0 | 0 | 0 | 0 | 49.8 |
| $Re_2O_7/SiO_2$ | 36 | 25.5 | 19.1 | 0.9 | 1.4 | 31.5 | 45.2 | 1.9 | 0 |
| $Re_2O_7+Sb_2O_3$ | 1 | 1.4 | 23.6 | 76.4 | 0 | 0 | 0 | 0 | 0 |
| $Sb_2O_3$ | 1 | 0 | 0 | 0 | 0 | 0 | 0 | 0 | 0 |
| $Sb_2O_4$ | 1 | 0 | 0 | 0 | 0 | 0 | 0 | 0 | 0 |
| $SbO_x^b$ | 56 | 1.3 | Trace | 26.4 | 0 | 0 | 18.0 | 0 | 55.5 |
| $Sb_2O_3/SiO_2$ | 35 | 9.6 | 23.9 | 49.3 | Trace | 0 | 7.0 | 7.8 | 11.6 |

[a] Catalyst weight=200mg; $He/O_2/MeOH=86.3/9.7/4.0$(mol%); GHSV=10000mL$h^{-1}$ $g^{-1}_{cat}$.

[b] Mainly $Sb_2O_5$.

Figure 1 shows the conversions and selectivities of the methanol reaction on the three Re-Sb-O compounds as a function of reaction temperature. The 100% conversion of methanol corresponds to a reaction rate of $16.4 \times 10^3$ mol$h^{-1}$ $g^{-1}_{cat}$. For $SbRe_2O_6$, the reaction rate and the methylal selectivity increased with increasing temperature up to 573 K, where the selectivity to methylal reached a maximum of 92.5%. The main byproduct was dimethyl ether, the formation of which decreased with increasing temperature. The lower the GHSV value, the higher the methanol conversion became, while the selectivity to methylal was almost independent of GHSV.

The catalytic reaction profiles for $SbOReO_4 \cdot 2H_2O$ and $Sb_4Re_2O_{13}$ were totally different from that for $SbRe_2O_6$. For $SbOReO_4 H_2O$, the reaction rate decreased with an increase in the temperature. The selectivity to methylal increased with reaction temperature and reached a maximum of about 60% at 533 K, then gradually decreased to zero at 593 K. In contrast to this, the selectivity to dimethyl ether decreased at first with reaction temperature below 533 K and then increased with an increase in the reaction temperature to be as high as 99% at 593 K. On the other hand, no catalytic activity was observed with $Sb_4Re_2O_{13}$ below 573 K and dimethyl ether was mainly produced above 573 K.

To examine the structures of the three crystalline Re-Sb-O samples relevant to the catalytic performances for the selective oxidation of methanol to methylal, they were characterized by means of XRD, XPS, Raman and SEM. Figure 2 shows the XRD patterns for $SbOReO_4 \cdot 2H_2O$, $Sb_4Re_2O_{13}$ and $SbRe_2O_6$ before and after methanol oxidation at 593 K. A dramatic change in the crystal structure was observed for $SbOReO_4 \cdot 2H_2O$ after selective methanol conversion, as shown in figure 2(a) and 2(b). The XRD pattern in figure 2(b) was similar to that for $Sb_4Re_2O_{13}$ (figure 2(d)). The XRD pattern for $Sb_4Re_2O_{13}$ did not change at 593 K as shown in figure 2(c) and 2(d). It was also found that XRD pattern

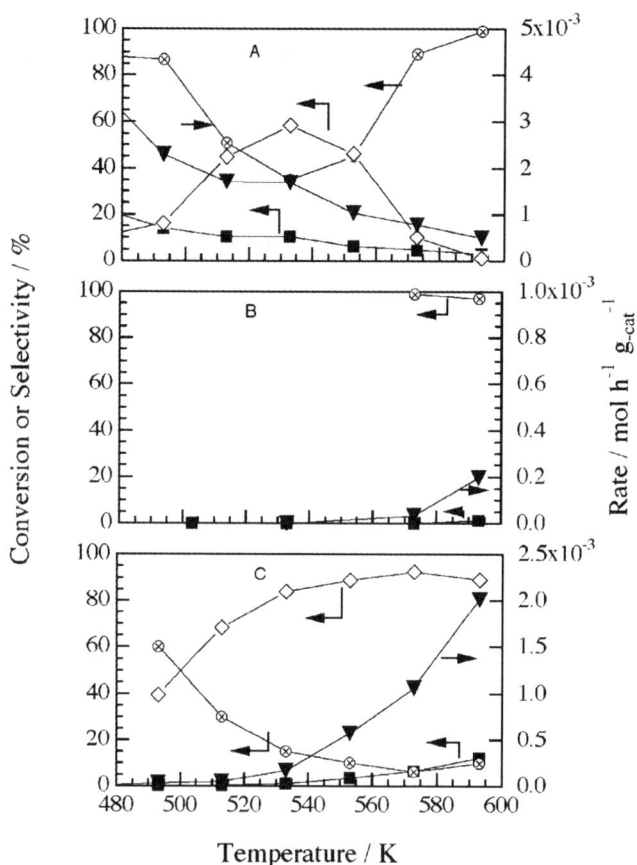

Figure 1　Catalytic methanol oxidation on(A)$SbOReO_4 \cdot 2H_2O$, (B)$Sb_4Re_2O_{13}$ and (C)$SbRe_2O_6$ as a function of reaction temperature(GHSV = 10 000 mLh$^{-1}$ g$_{-cat}^{-1}$; He/O$_2$/MeOH = 86. 3/9. 7/4. 0(mol%)): ■, methanol conversion; ◇, methylal selectivity;⊗, dimethyl ether selectivity;▼, reaction rate(molh$^{-1}$ g$_{-cat}^{-1}$).

for $SbRe_2O_6$ after the catalytic reaction was identical to that for the fresh sample(figure 2(e)and 2(f)).

　　Figure 3 shows Raman spectra for $SbOReO_4 \cdot 2H_2O$, $Sb_4Re_2O_{13}$ and $SbRe_2O_6$ before and after the methanol reaction at 593K(see table 2). The spectrum of $SbOReO_4 \cdot 2H_2O$ showed bands at 324,334, 341,361,459,854,919s,939 and 975vs(most intense peak)cm$^{-1}$. $Sb_4Re_2O_{13}$ exhibited peaks at 211,222, 236,304,315,330s,339s,350,358,403,421,464vw,859,874,899,914,929 and 971vs cm$^{-1}$. After methanol oxidation at 593 K, the Raman bands for $SbOReO_4 \cdot 2H_2O$ changed to similar ones to $Sb_4Re_2O_{13}$(figure 3(b)and 3(c)). The spectra for $Sb_4Re_2O_{13}$ remained unchanged by the reaction at 593 K. For $SbOReO_4 \cdot 2H_2O$ and $Sb_4Re_2O_{13}$,Raman bands at 970～975,914～919 and 330～334 cm$^{-1}$ were observed,which resemble the Raman bands for perrhenate ion in aqueous solution(971 cm$^{-1}$ for symmetric stretching mode; 916 cm$^{-1}$ for antisymmetric stretching mode; 332 cm$^{-1}$ for bending mode)[39]. Therefore, it is suggested that the surface rhenium species in $SbOReO_4 \cdot 2H_2O$ and $Sb_4Re_2O_{13}$ possess a [$ReO_4$] structure. Raman bands for fresh $SbRe_2O_6$ appeared at 152,171,227s, 239s,267,282,308,347,391,416,525 and 756cm$^{-1}$(figure 3(e)),which are entirely different from those for $SbOReO_4 \cdot 2H_2O$ and $Sb_4Re_2O_{13}$,although the assignments of the Raman bands for $SbRe_2O_6$ are not available currently. After methanol oxidation at 593 K,there was no significant change in the bands (figure 3(f)).

　　Figure 4 shows Re 4f XPS spectra for $SbOReO_4 \cdot 2H_2O$, $Sb_4Re_2O_{13}$ and $SbRe_2O_6$ before and after

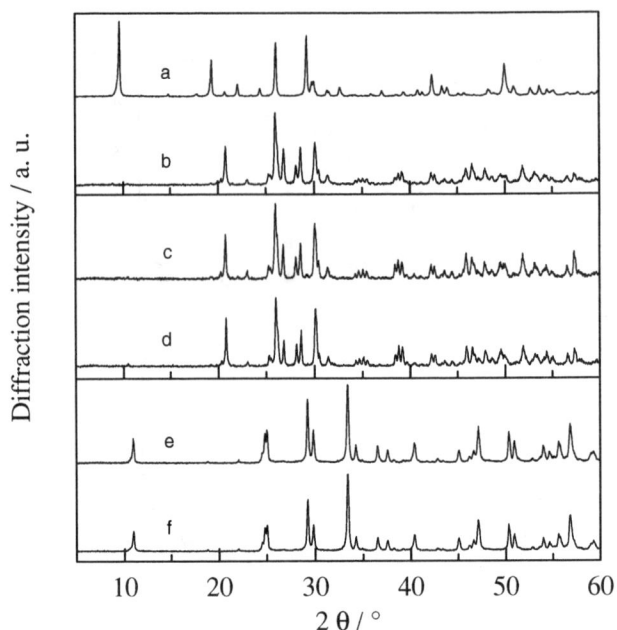

**Figure 2** XRD patterns for (a) fresh $SbOReO_4 \cdot 2H_2O$, (b) $SbOReO_4 \cdot 2H_2O$ after methanol reaction at 593 K, (c) fresh $Sb_4Re_2O_{13}$, (d) $Sb_4Re_2O_{13}$ after methanol reaction at 593 K, (e) fresh $SbRe_2O_6$, (f) $SbRe_2O_6$ after methanol reaction at 593 K.

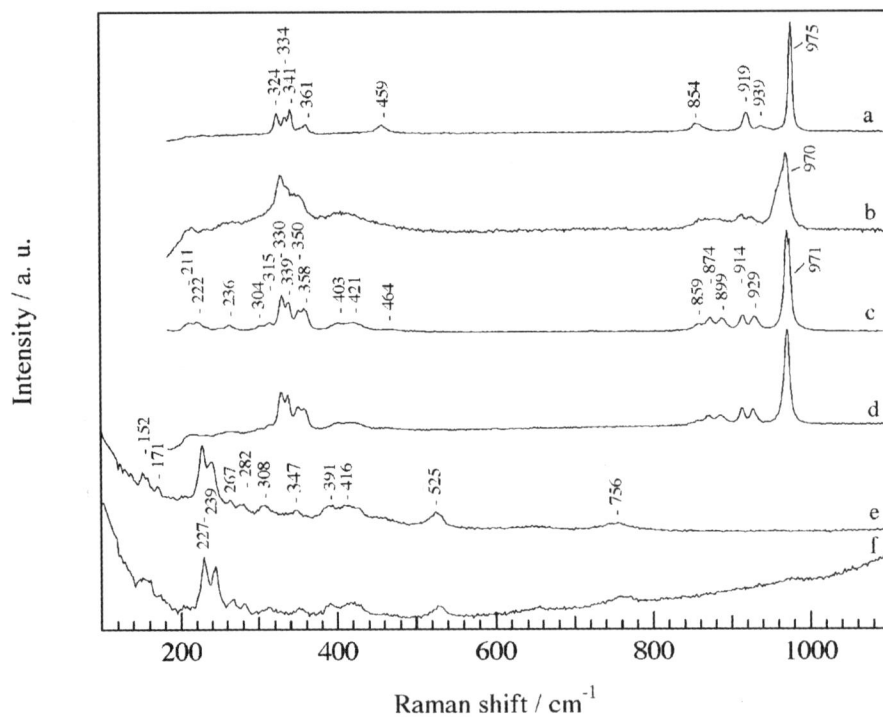

**Figure 3** Raman spectra for (a) fresh $SbOReO_4 \cdot 2H_2O$, (b) $SbOReO_4 \cdot 2H_2O$ after methanol reaction at 593 K, (c) fresh $Sb_4Re_2O_{13}$, (d) $Sb_4Re_2O_{13}$ after methanol reaction at 593 K, (e) fresh $SbRe_2O_6$, (f) $SbRe_2O_6$ after methanol reaction at 593 K.

methanol oxidation at 593 K. For these three samples before and after methanol oxidation, Sb 4d and Sb $3d_{3/2}$ bands were observed at binding energies of 34.4 and 539.7 eV, respectively, which are the values typical of $Sb^{3+}$. The Re 4f XPS spectra for fresh $SbOReO_4 \cdot 2H_2O$ and $Sb_4Re_2O_{13}$ were similar, exhibiting two peaks centered at 45.7 and 48.1 eV. The peaks are attributed to Re $4f_{7/2}$ and Re $4f_{5/2}$ levels of $Re^{7+}$ species, respectively. After methanol oxidation, the spectra did not change at all. In the case of $SbRe_2O_6$, three peaks were observed with both samples before and after the reaction. A small peak at 47.7 eV is tentatively assigned to Re $4f_{5/2}$ for $Re^{6+}$ species, while a definite peak at 42.3 eV is assigned to Re $4f_{7/2}$ for $Re^{4-5+}$ [40,41]. The observation of $Re^{4-5+}$ XPS peaks agrees with the chemical formula of the $SbRe_2O_6$ crystal containing Re ions with an oxidation state of $+4.5$ (mixed valency of $+4$ and $+5$) [36,37]. The most intense peak at about 44.7 eV is then considered as the sum of a Re $4f_{7/2}$ peak for $Re^{6+}$ species and a Re $4f_{5/2}$ peak for $Re^{4-5+}$ species as deconvoluted in figure 4(e) and 4(f). The $SbRe_2O_6$ sample exhibited no change in the Re 4f XPS spectra on methanol oxidation (figure 4(e) and 4(f)).

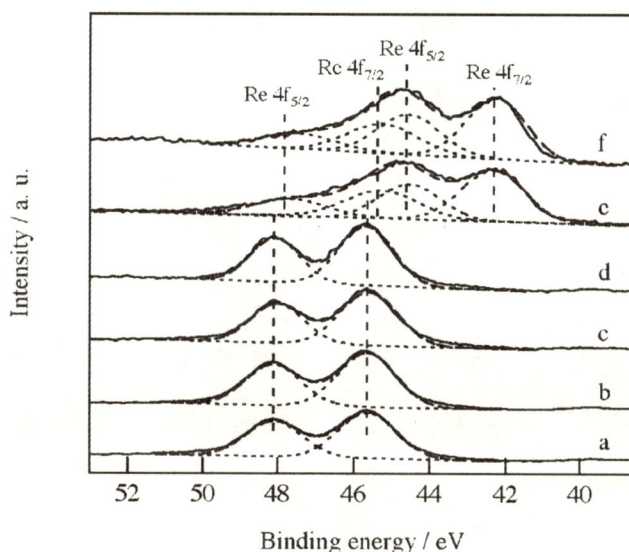

Figure 4  XPS spectra for (a) fresh $SbOReO_4 \cdot 2H_2O$, (b) $SbOReO_4 \cdot 2H_2O$ after methanol reaction at 593 K, (c) fresh $Sb_4Re_2O_{13}$, (d) $Sb_4Re_2O_{13}$ after methanol reaction at 593 K, (e) fresh $SbRe_2O_6$, (f) $SbRe_2O_6$ after methanol reaction at 593 K.

Table 2  Raman bands (in $cm^{-1}$) of three Re-Sb-O crystalline catalysts before and after catalytic methanol oxidation at 593 K.

| Catalyst | Fresh | After methanol oxidation |
| --- | --- | --- |
| $SbOReO_4 \cdot 2H_2O$ | 324,334,341,361,459,854,919s,939,975vs | 211,222,236,330s,339s,350,358,403,421,859, 874,899,914,929,970vs |
| $Sb_4Re_2O_{13}$ | 211,222,236,304,315,330s,339s,350,358, 403, 421, 464vw, 859, 874, 899, 914, 929, 971vs | Same bands as those for fresh catalyst |
| $SbRe_2O_6$ | 152,171,227s,239s,267,282,308,347,391, 416,525,756 | Same bands as those for fresh catalyst |

Figure 5 shows SEM images for $SbOReO_4 \cdot 2H_2O$, $Sb_4Re_2O_{13}$ and $SbRe_2O_6$ before and after methanol oxidation at 593 K. The fresh $SbOReO_4 \cdot 2H_2O$ was built up from an alum-like crystal with transparency, but it changed to chalk-like at the surface after the methanol reaction, with a morphology similar to that of $SbRe_2O_{13}$(figure 5(c)). The morphology of $SbRe_2O_{13}$ did not change after the reaction (figure 5(d)). For $SbRe_2O_6$, the fresh sample was composed of crystals possessing square basal faces of 0.5—3 $\mu$m in length and about 100 nm in thickness(figure 5(e)). The basal(100)faces were smooth and had sharp and regular edges. After selective oxidation of methanol to methylal at 593 K, the $SbRe_2O_6$ crystals exhibited almost the same morphology as that for the fresh sample(figure 5(f)).

**Figure 5  SEM images for(a)fresh $SbOReO_4 \cdot 2H_2O$, (b)$SbOReO_4 \cdot 2H_2O$ after methanol reaction at 593 K, (c)fresh $Sb_4Re_2O_{13}$, (d)$Sb_4Re_2O_{13}$ after methanol reaction at 593 K, (e)fresh $SbRe_2O_6$,(f)$SbRe_2O_6$ after methanol reaction at 593 K.**

Thus the results revealed no structural changes in the bulk and surface of $Sb_4Re_2O_{13}$ and $SbRe_2O_6$, while in the bulk and surface of $SbOReO_4 \cdot 2H_2O$ the structural transformation to $Sb_4Re_2O_{13}$ was observed after the methanol reaction at 593 K. The XPS spectra for $SbOReO_4 \cdot 2H_2O$ and $Sb_4Re_2O_{13}$ before and after the methanol reaction showed the oxidation states of Re and Sb to be $+7$ and $+3$, respectively. However, the catalytic properties of $SbOReO_4 \cdot 2H_2O$ and $Sb_4Re_2O_{13}$ were entirely different from each other (figure 1). The active $SbRe_2O_6$ catalyst exhibited $Re^{6+}$ species which are probably produced by oxidation of $Re^{4-5+}$ species at the surface. The three Re-Sb-O compounds have different crystalline structures. $SbRe_2O_6$ consists of alternate octahedral $(Re_2O_6)^{3-}$ and $(SbO)^+$ layers which are connected with each other making Re-O-Sb bonds[35]. The other two crystalline compounds, $SbOReO_4 \cdot 2H_2O$ and $Sb_4Re_2O_{13}$, are built up from tetrahedral $(ReO_4)^+$ anions and cationic $(SbO)$ layers[27,35]. The results indicate that the difference in the catalytic performances of the Re-Sb-O crystalline catalysts may be due not only to the difference in their surface Re oxidation states but also to the difference in their surface structures. It should be noted that $SbRe_2O_6$ is the first mixed oxide to show a high performance for the selective catalytic oxidation of methanol to methylal[34,35].

# 4  Conclusions

1. The catalysis of the oxidation of methanol to produce methylal is significantly related to the structure of Re-Sb-O compounds. Among the three Re-Sb-O catalysts studied, $SbRe_2O_6$ showed the highest activity and selectivity in the catalytic oxidation of methanol to methylal. The $SbRe_2O_6$ catalyst was stable during the catalytic reaction. Methylal can be synthesized from methanol and oxygen with a high selectivity of 92.5% at 6.5% conversion at GHSV$=$10 000 $mLh^{-1}g_{cat}^{-1}$ and 573K.

2. No structural change in the bulk and surface of $SbRe_2O_6$ and $Sb_4Re_2O_{13}$ occurred during methanol oxidation below 593 K, but $SbOReO_4 \cdot 2H_2O$ was transformed to $Sb_4Re_2O_{13}$ under the methanol oxidation conditions at 593 K.

3. The good performance of $SbRe_2O_6$ for the selective catalytic oxidation of methanol to methylal may be related to the octahedral $(Re_2O_6)^{3-}$ layer structure connecting with $(SbO)^+$ chains through Re-O-Sb bonds.

# Acknowledgements

This work has been supported by Core Research for Evolutional Science and Technology(CREST) of the Japan Science and Technology Corporation(JST) and the Natural Science Foundation of China (NSFC,Grant Nos 29873037,20023001 and 20021002).

# References

[1]J. M. Tatibouët,Appl. Catal. A:General **148**. (1997)213.

[2]G. Busca,A. S. Elmi and P. Forzatti,J. Phys. Chem. **91**(1987)5236.

[3]M. Ai,J. Catal. **77**(1982)279.

[4]N. Arora,G. Deo,I. E. Wachs and A. M. Hirt,J. Catal. **159**(1996)1.

[5]J. M. Tatibouët and J. E. Germain,Bull. Soc. Chim. Fr. I 9－10(1980)343.

[6](a)M. Fournier,A. Aouissi and C. Rocchiccioli-Deltchff,J. Chem. Soc. Chem. Commun. (1994) 307;(b) C. Rocchiccioli-Deltcheff, A. Aoussi, S. Launary and M. Fournier,J. Mol. Catal. A: Chemical. **114**(1996)331.

[7]I. J. Shannon, T. Maschmeyer, R. D. Oldroyd, G. Sankar, J. M. Thomas, H. Pernot, J. P. Baalikdjian and M. Che,J. Chem. Soc. Faraday Trans. **94**(1998)1495.

[8]For example,K. Otsuka and I. Yamanaka,Appl. Catal. **26**(1986)401.

[9]J. A. Moulijn and J. C. Mol, J. Mol. Catal. **46**(1988)1.

[10]J. C. Mol and J. A. Moulijn, in: Advance in Catalysis, eds D. D. Eley, H. Pines and P. B. Weisz, Vol. 24(Academic Press, New York, 1975)p. 131.

[11]I. E. Wachs, G. Deo, A. Andreini, M. A. Vuurman and M. de Boer, J. Catal. **160**(1996)322.

[12]J. -M. Jehng, H. Hu, X. Gao and I. E. Wachs, Catal. Today. **28**(1996)335.

[13]A. A. Olsthoorn and C. Boelhouwer, J. Catal. **44**(1976)207.

[14]S. M. Augustine and W. H. H. Sachtler, J. Catal. **116**(1989)184.

[15]T. A. Pecoraro and R. R. Chianelli, J. Catal. **67**(1981)430.

[16]R. Thomas, E. M. Van Oers, V. H. J. De Beer, J. Medema and J. A. Moulijn, J. Catal. **76**(1982) 241.

[17]J. Okai, L. Lepinski, L. Krajczky and M. Drozd, J. Catal. **188**(1999)140.

[18]J. Röty and T. A. Pakkanen, Catal. Lett. **65**(2000)276.

[19]L. Wang, R. Onishi and M. Ichikawa, J. Catal. **190**(2000)276.

[20]W. A. Herrmann, Angew. Chem. Int. Ed. Engl. **27**(1988)1297.

[21]J. Sundermeyer, Angew. Chem. **105**(1993)1195.

[22]R. K. Grasselli, Catal. Today. **49**(1999)141.

[23]S. Albonetti, F. Cavani and F. Trifirō, Catal. Rev. Sci. Eng. **38**(1996)413.

[24]G. Centi and F. Trifirō, Catal. Rev. Sci. Eng. **28**(1986)165.

[25]J. Nilsson, A. R. Land-Canovas, S. Hansen and A. Andersson, J. Catal. **186**(1999)442.

[26]F. J. Berry, Adv. Catal. **30**(1981)97.

[27]E. C. Gaigneaux, H. Liu, H. Imoto, T. Shido and Y. Iwasawa, Topics in Catal. **11－12**(2000) 185.

[28]H. Liu, E. C. Gaigneaux, H. Imoto, T. Shido and Y. Iwasawa, J. Phys. Chem. B. **104**(2000) 2033.

[29]H. Liu, E. C. Gaigneaux, H. Imoto, T. Shido and Y. Iwasawa, Appl. Catal. A: General. **202** (2000)251.

[30]H. Liu, T. Shido and Y. Iwasawa, Chem. Commun. (2000)1881.

[31]H. Liu, E. C. Gaigneaux, H. Imoto, T. Shido and Y. Iwasawa, Catal. Lett. **71**(2001)75.

[32]H. Liu, H. Imoto, T. Shido and Y. Iwasawa, J. Catal. **200**(2001)69.

[33]W. T. A. Harrison, A. V. P. McManus, M. P. Kaminsky and A. K. Cheetham, Chem. Mater. **3** (1993)1631.

[34]Y. Yuan, H. Liu, H. Imoto, T. Shido and Y. Iwasawa, Chem. Lett. (2000)674.

[35]Y. Yuan, H. Liu, H. Imoto, T. Shido and Y. Iwasawa, J. Catal. **195**(2000)51.

[36]H. Watanabe and H. Imoto, Inorg. Chem. **36**(1997)4610.

[37]H. Watanabe, H. Imoto and H. Tanaka, J. Solid State Chem. **138**(1998)245.

[38]R. Castillo, K. Dewaele, P. Ruiz and B. Delmon, Appl. Catal. A: General. **153**(1997)L1.

[39]F. Gonzalez-Vichez and W. P. Griffith, J. Chem. Soc. Dalton Trans. (1972)1416.

[40]A. Cimino, B. A. De Angelis, D. Gazzoli and M. Valigi, Z. Anorg. Allg. Chem. **460**(1980)86.

[41]E. S. Shpiro, V. I. Avaev, M. A. Ryashentseva and Kh. M. Minachev, J. Catal. **55**(1978)402.

■ **本文原载**:《物理化学学报》(自然科学版)第 19 卷第 11 期(2003 年)，第 1073~1077 页。

# 晶体中原子的平均范德华半径*

胡盛志① 周朝晖 蔡启瑞

(厦门大学化学系，物理化学研究所，厦门 361005)

**摘 要** 根据晶体中原子的平均体积数据提出包括全部金属元素的原子平均范德华半径值。与现有几个重要的范德华半径体系进行了初步的比较，指出范德华半径值在应用中值得注意的问题，简要提出了存关范德华半径今后研究的方向。

**关键词** 平均原子体积 范德华半径 次级键

化学在二十世纪的发展已深入到物质结构在原子和分子水平的定量研究阶段，几套有关原子大小的半径值——离子半径、共价半径、金属原子半径和范德华半径系统的建立就是一个重要标志。Pauling[1] 在上世纪三十年代首次提出晶体中原子的范德华半径(以下简称为 $r_w$)，至今已有多套 $r_w$ 值见诸文献。由于 $r_w$ 值与原子和分子所处的物态有关，导出 $r_w$ 值的原始实验数据质量的差异以及计算的原则和方法不同，不同系统之间的差异最大可超出 0.05 nm。我们知道，Allinger[2] 为分子力学进行能量优化计算建立了一套独立的 $r_w$ 值系统，它是对应于处在孤立状态下的原子，通常其数值比从晶体结构数据导出的结果要大 10%~30%。

除 Pauling 和 Allinger 这两套 $r_w$ 值外，目前影响较大的还有 Bondi[3] 和 Batsanov 等[4] 提出的 $r_w$ 系统。Bondi 系统的建立采用了不同物态的原始数据，即除晶体结构数据之外，还利用了原子的碰撞截面、临界体积和紫外光谱等不同来源的数据。尽管 Bondi 推出了包括若干金属原子在内的数据较为完整的 $r_w$ 值，并被广为引用，但其理论基础仍然值得商榷[5]。Batsanov 系统则包含了更多金属原子的 $r_w$ 值，但同样受到了质疑[6]。Zefirov[7] 指出，这两套 $r_w$ 系统都不是单纯源于晶体结构数据，即不是严格建立在分子间原子接触距离的基础上，因此，其实用范围尚有待于进一步的研究。可是到目前为止，Zefirov 等尚未提出全部源于晶体结构数据并足以取代上述 $r_w$ 值的完整系统，特别是周期表中很多重要的金属元素(包括镧系元素)原子的 $r_w$ 值仍属空白。

我们知道，从晶体结构数据对分子间原子接触距离进行检索是计算 $r_w$ 值简捷而又传统的方法。由于历史的原因，Pauling 当年建立 $r_w$ 系统的开创性工作仅涉及 H、惰性气体和 P 区元素。经过上世纪晶体结构数据的大量积累，现在完全依据晶体结构数据推导包括全部周期表元素的 $r_w$ 系统，不但迫切需要，而且完全可能了。然而检索分子间非金属原子 X 的范德华接触距离 X···X 比较容易，要对金属原子 M 检索 M···M 接触距离仍旧非常困难。为此，Batsanov[4] 曾试图从晶体结构中分子内的原子接触距离导出一些金属原子的 $r_w$ 值。另一方法则是基于络合物分子中各原子在晶体中的 Voronoi-Dirichlet 多面体(VDP)体积 $V_{VDP}$，令圆球体积等于 $V_{VDP}$ 后再从球半径计算原子的 $r_w$ 值[8]。这种方法虽适用于卤素原

---

\* 2003 年 5 月 16 日收到初稿。2003 年 6 月 19 日收到修改稿。

国家基础研究发展规划(CB001CB108906)和国家自然科学基金(29933040)资助项目。

① 联系人:胡盛志(E-mail:szhu@xmu.edu.cn;Tel:0592-2180297)。

子,但 H 原子的球半径为 0.142(7)nm,这个数值远大于 Pauling 推荐的 0.110 nm[9]。相反地,中心金属原子的球半径又显得过小,例如稀土金属原子仅在 0.16 nm 左右,这个数值大约只与它们的单键共价半径相当[2]。

与上述方法不同,我们试图从晶体中原子的平均体积建立复盖周期表全部元素的统一 $r_w$ 系统。由于在计算中既不计及原子配位数和氧化态的差异,也忽视原子或离子大小的各向异性这一重要特征[10,11],这样得到的原子作用势力范围和 $r_w$ 值显然具有统计平均性质。本文将它们与其他现有的几个重要的 $r_w$ 系统并列,以资对比和选用,并根据几个重要的实例对它们的适用性进行分析和初评。最后我们还简要指出范德华半径在实际应用和进一步研究中的若干问题。

# 1 方法与结果

本文提出从大量的晶体结构中统计得到的原子的平均体积来计算 $r_w$ 值,而晶体中原子体积的研究及相关数据的积累已有历史。众所周知,原子在晶体中的平均体积和有限的范德华半径数据都曾相当成功地用于晶体密度、单胞体积和单胞中分子数的估算。早在上世纪七十年代,Kempster 与 lipson[12] 仅选用 40 个有机晶体结构统计就得出有机化合物中非 H 原子平均体积为 0.018 nm³,这个有用的"魔数"是结构晶体学工作者非常熟悉的。随后有一系列的工作对它进行优化,直到 Mighell 等[13] 在上世纪八十年代成功地得到了周期表中全部元素在晶体中原子的平均体积。最近,Hofmann[14] 又从剑桥晶体结构数据库中选用 182 239 个有机和金属有机物晶体结构数据对 Mighell 等的工作进行修订,其中包含了热膨胀系数和部分的误差估计。Hofmann 修订原子平均体积的目的原本用于有机晶体密度的估算,而我们则用来计算 $r_w$ 值,即先令晶体中平均原子体积圆球化,再从其半径计算出平均 $r_w$ 值,计算结果见表 1. 对于 Hofmann 的数据中未涉及的 Ne 等 17 个稀有元素,我们采用 Mighell 等的数据进行计算。表 1 括号中的 $r_w$ 值即表示该计算数据结果源于参考文献[13]中的平均原子体积。

表 1　原子的范德华半径 $r_w$(nm)

Table 1　van der Waals radii $r_w$(nm)of the Elements

| Element | H | He | Li | Be | B | C | N | O | F | Ne |
|---|---|---|---|---|---|---|---|---|---|---|
| Pauling[1,9] | 0.110 | 0.140 | | | | 0.172 | 0.150 | 0.140 | 0.135 | 0.154 |
| Bondi[3] | 0.120 | 0.140 | 0.182 | | 0.213 | 0.170 | 0.155 | 0.152 | 0.147 | 0.154 |
| Batsanov[4] | 0.12 | 0.14 | 0.22 | 0.19 | 0.18 | 0.17 | 0.16 | 0.155 | 0.15 | 0.15 |
| Allinger[2] | 0.162 | 0.153 | 0.255 | 0.223 | 0.215 | 0.204 | 0.193 | 0.182 | 0.171 | 0.160 |
| Zefirov[7] | 0.116 | | 0.22 | 0.19 | 0.18 | 0.171 | 0.150 | 0.129 | 0.140 | |
| This work | 0.108 | (0.134)* | 0.175 | 0.205 | 0.147 | 0.149 | 0.141 | 0.140 | 0.139 | (0.168)* |
| Element | Na | Mg | Al | Si | P | S | Cl | Ar | K | Ca |
| Pauling[1,9] | | | | | 0.190 | 0.185 | 0.180 | 0.192 | | |
| Bondi[3] | 0.227 | 0.173 | 0.251 | 0.210 | 0.180 | 0.180 | 0.175 | 0.188 | 0.275 | |
| Batsanov[4] | 0.24 | 0.22 | 0.21 | 0.21 | 0.195 | 0.18 | 0.18 | 0.19 | 0.28 | 0.24 |
| Allinger[2] | 0.270 | 0.243 | 0.236 | 0.229 | 0.222 | 0.215 | 0.207 | 0.199 | 0.309 | 0.281 |
| Zefirov[7] | 0.23 | 0.20 | 0.20 | | 0.184 | 0.190 | | | 0.27 | 0.24 |
| This work | 0.184 | 0.205 | 0.211 | 0.207 | 0.192 | 0.182 | 0.183 | (0.193)* | 0.205 | 0.221 |

Con. Table 1

| Element | Sc | Ti | V | Cr | Mn | Fe | Co | Ni | Cu | Zn |
|---|---|---|---|---|---|---|---|---|---|---|
| Pauling[1,9] | | | | | | | | | | |
| Bondi[3] | | | | | | | | | | |
| Batsanov[4] | 0.23 | 0.215 | 0.205 | 0.205 | 0.205 | 0.205 | 0.20 | 0.20 | 0.20 | 0.21 |
| Allinger[2] | 0.261 | 0.239 | 0.229 | 0.225 | 0.224 | 0.223 | 0.223 | 0.2222 | 0.226 | 0.229 |
| Zefirov[7] | 0.22 | | | | | | | | 0.19 | 0.19 |
| This work | 0.216 | 0.187 | 0.179 | 0.189 | 0.197 | 0.194 | 0.192 | 0.184 | 0.186 | 0.210 |

| Element | Ga | Ge | As | Se | Br | Kr | Rb | Sr | Y | Zr |
|---|---|---|---|---|---|---|---|---|---|---|
| Pauling[1,9] | | | | | | | | | | |
| Bondi[3] | | | | | | | | | | |
| Batsanov[4] | 0.21 | 0.21 | 0.205 | 0.19 | 0.19 | 0.20 | 0.29 | 0.255 | 0.24 | 0.23 |
| Allinger[2] | 0.246 | 0.244 | 0.236 | 0.229 | 0.222 | 0.215 | 0.325 | 0.300 | 0.271 | 0.254 |
| Zefirov[7] | 0.21 | | | | 0.197 | | 0.28 | 0.26 | 0.23 | |
| This work | 0.208 | 0.215 | 0.206 | 0.193 | 0.198 | (0.212)* | 0.216 | 0.224 | 0.219 | 0.186 |

| Element | Nb | Mo | Tc | Ru | Rh | Pd | Ag | Cd | In | Sn |
|---|---|---|---|---|---|---|---|---|---|---|
| Pauling[1,9] | | | | | | | | | | |
| Bondi[3] | | | | | | 0.163 | 0.172 | 0.162 | 0.255 | 0.227 |
| Batsanov[4] | 0.215 | 0.21 | 0.205 | 0.205 | 0.20 | 0.205 | 0.21 | 0.22 | 0.22 | 0.225 |
| Allinger[2] | 0.243 | 0.239 | 0.236 | 0.234 | 0.234 | 0.237 | 0.243 | 0.250 | 0.264 | 0.259 |
| Zefirov[7] | | | | | | | 0.20 | 0.21 | 0.22 | |
| This work | 0.207 | 0.209 | 0.209 | 0.207 | 0.195 | 0.202 | 0.203 | 0.230 | 0.236 | 0.233 |

| Element | Sb | Te | I | Xe | Cs | Ba | La | Ce | Pr | Nd |
|---|---|---|---|---|---|---|---|---|---|---|
| Pauling[1,9] | 0.220 | 0.220 | 0.215 | 0.218 | | | | | | |
| Bondi[3] | 0.190 | 0.206 | 0.198 | 0.216 | | | | | | |
| Batsanov[4] | 0.22 | 0.21 | 0.21 | 0.22 | 0.30 | 0.27 | 0.25 | | | |
| Allinger[2] | 0.252 | 0.244 | 0.236 | 0.228 | 0.344 | 0.307 | 0.278 | 0.274 | 0.273 | 0.273 |
| Zefirov[7] | | | 0.214 | | 0.29 | 0.26 | 0.23 | | | |
| This work | 0.225 | 0.223 | 0.223 | 0.221 | 0.222 | 0.251 | 0.240 | 0.235 | 0.239 | 0.229 |

| Element | Pm | Sm | Eu | Gd | Tb | Dy | Ho | Er | Tm | Yb |
|---|---|---|---|---|---|---|---|---|---|---|
| Pauling[1,9] | | | | | | | | | | |
| Bondi[3] | | | | | | | | | | |
| Batsanov[4] | | | | | | | | | | |
| Allinger[2] | 0.272 | 0.271 | 0.294 | 0.271 | 0.270 | 0.269 | 0.267 | 0.267 | 0.267 | 0.279 |
| Zefirov[7] | | | | | | | | | | |
| This work | (0.236)* | 0.229 | 0.233 | 0.237 | 0.221 | 0.229 | 0.216 | 0.235 | 0.227 | 0.242 |

| Element | Lu | Hf | Ta | W | Re | Os | Ir | Pt | Au | Hg |
|---|---|---|---|---|---|---|---|---|---|---|
| Pauling[1,9] | | | | | | | | | | |
| Bondi[3] | | | | | | | | 0.175 | 0.166 | 0.170 |
| Batsanov[4] | | 0.225 | 0.22 | 0.21 | 0.205 | 0.20 | 0.20 | 0.205 | 0.21 | 0.205 |
| AHinger[2] | 0.265 | 0.253 | 0.243 | 0.239 | 0.237 | 0.235 | 0.236 | 0.239 | 0.243 | 0.253 |
| Zefirov[7] | | | | | | | | | 0.20 | 0.20 |
| This work | 0.221 | 0.212 | 0.217 | 0.210 | 0.217 | 0.216 | 0.202 | 0.209 | 0.217 | 0.209 |

Con. Table 1

| Element | T1 | Pb | Bi | Po | At | Rn | Fr | Ra | Ac | Tb |
|---|---|---|---|---|---|---|---|---|---|---|
| Pauling[1,9] | | | | | | | | | | |
| Bondi[3] | 0.196 | 0.202 | 0.187 | | | | | | | |
| Batsanov[4] | 0.22 | 0.23 | 0.23 | | | | | | | 0.24 |
| Allinger[2] | 0.259 | 0.274 | 0.266 | 0.259 | 0.251 | 0.243 | 0.364 | 0.327 | 0.308 | 0.274 |
| Zefirov[7] | 0.22 | | | | | | | | | |
| This work | 0.235 | 0.232 | 0.243 | (0.229)* | (0.236)* | (0.243)* | (0.256)* | (0.243)* | 0.260 | 0.237 |

| Element | Pa | U | Np | Pu | Am | Cm | Bk | Cf | Es | Fm |
|---|---|---|---|---|---|---|---|---|---|---|
| Pauling[1,9] | | | | | | | | | | |
| Bondi[3] | | | | | | | | | | |
| Batsanov[4] | | 0.23 | | | | | | | | |
| Allinger[2] | 0.264 | 0.252 | 0.252 | 0,252 | | | | | | |
| Zefimv[7] | | | | | | | | | | |
| This work | 0.243 | 0.240 | 0.221 | (0.256)* | (0.256)* | (0.256)* | (0.256)* | (0.256)* | (0.256)* | (0.256)* |

\* from Ref,[13]

# 2 讨 论

除我们的计算结果外,表1同时列出现有的几个重要的 $r_w$ 系统。它们将在实际工作中接受检验,在此我们提出以下几个问题作初步的分析和讨论。

(1)Allinger 推荐的分子力学程序中的计算参数与全部或主要基于晶体结构数据的 $r_w$ 系统推荐值差别很大。分子力学中的 $r_w$ 值显然自成独立系统,它在分子构型的能量优化计算中是相当成功的。但是它不能与其他系统混用,即使在其他的几个系列之间混用也要持谨慎的态度。例如 Iwaoka 等[15]对于蛋白质中的次级键 S…X(X=O、N、S)进行统计和理论研究时,C、N、O 和 S 原子的 $r_w$ 采用的是 Bondi 值,而 Fe 原子却采用基于 Allinger 值 0.223 nm 进行修改后的 0.170 nm。修改值比推荐值缩短 0.053 nm,但是原作者并未对修改的原因和方法作出具体说明。本文得到的 Fe 原子计算值为 0.194 nm,与 Allinger 值 0.223 nm 和上述修正值 0.170 nm 的平均值 0.196 nm 接近。有意思的是,Iwaoka 选定的 Fe…O 的混合范德华半径和为 0.170+0.152=0.322(nm),而用本文推荐值的范德华半径和为 0.194+0.140=0.334(nm),二者相当接近。

(2)Allinger 系统以外的几个 $r_w$ 系统虽然大同小异,但其差别在实际应用中可能也不容忽视。就以 H 原子为例,Bondi 和 Batsanov 系统采用的是 Pauling 最早提出的 0.120 nm[1],实际上 Pauling 本人早已将它修正为 0.110 nm[9]。后来 Iijima 等[16]和 Rowland 等[17]又从理论计算和实验数据统计分别推荐 H 原子的 $r_w$ 为 0.108 和 0.109 nm,与本文计算的 $r_w$ 值 0.108 nm 吻合,并与 Pauiing 的修正值非常接近。

近十年来,H 原子在气相和凝固态中 $r_w$ 值各向异性的研究相当活跃[6,18]。中子衍射数据显示 H 原子在 $C_{sp^3}$-H 键轴向的 $r_w$ 值仅为 0.101 mn,N 原子的 $r_w$ 椭球的短轴也只有 0.142 nm。实际上,不仅非金属原子的 $r_w$ 呈各向异性的椭球特征,金属原子的 $r_w$ 也不例外。它们可能源于原子之间的相互极化和次级键的特殊作用,具有重要的理论意义和应用价值。例如在固氮酶的 FeMo 辅因子还原作用机理的探讨过程中,我们就需要考究 H 原子等 $r_w$ 的可靠值及其各向异性特征,并会感到通常引用的 H 原子的 $r_w$ 为 0.120 nm 的确是太长了些。我们在此不对这个问题作进一步的讨论,只是指出:本文提出的 $r_w$ 值和一般文献和手册上列出的范德华半径全是它们的相当于各向同性的统计平均值。

（3）推出所有金属原子的 $r_w$ 值是本文的重要贡献，它们在次级键的界定和超分子化学等诸多化学领域的研究中具有重要的意义。本文推出的金属原子的 $r_w$ 值中，除了 $s$ 区元素外，其他金属元素的 $r_w$ 值与 Batsanov 系统和 Zefirov 系统相比，总的说来相当接近，与 Bondi 系统的差别则比较大。以 Ag 原子的 $r_w$ 值为例，Bondi 系统为 0.172 nm。据此，凡是 Ag⋯Ag 距离超过 0.344 nm 的次级键将不复存在，因两原子间距大于 $r_w$ 值之和一般被认为是没有任何相互作用的。假如按照本文推荐的 0.203 nm 考虑，我们就不难接受借助 Ag⋯Ag 次级键形成的一系列 Ag 原子簇多面体及其包合物结构的结构描述[19]。值得提出的是，本文首次推荐晶体中镧系元素 Ln 的 $r_w$ 值，填补了范德华半径数据库中的长期留下的空白。从表 1 可以看出，它们落在 0.22～0.24 nm 这一狭窄范围，这套数据对于确定诸如 Ln⋯O 等次级键具有参考价值[20]。而在此之前，由于大部分金属原子的 $r_w$ 值无从查询，在研究是否存在次级键时，Richardson[21] 曾建议了一个实用的 0.10 nm 判据，即原子短接触不超出两原子单键共价键长之和 0.10 nm 作为次级键存在的上限。这个经验判据意味着一个原子的 $r_w$ 大约比它的单键共价半径长 0.05 nm。这个数值比 Pauling[1] 当年为估计金属原子的 $r_w$ 值所提出的 0.08 nm[1] 要短很多。现在看来这个界定确是过于苛刻了，因为按此判据，一些原本客观存在的重要次级键将极有可能被排斥在我们的视野之外。但我们计算的 B、C 和 N 这三个非金属原子的 $r_w$ 值，相对于其他几个系统来说是太短了些。特别是 C 原子的 0.149 nm，它比 Pauling 的推荐值 0.172 nm 小 0.023 nm，而像对石墨这类无机层状结构中，C 原子的层间距为 0.34 nm，Pauling 的数据无疑是正确的。由于我们的计算完全源于剑桥数据库中存储的有机和金属有机化合物晶体结构数据，因此，这个 $r_w$ 系统应用于有机以及金属有机化合物领域有关的结构化学研究，将会有比较牢靠的理论基础。

我们认为，除本文介绍的几套 $r_w$ 值系统外，如何利用已经积累起来的大量宝贵的晶体结构数据，在总结更为精确而又适用的 $r_w$ 系统的同时，进一步探索其变化规律和它们与其他半径如单键共价半径和金属原子半径等的联系，是极有意义的基本理论研究课题。我们还认为，理论化学计算原子在不同配位数、不同氧化态在不同作用方向的 $r_w$ 值等方面都将是大有作为的[22]。这些研究不仅可以探索不同的半径系统之间可能存在的内在联系，并且对于作为超分子化学基础的次级化学键的界定和范德华键本质的深入探讨均不无裨益。

**致谢**：作者对于俄罗斯 S. S. Batsanov 教授提供的重要信息和有益的讨论深表谢忱。对审稿人提出的宝贵意见在此一并致谢。

# References

[1]Pauling, L. The nature of the chemical bond. NY：Cornell Univ. Press, 1939.

[2]Allinger, N. L. ；Zhou, X. ；Bergsma, J. J. Mol Struct., Theochem, 1994, **312**：69.

[3]Bondi, A. J. Phys. 1964, **8**：441.

[4]Batsanov, S. S. Runs. J. Inorg. Chem., 1991, **36**：1694；Inorg Mater., 2001, **37**：871.

[5]Zefirov, Y. V. Russ. J. Inorg. Chem., 2001, **46**：568.

[6]Zefirov, Y. V. Russ. J. Inorg. Chem., 2001, **46**：1211.

[7]Zefirov, Y. V. Russ. J. Inorg. Chem., 2000, **45**：1552.

[8]Blatova, O. A. ；Blatov, V, A. ；Serezhkin, V, N. Acta Cryst, 2001, **57**：261.

[9]Faulmg, L. ；Pauling, P. Chemistry. San Francisco：W. H. Freeman Company, 1975.

[10]Tsai, K. R. ；Harris, P. M. ；Lassettre, E. N. J. Phys. Chem, 1956, **60**：338.

[11]Shang,M. Y. ;Huang,J. L. ;Lu,J,X. Science in China B,1985,**28**:351〔商茂虞,黄金陵,卢嘉锡. 中国科学 B(Zhong guo Kexue B),1985,**9**:607〕.

[12]Kempster,C. J. E,;Lipson,H. Acta Cryst A,1972,**28**:3674.

[13]MigheH, A. D. ;Hubbard,C R. ;Stalick, J. K. ;Santoro,A,;Snyder,R. L. ;Holomany, M. ; Scidel,J. ;Lederman,S. National bureau of standards,Gaithersburg,MD 20899,USA,1987.

[14]Hofmann,D. W. M. Acta Cryst,2002,**58**:489.

[15]Twaoka,M,;Takemoto,S. ;Okada,M. ;Tomoda,S. Bull. Chem. Soc. Jpn,2002,**75**:1611.

[16]Iijima,H. ;Dunbar,J. B. ;Marshall,G,R,Proteins:Structure,Function,and Genetics,1987,**2**: 330.

[17]Rowland,R. S. ;Taylor,R. J. Phys. chem. ,1996,**100**:7384.

[18]Batsanov,S. S. Russ. J. Coord. Chem.,2001,**27**:890.

[19]Wang,Q. M. ;Mak,T. C. W. Angew. Chem,Int. Ed. ,2001,**40**:1130.

[20]Chen,M. D. ;Hu,S. Z. Acta Phys.,Chim. Sin,2002,**18**(12),1104〔陈明旦,胡盛志. 物理化学学报(Wuli Huaxue Xuebao),2002,**18**(12):1104〕;Hu,S. Z. Univ. Chem.,2001,**16**:6〔胡盛志. 大学化学(Daxue Huaxue),2001,**16**:6〕.

[21]Richardson, M. F,;Wujfsberg, G. ;Marlow, R. ;Zaghonni, S,;McCorlde, D. ;Shadid, K. ; Gagliardi,J. ;Farris,B. Inorg. Chem. ,1993,**32**:1913.

[22]Badenhoop,J. K. ;Weinhold,F. J. Chem. Phys.,1997,**107**:5422.

# Average van der Waals Radii of Atoms in Crystals[*]

Sheng-Zhi Hu, Zhao-Hui Zhou, Khi-Rui Tsai

(*Department of Chemistry and Institute of Physical Chemistry, Xiamen University, Xiamen* 361005)

**Abstract**  Provisional data of the atomic average van der Waals radii including all common metal elements in crystals are proposed,which is derived from the available data of the average atom volume in crystals. Average volume of elements at 298 K was obtained by a statistical analysis of the Cambridge Structural Database. The present van der Waals radii are compared with the other systems proposed previously by Pauling,Bondi,Batsanov,Allinger or Zefirov. The van der Waals radii of lanthanide and actmide elements are peculiar to the other systems since they have never appeared in literature. It is worthy to note that all the radii presented in this work are more suitable to organic and metal-organic compounds. Moreover,the characteristics of this system,some current problems of van der Waals radii application and further study in this area are also discussed.

**Key words**  Average atomic volumes  van der Waals radii  Secondary bonding

* The Project Supported by the Ministry of Science and Technology(CB001CB108906)and NSFC(29933040).

■ **本文原载**:《厦门大学学报》(自然科学版)第42卷第2期(2003年3月),第133～138页。

# 碳纳米管负载/促进 Cu-Cr 催化剂上甲醇分解制氢[*]

陈书贵[①]　周金梅　张鸿斌[②]　林国栋　蔡启瑞

(厦门大学化学化工学院　固体表面物理化学国家重点实验室,福建厦门　361005)

**摘　要**　用自行制备的碳纳米管(CNTs)作为载体,研制出一类高活性CNTs负载/促进甲醇分解制氢Cu-Cr/CNTs催化剂。实验结果显示,在0.1 MPa,503 K,$n(CH_3OH):n(Ar)=2:1$,GHSV=3 600 $h^{-1}$的反应条件下,27%$Cu_{10}Cr_1$/CNTs催化剂上$H_2$的时空产率达133 mmol-$H_2$ $h^{-1}$(g-catal.)$^{-1}$,是AC、$SiO_2$和$\gamma$-$Al_2O_3$负载相应参比催化剂(分别为:111、73.5、60.9 mmol-$H_2$ $h^{-1}$(g-catal.)$^{-1}$)的1.20、1.81和2.18倍。实验表征研究揭示,碳纳米管载体促使催化剂活性Cu表面积大为增加,并诱使Cu-Cr催化活性位上甲醇分子解离下来的吸附H物种向碳纳米管载体"溢流"、疏散、随后偶联成$H_2$(a)脱附,于是降低了副产物甲醛、甲酸甲酯的生成机率,有利于提高甲醇深度脱氢、生成$H_2$和CO的选择性。

**关键词**　多壁碳纳米管　碳纳米管载体　碳纳米管促进剂　Cu-Cr/CNTs　甲醇分解制氢

氢能是理想的洁净能源之一[1]。以氢燃料电池驱动电动机的氢能汽车是真正无污染的绿色汽车。氢能汽车的关键技术环节有二:氢的储运与氢燃料电池。基于安全的原因,利用液体燃料作为氢的载体,通过将其重整即时产生氢燃料电池所需要的燃料$H_2$,可有效地解决氢能利用中遇到的多种问题[2-4];而在诸多可重整制$H_2$的液体燃料中,甲醇以其能量密度高、易于储运处理、以及价格低廉等优点而占具优势。

甲醇催化分解制氢是甲醇制氢的3种途径之一[5]。该反应是合成气制甲醇的逆反应,用于甲醇合成的催化剂均可用于其催化分解,其中以Cu-基催化剂为主。文献[6,7]曾报道甲醇分解会伴随发生一些副反应,生成甲醛、甲酸甲酯、二甲醚等副产物,这需要对催化剂作一定改性来加以抑制。分解气中含有30%以上的CO,经低温水煤气变换可将其转化为$H_2$;变换气中尚含有1%～3%的CO,可通过低温选择氧化将其去除,最终得到CO含量低于$1\times10^{-4}$的高纯$H_2$。

碳纳米管是一类纳米级新型碳素材料,将其作为一种新型催化剂载体或促进剂,近年来已引起国内外催化学界日益注意[8,9]。新近,本研究组[10,11]报道一类碳纳米管促进高效甲醇合成Cu-ZnO-$Al_2O_3$催化剂,其催化合成气制甲醇的时空产率是不含碳纳米管的相应参比样的1.46倍,并将高的活性归因于碳

* 收稿日期:2002年12月4日。
　基金项目:国家自然科学基金(50072021),教育部科技基金(20010384002),福建省自然科学基金(2001H017)资助。
① 作者简介:陈书贵(1977— ),男,硕士研究生。
② Corresponding author 张鸿斌,男,教授,博士生导师。
　 E-mail:hbzhang@xmu.edu.cn。

纳米管促进剂对催化剂活性组分高的分散作用和对反应物之一 $H_2$ 强的吸附、活化、存储作用。

本文报道一类碳纳米管负载/促进 Cu-Cr 催化剂。初步工作显示,其对甲醇分解制氢显示出高的低温催化活性和对产物 $H_2$ 高的选择性;在常压、503 K 反应条件下,其 $H_2$ 的时空产率是三种常规载体 AC(活性炭)、$SiO_2$、$\gamma$-$Al_2O_3$ 分别负载参比样的 1.20、1.81、2.18 倍。本文结果对于增进对碳纳米管促进作用本质的理解,以及低温高效甲醇分解制氢催化剂的研究开发,有重要意义。

# 1 实验部分

碳纳米管(CNTs)的制备系在文献[12]报道的技术基础上经放大的操作方法进行。CNTs-负载 Cu-Cr 催化剂(标记为 W% $Cu_iCr_j$/CNTs)由等容浸渍法制备,即:将计量的 $Cu(NO_3)_2 \cdot 3H_2O$ 和 $Cr(NO_3)_3 \cdot 9H_2O$ 混合,加入计算量去离子水制成溶液,将其浸渍负载于计算量 CNTs 载体上,室温下过夜,后于 373 K 烘干 6 h,473 K 灼烧 4 h,冷至室温,即得氧化前驱态催化剂;经压片、破碎,筛选出 40～80 目试样供活性评价之用。作为对比研究用的三种常规载体 AC、$SiO_2$ 和 $\gamma$-$Al_2O_3$ 负载的相应催化剂,其制备操作及标记方式同上。催化剂对甲醇分解的活性评价在常压固定床连续流动微型反应器-GC 组合系统进行。每次试验催化剂用量为 1 000 mg(1.0 mL)。反应前,氧化前驱态催化剂先经低氢还原气(5% $H_2$+95% $N_2$)按一定升温程序进行原位预还原;而后,在反应温度下导入由载气 Ar 流经甲醇鼓泡蒸发器(恒温于 329 K)产生的 Ar 载甲醇原料气($n(CH_3OH):n(Ar)=2:1$),在 GHSV=3 600 $h^{-1}$ 的空速下进行甲醇分解反应。反应产物由配备 TC 检测器和双色谱柱的 102 GD 型气相色谱仪,以氩气为载气作在线分析。两支色谱柱分别填充 5A 分子筛和"有机担体-401"(均为上海试剂一厂产品),柱长均为 2 m;前者用于分离 $H_2$ 和 CO,后者用于分离 $CO_2$,$CH_3OH$,MF。甲醇转化率由内部归一化法计算。

$H_2$-TPR/TPD 测试在固定床连续流动反应器或吸附-脱附系统上进行。

# 2 结果和讨论

图 1 示出在 453～503 K 温度范围,Cu 负载量均为 25 w%、Cu/Cr 摩尔比有别的 4 种催化剂上甲醇分解反应活性的评价结果。4 种催化剂上甲醇转化率及 $H_2$ 的时空产率均随温度升高而上升,这与该反应系吸热反应的热力学性质相一致。4 种不同 Cu/Cr 摩尔比催化剂上甲醇分解活性高低顺序为:a) 27% $Cu_{10}Cr_1$/CNTs > b) 26.2% $Cu_{10}Cr_{0.6}$/CNTs > c) 29.1% $Cu_{10}Cr_2$/CNTs > d) 31.1% $Cu_{10}Cr_3$/CNTs,以 $n(Cu):n(Cr)=10:1$、相应之 $Cu_{10}Cr_1$ 总负载量为 27%(即负载 Cu、Cr 金属元素组分相对于 CNTs 载体的总质量百分数)的催化剂为最佳。

在优化 Cu/Cr 摩尔比的基础上,保持 $n(Cu):n(Cr)=10:1$ 不变,进而考察催化剂活性与金属元素组分总负载量的相关性,结

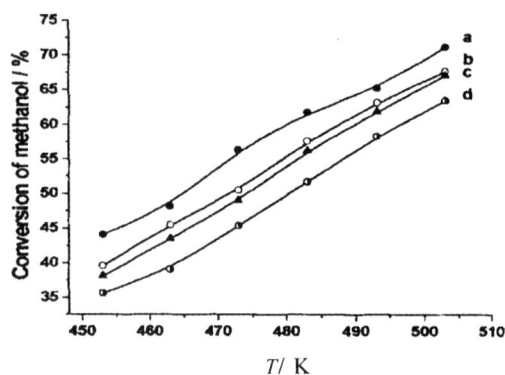

图 1 不同摩尔比(Cu 负载量均为 25%)催化剂上甲醇分解反应活性随温度的变化

Fig. 1 Change of reactivity of methanol decomposition with temperature over the catalysts:

a) 27.0% $Cu_{10}Cr_1$/CNTs;b) 26.2% $Cu_{10}Cr_{0.6}$/CNTs;
c) 29.1% $Cu_{10}Cr_2$/CNTs;d) 31.1% $Cu_{10}Cr_3$/CNTs
反应条件:0.1 MPa,$n(CH_3OH):n(Ar)=2:1$,GHSV=3600 $h^{-1}$

果表明,所调查 4 种催化剂的活性高低顺序为:$27\%Cu_{10}Cr_1/CNTs>22\%Cu_{10}Cr_1/CNTs>32\%Cu_{10}Cr_1/CNTs>16\%Cu_{10}Cr_1/CNTs$,以 $Cu_{10}Cr_1$ 总负载量为 27% 的催化剂为最佳。该催化剂在 503K、0.1 MPa,$n(CH_3OH):n(Ar)=2:1$,GHSV=3 600 $h^{-1}$ 的反应条件下,其甲醇转化率达 71.2%,明显地高于其他 3 种 Cu-Cr 负载量之催化剂(后 3 者甲醇转化率依次为 69.0%、66.8% 和 64.9%)。

为进一步了解碳纳米管载体对催化剂性能的影响,遂在相同条件下制备 3 种常规载体 AC、$SiO_2$ 和 $\gamma$-$Al_2O_3$ 分别负载之相应催化剂作为参比样,在相同反应条件下评价它们对甲醇分解的催化化学行为,并与 CNTs 负载催化剂作比较,结果示于图 2。由图 2 可见,载体能显著地影响催化剂的活性和选择性。在 0.1 MPa、503 K、$n(CH_3OH):n(Ar)=2:1$,GHSV=3600 $h^{-1}$ 的反应条件下,在 $27\%Cu_{10}Cr_1/CNTs$ 催化剂上,甲醇的转化率和 $H_2$ 的时空产率分别达 71.2% 和 133 mmol-$H_2$ $h^{-1}$ (g-catal.)$^{-1}$,分别是 AC、$SiO_2$ 和 $\gamma$-$Al_2O_3$ 负载参比样上甲醇转化率(分别为 64.3%、55.3%、50.4%)的 1.11、1.29、1.41 倍,$H_2$ 时空产率(分别为 111、73.5、60.9 mmol-$H_2$ $h^{-1}$ (g-catal.)$^{-1}$)的 1.20、1.81、2.18 倍。不同载体负载催化剂对甲醇分解的催化活性评价结果归纳示于表 1。

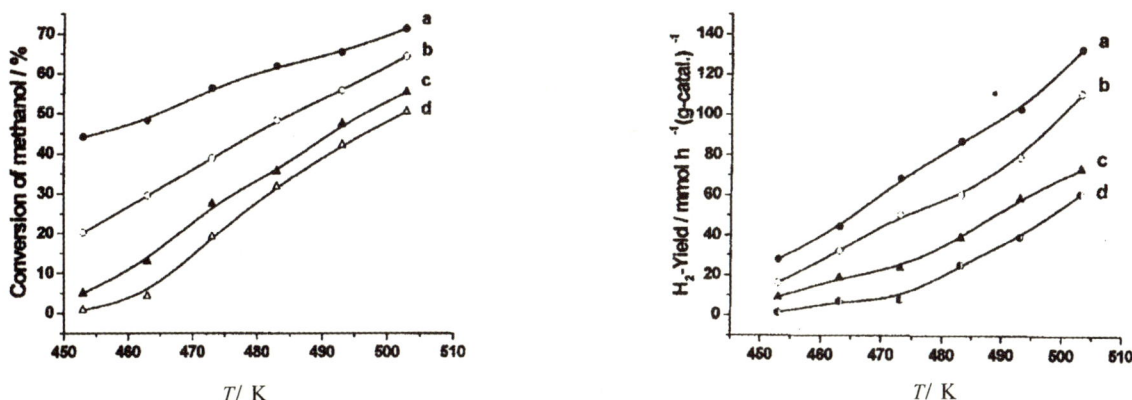

图 2 不同载体负载催化剂上甲醇分解反应活性随温度的变化

Fig. 2 Change of reactivity of methanol decomposition with temperature over the catalysts: a)$27\%Cu_{10}Cr_1/CNTs$;b)$27\%Cu_{10}Cr_1/AC$;c)$27\%Cu_{10}Cr_1/SiO_2$;d)$27\%Cu_{10}Cr_1/\gamma$-$Al_2O_3$ 反应条件:0.1 MPa,$n(CH_3OH):n(Ar)=2:1$,GHSV=3 600 $h^{-1}$

表 1 不同载体负载 Cu-Cr 催化剂上甲醇分解的反应活性

Tab. 1 Reactivity of decomposition of methanol over $Cu_{10}Cr_1/CNTs$(or AC,$SiO_2$,$\gamma$-$Al_2O_3$)

| 催化剂试样 | 反应温度/K | 甲醇转化率/% | 含碳产物选择性/% | | | 反应速度/$\mu$mol $CH_3OH s^{-1}$ $(m^2$-Cu surf.$)^{-1}$ | 氢选择性/% | 氢生成速度/$\mu$mol $H_2$ $s^{-1}$ $(m^2$-Cu surf.$)^{-1}$ | 氢时空产率/mmol $H_2$ $h^{-1}$ $(g$-catal.$)^{-1}$ |
|---|---|---|---|---|---|---|---|---|---|
| | | | CO | $CO_2$ | MF | | | | |
| $27\%Cu_{10}Cr_1$/CNTs | 453 | 44.1 | 30.2 | 0.8 | 69.0 | 0.38 | 29.6 | 0.22 | 27.9 |
| | 463 | 48.2 | 46.2 | 0.9 | 52.9 | 0.42 | 43.1 | 0.36 | 44.5 |
| | 473 | 56.3 | 58.6 | 0.9 | 40.5 | 0.49 | 56.8 | 0.55 | 68.4 |
| | 483 | 61.8 | 68.8 | 1.0 | 30.2 | 0.53 | 65.8 | 0.70 | 87.0 |
| | 493 | 65.3 | 76.6 | 1.0 | 22.4 | 0.56 | 73.6 | 0.83 | 103 |
| | 503 | 71.2 | 90.1 | 1.0 | 8.9 | 0.61 | 87.2 | 1.07 | 133 |
| $27\%Cu_{10}Cr_1$/AC | 453 | 20.2 | 38.0 | 2.6 | 59.4 | 0.21 | 37.6 | 0.15 | 16.2 |
| | 483 | 48.2 | 62.0 | 1.5 | 36.5 | 0.49 | 58.5 | 0.57 | 60.3 |
| | 493 | 55.8 | 72.9 | 1.4 | 25.7 | 0.57 | 66.3 | 0.75 | 79.2 |
| | 503 | 64.3 | 84.0 | 1.4 | 14.6 | 0.65 | 80.5 | 1.05 | 111 |

续表

| 催化剂试样 | 反应温度 /K | 甲醇转化率/% | 含碳产物选择性/% | | | 反应速度/μmol CH₃OHs⁻¹ (m²-Cu surf.)⁻¹ | 氢选择性 /% | 氢生成速度/μmol H₂ s⁻¹ (m²-Cu surf.)⁻¹ | 氢时空产率/mmol H₂ h⁻¹ (g-catal.)⁻¹ |
|---|---|---|---|---|---|---|---|---|---|
| | | | CO | CO₂ | MF | | | | |
| $27\%Cu_{10}Cr_1$ /SiO₂ | 463 | 10.9 | 12.4 | 87.6 | 0 | 0.17 | 80.2 | 0.26 | 18.7 |
| | 483 | 35.5 | 33.5 | 23.1 | 43.4 | 0.53 | 51.6 | 0.55 | 39.2 |
| | 493 | 47.3 | 40.1 | 21.9 | 38.0 | 0.70 | 58.0 | 0.82 | 58.7 |
| | 503 | 55.3 | 51.0 | 16.1 | 32.9 | 0.82 | 62.1 | 1.03 | 73.5 |
| $27\%Cu_{10}Cr_1$ /γ-Al₂O₃ | 473 | 19.1 | 11.8 | 11.2 | 77.0 | 0.31 | 19.3 | 0.12 | 7.9 |
| | 483 | 31.7 | 30.9 | 10.3 | 58.8 | 0.51 | 37.2 | 0.53 | 25.2 |
| | 493 | 42.1 | 38.5 | 9.2 | 52.3 | 0.68 | 43.7 | 0.60 | 39.4 |
| | 503 | 50.4 | 52.9 | 7.5 | 39.5 | 0.82 | 56.5 | 0.92 | 60.9 |

反应条件:0.1 MPa,$n(CH_3OH):n(Ar)=2:1$,GHSV=3 600 h⁻¹

图 3 示出 4 种不同载体负载催化剂氧化前驱态的 H₂-TPR 谱。CNTs 负载催化剂的 H₂-TPR 峰从 ~403 K 开始,在 466 K 达到峰顶,至 493 K 还原结束。相比之下,AC 负载催化剂的峰温位置虽与 CNTs 负载催化剂的相近,但其还原峰的面积强度只及碳纳米管负载催化剂的 52%;SiO₂ 和 γ-Al₂O₃ 负载催化剂 TPR 峰的面积强度虽分别达碳纳米管负载催化剂的 82% 和 99%,但它们的还原峰峰温(位于 503 K 和 520-535-550 K)却比碳纳米管负载之相应体系分别高约 37 和 54 度(K)。通过谱峰的峰温位置及相对面积强度的比较,可知这些试样的可还原性高低顺序为:$27\%Cu_{10}Cr_1$/CNTs > $27\%Cu_{10}Cr_1$/AC > $27\%Cu_{10}Cr_1$/SiO₂ > $27\%Cu_{10}Cr_1$/γ-Al₂O₃,这一顺序与它们对甲醇分解催化活性的高低顺序相一致。

若干负载型 Cu-Cr 催化剂的金属 Cu 表面积的测定结果示于表 2。碳纳米管负载催化剂的金属 Cu 表面积明显地高于三种常规载体负载之催化剂;其中,$27\%Cu_{10}Cr_1$/CNTs 的金属 Cu 表面积达到 34.6 m²(g-catal.)⁻¹,分别比 AC、SiO₂、γ-Al₂O₃ 负载相应参比样的 Cu 表面积(分别为:29.3、19.9 和 18.4 m²g⁻¹)高出 18%、74% 和 88%,尽管碳纳米管载体材料的 N₂-BET-比表面积(~130 m²g⁻¹)只为 AC、SiO₂、γ-Al₂O₃ 三种常规载体的 N₂-BET-比表面积(分别为:~590、~202 和 ~187 m²g⁻¹)分别的 22%、64% 和 70%。

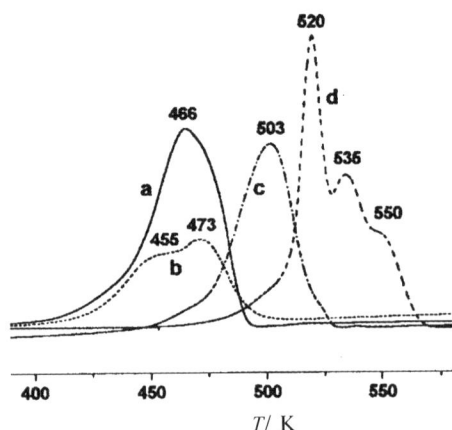

图 3 不同载体负载 Cu-Cr 催化剂的 H₂-TPR 谱
Fig. 3 H₂-TPR spectra of the oxidation precursor of supported catalysts:
(a)$27\%Cu_{10}Cr_1$/CNTs;(b)$27\%Cu_{10}Cr_1$/AC;
(c)$27\%Cu_{10}Cr_1$/SiO₂;(d)$27\%Cu_{10}Cr_1$/γ-Al₂O₃

表 2 若干催化剂的金属 Cu 表面积
Tab. 2 Metallic Cu-surface area of several catalysts

| 催化剂试样 | 金属 Cu 表面积(m²/g-catal.) |
|---|---|
| $16\%Cu_{10}Cr_1$/CNTs | 28.6 |
| $22\%Cu_{10}Cr_1$/CNTs | 31.7 |
| $32\%Cu_{10}Cr_1$/CNTs | 29.8 |
| $27\%Cu_{10}Cr_1$CNTs | 34.6 |
| $27\%Cu_{10}Cr_1$/AC | 29.3 |
| $27\%Cu_{10}Cr_1$/SiO₂ | 19.9 |
| $27\%Cu_{10}Cr_1$/γ-Al₂O₃ | 18.4 |

诚然，金属 Cu 表面的增加无疑将有助于提高催化剂的比活性，这是导致碳纳米管负载催化剂上甲醇分解反应活性明显高于其他三种载体负载催化剂的重要原因；这从这些催化剂上甲醇转化率的高低顺序（即：Cu-Cr/CNTs＞Cu-Cr/AC＞Cu-Cr/SiO₂＞Cu-Cr/γ-Al₂O₃）与这些催化剂的 Cu 表面积的大小顺序呈顺变关系，而单位 Cu 表面上甲醇分解的本征反应速度则彼此相当接近（见表 1），的结果可获得实验支持。然而，CNTs 负载催化剂上甲醇分解对 $H_2$ 的选择性则明显地高于其他 3 种载体负载之参比样，使得在大致相近的甲醇分解本征反应速度情形下，CNTs 负载体系上 $H_2$ 的时空产率明显地高于其他 3 种载体负载之催化剂；以 503 K 温度下的反应活性为例，CNTs 负载催化剂上 $H_2$ 的时空产率达到 133 mmol-$H_2$ h⁻¹(g-catal.)⁻¹，是 AC、SiO₂、γ-Al₂O₃ 负载之相应催化剂（分别为：111、73.5、60.9 mmol-$H_2$ h⁻¹(g-catal.)⁻¹）的 1.20、1.81、2.18 倍。差别如此之大的 $H_2$ 的选择生成显然无法从这些不同载体负载催化剂活性 Cu 表面积的差别得到合理解释。

在我们看来，碳纳米管负载 Cu-Cr 催化剂对甲醇分解所表现出的对 $H_2$ 优异的选择生成性能显然同这种载体材料特定的结构和性质密切相关。图 4 示出该类 CNTs 的 TEM 图象。多种表征技术联合观测证实，本文所制得 CNTs 为一类多壁碳纳米管，外管径在 15～50 nm 范围，内管径～3 nm，管长达数十至数百 μm，含碳量≥99 wt％，石墨状碳含量≥85 wt％。这类碳纳米管的 XRD 图（主强峰位于 $2\theta=26.1°$）与石墨的（主强峰位于 $2\theta=26.5°$）十分接近，但相应特征衍射峰强度明显减弱，峰形较为宽化，表明其管壁结构与石墨相近，但长程有序度不如石墨的高。HR TEM 观测进一步揭示，这类碳纳米管系由一片片具有类石墨片状结构的圆锥形面沿中空的管

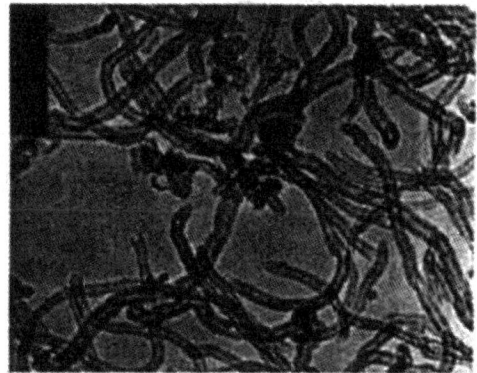

图 4　由 $CH_4$ 催化分解长成碳纳米管的 TEM 图($1\times10^5$)
Fig. 4　TEM image($1\times10^5$) of the CNTs grown catalytically from $CH_4$ decomposition

轴叠合而成，管壁纵截面呈所谓"鱼骨形"，类石墨结构圆锥形面与管空心轴倾斜[12]。

从化学催化的角度考虑，CNTs 具吸引力的特性，除显而易见高的机械强度，$sp^2$-C 构成的表面，类石墨结构的管壁，以及纳米级的内管腔之外，碳纳米管对氢强而可逆的吸附-脱附性能以及可期产生的促进"氢溢流"的作用也很值得注意。我们前已报道[13,14]，CNTs 在常温、常压下就能吸附可观量的 $H_2$；相应之吸附氢兼具非解离（即分子态）和解离（即原子态）两种形式；绝大部分（～99％）氢在 CNTs 上的吸附是可逆的，它们在从室温至 723 K 的加热升温过程中则以 $H_2$ 的形式脱附出来（所观测到的 $H_2$-TPD 谱主峰峰温为 400 K，并在 485 K 处有一肩峰，至 723 K 对应于该峰的氢吸附物种的脱附才完毕）。碳纳米管对氢这种既强而又可逆的吸附-脱附性能，使得作为载体的它、容易接受在 Cu-Cr 活性位上生成、溢流过来的氢吸附物种。于是可以预期，在 503 K 甲醇分解反应条件下，碳纳米管负载催化剂 Cu-Cr 活性位上甲醇分解生成的吸附氢将相当容易"溢流"、转移到碳纳米管上，并在碳纳米管上脱附为 $H_2$（CNTs 上的原子态吸附氢偶联成 $H_2$(a)随后脱附成 $H_2$(g)可期比 Cu-Cr 活性位上离子态吸附氢（H⁻ 和 H⁺ 或 $H^{\delta-}$ 和 $H^{\delta+}$）的偶联、脱附来得容易）。Cu-Cr 活性位上吸附氢得以更快地疏散、转移将有利于甲醇深度脱氢成 CO，减少脱氢中间产物甲醛、及副产物甲酸甲酯的生成机率，于是促进甲醇深度分解产物 $H_2$ 和 CO 选择性的提高。这一推断也可合理地解释碳纳米管负载催化剂上甲酸甲酯选择性何以比其他 3 种载体（尤其是 SiO₂ 和 γ-Al₂O₃）负载催化剂低得多的原因。

综上所述，在本文碳纳米管负载甲醇分解 Cu-Cr 催化剂体系中，碳纳米管不仅作为载体，同时也是 Cu-Cr 催化活性组分优良的分散剂，促进催化剂活性 Cu 表面积大幅度增加；在另一方面，碳纳米管也是

"氢溢流"的优良促进剂,帮助 Cu-Cr 催化活性位上甲醇分子解离下来的 H 物种"溢流"、疏散、转移至碳纳米管上,并随后偶联成 $H_2(a)$ 脱附,于是降低了副产物甲醛、甲酸甲酯的生成机率,有利于提高甲醇深度脱氢生成 $H_2$ 和 CO 的选择性。本工作也为碳纳米管可以起促进"氢溢流"的作用提供一典型实例。

## 参考文献

[1]Srinivasant S,Velve O A,Manko D J. High energy efficiency and high power density PEMAC-electrode kinetics and mass transport[J]. J. Power Sources,1991,36:299~304.

[2]Fierro J L G. Oxidative Methanol Reforming Reactions for the Production of Hydrogen[M]. Stud. Surf. Sci. Catal. 130,Eds. Corma A et al. Amsterdam:Elsevier,2000. 177~186.

[3]Cheng W H. Development of methanol decomposition catalysts for production of $H_2$ and CO[J]. Acc. Chem. Res.,1999,32:685~691.

[4]Appleby A J. 用于车辆的电化学发动机[J]. 科学(中译本),1999,10:34~39.

[5]Pena M A,Gomez J P,Fierro J L G. New catalytic routes for syngas and hydrogen production [J]. Appl. Catal. A:General,1996,144:7~57.

[6]Ai M. Dehydrogenation of methanol to methyl formate over copper-based catalysts[J]. Appl. Catal.,1984,11:259~270.

[7]Guerrero A,Rodriguez-Ramos I,Fierro J L G. Dehydrogenation of methanol to methyl formate over supported copper catalysts[J]. Appl. Catal.,1991,72:119~137.

[8]De Jong K P,Geus J W. Carbon nanofibers:catalytic synthesis and applications[J]. Catal. Rev. - Sci. Eng.,2000,42:481~510.

[9]Zhang Y,Zhang H B,Lin G D,et al. Preparation,characterization and catalytic hydroformylation properties of CNTs-supported Rh-phosphine catalyst[J]. Appl. Catal. A:General,1999,187:213~224.

[10]Zhang H B,Dong X,Lin G D,et al. Methanol synthesis from $H_2/CO/CO_2$ over CNTs-promoted Cu-ZnO-$Al_2O_3$ catalyst[J]. 223rd ACS National Mtg,Fuel Chem. Div. Preprints,2002,47(1),284~285.

[11]董鑫,张鸿斌,林国栋,等. 碳纳米管促进 Cu-基高效甲醇合成催化剂[J]. 厦门大学学报(自然科学版),2002,41(2):135~140.

[12]Chen P,Zhang H B,Lin G D,et al. Growth of carbon nanotubes by catalytic decomposition of $CH_4$ or CO on a Ni-MgO catalyst[J]. Carbon,1997,35(10-11):1 495~1 501.

[13]周振华,武小满,王毅,等. $H_2$ 在碳纳米管基材料上的吸-脱附特性[J]. 物理化学学报,2002,18(8):692~698.

[14]Zhang Hong-bin, Lin Guo-dong, Zhou Zhen-hua, et al. Raman Spectra of MWCNTs and MWCNTs-based $H_2$-Adsorbing Systems[J]. CARBON,2002,40(13),2 429~2 436.

# Production of $H_2$ from Decomposition of $CH_3OH$ over CNT-supported/promoted Cu-Cr Catalyst

Shu-Gui Chen, Jin-Mei Zhou, Hong-Bin Zhang*, Guo-Dong Lin, K. R. Tsai

(*Department of Chemistry & State Key Lab of Physical Chemistry for the Solid Surfaces, Xiamen University, Xiamen 361005, China*)

**Abstract** With a kind of multi-walled carbon nanotubes(CNTs)synthesized catalytically in-house and the nitrates of the corresponding metallic components,highly active CNT-supported and promoted catalysts, wt% $Cu_iCr_j$/CNTs, were prepared by an incipient wetness method, and their catalytic performance for production of $H_2$ from decomposition of $CH_3OH$ was investigated and compared with the corresponding those supported by AC, $SiO_2$ and $\gamma$-$Al_2O_3$. It was experimentally found that the support could significantly affect the catalyst activity for methanol decomposition. Over a 27% $Cu_{10}Cr_1$/CNTs catalyst,space-time-yield of $H_2$ reached 133 mmol-$H_2$ $h^{-1}$ (g-catal. )$^{-1}$ under reaction condition of 0. 1 MPa,503 K,$n(CH_3OH)$ : $n(Ar)$=2 : 1,GHSV=3 600 $h^{-1}$,which was 1. 20,1. 81 and 2. 18 times as high as those of the reference systems supported by AC, $SiO_2$ and $\gamma$-$Al_2O_3$ at the same reaction condition,respectively. The nature of promoter action by the CNTs was discussed together with the results of characterization study of the CNT-carrier and the CNT-supported catalysts by TEM, XRD, TPR, TPD and Cu exposed area. The results indicated that the CNTs served not only as a carrier,but also as an excellent dispersant for Cu-$Cr_2O_3$ components,which led to a significant increase of the Cu exposed area, generating more catalytically active Cu-sites at the catalyst surface for methanol decomposition. On the other hand,the CNTs could serve as an excellent promoter for "hydrogen spill-over" from the catalytically active sites of Cu-Cr to CNT-carrier,which would be conducive to scattering hydrogen-adspecies produced from dehydrogenation of methanol molecules at Cu-Cr sites and transferring them to the CNT carrier, followed by the coupling and desorption, thus favorable to reducing the probability of formation of the by-products, formaldehyde and methylformate, and enhancing the selectivity of deep dehydrogenation of methanol to form $H_2$ and CO.

**Key words** Multiwalled carbon nanotubes CNT-carrier CNT-promoter wt% $Cu_iCr_j$/CNTs Methanol decomposition $H_2$-production

■ **本文原载**：Inorganic Chemistry Communications 7 (2004)，pp. 169～172.

# Ammonium Barium Citrato Peroxotitanate(IV) Ba$_2$(NH$_4$)$_2$[Ti$_4$(O$_2$)$_4$(Hcit)$_2$(cit)$_2$] · 10H$_2$O: a Molecular Precursor of Stoichiometric BaTi$_2$O$_5$

Yuan-Fu Deng，Zhao-Hui Zhou[①]，Hui-Lin Wan，Khi-Rui Tsai

*(Department of Chemistry and State Key Laboratory for Physical Chemistry of Solid Surface, Xiamen University, Xiamen 361005, China)*

Received 21 September 2003；accepted 29 October 2003
Published online：2 December 2003

**Abstract**  Diammonium dibarium hydrogen citrato peroxotitanate(IV)，Ba$_2$(NH$_4$)$_2$[Ti$_4$(O$_2$)$_4$(Hcit)$_2$(cit)$_2$] · 10H$_2$O，is obtained directly from H$_2$O$_2$ oxidation of tricitrato titanate，or from the reaction of peroxotitanate with citric acid(H$_4$cit＝citric acid)，which is decomposed stoichiometrically into BaTi$_2$O$_5$ at the temperature of 800 ℃ in air.

**Key words**  Barium titanate  Citrate  Precursor  Pechini process  Crystal structure

Titanate containing materials have attracted special attention for their applications to the electronic industry because of its characteristic ferroelectric, piezoelectric and pyroelectric properties[1-3]. Previously, most titanates are prepared by reacting intimate mixtures of the constituent metal oxides or carbonates. The shortcomings of such an approach include the formation of relatively coarse particles leading to large and inhomogeneous porosity in sintered bodies, incomplete reaction of the oxide powders, and limitations on the ability to control composition[4]. For better controlling the chemical composition and homogeneity, phase and crystallinity of the fine multicomponent powders and ceramics, one way to closely approach the need requires utilization of a precursor that with the correct stoichiometry and homogeneity[5]. Some examples of the precursors which fix the stoichiometric ratios of barium and titanium in bimetallic ceramic oxide: such as barium titanyl oxlate Ba$_4$[Ti$_4$O$_4$(C$_2$O$_4$)$_8$] · 20H$_2$O[6], barium titanyl citrate Ba[Ti(H$_2$cit)$_3$] · $n$H$_2$O[7], barium citrato peroxotitanate Ba$_4$[Ti$_4$(O$_2$)$_4$(cit)$_4$] · 8H$_2$O[8]. In these precursors, only barium titanyl oxalate was structurally determined by single X-ray analysis[6]. In order to investigate the constitutions and structures of these molecular precursors, a lot of spectroscopic measurements such as IR, NMR, and Raman spectra were applied[7,8]. Despite two Ti(IV)-citrate complexes[9] and two Ti(IV)-peroxo-citrate complexes[10,11] for Ti-based ceram were isolated and well characterized, the spectroscopic and structural data for Ba-Ti-peroxo-citrate complexes are limited, which are important precursors for the electronic materials in the BaO-TiO$_2$ system. Only powder XRD

---

① Corresponding author. Tel.：+865922184531；fax：+865922183047. *E-mail address*：zhzhou@xmu.edu.cn(Z.-H. Zhou).

pattern is available for Ba$_4$[Ti$_4$(O$_2$)$_4$(cit)$_4$] · 8H$_2$O[8]. Here a reinvestigation of the Ti(IV)-peroxo-citrate system results in an isolation of the new tetranuclear species Ba$_2$(NH$_4$)$_2$[Ti$_4$(O$_2$)$_4$-(Hcit)$_2$-(cit)$_2$] · 10H$_2$O, which can be served as a molecular precursor for BaTi$_2$O$_5$. Moreover, the pH-dependent citrato peroxotitanate anions and the conversion of Ti(IV)-citrate and Ti(IV)-peroxo-citrate complexes were reported.

The reaction of TiCl$_4$-H$_2$O$_2$-H$_4$cit system with barium carbonate was followed by adjustment of pH with ammonia. The yellow crystals of compound **1** was produced in an ice bath. ① The formation of citrato peroxotitanate(IV)complexes depends mainly on the pH of the solution, which is similar to the results found in the citrato peroxovanadate(V)[12] and malato peroxovanadates(V)[13,14] complexes. At the low pH 1.5—3.0, it is shown that the protonated species of [Ti$_4$(O$_2$)$_4$(Hcit)$_2$(cit)$_2$]$^{6-}$ anion exists in the solution, whereas at the higher pH 4.0—7.0, the deprotoned product of [Ti$_4$(O$_2$)$_4$(cit)$_4$]$^{8-}$ anion is formed. The interconversions of the citrato peroxotitanate anions can be accomplished in an aqueous solution, which argues against the previous single composition over a broad range of pH 1—14[10].

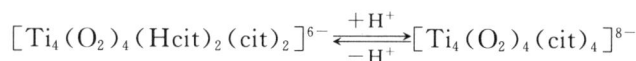

$$[Ti_4(O_2)_4(Hcit)_2(cit)_2]^{6-} \underset{-H^+}{\overset{+H^+}{\rightleftharpoons}} [Ti_4(O_2)_4(cit)_4]^{8-}$$

X-ray diffraction analysis reveals that **1** consists of ammonium, barium cations, citrato peroxotitanate anion, and lattice water molecules. ②Fig. 1 shows the structure and labeling scheme of the [Ti$_4$(O$_2$)$_4$(Hcit)$_2$(cit)$_2$]$^{6-}$ anion, which is composed of two dinuclear fragments interlinked through the bridging carboxyl group(C16-O16-O17) of one citrate ligand. In each dinuclear unit, there exists one [Ti$_2$O$_2$(O$_2$)$_2$]$^{2+}$ core, to which two citrate are bound. One of the citrate ligands is pentadentate and fully deprotonated, while the other is only tridentate, leaving the β-carboxylic acid group free. The

---

① Method 1: Titanium tetrachloride(1.9 g,10 mmol) was hydrolyzed in 15 ml of deioned water, the solution was added slowly with 10 ml 30% hydrogen peroxide and citric acid monohydrate(2.10 g,10 mmol). The mixture was further added with barium carbonate(0.99 g,5 mmol) with continuously stirring, the solution was adjusted to pH 2.0 with ammonia and filtered, and the yellow solids were collected and washed with deioned water three times and dried in vacuum oven. Yield:70%. Elemental analysis for Ba$_2$(NH$_4$)$_2$[Ti$_4$(O$_2$)$_4$(Hcit)$_2$(cit)$_2$] · 10H$_2$O, Found(Calcd. for C$_{24}$H$_{46}$O$_{46}$N$_2$Ba$_2$Ti$_4$): C 18.1%(18.4%); H 2.7%(3.0%); N 1.4%(1.8%). IR(KBr): $v$(COOH),1721.9$_m$; $v_{as}$(COO),1677.7$_m$,1596.7$_{vs}$, 1552.0$_s$; $v_s$(COO),1432.1$_s$,1397.1$_s$; $v$(O-O),884.5$_m$,880.8$_m$; $v$(Ti-O$_2$),622.8$_m$.

Method 2:Ba[Ti(H$_2$cit)$_3$] · 4H$_2$O(1.66 g,2 mmol)[9] was dissolved in 10 ml deioned water, and the solution was added slowly with titanium tetrachloride(0.38 g,2 mmol)and 10 ml 30% hydrogen peroxide. The solution was adjusted to pH 2.0 with ammonia and filtered, the resulting solution was kept in refrigerator for several days, and the yellow solids were collected and washed with deioned water three times and dried. Yield:65%.

Transformation of barium peroxo titanium citrate. In ice water, Ba$_2$(NH$_4$)$_2$[Ti$_4$(O$_2$)$_4$(Hcit)$_2$(cit)$_2$] · 10H$_2$O was added to the solution containing barium chloride. The mixture was adjusted to pH 6 with ammonia and stirred continuously. The yellow solids were collected and washed with deioned water three times and dried, which is confirmed as Ba$_4$[Ti$_4$(O$_2$)$_4$(cit)$_4$] · 8H$_2$O[8]. Yield: 60%. IR(KBr): $v_{as}$(COO),1686.4$_s$,1659.7$_s$,1581.3$_s$; $v_s$(COO),1425.1$_m$,1403.4$_s$; $v$(O-O),895.3$_m$,854.3$_m$; $v$(Ti-O$_2$), 598.2$_m$.

② Crystallographic data for **1**: crystal dimensions 0.07 × 0.12 × 0.13 mm, C$_{24}$H$_{46}$O$_{46}$N$_2$Ba$_2$Ti$_4$, $M_r$ = 1564.91, monoclinic, space group $P2_1/c$, $a$=10.2507(5)Å, $b$=27.026(1)Å, $c$=9.4169(4)Å, $\beta$=107.001(1)°, $V$=2494.8(2)Å$^3$, $Z$= 2, $D_c$=2.083 g cm$^{-3}$, $\mu$=2.29 mm$^{-1}$, 28741 reflections were collected, of which 5952 were unique ($R_{int}$=0.0659)$R$=0.053 and $wR$=0.117, GOF=1.179 for 398 parameters. The intensity data were measured on a Bruker Smart CCD diffractometer with graphite-monochromated Mo-K$\alpha$($\lambda$=0.7107 Å) at 296(2)K. Empirical absorption corrections were applied using the SADABS program. All calculation were performed using WinGX and SHELXL-97 program package[19,20]. The structure was solved by direct methods and refined by full-matrix least-squares methods. All non-hydrogen atoms were refined aniso-tropically, while all hydrogen atoms were located from differential Fourier maps. CCDC reference number is 220237.

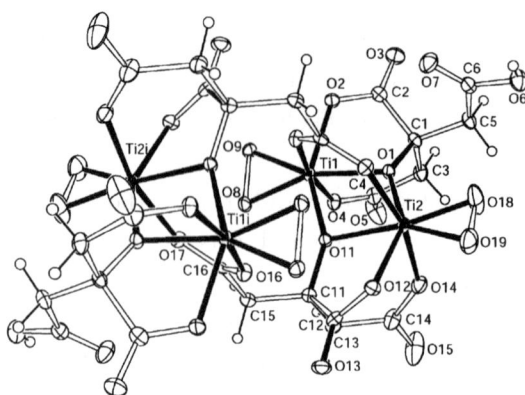

**Fig. 1** **ORTEP plot of the anion of for [Ti₄(O₂)₄(Hcit)₂(cit)₂]⁶⁻ at the 30% probability level.** **Selected bond lengths: Ti1-O1 2.048(3), Ti1-O2 2.056(3), Ti1-O4 2.001(3), Ti1-O16i 2.001 (3), Ti1-O8 1.873(3), Ti1-O9 1.884(3), Ti1-O11 2.003(3), Ti2-O1 2.001(3), Ti2-O11 2.077(3), Ti2-O12 2.055(3), Ti2-O14 2.014(4), Ti2-O17i 2.069(3), Ti2-O18 1.841(4), Ti2-O19 1.855(4), Ti1-Ti2 3.213(1) Å. Symmetry transformations: i 1 -x, 1-y, -z.**

coordination sphere around the four Ti(IV) ions is completed with peroxo groups, one on each Ti(IV) center. The peroxo group $O_2^{2-}$ binds to Ti(IV) in a side-on $\eta^2$-fashion, occupying two coordination sites in the equatorial plane, with the remaining three sites being taken by α-alkoxyl(O1) and α-carboxyl(O2) groups from one citrate, and one α-alkoxyl(O11) from another citrate ligand. The apical positions in the bipyramid are occupied with the terminal β-carboxyl oxygen(O4) of one citrate, and the other terminal β-carboxyl oxygen(O16i) of the third citrate ligand. As such, the coordination number around each Ti(IV) ion is seven, reflecting a distorted pentagonal bipyramidal coordination environment. The Ti-O bond distances in **1**[1.841(4)−2.077(3)Å] are similar to those found in related Ti(IV) oxygen containing complexes, like(NH₄)₈[Ti₄(O₂)₄(cit)₄]·8H₂O [1.863(1)−2.085(1)Å][10], (NH₄)₄[Ti₂(O₂)₂(cit)₂]₂·2H₂O [1.852(2)−2.085(2)Å][11], [Ti(ehpg)(H₂O)]·(11/3)H₂O [1.869(2)−2.091(2) Å], [{Ti(Hehpg)(H₂O)}₂O]·13H₂O [1.853(3)−2.095(3)Å][15], KMg₁/₂[Ti(H₂cit)₃]·6H₂O [1.865(1)−2.045(1)Å]and(NH₄)Mg₁/₂[Ti(H₂cit)₃]·6H₂O [1.866(2)−2.049(2)Å][9]. Moreover, the citrate ligands also coordinate to the barium ions. As shown in Fig. 2, the barium ions in **1** is in deca-coordination, surrounding by four water molecules, two peroxo groups(O8, O9 and O9ii), one oxygen atom (O4) from β-carboxyl group, two oxygen atoms (O12iii and O13iii) from α-carboxyl group. The average Ba-O bond distances of 2.867(3)Å agree well with the sum of ionic radii for oxygen and ten-coordinate barium ions[6,16].

Fig. 3 displays the results of differential thermal analysis (DTA) and thermogravimetric (TG) analysis for complex **1**. The sample loses water molecules with two endothermic peaks in the temperature at 78 and 127 ℃. The subsequent thermal decomposition of **1** is in the range 150∼600 ℃, which is very complicated. The DTA shows three main endothermic peas at 227, 427 and 528 ℃. A mass loss of ca. 61.2% between 25 and 800 ℃ corresponds to the theoretical weight loss(60.0%) of **1** to barium dititanate(BaTi₂O₅). The mode of decomposition of **1** is similar to that of Ba₄[Ti₄(O₂)₄(cit)₄]·8H₂O[8]. The powder annealed in air at 800 ℃ had XRD pattern typical for a well-crystallized BaTi₂O₅(Fig. 4) phase, which in excellent accord with the results reported[17,18].

The isolation and structural characterization of the complex **1** in the acidic solution show an instructional role in the synthesis the BaO-TiO₂ ceramic powders by employment of peroxide based

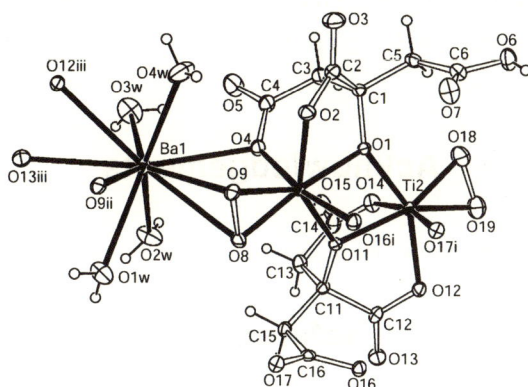

Fig. 2　ORTEP plot of the coordination mode of barium ion for 1 at the 30% probability level. Selected bond lengths: Ba1-O1w 2.757(4), Ba1-O2w 2.770(4), Ba1-O3w 2.831(5), Ba1-O4w 2.751(5), Ba1-O4 2.884(3), Ba1-O8 3.028(3), Ba1-O9 2.903(3), Ba1-O9ii 2.866(3), Ba1-O12iii 2.839(3), Ba1-O13iii 3.035(4)Å. Symmetry transformations: ii-$x$,1-$y$,-$z$; iii-1+$x$,$y$,$z$.

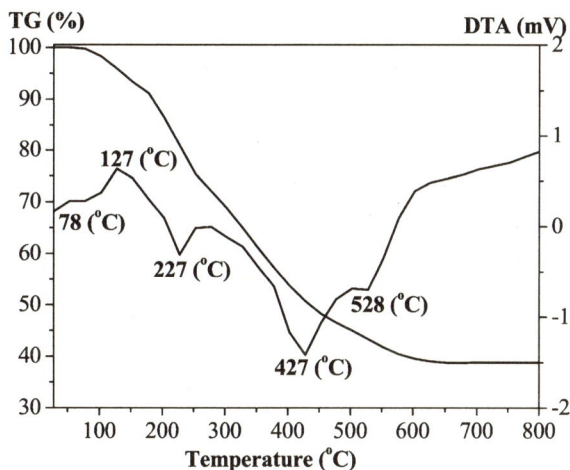

Fig. 3　TG-DTA data for 1 with heating rate of 5 ℃/min.

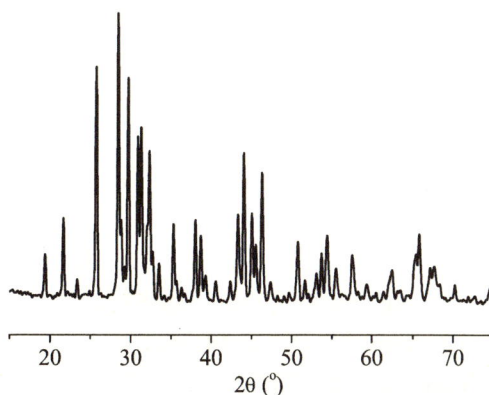

Fig. 4　X-ray powder diffraction pattern of BaTi$_2$O$_5$ obtained from complex 1 at 800 ℃, with holding in air for two hours.

route. Comparing to $Ba_4[Ti_4(O_2)_4(cit)_4] \cdot 8H_2O^{[8]}$ complex for $BaTiO_3$ synthesis, pH control is the key factor of acquiring the precursor $Ba_2(NH_4)_2[Ti_4(O_2)_4(Hcit)_2(cit)_2] \cdot 10H_2O$, which decompose into $Ba_2TiO_5$.

# Acknowledgements

This work is supported by the Ministry of Science and Technology(G1999022408).

# References

[1] J. W. R. Cook, H. Jaffe. Piezoeclectric Ceramics, Academic Press, New York, 1971.

[2] P. P. Phule, S. H. Risbud. J. Mater. Sci, 1990, **25**: 1169.

[3] M. A. Peńa, J. L. Fierro. Chem. Rev, 2001, **101**: 1981.

[4] W. G. Sluys, A. P. Sattleberger. Chem. Rev, 1990, **90**: 1027.

[5] (a) C. D. Chandler, C. Roger, M. J. Hampden-Smith. Chem. Rev, 1993, **93**: 1205;
　　(b) L. G. Hubert-Pfalzgraf. Inorg. Chem. Comm, 2003, **6**: 102.

[6] W. E. Rhine, R. B. Hallock, W. M. Davis, W. W. Ng. Chem. Mater, 1992, **4**: 1208.

[7] (a) M. P. Pechini, US Pat. No. 3231328, 1966;
　　　M. P. Pechini, US Pat. No. 3330697, 1967;
　　(b) B. J. Mulder. Am. Ceram. Bull, 1970, **49**: 990;
　　(c) D. Hennings, W. Maya. J. Solid State Chem, 1978, **26**: 329;
　　(d) G. A. Hutchins, G. H. Maher, S. D. Ross. Am. Ceram. Soc. Bull, 1987, **66**: 681;
　　(e) M. Arima, M. Kakihana, Y. Nakamura, M. Yashima, M. Yoshimura. J. Am. Ceram. Soc, 1996, **79**: 2847;
　　(f) M. Kakihana, M. Arima, Y. Nakamura. Chem. Mater, 1999, **11**: 438;
　　(g) J. D. Tsay, T. T. Fang. J. Am. Ceram. Soc, 1999, **82**: 1409.

[8] M. Tada, K. Tomita, V. Petrykin, M. Kakihana. Solid State Ionics, 2002, **151**: 293.

[9] Z. H. Zhou, Y. F. Deng, Y. Q. Jiang, H. L. Wan, S. W. Ng. Dalton Trans, 2003: 2636.

[10] M. Kakihana, M. Tada, M. Shiro, V. Petrykin, M. Osda, Y. Nakamura, Inorg. Chem. 2001, 40: 891.

[11] M. Dakanali, E. T. Kefalas, C. P. Raptopoulou, A. Terzis, G. Voyiatzis, I. Kyrikou, T. Mavromoustakos, A. Salifoglou. Inorg. Chem, 2003, **42**: 4632.

[12] M. Kaliva, E. Kyriakakis, A. Salifoglou. Inorg. Chem, 2002, **41**: 7015.

[13] C. Djordjevic, M. Lee-Renslo, E. Sinn. Inorg. Chem. Acta, 1995, **233**: 97.

[14] M. Kaliva, T. Giannadaki, A. Salifoglou. Inorg. Chem, 2001, **40**: 3711.

[15] M. Guo, H. Z. Sun, S. Bihari, J. A. Parkinson, R. O. Gould, S. Parsons, P. J. Sadler. Inorg. Chem, 2000, **39**: 206.

[16] R. D. Shannon, C. T. Prewitt. Acta Crystallogr. B, 1970, **26**: 1076.

[17] Y. B. Xu, G. H. Huang, H. Long. Mater. Lett, 2003, **57**: 3570.

[18] D. E. Rase, R. Roy. J. Am. Ceram. Soc, 1955, **38**: 102.

[19] L. J. Farrugia. J. Appl. Cryst, 1999, **32**: 837.

[20] G. M. Sheldrick, SHELX-97, Programs for Crystal Structure Analysis, University of Göttingen, Germany, 1997.

■ 本文原载:Journal of Inorganic Biochemistry 98(2004),pp. 1787～1794.

# Enantiomeric and Mesomeric Mandelate Complexes of Molybdenum-on Their Stereospecific Formations and Absolute Configurations

Zhao-Hui Zhou[1], Hong Zhao, Khi-Rui Tsai

(Department of Chemistry and State Key Laboratory of Physical Chemistry of Solid Surfaces, Xiamen University, Xiamen 361005, P.R. China )

Received 13 April 2004; received in revised form 22 July 2004; accepted 5 August 2004 Available online 11 September 2004

**Abstract** The stereospecific formation and absolute configuration of $R$-homocitrate coordinated FeMo-co in nitrogenase was mimicked through the structural analyses of a collection of enantiomeric and mesomeric mandelato molybdenum complexes, i.e., $(NH_4)_2[Mo_\Delta O_2(R\text{-mand})_2] \cdot 3H_2O$ (**1a**), $(NH_4)_2[Mo_\Lambda O_2(S\text{-mand})_2] \cdot 3H_2O$ (**1b**), $(NH_4)_4[Mo_\Lambda O_2(RS\text{-mand})_2][Mo_\Lambda O_2(RS\text{-mand})_2] \cdot 8H_2O$ (**2**), $(NH_4)_2[W_\Delta O_2(R\text{-mand})_2] \cdot 2H_2O$ (**3a**), $(NH_4)_2[W_\Lambda O_2(S\text{-mand})_2] \cdot 2H_2O$ (**3b**) ($H_2$mand = mandelic acid, $C_8H_8O_3$), which have been characterized by elemental analyses, optical rotation, circular dichroism, IR, NMR spectroscopes and X-ray single crystal studies. The $R$ and $S$ chiral mandelic acids induce the formations of the enantiomeric pair of chiral complexes, which are supported by the characterizations of optical rotation and circular dichroism. The configuration of the resulted metal center could be assigned as $\Delta$ or $\Lambda$. While the $RS$ racemic reagent yields only mesomeric compound. The $\Delta_{R,R}$-complexes **1a** and **3a** are enantiomers of $\Lambda_{s,s}$-**1b** and **3b**, respectively. Of the five complexes, Mo and W atoms are all hexa-coordinated by two $cis$-$oxo$ groups and two bidentate mandelate ligands through the deprotonated $\alpha$-alkoxyl and $\alpha$-carboxyl groups, forming a stable five-membered chelated rings. The average Mo(Ⅵ)-O bond distances with $\alpha$-alkoxyl and $\alpha$-carboxyl are 1.944 and 2.210 Å, respectively. Further comparison indicates that bonds of $\alpha$-alkoxyl groups in the hydroxycarboxylato molybdenum complexes are much sensitive to the change in the oxidation state of molybdenum, which support the possible Mo activation model in FeMo-co through the protonation and cleavage of $\alpha$-alkoxyl group in homocitrate ligand.

**Key words** Molybdate Tungstate Mandelic acid Mandelate Crystal structure Stereospecific FeMo-cofactor

---

① Corresponding author. Tel. :+865922184531;fax:+865922183047.
E-mail address:zhzhou@xmu. edu. cn(Z. -H. Zhou).

**1403**

# 1 Introduction

The essential role of molybdenum in biology has been known for decades, and molybdoenzymes are ubiquitous in living systems[1]. An example is the recent high resolution structural studies on the nitrogenase, in which the active site of FeMo-co (iron molybdenum cofactor) of the enzyme is revealed as a cage-like $MoFe_7S_9L$ bidentate $R$-homocitrate cluster[2,3]. The new and detailed analysis found a light atom inside the cavity of FeMo-co that makes the surrounding Fe atoms coordinated tetrahedrally like a terminal Fe. ENDOR and ESEEM observations show the light atom is an unexchangeable nitrogen strengthen the notion of molybdenum binding models[4-7]. Alternative polycarboxylic acids including citric, malic and citramalic acids participates in in vitro syntheses of FeMo-co are more than 10 times less active for $N_2$ reduction[8-11]. The absolute configuration of the FeMo-co is temporarily assigned as $\Delta_R$ based on the available protein data from the different sources of nitrogenases[12]. A hydroxycarboxylic acid is essential in the complex process of biological synthesis including the mobilization of molybdenum from the appropriate storage enzyme[10,13]. Moreover, high valent molybdenum oxotransferases are the enzymes that catalyze oxygen atom transfer from a substrate. In order to understand the chemistry and oxo-transfer properties of the enzymes, numerous dioxo-Mo(VI) complexes having a wide variety of ligands have been prepared and structurally characterized[14].

Within a broad interests in our investigations of molybdate with $\alpha$-hydroxycarboxylic acids[15-18], we report the interaction of molybdate(VI) and tungstate(VI) with mandelic acid of $R$ and $S$-configuration as well as $RS$-racemate, which results in the isolations and characterization of mandelato molybdates(VI) and tungstates(VI) in different configurations. This may supply a clue for the stereospecific formation of FeMo-co related to the coordination of homocitrate ligand.

The solution chemistry of molybdenum(VI) and tungsten(VI) with mandelic acid has been previously investigated[19-23]. On the base of the results of spectrophotometric, potentiometric, enthalpimetric titration and NMR studies, it is concluded that coordination behaviour of mandelic acid groups towards molybdenum(VI) in 1∶2 metal∶ligand complexes are much more stable than that of 1∶1 ratio. Our results show that $R$- or $S$-mandelate forms enantiomeric complexes with ammonium molybdate and tungstate in trihydrate $(NH_4)_2[MoO_2(mand)_2] \cdot 3H_2O$ (**1**) ($H_2mand$ = mandelic acid) and dihydrate $(NH_4)_2[WO_2(mand)_2] \cdot 2H_2O$ (**3**), while $RS$-mandelate molybdate(VI) in tetrahydrate $(NH_4)_2[MoO_2(RS\text{-mand})_2] \cdot 4H_2O$ (**2**) is formed preferentially for $RS$-mandelate.

# 2 Experimental

All chemicals were commercial analytical reagents and used without further purification and all reactions were carried out in air. Nanopure-quality water was used throughout this work.

## 2.1 Preparation of $(NH_4)_2[Mo_\Delta O_2(R\text{-mand})_2] \cdot 3H_2O$ (1a), $(NH_4)_2[Mo_\Delta O_2(S\text{-mand})_2] \cdot 3H_2O$ (1b), $(NH_4)_4[Mo_\Delta O_2(RS\text{-mand})_2][Mo_\Delta O_2(RS\text{-mand})_2] \cdot 8H_2O$ (2)

In a typical experiment, mandelic acid (4.4 mmol) dissolved in water (5 mL) was slowly added to a stirred solution of $(NH_4)_6[Mo_7O_{24}] \cdot 4H_2O$ (0.31 mmol) in water (3 mL). The pH value was adjusted to 6.0 with additions of 1.0 M ammonia hydroxide. The mixture was stirred for 8 h at room temperature, and kept in refrigerator for several days. The $R$-mandelato molybdate(VI) (**1a**) was isolated as colorless crystals (1.0 g, yield 89%). $[\alpha]_D^{20} = +24.9°$ ($c$ 1.04, $H_2O$) {$R$-(−)-mandelic acid, $[\alpha]_D^{20} = -153°$ ($c$ 2.5, $H_2O$)[24]}. Found: C, 36.8; H, 5.1; N, 4.7. Calc. for $C_{16}H_{26}MoN_2O_{11}$: C, 37.1; H, 5.1; N, 5.4. IR(KBr): $v_{as}$(C=O)$1642_{vs}$, $1614_{vs}$; $v_s$(C=O)$1475_s$, $1450_s$, $1400_s$, $1364_s$; $v$(Mo=O)$906_s$, $880_s$; $^1H$ NMR $\delta H$(500 MHz, $D_2O$) ppm, 7.43−7.53(m, 5H, Ph), 5.75(s, 1H, ≡CH); $^{13}C$ NMR $\delta c$($D_2O$) ppm, 186.65($CO_2$)$_a$; 143.36, 131.81, 131.52, 129.99(Ph); 89.14(HCO-). Similarly, $S$-mandelato molybdate (Ⅵ)(**1b**) was isolated as colorless crystals(0.96 g, yield 84%). $[\alpha]_D^{20} = -26.2°$ ($c$ 1.03, $H_2O$) {$S$-(-)-mandelic acid, $[\alpha]_D^{20} = +154°$ ($c$ 2.8, $H_2O$)[24]}. Found: C, 36.9; H, 5.1; N, 4.7. Calc. for $C_{16}H_{26}MoN_2O_{11}$: C, 37.1; H, 5.1; N, 5.4. IR(KBr): $v_{as}$(C=O)$1643_{vs}$, $1608_{vs}$; $v_s$(C=O)$1475_s$, $1449_s$, $1409_s$, $1364_s$; $v$(Mo=O)$910_s$, $882_s$; $^1H$ NMR $\delta_H$(500 MHz, $D_2O$) ppm, 7.42−7.53(m, 5H), 5.74(s, 1H, ≡CH); $^{13}C$ NMR $\delta c$($D_2O$) ppm, 186.65($CO_2$)$\alpha$; 143.33, 131.81, 131.49, 129.97(Ph); 89.14(HCO-). For the racemic mandelato molybdate(Ⅵ)**2**, the compound was isolated as colorless crystals(1.0 g, yield 90%). Found: C, 36.1; H, 5.1; N, 4.7. Calc. for $C_{16}H_{28}MoN_2O_{12}$: C, 35.8; H, 5.2; N, 5.2. IR(KBr): $v_{as}$(C=O)$1635_{vs}$; $v_s$(C=O)$1450_s$, $1400_{vs}$; $v$(Mo=O)$922_s$, $889_s$; $^1H$ NMR $\delta_H$(500 MHz, $D_2O$) ppm, 7.42−7.52(m, 5H), 5.73−5.82(m, 1H, ≡CH); $^{13}C$ NMR $\delta c$($D_2O$) ppm, 186.76($CO_2$)$\alpha$; 143.33, 131.81, 131.50, 130.86, 130.01(Ph); 89.20(HCO$^-$).

## 2.2 Preparation of $(NH_4)_2[W_\Delta O_2(R\text{-mand})_2] \cdot 2H_2O$(3a), $(NH_4)_2[W_\Delta O_2(S\text{-mand})_2] \cdot 2H_2O$(3b)

Mandelic acid(3 mmol) dissolved in water(5 mL) was slowly added to a stirred solution of $(NH_4)_5H_5[H_2(WO_4)_6] \cdot H_2O$(0.2 mmol) in water(3 mL). The pH value was adjusted to 6.0 with additions of 1.0 M ammonia hydroxide. The mixture was stirred for 8 h at room temperature, and kept in refrigerator for several days. The $R$-mandelato tungstate(Ⅵ)(**3a**) was isolated as colorless crystals (0.57 g, yield 80%). $[\alpha]_D^{20} = -16.4°$ ($c$ 1.02, water). Found: C, 32.5; H, 3.9; N, 4.3. Calc. for $C_{16}H_{24}WN_2O_{10}$: C, 32.7; H, 4.1; N, 4.8. IR(KBr): $v_{as}$(C=O)$1651_{vs}$; $v_s$(C=O)$1451_s$, $1402_{vs}$, $1355_s$; $v$(W=O)$927_{vs}$, $870_{vs}$; $^1H$ NMR $\delta_H$(500 MHz, $D_2O$) ppm, 7.47−7.66(m, 5H, Ph); 6.08, 5.93(d, 1H, =CH); $^{13}C$ NMR $\delta_C$($D_2O$) ppm, 187.71($CO_2$)$\alpha$; 143.32, 143.13, 131.99, 131.75, 131.58, 131.00, 130.06(Ph); 89.84, 88.53(HCO$^-$). Similarly, $S$-mandelato tung-state(Ⅵ)(**3b**) was isolated as colorless crystals (0.60 g, yield 85%). $[\alpha]_D^{20} = +15.5°$ ($c$ 1.06, water). Found: C, 32.3; H, 4.0; N, 4.9. Calc. for $C_{16}H_{24}WN_2O_{10}$: C, 32.7; H, 4.1; N, 4.8. IR(KBr): $v_{as}$(C=O)$1654_{vs}$; $v_s$(C=O)$1468_s$, $1402_s$, $1355_s$; $v$(W=O)$931_{vs}$, $872_{vs}$; $^1H$ NMR $\delta_H$(500 MHz, $D_2O$) ppm, 7.44−7.63(m, 5H); 6.04, 5.91(d, 1H, =CH); $^{13}C$ NMR $\delta c$($D_2O$) ppm, 187.59($CO_2$)$\alpha$; 143.30, 131.95, 131.73, 131.56, 130.95, 130.01(Ph); 89.77, 88.46(HCO$^-$).

Crystals of suitable quality for the subsequent X-ray diffraction studies were obtained as transparent rhombohedral blocks by slow evaporation of the related solution of compounds **1a-3b** at room temperature. The resulting crystals were sealed in capillary to prevent loss of water molecules.

## 2.3 Physical property determinations

Infrared spectra were recorded as Nujol mulls between KBr plates using a Nicolet 360 FT-IR spectrometer. Elemental analyses were performed using EA 1110 elemental analyzers. Optical rotations were measured with Perkin-Elmer 341 automatic polarimeter. CD spectra were measured on JASCO 810 spectrometer. $^1H$ NMR and $^{13}C$ NMR spectra were recorded on a Varian UNITY 500 MHz NMR spectrometer with $D_2O$ using DSS (sodium 2,2-dimethyl-2-silapentane-5-sulfonate) as internal reference, respectively.

## 2.4 X-ray data collections, structure solutions and refinements

Crystallographic data for the mandelato molybdates (Ⅵ) and tungstates (Ⅶ) are summarized in Table 1. Diffraction data were collected on a Bruker Smart Spex CCD diffractometer with graphite monochromated Mo Kα radiation at 296 K. A Lorentz polarization factor, anisotropic decay and empirical absorption corrections were applied. The structures were solved by Schlxs97 and refined by full-matrix least-squares procedures with anisotropic thermal parameters for all the non-hydrogen atoms. H atoms were located from difference Fourier map. All calculations were performed on a microcomputer using Schlxl97 program[25]. Selected atomic distances and bond angles are given in Table 2.

**Table 1    Crystal data**

|  | 1a | 1b | 2 | 3a | 3b |
|---|---|---|---|---|---|
| Empirical formula | $C_{16}H_{26}MoN_2O_{11}$ | $C_{16}H_{26}MoN_2O_{11}$ | $C_{16}H_{28}MoN_2O_{12}$ | $C_{16}H_{24}WN_2O_{10}$ | $C_{16}H_{24}WN_2O_{10}$ |
| Formula weight | 518.33 | 518.33 | 536.34 | 588.22 | 588.22 |
| Crystal color, habit |  |  | Colorless, block |  |  |
| Crystal dimmers(mm) | 0.16×0.20×0.27 | 0.15×0.25×0.50 | 0.15×0.19×0.26 | 0.23×0.09×0.05 | 0.17×0.13×0.09 |
| Crystal system |  |  | Monoclinic |  |  |
| *Cell constants* |  |  |  |  |  |
| $a(Å)$ | 10.3700(5) | 10.3633(2) | 14.5497(5) | 8.1532(6) | 8.1499(2) |
| $b(Å)$ | 10.2266(4) | 10.2295(3) | 9.3171(2) | 13.3360(9) | 13.3337(9) |
| $c(Å)$ | 10.7832(3) | 10.7874(3) | 18.0036(5) | 10.3372(7) | 10.3348(7) |
| $\beta(°)$ | 100.454(2) | 100.489(1) | 112.584(1) | 110.695(1) | 110.703(1) |
| $V(Å^3)$ | 1124.57(8) | 1124.48(5) | 2253.4(1) | 1051.4(1) | 1050.5(1) |
| Space group | $P2_1$ | $P2_1$ | $P2_1/n$ | $P2_1$ | $P2_1$ |
| *Formula units* | 2 | 2 | 4 | 2 | 2 |
| $D_{calc}(g\ cm^{-3})$ | 1.531 | 1.531 | 1.581 | 1.858 | 1.860 |
| $F(000)$ | 532 | 532 | 1104 | 576 | 576 |
| *Diffractometer* |  |  | *Smart Apex CCD* |  |  |
| *Radiation* |  |  | *Mo Kα(λ=0.7107 Å)* |  |  |
| *Temperature(℃)* |  |  | 23 |  |  |
| *Flack parameter*[26] | 0.04(4) | −0.01(3) |  | 0.05(5) | 0.01(2) |
| *GOF on $F^2$* | 1.028 | 1.104 | 1.062 | 1.106 | 1.184 |
| $R_1$(all data) | 0.037 | 0.028 | 0.068 | 0.067 | 0.044 |
| $wR$ indices[a] | 0.086 | 0.066 | 0.157 | 0.166 | 0.102 |

$^a R_1 = \sum\{|F_o|-|F_c|\}/\sum(|F_o|), wR_2^{-1/2} = \sum[w(F_o^2-F_c^2)^2]/\sum[w(F_o^2)^2].$

**Table 2　Selected bond distances(Å)and angles(°)for 1a-3b**

| Bond | 1a | 1b | 2 | Bond | 3a | 3b |
|---|---|---|---|---|---|---|
| Mo(1)—O(1) | 1.932(3) | 1.927(2) | 1.950(3) | W(1)—O(1) | 1.96(1) | 1.965(7) |
| Mo(1)—O(2) | 2.187(4) | 2.183(3) | 2.229(3) | W(1)—O(2) | 2.22(1) | 2.194(7) |
| Mo(1)—O(11) | 1.954(3) | 1.951(2) | 1.951(3) | W(1)—O(11) | 1.987(9) | 1.924(7) |
| Mo(1)—O(12) | 2.228(3) | 2.232(3) | 2.205(3) | W(1)—O(12) | 2.14(1) | 2.144(8) |
| Mo(1)—O(4) | 1.726(3) | 1.719(3) | 1.700(4) | W(1)—O(4) | 1.78(1) | 1.763(8) |
| Mo(1)—O(5) | 1.714(4) | 1.707(3) | 1.717(4) | W(1)—O(5) | 1.73(1) | 1.742(7) |
| Angle | 1a | 1b | 2 | Angle | 3a | 3b |
| O(1)—Mo(1)—O(2) | 75.0(1) | 75.2(1) | 74.8(1) | O(1)—W(1)—O(2) | 75.1(4) | 74.9(3) |
| O(1)—Mo(1)—O(11) | 152.0(1) | 152.3(1) | 152.3(1) | O(1)—W(1)—O(11) | 151.6(5) | 152.0(3) |
| O(1)—Mo(1)—O(12) | 84.9(1) | 85.09(9) | 85.1(1) | O(1)—W(1)—O(12) | 82.7(5) | 83.5(3) |
| O(1)—Mo(1)—O(4) | 92.7(2) | 92.6(1) | 92.6(2) | O(1)—W(1)—O(4) | 104.9(5) | 104.5(4) |
| O(1)—Mo(1)—O(5) | 104.9(2) | 104.6(1) | 103.1(2) | O(1)—W(1)—O(5) | 93.3(5) | 93.5(3) |
| O(2)—Mo(1)—O(11) | 82.3(2) | 82.5(1) | 82.5(1) | O(2)—W(1)—O(11) | 82.7(4) | 82.9(3) |
| O(2)—Mo(1)—O(12) | 78.3(2) | 78.1(1) | 77.8(1) | O(2)—W(1)—O(12) | 79.1(5) | 79.9(3) |
| O(2)—Mo(1)—O(4) | 162.0(2) | 162.0(1) | 161.5(2) | O(2)—W(1)—O(4) | 90.8(6) | 90.5(4) |
| O(2)—Mo(1)—O(5) | 92.6(2) | 92.4(1) | 91.7(2) | O(2)—W(1)—O(5) | 165.1(5) | 165.0(3) |
| O(11)—Mo(1)—O(12) | 74.5(1) | 74.30(9) | 74.9(1) | O(11)—W(1)—O(12) | 75.9(5) | 75.9(3) |
| O(11)—Mo(1)—O(4) | 104.9(2) | 104.7(1) | 105.2(2) | O(11)—W(1)—O(4) | 92.6(5) | 92.5(3) |
| O(11)—Mo(1)—O(5) | 92.1(2) | 92.4(1) | 92.9(2) | O(11)—W(1)—O(5) | 105.0(5) | 104.8(4) |
| O(12)—Mo(1)—O(4) | 87.7(2) | 87.9(1) | 87.8(2) | O(12)—W(1)—O(4) | 165.5(6) | 165.7(4) |
| O(12)—Mo(1)—O(5) | 164.6(2) | 164.4(1) | 164.6(2) | O(12)—W(1)—O(5) | 90.3(6) | 89.4(4) |
| O(4)—Mo(1)—O(5) | 103.5(2) | 103.6(2) | 104.6(2) | O(4)—W(1)—O(5) | 101.4(7) | 101.8(4) |

# 3　Results and discussion

## 3.1　Syntheses

The reactions of $R$-,$S$- and $RS$-mandelate with ammonium molybdate(2∶1 ratio)in water gave the desired $R$-,$S$- and $RS$-mandelato molybdate **1a**,**1b** and **2**,respectively,in good yields as shown in the equation below. In a similar manner,$R$-and $S$-mandelato tungstates(Ⅵ)**4a** and **4b** were prepared. The presence of a small excess mandelic acid in the reaction mixture was essential,since its deficiency caused in the isolation of homopolyacid of molybdate(Ⅴ)and tungstate(Ⅵ). The reaction proceeded most readily at the optimal pH range of 3.0－7.0,and the resulted cis-dioxo dimandelates were readily soluble in water.

$$(NH_4)_2MO_4+2(R\ or\ S\ or\ RS\text{-}H_2mand)\rightarrow(NH_4)_2[MO_2(R\ or\ S\ or\ RS\text{-}mand)_2]\,(M=Mo,W)$$

There is much information published on the reaction of molybdenum(Ⅵ) and tungsten(Ⅵ) with different α-hydroxycarboxylic acids(e. g., citric, malic and tartaric acid) in solution chemistry[27-33]. Previous works have shown that complexes with various metal-to-ligand ratios,e. g.,1∶1,2∶2 and 1∶2,can occur in preparations. For those ligands having only one α-hydroxyl and one α-carboxylic acid group, the preferred stoichiometry appears to be 1∶2[22,29]. The related mononuclear cis-dioxo complexes have been isolated in the solid state and their structures determined by single crystal X-ray analyses[12,16,18,34-37]. In all these investigations,both the molar ratio of the complexing agents and the pH of the solution are crucial. However,other factors such as possible tensions in chelate rings and steric constraints from the ligands may affect structures of various metal-to-ligand ratios[36].

## 3.2 Description of the crystal structures

As expected,compounds **1a** and **1b** are enantiomorphic to each other showing an equivalent but in opposite signs of optical rotation,which **2** gives no optical activity and a racemic mixture. Figs. 1 and 2 are the plots of **1a,1b** and **2**,respectively,obtained from the chiral and racemic monomeric molybdates. These show molybdenums(Ⅵ)are coordinated by R-, S-and RS-mandelate anions. The coordinations of tungstates **3a** and **3b** are in a similar way as shown in Figures S1 and S2. The structural analyses show the presences of monomeric anions having two chelated mandelato molecules coordinated to the central metal atom in the conventional cis disposition. As usual,for cis-dioxo complexes,the molybdenum and tungsten atoms are not at the center of the coordination octahedron but is shifted toward the terminal unshared oxygen atoms. Moreover, cis-MoO₂ and cis-WO₂ groups have the general properties of compressing those bond angles involving other atoms to less than the octahedral values of 90° or 180°.

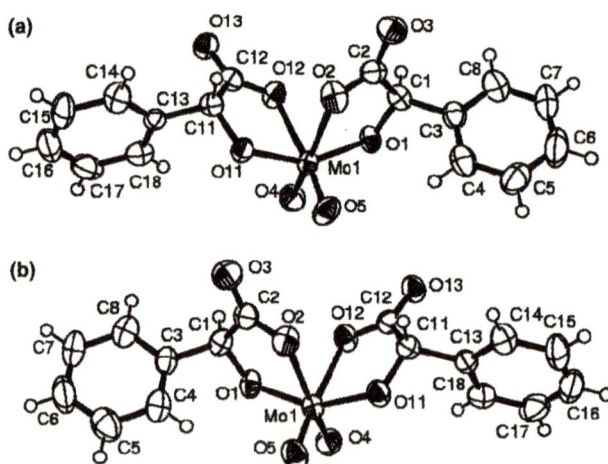

**Fig. 1** Perspective view of the enantiomeric anion structures of(NH₄)₂[Mo₄O₂(R-mand)₂]·3H₂O(1a) (a)and(NH₄)₂[Mo₄O₂(S-man-d)₂]·3H₂O(1b)(b).

Of the five complexes,the Mo-O and W-O distances in mandelato molybdates and tungstates vary systematically. Mo=O is in the range 1.700(3)-1.726(3)Å,while W=O is in the range 1.73(1)-1.78(1)Å,indicating that they are double bonds. The resulting O=Mo=O angles(103-105°) is considerably larger than the 90° regular octahedron value for cis groups. This is expected from the greater O⋯O repulsion between oxygen with short bonds to the metal atom. Owing to the tungsten

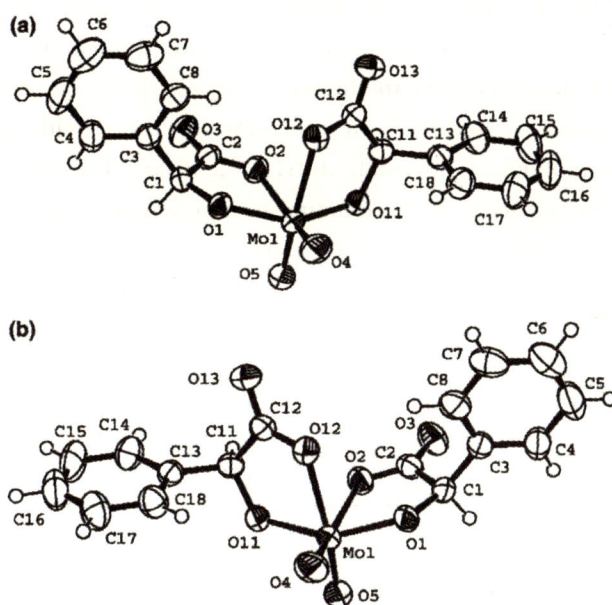

Fig. 2 Perspective view of anion structures of the mesomeric mandelato molybdate ( $NH_4$ )$_2$ [ $MoO_2$ ( $RS$-mand)$_2$ ] $\cdot$ 4H$_2$O(2) , $\Delta_{RS}$ (a) and $\Lambda_{RS}$ (b).

atom is larger than the molybdenum atom, the smaller O···O repulsion leads to the smaller $O = W = O$ angles(101. 4—101. 8°). In the complex $(NH_4)_2[MoO_2(OOCCO-Ph_2)_2] \cdot 2H_2O^{[36]}$, the replacement of a hydrogen atom by phenyl group on the ligand appears to have a little effect on the coordination geometry, it leads to the $O = Mo = O$ angle(102. 5°) smaller than those of the compounds **1a-2**. The single Mo-O bond distances vary from 1. 927 to 2. 232Å, and these variations may be ascribed to the different donor natures of α-alkoxyl and α-carboxyl oxygen atoms and *trans* effects of terminal oxo groups. So do the complexes **3a** and **3b**. The bond distances of Mo-O α-alkoxyl group [1. 927 ( 2 ) — 1. 954(3)Å]are slightly shorter than those of Mo-O α-carboxyl group *trans* to terminal oxide ligands [2. 183 ( 3 ) — 2. 232 ( 3 ) Å]. These Mo-O$_{carboxyl}$ distances are comparable with those found in the [MoO$_2$(OOCCOPh$_2$)2]$^{2-}$ anion$^{[36]}$. Deprotonated hydroxyl and carboxyl O atoms of the mandelate ligand give two five member rings. Another situations were found in the peroxo vanadium mandelate complexes like (Et$_4$N)$_2$[V$_2$O$_2$(O$_2$)$_2$(R-mand)$_2$] and (t-Bu$_4$N)$_4$[V$_2$O$_2$(O$_2$)$_2$(R-mand)$_2$][V$_2$O$_2$(O$_2$)$_2$ (S-mand)$_2$] $\cdot$ (R-H$_2$mand)(S-H$_2$mand)$^{[38,39]}$, in which the bond of V-O$_{carboxyl}$[1. 991(2)Å]is closed to the bond of V-O$_{alkoxyl}$[1. 979(2)Å].

## 3. 3  FT-IR and NMR spectroscopies

The FT-infrared spectra of the five complexes showed well-resolved strong and sharp absorption bands for the coordinated mandelates as shown in Figures S3 and S4, respectively. Antisymmetric stretching vibrations $v_{asym}$(COO$^-$)appeared between 1640 and 1580 cm$^{-1}$. The corresponding symmetric stretches $v_{sym}$(COO$^-$) appeared between 1455 and 1355 cm$^{-1}$. All of the carboxyl absorptions were shifted to lower frequencies with respect to those of free mandelic acid. This is in accord with a chelate mode by the mandelato ligand. Losses of the protons in compounds **1a-3b** reduce the number of bands and displace them to lower frequencies.

The $^1$H NMR spectra of **1a-3b** shows multiple lines at 7. 37—7. 66 ppm, assigned to the hydrogen atoms of the phenyl groups of mandelato molybdates( Ⅵ )and tungstate( Ⅵ ). At about 5. 73—6. 08 ppm, they show a group of double line, which can be assigned to the hydrogen atoms of the CH(O) groups. The compounds **1a-2,3a** and **3b** retain bound mandelate dianion ligands. In Fig. 3, large low-field shifts of some $^{13}C$ resonances of **1** in comparison to the free mandelic acids clearly show that both α-alkoxyl (Δδ 13. 49 ppm) and α-carboxyl groups(Δδ 7. 73 ppm) shift to lower fields and are coordinated. This is also true for *RS*-mandelato molybdates. A small shift(Δδ 2. 60 ppm) of the phenyl carbon bonded to the α-alkoxyl group indicates the influence of the bonding to molybdenum. As for the complexes **3a** and **3b**, a similar situation is also observed. But due to the ligand binding to different metal atom, it is found that the -COOH shifts on binding to two metals are larger for W(Ⅵ); the same occurs with the proton shifts of ═CHOH and-C₆H₅.

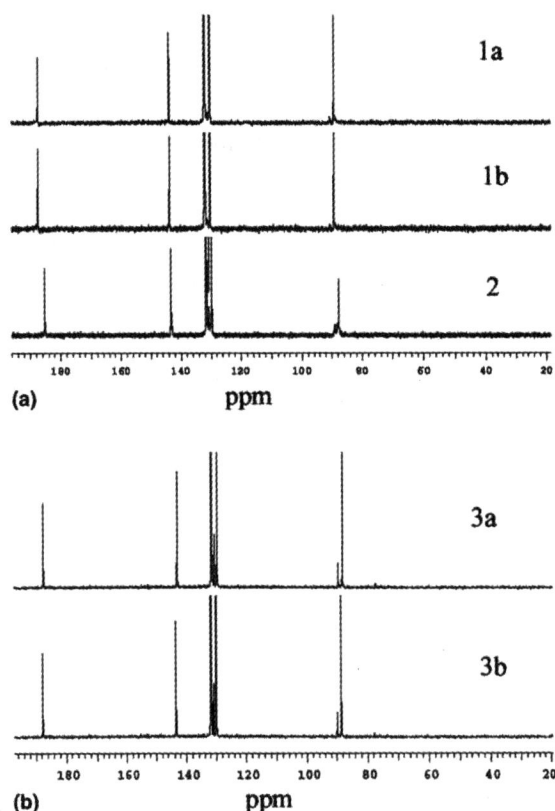

Fig. 3　$^{13}C$ NMR spectra of 1a-3b.

# 4　Biological relevance

For comparison, Table 3 shows the comparisons of Mo-O and W-O distances(Å) in bidentate hydroxycarboxylato molybdate ( Ⅵ ) like citric, malic, tartaric, lactic and glycolic acids; and the assignments of their absolute configurations. The bond distances of M-O$_{carboxyl}$ and M-O$_{alkoxyl}$ (M ═ Mo or W) are in the similar ranges for different organic acid complexes. Moreover, similar bond distances for tungstates and molybdates are due to the effect of lanthanide contraction. The bond distances of average Mo-O α-alkoxyl and α-carboxyl groups in the 12 molybdate(Ⅵ)complexes are 1. 944 and 2. 210 Å, respectively. There is no obvious relationship in the bond distance of Mo-O$_{carboxyl}$ with the change in oxidation state of molybdenum(Mo-O$_{carboxyl}$ 2. 162 Å in FeMo-co), while the bond distance of Mo(Ⅵ)-

$O_{alkoxyl}$ is shorter than that of homocitric acid to FeMo-co(2.212 Å). This shows that bond distance of $\alpha$-alkoxyl group to molybdenum increases with the decrease in the oxidation state of molybdenum. Strong reductions of mandelate and benzilato molybdates result in no coordination of hydroxyl group, which suggests a model for creation of a binding site for substrate at Mo by protonation and dissociation of the $\alpha$-alkoxyl group in $R$-homocitrate ligand[5−7,47−49] (see Fig. 4).

Table 3　Comparisons of Mo-O and W-O distances(Å)in new and known hydroxycarboxylato complexes($H_2$glyc = glycolic acid, $H_2$lact = lactic acid, $H_3$mal = malic acid; $H_4$tart = tartaric acid; $H_4$cit = citric acid)and the assignments of their absolute configurations

| Complex | M-O($\alpha$-alkoxyl) | M-O($\alpha$-carboxyl) | Configuration assignment | Ref. |
|---|---|---|---|---|
| $K_2[MoO_2(glyc)_2] \cdot H_2O$ | 1.929(3), 1.953(2) | 2.202(2), 2.204(2) | $\Delta/\Lambda$ | [18] |
| $\{Na_2[MoO_2(S\text{-}lact)_2]\} \cdot 13H_2O$ | 1.947(4), 1.969(5) | 2.181(4), 2.201(5) | $\Lambda_{S,S}$ | [18] |
| $(NH_4)_3[GdMo_6O_{15}(S\text{-}lact)_6]$ | 1.919(6) | 2.192(6) | $\Delta_{S,S}$ | [40] |
| $(NH_4)_2[MoO_2(R\text{-}mand)_2] \cdot 3H_2O(\mathbf{1a})$ | 1.932(3), 1.954(3) | 2.187(4), 2.228(3) | $\Delta_{R,R}$ | — * |
| $(NH_4)_2[MoO_2(S\text{-}mand)_2] \cdot 3H_2O(\mathbf{1b})$ | 1.927(2), 1.951(2) | 2.183(3), 2.232(3) | $\Lambda_{S,S}$ | — * |
| $(NH_4)_2[MoO_2(RS\text{-}mand)_2] \cdot 4H_2O(\mathbf{2})$ | 1.950(3), 1.951(3) | 2.205(3), 2.229(3) | $\Delta/\Lambda_{R,S}$ | — * |
| $(NH_4)[MoO_2(OOCCOPh_2)_2] \cdot 2H_2O$ | 1.966(2), 1.977(2) | 2.177(2), 2.166(3) | $\Delta/\Lambda$ | [36] |
| $Cs_2[MoO_2(Hmal)_2] \cdot H_2O$ | 1.939(8) | 2.243(9) | $\Delta_{S,S}$ | [34] |
| $Na_3[MoO_2H(mal)_2]$ | 1.935(2) | 2.182(3) | $\Lambda_{S,S}$ | [12] |
| $(NMe_4)[MoO_2(H_2tart)_2] \cdot EtOH \cdot 1.5H_2O$ | 1.94(1), 1.921(9) | 2.227(9), 2.226(8) | $\Lambda_{R,R}$ | [35] |
| $(NH_4)\{[Gd(H_2O)_6]Mo_2O_4(tart)_2\} \cdot 4H_2O$ | 1.944(4), 1.946(4) | 2.205(4), 2.227(4) | $\Delta_{R,R}$ | [41] |
| $Na_2[MoO_2(H_2cit)_2] \cdot 3H_2O$ | 1.944(4), 1.946(4) | 2.205(4), 2.227(4) | $\Delta/\Lambda$ | [16] |
| Average | 1.944 | 2.210 | | |
| $(NH_4)_2[W_\Delta O_2(R\text{-}mand)_2] \cdot 2H_2O(\mathbf{3a})$ | 1.96(1), 1.987(9) | 2.14(1), 2.22(1) | $\Delta_{R,R}$ | — * |
| $(NH_4)_2[W_\Lambda O_2(S\text{-}mand)_2] \cdot 2H_2O(\mathbf{3b})$ | 1.924(7), 1.965(7) | 2.144(8), 2.194(7) | $\Lambda_{S,S}$ | — * |
| $(NH_4)[WO_2(OOCCOPh_2)_2] \cdot 2H_2O$ | 1.962(3), 1.981(3) | 2.157(3), 2.160(3) | $\Delta/\Lambda$ | [42] |
| $Na_2[WO_2(Hmal)_2] \cdot 4H_2O$ | 1.95(3), 2.02(3) | 2.17(4), 2.21(3) | $\Delta_{S,S}$ | [43] |
| $Na_3[WO_2H(mal)_2]$ | 1.973(4) | 2.170(4) | $\Lambda_{S,S}$ | [44] |
| $Na_2[WO_2(H_2cit)_2] \cdot 3H_2O$ | 1.945(6), 1.968(7) | 2.189(8), 2.227(7) | $\Delta/\Lambda$ | [16] |
| $K_2[WO_2(C_7H_{10}O_6)_2] \cdot 1.5H_2O$ | 1.925(5), 1.938(4) | 2.174(5), 2.182(5) | $\Lambda_{R,R}$ | [37] |
| Average | 1.963 | 2.179 | | |
| $[Mo(\text{II})_2(R\text{-}Hmand)_4] \cdot 2THF$ | | $2.119(8)_{av}$ | | [45] |
| $[Mo(\text{II})_2(OOCCOPh_2)_4] \cdot 4THF$ | | $2.119(2)_{av}$ | | [46] |
| $[MoFe_7S_9N(R\text{-}homocitrate)(his)]$ in | 2.212 | 2.162 | $\Delta_R$ | [2] |
| Azotobacter vinelandii(Av1) | | | | |

* This work.

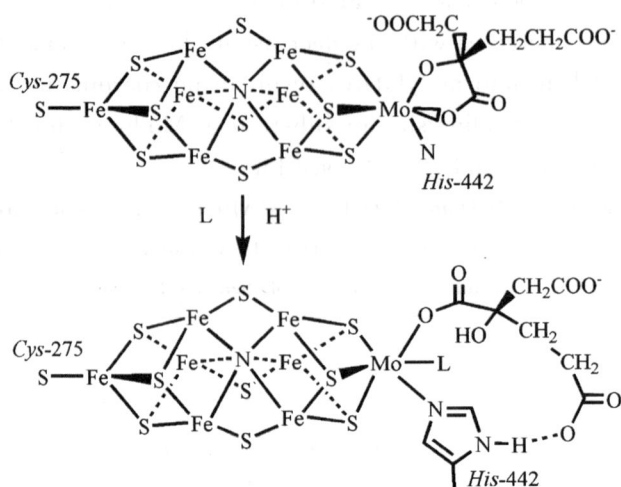

**Fig. 4** **Proposed model for binding substrate at Mo by protonation and dissociation of the α-alkoxyl group in *R*-homocitrate ligand.**

The chiral configuration of the metal center is determined by the chiral ligand and the noncentrosymmetric space group[50]. Based on the assignments of metal configurations of chiral mandelate complexes, it is deduced that there are some other possible diastereoisomers of the coordinated monomer, which result from the geometric and asymmetric octahedral coordination environments around molybdenum or tungsten. But only one of the favorable species is separated in **1a**, **1b**, **3a** and **3b**, respectively. The reasons may be attributed to the formation of much less energetically favorable species in the crystal growth process and the chiral effect of ligand and counterion cations. The $\Delta_{R,R}$ mandelato molybdate **1a** and tungstate **3a** are the enantiomers of $\Lambda_{S,S}$ mandelato molybdate **1b** and tungstate **3b**, respectively. This is further supported by the circular dichroism spectra of the two enantiomers shown in Fig. 5. In the mesomeric mandelato molybdate **2**, both *R*-and *S*-mandelate are coordinated simultaneously with molybdenum(Ⅵ) in the ammonium monomer **2**. Indeed, it is not yet known how molybdenum and *R*-homocitrate are inserted in the nitrogenase iron-sulfur cluster of *NifB*-co. As far as stereochemistry is concern, *R*-homocitrate is certainly biologically synthesized as a pure enantiomer and its stereospecific coordination to Mo must be governed by the proteins environment. Even citrate seems to coordinate in a single configuration[9]. However, the coordinations of different chiral ligands like *R*-and *S*-malic, citramalic and citroylformic acids influence the activities of substrate reduction in nitrogenase based on the natural *R*-homocitrate ligand[10]. The same situation may also be true for *S*-homocitrate ligand, which can be served as a probe for the substrate reduction.

## Acknowledgements

We thank the Ministry of Science and Technology (001CB108906) and the National Science Foundation of China(No. 20021002)for the generous supports of this research, and Professor Yuan L. Chow for stimulating discussions and many suggestions.

## Appendix A. Supporting information available

Tables of X-ray crystal structure refinement data, positional and thermal parameters for **1a-3b** are

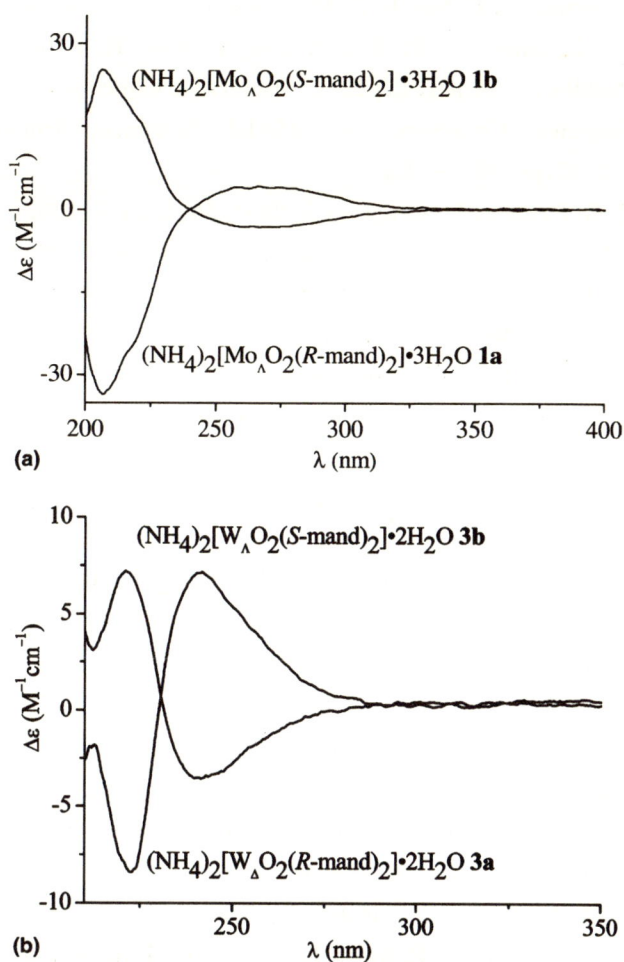

Fig. 5　CD spectra of 1a, 1b, 3a and 3b.

available via the Internet. CCDC reference numbers are 203217-203219,237819 and 237820. Supplementary data associated with this article can be found,in the online version,at doi:10. 1016/j. jinorgbio. 2004. 08. 003.

# References

[1]T. G. Spiro(Ed. ),Metal Ions in Biology,Molybdenum Enzymes,vol. 7,Wiley-Interscience Pub.,
1985.

[2]O. Einsle, F. A. Tezcan, S. L. A. Andrade, B. Schmid, M. Yoshida, J. B. Howard, D. C. Rees.
Science,2002,**297**:1696~1700.

[3]J. B. Howard,D. C. Rees. Chem. Rev,1996,**96**:2965~2982.

[4]H. I. Lee, P. M. Benton, M. Laryukhin, R. Y. Igarashi, D. R. Dean, L. C. Seefeldt. J. Am. Chem.
Soc,2003,**125**:5604~5605.

[5]P. L. Holland,in:J. A. McCleverty, T. J. Meyer(Eds. ),Comprehensive Coordination Chemisry
II,vol. **8**(L. Que,W. B. Tolman,Eds. ),Elsevier/Pergamon,2004,pp. 586~591.

[6]L. C. Seefeldt,I. G. Dance, D. R. Dean. Biochemistry,2004,**43**:1401~1409.

[7]M. C. Durrant. Biochemistry,2002,**41**:13934~13945.

[8]J. Imperial, T. R. Hoover, M. S. Madden, P. W. Ludden, V. K. Shah. Biochemistry, 1989, **28**:7796~
7799.

[9]S. M. Mayer,C. A. Gormal,B. E. Smith,D. M. Lawson. J. Biol. Chem,2002,**277**:35263~35266.

[10] P. W. Ludden, V. K. Shah, G. P. Roberts, M. Homer, R. Allen, T. Paustian, J. Roll, R. Chatterjee,M Madden, J. Allen, in：E. I. Stiefel, D. Coucouvanis, W. E. Newton（Eds.）, Molybdenum Enzymes, Cofactors and Model Systems, American Chemical Society, Washington,DC,1993,pp. 196~215.

.[11]G. N. Schrauzer,J. G. Palmer,Z. Naturforsch. 2001,**56**:1354~1359.

[12]Z. H. Zhou,W. B. Yan,H. L. Wan,K. R. Tsai. J. Inorg. Biochem,2002,**90**:137~143.

[13]B. K. Burgess,D. J. Lowe. Chem. Rev,1996,**96**:2983~3011.

[14]R. Hille. Chem. Rev,1996,**96**:2757~2816.

[15]Z. H. Zhou,H. L. Wan,K. R. Tsai. Polyhedron,1997,**16**:75~79.

[16]Z. H. Zhou,H. L. Wan,K. R. Tsai. J. Chem. Soc.,Dalton Trans,1999:4289~4299.

[17]Z. H. Zhou,H. L. Wan,K. R. Tsai. Inorg. Chem,2000,**39**:59~64.

[18]Z. H. Zhou,S. Y. Hou,Z. X. Cao,H. L. Wan,S. W. Ng. J. Inorg. Biochem,2004,**96**:1037~1044.

[19]K. S. Pakhomova,L. P. Volkova. J. Anal. Chem. USSR,1976,**31**:774~777.

[20]M. M. Caldeira,M. L. Ramos,V. M. S. Gil. Can. J. Chem,1987,**65**:827~832.

[21]M. Hlaïbi,S. Chapelle,M. Benaïssa,J. F. Verchère. Inorg. Chem,1995,**34**:4434~4440.

[22]J. J. Cruywagen,E. A. Rohwer. J. Chem. Soc.,Dalton Trans,1995:3433~3438.

[23]V. M. S. Gil. Pure Appl. Chem,1989,**5**:81~848.

[24]Aldrich,Handbook of Fine Chemicals and Laboratory Equipment,2001,p. 1038.

[25]G. M. Sheldrick,Shelxl97 and Shelxs97,University of Göttingen,Germany.

[26]H. D. Flack. Acta Cryst. A,1983,**39**:876~881.

[27]J. J. Cruywagen,J. B. B. Heyns,E. A. Rohwer. J. Chem. Soc.,Dalton Trans,1990:1951~1956.

[28]J. J. Cruywagen,L. Krüger,E. A. Rohwer. J. Chem. Soc.,Dalton Trans,1991:1727~1731.

[29]J. J. Cruywagen,L. Kruger,E. A. Rohwer. J. Chem. Soc.,Dalton Trans,1993:105~109.

[30]J. J. Cruywagen,E. A. Rohwer,G. F. S. Wessels. Polyhedron,1995,**14**:3481~3493.

[31]J. J. Cruywagen,E. A. Rohwer,R. F. van de Water,Polyhedron,1997,**16**:243~251.

[32]J. J. Cruywagen,L. Krüger,E. A. Rohwer. J. Chem. Soc.,Dalton Trans,1997:1925~1929.

[33]P. Lubal,J. Perutka. J. Havel,Chem. Papers,2000,**54**:289~295.

[34]C. B. Knobler, A. G. Wilson, R. N. Hider, I. W. Jensen, B. R. Penfold, W. T. Robinson, C. J. Wilkins. J. Chem. Soc.,Dalton Trans,1983:1299~1303.

[35]W. T. Robinson,C. J. Wikins. Trans. Metal Chem,1986,**11**:86~89.

[36]A. Cervilla, E. Llopis, A. Ribera, A. Doménech, A. J. P. White, D. J. Williams. J. Chem. Soc., Dalton Trans,1995:3891~3895.

[37]M. L. Ramos, M. M. Pereira, A. M. Beja, M. R. Silva, J. A. Paixāo, V. M. S. Gil. J. Chem. Soc., Dalton Trans,2002:2126~2131.

[38]I. K. Smatanová,J. Marek,P. Švancârek,P. Schwendt. Acta Crystallogr. C,2000,**56**:154~155.

[39]M. Ahmed,P. Schwendt,J. Marek,M. Sivák. Polyhedron,2004,**23**:655~663.

[40]C. D. Wu,C. Z. Lu,J. C. Liu,H. H. Zhuang,J. S. Huang. J. Chem. Soc.,Dalton Trans,2001：3202~3204.

[41]C. D. Wu,C. Z. Lu,X. Lin,D. M. Wu,S. F. Lu,H. H. Zhuang,J. S. Huang. Chem. Comm,2003：1284~1285.

[42]H. Zhao,Y. Q. Jiang,H. Zhang,Z. H. Zhou. Chinese J. Struct. Chem,2004,**23**:502~505.

[43] S. Y. Hou, W. B. Yan, Z. J. Ma, X. L. Liao, Z. H. Zhou, H. L. Wan. J. Coord. Chem, 2003, **56**: 133~139.

[44] Z. H. Zhou, G. F. Wang, S. Y. Hou, H. L. Wan, K. R. Tsai. Inorg. Chim. Acta, 2001, **314**: 184~188.

[45] F. A. Cotton, L. R. Falvello, C. A. Murillo. Inorg. Chem, 1983, **22**: 382~387.

[46] T. Liwporncharoenvong, T. Lu, R. L. Luck. Inorg. Chim. Acta, 2002, **329**: 51~58.

[47] A. Leigh. Science, 1995, **268**: 827~828.

[48] K. L. C. Grtönberg, C. A. Gormal, M. C. Durrant, B. E. Smith, R. A. Henderson. J. Am. Chem. Soc, 1998, **120**: 10613~10621.

[49] K. R. Tsai, H. L. Wan. J. Clust. Sci, 1995, **6**: 485~501.

[50] U. Knof, A. von Zelewsky. Angew. Chem. Int. Ed, 1999, **38**: 302~332.

蔡启瑞院士论文选集

■ 本文原载：Inorg. Chem. 43(2004)，pp. 923～930.

# Enzymatic Mechanism of Fe-Only Hydrogenase: Density Functional Study on H-H Making/Breaking at the Diiron Cluster with Concerted Proton and Electron Transfers

Tai-Jin Zhou,[①] Yi-Rong Mo,[a] Ai-Min Liu,[b] Zhao-Hui Zhou, K.R.Tsai

(Department of Chemistry and the State Key Laboratory for Physical Chemistry of the Solid Surface, Xiamen University, Xiamen 361005, People's Republic of China )

Received March 3, 2003

**Abstract**  The mechanism of the enzymatic hydrogen bond forming/breaking($2H^+ + 2e \leftrightarrows H_2$) and the plausible charge and spin states of the catalytic diiron subcluster $[FeFe]_H$ of the H cluster in Fe-only hydrogenases are probed computationally by the density functional theory. It is found that the active center $[FeFe]_H$ can be rationally simulated as $\{[H](CH_3S)(CO)(CN^-)Fe_p(CO_b)(\mu\text{-SRS})Fe_d(CO)(CN^-)L)\}$, where the monovalence $[H]$ stands for the $[4Fe4S]_H^{2+}$ subcluster bridged to the $[FeFe]_H$ moiety, $(CH_3S)$ represents a Cys-S, and $(CO_b)$ represents a bridging CO. L could be a CO, $H_2O, H^-, H_2$, or a vacant coordination site on $Fe_d$. Model structures of possible redox states are optimized and compared with the X-ray crystallographic structures and FTIR experimental data. On the basis of the optimal structures, we study the most favorable path of concerted proton transfer and electron transfer in $H_2$-forming/breaking reactions at $[FeFe]_H$. Previous mechanisms derived from quantum chemical computations of Fe-only hydrogenases (Cao, Z.; Hall, M. B. *J. Am. Chem. Soc.* **2001**, *123*, 3734; Fan, H.; Hall, M. B. *J. Am. Chem. Soc.* **2001**, *123*, 3828) involved an unidentified bridging residue ($\mu$-SRS), which is either a propanedithiolate or dithiomethylamine. Our proposed mechanism, however, does not require such a ligand but makes use of a shuttle of oxidation states of the iron atoms and a reaction site between the two iron atoms. Therefore, the hydride $H_b$ (bridged to $Fe_p$ and $Fe_d$) and $\eta^2\text{-}H_2$ at $Fe_p$ or $Fe_d$ most possibly play key roles in the dihydrogen reversible oxidation at the $[FeFe]_H$ active center. This suggested way of $H_2$ formation/splitting is reminiscent of the mechanism of $[NiFe]$ hydrogenases and therefore would unify the mechanisms of the two related enzymes.

①  Author to whom correspondence should be addressed. E-mail: tjzhou@ xmu. edu. cn. Fax: +86-592-2183795.

a Permanent address: Department of Chemistry, Western Michigan University, Kalamazoo, MI 49008.

b Permanent address: Department of Biochemistry, University of Mississippi Medical Center, Jackson, MS 39216.

## Introduction

Because hydrogen($H_2$) is regarded as a potential future energy resource, the efficient generation, storage, and oxidative conversion of $H_2$ have been the subject of intense studies. Before mankind has been able to design and synthesize catalysts for fast and economical production of $H_2$ or efficient conversion of $H_2$ into protons and electrons, nature has evolved a family of enzymes called hydrogenases (or hereafter $H_2$ ases), which are capable of catalyzing the reversible two-electron oxidation of $H_2$ as $H_2 \rightleftharpoons 2H^+ + 2e^-$. $H_2$ ases constitute the core of $H_2$ metabolism that is essential to many microorganisms of biotechnological interest[1,2]. The microorganisms can use $H_2$ as a source of electrons and protons or generate $H_2$ by reducing protons. In the organisms so far studied, $H_2$ ases as metalloenzymes are mainly grouped into two classes[3-6]: [NiFe]$H_2$ ases (which may include [NiFeSe]$H_2$ ase), which most often catalyze the forward reaction in which $H_2$ is consumed[7-14], and [FeFe]$H_2$ ases (iron-only $H_2$ ases), which catalyze the reduction of protons as terminal electron acceptors to yield $H_2$ and thus mainly function in $H_2$ production[15-24]. The [NiFe]$H_2$ ase in the bacterium *Desulfovibrio gigas* (DvH) has been crystallized in the air-oxidized form and its structure solved by Volbeda et al[7,8]. The crystal structure of [NiFe]$H_2$ ase reveals that the active center is composed of a [Ni] and a [Fe] bridged by two cysteinyl-S, as well as an oxygenic species, $\mu$-O, which is assumed to be removed during the reductive activation of the enzyme[25].

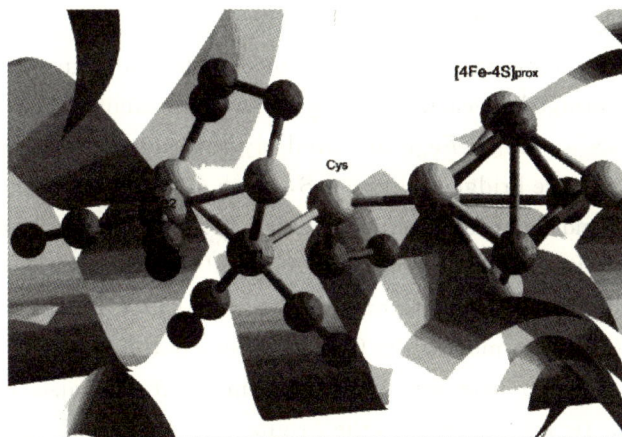

**Figure 1    Three-dimensional structure of the H cluster in the DdH enzyme and its proximal protein environment revealed by the X-ray crystallographic study.** [17]

For [FeFe]$H_2$ ases, recently two X-ray crystallographic structures of the enzymes from the bacteria *Clostridium pasterianum* (CpI)[15] and *Desulfovibrio desulfuricans* (DdH)[17] have been resolved. Compared with the [NiFe]$H_2$ ase, the [FeFe]$H_2$ ases are highly evolved catalysts. Under optimal conditions, each molecule of DdH can produce 9000 $H_2/s^{-1}$ at 30 ℃; for CpI, the figure is 6000 $s^{-1}$[1,26]. DdH contains three evenly spaced [4Fe-4S] clusters, which connect the active site with the molecular surface. The [6Fe-6S] cluster, termed H cluster(Figure 1), consists of a $[4Fe-4S]_H$ cuboidal cluster covalently bridged to an unusual $[FeFe]_H$ diiron cluster by a cysteinyl thiolate, S(Cys). The relative location of each of the metal clusters in the $H_2$ ase crystal structures depicts probable electron-transfer (ET) pathways. Peters et al[15,16]. and Nicolet et al[17-20]. have proposed a similar pathway of the proton transfer (PT) and ET to the H cluster based on their respective protein structures. The pathway consists of a few conserved amino acid side chains along with proteinbound water molecules. A potential channel for the delivery of $H_2$ molecules, which is hydrophobic and connects the active site to the molecular surface, has also been proposed[1,2,15-20].

The successive crystal structure determination of the above three $H_2$ ases and the discovery that the DvH(and later, the CpI and DdH as well) contains cyanide in its prosthetic group, as well as the technological significance of $H_2$ as a potential resource of clean fuel, have greatly piqued the interest of chemical, biochemical, and biophysical scientists. The prosthetic groups of these $H_2$ ases have been studied intensively with a wide variety of spectroscopic techniques such as EPR, ENDOR, Mössbauer, resonance Raman, and FTIR spectroscopy[26-29]. In particular, the FTIR spectra of the three redox states for DdH were investigated, and the results provided insights into the correlation between the geometrical structures and redox states in the H cluster. It is now generally believed that the $[4Fe-4S]_H$ always appears in the EPR-silent $2+$ state, or $[4Fe-4S]_H^{2+}$, and the $[FeFe]_H$ of DdH or CpI in both oxidized and reduced states are EPR-silent and may be better symbolized as $[Fe^{II}Fe^{II}]_{Hox}$($H_{ox}$ in short) and $[Fe^{I}Fe^{I}]_{H\,red}$(or $H_{red}$), respectively. In contrast, the semireduced diiron cluster, symbolized as $H_s$, in DdH or CpI is EPR-active $(S=1/2)$ and can be expressed as $[Fe^{II}Fe^{I}]_{Hs}$. These three states exhibit distinctive yet different IR bands for the CO ligands respectively(i. e., for CO in $H_{ox}$, 1847, 1983, and 2007 cm$^{-1}$; in $H_{red}$, 1894, 1916, 1940, and 1965 cm$^{-1}$; and in $H_s$, 1802, 1940, and 1965 cm$^{-1}$)[19,30]. Pickett et al. pointed out that the fully reduced $Fe^{I}Fe^{I}$ species favors the terminal CO conformation and the $Fe^{I}Fe^{II}$ transient bridging CO intermediate can be generated by the oneelectron oxidation of an $Fe^{I}Fe^{I}$ precursor bearing the thioether group. The observed IR spectra for CN and CO are consistent with the CO-inhibited paramagnetic center of $[FeFe]H_2$ ases[31,32].

Both of the cationic iron atoms in $[FeFe]_H$, assumed to be the active center, adopt an octahedral coordination mode and are bridged to each other by three nonprotein atoms, namely, two $\mu$-S and one putative $CO_b$. While $[4Fe-4S]_H$ and $[FeFe]_H$ are similarly associated by a cysteinyl-S in DdH, the two [Fe]'s of the $[FeFe]_H$ cluster are bridged by two $\mu$-S, which are linked together and initially modeled as $\mu$-propanedithiolate (PDT) and by an unspecified $\mu$-O[17]. Recently, on the basis of crystallographic and FTIR spectroscopic evidence from DdH, Nicolet et al. claimed that the previously observed $CO_b$ ligand in CpI (probably in the oxidized state) is terminally bound ($CO_t$) in the reduced DdH[18], and the transition of the $CO_b$ to $CO_t$ affects mostly the C atom, whereas the O atom essentially keeps the same position with respect to the two iron ions[19]. Furthermore, it has also been proposed that, from stereochemical and mechanistic considerations, the original assignment of the $\mu$-SRS to PDT(i. e., R= $(CH_2)_3$)might be incorrect. A dithiomethylamine (DTMA; i. e., R=$CH_2NHCH_2$) assignment, although still speculative, seems to make more sense because the latter would contribute a more basic, NH bridgehead that might be more effective for the heterolytic splitting of H-H. In addition, the distance between the NH of DTMA and a Cys178[19] residue can better rationalize the observed internuclear distance of 3. 2 Å than that between the central $CH_2$ of $\mu$-PDT and Cys178.

The $[FeFe]_H$ is well modeled by easily prepared and classical organometallic complexes, ($\mu$-SRS) $[Fe(CO)_2L]_2$ with L=CO, CN$^-$, PMe$_3$, etc., and R=-($CH_2$)$_3$-, or -($CH_2$NMeCH$_2$)-[31-53]. Important information about the charge, spin, and chemical states as well as the ligand exchange reactivity of $[FeFe]_H$ has been derived from the synthesis and characterization of ($\mu$-SRS)$[Fe(CO)_2L]_2$. It should be noted that known structures show the Fe-Fe internuclear distance of around 2. 5 Å[33,36,37]. well within the bonding range and in accordance with the 18-electron rule[38]. Particularly, the synthetic dinuclear iron models reported independently by Gloaguen et al. [41,44]. Zhao et al., [42] and Nehring et al[43]. present significant $H_2$ molecule formation via reduction/protonation and $H_2/D_2$ exchange activity.

Despite numerous discussions on the possible mechanisms for $H_2$ase[1,2,9,15-20,26,27,54], further quantitative analyses are pivotal for the future engineering of $H_2$ases with higher activity as well as the design of biomimetic materials to simulate the functions of $H_2$ases. In particular, computational studies on the structures and electronic states of the H cluster, transition state, and energy barrier, as well as vibrational frequencies of characteristic ligands, are indispensable to gain insights into the mechanism of enzymatic H-H bond making/breaking. Among various feasible methodologies, the hybrid density functional theory (DFT), which takes the electron correlation into account but retains the computational efficiency, has shown its reliability and popularity in modeling metalloenzymes[55,56]. Up to now, several groups[21-23,25,30,39,50,57-63] have attempted to model the [NiFe]$H_2$ase and [FeFe]$H_2$ases with the hydrid DFT method. On the basis of the DFT optimal structures for [FeFe]$_H$, the high correlation between the computed IR frequencies of the $CO_b$, CO, and CN ligands, and the known IR frequencies of these ligands in the [FeFe]$_H$ of DdH[22,30], strong theoretical supports have been garnered for a low-charge and low-spin diiron active center as previously inferred from EPR, ENDO, and Mössbauer spectroscopy[26-29]. A mechanism for the $\eta^2$-$H_2$ activation by side-on coordination at a vacant site of the $Fe^I Fe^{II}$ active center and the heterolytic splitting of the H-H bond assisted by an adjacent PDT-thiolate-$\mu$-S was proposed by Cao and Hall[22]. The mechanism was later refined by replacing the PDT with a DTMA based on further theoretical analyses[23,30]. Bruschi et al. pointed out that $H_2$ can bind to $Fe_p$ and its activation, involving both iron atoms and one of the bridging sulfur ligands, is associated with a very low activation energy, leading to intermediate species characterized by a $\mu$-H atom[63].

In view of the recent progresses in experimental and theoretical studies on the enzymatic mechanism of [FeFe]$H_2$ases and the fact that all of the DFT studies on $H_2$ases catalysis so far have been focused on the H-H bond breaking, it is desirable to explore the enzymatic mechanism of the H-H bond making in [FeFe]$H_2$ases with computational tools. Our major interest lies in the evolution of the spin/charge state of the diiron [FeFe]$_H$ cluster as well as the rational pathways of PT and ET in the process of $H_2$ production.

## Computational Details

Throughout the paper, all of the calculations were performed using the DFT program DMol[3] from Cerius[2] and Materials Studio software package[64]. DMol[3] utilizes a basis set of numeric atomic functions, which are in general more complete than a comparable set of linearly independent Gaussian functions, as most often adopted, and have been demonstrated to have negligible basis-set superposition errors[64b].

The geometries of various models have been optimized by local DFT methods, specifically with the Perdew-Wang (1992) functional. The local spin-density approximation can be used to accurately predict structures, vibrations, and relative energies of covalent systems. Stationary points on a hypersurface have been located by means of energy-gradient techniques. In Dmol[3], the second derivatives are numerically computed by finite differences of the first derivatives. A full vibrational analysis has been performed to further characterize each stationary point. The Dmol[3] uses fast convergent three-dimensional numerical integrations to calculate matrix elements occurring in the Ritz variation method. As for the basis sets, we adopt the double numerical with d functions (DND) for all of the atoms. The size of the DND basis set is comparable to the standard Gaussian 6-31G* basis set. However, the numerical basis sets, which are exact solutions to the Kohn-Sham equations for the atoms[64b], are much more accurate than a Gaussian basis set of the same size. The search for a transition structure (TS) comprises repeated steps of computations with the linear synchronous transit (LST)[65] and quadratic synchronous transit(QST)[66] methods. The LST procedure roughly locates a maximum along a path connecting two structures and provides a guess for the TS connecting them. A subsequent cycle of QST refinements involving a Brent line searches along the QST line, followed by an energy minimization of the maximum structure perpendicular to the QST line, and finally determines the TS.

## Results and Discussion

**Chemical Models of $[FeFe]_H$ for CpI and DdH.** The H cluster in the $H_2$ases has been experimentally studied in three different forms(see Figure 1 in ref 40): (1)the partially oxidized form, where one of the irons, putatively $[Fe_d]$, is weakly bound by a water molecule; (2)the reduced form, where the $H_2O$ is released from $Fe_d$ and replaced by a H(or $H_2$)or a vacant coordination site on $Fe_d$, which is generally considered to be the primary catalytic center[15-20]; (3)the deactivated form, where a CO displaces the terminally bound $H_2O$ around $Fe_d$. In the current paper, we intend to computationally explore the states of the H-cluster during the reaction process with cluster models, with the environmental effect inexplicitly taken into account. The models were built up based on the X-ray crystallographically determined structural coordinates of CpI and DdH[15,17,19] and by taking in the knowledge from the synthetic model complexes. In the crystal structures, the $[FeFe]_H$ consists of two octahedrally liganded Fe atoms that bind $\mu$-SRS and a total of two $CN^-$ and three CO ligands, of which one may be $CO_b$. The terminal N of each $CN^-$ may form a H bond with the adjacent amino acid residues, and the CO ligands occupy hydrophobic pockets. However, the dicyano dianion in the synthetic model complexes is stabilized by quaternary ammonium cations as counterions[38,40]. Thus, in $[FeFe]$ $H_2$ases, a dicyano anionic $[FeFe]_H$ may have charge compensation by H bonding to adjacent cationic residues. In other words, the chemical model can be envisaged as an anion. Darensbourg et al. [38,40] have pointed out that the CO ligands in $[FeFe]_H$ can be exchanged and stabilized by $CN^-\cdots H^+$(here the dotted line denotes a H bond). Consequently, there would be one $CN^-$ to each $[Fe]$ in $[FeFe]_H$[18], and the model $\{^-(NC)(CO)([H]SCH_3)Fe_p[(CO_b)(\mu-SRS)]Fe_d(CO)(CN)^-L\}$ can serve as an initial chemical model for the catalytic center $[FeFe]_H$ in iron-only $H_2$ases. In the above model, L can be a ligand such as $CO, H_2O, H^-, H_2$, or simply a vacant coordination site on $Fe_d$. The $\mu$-SRS in DdH was assigned initially as PDT[17] and later as DTMA by Nicolet et al. [19] It has been found that $[Ni(Et_2PCH_2NMeCH_2PEt_2)_2]^{2+}$ or $[Ni(PNP)_2]^{2+}$ catalyzes the electrochemical oxidation of $H_2$ and reacts with $H_2$ to form complexes in which the hydride ligand is associated with nickel and the proton is bound by nitrogen. Rapid exchange occurs between these two sites[48]. However, although the diiron azadithiolates $Fe_2[(SCH_2)_2NR](CO)_6$ and dicyanide complex$(NEt_4)_2[Fe_2\{(SCH_2)_2NMe\}(CO)_4(CN)_2]$ have been prepared[46,47], evidence is elusive for the participation of the nitrogen atom of$(SCH_2)_2NR$ in the $H_2$-forming/breaking reaction. Importantly, it is experimentally verified that the hydride $[Fe^{II}-H-Fe^{II}]$ can be produced via the protonation of the $Fe^IFe^I$ model complexes such as $Fe_2(S_2C_3H_6)(CO)_4(PMe_3)_2$(here we denote this complex as **M1**; thus, the protonation process is **M1**$+H^+\rightarrow$**M1H**$^+$), and subsequently $H_2$ is produced by the reaction **M1H**$^+ +H^+ +2e\rightarrow$**M1**$+H_2$[41-44]. Therefore, in this paper, we start from model $[FeFe]_{H red}=$**1**$[-2,0]$(as shown in Figure 2)determined experimentally, where PDT is recommended[17]. To make the description concise throughout the paper, we use label $x[x_1,x_2]$ to represent a model, where $x$ refers to a serial number and $x_1$ and $x_2$ are the charge and spin states of the model, respectively.

**Mechanism of $H_2$ Generation in $[FeFe]_H$.** We investigate the reaction process of $2H^+ + 2e + [FeFe]_{H red}\rightarrow\cdots\rightarrow[FeFe]_{H red}+H_2$ starting with the model **1**$[-2,0]$ as shown in Figure 2. Obviously, as a catalyst, $[FeFe]_H$ should be invariant in the cycle of $H_2$ generation. We want to address the following

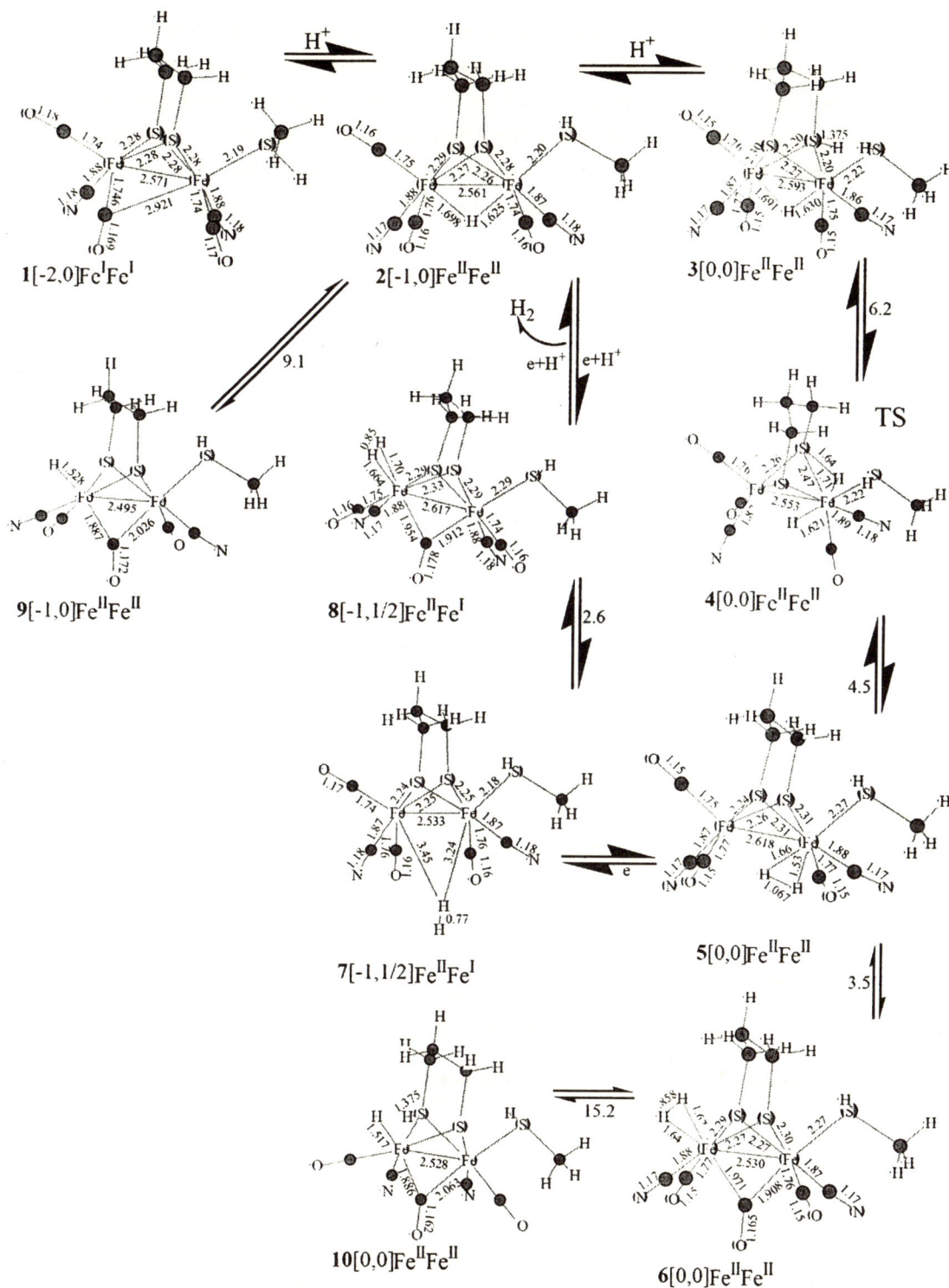

Figure 2   Optimal geometries of various probable models (1 − 10) in the process of $2H^+ + 2e \rightleftharpoons H_2$ at $[FeFe]_{H\,res}$.
The values on the arrowheads are the energies ( kcal/mol) at a temperature of 298 K. Highlighted
arrowheads represent the preferable paths.

issues: (i) What is the most likely structure of $[FeFe]_H$? (ii) What are the intermediates and reaction steps in the catalytic cycle? (iii) What insights can be garnered from the electronic structures of these redox states? First, excellent linear correlation between the computed and the observed CO frequencies for models **1—8** is found and shown in Figure 3[22,23]. This linear relationship is thus employed to predict the CO frequencies from calculated data. In fact, if we request the linear relationship to cross the origin, it would suggest a scaling factor of 0.9639 for the computed frequencies at the correlation factor $R^2 =$ 0.9863. The predicted frequencies are further compared with the observed bands from the experiment in Tables 1 and 2[19]. Table 1 listed the major structural parameters and stretching frequencies of $CO_b$ on **1** [−2,0]. The agreement between our data and others, particularly the experimental findings, legitimates **1**[−2,0] as a model for $[FeFe]_{H red}$. In Figure 2, the bond distances of Fe-CO and Fe-CN$^-$ are about 1.74—1.77 and 1.86—1.88 Å, respectively. The discrepancy between the ligands CO and CN$^-$ lies in the fact that CO is a better $\pi$-electron acceptor than CN$^{-[38]}$. Thus, the CO bond lengths in the models are a measurement of the $\pi$-donor ability or the redox state of the iron ion. For instance, the CO bond is longer in the Fe$^I$Fe$^I$ (**1**) and Fe$^{II}$Fe$^I$ (**7** and **8**) models than in the Fe$^{II}$Fe$^{II}$ models (**3—6**) and is longer in negative charge models **1,2,7,** and **8** than in neutral models **3—6**. Simultaneously, the CO bond lengths (in Figure 2) are closely correlated with the CO frequencies (in Table 2). In models **6** and **8**, the bond lengths of $CO_b$ are longer than those of $CO_t$; consequently, the $CO_b$ frequencies are lower than those of $CO_t$. Similar conclusions can be drawn by the comparison of models **3,5,** and **6** with models **2** and **8**.

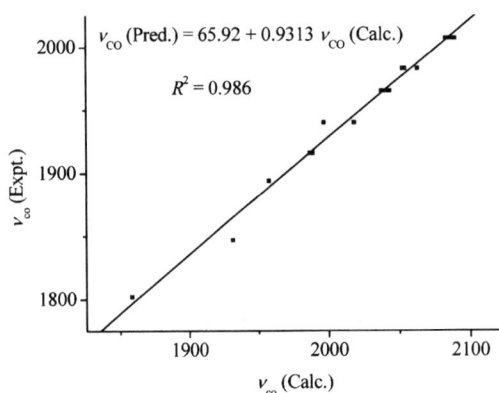

Figure 3    Linear fitting of the calculated CO stretching frequencies against the predicted data. The calculated CO stretching frequencies are taken from calculation results in models **1—3** and **5—8**.

Table 1    Comparison of Experimental Data and DFT-Optimized Results in the $CO_b$ Part of $[FeFe]_{H red}$

|  | model | | |
| --- | --- | --- | --- |
|  | a[a] | b[b] | **1**[−2,0] |
| $CO_b$-$Fe_p$ bond length(Å) | 2.40(2.56) |  | 2.921 |
| $CO_b$-$Fe_d$ bond length(Å) | 1.69(1.69) | 1.794 | 1.746 |
| $Fe_p$-$Fe_d$ bond length(Å) | 2.55(2.61) | 2.655 | 2.571 |
| C-O bond length(Å) |  | 1.185 | 1.169 |
| $v(CO_b)(cm^{-1})$ | 1894 | 1878 | 1889[c] |

[a] Experimental data are obtained from ref 19. [b] DFT calculation data are from model **26**[−3,0] in ref 22. [c] The predicted CO stretching frequency is calculated according to the linear-fitting formula in Figure 3.

**Table 2　Comparison of the Calculated, Predicted, and Observed CO Frequencies($cm^{-1}$)[a]**

| species | calculated | predicted[b] | observed |
|---------|-----------|--------------|----------|
| $1[-2,0]Fe^I Fe^I$ | *1957. 2* | *1889* | *1894*(b) |
| $2[0,0]Fe^{II} Fe^{II}$ | 1986. 7,1996. 7,2037. 2 | 1916,1926,1963 | 1916(b),1940(c,b),1965(c,b) |
| $3[0,0]Fe^{II} Fe^{II}$ | 2043. 0,2052. 2,2082. 5 | 1969,1977,2005 | 1965(c,b),1983(a),2007(a) |
| $5[0,0]Fe^{II} Fe^{II}$ | 2041. 9,2053. 2,2083. 9 | 1968,1978,2007 | 1965(c,b),1983(a),2007(a) |
| $6[0,0]Fe^{II} Fe^{II}$ | *1931. 8*,2062. 5,2087. 2 | *1865*,1987,2010 | *1847*(a),1983(a),2007(a) |
| $7[-1,1/2]Fe^{II} Fe^I$ | 2039. 4,2053. 6,2089. 1 | 1965,1979,2012 | 1965(b,c),1983(a),2007(a) |
| $8[-1,1/2]Fe^{II} Fe^I$ | *1858. 5*,1988. 7,2018. 0 | *1797*,1912,1945 | *1802*(c),1916(b),1940(b,c) |

[a] The observed data are taken from ref 19. The symbol in parentheses denotes the redox state of the species: (a) as isolated, $H_{ox}$, (b) after reduction, $H_{red}$, and (c) after reoxidation, $H_s$. The italicized data correspond to the $CO_b$ frequencies, and the others represent the $CO_t$ frequencies. [b] The predicted $v_{co}$ data are calculated according to the linear-fitting formula in Figure 3.

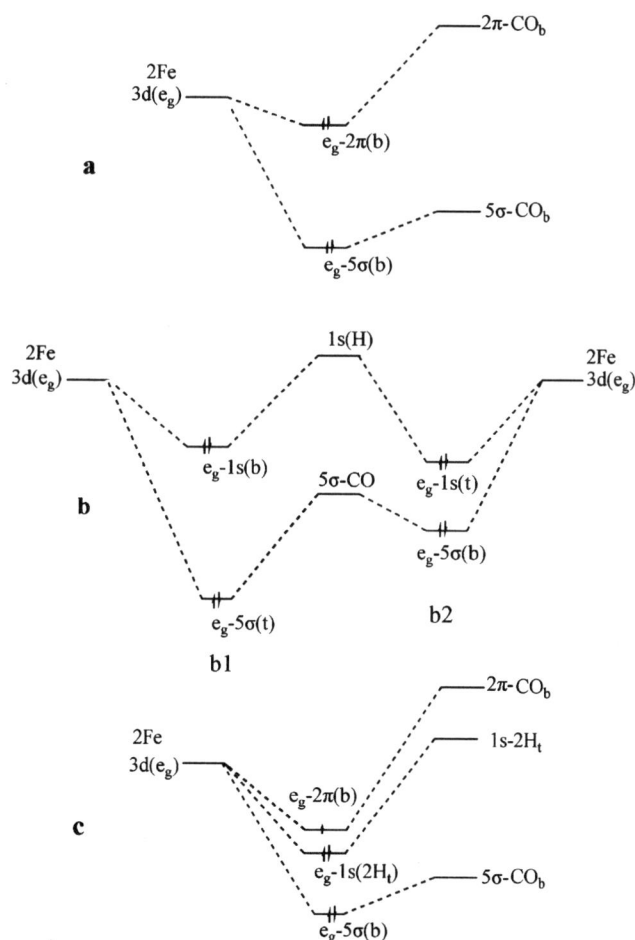

**Figure 4　Simplified molecular-orbital diagram showing the interaction among $CO_b$, $CO_t$, $H_b^-$, $H_t^-$, and Fe($3d$-$e_g$) (a) for model $1[-2,0]$; (b1) model $2[-1,0]$; (b2) model $9[-1,0]$; and (c) model $8[-1,^1/_2]$.**

Next, we attempt to build a flowchart for the formation of $H_2$ in $[FeFe]H_2$ases and investigate the evolution of the structures of $1[-2,0]$ combined with protons and electrons at various stages (Figure 2). Through the reaction of $1[-2,0]+H^+ \to 2[-1,0]$, the incoming proton takes the middle position between two irons, $Fe_p$ and $Fe_d$, and meanwhile, $CO_b$ is shifted to the $Fe_d$ side as a terminal $CO_t$ ligand. This process can be written as $H^+ + [Fe^I Fe^I](CO_b) \to [Fe^{II} Fe^{II}]H_b^- + CO_t$. Optimization results indicate that $2[-1,0]$ is energetically the most favorable state for the complex of system $1[-2,0]$ and $H^+$. In $2[-1,0]$, the bridging species is $H_b$, while the original $CO_b$ turns out to be a terminal $CO_t$. Population analysis demonstrated that the $H_b^-$ in $2[-1,0]$ essentially is $H_b^\delta$ (*Mulliken atomic charges* $\delta=0.143e$), and the bridging H is more like a hydrogen atom than an anion $H^-$. Thus, the electronic configuration of $[FeFe]$ in $2$ might be seen as a mixture of the three configurations $CO_b\text{-}H_{ox}$, $CO_b\text{-}H_s$, and $CO_b\text{-}H_{red}$. As a consequence, the $CO_t$ stretching frequencies (1916, 1926, and 1963 $cm^{-1}$) in $2$ can be seen as a mixture of 1916 $cm^{-1}$ ($H_s$) and 1940 and 1965 $cm^{-1}$ ($H_s$ and $H_{red}$) (see Table 2). There is another possible complex of $1[-2,0]$ plus $H^+$ where $CO_b$ is unchanged, but the incoming $H^+$ takes a terminal position (model $9[-1,0]$). However, the latter structure $9[-1,0]$ is unstable by 9.1 kcal/mol compared with $2[-1,0]$. The reaction $2[-1,0] \to 9[-1,0]$ occurs via ETs from $e_g-5\sigma(CO_t)$ and $e_g-1s(H_b)$ to $e_g-5\sigma(CO_b)$ and $e_g-1s(H_t)$ and is energetically unfavorable because the energy gap between $e_g-1s(H_t)$ and $e_g-1s(H_b)$ is smaller than that between $e_g-5\sigma(CO_t)$ and $e_g-5\sigma(CO_b)$ as shown in Figure 4b. As a matter of fact, these two cases are parallel to the structures $3[0,0]$ and $10[0,0]$ where the former is 13.4 kcal/mol more stable than the latter, as demonstrated in Figure 2. This implies

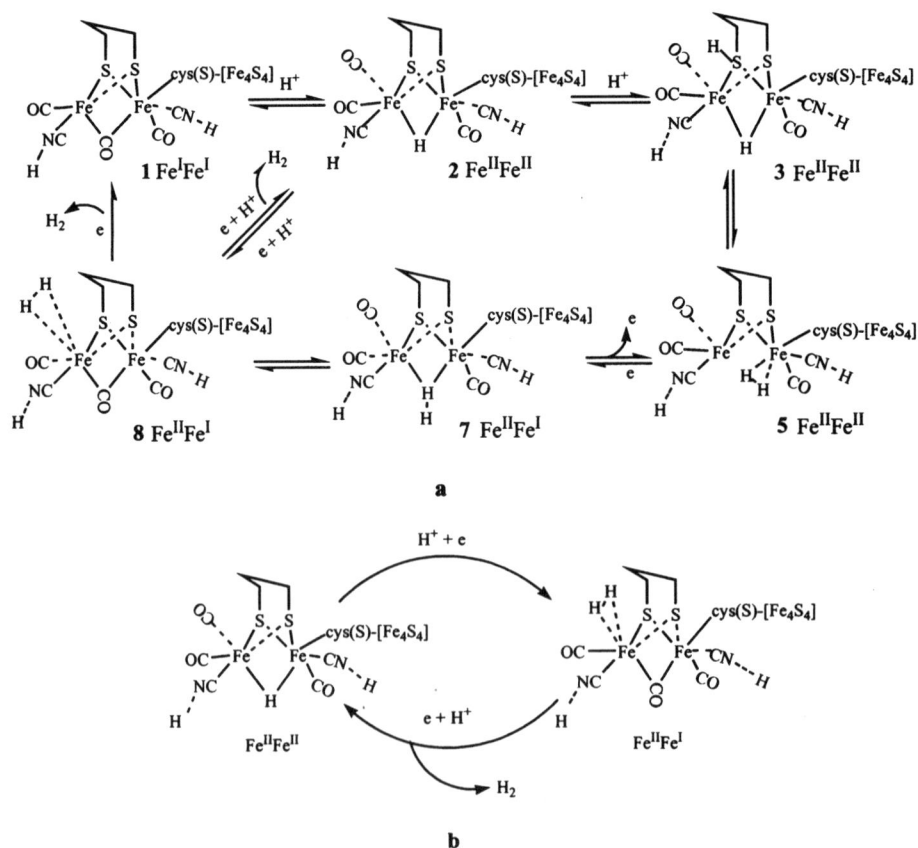

Figure. 5　(a) Catalytic cycle on $[FeFe]_H$ and (b) simplified key steps in the catalytic cycle.

that $H_t^-$ is less stable than $H_b^-$. Is there any possibility that model $2[-1,0]$ accepts an electron following the process $1[-2,0]+H^+ \rightarrow 2[-1,0]$? The answer might be no. This is due to the fact that all of the low-lying 3d bonding orbitals in $Fe_p$ and $Fe_d$ are almost fully occupied in $2[-1,0]$, and the energy level of LUMO is so high that the probability of the reaction of $2[-1,0]$ with another electron is very small. Therefore, $2[-1,0]$ is nucleophilic, and its protonation most possibly results in $3[0,0]$. Interestingly, although $\mu$-S is tetracoordinated in $3[0,0]$, the $(\mu$-S)-Fe bond remains strong; the bond lengths are basically unchanged. Because the negative charge of $H_b^{\delta-}$ is very small in $2[-1,0]$ or $3[0,0]$, $Hb^{\delta-}$ has a much smaller ion radius than the anion $H^-$. As a consequence, the relatively long distance of Fe-H(1.6 Å)implies that the $Fe_p \cdots H_b^{\delta-} \cdots Fe_d$ is a weak bond[67]. We further infer that the weakly bound $H^{\delta-}$ could attract the proton hanging on $\mu$-S. This results in the formation of $\eta^2$-$H_2$ at $Fe_p$ or model $5[0,0]$. The transformation from $3[0,0]$ to $5[0,0]$ needs to pass the transition state $4[0,0]$. The location of $4[0,0]$ shows that the dihydrogen formation or breakage on $Fe_p$ requires an activation energy of only 6.2—4.5 kcal/mol. Although the transformation from $5[0,0]$ to $6[0,0]$ is energetically favorable, it might be dynamically difficult because of the possible existence of nonclassical bonds $Fe_p$-$(\eta^2$-$H_2)$(model $5[0,0]$)and $Fe_d$-$(\eta^2$-$H_2)$(model $6[0,0]$), which are fairly stable in $[Fe^{II}Fe^{II}]$. The Fe-$(\eta^2$-$H_2)$ falls in the category of two-electrons-three-center(2e-3c) bonds[68]. Similar 2e-3c bonds have been found in $CH_5^+$[69,70], methane dehydrogenation in the absence of $O_2$[71], and the H-H activation by $[NiFe]H_2$ases[21,25,39,57,60]. When an electron enters $5[0,0]$ or $6[0,0]$, model $7[-1,^1/_2]$ or $8[-1,^1/_2]$ (active oxidized state $[FeFe]_{H_s}$)is generated. The comparison between Fe-$(\eta^2$-$H_2)$in $5[0,0]$ and $7[-1,^1/_2]$(or $6[0,0]$ and $8[-1,^1/_2]$)shows that the H-H distance in $7$(or $8$)becomes shorter than that in $5$(or $6$), while the distance between $\eta^2$-$H_2$ and $Fe_p$(or $Fe_d$)lengthens. This suggests that $Fe_p$-$(\eta^2$-$H_2)$ and $Fe_d-(\eta^2$-$H_2)$are more active in $[Fe^{II}Fe^I]$ than in $[Fe^{II}Fe^{II}]$, and ET is the driving force for the H-H bond making/breaking at the $Fe_p$(or $Fe_d$)active center. Therefore, the evolution from $7[-1,^1/_2]$ to $8[-1,^1/_2]$ is easier than that from $5[0,0]$ to $6[0,0]$. Because the structural data and predicted CO stretching frequencies listed in Figure 2 and Table 2 for models $2-8$ are in good agreement with the experimental findings, we propose that $2[-1,0]$, $7[-1,^1/_2]$, and $8[-1,^1/_2]$ would be the key structures in the catalytic cycle, whereas $3[0,0]$, $5[0,0]$, and $6[0,0]$ are possible intermediates of comparable stabilities. On the basis of our calculations and the above discussion, the mechanism of reversible $H_2$ oxidation is summarized in Figure 5. In particular, Figure 5b shows that an electron and a proton would simultaneously arrive at $[FeFe]_H$ to avoid costly charge separation.

Compared with the mechanism proposed by Hall and coworkers[22,23], our scheme highlights the importance of the hydride(Fe-$H_b$-Fe)in the reversible $H_2$ oxidation similar to the case of $[NiFe]$ $H_2$ase[21]. It is rational to compare the function of the cationic $[Ni]$ in $[NiFe]H_2$ases with the $[Fe_p]$ in $[FeFe]H_2$ases. The processes of dihydrogen production and dihydrogen uptake should be microscopically reversible. The existence of the hydride in $[FeFe]_H$ has been speculated[19]. Our mechanism also shows that a CO could rapidly alter from a bridging to an $Fe_d$ terminal position, as pointed out by Nicolet et al.[18,19] and Bruschi et al.[63]Cao and Hall suggested that the first step of the H-H bond breaking only removes a single proton to the neighboring base center $\mu$-S[22]. However, our computations reveal that the transformation from $6[0,0]$ to $10[0,0]$ is endothermic by 15.2 kcal/mol and thus does not seems energetically favorable, because the activation barrier must be higher than 15.2 kcal/mol.

In $[FeFe]_H$, both of the iron ions tend to satisfy the 18 electron rule. Thus, an Fe-Fe metal bond is formed if there is a surplus valence electron on each Fe ion in the valence state of $Fe^IFe^I$(model $1$). The short Fe-Fe distance indicates a strong metal-metal interaction. Liu and Hu[30] pointed out that the

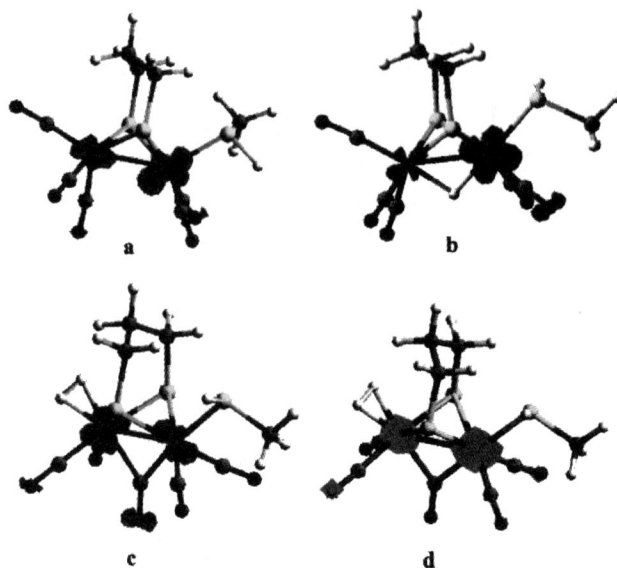

**Figure 6** Frontier molecular orbitals and the spin-density map: (a) HOMO of 1; (b) HOMO of 2; (c) HOMO of 6; (d) total spin density of 8.

interaction of $e_g(2Fe)$-$5\sigma(CO_b)$ and $e_g(Fe_d)$-$2\pi(CO_b)$ plays key roles in the stabilization of $[Fe^I Fe^I]CO_b$-$H_{red}$ (see Figure 4a). The full occupation of $e_g(Fe_d)$-$2\pi(CO_b)$ mainly strengthens the bonding between $CO_b$ and $Fe_d$ and leads to $CO_b$ shifting toward Fed in $1[-2, 0]$. The HOMO of model 1 is characterized by a large lobe underneath the Fe-$CO_b$-Fe frame representing the electron density of the metal-metal bond, as shown in part a of Figure 6. If $[FeFe]_H$ adopts an $Fe^{II} Fe^{II}$ valence state without surplus electrons, the metal-metal bond could not be formed, and the Fe-Fe distance ranges between 3.07 and 3.47 Å in some binuclear Fe(II) complexes[36]. However, we found that the Fe-Fe distances in models 2 and 6 of $Fe^{II} Fe^{II}$ remain short (2.55 — 2.62 Å). As depicted in parts b and c of Figure 5, the HOMOs of the models 2 and 6 involve $Fe(3d)$-$(H_b)$-$Fe(3d)$ and $Fe(3d)$-$2\pi(CO_b)$-$Fe(3d)$ interactions, respectively. This finding highlights the fact that the hydride and $CO_b$, which form a 2e-3c bond with two iron ions, are the key to stabilizing the Fe-Fe distance in $[Fe^{II} Fe^{II}]_H$. For $[Fe^{II} Fe^I]CO_b$-$H_s$ ($8[-1,^1/_2]$), the spin density is mainly located on the two iron ions as shown in Figure 6d (also see Figure 4c). This proves that $[Fe^{II} Fe^{II}]CO_b$-$H_{ox}$ has a low-lying unoccupied Fe-Fe front orbital and can be seen as an electron sink ($[Fe^{II} Fe^{II}]CO_b$-$H_{ox}$ + e → $[Fe^{II} Fe^I]CO_b$-$H_s$ + e → $[Fe^I Fe^I]CO_b$-$H_{red}$). However, $Fe_p$-$H_b^-$-$Fe_d$ is stable particularly in $[Fe^{II} Fe^{II}]$. Remarkable differences exist between the electronic configurations of $[Fe^{II} Fe^{II}]H_b^-$ and $[Fe^{II} Fe^{II}]CO_b$-$H_{ox}$.

## Conclusion

In this paper, we have explored the mechanism of the H-H bond making/breaking in the Fe-only hydrogenase. Model systems incorporating the chemical and biological characteristics of the enzyme are designed and investigated computationally at the DFT level with a basis set of the numeric atomic orbital. Results suggested that both $Fe_p$-$H_b^-$-$Fe_d$ and $Fe_d$-$(\eta^2$-$H_2)$ [also $Fe_p$-$(\eta^2$-$H_2)$] species exist in the catalytic cycle and may play important roles for the $H_2$ activation at the $[FeFe]_H$ active center. The electronic structures of $Fe_p$-$H_b^-$-$Fe_d$ and $Fe_p$-$CO_b$-$Fe_d$ are obviously different. While $Fe_p$-$H_b^-$-$Fe_d$ is stable in $[Fe^{II} Fe^{II}]$, $Fe_p$-$CO_b$-$Fe_d$ can exist in all of the three oxidation states ($H_{ox}$, $H_s$, and $H_{red}$). The

ET between $[Fe_4S_4]_H{}^{2+}$ and $[FeFe]_H$ would be the driving force for the H-H bond making/breaking. The $Fe^{II}Fe^{I}$ state is active for the dihydrogen reversible oxidation. Regarding the pathways of PT and ET, they are not necessarily the same but must be coupled to lower the overall energy barrier. The sequential routes of $H_2$ evolution with concerted PT and ET at the diiron active center are eventually presented and discussed.

# Acknowledgment

This work is supported by the National Science Foundation of China. We thank the reviewers for their very constructive suggestions to update the paper.

# Supporting Information Available

The optimized geometries of the models **1－10** in Cartesian coordinates. This material is available free of charge via the Internet at http://pubs. acs. org.

# References

[1]Cammack,R. Nature 1999,**397**,214.

[2]Collman,J. P. Nat. Struct. Biol. 1996,**3**,213.

[3]Holm,R. H. ;Kennepohl,P. ;Solomon,E. I. Chem. Rev. 1996,**96**,2239.

[4]Thauer,R. K. ;Klein,A. R. ;Hartmann,G. C. Chem. Rev. 1996,**96**,3031.

[5]Albracht,S. P. J. Biochim. Biophys. Acta 1994,**1118**,167.

[6]Frey,M. ChemBioChem 2002,**3**,153.

[7]Volbeda,A. ;Charon,M. -H. ;Plras,C. ;Hatchiklan,E. C. ;Frey,M. ;Fontecilla-Camps,J. C. Nature 1995,**373**,580.

[8]Volbeda,A. ;Garcin,E. ;Piras,C. ;de Lacey,A. L. ;Fernandez,V. M. ;Hatchikian,E. C. ;Frey, M. ;Fontecilla-Camps,J. C. J. Am. Chem. Soc. 1996,**118**,12989.

[9]Frey,M. Struct. Bonding 1998,**90**,97.

[10]Lenz,O. ;Friedrich,B. Proc. Natl. Acad. Sci. U. S. A. 1998,**98**,12474.

[11]Cammack,R. ,Reedijk,J. ,Ed. Bioinorganic Catalysis; Marcel Dekker:New York,1993;pp 189 －225.

[12]Sellmann,D. ;Geipel,F. ;Moll,M. Angew. Chem.,Int. Ed. 2000,**39**,561.

[13]Bertrand,P. ;Dole,F. ;Asso,M. ;Guigliarelli,B. J. Biol. Inorg. Chem. 2002,**5**,682.

[14]Fong,T. P. ;Forde,C. E. ;Lough,A. J. ;Morris,R. H. ;Rigo,P. ;Rocchini,E. ;Stephan,T. J. Chem. Soc.,Dalton Trans. 1999,4475.

[15]Peters,J. W. ;Lanzilotta,W. N. ;Lemon,B. J. ;Seefeldt,L. C. Science 1998,**282**,1853(Errata: 2002,**283**,35).

[16]Peters,J. W. Curr. Opin. Struct. Biol. 1999,**6**,670.

[17]Nicolet,Y. ;Piras,C. ;Legrand,P. ;Hatchikian,E. C. ;Fontecilla-Camps,J. C. Struct. Fold Des. 1999,**7**,13.

[18]Nicolet,Y. ;Lemon,B. J. ;Fontecilla-Camps,J. C. ;Peters,J. W. TIBS 2000,**25**,138.

[19]Nicolet,Y. ;Lacey,A. L. ;Vernede,X. M. ;Fernandez,V. M. ;Hatchikian,E. C. ;Fontecilla-

Camps, J. C. J. Am. Chem. Soc. 2001, **123**, 1596.

[20] Pierik, A. J. ; Hagen, W. R. ; Redeker, J. S. ; Wolbert, R. B. G. ; Boersma, M. ; Verhagen, M. F. J. M. ; Grande, H. J. ; Veeger, C. ; Mutsaers, P. H. A. ; Sand, R. H. ; Dunham, W. R. Eur. J. Biochem. 1992, **209**, 63.

[21] Niu, S. ; Thomson, L. M. ; Hall, M. B. J. Am. Chem. Soc. 1999, **121**, 4000.

[22] Cao, Z. ; Hall, M. B. J. Am. Chem. Soc. 2001, **123**, 3734.

[23] Fan, H. ; Hall, M. B. J. Am. Chem. Soc. 2001, **123**, 3828.

[24] Nicolet, Y. ; Cavazza, C. ; Fontecilla-Camps, J. C. J. Inorg. Biochem. 2002, **91**, 1.

[25] Pavlov, M. ; Siegbahn, P. E. M. ; Blomberg, M. R. A. ; Crabtree, R. H. J. Am. Chem. Soc. 1998, **120**, 548.

[26] Adams, M. W. W. Biochim. Biophys. Acta 1990, **1020**, 115.

[27] Adams, M. W. W. ; Stiefel, E. I. Curr. Opin. Chem. Biol. 2000, **4**, 214.

[28] Popescu, C. V. ; Munck, E. J. Am. Chem. Soc. 1999, **121**, 7877.

[29] Pereira, A. S. ; Tavares, P. ; Moura, I. ; Moura, J. G. ; Huynh, B. H. J. Am. Chem. Soc. 2001, **123**, 2771.

[30] Liu, Z. P. ; Hu, P. J. Am. Chem. Soc. 2002, **124**, 5175.

[31] (a) Razavet, M. ; Borg, S. J. ; George, S. J. ; Best, S. P. ; Fairhurst, S. A. ; Pickett, C. J. Chem. Commun. 2002, 700. (b) George, S. J. ; Cui, Z. ; Razavet, M. ; Pickett, C. J. Chem. ——Eur. J. 2002, **8**, 4037.

[32] Yang, X. ; Razavet, M. ; Wang, X. B. ; Pickett, C. J. ; Wang, L. S. J. Phys. Chem. A 2003, **107**, 4612.

[33] Erica, J. L. ; Georgakaki, I. P. ; Reibenspies, J. H. ; Darensbourg, M. Y. Angew. Chem. , Int. Ed. 1999, **38**, 3178.

[34] Alban, L. C. ; Stephen, P. B. ; Stacey, B. ; Sian, C. D. ; David, J. E. ; David, L. H. ; Christopher, J. P. Chem. Commun. 1999, 2285.

[35] Kaasjager, V. E. ; Henderson, R. K. ; Bouwman, E. ; Lutz, M. ; Spek, A. L. ; Reedijk, J. Angew. Chem. , Int. Ed. 1998, **37**, 1668.

[36] Liaw, W. F. ; Lee, N. H. ; Chen, C. H. ; Lee, C. M. ; Lee, G. H. ; Peng, S. M. J. Am. Chem. Soc. 2000, **122**, 488.

[37] Schmidt, M. ; Contakes, S. M. ; Rauchfuss, T. B. J. Am. Chem. Soc. 1999, **121**, 9736.

[38] Darensbourg, M. Y. ; Lyon, E. J. ; Smee, J. J. Coord. Chem. Rev. 2000, **206**, 533.

[39] Lawrence, J. D. ; Li, H. ; Rauchfuss, T. B. Chem. Commun. 2001, 1482.

[40] Lyon, E. J. ; Georgakaki, I. P. ; Reibenspies, J. H. ; Darensbourg, M. Y. J. Am. Chem. Soc. 2001, **123**, 3268.

[41] Gloaguen, F. ; Lawrence, J. D. ; Rauchfuss, T. B. J. Am. Chem. Soc. 2001, **123**, 9476.

[42] (a) Zhao, X. ; Georgakaki, I. P. ; Miller, M. L. ; Yarbrough, J. C. ; Darcensbourg, M. Y. J. Am. Chem. Soc. 2001, **123**, 9710. (b) Zhao, X. ; Georgakaki, I. P. ; Miller, M. L. ; Mejia-Rodriguez, R. ; Chiang, C. Y. ; Darcensbourg, M. Y. Inorg. Chem. 2002, **41**, 3917. (c) Zhao, X. ; Chiang, C. Y. ; Miller, M. L. ; Rampersad, M. V. ; Darcensbourg, M. Y. J. Am. Chem. Soc. 2003, **125**, 518.

[43] Nehring, J. L. ; Heinekey, D. M. Inorg. Chem. 2003, **42**, 4288.

[44] Gloaguen, F. ; Lawrence, J. D. ; Rauchfuss, T. B. ; Benard, M. ; Rohmer, M. -M. Inorg. Chem. 2002, **41**, 6573.

[45] Arabi, X. S. ; Matheu, R. ; Poilblanc, R. J. J. Organomet. Chem. 1979, **111**, 199.

[46] Lawrence, J. D. ; Li, H. ; Rauchfuss, T. B. ; Benard, M. ; Rohmer, M. -M. Angew. Chem. , Int.

Ed. 2001,**40**,1768.

[47]Li,H.；Rauchfuss,T. B. J. Am. Chem. Soc. 2001,**124**,726.

[48]Curtis,C. J.；Miedaner,A.；Ciancanelli,R.；Ellis,W. W.；Noll,B. C.；DuBois,M. R.；DuBois,
D. L. Inorg. Chem. 2003,**42**,216.

[49]Gloaguen,F.；Lawrence,J. D.；Schmidt,M.；Wilson,S. R.；Thomas,B.；Rauchfuss,T. B. J.
Am. Chem. Soc. 2001,**123**,12518.

[50]Georgakaki,I. P.；Thomson,L. M.；Lyon,E. J.；Hall,M. B.；Darensbourg,M. Y. Coord.
Chem. Rev. 2003,**238**,255.

[51]Lawrence,J. D.；Rauchfuss,T. B.；Wilson,S. R. Inorg. Chem. 2002,**41**,6193.

[52]Lawrence,J. D.；Rauchfuss,T. B.；Wilson,S. R. Inorg. Chem. 2002,**41**,6193.

[53]Darensbourg,M. Y.；Lyon,E. J.；Zhao,X.；Georgakaki,I. P. Proc. Natl. Acad. Sci. U. S. A.
2003,**100**,3683.

[54]Dole,F.；Fournel,A.；Magro,V.；Hatchikian,E. C.；Bertrand,P.；Guigliarelli,B. Biochemistry
1997,**36**,7847.

[55]Niu,S. Q.；Hall,M. B. Chem. Rev. 2000,**100**,353.

[56]Frenking,G.；Frohlich,N. Chem. Rev. 2000,**100**,717.

[57]Pavlov,M.；Blomberg,M. R. A.；Siegbahn,P. E. M. Int. J. Quantum Chem. 1999,**73**,197.

[58]Li,S.；Hall,M. B. Inorg. Chem. 2001,**40**,18.

[59]Siegbahn,P. E. M.；Eriksson,L.；Himo,F.；Pavlov,M. J. Phys. Chem. B 1998,**102**,10622.

[60]Siegbahn,P. E. M.；Margareta,R. A.；Blomberg,M. R. A.；Crabtree,P. R. H. J. Biol. Inorg.
Chem. 2001,**6**,460.

[61]De Gioia,L.；Fantucci,P.；Guigliarelli,B.；Bertrand,P. Int. J. Quantum Chem. 1999,**73**,187.

[62]De Gioia,L.；Fantucci,P.；Guigliarelli,B.；Bertrand,P. Inorg. Chem. 1999,**38**,2658.

[63](a)Bruschi,M.；Fantucci,P.；Gioia,L. Inorg. Chem. 2002,**41**,1421.

(b)Bruschi,M.；Fantucci,P.；Gioia,L. Inorg. Chem. 2003,**42**,4773.

[64](a)Delley,B. J. Chem. Phys. 1990,**92**,508. (b)Delley,B. In Density Functional Theory:A Tool
for Chemistry; Seminario, J. M., Politzer, P., Eds.；Elsevier：Amsterdam, The Netherlands,
1995.

[65]Halgren,T. A.；Lipscomb,W. N. Chem. Phys. Lett. 1977,**49**,225.

[66]Frisch, M. J.；Trucks, G. W.；Schlegel, H. B.；Gill, P. M. W.；Johnson,B. G.；Robb, M. A.；
Cheeseman, J. R.；Keith, T.；Petersson, G. A.；Montgomery, J. A.；Raghavachari, K.；Al-
Laham, M. A.；Zakrzewski, V. G.；Ortiz,J. V.；Foresman,J. B.；Peng, C. Y.；Ayala, P. Y.；
Chen,W.；Wong,M. W.；Andres,J. L.；Replogle,E. S.；Gomperts, R.；Martin, R. L.；Fox,D.
J.；Binkley,J. S.；Defrees, D. J.；Baker, J.；Stewart, J. P.；Head-Gordon, M.；Gonzalez, C.；
Pople,J. A. Gaussian 98,A. 3 ed.；Gaussian,Inc.：Pittsburgh,PA,1999.

[67]Bau,R.；Teller,R. G.；Kirtley,S. W.；Koetzle, T. F. Acc. Chem. Res. 1979,**12**,176.

[68]Kubas,G. J. Acc. Chem. Res. 1988,**21**,120.

[69]Scuseria,G. E. Nature 1993,**366**,512.

[70]Marx,D.；Parrinello,M. Nature 1995,**375**,216.

[71]Zhou,T.；Liu,A.；Mo,Y.；Zhang,H. J. Phys. Chem. A 2000,**104**,4505.

■ 本文原载:Journal of Inorganic Biochemistry 98(2004),pp. 1110~1116.

# Speciation and Transformation of Co(Ⅱ)/Ni(Ⅱ)-Citrate-Imidazole Ternary System-Synthesis, Spectroscopic and Structural Studies[*]

Yuan-Fu Deng, Zhao-Hui Zhou[①], Ze-Xing Cao, Khi-Rui Tsai

(Department of Chemistry, State Key Laboratory for Physical Chemistry
of Solid Surface, Xiamen University, Xiamen 361005, China )

Received 2 February 2004; received in revised form 19 March 2004;
accepted 23 March 2004 Available online 22 April 2004

**Abstract**   The cobalt and nickel imidazole citrate complexes [Co(Im)$_6$][Co (Im)$_3$(Hcit)] [Co (Im)$_3$(Hcit)] • 4H$_2$O (1) and [Ni(Im)$_6$][Ni (Im)$_3$ (Hcit)] [Ni (Im)$_3$ (Hcit)] • 4H$_2$O (2) (Im = imidazole, H$_4$cit = citric acid) were synthesized in aqueous solutions, which were isolated in pure crystalline forms and characterized spectroscopically. The X-ray structural analyses revealed that both compounds are isomorphous and consist of two types of metal ions in different coordination modes. One is octahedrally bound by six imidazole ligands, and the other is *quasi*-octahedrally coordinated by three imidazole and one citrate ligands, in which the citrate ion coordinates to the metal ions tridentately through α-hydroxyl, α-carboxyl and one of the β-carboxyl groups, leaving the other deprotonated β-carboxyl group free, which form intramolecular strong hydrogen bond with α-hydroxyl group. The co-existences of [M(Im)$_6$]$^{2+}$ and [M(Im)$_3$(Hcit)]$^-$ in the double salts(1) and(2)(M = Co$^{2+}$, Ni$^{2+}$) show an equilibrium between imidazole and imidazole-citrate species, which is further supported by the conversions of imidazole or citrate complexes to(1) and(2). The formations of the mixed ligand complexes present a case that the Co$^{2+}$ and Ni$^{2+}$ might interact competitively with monodentate imidazole and tridentate citrate ligands.

**Key words**   Cobalt   Nickel   Imidazole   Citrate

## 1   Introduction

The imidazole ring, as a histidine moiety, functions as a ligand towards transition metal ions in a number of biologically important molecules, e. g. one or more imidazole units are bound to metal ions in almost all copper-and zinc-metalloproteins or in nickel-containing urease, and thus have profound effects

---

* Supplementary data associated with this article can be found, in the online version, at doi:10. 1016/j. jinorgbio. 2004-03-009.

① Corresponding author. Tel. :+865922184531;fax:+865922183047.

*E-mail address*:zhzhou@xmu. edu. cn(Z. -H. Zhou).

on their biological actions[1-3]. Also,citric acid has diverse physiological roles in bacteria,as well as in higher organisms[4,5]. It is central to the citric acid cycle and forms complexes with many metal ions, which increases solubility and leads to enhanced bioavailability and subsequent absorption by biological issues. Such transition metal ions include manganese,iron,cobalt,nickel,copper and zinc et al. A wide variety of imidazole or citrate containing ligands have been investigated to mimic structural features of the relevant enzymes. Of these transition metal ions,cobalt(Ⅱ)ion is frequently used to substitute for zinc ion,and the cobalt(Ⅱ)-substituted enzymes often show about as much catalytic activity as the native zinc enzymes. This is a general characteristic since the coordination chemistry of cobalt(Ⅱ)is very similar to that of zinc(Ⅱ). Cobalt(Ⅱ)complexes with imidazole and carboxyl ligands have been studied as models for metalloproteins since they both contain functionalities in the side chain. The nickel(Ⅱ)ion is recognized as an essential trace element for bacteria, plants, animals and humans[6]. Nickel(Ⅱ)-imidazole and nickel(Ⅱ)-hydroxyl-carboxyl interactions are also importance in biological processes[7-10]. Apart from the important biological roles in the metalloproteins,imidazole and citrate, form a number of materials with interesting structural and magnetic properties[11-15].

Because of the co-existence of imidazole and citrate ligand in the biological system, like $nifV$ mutant iron-molybdenum cofactor in nitrogenase[16,17], and the limited structural information and the interactions of the model complexes containing the imidazole and citrate ligands, in our case, the behavior of cobalt(Ⅱ)and nickel(Ⅱ)ions in aqueous media,in the co-existence of imidazole and citrate ligands in the nearly physiological pH value range was investigated. Aimed at providing low-molecular-weight model complexes of cobalt(Ⅱ)and nickel(Ⅱ)ions containing the both biologically relevant ligands. Here, we report the syntheses, spectroscopic and structural characterizations as well as the transformations of cobalt and nickel imidazole citrate complexes: $[Co(Im)_6][Co_\Delta(Im)_3(Hcit)][Co_\Lambda(Im)_3(Hcit)] \cdot 4H_2O(1)$ and $[Ni(Im)_6][Ni_\Delta(Im)_3(Hcit)][Ni_\Lambda(Im)_3(Hcit)] \cdot 4H_2O(2)$ (Im= imidazole,$H_4$cit=citric acid).

# 2 Experimental

All experiments were carried out in open air. All chemicals were analytical reagents and used without further purification. $Co(Im)_6(NO_3)_2$, $Ni(Im)_6(NO_3)_2$, $(NH_4)_4[Co(Hcit)_2]$ and $(NH_4)_4[Ni(Hcit)_2] \cdot 2H_2O$ were prepared according to the procedures in the literatures[18-21]. Nanopure-quality water was used throughout this work. Infrared spectra were recorded as Nujol mulls between KBr plates using a Nicolet 360 FT-IR spectrometer. Electronic spectra were recorded on Cary 5000 UV-visible-NIR spectrophotometer. NMR spectra were recorded on a Varian UNITY 500 NMR spectrometer. Elemental analyses were performed using EA 1110 elemental analyzers.

## 2.1 Preparation of $[Co(Im)_6][Co_\Delta(Im)_3(Hcit)][Co_\Lambda(Im)_3(Hcit)] \cdot 4H_2O(1)$

$Co(NO_3)_2 \cdot 6H_2O(0.87 g,3 mmol)$and citric acid monohydrate(0.42 g,2 mmol)were dissolved in 5 mL of water with continuous stirring. The pH of the resulting solution was raised to ~7 with 2.0 M NaOH,and imidazole(0.82 g,12 mmol)was added to the above solution. The resulting solution was filtered and kept at room temperature for several days. Red crystals of$[Co(Im)_6][Co_\Delta(Im)_3(Hcit)][Co_\Lambda(Im)_3(Hcit)] \cdot 4H_2O(1)$were collected,washed with water and air-dried to give 1.16 g of product (yield 80%). Anal. found(Calc.)for(1):C,40.4%(39.9%);H,4.4%(4.6%).

## 2.2 Preparation of $[Ni(Im)_6][Ni_\Delta(Im)_3(Hcit)][Ni_\Lambda(Im)_3(Hcit)] \cdot 4H_2O(2)$

In a similar procedure as 2.1,$Ni(NO_3)_2 \cdot 6H_2O(0.87 g,3 mmol)$and citric acid monohydrate(0.42

g,2 mmol)were dissolved in 5 mL of water with continuous stirring. The pH of the solution was raised to ~7 with 2.0 M NaOH, and imidazole(0.82 g,12 mmol) was added to the above solution and blue precipitates deposited. The precipitates were heated at 40 ℃ with continuously stirring for half an hour and the precipitates redissolved. The solution was filtered and kept at room temperature for 2 weeks, blue crystals of $[Ni(Im)_6][Ni_\Delta(Im)_3(Hcit)][Ni_\Lambda(Im)_3(Hcit)] \cdot 4H_2O(\mathbf{2})$ were separated. These were collected, washed with water and air-dried to give 1.01 g of product(yield 70%). Anal. found(Calc.)for (**2**):C,40.5%(39.9%);H,4.5%(4.6%).

## 2.3 Transformation of $[Co(Im)_6](NO_3)_2$ to (1)

$[Co(Im)_6](NO_3)_2$(1.18 g,2 mmol)[18] was dissolved in 10 mL of water at 60 ℃ with continuous stirring, and $Na_3Hcit \cdot 2H_2O$(0.59 g,2 mmol) was added to the solution. The mixture was filtered and kept at room temperature for several days, red crystals of(**1**) were precipitated. These were collected, washed with water and air-dried to give 0.78 g of product(yield 80%). Positive identification was provided via IR spectrum.

## 2.4 Transformation of $[Ni(Im)_6](NO_3)_2$ to (2)

In a similar procedure as 2.3, $[Ni(Im)_6](NO_3)_2$[19] is converted into(**2**)with 75% yield. The products were characterized with IR spectrum.

## 2.5 Transformation of $(NH_4)_4[Co(Hcit)_2]$ to(1)

$(NH_4)_4[Co(Hcit)_2]$[20](1.53 g,3 mmol)was dissolved in 10 mL of water at 60 ℃ with continuous stirring, and imidazole(0.82 g,12 mmol)was added to the solution. The solution was filtered and kept at room temperature for a week, red crystals of(**1**)were separated, which were collected, washed with water and air-dried to give 0.73 g of product(yield 50%). Positive identification was provided via IR spectrum.

## 2.6 Transformation of $(NH_4)_4[Ni(Hcit)_2] \cdot 2H_2O$ to (2)

In a similar procedure as 2.5,$(NH_4)_4[Ni(Hcit)_2] \cdot 2H_2O$[21] is converted into(**2**)with 55% yield. The products were characterized with IR spectrum.

## 2.7 Transformation of (1) to $(NH_4)_4[Co(Hcit)_2]$

$[Co(Im)_6][Co_\Delta(Im)_3(Hcit)][Co_\Lambda(Im)_3(Hcit)] \cdot 4H_2O(\mathbf{1})$(2.89 g,2 mmol) was dissolved in 15 mL of water at 60 ℃ with continuous stirring,and $(NH_4)_3Hcit$(1.70 g,14 mmol) was added to the solution. The mixture was filtered and kept at room temperature for a week, red crystals of$(NH_4)_4[Co(Hcit)_2]$[20] were separated, which were collected, washed with water and air-dried to give 1.52 g of product(yield 48%). Positive identification was provided via IR spectrum.

## 2.8 Transformation of (2) to $(NH_4)_4[Ni(Hcit)_2] \cdot 2H_2O$

In a similar procedure as 2.7, imidazole citrato nickel(**2**)is converted into $(NH_4)_4[Ni(Hcit)_2] \cdot 2H_2O$[21] in 43% yield. The products were characterized with IR spectrum.

## 2.9 X-ray crystallographic study

X-ray diffraction data of(**1**)and(**2**)were collected on a Bruker Smart CCD diffractomerter with graphite monochromated Mo Kα(λ＝0.7107 Å) at 296(2)K. Empirical absorption corrections were applied using the SADABS program. All calculations were performed using the SHELXL-97 program package[22]. The structures were solved by direct methods and refined by full-matrix least-squares. All non-hydrogen atoms were refined anisotropically, while hydrogen atoms were generated geometrically or located from differential Fourier maps and refined isotropically. The crystallographic data of(1)and(2) are summarized in Table 1. Selected distances and bond angles are listed in Table 2.

**Table 1   Crystal data summaries of intensity data collection and structure refinement for(1)and(2)**

| Empirical formula | $C_{48}H_{66}O_{18}N_{24}Co_3$ (1) | $C_{48}H_{66}O_{18}N_{24}Ni_3$ (2) |
|---|---|---|
| Formula mass | 1444.04 | 1443.38 |
| Crystal color | Red | Blue |
| Crystal dimmers(mm) | 0.37×0.23×0.12 | 0.24×0.19×0.16 |
| Crystal system | Triclinic | Triclinic |
| Cell constants | $a$＝8.4123(4)Å | $a$＝8.3832(4)Å |
| | $b$＝9.3608(5)Å | $b$＝9.3096(4)Å |
| | $c$＝20.193(1)Å | $c$＝20.1881(9)Å |
| | $\alpha$＝91.858(1)° | $\alpha$＝91.929(1)° |
| | $\beta$＝91.835(1)° | $\beta$＝91.871(1)° |
| | $\gamma$＝92.987(1)° | $\gamma$＝92.902(1)° |
| | $V$＝1586.2(1)Å³ | $V$＝1571.7(1)Å³ |
| Space group | $P\bar{1}$ | $P\bar{1}$ |
| Formula units/unit cell | 1 | 1 |
| $D_{calc}$(gcm$^{-3}$) | 1.512 | 1.525 |
| $F_{000}$ | 747 | 750 |
| Diffractometer | Smart Apex CCD | |
| Radiation | Mo Kα (λ＝0.7107 Å) | |
| Temperature(℃) | 23 | |
| Reflections collected/ unique | 18,451/7317 [$R$(int)＝0.0903] | 18,297/7262 [$R$(int)＝0.0680] |
| Data/restraints/ parameters | 7309/7/436 | 7262/7/441 |
| $\theta$ range(°) | 1.01－28.31 | 1.01－28.32 |
| GOF on $F^2$ | 1.100 | 1.025 |
| $R_1$,$wR_2$[$I>2\sigma(I)$] | 0.038,0.105 | 0.042,0.096 |
| $R_1$,$wR_2$(all data) | 0.042,0.108 | 0.052,0.101 |
| Largest difference peak and hole(e Å$^{-3}$) | 0.446,－0.371 | 0.419,－0.388 |

**Table 2 Selected bond distances(Å) and angles(°) for(1) and(2)**

| Complex(**1**) | | Complex(**2**) | |
|---|---|---|---|
| Co(1)—N(1) | 2.162(1) | Ni(1)—N(1) | 2.118(2) |
| Co(1)—N(3) | 2.154(1) | Ni(1)—N(3) | 2.112(2) |
| Co(1)—N(5) | 2.162(2) | Ni(1)—N(5) | 2.121(2) |
| Co(2)—O(1) | 2.162(1) | Ni(2)—O(1) | 2.114(1) |
| Co(2)—O(2) | 2.085(1) | Ni(2)—O(2) | 2.061(1) |
| Co(2)—O(4) | 2.146(1) | Ni(2)—O(4) | 2.119(1) |
| Co(2)—N(7) | 2.134(1) | Ni(2)—N(7) | 2.088(2) |
| Co(2)—N(9) | 2.093(1) | Ni(2)—N(9) | 2.049(2) |
| Co(2)—N(11) | 2.102(2) | Ni(2)—N(11) | 2.061(2) |
| N(1)—Co(1)—N(3) | 90.41(5) | N(1)—Ni(1)—N(3) | 90.72(7) |
| N(1)—Co(1)—N(5) | 88.80(5) | N(1)—Ni(1)—N(5) | 88.83(7) |
| N(1)—Co(1)—N(1a) | 180.00(7) | N(1)—Ni(1)—N(1a) | 180.00(8) |
| N(1a)—Co(1)—N(3) | 89.59(5) | N(1a)—Ni(1)—N(3) | 89.28(7) |
| N(1a)—Co(1)—N(5) | 91.20(5) | N(1a)—Ni(1)—N(5) | 91.17(7) |
| N(3)—Co(1)—N(5) | 89.15(6) | N(3)—Ni(1)—N(5) | 88.83(7) |
| N(3)—Co(1)—N(3a) | 180.00(7) | N(3)—Ni(1)—N(3a) | 180.00(5) |
| N(3)—Co(1)—N(5a) | 90.85(6) | N(3)—Ni(1)—N(5a) | 91.17(7) |
| N(5)—Co(1)—N(5a) | 180.00(1) | N(5)—Ni(1)—N(5a) | 180.00(9) |
| O(1)—Co(2)—O(2) | 76.16(4) | O(1)—Ni(2)—O(2) | 77.48(5) |
| O(1)—Co(2)—O(4) | 84.01(4) | O(1)—Ni(2)—O(4) | 85.44(5) |
| O(1)—Co(2)—N(7) | 89.77(5) | O(1)—Ni(2)—N(7) | 89.68(6) |
| O(1)—Co(2)—N(9) | 170.18(5) | O(1)—Ni(2)—N(9) | 170.33(7) |
| O(1)—Co(2)—N(11) | 93.08(5) | O(1)—Ni(2)—N(11) | 93.17(6) |
| O(2)—Co(2)—O(4) | 86.66(5) | O(2)—Ni(2)—O(4) | 86.64(6) |
| O(2)—Co(2)—N(7) | 92.44(4) | O(2)—Ni(2)—N(7) | 92.61(6) |
| O(2)—Co(2)—N(9) | 94.72(5) | O(2)—Ni(2)—N(9) | 92.89(7) |
| O(2)—Co(2)—N(11) | 167.51(5) | O(2)—Ni(2)—N(11) | 168.56(6) |
| O(4)—Co(2)—N(7) | 173.75(5) | O(4)—Ni(2)—N(7) | 175.11(6) |
| O(4)—Co(2)—N(9) | 94.72(5) | O(4)—Ni(2)—N(9) | 93.17(6) |
| O(4)—Co(2)—N(11) | 85.90(5) | O(4)—Ni(2)—N(11) | 86.12(6) |
| N(7)—Co(2)—N(9) | 91.51(6) | N(7)—Ni(2)—N(9) | 91.69(7) |
| N(7)—Co(2)—N(11) | 93.86(6) | N(7)—Ni(2)—N(11) | 93.85(7) |
| N(9)—Co(2)—N(11) | 96.54(6) | N(9)—Ni(2)—N(11) | 96.35(7) |

[a]Symmetry transformations: $-x, -y+1; -z$.

# 3　Results and discussion

## 3.1　Synthesis

In an expedient synthetic procedure, $M(NO_3)_2 \cdot 6H_2O(M=Co, Ni)$, citric acid and imidazole were reacted in water solution with molar ratio of 3 : 2 : 12 at pH $\sim 7$. The preparations of the compounds (1)and(2)depend mainly on the pH of the solutions. In lower or higher pH(pH $<5$ or pH $> 9$),[M (Im)$_6$]$^{2+}$ (M=Co, Ni)complexes were isolated instead of the double salts(1)and(2). The effects of reaction temperature do not seem so crucial for the formation mixed-ligand products.

Previous solution studies[23,24] have suggested that [M(Hcit)]$^-$ species predominate at equimolor metal salt and citric acid with pH $> 5$. If citric acid is in excess,[M(Hcit)$_2$]$^{4-}$ species(M=Co, Ni)are the major components in the pH range $5-7$. The isolations and structural characterizations of the cobalt-citrate[20,25] and nickel-citrate[21,26] complexes in the corresponding pH range are in accord with the solution studies. In our cases,complexes(1)and(2)were isolated in the $M^{2+}$ (M=Co, Ni)-imidazole-citrate triad systems respectively. The complexes(1)and(2)can be also obtained from the reaction of [M (Im)$_6$](NO$_3$)$_2$ with citrate,or from the reaction of(NH$_4$)$_4$[M(Hcit)$_2$] with imidazole(M=Co, Ni). The competitive reactions of the $M^{2+}$ ions with imidazole and citrate in the $M^{2+}$ ions-imidazole-citrate triad systems are shown in Scheme 1,which suggests that mixed ligand complexes(1)and(2)are the dominant and the most stable species in the triad systems. The crystals of(1)and(2)were stable in air for a long time. Both complexes were soluble in aqueous solution at pH $\sim 7$,especially in hot water. NMR experiments show broad bands due to the paramagnetic effect of the metal ions.

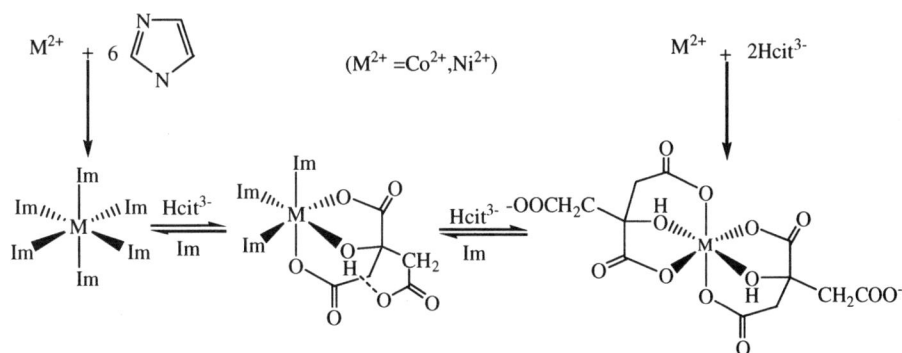

Scheme 1　Syntheses and transformation of Co( II )/Ni( II )-citrate-imidazole complexes.

## 3.2　Crystal structures

Complex(1)comprises one [Co(Im)$_6$]$^{2+}$ cation,one mesomeric[Co(Im)$_3$(Hcit)]$^-$ anions and four water molecules. Figs. 1 and 2 show the plots of [Co(Im)$_6$]$^{2+}$ cation,and the enantiomers of citrato imidazole cobalt( II )anions,[Co$_\Delta$(Im)$_3$(Hcit)]$^-$ and [Co$_\Lambda$(Im)$_3$(Hcit)]$^-$. The three cobalt( II )ions coordinated octahedrally in two different modes. The Co1 lies on a crystallographic symmetry center and Co2 occupies a general position. The triionized citrate ion serves as a tridenate ligand and coordinates with Co2 ion via its $\alpha$-hydroxyl,$\alpha$-carboxyl and one of the $\beta$-carboxyl groups,while the other $\beta$-carboxyl group does not participate in the coordination,dangling away from the complex. This typical coordination mode of citrate has been previously observed in the metal citrate complexes, such as

$(NH_4)_4[Co(Hcit)_2]^{[20]}$, $(NH_4)_4[Ni(Hcit)_2] \cdot 2H_2O^{[21]}$, $(NH_4)_4[Mn(Hcit)_2]^{[27]}$ and $(NH_4)_4[Zn(Hcit)_2]^{[28]}$. The six apices around the Co1 ion are completed by six imidazole molecules, and the geometry of Co1 ion is a normal octahedron. The six O and N apices around the $CO_2$ ion consist of three *fac*-oxygen atoms from the citrate ligand, and the remaining three points are occupied by nitrogen atoms from three imidazole ligands. The equatorial plane is defined by two oxygen atoms corresponding to the α-hydroxyl and α-carboxyl oxygen atoms (O1 and O2) and two nitrogen atoms (N9 and N11) coming from two imidazole ligands. None of the four atoms deviated more than 0.05 Å from the mean plane. The axial positions are occupied with the β-carboxyl oxygen atom (O4) and the other nitrogen atom (N7) of imidazole.

Nickel complex(2) is isostructural to cobalt complex(1), and the crystal structure of(2)comprises a discrete$[Ni(Im)_6]^{2+}$ cation, a mesomeric $[Ni(Im)_3(Hcit)]^-$ anion and four water molecules. The structures of $[Ni(Im)_6]^{2+}$ cation and mesomeric $[Ni(Im)_3(Hcit)]^-$ anion are in the same patterns like cobalt complex(1)in Figs. S1 and S2.

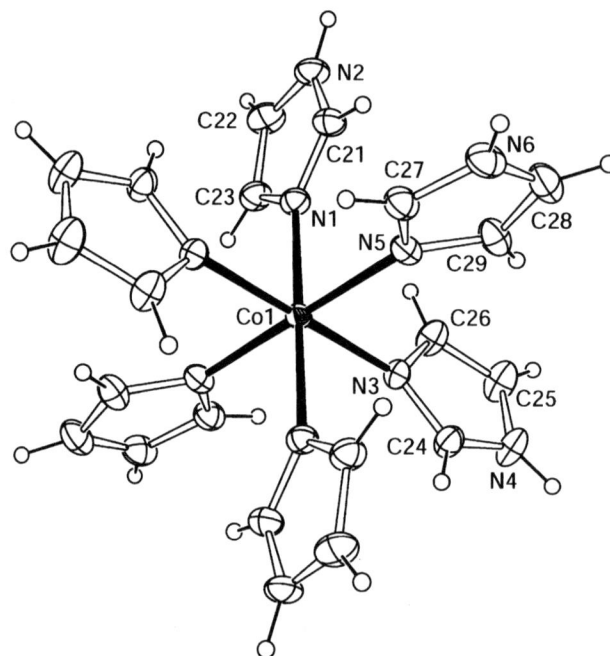

Fig. 1　Perspective view of the cation structure of $[Co(Im)_6]^{2+}$, showing the 30% thermal ellipsoids.

There is an extensive network of hydrogen-bonding interactions in the crystal structure of(1)and (2). The main interactions are given in Table 1. The most important intramolecular H-bonds in the two complexes are the strong hydrogen bonds of the α-hydroxyl to the deprotonated uncoordinated β-carboxyl moiety [O1···O7 = 2.635(2)Å for(1)and O1···O7 = 2.630(2)Å for(2)]. These are different from the intermolecular hydrogen bondings existing in similar citrato complexes, such as$(NH_4)_4[Co(Hcit)_2]^{[20]}$, $(NH_4)_4[Ni(Hcit)_2] \cdot 2H_2O^{[21]}$, $(NH_4)_4[Mn(Hcit)_2]^{[27]}$ and$(NH_4)_4[Zn(Hcit)_2]^{[28]}$. The hydrogen-bonding network is further extended by the involvement of the water molecules and the uncoordinated imidazole nitrogen atoms, which interact with both the coordinated and free carboxyl moieties, forming an extensive hydrogen bonds network. The assembly of hydrogen bonds is likely to be an important factor in the overall stability of the complexes. The long distance[O7···N7 = 3.724(2)Å for(1)and O7···N7 = 3.669(2)Å for(2)] between β-carboxyl group of citrate ligand and imidazole group shows no existence of intramolecular hydrogen bond in imidazole complexes. Based on the spacial consideration,

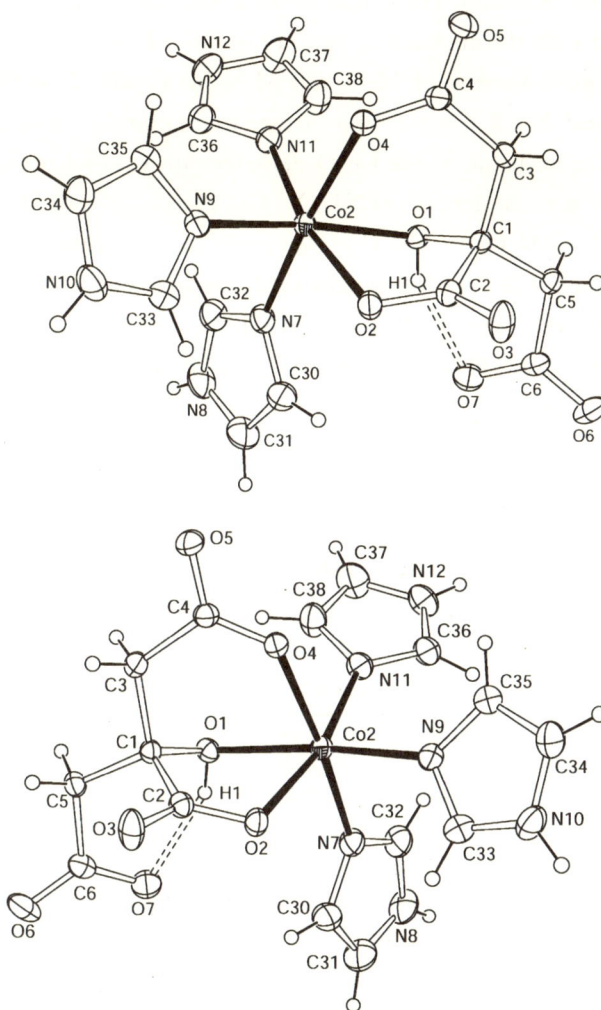

Fig. 2   Perspective view of the anion mesomeric structure of [Co$_\Delta$(Im)$_3$(Hcit)]$^-$(up)and[Co$_\Lambda$(Im)$_3$(Hcit)]$^-$(down),
showing the 30% thermal ellipsoids.

Table 3   Selected hydrogen bondings in(1)and(2)

| Interaction | H···A(Å) | D···A(Å) | D-H-A(°) | Symmetry transformations |
|---|---|---|---|---|
| Complex(**1**) | | | | |
| O1-H1···O7 | 1.91(2) | 2.635(2) | 142(2) | $x,y,z$ |
| O1W-H1W1···O4 | 1.917(9) | 2.762(2) | 176(2) | $x;y,z$ |
| O1W-H1W2···O5 | 1.85(1) | 2.664(2) | 164(2) | $-x,-y,1-z$ |
| O2W-H2W1···O5 | 1.926(9) | 2.782(2) | 178(2) | $-1+x,y,z$ |
| O2W-H2W2···O2 | 2.13(1) | 2.946(2) | 163(2) | $x,y,z$ |
| Complex(**2**) | | | | |
| O1-H1···O7 | 1.91(2) | 2.630(2) | 144(2) | $x,y,z$ |
| O1W-H1W1···O4 | 1.931(9) | 2.772(2) | 178(2) | $x,y,z$ |
| O1W-H1W2···O5 | 1.84(1) | 2.663(2) | 168(2) | $-x,-y,1-z$ |
| O2W-H2W1···O5 | 1.94(1) | 2.789(2) | 176(3) | $-1+x,y,z$ |
| O2W-H2W2···O2 | 2.10(1) | 2.940(2) | 170(3) | $x,y,z$ |

this supports the model that *R*-homocitrate ligand with one more methylene group might form an intramolecular hydrogen bond with the imidazole group in histidine of the FeMo-cofactor in nitrogenase[29,30].

Comparisons of the bond distances of M-O and M-N in(**1**)and(**2**)with corresponding structural parameters in the Co(Ⅱ)-and Ni(Ⅱ)-containing citrate or imidazole complexes are given in Tables 4 and 5,respectively. As shown in Table 4,the Co-O distances in(**1**)lies between 2. 085(1)and 2. 162(2) Å,which are comparable with those found in(NH$_4$)$_4$[Co(Hcit)$_2$][20],K$_2$[Co(Hcit)(H$_2$O)$_2$]$_2$ • 6H$_2$O and Na$_2$[Co(Hcit)(H$_2$O)$_2$]$_2$ • 6H$_2$O[25]. The Co-O distance of the α-hydroxyl group in(**1**)is the longest in the citrato cobalt complexes. The Ni-O distances in(**2**)are between 2. 061(1)and 2. 119(1)Å,similar distances of the Ni-O bonds were observed in K$_2$[Ni(Hcit)(H$_2$O)$_2$]$_2$ • 4H$_2$O[26],(NH$_4$)$_2$[Ni(Hcit) (H$_2$O)$_2$]$_2$ • 2H$_2$O and(NH$_4$)$_4$[Ni(Hcit)$_2$] • 2H$_2$O [21]. The Ni-O distance of α-carboxyl in(**2**)is the shortest in the citrato nickel complexes. The longer Co-O and Ni-O(hydroxyl)bonds[2. 162(1)Å for(**1**) and 2. 114(1)Å for(**2**)]indicate the protonation of hydroxyl group in citrate anion,which is different from the bridged α-alkoxyl group coordination in cobalt or nickel cluster citrate complexes, such as [(NMe$_4$)$_3$Na{Co$_4$(cit)$_4$[Co(H$_2$O)$_5$]$_2$}] • 11H$_2$O[11],(NMe$_4$)$_5$[Ni$_4$(cit)$_3$(OH)(H$_2$O)] • 18H$_2$O[31], (NMe$_4$)$_{14}$Na$_2$[Ni$_{21}$(cit)$_{12}$(OH)$_{10}$(H$_2$O)$_{10}$],(NMe$_4$)$_{16}$-[Ni$_{21}$(cit)$_{12}$(OH)$_{10}$(H$_2$O)$_{10}$][12],(NMe$_4$)$_8$Na$_8$ [Ni$_{21}$(cit)$_{12}$(OH)$_{10}$(H$_2$O)$_{10}$],(NMe$_4$)$_5$Na$_5$[Ni$_7$(cit)$_6$(H$_2$O)$_2$][12,13],(NMe$_4$)$_{10}$[Ni$_8$(cit)$_6$(OH)$_2$ (H$_2$O)$_2$]and(NMe$_4$)$_{10}$[Ni$_8$(cit)$_6$(OH)$_2$][14].

**Table 4    Bond distances of M(Ⅱ)-O bonds in M(Ⅱ)-citrate complexes(M=Co$^{2+}$,Ni$^{2+}$)**

| Complex | Co-O(Å)(hydroxyl) | Co-O(Å)(α-carboxyl) | Co-O(Å)(β-carboxyl) | Ref. |
|---|---|---|---|---|
| (NH$_4$)$_4$[Co(Hcit)$_2$] | 2. 157(2) | 2. 075(2) | 2. 051(2) | [20] |
| K$_2$[Co(Hcit)(H$_2$O)$_2$]$_2$ • 6H$_2$O | 2. 169(2) | 2. 044(2) | 2. 069(2) | [25] |
| Na$_2$[Co(Hcit)(H$_2$O)$_2$]$_2$ • 6H$_2$O | 2. 194(2) | 2. 065(2) | 2. 042(2) | [25] |
| [Co(Im)$_6$][Co(Im)$_3$(Hcit)]$_2$ • 4H$_2$O(**1**) | 2. 162(1) | 2. 085(1) | 2. 146(1) | This work |
| | Ni-O(Å)(hydroxyl) | Ni-O(Å)(α-carboxyl) | Ni-O(Å)(β-carboxyl) | Ref. |
| (NH$_4$)$_4$[Ni(Hcit)$_2$] • 2H$_2$O | 2. 021(3) | 2. 038(3) | 2. 072(3) | [21] |
| (NH$_4$)$_2$[Ni(Hcit)(H$_2$O)$_2$]$_2$ • 2H$_2$O | 2. 074(2) | 2. 020(2) | 2. 031(2) | [21] |
| K$_2$[Ni(Hcit)(H$_2$O)$_2$]$_2$ • 6H$_2$O | 2. 125(3) | 2. 054(3) | 2. 125(3) | [26] |
| [Ni(Im)$_6$][Ni(Im)$_3$(Hcit)]$_2$ • 4H$_2$O(**2**) | 2. 114(1) | 2. 061(1) | 2. 119(1) | This work |

**Table 5    Bond distances of M(Ⅱ)-N bonds in M(Ⅱ)-imidazole complexes(M=Co$^{2+}$,Ni$^{2+}$)**

| Complex | Co-N(Å) | Ref. |
|---|---|---|
| [Co(Im)$_6$](Bz)$_2$ | 2. 160−2. 186(1) | [32] |
| [Co(Im)$_6$](Mz)2 | 2. 169−2. 184(1) | [32] |
| [Co(Im)$_2$(Ac)$_2$] | 2. 013−2. 014(1) | [33,35,36] |
| [Co(Im)$_2$(C$_2$H$_5$COO)$_2$] | 2. 023−2. 037(1) | [34,35] |
| [Co(Im)$_6$][Co(Im)$_3$(Hcit)]$_2$ • 4H$_2$O(**1**) | 2. 093−2. 162(1) | This work |

续表

| Complex | Co-N(Å) | Ref. |
|---------|---------|------|
| | Ni-N(Å) | |
| [Ni(Im)$_6$](sal)$_2$ | 2. 120−2. 141(1) | [37] |
| [Ni(Im)$_6$](NO$_3$)$_2$ | 2. 129(2) | [19] |
| [Ni(Im)$_4$](Ac)$_2$ | 1. 986−2. 054(3) | [38] |
| [Ni(Im)$_6$](C$_4$H$_4$O$_5$) • C$_2$H$_5$OH | 2. 113−2. 139(1) | [39] |
| [Ni$_2$($\mu$-H$_2$O)($\mu$-Ac)$_2$(Im)$_4$(Ac)$_2$] | 2. 082−2. 095(3) | [40] |
| [Ni(Im)$_6$][Ni(Im)$_3$(Hcit)]$_2$ • 4H$_2$O(**2**) | 2. 049−2. 118(2) | This work |

The bond distances of Co-N found in(**1**)[2. 093−2. 162(1)Å] are in agreement with those reported for octahedral[Co(Im)$_6$]$^{2+}$ complexes, such as [Co(Im)$_6$](Mz)$_2$[32], but significantly longer than those of tetrahedral or distorted tetrahedral complexes, such as [Co(Im)$_2$(Ac)$_2$][33,35,36] and [Co(Im)$_2$(C$_2$H$_5$COO)$_2$][34,35], which have been reported repeatedly. Similarly, The Ni-N distances[2. 049−2. 118(2)Å] are comparable to those found in other octahedral Ni( Ⅱ )-complexes containing imidazole ligands, such as [Ni(Im)$_6$](sal)$_2$[37], [Ni(Im)$_6$](NO$_3$)$_2$[19], [Ni(Im)$_4$(Ac)$_2$][38], [Ni(Im)$_6$](C$_4$H$_4$O$_5$) • C$_2$H$_5$OH[39], and [Ni$_2$($\mu$-H$_2$O)($\mu$-Ac)$_2$(Im)$_4$(Ac)$_2$][40].

## 3. 3  Spectroscopic properties

The UV-Vis spectra of(**1**)and(**2**)(shown in(Fig. S3))were taken in water at pH $\sim$7 in the range from 300 to 800 nm. Complex(**1**)shows a broad peak between 460 and 620 nm with $\lambda_{max}$=518 nm($\varepsilon$= 11), which could be tentatively assigned as $^4T_{1g}$(F) $\rightarrow$ $^4T_{1g}$(P) transition. The observed absorption features for(**1**)are in consonance with those predicted for Co( Ⅱ )d$^7$ octahedral species[20,41]. Complex(**2**) shows an absorption at $\lambda_{max}$=380 nm($\varepsilon$=4. 7), and a weak absorption at 640 nm($\varepsilon$=3. 4). The band at 380 nm might be attributed to the $^3A_{2g} \rightarrow$ $^3T_{1g}$(P)transition, and a weak absorption at 640 nm is the $^3A_{2g} \rightarrow$ $^3T_{1g}$(F)transition[42]. The IR spectra of compounds(**1**)and(**2**)(shown in(Fig. S4))exhibit strong characteristic vibrations for the carbonyl groups of the citrate ligand. The asymmetric stretching vibrations $v_{as}$(COO$^-$) appear between 1661 and 1611 cm$^{-1}$ for both compounds and the symmetric stretching vibrations $v_s$(COO$^-$)are observed between 1418 and 1385 cm$^{-1}$ for(**1**)and(**2**), which are in agreement with previous results for deprotoned citrate complexes of various metals[27,28]. The two obvious peaks observed at 3387 and 3132 cm$^{-1}$ for both compounds are attributed to the N-H and O-H stretching vibration respectively, these bands are shifted to lower wave-number due to hydrogen-bondings. Especially, the strong intramolecular hydrogen bondings in both compounds result in big lower frequency shift of the O-H stretching vibration(see Table 3).

Crystallographic data for the structures reported in this paper have been deposited with the Cambridge Crystallographic Data Centre as Supplementary Publication Nos. CCDC-224847, 224848. Copies of the data can be obtained free of charge from the CCDC(12 Union Road, Cambridge CB2 1EZ, UK; Tel. : + 44-1223336-408; fax: + 44-1223-336-003; e-mail: deposit @ ccdc. cam. ac. uk; www: http://ccdc. cam. ac. uk).

## 3.4　Supporting information

The following data are available on the Elsevier web site. ①Perspective view of the cation structure of $[Ni(Im)_6]^{2+}$ (Fig. S1), perspective view of the anion mesomeric structure of $[Ni_\Lambda(Im)_3(Hcit)]^-$ and $[Ni_\Lambda(Im)_3(Hcit)]^-$ (Fig. S2), UV-Vis spectra of citrato imidazole complexes of (1) and (2) (Fig. S3), and IR spectra of citrato complexes of (1) and (2) (Fig. S4).

# Acknowledgements

Financial supports by the Ministry of Science and Technology (001CB108906) and the National Science Foundation of China(20021002) are gratefully acknowledged.

# References

[1] R. J. Sundberg, R. B. Martin, Chem. Rev. 1974, **74**: 471~517.

[2] J. J. R. Fraústo da Silva, R. J. P. Williams, The Biological Chemistry of the Elements: the Inorganic Chemistry of Life, Oxford University Press, London, 2001.

[3] I. Török, P. Surdy, A. Rockenbauer, L. Korecz Jr., G. J. Anthony, A. Koolhaas, T. Gajda, J. Inorg. Biochem. 1998, **71**: 7~14.

[4] J. P. Glusker, Acc. Chem. Res. 1980, **13**: 345~352.

[5] R. B. Martin, J. Inorg. Biochem. 1986, **28**: 181~187.

[6] M. A. Halcrow, G. Christou, Chem. Rev. 1994, **94**: 2421~2481.

[7] J. Lee, R. D. Reeves, R. P. Brooks, T. Jaffré, Phytochemistry 1977, **16**: 1503~1505.

[8] W. J. Kersten, R. R. Brooks, R. D. Reeves, T. Jaffré, Phytochemistry 1980, **19**: 1963~1965.

[9] F. A. Homer, R. R. Reeves, R. D. Brook, A. J. M. Baker, Phytochemistry 1991, **30**: 2141~2145.

[10] H. Siegel, A. Saha, N. Saha, P. Carloni, L. E. Kapinos, R. Griesser, J. Inorg. Biochem. 2000, **78**: 129~137.

[11] A. M. Atria, A. Vega, C. Contreras, J. Valenzuela, E. Spodine, Inorg. Chem. , 1999, **38**: 5681~5685.

[12] M. Murrie, S. J. Teat, H. Stoeckli-Evans, H. U. Güdel, Angew. Chem. , Int. Ed. 2003, **42**: 4653~4656.

[13] S. T. Ochsenbein, M. Murrie, E. Rusanov, H. Stoeckli-Evans, C. Sekine, H. U. Güdel, Inorg. Chem. 2002, **41**: 5133~5140.

[14] M. Murrie, H. Stoeckli-Evans, H. U. Güdel, Angew. Chem. , Int. Ed. 2001, **40**: 1957~1960.

[15] M. Murrie, D. Biner, H. Stoeckli-Evans, H. U. Güdel, Chem. Commun. 2003, 230~231.

[16] J. Liang, M. Madden, V. K. Shah, R. H. Burris, Biochemistry 1989, **29**: 8577~8581.

[17] S. M. Mayer, C. A. Gormal, B. E. Smith, D. M. Lawson, J. Biol. Chem. 2002, **277**: 35263~35266.

[18] E. Prince, A. D. Mighell, C. W. Reimann, A. Santoro, Cryst. Struct. Commun. 1972, **1**: 247~252.

[19] A. J. Finney, M. A. Hitchman, C. L. Raston, G. L. Rowbottom, A H. White, Aust. J. Chem. 1981, **34**: 2113~2123.

---

① Supplementary data is available in the online version of this paper.

[20] M. Matzapetakis, M. Dakanali, C. P. Raptopoulou, V. Tanglouis, A. Terzis, N. Moon, J. Giapintzakis, A. Salifoglou, J. Biol. Inorg. Chem. 2000, **5**: 469～474.

[21] Z. H. Zhou, Y. J. Lin, H. B. Zhang, G. D. Lin, K. R. Tsai, J. Coord. Chem. 1997, **42**: 131～141.

[22] G. M. Sheldrick, SHELX-97, Programs for Crystal Structure Analysis, University of Göttingen, Göttingen, Germany, 1997.

[23] G. R. Hedwig, J. R. Liddle, R. D. Reeves, Aust. J. Chem. 1980, **33**: 1685～1693, and references therein.

[24] E. Camp, G. Ostacoli, M. Meirone, G. Saini, J. Inorg. Nucl. Chem. 1964, **26**: 553～564.

[25] N. Kotsakis, C. P. Raptopoulou, V. Tangoulis, A. Terzis, J. Giapintzakis, T. Jakusch, T. Kiss, A. Salifoglou, Inorg. Chem. 2003, **42**: 22～31.

[26] E. N. Baker, H. M. Baker, B. Anderson, R. D. Reeves, Inorg. Chim. Acta 1983, **78**: 281～285.

[27] M. Matzapetakis, N. Karligiano, A. Bino, M. Dakanali, C. P. Raptopoulou, V. Tangoulis, A. Terzis, J. Gapintzakis, A. Salifoglou, Inorg. Chem. 2000, **39**: 4044～4051.

[28] R. Swanson, W. H. Ilsley, A. G. Stanislowski, J. Inorg. Biochem. 1983, **18**: 187～194.

[29] K. L. C. Grönberg, C. A. Gormal, M. C. Durrant, B. E. Smith, R. A. Henderson, J. Am. Chem. Soc. 1998, **120**: 10613～10621.

[30] K. R. Tsai, H. L. Wan, J. Clust. Sci. 1995, **6**: 485～501.

[31] J. Strouse, S. W. Layten, C. E. Strouse, J. Am. Chem. Soc. 1977, **99**: 562～572.

[32] Z. X. Wang, F. F. Jian, Y. R. Zhang, F. S. Li, H. K. Fun, K. Chinnakali, J. Chem. Crystallogr. 1999, **29**: 885～890.

[33] P. A. Gadet, Acta Cryst. B 1974, **30**: 349～353.

[34] W. D. Horrocks Jr., J. N. Ishley, B. Holmquist, J. S. Thompson, J. Inorg. Biochem. 1980, **12**: 131～141.

[35] W. D. Horrocks, J. N. Ishley, R. R. Whittle, Inorg. Chem. 1982, **21**: 3265～3269.

[36] X. M. Chen, B. H. Ye, X. C. Huang, Z. T. Xu, J. Chem. Soc., Dalton Trans. 1996, 3465～3468.

[37] F. F. Jian, Z. X. Wang, Z. P. Bai, X. Z. You, H. K. Fun, K. Chinnakali, J. Chem. Crystallogr. 1999, **29**: 359～363.

[38] P. Naumov, M. Ristova, M. G. B. Drew, S. W. Ng, Acta Cryst. C 2000, **56**: 372～373.

[39] M. Perec, R. Baggio, M. T. Garland, Acta Cryst. C 1999, **55**: 858～860.

[40] B. H. Ye, I. D. Williams, X. Y. Li, J. Inorg. Biochem. 2002, **92**: 128～136.

[41] F. A. Cotton, G. Wilkinson, Adv. Inorg. Chem, Wiley, New York, 1999, pp. 820～821.

[42] F. A. Cotton, G. Wilkinson, Adv. Inorg. Chem, Wiley, New York, 1999, pp. 838～839.

■ **本文原载**:《石油化工》第 33 卷增刊(2004 年),第 273～275 页。

# 高效甲醇合成催化剂及一次性通过过程研究[*]

沈炳顺[①]　张鸿斌　林国栋　董　鑫　蔡启瑞

(厦门大学化学化工学院固体表面物理化学国家重点实验室,福建　厦门 361005)

**摘　要**　以 Ni 修饰多壁碳纳米管复合材料($y\%$Ni/CNT)为促进剂,制备一类合成气高效制甲醇催化剂 $Cu_6Zn_3Al$-$x\%$($y\%$ Ni/CNT)。实验发现,在传统工业甲醇合成 Cu-ZnO-$Al_2O_3$ 催化剂中添加适量 $y\%$ Ni/CNT 能显著提高其对甲醇合成的催化活性,在 $Cu_6Zn_3Al$-12.5%(8% Ni/CNT)催化剂上,2.0 MPa,493 K,$H_2/CO/CO_2/N_2$=62/30/5/3,GHSV=2700 mL/(h·g)的反应条件下,CO 转化率达 34%,甲醇 STY 为 442 mg/(h·g),是非促进的参比催化剂 $Cu_6Zn_3Al$(330 mg/(h·g),在 513 K)的 1.34 倍。该类催化剂对于合成气一次性通过制甲醇过程有重要应用前景。

**关键词**　$Cu_6Zn_3Al$-$x\%$($y\%$ Ni/CNT)　合成气高效制甲醇　一次性通过甲醇合成过程

**中图分类号**　O 643.35　　　　　**文献标识码**　A

煤基燃料甲醇/二甲醚是基于能源化工原料多样化及煤炭利用洁净优化具战略意义的两大燃料化学品。现有甲醇合成工艺中,原料合成气只有一小部分(10%～15%)转化为甲醇,大部分未反应合成气需经分离后,反复、多次进行循环反应以提高原料合成气的利用率,工艺流程及设备比较复杂,并要额外消耗用于分离、循环的能量。新近提出的煤集成气化联合循环(CIGCC)发电联产甲醇系统,其优点之一就在于原料合成气"一次性通过"进行单程反应,未反应的合成气不再送回反应器,而是送入燃气/蒸汽联合循环电厂作动力燃料;这既利于简化甲醇生产流程,又可省去用于分离、循环的额外能耗。为使"一次性通过"工艺具有实用意义,现有甲醇合成单程转化率必须大大提高,其关键在于高转化率甲醇合成技术的开发;而寻找高效新型催化剂体系(包括新型载体及助剂),以降低反应温度,提高合成气一步合成甲醇的效率,则是值得注意的研究方向。本文报道一类 Ni 修饰多壁碳纳米管复合材料($y\%$ Ni/CNT)促进的甲醇合成 Cu 基催化剂,$Cu_6Zn_3Al$-$x\%$($y\%$ Ni/CNT),其催化转化合成气制甲醇兼具高的 CO 单程转化率和甲醇时空产率,展现出作为合成气"一次性通过"合成甲醇催化剂的良好前景。

## 1　实验

多壁碳纳米管(CNT)及其 Ni 修饰复合材料 $y\%$ Ni/CNT 参照文献[1,2]方法制备。$y\%$ Ni/CNT 促进的甲醇合成 Cu 基催化剂,$Cn_6Zn_3Al$-$x\%$($y\%$ Ni/CNT),由共沉淀法制备,具体操作大致为:在 353 K 恒定温度下,将计量 $Cu(NO_3)_2$·$3H_2O$,$Zn(NO_3)_2$·$6H_2O$,和 $Al(NO_3)_3$·$9H_2O$ 的混合水溶液与等离子浓度的 $Na_2CO_3$ 溶液(均为 AR 级)并流滴加入装有一定量 $y\%$Ni/CNT 的玻璃容器中,强烈搅拌并调节 pH=～7.0,5h 后停止加热,静置过夜;沉淀物经抽滤、洗涤、烘干、$N_2$ 气保护下 ～573 K 焙烧而得。

* 国家自然科学基金(50072021)、教育部科技基金(20010384002)和福建省自然科学基金(2001H017)资助项目。

① 沈炳顺(1977— ),男,福建省漳州市人,硕士生,电话 0592-2184591,电邮 hbzhang@ xmu.edu.cn。

不含 $y\%$ Ni/CNT 或单纯 CNT 促进的共沉淀型催化剂,$Cu_6Zn_3Al$ 或 $Cu_6Zn_3Al$-$x\%$ CNT,的制备程序与上述相仿。

合成气制甲醇反应活性评价在加压固定床连续流动反应器-GC 组合系统上进行。每次试验催化剂用量为 0.50 g,反应条件为:413～523 K,2.0 MPa,原料合成气组成为 $H_2/CO/CO_2/N_2 = 62/30/5/3$(体积比)。反应物和产物由一台在线双柱双气路气相色谱仪(GC-950 型)的 TCD 作现场分析;CO 转化率由 $N_2$-内标法计算,甲醇时空产率直接由外标法(即工作曲线法)测量。

# 2 结果与讨论

## 2.1 催化剂活性评价

图 1 示出共沉淀型非促进的和 CNT 或 $y\%$ Ni/CNT 促进的 $Cu_iZn_jAl_k$ 催化剂上甲醇合成反应活性的对比评价结果。$Cu_6Zn_3Al$-12.5%(8% Ni/CNT)催化剂在 ～413 K 就显示出可观活性:在 2.0 MPa,$H_2/CO/CO_2/N_2 = 62/30/5/3$,GHSV = 2700 mL/(h·g) 的反应条件下,所观测 CO 转化率为 4.8%,相应甲醇时空产率($STY_{CH_3OH}$)为 61.8 mg/(h·g)。最佳操作温度在 493 K 附近,CO 转化率达 34%,$STY_{CH_3OH}$ 为 442 mg/(h·g),比不含 $y\%$Ni/CNT 的参比样 $Cu_6Zn_3Al$(330 mg/(h·g),在 513 K)和单纯 CNT 促进催化剂 $Cu_6Zn_3Al$-12.5% CNT(378 mg/(h·g),在 503 K)分别增加 34% 和 17%。若干催化剂上 $CO/CO_2$ 加氢成甲醇的反应活性评价结果列于表 1。

表 1　若干催化剂上 $CO/CO_2$ 加氢制甲醇的反应活性*

| 催化剂组成 | 比面积(SSA)/$m^2 \cdot g^{-1}$ | 反应温度/K | CO 转化率/% | 甲醇时空产率 |
|---|---|---|---|---|
| $Cu_6Zn_3Al$-0%CNT | 50.1 | 503 | 26.4 | 320.3 |
| | | 513 | 29.7 | 330.7 |
| $Cu_6Zn_3Al$-12.5%CNT | 63.1 | 503 | 30.5 | 378.6 |
| | | 513 | 35.6 | 364.5 |
| $Cu_6Zn_3Al$-12.5%(5% Ni/CNT) | 70.3 | 503 | 25.7 | 354.0 |
| | | 513 | 30.8 | 379.0 |
| $Cu_6Zn_3Al$-12.5%(8% Ni/CNT) | 77.2 | 493 | 34.8 | 442.3 |
| | | 503 | 36.0 | 394.0 |
| | | 513 | 36.9 | 357.7 |
| $Cu_6Zn_3Al$-12.5%(16% Ni/CNT) | 59.2 | 503 | 24.3 | 311.0 |
| | | 513 | 30.8 | 320.0 |

\* 反应条件:2.0 MPa,$H_2/CO/CO_2/N_2 = 62/30//5/3$(体积比),GHSV = 2700 mL/(h·g);时空产率单位为:mg/(h·g)。

实验结果显示,在 ≤493 K 的反应温度下,在 $Cu_6Zn_3Al$-12.5%(8% Ni/CNT)催化剂上,$CO/CO_2$ 加氢成甲醇的选择性 ≥99.5%,反应尾气中主要副产物 $CH_4$ 的浓度在 GC 检测极限以下。当反应温度达到 503 K 时,始观测到可观量 $CH_4$ 的生成(见图 2b);此后,尽管 CO 转化率继续随反应温度升高而增加,但生成 $CH_4$ 的选择性快速上升,并检测到少量其他副产物(如乙醇、丁醇等)的生成,甲醇选择性相应下降,并导致其时空产率明显下降。从图 2 所示结果的比较,可见在这些催化剂上 $CO/CO_2$ 加氢生成

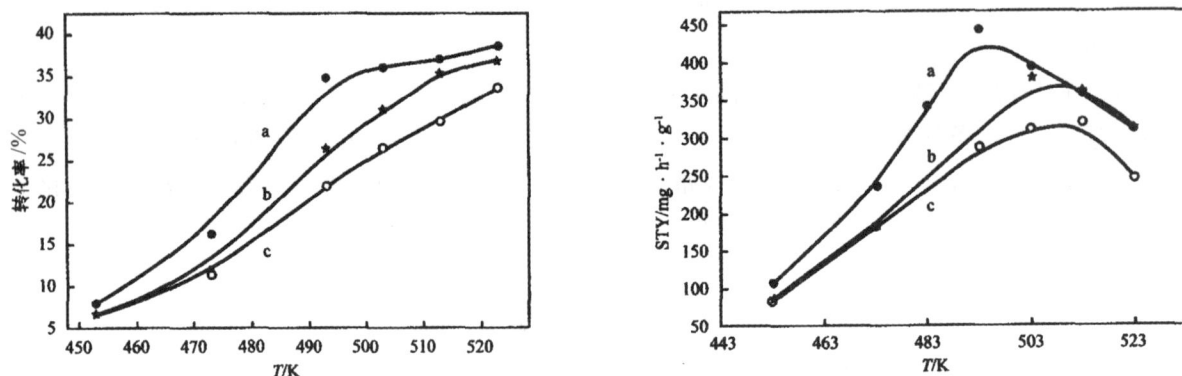

**图 1 y％Ni/CNT 促进的 $Cu_6Zn_3Al$ 催化剂及其参比体系上甲醇合成的反应活性**

a)$Cu_6Zn_3Al$-12.5％(8％Ni/CNT)；b)$Cu_6Zn_3Al$-12.5％GNT；c)$Cu_6Zn_3Al$

反应条件:2.0 MPa,$H_2CO/CO_2/N_2$=62/30/5/3(体积比),GHSV=2700 mL/(h·g)

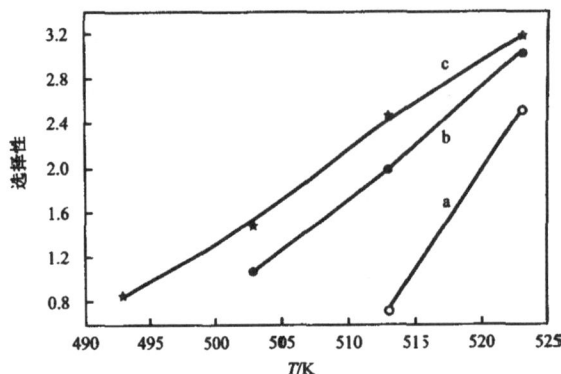

**图 2 若干催化剂上 $CO/CO_2$ 加氢成甲醇反应中副产 $CH_4$ 选择性**

a)$Cu_6Zn_3Al$-12.5％(5％Ni/CNT)；b)$Cu_6Zn_3Al$-12.5％(8％Ni/CNT)；c)$Cu_6Zn_3Al$-12.5％(16％Ni/CNT)

反应条件:2.0 MPa,$H_2/CO/CO_2/N_2$=62/30/5/3(体积比),GHSV=2700 mL/(h·g)

$CH_4$ 的选择性高低与促进剂(y％Ni/CNT)中 Ni 含量多寡存在相关性,其高低顺序为:$Cn_6Zn_3Al$-12.5％(5％Ni/CNT)＜$Cu_6Zn_3Al$-12.5％(8％Ni/CNT)＜$Cu_6Zn_3Al$-12.5％(16％Ni/CNT)。为兼获高的甲醇时空产率和选择性,催化剂组成看来以 $Cu_6Zn_3Al$-12.5％(8％Ni/CNT)为佳,反应温度以~493 K 为宜。

### 2.2 促进剂的作用本质

在我们看来,Ni-修饰 CNT 促进的 $Cu$-$ZnO$-$Al_2O_3$ 催化剂对甲醇合成所表现出的优异催化性能,很可能与单纯 CNT 的促进作用机制[3,4]相仿,即主要地与这种 Ni-修饰碳纳米材料对氢的吸附特性密切相关。图 3 示出 $H_2$ 在 $Cu_6Zn_3Al$-12.5％(8％Ni/CNT)及两个参比样上吸附的 TPD 谱。从总体上说,每一条谱线都包含一个低温峰(峰-I)和一高温峰(峰-II)。低温峰系源于弱吸附氢物种(很可能为分子态吸附氢,$H_2$(a))的脱附,高温峰可归属于强吸附氢物种(可能为解离化学吸附氢,H(a))的脱附。在所观测 3 个体系中,与峰-II 相关的解离吸附氢量以 8％Ni/CNT 促进的体系为最高。可以想象,在本文甲醇合成反应温度范围(453~533 K),与峰-I 相关的那些氢吸附物种的表面浓度可期非常低,工作态催化剂表面绝大多数氢吸附物种是同峰-II 相关的那些氢吸附物种。正是那些强吸附氢物种同甲醇合成反应活性密切相关。通过拟合可估算出图 3 所示 3 种试样的峰-II 相对面积强度比为:S($Cu_6Zn_3Al$-12.5％(8％Ni/CNT))/S($Cu_6Zn_3Al$-12.5％CNT)/S($Cu_6Zn_3Al$)=100/90/42。这个顺序正好与这些催化剂对甲醇合成催化活性的实验观测顺序相一致。

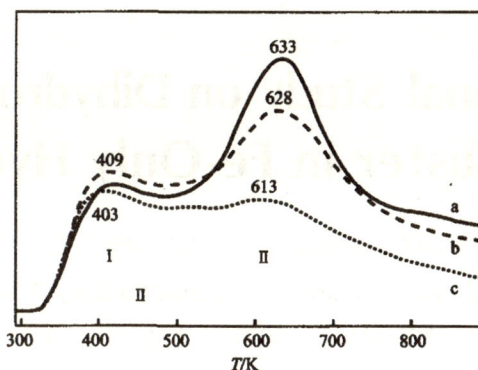

**图 3　8％ Ni/CNT 促进的 Cu-ZnO-Al₂O₃ 及其参比体系的 H₂-TPD 谱**
a)Cu₆Zn₃Al-12.5％(8％Ni/CNT);b)Cu₆Zn₃Al-12.5％ CNT;c)Cu₆Zn₃Al

　　基于上述几种催化剂上甲醇合成反应活性和相关谱学表征的比较研究,本文认为,在 Cu₆Zn₃Al-$x$％ ($y$％ Ni/CNT)催化剂上合成气制甲醇反应中,$y$％ Ni/CNT 促进剂的作用本质与单纯 CNT 的促进作用相仿;对 H₂ 具有强吸附活化能力的金属 Ni 组分对 CNT 的预修饰进一步提高了 CNT 对 H₂ 的解离吸附容量,并提高其所促进催化剂对 H₂ 的吸附活化能力,有利于在工作态催化剂上营造高稳态浓度的活泼氢吸附物种的表面氛围,在 $y$％Ni/CNT 上的这些活泼氢吸附物种通过"氢溢流"容易传输至 Cu⁰ 或 Cu⁺-ZnO 表面活性位,于是有助于提高 CO/CO₂ 加氢成甲醇反应过程中表面加氢反应的速率。

# 参考文献

[1]Chen P,Zhang H B,Lin G D,et al. [J]. Carbon,1997,**35**(10－11):1495～1501.

[2]沃尔夫冈·里德尔著.罗守福译.化学镀镍[M].上海:上海交通大学出版社,1996.27～65.

[3]Zhang H B,Dong X,Lin G D,et al. Methanol Synthesis from H₂/CO/CO₂ over CNT-Promoted Cu-ZnO-Al₂O₃ Catalyst. In:Liu C J,Mallinson R G,Aresta M,Eds. ACS Symp Ser No. 852:Utilization of Greenhouse Gases[C]. Washington DC:ACS,2003.195～209.

[4]Dong X,Zhang H B,Lin G D,et al. [J]. Catal Lett,2003,**85**(3－4):237～246.

■ 本文原载：Inorg. Chem. 44(2005),pp. 4941～4946.

# Density Functional Study on Dihydrogen Activation at the H Cluster in Fe-Only Hydrogenases

Tai-Jin Zhou[①,a], Yi-Rong Mo[①,b], Zhao-Hui Zhou[a], Khi-Rui Tsai[a]

(ᵃDepartment of Chemistry, State Key Laboratory for Physical Chemistry of Solid States, Xiamen University, Xiamen, Fujian 361005, P. R. China, and the ᵇDepartment of Chemistry, Western Michigan University, Kalamazoo, Michigan 49008)

Received November 2, 2004

Xiamen University.

Western Michigan University.

**Abstract**    Models simulating the catalytic diiron subcluster $[FeFe]_H$ in Fe-only hydrogenases have often been designed for computational exploration of the catalytic mechanism of the formation and cleavage of dihydrogen. In this work, we extended the above models by explicitly considering the electron reservoir $[4Fe\text{-}4S]_H$ which is linked to the diiron subcluster to form a whole H cluster($[6Fe\text{-}6S] = [4Fe\text{-}4S]_H + [FeFe]_H$). Large-scale density functional theory (DFT) computations on the complete H cluster, together with simplified models in which the $[4Fe\text{-}4S]_H$ subcluster is not directly involved in the reaction processes, have been performed to probe hydrogen activation on the Fe-only hydrogenases. A new intermediate state containing an $Fe_p\cdots H\cdots CN$ two-electron three-center bond is identified as a key player in the $H_2$ formation/cleavage processes.

## 1    Introduction

Hydrogenases(or hereafter $H_2$ases) reversibly catalyze hydrogen oxidation($H_2 \leftrightarrows 2H^+ + 2e^-$) and play a key role in hydrogen metabolism in many microorganisms, which use hydrogen as a source of electrons and protons or generate hydrogen as a way to reduce protons[1-27]. Recently, we systematically studied the mechanism of the enzymatic formation and cleavage of dihydrogen on the catalytic diiron subcluster, $[FeFe]_H$, of the H cluster in Fe-only $H_2$ases by simulating the active center, $[FeFe]_H$, with a simplified model, $\{[H](CH_3S)(CO)(CN^-)Fe_p(CO_b)(\mu\text{-}SRS)Fe_d(CO)(CN^-)L\}$, where $[H]$ stands for the $[Fe_4S_4]_H{}^{2+}$ subcluster bridged to the $[FeFe]_H$ moiety and L can be any hydrogen species such as $H^+$, $H^-$, or $H_2$. On the basis of the above model, structures of various possible redox states were optimized and compared with the X-ray crystallographic structures and other accessible experimental evidence[7,9,10,28-34]. On the basis of these calculations, the most probable pathway of the concerted proton transfer(PT)and electron transfer(ET)in the $H_2$ formation/cleavage reactions at $[FeFe]_H$ was

---

①    To whom correspondence should be addressed. Fax: +86-5922183795. E-mail: tjzhou@xmu. edu. cn(T. Z. ); ymo@wmich. edu(Y. M. ).

suggested and rationalized[35]. Our mechanism stressed that the proximal iron $Fe_p$ site is the activation center and that the hydride $H^-$ bridging to the two iron atoms, $Fe_p$ and $Fe_d$ (distal iron), takes a central position in the proposed $H_2$ formation/cleavage mechanism. In the present work, we significantly expanded the previous models and performed large-scale DFT computations on the whole H cluster ($[6Fe-6S] = [4Fe-4S]_H + [FeFe]_H$) by including the important $[4Fe-4S]_H$ subcluster which acts as an electron reservoir[10]. The electron transfer between $[Fe_4S_4]_H{}^{2+}$ and $[FeFe]_H$ has been regarded as the driving force for dihydrogen bond formation and cleavage. Simplified models[35], where the $[Fe_4S_4]_H{}^{2+}$ subcluster is not explicitly considered, were also employed whenever the subcluster is not actively involved in the reaction steps and acts as a spectator. In particular, we focused our investigation on the H-H bond formation and cleavage processes on $Fe_p$ and identified a new intermediate state containing an $Fe_p\cdots H\cdots CN$ two-electron three-center(2e-3c) bond which occurs at the initial stage for the H-H bond formation and final stage for the H-H bond cleavage.

## 2  Computational Details

Calculations were performed using the DFT program Dmol[3] from the Cerius[2] software package[36]. Geometries of the various models were optimized by unrestricted spin-polarized(or different orbital for different spin, DODS) local DFT methods with the Perdew-Wang(1992)functional[37]. For the basis sets, we adopted the double numerical basis set containing a $p$-polarization for H and $d$-polarization for all other atoms(abbreviated as DNP). The DNP basis sets are given numerically as cubic spline functions, and their quality and size are comparable to the standard Gaussian 6-31G(d, p) split-valence double-$\zeta$ plus polarization basis set. The numerical basis set, DNP, is the exact solution to the Kohn-Sham equations for atoms. By means of the potential energy surface(PES) scan, we searched the intermediate states with one imaginary frequency in the calculation of the second derivatives for the singlevariable models of proton transfer from one site to the other. Here, the intermediate state is defined as the structure with the highest energy between two optimal energy-minimum structures on a PES scan and may be better described as an approximate ( or quasi ) transition state ( TS ). Simultaneously, searches for a multivariable transition state were conducted progressively using the LST/QST method[38] in the Materials Studio software package[36].

## 3  Results and Discussion

**Optimal Structures at Various Redox and Charge States.** Because an iron atom with an oxidation state of $0, +1, +2,$ or $+3$ favors a high-spin state with its 3d shell partially occupied, there are significant spin polarization and coupling effects in many iron-sulfur protein enzymes[39]. As a result, ideally these systems should be described by a linear combination of various DODS wave functions ( or states ). Because the Dmol[3] program cannot deal with multiwave functions, the $[4Fe-4S]$ cluster is mainly imposed as a spectator during the $H_2$ formation and cleavage processes. Consequently, we assume that the spin-coupled effect within the $[4Fe-4S]$ cluster on the relative energy calculations of the $H_2$ cleavage may be insignificant, and in the following discussion, we only consider the low-spin state of the H cluster.

We started by considering four $[H cluster]_t$ models consisting of the $[FeFe]_H$ and $[4Fe-4S]_t$ moieties, where $[FeFe]_H([FeFe]_H = (CH_3S)(2CO)(CN)Fe_p(2H)(\mu-SRS)Fe_a(CO)(CN))$ contains two active hydrogen species($H^+ + H^-$ or activated $H_2$)and the type label,t(t=a,b,c,d),denotes the extent of the simplification of the [4Fe-4S] subcluster(i. e.,$[4Fe-4S]_a = Fe-[3(CH_3S-Fe)(3H)-4S]$,$[4Fe-4S]_b = Fe[3(CH_3S-Fe)(2H)-4S]$,$[4Fe-4S]_c = H$,and $[4Fe-4S]_d = null$). The symbol $x_t[y]_z$ is thus used to represent a model to make the following discussion concise and straightforward,where the serial number $x$ distinguishes the structure of $[FeFe]_H$,y is the total charge,and z represents the redox state(i. e.,z= ox is an oxidized state,$[Fe^{II}Fe^I]$,z=s is an EPR active semireduced state,$[Fe^{II}Fe^I](s=1/2)$,and z=red is a reduced state,$[Fe^IFe^I]$). Both experimental and computational studies have suggested a catalytic cycle as follows:$Fe^{II}Fe^{II} \leftrightarrow Fe^{II}Fe^I \leftrightarrow Fe^IFe^{I[13,14,19,20]}$. For simplicity and generality,we can further use the notation $x$ to define a structure and the symbol $x_z$ (e. g.,$1_ox$ or $1_s$) to refer to a structure at an assigned redox state.

Our previous investigation with the simplified model $[H\ cluster]_c$ has confirmed the significance of the hydride bridging the $Fe_p$ and $Fe_d$ sites in the hydrogen generation and consumption by Fe-only $H_2$ ases[35]. As our interest lies in the H-H bond formation and cleavage processes on $Fe_p$,we optimized four key models($1_a[1]_{ox}$,$1_b[0]_{ox}$,$1_c[0]_{ox}$,and $1_d[-1]_{ox}$) as shown in Figure 1. Among these four models,$1_a[1]_{ox}$ and $1_b[0]_{ox}$ fall into the category of complete H cluster models,and they are correlated as $1_a[1]_{ox} = 1_b[0]_{ox} + H^+$. The other two models,$1_c[0]_{ox}$ and $1_d[-1]_{ox}$,are simplified models,and $1_c[1]_{ox} = 1_d[0]_{ox} + H^+$. These four models($1_a[1]_{ox}$,$1_b[0]_{ox}$,$1_c[0]_{ox}$ and $1_d[-1]_{ox}$) can be uniformly denoted as $1_{ox}$. The addition of one electron to $1_{ox}$ results in a semireduced state,$1_s$,which also has four models,$1_a[0]_s$,$1_b[-1]_s$,$1_c[-1]_s$,and $1_d[-2]_s$,and all eight models($1_a[1]_{ox}$,$1_b[0]_{ox}$,$1_c[0]_{ox}$,$1_d[-1]_{ox}$,$1_a[0]_s$,$1_b[-1]_s$,$1_c[-1]_s$,and $1_d[-2]_s$)are collectively referred to as structure 1. This class of structures shows energy-minimum states that were not previously determined. The unique characteristic for structure 1 is that there is one hydride bridging iron ions $Fe_p$ and $Fe_d$ and one proton participating in an $Fe_p \cdots H \cdots CN$ two-electron three-center bond. This seems to be in accordance with Bruschi's note[40]

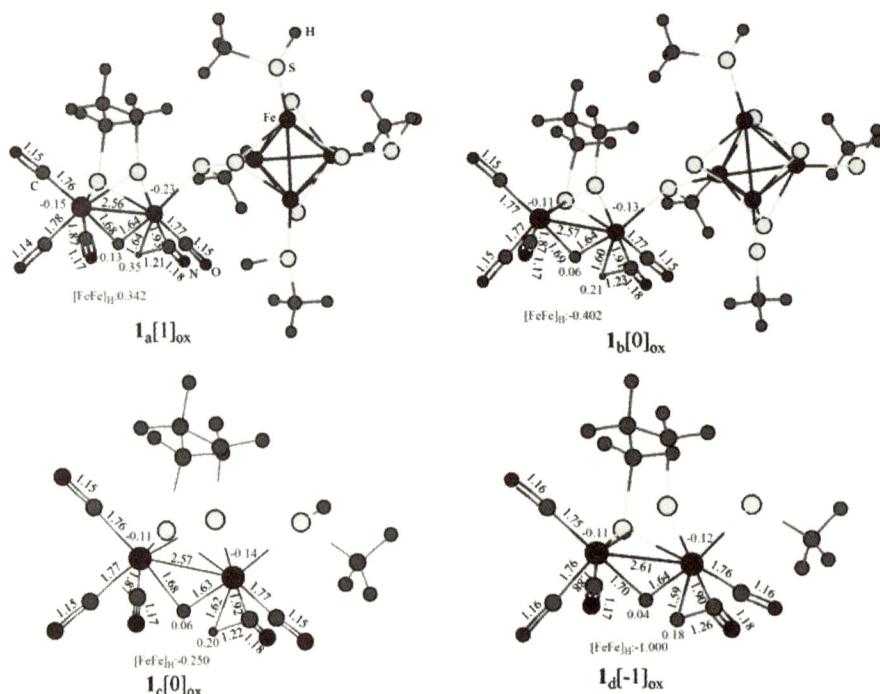

Figure 1　Four key models($1_a[1]_{ox}$,$1_b[0]_{ox}$,$1_c[0]_{ox}$,and $1_d[-1]_{ox}$)with optimal structural parameters(in black)and Mulliken charges(in red,also listed is the total charge for the $[FeFe]_H$ moiety). The atoms are symbolized with colors(cyan for H,gray for C,red for O,blue for N,yellow for S,and brown for Fe).

which states that the release of HCN from the optimized transition state in the proton transfer from the $\mu$-SRS site to the $Fe_p$ site can often be observed. Starting from structure **1**, we attempted to take the H-H distance as the reaction coordinate and derive energy profiles to find the energy-minimal structures, **2**, containing an $Fe_p \cdots \eta^2$-$H_2$ bond. However, we found that only for models $\mathbf{1_b}[-1]_s$, $\mathbf{1_c}[-1]_s$, $\mathbf{1_d}[-1]_{ox}$, and $\mathbf{1_d}[-2]_s$, can both the transition and product states on the energy profiles be successfully located. For models $\mathbf{1_a}[1]_{ox}$, $\mathbf{1_a}[0]_s$, $\mathbf{1_b}[0]_{ox}$, and $\mathbf{1_c}[0]_{ox}$, the molecular energy monotonically increases as the H-H bond distance decreases in the PES scan; thus, we were unable to locate structures $\mathbf{2_a}[1]_{ox}$, $\mathbf{2_a}[0]_s$, $\mathbf{2_b}[0]_{ox}$, and $\mathbf{2_c}[0]_{ox}$ because the optimizations uniformly converged to structure **1** (i. e., $\mathbf{1_a}[1]_{ox}$, $\mathbf{1_a}[0]_s$, $\mathbf{1_b}[0]_{ox}$, and $\mathbf{1_c}[0]_{ox}$). Model $\mathbf{3_{ox}}$ comes from a proton transfer from the $Fe_p \cdots H \cdots CN$ site in $\mathbf{1_{ox}}$ to the Cys-S site. The hopping of this proton in either $\mathbf{1_{ox}}$ or $\mathbf{3_{ox}}$ leads to structure $\mathbf{4_{ox}}$, in which the proton is bonded to $\mu$-SRS. Obviously, for type c where $[4Fe\text{-}4S]_c = H$, we cannot put two H atoms on $CH_3$-S, with one H representing the $[4Fe\text{-}4S]$ cluster and one H representing the transferable proton. Therefore, it is impossible to locate the structure **3** for type c. The transition state from $\mathbf{1_s}$ to $\mathbf{2_s}$ is defined as structure $\mathbf{5_s}$, and structure $\mathbf{6_{ox}}$ is the transition state from $\mathbf{3_{ox}}$ to $\mathbf{4_{ox}}$. For the process $\mathbf{1_{ox}} \leftrightarrows \mathbf{4_{ox}}$, the transition state is $\mathbf{7_{ox}}$. However, in the optimal geometry of model $\mathbf{6_b}[0]_{ox}$, as shown in Figure 2, the migrating proton is simultaneously in close contact with $Fe_p$(1. 67 Å), Cys-S(1. 89 Å), and $\mu$-SRS(1. 57 Å). Thus, $\mathbf{6_b}[0]_{ox}$ seems to be not only the transition state for the $\mathbf{3_b}[0]_{ox} \leftrightarrows \mathbf{4_b}[0]_{ox}$ process, but also very close to the transition states for the other two processes, $\mathbf{3_b}[0]_{ox} \leftrightarrows \mathbf{1_b}[0]_{ox}$ and $\mathbf{1_b}[0]_{ox} \leftrightarrows \mathbf{4_b}[0]_{ox}$. A reaction flowchart and the associated energy changes for the proton and electron transfers are presented in Figure 3.

To differentiate the models with the same structure, **x**, and different types, t, we computed the sum of Mulliken atom charges in the $[FeFe]_H$ part (Figures 1 and 2). We found that the CO, CN, Fe-Fe bond length, and, particularly, the C-H distances in the $Fe_p \cdots H \cdots CN$ two-electron three-center bond of structure **1** shown in Figures 1 and 2 and the reaction energies in Figure 3 from structures **1** to **2** are closely related to the total Mulliken atomic charges in the $[FeFe]_H$ part and the acid-base characters of the models, as shown in Figure 4. The $[4Fe\text{-}4S]$ part imposes its role of "spectator" via electrostatic interactions with $[FeFe]_H$. Structure $\mathbf{1_a}[1]_{ox}$ with a positive Mulliken charge(0. 342 e) in the $[FeFe]_H$ part is assumed to be a weak acid model and favorable for $Fe_p \cdots H \cdots CN$ bond formation on the basis of the structural data in Figure 1. In fact, we found that the higher the positive charge on $[FeFe]_H$, the stronger the $Fe_p \cdots H \cdots CN$ bond. Models of type a are not active for H-H bond formation in the $Fe_P$ site, and thus, $\mathbf{1_a}[1]_{ox}$ could be taken as a model for the isolated oxidized state. In contrast, weak basic models of types b and c favor the H-H bond formation and cleavage reactions, and models of type d are stronger bases than those of type c; therefore, they are more favorable for the formation of the H-H bond. Particularly, model $\mathbf{1_b}[-1]_s$, a weak base, is active for both the H-H bond formation and cleavage processes with very low activation barriers. On the basis of these results, discussed above, we conclude that the models of type b are the most rational models with which to study the reversible hydrogen oxidation.

For the three identical reactions $\mathbf{3_a}[1]_{ox} - 1. 4\ kcal/mol \leftrightarrows \mathbf{4_a}[1]_{ox}$, $\mathbf{3_b}[0]_{ox} + 2. 0\ kcal/mol \leftrightarrows \mathbf{4_b}[0]_{ox}$ and $\mathbf{3_d}[-1]_{ox} + 15. 3\ kcal/mol \leftrightarrows \mathbf{4_d}[-1]_{ox}$, we found a remarkable discrepancy among their reaction energies, which once again confirms that the simplified models cannot rationally predict the proton transfer from Cys-S to other parts of the $[FeFe]_H$ cluster and highlights the importance of the explicit consideration of the $[4Fe\text{-}4S]$ part. In the optimal structures of $\mathbf{1_d}[-2]_s$ and $\mathbf{2_d}[-2]_s$, the distances of the Fe-Fe, CO, and CN bonds are unanimously much larger than the experimental values(Figure 2); this suggests that the simplified models of type d with large negative charges are unreliable. However, by

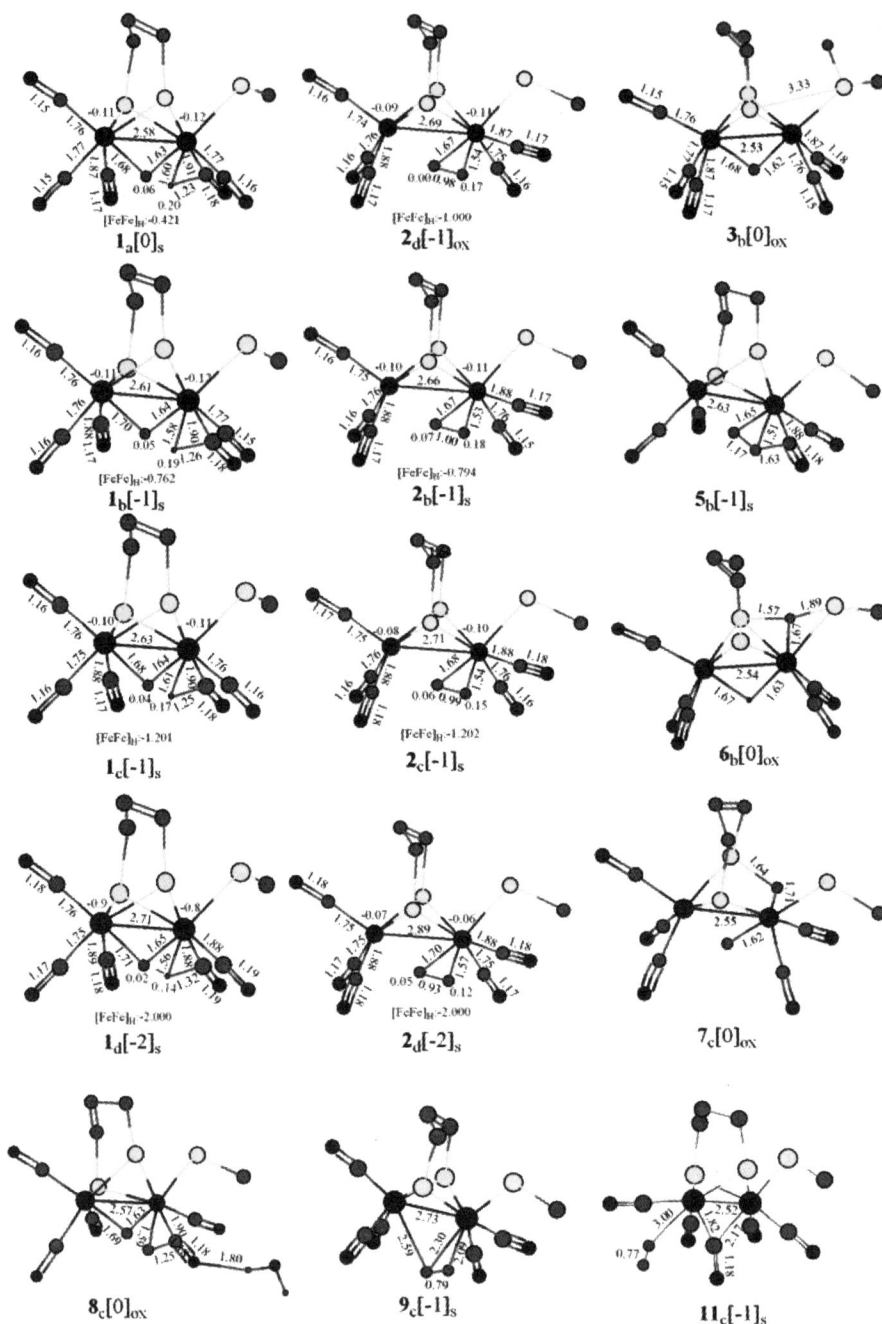

Figure 2    Optimal geometries and Mulliken charges of various important models. For simplicity, the [4Fe-4S]$_t$ part and hydrogen atoms in PDT and -CH$_3$ are not shown.

comparing the $1_b[0]_{ox}+3.2$ kcal/mol$\leftrightarrows 4_b[0]_{ox}$, $1_b[-1]_s+0.4$ kcal/mol$\leftrightarrows 2_b[-1]_s$ and $1_c[0]_{ox}+2.4$ kcal/mol$\leftrightarrows 4_c[0]_{ox}$, $1_c[-1]_s- 0.4$ kcal/mol$\leftrightarrows 2_c[-1]_s$ reactions, we found that their reaction energies (3.2 vs 2.4 kcal/mol and 0.4 vs. -0.4 kcal/mol) are very close. In addition, model $1_c[-1]_s$ is also a weak base, and the Mulliken charges of the same structures of b and c are similar(see Figures 1 and 2). As a result, simplified model structures of type c are also used in our subsequent calculations and discussion when the $[Fe_4S_4]_H{}^{2+}$ subcluster is not directly involved in the reactions.

Because our computations are conducted in the gas phase, one question related to the new

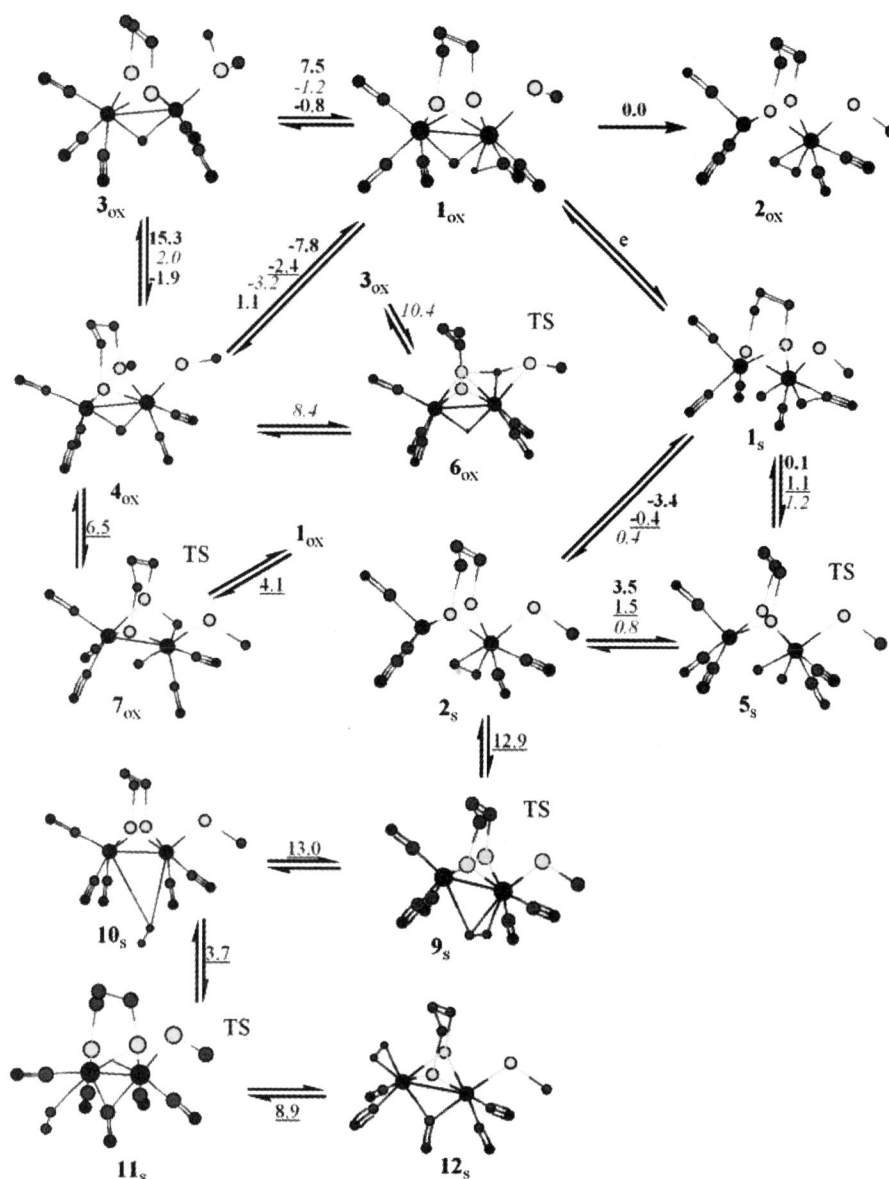

**Figure 3** Proton transfer(PT)reaction flowchart where the [4Fe-4S]$_t$ part and hydrogen atoms in PDT and -CH$_3$ are not shown. $x_z$ corresponds to a structure at the assigned redox state, and the data on the arrowheads are the reaction energies in kcal/mol(black plain text for model type a, green italic text for type b, blue underlined text for type c, and red bold text for type d).

intermediate state, **1**, is whether the Fe$_p$ ⋯ H ⋯ CN two-electron three-center bond could exist in a protein environment and aqueous solution, as the surrounding amino acid residues or water molecules could form hydrogen bonds with the CN$^-$ ligands in the [FeFe]$_H$ which would subsequently disfavor the two-electron three-center bond. To confirm the existence of such an Fe$_p$ ⋯ H ⋯ CN bond, we put water molecules around CN$^-$, fixed the N⋯H distance at 1.8 Å, and optimized the resulting structure, **8**$_c$[0]$_{ox}$ = **1**$_c$[0]$_{ox}$ + H$_2$O. Figure 2 shows the optimal geometry of **8**$_c$[0]$_{ox}$. We found that the Fe$_p$ ⋯ H ⋯ CN bond in model **8**$_c$[0]$_{ox}$ is only slightly weaker than that in **1**$_c$[0]$_{ox}$. Thus, we surmise that the possible hydrogen bond between the CN$^-$ ligand and the surroundings imposes a negligible effect on the unique Fe$_p$ ⋯ H ⋯ CN bond in intermediate structure **1**.

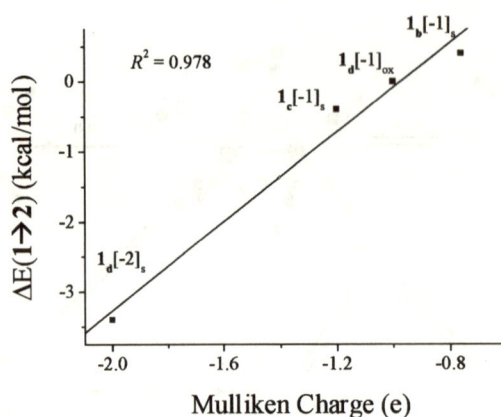

Figure 4　Correlation between the reaction energies(kcal/mol)from **1** to **2** and the overall Mulliken charges for the [FeFe]$_H$ part.

**H-H Bond Formation and Cleavage on [FeFe]$_H$.** The conversion between **1**$_s$ and **2**$_s$ requires very little energy(**1**$_b$[−1]$_s$+0.4 kcal/mol⇆**2**$_b$[−1]$_s$ or **1**$_c$[−1]$_s$-0.4 kcal/mol⇆**2**$_c$[−1]$_s$) with a very low activation barrier(**1**$_b$[−1]$_s$+1.2 kcal/mol⇆**5**$_b$[−1]$_s$(TS)and **5**$_b$[−1]$_s$-0.8 kcal/mol⇆**2**$_b$[−1]$_s$ or **1**$_c$[−1]$_s$+1.1 kcal/mol⇆**5**$_c$[−1]$_s$(TS)and **5**$_c$[−1]$_s$-1.5 kcal/mol⇆**2**$_c$[−1]$_s$) as shown in Figure 3. These data are in agreement with the experimental findings and confirm that Fe-H$_2$ases can catalyze both the H-H formation and cleavage reactions efficiently. Because the H-H formation and cleavage directions are closely related to the total charge in the[FeFe]$_H$ part, as shown in Figure 4, the ET between [Fe$_4$S$_4$]$_H^{2+}$ and [FeFe]$_H$ is regarded as the driving force for the generation and consumption of hydrogen. The process from **2** to **1** corresponds to the splitting of the H-H bond as shown above.

Hall and co-workers[19,20] considered that a hydride is bound to the terminal Fe$_d$ in [FeFe]$_H$. Spectroscopically, such an intermediate remains undetected in enzyme systems, and no synthetic diiron thiolate models with terminally bound hydrides have ever been characterized[4]. In contrast, model complexes n-H$^+$=[Fe$^{II}$($\mu$-H)Fe$^{II}$] containing a bridge hydride[28-33] can be easily produced via the protonation of n=[Fe$^I$Fe$^I$] (n+H$^+$⇆n-H$^+$), and subsequently, H$_2$ is generated through the reaction n-H$^+$+2e+H$^+$⇆n+H$_2$[28,29]. The position of the bridging hydride in n-H$^+$ has been located by NMR spectra[29]. Thus, the probable existence of the structures of **2** and **3** proposed in this work could be experimentally verified. Zhao et al. pointed out that the formation of H$_2$ must be preceded by a reduction to the Fe$^{II}$Fe$^I$ redox state[30]; our calculations on models of types b and c, shown in Figure 3, supported this assumption. As structure **1** is the key to the present mechanism, experimental verifications of the existence of the Fe$_p$···H···CN two-electron three-center bond in Fe-H$_2$ases or related model complexes are vital for endorsing our mechanism. From the structural point of view, the Fe$_p$-CN bond in **1** is about 0.02−0.07 Å longer than those in models **2** and **3**, while the Fe$_p$C-N bond changes negligibly on the basis of the optimized geometries of models of types a and b, as compiled in Table 1; thus, the change of the Fe$_p$-CN bond can be an indicator for the existence of the two-electron three-center bond.

In addition, the **2**$_s$⇆**9**$_s$(TS)⇆**10**$_s$ and **10**$_s$⇆**11**$_s$(TS)⇆**12**$_s$ processes have relatively low reaction energies, from our previous work[35], and in $10_s$, the dihydrogen can practically be seen as a free $H_2$ molecule. The multivariable transition states $9_c[-1]_s$ and $11_c[-1]_s$ were solved using the LST/QST/CG method[38] as illustrated in Figure 5. Therefore, the three states ($2_s$, $10_s$, and $12_s$) could easily interchange from one to another with very low activation barriers. The $10_s \leftrightarrows 12_s$ process refers to the hopping of CO from an $Fe_d$ terminal site to a bridging site[7] ($Fe_p \cdots (\eta^2 - H_2) \cdots Fe_d + Fe_d\ CO \leftrightarrows Fe_p \cdots (CO) \cdots Fe_d + Fe_d - (\eta^2 - H_2)$). Finally, $H_2$ evolves from $Fe_d$[3,41].

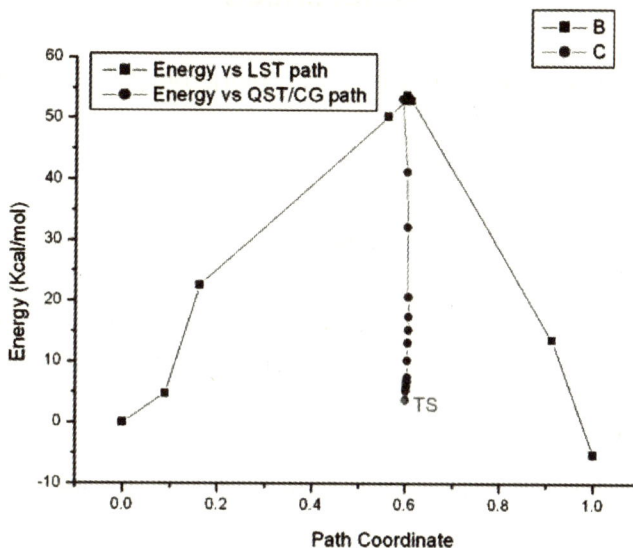

Figure 5   Search of transition state $11_c[-1]_s$ from $10_c[-1]_s$ to $12_c[-1]_s$ by means of the LST/QST/CG strategy.

Table 1   Optimal $Fe_P$-CN and $Fe_P$C-N Bond Lengths in Structures 1, 2, and 3(Å)

|  | $1_a[1]_{ox}$ | $3_a[1]_{ox}$ | $1_b[0]_{ox}$ | $3_b[0]_{ox}$ | $1_a[0]_s$ | $1_b[-1]_s$ | $2_b[-1]_s$ |
|---|---|---|---|---|---|---|---|
| $Fe_P$-CN | 1.926 | 1.862 | 1.912 | 1.865 | 1.912 | 1.904 | 1.876 |
| $Fe_P$C-N | 1.176 | 1.174 | 1.177 | 1.175 | 1.178 | 1.179 | 1.173 |

# 4   Conclusion

On the basis of the extended model studies of Fe-only hydrogenases with density functional theory, we identified a new intermediate state with an $Fe_p \cdots H \cdots CN$ two-electron three-center bond. This intermediate state is assumed to play an important role in the proton and electron transfers and, subsequently, in the reversible hydrogen oxidation in hy-drogenases, which are essential for the elucidation of the potent catalytic power of this type of enzymes to generate and consume hydrogen.

# Acknowledgment

This work has been supported by State Key Laboratory for Physical Chemistry of Solid States at Xiamen University. Partial support from Western Michigan University is gratefully acknowledged(Y. M.). We thank Dr Kai Tan and Dr Ze-Xing Cao for their help with the calculations of the transition state using the Materials Studio software package.

## Supporting Information Available

Cartesian coordinates for the optimal geometries of various models. This material is available free of charge via the Internet at http://pubs. acs. org.

## References

[1]Nicolet Y.,Cavazza C.,Fontecilla-Camps J. C. J. Inorg. Biochem. 2002,**91**,1.

[2]Frey M. ChemBioChem 2002,**3**,153.

[3]Cammack R. Nature1999,**397**,214.

[4]Evans D. J.,Pickett C. J. Chem. Soc. Rev. 2003,**32**,268.

[5]Rees D. C.,Howard J. B. Science 2003,**300**,929.

[6]Collman J. P. Nature Struct. Biol. 1996,**3**,213.

[7]Nicolet Y.,Piras C.,Legrand P.,Hatchikian E. C.,Fontecilla-Camps J. C. Structure 1999,**7**,13.

[8]Nicolet Y.,Lemon B. J.,Fontecilla-Camps J. C.,Peters J. W. TIBS 2000,**25**,138.

[9]Nicolet Y.,de Lacey A. L.,Vernede X.,Fernandez V. M.,Hatchikian E. C.,Fontecilla-Camps J. C. J. Am. Chem. Soc. 2001,**123**,1596.

[10]Peters J. W.,Lanzilotta W. N.,Lemon B. J.,Seefeldt L. C. Science 1998,**282**,1853(Errata:283, 35;283,2102).

[11]Peters J. W. Curr. Opin. Struct. Biol. 1999,**6**,670.

[12]Holm R. H.,Kennepohl P.,Solomon E. I. Chem. Rev. 1996,**96**,2239.

[13]Darensbourg M. Y.,Lyon E. J.,Smee,J. J. Coord. Chem. Rev. 2000,**206**,533.

[14]Darensbourg M. Y.,Lyon E. J.,Zhao X.,Georgakaki I. P. Proc. Natl. Acad. Sci. U. S. A. 2003, **100**,3683.

[15]Thauer R. K.,Klein A. R.,Hartmann G. C. Chem. Rev. 1996,**96**,3031.

[16]Stein M.,Lubitz W. J. Inorg. Biochem. 2004,**98**,862.

[17]Niu S.,Thomson L. M.,Hall M. B. J. Am. Chem. Soc. 1999,**121**,4000.

[18]Pavlov M.,Siegbahn P. E. M.,Blomberg M. R. A.,Crabtree R. H. J. Am. Chem. Soc. 1998,**120**, 548.

[19]Cao Z.,Hall M. B. J. Am. Chem. Soc. 2001,**123**,3734.

[20]Fan H.,Hall M. B. J. Am. Chem. Soc. 2001,**123**,3828.

[21]Liu Z. P.,Hu P. J. Am. Chem. Soc. 2002,**124**,5175.

[22]Adams M. W. W.,Stiefel E. I. Curr. Opin. Chem. Biol. 2000,**4**. 214.

[23]Pereira A. S.,Tavares P.,Moura I.,Moura J. G.,Huynh B. H. J. Am. Chem. Soc. 2001,**123**, 2771.

[24]Razavet M.,Borg S. J.,George S. J.,Best S. P.,Fairhurst S. A. Pickett,C. J. Chem. Commun. 2002,700.

[25]George S. J.,Cui Z.,Razavet M.,Pickett C. J. Chem. -Eur. J. 2002,**8**,4037.

[26]Dole F.,Fournel A.,Magro V.,Hatchikian E. C.,Bertrand,P Guigliarelli,B. Biochem. 1997, **36**,7847.

[27]Tard C. D.,Liu X.,Kalbrahim S.,Bruschi M.,Gioia L. D.,Davies S. C.,Yang X.,Wang L. -S., Sawers G.,Pickett C. J. Nature 2005,**433**,610.

[28] Gloaguen F., Lawrence J. D., Rauchfuss T. B. J. Am. Chem. Soc. 2001, **123**, 9476.

[29] Gloaguen F., Lawrence J. D., Rauchfuss T. B., Benard M., Rohmer M. -M. Inorg. Chem. 2002, **41**, 6573.

[30] Zhao X., Georgakaki I. P., Miller M. L., Mejia-Rodriguez R., Chiang C. Y., Darcensbourg M. Y. Inorg. Chem. 2002, **41**, 3917.

[31] Zhao X., Chiang C. -Y., Miller M. L., Rampersad M. V., Darens-bourg M. Y. J. Am. Chem. Soc. 2003, **125**, 518.

[32] Zhao X., Georgakaki I. P., Miller M. L., Yarbrough J. C., Darcensbourg M. Y. J. Am. Chem. Soc. 2001, **123**, 9710.

[33] Nehring J. L., Heinekey D. M. Inorg. Chem. 2003, **42**, 4288.

[34] Fiedler A. T., Brunold T. C. Inorg. Chem. 2005, **44**, 1794.

[35] Zhou T., Mo Y., Liu A., Zhou Z., Tsai K. Inorg. Chem. 2004, **43**, 923.

[36] (a) Delley B. J. Chem. Phys. 1990, **92**, 508. (b) Delley B. J. Chem. Phys. 1991, **94**, 7245. (c) Delley B. J. Chem. Phys. 2000, **113**, 7756. (d) Delley B. In Density Functional Theory: A Tool for Chemistry; Seminario J. M., Politzer, P., Eds. ; Elsevier: Amsterdam, The Netherlands, 1995.

[37] Perdew J. P., Chevary J. A., Vosko S. H., Jackson K. A., Pederson M. R., Fiolhais C. Phys. Rev. B 1992, **46**, 6671.

[38] (a) Halgren T. A., Lipscomb W. N. Chem. Phys. Lett. 1977, **49**, 225. (b) Ayala P. Y., Schlegel H. B. J. Chem. Phys. 1997, **107**, 375.

[39] (a) Noodleman L., Lovell T., Han W. -G., Li J., Himo F. Chem. Rev. 2004, **104**, 459. (b) Lovell T., Himo F., Han W. -G., Noodleman L. Coord. Chem. Rev. 2003, **238**, 211. (c) Noodleman L., Peng C. Y., Case D. A., Mouesa J. -M. Coord. Chem. Rev. 1995, **144**, 199.

[40] (a) Bruschi M., Fantucci P., Gioia L. Inorg. Chem. 2002, **41**, 1421.
(b) Bruschi M., Fantucci P., Gioia L. Inorg. Chem. 2003, **42**, 4773.
(c) Bruschi M., Fantucci P., Gioia L. Inorg. Chem. 2004, **43**, 3733.

[41] Igarashi R. Y., Laryukhin M., Dos Santos P. C., Lee H. -I., Dean D. R., Seefeldt L. C., Hoffman B. M. J. Am. Chem. Soc. 2005, **127**, 6231.

■ 本文原载：Polyhedron 25(2006)，pp. 1909～1914.

# pH Sependent Transformation of Nitrilotriacetato Molybdates(VI)-Synthesis, Spectral and Structural Characterization

Zhao-Hui Zhou[①], Jing Lin, Wen-Bin Yan, Hui Zhang, Khi-Rui Tsai

(*Department of Chemistry, College of Chemistry and Chemical Engineering and State Key Laboratory of Physical Chemistry of Solid Surfaces, Xiamen University, Xiamen 361005, China* )

Received 10 July 2005; accepted 5 December 2005 Available online 21 February 2006

**Abstract**  Investigation on the aqueous coordination chemistry of nitrilotriacetato molybdate(VI) resulted in the isolation of molybdenum (VI) nitrilotriacetato complexes $Na_{6n}\{[Mo_2O_5(ntaH)_2][(nta)_2Mo_2O_5]\}_n \cdot 12nH_2O$ (**1**) and $(NH_4)_{3n}[Mo_2O_5H(nta)_2]_n \cdot 4nH_2O$ (**2**) ($H_3nta =$ nitrilotriacetic acid). The complexes have been characterized by elemental analyses, IR, NMR spectroscopies and X-ray structural analyses. The anions of complexes **1** and **2** contain a linear $(O_2Mo)O(MoO_2)$ core with the bridging oxo group lying at a center of symmetry. Each molybdenum atom is octahedrally coordinated by two terminal oxo groups, a bridging oxygen and a tridentate nitrilotriacetate ligand, which binds through the nitrogen atom and two $\mu$-O carboxylates, while the other carboxylate remains free. Strong hydrogen bonds are found to exist in the two complexes. This is supported by IR and NMR spectra, and revealed by single crystal X-ray analysis. The complexes **1** and **2** show obvious decomposition in solution based on $^{13}$C NMR observations.

© 2006 Published by Elsevier Ltd.

**Key words**  Molybdenum  Molybdate  Nitrilotriacetic acid  Nitrilotriacetate  X-ray crystal structures  Hydrogen bonding

## 1  Introduction

The deprotonated trianion of nitrilotriacetic acid forms water-soluble salts with a plethora of polyvalent metal ions like Ti[1], V[2,3], Nb[4], Cr[5,6], Fe[7], Co[8,9] and Cu[10], in which the nitrilotriacetato ligand behaves as a tetradentate chelate that binds through the amino nitrogen and three carboxyl oxygen atoms. Among these complexes, several oxomolybdates feature tridentate chelation instead of tetradentate chelation[11-14]. Complex formation between molybdate and nitriotriacetate has been studied by UV, IR, EPR, potentiometric, enthalpimetric and NMR measurements[15-19]. Previous

---

①  Corresponding author. Tel.：+86 592 2184531；fax：+86 592 2183047.

*E-mail address*：zhzhou@xmu.edu.cn(Z.-H. Zhou).

studies revealed the existence of several complexes with Mo:nta molar/ratios of $1:1, 2:2$ and $2:1$ in solution, of which three are mononuclear and four are dinuclear complexes[15]. Up until now, two kinds of species have been determined by X-ray structure analysis as $K_3[MoO_3(nta)] \cdot H_2O^{[11]}$ and $M_2[Mo_2O_5(Hnta)_2] \cdot xH_2O$ $[M=Na, x=8; M=pyH, x=0; M=(n-C_4H_9)_4N, x=1]^{[12-14]}$. However, the other pH-dependant species remain unsolved.

For a better understanding of the Mo-nta system, here we report the syntheses, spectroscopic properties and structures of the molybdenum(Ⅵ) hydrogen nitrilotriacetato complexes $Na_{6n}\{[Mo_2O_5(ntaH)_2][(nta)_2Mo_2O_5]\}_n \cdot 12nH_2O$ (1) and $(NH_4)_{3n}[Mo_2O_5H(nta)_2]_n \cdot 4nH_2O$ (2), and their transformation at pH 2—7. Complexes 1 and 2 represent an adduct or an intermediate between the fully deprotonated and protonated nitrilotriacetato molybdates(Ⅵ) moieties. Strong hydrogen bonds are found to exist in the two complexes.

## 2　Experimental

### 2.1　Preparation of $Na_{6n}\{[Mo_2O_5(ntaH)_2][(nta)_2Mo_2O_5]\}_n \cdot 12nH_2O(1)$

Sodium molybdate(20 mmol) was added to a mixture of nitrilotriacetic acid(40 mmol) and sodium hydroxide(20 mmol). The solution was stirred in a water bath at 90 ℃ for 8 h. The mixture(pH 3.5) was filtered and evaporated to about 20 ml. The colorless crystals obtained after standing for two days were collected and washed with 50% ethanol and 95% ethanol. Yield:4.7 g(58%). Anal. Calc. for $C_{24}H_{50}Mo_4N_4Na_6O_{46}$: C, 17.4; H, 3.1; N, 3.4. Found: C, 17.4; H, 2.8; N, 3.3%. IR(KBr): $v$(COOH) $1735_w$; $v_{as}$(C＝O)$1652_{vs}$; $v_s$(C＝O)$1446_m$, $1383_{vs}$, $1340_s$; $v$(Mo＝O)$920_{vs}$, $885_b$; $v_{as}$(MoO$_b$Mo)$779_{vs}$; $v_s$(MoO$_b$Mo)$685_m$. $^1$H NMR $\delta_H$(500 MHz, D$_2$O, ppm):4.137(s). $^{13}$C NMR $\delta_C$(D$_2$O, ppm):180.2, 179.0 and 177.8(CO$_2$)$_\beta$, 65.5(CH$_2$).

### 2.2　Preparation of $(NH_4)_{3n}[Mo_2O_5H(nta)_2]_n \cdot 4nH_2O(2)$

Ammonium heptamolybdate(5 mmol) and nitrilotriacetic acid(35 mmol) were added to dilute ammonia(5 M, 40 ml). The solution(pH 4.0) was stirred until the color changed into light green and then it was evaporated to 20 ml. The colorless crystals obtained after standing overnight were collected and washed by 50% ethanol and 95% ethanol. Yield:8.4 g(62%). Anal. Calc. for $C_{12}H_{33}Mo_2N_5O_{21}$: C, 18.6; H, 4.3; N, 9.0. Found: C, 18.2; H, 4.0; N, 8.0%. IR(KBr): $v$(COOH)$1719_w$; $v_{as}$(C＝O)$1639_{vs}$; $v_s$(C＝O)$1448_s$, $1399_{vs}$, $1312_w$; $v$(Mo＝O)$929_{vs}$, $914_{vs}$; $v_{as}$(MoO$_b$Mo)$883_s$; $v_s$(MoO$_b$Mo)$690_m$. $^1$H NMR $\delta_H$(500 MHz, D$_2$O, ppm):3.976(s). $^{13}$C NMR $\delta_C$(D$_2$O, ppm):179.0(CO$_2$)$_\beta$, 65.7(CH$_2$).

Crystals of suitable quality for the subsequent X-ray diffraction studies were obtained as transparent prisms or rhombohedral blocks by slow evaporation of the related solution for one week at room temperature. The resulting crystals were sealed in a capillary to prevent loss of water molecules.

### 2.3　Transformation of 1-3 and 4

To the solution of $Na_{6n}\{[Mo_2O_5(ntaH)_2][(nta)_2Mo_2O_5]\}n \cdot 12nH_2O(1)$(pH 3.5), an appropriate amount of hydrochloric acid was added and the pH was adjusted to 2.5. The solution was then filtered and evaporated slowly. The crystals formed were collected and washed. The structural analysis shows the product as $Na_2[Mo_2O_5(Hnta)_2]$(3). Anal. Calc. for $C_{12}H_{14}Mo_2N_2Na_2O_{17}$: C, 20.7; H, 2.0; N, 4.0. Found: C, 20.0; H, 2.4; N, 3.8%. IR(KBr): $v$(COOH)$1750_s$, $v_{vs}$(C＝O)$1673_{vs}$, $1600_v$; $v_s$(C＝O)

$1428_m$,$1401_s$,$1339_s$,$1318_s$;$v$($Mo=O$)$982_m$,$947_s$,$912_{vs}$,$883_s$;$v_{as}$($MoO_bMo$)$785_{vs}$;$v_s$($MoO_bMo$)$695_m$.$^1$H NMR $\delta_H$(500 MHz,$D_2O$,ppm):4. 119(s).$^{13}$C NMR $\delta_c$($D_2O$,ppm):174. 8($CO_2$)$_\beta$,62. 5($CH_2$).

In another experiment,sodium hydroxide was added and the pH was adjusted to 7. 0. The solution was evaporated slowly. The crystals formed were collected and washed. The structural analysis shows the product as $Na_3$[$MoO_3$($nta$)] $\cdot$ $4H_2O$(**4**). *Anal.* Calc. for $C_6H_{14}MoNNa_3O_{13}$:C,15. 2;H,3. 0;N, 3. 0. Found:C,15. 1;H,2. 8;N,2. 9%. IR(KBr):$v_{vs}$($C=O$)1644$_s$,$1602_{vs}$($C=O$)1439$_m$,1409$_s$;$v$($Mo=O$)939$_{ms}$,897$_s$,857$_{vs}$.$^1$H NMR $\delta_H$(500 MHz,$D_2O$,ppm):3. 982(s).$^{13}$C NMR $\delta_c$($D_2O$,ppm):178. 1($CO_2$)$_\beta$,64. 6($CH_2$).

## 2. 4 Physical measurements

Infrared spectra were recorded as Nujol mulls between KBr plates using a Nicolet 360 FT-IR spectrometer. Elemental analyses were performed using EA 1110 elemental analyzers. $^1$H NMR and $^{13}$C NMR spectra were recorded on a Varian UNITY 500 NMR spectrometer with $D_2O$ using DSS(sodium 2,2-dimethyl-2-silapentane-5-sulfonate)as internal reference.

## 2. 5 X-ray data collection,structure solution and refinement

Crystallographic data for the nitrilotriacetato molybdates of **1 — 4** are summarized in Table 1. Diffraction data were collected on Bruker Smart CCD diffractometer with graphite monochromated Mo Kα(λ=0. 71073 Å)radiation at 296 K. Empirical absorption corrections were applied using the SADABS program[20]. The structures were primarily solved by the WINGX program[21] and refined by full-matrix least-squares procedures with anisotropic thermal parameters for all of the non-hydrogen atoms with SHELXL-97[22]. Hydrogen atoms were located from the difference Fourier map and refined isotropically.

**Table 1    Crystallographic data for $Na_{6n}${[$Mo_2O_5$($ntaH$)$_2$][($nta$)$_2Mo_2O_5$]}$_n$ $\cdot$ $12nH_2O$(1),**
**($NH_4$)$_{3n}$[$Mo_2O_5$H($nta$)$_2$]$_n$ $\cdot$ $4nH_2O$(2),$Na_2$[$Mo_2O_5$($Hnta$)$_2$](3)and $Na_3$[$MoO_3$($nta$)] $\cdot$ $4H_2O$(4)**

| Complex | 1 | 2 | 3 | 4 |
|---|---|---|---|---|
| Empirical formula | $C_{24}H_{50}Mo_4N_4Na_6O_{46}$ | $C_{12}H_{33}Mo_2N_5O_{21}$ | $C_{12}H_{14}Mo_2N_2Na_2O_{17}$ | $C_6H_{14}MoNNa_3O_{13}$ |
| Formula weight | 1652. 38 | 775. 31 | 696. 11 | 473. 09 |
| Crystal system/space group | triclinic,$P\bar{1}$ | triclinic,$P\bar{1}$ | orthorhombic,$Pbca$ | triclinic,$P\bar{1}$ |
| $a$(Å) | 10. 0630(6) | 6. 7850(4) | 13. 0373(9) | 8. 0291(5) |
| $b$(Å) | 11. 6450(7) | 9. 6462(5) | 11. 2693(8) | 9. 4837(6) |
| $c$(Å) | 12. 9230(8) | 10. 0768(6) | 14. 257(1) | 10. 1010(6) |
| $\alpha$(°) | 80. 381(1) | 85. 571(1) | | 84. 267(1) |
| $\beta$(°) | 85. 258(1) | 82. 124(1) | | 87. 065(1) |
| $\gamma$(°) | 66. 936(1) | 82. 994(1) | | 77. 779(1) |
| $V$(Å$^3$) | 1373. 5(1) | 647. 19(6) | 2094. 7(3) | 747. 60(8) |
| $Z$ | 1 | 1 | 4 | 2 |
| $F$(000) | 824 | 392 | 1368 | 472 |
| $D_c$(g cm$^{-3}$) | 1. 998 | 1. 989 | 2. 207 | 2. 102 |
| $\mu$(mm$^{-1}$) | 1. 061 | 1. 072 | 1. 331 | 1. 035 |

续表

| Complex | 1 | 2 | 3 | 4 |
|---|---|---|---|---|
| Crystal size(mm³) | 0. 12×0. 10×0. 05 | 0. 16×0. 13×0. 06 | 0. 15×0. 10×0. 10 | 0. 18×0. 12×0. 10 |
| $\theta$ Range(°) | 1. 60—28. 34 | 2. 04—26. 00 | 2. 78—28. 39 | 2. 03—28. 47 |
| Radiation($\lambda$/Å) | Mo K$\alpha$(0. 7107) | | | |
| $T$(K) | 296(2) | | | |
| GOF on $F^2$ | 0. 895 | 1. 083 | 1. 200 | 1. 027 |
| $R_1[I>2\sigma(I)]$ | 0. 0463 | 0. 0367 | 0. 0767 | 0. 0506 |
| $wR_2[I>2\sigma(I)]$ | 0. 0706 | 0. 0934 | 0. 1403 | 0. 1383 |
| $R_1$(all data) | 0. 0718 | 0. 0389 | 0. 1103 | 0. 0552 |
| $wR_2$(all data) | 0. 0760 | 0. 0947 | 0. 1534 | 0. 1421 |
| Maximum and minimum transmission $\Delta\rho$(e Å³) | 0. 788 and -0. 562 | 0. 677 and -0. 794 | 1. 040 and -1. 673 | 1. 932 and -1. 341 |

# 3  Results and discussion

## 3. 1  Synthesis

Nitrilotriacetic acid is able to provide tetradentate chelation and different protonated forms,which might be expected to produce various types of anion complexes with molybdenum (Ⅵ). Previous potentiometric,spectrophotometric and enthalpimetric titrations in the molybdate(Ⅵ)-nitrilotriacetate system have been proceeded in the pH range 1. 0—9. 0 in solution[15]. The formation of the products may be represented by the general equation

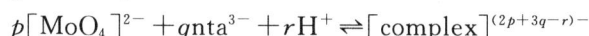

$$p[MoO_4]^{2-}+q\text{nta}^{3-}+r\text{H}^+\rightleftharpoons[\text{complex}]^{(2p+3q-r)-}$$

The major species are proposed to have Mo:nta:H ratios of 1:1:2,1:1:3,1:1:4,2:2:7,2:2:8,2:1:5 and 2:1:6 in solution,depending mainly on the pH[15]. In our experiment,only the 1:1:2,2:2:6,2:2:7 and 2:2:8 species of nitrilotriacetato molybdates are separated in the acidic,weak acidic and neutral aqueous media. The transformation and the isolation of nitrilotriacetate complexes in different protonated forms have proven that 1:1 and 2:2 ratios (Mo:nta) of nitrilotriacetato molybdates are the major species. The 2:1 species is unfavorable due to the competitive formation of homopolyacid.

In an acidic solutions of pH 2. 0—3. 0,nitrilotriacetic acid can react with a Mo(Ⅵ)complex to form the protonated anion [Mo₂O₅(Hnta)]²⁻, while in neutral solutions of pH 5. 0—7. 0, mononuclear [MoO₃(nta)]³⁻ is formed. The adduct of protonated and deprotonated species Na₆ₙ-{[Mo₂O₅(ntaH)₂][(nta)₂Mo₂O₅]}ₙ • 12nH₂O(1) and hydrogen nitrilotriacetato molybdate(NH₄)₃ₙ[Mo₂O₅H(nta)₂]ₙ • 4nH₂O(2)were isolated in weak acidic solutions of pH 3. 5—4. 0. The later represents an intermediate between protonated [Mo₂O₅(Hnta)₂]²⁻ and deprotonated [Mo₂O₅(nta)₂]⁴⁻ molybdenum (Ⅵ)-nitrilotriacetate complexes. In compound 1,the two uncoordinated β-carboxyl groups of nitrilotriacetate in one dimer are protonated,while those in the other are deprotonated. This is supported by a downshift of the $v$(COOH)vibration from 1750 in 3 to 1735 cm⁻¹ in 1. Transformations between 1,3 and 4 can be accomplished by adjusting the pH. The pH values in the reactions are controlled easily by hydrochloric acid or sodium hydroxide as shown in Scheme 1. The transformations of monomeric and dimeric nitrilotriacetato molybdate ( VI ) are different from that described previously, in which only the monomeric species is considered in the precursor of a NiMo/SiO$_2$ hydrotreating catalyst[23].

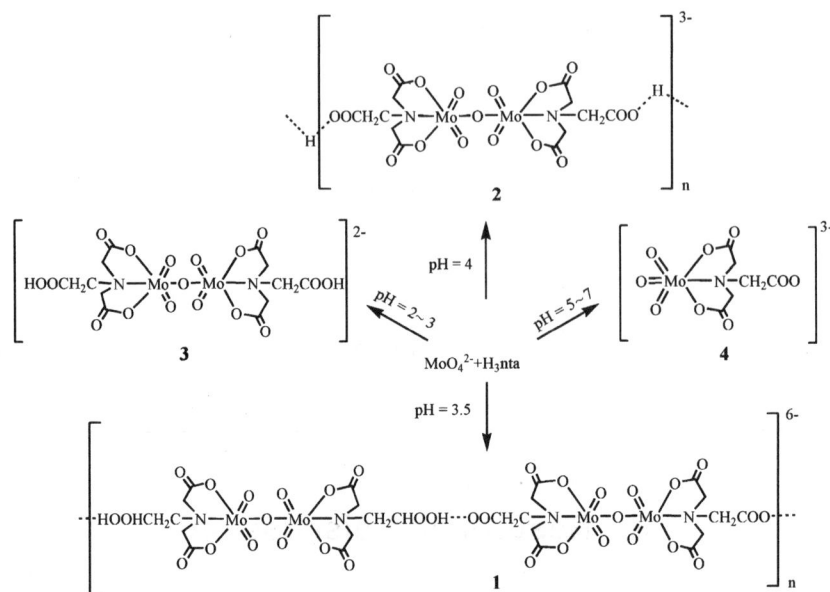

**Scheme 1   Synthesis and transformation of nitrilotriacetato molybdates( VI ).**

## 3.2   Structure descriptions of complexes 1 and 2

Complex **1** consists of a molybdenum-nitrilotriacetate adduct in a 2 : 2 stoichiometric ratio. The perspective view of the anion is presented in Fig. 1. Each molybdenum atom is octahedrally surrounded by two terminal oxygen atoms, one bridging oxygen atom and a tridentate nitrilotriacetate ligand, which binds through two β-carboxyl oxygen atoms and one nitrogen atom. The non-coordi-nated β-carboxyl oxygen O15 is protonated. This is supported by the relatively long C16-O15 bond, 1.307(9) Å, as compared with C16-O16, 1.214(9) Å. However, the noncoordinated carboxyl groups of the other do not show a distinct difference. The two neighboring carboxylate oxygen atoms are linked by a strong

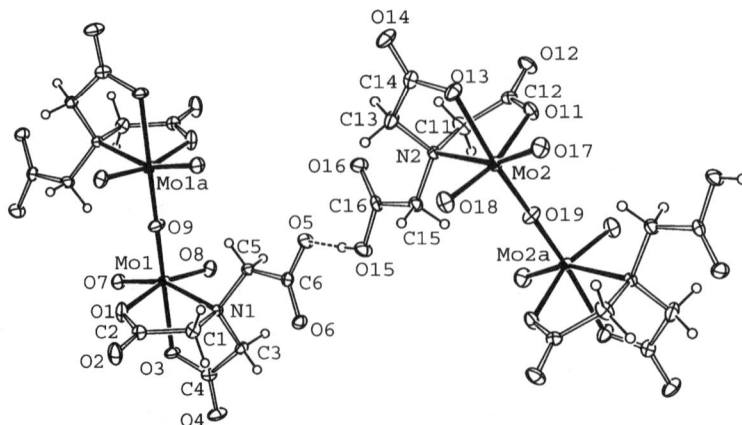

**Fig. 1   ORTEP plot of the anion in Na$_{6n}$ { [ Mo$_2$O$_5$ ( ntaH )$_2$ ] [ ( nta )$_2$ Mo$_2$O$_5$ ] }$_n$ · 12$n$H$_2$O( 1 ) at the 30% probability level.**

hydrogen bond[O5···H-O15 2.481(3)Å], generating a one-dimensional coordination polymer through the aggregation of fully protonated species [Mo$_2$O$_5$(Hnta)$_2$]$^{2-}$ and deprotonated units [Mo$_2$O$_5$ (nta)$_2$]$^{4-}$.

Fig. 2 shows a perspective view of the anion in complex 2. The crystal structure comprises ammonium cations, hydrogen oxo-$\mu$-bis[*trans*-nitrilotriacetato-*cis*-dioxomolybdate] anions and lattice water molecules. The dimeric anion consists of MoO$_2$(Hnta) and MoO$_2$(nta) units bridged by an oxygen linkage[Mo-O-Mo = 180.00(3)°]. The two oxo atoms are *cis* to each other in the coordination octahedron. The two *cis*-dioxo groups are in a *trans*-configuration, which is the same as that found in tripotassium oxo-$\mu$-bis[*trans*-nitrilotriacetato-*cis*-dioxotungstate] K$_3$[W$_2$O$_5$H(nta)$_2$] · 2H$_2$O[W-O-W = 161.9(7)°][24]. The other three sites are occupied by the nta entity. Adjacent Mo$_2$O$_5$H(nta)$_2$ units are linked into an infinite zigzag chain through the acidic hydrogen[O6···H···O6b=2.486(3)Å]. The space between the chains is occupied by the ammonium cations and the lattice water molecules.

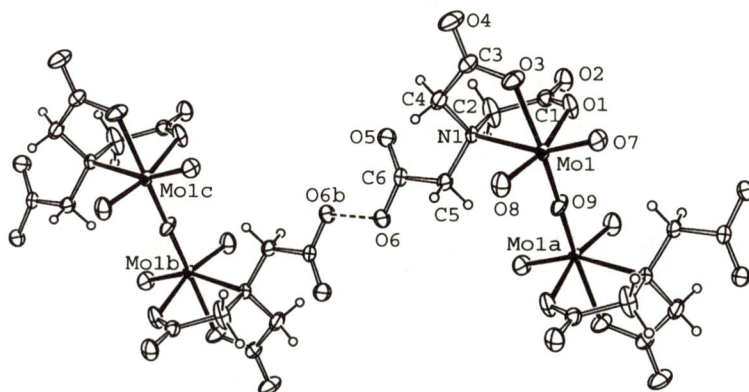

Fig. 2  ORTEP plot of the anion in(NH$_4$)$_{3n}$[Mo$_2$O$_5$H(nta)$_2$]$_n$ · 4$n$H$_2$O(2)at the 30% probability level.

The anion structures of compounds 3 and 4 are similar to the previous reported structures[11-14]. Evidently, the uncoordinated $\beta$-carboxyl groups in complex 3 are fully protonated, this is not only showed by the visibility of hydrogen atoms in difference maps, but also by the difference between the C-O distances of the terminal carboxylates, as well as from the consideration of charge balance. The conclusion of full deprotonation of the monomer 4 can be drawn from the observed carbon-oxygen bond distances of carboxyl groups that are equivalent [1.259(5),1.251(5)Å](see Figs. 3 and 4).

As shown in Table 2, the Mo-O distances in nitrilotriacetato molybdates

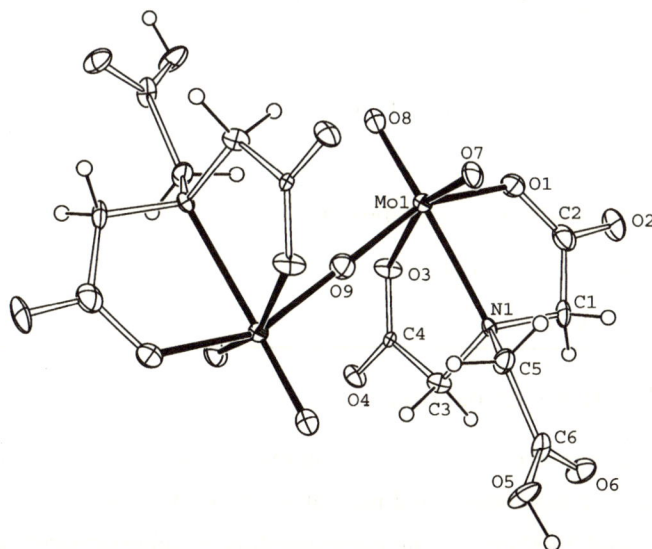

Fig. 3  ORTEP plot of the anion in Na$_2$[Mo$_2$O$_5$(Hnta)$_2$] (3)at the 30% probability level.

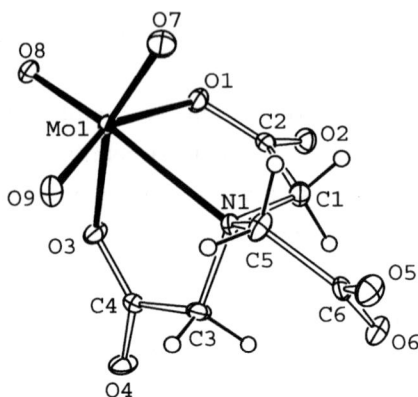

**Fig. 4  ORTEP plot of the anion in Na$_3$[MoO$_3$(nta)] · 4H$_2$O(4) at the 30% probability level.**

vary systematically. The Mo=O bond lengths lie in the range 1.694—1.737 Å, indicating that they are double bonds. The observed O=Mo=O angles of 106.6(3), 106.1(3), 106.4(2), 105.8(2), 106.7(2), 104.9(2) and 105.8(1)° in compounds 1—4 are consistent with the large repulsion between oxo groups with short bonds to the metal atom. The bridging distances Mo-O$_b$-Mo are 1.8846(6), 1.8774(7), 1.8714(7) and 1.8724(3) Å. The Mo-N bonds are similar in all four structures, being in the range 2.37—2.46 Å.

**Table 2  Comparisons of the Mo-O and Mo-N distances(Å) in the complexes of nitrilotriacetato molybdates(VI)**

| Compound | Mo-O$_t$ | Mo-O$_b$ | Mo-O$_{c1}$ | Mo-O$_{c2}$ | Mo-N | Ref. |
|---|---|---|---|---|---|---|
| K$_3$[MoO$_3$(nta)] · H$_2$O | 1.737(5) | | 2.176(4) | 2.245(3) | 2.388(5) | [11] |
| Na$_2$[Mo$_2$O$_5$(Hnta)$_2$] · 8H$_2$O | 1.695(6) | 1.880(1) | 2.069(5) | 2.134(5) | 2.410(6) | [12] |
| (pyH)$_2$[Mo$_2$O$_5$(Hnta)$_2$] | 1.695(3) | 1.870(3) | 2.071(3) | 2.145(3) | 2.413(3) | [13] |
| [(n-C$_4$H$_9$)N]$_2$[Mo$_2$O$_5$(Hnta)$_2$] · H$_2$O | 1.702(7) | 1.854(7) | 2.077(7) | 2.144(6) | 2.457(7) | [14] |
| Na$_2$[Mo$_2$O$_5$(Hnta)$_2$] | 1.703(5) | 1.8846(6) | 2.029(5) | 2.187(5) | 2.394(6) | |
| Na$_{6n}$\{[Mo$_2$O$_5$(ntaH)$_2$][(nta)$_2$Mo$_2$O$_5$]\}$_n$ · 12nH$_2$O | 1.697(6) | 1.8774(7) | 2.083(5) | 2.110(5) | 2.436(6) | |
| | 1.698(5) | 1.8714(7) | 2.074(5) | 2.115(5) | 2.377(6) | this work |
| (NH$_4$)$_{3n}$[Mo$_2$O$_5$H(nta)$_2$]$_n$ · 4nH$_2$O | 1.695(3) | 1.8724(3) | 2.059(3) | 2.162(3) | 2.405(3) | |
| Na$_3$[MoO$_3$(nta)] · 4H$_2$O | 1.728(3) | | 2.186(3) | 2.214(3) | 2.395(3) | |
| Average(Å) | 1.706 | 1.873 | 2.092 | 2.162 | 2.408 | |

## 3.3  IR spectroscopy

The infrared spectra of complexes 1—4 show strong and sharp absorption bands for the carboxylate groups of the coordinated nitrilotriacetate. Asymmetric stretching vibrations $v_{as}$(COO$^-$) appear between 1673 and 1600 cm$^{-1}$. The corresponding symmetric stretches $v_s$(COO$^-$) appear between 1448 and 1318 cm$^{-1}$. Compound 3 gives a typical band of a non-bonded and undissociated carboxylic acid group at 1750 cm$^{-1}$. The weak absorption bands at 1735 and 1719 cm$^{-1}$ for compounds 1 and 2 indicate strong H-bonds between the free carboxylic acid and carboxyl groups. All of the carboxyl absorptions are shifted to lower frequencies with respect to those of the free nitrilotriacetic acid due to the influence of hydrogen bonds and counterions. These observations are in accord with the formation of a chelate ring and the bridging mode of the nitrilotriacetate ligand. Loss of the proton in compounds **1** and **2** and the absence of a oxo-bridge group in compound **4** reduce the number of bands and displace them to lower frequencies. In the region between 982 and 857 $cm^{-1}$, these complexes show several bands that result from the presence of *cis*-dioxo cores in two different environments. The low frequency symmetric $MoO_2$ stretching may be explained by intramolecular hydrogen bonding and the coordination of a sodium or ammonium cation. The strong IR band only observed for the dimers around 700 $cm^{-1}$ may be attributed to the $Mo-O_b-Mo$ bridges. The absence of IR bands between $1740-1700$ $cm^{-1}$ and $790-775$ $cm^{-1}$ indicate full deprotonation of β-carboxyl group in compound **4**[25].

## 3.4 Solution NMR spectroscopy

The solution $^{13}C$ NMR spectra of complexes **1—4** were measured in $D_2O$. The $^{13}C$ NMR spectra of **1** and **2** in Fig. 5 show two sets of resonances, which can be attributed to the nitrilotriacetate ligands in the molybdenum complexes and the decomposed free nitrilotriacetate [$Na_2Hnta$: $^{13}C$ NMR $\delta_C$ ($D_2O$) 171. 3($CO_2$)$_\beta$, 58. 1($CH_2$)]. Complex **1** shows one resonance in the high-field region(65. 5 ppm) for the methylene groups of the nitrilotriacetate ligands. The resonances at 180. 2, 179. 0 and 177. 8 ppm in the lower-field region can be assigned to the β-carboxylate carbons coordinated to the Mo( Ⅵ )central ion. The $^{13}C$ NMR spectra of **2—4** resemble one another. Large downfield shifts of the β-carboxylate($\Delta\delta$ 7. 7, 4. 4, 6. 8) and methylene($\Delta\delta$ 7. 6, 3. 5, 6. 5) carbons compared with the free nitrilotriacetate are observed for **2—4**, respectively. The downfield shifts of the β-carboxylate and $CH_2$ groups are in agreement with those in earlier studies[16,26]. The $^1H$ NMR spectra of all four complexes show one wide peak for the methylene protons, which may arise from the similar magnetic environment in solution. The similarities of the NMR spectra of the isolated Mo( Ⅵ )complexes suggest that the coordination mode of the nitrilotriacetate ligand is similar not only in the solid state but also in solution. The very strong hydrogen bonds in **1** and **2** may be related to the downshift of the β-carboxyl group from 177. 8 to 179. 0 ppm.

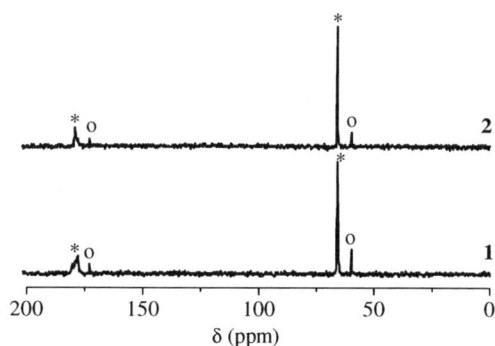

Fig. 5 $^{13}C$ NMR spectra of the complexes: (bottom)$Na_{6n}\{[Mo_2O_5(ntaH)_2][(nta)_2Mo_2O_5]\}_n \cdot 12nH_2O(1)$; (top)$(NH_4)_{3n}[Mo_2O_5H(nta)_2]_n \cdot 4nH_2O(2)$. The symbol ' * ' indicates resonance signals of the coordinated ligand, 'o' of the free ligand.

# Acknowledgements

We thank the Ministry of Science and Technology (001CB108906 and 2005CB221408) and the National Science Foundation of China (20571061, 20021002 and 20423002) for their generous financial support.

# Appendix A. Supplementary data

Crystallographic data are available from the CCDC, 12 Union Road, Cambridge CB2 1EZ, UK on request, quoting the Deposition Nos. 271976 — 271979 for compounds **1 — 4**. Supplementary data associated with this article can be found, in the online version, at doi:10.1016/j. poly. 2005.12.017.

# References

[1]K. Wieghart, U. Quilitzsch, J. Weiss, B. Nuber, Inorg. Chem. **19**(1980)2514.

[2]K. Okamoto, J. Hidaka, M. Fukagawa, K. Kanamori, Acta Crystallogr., Sect. C **48**(1992)1025.

[3]Q. Z. Zhang, C. Z. Lu, W. B. Yang, S. M. Chen, Y. Q. Yu, X. He, Y. Yan, J. H. Liu, X. J. Xu, C. K. Xia, L. J. Chen, X. Y. Wu, Polyhedron **23**(2004)1975.

[4]M. Sokolov, H. Imoto, T. Saito, Inorg. Chem. Commun. **3**(2000)96.

[5]H. G. Visser, W. Purcell, S. S. Basson, Polyhedron **18**(1999)2795.

[6]G. Novitchi, J. P. Costes, V. Ciornea, S. Shova, I. Filippova, Y. A. Simonov, A. Gulea, Eur. J. Inorg. Chem. (2005)929.

[7]S. L. Heath, A. K. Powell, H. L. Utting, M. Helliwell, J. Chem. Soc., Dalton Trans. (1992)305.

[8]H. G. Visser, W. Purcell, S. S. Basson, Polyhedron **20**(2001)185.

[9]N. R. Lien, M. A. Timmons, G. J. H. Belkin, J. R. Holst, M. L. Janzen, R. Kanthasamy, W. J. Lin, A. Mubayi, M. I. Perring, L. M. Rupert, A. Saha, N. J. Schoenfeldt, A. N. Sokolov, J. R. Telford, Inorg. Chim. Acta **358**(2005)1284.

[10]X. Q. Lu, J. J. Jiang, C. L. Chen, B. S. Kang, C. Y. Su, Inorg. Chem. **44**(2005)4515.

[11]R. J. Butcher, B. R. Penfold, Cryst. Mol. Struct. **6**(1976)13.

[12]C. B. Knobler, W. T. Robinson, C. J. Wilkins, A. J. Wilson, Acta Crystallogr., Sect. C **39**(1983) 443.

[13]K. Matsumoto, Y. Marutani, S. Ooi, Bull. Chem. Soc. Jpn. **57**(1984)2671.

[14]S. C. Liu, L. D. Ma, D. McGowty, J. Zubieta, Polyhedron **9**(1990)1541.

[15]J. J. Cruywagen, J. B. B. Heyns, E. A. Rohwer, J. Chem. Soc., Dalton Trans. (1994)45.

[16]K. F. Miller, R. A. D. Wentworth, Inorg. Chem. **17**(1978)2679.

[17]S. Funahashi, Y. Kato, M. Nakayama, M. Thanaka, Inorg. Chem. **20**(1981)1752.

[18]S. I. Chan, R. J. Kula, D. T. Sawyer, J. Am. Chem. Soc. **86**(1964)377.

[19]K. Majlesi, K. Zare, F. Teimouri, J. Chem. Eng. Data **48**(2003)680.

[20]SADABS, University of Göttingen, Germany, 1997.

[21]L. J. Farrugia, J. Appl. Crystallogr. **32**(1999)837.

[22]G. M. Sheldrick, SHELXL-97 and SHELXL-97, University of Göttingen, Göttingen, Germany, 1997.

[23]L. Medici, R. Prins, J. Catal. **163**(1996)28.

[24]Z. H. Zhou, S. Y. Hou, Z. J. Ma, H. L. Wan, K. R. Tsai, S. W. Ng, Inorg. Chem. Commun. **5** (2002)388.

[25]N. Kazuo, Infrared and Raman Spectra of Inorganic and Coordination Compounds, Wiley, New York, 1997, pp. 59—62;168—169.

[26]M. A. Freeman, F. A. Schultz, C. N. Reilley, Inorg. Chem. **21**(1982)567.

■ 本文原载：Inorg. Chem. 2006,45,pp. 8447～8451.

# Syntheses, Spectroscopies and Structures of Molybdenum(Ⅵ) Complexes with Homocitrate

Zhao-Hui Zhou①, Shu-Ya Hou, Ze-Xing Cao, Khi-Rui Tsai, Yuan L.Chow

(Department of Chemistry, College of Chemistry and
Chemical Engineering and State Key Laboratory of Physical Chemistry of
Solid Surfaces, Xiamen University, Xiamen 361005, China )

Received July 29,2006

**Abstract**    Initial investigations into the possible role of homocitric acid in iron molybdenum cofactor (FeMo-co) of nitrogenase lead us to isolate and characterize two tetrameric molybdate(Ⅵ) species. The complexes $K_2(NH_4)_2[(MoO_2)_4O_3(R,S\text{-}Hhomocit)_2] \cdot 6H_2O(1)$ and $K_5[(MoO_2)_4O_3(R,S\text{-}Hhomocit)_2]$ $Cl \cdot 5H_2O(2)$ (homocitric acid = $H_4$ homocit, $C_7H_{10}O_7$) are prepared from the reactions of acyclic homocitric acid and molybdates, which represent the first synthetic structural examples of molybdenum homocitrate complexes. The homocitrate ligand trapped by tetranuclear molybdate coordinates to the molybdenum (Ⅵ) atom through $\alpha$-alkoxy and $\alpha$-, $\beta$-carboxy groups. The physical properties, structural parameters, and their possible biological relevances are discussed.

## 1　Introduction

Homocitric acid, which exists in the $\gamma$-lactone form in solution[1−10], is an integral part of iron molybdenum cofactor (FeMo-co) in nitrogenase[11−13]. While the precise role of homocitrate in the dinitrogen reduction pathway is poorly understood, the discovery and elucidation of the interactions between molybdenum and this tricarboxylic acid are needed in nitrogenase biochemistry[14,15]. Homocitrate is believed to be a key factor during biological nitrogen fixation as contrasted to the other alternative hydroxycarboxylates, such as citrate and malate, which greatly reduce the activities of the dinitrogen reduction[16−18]. The homocitrato entity in FeMo-co uses its oxygen atoms of $\alpha$-alkoxy and $\alpha$-carboxy groups to chelate the molybdenum atom, forming a giant cluster $MoFe_7S_9X(S\text{-}cys)(N\text{-}His)$ (homocit)[19−27]. It is proposed that homocitrate may form an intramolecular hydrogen bond with the imidazole group of histidine[14,28]. It may facilitate the binding of dinitrogen molecule through the dissociation of either the bound $\alpha$-carboxy or $\alpha$-alkoxy group from the molybdenum atom[29−34]. In vitro biosynthesis experiments lacking homocitrate show no[99] Mo incorporation in any proteins other than the Mo-storage enzyme[35].

①  To whom correspondence should be addressed. E-mail：zhzhou@ xmu. edu. cn. Tel：+86-592-2184531. Fax：+86-592-2183047.

However, the isolation of molybdenum homocitrate complexes in a crystalline form suitable for X-ray structural analysis has been proved to be difficult[36], leaving the interaction between molybdate and homocitrate poorly understood and requiring further attention. Taking into consideration the syntheses of homocitrato lactone and homocitrato vanadate[3,37], two yet undocumented tetrameric homocitrato molybdates($\text{VI}$), $[(MoO_2)_4O_3(R,S\text{-Hhomocith})_2]^{4-}$, have been reported.

# 2  Experimental Section

All experiments were carried out in the open air. All chemicals were analytical reagents and used without further purification.

**Preparation of $K_2(NH_4)_2[(MoO_2)_4O_3(R,S\text{-Hhomocit})_2] \cdot 6H_2O(1)$ and $K_5[(MoO_2)_4O_3(R,S\text{-Hhomocit})_2]Cl \cdot 5H_2O(2)$.** A prepared racemic homocitrato γ-lactone acid[3] (0.14 g, 0.7 mmol) was dissolved in a minimal amount of water. The pH value was adjusted by potassium hydroxide to 11 to generate the acyclic homocitrate as monitored with HPLC. After hydrolysis, ammonium paramolybdate (0.26 g, 0.2 mmol) was added in small portions over a few minutes. Upon cooling in an ice bath for 15 min, the pH value of the solution was adjusted to 2 to induce complex formation. The salts of chloride were filtrated with precipitation. Complex **1** was formed after sitting at room temperature for a few days. Yield: 0.49 g(56%). Found(calcd for $C_{14}H_{34}N_2O_{31}K_2Mo_4$): C, 14.0(14.1); H, 3.0(2.9). IR(KBr, cm$^{-1}$): $v_{as(COOH)}$ 1702$_s$, $v_{as(coo)}$ 1629$_s$, 1578$_s$; $v_{s(coo)}$ 1422$_s$, 1392$_{vs}$; $v_{s(Mo=O)}$ 926$_s$, 865$_{vs}$ cm$^{-1}$. $^1$H NMR (500 MHz, D$_2$O): $\delta_H$ 2.937(d, $J = 18.0$ Hz, CH$_2$), 2.860(d, $J = 18.5$ Hz, CH$_2$), 2.511~2.478, 2.378~2.340, 2.169~2.123, 1.892~1.860(m, CH$_2$). $^{13}$C NMR in D$_2$O: see Table 1.

**Table 1  $^{13}$C NMR Spectral Data(in ppm)of Complexes 1 and 2 and Homocitrato γ-Lactone Acid$^a$**

| compound | ($\equiv$CO) | (CO$_2$)$_\alpha$ | (CO$_2$)$_\beta$ | (CO$_2$)$_\gamma$ | ($=$CH$_2$) |
|---|---|---|---|---|---|
| homocitrato γ-lactone acid | 83.5 | 176.5 | 172.5 | 170.5 | 41.7, 31.9, 28.3 |
| **1** | 90.0(6.5) | 186.5(10.0) | 181.8(9.3) | 180.7(10.2) | 47.7(6.0), 37.1(5.2)31.7(3.4) |
| **2** | 90.2(6.7) | 186.6(10.1) | 182.0(9.5) | 181.1(10.6) | 47.8(6.1), 37.3(5.4), 31.9(3.6) |

$^a$ $\Delta\delta$ values are given in brackets.

The same procedure was applied to prepare complex **2**, and potassium molybdate(0.95 g, 4.0 mmol) was allowed to react with homocitric acid(0.48 g, 2.0 mmol). Yield 0.45 g(34%). Found(calcd for $C_{14}H_{24}O_{30}ClK_5Mo_4$): C, 12.6(13.1); H, 2.0(1.9). IR(KBr, cm$^{-1}$): $v_{as(COOH)}$ 1716$_s$, $v_{as(COO)}$ 1638$_{vs}$; $v_{s(COO)}$ 1422$_s$, 1401$_{vs}$; $v_{s(Mo=O)}$ 920$_s$, 892$_{vs}$ cm$^{-1}$. $^1$H NMR(500 MHz, D$_2$O): $\delta_H$ 2.81~2.99, 2.60~2.76, 2.32~2.42, 2.267, 2.12~2.24, 1.84~1.94(CH$_2$). $^{13}$C NMR in D$_2$O: see Table 1.

**Physical Measurements.** Infrared spectra were recorded as Nujol mulls between KBr plates using a Nicolet 360 FT-IR spectrometer. $^1$H and $^{13}$C NMR spectra were recorded in D$_2$O on a Varian UNITY 500 NMR spectrometer or a Bruker AV400 NMR spectrometers using DSS(sodium 2,2-dimethyl-2-silapentane-5-sulfonate)as an internal reference. Elemental analyses were performed using an EA 1110 element analyzer.

**X-ray Structure Determination.** Diffraction data were collected on a Bruker Smart Apex CCD diffractometer with graphitemonochromated Mo Kα radiation at 296 K. The structures were solved by SHELXS-97 and refined by full-matrix least-squares procedures with anisotropic thermal parameters for all of the nonhydrogen atoms. Hydrogen atoms were located from a difference Fourier map. All calculations were performed on a microcomputer using SHELXL-97 and SHELXS-97 programs[38,39]. Crystallographic data for homocitrato molybdates **1** and **2** are summarized in Table 2. The large differential peaks and holes may be related to the heavy molybdenum atom. Selected bond distances and angles are given in Table 3.

**Table 2  Crystal Data and Structure Refinements for $K_2(NH_4)_2[(MoO_2)_4O_3(R,S\text{-Hhomo})_2] \cdot 6H_2O(1)$ and $K_5[(MoO_2)_4O_3(R,S\text{-Hhomo})_2]Cl \cdot 5H_2O(2)$**

| | | |
|---|---|---|
| empirical formula | $C_{14}H_{34}N_2O_{31}K_2Mo_4$ | $C_{14}H_{24}O_{30}ClK_5Mo_4$ |
| fw | 1188. 39 | 1287. 04 |
| temp(℃) | 23 | |
| radiation | Mo K$\alpha$($\lambda$=0. 7107 Å) | |
| cryst color | colorless | |
| cryst syst | monoclinic | triclinic |
| space group | $P2_1/c$ | $P\bar{1}$ |
| formula units/ unit cell | 4 | 2 |
| diffractometer | Smart Apex CCD | |
| crystal size(mm³) | 0. 35×0. 17×0. 14 | 0. 28×0. 16×0. 07 |
| cell constants | | |
| $a$(Å) | 11. 6965(5) | 11. 1529(4) |
| $b$(Å) | 15. 2089(6) | 12. 1790(4) |
| $c$(Å) | 22. 1252(9) | 15. 2720(5) |
| $\alpha$(°) | | 86. 640(1) |
| $\beta$(°) | 101. 424(1) | 72. 297(1) |
| $\gamma$(°) | | 67. 387(1) |
| $V$(Å³) | 3857. 9(3) | 1820. 2(1) |
| $D_{calc}$(g,cm⁻³) | 2. 046 | 2. 348 |
| abs coeff(mm⁻¹) | 1. 589 | 2. 096 |
| $F000$ | 2344 | 1256 |
| $\theta$ range(deg) | 1. 64~28. 37 | 1. 40~25. 00 |
| reflns collected/ unique | 44 039/9215 | 17 542/6394 |
| | [$R$(int)=0. 0837] | [$R$(int)=0. 1224] |
| data/restraints/params | 9215/34/568 | 6394/17/523 |
| GOF on $F^2$ | 0. 959 | 1. 029 |
| final $R$ indices [$I>2\sigma(I)$]$^a$ | R1=0. 0436 | R1=0. 0375 |
| | wR2=0. 1398 | wR2=0. 0941 |
| $R$ indices(all data)$^a$ | R1=0. 0481 | R1=0. 0416 |
| | wR2=0. 1434 | wR2=0. 0962 |
| largest diff. peak and hole(e · Å⁻³) | 1. 413 and $-$2. 147 | 1. 310 and $-$0. 776 |

$^a$ $R1=\sum||F_o|-|F_c||/\sum(|F_o|)$,$wR2=\sum[w(F_o^2-F_c^2)^2]/\sum[w(F_o^2)^2]^{1/2}$.

### Table 3　Selected Bond Distances(Å)and Angles(deg)for 1 and 2

| | 1 | 2 | | 1 | 2 |
|---|---|---|---|---|---|
| Mo(1)—O(1)$_{\alpha\text{-alkoxy}}$ | 1.929(4) | 1.947(3) | Mo(1)—O(2)$_{\alpha\text{-carboxy}}$ | 2.194(4) | 2.193(3) |
| Mo(1)—O(4)$_{\beta\text{-carboxy}}$ | 2.327(4) | 2.312(3) | Mo(1)—O(8) | 1.708(4) | 1.701(3) |
| Mo(1)—O(9) | 1.731(4) | 1.701(3) | Mo(1)—O(10) | 1.879(4) | 1.920(3) |
| Mo(2)—O(4)$_{\beta\text{-carboxy}}$ | 2.288(4) | 2.265(3) | Mo(2)—O(10) | 1.950(4) | 1.948(3) |
| Mo(2)—O(15)$_{\beta\text{-carboxy}}$ | 2.316(4) | 2.332(3) | Mo(2)—O(18) | 1.701(5) | 1.718(3) |
| Mo(2)—O(19) | 1.696(5) | 1.701(3) | Mo(2)—O(20) | 1.910(4) | 1.897(3) |
| Mo(3)—O(5)$_{\beta\text{-carboxy}}$ | 2.318(4) | 2.262(3) | Mo(3)—O(14)$_{\beta\text{-carboxy}}$ | 2.292(4) | 2.313(3) |
| Mo(3)—O(20) | 1.886(4) | 1.929(3) | Mo(3)—O(21) | 1.974(4) | 1.928(3) |
| Mo(3)—O(22) | 1.698(5) | 1.694(3) | Mo(3)—O(23) | 1.706(5) | 1.706(3) |
| Mo(4)—O(11)$_{\alpha\text{-alkoxy}}$ | 1.955(4) | 1.931(3) | Mo(4)—O(12)$_{\alpha\text{-carboxy}}$ | 2.181(4) | 2.200(3) |
| Mo(4)—O(14)$_{\beta\text{-carboxy}}$ | 2.317(4) | 2.296(3) | Mo(4)—O(21) | 1.871(4) | 1.903(3) |
| Mo(4)—O(24) | 1.707(4) | 1.691(3) | Mo(4)—O(25) | 1.727(4) | 1.728(3) |
| C(7)—O(6) | 1.316(8) | 1.34(1) | C(7)—O(7) | 1.222(8) | 1.20(1) |
| O(1)—Mo(1)—O(2) | 74.7(2) | 75.2(1) | O(1)—Mo(1)—O(4) | 77.9(1) | 77.6(1) |
| O(1)—Mo(1)—O(8) | 106.8(2) | 91.5(1) | O(1)—Mo(1)—O(9) | 93.5(2) | 106.9(2) |
| O(1)—Mo(1)—O(10) | 145.8(2) | 146.4(1) | O(2)—Mo(1)—O(4) | 77.0(1) | 76.5(1) |
| O(2)—Mo(1)—O(8) | 92.2(2) | 163.7(1) | O(2)—Mo(1)—O(9) | 163.3(2) | 89.1(2) |
| O(2)—Mo(1)—O(10) | 83.6(2) | 86.2(1) | O(4)—Mo(1)—O(8) | 166.8(2) | 91.7(1) |
| O(4)—Mo(1)—O(9) | 89.1(2) | 163.4(1) | O(4)—Mo(1)—O(10) | 71.6(2) | 70.9(1) |
| O(8)—Mo(1)—O(9) | 102.7(2) | 104.0(2) | O(8)—Mo(1)—O(10) | 99.9(2) | 100.6(2) |
| O(9)—Mo(1)—O(10) | 101.0(2) | 100.5(1) | O(4)—Mo(2)—O(10) | 71.4(1) | 71.5(1) |
| O(4)—Mo(2)—O(15) | 78.8(1) | 80.3(1) | O(4)—Mo(2)—O(18) | 89.0(2) | 163.0(1) |
| O(4)—Mo(2)—O(19) | 164.3(2) | 92.1(1) | O(4)—Mo(2)—O(20) | 82.7(2) | 82.8(1) |
| O(10)—Mo(2)—O(15) | 77.8(2) | 77.1(1) | O(10)—Mo(2)—O(18) | 98.0(2) | 98.8(1) |
| O(10)—Mo(2)—O(19) | 98.8(2) | 98.9(1) | O(10)—Mo(2)—O(20) | 149.5(2) | 148.9(1) |
| O(15)—Mo(2)—O(18) | 167.9(2) | 83.9(1) | O(15)—Mo(2)—O(19) | 87.3(2) | 172.1(1) |
| O(15)—Mo(2)—O(20) | 81.6(2) | 81.8(1) | O(18)—Mo(2)—O(19) | 104.7(3) | 103.5(2) |
| O(18)—Mo(2)—O(20) | 97.4(2) | 101.3(1) | O(19)—Mo(2)—O(20) | 102.5(2) | 99.2(1) |
| O(5)—Mo(3)—O(14) | 79.3(1) | 79.6(1) | O(5)—Mo(3)—O(20) | 81.2(2) | 81.6(1) |
| O(5)—Mo(3)—O(21) | 76.5(2) | 79.7(1) | O(5)—Mo(3)—O(22) | 169.4(2) | 88.4(1) |
| O(5)—Mo(3)—O(23) | 86.4(2) | 168.2(1) | O(14)—Mo(3)—O(20) | 83.5(2) | 83.3(1) |
| O(14)—Mo(3)—O(21) | 70.7(1) | 71.5(1) | O(14)—Mo(3)—O(22) | 90.3(2) | 166.5(1) |
| O(14)—Mo(3)—O(23) | 163.7(2) | 88.9(1) | O(20)—Mo(3)—O(21) | 148.4(2) | 150.9(1) |
| O(20)—Mo(3)—O(22) | 99.6(2) | 101.1(1) | O(20)—Mo(3)—O(23) | 102.1(2) | 94.7(1) |
| O(21)—Mo(3)—O(22) | 98.5(2) | 100.5(1) | O(21)—Mo(3)—O(23) | 98.5(2) | 99.1(1) |
| O(22)—Mo(3)—O(23) | 103.6(2) | 103.4(2) | O(11)—Mo(4)—O(12) | 74.4(2) | 75.0(1) |
| O(11)—Mo(4)—O(14) | 77.2(1) | 79.0(1) | O(11)—Mo(4)—O(21) | 146.2(2) | 147.2(1) |
| O(11)—Mo(4)—O(24) | 106.0(2) | 104.7(1) | O(11)—Mo(4)—O(25) | 90.4(2) | 94.0(1) |
| O(12)—Mo(4)—O(14) | 77.3(2) | 76.6(1) | O(12)—Mo(4)—O(21) | 86.0(2) | 83.1(1) |
| O(12)—Mo(4)—O(24) | 90.0(2) | 91.3(1) | O(12)—Mo(4)—O(25) | 162.2(2) | 163.3(1) |
| O(14)—Mo(4)—O(21) | 71.7(2) | 72.3(1) | O(14)—Mo(4)—O(24) | 165.7(2) | 166.1(2) |
| O(14)—Mo(4)—O(25) | 90.6(2) | 89.2(1) | O(21)—Mo(4)—O(24) | 101.1(2) | 99.8(1) |
| O(21)—Mo(4)—O(25) | 102.7(2) | 101.1(1) | O(24)—Mo(4)—O(25) | 103.2(2) | 103.7(2) |

# 3 Results and Discussions

$^{13}$C NMR chemical shifts observed on complex **2** are shown in Figure 1. This is a clear indication for the coordination of the α-alkoxy, α-carboxy, and β-carboxy groups in the complex. In comparison with free ligand under comparable conditions, the title complexes show generally large downfield shifts of the corresponding $^{13}$C resonance as in Table 1. For example, both α-and β-carboxy carbons show large downfield shifts of $\Delta\delta$ 10.0, 9.3 ppm for **1** and 10.1, 9.5 ppm for **2**, respectively, in each case arising from the coordination. However, $^{13}$C resonance shifts of γ-car-boxylic acid in the open chain

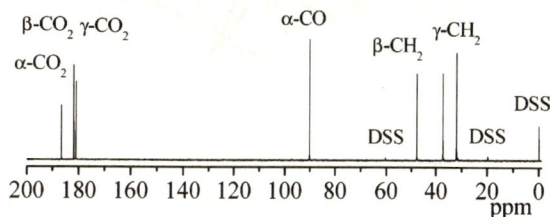

**Figure 1** $^{13}$C NMR spectra of molybdate(VI)

$K_5[(MoO_2)_4O_3(R,S\text{-}Hhomo)_2]Cl \cdot 5H_2O(2)$.

of complexes **1** and **2** cannot be compared directly with free homocitric acid. The later forms a five-membered ring lactone by γ-carboxy and α-hydroxy groups. The α-alkoxy carbons have somewhat small downfield shifts $\Delta\delta$ of 6.5 and 6.7 ppm for **1** and **2**, respectively. This is also related to the formation of lactone. Other peaks also showed downfield shifts in general for **1** and **2** and can be easily assigned to the remaining methylene carbons. The $^{13}$C NMR spectra of the two complexes in solution show no obvious decomposition for 1 week. The tetrameric species are believed to be stable in solution.

The FT-infared spectra of the two title complexes display the characteristic features of the coordinated homocitrato ligand and oxo groups. The antisymmetric stretching carboxy vibrations of the two complexes all shift to lower values with respect to that of uncoordinated ligand. The bands at 1440~1390 cm$^{-1}$ correspond to symmetric stretching vibrations, $v_{s(COO)}$, for the two compounds. The remaining $v(COOH)$ of γ-carboxylic acidic group appears at 1702 cm$^{-1}$ for **1** and 1716 cm$^{-1}$ for **2**. In the region around 900 cm$^{-1}$, both complexes show several bands that result from the presence of cis-dioxo molybdenum.

The ORTEP plots of the anions of complexes **1** and **2** are shown in Figure 2. The crystal structure of **1** comprises potassium cations, ammonium cations, lattice water molecules, and the tetrameric homocitrato molybdate(VI) anion. Each Mo atom contains a cis-dioxo-MoO$_2$ unit and exists in an approximately octahedral geometry. The two homocitrate ligands coordinate in a similar fashion with the molybdenum center. One homocitrate ligand coordinates to Mo1 with O1 of α-alkoxy, O2 of α-carboxy, and O4 of β-carboxy groups, as well as Mo2 and Mo3 with O4 and O5 of the β-carboxy group, respectively, leaving the γ-carboxylic acidic group free. The charge balance and the difference of C-O distances suggest that this γ-carboxy group is protonated. Thus, homocitrates act as tridentate ligands in this complex.

The structure analyses of complex **2** reveal that it has a similar anion with **1** except for the conformation of the pendant CH$_2$CH$_2$CO$_2$ arm of homocitrate. The latter is sufficiently long and flexible that its γ-carboxy group can orient differently in the two complexes. In complex **1**, the γ-carboxy group (C7) curls back toward the α-carboxy group(C2). However, it keeps away from α-carboxy group in complex **2**. The difference of the conformation may be caused by the cationic partners.

It is useful to examine some bond distances in Table 4 to evaluate the Mo-O bond strength. The α-and β-carboxy oxygen atoms of homocitrate orient opposite to the cis-dioxo molybdenum core in both complexes. Two sets of Mo-O(β-carboxy) distances in the two homocitrate complexes are averaged to

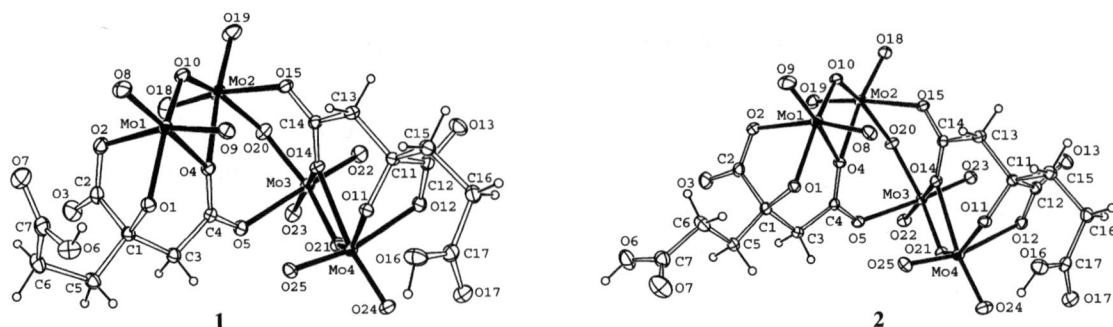

**Figure 2** Perspective view of the anion structures of $K_2(NH_4)_2[(MoO_2)_4O_3(R,S\text{-}Hhomo)_2] \cdot 6H_2O$ (**1**, left) and $K_5[(MoO_2)_4O_3(R,S\text{-}Hhomo)_2]Cl \cdot 5H_2O$ (**2**, right) in $\Lambda_R\Lambda\Lambda\Lambda_R$ configurations.

give 2.311(4) and 2.309(4) Å for **1** and 2.280(3) and 2.314(3) Å for **2**. These figures are longer than the Mo-O($\alpha$-carboxy) bond distances of $>2.20$ Å in both complexes. That might arise from the strong trans influence of the dioxo groups. As shown in Table 3, the Mo-O($\alpha$-alkoxy) distances are shorter, implying the deprotonation of $\alpha$-hydroxy group in homocitrate. The bond strength of Mo-O in homocitrate to molybdenum is in the order of $\alpha$-alkoxy $>$ $\alpha$-carboxy $>$ $\beta$-carboxy $>$ $\gamma$-carboxy groups. The strong coordination of $\alpha$-alkoxy and $\alpha$-carboxy groups to molybdenum has been found in protein-bound FeMo-co.[25,26]

**Table 4** Mo-O Distances(Å) and Absolute Configuration Assignments in Homocitrato and Citrato Molybdates

| complex | Mo-O$_{\alpha\text{-alkoxy}}$ | Mo-O$_{\alpha\text{-carboxy}}$ | Mo-O$_{\beta\text{-carboxy}}$ | configuration assignments | ref |
|---|---|---|---|---|---|
| $K_2(NH_4)_2[(MoO_2)_4O_3(R,S\text{-}Hhomocit}_2] \cdot 6H_2O$(**1**) | 1.929(4),1.955(4) | 2.194(4),2.181(4) | 2.311$_{av}$(4),2.309(4) | $\Lambda_R\Lambda\Lambda\Lambda_R/\Delta_S\Delta\Delta\Delta_S$ | this work |
| $K_5[(MoO_2)_4O_3(R,S\text{-}Hhomocit})_2]Cl \cdot 5H_2O$(**2**) | 1.947(3)1.931(3) | 2.193(3),2.200(3) | 2.280$_{av}$(3),2.314$_{av}$(3) | | |
| $K_4[(MoO_2)_4O_3(Hcit)_2] \cdot 5H_2O$ | 1.976(5),1.968(5) | 2.185(5),2.211(5) | 2.310$_{av}$(5),2.357$_{av}$(5) | | [40] |
| $[Me_3N(CH_2)_6NMe_3]_2[(MoO_2)_4O_3(Hcit)_2] \cdot 12H_2O$ | 1.970(6) | 2.217(6) | 2.328$_{av}$(5) | | [41] |
| $MoFe_7S_9X(S\text{-}cys)(N\text{-}His)(homocit)$ | | | | | |
| *Azotobacter vinelandii*(*Av*1,1992) | 1.996 | 2.167 | | $\Delta_R$ | [19] |
| *Clostridium pasturianum*(*Cp*1) | 2.035 | 2.206 | | | [20] |
| *Klebsiella pneumoniae*(*Kp*1) | 2.35(2)$_{re}$,2.35(2)$_{ox}$ | 2.29(1)$_{re}$,2.29(2)$_{ox}$ | | | [25] |
| *Azotobacter vinelandii*(*Av*1,2002) | 2.212 | 2.162 | | | [36] |

For comparison, some related Mo-O bond distances and absolute configurations of citrato and homocitrato molybdates are listed in Table 4. In tetranuclear citrato molybdates, the bond distances of $\alpha$-alkoxy to molybdenum are shorter than those of $\alpha$-carboxy group to molybdenum of complexes **1** and **2**. However, the difference of Mo-O ($\alpha$-carboxy) and Mo-O ($\beta$-carboxy) distances is smaller, which demonstrates a $\beta$-carboxy group has comparable coordination ability. In this point, this implies the possible substitution of imidazole group by $\beta$-carboxy group of homocitrate in NMF extraction of the cofactor of nitrogenase.[14]

The reaction of a racemic mixture of homocitrate or citrate results in the isolation of racemic mixture of $\Lambda$- and $\Delta$-complexes. The chiral configuration of molybdenum center in wild-type FeMo-co is $\Delta$, which might be induced by the coordination of the chiral $R$-homocitrate. However, recent crystallographic structural analysis shows that ( FeMo-co )* from a *nifV* mutant of *Klebsiella pneumoniae*, although achiral citrate acts as the ligand to the molybdenum, also possesses the same molybdenum configuration with the wild-type FeMo-co. This is the further manifestation that the asymmetric coordination environment can induce chiral effects in the biosynthesis of FeMo-co.

## Acknowledgment

Financially support by the Ministry of Science and Technology( 001CB108906 ) and the National Science Foundation of China( 20571061, 20021002, 20423002 ) is gratefully acknowledged. We thank the reviewers for their very constructive suggestions to update the paper.

## Supporting Information Available

X-ray crystallographic files in CIF format, IR ( Figure S1 ) and NMR spectra ( Figure S2 ). This material is available free of charge via the Internet at http://pubs. acs. org. IC061429F

## References and notes

[1]Rodríguez,G. H. ;Biellmann,J. F. J. Org. Chem. 1996,**61**,1822~1824.

[2]Ancliff,R. A. ;Russell,A. T. ;Sanderson,A. J. Tetrahedron Asym. 1997,**8**,3379~3382.

[3]Li,Z. C. ;Xu,J. Q. Molecules1998,**3**,31~34.

[4]Ma,G. X. ;Palmer,D. R. J. Tetrahedron Lett. 2000,**41**,9209~9212.

[5]Paju,A. ;Kanger,T. ;Pehk,T. ;Eek,M. ;Lopp,M. Tetrahedron 2004,**60**,9081~9084.

[6]Tavassoli,A. ;Duffy,J. E. S. ;Young,D. W. Tetrahedron Lett. 2005,**46**,2093~2096.

[7]Xu,P. F. ;Matsumoto,T. ;Ohki,Y. ;Tatsumi,K. Tetrahedron Lett. 2005,**46**,3815~3818.

[8]Huang,P. Q. ;Li,Z. Y. Tetrahedron Asym. 2005,**46**,2093~2096.

[9]Jia,Y. H. ;Palmer,D. R. J. ;Quail,J. W. Acta Crystallogr. 2005,**E61**,o4034~o4036.

[10]Tavassoli,A. ;Duffy,J. E. S. ;Young,D. W. Org. Biomol. Chem. 2006,**4**,569~580.

[11]Shah,V. K. ;Brill,W. J. Proc. Natl. Acad. Sci. U. S. A. 1977,**74**,3249~3253.

[12] Hoover, T. R. ; Robertson, A. D. ; Cerny, R. L. ; Hayes, R. N. ; Imperial, J. ; Shah, V. K. ; Ludden,P. W. Nature(London)1987,**329**,855~857.

[13]Hoover,T. R. ;Imperial,J. ;Ludden,P. W. ;Shah,V. K. Biochemistry 1989,**28**,2768~2771.

[14]Grönberg, K. L. C. ; Gormal, C. A. ; Durrant, M. C. ; Smith, B. E. ; Henderson, R. A. J. Am. Chem. Soc. 1998,**120**,10613~10621.

[15]Burgess,B. K. ;Lowe,D. J. Chem. Rev. 1996,**96**,2983~3011.

[16]Ludden, P. W. ; Shah, V. K. ; Roberts, G. P. ; Homer, M. ; Allen, R. ; Paustian, T. ; Roll, J. ; Chatterjee,R. ;Madden,M. ;Allen,J. In Molybdenm Enzymes,Cofactors and Model Systems; Stiefel, E. I., Coucouvanis, D., Newton W. E., Eds. ; American Chemical Society: Washington, DC,1993;pp 196~215.

[17]Mayer, S. M. ; Gormal, C. A. ; Smith, B. E. ; Lawson, D. M. J. Biol. Chem. 2002,**277**,35263~ 35266.

[18]Palmer,J. G. ;Doemeny,P. A. ;Schrauzer,G. N. Z. Naturforsch. 2001,**56b**,386~393.

[19]Kim,J. ;Rees,D. C. Science1992,**257**,1677~1682.

[20]Chan,M. K. ;Kim,J. ;Rees,D. C. Science1993,**260**,792~794.

[21]Bolin,J. T. ;Ronco, A. E. ;Morgan, T. V. ;Mortenson, L. E. ;Xuong, N. H. Proc. Natl. Acad. Sci. U. S. A. 1993,**90**,1078~1082.

[22]Howard,J. B. ;Rees,D. C. Chem. Rev. 1996,**96**,2965~2982.

[23]Peters,J. W. ;Stowell, M. H. B. ;Soltis, S. M. ;Finnegan, M. G. ;Johnson, M. K. ;Rees,D. C. Biochemistry 1997,**36**,1181~1187.

[24]Schindelin, H. ;Kisker, C. ;Schlessman, J. L. ;Howard, J. B. ;Rees, D. C. Nature(London) 1997,**387**,370~376.

[25]Mayer,S. M. ;Lawson,D. M. ;Gormal,C. A. ;Roe,S. M. ;Smith,B. E. J. Mol. Biol. 1999,**292**, 871~891.

[26]Einsle, O. ;Tezcan, F. A. ;Andrade, S. L. A. ;Schmid, B. ;Yoshida, M. ;Howard,J. B. ;Rees, D. C. Science 2002,**297**,1696~1700.

[27]Yang,T. C. ;Maeser, N. K. ;Laryukhin, M. ;Lee, H. I. ;Dean,D. R. ;Seefeldt, L. C. ;Hoffman, B. M. J. Am. Chem. Soc. 2005,**127**,12804~12805.

[28]Tsai, K. R. ;Wan,H. L. J. Clust. Sci. 1995,**6**,485~501.

[29]Leigh,G. J. Science 2003,**301**,55~56.

[30]Leigh,G. J. Science 1995,**268**,827~828.

[31]Holland, P. L. In Comprehensive Coordination Chemistry II; Mc-Cleverty J. A.,Meyer T. J., Eds. ;Elsevier Paragamon:New York,2004; Vol. 8,(Que, L., Tolman, W. B.,Eds. ;pp 586~ 591.

[32]Durrant, M. C. Biochemistry 2002,**41**,13934~13945.

[33]Seefeldt, L. C. ;Dance,I. G. ;Dean,D. R. Biochemistry 2004,**43**,1401~1409.

[34]Zhou,Z. H. ;Zhao,H. ;Tsai, K. R. J. Inorg. Biochem. 2004,**98**,1795~1802.

[35]Allen, R. M. ;Roll,J. T. ;Rangaraj, P. ;Shah, V. K. ;Roberts, G. P. ;Ludden, P. W. J. Biol. Chem. 1999,**274**,15869~15874.

[36]Li,D. M. ;Xing, Y. H. ;Li,Z. C. ;Xu,J. Q. ;Song, W. B. ;Wang, T. G. ;Yang, G. D. ;Hu,N. H. ;Jia,H. Q. ;Zhang,H. M. J. Inorg. Biochem. 2005,**99**,1602~1610.

[37]Wright,D. W. ;Chang,R. T. ;Mandal,S. K. ;Armstrong,W. H. ;Orme-Johnson, W. H. J. Biol. Inorg. Chem. 1996,**1**,143~151.

[38]Farrugia,L. J. J. Appl. Crystallogr. 1999,**32**,837~838.

[39]Sheldrick, G. M. SHELXS-97 and SHELXL-97, Programs for Solution and Refinement of Crystal Structures;University of Göttingen:Göttingen,Germany,1997.

[40]Alcock, N. W. ;Dudek, M. ;Grybooe, R. ;Hodorwicz, E. ;Kanas, A. ;Samotus, A. J. Chem. Soc.,Dalton. Trans. 1990,707~711.

[41]Nassimbeni,L. R. ;Niven, M. L. ;Cruywagen, J. J. ;Heyns, J. B. B. J. Crystallogr. Spectrosc. Res. 1987,**17**,373~382.

■ 本文原载：Tetrahedron 2007,63,pp. 2148~2152.

# Expeditious Biomimetically-Inspired Approaches to Racemic Homocitric Acid Lactone and Per-Homocitrate

Hong-Bin Chen, Ling-Yan Chen, Pei-Qiang Huang[①], Hong-Kui Zhang, Zhao-Hui Zhou, Khi-Rui Tsai

(Department of Chemistry, Key Laboratory for Chemical Biology of Fujian Province and State Key Laboratory for Physical Chemistry of Solid Surfaces, College of Chemistry and Chemical Engineering, Xiamen University, Xiamen 361005, P.R. China)

Received 30 October 2006; revised 28 December 2006; accepted 28 December 2006 Available online 4 January 2007

**Abstract** Two concise and flexible biomimetically-inspired approaches to homocitric acid lactone(**3**) and its higher homolog, triethyl per-homocitrate(**12**), are presented herein. The key steps include an efficient indium metal-mediated allylation-oxidative cleavage procedure and a one-step ethoxycarbonylmethylation of $\alpha$-oxo-diesters.

**Key words** Homocitric acid   Homocitric acid lactone   Per-homocitric acid   Indium   Allylation.

## 1  Introduction

The mechanism of biological nitrogen fixation has been extensively studied in many laboratories over the world as an important interdisciplinary research. It has been shown that nitrogenases, mainly MoFe-nitrogenase, catalyze the ATP-dependent reduction of $N_2$ to ammonia with obligatory evolution of at least one $H_2$ at ambient temperature[1]. In addition, this enzyme can also catalyze the reduction of several types of substrates besides $N_2$ and $H^+$. Recent progress in mechanistic study of this complex metalloenzyme revealed that, (R)-homocitric acid (**1**) as a chelating ligand bound to $Mo^{IV/III}$ of FeMo-cofactor plays critical roles in nitrogen fixation[2]. Moreover, (R)-homocitric acid is also a key intermediate in the biosynthetic pathway to the essential amino acid lysine[3] in fungi and euglenids. Because this pathway is absent in plants and mammalians, (R)-homocitric acid and its derivatives are considered to be promising candidates for potential anti-fungal therapy in medicine, and anti-fungal agents for crop protection[4]. For further studies in both areas, it requires convenient methods to synthesize homocitric acid. Although several methods have been developed to synthesize homocitric acid lactone, a more stable version of homocitric acid[5-7], almost all of them are either lengthy, low yielding,

---

①　Corresponding author. Tel. :+86 592 2180992;fax:+86 592 2186405;e-mail:pqhuang@xmu.edu.cn

expensive, unscalable, or involving toxic reagents. As a consequence, current supply of homocitric acid is quite limited and very costly. In addition, the higher homolog of homocitric acid, namely per-homocitric acid is also important for elucidating the mechanism of nitrogenase-catalyzed biological nitrogen fixation. However, the method for the synthesis of per-homocitric acid is still not available (Chart 1). [8]

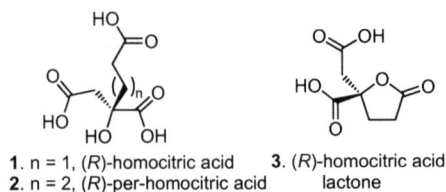

**1**. n = 1, (*R*)-homocitric acid
**2**. n = 2, (*R*)-per-homocitric acid
**3**. (*R*)-homocitric acid lactone

**Chart 1    Structure of homocitric acid and per-homocitric acid.**

In continuation of our studies on the nitrogenase[9,7e], we have turned our attention to explore the roles that homocitrate played in the process of biological nitrogen fixation. To this end, an efficient, simple, and scalable route to homocitric acid and per-homocitric acid is required. We describe herein two expeditious, practical, and flexible approaches to both racemic homocitric acid lactone and triethyl per-homocitrate.

Inspired by the biosynthesis of homocitrate (Fig. 1)[6b,10,3], our retrosynthetic analysis of homocitric acid and its higher homologs (Scheme 1) implies the installation of the C-2 side chain to α-oxo-dicarboxylic acids or their diesters. Although such a $C_2^d$ synthon has recently been shown to be introducible, albeit in modest yields (54 — 58%)[11], via silyl ketene acetate, to facilitate the product isolation and purification, we elected to use firstly the allyl group as a synthetic equivalent[12] of the requisite $C_2^d$ synthon, and to use α-ketodi-esters[13] as the substrates.

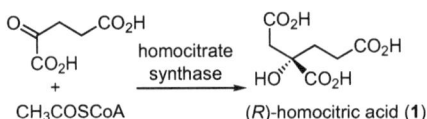

**Figure 1    Biosynthetic pathway for homocitric acid.**

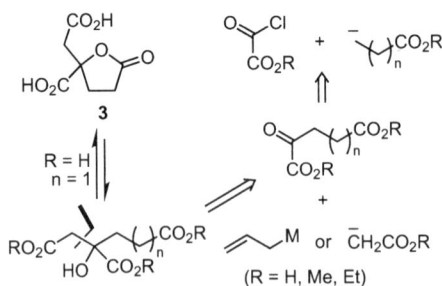

**Scheme 1    Retrosynthetic analysis of homocitric acid and per-homocitric acid.**

## 2    Results and discussion

The synthesis of homocitric acid lactone (**3**) started from the di-esterification of commercially available α-oxo-glutaric acid (**4**) (Scheme 2). Thus treatment of α-oxo-glutaric acid with acetyl chloride containing absolute ethanol gave diethyl α-oxo-glutarate **5** in 91% yield. The efficient allylation of **5** was the key step for our strategy. Among numerous allylation methods[14], modern indium metal-mediated Barbier-type allylation[15] appeared to be the most attractive in terms of its simplicity and low toxicity. Indeed, simply by stirring a mixture of diethyl α-oxo-glutarate **5**, allyl bromide, indium metal, and sodium iodide in a mixed solvent system($H_2O$-ethanol 6/1) at rt, the desired carbinol **6** was obtained in 86% yield[16].

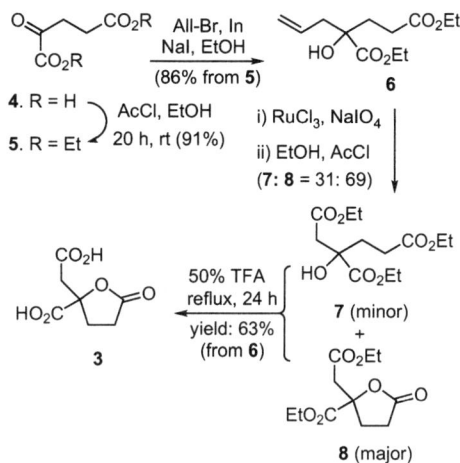

**Scheme 2**

Next, we proceeded to investigate the oxidative cleavage of the double bond of the allyl group into a carboxyl group using ruthenium tetraoxide as an oxidant. Although subjection of **6** to the standard conditions established by Sharpless[12] (2.2% $RuCl_3$, 4.1 mol equiv $NaIO_4$ in $CCl_4$-MeCN-$H_2O$, 2/2/3) led to the desired product in satisfied yield, a recent report on replacing environmentally harmful $CCl_4$ by EtOAc[17] drew our attention. To our delight, mixed solvent system(EtOAc-MeCN-$H_2O$, 2/2/3, v/v)[17] worked similarly well(rt, 1 h). To facilitate the product isolation, the crude acid-diester was transformed into the corresponding triester by treating with $CH_3COCl$-EtOH at rt, which gave a separable mixture of **7** and **8** in a ratio of 31/69. The mixture of triester-lactone **7/8** was treated with a 50% solution of aqueous trifluoroacetic acid to give homocitric acid lactone(**3**) in an overall yield of 63% from **6**. It is worth to mention that only one flash chromatography purification and two simple filtrations through silica gel were required for the synthesis of homocitric acid lactone **3** from **4**.

The method was extended to the synthesis of per-homocitric acid triester(**12**). For this purpose, zinc-copper reagent **9**, prepared by the known procedure[18], was treated with ethyl oxalyl chloride at −30 ℃ to afford the desired diethyl α-oxo-adipate **10** in 70% yield(Scheme 3). Indium-mediated allylation of α-oxo-adipate **10**, followed by $RuO_4$-mediated oxidative cleavage of the C = C bond under improved conditions(2.2% $RuCl_3$, 6.0 mol equiv $NaIO_4$ in EtOAc-MeCN-$H_2O$, 2/2/3, v/v)[17], and esterification furnished compound **12** in an overall yield of 79% from **10**.

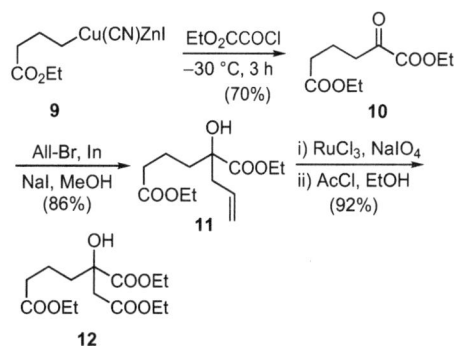

**Scheme 3**

We next turned our attention to investigate a biomimetically-inspired one-step approach to triethyl homocitrate **7** and triethyl per-homocitrate **12** by using ethyl lithioacetate as a $C_2^d$ synthon. Treatment of diethyl α-oxo-glutarate **5** with ethyl lithioacetate, in situ generated from ethyl acetate and LiHMDS, afforded **7** in 73% yield(Scheme 4).[19] Similarly, subjection of diethyl α-oxo-adipate **10** to react with ethyl lithioacetate provided **12** in 72% yield.

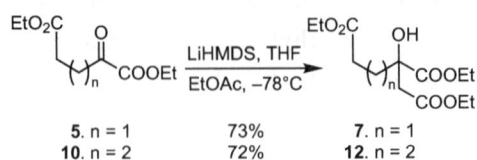

EtO$_2$C O → LiHMDS, THF / EtOAc, −78°C → EtO$_2$C OH COOEt COOEt

**5**. n = 1  73%  **7**. n = 1
**10**. n = 2  72%  **12**. n = 2

**Scheme 4**

# 3 Conclusion

In conclusion, we have developed a simple, convenient, and flexible synthesis of homocitric acid lactone(**3**) and its higher homolog(as triester **12**) starting from two α-oxo-diesters. The use of Barbier-type indium-mediated allylation reaction, and substitution of harmful carbon tetrachloride by ethyl acetate in the RuO$_4$-mediated oxidative olefin cleavage, combined with simple purification procedure for the synthesis of homocitric acid lactone or per-homocitrate, render the present method more environmentally benign. Moreover, an even simpler one-step synthesis of triethyl homocitrate **7** and its higher homolog, triethyl per-homocitrate **12**, has also been achieved. The easy access to homo-citric acid lactone and its higher homologs, will contribute, in its own right, to the studies of the mechanism of biological nitrogen fixation.

# 4 Experimental section

## 4.1 General

Melting points were determined on a Yanaco MP-500 micro melting point apparatus and were uncorrected. Infrared spectra were measured with a Nicolet Avatar 330 FTIR spectrometer using film KBr pellet technique. NMR spectra were recorded in CDCl$_3$ on a Bruker AV400 spectrometer with tetramethylsilane as an internal standard. Chemical shifts are expressed in δ(parts per million)units downfield from TMS. Mass spectra were recorded by Bruker Dalton Esquire 3000 plus LC-MS apparatus. Flash column chromatography was carried out on silica gel(300−400 mesh). THF was distilled over sodium.

**4.1.1  Diethyl 2-allyl-2-hydroxypentanedioate(6)**. To an ethanolic solution(20 mL)of 5(4.04 g, 20.0 mmol)were successively added an aqueous solution(120 mL)of sodium iodide(0.30 g,2.0 mmol), allyl bromide(4.84 g,3.5 mL,40.0 mmol),and indium metal(2.76 g,24.0 mmol). The reaction mixture was stirred at room temperature until indium metal completely dissolved. The mixture was diluted with 4 N HCl(15 mL) and extracted with ethyl ether(5×50 mL). The combined organic phases were successively washed with saturated aqueous NaHCO$_3$(10 mL)and brine(10 mL),dried over anhydrous Na$_2$SO$_4$,filtered,and concentrated. The residue was purified by flash chromatography on silica gel (EtOAc-PE 1/8,$R_f$=0.35)to afford **6**(4.20 g,17.2 mmol,yield:86%)as a pale yellow oil. IR(film) $v_{max}$:3513,3077,1735,1644,1223,1181;$^1$H NMR(CDCl$_3$,400 MHz)δ:5.82−5.70(m,1H),5.15−5.07 (m,2H),4.29−4.18(m,2H),4.12(q,$J$=7.0 Hz,2H),3.31(s,1H,*OH*,D$_2$O exchangeable),2.52−

2.38(m,3H),2.25−2.15(m,1H),2.15−1.99(m,2H),1.30(t,$J$=7.0 Hz,3H),1.25(t,$J$=7.0 Hz,
3H);$^{13}$C NMR(CDCl$_3$,100 MHz)δ:175.5,173.2,131.9,119.2,76.3,62.1,60.5,43.8,33.5,28.8,
14.3,14.2;MS(ESI,$m/z$):244(M+H)$^+$,266(M+Na)$^+$. Anal. Calcd for C$_{12}$H$_{20}$O$_5$:C,59.00;H,
8.25. Found:C,58.93;H,8.19.

**4.1.2 Homocitric acid lactone(3).** To a solution of **6**(2.44 g,10.0 mmol)in EtOAc(20 mL)and
MeCN(20 mL) were successively added H$_2$O(30 mL),NaIO$_4$ (12.8 g,60.0 mmol),and ruthenium
trichloride hydrate(0.05 N in H$_2$O,4.4 mL,2.2% mol equiv). The dark brown suspension was stirred
vigorously at rt for ca. 1 h. Then the reaction mixture was quenched with *iso*-propanol and filtered to
remove the insoluble solids. The filtrate was concentrated under reduced pressure. The remaining
aqueous solution was extracted with EtOAc(5×40 mL). The combined organic phases were washed
with brine(10 mL),dried over anhydrous Na$_2$SO$_4$,filtered,and concentrated under reduced pressure.
The oily residue was dissolved in 20 mL of absolute ethanol and cooled to 0 ℃ before acetyl chloride
(0.39 g,0.36 mL,5.0 mmol)was added dropwise. The resulting mixture was warmed to rt and stirred
overnight. Ethanol was removed under reduced pressure and the residue was passed through a short
silica gel column eluting with EtOAc-PE(60−90 ℃,1/1)to give a residue,which is a mixture of **7** and **8**
(*method 1*). An analytical sample was obtained by flash chromatographic purification and the crude
mixture was used in the next step as it was.

The crude mixture of **7/8** was dissolved in 50% trifluoroacetic acid(10 mL)and refluxed for 24 h.
After cooling,the solvent was removed under reduced pressure. The residue was washed with anhydrous
diethyl ether to give **3**(1.18 g,6.3 mmol;overall yield from **6**:63%)as colorless crystals. Mp:166−168
℃(Et$_2$O)(lit.$^{6b}$ 161−162 ℃). IR(film)$v_{max}$:3421,1787,1730,1703,1222,1200,1177,1064 cm$^{-1}$;$^1$H
NMR(methanol-$d_4$,400 MHz)δ:2.93(d,$J$=17.0 Hz,1H),2.68(d,$J$=17.0 Hz,1H),2.45−2.31(m,
2H),2.31−2.20(m,1H),2.20−2.09(m,1H);$^{13}$C NMR(methanol-$d_4$,100 MHz)δ:178.7,174.1,
172.4,84.7,42.1,32.3,28.7;HRMS calcd for [C$_7$H$_8$O$_6$-1]$^-$:187.0240,found:187.0237.

**4.1.3 Compound 8.** Pale yellow oil. IR(film)$v_{max}$:2981,1793,1739,1376,1178,1070;$^1$H NMR
(CDCl$_3$,400 MHz)δ:4.28(q,$J$=7.1 Hz,2H),4.16(q,$J$=7.1 Hz,2H),3.11(d,$J$=16.7 Hz,1H),
2.96(d,$J$=16.7 Hz,1H),2.74−2.52(m,3H),2.40−2.30(m,1H),1.32(t,$J$=7.1 Hz,3H),1.26(t,
$J$=7.1 Hz,3H);$^{13}$C NMR(CDCl$_3$,100 MHz)δ:175.5,170.5,168.5,82.9,62.4,61.2,41.6,31.2,27.8,
14.1,14.0;MS(ESI,$m/z$):244(M+H$^+$),266(M+Na$^+$). Anal. Calcd for C$_{11}$H$_{16}$O$_6$:C,54.09;H,
6.60. Found:C,54.13;H,6.60.

**4.1.4 Triethyl homocitrate(7).**
**4.1.4.1.** *Method 2*（via enolate addition to **5**）. To a solution of HMDS(0.63 g,0.83 mL,3.90
mmol)in 2.6 mL of anhydrous THF at 0 ℃ was added dropwise *n*-BuLi(2.5 M solution in *n*-hexane,
2.93 mmol,1.17 mL). After stirring for about 30 min,the mixture was cooled to−78 ℃ and to which
was added EtOAc(0.26 g,0.76 mL,2.93 mmol). After stirring at−78 ℃ for 30 min,a THF solution
(5.50 mL)of **5**(0.39 g,1.95 mmol)was added dropwise. The mixture was stirred at−78 ℃ for another
3 h and then quenched with saturated aqueous NH$_4$Cl. The mixture was extracted with ethyl ether(3×
8 mL). The combined organic layers were dried over anhydrous Na$_2$SO$_4$,filtered,and concentrated under
reduced pressure. The residue was purified by flash chromatography on silica gel(eluent:EtOAc-PE 1/
6)to afford **7**(0.28 g,0.98 mmol)in 73% yield. Pale yellow oil. IR(film)$v_{max}$:3505,2982,2938,1736,
1446,1373,1191 cm$^{-1}$;$^1$H NMR(CDCl$_3$,400 MHz)δ:4.22−4.31(m,2H),4.13(2q,overlapped,each $J$
=7.1 Hz,4H),3.77(s,1H,*OH*,D$_2$O exchangeable),2.94(d,$J$=16.2 Hz,1H),2.68(d,$J$=16.2 Hz,

1H),2.46—2.54(m,1H),2.21—2.29(m,1H),2.02—2.07(m,2H),1.31(t,$J=7.1$ Hz,3H),1.25(2t,overlapped,each $J=7.1$ Hz,6H);$^{13}$C NMR(CDCl$_3$,100 MHz)$\delta$:174.6,172.9,170.5,74.2,62.2,60.9,60.6,43.5,33.9,28.3,14.1;MS(ESI,$m/z$):290(M$+$H$^+$),312(M$+$Na$^+$). Anal. Calcd for C$_{13}$H$_{22}$O$_7$:C,53.78;H,7.64. Found:C,53.62;H,7.60.

### 4.1.5 Triethyl per-homocitrate(12).

**4.1.5.1 *Method 1*(via indium-mediated allylation of 10 followed by subsequent oxidative cleavage and esterification).** Following the procedure described for the allylation of **5**,**11** was synthesized in 86% yield from **10**.

Compound **11**: pale yellow oil. IR(film)$v_{max}$:3517,3073,2980,1735,1637,1222,1175 cm$^{-1}$;$^1$H NMR(CDCl$_3$,400 MHz)$\delta$:5.83—5.70(m,1H),5.16—5.08(m,2H),4.30—4.21(m,2H),4.13(q,$J=7.1$ Hz,2H),3.25(s,1H,*OH*,D$_2$O exchangeable),2.48—2.35(m,2H),2.35—2.24(m,2H),1.88—1.77(m,2H),1.74—1.65(m,1H),1.55—1.45(m,1H),1.31(t,$J=7.1$ Hz,3H),1.26(t,$J=7.1$ Hz,3H);$^{13}$C NMR(CDCl$_3$,100 MHz)$\delta$:175.8,173.2,132.2,119.0,76.7,62.0,60.3,43.9,38.0,34.2,19.1,14.3,14.2. MS(ESI,$m/z$):259(M$+$H)$^+$,281(M$+$Na)$^+$. Anal. Calcd for C$_{13}$H$_{22}$O$_5$:C,60.45;H,8.58. Found:C,60.51;H,8.49.

Compound **12** was prepared in 92% yield from **11** by following the procedure described for the synthesis of **7/8**.

**4.1.5.2 *Method 2*(via enolate addition to 10).** To a solution of HMDS(0.500 g,0.65 mL,3.08 mmol)in 2.0 mL anhydrous THF at 0 ℃ was added dropwise $n$-BuLi(2.5 M solution in $n$-hexane,2.31 mmol,0.90 mL). After stirring for about 30 min,the mixture was cooled to$-78$ ℃. To which was added EtOAc(0.65 mL,2.51 mmol)and the stirring was continued at$-78$ ℃ for about 30 min. To the resulting mixture was added dropwise **10**(0.33 g,1.54 mmol)in 4.50 mL anhydrous THF. The reaction mixture was stirred at$-78$ ℃ for another 3 h and then quenched with saturated NH$_4$Cl. The resulting mixture was extracted with diethyl ether(3$\times$5 mL). The combined organic layers were dried over anhydrous Na$_2$SO$_4$,filtered,and concentrated under reduced pressure. The residue was purified by flash chromatography on silica gel(eluent:EtOAc-PE 1/6)to afford **12**(0.34 g,1.12 mmol,yield:72%)as a pale yellow oil. IR(film)$v_{max}$:3517,2980,1735,1641,1446,1372,1222,1175,1025 cm$^{-1}$;IR(film)$v_{max}$:3508,2982,2938,1739,1736,1732,1374,1183,1096,1030 cm$^{-1}$;$^1$H NMR(CDCl$_3$,400 MHz)$\delta$:4.32—4.21(m,2H),4.12(2q,overlapped,each $J=7.0$ Hz,4H),3.75(s,1H,*OH*,D$_2$O exchangeable),2.91(d,$J=16.2$ Hz,1H),2.67(d,$J=16.2$ Hz,1H),2.36—2.23(m,2H),1.84—1.58(m,4H),1.30(t,$J=7.0$ Hz,3H),1.24(2t,overlapped,each $J=7.0$ Hz,6H);$^{13}$C NMR(100 MHz,CDCl$_3$)$\delta$:174.8,173.0,170.6,74.8,62.0,60.8,60.3,43.5,38.4,34.0,18.8,14.2,14.1,14.0;MS(ESI,$m/z$):305(M$+$H)$^+$,327(M$+$Na)$^+$. Anal. Calcd for C$_{14}$H$_{24}$O$_7$:C,55.25;H,7.95. Found:C,55.15;H,8.04.

# Acknowledgements

We thank the NSF of China for financial support. Partial support from the program for Innovative Research Team in Science and Technology in Fujian Province University is acknowledged.

# References and notes

[1]Burgess,B. K.;Lowe,D. J. Chem. Rev. 1996,**96**,2983~3011.

[2](a) Hoover,T. R.;Robertson,A. D.;Cerny,R. L.;Hayes,R. N.;Imperial,J.;Shah,V. K.;

Ludden, P. W. Nature 1987, **329**, 855～857; (b) Hoover, T. R.; Imperial, J. P.; Ludden, W.;
Shah, V. K. Biochemistry 1989, **28**, 2768～2771; (c) Kim, J.; Rees, D. C. Science 1992, **257**,
1677～1682.

[3] Zabriskie, T. M.; Jackson, M. D. Nat. Prod. Rep. 2000, **17**, 85～97.

[4] (a) Suvarna, K.; Seah, L.; Bhattacharjee, V.; Bhattacharjee, J. K. Curr. Genet. 1998, **33**, 268～
275; (b) Palmer, D. R. J.; Balogh, H.; Ma, G.; Zhou, X.; Marko, M.; Kaminskyj, S. G. W.
Pharmazie 2004, **59**, 93～98.

[5] For racemic synthesis of homocitric acid lactone, see: (a) Maragoudakis, M. E.; Strassman, M. J.
Biol. Chem. 1966, **241**, 695～699; (b) Tucci, A. F.; Ceci, L. N.; Bhattacharjee, J. K. Methods
Enzymol. 1969, **13**, 619～623; (c) Li, Z. -C.; Xu, J. -Q. Molecules 1998, **3**, 31～34.

[6] For a racemic synthesis followed by resolution, see: (a) Ancliff, R. A.; Russell, A. T.;
Sanderson, A. J. Tetrahedron: Asymmetry 1997, **8**, 3379～3382; (b) For chiral synthesis, see:
Thomas, U.; Kalyanpur, M. G.; Stevens, C. M. Biochemistry 1966, **5**, 2513～2516; See also: (c)
Tavassoli, A.; Duffy, J. E. S.; Young, D. W. Tetrahedron Lett. 2005, **46**, 2093～2096.

[7] (a) For asymmetric synthesis of homocitric acid lactone, see: Rodriguez, G. H. R.; Biellmann, J.
F. J. Org. Chem. 1996, **61**, 1822～1824; (b) Ma, G.; Palmer, D. R. J. Tetrahedron Lett. 2000, **41**,
9209～9212; (c) Xu, P. -F.; Matsumoto, T.; Ohki, Y.; Tatsumi, K. Tetrahedron Lett. 2005, **46**,
3815～3818; (d) Paju, A.; Kanger, T.; Pehk, T.; Eek, M.; Lopp, M. Tetrahedron 2004, **60**,
9081～9084; (e) Huang, P. -Q.; Li, Z. -Y. Tetrahedron: Asymmetry 2005, **16**, 3367～3370.

[8] For an asymmetric synthesis of a precursor of (S)-per-homocitric acid, see Ref. 7e.

[9] (a) Tsai, K. R.; Wan, H. L. J. Cluster Sci. 1995, **6**, 485～501; (b) Zhou, Z. H.; Wan, H. L.; Tsai,
K. R. Inorg. Chem. 2000, **39**, 59～64; (c) Zhou, Z. H.; Hou, S. Y.; Cao, Z. X.; Tsai, K. R.;
Chow, Y. L. Inorg. Chem. 2006, **45**, 8447～8451.

[10] (a) Strassman, M.; Ceci, L. N. Biochem. Biophys. Res. Commun. 1964, **14**, 262～267; (b)
Strassman, M.; Ceci, L. N. J. Biol. Chem. 1965, **240**, 4357～4361; (c) Hogg, R. W.; Broquist,
H. P. J. Biol. Chem. 1968, **243**, 1839～1845.

[11] (a) Loh, T. -P.; Huang, J. -M.; Goh, S. -H.; Vittal, J. J. Org. Lett. 2000, **2**, 1291～1294; (b)
Nakamura, S.; Sato, H.; Hirata, Y.; Watanabe, N.; Hashimoto, S. Tetrahedron 2005, **61**, 11078
～11106; (c) Roers, R.; Verdine, G. L. Tetrahedron Lett. 2001, **42**, 3563～3565.

[12] Carlsen, P. H. J.; Katsuki, T.; Martin, V. S.; Sharpless, K. B. J. Org. Chem. 1981, **46**, 3936～
3938.

[13] For allyl metal addition to α-keto-esters or α-keto-amides, see: (a) Soai, K.; Ishizaki, M. J. Org.
Chem. 1986, **51** 3290～3295; (b) Kiegiel, K.; Jurczak, J. Tetrahedron Lett. 1999, **40**, 1009～
1012; (c) Loh, T. -P.; Huang, J. -M.; Xu, K. -C.; Goh, S. -H.; Vittal, J. J. Tetrahedron Lett.
2000, **41**, 6511～6515; (d) Shin, J. A.; Cha, J. H.; Pae, A. N.; Choi, K. I.; Koh, H. Y.; Kang,
H. -Y.; Cho, Y. S. Tetrahedron Lett. 2001, **42**, 5489～5492; (e) Basavaiah, D.; Sreenivasulu, B
Tetrahedron Lett. 2002, **43**, 2987～2990; (f) Kaur, P.; Singh, P.; Kumar, S. Tetrahedron 2005,
**61**, 8231～8240; (g) Chen, J. H.; Venkatesham, U.; Lee, L. C.; Chen, K. M. Tetrahedron 2006,
**62**, 887～893.

[14] For a review on selective reactions using allylic metals, see: Yamamoto, Y. Chem. Rev. 1993,
**93**, 2207～2293.

[15] For reviews, see: (a) Li, C. -J.; Chan, T. -K. Tetrahedron 1999, **55**, 1149～1176; (b) Ranu, B. C.
Eur. J. Org. Chem. 2000, 2347～2356; (c) Pae, A. M.; Cho, Y. -S. Curr. Org. Chem. 2002, **6**,
715～737; (d) Podlech, J.; Maier, T. C. Synthesis 2003, 633～655; (e) Nair, V.; Ros, S.; Jayan,

C. N. ;Pillai,B. Tetrahedron 2004,**60**,1959～1982;(f)Li,C. -J. Chem. Rev. 2005,**105**,3095～3165;(g)Loh,T. P. ;Chua,G. L. J. Synth. Org. Chem. Jpn. 2005,**63**,1137～1146.

[16]During the progress of this work,an interesting indium-mediated allylation of α-oxo-glutaric acid leading to lactone was reported, see：Singh, P. ; Mittal, A. ; Kaur, P. ; Kumar, S. Tetrahedron 2006,**62**,1063～1068.

[17]Zimmermann,F. ;Mend,E. ;Oget,N. Tetrahedron Lett. 2005,**46**,3201～3203.

[18]Yeh, M. -C. ; Sheu, B. -A. ; Fu, H. -W. ; Tau, S. -I. ; Chuang, L. -W. J. Am. Chem. Soc. 1993, **115**,5941～5952.

[19]While this work is in progress,a concise Reformatsky reaction-based synthesis followed by optical resolution of homocitric acid lactone was described,in which harmful benzene was used as a solvent：Jia, Y. -H. ; Palmer, D. R. J. ; Quail, J. W. Acta Crystallogr. 2005, **E61**, o4034～o4036.

■ 本文原载：Inorg. Chem. 2008，47，pp. 8714～8720.

# Formations of Mixed-Valence Oxovanadium$^{V, IV}$ Citrates and Homocitrate with N-Heterocycle Chelated Ligand

Can-Yu Chen， Zhao-Hui Zhou①， Hong-Bin Chen，
Pei-Qiang Huang， Khi-Rui Tsai， Yuan L. Chow

(State Key Laboratory of Physical Chemistry of Solid Surfaces and
Department of Chemistry, College of Chemistry and Chemical Engineering,
Xiamen University, Xiamen, 361005, China )
Received March 27, 2008

**Abstract** Dimeric mixed-valence oxovanadium citrate $[V_2O_3(phen)_3(Hcit)] \cdot 5H_2O$ **(1)** ($H_4$cit = citric acid, phen = 1, 10-phenanthroline) was isolated from a weak acidic medium. It could be converted quantitatively into a tetrameric oxovanadium citrate adduct of 1, 10-phenanthroline $[V_2O_3(phen)_3(Hcit)_2(phen)_3O_3V_2] \cdot 12H_2O$ **(2)**. This was supported by the trace of infrared spectra and X-ray diffraction patterns. The two compounds feature a bidentate citrate group that chelates only to one vanadium center through their negatively charged $\alpha$-alkoxy and $\alpha$-carboxy oxygen atoms, while the other $\beta$-carboxy and $\beta$-carboxylic acid groups are free to participate in strong intramolecular and intermolecular hydrogen bonding [2. 45(1) in **1** and 2. 487(2) Å in **2**], respectively. This is also the case of homocitrato vanadate(V/IV) $[V_2O_3(phen)_3(R, S-H_2homocit)]Cl \cdot 6H_2O$ **(3)** ($H_4$homocit = homocitric acid), which features a binding mode similar to that found in the $R$-homocitrato iron molybdenum cofactor of Mo-nitrogenase. Moreover, the homocitrato vanadate (V) $[VO_2(phen)_2]_2$ $[V_2O_4(R, S-H_2homocit)_2] \cdot 4H_2O \cdot 2C_2H_5OH$ **(4)** is isolated as a molecular precursor for the formation of mixed-valence complex **3**. The V-$O_{\alpha\text{-alkoxy}}$ and V-$O_{\alpha\text{-carboxy}}$ bond distances of homocitrate complexes **3** and **4** are 1. 858(4) and 1. 968(6)$_{av}$ and 2. 085(4) and 1. 937(5) Å, respectively. They are shorter than those of homocitrate to FeVco(2. 15 Å). The $\gamma$-carboxy groups of coordinated homocitrato complexes **3** and **4**, and the free homocitrate salt $Na_3(Hhomocit) \cdot H_2O$ **(5)**, form strong hydrogen bonds with the chloride ion and the water molecule [2. 982(5) in **3**, 2. 562(9) in **4**, and 2. 763(1) Å in **5**], respectively.

## 1 Introduction

Detailed crystallographic analysis of nitrogenase has revealed a previously unrecognized light atom in the center of six iron atoms of the $R$-homocitrato-MoFe$_7$S$_9$X (FeMoco) cluster[1], which is supported

---

① Author to whom correspondence should be addressed. Phone：+86 592 2184531. Fax：+86 592 2183047. E-mail address：zhzhou@xmu. edu. cn.

by the theoretical study and model compound synthesis of iron sulfur clusters with similar cores of $Fe_8S_9$ and $Fe_8S_8N$, as well as analyses of $^{57}Fe$ nuclear resonance vibrational spectroscopy data[2-4]. This idea has been expanded, and it is now believed that iron-centered vanadium cofactor(FeVco) has a very similar chemical environment to that of FeMoco[5]. On the other hand, nitrogenase having a citrate in the place of homocitrate manifests itself with a weaker reduction of $N_2$, while the citrate analogue retains $C_2H_2$ reduction activity comparable to that of the wild type[6-8]. It is proposed that homocitrate may facilitate the binding of the dinitrogen molecule through the dissociation of the bound $\alpha$-carboxy or $\alpha$-alkoxy groups from the Mo atom[9,10], and also the formation of an intramolecular hydrogen bond[11]. In the previous study, our investigations have addressed the coordination chemistry of citrato or homocitrato vanadium and molybdenum complexes[12-16]. In order to understand the interaction of tricarboxylate ligands with vanadium in detail, the binding of citrates is now extended to mixed-valence citrate and homocitrate vanadium systems; this is reported in this paper.

# 2　Experimental Section

**Physical Measurements.** The pH value was measured by a potentiometric method with a digital PHB-8 pH meter. Infrared spectra were recorded as Nujol mulls in KBr plates on a Nicolet 360 FT-IR spectrometer. Elemental analyses were performed with an EA 1100 elemental analyzer. The electronic spectra were recorded on a Shimadzu UV 2501 spectrophotometer with an integrating sphere for reflectance spectroscopy. X-ray photoelectron spectra(XPS) were recorded on a Quantum 2000 Scanning ESCA Microprobe electron spectrometer using Al K$\alpha$ radiation with a pass energy of 46.95 eV. The binding energy(BE) scale was regulated by setting the C 1s transition at 284.6 eV(accuracy of BE was $\pm0.1$ eV). Electron paramagnetic resonance(EPR) spectra of compounds were collected at different temperatures on a Bruker EMX-10/12 spectrometer. X-ray diffraction(XRD) patterns were recorded on a Philips X'Pert Pro Super X-ray diffractometer equipped with X'Celerator and Xe detection systems. $^{13}C$ NMR spectra were recorded in $D_2O$ on a Bruker AV400 NMR spectrometer using sodium 2,2-dimethyl-2-silapentane-5-sulfonate as an internal reference.

**Preparations of** $[V_2O_3(phen)_3(Hcit)]\cdot5H_2O(1)$ **and** $[V_2O_3(phen)_3(Hcit)_2(phen)_3O_3V_2]\cdot12H_2O$ **(2).** Vanadium pentoxide(0.55 g, 3.0 mmol) dissolved in 2 mol$\cdot$L$^{-1}$ of potassium hydroxide(5.0 mL) was mixed with citric acid monohydrate(1.91 g, 9.1 mmol) solution in 100 mL of water. 1,10-Phenanthroline monohydrate(0.61 g, 3.1 mmol) in 95% ethanol(50 mL) was added, and the resulting mixture was brought to pH 4.0 with a potassium hydroxide solution. The dark green solution was set aside for two weeks to deposit crystals that were collected and washed with ethanol to afford **1**(0.81 g, 81% yield based on phenanthroline). Found(calcd for $C_{42}H_{39}N_6O_{15}V_2$): C, 51.7(52.0); H, 4.2(4.1); N, 8.5(8.7%). IR(KBr, cm$^{-1}$): $v_{as}$(COO), 1627$_{vs}$; $v_s$(COO), 1427$_s$, 1385$_w$, 1342$_w$; $v$(V=O), 965$_m$, 936$_m$; $v$(V-O-V), 851$_s$. A similar procedure was used except that the final pH was adjusted to 5.1 to obtain **2**. Vanadium pentoxide(90.0 mg, 0.50 mmol) dissolved in 2 mol$\cdot$L$^{-1}$ of potassium hydroxide solution was mixed with a solution of citric acid monohydrate(0.22 g, 1.02 mmol) and 1,10-phenanthroline monohydrate(0.20 g, 1.02 mmol) in 25 mL of water-ethanol(2:1 by volume). The mixture was brought to pH 5.1 with a potassium hydroxide solution. The dark green crystals were isolated as **2**(0.13 g, 50% yield based on phenanthroline). Found(calcd for $C_{84}H_{82}N_{12}O_{32}V_4$): C, 51.3(51.1); H, 4.3(4.2); N, 8.6(8.5%). IR(KBr, cm$^{-1}$): $v_{as}$(COO), 1625$_{vs}$; $v_s$(COO), 1426$_s$, 1385$_w$, 1344$_w$; $v$(V=O), 961$_m$, 936$_m$; $v$(V-O-V), 853$_s$.

**Preparation of** $[V_2O_3(phen)_3(R,S-H_2homocit)]Cl\cdot6H_2O(3)$. Homocitric acid $\gamma$-lactone(38.2 mg,

0. 20 mmol) prepared according to the published method[17] was dissolved in 4 mL of a potassium hydroxide solution(1. 0 mol $\cdot$ L$^{-1}$)at pH 11 with stirring, to which a solution(4 mL) of potassium vanadate(27. 6 mg, 0. 20 mmol)and an ethanol solution(3 mL)of 1, 10-phenanthroline monohydrate (74. 5 mg, 0. 37 mmol)were added. The pH value of the mixture was brought to 2. 5 by adding diluted HCl. The color changed to light yellow-green. The mixture was set aside to evaporate slowly for two weeks. The precipitate was filtered and washed with ethanol to afford dark green crystals(54. 0 mg, 51% based on vanadium). Found(calcd for $C_{43}H_{44}Cl_1N_6O_{16}V_2$): C, 50. 0(49. 7); H, 4. 1(4. 3); N, 8. 0(8. 1%). IR(KBr, cm$^{-1}$): $v$(COOH), 1720$_s$; $v_{as}$(COO), 1607$_{vs}$; $v_s$(COO), 1425$_s$, 1383$_w$, 1346$_{vw}$; $v$(V=O), 974$_m$, 941$_m$; $v$(V-O-V), 851$_s$. In the synthesis of **3**, $[VO_2(phen)_2]_2[V_2O_4(R,S-H_2homocit)_2]$ $\cdot$ 4H$_2$O $\cdot$ 2C$_2$H$_5$OH(**4**) could be obtained in small amounts(yield ~5%)as a yellow solid, which redissolved to form **3** as the final product.

**Preparation of Racemic Homocitrate Salt Na$_3$(Hhomocit) $\cdot$ H$_2$O(5).** Racemic homocitric acid $\gamma$-lactone (107 mg, 0. 57 mmol)prepared according to the published method[17] was dissolved in 2. 0 mol $\cdot$ L$^{-1}$ of sodium hydroxide solution to a pH of 10. 6 with stirring. The solution was hydrolyzed for 2 days at room temperature. The colorless product of racemic homocitrate salt **5** was obtained after one week of standing in a refrigerator. The yield of **5** was 85 mg(51%). Found(calcd for $C_7H_9O_8Na_3$): C, 28. 8 (29. 0); H, 3. 0(3. 1%). IR(KBr, cm$^{-1}$): $v_{as}$(COO), 1675, 1598, 1571, $v_s$(COO), 1403. $^{13}$C NMR(D$_2$O, ppm): 185. 8 ($\alpha$-CO$_2$), 184. 4 ($\beta$-CO$_2$), 182. 3 ($\gamma$-CO$_2$), 79. 3 ($\equiv$COH), 48. 8 ($-$CH$_2$CO$_2$), 38. 1 (CH$_2-$CH$_2$CO$_2$), 35. 0($-$CH$_2$CH$_2$CO$_2$)ppm.

**Transformations of 1 to 2.** Compound **1**(0. 100 g) was suspended in a water-ethanol(1 : 2 in volume)solution. The pH of the mixture was 5. 1. Dark green crystals separated from the green solution when the solvent was slowly evaporated over two weeks to give **2** in quantitative yield. The compound was identified by IR spectra, XRD patterns, and also unit cell dimensions.

**X-Ray Structure Determination.** Crystals of **1**~**5** in oil were analyzed with a Bruker Smart Apex CCD area detector diffractometer or an Oxford Gemini S Ultra system with graphite monochromate Cu K$\alpha$ radiation ($\lambda$=1. 54180 Å)or Mo K$\alpha$ ($\lambda$=0. 71073 Å)radiation at 173 K. The data were corrected for absorption by using the SADABS program[18]. The structures were primarily solved by SHELXS in the WinGX program[19] and refined by full-matrix least-squares procedures with anisotropic thermal parameters for all of the nonhydrogen atoms with SHELX-97[20]. Hydrogen atoms were located from a difference Fourier map and refined isotropically; they were omitted for clarity in the displayed figures, except for **5**.

Crystal data for **1**: $C_{42}H_{39}N_6O_{15}V_2$, $M$ = 969. 67, triclinic, space group P$\bar{1}$, $a$ = 12. 5702(4), $b$ = 13. 5289(6), $c$ = 14. 1029(6) Å, $\alpha$ = 73. 634(4), $\beta$ = 82. 865(3), $\gamma$ = 72. 525(4)°, $V$ = 2192. 9(2)Å$^3$, $D_c$ = 1. 469 g/cm$^3$, $Z$ = 2, $R_1$ = 0. 071, $wR_2$ = 0. 239. Crystal data for **2**: $C_{84}H_{82}N_{12}O_{32}V_4$, $M$ = 1975. 38, triclinic, space group P$\bar{1}$, $a$ = 13. 0831(5), $b$ = 13. 0996(4), $c$ = 14. 8791(5)Å, $\alpha$ = 101. 442(3), $\beta$ = 105. 709(3), $\gamma$ = 115. 186(3)°, $V$ = 2072. 5(1)Å$^3$, $D_c$ = 1. 583 g/cm$^3$, $Z$ = 1, $R_1$ = 0. 031, $wR_2$ = 0. 096. Crystal data for **3**: $C_{43}H_{44}Cl_1N_6O_{16}V_2$, $M$ = 1038. 17, triclinic, space group P$\bar{1}$, $a$ = 12. 9695(8), $b$ = 13. 6195(8), $c$ = 15. 4979(9) Å, $\alpha$ = 71. 586(5), $\beta$ = 78. 905(5), $\gamma$ = 63. 745(6)°, $V$ = 2324. 9(2)Å$^3$, $D_c$ = 1. 483 g/cm$^3$, $Z$ = 2, $R_1$ = 0. 059, $wR_2$ = 0. 177. Crystal data for **4**: $C_{66}H_{68}N_8O_{28}V_4$, $M$ = 1625. 04, triclinic, space group P$\bar{1}$, $a$ = 8. 0818(9), $b$ = 13. 932(2), $c$ = 15. 770(2)Å, $\alpha$ = 113. 305(2), $\beta$ = 93. 113(2), $\gamma$ = 91. 559(2)°, $V$ = 1626. 0(3)Å$^3$, $D_c$ = 1. 660 g/cm$^3$, $Z$ = 1, $R_1$ = 0. 94, $wR_2$ = 0. 198. Crystal data for **5**: $C_7H_9O_8Na_3$, $M$ = 290. 11, triclinic, space group P$\bar{1}$, $a$ = 5. 8041(3), $b$ = 8. 0920(5), $c$ = 10. 9032(8) Å, $\alpha$ = 80. 153(6), $\beta$ = 82. 600(5), $\gamma$ = 87. 431 (5)°, $V$ = 500. 21(5)Å$^3$, $D_c$ = 1. 926 g/cm$^3$, $Z$ = 2, $R_1$ = 0. 030, $wR_2$ = 0. 069.

# 3 Results and Discussion

The reaction of vanadium pentoxide with citric or homocitric acid in the presence of 1, 10-phenathroline in an aqueous ethanol solution is delicately sensitive to pH adjustment, as demonstrated

by a digital pH monitor. Excess citrate and a lesser amount of 1, 10-phenathroline are needed for the completeness of the reaction. At pH 4.0, a dimeric mixed-valence complex $[V_2O_3(phen)_3(Hcit)] \cdot 5H_2O$ (1) was isolated as dark green crystals in high yield. At pH $5.0 \sim 6.0$, however, a tetrameric complex, $[V_2O_3(phen)_3(Hcit)_2(phen)_3O_3V_2] \cdot 12H_2O$ (2), was obtained as dark green crystals. The relation was demonstrated by the conversion of 1, dissolving in a water-ethanol solution from which 2 was deposited quantitatively. As shown in Figure 1, the IR patterns of 1 and 2 are very similar. However, the frequency of V = O vibration around 950 cm$^{-1}$ could be served as a probe for the reaction. IR tracing of the reaction product from the solution shows that the conversion takes place in the early stage. The switch from

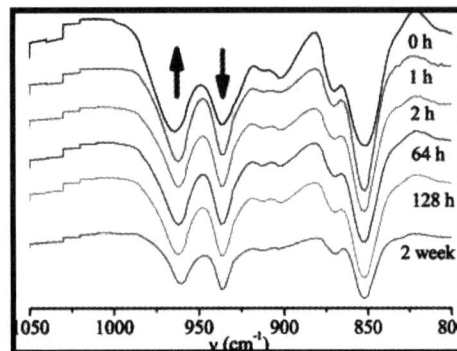

Figure 1    Qualitative IR tracing of the conversion from $[V_2O_3(phen)_3(Hcit)] \cdot 5H_2O(1)$ to $[V_2O_3(phen)_3(Hcit)_2(phen)_3O_3V_2] \cdot 12H_2O(2)$ at different times.

an intramolecular bond in 1 to an intermolecular hydrogen bond of the citrate ion in 2 is interesting. Complex 1 could be quantitatively changed into 2 after two weeks in a water-ethanol solution, as further proved by bulk XRD measurements in Figure S1 (Supporting Information). This, in turn, reveals the stability of tetramer 2, which is experimentally demonstrated in a solid.

Complex 1 is shown to be stabilized by strong intramolecular hydrogen bonding (Figure 2), whereas the formation of tetramer 2 is enforced by strong intermolecular hydrogen bonding (Figure 3). This is eventually substantiated by X-ray structures (vide infra). The detail crystallization processes is not clear at this stage, but it is the fact that, at pH 4.0, monomer 1 crystallized faster.

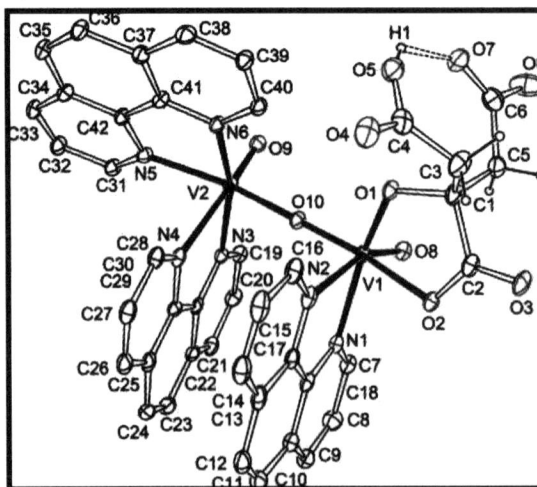

Figure 2    The ORTEP plot of $[V_2O_3(phen)_3(Hcit)] \cdot 5H_2O(1)$ at the 20% probability level. Hydrogen atoms of phenanthroline groups were omitted for clarity.

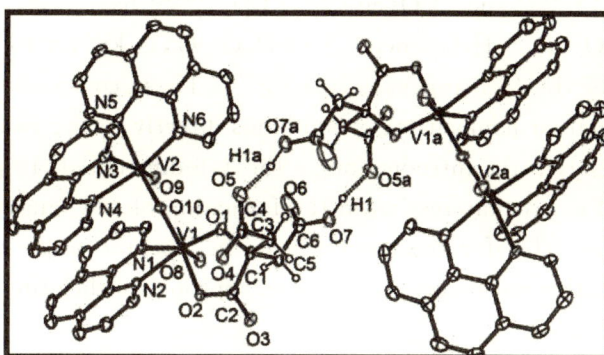

**Figure 3** The ORTEP plot of $[V_2O_3(phen)_3(Hcit)_2(phen)_3O_3V_2] \cdot 12H_2O(2)$ at the 30% probability level. Hydrogen atoms of phenanthroline groups were omitted for clarity.

The reaction of vanadate and racemic homocitrate, both as potassium salts, in the presence of 1, 10-phenanthroline at pH 2.5 gave protonated homocitrato vanadate $(V/IV)[V_2O_3(phen)_3(R,S-H_2homocit)]Cl \cdot 6H_2O(3)$ as dark green crystals (Scheme 1). In the synthesis of **3**, yellow solid **4** could be obtained in a small amount as an intermediate, which redissolved to form the dark green solid **3** as the final product. X-ray structural analysis shows that the precursor $[VO_2(phen)_2]_2[V_2O_4(R,S-H_2homocit)_2] \cdot 4H_2O \cdot 2C_2H_5OH$ (**4**) is an adduct of the diphennanthroline dioxovanadium cation and dihydrogen homocitrato dioxovanadium anion. The cation diphenanthroline dioxovanadate (V) $[V^VO_2(phen)_2]^+$ is a direct reaction product of vanadate (V) and the phenanthroline ligand, which is easily reduced to $[V^{IV}O_2(phen)_2]^{[21]}$. The reduced $[V^{IV}O_2(phen)_2]$ unit might be used as an reactant for the further formation of mixed-valence complex **3**, where the $V_2O_3^{3+}$ core can be viewed as a complex between the donor $V^VO_2^+$ and the acceptor $V^{IV}O^{2+}$. The anion structure $[V_2O_4(R,S-H_2homocit)_2]^{2-}$ of **4** is similar to that found in $K_2[V_2O_4(R,S-H_2homocit)_2] \cdot 6H_2O$, as shown in Figure S2 (Supporting Information)$^{[22]}$, which is a reaction product of vanadate and homocitrate containing a $V_2O_4$ unit. The homocitrate dianion in **4** bidentately coordinates to the vanadium atom through $\alpha$-carboxy and the $\alpha$-alkoxy groups, where the latter acts as a bridging group. A detailed reaction mechanism for the formation of mixed-valence complex **3** is under further study.

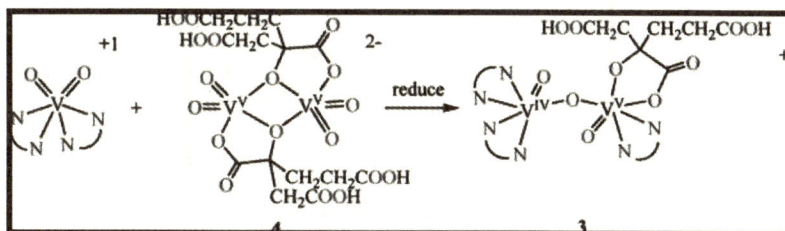

**Scheme 1** Transformation of Homocitrato Vanadate (V) to Mixed-Valence Homocitrato Vanadium Complex

A recemic homocitrate solution was obtained by dissolving the known homocitric acid lactone in aqueous potassium hydroxide and had a pH 10.6. This is proved by the isolation of sodium homocitrate $Na_3(Hhomocit) \cdot H_2O$. All of the isolated citrato and homocitrato mixed-valance vanadium (V/IV) complexes **1~3** are insoluble in water and ethanol and slightly soluble in the mixed solvent of water-ethanol.

Vanadium citrate complexes such as the dioxovanadates (V) $[VO_2(H_2cit)]_2^{2-}$ (**6**)$^{[13,23,24]}$, $[VO_2(Hcit)]_2^{4-}$ (**7**)$^{[14,25,26]}$, and $[VO_2(cit)]_2^{6-}$ (**8**)$^{[12,26,27]}$ and vanadyl citrate complexes of

$[V_2O_2(Hcit)(cit)]^{3-}$ (**10,11**)[28,29] and $[VO(cit)]_2^{4-}$ (**12**)[12,30,31] have been reported. The homocitrato dioxovanadate(V), $K_2[V_2O_4(R,S-H_2homocit)_2] \cdot 6H_2O$ (**9**), is known to mimic an early mobilized precursor in the synthesis of the V-nitrogenase cofactor[22]. These compounds have $V_2(V)O_4$ and $V_2(IV)O_2$ cores, while the citrate or homocitrate group binds directly to the two vanadium atoms through its α-alkoxy group. Previously, in the introduction of the 2,9-dimethyl-1,10-phenanthroline ligand, the N-heterocycle group served as a countercation in the formation of the vanadyl citrate complex[29]. The methyl substituents are unsuitable for the coordination. In fact, the present study shows that 1,10-phenanthroline without a methyl substituent group is an appropriate chelating entity for the synthesis of the mixed ligand complexes **1~3**.

The crystal structures of **1** and **2** show a *cis*-oxo-$V_2O_3$ unit along with a bent oxo bridge [172.2(3) and 160.3(1)°] in Figures 1 and 2, respectively. The V1···V2 distances are on the order of 3.5 Å, a distance that implicates a weak interaction between two metal centers. The vanadium atom is six-coordinate in a distorted octahedral geometry. The citrate group chelates to one vanadium atom as a bidentate ligand via its α-alkoxy and carboxy oxygen atoms, whereas the other β-carboxy and β-carboxylic acid groups are free. The free groups participate in a very strong intramolecular hydrogen bond [**1**, O5···O7 = 2.46(1) Å] and intermolecular hydrogen bond [**2**, O5a···O7 = 2.487(4) Å]. The O···O distances are short in comparison with the bond found in, for example, dimeric acetic acid (2.68 Å)[32]. Here, the β-carboxy and β-carboxylic acid groups are involved in the formation of an intramolecular hydrogen bond in **1**. In a previous quantum-mechanical computation and a model suggestion, the γ-carboxy group of R-homocitrate is proposed as undergoing intramolecular hydrogen bonding with the imidazole ligand on Mo, while the β-carboxy group of the citrate ligand is not[11].

Compound **1** exhibits photoinstability under UV radiation, as shown in Figure S3 (Supporting Information). This is inconsistent with the strong absorption of solid **1** in the UV diffused reflectance spectra in Figure 4. Bands in the 550−800 nm region are assigned to a d-d transition, and the bands in the 350−550 nm region can be assigned to metal-to-ligand charge transfer. The bands below 350 nm region should be the π-π transition[33].

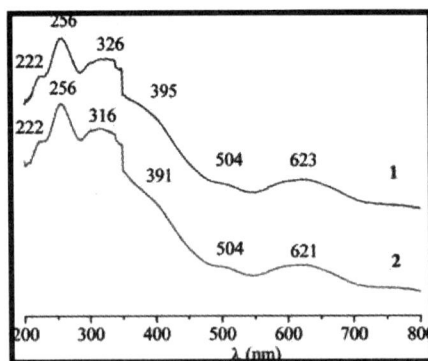

Figure 4   Diffuse reflectance spectra of solid $[V_2O_3(phen)_3(Hcit)] \cdot 5H_2O$(**1**) and $[V_2O_3(phen)_3(Hcit)_2(phen)_3O_3V_2] \cdot 12H_2O$(**2**).

The V-O distances in **1** and **2** vary systematically according to the bond types in Table S1 (Supporting Information); the short V=O distances are in agreement with the double-bond character. The bond distances for the VOV bridge are unequal but fall within the range found for the oxo-bridged, mixed-valence complex, $(NH_4)_3[V_2O_3(nta)_2] \cdot 3H_2O$[34]. Bond valence calculations give valences of 5.0 and 4.3 for **1** and 5.0 and 4.3 for **2**[35]. The assignments are supported by an XPS analysis of V2p as shown in Figure S4 of the Supporting Information. The peaks at 516 and 523 eV correspond to vanadium

$2p_{3/2}$ and $2p_{1/2}$ in the oxidation states of V and IV[36]. Moreover, the X-band EPR spectra of VO(IV/V) mixed-valence complexes **1** and **2** were recorded in the solid state at 100, 150, 200, and 250 K, as shown in Figures S5a and S5b(Supporting Information). The EPR spectra of **1** and **2** exhibit one broad band centered on $g = 1.991$ and $g = 1.995$, respectively, without a resolved hyperfine structure. In particular, the hyperfine coupling around the $^{51}V(I = 7/2, S = 1/2)$ nucleus is not observable. The absence of vanadium's hyperfine coupling is common in the solid state and is attributed to the simultaneous flipping of neighboring electron spin or is due to strong exchange interactions, which average out the interaction with the nuclei[37]. In homocitrato complex $[V_2O_3(phen)_3(R,S-H_2homocit)]Cl \cdot 6H_2O(3)$, the molecular structure is similar to those of **1** and **2**, see Figure 5. The V(1)-O(10) distance [1.713(4) Å] is much shorter than the V(2)-O(10) distance [1.884(4) Å], which implies an unsymmetric configuration. The bond valence summaries are 5.0 and 4.4 for V1 and V2, respectively[35]. The homocitrate ion acts as a bidentate ligand through its $\alpha$-alkoxy and $\alpha$-carboxy groups, leaving the $\beta$-carboxylic and $\gamma$-carboxylic acid groups free. The V-O distance[$\alpha$-alkoxy 1.858(4) Å] is comparable to those of **1** and **2** but much shorter than that of $K_2[V_2O_4(R,S-H_2homocit)_2] \cdot 6H_2O(9)[1.980(8)_{av}$ Å][22], while the V-O distance [$\alpha$-carboxy, 2.085(4) Å] is longer than that of **9** [1.959(8)$_{av}$ Å]. This should be the result of the strong trans influence of the short V(1)-O(10) bond. Moreover, the discrete structure of complex **3** is linked by the weak hydrogen bonding between O(5)-Cl(1) and O(7)-Cl(1) [3.004(5), 2.982(5) Å]; the latter is from the $\gamma$-carboxy group of the homocitrate ligand. The hydrogen bond from the $\gamma$-carboxy group could also be seen from the structural analysis of homocitrate salt **5**, which forms a strong hydrogen bond between the $\gamma$-carboxy group and the water molecule [2.763(1) Å], as shown in Figure 6. Moreover, the anion bridged core of **4** is defined by the O(1)-V(1)-O(1a) [71.2(3)°] and the V(1)-O(1)-V(1a)[108.8(3)°] angles. This is similar to the 71.6(3)° and 108.4(3)° given for **9**. The V-O distances [$V-O_{alkoxy} = 1.968(6)_{av}$, $V-O_{carboxy} = 1.937(5)$ Å] of complex **4** are the shortest bonds among the citrate and homocitrato vanadates(V). The hydrogen bond between the $\gamma$-carboxy group and the water molecule [2.562(9) Å] in **4** is the most strongest in the reported homocitrato complexes[16,22]. The transformation of **4** to **3** shows that bulky conjugated phenanthroline is useful for the formation of bidentate homocitrate coordinated in only one vanadium center, which avoids the formation of a bridging $\alpha$-alkoxy group.

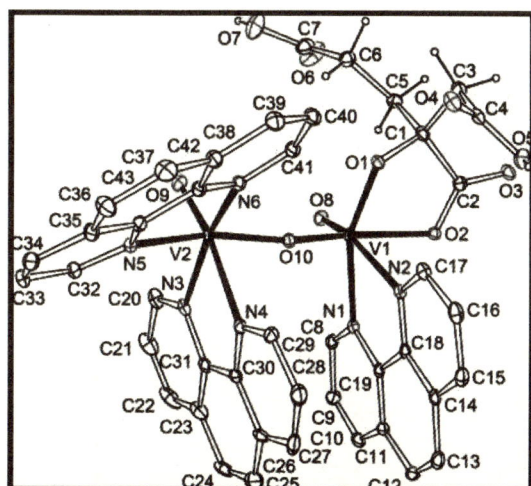

Figure 5 The ORTEP plot of $[V_2O_3(phen)_3(R,S-H_2homocit)]Cl \cdot 6H_2O(3)$ at the 30% probability level. Hydrogen atoms of phenanthroline groups were omitted for clarity.

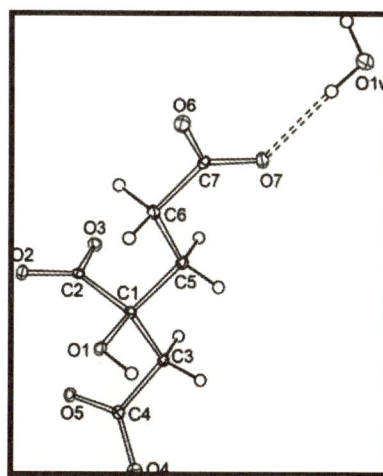

Figure 6 The ORTEP plot of $Na_3(Hhomocit) \cdot H_2O(5)$ at the 30% probability level.

Obvious downfield shifts of $^{13}$C NMR spectra in **5** are observed compared with those of homocitric acid $\gamma$-lactone, except for that of $\alpha$-hydroxy carbon in $D_2O$. For example, both $\beta$-and $\gamma$-carboxy carbons of **5** show large downfield shifts of $\Delta\delta$ 6. 9 and 6. 8 ppm, respectively, arising from the basic medium and delactonization. The $\alpha$-carboxy carbon has somewhat small downfield shifts of $\Delta\delta$ 3. 1 ppm. Other peaks of methylene carbons also show downfield shifts in general. This is a clear indication for the hydrolysis of the $\gamma$-lactone. Moreover, the downfield shifts (2~3 ppm) of $^{13}$C NMR spectra in the strong polar solvent $D_2O$ are observed compared with that of $CD_3OD$. Table 1 gives a $^{13}$C NMR data comparison between homocitric acid $\gamma$-lactone and sodium homocitrate **4** in $D_2O$ and $CD_3OD$ solvents. The $^{13}$C spectra of homocitrato acid $\gamma$-lactone and sodium homocitrate **4** are shown in Figures S6 and S7 (Supporting Information).

**Table 1** $^{13}$C NMR Spectral Data(in ppm, $D_2O$ or $CD_3OD$)of Homocitrato Acid $\gamma$-Lactone and Homocitrate Na$_3$(Hhomocit) $\cdot$ H$_2$O(5)$^a$

| compound | ('CO) | (CO$_2$)$_\alpha$ | (CO$_2$)$_\beta$ | (CO$_2$)$_\gamma$ | ( =CH$_2$) |
|---|---|---|---|---|---|
| homocitric acid $\gamma$-lactone($D_2O$) | 87. 0 | 182. 7 | 177. 5 | 175. 8 | 44. 0, 33. 7, 30. 3 |
| homocitrate **5**($D_2O$) | 79. 3(−7. 7) | 185. 8(3. 1) | 184. 4(6. 9) | 182. 3(6. 8) | 48. 8(4. 8), 38. 1(4. 4), 35. 0(4. 7) |
| homocitric acid $\gamma$-lactone($CD_3OD$)$^{17}$ | 84. 7 | 178. 77 | 174. 1 | 172. 48 | 42. 1, 32. 3, 28. 7 |

$^a\Delta\delta$ values are given in parentheses.

Furthermore, attempts to isolate a similar citrato complex of $[V_2O_3(\text{phen})_3(H_2\text{cit})] \cdot Cl$ were unsucessful. This may be related to the formations of the species with very strong hydrogen bonds in **1** and **2**. On the other hand, no isolation of $[V_2O_3(\text{phen})_3(\text{Hhomocit})]$ is attributed temporarily to the equilibrium between homocitric acid $\gamma$-lactone and homocitrate in a weak acidic solution and the dissociation of homocitrato vanadate in solution. The present isolations of complexes **1~3** give a new class of citrato and homocitrato vandates in different oxidation states, as shown in Scheme 2.

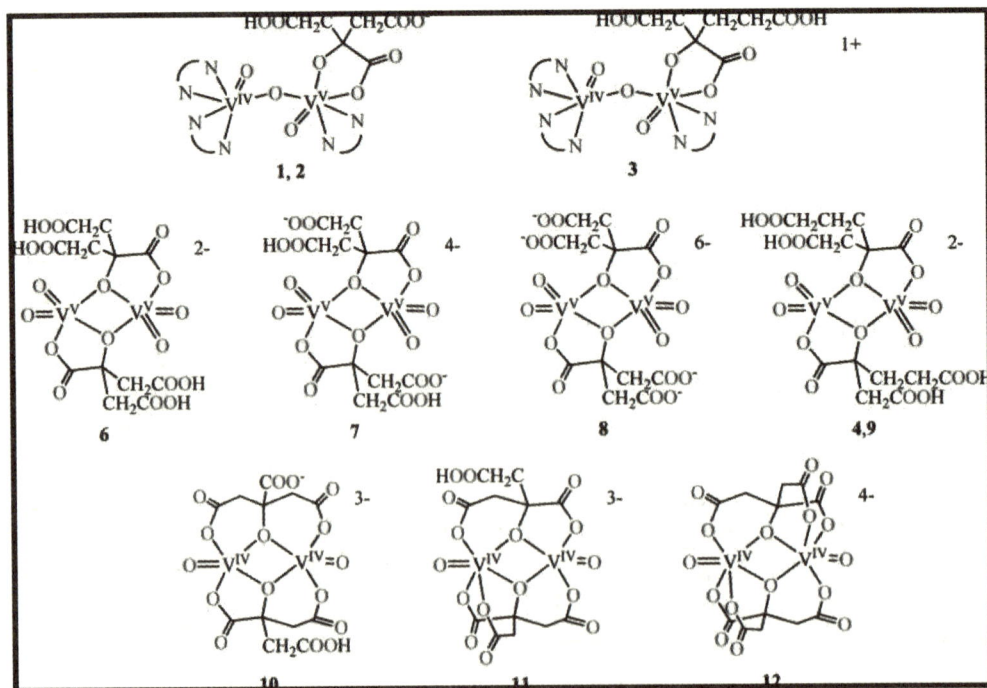

**Scheme 2** Typical Coordination Modes of Citrate and Homocitrate with Vanadium ( V , IV ): Mixed-Valence Citrato and Homocitrato Vanadate ( V/IV ), **1~3**; Citrato and Homocitrato Vanadate ( V ), **6~9**[12~14,23~27]; and Vanadyl Citrate, **10~12**[12,28~31]

In the mixed-valence dimeric citrato vanadates(V/IV), the bond distances of $\alpha$-alkoxy to vanadium are shorter than those of vanadates(V) and vanadyl citrates, see Table 2. The latter ones are strongly influenced by the bridging mode coordination in the $V_2O_4$ or $V_2O_2$ units. However, the differences between V-O($\alpha$-carboxy) distances(1. 979～2. 137 Å) are smaller, being close to the V-O distance(2. 15 Å) in homocitrato FeVco[38]. The short distances of V-O($\alpha$-alkoxy) in compounds 1～3 are worthy of note, which indicates strong binding of the $\alpha$-alkoxy group to the vanadium atom in the high oxidation state.

Table 2　Comparions of V-O Distances in Citrato and Homocitrato Vanadium Complexes
(neo=2,9-Dimethyl-1,10-phenanthroline)

| complex | V-O(Å)($\alpha$-alkoxy) | V-O(Å)($\alpha$-carboxy) | ref |
|---|---|---|---|
| $[V_2O_3(phen)_3(Hcit)]\cdot 5H_2O$(1) | 1. 851(4) | 2. 082(4) | this work |
| $[V_2O_3(phen)_3(Hcit)_2(phen)_3O_3V_2]\cdot 12H_2O$(2) | 1. 858(1) | 2. 072(1) | this work |
| $[V_2O_3(phen)_3(R,S\text{-}H_2homocit)]Cl\cdot 6H_2O$(3) | 1. 858(4) | 2. 085(4) | this work |
| Average | **1. 856(4)** | **2. 080(4)** | |
| $K_2[V_2O_4(H_2cit)_2]\cdot 4H_2O$(6) | 1. 986(2)$_{av}$ | 1. 980(3) | 13,23 |
| $Na_2K_2[V_2O_4(Hcit)_2]\cdot 9H_2O$(7) | 1. 984(4)$_{av}$ | 1. 979(4)$_{av}$ | 14,24—26 |
| $K_2(NH_4)_4[V_2O_4(cit)_2]\cdot 6H_2O$(8) | 1. 983(2)$_{av}$ | 1. 981(2) | 12,26,27 |
| $K_2[V_2O_4(R,S\text{-}H_2homocit)_2]\cdot 6H_2O$(9) | 1. 980(8)$_{av}$ | 1. 959(8)$_{av}$ | 22 |
| $[VO_2(phen)_2]_2[V_2O_4(R,S\text{-}H_2homocit)_2]\cdot 4H_2O\cdot 2C_2H_5OH$(4) | 1. 968(6)$_{av}$ | 1. 937(5) | this work |
| Average | **1. 980(8)** | **1. 967(8)** | |
| $K_3[V_2O_2H(cit)_2]\cdot 7H_2O$(10) | 1. 976(5)$_{av}$ | 1. 981(6) | 28 |
| $(Hneo)_3[V_2O_2H(cit)_2]\cdot 4H_2O$(11) | 1. 984(4)$_{av}$ | 2. 137(4)$_{av}$ | 29 |
| $Na_4[V_2O_2(cit)_2]\cdot 6H_2O$(12) | 2. 089(2)$_{av}$ | 2. 038(2) | 12,30,31 |
| Average | **2. 016(5)** | **2. 052(6)** | |
| $VFe_7S_9X(homocit)(S\text{-}cys)(N\text{-}His)$ | 2. 15 | | 38 |

# 4　Conclusions

There is sufficient evidence to show that the vanadyl citrate engages in tridentate and tetradentate coordination via the $\alpha$-alkoxy, $\alpha$-carboxy, and $\beta$-carboxy groups in the complexes $K_3[V_2O_2H(cit)_2]\cdot 7H_2O$(10), $(Hneo)_3[V_2O_2H(cit)_2]\cdot 4H_2O$(11), and $Na_4[V_2O_2(cit)_2]\cdot 6H_2O$(12)[12,28-31]. The present formation of symmetric and unsymmetric bidentate citrato and homocitrato vanadate 1～4 could be related to the bidentate homocitrato FeVco in nitrogenase, which shows a resemblance in coordination mode. Although, a detailed structure of the homocitrate vanadate of FeVco is unknown[38]. It is also noted that a large gap exists between the homocitrate vanadate in the natural cofactor and in the current model molecules. The present structural examples show that citrate and homocitrate mixed-valence complexes feature bidentate citrato or homocitrato groups that chelate to one vanadium atom through its $\alpha$-alkoxy and $\alpha$-carboxy oxygen atoms, while the other $\beta$- or $\gamma$-carboxylic acid groups are free. The bond distances $[2. 080(4)_{av}$ Å$]$ of $\alpha$-carboxy groups to vanadium in 1～3 are close to that of homocitrate to FeVco(2. 15 Å), while those $[1. 856(4)_{av}$ Å$]$ of $\alpha$-alkoxy groups to vanadium are shorter(2. 15 Å). This is attributed to a strong coordination of vanadium in the mixed-valence citrate and homocitrate vanadium[V/IV] complexes. The chemical conversion of mixed-valence vanadium citrate complexes suggests that the citrate complex **1** with an intramolecular hydrogen bond could be converted to the tetrameric form **2** with strongly interacting hydrogen bonds in the solid. Moreover, the $\beta$-carboxy group in the citrate ligand is also effective and should be noted in the further consideration of intramolecular hydrogen bonds in the vanadium citrate complex, including the biomolecular chemistry of the homocitrato-FeVco in nitrogenase.

## Acknowledgment

Financial support provided by the National Science Foundation of China (20571061) and the Ministry of Science & Technology (2005CB221408) is gratefully acknowledged. We thank Professor S. W. Ng for stimulating discussion and the reviewers for their very constructive suggestions to update the paper.

## Supporting Information Available

IR, XPS, EPR, XRD, $^{13}$C NMR, and the image of the conversion between **1** and **2**; tables of crystal and refinement data; atomic positions and displacement parameters; anisotropic displacement parameters; and bond lengths and angles in CIF format. This material is available free of charge via the Internet at http://pubs. acs. org. IC800553P

## References

[1] (a) Einsle, O. ; Tezcan, F. A. ; Andrade, S. L. A. ; Schmid, B. ; Yoshida, M. ; Howard, J. B. ; Rees, D. C. Science 2002, **297**, 1696~1700. (b) Howard, J. B. ; Rees, D. C. Proc. Natl. Acad. Sci. U. S. A. 2006, **103**, 17088~17093.

[2] Hinnemann, B. ; Nørskov, J. K. J. Am. Chem. Soc. 2003, **125**, 1466~1467.

[3] Ohhi, Y. ; Ikagama, Y. ; Tatsumi, K. J. Am. Chem. Soc. 2007, **297**, 10457~10465.

[4] Xiao, Y. ; Fischer, K. ; Smith, M. C. ; Newton, W. ; Case, D. A. ; George, S. J. ; Wang H. ; Sturhahn, W. ; Alp, E. E. ; Zhao, J. ; Yoda, Y. ; Cramer, S. P. J. Am. Chem. Soc. 2006, **128**, 7608~7612.

[5] (a) Hales, B. J. ; Case, E. E. ; Morningstar, J. E. ; Dzeda, M. F. ; Mauterer, L. A. Biochemistry 1986, **25**, 7251 ~ 7264. (b) Strange, R. W. ; Eady, R. R. ; Lawson, D. ; Hasnain, S. S. J. Synchrotron Radiat. 2003, **10**, 71~75.

[6] (a) Hoover, T. R. ; Robertson, A. D. ; Cerny, R. L. ; Hayes, R. N. ; Imperial, J. ; Shah, V. K. ; Ludden, P. W. Nature 1987, **329**, 855 ~ 857. (b) Hoover, T. R. ; Imperial, J. ; Ludden, P. W. ; Shah, V. K. Biochemistry 1989, **28**, 2768~2771.

[7] (a) Imperial, J. ; Hoover, T. R. ; Madden, M. S. ; Ludden, P. W. ; Shah, V. K. Biochemistry 1989, **28**, 7796~7799. (b) Mayer, S. M. ; Gormal, C. A. ; Smith, B. E. ; Lawson, D. M. J. Biol. Chem. 2002, 277, 35263~35266.

[8] (a) Burgess, B. K. Chem. Rev. 1990, **90**, 1377~1406. (b) Ludden, P. W. ; Shah, V. K. ; Roberts, G. P. ; Homer, M. ; Allen, R. ; Paustian, T. ; Roll, J. ; Chatterjee, R. ; Madden, M. ; Allen, J. ACS SymP. Ser. 1993, **535**, 196~215.

[9] Holland, P. L. In Comprehensive Coordination Chemistry II; Mc-Cleverty, J. A., Meyer, T. J., Eds. ; Elsevier Pergamon; New York, 2004, Vol. 8; pp 586~591.

[10] (a) Durrant, M. C. Biochemistry 2002, **41**, 13934 ~ 13945. (b) Seefeldt, L. C. ; Dance, I. G. ; Dean, D. R. Biochemistry 2004, **43**, 1401~1409. (c) Zhou, Z. H. ; Zhao, H. ; Tsai, K. R. J. Inorg. Biochem. 2004, **98**, 1787~1794.

[11] (a) Grönberg, K. L. C. ; Gormal, C. A. ; Durrant, M. C. ; Smith, B. E. ; Henderson, R. A. J. Am. Chem. Soc. 1998, **120**, 10613~10621. (b) Tsai, K. R. ; Wan, H. L. J. Cluster Sci. 1995, **6**, 485~ 501.

[12] Zhou, Z. H. ; Wan, H. L. ; Hu, S. Z. ; Tsai, K. R. Inorg. Chim. Acta 1995, **237**, 193~197.

[13] Zhou, Z. H. ; Yan, W. B. ; Wan, H. L. ; Tsai, K. R. ; Wang, J. Z. ; Hu, S. Z. J. Chem. Crystallogr. 1995, **25**, 807~811.

[14] Zhou, Z. H. ; Zhang, H. ; Jiang, Y. Q. ; Lin, D. H. ; Wan, H. L. ; Tsai, K. R. Trans. Met. Chem. 1999, **24**, 605~609.

[15] Zhou, Z. H. ; Deng, Y. F. ; Cao, Z. X. ; Zhang, R. H. ; Chow, Y. L. Inorg. Chem. 2005, **44**, 6912~6914.

[16] Zhou, Z. H. ; Hou, S. Y. ; Cao, Z. X. ; Tsai, K. R. ; Chow, Y. L. Inorg. Chem. 2006, **46**, 8447~ 8451.

[17] Chen, H. B. ; Chen, L. Y. ; Huang, P. Q. ; Zhang, H. K. ; Zhou, Z. H. ; Tsai, K. R. Tetrahedron 2007, **63**, 2148~2152.

[18] SADABS; University of Göttingen; Göttingen, Germany, 1997.

[19] Farragia, L. J. J. Appl. Crystallogr. 1999, **32**, 837~838.

[20] Sheldrick, G. M. SHELXL97; SHELXS97; University of Göttingen, Göttingen, Germany, 1997.

[21] Qi, Y. J. ; Yang, Y. L. ; Cao, M. H. ; Hu, C. W. ; Wang, E. B. ; Hu, N. H. ; Jia, H. Q. J. Mol. Struct. 2003, **648**, 191~201.

[22] Wright, D. W. ; Chang, R. T. ; Mandal, S. K. ; Armstrong, W. H. ; Orme-Johnson, W. H. J. Biol. Inorg. Chem. 1996, **1**, 143~151.

[23] Wright, D. W. ; Humiston, P. A. ; Orme-Johnson, W. H. ; Davis, W. M. Inorg. Chem. 1995, **34**, 4194~4197.

[24] Tsaramyrsi, M. ; Kavousanaki, D. ; Raptopoulou, C. P. ; Terzis, A. ; Salifoglou, A. Inorg. Chim. Acta 2001, **320**, 47~59.

[25] Kaliva, M. ; Giannadaki, T. ; Salifoglou, A. ; Raptopoulou, C. P. ; Terzis, A. Inorg. Chem. 2002, **41**, 3850~3858.

[26] Kaliva, M. ; Raptopoulou, C. P. ; Terzis, A. ; Salifoglou, A. J. Inorg. Biochem. 2003, **93**, 161~ 173.

[27] Aureliano, M. ; Tiago, T. ; Gândara, R. M. C. ; Sousa, A. ; Moderno, A. ; Kaliva, M. ; Salifoglou, A. ; Duarte, R. O. ; Moura, J. J. G. J. Inorg. Biochem. 2005, **99**, 2355~2361.

[28] Tsaramyrsi, M. ; Kaliva, M. ; Salifoglou, A. ; Raptopoulou, C. P. ; Terzis, A. ; Tangoulis, V. ; Giapintzakis, J. Inorg. Chem 2001, **40**, 5772~5779.

[29] Burojevic, S. ; Shweky, I. ; Bino, A. ; Summers, D. A. ; Thompson, R. C. Inorg. Chim. Acta 1996, **251**, 75~79.

[30] Rehder, D. ; Pessoa, J. C. ; Geraldes, C. F. G. C. ; Castro, M. M. C. A. ; Kabanos, T. ; Kiss, T. ; Meier, B. ; Micera, G. ; Pettersson, L. ; Rangel, M. ; Salifoglou, A. ; Turel, I. ; Wang, D. R. J. Biol. Inorg. Chem. 2002, **7**, 384~396.

［31］Velayutham，M.；Varghese，B.；Subramanian，S. Inorg. Chem. 1998，**37**，1336～1340.

［32］Rodríguez-Cuamatzi, P.；Arillo-Flores, O. I.；Bernal-Uruchurtu, M. I.；Höpfl, H. Cryst. Growth Des. 2005，**5**，167～175.

［33］Kovalevsky, A. Y.；Gembicky, M.；Novozhilova, I. V.；Coppens, P. Inorg. Chem. 2003，**42**，8794～8802.

［34］Nishizawa，M.；Hirotsu，K.；Ooi，S.；Saito，K. Chem. Comm. 1979，707.

［35］Brown，I. D.；Altermatt，D. Acta Crystallogr.，Sect. B 1985，**41**，244～247.

［36］Silversmit，G.；Depla，D.；Poelman，H.；Marin，G. B.；De Gryse，R. J. Electron Spectrosc. Relat. Phenom. 1991，**57**，189～197.

［37］Bencini，A.；Gatteschi，D. EPR of Exchange Coupled Systems；Springer Verlag：New York，1990.

［38］George，G. N.；Coyle，C. L.；Hales，B. J.；Cramer，S. P. J. Am. Chem. Soc. 1988，**110**，4057～4059.

■ 本文原载：Dalton Trans.,2008,pp. 2475～2479.

# N-Heterocycle Chelated Oxomolybdenum(VI and V) Complexes with Bidentate Citrate*

Zhao-Hui Zhou①, Can-Yu Chen, Ze-Xing Cao, Khi-Rui Tsai, Yuan L. Chow

Received 12th November 2007, Accepted 1st February 2008
First published as an Advance Article on the web 20th March 2008
DOI: 10.1039/b717452g

**Abstract**   A 1,10-phenanthroline(phen)chelated molybdenum(VI) citrate, $[(MoO_2)_2O(H_2cit)(phen)(H_2O)_2] \cdot H_2O(\mathbf{1})(H_4cit = $ citric acid), is isolated from the reaction of citric acid, ammonium molybdate and phen in acidic media (pH 0.5 - 1.0). A citrato oxomolybdenum(V) complex, $[(MoO)_2O(H_2cit)_2(bpy)_2] \cdot 4H_2O(\mathbf{2})$, is synthesized by the reduction of citrato molybdate with hydrazine hydrochloride in the presence of 2,2′-bipyridine(bpy), and a monomeric molybdenum(VI) citrate $[MoO_2(H_2cit)(bpy)] \cdot H_2O(\mathbf{6})$ is also isolated and characterized structurally. The citrate ligand in the three neutral compounds uses the $\alpha$-alkoxy and $\alpha$-carboxy groups to chelate as a bidentate leaving the two $\beta$-carboxylic acid groups free, that is different from the tridentate chelated mode in the citrato molybdate(VI and V)complexes. **1** and **2** in solution show obvious dissociation based on $^{13}C$ NMR studies.

## 1   Introduction

The undefined X atom in the bidentately coordinated $R$-homocitrato-MoFe$_7$S$_9$X cluster of FeMoco in nitrogenase has made the six central Fe atoms coordinately saturated[1-3], which is supported by the model compound synthesis of iron sulfur clusters with a similar core of Fe$_8$S$_9$ and Fe$_8$S$_8$N[4]. Despite intensive research and knowledge of the macromolecular structures of the nitrogenase component proteins from several different sources, the precise location of dinitrogen coordination at FeMoco remains an open question[5-7]. Moreover, it is proposed that homocitrate and molybdenum are inserted into the precursor of FeMoco in the final step[8]. This stimulates our interests in the coordination chemistry of molybdenum complexes with homocitrate and citrate, both are homologous to hydroxy-tricarboxylate ligands[9]. In our former studies on the molybdenum citrate system, we have attempted to show the reactivity of the Mo-tricarboxylate by isolating citrato molybdate under highly acidic

---

①   State Key Laboratory for Physical Chemistry of Solid Surfaces and Department of Chemistry, College of Chemistry and Chemical Engineering, Xiamen University, Xiamen, 361005, China. E-mail: zhzhou@xmu.edu.cn; Fax: + 86-592-2183047; Tel: + 86-592-2184531.

*   Electronic supplementary information (ESI) available: Experimental details. CCDC reference numbers 673570-673572. For crystallographic data in CIF or other electronic format See DOI:10.1039/b717452g.

conditions, where the coordinated β-carboxylic acid group remains protonated[10]. The present results show that, with the aid of N-heterocycle ligands, two novel dinuclear bidentate citrate molybdenum(VI and V)complexes $[(MoO_2)_2O(H_2cit)(phen)(H_2O)_2] \cdot H_2O(1)$, $[(MoO)_2O(H_2cit)_2(bpy)_2] \cdot 4H_2O$ (2)and a monomeric citrate molybdenum(VI)$[MoO_2(H_2cit)(bpy)] \cdot H_2O(6)$are prepared.

# 2　Experimental

## Physical measurements

Infrared spectra were recorded as Nujol mulls between KBr plates on a Nicolet 360 FT-IR spectrometer. Elemental analyses were performed with an EA 1100 elemental analyzer. Electronic spectra were recorded on a UV 2501 spectrophotometer with an intergrating sphere for reflectance spectroscopy. $^1H$ and $^{13}C$ NMR spectra were recorded in DMSO solution at 298 K on a Bruker BIOSPIN AV400 spectrometer. $^1H$ and $^{13}C$ shifts are referenced to internal solvent resonances and reported relative to 3-(trimethylsilyl)-1-propanesulfonic acid sodium salt(DSS). XRD patterns were recorded on a Philips X′Pert Pro Super X-ray diffractometer equipped with X′Celerator and Xe detection systems.

## Preparations of $[(MoO_2)_2O(H_2cit)(phen)(H_2O)_2] \cdot H_2O(1)$

A mixture of $(NH_4)_6Mo_7O_{24} \cdot 4H_2O(0.18$ g, 0.14 mmol), citric acid monohydrate(0.66 g, 3.15 mmol), 1,10-phenanthroline monohydrate(0.10 g, 0.52 mmol)in aqueous ethanol solution(1∶1v/v)and a few drops of concentrated hydrochloric acid was heated at 100 ℃ for 3 h and then at 86 ℃ for 7.5 h and cooled to rt. The milky solution treated with hydrothermal condition changed to a light blue cloudy solution after one week at rt, product **1** was isolated as light yellow-brown spicules. 43% yield(0.145 g) based on molybdenum. C, N and H elemental analyses. Found(calcd for $C_{18}H_{20}N_2O_{15}Mo_2$): C, 30.7 (31.1); H, 3.1(2.9); N, 4.0(4.0). IR(KBr, cm$^{-1}$): $v(\beta$-$CO_2H)1721_s$; $v_{as}(CO_2)1691_s, 1634_{vs}$; $v_s(CO_2)$ $1431m, 1392m, 1357m$ $v_s(Mo=O)944_s, 906_{vs}$; $v_{as}(Mo$-$O$-$Mo)806_{vs}$. $^1H$ NMR $\delta_H$(400 MHz, DMSO-d$_6$): $\delta_H$ 9.498(d, J 4.0 Hz, CH, 2H), 8.892(d, J 8.0 Hz, CH, 2H), 8.450(s, CH, 1H), 8.238(s, CH, 1H), 8.150(dd, J 8.0, 4.0 Hz, CH, 2H), 2.752(d, J 16.8 Hz, CH$_2$, 2H), 2.576(d, J 16.8 Hz, CH$_2$, 2H). $^{13}C$ NMR(DMSO-d$_6$): $\delta_c$ 178.9($\alpha$-$CO_2$), 171.0($\beta$-$CO_2$), 153.1, 141.8, 140.8, 129.5, 127.4, 126.4(phen), 81.0($\equiv$CO), 43.3(=CH$_2$)ppm.

## Preparations of $[(MoO)_2O(H_2cit)_2(bpy)_2] \cdot 4H_2O(2)$

Na$_2$MoO$_4 \cdot 2H_2O(0.24$ g, 1.0 mmol), citric acid monohydrate(0.46 g, 2.2 mmol)and 2,2′-bipyridine(0.16 g, 1.01 mmol)were mixed in 100 ml aqueous ethanol solution(1∶2 v/v). The reactants were dissolved by continuous stirring, then hydrazine dihydrochloride(0.44 g, 4.2 mmol)was added. The solution turned carmine. The pH value was adjusted to 3.0 and the solution turned dark red. The product **2** was isolated as dark violet after three days. 64% yield(0.319 g)based on molybdenum. C, N and H elemental analyses for $C_{32}H_{36}N_4O_{21}Mo_2$: found(calc.): C, 37.8(38.2); N, 5.5(5.6); H, 3.8 (3.6). IR(KBr): $v(\beta$-$CO_2H)1725_{vs}$; $v_{as}(CO_2)1644_{vs}$; $v_s(CO_2)1475_m, 1442_m, 1403_m$; $v_s(Mo=O)949_s$, $902_w$; $v_{as}(Mo$-$O$-$Mo)851_m$ cm$^{-1}$. $^1H$ NMR $\delta_H$(400 MHz, DMSO-d$_6$): $\delta_H$ 8.689(d, J 4.8 Hz, CH, 2H), 8.387(d, J 8.0 Hz, CH, 2H), 7.956(t, J 15.2, 7.6 Hz, CH, 2H), 7.461(t, J 12.4, 6.0 Hz, CH, 2H), 2.714(dd, J 27.2, 15.4 Hz, CH$_2$, 4H). $^{13}C$ NMR(DMSO-d$_6$): $\delta_c$ 180.9($\alpha$-$CO_2$), 173.1($\beta$-$CO_2$), 157.0, 151.2, 139.3, 126.1, 122.3(bpy), 82.8($\equiv$C—O), 44.4(=CH$_2$)ppm.

## Preparations of $[MoO_2(H_2cit)(bpy)]\cdot H_2O(6)$

A mixture of $(NH_4)_6Mo_7O_{24}\cdot 4H_2O(0.18\ g,0.14\ mmol)$, citric acid monohydrate $(0.64\ g,3.02$ mmol$),2,2'$-bipyridine$(0.18\ g,1.14mmol)$in water and a few drops concentrated HCl was heated at 100 ℃ for 8 hours,then cooled to rt at 5 ℃/h. The milky cloudy solution was turned to pink solution. The colorless crystal **6** was isolated from the white precipitate after one month. 5% yield$(0.016\ g)$is based on bipyridine. C,N and H elemental analyses. Found(calcd for $C_{16}H_{16}N_2O_{10}Mo_1$):C,38.7(39.0);N,5.6 (5.7);H,3.2(3.3). IR$(KBr,cm^{-1})$:$v(\beta\text{-}CO_2H)1748_s$;$v_{as}(CO_2)1658_s,1603_{vs}$;$v_s(CO_2)1444_m,1400_m$, $1339_m$;$v_s(Mo=O)937_s,909_{vs}$.

### X-Ray structure determination

Crystals of **1,2** and **6** were measured on a Bruker Smart Apex CCD area detector diffractometer with graphite monochromatic Mo-Kα radiation $(\lambda=0.71073\ Å)$at $-50$ ℃ for **1** and 23 ℃ for **2** and **6**. Data were corrected for absorption by using the SADABS program[11]. The structures were primarily solved by SHELXS in the WinGX program[12] and refined by full-matrix least-squares procedures with anisotropic thermal parameters for all of the nonhydrogen atoms with SHELX-97[13]. Hydrogen atoms were located from a difference Fourier map and refined isotropically.

**Crystal data for 1.** $C_{18}H_{20}N_2O_{15}Mo_2$,$M=696.24$,triclinic,space group $P\text{-}1,a=7.6946(6),b=$ $11.768(1)c=13.741(1)Å,\alpha=87.550(2),\beta=76.478(1),\gamma=71.122(2)°,V=1143.9(2)Å^3,D_c=2.021$ $g/cm^3,Z=2$,reflections collected/unique/observed$=7447/4873/3608,R_{int}=0.1152,R1=0.066,wR2=$ 0.140.

**Crystal data for 2.** $C_{32}H_{36}N_4O_{21}Mo_2$,$M=1004.53$,triclinic,space group $P\text{-}1,a=10.306(3),b=$ $13.089(5)c=16.049(9)Å,\alpha=103.71(4),\beta=107.11(4),\gamma=94.48(3)°,V=1985(2)Å^3,D_c=1.681\ g/$ $cm^3,Z=2$,reflections collected/unique/observed$=24574/12584/5116,R_{int}=0.0619,R=0.055,wR2=$ 0.125.

**Crystal data for 6.** $C_{16}H_{16}N_2O_{10}Mo_1$,$M=492.25$,monoclinic,space group $P\ 2_1/n,a=14.688(1),b$ $=7.7377(8),c=16.084(2)Å,\beta=100.151(2)°,V=1799.3(3)Å^3,D_c=1.817\ g/cm^3,Z=4$,reflections collected/unique/observed$=10660/3167/2091,R_{int}=0.1278,R=0.085,wR2=0.160$.

# 3　Results and discussion

The preparations of **1** and **2** are outlined in Scheme 1. The solution of $Mo:H_4cit(1:1)$ was investigated by $^{13}C$ NMR spectra in the pH range of $0.5-4.0$, showing that dimeric species, K $[(MoO_2)_2(OH)(H_2cit)_2]\cdot 4H_2O(3)^{[10]}$ and $K_4[(MoO_2)_2O(Hcit)_2]\cdot 4H_2O(4)^{[14]}$,are in equilibrium with the free ligand:this led to the isolation of the three compounds. An attempt at synthesizing $K_2$ $[(MoO_2)_2O(H_2cit)_2(H_2O)_2](5)$ has been unsuccessful owing to the fact that **3** is a stable,hydrogen-bonded species and resists further solvation[10]. It takes the assistance of strongly chelating 1,10-phenanthroline to displace the tridentate citrate with water to give neutral compound **1**, which is stabilized by the intramolecular hydrogen bonding of the coordinated water. Reduction of citrato molybdate **4** in the presence of 2,2'-bipyridine gives the dimeric bipyridine hydrogen citrato oxomolybdenum(Ⅴ)$[(MoO)_2O(H_2cit)_2(bpy)_2]\cdot 4H_2O(2)$in 64% yield. When bipyridine is used instead of phen in the reaction of **3**, a mononuclear bipyridine citrato molybdate(Ⅵ),$[MoO_2(H_2cit)$ $(bpy)]\cdot H_2O(6)$has been isolated in a very low yield;this has not been reproduced at present.

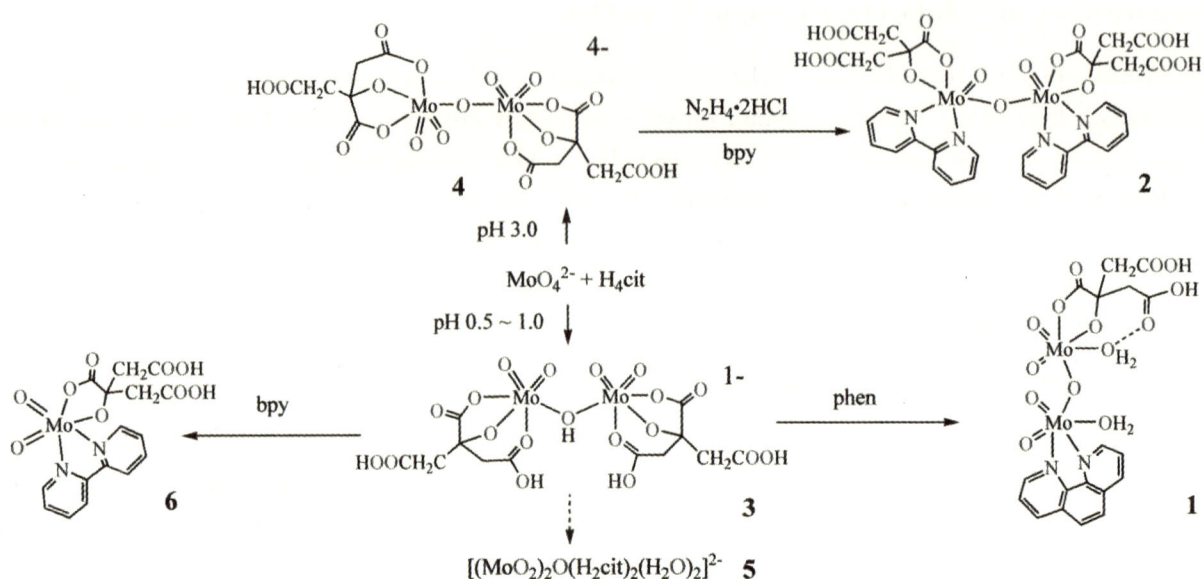

**Scheme 1**  Synthesis and transformation of molybdenum(Ⅵ, Ⅴ)citrate complexes in the presence of 1, 10-phenanthroline and 2,2′-bipyridine. [10]

The X-ray structure of **1**(Fig. 1)shows two *cis*-dioxo-Mo units oriented in a staggered *syn*-mode with respect to the oxo bridge, and both molybdenum atoms possessing six-coordinate, octahedral geometries. The citrate moiety coordinates to one molybdenum atom as a bidentate ligand *via* its α-alkoxy and α-carboxy groups leaving both β-carboxylic acid groups free to participate in intramolecular and intermolecular hydrogen bonding [O4···O1w 2.744(8),O7···O3a 2.604(9)Å;*a*,−1+*x*,*y*,*z*]. The intramolecular hydrogen bond in FeMoco,when investigated theoretically using a model and quantum-mechanical computation,suggests that the long arm of *R*-homocitrate(γ-carboxy group)but not its free β-carboxy group is capable of undergoing intramolecular hydrogen bonding with the imidazole ligand on Mo[15,16].

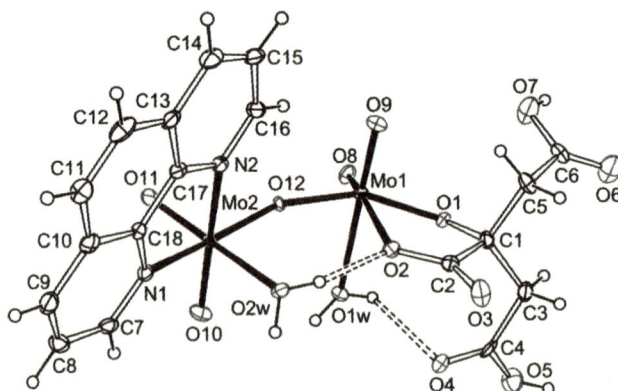

**Fig. 1**  The ORTEP plot of the molecular structure of $[(MoO_2)_2O(H_2cit)(phen)(H_2O)_2] \cdot H_2O(1)$ at the 30% probability levels.

Selected bond distances(Å)and angles(°)of complexes **1,2** and **6** are shown in Table 1. The Mo-O distances in complex **1** vary systematically according to its bond types. The Mo=O distances ranged 1.667(6)to 1.710(5)Å are in agreement with the double bond character. The Mo-O bridge is necessarily unsymmetrical exhibiting a shorter bond distance of Mo(2)-O(12)[1.840(4)Å] than that of Mo(1)-O

(12)[1.932(4)Å] arising from the *trans*-effect of the phenanthroline ligand. The Mo-O($\alpha$-alkoxy)and Mo-O($\alpha$-carboxy)distances observed in **1** are 1.941(4)and 2.209(5)Å respectively,and the coordinated water molecules are longer [2.286(5)and 2.307(5)Å] than those coordinated in trinuclear molybdenum (IV)complexes such as [Mo₃(CH₃CO₂)₃Br₃(H₂O)₃(CCH₃)]·ClO₄·4H₂O and(Ph₄P)₂[Mo₃($\mu_3$-S) ($\mu_2$-S)₃(C₂O₄)₃(H₂O)₃]·11H₂O [2.211(9)$_{av}$,2.15—2.22 Å][17,18]. Thus insertion of a water molecule to form Mo-O$_w$ bonds in **1** may be justifiable in view of the more weakly coordinated Mo-O$_{\beta\text{-carboxy}}$[2.463 (6)Å] in **3**[10].

Table 1　Selected bond distances(Å)and angles(°)for [(MoO₂)₂O(H₂cit)(phen)(H₂O)₂]·H₂O(**1**), [(MoO)₂O(H₂cit)₂(bpy)₂]·4H₂O(**2**)and [MoO₂(H₂cit)(bpy)]·H₂O(**6**)

| **1** | | | |
| --- | --- | --- | --- |
| Mo(1)—O(1) | 1.941(4) | Mo(2)—O(10) | 1.698(4) |
| Mo(1)—O(2) | 2.209(5) | Mo(2)—O(11) | 1.704(5) |
| Mo(1)—O(8) | 1.710(5) | Mo(2)—O(12) | 1.840(4) |
| Mo(1)—O(9) | 1.695(5) | Mo(2)—O(2w) | 2.286(5) |
| Mo(1)—O(12) | 1.932(4) | Mo(2)—N(1) | 2.210(6) |
| Mo(1)—O(1w) | 2.307(5) | Mo(2)—N(2) | 2.294(5) |
| Mo(1)—O(12)—Mo(2) | 160.8(3) | O(1)—Mo—O(2) | 75.2(2) |
| **2** | | | |
| Mo(1)—O(1) | 1.987(3) | Mo(2)—O(9) | 1.888(3) |
| Mo(1)—O(2) | 2.075(3) | Mo(2)—O(11) | 1.976(3) |
| Mo(1)—O(8) | 1.676(4) | Mo(2)—O(12) | 2.077(3) |
| Mo(1)—O(9) | 1.875(3) | Mo(2)—O(18) | 1.686(3) |
| Mo(1)—N(1) | 2.277(4) | Mo(2)—N(11) | 2.257(4) |
| Mo(1)—N(2) | 2.351(4) | Mo(2)—N(12) | 2.342(4) |
| Mo(1)—O(9)—Mo(2) | 169.9(2) | O(1)—Mo(1)—O(2) | 78.3(1) |
| **6** | | | |
| Mo(1)—O(1) | 1.907(6) | Mo(1)—O(9) | 1.667(6) |
| Mo(1)—O(2) | 2.144(6) | Mo(1)—N(1) | 2.189(8) |
| Mo(1)—O(8) | 2.351(4) | Mo(1)—N(2) | 2.261(7) |
| O(1)—Mo(1)—O(2) | 75.8(3) | O(8)—Mo(1)—O(9) | 104.5(4) |

As shown by the X-ray structure(Fig. 2),**2** is a dimeric bipyridine citrate molybdenum(V)complex consisting of a core of a *anti*-oxo-Mo entity with respect to the angled oxo-bridge with molybdenum in a distorted octahedral geometry. The Mo=O distances[1.676(4)and 1.686(3)Å] are in a shorter range compared to those of **1** and other dimeric citrate molybdenum(V, VI)complexes.[10,14,19] The Mo(1)-O (9)-Mo(2)bridge [169.9(2)°] is bent and has approximately symmetric [1.875(3)and 1.888(3)Å] configuration. It is noteworthy that two coordinated bipyridine molecules are structurally non-equivalent [2.257(4)and 2.277(4);2.342(4)and 2.351(4)Å] owing to the *trans* effect of the terminal oxo group.

Complex **6** is a monomeric bipyridine citrate molybdenum(VI) complex as shown in Fig. 3, consisting of a core of a *cis*-dioxo-Mo entity in a distorted octahedral geometry. The Mo=O distances [1.667(6) and 1.670(7)Å] are in a shorter range compared to those of **1** and **2**. Both free $\beta$-carboxylic

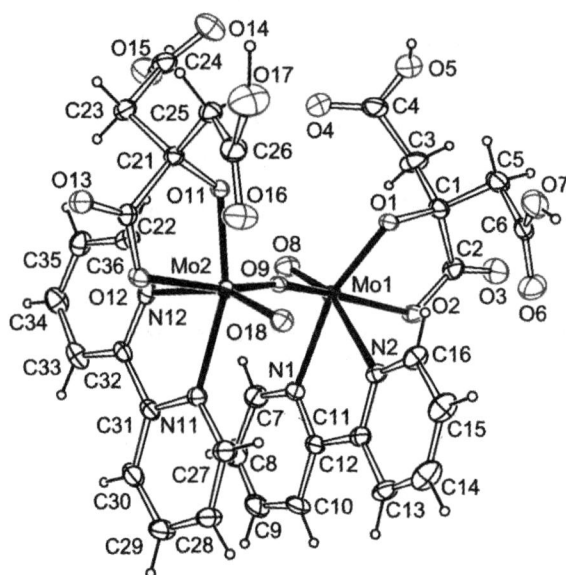

Fig. 2　The ORTEP plot of the molecular structure of $[(MoO)_2O(H_2cit)_2(bpy)_2] \cdot 4H_2O(2)$ at the 15% probability levels.

acid groups participate in hydrogen bonding with water molecule $[O5a \cdots O1w \, 2.67(1), O7b \cdots O1w \, 2.61(1) Å; a, 2-x, -y, 2-z; b, 1.5-x, -0.5-y, 1.5-z]$.

The citrate ligand in **2** and **6** coordinates bidentately through α-alkoxy and α-carboxy groups, leaving all the β-carboxylic acid groups free. In comparison to **1** and **6**, the reduced state of molybdate (V) in **2** results in longer Mo-O bond distances for α-alkoxy groups [1.941(4) and 1.907(6) Å for **1** and **6** respectively, 1.976(3) and 1.987(3) Å for **2**] and shorter Mo-O distances for α-carboxy groups [2.209(5) and 2.144(6) Å for **1** and **6** respectively, 2.075(3), 2.077(3) Å for **2**]; this situation is comparable to the coordination of FeMoco in nitrogenase(Av, 2.212, 2.162 Å; Kp-*nifV*, 2.253, 2.269 Å for α-alkoxy and α-carboxy groups)[1,9].

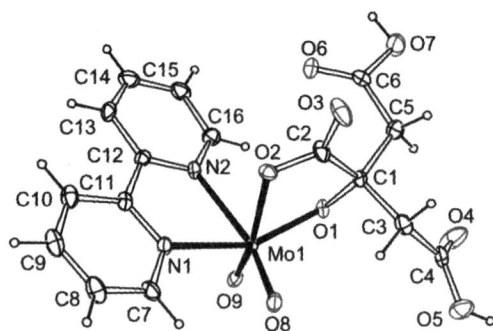

Fig. 3　The ORTEP plot of the molecular structure of $[MoO_2(bpy)(H_2cit)] \cdot H_2O(6)$ at the 15% probability levels.

Complexes **1** and **2** are insoluble in water and ethanol but sparingly soluble in DMSO. [13]C NMR spectra in deuterated DMSO are shown in Fig. 4 and 5 with long time superposition. Two sets of signals are attributed to the coordinated and free citrate species: the chemical shifts for the latter free citrate are comparable with those for $KH_3cit$ ion.

In particular, the carbon atoms of coordinated α-alkoxy and α-carboxy groups show distinct downfield shifts Δδ of 4.3—4.6 and 8.6 ppm, respectively, which are a clear indication of the bidentate coordination. These are in agreement with the state of an equilibrium in spectral solution involving the dissociation of both α-carboxy and a-alkoxy groups. Further comparison of the [13]C NMR pattern of molybdate and citrate(1 : 1 molar ratio)in the pH range of 0.5—4.0 indicates that citrate molybdenum (Ⅵ)compounds **3** and **4** also show obvious dissociation in these acidic solutions. The XRD pattern of solid complex **1** is consistent in comparison with a theoretically simulated pattern. The solid species

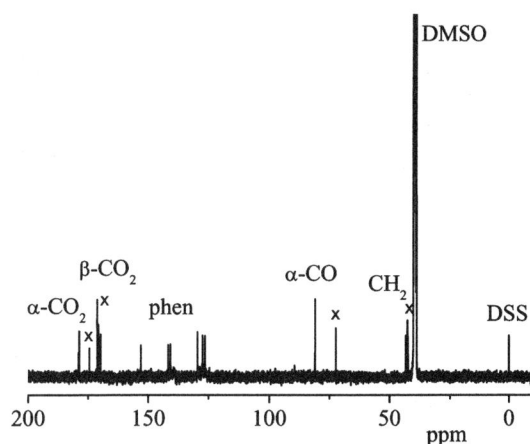

Fig. 4  The $^{13}$C NMR spectrum of $[(MoO_2)_2O(H_2cit)(phen)(H_2O)_2]\cdot H_2O$
(1)in DMSO-$d_6$. ($\times$)denotes free citrate($H_3cit^-$).

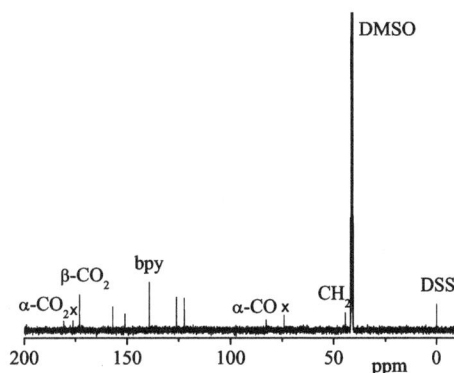

Fig. 5  The $^{13}$C NMR spectrum of $[(MoO)_2O(H_2cit)_2(bpy)_2]\cdot 4H_2O(2)$ in
DMSO-$d_6$. ($\times$)denotes free citrate($H_3cit^-$).

isolated in macroscale is the same as the sample for X-ray structural analysis. This is supported by the theoretical simulation of XRD pattern shown in Fig. S4. †

# 4  Conclusion

There is sufficient evidence to show that the citrate group engages in tridentate coordination *via* the α-alkoxy, α-carboxy and β-carboxy groups in complexes $K[(MoO_2)_2(OH)(H_2cit)_2]\cdot 4H_2O^{[10]}$, $K_4$ $[(MoO_2)_2O(Hcit)_2]\cdot 4H_2O^{[14]}$, and $K_4Na_2[(MoO_2)_2O(cit)_2]\cdot 5H_2O^{[19]}$. In fact, direct substitution of $K_4[MoO_3(cit)]\cdot 2H_2O^{[14]}$ in neutral conditions with a monodentate ligand has been largely unsuccessful because of the full dissociation of the citrate ligand. Even under highly acidic solutions, protonated species $[(MoO_2)_2(OH)(H_2cit)_2]^-$ (4)[10] can form without any direct monodentate substitution. The present formation of symmetric and unsymmetric bidentate citrato molybdates 1,2 and 6 could be related to the assembly of homocitrato FeMoco in nitrogenase, that shows certain resemblance to a monodentate subtitution of β-carboxy group with a coordinated water molecule. Although the structure and coordination of homocitrate molybdate precursor of FeMoco is unknown in the final process of biosynthesis[8,20,21]. It is also noted that a large gap exists between the homocitrate molybdate in natural cofactor assembly and current model molecules, homocitrate and citrate, especially the very low pH required to induce bidentate coordination.

# Acknowledgements

Financial supports provided by National Natural Science Foundation of China(20571061,20673087) and the Ministry of Science & Technology (2005CB221408) are gratefully acknowledged. We thank Professor G. M. Blackburn for stimulating discussions.

# References

[1] O. Einsle, F. A. Tezcan, S. L. A. Andrade, B. Schmid, M. Yoshida, J. B. Howard and D. C. Rees, Science, 2002,**297**,1696.

[2] T. C. Yang, N. K. Maeser, M. Laryukhin, H. I. Lee, D. R. Dean, L. C. Seefeldt and B. M. Hoffman, J. Am. Chem. Soc., 2005,**127**,12804.

[3] J. B. Howard and D. C. Rees, Proc. Natl. Acad. Sci. U. S. A., 2006,**103**,17088.

[4] Y. Ohki, Y. Ikagawa and K. Tatsumi, J. Am. Chem. Soc., 2007,**129**,10457.

[5] F. Barrière, Coord. Chem. Rev., 2003,**236**,71.

[6] L. C. Seefeldt, I. G. Dance and D. R. Dean, Biochemistry, 2004,**43**,1401.

[7] B. M. Barney, H. I. Lee, P. C. D. Santos, B. M. Hoffman, D. R. Dean and L. C. Seefeldt, Dalton Trans., 2006,2277.

[8] M. C. Corbett, Y, L. Hu, A. W Fay, M. W Ribbe, B. Hedman and K. Hodgson, Proc. Natl. Acad. Sci. U. S. A., 2006,**103**,1238.

[9] S. M. Mayer, C. A. Gormal, B. E. Smith and D. M. Lawson, J. Biol. Chem., 2002, **277**,35263.

[10] Z. H. Zhou, Y F. Deng, Z. X. Cao, R. H. Zhang and Y L. Chow, Inorg. Chem., 2005,**44**,6912.

[11] SADABS, University of Göttingen, Göttingen, Germany,1997.

[12] L. J. Farragia, J. Appl. Crystallogr.,1999,**32**,837.

[13] G. M. Sheldrick, SHELX-97, Programs for Crystal Structure Analysis, University of Göttingen, Germany,1997.

[14] Z. H. Zhou, H. L. Wan and K. R. Tsai, Inorg. Chem.,2000,**39**,59.

[15] K. L. C. Grönberg, C. A. Gormal, M. C. Durrant, B. E. Smith and R. A. Henderson, J. Am. Chem. Soc., 1998,**120**,10613.

[16] K. R. Tsai and H. L. Wan, J. Clust. Sci., 1995,**6**,485.

[17] A. Birnbaum, F. A. Cotton, Z. Dori, M. Kapon, D. Marler, G. M. Reisner and W Schwotzer, J. Am. Chem. Soc., 1985,**107**,2405.

[18] M. N. Sokolov, A. L. Gushchin, D. Y. Naumov, O. A. Gerasko and V P. Fedin, Inorg. Chem., 2005,**44**,2431.

[19] Z. H. Zhou, H. L. Wan and K. R. Tsai, Polyhedron,1997,**16**,75.

[20] P. C. D. Santos, D. R. Dean, Y Hu and M. W Ribbe, Chem. Rev.,2004,**104**,1159.

[21] S. J. George, R. Y. Igarashi, C. Piamonteze, B. Soboh, S. P. Cramer, L. M. Rubio and M. Luis, J. Am. Chem. Soc., 2007,**129**,3060.

▧▧▧▧▧▧▧▧▧▧▧▧▧▧▧▧▧▧▧▧▧▧▧▧▧

■ 本文原载:Chinese Journal of Catalysis,2010,31:pp. 153~155.

# Deuterium Inverse Isotopic Effect in Ammonia Synthesis over Ru-Based and Fe-Based Catalysts

Jing-Dong Lin①, Dai-Wei Liao②, Hong-Bin Zhang, Hui-Lin Wan, Khi-Rui Tsai
(National Engineering Laboratory for Green Chemical Productions of
Alcohols-Ethers-Esters, State Key Laboratory of Physical
Chemistry of Solid Surfaces, Institute of Physical Chemistry,
Department of Chemistry, College of Chemistry and Chemical Engineering,
Xiamen University, Xiamen 361005, Fujian, China )

**Abstract**　　The deuterium inverse isotopic effect(DIIE)is important for the clarification of the ammonia synthesis mechanism over transition metal catalysts. We investigated the DIIE over Ru-based and Fe-based catalysts including pure Fe, multi-promoted fused Fe, pure Ru, $Ru/\gamma$—$Al_2O_3$, K—$Ru/\gamma$—$Al_2O_3$, $Ru/MgO$, K—$Ru/MgO$, and Ba—$Ru/MgO$ under the reaction conditions of $N_2/H_2$(or $D_2$) volume ratio$=1/3$, 0.2 MPa, 633~733 K, and GHSV$=24\,000$ and $12\,000\ h^{-1}$. A strong DIIE($r_D/r_H$ of about 2)was observed over the above catalysts. The trend of the DIIE with catalysts and temperature indicated a strong competition between the dynamic and thermodynamic factors.

**Key words**　　ammonia synthesis　deuterium inverse isotopic effect　Ru-based catalysts　Fe-based catalysts　hydrogen-assisted dissociation　dissociative mechanism　associative mechanism

The catalytic hydrogenation of nitrogen into ammonia, $N_2 + 3H_2 \rightarrow 2NH_3$, is a simple reaction without any byproduct. Research achievements in both experiment and theory on this catalytic process in the past hundred years have greatly promoted development in heterogeneous catalysis and related fields such as surface science[1]. However, there are still different opinions on the catalytic mechanism, whether it is the dissociative mechanism or associative mechanism[2,3].

In the dissociative mechanism, the dissociative chemisorption of molecular nitrogen($N_2 \rightarrow 2N_{ad}$) is the rate-determining step(RDS). This means that hydrogen is not involved in the RDS. Therefore, the deuterium inverse isotopic effect(DIIE)was explained by the thermodynamic reason that the reaction equilibrium constants are different for the hydrogenation and deuteration of the most abundant species chemisorbed on the catalysts[4]. However, in the associative mechanism, hydrogen-assisted dissociation of chemisorbed molecular nitrogen($N_2H_{ad} \rightarrow N_{ad} + NH_{ad}$ or $N_2H_{2ad} \rightarrow 2NH_{ad}$)is the RDS. This means that hydrogen is involved in the RDS. In this case, the dynamic reason of a different reaction energy barrier

---

①　**Corresponding author.** Tel: +86-592-2183045; Fax: +86-592-2183043; E-mail: jdlin@xmu. edu. cn

②　**Corresponding author.** Tel: +86-592-2183045; Fax: +86-592-2183043; E-mail: dwliao@xmu. edu. cn

**Foundation item**: Supported by the National Natural Science Forndation of China(20673089)and the National Basic Research Program of China(973 Program,2009CB939804).

due to the different zero point energies of the adsorbed hydrogen-species and deuterium-species was used to explain the DIIE.

The DIIE in ammonia synthesis over a doubly-promoted iron catalyst was reported first by Ozaki and co-workers[4] in 1960. It is an important experimental evidence for the ammonia synthesis mechanism over transition metal catalysts, and in the past 50 years, both experimental and theoretical research on the DIIE over Fe-based catalysts, Ru-based catalysts, and other transition-metal catalysts has been continuously reported[5-13]. Unfortunately, the results and explanations were inconclusive. Therefore, a new systematic experimental re-examination of the DIIE is necessary.

In this study, the precursor of Fe was $Fe_2O_3$ (99.999%) supplied by STREM Chemicals. The multi-promoted fused Fe(commercial A110-3) catalyst was supplied by Fuzhou University and was composed of 68%~70% Fe, 1.8%~2.4% $Al_2O_3$, 1.2%~1.6% CaO, and 0.5%~0.7% $K_2O$. The Ru catalyst was Ru powder of 99.9% and 200 mesh supplied by Acros Company. ①

The $Ru/MgO$ and $Ru/\gamma-Al_2O_3$ catalyst precursors were prepared by impregnation, using MgO obtained by calcining basic magnesium carbonate at 973 K for 5 h or $\gamma-Al_2O_3$ supplied by Alfa Aesar. $RuCl_3 \cdot nH_2O$(37% Ru, Heraeus Materials Technology Shanghai Ltd.), acetone(100 ml), and MgO or $\gamma-Al_2O_3$ support (Ru/support mass ratio of 4/100) were mixed for 6 h, evaporated to dryness, and calcined at 383 K overnight to obtain the precursor. The precursors of $K-Ru/MgO$ and $K-Ru/\gamma-Al_2O_3$ were prepared by impregnation. The above precursor of $Ru/MgO$ or $Ru/\gamma-Al_2O_3$ was reduced by $N_2/H_2$ (volume ratio = 1/3) at 723 K for 6 h, followed by impregnation with 50% ethanol aqueous solution at a K/Ru ratio of 4/1 and room temperature for 4 h, evaporating to dryness and drying at 383 K overnight to obtain the precursor. The precursor of $Ba-Ru/MgO$ was prepared by impregnation. $Ru/MgO$ was prepared by above same treatment, followed by impregnation with 50% ethanol aqueous solution at a Ba/Ru ratio of 1/1 and room temperature for 4 h, evaporating to dryness and calcining at 383 K overnight to obtain the precursor.

The ammonia synthesis activity of the catalysts was evaluated over 0.2 or 0.4 ml of catalyst powder in a fixed-bed quartz reactor. Before entering the catalyst bed, the reactant gases($N_2/H_2$ volume ratio = 1/3, 99.999%, supplied by Linde Company or $N_2/D_2$ volume ratio = 1/3, 99.50%, supplied by the 718 Institute of CSIC) were passed through a HON—HG—V—M0.3/1.0 high purity gas purification unit(Dalian Samat Chemicals Co. Ltd) to remove traces of oxygen and water. All catalysts were activated by a $N_2/H_2$ flow(volume ratio = 1/3) of 30 ml/min under atmospheric pressure using a programmed temperature increase. The reactant gas(0.2 MPa, GHSV = 24 000 or 12 000 $h^{-1}$) was switched into the reactor after the activation of catalyst was completed. When the activity reached the steady state, the amount of ammonia, $NH_3$ and $ND_3$, produced at different reaction temperatures (633~733 K) were determined by a chemical titration method and denoted $r_H$ and $r_D$ (ml/(g·h)), respectively. The ratio $r_D/r_H$ under the same reaction conditions was the DIIE value.

We carried out the evaluation of the DIIE values in ammonia synthesis over all the Fe-based catalysts and Ru-based catalysts prepared above. The results are listed in Table 1. A strong DIIE, $r_D/r_H \approx 2$, was observed for all the catalysts. The values of $r_D/r_H$ were dependent on the reaction

① Received date: 8 December 2009.

Foundation item: Supported by the National Natural Science Foundation of China(20673089) and the National Basic Research Program of China(973 Program, 2009CB939804).

DOI: 10.1016/S1872-2067(09)60040-1.

temperatures.

Two different opinions, namely, the thermodynamic effect and dynamic effect, have been proposed to explain the DIIE in ammonia synthesis. In order to be in accord with the dissociative mechanism of ammonia synthesis, Ozaki and co-workers[4,5] suggested that the most abundant chemisorbed species was $N_{ad}$ on unpromoted, singly-promoted, and doubly-promoted (at high temperature) iron catalysts, and $NH_{ad}$ on doubly-promoted iron catalyst, and explained the DIIE by the thermodynamic reason of a difference in the equilibrium constants for the hydrogenation and deuteration reactions.

Table 1   Deuterium inverse isotopic effect($r_D/r_H$) in ammonia synthesis over Fe-based and
Ru-based catalysts at different reaction temperatures

| Catalyst | $r_D/r_H$ | | | | | |
|---|---|---|---|---|---|---|
| | 633 K | 653 K | 673 K | 693 K | 713 K | 733 K |
| Fe * | 1. 79 | 1. 89 | 1. 97 | 2. 05 | 2. 16 | 2. 36 |
| A110—3 * | 1. 93 | 1. 96 | 2. 07 | 2. 13 | 2. 25 | 2. 44 |
| Ru | 1. 53 | 1. 70 | 1. 40 | 1. 29 | 1. 60 | 1. 63 |
| Ru/Y—Al₂O₃ | 1. 12 | 1. 44 | 1. 40 | 1. 65 | 2. 03 | 1. 42 |
| K—Ru/Y—Al₂O₃ | 1. 72 | 2. 02 | 2. 04 | 2. 11 | 2. 39 | 2. 46 |
| Ru/MgO | 1. 53 | 1. 96 | 1. 71 | 1. 69 | 1. 68 | 1. 77 |
| K—Ru/MgO | 2. 04 | 1. 92 | 1. 76 | 1. 80 | 1. 83 | 1. 88 |
| Ba—Ru/MgO | 2. 56 | 2. 10 | 1. 87 | 1. 84 | 1. 77 | 1. 79 |

Reaction conditions: 0. 2 MPa, GHSV = 12000 $h^{-1}$ ( * 24000 $h^{-1}$).

In the second explanation based on the associative mechanism, Tsai et al.[10] proposed that the DIIE was due to the dynamic reason of a difference in the zero point energies. This was similar to the successful explanation of the DIIE in the hydrogenation of ethene at low temperature by Kokes[14]. It is to be noted that the ethene was not dissociated. Tsai et al.[10] also suggested that the DIIE over iron catalysts for ammonia synthesis may be due to a combination of the dynamic effect and thermodynamic effect. Using the associative mechanism, Huang[11] deduced a rate equation similar to the Tem- kin-Pizhev equation, and then by using a calculation performed with the experimental kinetic data reported by Ozaki and co-workers[4], he explained the DIIE as due to a combined effect of both the dynamic and thermodynamic factors. Liao[12,13] also deduced similar rate equations based on the associative route and used experimental molecular spectroscopic data[10,15] to explain the DIIE in ammonia synthesis by the dynamic reason of a zero point energy difference.

For the thermodynamic explanation to work, an atomic nitrogen adspecies has to be the most abundant adsorbed species. However, this may be not true. In situ FT-IR and laser Raman spectroscopy studies showed that the dominant chemisorbed species under reaction conditions was molecular nitrogen[12,15,16]. Takezawa[8] indicated that only L-type chemisorbed nitrogen, which would be undissociated dinitrogen, can exhibit the DIIE, but H-type chemisorbed nitrogen, which would be dissociated nitrogen, only exhibited the normal deuterium isotopic effect. Vibrational signals in the range of 2 268~1 910 $cm^{-1}$ that were assigned to the dinitrogen chemisorbed species on the Fe-based catalysts and Ru-based catalysts were observed in our experiments[10,12,15-17]. This experimental finding supported the argument that the most abundant adsorbed species was IR and Raman active dinitrogen adspecies, and these chemisorbed dinitrogen species were activated to react directly with hydrogen to ammonia. Therefore, the dynamic reason invoking the hydrogenation of molecular nitrogen adspecies would be the explanation of the DIIE.

A statistical expression for the dynamic isotopic effect can be deduced from the transition state

**1503**

theory and statistical mechanics. The expression can be approximated by a simple expression by including only one term of the zero point energy. When this is treated as a first order isotopic effect of hydrogen, we can write the expression as $k_D/k_H = \exp\{(1.4388/T)[(\omega_{FeD} - \omega_{FeH}) - (\omega_{ND} - \omega_{NH})]\}$. By substitution with the bond vibration frequences ($\omega$) of the transition state and temperature ($T$), we can calculate that the ratio of rate constants, $k_D/k_H$, is 2.03 for the transition states of $H_2N_2/Fe$ and $D_2N_2/Fe$. This is in good agreement with the experimental value of 2.07 in Table 1. We can also calculate the value of 1.42 for the transition state of $HN_2/Fe$ and $DN_2/Fe$ at 673 K over the multi-promoted iron catalyst[13].

According to the above approximate expression, the values of the dynamic isotopic effect change with both the vibrational frequencies of the transition state and the temperature. Our experimental results also showed the change of the $r_D/r_H$ values with increased temperatures. The trend change indicated a strong competition between the dynamic isotopic effect and thermodynamic isotopic effect, which was influenced by both catalyst and reaction conditions. The infrared-active dinitrogen adspecies, for example, was observed at $2\,170 \sim 2\,150$ cm$^{-1}$ over MgO but not over $Al_2O_3$[18]. This indicated that the MgO support gave the adsorption and activation of molecular nitrogen, but this was not the case with $Al_2O_3$. Therefore, the extent of the competition between the dynamic factor and thermodynamic factor over MgO-supported catalysts and $Al_2O_3$-supported catalysts would be different.

In conclusion, strong deuterium inverse isotopic effects, $r_D/r_H$ of about 2, were observed over both Ru-based and Fe-based catalysts. This was explained as due to a strong competition between the dynamic isotopic effect and thermodynamic isotopic effect. This was a reliable and important experimental evidence that hydrogen was involved in the rate determining step. This result supported the associative mechanism of ammonia synthesis, in which the associative route was the major pathway, the dissociative route was the minor pathway, and the two routes strongly competed with each other.

# References

[1]Ertl G. Angew Chem, Int Ed, 2008, **47**:3524.

[2]Schlogl R. Angew Chem, Int Ed, 2003, **42**:2004.

[3]Boudart M. CatalRev-Sci Eng, 1981, **23**:1.

[4]Ozaki A, Taylor H, Boudart M. Proc R Soc London Ser A, 1960, **258**:47.

[5]Aika K, Ozaki A. J Catal, 1970, **16**:97.

[6]Aika K, Ozaki A. J Catal, 1969, **13**:232.

[7]Logan S R, Philp J. J Catal, 1968, **11**:1.

[8]Takezawa N. J Catal, 1972, **24**:417.

[9]Asscher M, Carrazza J, Khan M M, Lewis K B, Somorjai G A. J Catal, 1986, **98**:277.

[10]Liao D W, Zhang H B, Wang Z Q, Tsai K R. Sci Sin(Ser B), 1987, **30**:246.

[11]Huang K H. J Xiamen Univ(Natur Sci), 1978, **17**:130.

[12]Liao D W. [PhD Dissertation]. Xiamen: Xiamen Univ, 1985.

[13]Liao D W. J Xiamen Univ(Natur Sci), 1995, **34**:204.

[14]Kokes R J. Catal Rev, 1972, **6**:1.

[15]Liao D W, Wang Zh Q, Zhang H B, Tsai K R. J Xiamen Univ(Natur Sci), 1982, **21**:100.

[16]Zhang H B, Schrader G L. J Catal, 1986, **99**:461.

[17]Kubata J, Aika K. JPhys Chem, 1994, **98**:11293.

[18]Aika K, Midorikawa H, Ozaki A. J Catal, 1982, **78**:147.

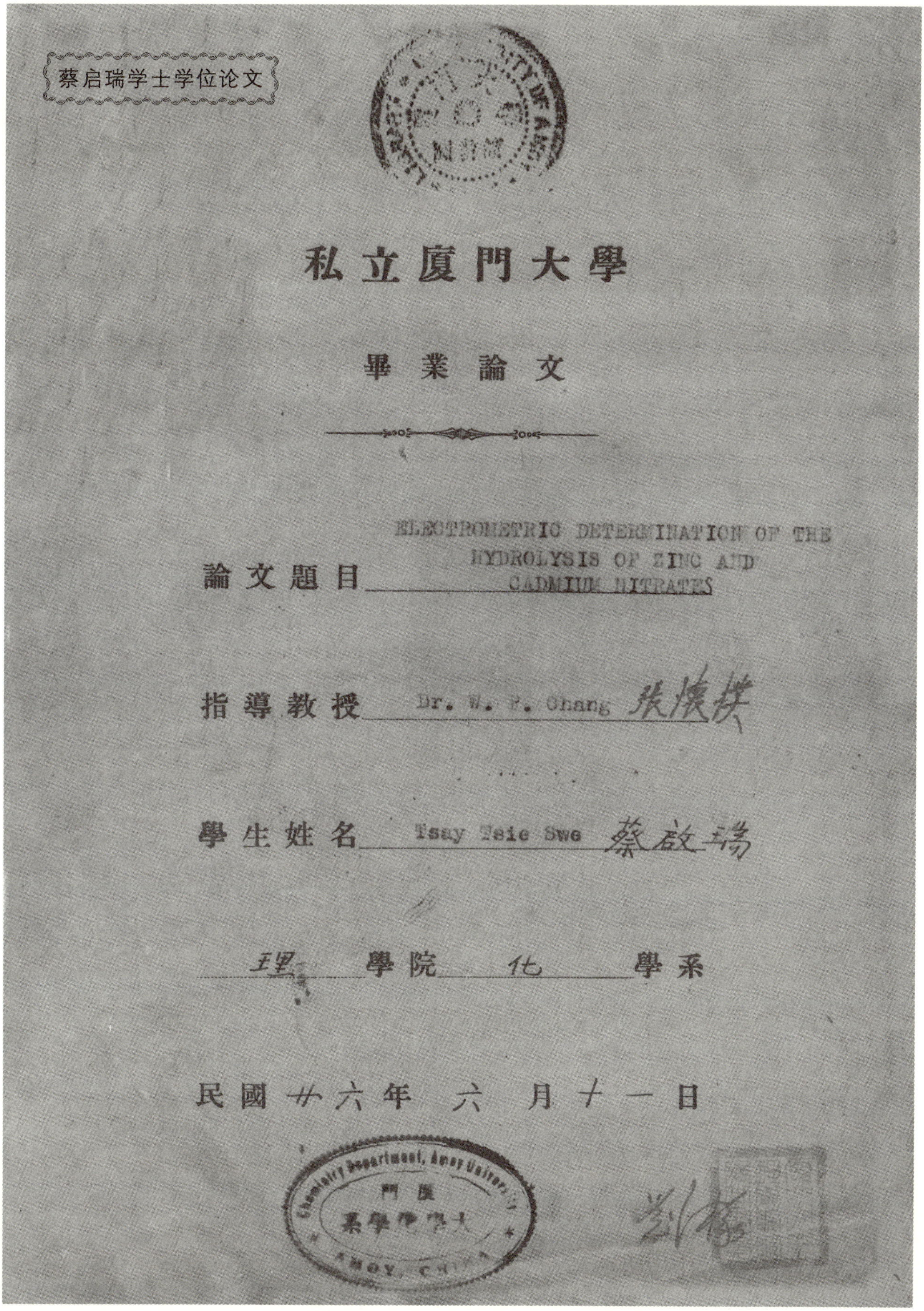

蔡启瑞学士学位论文

私立廈門大學

畢業論文

論文題目 ELECTROMETRIC DETERMINATION OF THE HYDROLYSIS OF ZINC AND CADMIUM NITRATES

指導教授 Dr. W. P. Chang 張懷樸

學生姓名 Tsay Tsie Swe 蔡啟瑞

理 學院 化 學系

民國廿六年 六 月十一日

# ELECTROMETRIC DETERMINATION OF THE HYDROLYSIS

## OF ZINC AND CADMIUM NITRATES

by

Tsay Tsie Swe

A Thesis Submitted to the Faculty of

the College of Science

University of Amoy

In Partial Fulfillment of the Requirements

for the Degree of Bachelor of Science

Approved by:

*In Charge of Major Work*

*Head of Department*

*Dean of College of Science*

University of Amoy

Jun., 1937.

## ACKNOWLEDGMENT

The auther whishes to express his gratitude to Dr. W. P. Chang for his unstinted help and valuable suggestions in developing the theoretical aspects of the problem.

## CONTENTS

--- 1 ---

## THEORETICAL INTRODUCTION

When a normal salt is dissolved in water, the pH of the resulting solution is seldom that of exact "neutrality" and varies with the chemical nature of the salt. If the salt is formed by a strong acid and a weak base, the salt solution will show acid reaction; conversely, if the basic component is "stronger", it will react alkaline. Only when the acid and the basic component are of the same strength, will the pH be that of exact neutrality. This obscure phenomenum is due to a process, called "Hydrolysis".

In the solution of a uni-univalent electrolyte with either or both redicals weak, there are involved several dissociation equilibria:

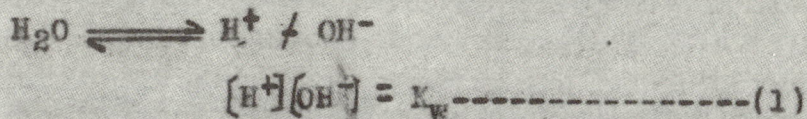

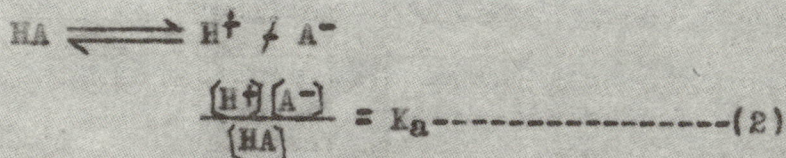

$$H_2O \rightleftharpoons H^+ + OH^-$$

$$[H^+][OH^-] = K_w \text{------------(1)}$$

$$HA \rightleftharpoons H^+ + A^-$$

$$\frac{[H^+][A^-]}{[HA]} = K_a \text{------------(2)}$$

--- 2 ---

$$MOH \rightleftharpoons M^+ + OH^-$$

$$\frac{[M^+][OH^-]}{[BOH]} = K_b \text{-------------(3)}$$

$$M^+A^- + H_2O \rightleftharpoons HA + MOH$$

$$\frac{[HA][MOH]}{[B^+][A^-][H_2O]} = K_h \text{------(4)}$$

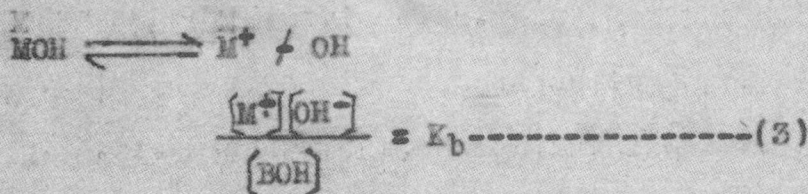

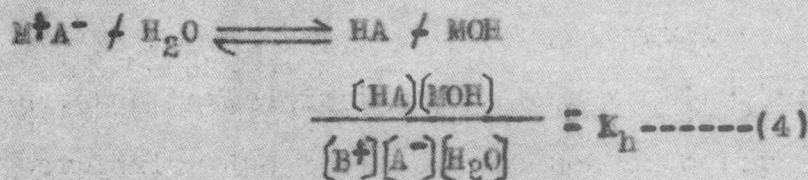

the last one being the hydrolytic dissociation
or simple called "hydrolysis". Thus hydrolysis
is the reverse of neutralization. If the acid
and base set free by hydrolysis do not dissociate to the same extent, a greater or less
hydrogen ion concentration than that of the
hydroxyl ion results, and the solution is no
longer "neutral".

Now the mechanism of $_{\wedge}$ equations (1), (2),
and (3): *equation (4) may be represented by*

$$[H^+][OH^-]\cdot\frac{[HA]}{[H^+][A^-]}\cdot\frac{[MOH]}{[M^+][OH^-]} = \frac{[HA][MOH]}{[M^+][A^-]} = \frac{K_w}{K_a K_b} = K_h \text{----(5)}$$

In dealing with a salt of which the acid
redical is "strong", we may regard the acid set

--- 3 ---

free by hydrolysis as completely dissociated;
then we derive.

$$\frac{[H^+][MOH]}{[M^+]} = \frac{K_W}{K_b} = K_h \text{-----------------} (6)$$

Conversely, if the basic component is
strong, we have

$$\frac{[OH^-][HA]}{[A^-]} = \frac{K_W}{K_a} = K_h \text{-------------------} (7)$$

Now, if we denote the degree of hydrolysis
by h, the total salt concentration by c, and
assume, for the sake of simplicity, that the
strong electrolytes are completely dissociated,
and the the activities of the ions and of the
undissociated molecules can be substituted by
their respective concentrations, equations (5),
(6), and (7) may be changed into the forms

$$\frac{[HA][MOH]}{[A^-][M^+]} = \frac{ch \cdot ch}{c(1-h)\cdot c(1-h)} = \frac{h^2}{(1-h)^2} = \frac{K_W}{K_a K_b} = K_h \text{----}(5)'.$$

$$\frac{[M^+][MOH]}{[M^+]} = \frac{ch \cdot ch}{c(1-h)} = \frac{ch^2}{1-h} = \frac{K_W}{K_b} = K_h \text{--------}(6)'.$$

$$\frac{[OH^-][HA]}{[A^-]} = \frac{ch \cdot ch}{c(1-h)} = \frac{ch^2}{1-h} = \frac{K_W}{K_a} = K_h \text{----------} (7)'.$$

— 4 —

Measurements of h, of $[H^+]$, or of $[OH^-]$, at several concentrations, enable us to calculate $K_h$, and furnish us a means to ascertain whether the Mass Action Law, is applicable to the solution investigated and also to examine the deviations arising from the substitution of concentrations for activities.

From equation (6)' or (7)', the ionisation constant $K_a$, or $K_b$, may be calculated if $K_h$ and $K_w$ are known. Conversely, if $K_h$ and $K_a$ (or $K_b$) have been determined in advance, $K_w$ may be calculated. In fact many important data for the ionization of water at various temperatures have been obtained from such hydrolysis measurements.

Finally, if $K_h$ at several temperatures has been determined, the heat of hydrolysis, $\Delta H_h$, may be calculated from the well known equation of Vant Hoff

$$\frac{d.lu.k}{dT} = \frac{\Delta H}{RT^2},$$

or

$$\Delta H_h = \frac{-R\,d\,lu\,k_h}{d\left(\frac{1}{T}\right)} \quad\text{------(8)}$$

--- 5 ---

It follows (at once) from equation (8), that, if the values of $-(R\ln K_h \neq \text{constant})$ are plotted against that of $(-\frac{1}{T})$, the slope at every point of the curve gives directly the value of $H_h$ at the corresponding absolute temperature.[9]

For the accurate measurements of the amount of hydrolysis in salt solutions three methods are gererally employed; namely, the conductance, the kinetic, and the E. M. F. method. The conductance method consists in measuring the increase or decrease in conductance due to hydrolysis, and in calculating the degree of hydrolysis from such data, with the aid of some previous knowlege about the equivalent conductance of the ions invalved. The kinetic method depends upon the determination of the velocity constant for the hydrolysis of some organic esters, or for the inversion of cane sugar, catalysed by the salt solution, or more properly spoken, by the H ions set free by hydrolysis of the salt. The chelf defect of this method lies in that it involves

--- 6 ---

too many unknown intricacies. By fax the most reliable method is that depending upon the E. M. F. measurement, which furnishes us a means for the direct determination of hydrogen ion activity. It has an advantage over the conductance method in that it does not involve the perplexity due to the presence of other ions. Formerly the hydrogen electrode was used exclusively in the potentiometric determination of hydrogen ions. But it has been discovered later that this electrode is not suitable for that purpose in case the salt solution contains such easily reducible anions as $NO_3^-$ and $SO_4^=$, or such anions as are derived from metals less "noble" than hydrogen. Quinhydrone electrode has been purposely employed in such cases, to take the place of hydrogen electrode; and has been found to be of satisfactory service in measuring the pH of any acidic solution.

--- 7 ---

The hydrolysis of many uni-univalent salts
has been studied in detail by the aforesaid three
methods. Among the more important works, that
of Noyes[1] and that of Shields[2] may be mentioned.
Shields determined the hydrolysis of sodium
acetate, and Noyes determined the hydrolysis
of ammonium acetate by the conductance method.
Both of them have calculated the ionization constant of water at 25°C., and the two values of
$K_W$ have checked remarkably well with each other.

The Hydrolysis of Several Salts other
than the uni-univalent has also been studied
by the kinetic and the potentiometric methods.
That of zinc and cadmium salts will be mentioned
here. Carl Kullgren[5] studied the hydrolysis of
$ZnCl_2$, zinc nitrate, zinc sulfate, and cadium
chloride by the inversion method, and obtained
the value 0.094% for the degree of hydrolysis of
a $1/64$ molar solution of zinc nitrate at 100°C.
Denham and Marries[3], in the study of hydrolysis
of E. M. F. measurements, a number of salt solutions by means obtained with a hydrogen electrode, found that the E. M. F. of

the cell, using zinc salts as electrolytes,
varied most irregularly from day to day over
long periods. They ascribed this abnormal
behavior to the disturbance of some/ heterogeneous equilibria of the colloidal type. The same
authors [4], however, have succeeded later in determining electrometrically with a quinhydrone electrode, the hydrolysis constant of cadmium sulfate
solution. Similar results have been obtained electrometrically by V. Cupr, [6] also with a
quinhydrone electrode.

So far, the electrometric study of the hydrolysis of zinc and cadmium nitrates has never
been discussed in literature. Nor has the
hydrolysis of the later salt been studied by any
other method.

In the present investigation the hydrolysis
of these two salts, zinc nitrate and cadmium
nitrate, in aqueous solutions, will be studied
electrometrically with a quinhydrone electrode.
The following cell.

--- 9 ---

Mg / MgCl, KCl (sat'd) / quinhydrone solution / Pt
will be used for the investigation, on account
of its small temperature coefficient, and of
its accommodation of the use of a combined
electrode.

--- 10 ---

## II. PREPARATION OF MATERIALS

Water---Distilled water was redistilled in glass apparatus, over small amount of potassium acid sulphate.. The middle portion of the distillate, often being washed with a stream of carbon-dioxide-free air for about thirty minutes, had an average specific conductance of $1.2 \times 10^{-6}$ mhos at $18^{\circ}C$.

Zinc nitrate---Merck's pure zinc nitrate was recrystallized four times from conductivity water. The salt solution was saturated at $60-70^{\circ}C$, and the crystallization was allowed to take place around $-5^{\circ}C$. The purified salt was drained dry on the suction filter, then it was pressed between filter paper, and dried over anhydrous calcoum chloride for twenty four hours.

By Analysis:---%Zn in the nitrate: 21.89.

% Zn in $Zn(NO_3)_2.6H_2O$: 21.98.

Hence the purified salt was in the 6-hydrated form, $Zn(NO_3)_2.6H_2O$

--- 11 ---

Cadmium Nitrate---Merck's extra pure fused salt, $Cd(NO_3)_2 \neq 2H_2O$, was recrystallized three times from conductivity water, in a manner similar to that for purifiying zinc nitrate. The crystallized salt was also dried over anhydrous calcium chloride

By Analysis: % Cd in the nitrate          36.50.

% Cd in $Cd(NO_3)_2 \cdot 4H_2O$                          36.42.

Hence the recrystallized salt was in the form, $Cd(NO_3)_2 \cdot 4H_2O$.

Potassium Chloride---Merck's extra pure crystals of potassium chloride were recrystallized two times from conductivity water. The recrystallized salt was dried in oven at $160^\circ C$. for a few hours. To be used for ~~pp~~ preparing buffer solutions, it was further ignited.

Mercury---Merck's extra pure mercury was distilled under diminished pressure in a current of air. The middle portion of the distillate was filtered through a series of perforated harden filter paper, to remove any oxide film that night be carried over.

--- 12 ---

Calomel Paste---Calomel paste was prepared electrolytically from redistilled Hg and N HCl after the modified method of Lips-Comb and Hulett.[8] The calomel paste so prepared was heavily laden with finely divided mercury. It was wasted with repeated changes of saturated potassium chloride solution before it was placed in the cell.

Quinhydrone---Prepared from ferric ammonium sulfate and Chemically pure hydroquinone in the manner described by Kolthoff and Furman.[7] It was dried over anhydrous calcium chloride and kept in a colored bottle. Both the crude and the recrystallized product were used; the results were not different.

Potassium Acid Phthalate---The preparation followed the method of Clark,[8] extra pure potassium hydroxide and resublimed ortho-Phthalic anhydride being started with. The salt was dried at 110° to constant weight.

Salt solutions---Seven different molar concentrations for each salt were prepared

--- 13 ---

were prepared from one molar solution. These
were kept in steamed-out Pyrex glass flasks
fitted with soda-lime tube to guard against
the entrance of Carbon dioxide.

Buffer Solutions---Several "phthalate
buffers"[5] were prepared according to the directions of Clark's Determination of Hydrogen Ions.
For daily checking of the cell, a carefully
prepared $\dfrac{M}{20}$ potassium acid phthalate was
used.

--- 14 ---

### III. APPARATUS

Thermostat---A tall iron cylindric of
four liters capacity was placed in a water thermostat holding about 20 liters and provided with
a 110 V. 250 W. Roberston (Griffen and Tatlock, *lamp, an efficient stirrer, a hot* 
*relay*
L 38328), and *a* toluene regulator sensitive
within 0.02°C. This iron cylinder served as
an air thermostat; a piece of thick paper, laden
with a layer of cotton and perforated for the
binding posts on the combined electrode, made
the cover.

Electrodes--A combined electrode, L and N
hydrogen (quinhydrone)--Calomel electrode No.
7700-A-1, was used. The bright pt-wire electrode was used in this case without being
platinized. The filling of the calomel half
cell was made according to the directions
accompanying the electrode.

--- 15 ---

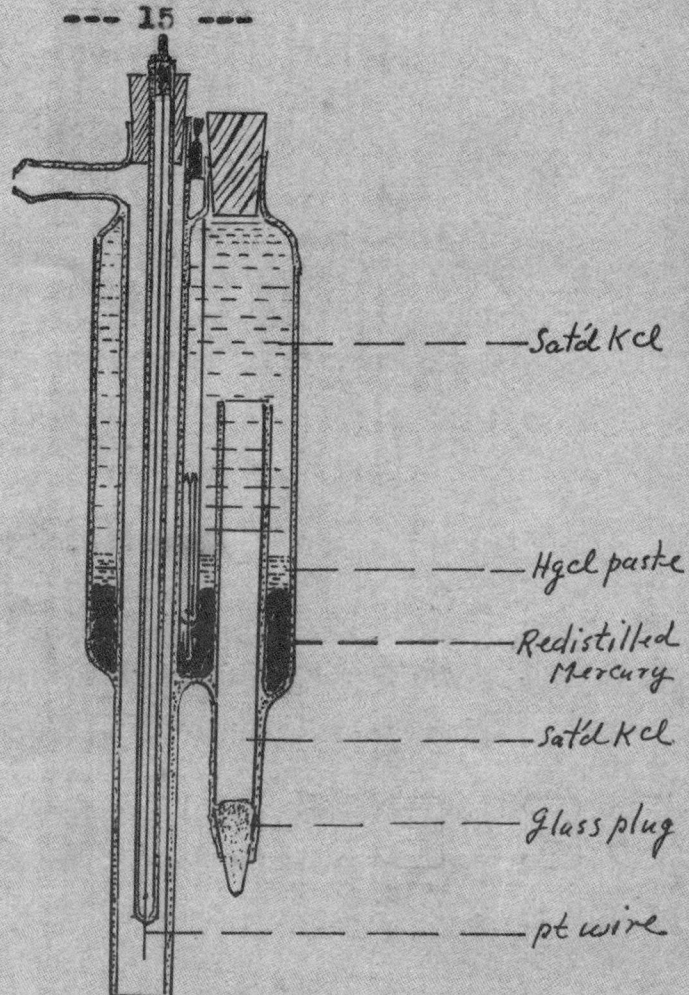

Sat'd KCl

Hgcl paste

Redistilled
Mercury

Sat'd KCl

Glass plug

pt wire

Potentiometer Assembly---A Type-K potentiometer (L and N, No. 85129); a portable
galvanmeter (L and N, No. 243538), provided
with lamp and scale, and capable of measuring
0.01 milivolt with accuracy.

Standard cell---Working cell Weston Standard Cell, No. 5791; cell for checking, Weston
Standard Cell No. 8479.

--- 16 ---

### IV. MEASUREMENTS

The old Weston Standard Cell, No.5791, was standardized against a new one, No. 8479.

The sat'd calomel half cell was standardized daily with "4.0" phthalate Buffer and $\frac{M}{20}$ potassium acid phthalate. In case the E. M. F. was found to change, the cell materials were renewed. It must be noted that throughout the working period of the sat'd calomel half cell, the pH measured for the $\frac{M}{20}$ potassium phthalate have invariably checked with that published by Clark and Lubs, namely "3.97".

The calomel half cell having been checked, some preliminary measurements of the pH of the salt solutions were carried out under room temperature, which was quite close to 25°C. About 60 cc. of the salt solution were siphoned into a small conical flask; it was shaken with 0.2-0.3 gm. of quinhydrone for about 20 seconds. Then it was poured into a 100 cc. lipless beaker, and the bright Pt electrode and the freshly-flashed salt bridge, both carefully rinced

--- 17 ---

with the same solution, were immersed into it.
Then the potentiometer was balanced and the
record taken down.

These measurements were repeated for several
days; the E. M. F. was found to change irregularly with time.

New solutions were carefully prepared from
one molar stock solutions. These were allowed
to stand for four days at room temperature
around 25°C. Then several "time experiments"
on the E. M. F. measurement were carried out.
Fifty cc. of the solutions were carefully shaken
with 0.3 g. of quinhydrone; the whole cell was
placed in the air thermostat described above,
and allowed to take the temperature of the
latter, which was maintained at 25 ± 0.01°C.
The measurement was extended over one or two
hours, and the value of E. M. F. was taken
downwith time at intervals of 2, 5, or 10
minutes.

Iregular variations of the E. M. F. were

--- 18 ---

again observed.

At first it was doubted that polorization occurred, due to the reduction of $NO_3^-$. The E. M. F. of a cell with 0.04 M nitric acid as electrolyte was measured. No change with time was observed; the E. M. F. remained constant around 0.3090 volt, over several hours, with small variations within 0.1 mili-volt.

--- 19 ---

DATA, CALCULATIONS, AND RESULTS

Data:                                              May 25th, 1937.

Zinc Nitrate Solution

| Molarity | time | E. M. F. 25°C. |
|----------|------|----------------|
| 0.05 | $3°42'$ | 0.14617 |
| | $3°43'$ | 0.14617 |
| | $3°45'$ | 0.14450 |
| | $3°50'$ | 0.14518 |
| | $3°55'$ | 0.14780 |
| | $4°0'$ | 0.14780 |
| | $4°5'$ | 0.14805 |
| | $4°10'$ | 0.14850 |
| | $4°15'$ | 0.14880 |
| | $4°20'$ | 0.14910 |
| | $4°25'$ | 0.14918 |
| | $4°30'$ | 0.14923 |
| | $4°34$ | 0.14937 |
| | $4°42'$ | 0.15050 |
| | $4°49'$ | 0.15233 |
| | $5°7'$ | 0.15275 |
| | $5°8'$ | 0.15271 |
| | $5°14'$ | 0.15307 |

--- 20 ---

| | |
|---|---|
| $5^{o}25'$ | 0.15340 |
| $5^{o}32'$ | 0.15337 |
| $5^{o}38'$ | 0.15107 |
| $5^{o}45'$ | 0.15027 |
| $5^{o}50'$ | 0.16012 |
| $5^{o}55'$ | 0.15022 |
| $6^{o}2'$ | 0.15005 |
| $6^{o}7'$ | 0.15005 |
| $6^{o}10'$ | 0.14984 |
| $6^{o}16'$ | 0.14984 |
| $6^{o}21'$ | 0.14984 |
| $6^{o}30'$ | 0.14983 |
| $7^{o}5'$ | 0.14914 |
| $7^{o}0'$ | 0.14884 |

Most probable value 0.1498

---

May 25th, 1937.

<u>Zinc Nitrate Solution</u>

| Molarity | time | E. M. F. $25^{o}C$. |
|---|---|---|
| 0.005 | $9^{o}20'$ | 0.13525 |
| | $9^{o}25'$ | 0.13420 |
| | $9^{o}30'$ | 0.13550 |
| | $9^{o}4$ | |

--- 21 ---

|  |  |
|---|---|
| 9°40' | 0.13590 |
| 9°45 | 0.13700 |
| 9°50' | 0.13830 |
| 9°57' | 0.13849 |
| 10°5' | 0.13867 |
| 10°10' | 0.13870 |
| 10°20' | 0.13870 |
| 10°30' | 0.13865 |
| 10°40' | 0.13872 |
| 10°50' | 0.13875 |
| 11°0' | 0.13871 |
| 11°10' | 0.13818 |
| Most probable value | 0.1387 |

May 26th, 1937.

### Zinc Nitrate Solution

| Molarity | time | E. M. F. 25°C. |
|---|---|---|
| 0.01 | 5°14' | 0.15563 |
|  | 5°20' | 0.15017 |
|  | 5°53' | 0.14944 |
|  | 6°8' | 0.15002 |
|  | 6°17' | 0.15015 |
|  | 6°18' | 0.15019 |
| Most probable value | | 0.1501 |

--- 22 ---

May 26th, 1937.

### Zinc Nitrate Solution

| Molarity | time | E. M. F. 25°C. |
|---|---|---|
| 0.01 | 6°50' | 0.14814 |
| | 6°55' | 0.14848 |
| | 7°3' | 0.14845 |
| | 7°14' | 0.15010 |
| | 7°17' | 0.14982 |
| | 7°24' | 0.14930 |
| | 7°30' | 0.14860 |
| | 7°39' | 0.14847 |
| | 8°30' | 0.15201 |
| | 8°45' | 0.15103 |
| | 8°50' | 0.15095 |
| | 8°55' | 0.15092 |
| | 9°0' | 0.15087 |
| | 9°5' | 0.15083 |
| | 9°10' | 0.15082 |
| Most probable value | | 0.1509 |

--- 23 ---

May 24th, 1937.

### Cadmium Nitrate Solution

| Molarity | time | E. M. F. 25°C. |
|----------|------|----------------|
| 0.005 | 2' 30" | 0.15787 |
| | 2' 35" | 0.15786 |
| | 2' 40 | 0.15736 |
| | 2°44 | 0.15963 |
| | 2°46 | 0.15960 |
| | 2°55 | 0.15870 |

--- 24 ---

### V. Calculations and Results

Kolthoff represents the transformation of quinone into hydroquinone by the electochemical equation:

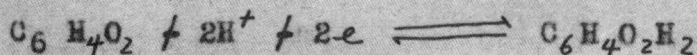

$$C_6H_4O_2 + 2H^+ + 2e \rightleftharpoons C_6H_4O_2H_2$$

As quinhydrone is an equimolecular compound of quinone and hydroquinone, we have

$$[C_6H_4O_2] = [C_6H_4O_2H_2]$$

and the potential of the electrode will therefore be

$$*E = \varepsilon_0 - \frac{RT}{nF} \ln \frac{[C_6H_4O_2H_2]}{[C_6H_4O_2][H^+]^2}$$

$$= \varepsilon_0 + [0.0577 + 0.0002(t-18)] \log[H^+]$$

$$= \varepsilon_0 + 0.05912 \log[H^+] \qquad (25°)$$

According to E. Bulmann and his coworkers, between 0° and 37°C.

$$\varepsilon_0 = 0.7044 - 0.00074(t-18).$$

The potential of saturated calomel electrode has been given in Clark's book

$$E_{s.c.} = 0.2504 - 0.00065(t-18)$$

6

* Size of electrode potential is according to the European Convention.

--- 25 ---

Hence for the (combination) cell

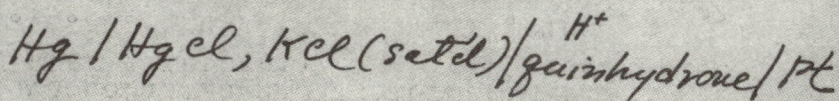

$$Hg \,/\, HgCl, \, KCl\,(setd)\,/\, \overset{H^+}{quinhydrone}\,/\,Pt$$

$$E = E_Q - E_{S.C.} = E_0 + [0.0577 + 0.0002(t-18)]\log[H^+]$$
$$\qquad\qquad - 0.2504 + 0.00065(t-18)$$

$$= 0.7044 - 0.00074(t-18)$$
$$\quad - 0.2504 + 0.00065(t-18)$$
$$\quad + [0.0577 + 0.0002(t-18)]\log[H^+]$$

Hence

$$[H^+] = -antilog\left[\frac{*(0.4534 - E)}{0.05912}\right] \text{ (at } 25°C)$$

$$\text{- - - - - - - - - - - - - - - - - - } (9)$$

Equation (9) will be used for calculating the pH of the salts solution.

In aqueous solution the hydrolysis of a Bi--univalent salt, like $Zn(NO_3)_2$, occurs progressively in two stages:

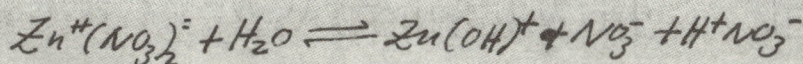

$$Zn^{++}(NO_3)_2^= + H_2O \rightleftharpoons Zn(OH)^+ + NO_3^- + H^+ NO_3^-$$

$$\frac{[Zn(OH)^+][H^+]}{[Zn^{++}]} = K_A' \text{ - - - - - - - - } (10),$$

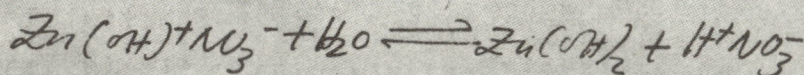

$$Zn(OH)^+ NO_3^- + H_2O \rightleftharpoons Zn(OH)_2 + H^+ NO_3^-$$

$$\frac{[Zn(OH)_2][H^+]}{[Zn(OH)^+]} = K_A'' \text{ - - - - - - - } (11)$$

--- 2b ---

The mechanism of equations (10) and (11) may be represented by the three ionic equations

$$Zn(OH)_2 \rightleftharpoons Zn(OH)^+ + OH^-$$

$$\frac{[Zn(OH)^+][OH^-]}{[Zn(OH)_2]} = K_1 \quad \cdots \cdots \cdots \cdots \quad (12)$$

$$Zn(OH)^+ \rightleftharpoons Zn^{++} + OH^-$$

$$\frac{[Zn^{++}][OH^-]}{[Zn(OH)^+]} = K_2 \quad -------- \quad (13)$$

$$H_2O \rightleftharpoons H^+ + OH^-$$

$$[H^+][OH^-] = K_w$$

Thus

$$\frac{[Zn(OH)^+]}{[Zn^{++}][OH^-]}[H^+][OH^-] = \frac{[Zn(OH)^+][H^+]}{[Zn^{++}]}$$

$$= \frac{K_w}{K_2} = K_h' \quad ----------- \quad (14)$$

$$\frac{[Zn(OH)_2]}{[Zn(OH)^+][OH^-]} \cdot [H^+][OH^-] = \frac{[Zn(OH)_2][H^+]}{[Zn(OH)^+]}$$

$$= \frac{K_w}{K_1} = K_h'' \quad ---------- \quad (15)$$

In general the amount of second-step hydrolysis is negligible in comparison with that of the first-step; practically all of the free acid comes from the first step hydrolysis.

s

--- 227 ---

Starting with this as potulate we may calculate $K_h'$ if c and $[H^+]$ are known. Knowing c, $[H^+]$ and $K_h'$ we may proceed to calculate $K_h''$. The calculation is as follows:

From eq (14)

$$K_h'[Zn^{++}] - [H^+][Zn(OH)^+] = 0 \quad ----- (i)$$

By the rule of electrical neutrality

$$[H^+] + 2[Zn^{++}] + [Zn(OH)^+] = [NO_3^-] + [OH^-]$$

of

$$2[Zn^{++}] + [Zn(OH)^+] + \{[H^+] - 2c - [OH^-]\} = 0 \quad \cdots (ii)$$

Since total zinc = total salt = c

$$[Zn^{++}] + [Zn(OH)^+] + Zn(OH)_2 - c = 0 \quad \cdots \cdots (iii)$$

Solving the three simutaneous equations we get

$$[Zn(OH)^+] = - \frac{[H^+]\{[H^+] - 2c - [OH^-]\}}{K_h' + 2[H^+]}$$

$$[Zn(OH)_2] = \frac{[H^+]^2 - K_1\{c + [OH^-] - [H^+]\} - K_w}{K_h' + 2[H^+]}$$

and

$$K_h'' = \frac{K_w}{K_1} = \frac{[Zn(OH)_2][H^+]}{[Zn(OH)^+]}$$

$$= \frac{[H^+]^2 - K_h'\{c + [OH^-] - [H^+]\} - K_w}{2c + [OH^-] - [H^+]}$$

$$K_1 = \frac{K_w}{K_h''}$$

$$K_2 = \frac{K_w}{K_h'}$$

--- 24 ---

## Zinc Nitrate Solution

| M | E. M. F. | $H^- \cdot X10^{-6}$ | $K_h' \times 10^{-9}$ | $K_h \times 10^{-12}$ |
|---|---|---|---|---|
| 0.005 | 0.1498 | 7.311 | 1.069 | 6.8 |
| 0.005 | 0.1387 | 4.753 | 4.519 | 1.2 |
| 0.001 | 0.1501 | 7.413 | 5.495 | 1.5 |
| 0.001 | 0.1509 | 7.638 | 5.834 | 1.7 |

## Ionization Constants of Zinc Hydroxide calculated from $K_w$ $K_h'$ and $K_h''$

| Molarity of $Zn(NO_3)_2$ | $K_1 = \dfrac{K_w}{K_h''}$ | $K_2 = \dfrac{K_w}{K_h'}$ |
|---|---|---|
| 0.005 | $1.47 \times 10^{-3}$ | $9.37 \times 10^{-7}$ |
| 0.005 | $8.33 \times 10^{-3}$ | $2.21 \times 10^{-6}$ |
| 0.001 | $6.67 \times 10^{-3}$ | $1.82 \times 10^{-6}$ |
| 0.001 | $5.37 \times 10^{-3}$ | $1.72 \times 10^{-6}$ |

--- 27 ---

## SUMMARY AND CONCLUSION

I.  The theoretical significance of hydrolysis has been discussed.

II.  The hydrolysis of zinc and cadmium nitrates has been studied electrometrically with a quinhydrone electrode.  The E. M. F. of the cell with either zinc nitrate solution or cadmium nitrate solutions as electrolyte has been found to change irregularly with time.

III.  An equation for the calculation of hydrogen ion activity from the E. M. F. of the cell Quinhydrone electrode---Saturated calomel electrode, has been derived.

IV.  An equations for the calculation of the first and second hydrolytic dissociation constants of zinc nitrate or cadmium nitrate, from the hydrogen ion activity of the solution, has also been derived.

--- 30 ---

V. The first and the second hydrolytic
hydrolytic dissociation constants $K_h'$ and $K_h''$ of
the two salts [ZnNO₃] were calculated from E. M. F. data.
These do not appear to be constant.

VI. More information about the nature of
the two salts on hydrolysis will be sought in
the future, as potentiometric work requires
special technique and much experiance, which
can not be obtained in such a short time. At
present it can be said that the hydrolysis of
the two salts, zinc nitrate and cadmium nitrate
as determined electromotrically with a quinhydrone electrode, appears to be abnormal, and
that this abnormality is not due to ⟨the salt as a whole rather than⟩ the reducible anion $NO_3^-$.

--- 31 ---

# REFERENCE

(1) A. A. Noyes:  Journ. Am. Chem. Soc.,

  32, 159 (1910).

(2) Shields: Zeit. Phys. Chem. 12, 167 (1893).

  Cited in Getman Theoretical Chemistry.

(3) H. G. Denham: Z. Anorg. Chem. 57, 378-94

  C. A. 2 : $2052^5$    (1907).

(4) H. G. Denham and N. A. Marris: Trans.

  Faraday Soc. 24, 510-5 (1928).

  C. A.  8 : $608^9$

(5) Carl Kullgen: Zeit Physik. Chem.  85, 466-80.

  C. A.  8:6089.

(6) V. Cupr: Collection Czechoslov.

  Chem. Comm. 1, 467-76.

  C. A.  4766.
        + Furman
(7) Kolthoff: Potentiometric Titration.
     ( 2nd Eddition )
(8) Clark: Determination of Hydrogen Ions

  (3rd. Eddition)

(9) Lewis and Randall: Thermodynamics.

(10) Getman : Theoretical Chemistry

蔡启瑞博士学位论文

A STUDY OF MACRO-RING CLOSURE IN HETEROGENEOUS REACTIONS;
SURFACE FILMS OF HIGH POLYMETHYLENE DICARBOXYLIC
ACIDS AND GLYCOLS.

DISSERTATION

Presented in Partial Fulfillment of the Requirement for
the Degree of Doctor of Philosophy in the
Graduate School of the Ohio State University

By

KHI-RUEY TSAI, B. S.
THE OHIO STATE UNIVERSITY
1950

Approved by:

_P.M. Harris_ _for E. Mack, Jr._
Advisor

-i-

# TABLE OF CONTENTS

-1-

STUDY OF MACRO-RING CLOSURE IN HETEROGENEOUS REACTIONS:

SURFACE FILMS OF HIGH POLYMETHYLENE DICARBOXYLIC

ACIDS AND GLYCOLS.

The purpose of the present work is (1) to devise a method for the synthesis of high polymethylene dicarboxylic acids and glycols, and (2) to study the surface films of such compounds on water. The theoretical aspects and experimental part of (1) are presented in Chapters II, III and IV. The object of the surface film measurement (Chapter V) is to ascertain the proximity of the two end-groups when such compounds are spread on water; this will be of important bearing to macro-ring closure by heterogenous reactions, a theoretical analysis of which forms the introduction for this dissertation.

## I.  INTRODUCTION

A reaction which leads to ring-closure may also lead to the formation of polymers when the molecules react intermolecularly instead of intramolecularly, the course of the reaction depends upon the steric factor and the experimental conditions. In the case of closure of 5- and

-2-

6-membered rings, the steric factor is so favorable to cyclization that the polymerization reaction hardly ever has a chance to show up. On the other hand, the 9-, 10-, 11- and 12-membered rings are known to be very difficult to form because of the congestion of hydrogen atoms in the ring space as demonstrated with models by M. Stoll (6); here even the best experimental procedure fails to make the cyclization reaction prevail, and the yields are always very low.

With increasing size of the ring, the steric factor becomes less important, and the choice of experimental conditions begins to play the major role in determining the course of the reaction. Since the polymerization reaction is kinetically of higher order than the first, one way of suppressing this undesirable side-reaction is to carry out the cyclization reaction in highly dilute solution. This principle was first successfully employed by Ziegler (7), (8) in the cyclization of dinitriles with sodio derivative of a secondary amine as the condensing agent.

A new and improved method of ring-closure has recently been described by Stoll and Rouve (5). It is based upon the condensation of dicarboxylic esters by means of molten sodium suspended in boiling xylene solution. This acyloin condensation method gives much better yield of cyclic

-3-

product than either Ruzicka's (4) or Ziegler's method and
does not necessitate the use of highly dilute solution.
The explanation for this brilliant success is that the reaction probably takes place heterogeneously:  the molten
sodium metal attacks the carbonyl oxygen atoms of the ester
groups with the formation of quasi-salt linkages, thus
leaving the carbonyl carbon atoms partially unsaturated and
at the same time binding the reactive end-groups in the
metal-xylene interface;  impacts by foreign molecules will
tend to throw the two reactive end-groups of the same chain
together, thus giving them a better chance to react intramolecularly.

The mechanism of Ruzicka's method, which is based upon
the pyrolysis of the thorium salts of dicarboxylic acids, is
still obscure.  The fact that polyvalent metal salts of carboxylic acids give ketones on pyrolysis while the corresponding alkali salts give only hydrocarbons leaves little
doubt that the capacity of the polyvalent metals in binding
the carboxyl groups together is essential for the subsequent formation of ketones.  Following this reasoning, it
seems logical to suppose that ketone formation by pyrolysis of metal carboxylates can hardly take place when the
carboxyl groups are attached to different metal atoms, and
as a further deduction, that a cyclic ketone can be obtained

-4-

by pyrolysis of polyvalent metal dicarboxylate only when the sample contains cyclic salts in which the two reactive end-groups are bound to the same metal atom. However, Carothers(2) suggested that polyanhydrides might form first during the pyrolysis of polyvalent metal dicarboxylates through the thermal decomposition of linear polymeric salts, these polyanhydrides might then undergo interchain reaction on subsequent cracking to form cyclic ketones. His experimental evidence is indirect and not conclusive. Furthermore, his theory fails to explain why the quadri-valent salts (Th and Ce) should give better yields of cyclic ketones than the corresponding bivalent salts (Ca and Ba).

By the principle of steric hindrance, it is possible to give a satisfactory answer to the last question if we start with the assumption that the formation of cyclic salts is the prerequisite step towards the subsequent formation of cyclic ketones by pyrolysis. The procedure recommended by Ruzicka (4) for carrying the precipitation of the polyvalent metal salts of dicarboxylic acids is based upon homogeneous double-decomposition of the alkali dicarboxylates with the chlorides of the polyvalent metals. The small-sized quadri-valent Th or Ce ion (ionic radius: Ce $1.04^{\circ}$A) requires four carboxylate end-groups (anions) for complete charge neutralization. When two carboxylate end-

-5-

groups from separate chains become attached to the metal
ion, the two chains are still free to vibrate; consequently,
there is a better chance for one of the two free ends to
wind around and approach the central metal ion than for a
foreign end-group, with its long tail, to fight its way
through the double hindrance set up by the two vibrating
chains. Therefore, there is a good chance of obtaining some
cyclic salts in this way. In the case of precipitation of
the corresponding bivalent salts, however, the cation $Ca^{++}$
or $Ba^{++}$ requires only two carboxylate anions for complete
charge neutralization; when one carboxylate ion is
attached to the bivalent metal ion (ionic radius: $Ca^{++}$
$0.99\overset{o}{A}$), the latter becomes only singly hindered and is still
open to attack by foreign end-groups; hence the chance of
getting cyclic salts is small unless the steric factor is
very favorable. If cyclic salts are the necessary intermediates in the formation of cyclic ketones by pyrolysis, it
is then easy to understand why the quadri-valent metal
salts should, in general, give better yields of cyclic
ketones than the corresponding bivalent metal salts of dicarboxylic acids. Since the steric hindrance set up by
a chain usually does not increase appreciably with further
increase in chain length beyond 5 or 6 carbons atoms whereas
the average distance between the free carboxylate end-group

-6-

and the one attached to the metal ion does increase proportionally with the square root of the chain length, it can be anticipated that with further increase in chain length beyond certain limits, the yields of cyclic ketones obtained with Ruzicka's method should begin to fall off again. Finally, if pure cyclic thorium salts of dicarboxylic acids should be obtainable and subject to pyrolysis, they would still give considerable proportion of side-products such as cyclic diketones, whereas with pure cyclic calcium salts the yields of cyclic mono-ketones should be higher, because a cyclic calcium salt carries only one chain per metal atom whereas a cyclic thorium salt carries two chains. These last two deductions are also supported by experimental facts: the yields with Ruzicka's method begin to fall off slightly when the ring of the ketone gets larger than the 17 or 18-membered (7); and calcium adipate, which probably contains a very high percentage of cyclic salt, actually gives better yield of cyclopentanone than does the corresponding thorium salt (2). The yields of cyclic ketones by different methods are shown in Figure I.

Thus Ruzicka's method may be regarded as belonging to the same category as the acyloin condensation method in the sense that both methods enhance the chance of ring-closure by bringing the two reactive end-groups into proximity.

-7-

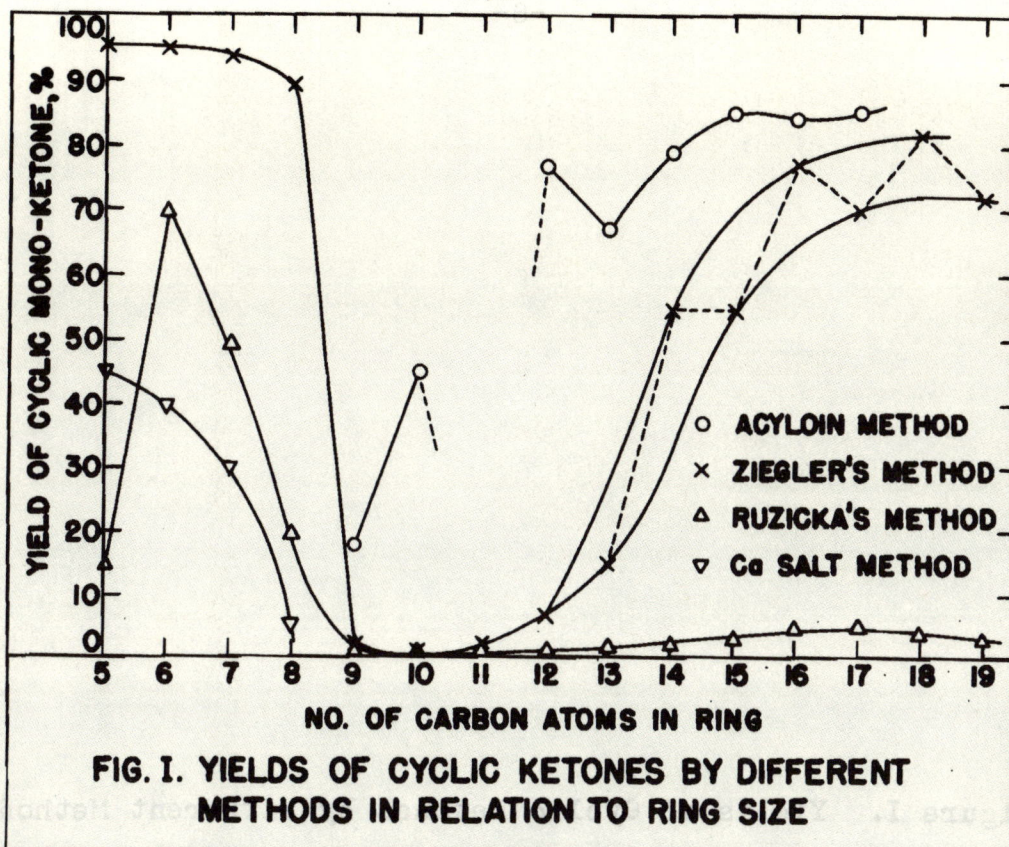

FIG. I. YIELDS OF CYCLIC KETONES BY DIFFERENT
METHODS IN RELATION TO RING SIZE

-8-

Figure I.  Yields of Cyclic Ketones by Different Methods
as  Functions of Chain Length.

But the advantage of the acyloin condensation method is
obvious because it does not have to deal with substance in
the solid state.

The brilliant success of the acyloin condensation
method also illuminates other possibilities of exploiting
heterogeneous reactions to favor ring-closure.  Any cyclization reactions which involve two polar end-groups separated by a long chain, such as reductive dehalogenation of
polymethylene dihalides with dilute metal amalgam, the
formation of higher lactones, intramolecular pinacol reduc-

-9-

tion, intramolecular condensation of alkali di-acetylides or di-malonates (provided that the melting-points of such pseudo salts are not too high), intramolecular Claisen condensation of diesters, and precipitation of polyvalent metal dicarboxylates may be carried out heterogeneously, the last two types of reactions being distinct possibilities. It may not be necessary, nor always practicable, to employ oil-water interface; other immiscible solvent pairs, such as oil-mercury (dilute amalgam), oil-liquid-sulphur-dioxide, oil-glycerol, oil-ethyleneglycol, oil-ethylene-diamide, and oil-formamide, are possible media. The essential idea is that, by carrying out the reaction heterogeneously, the two reactive end-groups are restricted to move mostly in the two-dimensional interface instead of in the three-dimensional solution bulk; furthermore, because of the tendency to minimize surface free-energy, also because of the impacts by foreign molecules, the two endgroups of the same chain are likely to be close together; accordingly, there is a better chance for ring-closure to take place.

The whole weight of the argument for the advantage of cyclization by heterogeneous reactions rests upon the assumption that the two polar end-groups of the same chain are close together when such molecules are spread in a

-10-

suitable interface. If contrary to expectation, the molecules should lie flat with intermolecular association, then,
obviously there would be even less chance of effecting
cyclization by heterogeneous reactions than by homogeneous
reactions. In view of this point, it is very desirable to
make a study of the surface films of the high polymethylene
dicarboxylic acids and glycols, for this appears to be the
only effective means of ascertaining the proximity of the
two polar end-groups when the molecules are spread on aqueous
surface.

If the hydrocarbon chains separating the two polar
end-groups are long enough to overcome the solubility effect
of the latter, and also long enough to bend back and form
double-chain without encountering appreciable steric hindrance, then, by the principle of like attracts like, the
molecules are expected to form condensed films at ordinary
temperature, with the polar end-groups sticking together and
the hydrocarbon residue bent in the form of parallel double-
chains. At ordinary temperature, there will not be appreciable free rotation of the double-chains about their respective axes of symmetry, hence a close packing of the
chains should be possible and the surface condensed area for
each branch of the chain should be about the same as that
for a long normal fatty acid molecule in the condensed-

-11-

film state; namely, about 20.5 $\overset{o_2}{A}$. As a matter of fact, this has been found to be the case with the high polymethylene dicarboxylic esters (1). Furthermore, this condensed area should be practically independent of the chain length whereas the thickness of the monolayer should take on definite increament for each additional $CH_2$ unit in the chain. Conversely, if the chains should be interlocked, or if the molecules should lie flat on the aqueous surface, then the condensed area per molecule should increase with the chain length. Thus from the study of the surface films of such dicarboxylic acids and glycols, valuable informations can be obtained regarding the proximity of the end-groups and the arrangement of the molecules on the aqueous surface.

At higher temperature, the polymethylene dicarboxylic acid or glycol molecules may form expanded films on aqueous surface; with rapid interchange of kinetic and potential energies, the molecules are expected to be oscillating between lying-flat and standing-erect through all intermediate positions like the opening and closing of a standing zipper. If the polar end-groups are fully solvated, they may partake the motion of the underlying water molecules as if gliding on an equipotential surface. But any partial opening of the double-chain will be attended by an energy change, $e_1$, corresponding to (1) partial separation of a number, i, of the $CH_2$ units from close packing against their cohesive force, and (2) replacement of an air-water interface by an equivalent water-hydrocarbon interface, $s_1$. A knowledge of the heat of vaporization of hydrocarbons per $CH_2$ unit and the difference between the two interfacial energies should enable us to estimate theoretically the statistical weight for each degree of opening of the double-chain, at that particular temperature, by means of Boltzmann partition function. Since we are only interested in getting a qualitative picture, we shall merely point out that, in the state of expanded films, the intermolecular spacings are large, and that, even though some of the molecules may transitorily lie flat, the frequency of collision of the two end-groups is still expected to be high because of the dynamic equilibrium between thermal agitations and molecular cohesions and adhesions. Since with greater intermolecular spacings there is less chance for intermolecular reaction, the expanded-film state should be even more favorable than the condensed-film state for cyclization by heterogeneous reactions.

With the polymethylene dicarboxylic acids, glycols, or similar types of polymethylene compounds spread between oil-water interface instead of between air-water interface, the situation is not so favorable. If the oil is a satur-

-13-

ated hydrocarbon, then the displacement of the oil molecules from the interface by the polymethylene chains will
result in no increase in surface energy; hence, except
for the impacts by foreign molecules, most of the dicarboxylic acid or glycol molecules are expected to lie flat
in the interface. However, if the oil employed is a higher fatty alcohol, or a phenolic compound, which has an
interfacial tension with water considerably lower than the
hydrocarbon-water interfacial tension, then the oil molecules will tend to squeeze the hydrocarbon chains into a
standing-up position. This consideration may serve as a
guide in the choice of suitable immiscible solvents for
cyclization experiments.

A direct experimental test of the advantage of cyclization by heterogeneous reactions can be made by precipita-
ting calcium or barium dicarboxylates from oil-water emul-
sions containing the high polymethylene dicarboxylic acids,
through the addition of lime or baryta water provided
that these cyclic salts are stable in the presence of polar
solvents and that their transition temperatures are sufficiently high. The physical characteristics of the samples
(such as x-ray diffraction patterns, densities, and wetting
property) and their yields of cyclic ketones on pyrolysis
can then be compared with the characteristics of, and the

-14-

yields of cyclic ketones from, the precipitates obtained
in the usual way through double-decomposition in homogeneous solutions.  Incidentally, if such cyclic salts can be
prepared, one should be able to obtain an answer to the
question   whether they are essential for cyclic-ketone
formation.

Since both the surface-film and the cyclization experiments require the use of high polymethylene dicarboxylic
acids and derivatives, for which no satisfactory general
method of preparation is yet available, a part of this
dissertation will be devoted to the description of a new
method for synthesizing such compounds. This new method
also appears to be quite general and useful for the
lengthening of hydrocarbon chains in the synthesis of aliphatic mono-carboxylic acids and hydrocarbons.

Since one of the essential steps in the new method of
synthesis is based upon preferential esterification of the
primary carboxyl groups in polycarboxylic acids carrying
primary and secondary carboxyl groups, the kinetics of
acid-catalyzed esterification of such acids as well as a
few polymethylene dicarboxylic acids will also be studied
so as to establish the ratios of rate constants for the
esterification of the primary and secondary carboxyl groups
and to find the optimum condition for carrying out the pre-

-15-

ferential esterification. This kinetic study also furnishes additional experimental support for the general
validity of Goldschmidt equation and his proposed mechanism
of hydrion-catalyzed esterification of carboxylic acids in
alcoholic solutions.

For the synthesis of *even-membered* polymethylene dicarboxylic acids below $C_{30}$, the well-known electrolytic method
based upon anodic decarboxylation and condensation of ester
salts, is more convenient. The need of considerable amounts
of the starting half-esters has led us to the development
of a new and much improved method for making such half-
esters (sebacic and adipic half-esters). The method is
based upon simultaneous partial esterification of the dicarboxylic acids and extraction of the half-esters from
the aqueous-alcoholic phase by a hydrocarbon solvent
in which the dicarboxylic acids are practically insoluble.
A fairly thorough theoretical study of the kinetics of
catalyzed esterification involving phase equilibria has
paved the way for the development of the optimum procedure.
The practical significance and further application of the
method will be discussed later.

Since this dissertation consists of several topics,
each one of which, though developed in related sequence,
can be considered as complete in itself, it is more conven-

-16-

lent for presentation to treat each topic in a separate chapter, together with the corresponding literature review, and discussion. These topics will be presented in the following order:

II. Kinetics of Acid-Catalyzed Esterification of Polycarboxylic Acids in Methanol Solutions; Preferential Esterification of Primary Carboxyl Groups in Polycarboxylic Acids.

III. Partial Esterification of Dicarboxylic Acids by Simultaneous Extraction of Half-Esters. A New Method for Preparing Half-Esters.

IV. Synthesis of High Polymethylene Dicarboxylic Acids and Glycohols. A new and General Method for the Lengthening of Hydrocarbon Chain in the Synthesis of Aliphatic Acids and Hydrocarbons.

V. Surface Films of High Polymethylene Dicarboxylic Acids and Glycols.

-17-

## Literature

(1) Adam, N. K., and Jessop, Proc. Roy. Soc. A, <u>120</u> 473 (1928).

(2) Carothers, W. H., and Hill, J. W., Journ. Amer. Chem. Soc., <u>55</u>, 5043 (1933).

(3) Gilman, H., "Organic Chemistry" 2nd Ed., John Wiley and Sons. Vol. 1, 79 (1943).

(4) Ruzicka, L., Stoll, M., and Schinz, H., Helv. Chem. Acta. <u>9</u>, 249-263 (1926).

(5) Stoll, M., and Rouve, A., ibid., <u>30</u>, 1822 (1947).

(6) Stoll, M., and Stoll-Comte, ibid., <u>13</u>, 1190 (1930).

(7) Ziegler, K., Ber. A, <u>67</u>, 139 (1934).

(8) Ziegler, K., Eberle, H., and Phlinger, H., Annalen <u>504</u>, 94 (1933).

-18-

## II. KINETICS OF HYDRION-CATALYZED ESTERIFICATION OF POLYCARBOXYLIC ACIDS IN METHANOL; PREFERENTIAL ESTERIFICATION OF PRIMARY CARBOXYL GROUPS IN POLYCARBOXYLIC ACIDS.

The kinetics of acid-catalyzed esterification of aliphatic acids has been the subject of extensive investigation.
The work of Goldschmidt (1), (2) is particularly important
in that it throws some light on the reaction mechanism.
Considering the presence of a protolytic exchange equilibrium between proton-alcohol, or proton-organic-acid, complex and water molecules, Goldschmidt was able to show that
the two possible reaction mechanisms,

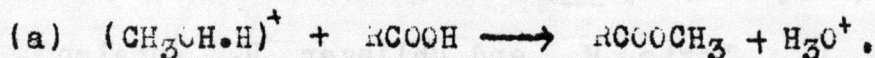

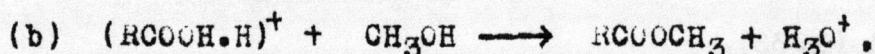

(a) $(CH_3OH \cdot H)^+ + RCOOH \longrightarrow RCOOCH_3 + H_3O^+$.

(b) $(RCOOH \cdot H)^+ + CH_3OH \longrightarrow RCOOCH_3 + H_3O^+$.

for hydrion-catalyzed esterification of carboxylic acids
in methanol solutions have different kinetic expressions,
and that experimental kinetic evidence supports mechanism
(a) as representing the rate-determining step. Later,
Hinshelwood showed that the esterification is also catalyzed
by the undissociated acid (5), but that this becomes insignificant in the presence of strong mineral acid catalyst
(10). More recently, a systematic study of the steric
effect on esterification rate constants of normal and
branched aliphatic acids has been reported by H. A. Smith

and his coworkers (6), (7), (8), (9);  their results also substantiate the validity of Goldschmidt equation.

However, no work has been done on the catalyzed esterification of aliphatic polycarboxylic acids carrying both secondary and primary carboxyl groups.  Although from known data of rate constants for mono-n-alkyl and di-n-alkyl acetic acids, it is possible to estimate the relative magnitudes of esterification rate constants for the primary and secondary carboxyl groups in a polycarboxylic acid, yet direct measurement of these relative magnitudes is desirable as this will enable us to determine more accurately whether preferential esterification of the primary carboxyl groups is practicable. This can be done by determining the initial rates of esterification of a polycarboxylic acid and its partial ester in which the primary carboxyl groups are fully esterified.

In the present investigation, the HCl-catalyzed esterification of (a) heneicosanetricarboxylic-acid-1,11,21 $HOOC(CH_2)_{10}CH(COOH)(CH_2)_{10}COOH$, m.p. 90-91°C., (b) dimethyl ester, $CH_3OOC(CH_2)_{10}CH(COOH)(CH_2)_{10}COOCH_3$, m.p. 62.5-63.5°C, (c) dotricontanetetracarboxylic-acid-1,11,22,32, $HOOC(CH_2)_{10}CH(COOH)(CH_2)_{10}CH(COOH)(CH_2)_{10}COOH$, m.p. 104-106°C, and (d) dimethyl ester, $CH_3OOC(CH_2)_{10}CH(COOH)$ $(CH_2)_{10}CH(COOH)(CH_2)_{10}COOCH_3$, m.p. 108-109.5 C, in methanol

-20-

solutions was studied, together with the catalyzed esterification of sebacic, and decamethylene-dicarboxylic acids.
These acids are chosen for study because they represent
three types of polycarboxylic acids in which the carboxyl
groups are spaced sufficiently apart. The purity of each
acid studied was established from its neutralization
equivalent weight. The synthesis and proof of structures
of the tri-acid, the tetra-acid, and their dimethyl esters
will be described in a later chapter. The sebacic acid
used was Eastman-Kodak pure product, twice recrystallized
from benzene, m.p. 141-142°C. The decamethylene-dicarboxylic acid was prepared in 65 % yield from ω-bromo
undecanoic acid through the nitrile synthesis, m.p. 129-130°C.
The methanol used was Eastman-Kodak reagent-grade product
carefully fractionated with a 3 ft. packed column.

The progress of the esterification was followed by
titrating pipetted samples with $CO_2$-free standard NaOH in
isopropanol-water (3:1) solution, using phenolphthalein as
indicator. The samples were pipetted into cold isopropanol-water solutions containing small amounts of the standard alkali in slight excess over the amount of acid catalyst
present in the samples. In some cases, inaccuracy in
locating the initial time (This may cause a systematic
drifting in the calculated values of $k_c$.) was avoided by

-21-

taking the first time and titration readings as $t_o$ and $C_o$ (organic acid catalyst), the extent of esterification that had occurred during the mixing of the organic acid solution and the catalyst acid solution was then calculated from $C_o$ and the true initial total acid concentration; this gave the amount of water produced, $\underline{n}$; values of $k_G$ were calculated from the Goldschmidt equation.

$$k_G = (1 + \frac{n+a}{r}) \frac{2.303 \log (a/C)}{(HCl)(t-t_o)} - \frac{(a-C)}{r(HCl)(t-t_o)} .$$

where $\underline{n}$ is the initial concentration of water in the methanol, $\underline{a}$ the initial equivalent concentration of the carboxylic acid, $\underline{c}$ the concentration at time $\underline{t}$, and $\underline{r}$ is the protolytic exchange "mass-action constant" for the system alcohol-$H_2O$-proton; the values of $\underline{r}$ at different temperatures are 0.11 at $0^o$, 0.22 at $22^o$, and 0.42 at $50^o C$, these being taken from Smith's paper (8).

Rate constants for each acid investigated were calculated both from the simple first-order rate equation and from Goldschmidt equation; in the case of the triacid and the tetra-acid, these calculated values were extrapolated to zero degree of esterification where $k_G$ and $(k)_{1st-order}$ converge to one value, indicating that the methanol used was free from water. For the tri-acid, this initial rate constant is equal to $(k_1 + \frac{1}{2}k_2)$ where $k_1$ and $k_2$ denote the

-22-

rate constants for the primary and the secondary carboxyl groups, respectively. For the tetra-acid, the initial rate constant thus determined by extrapolation is equal to $(k_1 + k_2)$. The average values of $k_G$ for the dimethyl esters of the tri-acid and the tetra-acid give $k_2$ for the corresponding acids.

In the same way, the esterification rate constants of the tricarboxylic acid and its dimethyl ester at $0^{\circ}$ (ice temperature) and $50^{\circ}$C. were determined and tabulated in Table VIII with the rate constants at $25^{\circ}$C. From these values the activation energies for the catalyzed esterification of the primary and secondary carboxyl groups were determined graphically by plotting values of $-\log k$ against $1/T$. In each case, the three points fall nicely on a straight line.

**Table I. HCl-catalyzed Esterification of Sebacic Acid in Methanol.**
**Sebacic Acid (0.1000 M), HCl (0.0100 M).**
**Temperature: 25.00 _ 0.05°C**

| Time, t (seconds) | Total Acid Concentration (Equiv. per L.) | Conc'n Organic Acid, C | $k \times 10^2$ (L. per Equiv. per second) (1st-order) | $k_G \times 10^2$ (L. per Equiv. per second) (Goldschmidt) |
|---|---|---|---|---|
| 0 | 0.2100 | 0.2000 | | |
| 482 ($t_o$) | 0.1822 | 0.1722 (a) | | |
| 1062 | 0.1584 | 0.1484 | 2.56 | 3.03 |
| 1543 | 0.1426 | 0.1326 | 2.46 | 3.00 |
| 2005 | 0.1293 | 0.1193 | 2.41 | 3.03 |
| 2550 | 0.1162 | 0.1062 | 2.33 | 3.01 |
| 3283 | 0.1020 | 0.0920 | 2.24 | 2.98 |
| 4105 | 0.0885 | 0.0785 | 2.17 | 2.97 |

Average: $(k_1)_G = 3.00 \times 10^2$ L. per Equiv. per second.

Table II. Esterification of Decamethylene Dicarboxylic Acid
in Methanol at 25.00 _ 0.05°C.
Catalyst: HCl(0.00947 M).

| Time, t (seconds) | Conc'n Organic Acid, C Equiv. per L.) | $k = \dfrac{2.303 \log(a/C)}{[HCl](t - t_o)}$ | $k_G = (1+\dfrac{n+a}{r})k$ $- \dfrac{(a - C)}{[HCl]/r(t-t_o)}$ |
|---|---|---|---|
| 0 | 0.0924 (n a) | | |
| 609($t_o$) | 0.0766 (a) | | |
| 1198 | 0.0654 | $2.83 \times 10^{-2}$ | $3.10 \times 10^{-2}$ |
| 1409 | 0.0622 | 2.75 | 3.03 |
| 2150 | 0.0524 | 2.60 | 2.93 |
| 3524 | 0.0385 | 2.49 | 2.91 |

Average: $(k_1)_G$ 2.99 × 10$^{-2}$
L. per Equiv. per second.

Table III. Ester'n of Heneicosanetricarboxylic-acid Dimethyl Ester
in Methanol at 25.00 _ 0.05°C.
$C_o = (C_2)_o = a_2 = 0.02000N$.
Catalyst: HCl, c = 0.0100N.*

| Time, t (seconds) | Conc'n Organic Acid $C_2$ (Equiv. per L.) | $(k_2)_{1st-order}$ | $(k_2)_{Goldschmidt}$ |
|---|---|---|---|
| 0 | 0.02000 ($a_2$) | | |
| 38940 | 0.0172 | $3.87 \times 10^{-4}$ | $3.90 \times 10^{-4}$ |
| 51060 | 0.0164 | 3.89 | 3.93 |
| 58020 | 0.0159 | 3.95 | 3.99 |
| 70800 | 0.0153 | 3.81 | 3.85 |

Average: $(k_2)_G = 3.92 \times 10^{-4} \_ k_2$
L. per mol per second.

*Note: Dilute HCl in methanol at 25° was found to decrease in strength by about 0.8% in 12 × 3600 seconds. Corrections have been applied in calculating $C_2$ from the total acid concentrations.

## Table IV. Ester'n of Heneicosanetricarboxylic-acid-1,11,21 in Methanol at 25.00 ± 0.05°C.
### $C_0 = 0.1000M$; /HCl/ = c = M/100.3

| Time, t (seconds) | Conc'n Org. Acid, C (Equiv./L.) | Conc'n Primary Carboxyl, $C_1$, Assumed Secondary Carboxyl Unester'd | $k_1 \times 10^2$ (1st-order) | $(k_1)_G \times 10^2$ (Goldschmidt) |
|---|---|---|---|---|
| 0 | 0.3000 | 0.2000 ($a_1$) | | |
| 523 | 0.2734 | 0.1734 | 2.73 | 2.91 |
| 1070 | 0.2518 | 0.1518 | 2.58 | 2.88 |
| 1620 | 0.2330 | 0.1330 | 2.51 | 2.93 |
| 2200 | 0.2192 | 0.1192 | 2.35 | 2.83 |
| 3104 | 0.1984 | 0.0984 | 2.29 | 2.88 |
| 4298 | 0.1786 | 0.0786 | 2.17 | 2.87 |

By extrapolation, $(k_1)_G$, obs. = $2.89 \times 10^{-2}$ L. per equiv. per second.

From Tables III and IV, $k_1 = (k_1)_o$, obs. $-\frac{1}{2}k_2$
$= 2.87 \times 10^{-2}$ L. per Equiv. per second.

## Table V. Dotricontanetetracarboxylic-acid-1,11,22,32 Dimethyl Ester HCl-catalyzed Esterification in Methanol.
### Temperature: 25 ± 0.05°C. /HCl/ = M/100.

| Time, t (seconds) | Total Acid (Equiv./L.) | /HCl/ = c | Conc'n Org. Acid $C_2$ (Equiv./L.) | $\frac{/HCl/}{c}$ av. | $(k_2)$ 1st.ord $\times 10^4$ | $(k_2)_G \times 10^4$ |
|---|---|---|---|---|---|---|
| 0 | 0.0206 | 0.0100 | 0.0106 | 0.0100 | | |
| 32940 | 0.0193 | 0.0094 | 0.0094 | 0.0100 | 3.76 | 3.78 |
| 72800 | 0.0180 | 0.0098 | 0.0082 | 0.0099 | 3.65 | 3.67 |
| 179400 | 0.0162 | 0.0097 | 0.0065 | 0.0098 | (2.78) | (2.81)* |

Mean: $(k_2)_G = 3.73 \times 10^{-4}$

*Note: This low value probably due to partial hydrolysis of primary ester groups.

-26-

### Table VI. Ester'n of Dotricontanetetracarboxylic-acid-1,11,22,32 in Methanol at 25.00 _ 0.05°C. Catalyst: HCl, c = 0.0100 N.

| Time, t (seconds) | Conc'n Org. Acid (Equiv./L.) | Conc'n Primary Carboxyl, $C_1$ (Assumed Secondary Carboxyls Unesterified) | $k_1 \times 10^2$ (1st. Ord.) | $(k_1)_{G,obs} \times 10^2$ |
|---|---|---|---|---|
| 0 | 0.0976 | 0.0488 ($a_1$) | | |
| 676 | 0.0890 | 0.0402 | 2.85 | 2.91 |
| 1222 | 0.0832 | 0.0344 | 2.83 | 2.93 |
| 1875 | 0.0775 | 0.0287 | 2.80 | 2.94 |
| 2768 | 0.0711 | 0.0223 | 2.80 | 2.99 |

By extrapolation, $(k_1)_{o,obs.} = 2.88 \times 10^{-2}$ L. per mol per second.

From Tables V and VI, $k_1 = (k_1)_{o,obs.}$ $-k_2 = 2.84 \times 10^{-2}$ L. per mol per second.

### Table VII. Specific Rates and Activation Energies for Hydrogen-ion-catalysed Esterification of Heneicosanetricarboxylic-acid-1,11,21 in Methanol Solution.

| | Temperature °C. | Specific Rate (L. per Equiv. per Second) | Activation Energy, Kilo-calories per Equivalent |
|---|---|---|---|
| Primary Carboxyl Groups | 0° | $5.78 \times 10^{-3}$ ($k_1$) | |
| | 25.00 $\pm$ 0.05 | $2.87 \times 10^{-2}$ ($k_1$) | 10.4 |
| | 50.0 $\pm$ 0.1° | $1.14 \times 10^{-1}$ ($k_1$) | |
| Secondary Carboxyl Group | 0° | $4.81 \times 10^{-5}$ ($k_2$) | |
| | 25.00 $\pm$ 0.05 | $3.92 \times 10^{-4}$ ($k_2$) | 13.8 |
| | 50.0 $\pm$ 0.1° | $2.40 \times 10^{-3}$ ($k_2$) | |

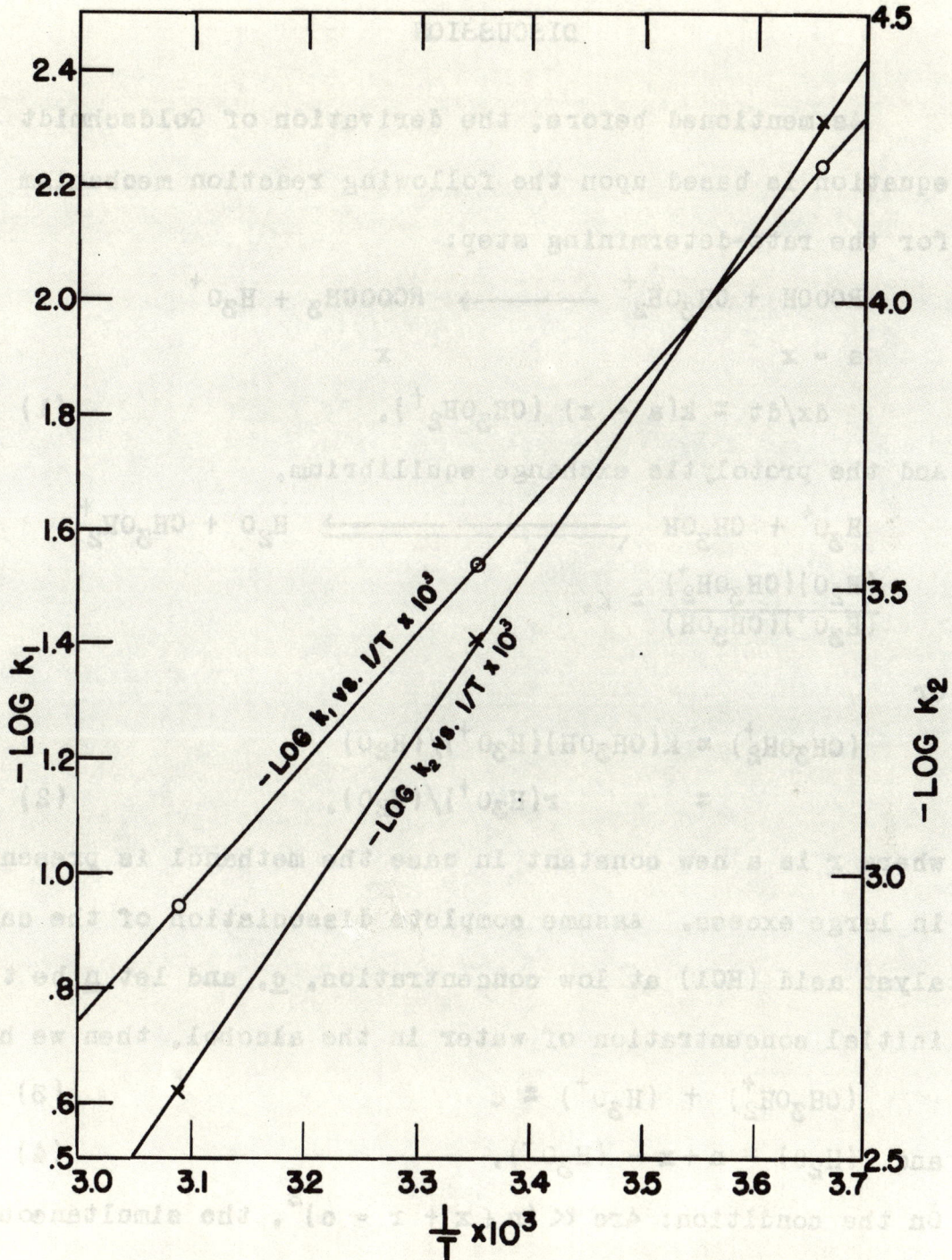

FIG. II. CHANGE IN RATE CONSTANTS WITH TEMPERATURE
FOR HYDROGEN ION CATALYZED ESTERIFICATION OF
THE PRIMARY AND SECONDARY CARBOXYL GROUPS
IN HENEICOSANETRICARBOXYLIC ACID −1,11,21

-28-

## DISCUSSION

As mentioned before, the derivation of Goldschmidt equation is based upon the following reaction mechanism for the rate-determining step:

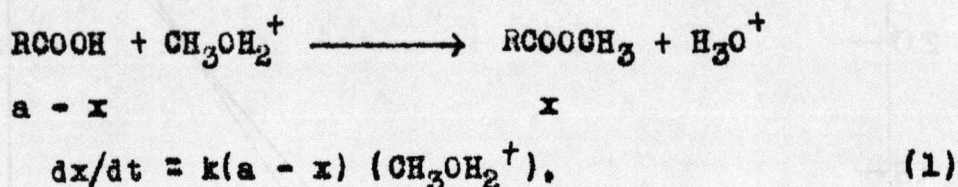

$$RCOOH + CH_3OH_2^+ \longrightarrow RCOOCH_3 + H_3O^+$$

$$a - x \qquad\qquad\qquad x$$

$$dx/dt = k(a - x)(CH_3OH_2^+), \qquad\qquad (1)$$

And the protolytic exchange equilibrium,

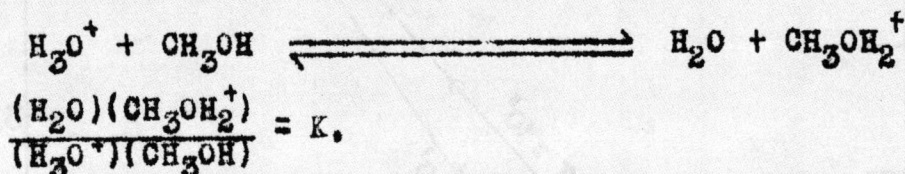

$$H_3O^+ + CH_3OH \rightleftharpoons H_2O + CH_3OH_2^+$$

$$\frac{(H_2O)(CH_3OH_2^+)}{(H_3O^+)(CH_3OH)} = K,$$

or

$$(CH_3OH_2^+) = K(CH_3OH)(H_3O^+)/(H_2O)$$

$$= r(H_3O^+)/(H_2O). \qquad\qquad (2)$$

where $r$ is a new constant in case the methanol is present in large excess. Assume complete dissociation of the catalyst acid (HCl) at low concentration, $c$, and let $n$ be the initial concentration of water in the alcohol, then we have

$$(CH_3OH_2^+) + (H_3O^+) = c \qquad\qquad (3)$$

and $(H_2O) = n + x - (H_3O^+).$ $\qquad\qquad (4)$

On the condition: $4rc \ll (n + x + r - c)^2$, the simultaneous equations (2), (3), and (4) give

$$(CH_3OH_2^+) \approx rc/(n + x + r - c). \approx rc/(n + x + r). \qquad (5)$$

-29-

Whence from (1) and (5)

$$dx/dt = k(a - x)rc/(r + n + x) \tag{1}'$$

$$k_G = (1 + \frac{n+a}{r}) \frac{2.303 \log}{ct} \left(\frac{a}{a-x}\right) - \frac{x}{rct} \tag{6}$$

r can be calculated from two sets of values of x and t. Its
value has been found by previous investigators (1), (8),(9),
to be reasonably constant for different carboxylic acids;
this is a strong support for mechanism (a). It is well
known that HCl is a strong electrolyte in methanol solution,
but a very weak electrolyte in glacial acetic acid (3),(4),
hence it is reasonable to assume that, in the presence of
small amount of carboxylic acid and large amount of alcohol,
the HCl is almost entirely tied up with the alcohol and
that the proton-alcohol complex should play an important
role in the esterification catalysis. However, as pointed
out by Hinshelwood from collision frequency and probability
considerations (5), the observed magnitudes of rate constant
and activation energy for HCl-catalyzed esterification of
acetic acid in methanol do not exclude the possibility of
a termolecular mechanism involving an unprotonated alcohol
molecule as well as the proton-alcohol complex and the
organic acid molecule.

By rewriting the Goldschmidt equation as

$$k_G = \left(1 + \frac{n}{r}\right) \frac{1}{ct} \log_e \left[a/(a-x)\right] = \frac{-a \log_e(1 - x/a) - x}{rct}$$

-30-

and expanding the right-hand side into power series of $x$,
it is readily seen that

$$k_G - (1 + n/r) k_{1st\ order} \approx 0$$

as $x$ approaches zero, since $dx/dt$ remains finite. Hence
if $n$ is zero, $k_G$ and $k_{1st\ order}$ should converge to one
value.

In the calculation of the rate constants for the polycarboxylic acids, it was tacitly assumed that the two primary carboxyl groups in each of the four acids studied have
practically the same rate constant, $k_1 \approx k_1'$. The fact
that the observed rate constants for the dicarboxylic acids
show only very slight drift with $x$ justified this assumption. Self-catalysis of the undissociated-acid catlysis
type, due to intra-molecular collisions of the two carboxyl
groups, may tend to make $k_1$ (first step) larger than $k_1'$
(second step), but the effect appears to be negligible.
However, the effect may become considerable in the case of
succinic and malonic acids.

To consider the effect of hydrolysis on the observed
rate of esterification, we have to add one more term to
the right-hand side of equation $(1)'$

$$dx/dt = [k(a - x)\ rc/(r+n+x)] - [k_h x(n + x)\ c/(r+n+x)], \quad (1)''$$

where $k_h$ is the specific rate of hydrolysis,

$$ROOOCH_3 + H_3O^+ \xrightarrow{\quad k_h \quad} ROOOH + CH_3OH_2^+.$$
$$(x) \qquad\qquad\qquad\qquad (a-x)$$

-31-

The $k_h$ term is readily obtained since $(H_3O^+) = c - (CH_3OH_2^+)$

$$= c - \frac{rc}{r+n+x}$$

$$= \frac{c(n+x)}{r+n+x}$$

It can be shown that

$$k/k_h = k_e/k = k_e (CH_3OH)/r = 4 \times 25 \div 0.22 = 4.5 \times 10^2$$

where $k_e$ is the esterification equilibrium constant whose value is known to be about 4 for most carboxylic acids and alcohols (4.5 for adipic acid and methanol at 73°C, and for sebacic acid and methanol at 75°C, as shown in the next chapter). Hence the $k_h$ term in (1)'' is negligible unless $x(n+x)/(a - x)r$ becomes very large. In dry methanol where n is zero, the error due to hydrolysis will still be less than 2 % even up to $x/(a - x) = 9$, or 90% esterification of a 0.2 N solution. However, in the case of determining $k_2$ with the ester-acids, $k_2$ is only about 6 times as large as $k_h$ of the primary ester groups; if the kinetic measurement is extended to high degree of esterification, or if the alcohol used is not sufficiently dry, the primary ester groups will be hydrolyzed to some extent.

The observed esterification rate constants of the primary carboxyl groups in the tri- and tetra-carboxylic acids are seen to be about the same as that of a higher normal fatty acid (e.g. lauric acid, $k = 3.05 \times 10^{-2}$ at 25°). Energy of activation for the primary carboxyl groups in the tricarboxylic acid ($C_{24}$) is about 4% higher than the average value for a normal fatty acid (8). Thus the presence of a secondary carboxyl group in the middle of a long chain has very little effect on the rate constant of the primary carboxyl groups. The rate constant of the secondary carboxyl groups in a polycarboxylic acid is also seen to be about the same as that of a di-n- alkyl acetic acid (e. g. dipropyl acetic acid, $k_{25} = 4.25 \times 10^{-4}$) (7). The observed activation energy for the secondary carboxyl group in the tri-acid is considerably higher than that for dibutyl acetic acid (12.9 k cals per mole) reported by Smith (6).

For the tricarboxylic acid, the ratio of $k_1$ and $k_2$ is 73:1; for the tetracarboxylic acid, the ratio is 77:1. It can be assumed that for other polycarboxylic of the same type, $k_1/k_2$ will be above 70. From the simultaneous rate equations

$$dx_1/dt = k_1(a_1 - x_1) \cdot \frac{r c}{(r + n + x_1 + x_2)}$$

$$dx_2/dt = k_2(a_2 - x_2) \cdot \frac{r c}{(r + n + x_1 + x_2)},$$

we have

$$\frac{\log(1-x_1/a_1)}{\log(1-x_2/a_2)} = \frac{\log(1 - f_1)}{\log(1 - f_2)} = K_1/k_2 > 70;$$

it can readily be seen that when the primary carboxyl groups are 90% esterified, ($x_1/a_1 = f_1 = 0.9$), the secondary ones are only about 3% reacted, ($x_2/a_2 = f_2 \cong 0.03$); hence it

is practicable to have the primary carboxyl groups preferentially esterified. In practice, this is greatly assisted
by low solubilities of the partial esters in methanol. In
the case of the tricarboxylic acid ($C_{24}$) and the tetracarboxylic acid ($C_{36}$) studied, the dimethyl esters precipitated from the solution during the course of the preferential
esterification if  fairly concentrated solutions of the
acids are started with. Since polycarboxylic acids of
these types, or dicarboxylic acids with a secondary carboxyl
group, can be prepared by malonic ester synthesis, and
since the unesterified secondary carboxyl groups can be
readily removed by means of the bromine-silver-salt reaction, the possibility of preferentially esterifying the
primary carboxyl groups opens a new way of lengthening the
hydrocarbon chains in the synthesis of aliphatic acids. This
will be illustrated in a later chapter.

## SUMMARY

The kinetics of HCl-catalyzed esterification of
sebacic, decamethylene dicarboxylic acid, heneicosanetricar-
boxylic-acid-1,11,21, dotricontanetetracarboxylic-acid,1,
11,22,32, and the dimethyl esters of the latter two acids
in methanol have been studied. The rate constants for the
primary and secondary carboxyl groups in the tri- and
tetra-carboxylic acids have been determined by means of

-34-

Goldschmidt equation; these values have been found to be $k_1 = 2.87 \times 10^{-2}$, $k_2 = 3.92 \times 10^{-4}$, for the tri-acid; and $k_1 = 2.84 \times 10^{-2}$, $k_2 = 3.73 \times 10^{-4}$, for the tetra-acid. The practicability of preferential esterification and its application in the synthesis of long aliphatic acids have been discussed.

LITERATURE

(1) Goldschmidt, H. and Udby, O., Zeit. Physik. Chem., 60, 728 (1907).

(2) Goldschmidt, H. and Theusen, A., ibid., 81, 30 (1912).

(3) Heston, B. O., and Hall, N. F., Journ. Am. Chem. Soc., 56, 1462 (1934).

(4) Kolthoff, I. M., and William, A., ibid., 56, 1007. (1934).

(5) Rolfe, A. C., and Hinshelwood, C. N., Trans. Farad. Soc., 30, 935, (1934).

(6) Smith, H. A., Journ. Amer. Chem. Soc., 61, 254, (1939), 62 1136 (1940).

(7) Smith, H. A. and Burn, J., ibid., 66 (1494, (1944).

(8) Smith, H. A., and Reichardt, C. H., ibid., 63, 605 (1941).

(9) Smith, H. A., and Steele, J. H., ibid., 63, 3466 (1941).

(10) Williamson, A. T., and Hinshelwood, C. N., Trans. Farad. Soc. 30 1145 (1934).

-55-

## III: PARTIAL ESTERIFICATION OF DICARBOXYLIC ACIDS BY SIMULTANEOUS EXTRACTION OF HALF ESTERS: A NEW METHOD FOR PREPARING HALF-ESTERS.

In the synthesis of high polymethylene dicarboxylic acids by either the electrolytic method or the new method to be described in the following chapter, an essential step consists of converting a lower dicarboxylic acid into the half-ester which can then be electrolyzed in the form of ester-salt or be converted into the ω-bromo ester for the malonic ester condensation. Because of other uses of the half-esters in organic synthesis, numerous methods have been described in literature for converting dicarboxylic acids or diesters into the half-esters; these are based upon (a) partial saponification of the di-esters (1), (5), (7), (14), or (b) partial esterification of the di-acids, or (c) half-esterification of the anhydrides of the dicarboxylic acids. Method (c) generally gives excellent yields (2), (9) but its application is limited to a few dicarboxylic acids which readily form anhydrides. Method (a) works very well with esters of the higher dicarboxylic acids ($C_{12}$ or higher), by the use of barium hydroxide in hot methanol; but gives low yields with sebatic and adipic esters (6). For method (b), several empirical procedures have been described in literature (1), (2), (6), (12), (13);

these consist of esterifying the acids with limited amounts of alcohols, or with a mixture of alcohol and concentrated aqueous hydrochloric acid, and give around 35% yield of the half-esters (7), although by recirculation of the full esters obtained as side-products the yields of the half-esters can be considerably increased (13). Separation of the reaction products by method (b) is commonly done by fractionation under reduced pressure; this becomes very laborious for the higher-membered acids, like sebacic and azelaic acids, for the boiling points of the reaction products are too high and too close and further interaction may occur during the fractionation.

In this chapter, a new method for making mono-methyl sebacate and adipate will be described; this is based upon catalyzed esterification of the acids in aqueous methanol in the presence of an immiscible solvent which continuously extracts the half-esters from the aqueous-alcoholic phase. The extracting solvent used is skelly-solve C and tuluene, in which the lower dicarboxylic acids are practically insoluble even in the hot, while the half-esters and full esters are soluble; skellysolve C has a limited miscibility with methanol; so most of the methanol remains in the aqueous phase, together with prac-tically the entire amount of the catayst acid, HCl. With

-37-

suitable proportions of the reacting components, a 63-67% yield of pure mono-methyl sebacate can be obtained in one operation. The half-ester readily crystallizes out when the concentrated hydrocarbon layer is cooled in ice; so the process can be made continuous if necessary, in which case, the half-ester can be maintained at low concentration and the yield is expected to be even better. For mono-methyl adipate, toluene-skellysolve (7:1) is used as the extracting solvent and a 70-75% can be easily obtained by saturating the aqueous layer with NaCl. The operation is rate-controlled, but slight overstepping causes no harm, since the concentration of the half-ester in the aqueous phase is low.

The principle involved in the rate-controlled partial esterification with or without the extracting solvent is of theoretical interest, and will be considered in some detail.

## THEORETICAL

As shown in the preceding chapter, the kinetics of catalyzed esterification of polycarboxylic acids follows Goldschmidt equation (3). For the polymethylene dicarboxylic acids, the consecutive reactions are:

$$A\text{-}A + CH_3OH_2^+ \longrightarrow A\text{-}E + H_3O^+$$
$$(a\text{-}x\text{-}y) \qquad\qquad (x)$$

-38-

$$a-E + CH_3OH_2^+ \xrightarrow{k'} E-a + H_3O^+$$

(x)                          (y)

$$dx/dt = 2k\ (a-x-y)\ (CH_3OH_2^+) - k'x\ (CH_3OH_2^+)$$

$$= \left[2k\ (a-x-y)\ rc\ /\ (r+n+x+2y)\right]-\left[k'xrc\ /(r+n+x+2y)\right].$$

(1)

$$dy/dt = k'xrc/(r+n+x+2y).$$ (2)

where $\underline{r}$ is the protolytic exchange 'mass-action constant' implying the constant alcohol concentration, $\underline{n}$ is the initial molar concentration of water in the alcohol, and $\underline{c}$ is the concentration of the catalyst acid. The derivation follows equations (2), (3), (4), (5) of chapter II. It is possible to obtain a solution in series for these simultaneous differential equations, but this is not necessary if we are only interested in knowing the maximum yield of the ester-acid, a-E, permitted by theory. To indicate the generality of the case, equations (1) and (2) can be written as

$$dx/dt = 2k(a-x-y)\ f(x,y,c),\ -kxf(x,y,c),$$ (1)'

and

$$dy/dt = k'\ xf(x,y,c).$$ (2)'

where $f(x,y,c)$ is by its physical significance a non-vanishing function.

Whence

$$dx/dy = (2k/k')(a-y)/x - \left[(2k/k') + 1\right]$$

$$= n(a-y)/x \qquad - (R + 1)$$ (3)

-39-

where $R = 2k/k'$. Integrating equation (3) after making the substitution

$$x = (a-y)v, \qquad (4)$$

we have

$$\log_e \{a/(a-y)\} = \{1/(R-1)\} \{R \log_e (R-v)/R - \log_e(1-v)\}, \qquad (5)$$

using the boundary condition: $x = 0$, $y = 0$, and $v = 0$, at $t = 0$, in determining the integration constant.

The condition for maximum $\underline{x}$ is $dx/dt = 0$; hence by equation (1)',

$$(a-x-y)/x = 1/R \qquad \bigg|_{dx/dt = 0},$$

or in view of equation (4),

$$v = x/(a-y) = R/(R+1) \qquad \bigg|_{dx/dt = 0}. \qquad (6)$$

Hence from equations (5) and (6) we have, for maximum $\underline{x}$:

$$\frac{y}{a} = 1 - \frac{1}{R^{\frac{R}{(R-1)}}}, \qquad (7)$$

and

$$\frac{x}{a} = \frac{1}{R^{(\frac{1}{R-1})}}. \qquad (8)$$

Now, $(x/a) \times 100\%$ is the percentage yield of the half-ester, a-E, and $(x+2y)/2a$ is the total degree of esterifi-

-40-

cation, hence the maximum yield of the half-ester and the corresponding degree of esterification can be calculated from the ratio of the rate constants, $\underline{k}$ and $\underline{k}'$, for the first and second steps of catalyzed esterification of the dicarboxylic acid. These are seen to be independent of the time variable and the nature of the function $f(x,y,c)$. It is obvious that this theoretical calculation can be applied just as well to the hydrolysis or saponification of diesters, or to any other stepwise reactions involving a bi-functional molecule (symmetrical); for the function may represent the concentration of hydronium ions, of hydroxyl ions, or of any other reagents. The calculated values of $(x/a)_{max.}$ x 100% and $(x+2y)/2a$ for different values of $R = 2k/k'$ are shown in table I.

Table I. Maximum Yields of Half-Esters (or Ester-Salts) in Relation to Ratios of Rate Constants for the First and Second Steps of Esterification of Dicarboxylic Acids (or Hydrolysis, or Saponification of Dicarboxylic Esters).

| Ratio of Rate Constants, $R = 2k/k'$ | 2 | 4 | 6 | 8 | 10 | 20 | 100 |
|---|---|---|---|---|---|---|---|
| Maximum Yield of Half-Ester, % | 50 | 63 | 70 | 74 | 77 | 85 | 95 |
| Optimum Degree of Esterification | .50 | .53 | .53 | .53 | .54 | .53 | .51 |

-41-

The above relation between maximum x/a and k/k' is in accord with probability considerations, for the function f(x,y,c) may be regarded as equivalent to any mechanical operation which is repeated in the second step of the consecutive transformations.  Hence the maximum concentration of the intermediate is dependent only on its relative <u>a priori</u> probability in undergoing the transformation.  The result for <u>k</u> = <u>k</u>' is obvious by such simple reasoning.

Direct verification of the above relation can be obtained by considering one of the simplest cases of consecutive reactions:  the esterification of a simple dicarboxylic acid in <u>dilute dry</u> methanol solution.  For very low concentration of the dicarboxylic acid, the Goldschmidt's correction for the water produced may be neglected, and the kinetic differential equations are first-order in which the constant concentration of the catalyst acid may be incorporated in the two rate constants, k and k':

$$dx/dt = 2k(a - x - y) - k'x, \qquad (9),$$

and

$$dy/dt = k'x, \qquad (10).$$

The solution of these equations is exactly the same as that derived by Meyer (10) for the hydrolysis of dicarboxylic esters in excess water, but it can be more readily obtained by differentiating the first equation with <u>t</u>.

eliminating the dy/dt term by means of the second equation, solving the resulting second-order differential equation with constant coefficients, and substituting the boundary conditions; $x = 0$, $y = 0$, at $t = 0$; and $x = 0$, $y = a$, at $t = $ infinity. The result is

$$x = 2a \ (e^{-k't} - e^{-2kt}),$$

and

$$y = a \ (1 + e^{-2kt}) - 2 \ a \ e^{-k't}$$

From these equations, relation (7) and (8) can be readily verified. Note that $\underline{2k}$ here is equivalent to $\underline{k}$ commonly reported in literature since we have regarded each carboxyl group separatedly for the convenience in developing the kinetic expressions.

It is interesting to note that the optimum degree of overall reaction lies very close to 0.50, for all ratios of rate constants ($2k/k' \gtreqless 2$). In the case of rate-controlled partial esterification, or hydrolysis, it is therefore possible to use aqueous-alcohol solution of fairly wide composition range, so long as this does not allow an appreciable extent of the reverse reaction to take place before the optimum degree of reaction is reached. In the case of saponification, which is an irreversible reaction, the problem is very simple: no rate control is necessary; just mix the diester with an amount of alkali

solution as required for the optimum degree of overall
saponification, and allow the reaction to proceed until the
solution is neutral to phenolphthalein. However, it is
essential to success to add enough solvent (alcohol) so as
to keep the alkali and the diester in one phase. Otherwise,
the ester-salt being soluble in water will be further
attacked by the alkali.

It is to be noted also that if the ratios of rate constants, k/k', both for esterification and for hydrolysis
are equal to unity, then the optimum degrees of reaction
are the same, namely 0.50, and the sum is unity; under
this condition, the maximum yields of half-esters permitted
by theory are the same both for rate-controlled methods
and for equilibrium methods. This condition is generally
met with in the case of polymethylene dicarboxylic acids
higher than $C_3$; for these acids, k/k' are all very close
to unity (e.g., 2k/k' = 1.98 and 1.90 for Me and Et succi-
nate respectively). Hence a 50% yield of half-esters is
about the best result obtainable with both methods, provided that no phase separation takes place. In practice,
when a dicarboxylic acid is esterified with limited amount
of alcohol (Since the optimum degree of esterification is
about 0.50, a simple calculation will show that 4/3 to 5/4
mols of alcohol to each mol of diacid is theoretically the

best ratio, assuming that the esterification equilibrium constant lies between 4 and 5 and that the reaction is homogeneous), separation of an aqueous, or an ester phase usually takes place, especially with the higher di-acids; unless a suitable inert solvent is added (13). With prolonged reflux, phase separation tends to produce more di-ester at the expense of the di-acid and the half-ester.

The ratios of the first and second saponification rate constants for the esters of the lower polymethylene dicarboxylic acids are very large, and therefore favorable for stepwise saponification. This is due to the internally propagated polar effect and externally propagated electrostatic effect (Debye-Hueckel type) of the carboxylate anion, produced in the first step of saponification, on the hydroxyl ions approaching the remaining ester group. The following data for the ratios of saponification rate constants are taken from Ingold's paper (8), which contains a good collection of such data from several different sources. Comparison of Tables I and II will show that the method of partial saponification of the di-esters with calculated amounts of alkali should offer distinct advantage over the method of partial esterification, for the preparation of half-esters or ester-salts of the lower dicarboxylic acids. By the method of partial saponification,

-45-

a 70% (glutarate) to an almost quantitative yield (oxalate)

of half-esters can be expected, in view of Tables I. and

Table II. Ratios of First and Second Rate Constants for the
Saponification of Polymethylene Dicarboxylic Esters.

| Ester Saponified | | Ref. | k | 2k/k' |
|---|---|---|---|---|
| Oxalate | Me | (d) | $10^6$ | 104 |
| | Et | (e) | $10^{5.5}$ | |
| Malonate | Me | (c) | 170 | 91 |
| | Et | (b) | 112 | |
| | | (c) | 136 | |
| Succinate | Me | (c) | 20.5 | 9.7 |
| | | (f) | 26.0 | |
| | Et | (a) | 13.8 | |
| | | (c) | 16.5 | |
| Glurate | Me | (f) | 21.6 | 6.4 |
| | Et | (f) | 10.0 | |
| Suberate | Me | (f) | 1.53 | 3.1 |
| Azelate | Me | (f) | 1.24 | 2.95 |
| Sebacate | Me | (f) | 1.20 | 2.80 |

Original Sources of Reference:

(a) Knoblauch, Zeit. physik, Chem. 67, 257 (1909).

(b) Goldschmidt and Scholz, Ber. 26, 1333 (1903).

(c) Meyer, J., Zeit. Physik. Chem., 67, 257 (1909).

(d) Skrabal, Monatsch., 38., 29 (1917).

(e) Skrabal and Matievic, Ibid., 39, 765 (1918).

(f) Skrabal and Singer, Ibid., 41, 339 (1920).

II.  It should not be difficult to verify this prediction experimentally by successive extractions of the full ester and half-ester (after liberation with dilute acid) from the partially saponified mixture, with ether and petroleum ether, and determining the amount of each component.  The lower dicarboxylic acids are expected to be completely insoluble in petroleum ether.

For the higher dicarboxylic acids, the ratios of the first and second esterification, hydrolysis (10) or saponification constants (8) are all very close to unity, hence good results cannot be expected with ordinary procedures. An improved procedure consists of adding diester to the starting mixture together with some dibutyl ether which is added to prevent the separation of two phases.  In this way, Swann and coworkers claimed to have prepared sebacic and adipic mono-ethyl esters in 60-65% and 71-75% yields, respectively (13).  Only the first datum has been checked by Smith and Clegg (13).  It is pertinent here to make a critical analysis of such a procedure.  For adipic and sebacic acids, k and k' are practically equal, hence $R = 2$, and by equations (7) and (8), the maximum yield of the half-esters corresponds to $x/a = 0.50$, $y/a = 0.25$, and $(a-x-y)/a = 0.25$.  These represent the mol fractions of the half-ester, the full ester, and the diacid, respectively, in the equil-

-47-

ibrium mixture. (The proportions of alcohol and water
should be so adjusted that their concentration ratio at
this point is n/m * $K_e$ since the optimum degree of esterification is 0.50.) It is therefore obvious that if the
entire amount of the di-ester added is to be recovered, then
the mol ratio of the diester and di-acid in the starting
mixture should be 1:3, and that the maximum yield of the
half-ester predicted by theory is only 66.7%. The addition
of diester may prove to be advantageous for semi-continuous
process, but not for single-batch operation since the diester
has to be made separately and its addition also increases
the duty of final separation.

Thus we have seen that the partial saponification
methods are more promising to give good yields in the case
of $C_2$ to $C_5$ diacids, and that the 'normal baryta' method
takes care of the di-acids above $C_{10}$; but for the gap from
$C_6$ to $C_{10}$, which includes two technically important di-
acids, none of the numerous existing methods of half-esterification is entirely satisfactory.

The method to be described in this chapter is designed
to fill this gap; it aims at increasing the yields of the
half-esters and at simplying the procedure for separating
the reaction products in the case of the higher di-acids
where fractionation becomes difficult. In this method, the

-48-

esterification is carried out in the presence of an immiscible solvent, which continuously and preferentially extracts the half-ester and the ester from the aqueous-alcoholic phase in which the esterification is taking place, thus increasing the yield of the half-ester by lowering its concentration in the reaction mixture. A more precise theoretical analysis of this method is given below.

When the aqueous solution of an organic substance is extracted with an immiscible organic solvent, the concentration of the substances remaining in the aqueous phase after phase equilibrium has been established is related to the initial concentration $\underline{C}$, the relative volumes of the two phases, $V_{org} : V_W$ and the distribution constant, $K = C_{org}/C_w$, by the following equation

$$C_W = \frac{V_W}{V_W + K\,V_{org}} \cdot C = \frac{1}{1 + K(V_{org}/V_W)} \cdot C$$
$$= p \cdot C,$$

where $\underline{p}$ is the new partition coefficient involving the constant volume ratio. The above equation follows readily from a material balance of the organic substance before and after extraction. Hence if the esterification takes place in the presence of an immiscible organic solvent, the concentrations of the half-ester, the full ester, and the diacid in the aqueous phase at any instant $\underline{t}$ are, respectively.

-49-

given by the following equations:

$$x_w = p_1 x,$$

$$y_w = p_2 y,$$

and

$$(a-x-y)_w = p_0 (a-x-y).  \tag{11}$$

where $p_1$, $p_2$, and $p_0$ are the partition coefficients for the three compounds, and $x$, $y$, and $a$ follow the same notations as in equation (1). Since practically the entire amount of the dissociated mineral acid catalyst will remain in the aqueous phase, the esterification will also be practically confined to this phase. Hence the total rate of esterification is virtually given by the following differential equations:

$$dx/dt = 2k \, p_0(a-x-y) \cdot f(x,y,c)_w - k'p_1 x \cdot f(x,y,c)_w.  \tag{1''}$$

$$dy/dt = k'p_1 x \cdot f(x,y,c)_w.  \tag{2''}$$

Thus the original forms of the differential equations (1)' and (2)' remain unchanged: we merely replace $2k$ and $k'$ by two new constants $(2kp_0)$ and $(k'p_1)$; and equations (7) and (8) for maximum $x/a$ still hold except that $R$ is to be replaced by the new ratio $R'$:

$$R' = (2kp_0)/(k'p_1) = R(p_0/p_1).  \tag{12}$$

It is generally observed that the greater the number of polar groups an organic compound carries, the greater will be its water solubility and the lower its solubility in

-50-

non-polar solvents. Hence with suitable immiscible solvents, $p_1$ and $p_2$ can be made small in comparison with $p_0$,
and it is therefore possible to obtain in this way an
apparent increase in the ratio of the rate constants which
cannot be changed otherwise. In formulating the partition
equation (11), we have tacitly assumed that the acid and
the half-ester are not appreciably associated in the organic phase and that their concentrations in the two phases
will not reach such heights as to modify the solubilizing
powers of the two immiscible solvents. These of course are
only approximations. Actually, in saturated hydrocarbon
layer, the half-ester (th di-acid will remain almost entirely in the aqueous phase, i.e. $p_0$ will be close to unity if
a saturated hydrocarbon solvent is used.) will be associated
to a considerable extent even in the presence of some dissolved alcohol which has come over from the aqueous layer;
hence in this case, $dy/dt$ will be proportional, not to the
first power, but to a fractional power (between 1 and 0.5)
of $x$; but this is even more favorable to the maximum
yield of the half-ester. For the lower dicarboxylic acids
($C_6$, $C_7$) whose solubilities in hot water are high, it is
possible to saturate the aqueous layer with NaCl so as to
decrease the solubilities of the half-esters and make $p_1$
very small; in this case the yields of the half-ester can

be further increased.

In addition to giving a favorable ratio of the partition coefficients, $p_o/p_1$, the immiscible solvent must also
possess the property of having limited miscibility with the
alcohol so that it will not extract too much of the alcohol
from the aqueous phase. A survey of the International
Critical Tables shows that the saturated hydrocarbon solvents offer the best choice. Skellysolve C, a commercial
solvent of this nature, is therefore used in the present
investigation; and it has been found to be suitable for
the preferential extraction of mono-methyl sebacate. Mono-
methyl adipate, however, is only slightly soluble in this
solvent, hence a favorable value of $p_o/p_1$ cannot be obtained
with this solvent; but when used with a predominant proportion of toluene, it has been found to be very effective
in reducing the solubility of adipic acid in the toluene
layer containing considerable amount of the half-ester.

The choice of the proportions of carboxylic acid,
alcohol, and water, and organic solvent depends upon a
knowledge of the partition coefficients and the esterification equilibrium, $K_e$. Whenever practical, the amount of
alcohol used must be enough to hold the dicarboxylic acid
in solution so that there will be no separation of a middle
layer as is observed in the case of sebacic acid. But if

-52-

too large a proportion of alcohol is used, it might level the difference in the partition coefficients, $p_0$ and $p_1$, by modifying the solvent power of the two layers. The effect of hydrolysis must also be considered since it affects the half-ester but does not appreciably affect the full ester whose solubility in the aqueous layer is expected to be low ($p_2$   $p_1$   $p_0$). It can be shown that the more general form of Goldschmidt equation, for the case in which the concentration of water is not small in comparison with that of alcohol and hydrolysis is not negligible, should be, for a simple carboxylic acid,

$$dx/dt = k c (a - x)m.K/ (m.K + n + x)$$
$$-k_h c\, x(n + x)/(m.K+n+x), \qquad (1)'''$$

where $m$ and $n$ are the molar concentrations of alcohol and water, respectively, and $K$ is the equilibrium constant for the protolytic exchange equilibrium; this equation follows from equation $(1)'''$ of Chapter II by replacing $r$ with $m.K$. $K$ can be calculated from given or extrapolated value of $r$ by means of the relation: $K(CH_3OH) = r$. $k/k_h$ as before is equal to $K_e/K$, where $K_e$ is the esterification equilibrium constant whose value for most carboxylic acids is known to be around 4 and does not change much with temperature (4.5 for both sebacic and adipic acids at 75 C$^0$ as found in the present investigation.). In the presence of an extract-

-53-

ing solvent, $x$ and $(a - x)$ in equation (1)''' should be replaced by $x_w = p_1 x$ and $(a-x)_w = p_0(a - x)$. In order that the term $k_h$ in equation (1)''' shall be less than 10% of the $k$ term near the 50%-esterification point where $x/(a-x) = 1$, we must have

$$\frac{k \cdot c p_0 (a - x) \, m \, K}{k_h e \cdot p_1 x \cdot (n + x)} \approx \frac{k p_0 (a - x) \, m \, K}{k_h \cdot p_1 x \cdot \text{ℝ}} > 10,$$

or

$$(p_0/p_1)(m/n) > 10 \ /x/(a-x)/ \ /k_h/k.K/ \approx 10/K_e = 2.2,$$

that is, the concentration ratio of alcohol and water, $m/n$, near the 50%-esterification point should be greater than two times the ratio of $p_1$ over $p_0$ in order that the rate of hydrolysis shall be less than 10% of the rate of esterification at this point. This condition can usually be met with; so, for approximate estimation of possible yield, the $k_h$ term can be neglected.

By neglecting the $k_h$ term in equation (1)''' and substituting the extrapolated values of $k$ and $K$ (using the curves of $-\log(k)$ and $-\log(K)$ against $1/T$) in the first term, the rate of esterification can also be estimated. For a hypothetical case: m:n ≈ 1:2, at $75^{\circ}$C., a 50% esterification will be reached in about 20 minutes if the hydrogen ion concentration is about 0.1 M. Taking the values of $k$ and $K$ at $75^{\circ}$ to be 0.4 and 0.03, respectively

-54-

(roughly extrapolated from data listed in Chapter II):
we have

$$t_{1/2} = \frac{0.69}{k \cdot c \cdot (mK/n)} = \frac{1.2}{c} \cdot 10^2 \text{ seconds.}$$

This is a very rough estimation based on the assumption
that Goldschmidt equation still holds for aqueous-alcoholic
solution. In practice, a 50% esterification is reached in
about 30-40 minutes if there is no separation of a middle
layer (as in the case of adipic acid), but the reaction
should be allowed to proceed until the acid concentration
in the organic layer has reached its maximum. However,
the required duration of reflux is fairly reproducible,
and an overstepping of 20-30 minutes causes no appreciable
decrease in the yield of the half-ester, but produces more
di-ester at the expense of the di-acid.

## EXPERIMENTAL

(A) Preparation of Mono-methyl Sebacate.

In a three-necked round-bottomed flask provided
with a glass stirrer and a reflux condenser were placed
50.5 g. (0.25 mol) of sebacic acid (E. K. 612), 44.5 cc.
(1.1 mol) of pure methanol, 50 cc. of water, and 1 g. of
NaCl. The mixture was heated on the steam-bath until the
solid had dissolved or melted. 400 cc. of skellysolve C
were then added slowly. (The boiling-point of the mixture

was about 75°C. and the acid concentration of the skellysolve layer was found by titration to be 0.009 N (hence
$p_o = 0.99$)). 5.0 cc. of dilute HCl (3.0 N) were then added
to the boiling mixture under rapid stirring, and the reflux
continued for 5.5 to 6 hours.  The middle layer disappeared
after about 5 hours of reflux, and the acid concentration
in the organic layer reached its maximum of 0.36 N after
5.5 hours of reflux and stayed around that value for about
half an hour before it began to drop appreciably.  The two
layers were separated while still hot, and the organic layer
was washed once with 50 cc. of hot brine.  The washing and
a small amount of middle layer were combined with the aqueous
layer, this was then saturated with common salt and extracted twice with 50-cc. portions of hot petroleum ether (b.p.
range 65-110°).  By cooling the skellysolve layer (total
volume at room temperature: 420 cc.; acid concentration
or neutration value: 0.352 N; and total ester concentration or saponification value: 0.66 N, determined by sapon-
ifying a neutralized sample with excess alkali on steam-
bath for about an hour and back titrating the excess alkali
with standard hydrochloric acid.) in ice-salt mixture, 26.3
g. of white crystals were obtained, m.p. 39-40°.  A small
amount of petroleum ether washings was combined with the
petroleum ether extract from the aqueous layer, and evapor-

-56-

ated to about 50 cc. from which by cooling, an additional
amount of white crystals. 4.0 g., was obtained. The skellysolve mother-liquor still contained about 8 g. of half-ester,
which was extracted with 10 g. of solid potassium carbonate, reliberated with cold dilute hydrochloric acid and extracted with petroleum ether. This gave an additional 6.3
g. of half-ester, m.p. 37-39°. Total yield: 36.6 g. (68%
of the theoretical). (The first and second crops are sufficiently pure for most practical purposes, but may be recrystallized from petroleum ether with about 5% loss in the
mother-liquor. The yield based upon purified product,
m.p. 40-41°C., is about 63-64%. This estimation is based
upon experience in many previous runs.) About 1 g. of
sebacic acid was recovered from the cooled aqueous layer;
From the skellysolve mother-liquor, 13.7 g. of crude
dimethyl sebacate were obtained, this was combined with
that obtained in previous runs and distilled at 141°C. and
5 mm. Hg. The crude ester was found to be practically pure,
containing only a very small amount of unextracted half-
ester. The total recovery of material was better than
95%. The total degree of esterification was about $(0.66 \times 420)/500 = 0.56$.

The partition coefficient of pure mono-methyl sebacate, 1.00 cc. (4.63 m.e. determined by titration), between
two phases composed of 5.5 cc. $H_2O$, 4.5 cc. methanol, 0.1

-57-

g. NaCl, and 40 cc. skellysolve C, was found to be $p_1 = 0.32$.
For 2.00 cc. between two phases of the same over-all compositions as above, the observed partition coefficient became $p_1 = 0.24$. The average value of $p_1$ during the esterification run described above must be closer to 0.24, because of increasing concentration of the half-ester and decreasing concentration of the methanol. Hence $p_0/p_1$ is about 4 or slightly greater, and $\alpha'$ about 8, and a 74% yield is roughly predicted by theory.

The duration of reflux can be shortened to two hours, with slight decrease in yield, by using 20 cc. of methanol, 20 cc. of 5% aqueous NaCl, 5 cc. of 1.2 N HCl, and 80 cc. of skellysolve per 10.1 g. (0.05 mol) of sebacic acid. In this case, the middle layer disappeared in about 1.5 hours, and the maximum acid concentration in the organic layer reached 0.34 N, indicating a 62-65% yield of half-ester. The yield actually obtained was around 55%.

Other procedures have been tried: (a) 50.5 g. sebacic acid, 60 cc. methanol, 90 cc. water, 10 cc. HCl (2 N), 350 cc. skellysolve C, and 6 to 6.5 hours of vigorous reflux with slow stirring. The maximum acid concentration attained was 0.58 N. (b) 50.5 g. of sebacic acid, 44.4 cc. of methanol, 50 cc. of water, 1 g. of NaCl, 300 cc. of skellysolve C, 100 cc. of toluene, and 4 cc. of 3N HCl. After

3.75 hours of reflux, the acid concentration of the organic layer reached 0.44 N, of which 0.06 N was due to dissolved sebacic acid. Subsequent work-up gave 35.5 g. of half-ester (62% yield) and 9.0 g. of sebacic acid; no attempt being made to purify the crude di-ester. It is quite possible that by prolonging the reflux to 4.5 or 5 hours, the yield of half-ester by this procedure may be substantially increased, probably to 70%, but the greater solubility of the di-acid in the skellysolve layer in the presence of considerable amount of toluene and half-ester makes this addition of toluene unsuitable for continuous operation. For single-batch operation, however, the addition of some toluene (10 - 15% by volume in the skelly-solve) may prove profitable.

(B)   Preparation of Mono-methyl adipate.

(i)   36.5 g. of adipic acid (0.25 mol), 21 cc. of methanol, 42 cc. of water, 14 g. of NaCl, 250 cc. of toluene, 50 cc. of skellysolve C, and 2.0 cc. of 3N HCl were refluxed (boiling-point of mixture = $85^{\circ}$) for two hours. The acid concentration in the toluene layer reached a maximum of 0.458 N. 20 minutes later the acid concentration became 0.452 N and the saponification value 0.60. The toluene layer (450 cc.) was separated from the aqueous layer (58 cc., acid 0.59 N, saponification value = 0.11)

and washed twice with 30 cc. portions of warm conc'd brine.
The solvent was distilled off, and the residual liquid
(40.1 g.) fractionated at 4 mm. Hg.  The following fractions
were collected:  90-110°, 4.0 g. 110-130°, 2 g.; 130-138°,
1 g.;  138-141° (constant), 28 g.; 141-150°, a few drops.
Residue, about 2 cc., solidified on cooling to room temperature.  The pure fraction constituted 70% of the theoretical yield, and possessed a neutralization equivalent of
158.  The total yield is estimated at about 72-75%.

(ii)  The yield may be improved by using 5% less of
the amounts of alcohol and water and 1 g. more of NaCl. The
maximum acid concentration in the toluene layer was 0.502 N,
reached in 2.33 hours of reflux, and the saponification
value was 0.64 N.  A 78-80% yield was therefore indicated;
but the reaction mixture was not worked up to isolate the
products.  The total degree of esterification after $2\frac{1}{4}$ hours
of reflux was about $(440 \times 0.64)/500 = 0.56$.  At the beginning of the experiment, before adding the catalyst acid,
the acid concentration of the toluene layer at the boiling-
point of the mixture, 77°, was found to be 0.042 N;  hence
the partition coefficient of the di-acid was about 0.95
($p_0$).  The partition coefficient, $p_1$, of pure mono-methyl
adipate, 1.00 cc. ( $\equiv$ 67.85 cc. N/10 NaOH), between two
phases composed of 4.0 cc. methanol, 4.2 cc. water, 1.5 g.

-60-

NaCl, 35 cc. toluene, and 5 cc. skellysolve C was found to be about 0.07. Hence $p_0/p_1 = 14$. However, the solubility of the di-acid was found to increase considerably with increasing amount of the half-ester in the toluene layer.

(iii) Two other runs with 350 cc. of toluene and 10 g. of NaCl, 1.0 cc. of 12N HCl and one hour of reflux also indicated about 70% yield of the half-ester. If the reflux had been prolonged to 1½ hours, the yield probably would have been better. However, Procedure (ii) is recommended.

For laboratory purpose, the di-acid and the half-ester in the reaction mixture may be extracted with cold aqueous potassium carbonate, reliberated with ice-cooled dilute hydrochloric acid, and extracted with successive portions of ether after saturating the aqueous solution with common salt. The crude half-ester may then be purified by distillation under reduced pressure. The full ester may be obtained practically pure by distilling off the solvent from the alkali-extracted toluene solution. The half-ester boils constant at 146°C. and 6 mm Hg., and the full ester boils at 105° and 6 mm. Hg.

With skellysolve as the extracting solvent and a greater proportion of methanol and longer period of reflux, the procedure can be modified to give over 70% yield of the dimethyl ester.

-61-

## DISCUSSIONS

By varying the concentration of salt and the proportions of toluene and skellysolve, or by using other extracting solvents, it should be possible to modify the procedure for the half-esterification of other dicarboxylic acids between the $C_5$ the $C_{11}$ range. In some cases, monoethyl esters may also be prepared, but the extracting solvent probably will be limited to the saturated hydrocarbon type. Theoretically, the greater proportion of the extracting solvent, the higher will be the yield; in practice, however, too high a proportion of extracting solvent will not be convenient.

A method of half-esterification due to Contzen and Crowet (3) consists of removing the water formed by distilling the reaction mixture with a large amount of carbon tetrachloride and stopping the reaction when approximately one mole of water per mole of di-acid being esterified has distilled over. This is roughly a rate-controlled process; but there is no phase separation since absolute alcohols are used. The reported yields, recalculated to percent of the theoretical are: for mono-ethyl oxalate, (55/118).100% = 46.6%; and for mono-isopropyl oxalate, (78/132).100% = 59%; for succinic and higher, below 40% yield. This is interesting from the stand-point of steric hindrance and

-62-

our present theory. Because of the proximity of the ester
alkyls to the free carboxyl in the case of oxalic half-
esters, some hindrance is expected to make k' smaller than
k;  the relative magnitudes of k and k' for acid-catalyzed
esterification of oxalic acid in ethyl alcohol is probably
in about the same ratio as that of the rate constants for
acetic and butyric acids, i.e., about 2 : 1;  and for esterification in isopropyl alcohol, k : k' is probably about
3 : 1, as is the ratio of rate constants for acetic and
isobutyric acids.  Hence for rate-controlled process, the
maximum yields of half-esters predicted from Table I are
63% and 70%, respectively, for the mono-ethyl and mono-
isopropyl oxalates.  For the half-esters of higher dicarboxylic acids, the ester alkyls do not exert appreciable
hindrance on the free carboxyl group, hence k and k' will
be practically equal and the maximum yield of half-esters
obtainable by homogeneous reactions is only 50%, as explained before.

One potential application of the heterogeneous method
is for making methyl esters of water-soluble carboxylic
acids, or of any carboxylic acids whose water solubilities
are considerably higher than that of their methyl esters.
For the esterification equilibrium is established only
when the ester has reached a concentration such that

-63-

$$\frac{(Ester)_W \ (H_2O)_W}{(Acid)_W \ (Alc.)_W} = \frac{p_0(x) \cdot n}{p_1(a-x) \cdot m} = K_e = 4,$$

or

$$x/(a-x) = 4(m/n)(p_1/p_0).$$

Hence if $(p_1/p_0) = 0.1$, a 90% esterification can be obtained by the heterogeneous method with the alcohol and water in the mol ratio of 1 : 4. This method will be particularly useful for making methyl esters of hygroscopic acids, like lactic acid, or some other hydroxyl acids; since aqueous solutions of the acids can be used directly with the addition of suitable proportions of methanol and extracting solvents, and salt and acid catalyst.

## SUMMARY

A theoretical analysis of the problem of half-esterification of dicarboxylic acids, or half-saponification of dicarboxylic esters, has been made. Equations have been derived which relate the maximum yields of half-esters, or ester-salts, obtainable, and the optimum total degrees of esterification to the ratios of the first and second rate constants. It is concluded that, for making half-esters or ester-salts of polymethylene dicarboxylic acids below $C_6$ and above $C_{10}$, the half-saponification method will give better yields if properly conducted. For the half-esterification of the dicarboxylic acids in the range

-64-

$C_6$ to $C_{10}$ particularly, a new and improved method has been devised which is based upon preferential extraction of the half-esters and full esters being formed in the aqueous-alcoholic phase by means of an immiscible solvent with or without the addition of salt to the aqueous phase. The method has been successfully applied to the preparation of mono-methyl adipate (72-75% yield) and mono-methyl seba-cate (62-68% yield), with the use of toluene and skelly-solve C as the extracting solvents. The method of preferential extraction also greatly simplifies the isolation of the mono-methyl sebacate from the unreacted di-acid. The theoretical aspects of the heterogeneous method and its possible applications to the esterification if water-soluble acids in general have also been analyzed.

## LITERATURE

(1)  Blaise, E. E. and Koehler, A., Bull. soc. chim., /4/ 7, 218 (1910).

(2)  Carson, J. Organic Synthesis, 25, 19 (1945).

(3)  Contzen-Crowet, Bull. soc. chim. Belg., 35, 180 (1926); Chem. Zentr. 1926 II, 1126 (1926).

(4)  Goldschmidt, H. and Udby, O., Zeit. physik. Chem., 60, 728 (1907).

(5)  Grun, A. and Wirth, T., Ber., 55, 2216 (1922).

-65-

(6)  Fourneau, E. and Sabetay, S., Bull. soc. chim. (4),
     43, 859 (1928).  Ibid., (4), 45, 834 (1929).

(7)  Hunsdiecker, H. and Hunsdiecker, C., Ber., 75, 296
     (1942).

(8)  Ingold, C. K., Journ. Chem. Soc., 1930, 1375 (1930).

(9)  Kenyon, J., Organic Synthesis, Coll. Vol. I. 418
     (1941).

(10) Meyer, J., Zeit. physik. Chem., 66, 81 (1909).

(11) Meyer, J., Ibid., 67, 257 (1909).

(12) Neisen, E., Journ. Chem. Soc., 29, 319 (1876).

(13) Swann, S., Oehler, R., and Buswell, R. J., Org. Syn.,
     17, 45 (1939).

(14) Walker, J., Journ. Chem. Soc., 61, 713 (1892).

## IV. SYNTHESIS OF HIGH POLYMETHYLENE DICARBOXYLIC ACIDS AND GLYCOHOLS. A NEW AND GENERAL METHOD FOR THE LENGTHENING OF HYDROCARBON CHAIN IN THE SYNTHESIS OF ALIPHATIC ACIDS AND HYDROCARBONS.

Polymethylene dicarboxylic acids may be synthesized in the stepwise fashion by converting the lower diesters into glycohols, dibromides, then into dicyanides or bis-malonic-esters, thus stepping up the chains by 2 or 4 carbon atoms; the process can be repeated until the desired chain-length is obtained. In this way, Chuit and his co-workers (5) have prepared a series of polymethylene dicarboxylic acids ranging from $C_{11}$ to $C_{21}$. The synthesis is quite straight-forward and generally gives good yield in each step, but the work involved is almost prohibitive when it is required to repeat the process several times in order to obtain a chain-length of more than 20 carbon atoms.

The most important general method so far reported in literature for the synthesis of even-membered dicarboxylic acids is the electrolytic method based upon the following anodic reaction:

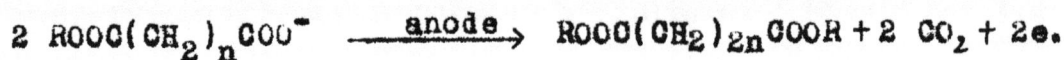

$$2\ ROOC(CH_2)_nCOO^- \xrightarrow{\text{anode}} ROOC(CH_2)_{2n}COOR + 2\ CO_2 + 2e.$$

The method was first employed by Brown and Walker (4) for the synthesis of the $C_{14}$ and $C_{18}$ diacids, and later modified and improved by Ruzicka, Fairweather (6), (7), (13),

-67-

and Ziegler (14) for the synthesis of the higher-membered diacids.

In the electrosynthesis of the higher-membered diacids, the following difficulties are encountered: (1) the ester-salts become increasingly soapy with increasing length of the chains; and (2) the crude esters tend to solidify on the electrode and thus stop the passage of current. In 1925, Fairweather (6),(7) obviated some of these difficulties by conducting the electrolysis in aqueous-alcoholic solutions of the ester-salts at temperatures above the melting-points of the crude di-esters expected. He obtained the diethyl ester of the $C_{22}$ dicarboxylic acid in 41% yield and the di-ethyl esters of the $C_{26}$ and $C_{30}$ di-acids in about 5% yields. In each case, the main product was found to be the $\omega$-unsaturated ester with one carbon atom less than the starting ester-salt. He also reported the electro-synthesis of the diethyl ester of $C_{34}$ diacid in about 12% yield.

In 1928, Ruzicka (13) prepared the $C_{30}$ dicarboxylic acid in 11% yield by electrolyzing potassium methyl tetra-decamethylene-dicarboxylate in methanol solution. However, the acid he obtained melted much too low (110°C) even after elaborate purification.

In 1937, Ziegler (14) was able to improve the yields

of the dimethyl esters of the $C_{22}$, $C_{26}$, and $C_{30}$ di-acids
to about 40%, 50%, and 30% respectively, by using platinum
wire-gauze anode to improve circulation of the methanol
solution of the ester-salt being electrolyzed. He also
carried out the electrosynthesis of the $C_{34}$ di-acid, but
reported that great difficulties in the purification of the
crude diester were encountered; only after elaborate
fractionation and repeated recrystallization was the ester
finally obtained pure (m.p. 87-88°; corresponding di-acid
m.p. 126-127°C. compared with that prepared by Fairweather
which melted at 123°C.), and the yield was probably very
low; he did not report the amount of the ester-salt employed in this case.

The electrolytic method is practical only for the
even-membered di-acids though Ruzicka (12) did carry out
the electrosynthesis of the $C_{21}$ diacid by a very involved
and laborious scheme.

In the electrosynthesis of the dicarboxylic esters,
the anodic reaction probably involves carboxyl and alkyl
free-radicals as intermediates, the final step being the
condensation of two alkyl free-radicals (with ester function at the other end of the chain) to give the desired
product. If there is local deficiency of carboxylate ions
near the anode, due to low ionic mobility or ineffective

circulation of the electrolyte, then the alkyl free-radicals produced, being of unfavorably low concentration for bimolecular condensation, will undergo further anodic decomposition to give the $\omega$-unsaturated ester or disproportionation products; this explains why improved circulation of the electrolyte will improve the yields of the diesters. However, the alkyl free-radical may also couple with the carboxyl free radical to form ester-linkage, since this ester formation side-reaction is also bimolecular, it will not be eliminated by improved circulation of the electrolyte.

The $C_{34}$ dicarboxylic acid appears to be the limit for the electrolytic method. With higher-membered di-acids, the difficulty of preventing the crude di-esters from solidifying on the electrode may become a very serious problem. But the main trouble lies in the increasing soapy nature of the ester-salts as this indicates the increasing tendency of the long-chained carboxylate anions to form large ionic aggregates, the so-called "micelles". With such highly associated anions, the complication of the anodic side-reactions can well be imagined.

There are methods of increasing the chain-length which may be useful in some special cases. Among these may be mentioned the method based upon the alkylation the acetylenes (2), (3), (11):

$$RBr + NaC \equiv CH \xrightarrow{\text{liq. } NH_3} RC \equiv CH + NaBr.$$

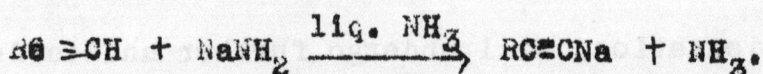

$$RC \equiv CH + NaNH_2 \xrightarrow{\text{liq. } NH_3} RC \equiv CNa + NH_3.$$

$$RC \equiv CNa + p\text{-}CH_3C_6H_4SO_3(CH_2)_nCl \longrightarrow RC \equiv C(CH_2)_nCl + p\text{-}CH_3C_6H_4SO_3Na$$

or

$$RC \equiv CNa + Br(CH_2)_nCl \longrightarrow RC \equiv C(CH_2)_nCl + NaBr.$$

The acetylenic chloride may then be converted into iodide
and cyanide and finally into carboxylic ester and acid. Obviously, the method may be modified for the synthesis of
dicarboxylic acids. However, the alkylation of acetylenes
usually does not give good yields, especially with llx of
higher molecular weights (3), probably because of the low
solubility of the alkyl halides in liquid ammonia and the
occurrance of amination as a side-reaction (8). The polymethylene mixed halides and p-tosyl-ester omega-chloride are
not readily available; but there is a possibility of using,
instead, the more easily prepared ω-bromo acids, in the
form of the sodium or potassium salts, provided that these
are not too soapy; this will eliminate several steps in
the synthesis; furthermore, the salts will be much more
soluble in liquid ammonia than do the mixed halides. At low
temperature, there is little danger of having the carboxylate ions involved in the nuclear-phillic replacement of
the bromine atom. However, the alkali salts of ω-bromo

undecanoic acid are laready quite soapy in aqueou solution, so the above scheme of synthesis probably will be
limited for making di-acids below $C_{20}$. A possible practical application is the synthesis of $\triangle^{9,10}$ octadecen-dioic
acid for making synthetic civetone.

Based upon preferential esterification of the primary
carboxyl groups in polycarboxylic acids, it is possible
to develop a new and general method for the lengthening of
carbon chains in the synthesis of high polymethylene dicarboxylic acids and

The method to be described in this Chapter consists
of the following steps: (1) malonic ester synthesis of
polycarboxylic acids from omega-bromo carboxylic ester and
partial decarboxylation of the substituted malonic acids
by pyrolysis; (2) preferential esterification of the end,
or 'primary', carboxyl groups; (3) replacement of the
'secondary' carboxyl groups by bromine by means of the bromine-silver-salt reaction; (4) reductive dehalogenation
of the bromo-esters by means of zinc dust in boiling glacial
acetic acid. These are illustrated in the following diagrams for the synthesis of (a) heneicosamethylene dicarboxylic and (b) dotricontamethylene dicarboxylic acid.

-72-

(a) Synthesis of Heneicosamethylene Dicarboxylic Acid, ($C_{23}$).

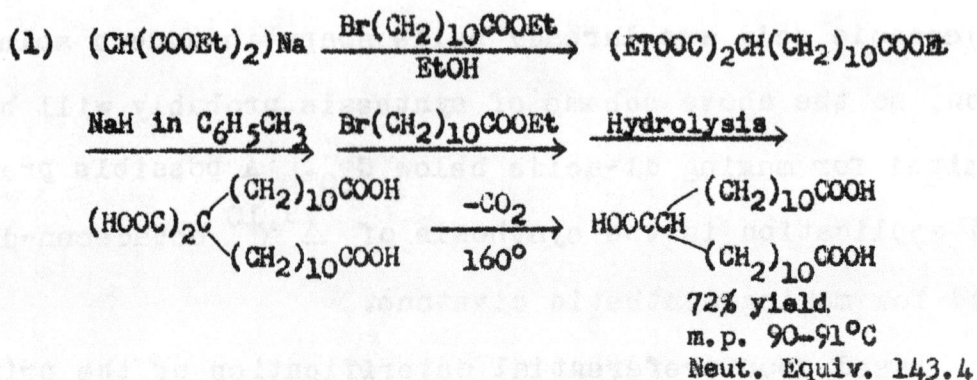

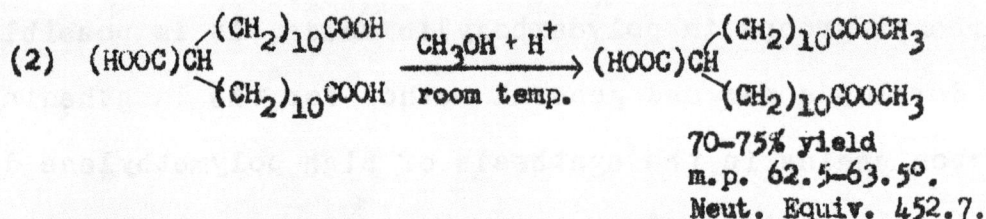

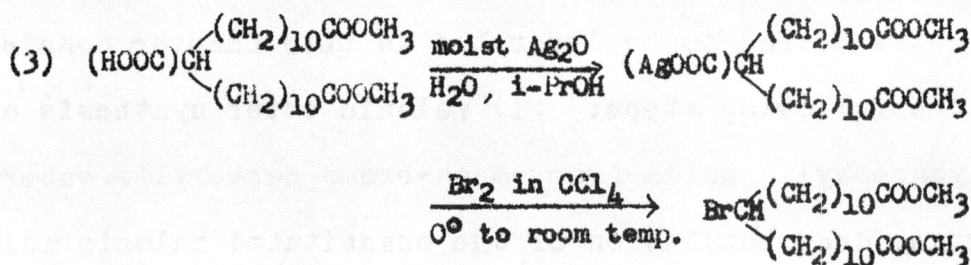

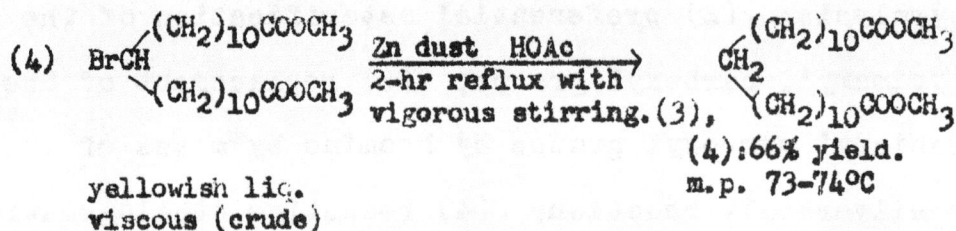

(1) $(CH(COOEt)_2)Na \xrightarrow[\text{EtOH}]{Br(CH_2)_{10}COOEt} (ETOOC)_2CH(CH_2)_{10}COOEt$

$\xrightarrow{NaH \text{ in } C_6H_5CH_3} \xrightarrow{Br(CH_2)_{10}COOEt} \xrightarrow{\text{Hydrolysis}}$

$(HOOC)_2C \Big\langle {}^{(CH_2)_{10}COOH}_{(CH_2)_{10}COOH} \xrightarrow[160°]{-CO_2} HOOCCH \Big\langle {}^{(CH_2)_{10}COOH}_{(CH_2)_{10}COOH}$

72% yield.
m.p. 90-91°C
Neut. Equiv. 143.4

(2) $(HOOC)CH \Big\langle {}^{(CH_2)_{10}COOH}_{(CH_2)_{10}COOH} \xrightarrow[\text{room temp.}]{CH_3OH + H^+} (HOOC)CH \Big\langle {}^{(CH_2)_{10}COOCH_3}_{(CH_2)_{10}COOCH_3}$

70-75% yield
m.p. 62.5-63.5°.
Neut. Equiv. 452.7.

(3) $(HOOC)CH \Big\langle {}^{(CH_2)_{10}COOCH_3}_{(CH_2)_{10}COOCH_3} \xrightarrow[H_2O, \text{ i-PrOH}]{\text{moist } Ag_2O} (AgOOC)CH \Big\langle {}^{(CH_2)_{10}COOCH_3}_{(CH_2)_{10}COOCH_3}$

$\xrightarrow[\text{0° to room temp.}]{Br_2 \text{ in } CCl_4} BrCH \Big\langle {}^{(CH_2)_{10}COOCH_3}_{(CH_2)_{10}COOCH_3}$

(4) $BrCH \Big\langle {}^{(CH_2)_{10}COOCH_3}_{(CH_2)_{10}COOCH_3} \xrightarrow[\substack{\text{2-hr reflux with} \\ \text{vigorous stirring.(3),}}]{Zn \text{ dust } HOAc} CH_2 \Big\langle {}^{(CH_2)_{10}COOCH_3}_{(CH_2)_{10}COOCH_3}$

yellowish liq.
viscous (crude)

(4):66% yield.
m.p. 73-74°C

(b) Synthesis of Dotricontamethylene Dicarboxylic Acid ($C_{34}$).

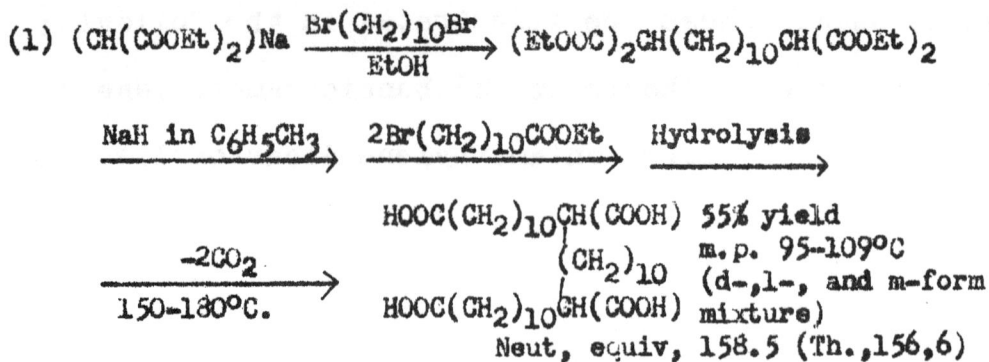

(1) $(CH(COOC)_2)Na \xrightarrow[\text{EtOH}]{Br(CH_2)_{10}Br} (EtOOC)_2CH(CH_2)_{10}CH(COOEt)_2$

$\xrightarrow{NaH \text{ in } C_6H_5CH_3} \xrightarrow{2Br(CH_2)_{10}COOEt} \xrightarrow{\text{Hydrolysis}}$

$\xrightarrow[\text{150-180°C.}]{-2CO_2}$

$HOOC(CH_2)_{10}CH(COOH)$
$\qquad\qquad\quad (CH_2)_{10}$
$HOOC(CH_2)_{10}CH(COOH)$

55% yield
m.p. 95-109°C
(d-,l-, and m-form
mixture)

Neut. equiv. 158.5 (Th.,156,6)

(2) $\xrightarrow[\text{room temp.}]{\text{MeOH + H}^+}$ 

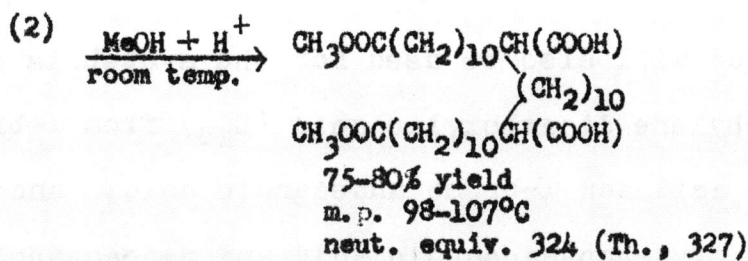

$CH_3OOC(CH_2)_{10}CH(COOH)$
　　　　　　　　　　　$(CH_2)_{10}$
$CH_3OOC(CH_2)_{10}CH(COOH)$
75-80% yield
m. p. 98-107°C
neut. equiv. 324 (Th., 327)

(3) $\xrightarrow[\text{H}_2\text{O, i-PrOH, (i-Pr)}_2\text{O}]{\text{Ag}_2\text{O, moist}}$ $\xrightarrow[\text{0° to room temp.}]{\text{Br}_2 \text{ in CCl}_4}$

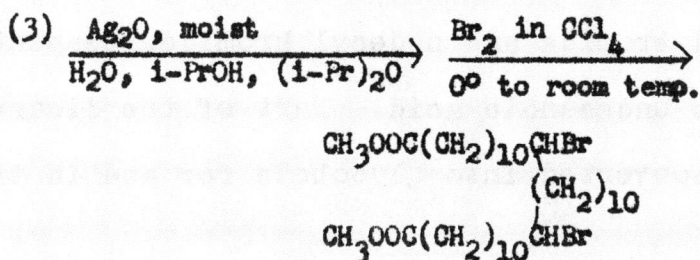

$CH_3OOC(CH_2)_{10}CHBr$
　　　　　　　　　$(CH_2)_{10}$
$CH_3OOC(CH_2)_{10}CHBr$

(4) $\xrightarrow{\text{Zinc dust in boiling HOAc}}$

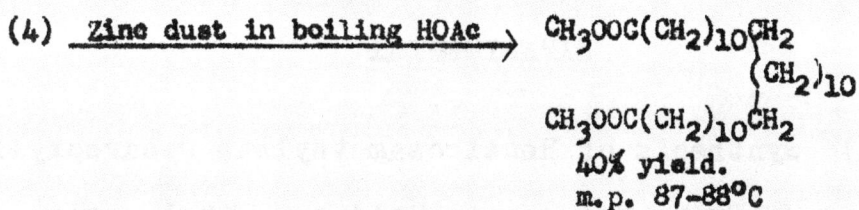

$CH_3OOC(CH_2)_{10}CH_2$
　　　　　　　　　　$(CH_2)_{10}$
$CH_3OOC(CH_2)_{10}CH_2$
40% yield.
m. p. 87-88°C

$\xrightarrow{\text{Alc. KOH}}$ $\xrightarrow{\text{dil. HCl}}$ $HOOC(CH_2)_{32}COOH$
m. p. 127-128°C.

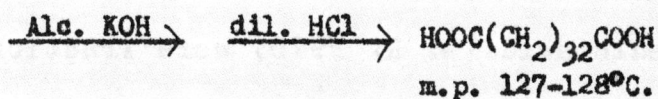

-74-

The method will also be used for the synthesis of heptaicosamethylene dicarboxylic acid ($C_{29}$) from ω-bromo heptadecanoic acid and ω-bromo undecanoic acid; and for the synthesis of n-heptadecanoic acid and n-docosanoic acid from n-amyl bromide and n-decyl bromide, respectively, and ω-bromo undecanoic acid. Part of the dicarboxylic esters will be converted into glycohols for use in the surface-film work.

EXPERIMENTAL

(a) Synthesis of Heneicosamethylene Dicarboxylic Acid.

ω-Bromo Undecanoic Acid and Ethyl Ester.---250 cc. of undecylenic acid (E. K. P975) were fractionated at 4-5 mm. pressure; the middle portion weighing 146 g., m.p. 21-22.5° (b.p.r. 160-162°) was dissolved in about 1400 cc. of purified petroleum ether (b.p.r. 65-80°) and saturated at 0-15° with HBr from gas cylinder in the course of about 45 minutes. The reaction mixture was washed with salt water and cooled to about -5° and the crystals filtered. The crude product was recrystallized from petroleum ether. Yield: 142 g. (67%), white crystals, m.p. 50-52°C.

(The procedure is essentially that described by Ashton and Smith (1), except that here more dilute solution was used.  The washing of the HBr-saturated solution before crystallization of the reaction product was also found to simplify the purification process.

The experiment was repeated more than 15 times using samples of different degrees of purity and under varying conditions.  But the yield never got beyond 70%.  It was found that if the undecylenic acid solution was added slowly, with stirring, into an excess of cold HBr solution in petroleum ether, the yield of $\omega$-bromo acid was below 20%. The use of toluene as solvent was found to give somewhat better yield, but toluene had to be distilled off and replaced by petroleum ether before crystallization could be done.)

105 g. of the $\omega$-bromo acid were treated with 80 cc. of redistilled thionyl chloride and the reaction mixture poured into 150 cc. of alcohol under cooling.  Yield of ethyl ester:  107.6 g. (93%), b.p. 139.5°.

Heneicosanetetracarboxylic Acid and Tricarboxylic Acid. --- 9.0 g. (0.39 mol) of metallic sodium were dissolved in 250 cc. of absolute alcohol;  80 cc. (about 25% excess) of pure malonic ester were then added followed by 115 g. (0.39 mol) of ethyl -bromo undecanoate.  The mixture was refluxed for 5 or 6 hours.  The alcohol was distilled off on steam-bath and the residue taken up with

ether, washed with brine, dried with unhydrous sodium sulfate. The volatile solvent was distilled off, and the
crude product was distilled on oil-bath maintained at about
120°. The excess malonic ester distilled over at 75° and
2.3 mm. and measured 20.6 cc. The clear yellowish liquid
left in the distilling flask weighed 135.5 g.

114 g. (0.306 mol) of the crude substituted malonic
ester obtained above were added with stirring to 7.3 g.
(0.304 mol) of sodium hydride in 300 cc. of hot toluene.
After foaming subsided, 97 g. (0.33 mol) of ethyl $\omega$-bromo
undecanoate were added slowly in the course of about 1/2
hour. The mixture was refluxed for 40 hours. At the end
of this period, titration of a pipetted sample with alkali
and methyl orange indicator show that the reaction was
practically complete (97%). The reaction mixture was taken
up with ether, washed with water, dilute aqueous acetic
acid (1%), and brine. The solvent was distilled off, and
the residue saponified, first with 10 g. of NaOH in 250 cc.
of 95% ethyl alcohol for about an hour under reflux, then
with excess of 6% aqueous sodium hydroxide on the steam-
bath, the alcohol being allowed to distill off through the
take-off condenser. After twenty four hours of heating,
the alkaline solution was partially neutralized with dilute
HCl the addition was stopped when the precipitate formed

by local excess of acid barely redissolved upon shaking the mixture), and extracted twice with benzene to remove most of the excess undecanoic ester. Excess of dilute HCl was added (Note: this order of addition was later found to be wrong; the alkaline solution should be <u>added</u> <u>slowly</u> into the acid solution with stirring to prevent occlusion of the soap by the insoluble organic acid being liberated.) The solid precipitate (white granular) was filtered, digested twice with 2% HCl.aq, twice with benzene, and finally washed with a little ether which bleached out most of the coloring matter. After drying in vacuum over conc. $H_2SO_4$ for about 12 hours, the solid (crystalline) melted at $116-118^{o}C.$ and had a neutralization equivalent of 140 (118.5) for pure tetra-carboxylic acid, $C_{26}$), probably stilled contaminated with some occluded salt. 100 g. of the crude product decarboxylated at $160^{o}C.$ for about 85 minutes gave 89.5 g. of crude tri-acid, m.p. $85-88^{o}$, neut. equiv. 149 (Th., 142.7). This was dissolved in aqueous NaOH and reprecipitated with dilute HCl: The precipitate was washed with dilute HCl, and small amounts of ether and acetone, then dried over conc. $H_2SO_4$. 88 g. were obtained, including 7.0 g. recovered from the washings by concentration and recrystallization. (72%). m.p. $89-90.5^{o}C.$, neut. equiv. 143.4 (Th., 142.7). Recrystallization from acetone yielded a product which melted at

## 78

90-91°C. With precaution, it seems possible to step up the yield to over 80%.

Heneicosanetricarboxylic-acid-1,11,21 Dimethyl Ester. 21 g. of the triacid were dissolved in about 400 cc. of dry methanol and 30 cc. of 0.381-N HCl-MeOH added. After standing at room temperature for an hour, the mixture with its container was put in ice-box and allowed to stand overnight. Yield of dried crystals: 16.0 g., m.p. 62.5-64°C. An additional crop was obtained from the mother liquid after 4 more days' standing in the ice-box. This was combined with that obtained in a previous run, total weighing 36 g., and recrystallized from acetone. 28 g. of pure crystals, m.p. 62.5-63.5° (corr.).

Analysis*: C 68.43%, H 10.33%

Theoretical: C 68.5% H 10.6%

as the first crop, the mother liquid being reserved for next run. Yield: 70-75%. The amount remaining may be quantitatively recovered in the form of the triacid, so there is practically no material loss.

11-Bromo Heneicosamethylene Dicarbocarxylic Acid Dimethyl Ester. 20.0 g. of the pure tricarboxylic-acid dimethyl ester were heated with stirring with a slight excess

---

* Note: All the micro-analyses were kindly done by Mrs. E. H. Klotz.

of silver oxide, (freshed precipitated from 8.0 g. of AgNO₃ and 2.5 g. of NaOH in water), in the presence of about 100 cc. of water and 10 cc. of ether, under reflux, for an hour. Some CCl₄ was added and the ether allowed to evaporate. Methanol was added to coagulate the silver salt, and the brownish cake was washed hot methanol, dried in vacuum at 60-70°. Grey granular solid, creepy, weighing 24.0 g. was treated with 5.0 cc. of bromine in dried CCl₄. The reaction mixture was filtered, and the filtrate washed with cold aqueous K₂CO₃, and brine, dried and heated under vacuum to distill off the CCl₄. The light yellowish liquid set to a soft waxy solid at room temperature.

Heneicosa-methylene Dicarboxylic Acid;   Reductive
Dehalogenation of the Secondary Bromo-Ester (Crude).  The
crude bromo ester was heated under reflux with 150 cc. of glacial acetic acid and 15-20 g. of zinc dust, an efficient glass-stirrer being used to prevent the zinc dust from settling down. After two hours of heating, the liquid was decanted into ice-water, and the solid residue leached three times with small amounts of hot acetic acid, and finally with a little water. White flocculent precipitate. Recrystallized from methanol, shiny white flakes, m.p. 73-74°C., 12 g. (66% yield, based upon the 20.0 g. of

pure tricarboxylic-acid dimethyl ester). From the methanol mother liquid by evaporation, 5 g. of soft waxy solid was obtained.

The free dicarboxylic acid was by alkaline hydrolysis of the dimethyl ester and recrystallization from benzen-methanol solution. m.p. 127-128°C. Neut. equiv. 193. This [cale 191.5] acid has been isolated from Japan work [Wax] by Flaschentrager Halle, m.p. 127.5°C (7).

4.0 g. of the dimethyl ester were treated with excess LiAlH$_4$ in ether, and the reaction mixture allowed to reflux overnight. Yield of glycohol after recrystallization from benzene-methanol, 2.5 g. (73%); silvery white flakes, m.p. 109°-110°C.

Analysis: C 77.66, H 13.48

Theoretical: C 77.5%, H 13.67.

(b) Synthesis of Dotricontamethylene Dicarboxylic Acid.

Dotricontanetetracarboxylic-acid-1,11,22,32. 18.4 g. (0.80 mol) of metallic sodium were dissolved in 400 cc. of absolute alcohol 160 g. (1.0 mol) of diethyl malonate were added, followed by 120 g. (.40 mol) of decamethylene dibromide in 200 cc. of absolute alcohol and a small amount of dried benzene. The mixture was refluxed for about 8 hours. Subsequent work-up gave 214 g. of crude

-81-

product, containing the excess malonic ester.  This excess
malonic ester was removed by vacuum distillation as described in (a), (the greater portion of the crude product
was used for the preparation of dodecamethylene dicarboxylic acid,) 37.5 g. of this crude bis-malonic ester treated with 4.4 g. of NaH in toluene and 58 g. of  -bromo undecanoic ester in the same way as described in (a), gave,
after hydrolysis and decarboxylation (at 160° -180), 39 g.
of crude tetra-acid, spoonzy white solid.  (The acid liberated from alkaline solution by dilute mineral acid had
a very troublesome tendency to occlude some salt.)  Recrystallization from acetone plus a few cc. of conc. HCl
gave 27.4 g. (55%) of white powder, m.p. 95-100°C., neut.
equiv. 158.5 (Th. 156.6).  A small portion of the product,
upon repeated recrystallization from acetone, was resolved
into two portions, one melting at 108-109.5° (complete
clearance of melt at about 112°), the other melting at 95°-
105°. However, they were found to have practically the
same neutralization equivalent (all titrations done in
$CO_2$-free 75% aqueous isopropanol) and esterify at the same
rate in subsequent rate experiment reported in Chapter II.)
Hence it was concluded that they were  a mixture of the
meso-form and recamic-form.

Dotricontanetetracarboxylic-acid Dimethyl Ester.--6.26
g. (0.01 mol) of the tetraacid in 300 cc. dry methanol

were catalytically esterified with 0.01 N HCl in the methanol for three hours at room temperature, then the mixture, containing a lot of precipitated partial ester, was allowed to stand in an ice-box overnight. 5.7 g. of nice white crystals were obtained, m.p. 105-110° (complete clearance at about 116°). Recrystallized from acetone, m.p. 106-110° (complete clearance at about 115°. By slow fraction crystallization, a fraction with m.p. 108-109.5° were obtained. Yield of pure dimethyl ester, neut. equiv. 324 (Th., 327) was about 75-80% from two runs. The mother liquid was reserved for recovery of the tetra-acid.

**11,22-Dibromo-Dotricontamethylene Dicarboxylic Acid Diethyl Ester.**---5.4 g. of the half ester were heated under reflux for four hours and with stirring with 50 cc. isopropyl alcohol, an equivalent amount of moist silver oxide, 10 cc. water, and 10 cc. of di-isopropyl ether (this was intended to be a mild complex-former for the silver salt so as to reduce the degree of occlusion and premature precipitation of the mono-silver salt. Other ways of forming the silver salt had been tried, but this appeared to give the best result in subsequent operation). The volatile solvent was allowed to distill off, and the precipitate washed with hot methanol. 6.5 g. of soft fine powder (greyish) were obtained.

The powdered silver salt was added to 1.2 cc. of dried bromine in 50 cc. of dried $CCl_4$. Evolution of gas occurred when solid hit the $CCl_4$ solution; shaking decreased gas evolution. The reaction temperature was about $5^{\circ}C$. After about half an hour, the excess of bromine was distilled off on steam-bath under water-pump suction. The residue was taken up with petroleum ether, filtered, and the filtrate treated with solid sodium carbonate. The solvent was driven off and the crude ester (low-melting solid, pale-yellow to greenish) heated on steam-bath excess of zinc dust and 75 cc. glacial acetic acid. Within five minutes, white crystalline solid began to appear on the surface of zinc dust; (this was later found to very soluble in water and give positive test with silver nitrate acidlfied with dil. $HNO_3$; hence it was probably $ZnBr_2$) After about two hours, the solution from the reaction mixture was decanted into cold water, the solid residue leached with hot acetic acid. 1.6 g. of powdery solid was obtained after recrystallization from petroleum ether, m.p. $86.5-88^{\circ}$. A small portion recrystallized again from petroleum ether plus about 20% of methanol gave fine crystalline product, m.p. $87-88^{\circ}C$. The yield was about 40% based on the half-ester.

Dotricontamethylene Dicarboxylic Acid and the $C_{34}$

-84-

Glycohol.---The free dicarboxylic acid was obtained from
the dimethyl ester by alkaline hydrolysis (sodium salt
very insoluble in alcohol and only slightly soluble in
water.) and recrystallisation from benzene-methanol and
petroleum-ether-methanol, m.p. 126.5-127.5°C.

0.40 g. of the dimethyl ester was refluxed with excess LiAlH$_4$ in ether-petroleum-ether (about 100 cc. :
20 cc., the petroleum ether was added to increase the solubility of the diester.) overnight. After extraction
with hot alkali, the crude product melted at 110-118°.
Repeated recrystallization from petroleum-methanol yielded 0.25 g. of white powder, m.p. 117-118.5°C.

Analysis[*]: H 13.73%, C 79.95% (H 13.82%; C
79.92%, required by theory).

## DISCUSSIONS

In case the disubstituted malonic acid obtained in
the first step can be easily recrystallized (preferably
from benzene-acetone), it may be profitable to proceed
with the preferential esterification of end carboxyl groups
before pyrolysis of the substituted malonic acid, because
the carboxyl groups on the malonic nucleus should be even
harder to esterify. Furthermore, the purification of the
partial ester thus formed is expected to be much simpler,

-85-

because of the absence of stereo-isomers. With one more
carboxyl group on the hydrocarbon chain, the methanol
solubility of the partial ester is also expected to be
higher. This means less danger of incomplete preferential esterification due to premature precipitation of the
mono-methyl ester. In this way, it should be possible to
extend the synthesis to very high-membered di-acids.

Synthesis through the tetra-carboxylic-acid-type
intermediate by the use of polymethylene bis-malonic-
esters as in (b) is obviously not a good procedure, because
of the low yield of the tetra-carboxylic step (1), and
low yield of dibromo-ester in step (3). This way of synthesis should be discarded in favor of the tri-acid-type,
or mono-malonic-acid type synthesis. For instance, the
$C_{34}$ diacid could well have been prepared in much better
yield from ω-bromo-docosanoic acid (ethyl ester) and
ω-bromo-undecanoic acid (ethyl ester).

The method obviously can be used for the synthesis
of higher normal, or branched, aliphatic mono-carboxylic
acids, by the use of different alkyl bromides, or iodides.

It also can be modified, with the elimination of the
preferential esterification step, for use in the synthesis
of aliphatic hydrocarbons, by using two mols of RX in the
first step of synthesis. There is an obvious advantage

-86-

for this type of chain-lengthening, for the intermediate
product in each step is easy to purify, by alkali extraction or otherwise.

　　Thus it is seen that the method provides a general
way for lengthening the hydrocarbon chains by any number
of carbon atoms so long as the product or the intermediates
do not become too insoluble to be conveniently handled.
In principle, it is reminiscent of the coppersmith's technique of joining two chains together and then subsequently
smoothing out the juncture by removing the excess solder

## SUMMARY

　　A new method has been described for the synthesis of
high polymethylene dicarboxylic acids.  It is based upon
(1) malonic ester synthesis of tri- or tetra-carboxylic
acids from ω-bromo carboxylic esters, (2) preferential
esterification of the primary carboxyl groups and (3) removal of the secondary carboxyl group or groups through the
bromine-silver-salt reaction followed by reductive debrom-
ination of the resulting bromo-ester by means of zinc and
acetic acid.  The method has been applied to the synthesis
of heneicosamethylene dicarboxylic acid ($C_{23}$) and dohi-
cosamethylene dicarboxylic acid ($C_{34}$). Its extended application as a general method for the lengthening of hydrocarbon chains in the synthesis of aliphatic acids and

-87-

hydrocarbons has been discussed.

## LITERATURE

(1) Ashton, R. and Smith, J. C.    Journ. Chem. Soc. <u>1934</u>, 435 (1934).

(2) Bried, E. A. and Henion, G. F.    Journ. Am. Chem. Soc. 1310 (1937).

(3) Bried, E. A. and Henion, G. F.    Ibid., <u>60</u>, 1717 (1938).

(4) Brown, A. C. and Walker, J.    Ann. <u>261</u>, 107 (1891).

(5) Chiut, P.    Helv. Chem. Acta <u>9</u>, 264 (1926).
    Chiut, Baelsing and Malet, Ibid <u>12</u>, 1096 (1929).
    Chiut and Hauser, Ibid <u>12</u>, 657 (1929).

(6) Fairweather, D. A., Proc. Roy. Soc. Edinburgh <u>45</u>, 23 (1925).

(7) Fairweather, D. A., Ibid, <u>45</u>, 283 (1925).

(8) Flanchentrager, B. and Halle, F.    Zeit. physiol. Chem. 190, 120 (1930).

(9) Henne, A. L. and Greenlee, K. W.    Journ. Am. Chem. Soc. 484 (1945).

(10) Hunsdiecker, H. and Hunsdiecker, C., Ber. <u>75</u>, 291 (1942).

(11) Newman, M. S. and Wotiz, J. H.,    Journ. Am Chem. Soc. <u>71</u>, 1292 (1949).

-88-

(12) Ruzicka, L., Stoll, M. and Schinz, Helv. Chem. Acta.
(11), 681 (1928).

(13) Ruzicka, L., Stoll, M. and Schinz, Ibid 11. 1179
(1928).

(14) Ziegler, K., Hechelhammer, W., Ann. 528. 1177 (1937).

## V. SURFACE FILMS OF HIGH POLYMETHYLENE DICARBOXYLIC ACIDS AND GLYCOLS.

The classical pioneer work of Pockels (5), Rayleigh (6), and Langmuir(4) has laid the foundation for our understanding of the physics and chemistry of surface films.   Later systematic work of Adam and collaborators has shown that four different types of insoluble films can exist with the long-chain aliphatic substances. These are classified as (1) condensed films, (2) liquid-expanded films, (3) vapor-expanded films, and (4) gaseous or vapor films, according to different degrees of lateral adhesion between the molecules.

For the condensed films of straight-chain aliphatic compounds with various types of head-groups the film area per molecule has been found to be practically the same, within about 1 sq. $\overset{o}{A}$ of 20.5 sq. $\overset{o}{A}$ at no compression, indicating that the hydrocarbon chains are close-packed and vertical, or tilted to the same angle (7).

The liquid-expanded films of simple aliphatic substances with various types of water-attracting head-groups also tend to a definite area at low compressions;  this being about 48 sq. $\overset{o}{A}$ per molecule for the single-chain aliphatic compounds;  83 sq. $\overset{o}{A}$ per molecule for the double-chain compounds, like glycol dilaurate (2), octyl palmitate (2),

and hexadecanecarboxylic acid-6 (8); and about 75 to 80%
of the area of three separate single-chain molecules for
each triple-chain molecule like that of a triglyceride (1).

It is now interesting to see how a compound with two
polar head-groups separated by a long hydrocarbon chain
will behave when it is spread on water surface. However,
very little work has been done on the surface films of such
compounds, and the only paper we can quote is that of Adam
and Jessop published in 1926 (3). These authors found
that the diethyl esters of the $C_{22}$ and $C_{34}$ dicarboxylic
acids form fairly stable films on water surface at room
temperature, and that the condensed area per molecule is
about 20.5 sq. $\overset{o}{A}$, indicating that only one of the two
ester head-groups is attached to the water surface. They
did point out, however, that at low compression of about
2 dynes per cm., there was some indication that some of the
chains might form vertical loops with the two head-groups
attached to the aqueous surface. They also found that the
$C_{22}$ and $C_{34}$ dicarboxylic acids form very unstable films
with a great deal of contraction and variable condensed
area between 6 and 18 sq. $\overset{o}{A}$ per molecule, i.e. about 1/7
to 3/7 of the normal area for two head-groups on separate
molecules. Adam and Jessop did not suppose that this indicated the existence of multi-molecular layers on the

aqueous surface; they suggested that some of the material might have been squeezed out of the aqueous surface.

Since the result obtained by Adam and Jessop on the polymethylene dicarboxylic acids does not appear to be conclusive, and since the behaviors of such polymethylene compounds in the surface-film state will have important bearing on cyclization by heterogeneous reactions, the surface films of the $C_{23}$ and $C_{34}$ polymethylene dicarboxylic acids and glycols and the dimethyl esters of the two acids has been studied in the present investigation. The preparation of the compounds has been described in the preceding chapter. The measurements were made with an ordinary Langmuir-Adam type hydrophile balance. Solutions of the esters were prepared from benzene; that of the glycols and dicarboxylic acids from benzene-methanol (4 : 1). It was found that the nature of the solvent and the extent of the dilution were very important in handling the glycols and the dicarboxylic acids. Without methanol, or with insufficient dilution, the film area tended to be too small and the results were inconsistent. This is probably due to the fact that these compounds may form association polymers with attending high solution viscosity and greater difficulty to spread. The actual concentration used for each compound was about 2 to 3 x $10^{-8}$ mol per cc. With such

-92-

concentration fairly reproducible results could be obtained. The time allowed to elapse before taking each pressure reading was about 40 seconds to one minute.

Typical F-A curves are shown in Figures III and IV. The results will be briefly discussed as follows.

(i) The dimethyl esters of the $C_{23}$ and $C_{34}$ dicarboxylic acids form condensed films with zero-compression area about 22 sq. Å per molecule, indicating only one of the two end-groups being attached to the aqueous surface as in the case of the diethyl esters of the $C_{22}$ and $C_{34}$ diacids (3). Evidently, this is due to the fact that the ester groups are only weakly hydrophillic, hence one of the two end-groups can be easily squeezed out of the surface by compression.

(ii) The $C_{23}$ dicarboxylic acid appears to form expanded film with two end-groups attached to the aqueous surface. This is based upon the following considerations: (1) The area per molecule under low compression is very close to that of liquid-expanded films of double-chain aliphatic compounds, this being about 83 sq. Å. (2) It is highly improbable for the $C_{23}$ diacid molecules to exist in the expanded-film state with only one of the two end-groups attached to the aqueous surface and the chain straight, since the lateral adhesion between the $C_{21}$ poly-

92-a

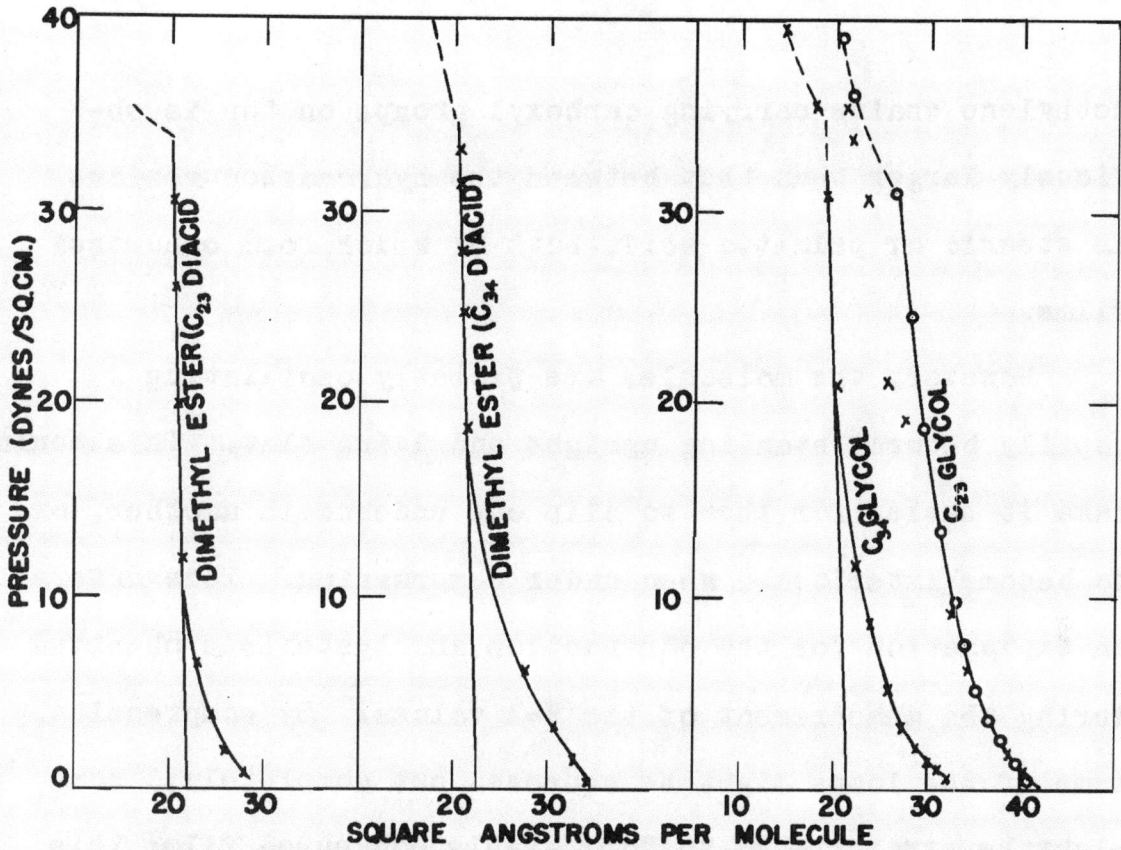

FIG. III   F-A CURVES FOR DIESTERS AND GLYCOLS AT 25°C

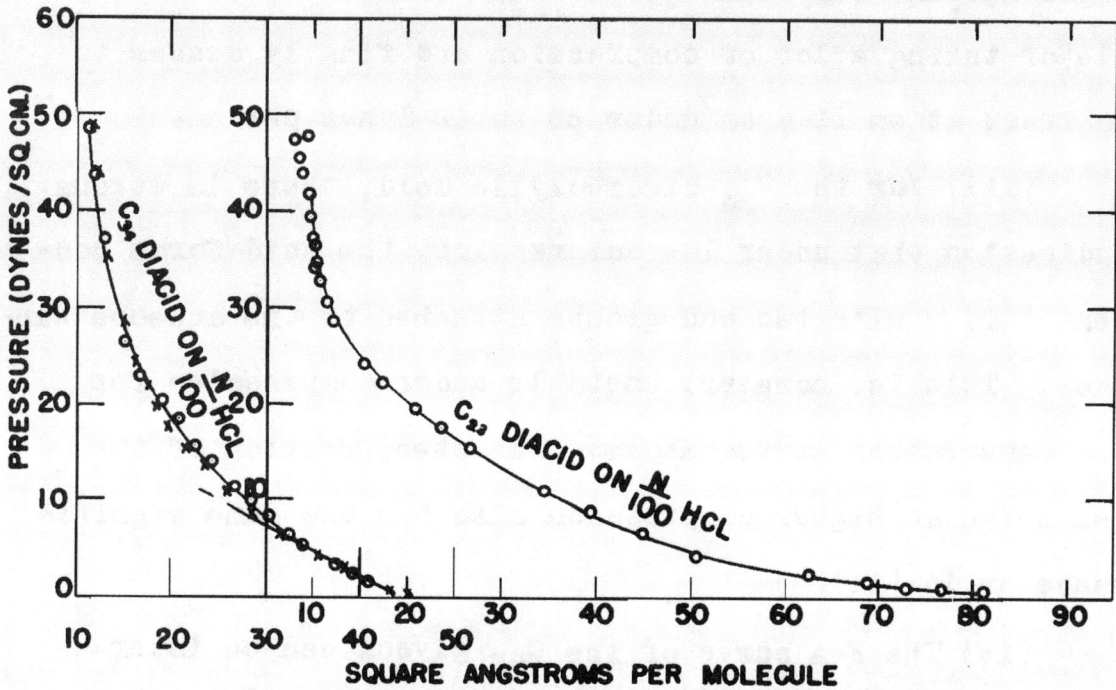

FIG. IV   F-A CURVES FOR POLYMETHYLENE DICARBOXYLIC ACID
AT 25°C

-93-

methylene chains carrying carboxyl groups on top is obviously larger than that between the hydrocarbon residue in stearic or palmitic acid, both of which form condensed films.

However, the molecules are probably oscillating rapidly between standing upright and lying flat. This should make it easier for them to slip one underneath another, or to become interlocked when under compression. This offers an explanation for the contraction and hysterisis observed during the measurement of the F-A values. By compression, some of the loops might be squeezed out completely; some might be straightened to form stable condensed film; this might account for the observed fact that the film is capable of taking a lot of compression and finally ceases to contract at small area under 35 to 45 dynes per cm.

(iii) For the $C_{34}$ dicarboxylic acid, there is strong indication that under low compression, the acid forms condensed film with two end-groups attached to the aqueous surface. This is, however, unstable under compression for the same reason stated above. The steep portion of the F-A curve at higher compression also has the same significance as in (ii).

(iv) The F-A curve of the $C_{23}$ glycol can be interpreted as indicating the presence of both loop-form and

-94-

the straight-form molecules in the condensed-film state. Again there is some indication that at low compression the film molecules probably exist mainly in the loop-form. The film collapsed at about 31 dynes per cm.

(v) The $C_{34}$ glycol behave like the dimethyl ester in forming a condensed film with one end-group attached to the aqueous surface. However, the solution employed in this particular case happened to be about four times as concentrated as that of the $C_{23}$ glycol. Probably at high dilution and low compression, both glycols will behave the same in forming co condensed films with both end-groups attached to the aqueous surface.

The orientation of the molecules in the glycol and diacid films may be decided by measuring the film dipole moments, or by depositing, if possible, the multi-layers on plates and examining the wetting property of the alternate layers.

## LITERATURE

(1) Adam, N. K., Proc. Roy. Soc., A. <u>101</u>, 516 (1922).

(2) Adam, N. K., Ibid., <u>126</u>, 366 (1930).

(3) Adam and Jessop, Ibid., <u>112</u>, 376 (1926).

(4) Langmuir, I., Journ. Am. Chem. Soc. <u>39</u>, 1848 (1917).

(5) Pockel, Nature, <u>43</u>, 437 (1891).

(6) Rayleigh, Phil. Mag., <u>48</u>, 337 (1899).

(7) Rideal, Proc. Roy. Soc., A <u>124</u>, 333 (1929).

(8) Stenhagen, E., Trans. Farad. Soc., <u>36</u>, 597 (1940).

-95-

# ACKNOWLEDGEMENT

This work was done under the direction of Professors E. Mack, Jr. and P. M. Harris;  I wish to acknowledge their valuable advice and constant encouragement;  and their kind interest and help, not only in my work, but also in my personal welfare.  I wish also to express my gratitude to Professor M. S. Newman for his helpful criticism and his inspiring lectures in steric hindrance;  to the United State Department of State, the Graduate School and the Chemistry Department of the Ohio State University for the grants of fellowship and scholarship;  to the National University of Amoy (China) for the grant of one-year furlough;  and to the late Dr. A. P. T. Sah of the National University of Amoy and Academia Sinica for his constant encouragement during the past years.

-96-

## AUTOBIOGRAPHY

I, Khi-Ruey Tsai, was born in Tung-An, Fukien, China, December 1, 1913. I received my secondary school education in Yu Min Grade School and Chip-Bee High School, both of Tung-An, Fukien; my undergraduate education at the University of Amoy, Amoy, China, from which I obtained the degree Bachelor of Science in 1937. While in college, I was able to get the University scholarship to support myself. After my graduation, I received from the University of Amoy an appointment as assistant in chemistry in the Department of Chemistry, which position I held for four years. After that I was promoted to lecturership in analytical and elementary physical chemistry in the same Department, where I remained teaching until the spring of 1947, when I obtained through the recommendation of the University of Amoy, a fellowship kindly offered by the United States Department of State. I came over here in the spring of 1947 and enrolled in the Graduate School of the Ohio State University for postgraduate work in Chemistry. After the expiration of the State Department Fellowship in March 1949, I have held the appointment as research fellow in the Department of Chemistry of The Ohio State University.

# 蔡启瑞院士论文总目录

## （截至 2013 年）

11. 多相催化理论的进展

蔡启瑞.

《1959 年全国催化研究工作报告会会刊》中国科学院石油研究所编，

科学出版社，1962, pp. 119～123

12. 离子晶体晶格能的计算——Ⅱ. 排斥指数的估计

蔡启瑞.

厦门大学学报（自然科学版），1962, **9**(1): 1～12

13. 钛酸钡晶体的天然极化、极化能和晶格能

林建新，蔡启瑞.

厦门大学学报（自然科学版），1962, **9**(2): 79～86

14. α-TiCl₃ 晶体的极化电场与 α-烯烃在 Ziegler-Natta 催化剂上定向聚合的机理

蔡启瑞.

厦门大学学报（自然科学版），1963, **10**(1): 85～86

15. α-TiCl₃ 晶体的极化能和晶格能

周泰锦，蔡启瑞.

厦门大学学报（自然科学版），1964, **11**(1): 1～10

16. α-TiCl₃ 电子能级的晶体场分裂

周泰锦，万惠霖，蔡启瑞.

厦门大学学报（自然科学版），1964, **11**(2-3): 1～8

17. 络合活化催化作用

蔡启瑞.

厦门大学学报（自然科学版），1964, **11**(2): 23～40

全国高等学校学报化学化工版（试刊），1965: 486

18. Estimation of repulsive exponents in the calculation of lattice energies of ionic crystals

Tsai, K. R.,

Scientia Sinica (English Edition), 1964, **13**(1): 47～60

19. 过渡金属化合物催化剂络合活化催化作用(I)-附载型氧化铬和氧化铌催化剂的研究与炔类环聚芳构化催化反应机理

厦门大学化学系催化教研室.

中国科学，1973, **16**(4): 373～388

20. 关于固氮酶的作用机理和活性中心结构

厦门大学化学系固氮研究组.

厦门大学学报（自然科学版），1974, **13**(1): 111～126

21. 络合活化催化作用——Ⅱ. 乙炔气相水合制乙醛锌系催化剂的研究

厦门冰醋酸厂，厦门橡胶厂，厦门大学化学系.

化学学报，1975, **33**(2): 113～124

22. 固氮酶的活性中心结构和化学模拟生物固氮的络合催化问题

厦门大学化学系催化教研室固氮小组.

《化学模拟生物固氮进展》（第 2 集），科学出版社，1976, pp. 163～209

23. 固氮酶的活性中心模型和催化作用机理
    厦门大学化学系催化教研室固氮研究组.
    中国科学,1976,**34**(5):479~491

24. A model of nitrogenase active-centre and mechanism of nitrogenase catalysis
    Nitrogen-Fixation Research Group.
    Scientia Sinica, 1976,**19**(4):460~474

25. 热泪盈眶洒像台
    蔡启瑞.
    厦门大学学报(哲学社会科学版),1977,(1):28

26. 生物固氮与络合催化
    蔡启瑞.
    化学通报,1978,(2):5~6

27. 固氨酶活性中心模型的演进和酶催化机理
    蔡启瑞,林硕田,万惠霖.
    厦门大学学报(自然科学版),1979,**18**(2):30~44

28. 化学模拟生物固氮的新里程
    蔡启瑞.
    化学通报,1979,(5):21~24

29. 化学模拟生物固氮——Ⅵ.铁钼辅基模型化合物的合成及其催化性能
    许志文,颜翠竹,丁马太,张藩贤,林硕田,许良树,蔡启瑞.
    厦门大学学报(自然科学版),1980,**19**(2):41~49

30. 化学模拟生物固氮——Ⅶ.乙炔选择性还原成乙烯作为原子簇活性中心多核络合活化底物的一种判据
    张藩贤,林正忠,许志文,林国栋,蔡启瑞.
    厦门大学学报(自然科学版),1980,**19**(2):50~56

31. 化学模拟生物固氮——Ⅸ 铁钼辅基模型化合物的合成和性能表征
    许志文,林国栋,林硕田,颜翠竹,丁马太,林培三,韩国彬,张藩贤,许良树,蔡启瑞.
    厦门大学学报(自然科学版),1980,**19**(4):67~73

32. A novel synthesis of 1,21-heneicosanedioic acid
    Tsai, K. R.; Newman, M. S.,
    Journal of Organic Chemistry, 1980,**45**:4785~4786

33. Correlation between chemisorption and coordination; Cluster approach to the nature of active sites on ammonia synthesis catalysts
    Tsai, K. R.,
    7$^{th}$ ICC Post-Congress Symposiumon Nitrogen Fixation (Tokyo),1980,Invited lecture

34. Development of a model of nitrogenase active-center and mechanism of nitrogenase catalysis
    Tsai, K. R.,
    Nitrogen Fixation, Newton, W. E.; Orme-Johnson, W. H. Eds.,
    Univ. Park Press, Baltimore, USA, 1980,**1**, pp. 373~387

35. 催化科学的新成就和发展动向

蔡启瑞.

《中国化学会 1978 年年会学术报告集》中国化学会编,科学出版社,1981,pp. 79～92

36. 化学模拟生物固氮——ⅩⅣ EHMO 法研究环丙烯等固氮酶底物的 $\mu_3\eta^2$ 型络合方式和分子氮还原加氢中间态

万惠霖,蔡启瑞.

厦门大学学报(自然科学版),1981,**20**(1):62～73

37. Studies on the mechanism and chemical modeling of nitrogenase catalysis

Tsai, K. R.; Wan, H. L.; Lin, S. T.; Lin, G. D.; Lai, W. J.; Zeng, D.; Ding, M. T.,

Current Perspectives in Nitrogen Fixation, Gibson, A. H.; Newton, W. E. Eds.,

Australian Academy Press, Camberra, Australia, 1981, p. 344

38. 氨合成铁催化剂上氮吸附态的研究——Ⅰ.氨合成铁催化剂表面上吸附氮的激光 Raman 光谱和红外光谱

廖代伟,王仲权,张鸿斌,蔡启瑞.

厦门大学学报(自然科学版),1982,**21**(1):100～103

39. $Fe_4S_4^*$ 原子簇与 ATP 的络合及电子传递与 ATP 水解的偶联

陈鸿博,林硕田,林国栋,蔡启瑞.

厦门大学学报(自然科学版),1982,**21**(1):104～106

40. 酶催化与非酶催化固氮成氨

厦门大学固氮研究组,物理化学研究所.

厦门大学学报(自然科学版),1982,**21**(4):424～442

41. 我国催化研究五十年

张大煜,蔡启瑞,余祖熙,闵恩泽.

自然杂志,1982,**5**(11):817～821

42. Coordination catalysis by transition metal complexes: nitrogenase catalysis and its chemical modeling

Tsai, K. R.; Wan, H. L.,

Fundamental Research in Organometallic Chemistry,

Tsutsui, M.; Ishii Y.; Huang, Y. Z. Eds., Unvi Park Press, Baltimore, USA, 1982, pp. 1～12

43. Synthesis and catalytic activities of FeMo-co Modeling compounds

Tsai, K. R.; Xu, Z. W.; Lin, S. T.; Lin, G. D.; Ding, M. T.; Zeng, D.,

Fundamental Research in Organometallic Chemistry, Tsutsui, M.; Ishi Y.; Huang, Y. Z. Eds.,

Unvi Park Press, Baltimore, USA, 1982, pp. 603～611

44. 固氮酶铁钼辅因子模型化合物合成的研究——Ⅰ、光谱法研究合成方法的设计方案

刘敏敦,张鸿图,林国栋,廖远琰,林正中,蔡启瑞.

厦门大学学报(自然科学版),1983,**22**(1):38～44

45. 金属铝促进的氨合成铁系催化剂研究

厦门大学固氮(Ⅰ)研究组,物理化学研究所.

化肥与催化,1983,**22**(2):15～20

46. 要注意多培养跨学科人才

蔡启瑞.

中国高等教育，1983，(8)：4～6

47. Laser Raman-spectra of chemisorbed species on ammonia-synthesis iron catalysts
Liao, D. W.; Wang, Z. C.; Zhang, H. B.; Tsai, K. R.,
Abstracts of Papers of the American Chemical Society, 1983, **186**，(AUG)，96-COLL

48. Studies on supported transition-metal catalysts for syngas conversion to ethanol on Rh-Nb$_2$O$_5$/SiO$_2$ catalysts
Yang, Y. C.; Liu, J. P.; Chen, D. A.; Tsai, K. R.,
Abstracts of Papers of the American Chemical Society, 1983, **186**，(AUG)，62-COLL

49. Study of ATP-driven electron transport in nitrogenase reactions with model systems
Lin, G. D.; Zhang, H. T.; Chen, H. B.; Wu, Y. F.; Tsai, K. R.,
Abstracts of Papers of the American Chemical Society, 1983, **186**，(AUG)，16-INOR

50. Unified elucidation of N$_2$-ase-catalyzed H$_2$-evolution reactions with edge-sharing twin-cubanes model and further-studies on synthesis of FeMo-Co modeling compounds
Tsai, K. R.; Zhang, H. T.; Lin, G. D.; Wu, M. G.; Han, G. B.; Yang, H. H.; Lai, W. J.; Liao, D. W.,
Abstracts of Papers of the American Chemical Society, 1983, **186**，(AUG)，95-INOR

51. 簇结构敏感型的过渡金属催化作用及其与原子簇络合物催化作用的关联
万惠霖，张鸿斌，廖代伟，周泰锦，蔡启瑞.
厦门大学学报（自然科学版），1984，**23**(1)：61～74

52. 乙烯光电催化氧化的研究—金属/n-GaP 光阳极
邬正伟，庄启星，蔡启瑞.
感光科学与光化学，1984，(3)：49～55

53. Advances in the studies on nitrogen fixation catalyzed by nitrogenase and by iron catalysts
Tsai, K. R.; Lin, G. D.; Zhang, H. T.; Zhang, H. B.; Liao, D. W.,
Abstracts of Papers, The International Chemical Congress of Pacific Basin Societies (Honolulu, Hawii, USA), 1984, 05082

54. Coordination and catalysis on ammonia synthesis iron catalysts
Tsai, K. R.; Liao, D. W.; Wang, Z. Q.; Zhang, H. B.,
Book of Abstracts of 23$^{rd}$ International Conference Coordination Chemistry (ICCC, Colorado), 1984, p. 277

55. Coordination catalysis by cluster complexes (Ⅰ) Synthesis and catalytic properties of FeMo-cofactor modelling compounds
Xu, Z. W.; Lin, G. D.; Lin, S. T.; Zhang, H. T.; Yan, C. Z.; Xu, L. S.; Wu, M. G.; Tsai, K. R.,
New Frontiers in Organometallic and Inorganic Chemistry, Huang, Y. Z.; Yamamoto, A.; Teo, B. K. Eds., Science Press, Beijing, China, 1984, p. 9

56. Coordination catalysis by cluster complexes (Ⅱ) Coupling of ATP hydrolysis with electron transfer by [Fe$_4$S$_4^*$L$_4$]$^{2-}$ cluster complexes
Chen, H. B.; Lin, S. T.; Lin, G. D.; Tsai, K. R.,
New Frontiers in Organometallic and Inorganic Chemistry, Huang, Y. Z.; Yamamoto, A.; Teo, B. K. Eds., Science Press, Beijing, China, 1984, p. 11

57. Spectroscopic studies of SMPI in niobia-promoted rhodium and palladium catalysis for syngas conversion to alcohols

Gu, G. X.；Liu, J. P.；Tsai, K. R.,

The International Chemical Congress of Pacific Basin Societies（Hawii,USA）,1984,O3G28

58. 催化作用和化学动力学

张大煜,蔡启瑞,余祖熙,闵恩泽.

《中国化学50年》（中国化学会专论）,科学出版社,1985,pp.123～140

59. 固氮酶反应中 ATP 驱动电子传递的化学模拟——Ⅰ.[Fe$_4$S$_4$(SR)$_4$]$^{2-}$原子簇与 ATP 络合的极谱及电子吸收光谱的研究

陈鸿博,张鸿图,林国栋,蔡启瑞.

厦门大学学报（自然科学版）,1985,**24**(4)：448～456

60. 合成气制乙醇 Rh-Nb$_2$O$_5$/SiO$_2$ 催化剂中的 SMPI 和助催剂作用本质的研究

顾桂松,刘金波,杨意泉,陈德安,林建毅,蔡启瑞,郭可珍.

物理化学学报,1985,**1**(2)：177～185

61. 化学法 MgCl$_2$—n-BuOH—SiCl$_4$—TiCl$_4$ 体系丙烯等规聚合催化剂中 n-Bu 含量对催化行为的影响

王耀华,刘金波,陈德安,曾金龙,郑荣辉,蔡启瑞.

合成树脂与塑料,1985,(3)：18～24

62. 乙苯脱氢制苯乙烯氧化铁系催化剂的研究——晶格氧的作用

陈慧贞,何淡云,肖漳龄,蔡启瑞.

高等学校化学学报,1985,**6**(5)：433～440

63. A S$_2$-containing edge-sharing twin-cubane model of nitrogenase active center and synthesis of FeMo-co modeling compounds

Zhang, H. T.；Song, Y.；Lin, G. D.；Xu, L. S.；Liao, Y. Y.；Chen, J. F.；Zhang, F. Z.；Ding, M. T.；Wan, H. L.；Tsai, K. R.,

Book of Abstracts,6$^{th}$ International Symposium on Nitrogen Fixation（Corvallis, Oregen, USA）,1985

64. Chemical modeling of ATP-driven electron transport in nitrogenase reactions — Partial hydrolysis of ATP catalyzed by redox reaction between Fe$_4$S$_4$L$_4$-ATP complex and H$_2$O$_2$

Wu, Y. F.；Lin, G. D.；Yu, X. X.；Xu, L. S.；Tsai, K. R.,

Book of Abstracts,6$^{th}$ International Symposium on Nitrogen Fixation（Corvallis, Oregen, USA）,1985

65. In-situ FT-IR study of chemisorbed species on ammonia synthesis iron catalyst — a complementary work to laser Raman studies

Liao, D. W.；Tsai, K. R.,

A Scientific paper presented at 2$^{nd}$ China-Japan-USA Symposium on Heterogeneous Catalysis（Berkeley）,1985

66. Molecular catalysis in hydrogenation of N$_2$ and of CO over metal catalysts

Tsai, K. R.；Zhang, H. B.；Wan, H. L.；Guo, X. X.；Lin, L. W.；Jiang, B. N.,

2$^{nd}$ China-Japan-USA Symposium on Heterogeneous Catalysis（Berkeley）,1985,Plenary presentation

67. 氨合成铁催化剂上化学吸附物种的 Raman 光谱

廖代伟,张鸿斌,王仲权,蔡启瑞.

中国科学 B,1986,(7)：673～680

68. 丙烯定向聚合高效负载型 Ziegler-Natta 催化剂的研究

翁维正,万惠霖,蔡启瑞.

厦门大学学报(自然科学版),1986,**25**(3):304～314

69. 丙烯腈的等离子体聚合研究

许颂临,吴丽云,庄启星,伍振尧,曹守镜,张光辉,蔡启瑞.

厦门大学学报(自然科学版),1986,**25**(3):321～327

70. F-T 合成铁催化剂上的配位催化作用

张鸿斌,蔡启瑞.

厦门大学学报(自然科学版),1986,**25**(6):658～665

71. XPS 研究合成气制醇的 Rh-$Nb_2O_5$/$SiO_2$ 催化剂的金属-助催化剂-载体的相互作用

林建毅,顾桂松,刘金波,蔡启瑞,郭可珍.

催化学报,1986,**7**(2):118～123

72. Coordination and catalysis in hydrogenation of $N_2$ and of CO over transition-metal catalysts

Zhang, H. B.; Wan, H. L.; Tsai, K. R.,

Book of Abstracts, 24$^{th}$ ICCC (Athens, Greece), 1986, p. 636

73. Hydrogenation of CO to ethanol over rare earth oxide promoted Rh catalysts

Du, Y. H.; Chen, D. A.; Tsai, K. R.,

Proceedings of China-Japan Bilateral Symposium on Utilization of CO and $CO_2$ (Lanzhou), 1986, p 3

74. On the complementary utilization of petroleum, natural gas, and coal resources and coordinative development of synthetic fuels & chemical industry

Tsai, K. R.,

Fujian Provincial Association of Science and Technology, Scientists' & Engineers' Proposals, Circ. 1986, No. 41986

75. 固氮酶活性中心模型的 EHMO 研究

周泰锦,万惠霖,王南钦,廖代伟,蔡启瑞.

厦门大学学报(自然科学版),1987,**26**(2):195～204

76. Cluster catalysis in fixation of nitrogen to ammonia catalyzed by nitrogenase and by iron catalysts

Cai, Q. R.; Zhang, H. B.; Lin, G. D.,

Advances in Science of China-Chemistry, 1987, **2**:125

77. Coordination and catalysis in coupled electron-transport & ATP hydrolysis by transition metal complexes

Wu, Y. F.; Chen, H. B.; Lin, G. D.; Yu, X. S.; Zhang, H. T.; Wan, H. L.; Tsai, K. R.,

25$^{th}$ ICCC (Nanjing, China), 1987

78. Nature of active site & mechanism of synergistic catalysis in methanol synthesis over Cu-ZnO-$M_2O_3$ catalysts

Chen, H. B.; Wang, S. J.; Liao, Y. Y.; Cai, J. X.; Zhang, H. B.; Tsai, K. R.,

Proceedings of 3$^{rd}$ China-Japan-USA Trilateral Symposium on Catalysis (Xiamen), B-66 (Oral presentation), 1987, p. 97

79. Promoter action of rare-earth-oxides in rhodium silica catalysts for the conversion of syngas to ethanol

Du, Y. H.; Chen, D. A.; Tsai, K. R.,

Applied Catalysis，1987，**35**(1)：77～92

80. Raman-spectra of chemisorbed species on ammonia-synthesis iron catalysts

Liao, D. W.；Zhang, H. B.；Wang, Z. Q.；Cai, Q. R.，

Scientia Sinica Series B-Chemical Biological Agricultural Medical & Earth Sciences，1987，**30**(3)：246～255

81. Reaction mechanisms and promoter action in syngas conversion to ethanol over supported rhodium catalysts

Liu, J. P.；Fu, J. K.；Wang, H. Y.；Tsai, K. R.，

Proceedings of 3$^{rd}$ China-Japan-USA Trilateral Symposium on Catalysis (Xiamen)，B-67 (Oral presentation)，1987，p. 99

82. Trends of development in chemical catalysis

Tsai, K. R.，

New Asia Life (a monthly periodical of the New Asia College, Chinese Univ. of Hong Kong)，1987，pp. 2570～2574

83. 丙烯酸酯类氢硅化中的基团效应

林旭，洪满水，蔡启瑞.

厦门大学学报(自然科学版)，1988，**27**(5)：558～561

84. 光电子能谱研究甲酸和乙酸在预氧化的铁表面上的吸附和分解

张兆龙，蒋安北，区泽棠，蔡启瑞.

分子催化，1988，**2**(1)：56～59

85. Carbene-ketene mechanism for Rh-catalyzed ethanol synthesis studied with model reaction involving supported ($\mu_2$-CH$_2$) or ($\mu_3$-CCO) metal clusters

Gao, J. X.；Zhou, Z. H.；Zheng, L. S.；Tsai, K. R.，

Post Congress Symposium of 9$^{th}$ International Congress on Catalysis：The Use of Metal Complexes in the Preparation of Catalysts (Quebec, Canada)，1988，B-6

86. Cluster-complex mediated electron-transfer and ATP hydrolysis

Wu, Y. H.；Chen, H. B.；Lin, G. D.；Yu, X. S.；Zhang, H. T.；Wan, H. L.；Tsai, K. R.，

Pure and Applied Chemistry，1988，**60**(8)：1291～1298

87. Coordination and catalysis in syngas conversion to methanol and ethanol over metal-oxide promoted metal catalysts

Chen, D. A.；Zhang, H. B.；Liu, J. P.；Tsai, K. R.，

Proceedings of 2$^{nd}$ Japan-China Bilateral Symposium on Utilization of CO and CO$_2$ (Osaka, Japan)，1988，p. 61

88. In-situ chemical trapping of ketene intermediate in syngas conversion to ethanol over promoted rhodium catalysts

Liu, J. P.；Wang, H. Y.；Fu, J. K.；Li, Y. G.；Tsai, K. R.，

Proceedings of 9$^{th}$ International Congress on Catalysis (Calgary, Canada)，1988，**2**：735

89. Mechanism of synergistic catalysis by Cu-ZnO-M$_2$O$_3$ catalysts in methanol synthesis

Chen, H. B.；Wang, S. J.；Liao, Y. Y.；Cai, J. X.；Zhang, H. B.；Tsai, K. R.，

Proceedings of 9$^{th}$ International Congress on Catalysis (Calgary, Canada)，1988，**2**：537

90. 催化作用中的某些结构化学问题

万惠霖, 蔡启瑞.

结构化学, 1989, **8**(5): 349~356

91. 电子光谱法研究不饱和酯类硅氢化的催化作用机理

林旭, 方钦和, 洪满水, 蔡启瑞.

厦门大学学报(自然科学版), 1989, **28**(2): 158~162

92. 光电子能谱研究甲酸和乙酸在铁表面上的吸附和分解

张兆龙, 蒋安北, 区泽棠, 蔡启瑞.

催化学报, 1989, **10**(1): 68~70

93. 甲烷氧化偶联 $K_2CO_3/BaCO_3$ 催化剂的表征

张兆龙, 黄文秀, 区泽棠, 蔡启瑞.

催化学报, 1989, **10**(4): 340~345

94. 甲烷在 $MnO_x/SiO_2$ 催化剂上氧化偶联反应的研究

张兆龙, 于新生, 区泽棠, 蔡启瑞.

分子催化, 1989, **3**(2): 104~109

95. LR spectroscopic study of chemisorbed dinitrogen species on ammonia synthesis iron catalysts

Zhang, H. B.; Tsai, K. R.,

Catalysis Letters, 1989, **3**(2): 129~141

96. Mechanistic analogy of ammonia synthesis & methanation over metal catalysts

Zhang, H. B.; Wan, H. L.; Tsai, K. R.,

Symposium on Activation, Conversion & Utilization of Methane, Pacific Chem '89 Congress (Honolulu), 1989, Preprints 3B

97. Model reactions of supported $\mu_2$-CH$_2$ or $\mu_3$-CCO metal carbonyls with CD$_3$OD and syngas for mechanism study of ethanol synthesis

Zhou, Z. H.; Gao, J. X.; Li, Y. G.; Fu, J. K.; Wang, H. Y.; Tsai, K. R.,

Abstract Volume of the 5th China-Japan-USA Symposium on Organometallic Chemistry, 1989, p. 70

98. Reactivities of supported iron-carbonyl cluster containing ($\mu_2$-CH$_2$) or ($\mu_3$-CCO) ligand with syngas and CD$_3$OD

Zhou, Z. H.; Gao, J. X.; Li, Y. G.; Fu, J. K.; Wang, H. Y.; Tsai, K. R.,

The 1989 International Chemical Congress of Pacific Basin (Honolulu, Hawaii, USA), 1989, Inor 467

99. Spectroscopic characterization of Mo-based cluster complexes-derived catalysts for desulfurization and denitrification

Zhang, H. B.; Lin, G. D.; Yang, Y. Q.; Liu, Y. D.; Tsai, K. R.,

27th International Conference Coordination Chemistry (Broadbeach, Australia), 1989, M51

100. Studies of methane oxidative coupling over LiCl-MnO$_x$-TiO$_2$ catalysts

Wang, F. C.; Zhang, Z. L.; Au, C. T.; Tsai, K. R.,

Symposium on Activation, Conversion & Utilization of Methane, Pacific Chem '89 Congress (Honolulu), 1989, Preprints 3B

101. 低压铜基甲醇合成催化剂活性表面的 XPS 和 TPD 表征

陈鸿博, 蔡俊修, 张鸿斌, 蔡启瑞.

厦门大学学报(自然科学版),1990,**29**(4):411~415

102. ESCA 研究 $CH_4$-$O_2$ 在铁和锰表面上的化学行为.

张兆龙,王水菊,区泽棠,蔡启瑞.

分子催化,1990,**4**(3):194~199

103. 福建省发展特殊精细石油化工产品的前景和重要性

蔡启瑞.

福建科技特刊,1990

104. $FeCl_2$-$(NH_4)_3VS_4$ 体系催化还原乙炔为乙烯

林国栋,周朝晖,张鸿图,蔡启瑞.

厦门大学学报(自然科学版),1990,**29**(5):542~545

105. 含卡宾或烯酮基铁簇合物作为乙醇合成机理模型的研究

周朝晖,高景星,傅锦坤,汪海有,李玉桂,蔡启瑞.

厦门大学学报(自然科学版),1990,**29**(3):286~290

106. 可溶于水的钌原子簇新体系催化丙烯和乙烯的醛化

高景星,区泽棠,王水菊,殷传光,蔡启瑞.

分子催化,1990,**4**(1):68~74

107. 重氧水和合成气与卡宾簇合物的模型反应研究铑催化乙醇合成机理

周朝晖,高景星,李玉桂,汪海有,蔡启瑞.

分子催化,1990,**4**(4):257~262

108. Methane oxidative coupling to $C_2$ hydrocarbons over lanthanum promoted barium catalysts

Zhang, Z. L.; Au, C. T.; Tsai, K. R.,

Applied Catalysis, 1990, **62**(2):L29~L33

109. ATP 与 $Fe_4S_4^*$ 络合的 $^{31}$P-NMR 研究

吴也凡,李春芳,曾定,洪亮,林国栋,蔡启瑞.

物理化学学报,1991,**7**(4):400~403

110. ATP 与 $MoFe_3S_4^*$ 原子簇的络合

袁友珠,吴也凡,曾定,洪亮,林国栋,蔡启瑞.

科学通报,1991,(15):1199

111. AHTD 法铜基催化剂合成甲醇的研究Ⅰ.催化剂的催化性能

胡云行,蔡俊修,万惠霖,蔡启瑞.

燃料化学学报,1991,**19**(2):181~184

112. 电子传递促进的 ATP 水解

吴也凡,曾定,林国栋,洪亮,蔡启瑞.

生物化学与生物物理进展,1991,**18**(5):374~375

113. 分子氮的电催化还原

吴也凡,王水菊,洪亮,林国栋,袁友珠,蔡启瑞.

高等学校化学学报,1991,**12**(9):1251~1252

114. 合成气转化为乙醇的反应机理

汪海有,刘金波,傅锦坤,蔡启瑞.

物理化学学报,1991,**7**(6):681~687

115. 磺化聚苯硫醚的制备与表征

　　吴也凡，林国栋，洪亮，蔡启瑞.

　　化学通报，1991，(5)：25～27

116. 磺化聚苯硫醚铑配合物催化剂初探

　　吴也凡，洪亮，林国栋，蔡启瑞.

　　化学研究与应用，1991，**3**(2)：98～99

117. 雾化高温分解法铜基催化剂合成甲醇的研究

　　胡云行，蔡俊修，万惠霖，蔡启瑞.

　　科学通报，1991，**36**(6)：476～477

118. 腺苷酸化合物与固氮酶组分结合的化学模拟

　　吴也凡，曾定，林国栋，洪亮，蔡启瑞.

　　生物化学与生物物理进展，1991，**18**(5)：375～376

119. 腺苷酸化合物与$[Mo_2Fe_6S_8(SPh)_6(MeO)_3]^{3-}$原子簇的络合

　　吴也凡，曾定，林国栋，洪亮，蔡启瑞.

　　厦门大学学报(自然科学版)，1991，**30**(4)：428～432

120. 烯烃醛化和羰化的新催化剂

　　高景星，区泽棠，万惠霖，蔡启瑞.

　　厦门大学学报(自然科学版)，1991，**30**(5)：486～491

121. 用同位素研究合成气制乙醇的反应机理

　　汪海有，刘金波，傅锦坤，李玉桂，蔡启瑞.

　　分子催化，1991，**5**(1)：16～23

122. Methane activation over titanium and titanium dioxide

　　Wang, F. C.; Wang, S. J.; Xu, F. C.; Au, C. T.; Wan, H. L.; Tsai, K. R.,

　　5th China-Japan-USA Trilateral Symposium on Catalysis (Evanston, Illinois), 1991, p. 20

123. Nature of composite active-sites in CO hydrogenation to methanol and to ethanol over metal/metal-oxide catalysts

　　Wang, H. Y.; Liu, J. P.; Cai, J. X.; Chen, H. B.; Zhang, H. B.; Wan, H. L.; Tsai, K. R.,

　　5th China-Japan-USA Trilateral Symposium on Catalysis (Evanston, Illinois), 1991, p. 33

124. New catalyst for hydroformylation and carbonylation of olefins

　　Gao, J. X.; Wan, H. L.; Tsai, K. R.,

　　Proceedings of 4th Asia Chemical Congress (August, Beijing), 1991, p. 906

125. Oxidative coupling of methane over $Na_2CO_3$ doped zirconium dioxide

　　Gong, Y. Q.; Wang, F. C.; Au, C. T.; Wan, H. L.; Tsai, K. R.,

　　Chinese Chemical Letters, 1991, **2**(12)：967～970

126. Oxidative coupling of methane over $Na_2CO_3$ doped zirconium dioxide

　　Gong, Y. Q.; Wang, F. C.; Au, C. T.; Wan, H. L.; Tsai, K. R.,

　　Proceedings of 4th Asia Chemical Congress (August, Beijing), 1991, p. 907

127. Promoter effect on hydroformylation, carbonylation and hydroesterification of olefins catalyzed by water-soluble ruthenium-phosphine-cluster complex

　　Gao, J. X.; Wan, H. L.; Au, C. T.; Tsai, K. R.,

China-Japan Bilateral Symposium on Effective Utilization of Carbon Resource，1991

128. Study on methanol synthesis over Cu-based catalysts prepared by AHTD
Hu, Y. H.; Cai, J. X.; Wan, H. L.; Tsai, K. R.,
China-Japan Bilateral Symposium on Effective Utilization of Carbon Resource，1991，pp. 14~15

129. Study on the mechanism of ethanol synthesis from syngas by chemical trapping and isotopic exchange reactions
Wang, H. Y.; Liu, J. P.; Fu, J. K.; Tsai, K. R.,
5th China-Japan-USA Trilateral Symposium on Catalysis（Evanston, Illinois），1991，p. 18

130. The effects of promoter to the oxidative coupling of methane over $TiO_2$
Wang, F. C.; Wan, H. L.; Tsai, K. R.,
Proceedings of 4th Asia Chemical Congress（August，Beijing），1991，p. 908

131. AHTD法铜基催化剂合成甲醇的研究 II. 催化剂氧化态前驱体的性质
胡云行，蔡俊修，万惠霖，蔡启瑞.
燃料化学学报，1992，**20**(2)：131~137

132. 合成气制乙醇铑催化剂中助剂的作用
傅锦坤，刘金波，许金来，汪海有，蔡启瑞.
厦门大学学报（自然科学版），1992，**31**(4)：387~391

133. 壳聚糖富氧膜的研究（Ⅰ）
丁俊琪，邹伟，何旭敏，夏海平，丁马太，蔡启瑞.
高等学校化学学报，1992，**13**(7)：985~986

134. 壳聚糖富氧膜的研究（Ⅱ）
丁俊琪，何旭敏，邹伟，夏海平，丁马太，蔡启瑞.
高等学校化学学报，1992，**13**(8)：1126~1127

135. 铑催化合成气转化为乙醇反应中甲酰基中间体的化学捕获
汪海有，刘金波，胡奕明，傅锦坤，蔡启瑞.
分子催化，1992，**6**(5)：346~351

136. 铑基催化剂上 $CO/H_2$ 合成甲醇及乙醇反应中的氘逆同位素效应
汪海有，刘金波，傅锦坤，蔡启瑞.
分子催化，1992，**6**(2)：156~159

137. 若干中性配体对 Mo-Fe-S 簇合物自兜的影响
张鸿图，林国栋，杨如，宋岩，蔡启瑞.
高等学校化学学报，1992，**13**(3)：362~365

138. ATP binding to nitrogenase and ATP-driven electron transfer in nitrogen fixation
You, C. B.; Song, W.; Zeng, D.; Tsai, K. R.,
The Nitrogen Fixation and its Research in China, Hong, G. F. Ed., Springer-Verlag & Shanghai Scientific & Technical Publishers, 1992：119~150

139. Characteristic studies of $MoS_x/K^+-SiO_2$ catalysts for synthesis of mixed alcohols from syngas
Zhang, H. B.; Huang, H. P.; Lin, G. D.; Tsai, K. R.,
Journal of Natural Gas Chemistry（China），1992，**1**(1)：1~10

140. Electron spectroscopic study of adsorption of methane on titanium dioxide surfaces

Wang, F. C.; Wan, H. L.; Tsai, K. R.; Wang, S. J.; Xu, F. C.,

Catalysis Letters, 1992, **12**(1-3): 319~326

141. Hydroformylation of higher olefins catalyzed by soluble polymer-bound rhodium complexes in homogeneous and heterogeneous systems

Yuan, Y. Z.; Zhang, H. T.; Zhang, H. B.; Tsai, K. R.,

7th International Symposium on Relations between Homogeneous and Heterogeneous Catalysis (Tokyo, Japan), 1992, pp. 108~114

142. Oxidative coupling of methane over rare-earth-nonreducible composite-metal oxides catalysts

Liu, Y. D.; Lin, G. D.; Zhang, H. B.; Cai, J. X.; Wan, H. L.; Tsai, K. R.,

203rd ACS National Meeting (San Francisco), 1992, Oral presentation 058-CATL Part 3; Abstracts of papers of ACS National Meeting, 1992, **203**, 53-PHYS

143. Studies on the mechanism of nitrogenase catalysis-substrates-cluster-coordination-chemistry approach

Tsai, K. R.; Wan, H. L.; Zhang, H. T.; Xu, L. S.,

The Nitrogen Fixation and its Research in China, Hong, G. F. Ed., Springer-Verlag & Shanghai Scientific & Technical Publishers, 1992: 87~117

144. Study on methanol synthesis over Cu-based catalysts by AHTD

Hu, Y. H.; Cai, J. X.; Wan, H. L.; Tsai, K. R.,

Chinese Science Bulletin, 1992, **37**(3): 262~263

145. Study on the mechanism of ethanol synthesis from syngas by in-situ chemical trapping and isotopic exchange reactions

Wang, H. Y.; Liu, J. P.; Fu, J. K.; Wan, H. L.; Tsai, K. R.,

Catalysis Letters, 1992, **12**(1-3): 87~96

146. Study on the method of chemical trapping for formyl intermediates

Wang, H. Y.; Liu, J. P.; Fu, J. K.; Zhang, H. B.; Tsai, K. R.,

Research on Chemical Intermediates, 1992, **17**(3): 233~242

147. AHTD法铜基催化剂中氧化铝的作用

胡云行, 万惠霖, 蔡启瑞.

高等学校化学学报, 1993, **14**(1): 106~108

148. 丙烯选择氧化铋钼铁复氧化物催化剂组成、结构及性能的研究

翁维正, 万惠霖, 戴深峻, 蔡俊修, 蔡启瑞.

分子催化, 1993, **7**(5): 339~346

149. 催化合成气合成乙醇的铑基催化剂中助剂锰的作用本质研究

汪海有, 刘金波, 傅锦坤, 蔡启瑞.

分子催化, 1993, **7**(4): 252~260

150. 纯稀土基丙烷氧化脱氢制丙烯催化剂的研究

张伟德, 万惠霖, 蔡启瑞.

高等学校化学学报, 1993, **14**(4): 566~567

151. 程序升温脱附导数谱

胡云行, 万惠霖, 蔡启瑞.

高等学校化学学报，1993，**14**（2）：238～243

152. 担载型水溶性膦铑配合物催化剂研究
袁友珠，杨意泉，张鸿斌，蔡启瑞.
高等学校化学学报，1993，**14**（6）：863～865

153. 钒促铑基催化剂上合成气反应中的 $H_2/D_2$ 同位素效应
汪海有，刘金波，傅锦坤，蔡启瑞.
高等学校化学学报，1993，**14**（8）：1157～1158

154. 固氮酶活性中心结构模型的 EHMO 研究
刘爱民，周泰锦，万惠霖，蔡启瑞.
高等学校化学学报，1993，**14**（7）：996～999

155. 合成气制乙醇的铑基催化剂中助剂铁、锂的作用本质研究
汪海有，刘金波，傅锦坤，蔡启瑞.
燃料化学学报，1993，**21**（4）：337～343

156. 甲烷氧化偶联制 $C_2$ 烃 $La_2O_3$ 基催化剂的研究
林国栋，王泉明，苏巧娟，刘玉达，张鸿斌，蔡启瑞.
厦门大学学报（自然科学版），1993，**32**（3）：312～316

157. 铑催化剂上 CO 的吸附态及其加氢反应
傅锦坤，许金来，刘金波，汪海有，蔡启瑞.
厦门大学学报（自然科学版），1993，**32**（5）：604～608

158. Raman 光谱方法在催化研究中应用实例枚举
张鸿斌，林国栋，廖远琰，刘玉达，蔡启瑞.
《原位技术在催化研究中的应用》（辛勤主编），北京大学出版社，1993，pp.253～265

159. $SiO_2$ 负载的磺化三苯膦铑配合物催化高碳烯氢甲酰化
袁友珠，陈鸿博，蔡启瑞.
应用化学，1993，**10**（4）：13～17

160. $SiO_2$ 负载的磺化三苯膦铑配合物催化高碳烯氢甲酰化及反应中的氘逆同位素效应
袁友珠，刘爱民，杨意泉，许金来，张鸿斌，蔡启瑞.
分子催化，1993，**7**（6）：439～445

161. 雾化高温分解法铜基甲醇合成催化剂的活性位
胡云行，黄爱民，蔡俊修，万惠霖，蔡启瑞.
催化学报，1993，**14**（6）：415～419

162. 稀土基复合物丙烷氧化脱氢制丙烯催化剂的研究
张伟德，周小平，蔡俊修，王水菊，万惠霖，蔡启瑞.
厦门大学学报（自然科学版），1993，**32**（4）：453～456

163. 载体促进输送膜进展
周水琴，丁马太，杨建灵，蔡启瑞.
材料科学与工程，1993，**11**（3）：32～36

164. Activation of $O_2$ and catalytic properties of $CeO_2/CaF_2$ catalysts for methane oxidative coupling
Zhou, X. P.; Wang, S. J.; Weng, W. Z.; Wan, H. L.; Tsai, K. R.,
Journal of Natural Gas Chemistry(China), 1993, **2**(4)：280～289

165. Beneficial effect of F⁻ on Sr/La oxide catalysts for the oxidative coupling of methane
Zhou, S. Q.; Zhou, X. P.; Weng, W. Z.; Huang, Y. P.; Wan, H. L.; Tsai, K. R.,
3$^{rd}$ China-France Symposium on Catalysis (Dalian, China), 1993

166. Catalyst design in oxidative dehydrogenation of C₁-C₄ alkanes to alkenes over rare-earth-based catalysts
Tsai, K. R.; Wan, H. L.; Zhang, H. B.; Lin, G. D.,
Proceedings of 34$^{th}$ IUPAC Congress (Beijing, China), 1993, Invited lecture, 5005

167. Catalytic mechanism of olefin hydroformylation over supported aqueous-phase catalysts
Yuan, Y. Z.; Xu, J. L.; Zhang, H. B.; Tsai, K. R.,
3$^{rd}$ China-France Symposium on Catalysis (Dalian, China), 1993, Oral presentation, O28

168. Characteristic of derivative temperature-programmed desorption spectra
Hu, Y. H.; Wan, H. L.; Tsai, K. R.,
6$^{th}$ China-Japan-USA Symposium on Catalysis (Beijing, China), 1993

169. Characterization of $MoS_x$-K⁺/$SiO_2$ catalysts for synthesis of mixed alcohols from syngas
Zhang, H. B.; Yang, Y. Q.; Huang, H. P.; Lin, G. D.; Tsai, K. R.; Nam, I. S.; Sermon, D. A.; Portela, L.,
Studies in Surface Science and Catalysis, 1993, **75**: 1493~1505

170. Effects of formic acid and $CO_2$ in CO hydrogenation to methanol over copper-based catalysts and nature of active sites
Cai, J. X.; Liao, Y. Y.; Chen, H. B.; Tsai, K. R.,
Studies in Surface Science and Catalysis, 1993, **75**: 2769~2772

171. Investigation on high efficient catalysts for the oxidative dehydrogenation of ethane
Zhou, X. P.; Zhou, S. Q.; Xu, F. C.; Wang, S. J.; Wan, H. L.; Tsai, K. R.,
6$^{th}$ USA-Japan-China Symposium on Catalysis (Beijing, China), 1993

172. Investigation on high efficient catalysts for the oxidative dehydrogenation of ethane
Zhou, X. P.; Zhou, S. Q.; Xu, F. C.; Wang, S. J.; Weng, W. Z.; Wan, H. L.; Tsai, K. R.,
Chemical Research in Chinese Universities, 1993, **9**(3): 269~272

173. Ligand-assisted self-assembly of FeMoco model compounds
Liu, A. M.; Zhang, H. T.; Zhou, M. Y.; Yang, R.; Wan, H. L.; Tsai, K. R.,
Proceedings of 34$^{th}$ IUPAC Congress (Beijing, China), 1993, p. 483

174. Low temperature catalysts modified by F⁻ for the oxidative coupling of methane
Zhou, X. P.; Zhou, S. Q.; Wang, S. J.; Wan, H. L.; Tsai, K. R.,
Proceedings of 34$^{th}$ IUPAC Congress (Beijing, China), Oral presentation, 1993, p. 687

175. Methane monooxygenase structure and mechanism - Where to Now?
Liu, A. M.; Li, S. B.; Wan, H. L.; Tsai, K. R.,
Abstracts of Papers of the American Chemical Society, 1993, **205**, 454-INOR

176. Methane oxidative coupling over alkaline-earth fluoro-oxide catalysts
Zhou, X. P.; Zhang, W. D.; Wang, S. J.; Wan, H. L.; Tsai, K. R.,
Journal of Natural Gas Chemistry(China), 1993, **2**(4): 344~347

177. Methane oxidative coupling over a new type of catalysts containing fluorides
Zhou, X. P.; Zhou, S. Q.; Wang, S. J.; Cai, J. X.; Wan, H. L.; Tsai, K. R.,

Abstracts of Papers of the American Chemical Society，1993，**205**，278-COLL

178. Methane oxidative coupling over fluoride-promoted cerium oxide catalysts

    Zhou，X. P.；Zhou，S. Q.；Wang，S. J.；Cai，J. X.；Weng，W. Z.；Wan，H. L.；Tsai，K. R.，

    Chemical Research in Chinese Universities，1993，**9**(3)：264～268

179. Methane oxidative coupling over fluorides promoted cerium oxide catalysts

    Zhou，X. P.；Zhou，S. Q.；Wang，S. J.；Cai，J. X.；Wan，H. L.；Tsai，K. R.，

    6ᵗʰ USA-Japan-China Symposium on Catalysis (Beijing，China)，1993

180. Methane oxidative coupling over fluoro-oxide catalysts

    Zhou，X. P.；Zhang，W. D.；Wan，H. L.；Tsai，K. R.，

    Catalysis Letters，1993，**21**(1-2)：113～122

181. Methane oxidative coupling over metal oxyfluoride catalysts

    Zhou，X. P.；Zhang，W. D.；Wan，H. L.；Tsai，K. R.，

    Chinese Chemical Letters，1993，**4**(7)：603～604

182. Mechanism of methane monooxygenase enzymatic chemical reactions

    Liu，A. M.；Li，S. B.；Wan，H. L.；Tsai，K. R.，

    Abstracts of Papers of the American Chemical Society，1993，**205**

183. Nature of active oxygen species for selective methane-oxidative coupling reactions

    Liu，Y. D.；Lin，G. D.；Zhang，H. B.；Tsai，K. R.，

    6ᵗʰ China-Japan-USA Symposium on Catalysis (Beijing)，1993，Oral presentation

184. New evidence for co-existence of ketene and acetyl intermediates in syngas conversion to ethanol over rhodium-based catalysts

    Wang，H. Y.；Liu，J. P.；Fu，J. K.；Cai，J. X.；Zhang，H. B.；Tsai，K. R.，

    Journal of Natural Gas Chemistry (China)，1993，**2**(1)：13～18

185. Oxidative coupling of methane over $BaF_2$-$TiO_2$ catalysts

    Zhou，S. Q.；Zhou，X. P.；Wan，H. L.；Tsai，K. R.，

    Catalysis Letters，1993，**20**(3-4)：179～183

186. Oxidative coupling of methane over new anion modified catalysts

    Zhou，S. Q.；Zhou，X. P.；Wan，H. L.；Tsai，K. R.，

    4ᵗʰ Japan-China Symposium on Coal and $C_1$ Chemistry (Osaka，Japan)，1993，Oral presentation

187. Preparation and performance of $ThO_2$-based methane-oxidative-coupling catalysts

    Liu，Y. D.；Lin，G. D.；Zhang，H. B.；Tsai，K. R.，

    Proceedings of 34ᵗʰ IUPAC Congress (Beijing，China)，1993，Poster presentation

188. Raman spectra of hydrogen adspecies on ammonia synthesis iron catalyst

    Chen，H. B.；Liao，Y. Y.；Zhang，H. B.；Tsai，K. R.，

    Chinese Chemical Letters，1993，**4**(5)：457～458

189. Studies of chemical modeling of nitrogenase catalysis

    Liu，A. M.；Zhang，F. Z.；Zhang，H. T.；Wan，H. L.；Tsai，K. R.，

    Proceedings of 34ᵗʰ IUPAC Congress (Beijing，China)，1993，p. 175

190. Studies on methane-oxidative-coupling (MOC) catalysis Ⅱ. Design，preparation and performance of rare-earth-based MOC catalysts

Liu, Y. D.; Lin, G. D.; Zhang, H. B.; Tsai, K. R.,

International Natural Gas Conversion Symposium (Sydney), 1993, Oral presentation

191. Study on the mechanism for clearage of C-O bond of CO and the rate-determining step of ethanol formation in hydrogenation of carbon monooxide on rhodium catalysts

Wang, H. Y.; Liu, J. P.; Xu, J. L.; Fu, J. K.; Lin, Z. Y.; Tsai, K. R.,

Proceedings of 34$^{th}$ IUPAC Congress (Beijing, China), 1993

192. Sulfonated polyphenylene-sulfide as ligands in Rh-complexes catalyzed hydroformylation

Yuan, Y. Z.; Zhang, H. T.; Zhang, H. B.; Tsai, K. R.,

Chinese Chemical Letters, 1993, **4**(2): 163～166

193. Supported aqueous-phase catalysts for hydroformylation: inverse deuterium effects and in-situ IR study

Yuan, Y. Z.; Xu, J. L.; Yang, Y. Q.; Zhang, H. B.;

Proceeding of 1$^{st}$ China Postdoctoral Academic Congress, 国防工业出版社, 北京, 1993, pp. 1423～1426

194. The investigation of methane oxidative coupling reaction on $CeO_2/BaF_2$ catalysts

Zhou, X. P.; Chao, Z. S.; Wang, S. J.; Weng, W. Z.; Wan, H. L.; Tsai, K. R.,

4$^{th}$ China-Japan Bilateral Symposium on Effective Utilization of Carbon Resources (Dalian, China), 1993, Oral presentation, p. 37

195. The promoting effect of fluoride in OCM catalyst $ZrO_2$-$BaF_2$

Chao, Z. S.; Zhou, X. P.; Weng, W. Z.; Wang, S. J.; Wan, H. L.; Tsai, K. R.,

4$^{th}$ China-Japan Bilateral Symposium on Effective Utilization of Carbon Resources (Dalian, China), 1993, Oral presentation, p. 129

196. α-钼酸铋的表面和体相组成与催化性能的关系

戴深峻, 翁维正, 祝以湘, 万惠霖, 蔡俊修, 蔡启瑞.

厦门大学学报(自然科学版), 1994, **33**(sup.): 198～203

197. 固氮酶底物络合活化模式的量子化学计算

刘爱民, 周泰锦, 张鸿图, 万惠霖, 蔡启瑞.

生物化学与生物物理进展, 1994, **21**(2): 171～172

198. 改进型铜基甲醇合成催化剂的制备研究

杨意泉, 张鸿斌, 林国栋, 陈汉忠, 袁友珠, 蔡启瑞.

厦门大学学报(自然科学版), 1994, **33**(4): 477～480

199. 合成气转化为甲醇和乙醇反应的速控步骤研究

汪海有, 夏文生, 刘金波, 张鸿斌, 蔡启瑞.

厦门大学学报(自然科学版), 1994, **33**(1): 58～62

200. 合成气制乙醇催化反应机理述评

汪海有, 刘金波, 蔡启瑞.

分子催化, 1994, **8**(6): 472～480

201. 含 Si-H 功能键聚硅氧烷/聚芳砜嵌段型高分子配体的合成与表征

周水琴, 杨建灵, 丁马太, 丁俊琪, 夏海平, 蔡启瑞.

厦门大学学报(自然科学版), 1994, **33**(5): 656～661

202. 金属氧化物催化剂分子设计的物理化学判则和数学模型

廖代伟,龚文华,李堂秋,林银钟,郭峰,蔡俊修,万惠霖,蔡启瑞.

厦门大学学报(自然科学版),1994,**33**(sup.):190~197

203. 甲烷氧化偶联 Ba-ZrO$_2$ 催化剂的研究

王泉明,林国栋,刘玉达,苏巧娟,张鸿斌,蔡启瑞.

厦门大学学报(自然科学版),1994,**33**(1):51~57

204. 几种具有缺陷 CaF$_2$ 型结构催化剂在甲烷氧化偶联反应中的行为

刘玉达,张鸿斌,林国栋,蔡启瑞.

厦门大学学报(自然科学版),1994,**33**(sup.):214~219

205. 铑催化合成气制乙醇反应中 CO 断键途径的研究

汪海有,刘金波,许金来,傅锦坤,林种玉,张鸿斌,蔡启瑞.

分子催化,1994,**8**(2):111~116

206. La$_2$O$_3$ 基催化剂表面的某些特征对其甲烷氧化偶联反应行为的影响

刘玉达,张鸿斌,林国栋,蔡启瑞.

厦门大学学报(自然科学版),1994,**33**(sup.):220~224

207. MgO(100)面上甲烷氧化偶联反应的量子化学研究

林银钟,廖代伟,郭峰,龚文华,万惠霖,蔡启瑞.

厦门大学学报(自然科学版),1994,**33**(sup.):351~354

208. 双齿配体 DPPE 和 DPPM 对 Mo-Fe-S 簇合物自兜合成的影响

刘爱民,袁友珠,张鸿图,周明玉,杨如,万惠霖,蔡启瑞.

厦门大学学报(自然科学版),1994,**33**(6):809~813

209. 铜基甲醇合成催化剂 TPR 导数谱

胡云行,万惠霖,蔡启瑞.

高等学校化学学报,1994,**15**(10):1550~1552

210. 优化固氮活性模型信息研究——双齿配体 DPPE 和 DPPM 对固氮酶促反应的影响

刘爱民,张凤章,张鸿图,袁友珠,许良树,万惠霖,蔡启瑞.

分子催化,1994,**8**(2):81~85

211. 载体酸碱性对丙烷氧化脱氢钒基催化剂性能的影响

方智敏,张伟德,刘为东,蔡俊修,万惠霖,蔡启瑞.

厦门大学学报(自然科学版),1994,**33**(sup.):225~228

212. A novel metal fluoride promoted metal-oxide catalyst system for methane and light alkane(C$_2$-C$_4$) conversion in presence of oxygen

Zhou, X. P.; Zhou, S. Q.; Zhang, W. D.; Chao, Z. S.; Weng, W. Z.; Long, R. Q.; Wang, S. J.; Cai, J. X.; Wan, H. L.; Tsai, K. R.,

Abstracts of Papers of the American Chemical Society, 1994, **207**, 13-CATL

213. Enhanced C$_2$ yield of methane oxidative coupling by means of a double layered catalyst bed

Liu, Y. D.; Zhang, H. B.; Su, Q. J.; Lin, G. D.; Tsai, K. R.,

Chinese Chemical Letters, 1994, **5**(10):863~864

214. New perspectives on the structures and functions of nitrogenase M-cluster and P-cluster pair

Tsai, K. R.; Wan, H. L.,

Bioinorganic Chemistry, 1994

215. In-situ Raman-spectra of oxygen-containing adspecies on Th-La-O$_x$/BaCO$_3$ Catalysts for Methane Oxidative Coupling

Zhang, H. B.; Liu, Y. D.; Lin, G. D.; Liao, Y. Y.; Tsai, K. R.,

Abstracts of Papers of the American Chemical Society, 1994, **207**, 7-COLL (1994)

216. In-situ Raman spectroscopic study of oxygen adspecies on a Th-La-O$_x$ catalyst for methane oxidative coupling reaction

Liu, Y. D.; Zhang, H. B.; Lin, G. D.; Liao, Y. Y.; Tsai, K. R.,

Journal of the Chemical Society-Chemical Communications, 1994, (16): 1871~1872

217. Oxidative dehydrogenation of isobutane to isobutylene over F- modified rare earth metal oxide catalysts

Zhang, W. D.; Tang, D. L.; Lai, W. Q.; Wan, H. L.; Tsai, K. R.,

Chemical Research in Chinese Universities, 1994, **10**(4): 381~383

218. Oxidative dehydrogenation of propane over fluorine promoted rare earth-based catalysts

Zhang, W. D.; Zhou, X. P.; Tang, D. L.; Wan, H. L.; Tsai, K.,

Catalysis Letters, 1994, **23**(1-2): 103~106

219. Rate-determining step in olefin hydroformylation over supported aqueous-phase catalysts

Yuan, Y. Z.; Yang, Y. Q.; Xu, J. L.; Zhang, H. B.; Tsai, K. R.,

Chinese Chemical Letters, 1994, **5**(4): 291~294

220. Selective oxidative dehydrogenation of isobutane over a Y$_2$O$_3$-CeF$_3$ catalyst

Zhang, W. D.; Tang, D. L.; Zhou, X. P.; Wan, H. L.; Tsai, K. R.,

Journal of the Chemical Society-Chemical Communications, 1994, (6): 771

221. Studies on methane-oxidative-coupling (MOC) catalysts . II. design, preparation and characterization of rare-earth-based MOC catalysts

Liu, Y. D.; Lin, G. D.; Zhang, H. B.; Tsai, K. R.,

In Natural Gas Conversion II, Elsevier Science Publ B V: Amsterdam; 1994, **81**, pp. 131~136

222. The beneficial effect of alkali-metal salt on supported aqueous-phase catalysts for olefin hydroformylation

Yuan, Y. Z.; Xu, J. L.; Zhang, H. B.; Tsai, K. R.,

Catalysis Letters, 1994, **29**(3-4): 387~395

223. The oxidative coupling of methane and the activation of molecular O$_2$ on CeO$_2$/BaF$_2$

Zhou, X. P.; Chao, Z. S.; Weng, W. Z.; Zhang, W. D.; Wang, S. J.; Wan, H. L.; Tsai, K. R.,

Catalysis Letters, 1994, **29**(1-2): 177~188

224. The promoting effect of fluoride to OCM catalyst ZrO$_2$-BaF$_2$

Chao, Z. S.; Zhou, X. P.; Wang, S. J.; Xu, F. C.; Wan, H. L.; Tsai, K. R.,

Chinese Chemical Letters, 1994, **5**(8): 685~686

225. 催化与环境保护

高利珍, 张伟德, 万惠霖, 蔡启瑞.

大自然探索, 1994, **14**(52): 6~9

226. 低温催化裂解烷烃法制备碳纳米管

陈萍,王培峰,林国栋,张鸿斌,蔡启瑞.

高等学校化学学报,1995,**16**(11):1783～1784

227. $Fe_4S_4$ 簇合物中 $\mu_3$-S 的酸不稳定性研究

黄静伟,张鸿图,周明玉,杨如,万惠霖,蔡启瑞.

厦门大学学报(自然科学版),1995,**34**(5):737～740

228. 氟化锶/氧化钕催化剂的甲烷氧化偶联性能及其吸附氧物种的原位 FTIR 光谱研究

龙瑞强,万惠霖,赖华龙,蔡启瑞.

高等学校化学学报,1995,**16**(11):1796～1797

229. 固氮酶及合成氨 Fe 催化剂中 $N_2$ 的络合位

黄静伟,张凤章,许良树,张鸿图,万惠霖,蔡启瑞.

高等学校化学学报,1995,**16**(6):920～923

230. 固氮酶中 $N_2$ 键合位的研究

黄静伟,张凤章,许良树,张鸿图,万惠霖,蔡启瑞.

厦门大学学报(自然科学版),1995,**34**(3):378～381

231. 含 Si-H 功能键聚硅氧烷/聚芳砜嵌段共聚物膜性能研究

周水琴,丁马太,夏海平,杨建灵,丁俊琪,蔡启瑞.

厦门大学学报(自然科学版),1995,**34**(1):71～75

232. 甲烷氧化偶联 $CaF_2/Sm_2O_3$ 催化剂的研究

周水琴,龙瑞强,黄亚萍,万惠霖,蔡启瑞.

高等学校化学学报,1995,**16**(2):290～292

233. 甲烷氧化偶联的分子催化

刘玉达,张鸿斌,林国栋,蔡启瑞.

《碳一化学中的催化作用》(蔡启瑞,彭少逸等编著),化学工业出版社,1995,pp.47～78

234. 碱土金属氟化物/三氧化二钐催化剂的甲烷氧化偶联性能

周水琴,龙瑞强,翁维正,万惠霖,蔡启瑞.

厦门大学学报(自然科学版),1995,**34**(3):382～386

235. 模拟固氮酶中 Mo 微环境的化学探测

黄静伟,张鸿图,杨如,周明玉,万惠霖,蔡启瑞.

高等学校化学学报,1995,**16**(4):632～634

236. 柠檬酸氧钒(V)配合物 $Na_2(NH_4)_4$-$[VO_2(cit)_2]$·$6H_2O$ 的合成和晶体结构

周朝晖,万惠霖,蔡启瑞.

科学通报,1995,**40**(4):321～323

237. 脱附过程中表面物种的迁移效应

胡云行,万惠霖,蔡启瑞.

中国科学 B 辑,1995,**25**(5):465～471

238. Biphasic synergy catalysis of the oxidative dehydrogenation of propane over VMgO catalysts

Fang, Z. M.; Weng, W. Z.; Wan, H. L.; Tsai, K. R.,

Journal of Molecular Catalysis (China), 1995, **9**(6):401～410

239. Bridge theory and physical-chemistry in molecular design of catalyst

Liao, D. W.; Huang, Z. N.; Wan, H. L.; Tsai, K. R.,

Abstracts of Papers of the American Chemical Society, 1995, **209**, 243-PHYS

240. Effects of surface migration of adsorbed species on desorption

Hu, Y. H.; Wan, H. L.; Tsai, K. R.,

Science in China Series B-Chemistry Life Sciences & Earth Sciences, 1995, **38**(8): 903~911

241. EHMO studies of chemisorbed dioxygen species on $Na_2O$ and $K_2O$ and of their interaction with $CH_4$ and $CH_3$-radical

Su, Q. J.; Zhang, H. B.; Zhou, T. J.; Liu, Y. D.; Lin, G. D.; Tsai, K. R.,

Chemical Research in Chinese Universities, 1995, **11**(1): 50~57

242. In situ FTIR spectral study of the oxygen adspecies on $SrF_2/La_2O_3$ catalyst during the oxidative coupling of methane

Long, R. Q.; Zhou, S. Q.; Huang, Y. P.; Wang, H. Y.; Wan, H. L.; Tsai, K. R.,

Chinese Chemical Letters, 1995, **6**(8): 727~730

243. In situ Raman study of oxygen species of $BaF_2$-LaOF catalysts

Chao, Z. S.; Zhou, X. P.; Wan, H. L.; Tsai, K. R.,

Chinese Chemical Letters, 1995, **6**(3): 239~242

244. Metal-hydroxycarboxylate interactions: syntheses and structures of $K_2[VO_2(C_6H_6O_7)]_2 \cdot 4H_2O$ and $(NH_4)_2[VO_2(C_6H_6O_7)]_2 \cdot 2H_2O$

Zhou, Z. H.; Yan, W. B.; Wan, H. L.; Tsai, K. R.; Wang, J. Z.; Hu, S. Z.,

Journal of Chemical Crystallography, 1995, **25**(12): 807~811

245. Methane oxidative coupling on $BaF_2$/LaOF catalyst

Chao, Z. S.; Zhou, X. P.; Wan, H. L.; Tsai, K. R.,

Applied Catalysis A-General, 1995, **130**(2): 127~133

246. Molecular recognition in nitrogenase catalysis

Tsai, K. R.; Wan, H. L.; Huang, J. W.; Zhang, H. Z.; Xu, L. S.; Zhang, H. T.,

Faseb Journal, 1995, **9**(6): A1460

247. On the structure-function relationship of nitrogenase M-cluster and P-cluster pairs

Tsai, K. R.; Wan, H. L.,

Journal of Cluster Science, 1995, **6**(4): 485~501

248. Oxidative dehydrogenation of ethane over $BaF_2$-LaOF catalysts

Zhou, X. P.; Chao, Z. S.; Luo, J. Z.; Wan, H. L.; Tsai, K. R.,

Applied Catalysis A-General, 1995, **133**(2): 263~268

249. Oxidative dehydrogenation of ethane(ODE) over LaOF catalyst promoted with $BaF_2$

Zhou, X. P.; Chao, Z. S.; Wan, H. L.; Tsai, K. R.,

Chinese Chemical Letters, 1995, **6**(4): 347~348

250. Promoting effect of $F^-$ on Sr/La oxide catalysts for the oxidative coupling of methane

Long, R. Q.; Zhou, S. Q.; Huang, Y. P.; Weng, W. Z.; Wan, H. L.; Tsai, K. R.,

Applied Catalysis A-General, 1995, **133**(2): 269~280

251. Studies of catalysis in partial oxidation of methane to syngas(II) - Chemisorbed species，energetics and mechanism

Chen，P.；Zhang，H. B.；Lin，G. D.；Tsai，K. R.，

Chemical Research in Chinese Universities，1995，**11**(4)：323～335

252. Synthesis and crystal-structure of sodium ammonium dimeric（citrato）dioxovanadium（V）Na₂ (NH₄)₄[VO₂(cit)]₂·6H₂O

Zhou，Z. H.；Wan，H. L.；Tsai，K. R.，

Chinese Science Bulletin，1995，**40**(9)：749～752

253. Syntheses and structures of the potassium-ammonium dioxocitratovanadate(V) and sodium oxoci-tratovanadate(IV) dimers

Zhou，Z. H.；Wan，H. L.；Hu，S. Z.；Tsai，K. R.，

Inorganica Chimica Acta，1995，**237**(1-2)：193～197

254. The investigation of methane oxidative coupling on BaF₂-LaOF catalyst

Chao，Z. S.；Zhou，X. P.；Wan，H. L.；Tsai，K. R.，

Chinese Chemical Letters，1995，**6**(8)：715～718

255. The investigation of oxygen absorption over LaOF by means of Raman-spectroscopy

Zhou，X. P.；Chao，Z. S.；Wang，S. J.；Wan，H. L.；Tsai，K. R.，

Chemical Research in Chinese Universities ，1995，**11**(1)：84～86

256. 氨合成铁催化剂上化学吸附物种的 in-situ FTIR 谱

廖代伟，林种玉，蔡启瑞.

厦门大学学报(自然科学版)，1996，**35**(5)：734～738

257. 丙烷氧化脱氢 VMgO 催化剂双相协同催化作用和活性位的研究

方智敏，翁维正，万惠霖，蔡启瑞.

科学通报，1996，**41**(20)：1919～1920

258. 催化剂分子设计专家系统 ESMDC 及其在甲烷氧化偶联催化剂设计中的应用

黄遵楠，廖代伟，张鸿斌，万惠霖，蔡启瑞.

计算机与应用化学，1996，**13**(3)：167～171

259. 催化剂分子设计专家系统推理机的设计

黄遵楠，廖代伟，林银钟，张鸿斌，万惠霖，蔡启瑞.

厦门大学学报(自然科学版)，1996，**35**(3)：389～392

260. 催化剂分子设计专家系统中数据库和知识库的建立

黄遵楠，廖代伟，林银钟，张鸿斌，万惠霖，蔡启瑞.

厦门大学学报(自然科学版)，1996，**35**(5)：739～744

261. 低温催化裂解烷烃法制备碳纳米管

陈萍，张鸿斌，林国栋，蔡启瑞，翟和生.

厦门大学学报(自然科学版)，1996，**35**(1)：61～66

262. 负载型水溶性膦铑配合物催化剂上气、液态烯烃氢甲酰化

袁友珠，杨意泉，林国栋，张鸿斌，蔡启瑞.

《羰基合成化学》(殷元骐主编)，化学工业出版社，1996，pp.64～91

263. 负载型水相膦铑配合物催化剂上丙烯氢甲酰化制丁醛
袁友珠,张宇,杨意泉,林国栋,张鸿斌,蔡启瑞.
厦门大学学报(自然科学版),1996,**35**(2):220～225

264. 固氮酶中的电子传递
黄静伟,张鸿图,万惠霖,蔡启瑞.
生物化学与生物物理进展,1996,**23**(1):18～20

265. 化学探针方法研究固氮酶 M-簇和 P-簇对的结构与功能关系
万惠霖,黄静伟,张凤章,周朝晖,张鸿图,许良树,蔡启瑞.
厦门大学学报(自然科学版),1996,**35**(6):890～899

266. 甲烷氧化偶联反应催化剂分子设计专家系统的建立
黄遵楠,廖代伟,张鸿斌,万惠霖,蔡启瑞.
福建电脑,1996,(2):1～8

267. 甲烷氧化偶联金属氧化物催化剂组分分子设计程序
廖代伟,黄遵楠,林银钟,龚文华,万惠霖,张鸿斌,蔡启瑞.
厦门大学学报(自然科学版),1996,**35**(1):133～136

268. $Mn^{2+}$ 和 $La^{3+}$ 对提高六铝酸盐热稳定性燃烧活性的作用
陈笃慧,毛通双,杨乐夫,蔡启瑞.
天然气化工(C1 化学与化工),1996,**21**(4):24～27

269. 强酸促进的 Mo/HZSM-5 基催化剂上甲烷脱氢芳构化研究
曾金龙,熊智涛,林国栋,张鸿斌,蔡启瑞.
厦门大学学报(自然科学版),1996,**35**(6):900～906

270. Computer-aided molecular design of catalysts based on mechanism and structure
Liao, D. W.; Huang, Z. N.; Lin, Y. Z.; Wan, H. L.; Zhang, H. B.; Tsai, K. R.,
Journal of Chemical Information and Computer Sciences, 1996, **36**(6): 1178～1182

271. Computer simulation of derivative TPD
Hu, Y. H.; Wan, H. L.; Tsai, K. R.; Au, C. T.,
Thermochimica Acta, 1996, **274**: 289～301

272. Constituent selection and performance characterization of catalysts for oxidative coupling of methane and oxidative dehydrogenation of ethane
Wan, H. L.; Chao, Z. S.; Weng, W. Z.; Zhou, X. P.; Cai, J. X.; Tsai, K. R.,
Catalysis Today, 1996, **30**(1-3): 67～76

273. Preparation of supported gold catalysts from gold complexes and their catalytic activities for CO oxidation
Yuan, Y. Z.; Asakura, K.; Wan, H. L.; Tsai, K.; Iwasawa, Y.,
Catalysis Letters, 1996, **42**(1-2): 15～20

274. Structure and catalysis of a $SiO_2$-supported gold-platinum cluster $[(PPh_3)Pt(PPh_3Au)_6](NO_3)_2$
Yuan, Y. Z.; Asakura, K.; Wan, H. L.; Tsai, K.; Iwasawa, Y.,
Chemistry Letters, 1996, (2): 129～130

275. Studies on catalysis in partial oxidation of methane to syngas（Ⅰ）—Performance and characterization of Ni-based catalysts

Chen, P.; Zhang, H. B.; Lin, G. D.; Tsai, K. R.,
Chemical Research in Chinese Universities, 1996, **12**(1): 70～80

276. Study on catalytic synthesis of methanol from syngas via methylformate in heterogeneous "one-pot" reactions
Zhang, H. B.; Li, H. Y.; Lin, G. D.; Liu, Y.; Tsai, K. R.,
In 11<sup>th</sup> International Congress on Catalysis - 40<sup>th</sup> Anniversary, Pts a and B, Elsevier Science Publ B V: Amsterdam; 1996, **101**, pp 1369～1378

277. Supported gold catalysts derived from gold complexes and as-precipitated metal hydroxides, highly active for low-temperature CO oxidation
Yuan, Y. Z.; Asakura, K.; Wan, H. L.; Tsai, K. R.; Iwasawa, Y.,
Chemistry Letters, 1996, (9): 755～756

278. The performance and structure of rare earth oxides modified by strontium fluoride for methane oxidative coupling
Long, R. Q.; Huang, Y. P.; Weng, W. Z.; Wan, H. L.; Tsai, K. R.,
Catalysis Today, 1996, **30**(1-3): 59～65

279. 氨合成反应的 BOC-MP 法研究
黑美军, 陈鸿博, 林贻基, 洪琦, 林银钟, 易军, 廖代伟, 蔡启瑞.
厦门大学学报(自然科学版), 1997, **36**(6): 879～884

280. 促进型甲酸甲酯氢解制甲醇铜基催化剂的研究
李海燕, 张鸿斌, 林国栋, 杨意泉, 蔡启瑞.
厦门大学学报(自然科学版), 1997, **36**(3): 381～387

281. CH₄-CO₂ 重整制合成气 Ni 基催化剂的失活研究
陈萍, 张鸿斌, 林国栋, 严程, 蔡启瑞.
天然气化工(C1 化学与化工), 1997, **22**(1): 12～15

282. 过渡金属配合物催化剂及其分子设计构思的发展与相互作用
万惠霖, 袁友珠, 高景星, 张鸿斌, 蔡启瑞.
高等学校化学学报, 1997, **18**(7): 1185～1193

283. 计算机辅助催化剂组分分子设计的思想和实现
黄遵楠, 廖代伟, 万惠霖, 张鸿斌, 蔡启瑞.
计算机与应用化学, 1997, **14**(1): 12～16

284. 膜反应器——化学、化工、材料科学发展中的机遇
高利珍, 李基涛, 万惠霖, 蔡启瑞.
科技导报, 1997, (3): 35～37

285. 手性多齿胺膦钌配合物的设计合成和在芳香酮不对称氢化反应中的应用
高景星, 许翩翩, 黄培强, 万惠霖, 蔡启瑞, 碇屋隆雄, 野依良治.
分子催化, 1997, **11**(6): 413～416

286. 碳纳米管负载铑催化剂上丙烯氢甲酰化
张宇, 吴汜昕, 张鸿斌, 林国栋, 袁友珠, 蔡启瑞.
物理化学学报, 1997, **13**(12): 1057～1060

287. 同位素方法在催化反应机理研究中的应用
汪海有,蔡启瑞.
化学通报,1997,(9):1～6

288. 新的胺膦钌配合物的合成及在氢转移氢化反应中的应用
许翩翩,高景星,王文国,陈忠,黄培强,万惠霖,蔡启瑞.
应用化学,1997,**14**(5):19～22

289. 新的双胺双膦钌配合物的合成、表征和催化性能
许翩翩,高景星,王文国,陈忠,黄培强,万惠霖,蔡启瑞.
物理化学学报,1997,**13**(6):484～488

290. 重要化工原料——异丁醛的制备方法
陈笃慧,袁友珠,蔡启瑞.
天然气化工(C1 化学与化工),1997,**22**(1):41～45

291. Biphasic synergy catalysis and active sites of VMgO catalysts for oxidative dehydrogenation of propane
Fang, Z. M.; Weng, W. H.; Wan, H. L.; Cai, Q. R.,
Chinese Science Bulletin, 1997, **42**(2):172～174

292. Characterization and catalysis of a $SiO_2$-supported $[Au_6Pt]$ cluster $[(AuPPh_3)_6Pt(PPh_3)]^{2+}/SiO_2$
Yuan, Y. Z.; Asakura, K.; Wan, H. L.; Tsai, K. R.; Iwasawa, Y.,
Journal of Molecular Catalysis A-Chemical, 1997, **122**(2-3):147～157

293. Coke-resistant Ni-based catalyst for partial oxidation and $CO_2$-reforming of methane to syngas
Chen, P.; Zhang, H. B.; Lin, G. D.; Guo, Z. P.; Tsai, K. R.,
Chemical Research in Chinese Universities, 1997, **13**(1):83～85

294. Growth of carbon nanotubes by catalytic decomposition of $CH_4$ or CO on a Ni-MgO catalyst.
Chen, P.; Zhang, H. B.; Lin, G. D.; Hong, Q.; Tsai, K. R.,
Carbon, 1997, **35**(10-11):1495～1501

295. Molybdenum(VI) complex with citric acid: Synthesis and structural characterization of 1:1 ratio citrate molybdate $K_2Na_4[(MoO_2)_2O(cit)_2] \cdot 5H_2O$
Zhou, Z. H.; Wan, H. L.; Tsai, K. R.,
Polyhedron, 1997, **16**(1):75～79

296. Supported Au catalysts prepared from Au phosphine complexes and As-precipitated metal hydroxides: Characterization and low-temperature CO oxidation
Yuan, Y. Z.; Kozlova, A. P.; Asakura, K.; Wan, H. L.; Tsai, K.; Iwasawa, Y.,
Journal of Catalysis, 1997, **170**(1):191～199

297. Syntheses, structures and spectroscopic properties of nickel(II) citrato complexes, $(NH_4)_2[Ni(Hcit)(H_2O)_2]_2 \cdot 2H_2O$ and $(NH_4)_4[Ni(Hcit)_2] \cdot 2H_2O$
Zhou, Z. H.; Lin, Y. J.; Zhang, H. B.; Lin, G. D.; Tsai, K. R.,
Journal of Coordination Chemistry, 1997, **42**(1-2):131～141

298. 丙烷氧化脱氢 VMgO 催化剂活性位的研究
方智敏,翁维正,万惠霖,蔡启瑞.
厦门大学学报(自然科学版),1998,**37**(4):525～531

299. $Cr_2O_3$ 在铜基甲醇合成催化剂中的作用

陈鸿博,于腊佳,廖代伟,林国栋,张鸿斌,蔡启瑞.

物理化学学报,1998,**14**(6):534～539

300. 催化裂解 $CH_4$ 或 $CO$ 制碳纳米管结构性能的谱学表征

陈萍,张鸿斌,林国栋,蔡启瑞.

高等学校化学学报,1998,**19**(5):765～769

301. 负载型水溶性铑膦配合物催化剂的结构和性能

袁友珠,张宇,陈忠,张鸿斌,蔡启瑞.

物理化学学报,1998,**14**(11):1013～1019

302. 甲醇在过渡金属上反应机理的 BOC-MP 法研究

黑美军,陈鸿博,林贻基,洪琦,林银钟,易军,廖代伟,蔡启瑞.

化学物理学报,1998,**11**(2):166～173

303. 清洁及氧修饰 $Cu(100)$ 表面上水煤气变换反应的能量学

吴廷华,夏文生,汪海有,王仲权,张鸿斌,万惠霖,蔡启瑞.

化学物理学报,1998,**11**(1):69～73

304. $SiO_2$ 气凝胶的制备及微孔分布

高利珍,汤皎宁,李基涛,万惠霖,蔡启瑞,严前古,储伟,于作龙.

无机化学学报,1998,**14**(3):292～297

305. 铜基催化剂表面缺陷对催化性能的影响

陈鸿博,于腊佳,廖代伟,林国栋,张鸿斌,蔡启瑞.

化学物理学报,1998,**11**(5):456～460

306. 推进化工生产可持续发展的途径——绿色化学与技术

闵恩泽,陈家镛,蔡启瑞,沈家聪,戴立信,胡英.

中国科学院院刊,1998,(6):413～415

307. VMgO 催化剂上丙烷氧化脱氢反应的原位 Raman 谱学研究

方智敏,翁维正,万惠霖,蔡启瑞.

分子催化,1998,**12**(3):207～213

308. 新的膦钯配合物的合成、表征及其催化氢化性能

许翩翩,高景星,丁开宁,林东海,阮源萍,陈守正,黄培强,万惠霖,蔡启瑞.

厦门大学学报(自然科学版),1998,**37**(1):52～57

309. 稀土氟氧化物在丙烷、异丁烷氧化脱氢中的催化作用

张伟德,洪碧凤,古萍英,万惠霖,蔡启瑞.

应用化学,1998,**15**(3):76～78

310. 载体酸碱性对钒基催化剂丙烷氧化脱氢性能的影响及其 IR 和 EPR 研究

方智敏,翁维正,万惠霖,蔡启瑞.

天然气化工(C1 化学与化工),1998,**23**(2):25～29

311. 在 $YBa_2Cu_3O_{6～7}$ 超导催化剂上 $CO_2$ 加氢制醇的研究

高利珍,严前古,于作龙,李基涛,万惠霖,蔡启瑞.

化学学报,1998,**56**(10):1015～1020

312. 正丁基苯甲酸选择加氢催化剂的制备及其催化性能

陈鸿博,廖代伟,黑美军,林贻基,洪琦,于腊佳,蔡启瑞.

厦门大学学报(自然科学版),1998,**37**(2):306～308

313. Characterization and propene hydroformylation of rhodium-phosphine complex supported on MCM-41 mesoporous materials

Yuan, Y. Z.; Fu, Q. J.; Tsai, K. R.,

Abstracts of Papers of the American Chemical Society, 1998, **216**, 199-COLL

314. Development of coking-resistant Ni-based catalyst for partial oxidation and $CO_2$-reforming of methane to syngas

Chen, P.; Zhang, H. B.; Lin, G. D.; Tsai, K. R.,

Applied Catalysis A-General, 1998, **166**(2):343～350

315. Enantioselective transfer hydrogenation of aromatic ketones catalyzed by new diaminodiphosphine Ru(II) complexes

Xu, P. P.; Gao, J. X.; Yi, X. D.; Huang, Y. Q.; Zhang, H.; Wan, H. L.; Tsai, K. R.; Takao, I.,

Chemical Research in Chinese Universities, 1998, **14**(3):340～343

316. Halide promoted catalysts with defective fluorite structure for selective oxidative dehydrogenation of ethane to ethene

Wan, H. L.; Weng, W. Z.; Chen, M. S.; Zhu, Y. M.; Chen, T.; Tsai, K. R.,

Abstracts of Papers of the American Chemical Society, 1998, **216**, 200-COLL

317. Molecular recognition in nitrogenase catalysis and two proton-relay pathways from P-cluster to M-center

Wan, H. L.; Huang, J. W.; Zhang, F. Z.; Wu, Y.; Xu, L. S.; Li, J. L.; Tsai, K. R.,

Current Plant Science and Biotechnology in Agriculture, 31, (Biological Nitrogen Fixation for the 21st Century), 1998:78～79

318. Nonoxidative dehydrogenation and aromatization of methane over W/HZSM-5-based catalysts

Zeng, J. L.; Xiong, Z. T.; Zhang, H. B.; Lin, G. D.; Tsai, K. R.,

Catalysis Letters, 1998, **53**(1-2):119～124

319. Structural properties of $[(AuPH_3)_6Pt(H_2)(PH_3)]^{2+}$: theoretical study of dihydrogen activation

Xu, X.; Yuan, Y. Z.; Asakura, K.; Iwasawa, Y.; Wan, H. L.; Tsai, K. R.,

Chemical Physics Letters, 1998, **286**(1-2):163～170

320. Supported gold catalysis derived from the interaction of a Au-phosphine complex with as-precipitated titanium hydroxide and titanium oxide

Yuan, Y. Z.; Asakura, K.; Kozlova, A. P.; Wan, H. L.; Tsai, K. R.; Iwasawa, Y.,

Catalysis Today, 1998, **44**(1-4):333～342

321. The catalytic properties of modified methanol synthesis catalyst NC208

Yang, Y. Q.; Dai, S. J.; Yuan, Y. Z.; Zhang, H. B.; Tsai, K. R.,

In: Advances of alcohol fuels in the world (Eds. Zhu Qiming et al.) Tsinghua Univ. Press, Beijing, 1998, pp. 14～20

322. The effects of $M_2O_3$ on stabilizing monocopper over the surface of Cu-ZnO-$M_2O_3$ catalysts for methanol synthesis

Chen, H. B.; Liao, D. W.; Yu, L. J.; Yi, J.; Zhang, H. B.; Tsai, K. R.,

Journal of the Chinese Chemical Society, 1998, **45**(5): 673～678

323. The long-neglected, active surface-oxygen species, (O)under-bar(3)(2-), in oxidative coupling of methane

Tsai, K. R.; Zhang, H. B.; Liu, Y. D.; Lin, G. D.; Wan, H. L.; Weng, W. Z.,

Abstracts of Papers of the American Chemical Society, 1998, **216**, U286

324. 氨合成催化剂及其催化反应机理研究进展

王丽华, 陈守正, 廖代伟, 蔡启瑞.

化学进展, 1999, **11**(4): 376～384

325. 丙烷在负载型 $V_2O_5$ / $Zr_3(PO_4)_4$ 催化剂上的氧化脱氢

张伟德, 沙开清, 于腊佳, 万惠霖, 蔡启瑞.

应用化学, 1999, **16**(1): 34～37

326. 负载型油溶性铑膦配合物的结构及其氢甲酰化催化性能

袁友珠, 张宇, 蔡阳, 杨意泉, 张鸿斌, 蔡启瑞.

天然气化工(C1 化学与化工), 1999, **24**(3): 16～21

327. 高分辨 NMR 研究金属盐对水溶性铑膦配合物分子结构的影响

袁友珠, 张宇, 叶剑良, 廖新丽, 姚春香, 杨意泉, 蔡启瑞.

高等学校化学学报, 1999, **20**(6): 914～917

328. 甲烷脱氢芳构化 W/HZSM-5 基催化剂的制备

熊智涛, 曾金龙, 张鸿斌, 林国栋, 蔡启瑞.

催化学报, 1999, **20**(1): 35～40

329. $Nb_2O_5$ 对丙烷氧化脱氢催化剂 $YVO_4$ 的助催化作用

张伟德, 沙开清, 李基涛, 古萍英, 万惠霖, 蔡启瑞.

天然气化工(C1 化学与化工), 1999, **24**(1): 24～27

330. 酸碱存在下水溶性铑膦配合物的 NMR 表征及氢甲酰化性能研究

张宇, 袁友珠, 廖新丽, 叶剑良, 姚春香, 蔡启瑞.

高等学校化学学报, 1999, **20**(10): 1589～1594

331. 重水中固氮酶催化还原乙炔产物的 $^1H$ NMR 研究

陈忠, 林国兴, 蔡淑惠, 徐昕, 黄静伟, 万惠霖, 蔡启瑞.

化学学报, 1999, **57**(8): 907～913

332. Asymmetric transfer hydrogenation of prochiral ketones catalyzed by chiral ruthenium complexes with aminophosphine ligands

Gao, J. X.; Xu, P. P.; Yi, X. D.; Yang, C. B.; Zhang, H.; Cheng, S. H.; Wan, H. L.; Tsai, K. R.; Ikariya, T.,

Journal of Molecular Catalysis A-Chemical, 1999, **147**(1-2): 105～111

333. Bidentate citrate with free terminal carboxyl groups, syntheses and characterization of citrato oxo-molybdate(VI) and oxotungstate(VI), $\Delta/\Lambda$-$Na_2[MO_2(H_2cit)_2] \cdot 3H_2O$ (M = Mo or W)

Zhou, Z. H.; Wan, H. L.; Tsai, K. R.,

Journal of the Chemical Society-Dalton Transactions, 1999, (24): 4289～4290

334. Characterization of CO- and $H_2$-adsorbed $Au_6Pt$-phosphine clusters supported on $SiO_2$ by EXAFS, TPD, and FTIR
Yuan, Y. Z.; Asakura, K.; Wan, H. L.; Tsai, K.; Iwasawa, Y.,
Bulletin of the Chemical Society of Japan, 1999, **72**(12): 2643~2653

335. Complexation between vanadium(V) and citrate: spectroscopic and structural characterization of a dinuclear vanadium(V) complex
Zhou, Z. H.; Zhang, H.; Jiang, Y. Q.; Lin, D. H.; Wan, H. L.; Tsai, K. R.,
Transition Metal Chemistry, 1999, **24**(5): 605~609

336. Diluted hydrogen activation of modified copper-based catalyst—NC208 for methanol synthesis
Yang, Y. Q.; Dai, S. J.; Yuan, Y. Z.; Lin, R. C.; Zhang, H. B.; Tsai, K. R.,
Journal of Natural Gas Chemistry, 1999, **8**(3): 223~230

337. Forty years of applied catalysis research at Xiamen University and its interaction with fundamental catalysis research
Tsai, K. R.; Chen, D. A.; Wan, H. L.; Zhang, H. B.; Lin, G. D.; Zhang, P. X.,
Catalysis Today, 1999, **51**(1): 3~23

338. Hydrogenation and hydroformylation of olefins with water-soluble $Ru_3(CO)_9(TPPMS)_3$ catalyst
Gao, J. X.; Xu, P. P.; Yi, X. D.; Wan, H. L.; Tsai, K. R.,
Journal of Molecular Catalysis A-Chemical, 1999, **147**(1-2): 99~104

339. Influence of trivalent metal ions on the surface structure of a copper-based catalyst for methanol synthesis
Chen, H. B.; Liao, D. W.; Yu, L. J.; Lin, Y. J.; Yi, J.; Zhang, H. B.; Tsai, K. R.,
Applied Surface Science, 1999, **147**(1-4): 85~93

340. Preparation, characterization and catalytic hydroformylation properties of carbon nanotubes-supported Rh-phosphine catalyst
Zhang, Y.; Zhang, H. B.; Lin, G. D.; Chen, P.; Yuan, Y. Z.; Tsai, K. R.,
Applied Catalysis A-General, 1999, **187**(2): 213~224

341. 高稳定度 $CH_4/CO_2$ 重整 Ni/MgO 催化剂的研究
李基涛，陈明旦，严前古，万惠霖，蔡启瑞.
高等学校化学学报，2000，**21**(9): 1445~1447

342. 介孔分子筛负载型铑膦配合物催化剂的制备
蔡阳，袁友珠，傅琪佳，杨意泉，蔡启瑞.
厦门大学学报(自然科学版)，2000，**39**(1): 128~131

343. 抗积炭 $CH_4/CO_2$ 重整用 $Ni/Al_2O_3$ 催化剂
李基涛，严前古，陈明旦，万惠霖，蔡启瑞.
应用化学，2000，**17**(5): 530~532

344. 载体对负载型水溶性铑膦配合物结构和氢甲酰化性能的影响
袁友珠，张宇，蔡阳，杨意泉，张鸿斌，蔡启瑞.
分子催化，2000，**14**(1): 20~24

345. Carbon nanotubes-supported Rh-phosphine complex catalysts for propene hydroformylation
Zhang, H. B.; Zhang, Y.; Lin, G. D.; Yuan, Y. Z.; Tsai, K. R.,

Studies in Surface Science and Catalysis, 130D, (International Congress on Catalysis, 2000, Pt. D), 2000: 3885~3890

346. New chiral catalysts for reduction of ketones
Gao, J. X.; Zhang, H.; Yi, X. D.; Xu, P. P.; Tang, C. L.; Wan, H. L.; Tsai, K. R.; Ikariya, T.,
Chirality, 2000, **12**(5-6): 383~388

347. Syntheses and spectroscopic and structural characterization of molybdenum(VI) citrato monomeric raceme and dimer, $K_4[MoO_3(cit)] \cdot 2H_2O$ and $K_4[(MoO_2)_2O(Hcit)_2] \cdot 4H_2O$
Zhou, Z. H.; Wan, H. L.; Tsai, K. R.,
Inorganic Chemistry, 2000, **39**(1): 59~64

348. 固氮酶催化作用机理及其化学模拟
周朝晖,颜文斌,张凤章,万惠霖,蔡启瑞.
厦门大学学报(自然科学版),2001,**40**(2):320~329

349. 苹果酸钼外消旋体的合成、光谱性质和结构表征
颜文斌,周朝晖,章慧,万惠霖,蔡启瑞.
高等学校化学学报,2001,**22**(12):1967~1970

350. 水溶性铑膦配合物催化剂的制备、结构和性能
袁友珠,杨意泉,林国栋,张鸿斌,蔡启瑞.
厦门大学学报(自然科学版),2001,**40**(2):407~417

351. 碳纳米管的催化合成、结构表征及应用研究
张鸿斌,林国栋,蔡启瑞.
厦门大学学报(自然科学版),2001,**40**(2):387~397

352. Active-oxygen species on non-reducible rare-earth-oxide-based catalysts in oxidative coupling of methane
Zhang, H. B.; Lin, G. D.; Wan, H. L.; Liu, Y. D.; Weng, W. Z.; Cai, J. X.; Shen, Y. F.; Tsai, K. R.,
Catalysis Letters, 2001, **73**(2-4): 141~147

353. Tungsten-malate interaction. Synthesis, spectroscopic and structural studies of homochiral S-malato tungstate(VI), $\Lambda$-$Na_3[WO_2H(S\text{-}mal)_2]$
Zhou, Z. H.; Wang, G. F.; Hou, S. Y.; Wan, H. L.; Tsai, K. R.,
Inorganica Chimica Acta, 2001, **314**(1-2): 184~188

354. 负载型铼催化剂体系与甲醇选择氧化性能的关系
袁友珠,曹为,蔡启瑞,岩泽康裕.
高等学校化学学报,2002,**23**(5):902~905

355. 碳纳米管促进Cu-基高效甲醇合成催化剂
董鑫,张鸿斌,林国栋,袁友珠,蔡启瑞.
厦门大学学报(自然科学版),2002,**41**(2):135~140

356. A hydrogen-bonded oxo-$\mu$-bis[trans-nitrilotriacetato-cis-dioxotungstate(VI)] synthesized directly from an alkali tungstate and nitrilotriacetic acid
Zhou, Z. H.; Hou, S. Y.; Ma, Z. J.; Wan, H. L.; Tsai, K. R.; Ng, S. W.,
Inorganic Chemistry Communications, 2002, **5**(6): 388~390

357. Energy Policy restructuring and a scheme of clean coal technologies
Tsai, K. R.; Zhang, H. B.; Yuan, Y. Z.,
Proceeding of 2002' China International Hi-tech Symposium on Coal Chemical Industry and Conversion (Beijing, 2002, 11, 6-8), Oral presentation. 煤化工, 2002, **30**(Sup.): 177~179

358. Methanol synthesis from $H_2/CO/CO$, over CNTs-promoted Cu-ZnO-Al$_2$O$_3$ catalyst
Zhang, H. B.; Dong, X.; Lin, G. D.; Yuan, Y. Z.; Tsai, K. R.,
Abstracts of Papers of the American Chemical Society, 2002, **223**, U580

359. Structure aspects and hydroformylation performance of water-soluble HRh(CO)[P(m-C$_6$H$_4$SO$_3$Na)$_3$]$_3$ complex supported on SiO$_2$
Yuan, Y. Z.; Zhang, H. B.; Yang, Y. Q.; Zhang, Y.; Tsai, K. R.,
Catalysis Today, 2002, **74**(1-2): 5~13

360. Study on highly active catalyst and once-through process for methanol synthesis from syngas
Zhang, H. B.; Dong, X.; Lin, G. D.; Yuan, Y. Z.; Tsai, K. R.,
Proceeding of 2002' China International Hi-tech Symposium on Coal Chemical Industry and Conversion (Beijing, 2002, 11, 6-8), Oral presentation. 煤化工, 2002, **30**(Sup.): 173~176

361. Synthesis and characterization of homochiral polymeric S-malato molybdate(VI): toward the potentially stereospecific formation and absolute configuration of iron-molybdenum cofactor in nitrogenase
Zhou, Z. H.; Yan, W. B.; Wan, H. L.; Tsai, K. R.,
Journal of Inorganic Biochemistry, 2002, **90**(3-4): 137~143

362. 晶体中原子的平均范德华半径
胡盛志, 周朝晖, 蔡启瑞.
物理化学学报, 2003, **19**(11): 1073~1077

363. 碳纳米管负载/促进 Cu-Cr 催化剂上甲醇分解制氢
陈书贵, 周金梅, 张鸿斌, 林国栋, 蔡启瑞.
厦门大学学报(自然科学版), 2003, **42**(2): 133~138

364. Highly active CNT-promoted Cu-ZnO-Al$_2$O$_3$ catalyst for methanol synthesis from $H_2/CO/CO_2$
Dong, X.; Zhang, H. B.; Lin, G. D.; Yuan, Y. Z.; Tsai, K. R.,
Catalysis Letters, 2003, **85**(3-4): 237~246

365. Methanol synthesis from $H_2/CO/CO_2$ over CNT-promoted Cu-ZnO-Al$_2$O$_3$ catalyst
Zhang, H. B.; Dong, X.; Lin, G. D.; Yuan, Y. Z.; Zhang, P.; Tsai, K. R.,
In Utilization of Greenhouse Gases, Amer Chemical Soc: Washington; 2003, **852**, pp 195~209

366. Selective methanol conversion to methylal on Re-Sb-O crystalline catalysts: catalytic properties and structural behavior
Yuan, Y. Z.; Tsai, K. R.; Liu, H. H.; Iwasawa, Y.,
Topics in Catalysis, 2003, **22**(1-2): 9~15

367. 高效甲醇合成催化剂及一次性通过过程研究
沈炳顺, 张鸿斌, 林国栋, 董鑫, 蔡启瑞.
石油化工, 2004, **33**(增刊): 273~275

368. Ammonium barium citrato peroxotitanate(IV) Ba$_2$(NH$_4$)$_2$[Ti$_4$(O$_2$)$_4$(Hcit)$_2$(cit)$_2$]·10H$_2$O: a molecular precursor of stoichiometric BaTi$_2$O$_5$

Deng, Y. F.; Zhou, Z. H.; Wan, H. L.; Tsai, K. R.,
Inorganic Chemistry Communications, 2004, **7**(2): 169~172

369. Enantiomeric and mesomeric mandelate complexes of molybdenum — on their stereospecific formations and absolute configurations
Zhou, Z. H.; Zhao, H.; Tsai, K. R.,
Journal of Inorganic Biochemistry, 2004, **98**(11): 1787~1794

370. Enzymatic mechanism of Fe-only hydrogenase: density functional study on H-H making/breaking at the diiron cluster with concerted proton and electron transfers
Zhou, T. J.; Mo, Y. R.; Liu, A. M.; Zhou, Z. H.; Tsai, K. R.,
Inorganic Chemistry, 2004, **43**(3): 923~930

371. Speciation and transformation of Co(II)/Ni(II)-citrate-imidazole ternary system — synthesis, spectroscopic and structural studies
Deng, Y. F.; Zhou, Z. H.; Cao, Z. X.; Tsai, K. R.,
Journal of Inorganic Biochemistry, 2004, **98**(6): 1110~1116

372. Activation of homocitrate ligand of femo-cofactor as mimicked by the coordination of homocitrato and its homologues
Zhou, Z. H.; Hou, S. Y.; Cao, Z. X.; Deng, Y. F.; Wan, H. L.; Tsai, K. R.,
In Biological Nitrogen Fixation, Sustainable Agriculture and the Environment; 2005, **41**, pp. 43~45

373. Density functional study on dihydrogen activation at the H cluster in Fe-only hydrogenases
Zhou, T. J.; Mo, Y. R.; Zhou, Z. H.; Tsai, K. R.,
Inorganic Chemistry, 2005, **44**(14): 4941~4946

374. pH dependent transformation of nitrilotriacetato molybdates (VI) — Synthesis, spectral and structural characterization
Zhou, Z. H.; Lin, J.; Yan, W. B.; Zhang, H.; Tsai, K. R.,
Polyhedron, 2006, **25**(9): 1909~1914

375. Syntheses, spectroscopies and structures of molybdenum(VI) complexes with homocitrate
Zhou, Z. H.; Hou, S. Y.; Cao, Z. X.; Tsai, K. R.; Chow, Y. L.,
Inorganic Chemistry, 2006, **45**(20): 8447~8451

376. Expeditious biomimetically-inspired approaches to racemic homocitric acid lactone and per-homocitrate
Chen, H. B.; Chen, L. Y.; Huang, P. Q.; Zhang, H. K.; Zhou, Z. H.; Tsai, K. R.,
Tetrahedron, 2007, **63**(10): 2148~2152

377. Formations of mixed-valence oxovanadium$^{V,IV}$ citrates and homocitrate with N-heterocycle chelated ligand
Chen, C. Y.; Zhou, Z. H.; Chen, H. B.; Huang, P. Q.; Tsai, K. R.; Chow, Y. L.,
Inorganic Chemistry, 2008, **47**(19): 8714~8720

378. N-heterocycle chelated oxomolybdenum(VI and V) complexes with bidentate citrate
Zhou, Z. H.; Chen, C. Y.; Cao, Z. X.; Tsai, K. R.; Chow, Y. L.,
Dalton Transactions, 2008, (18): 2475~2479

379. Mechanism and isotope effect of ammonia synthesis over Fe and Ru catalysts

Lin, J. D.; Liao, D. W.; Zhang, H. B.; Wan, H. L.; Tsai, K. R.,

Abstracts of Papers of the American Chemical Society, 2009, **237**: 834～834

380. Deuterium Inverse Isotopic Effect in Ammonia Synthesis over Ru-Based and Fe-Based Catalysts

Lin, J. D.; Liao, D. W.; Zhang, H. B.; Wan, H. L.; Tsai, K. R.,

Chinese Journal of Catalysis, 2010, **31**(2): 153～155

# 发明专利目录

| 序号 | 发表专利名称 | 申请号 | 专利号/公开号 | 当前法律状态及状态日 | 专利权人/申请人 | 发明人 |
|---|---|---|---|---|---|---|
| 1 | 甲烷氧化偶联制乙烯催化剂 | 91110922.6 | 专利号：ZL 91110922.6 | 授权,授权日：1995年10月4日 | 厦门大学 | 刘玉达;林国栋;张鸿斌;蔡俊修;万惠霖;蔡启瑞 |
| 2 | 氧化钍-氧化钙-碳酸钡系甲烷氧化偶联制乙烯催化剂 | 92100377.3 | 公开号：CN 1074391A | 公开,公开日：1993年7月21日 | 厦门大学 | 刘玉达;林国栋;张鸿斌;蔡俊修;万惠霖;蔡启瑞 |
| 3 | 甲烷氧化偶联制碳二以上烃催化剂 | 92105258.8 | 专利号：ZL 92105258.8 | 授权,授权日：1995年10月25日 | 厦门大学 | 周小平;万惠霖;蔡俊修;蔡启瑞 |
| 4 | 钍系甲烷氧化偶联制乙烯催化剂 | 92109209.1 | 公开号：CN 1081627A | 公开,公开日：1994年2月9日 | 厦门大学 | 刘玉达;林国栋;张鸿斌;蔡俊修;万惠霖;蔡启瑞 |
| 5 | 乙烷氧化脱氢制乙烯催化剂 | 92110008.6 | 专利号：ZL 92110008.6 | 授权,授权日：1995年11月29日 | 厦门大学 | 周小平;万惠霖;蔡启瑞 |
| 6 | 丙烷氧化脱氢制丙烯催化剂 | 92111518.0 | 专利号：ZL 92111518.0 | 授权,授权日：1995年10月4日 | 厦门大学 | 张伟德;周小平;万惠霖;蔡启瑞 |
| 7 | 高碳数端烯氢甲酰化制高碳醛负载型水溶性催化剂 | 93100802.6 | 专利号：ZL 93100802.6 | 授权,授权日：1997年8月6日 | 厦门大学 | 袁友珠;蔡启瑞;张鸿斌;洪亮 |
| 8 | 异丁烷氧化脱氢制异丁烯催化剂及其制造方法 | 93115308.5 | 专利号：ZL 93115308.5 | 授权,授权日：1999年8月25日 | 厦门大学 | 张伟德;汤丁亮;万惠霖;蔡启瑞 |
| 9 | 甲烷氧化偶联制碳二烃催化剂 | 94107757.8 | 专利号：ZL 94107757.8 | 授权,授权日：1999年8月25日 | 厦门大学 | 晁自胜;周小平;万惠霖;蔡启瑞 |
| 10 | 无死体积流量调节阀 | 94216485.7 | 专利号：ZL 94216485.7 | 授权,授权日：1996年1月31日 | 厦门大学 | 方智敏;万惠霖;蔡启瑞 |
| 11 | 一种用于气相色谱仪的单柱化双柱 | 94216622.1 | 专利号：ZL 94216622.1 | 授权,授权日：1996年6月19日 | 厦门大学 | 方智敏;万惠霖;蔡启瑞 |

续表

| 序号 | 发表专利名称 | 申请号 | 专利号/公开号 | 当前法律状态及状态日 | 专利权人/申请人 | 发明人 |
|---|---|---|---|---|---|---|
| 12 | 抗积炭甲烷部分氧化制合成气催化剂及其制造方法 | 96101766.X | 专利号：ZL 96101766.X | 授权,授权日：1999年10月27日 | 厦门大学 | 陈萍;张鸿斌;林国栋;蔡启瑞 |
| 13 | 负载型强碱液膜甲醇羰化制甲酸甲脂催化剂及其制备方法 | 96103148.4 | 专利号：ZL 96103148.4 | 授权,授权日：2000年1月26日 | 厦门大学 | 林国栋;李海燕;周金海;张鸿斌;刘玉达;蔡启瑞 |
| 14 | 过渡金属催化剂及用于制备均匀管径碳纳米管的方法 | 96110252.7 | 专利号：ZL 96110252.7 | 授权,授权日：2000年11月1日 | 厦门大学 | 陈萍;张鸿斌;林国栋;蔡启瑞 |
| 15 | 合成气制低碳醇的铑基催化剂及其制备方法 | 96112685.X | 公开号：CN 1179993A | 公开,公开日：1998年4月29日 | 厦门大学 | 汪海有;蔡启瑞;刘金波;傅锦坤;高景星 |
| 16 | 手性胺膦金属配合物及其制备方法和在不对称催化氢化的应用 | 97112606.2 | 专利号：ZL 97112606.2 | 授权,授权日：1999年12月22日 | 厦门大学 | 高景星;许翩翩;黄培强;万惠霖;蔡启瑞 |
| 17 | 非氧化条件下甲烷脱氢芳构化催化剂 | 97100978.3 | 专利号：ZL 97100978.3 | 授权,授权日：2000年11月29日 | 厦门大学 | 曾金龙;张鸿斌;林国栋;熊智涛;蔡启瑞 |
| 18 | 一种烯烃氢甲酰化制醛的淤浆型催化剂 | 01132718.9 | 专利号：ZL 01132718.9 | 授权,授权日：2003年7月16日 | 厦门大学 | 袁友珠;李志华;彭庆蓉;蔡启瑞 |
| 19 | 碳纳米管促进铜-基甲醇合成催化剂及其制备方法 | 02102608.4 | 专利号：ZL 02102608.4 | 授权,授权日：2003年10月22日 | 厦门大学 | 张鸿斌;董鑫;林国栋;蔡启瑞 |
| 20 | 氨合成催化剂及其制备方法 | 02142327.X | 专利号：ZL 02142327.X | 授权,授权日：2005年3月23日 | 厦门大学 | 廖代伟;林敬东;王欣莹;陈鸿博;蔡启瑞 |
| 21 | 外消旋高柠檬酸内酯制备方法 | 200510108100.7 | 专利号：ZL 200510108100.7 | 授权,授权日：2009年5月13日 | 厦门大学 | 蔡启瑞;陈洪斌;黄培强;周朝晖 |

# 蔡启瑞院士主要活动年表

## (1) 主要经历

| | |
|---|---|
| 1913.12 | 出生于福建省同安县(今厦门市翔安区)马巷镇番薯市五甲尾的华侨店员家庭,祖籍金门琼林 |
| 1921.2 —1925.7 | 辗转就读于马巷镇番薯市礼拜堂小学、马巷镇牖民小学、同安丙洲砥江小学 |
| 1925.9 —1926.1 | 中华中学学生 |
| 1926.2 —1929.7 | 集美中学学生 |
| 1929.9 —1931.7 | 厦门大学预科化学组学生 |
| 1931.9 —1937.7 | 厦门大学化学系本科学生,期间因病休学两年,获厦门大学理学学士学位 |
| 1937.9 —1940.7 | 厦门大学化学系助教 |
| 1940.8 —1947.2 | 厦门大学化学系讲师 |
| 1947.3 —1950.3 | 美国俄亥俄州立大学研究生,师从马克(E. Mack, Jr)、哈里斯(P. M. Harris)和纽曼(M. S. Newman)三位教授,从事多亚甲基长链二醇及二羧酸的 L-B 膜行为的研究,获俄亥俄州立大学哲学博士学位(化学领域) |
| 1950.4 —1951 | 美国俄亥俄州立大学结构化学博士后,从事铯氧化物晶体结构测定的研究 |
| 1951 —1956.3 | 美国俄亥俄州立大学化学系副研究员,从事无机结构化学、酶反应动力学方面的研究 |
| 1956.8 — | 厦门大学化学系教授 |
| 1980.11 — | 中国科学院化学部学部委员(院士) |
| 1981.11 — | 经国务院学位委员会审批,成为首批物理化学学科博士生指导教师 |
| 1986.6 — | 博士后合作教授 |
| 1997.3 | 获香港浸会大学荣誉理学博士学位 |
| 1998.7 — | 首批"资深院士" |

## (2) 主要科研学术活动

| | |
|---|---|
| 1955.9 | 在美国化学学会第 128 届年会上,宣读有关离子晶体极化现象的论文 |
| 1958.3—4 | 作为中国科学代表团团长,赴莫斯科参加全苏催化工作会议并参观访问;回北京后参加催化工作协调会议 |
| 1958.9 | 在厦门大学创建中国高校第一个催化教研室,开创中国催化科学领域的教学与研究基地 |
| 1962 —1965 | 作为国家科委化学学科组成员,参与制订国家基础学科发展规划和国家重点 |

科研项目研究规划

| 1964 | 承担国家重点科学研究项目第 29 项——催化和化学动力学研究 |
|---|---|
| 1964.12—1966 | 受国家高教部委托,带领厦门大学催化教研室承办全国性的催化学术讨论班 |
| 1972 | 赴长春,与卢嘉锡教授和唐敖庆教授一起进行化学模拟生物固氮跨学科大协作研究的项目规划工作 |
| 1973.2 | 在鼓浪屿召开由厦门大学承办的全国固氮会议上作报告,提出国际上较早的固氮活性中心模型——"厦门模型"以及已知固氮底物的多核配位活化模式 |
| 1977.10 | 参观美国密执安大学,将三十烷醇研究引入厦门大学 |
| 1977.11—12 | 参加中国理工科高教代表团访美,考察美国高等教育并进行学术交流 |
| 1978.6 | 赴美国威斯康辛大学参加第三届国际固氮会议,交流论文《固氮酶活性中心模型的演进和催化机理》 |
| 1979.10—1980.2 | 带领厦门大学催化团队举办催化进修班和现代催化研究方法研讨班 |
| 1980.6 | 赴北京参加中、日、美三国金属有机化学讨论会 |
| 1980.7 | 赴日本东京参加第七届国际催化会议,在固氮专题讨论会上作特邀报告《化学吸附与配位络合之间的关联—氨合成催化剂活性位本质的原子簇方法》 |
| 1980.12 | 赴澳大利亚堪培拉参加第四届国际固氮会议,会上对厦门大学化学模拟生物固氮协作组具有明显生物重组活性的模型化合物给予相当高的评价和重视 |
| 1981.2—4 | 率厦门大学催化-固氮教育和科学研究考察团访问日本、美国,共考察 11 所高校和 2 个工业研究所 |
| 1981 | "催化和固氮"的研究课题获联合国科教文组织开发计划署 40 万美元的资助 |
| 1983.5 | 赴上海参加中澳生物固氮学术讨论会 |
| 1983.8—9 | 赴华盛顿参加第 186 届美国化学会年会;后带助手林国栋赴荷兰、西德、比利时、瑞士、法国、英国进行催化科学考察和学术交流 |
| 1983 | 在福州举行的国际分子结构学术讨论会上作学术报告 |
| 1984.7 | 赴联邦德国西柏林参加第八届国际催化会议及会后专题讨论会 |
| 1984.10 | 带领厦门大学催化学科承办第二届全国催化会议 |
| 1985.8 | 赴美国加州大学伯克利分校参加第二届中、日、美催化学术讨论会,并作专题报告 |
| 1986.9 | 赴香港中文大学讲学,为时三周 |
| 1986.11 | 由中国化学会等七个单位联合组织在厦门大学举行"祝贺卢嘉锡、蔡启瑞从事化学工作五十年"学术讨论会 |
| 1986 | 在联合国 UNDP 赞助举行的电催化、光催化和金属仿生催化国际学术讨论会上作学术报告 |
| 1987 | 在厦门召开的第三届中、日、美催化会议上作学术报告 |
| 1987—1992 | 与中国科学院山西煤炭化学研究所彭少逸院士共同主持国家"七五"自然科学基金重大项目"碳一化学基础研究" |
| 1988.8 | 赴香港参加国际精细化工学术讨论会并作专题报告 |
| 1988.8 | 赴加拿大参加第九届国际催化会议并宣读论文 |
| 1989.3 | 应邀赴比利时作催化与固氮的专题讲学 |
| 1989.11 | 出席 1989 年太平洋化学会议并作专题报告 |

| 1990.5 | 向福建省教委提交"关于1990年省联合办学增招石油加工专业40名本科生的报告",推动翌年厦门大学设立化工系 |
| 1991.3 | 应邀出访香港并作固氮专题讲学 |
| 1991.5 | 赴美国伊利诺伊州参加第五届中、日、美催化会议 |
| 1991.7 | 赴苏联参加学术会议 |
| 1992.4 | 赴美国旧金山参加第203届美国化学会年会 |
| 1993.8 | 赴北京参加国际纯粹与应用化学联合会第34届学术大会 |
| 1993 —1997 | 协同主持国家"八五"973项目"煤炭、石油、天然气优化利用的催化基础" |
| 1994.1 | 蔡启瑞八秩华诞祝寿会暨学术研讨会召开 |
| 1994.3 | 赴美国圣地亚哥参加第207届美国化学会年会 |
| 1996.10 | 带领厦门大学催化学科承办第八届全国催化会议 |
| 1998 | 作为中国科学院化学学部"绿色化学与科技"咨询组的六名成员之一,参加中国科学院向国务院提交咨询报告的工作 |
| 2000.6 | 为推动两岸学术交流,赴台参加在台北举行的催化学科学术会议 |
| 2000.10 | 和助手张鸿斌教授赴北京参加由国家科委主办的21世纪新一代煤化工技术发展研讨会,作题为《煤洁净发电联产甲醇燃料化工发展甲醇汽车及混合动力汽车》的发言 |

# (3)主要任职

| 1945.7 | 当选为厦门大学首届教师会的九位理事之一 |
| 1960.5 | 任与中国科学院福建分院合办的厦门大学化学研究所所长 |
| 1971.12 —1978 | 任厦门大学革委会副主任 |
| 1978.4 —1984.8 | 任厦门大学副校长 |
| 1978.5 —1984 | 任厦门大学自然科学学术委员会主任 |
| 1978.5 | 任厦门大学自然辩证法研究会理事会会长兼理事 |
| 1980.4 | 当选厦门大学校友总会理事长 |
| 1982.3 | 当选厦门大学学位评定委员会第一届主席 |
| 1983 | 任"物理化学研究所(厦门大学)"首任所长 |
| 1984 | 任厦门大学理科学衔委员会主任 |
| 1987.4 | 任厦门大学校务委员会委员 |
| 1987.5 —1996 | 任"固体表面物理化学国家重点实验室(厦门大学)"首届学术委员会主任 |
| 1990.12 —1999 | 被聘为厦门大学侨联第四、五届委员会顾问 |
| 1992.3 | 当选厦门大学校友总会理事会顾问 |
| 1994.10 | 任厦门大学第四届学位评定委员会委员 |
| 1996.10—1999 | 被聘为"固体表面物理化学国家重点实验室(厦门大学)"第二届学术委员会名誉主任 |
| 2002.4 | 被聘为厦门大学纳米科技中心第一届学术委员会委员 |
| 2009 —2012 | 被聘为"醇醚酯化工清洁生产国家工程实验室(厦门大学)"第一届技术委员会名誉主任 |

**国际：**

| | |
|---|---|
| 1984.7—1988 | 任第八届国际催化大会理事会理事 |

**全国：**

| | |
|---|---|
| 1962 —1965 | 任国家科委化学学科组成员 |
| 1962 | 任第三届全国政协特邀代表 |
| 1963 —1990 | 当选为中国化学会第二十、二十一、二十二届理事会理事 |
| 1964、1975、1978 | 任第三、四、五届全国人民代表大会代表 |
| 1978 —1982 | 任国家科委化学学科组成员 |
| 1981 —1985 | 任国务院学位委员会委员及理科评议组成员 |
| 1982 —1986 | 任国务院第一届学位委员会委员 |

**福建省：**

| | |
|---|---|
| 1958.7 | 省青联代表大会上当选为福建省民主青年联合会副主席 |
| 1985 | 任福建化学学会名誉理事长 |
| 1988.8、1992.7 | 被聘为福建省工程咨询总公司第一、二届专家委员会顾问 |
| 1992.8、1996.11、2000 | 当选福建省留学生同学会名誉会长 |
| 2009.2 —2012 | 被聘为福建省人民政府第三届经济社会发展顾问 |

**厦门市：**

| | |
|---|---|
| 1980.10 | 当选集美学校校友会第一届理事会名誉理事长;之后连任八届名誉理事长 |
| 1981 —1991 | 任厦门市科协主席 |
| 1992.7 | 当选厦门市科协名誉主席 |
| 1993.12—1999 | 当选厦门市侨界知识分子联谊会第一、二届名誉理事长 |
| 1994.11 | 被聘为政协厦门市第八届委员会顾问 |
| 1994.11 | 被厦门市人民政府聘为厦门市科技中学名誉校长 |
| 1995.4、2005.5、2011.5 | 被聘为厦门市老教授协会第一、三、四届名誉会长 |
| 1995.7 | 被聘为厦门市专家协会顾问 |
| 1995.8 | 被聘为厦门市工程咨询专家委员会顾问 |
| 2001.12 | 被厦门市政府聘为厦门市科学技术顾问 |
| 2003.2 | 被聘请为厦门市知识创新与知识产权保护协会高级顾问 |

**其他地区：**

| | |
|---|---|
| 1962.12—1964.4 | 任华东物质结构研究所催化电化研究室主任 |
| 1995.7 | 被国家碳一化学工程技术研究中心、化工部西南化工研究院聘为"国家碳一化学工程技术研究中心"工程技术委员会委员 |
| 1996.11—1999.11 | 被中国科学技术大学聘为兼职教授 |
| 1997.3 —2000 | 被中科院兰州化学物理研究所聘为羰基合成与选择氧化国家重点实验室第二届学术委员会名誉主任 |

| | |
|---|---|
| 2001.7 | 被武夷山市人民政府聘为第十三届政府科技顾问 |

**刊物：**

| | |
|---|---|
| 1986.6、1990.3 | 被中科院兰州化学物理研究所聘为《分子催化》期刊第一、二届编辑委员会顾问 |
| 1991.12、1994.11 | 被中科院成都有机化学研究所聘为《天然气化学》期刊(英文版)顾问 |
| 1991.12、1996.12<br>2000.9、2007.1 | 被中科院福建物质结构研究所依次聘为《结构化学》第三届至第六届编辑委员会顾问 |
| 1993.9 —1998 | 被国家教委科技司聘为《高等学校化学学报》第三届编委顾问 |
| 2002.1 —2006 | 被中国物理学会光散射专业委员会聘为《光散射学报》顾问 |
| 2006.11 | 被福建省科学技术协会聘请担任《海峡科学》杂志顾问 |
| 2007.5 | 被聘为《工业催化》编辑委员会特邀顾问 |

# (4)主要奖项

| | |
|---|---|
| 1977.5 | 全国工业学大庆会议上被授予积极分子光荣册 |
| 1977 | 获全国劳动模范称号 |
| 1978.3 | 全国科学大会上被评为"先进工作者"；《络合催化理论与化学模拟生物固氮》、《石油化学中新型催化剂的研究》、《乙炔催化加合聚合新型催化剂的研究》、《高密度聚乙烯》获全国科学大会奖 |
| 1978 | 获福建省教育战线先进工作者光荣册 |
| 1978 | 获全国劳动模范称号 |
| 1979.12 | 获全国劳动模范称号 |
| 1979.12 | 获福建省劳动模范称号和金质奖章 |
| 1979 | 《络合催化理论的研究》(1960—1977)、《化学模拟生物固氮》(1972—1977)、《聚乙烯高效催化剂的研究》(1971—1977)、《乙炔水合制乙醛磷酸镉钙催化剂》(1960—1966)、《乙炔合成苯氧化铌催化剂的研究》、《乙炔气相水合制乙醛氧化锌催化剂的研究》(1972—1977)获 1979 年福建省科学大会奖 |
| 1980.4 | 《化学模拟生物固氮研究——固氮酶活性中心模型和酶催化机理及模拟体的合成活性和选择性》获 1979 年福建省科技成果奖二等奖 |
| 1981 | 《络合催化理论的研究》获国家教委科技进步奖二等奖(第 1 完成人) |
| 1981 | 评为厦门市特等劳动模范 |
| 1982.3 | 评为厦门市劳动模范 |
| 1982.7 | 《络合催化理论的研究》获国家自然科学奖三等奖(第 1 完成人) |
| 1982 | 获福建省劳动模范称号 |
| 1983 | 评为厦门市劳动模范 |
| 1984.4 | 评为厦门市劳动模范 |
| 1984 | 获全国劳动模范称号 |
| 1986.5 | 《在固氮酶作用下和铁催化剂作用下固氮成氨的研究》获 1985 年国家教委科 |

技进步奖二等奖（第 1 完成人）

| | |
|---|---|
| 1988.8 | 《在固氮酶作用下和铁催化剂作用下固氮成氨的研究》获 1987 年国家自然科学奖三等奖（第 1 完成人） |
| 1989.12 | 评为全国优秀归侨、侨眷知识分子 |
| 1990.12 | 获全国高等学校先进科技工作者称号 |
| 1994.6 | 在为实现"八五"计划和十年规划做贡献活动中,成绩突出,被中华全国归国华侨联合会评为先进个人 |
| 1995.5 | 《合成气制乙醇催化反应机理的研究》获国家教委科技进步奖一等奖（第 1 完成人） |
| 1995.5 | 《群表示约化的方法、程序与应用》获国家教委科技进步奖二等奖（第 7 完成人） |
| 1995.6 | 获 1995 年度"集友科技成就奖" |
| 1995.11 | 《合成气制乙醇催化反应机理的研究》获福建省第六届王丹萍科学技术奖三等奖 |
| 1995.12 | 《合成气制乙醇催化机理的研究》获国家自然科学奖三等奖（第 1 完成人） |
| 1997.5 | 报送的信息《发展煤气化综合洁净利用、发展甲醇汽车和甲醇燃料电池》被评为 1996 年度厦门市政协优秀信息 |
| 1997.10 | 《NC208 型甲醇合成催化剂》获厦门市科技进步奖一等奖 |
| 1997.12 | 获 1997 年度光华科技基金奖二等奖 |
| 1997.12 | 获福建省优秀归侨、侨眷知识分子称号 |
| 1999.1 | 《甲烷氧化偶联含氟稀土基催化剂的研究》获教育部科技进步奖一等奖（第 5 完成人） |
| 1999.1 | 《铁催化剂上的合成氨反应机理研究》获教育部科技进步奖三等奖（第 2 完成人） |
| 1999.10 | 获 1999 年度何梁何利基金科学与技术进步奖 |
| 1999.12 | 陈村牧基金首届获奖者 |
| 2001.5 | 《烯烃氢甲酰化负载型铑配合物催化剂的研究》获 2000 年中国高校科技进步奖二等奖（第 5 完成人） |
| 2001 | 《铜基甲醇合成催化剂各组分的协同催化作用机理研究》获福建省科技进步奖三等奖（第 2 完成人） |
| 2002.6 | 获"福建省优秀专家"称号 |
| 2004 | 评为"厦门市十佳教育之家" |
| 2006.9 | 获 2005 年度福建省科学技术重大贡献奖 |
| 2007.12 | 获福建省老科学技术工作者协会颁发的"突出贡献奖" |
| 2011.12 | 获厦门经济特区建设 30 周年杰出建设者称号 |
| 2013.4 | 获厦门大学首届"南强杰出贡献奖" |

# 蔡启瑞所指导的研究生及其
# 研究方向或学位论文题目

| 姓　名 | 入学至毕业/答辩时间 | 学位论文题目或研究方向、内容 | 学位或备注 |
|---|---|---|---|
| 黄开辉 | 1955—1958 | 乙炔二聚做乙烯基乙炔<br>乙烯基乙炔选择加氢做丁二烯 | 卢嘉锡调整来 |
| 施彼得 | 1955—1958 | 硫化物晶体结构 | 卢嘉锡调整来 |
| 陈德安 | 1957—1961 | 醇醛缩合催化研究——负载型氧化物催化剂 | 招收的第一位 |
| 肖漳龄 | 1959— | 醇醛缩合催化研究——离子交换树脂催化剂 | 在职研究生 |
| 陈祖炳 | 1961—1963 | 丙烯氨氧化做丙烯腈 | 1963 年转筹备讨论班工作 |
| 林国栋 | 1961—1966 | 乙炔水合 | |
| 陈守正 | 1961—1966 | 乙炔水合 | |
| 王秀丽 | 1961—1963 | 乙炔二聚 | 1963 年转催化电化研究室工作 |
| 万惠霖 | 1962—1966 | 快速反应动力学 | |
| 张鸿斌 | 1962—1966 | 烯烃聚合催化剂 | |
| 高沈林 | 1964— | | 因 1966 年"文革"无结业 |
| 林去存 | 1964— | | 因 1966 年"文革"无结业 |
| 王镜和 | 1964— | | 因 1966 年"文革"无结业 |

### 硕士研究生

| 姓　名 | 入学至毕业/答辩时间 | 学位论文题目或研究方向、内容 | 学位或备注 |
|---|---|---|---|
| 廖代伟 | 1978.10—1981.10.09 | 氨合成铁催化剂上氮吸附态的研究 | 理学硕士学位 |
| 陆维敏 | 1978.10—1981.10.09 | 负载型聚丙烯高效催化剂的研究 | 理学硕士学位 |
| 陈鸿博 | 1978.10—1981.10.09 | $Fe_4S_4{}^*$ 原子簇与 ATP 的络合及电子传递与 ATP 水解的偶联 | 理学硕士学位 |
| 林建毅 | 1978.10— | | 后赴美留学,无结业 |
| 曾晓鸣 | 1978.10—1981.10.09 | 铁催化剂上氢的升温脱附谱和活化吸附态的研究 | 理学硕士学位 |

续表

| 姓　名 | 入学至毕业/答辩时间 | 学位论文题目或研究方向、内容 | 学位或备注 |
|---|---|---|---|
| 陈慧贞 | 1978.10—1982.10.22 | 乙苯脱氢制苯乙烯 11# 及无铬 210 催化剂的研究——晶格氧的作用 | 理学硕士学位 |
| 郇正伟 | 1979.09—1983.04.09 | 乙烯光电催化氧化的研究 | 理学硕士学位 |
| 吴也凡 | 1980.09—1983.10.21 | $Fe_4 S_4^*$ 原子簇与 ATP 的络合 | 理学硕士学位 |
| 刘敏敦 | 1980.03—1983.10.22 | 固氮酶活性中心模型化合物合成方法的探索——($K_2 MoS_4$-$FeCl_2$-KOR-NaHS)体系的电子光谱及其催化固氮酶的某些底物还原的性能 | 理学硕士学位 |
| 杨意泉 | 1980.09—1983.10.25 | 合成气制乙醇负载型铑催化剂——Rh-$Nb_2 O_5$/$SiO_2$ 催化剂活性位组成的研究 | 理学硕士学位 |
| 王耀华 | 1980.09—1983.11.02 | 反应法制备的聚丙烯负载型催化剂的研究 | 理学硕士学位 |
| 顾桂松 | 1982.02—1984.12.13 | 合成气制乙醇 Rh-$Nb_2 O_5$/$SiO_2$ 催化剂金属-助催剂-载体相互作用机理的研究 | 理学硕士学位 |
| 陈建平 | 1982.02—1984.12.13 | 氧化铁系催化剂上乙苯脱氢反应机理的研究 | 理学硕士学位 |
| 宋　岩 | 1982.02—1985.01.28 | 固氮酶活性中心化学模拟合成体系研究 | 理学硕士学位 |
| 许颂临 | 1982.03—1985.01.28 | 等离子体聚合修饰半导体光电极的研究 | 理学硕士学位 |
| 洪　亮 | 1982.03—1985.01.31 | 负载型锰氧化物引发的丙二酸酯与端烯自由基加成反应的研究 | 理学硕士学位 |
| 钟传建 | 1982.03—1985.02.01 | 乙炔在铜/石墨阴极上的电催化还原加氢和电引发聚合 | 理学硕士学位 |
| 林　旭 | 1982.09—1985.06.24 | 不饱和酯类催化氢硅化的研究 | 理学硕士学位 |
| 翁维正 | 1982.09—1985.12.02 | 丙烯定向聚合负载型高效催化剂的研究 | 理学硕士学位 |
| 林建生 | 1982.09—1985.09.12 | 硝基苯与一氧化碳反应的催化剂与催化作用机理研究 | 理学硕士学位 |
| 王伟斌 | 1982.09—1985.11.04 | 在可见光照射下乙醇——水体系光催化脱氢的研究 | 理学硕士学位 |
| 杜玉华 | 1983.09—1986.08.04 | 合成气制乙醇负载型铑催化剂稀土金属氧化物助催剂的促进作用研究 | 理学硕士学位 |
| 林　珊 | 1983.09—1986.08.27 | 光电极表面保护与催化双功能膜的研制 | 理学硕士学位 |
| 张兆龙 | 1983.09—1986.09.01 | 蒽醌磺酸盐-聚吡咯膜电极的制备、电化学性质及其应用的研究 | 理学硕士学位 |
| 朱爱民 | 1983.09—1986.09.04 | 负载型铑催化剂中金属——促进剂的相互作用及催化合成气制乙醇的研究 | 理学硕士学位 |

续表

| 姓　名 | 入学至毕业/答辩时间 | 学位论文题目或研究方向、内容 | 学位或备注 |
| --- | --- | --- | --- |
| 林　俊 | 1984.09—1987.06.24 | 一步法合成甲苯二异氰酸酯催化剂及催化作用机理的研究 | 理学硕士学位 |
| 林淑琼 | 1984.09—1987.07.06 | 钼系、钼铋系和三元钼铋基氧化物上氧吸附物种的 EPR 研究 | 理学硕士学位 |
| 何基良 | 1984.09—1987.09.18 | $CO/H_2/NH_3$ 合成 $CH_3CN$ 钼基催化剂的 XPS 表征和催化活性 | 理学硕士学位 |
| 汪海有 | 1985.09—1988.10.29 | Rh 系催化剂上合成气转化为乙醇的反应机理、助剂作用及本质的研究 | 理学硕士学位 |
| 王京华 | 1986.09—1989.08.09 | 还原羰基化制备苯氨基甲酸乙酯钯催化剂研究 | 理学硕士学位 |
| 潘填元 | 1986.09—1989.09.07 | 醋酸甲酯羰基化制醋酐均相催化体系的助剂及其机理研究 | 理学硕士学位 |
| 周小平 | 1986.09—1989.09.13 | $[NH_4]_2[FeMo-S_4Cl_2]$ 在 ATP 作用下电子传递的研究<br>固氮酶 FeMo-Co 模拟体系的活性评价和合成尝试 | 理学硕士学位 |
| 余云帆 | 1986.09—1989.09.23 | 丙烯选择（氨）氧化钼酸铋催化剂性能控制因素的研究 | 理学硕士学位 |
| 李　勇 | 1986.09—1989.09.29 | 负载型乙腈合成钼基催化剂研究——第二金属添加组分的作用及工作态催化剂的 ESR、XPS 和 LRS 表征 | 理学硕士学位 |
| 王凡成 | 1986.09—1989.11.03 | 甲烷氧化偶联制乙烯的研究 | 理学硕士学位 |
| 林玉琴 | 1986.09—1989.11.14 | 制甲醛新型催化剂的研制及其表征 | 理学硕士学位 |
| 龚永强 | 1987.09—1990.08.07 | $ZrO_2$ 体系上甲烷氧化偶联制乙烯乙烷及其机理的研究 | 理学硕士学位 |
| 李文莹 | 1987.09—1990.08.07 | 脉冲激光离子源直线式双电场飞行时间质谱计的研制及碳原子簇质谱研究 | 理学硕士学位 |
| 黄世转 | 1987.09—1990.09.11 | 铑/碘甲烷催化醋酸甲酯羰基化制醋酐的研究 | 理学硕士学位 |
| 游晨涛 | 1987.09—1990.10.08 | 甲醇在负载型氧化物催化剂上脱氢制甲醛的研究 | 理学硕士学位 |
| 邱育南 | 1987.09—1990.11.02 | 甲苯的电催化氧化 | 理学硕士学位 |
| 黄浩平 | 1987.09—1990.11.15 | 合成气制低碳混合醇硫化钼基催化剂研究 | 理学硕士学位 |
| 周水琴 | 1988.09—1991.09.11 | 固定载体高分子富氧膜研究——新型膜材料的合成与性能研究 | 理学硕士学位 |

续表

| 姓　名 | 入学至毕业/答辩时间 | 学位论文题目或研究方向、内容 | 学位或备注 |
|---|---|---|---|
| 王　炜 | 1988.09—1991.09.12 | 硝基苯还原羰基化合成苯氨基甲酸乙酯催化剂及其作用机理研究 | 理学硕士学位 |
| 马　坚 | 1989.09—1992.07.07 | 甲醇催化脱氢制无水甲醛的研究 | 理学硕士学位 |
| 王泉明 | 1989.09—1992.08.25 | 甲烷氧化偶联催化剂的研究 | 理学硕士学位 |
| 何小龙 | 1990.09—1993.08.22 | 铜锰复氧化物催化剂对芳烃的深度氧化作用 | 理学硕士学位 |

**博士研究生**

| 姓　名 | 入学至毕业/答辩时间 | 学位论文题目或研究方向、内容 | 学位或备注 |
|---|---|---|---|
| 廖代伟 | 1982.04—1985.04.20 | 铁催化剂上的化学吸附物种和固氮成氨 | 理学博士学位 |
| 陈鸿博 | 1982.04—1986.12.09 | 金属—金属氧化物助剂协合催化作用的研究——Cu/ZnO/M$_2$O$_3$甲醇合成催化剂的活性中心本质 | 理学博士学位 |
| 沈雁飞 | 1985.09—1988.06.14 | A STUDY OF PROMOTER EFFECTS ON THE ACTIVE CENTERS AND THE MECHANISMS OF CO HYDROGENATION OVER SUPPORTED PALLADIUM CATALYSIS | 理学博士学位 |
| 吴也凡 | 1984.03—1988.09.08 | 铁硫立方烷簇合物及聚硫醚金属配合物催化剂仿生配位催化研究 | 理学博士学位 |
| 洪　亮 | 1985.03—1988.12.27 | 麝香酮及环十五酮环合前体合成法的研究——过渡金属和元素有机化合物在合成中的应用 | 理学博士学位 |
| 陈建平 | 1985.03—1994.05.11 | 合成气在第Ⅷ族金属催化剂上的转化及其一些相关的基础问题的研究 | 理学博士学位 |
| 张兆龙 | 1986.09—1989.09.28 | 关于甲烷氧化偶联的表面科学与催化作用基础研究 | 理学博士学位 |
| 翁维正 | 1986.09—1989.11.26 | 钼铋铁系复氧化物催化剂组成、结构与性能关系和丙烯选择氧化、氨氧化反应主、副产物形成机理的研究 | 理学博士学位 |
| 周朝晖 | 1986.09—1989.12.08 | 有关固氮酶模拟反应和铑催化乙醇合成模型反应的簇合物反应性能研究 | 理学博士学位 |
| 胡云行 | 1987.09—1990.12.24 | AHTD法铜基催化剂合成甲醇的研究 | 理学博士学位 |
| 汪海有 | 1988.09—1991.10.10 | 负载型铑基催化剂上合成气转化为乙醇的反应机理 | 理学博士学位 |

续表

| 姓　名 | 入学至毕业/答辩时间 | 学位论文题目或研究方向、内容 | 学位或备注 |
|---|---|---|---|
| 刘玉达 | 1989.09—1992.12.29 | CATALYST DESIGN AND PREPARATION FOR METHANE OXIDATIVE COUPLING (MOC) REACTIONS | 理学博士学位 |
| 周小平 | 1990.09—1993.12.23 | 氟离子调变的甲烷氧化偶联及乙烷氧化脱氢催化剂的研究 | 理学博士学位 |
| 黄静伟 | 1990.09—1994.04.11 | THE CHEMICAL SIMULATION OF BIOLOGICAL NITROGEN FIXATION | 理学博士学位 |
| 李海燕 | 1990.09—1995.10.24 | 合成气经甲酸甲酯制甲醇新催化过程和催化剂研究 | 理学博士学位 |
| 晁自胜 | 1992.09—1995.10.24 | 甲烷氧化偶联制碳二烃及乙烷氧化脱氢制乙烯 | 理学博士学位 |
| 方智敏 | 1990.09—1996.08.28 | 丙烷氧化脱氢 VMgO 催化剂双相协同催化作用和活性位的研究 | 理学博士学位 |
| 林海强 | 1996.09—1999.12.09 | 分子筛膜及纳米分子筛合成研究 | 理学博士学位 |
| 侯书雅 | 2001.09—2004.06.30 | 高柠檬酸钼及其同系物的研究 | 理学博士学位 |
| 陈洪斌 | 2002.09—2007.01.12 | 高柠檬酸同系物的合成及固氮酶催化反应中的质子传递研究 | 理学博士学位 |

博士后

| 姓名 | 时间 | 研究方向 | 备注 |
|---|---|---|---|
| 郑兰荪 | 1986.08—1988.08 | 铁原子簇羰基化合物的催化作用机理研究 | 博士后 |
| 杨建灵 | 1988.10—1990.10 | 配位络合分离膜和配位催化功能膜的研究 | 博士后 |
| 袁友珠 | 1990.08—1992.12 | 均相配位催化剂多相化 | 博士后 |
| 张伟德 | 1991.08—1993.10 | 低碳烷烃催化氧化脱氢制烯烃的研究 | 博士后 |
| 刘爱民 | 1992.02—1993.10 | 生物固氮及其化学模拟 | 博士后 |
| 高利珍 | 1994.08—1996.10.30 | 特殊聚集态催化剂的制备 | 博士后 |

**图书在版编目(CIP)数据**

蔡启瑞院士论文选集/蔡启瑞著. —厦门:厦门大学出版社,2013.11
ISBN 978-7-5615-3126-6

Ⅰ.①蔡… Ⅱ.①蔡… Ⅲ.①催化-文集 Ⅳ.①O643.3-53

中国版本图书馆 CIP 数据核字(2013)第 258398 号

厦门大学出版社出版发行

(地址:厦门市软件园二期望海路 39 号 邮编:361008)

http://www.xmupress.com

xmup @ xmupress.com

厦门集大印刷厂印刷

2013 年 11 月第 1 版 2013 年 11 月第 1 次印刷

开本:889×1194 1/16 印张:108.25

插页:12 字数:2980 千字

定价:290.00 元(上、下册)

本书如有印装质量问题请直接寄承印厂调换